FACTORIAL NOTATION

$$n! = n(n-1)(n-2)\cdots 4\cdot 3\cdot 2\cdot 1$$
$$n! = n(n-1)!$$
$$0! = 1$$

PERMUTATIONS

$$P_{n,r} = \frac{n!}{(n-r)!}$$

COMBINATIONS

$$C_{n,r} = \frac{n!}{r!(n-r)!}$$

PROBABILITY

S = sample space $\quad E$ = event $\quad E'$ = complement of E

$\varnothing$ = empty set

$$P(\varnothing) = 0 \qquad P(S) = 1$$
$$0 \le P(E) \le 1$$
$$P(E') = 1 - P(E)$$

INTEREST FORMULAS

SIMPLE INTEREST

$$I = Prt$$

$$S = P(1 + rt) \qquad P = \frac{S}{1 + rt}$$

SIMPLE DISCOUNT

$$D = Sdt$$

$$S = \frac{P}{1 - dt} \qquad P = S(1 - dt)$$

COMPOUND INTEREST

$$S = P(1+i)^n \qquad P = S(1+i)^{-n}$$

CONTINUOUS COMPOUND INTEREST

$$S = Pe^{jt} \qquad P = Se^{-jt}$$

MATHEMATICS WITH APPLICATIONS FOR THE MANAGEMENT, LIFE, AND SOCIAL SCIENCES

FOURTH EDITION

BERNARD KOLMAN
Drexel University

HOWARD ANTON
Drexel University

BONNIE AVERBACH
Temple University

with the assistance of
Charles G. Denlinger
Millersville University

SAUNDERS COLLEGE PUBLISHING
A Harcourt Brace Jovanovich College Publisher

FORT WORTH PHILADELPHIA SAN DIEGO NEW YORK ORLANDO AUSTIN SAN ANTONIO
TORONTO MONTREAL LONDON SYDNEY TOKYO

In memory of Shirley M. Anton.
H. A.

In memory of my mother, Eva.
B. K.

To Rachelle, Wayne, Robert, Stephen, Debra, Gary, and Jacob.
B. A.

Cover art: Victor Vasarely, *Torony,* 1970. Used with permission.

ISBN: 0-15-555228-7

Library of Congress Catalog Card Number: 91-73224

Printed in the United States of America

Preface

This book presents the fundamentals of finite mathematics and calculus in a style tailored for beginners, but at the same time covers the subject matter in sufficient depth so that the student can see a rich variety of realistic and relevant applications. Since many students in this course have a minimal mathematics background, we have devoted considerable effort to the pedagogical aspects of this book—examples and illustrations abound. We have avoided complicated mathematical notation and have painstakingly worked to keep technical difficulties and tedious algebraic manipulations from hiding otherwise simple ideas. We have, however, included an algebra review section following Chapter 19. This review affords the instructor a choice of covering the material or not, as time and students' levels permit.

We have been very pleased with the wide acceptance of this book in its three earlier editions and gratefully acknowledge the many comments from students and faculty.

The changes we have made in preparing this edition are in keeping with our main objective: *To produce a textbook that the student will find readable and valuable.* The writing style, illustrative examples, exercises, and applications have been designed with this in mind.

Where appropriate, each exercise set begins with basic computational "drill" problems and then progresses to problems with more substance. Each chapter concludes with a set of Supplementary Exercises and a Chapter Test.

Since there is much more material available than can be included in a single reasonably sized text, it was necessary for us to be selective in the choice of material. We have tried to select those topics that we believe are *most likely to prove useful* to the majority of readers.

In keeping with the title, *Mathematics with Applications*, we have included a host of applications. They range from artificial "applications" which are designed to point out situations in which the material might be used, all the way to bona fide relevant applications based on "live" data and actual research papers. We have tried to include a balanced sampling from business, finance, biology, behavioral sciences, and social sciences.

There is enough material in this book so that each instructor can select the topics that best fit the needs of the class. To help in this selection, we have included a discussion of the structure of the book and a flow chart suggesting possible organizations of the material.

CHAPTER 1 gives an introduction to Cartesian coordinate systems and graphs. Equations of straight lines are discussed and applications are given to problems in simple interest, linear depreciation, and prediction. We also consider systems of two linear equations in two unknowns, the least squares method for fitting a straight line to empirical data, and material on linear inequalities that will be needed for linear programming. Portions of this chapter may be familiar to some students, in which case the instructor can review this material quickly.

CHAPTER 2 discusses basic material on matrices, the solution of linear systems, and applications. Many of the ideas here are used in later sections.

CHAPTER 3 is devoted to an elementary introduction to linear programming from a geometric point of view. A more extensive discussion of linear programming, including the simplex method, appears in Chapter 4. Since Chapter 4 is technically more difficult, some instructors may choose to limit their treatment of linear programming entirely to Chapter 3, omitting Chapter 4.

CHAPTER 4 gives an elementary presentation of the simplex method for solving linear programming problems. Although our treatment is as elementary as possible, the material is intrinsically technical, so that some instructors may choose to omit this chapter.

CHAPTER 5 discusses the elementary set theory needed in the next chapter.

CHAPTER 6 discusses permutations, combinations, and counting methods that are needed in the probability chapter.

CHAPTER 7 introduces probability for finite sample spaces. We carefully explain the nature of a probability model so that the student understands the relationship between the model and the corresponding real-world problem. Section 7.6, which is optional, deals with Bayes' Formula and stochastic processes.

CHAPTER 8 discusses basic concepts in statistics. Binomial random variables as well as the normal approximation to the binomial are considered. Section 8.8 introduces hypothesis testing by means of the chi-square test, thereby exposing the student to some realistic statistical applications. Section 8.4 on Chebyshev's inequality is included because it helps give the student a better feel for the notions of mean and variance. It can be omitted from the chapter without loss of continuity.

CHAPTER 9 covers a number of topics in the mathematics of finance. It includes the time value of money and equations of value, simple interest, compound interest, and bank discount. The chapter is self contained.

CHAPTER 10 is intended to give the student some solid, realistic applications of the material. The topics in this chapter are drawn from a variety of fields so that the instructor can select those sections that best fit the needs and interests of the class. It draws on the material on matrices, probability, and mathematics of finance.

CHAPTER 11 provides an introduction to functions with special emphasis on polynomials.

CHAPTERS 12–19 introduce the elements of calculus and its applications. The approach is intuitive with emphasis on applications rather than theory.

APPENDIX A: ALGEBRA REVIEW is designed to review those algebraic skills needed in this text. Some instructors may prefer to start the course with this material, while others may prefer to cover it section by section as needed.

APPENDIX B: LOGIC presents a very elementary introduction to the foundations of logic.

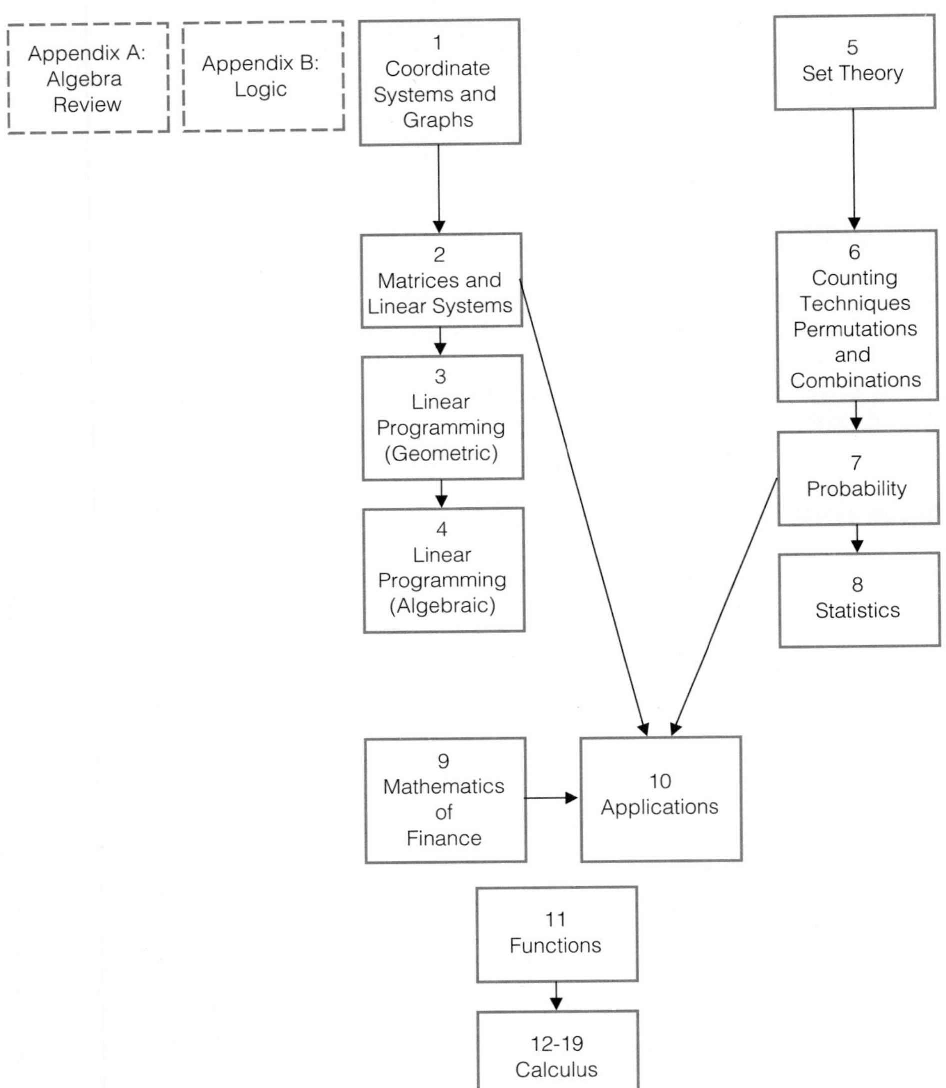
Appendix A: Algebra Review
Appendix B: Logic
1 Coordinate Systems and Graphs
5 Set Theory
2 Matrices and Linear Systems
6 Counting Techniques Permutations and Combinations
3 Linear Programming (Geometric)
7 Probability
4 Linear Programming (Algebraic)
8 Statistics
9 Mathematics of Finance
10 Applications
11 Functions
12-19 Calculus

Supplementary Materials

STUDENT SOLUTIONS MANUAL

Contains detailed solutions to selected odd-numbered exercises.

INSTRUCTOR'S MANUAL WITH TESTS

Contains answers to all even-numbered exercises and a test bank.

COMPUTERIZED TEST BANK

The Micro-Pac Genie Testing System allows the user to select, edit, and add or delete questions. It is available for Apple II and IBM or IBM-compatible computers with two disk drives (5 1/4″ or 3 1/2″) or one disk drive and a hard disk, 512K memory, DOS 2.0 or higher, color graphics card, graphics monitor, and graphics printer.

COMPUTER SOFTWARE

- INTERACTIVE FINITE MATH™ from Mathlab is available for Apple II and IBM-PC or IBM-compatible computers with two disk drives (5 1/4″ or 3 1/2″), 256K memory, and DOS 2.0 or higher.
- MathPath software by George W. Bergeman provides computational and graphical support for topics in this text. Computational routines serve as a "super calculator" to perform otherwise cumbersome operations and free students to concentrate on concepts and underlying theory. Graphical displays help students visualize, explore, and understand a variety of topics. Modules include Least Squares Linear Regression, Matrix Operations, Linear Programming via the Graphical Method, Simplex Tableau Evaluation, Function Graphing, Finance Calculations, Descriptive Statistics, Permutations and Combinations, Normal Curve Calculations, and Numerical Integration.
- Graph 2D/3D software by George W. Bergeman graphs functions in one variable and surfaces of functions in two variables. It provides an easy-to-use, basic tool to help users visualize, explore, and understand a variety of important topics.

  For graphing $y = f(x)$, 2D/3D can zoom-in, zoom-out, display function values, graph several functions simultaneously, compute areas under curve, and save set-ups to disk.

  For graphing $z = f(x, y)$, 2D/3D can rotate and raise or lower surfaces to view from varying vantage points, display function values, and save set-ups to disk.

- Both MathPath and Graph 2D/3D are available for IBM-PC or IBM-compatible computers with one disk drive, two disk drives, or a hard disk, 512K memory, DOS 2.1 or higher, color graphics card, graphics writers, and graphics printer.

New Features in the Fourth Edition

- The chapters have been reordered. Chapters 1 and 2 now cover the algebraic and geometric ideas and techniques needed for the study of linear programming, which is presented in Chapters 3 and 4. In addition, the student now has more time to improve his or her algebraic skills. The instructor can determine early in the semester how much of Appendix A, Algebra Review, needs to be covered. Chapters 5 and 6 cover sets and counting techniques, which are needed for the study of probability and statistics in Chapters 7 and 8.
- Appendix B, Logic, is new to this edition.
- More exercises and figures have been added.
- The clarity of writing and pedagogy have been improved in many places.
- A chapter on Functions (Chapter 11) now introduces the calculus portion of the book.

Acknowledgments

We gratefully acknowledge the contributions of the following people whose comments, criticisms, and assistance greatly improved the entire manuscript. For the first three editions:

Robert E. Beck, *Villanova University*
Elizabeth Berman, *Rockhurst College*
Alan I. Brooks, *UNISYS*
David Cochener, *Angelo State University*
Ronald Davis, *Northern Virginia Community College*
Jerry Ferry, *Christopher Newport College*
Mary W. Gray, *American University*
Beryl M. Green, *Oregon College of Education*
Albert J. Herr, *Drexel University*
Robert L. Higgins, *QUANTICS, Inc.*
David Hinde, *Rock Valley College*
Joe Hoffman, *Tallahassee Community College*
Leo W. Lampone, *METRON, Inc.*
Stanley Lukawecki, *Clemson University*
Dan Madden, *University of Arizona*
Lawrence P. Maher, *North Texas State University*
Daniel P. Maki, *Indiana University*
Samuel L. Marateck, *New York University*
Daniel Martinez, *California State University, Long Beach*
Curtis McKnight, *University of Oklahoma*
J. S. Mehta, *Temple University*
Robert Moreland, *Texas Tech University*
J. A. Moreno, *San Diego City College*
Donald E. Myers, *University of Arizona*
Robert Pruitt, *San Jose State University*
John Quigg, *Arizona State University*

Ellen Reed, *University of Massachusetts at Amherst*
Bernard Schroeder, *University of Wisconsin at Platteville*
Bob Simpson, *Bryan College*
James Snow, *Lane Community College*
Ward Soper, *Walla Walla College*
Leon Steinberg, *Temple University*
Paul Vicknair, *California State College*
William H. Wheeler, *Indiana State University*
Howard Wilson, *American University*

For the fourth edition:

Charles G. Denlinger, *Millersville University*
Arnaldo Horta, *University of Miami*
Donald E. Myers, *University of Arizona*
Robert Moreland, *Texas Tech University*
Max Plager, *Roosevelt University*
Brenda Roberts, *University of Tulsa*
Gayle Smith, *Eastern Washington University*
W. Vance Underhill, *East Texas State University*

We thank Ron Perline, Drexel University, for helpful information on the mathematics of CAT scans.

Our thanks go to Neil Soiffer of Tektronix, Inc. for the computer-generated graphs appearing in Chapter 19.

We thank Gail Edinger, Fisher College, for reading galleys and pages, Enid Savitz, Temple University, for preparing the *Instructor's Manual*, Susan Friedman, Bernard M. Baruch College of CUNY, for preparing the *Student Solutions Manual*, and Stephen M. Kolman for preparing the index.

We also thank Elizabeth van Dusen, Enid Savitz, and David Zitarelli, all of Temple University, and Susan Friedman for calling our attention to several errors.

Finally, we wish to thank our editor, Pamela Whiting; our production editor, Mary Douglas; and the entire staff of Harcourt Brace Jovanovich for their enthusiastic cooperation and support throughout the many phases of this project.

Contents

CHAPTER 4 LINEAR PROGRAMMING (AN ALGEBRAIC APPROACH) 124

CHAPTER 5 SET THEORY 163

CHAPTER 6 COUNTING TECHNIQUES; PERMUTATIONS AND COMBINATIONS 188

CHAPTER 7 PROBABILITY 232

CHAPTER 8 STATISTICS AND PROBABILITY 294

1

Coordinate Systems and Graphs

In many physical problems we are interested in relationships between variable quantities. For example, an astronomer may be interested in the relationship between the size of a star and its brightness, an economist in the relationship between the cost of a product and the quantity available, and a sociologist in the relationship between the increase in crime and the increase in population in U.S. cities. In this chapter we discuss two methods for describing such relationships: analytically, by means of formulas, and geometrically, by means of graphs.

Although you may find some of the topics in this chapter familiar, others are likely to be new. Because the ideas we develop here are crucial to our later work, we discuss graphs and functions from the very beginning to establish a common starting point. You may wish to skim the more familiar material in this chapter.

1.1 COORDINATE SYSTEMS

In this book we use the language of sets from time to time. A **set** is simply a collection of objects; the objects are called the **elements** or **members** of the set. Chapter 5 is devoted to a more careful study of sets.

Real Numbers

The set of numbers that we all learn first is the set of **natural numbers**, also called **positive integers** or **counting numbers**:

$$1, 2, 3, 4, 5, \ldots$$

This set is adequate for keeping account of "how many."

To keep track of what one person owes another, we need negative numbers (deficits) and zero. This gives rise to the set of integers

$$\ldots, -5, -4, -3, -2, -1, 0, 1, 2, 3, 4, 5, \ldots$$

consisting of the positive integers, the negative integers, and zero. In this section we examine the composition and some of the properties of the set of all real numbers.

To divide a number of objects equally among a group of people (e.g., 5 apples among 20 people), we are led to a new kind of number: a rational number, or fraction. When dividing m units equally among n people, each person gets m/n units. The quotient m/n, where m and n are integers and $n \neq 0$, is called a **rational number**. Recall that division by zero is not defined; therefore, the denominator of a rational number cannot be zero.

Decimals provide an alternate way of writing numbers, for example,

$$\frac{1}{4} = 0.25 \qquad \frac{19}{8} = 2.375 \qquad -\frac{1}{1000} = -0.001$$

Some decimals, such as $\frac{1}{3} = 0.3333\ldots$ and $\frac{3}{11} = 0.272727\ldots$, are **nonterminating** in the sense that they require infinitely many decimal places for their complete expression. If a nonterminating decimal has a block of one or more digits that repeats, then that decimal is called a **repeating decimal**. For example,

$$\frac{2}{3} = 0.666666\ldots \qquad \text{the "6" repeats}$$

$$\frac{1}{11} = 0.090909\ldots \qquad \text{the block "09" repeats}$$

$$\frac{8}{13} = 0.615384615384\ldots \qquad \text{the block "615384" repeats}$$

Terminating decimals can be regarded as repeating decimals as well. For example,

$$\frac{1}{2} = 0.50000\ldots \qquad \text{the zero repeats}$$

$$\frac{13}{8} = 1.6250000\ldots \qquad \text{the zero repeats}$$

It has become a standard convention to denote repeating decimals by listing the block of repeating digits only once but with a bar over the block to indicate that it repeats. Thus, we write

$$\frac{2}{3} = 0.\overline{6} \qquad \frac{1}{11} = 0.\overline{09} \qquad \frac{8}{13} = 0.\overline{615384}$$

The following fundamental result can be established.

Every rational number is represented by a repeating decimal, and vice versa.

It follows from this result that there exist numbers that are not quotients of integers, namely, those numbers that are represented by nonterminating, nonrepeating decimals. For example,

$$0.101001000100001000001\ldots$$

is not a quotient of integers because no block of digits repeats and the decimal is nonterminating. Such numbers are called **irrational numbers**. For example, it can be shown that $\pi = 3.141592653\ldots$ and $\sqrt{2} = 1.41421356\ldots$ do not terminate or repeat from some point on; therefore, π and $\sqrt{2}$ are irrational numbers.

The rational and irrational numbers together constitute the set of **real numbers** (Figure 1).

FIGURE 1

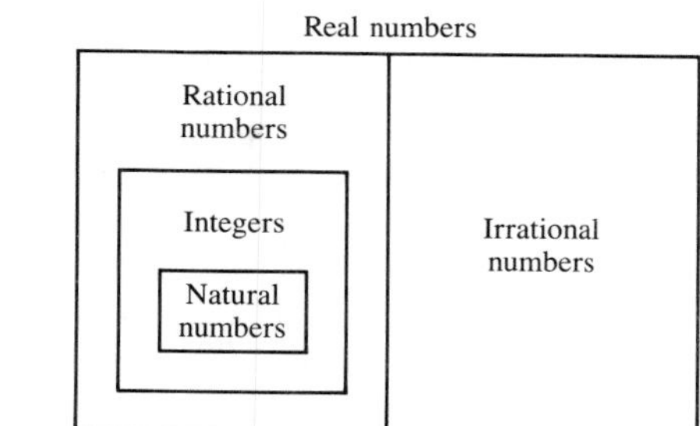

Calculator and Computer Computation

In most applications, complicated calculations are performed with calculators or computers. Most calculators and computers can display up to 10 digits, but this varies among machines. On a machine that displays 10 digits, the numbers $\frac{1}{3}$, $\sqrt{2}$, π, $\frac{115}{7}$, and $\frac{2}{17{,}635}$ are displayed as shown in Table 1.

TABLE 1

Number	Machine Display	Rounded to Four Decimal Places
$\frac{1}{3}$	0.333333333	0.3333
$\sqrt{2}$	1.414213562	1.4142
π	3.141592654	3.1416
$\frac{115}{7}$	16.42857142	16.4286
$\frac{2}{17{,}635}$	0.000113410	0.0001

Frequently, a calculator or computer supplies more digits than are needed, and it is necessary to eliminate the excess digits. One procedure for doing this is called **rounding off**. A number is said to be **rounded to *n* decimal places** if it is replaced by the closest number with n digits after the decimal point. One method for rounding is as follows.

Rounding to *n* Decimal Places

1. If the $(n + 1)$th digit is less than 5, then discard all digits after the nth.
2. If the $(n + 1)$th digit is 5 or greater, then increase the nth digit by 1 and discard all digits after the nth.

EXAMPLE 1 Round the numbers in Table 1 to four decimal places.

Solution The symbol $\cong$ (read "is approximately equal to") is used to denote approximations. Thus, we write $\sqrt{2} \cong 1.414$ to emphasize that 1.414 is only an approximation of $\sqrt{2}$.

Remark If we round 2.71625 to four decimal places using the method above, we obtain 2.7163. However, the number 2.7162 is equally close to 2.71625 and is just as good an approximation as 2.7163. This situation occurs whenever we round to n decimal places and the $(n + 1)$th decimal digit is 5 and all subsequent decimal digits are zero. In this ambiguous case our method selects the larger of the numbers. This is called **rounding up**. Other possible methods (which we do not use) call for **rounding down** or for rounding up in half the computations and down in the other half.

The Real Number Line

The **real number line** provides a geometric description of real numbers. Starting with a horizontal line, select a point on this line to serve as a reference and call it the **origin**. Next, choose one direction from the origin as the **positive direction**, and let the other direction be called the **negative direction**. It is common to choose the right side of the origin as the positive direction and to mark it with an arrowhead, as shown in Figure 2a. Finally, select a unit of length for measuring distances. With each real number, we can now associate a point on the real number line in the following way.

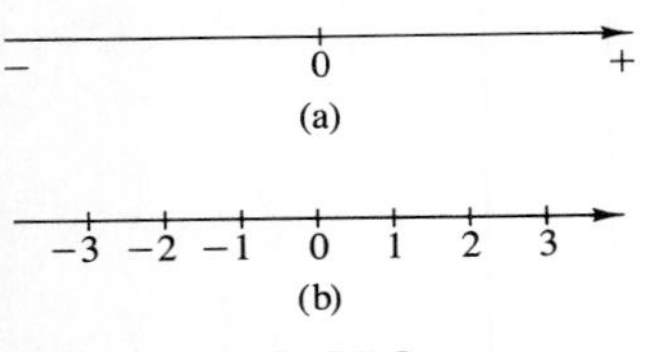

FIGURE 2

(a) With each positive number r, associate the point that is a distance of r units from the origin in the positive direction.

(b) With each negative number $-r$, associate the point that is a distance of r units from the origin in the negative direction.

(c) Associate the origin with the real number 0.

In Figure 2b we have marked on the real number line the points that are associated with some of the integers.

The real number corresponding to a point on the real number line is called the **coordinate** of the point.

EXAMPLE 2 In Figure 3 we have marked the points whose coordinates are $-\sqrt{2}$, $-\frac{1}{2}$, 1.25, and π.

FIGURE 3

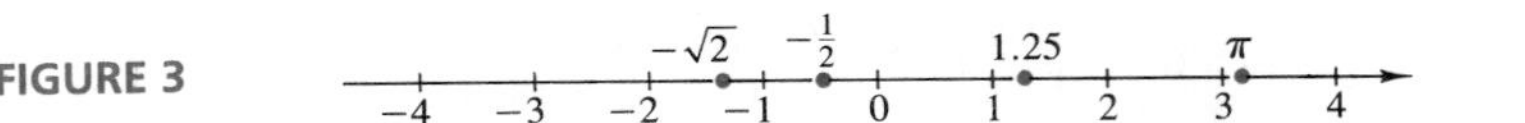

It is evident from the way in which real numbers and points on the real number line are related that each real number corresponds to a single point and each point corresponds to a single real number. We describe this by stating that the set of real numbers and the set of points on the real line are in **one-to-one correspondence**.

Order The real number line makes it possible to visualize order relationships between real numbers. If a is less than b, then the point with coordinate a is to the left of the point with coordinate b; we write $a < b$ (read "a is less than b"). If either $a < b$ or $a = b$, then we may write $a \leq b$ (read "a is less than or equal to b"). Geometrically, this means that the point with coordinate a is to the left of the point with coordinate b or that the two points coincide. The inequality $a < b$ can be written in the alternate form $b > a$ (read "b is greater than a"), and the inequality $a \leq b$ can be written as $b \geq a$ (read "b is greater than or equal to a"). Table 2 lists the basic inequalities and their geometric interpretations.

TABLE 2

Inequality	Geometric Interpretation	Examples
$a < b$	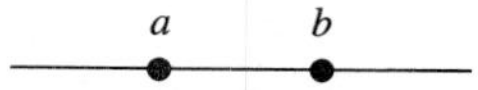 a b (points on line, a left of b)	$3 < 5$ $-6 < -3$
$a > b$	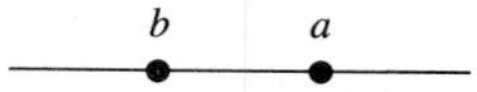 b a (points on line, b left of a)	$8 > 4$ $-2 > -5$
$a \leq b$	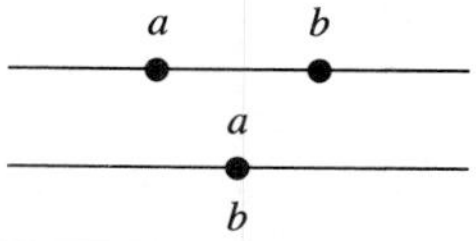 a b a / b (same point)	$-6 \leq 8$ $2 \leq 2$
$a \geq b$	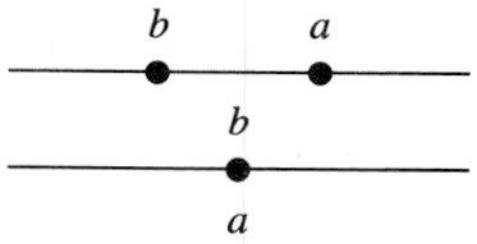 b a b / a (same point)	$4 \geq 1$ $-9 \geq -9$
$a < 0$	 a 0	$-5 < 0$
$a > 0$	 0 a	$9 > 0$

Sometimes two or more inequalities are combined into a single expression. For example, the expression

$$a < b < c$$

is equivalent to the pair of inequalities

$$a < b \quad \text{and} \quad b < c$$

(Figure 4). This notation implies that b is between a and c on the real number line.

FIGURE 4

As a matter of terminology a real number a is called **positive** if $a > 0$, **negative** if $a < 0$, and **nonnegative** if $a \geq 0$. Thus, the nonnegative real numbers consist of the positive real numbers and zero. (For the algebraic rules for operating with inequalities, see Section A.4 of Appendix A.)

THE DEVELOPMENT OF COORDINATE GEOMETRY

In 1637, René Descartes* published a philosophical work called *Discourse on the Method of Rightly Conducting the Reason.* In the back of that book there were three appendixes that purported to show how the "method" could be applied to concrete examples. The first two appendixes were minor works that endeavored to explain the behavior of lenses and the movement of shooting stars. The third appendix, however, was a landmark stroke of genius. It was described by the nineteenth century British philosopher John Stuart Mill as "the greatest single step ever made in the progress of the exact sciences."

In that appendix René Descartes linked together two branches of mathematics: algebra and geometry. His work evolved into a new subject called *analytic geometry;* it gave a way of describing algebraic formulas by means of geometric curves and a way of describing geometric curves by algebraic formulas. In this section we examine some of these ideas.

Although Descartes is widely credited with the creation of the Cartesian coordinate system and the resulting field of analytic geometry, it appears that the same results were obtained independently by Pierre Fermat† as early as 1629. Descartes receives more credit for the discovery because Fermat's work was not published in book form until 1679 and because Descartes' book was an instant success when published in 1637.

* *René Descartes* (1596–1650) was the son of a government official. He graduated from the University of Poitiers at age 20 with a law degree, after which he went to Paris. There, he lived a dissipative life as a man of fashion. In 1618 he joined the army of the Prince of Orange, for whom he worked as a military engineer. Descartes was a genius of the first magnitude who made major contributions in philosophy, mathematics, physiology, and science in general. His work in mathematics gave new direction to science; along with William Harvey, he is considered a founder of modern physiology.

† *Pierre Fermat* (1609–1665), born in France, was the son of a leather merchant. He studied law at the University of Toulouse and lived a quiet life as a king's councillor. Although he worked in mathematics in his spare time, he was a first-rate mathematician whose contributions are of major importance.

Coordinate Systems

We have seen how points on a line can be placed in a one-to-one correspondence with real numbers. We now present an analogous procedure for representing points in the plane.

Construct two perpendicular real number lines. For convenience, make one of the lines horizontal with its positive direction to the right, make the other line vertical with its positive direction upward, and locate the lines so that they intersect at their origins (Figure 5a). The two lines are called **coordinate axes**; the horizontal line is called the **x-axis**, the vertical line the **y-axis**. The coordinate axes together form what is called a **Cartesian* coordinate system** or sometimes a

*After René Descartes.

rectangular coordinate system. The point of intersection of the coordinate axes is denoted by O and is called the **origin** of the coordinate system (Figure 5b).

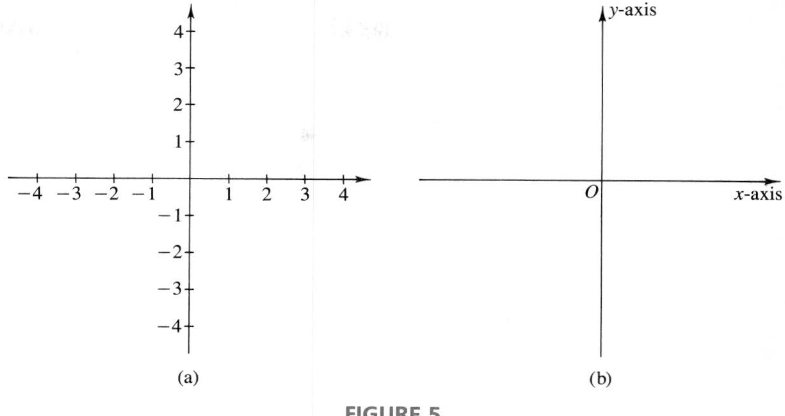

FIGURE 5

If P is a point in the plane, draw two lines through P, one parallel to the y-axis and one parallel to the x-axis. If the first line intersects the x-axis at the point whose coordinate is a and the second intersects the y-axis at the point whose coordinate is b, then we associate the **ordered pair** (a, b) with the point P (Figure 6). The number a is called the **x-coordinate** or **abscissa** of P, and the number b is called the **y-coordinate** or **ordinate** of P; we say that P is the point with **coordinates** (a, b) and denote the point by $P(a, b)$, read "P of a, b."

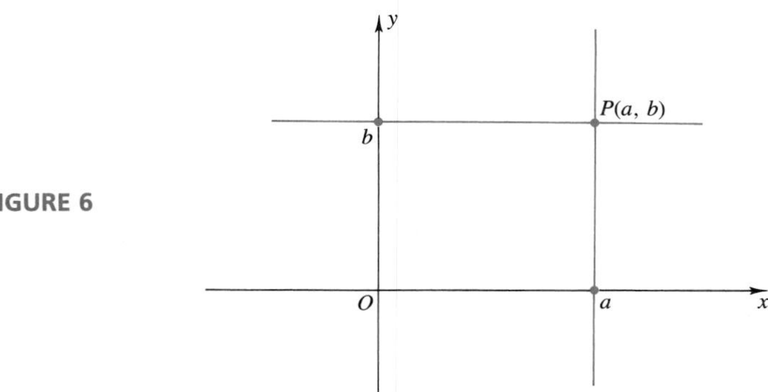

FIGURE 6

The order of listing the elements in an ordered pair is important. The ordered pairs (2, 3) and (3, 2) are different. If $(2, 3) = (a, b)$, then $a = 2$ and $b = 3$.

When two points are under discussion, we sometimes denote them as $P_1(x_1, y_1)$ and $P_2(x_2, y_2)$, where the subscript 1 denotes the coordinates of the first point and the subscript 2 denotes the coordinates of the second point. (We read x_1 as "x one" or as "x sub one.")

EXAMPLE 3 In Figure 7 we have located the points $P(4, 3)$, $Q(-2, 5)$, $R(-4, -2)$, $S(2, -5)$, and $T(-5, 2)$.

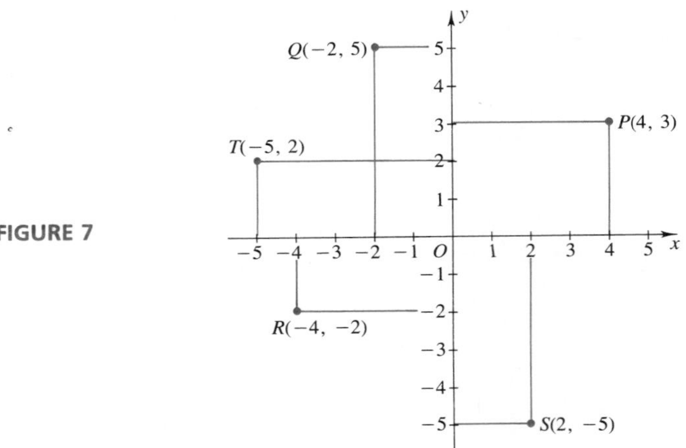

FIGURE 7

The coordinate axes divide the plane into four parts called **quadrants**. These quadrants are numbered from one to four, as shown in Figure 8a. As shown in Figure 8b, it is easy to determine the quadrant in which a point lies from the signs of its coordinates. A point with two positive coordinates $(+, +)$ lies in quadrant I, a point with a negative x-coordinate and a positive y-coordinate $(-, +)$ lies in quadrant II, and so on. A point with a zero coordinate lies on one of the axes and is not in any quadrant. The origin has coordinates $(0, 0)$.

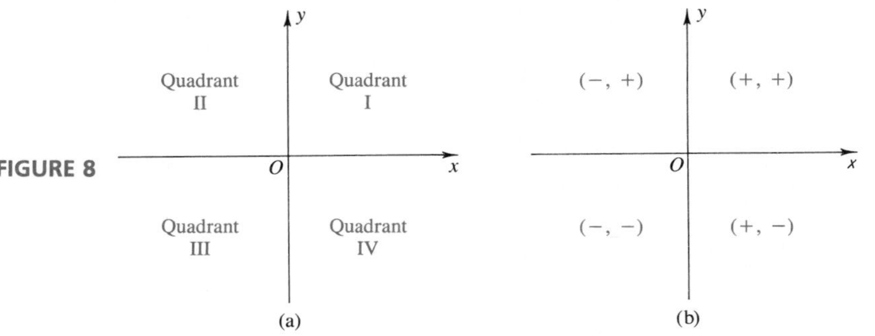

FIGURE 8

EXERCISE SET 1.1

1. Which of the following are (or can be represented by) terminating decimals? Repeating decimals? Neither? Which of these numbers are rational?
   (a) 5.02 (b) 5.020202 . . .
   (c) 5.02002000200002 . . .
   (d) 0.716543962121212121 . . .
   (e) 0.125

2. Apply the directions of Exercise 1 to the following numbers.
   (a) 3.146
   (b) 0.0000007
   (c) $\frac{\pi}{2} = 1.57079632\ldots$
   (d) $\frac{1}{7} = 0.1428571\ldots$
   (e) 2.01001000100001 . . .

3. Draw a real number line and plot the points whose coordinates are
(a) 6 (b) -4 (c) 2.5 (d) -1.7
(e) $\sqrt{3}$ (f) 0

4. Draw a real number line and plot the points whose coordinates are
(a) 4 (b) -2 (c) $\frac{5}{2}$ (d) -3.5
(e) 2.25 (f) $\sqrt{5}$

In Exercises 5–10 indicate which of the two given numbers appears first, viewed from left to right, on the real number line.

5. $4, 6$ **6.** $\frac{1}{2}, 0$ **7.** $-2, \frac{3}{4}$

8. $0, -4$ **9.** $-5, -\frac{2}{3}$ **10.** $4, -5$

In Exercises 11–14 show the given sets of numbers on a real number line.

11. the natural numbers less than 8

12. the integers greater than 4 and less than 10

13. the integers greater than 2 and less than 7

14. the integers greater than -5 and less than or equal to 1

In Exercises 15–24 express the statement as an inequality.

15. 10 is greater than 9.99

16. -6 is less than -2

17. a is nonnegative

18. b is negative

19. x is positive

20. a is between 3 and 7

21. b is less than or equal to -4

22. a is between $\frac{1}{2}$ and $\frac{1}{4}$

23. b is greater than or equal to 5

24. x is negative

In Exercises 25–32 replace the empty square by the symbol $<$ or $>$ to make a true statement.

25. $3 \square 5$ **26.** $4 \square -3$

27. $-3 \square -2$ **28.** $-\frac{1}{2} \square \frac{1}{3}$

29. $\frac{1}{2} \square -\frac{1}{4}$ **30.** $-\frac{1}{5} \square -\frac{1}{3}$

31. $-3.4 \square 4.2$ **32.** $-5.4 \square -6.1$

In Exercises 33–38 refer to Figure 9 and replace the square by the symbol $<$ or $>$ to make a true statement.

FIGURE 9 f c a 0 b d e

33. $a \square 0$ **34.** $b \square a$ **35.** $e \square f$

36. $d \square c$ **37.** $0 \square e$ **38.** $d \square a$

39. In Table 3 use an x to indicate the quadrant in which the point lies. If the point does not lie in any quadrant, make no mark.

TABLE 3

	Quadrant			
Coordinates	**I**	**II**	**III**	**IV**
$(-1, 3)$				
$(2, 7)$				
$(4, -8)$				
$(0, 2)$				
$(-\sqrt{2}, -\pi)$				
$(4, 0)$				

40. Follow the directions of Exercise 39 in Table 4.

TABLE 4

	Quadrant			
Coordinates	**I**	**II**	**III**	**IV**
$(0, 0)$				
$(-\frac{1}{3}, \frac{7}{5})$				
$(0.2, 0.13)$				
$(5, -1)$				
$(\sqrt{5}, 0)$				
$(-3.0, -5.2)$				

41. Draw a Cartesian coordinate system and plot the points whose coordinates are
(a) $(2, 3)$ (b) $(-4, 6)$ (c) $(-3, -5)$
(d) $(7, -3)$

42. Draw a Cartesian coordinate system and plot the points whose coordinates are
(a) $(3, 0)$ (b) $(0, -3)$ (c) $(-2.5, 3.5)$
(d) $(\frac{1}{2}, -\frac{3}{4})$

43. Sketch the set of all points in the plane whose x-coordinates are 2.

44. Sketch the set of all points in the plane whose y-coordinates are -3.

45. Find the fourth corner of the rectangle, three of whose corners are (8, 3), (−2, 3), and (8, 9).

46. Find the coordinates of the point that lies 3 units to the left of and 4 units below the point (2, −1).

1.2 GRAPHS OF EQUATIONS

Cartesian coordinate systems are helpful for giving a geometric description of the set of solutions of an equation involving two variables. We will assume in our discussion that a Cartesian coordinate system has been constructed and that we are given an equation involving only two variables, x and y, such as $3x + 3y = 4$, $x^2 + y^2 = 1$, or $y = 1/x$.

> Given an equation involving only the variables x and y, we call an ordered pair of real numbers (a, b) a **solution of the equation** if the equation is satisfied when we substitute
> $$x = a \qquad y = b$$

EXAMPLE 1 The ordered pair (4, 5) is a solution of the equation

$$3x - 2y = 2$$

since the equation is satisfied when we substitute $x = 4$, $y = 5$. Thus

$$\begin{aligned} 3(4) - 2(5) &= 2 \\ 2 &= 2 \end{aligned}$$

and the equation is satisfied. However, the ordered pair (2, 1) is not a solution of the equation since the equation is not satisfied when we substitute $x = 2$ and $y = 1$:

$$3(2) - 2(1) = 4 \neq 2$$

The set of all solutions of an equation is called its **solution set**.

EXAMPLE 2 Like many equations in x and y, the solution set of

$$y = x^2 + 1 \tag{1}$$

has infinitely many members, so it is impossible to list them all. However, some sample members of the solution set can be obtained by substituting *arbitrary* values for x into the right-hand side of (1) and solving for the associated values of y. Some typical computations are given in Table 1.

TABLE 1

x	0	1	2	3	−1	−2	−3
$y = x^2 + 1$	1	2	5	10	2	5	10
Solution	(0, 1)	(1, 2)	(2, 5)	(3, 10)	(−1, 2)	(−2, 5)	(−3, 10)

If we plot the points corresponding to the ordered pairs in the solution set of an equation in x and y, we obtain a geometric picture of the equation. This picture is called the **graph** of the equation.

EXAMPLE 3 Sketch the graph of the equation $y = x^2 + 1$ given in Example 2.

Solution Since there are infinitely many points on the graph of $y = x^2 + 1$, it is impossible to plot them all. The general procedure for sketching the graph of an equation is to plot enough points on the graph to indicate its general shape. Then the graph is approximated by drawing a smooth curve through the plotted points.

In Figure 1 we have sketched the graph of $y = x^2 + 1$ by plotting the points in Table 1 and connecting them with a smooth curve.

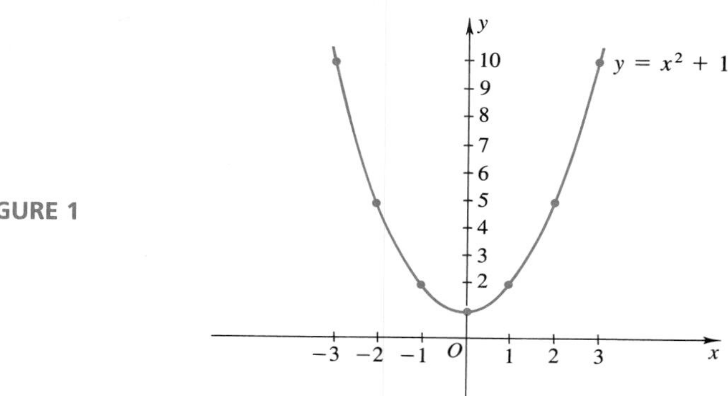

FIGURE 1

EXAMPLE 4 Sketch the graph of the equation $y - 3x = 1$.

Solution We first rewrite this equation as

$$y = 3x + 1$$

and substitute the values of x indicated in Table 2. Plotting the points in this table, we obtain the graph in Figure 2.

TABLE 2

x	0	1	2	−1	−2
$y = 3x + 1$	1	4	7	−2	−5
Solution	(0, 1)	(1, 4)	(2, 7)	(−1, −2)	(−2, −5)

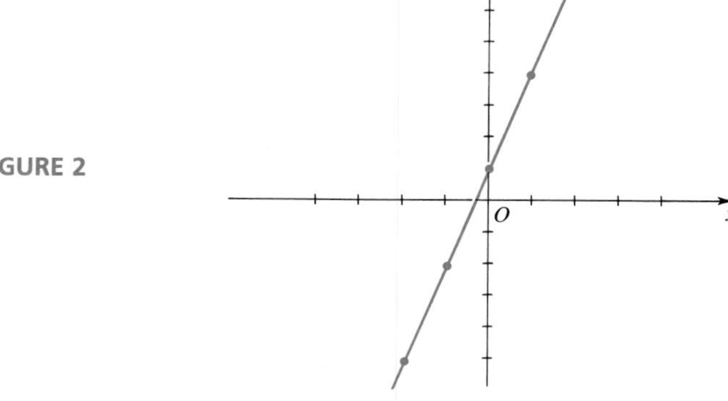

FIGURE 2

EXAMPLE 5 Sketch the graph of the equation

$$x = y^2 - 1$$

Solution We substitute the values of y indicated in Table 3 and use the given equation to compute the corresponding values of x. Plotting the points in Table 3, we obtain the graph shown in Figure 3.

TABLE 3

y	-3	-2	-1	0	1	2	3
$x = y^2 - 1$	8	3	0	-1	0	3	8
Solution	$(8, -3)$	$(3, -2)$	$(0, -1)$	$(-1, 0)$	$(0, 1)$	$(3, 2)$	$(8, 3)$

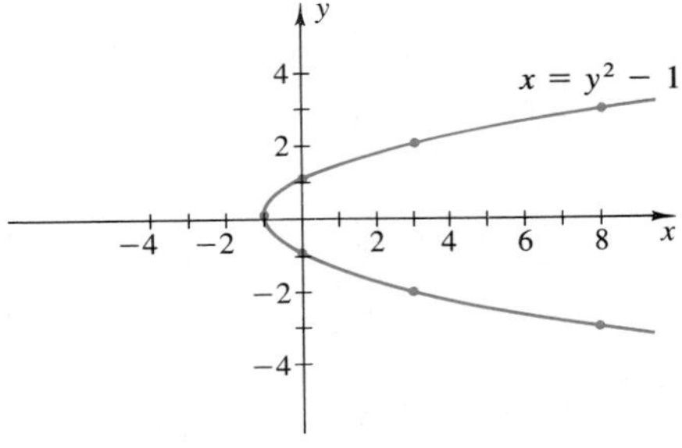

FIGURE 3

Remark In general, to graph an equation you can arbitrarily choose x-values and then determine y from the equation, or vice versa. Usually, you should choose the method that is easier for the particular equation.

There is one flaw in the graphing technique we have just discussed. To see it, consider the graph in Figure 2. Although it appears from our sketch that the points on the graph form a straight line, this conclusion is just guesswork based on plotting a few points. It is conceivable that the actual graph oscillates between the points we have plotted or begins to curve off once we pass the points we have plotted. In fact, based on the data given in Table 2, it is logically conceivable that the graph of $y = 3x + 1$ might look like the curve in color in Figure 4.

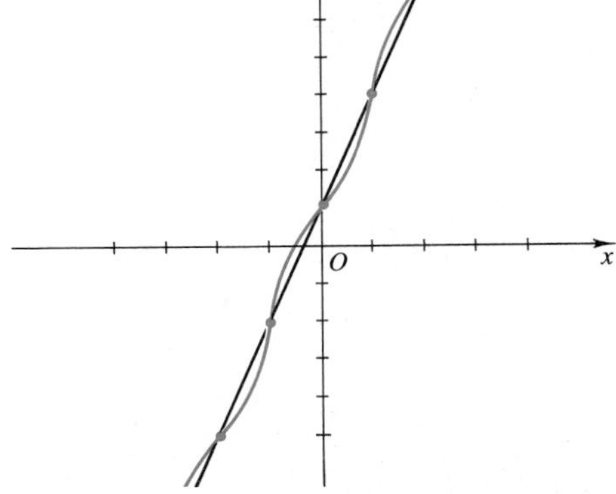

FIGURE 4

Although we show in the next section that the graph of $y = 3x + 1$ is actually a straight line, we are emphasizing that the graphing technique described here provides us only with a reasonable *guess* about the shape of the graph. There are more powerful graphic techniques, based on calculus, which allow us to picture with more certainty the graph of an equation.

Intercepts The x-coordinate of a point at which the graph of an equation intersects the x-axis is called an **x-intercept** of the graph. The y-coordinate of a point at which the graph intersects the y-axis is called a **y-intercept** of the graph. The x-intercepts are found by setting $y = 0$ and solving the equation for x; the y-intercepts are found by letting $x = 0$ and solving for y.

EXAMPLE 6 Find the x- and y-intercepts of the graph of the equation $y = x^2 - 1$.

Solution To find the x-intercepts we set $y = 0$ and solve for x. We obtain

$$x^2 - 1 = 0 \quad \text{or} \quad x^2 = 1$$

so the x-intercepts are $x = 1$ and $x = -1$. To find the y-intercepts, we set $x = 0$ and solve for y. This yields $y = -1$ as the only y-intercept. The graph of $y = x^2 - 1$ is shown in Figure 5.

FIGURE 5

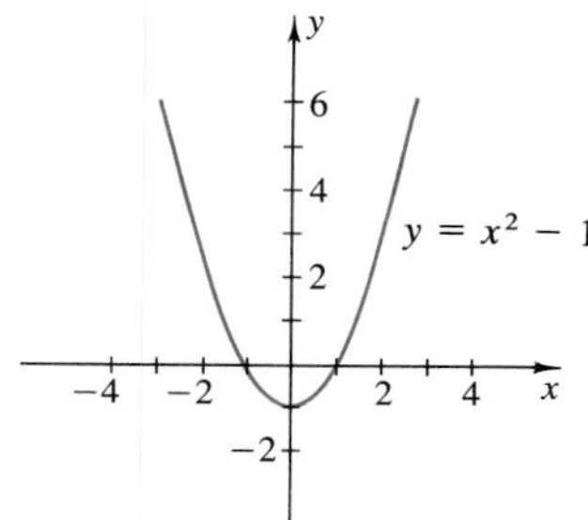

EXAMPLE 7 Find the x- and y-intercepts of the graph of the equation $y = x^2 + 1$.

Solution To find the x-intercepts we set $y = 0$ and solve for x, obtaining

$$x^2 + 1 = 0 \quad \text{or} \quad x^2 = -1$$

Since there is no real number whose square is negative, we conclude that the graph has no x-intercepts. To find the y-intercepts, we set $x = 0$ and solve for y, obtaining $y = 1$. The graph of $y = x^2 + 1$ is shown in Figure 6.

FIGURE 6

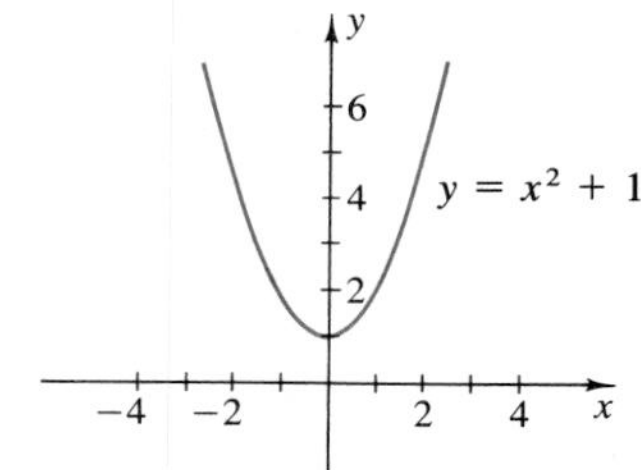

EXERCISE SET 1.2

1. Which of the following are solutions of $2x + 3y = 1$?
(a) $(0, \frac{1}{3})$ (b) $(\frac{1}{2}, 0)$
(c) $(-2, 3)$ (d) $(-1, 1)$

2. Which of the following are solutions of $4x^2 + 9y = 36$?
(a) $(0, 4)$ (b) $(3, 0)$
(c) $(-1, 1)$ (d) $(\frac{1}{2}, \frac{35}{9})$

3. In each part fill in the blank so that the ordered pair is a solution of the equation $3x - 2y = 3$.
(a) $(\_\_, 2)$ (b) $(-1, \_\_)$
(c) $(0, \_\_)$ (d) $(\_\_, 3)$

4. Fill in Table 4, plot the corresponding points, and sketch the graph of $y = 2x^2 + 1$.

TABLE 4

x	-3	-2	-1	0	1	2	3
$y = 2x^2 + 1$	19						
Solution	$(-3, 19)$						

5. Fill in Table 5, plot the corresponding points, and sketch the graph of $y = 1 - x^2$.

TABLE 5

x	-3	-2	-1	0	1	2	3
$y = 1 - x^2$	-8						
Solution	$(-3, -8)$						

In Exercises 6–15 determine the intercepts of the graphs of the given equation.

6. $y = 2x - 3$
7. $x = 3y + 4$
8. $y = x^2 - 4$
9. $y = x^3 + 1$
10. $y = x^3 - 1$
11. $x = y^3 + 1$
12. $y = \dfrac{1}{x}$
13. $y = \dfrac{1}{x + 1}$
14. $x^2 + y^2 = 16$
15. $x^2 - y^2 = 16$

In Exercises 16–25 find the intercepts and sketch the graph of the given equation.

16. $y = 3x + 2$
17. $y = -2x - 3$
18. $y = x^3 + 1$
19. $2x + 3y = 5$
20. $y = x^2 + 2$
21. $x = y^3 - 1$
22. $x = y^2 - 4$
23. $4x^2 + y = 3$
24. $y = \dfrac{1}{x^2 + 1}$
25. $xy = 3$

26. Find the value of k if the curve $y = 3x + k$ passes through the origin.

27. Find the value of k if the curve $kx^2 + y = 7$ passes through the point $(2, 3)$.

28. **(Temperature conversion)** The formula for changing from centigrade (Celsius) to Fahrenheit temperature readings is

$$F = \frac{9}{5}C + 32$$

(a) Draw a Cartesian coordinate system with vertical axis F and horizontal axis C, and sketch the graph of the above equation.
(b) If the temperature is 20°C, what is the Fahrenheit temperature?
(c) If the temperature is 70°F, what is the centigrade temperature?

29. **(Newton's laws of motion)** It follows from Newton's laws that the distance s traveled after t seconds by a body falling from rest is approximately given by

$$s = 16t^2$$

where t is in seconds and s is in feet. Draw a Cartesian coordinate system with vertical axis s and horizontal axis t. Sketch the graph of the above equation for $t \geq 0$.

30. **(Investment)** If \$10 is invested at an annual simple interest rate of 10%, then the value S of the investment after t years is

$$S = 10(1 + .1t)$$

Draw a Cartesian coordinate system with vertical axis S and horizontal axis t. Sketch the graph of the above equation for $t \geq 0$.

1.3 THE STRAIGHT LINE

In this section we discuss equations and graphs of straight lines. Among other things, you will learn to recognize those equations whose graphs are straight lines. This material has many important applications, which we explore in later sections.

To begin, consider a particle moving along a *nonvertical* line L from a point $P_1(x_1, y_1)$ to a point $P_2(x_2, y_2)$. As shown in Figure 1, the particle moves $y_2 - y_1$ units in the y direction as it travels $x_2 - x_1$ units in the x direction. Note that $x_2 - x_1$ can be positive or negative, while $y_2 - y_1$ can be positive, negative, or zero. The vertical change $y_2 - y_1$ is called the **rise** and the horizontal change $x_2 - x_1$ the **run** (Figure 1).

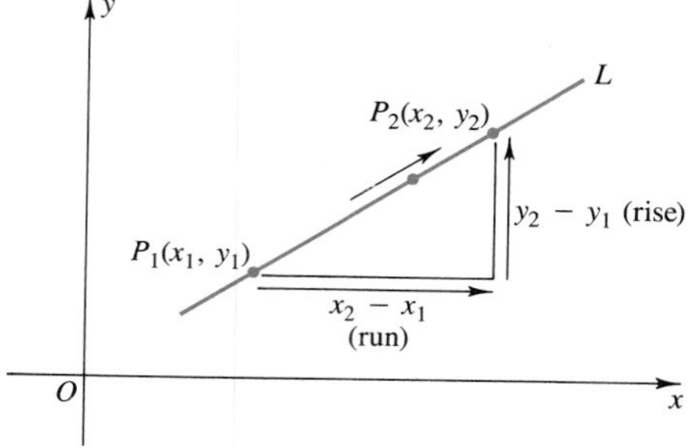

FIGURE 1

The ratio of the rise to the run is called the **slope** of the line and is denoted by m.

If $P_1(x_1, y_1)$ and $P_2(x_2, y_2)$ are distinct points on a nonvertical line, then the slope m of the line is given by

$$m = \frac{\text{rise}}{\text{run}} = \frac{y_2 - y_1}{x_2 - x_1} \tag{1}$$

EXAMPLE 1 Let L be the line determined by the points $P_1(4, 1)$ and $P_2(6, 9)$. From Formula (1), the slope of L is

$$m = \frac{9 - 1}{6 - 4} = 4$$

Similarly, the slope of the line determined by the points (1, 11) and (4, 5) can be obtained as follows: Letting P_1 be the point (1, 11) and P_2 the point (4, 5), it follows from Formula (1) that

$$m = \frac{5 - 11}{4 - 1} = \frac{-6}{3} = -2$$

Note that it does not matter which point is called P_1 and which is called P_2. Thus, had we chosen P_1 as (4, 5) and P_2 as (1, 11), we would have obtained

$$m = \frac{11 - 5}{1 - 4} = \frac{6}{-3} = -2$$

which agrees with the earlier result.

Remark We emphasize that Formula (1) does not apply to vertical lines. For such lines we would have $x_2 = x_1$. (Can you see why?) Thus the right-hand side of Formula (1) would have a zero denominator, which is not allowed. Lines parallel to the y-axis are said to have **undefined slope**.

Remark Using plane geometry, it can be shown that the value of m is the same no matter which two points P_1 and P_2 on a straight line L are used (see Exercise 27 of this section). Thus, to find the slope of a line L, we can apply Formula (1) to any two points on L.

EXAMPLE 2 Show that a line parallel to the x-axis has slope $m = 0$.

Solution If L is parallel to the x-axis, then all points on L have equal y-coordinates (Figure 2). Thus, if $P_1(x_1, y_1)$ and $P_2(x_2, y_2)$ are points on L, we must have $y_2 = y_1$. Therefore, m has a zero numerator and nonzero denominator in Formula (1). Thus, $m = 0$.

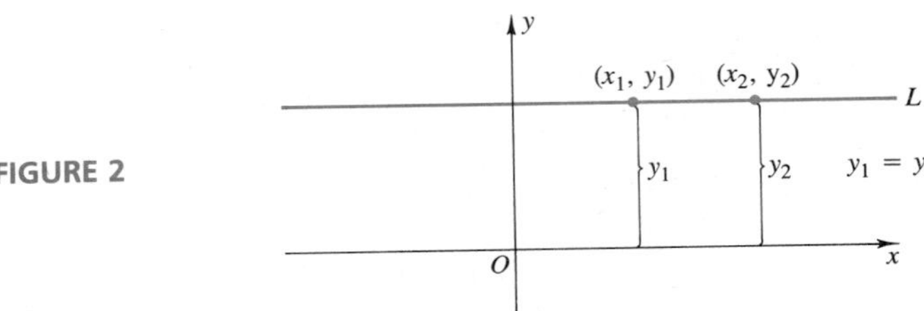

FIGURE 2

The slope of a line has a useful physical interpretation. If a particle moves left to right along a line L from a point $P_1(x_1, y_1)$ to a point $P_2(x_2, y_2)$, then it follows from Formula (1) that the rise and run are related by

$$y_2 - y_1 = m(x_2 - x_1) \tag{2}$$

Equation (2) states that the rise (or change in height) is proportional to the run (or change in horizontal distance), and the slope m is the constant of proportionality. For this reason, m is said to measure the **rate** at which y changes with x along the line L.

EXAMPLE 3 In Example 1, we show that the line determined by the points $P_1(4, 1)$ and $P_2(6, 9)$ has slope $m = 4$. This means that a particle traveling left to right along this line will gain 4 units in height for every unit it moves in the positive x direction (Figure 3).

Example 1 also shows that the line determined by the points $P_1(1, 11)$ and $P_2(4, 5)$ has slope $m = -2$. This means that a particle traveling left to right along this line will lose 2 units in height for every unit it moves in the positive x direction (Figure 4).

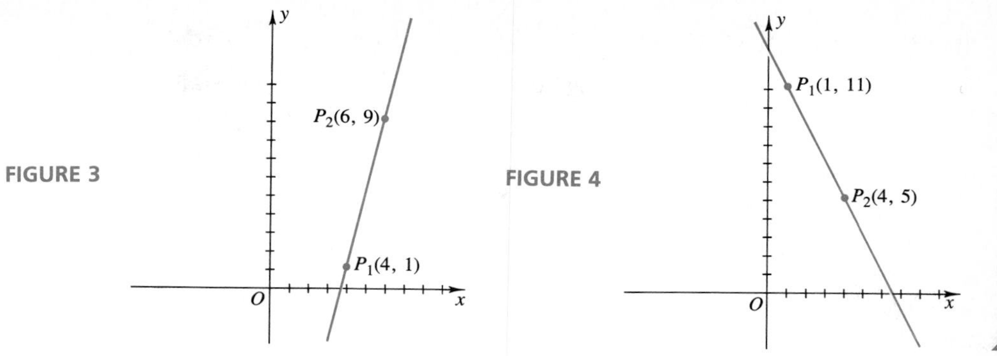

FIGURE 3

FIGURE 4

In Figure 5 we have sketched several lines with varying slopes. Observe that lines with positive slope are inclined upward to the right, whereas lines with negative slope are inclined downward to the right.

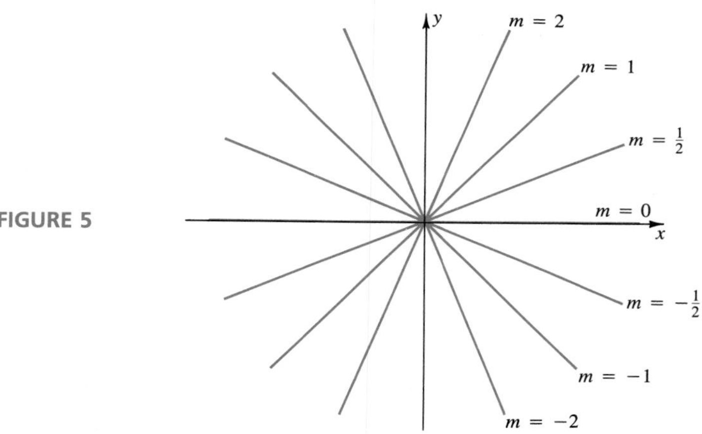

FIGURE 5

It is obvious from Figure 6 that for parallel lines, equal runs produce the same rise. This suggests the following result, which we state without proof.

Nonvertical parallel lines have the same slope, and, conversely, nonvertical lines with the same slope are parallel or coincident.

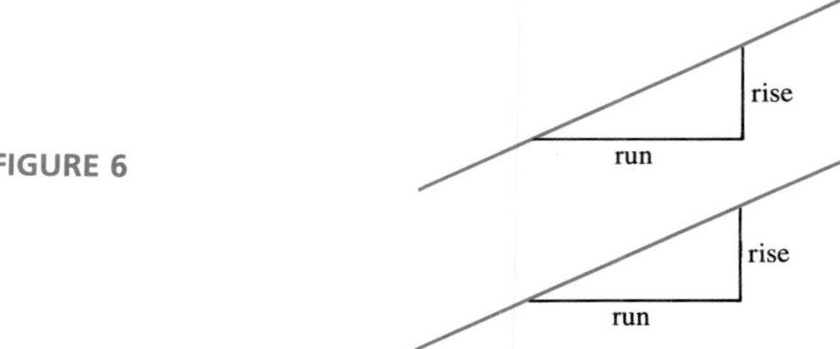

FIGURE 6

EXAMPLE 4 Let L be the line determined by the points $P_1(3, 4)$ and $P_2(4, 1)$, and let L' be the line determined by the points $P_1'(5, 6)$ and $P_2'(7, 0)$. Determine whether L and L' are parallel.

Solution The slope m of the line L is

$$m = \frac{1 - 4}{4 - 3} = -3$$

and the slope m' of the line L' is

$$m' = \frac{0 - 6}{7 - 5} = -3$$

Since the lines have the same slope, they are parallel. These lines are shown in Figure 7.

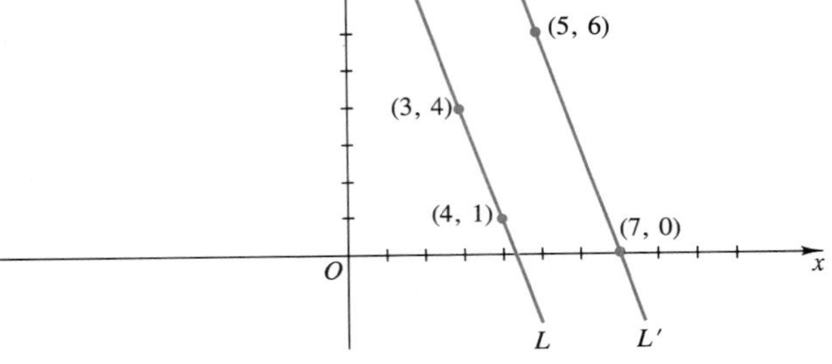

FIGURE 7

We now consider the problem of finding an equation for a line when we have been given information about the line. For example, a line can be specified by giving one of the following:

(a) its slope and any point on the line

(b) two points on the line

(c) its slope and the point where the line intersects the y-axis

First, let us consider the problem of finding an equation for the line passing through a given point $P_1(x_1, y_1)$ and having slope m. If (x, y) is any point on the line different from (x_1, y_1), then we can calculate the slope of the line from the points $P(x, y)$ and $P_1(x_1, y_1)$ (see Figure 8).

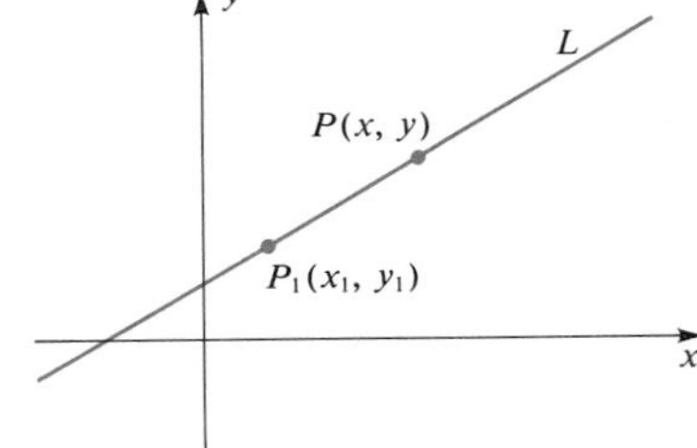

FIGURE 8

Using Formula (1), we obtain

$$m = \frac{y - y_1}{x - x_1}$$

If we multiply both sides of this equation by $x - x_1$, we obtain Equation (3), which is called the **point–slope** form of the line.

Point–Slope Form The equation of the line that has slope m and passes through the point (x_1, y_1) is

$$y - y_1 = m(x - x_1) \tag{3}$$

EXAMPLE 5 Find an equation for the line passing through the point $P_1(3, -2)$ and having slope $m = 5$.

Solution Substituting the values $x_1 = 3$, $y_1 = -2$, and $m = 5$ in the point–slope form (3) yields

$$y - (-2) = 5(x - 3) \quad \text{or} \quad y + 2 = 5(x - 3)$$

EXAMPLE 6 Find an equation for the line passing through the points $P_1(1, 3)$ and $P_2(2, 7)$.

Solution We can use the given points to calculate the slope m and then substitute this slope and the coordinates of either given point in Equation (3). The slope is

$$m = \frac{7 - 3}{2 - 1} = 4$$

Substituting this value for m and the coordinates of P_1 in Equation (3) yields

$$y - 3 = 4(x - 1)$$

Solving this equation for y we obtain

$$y = 4x - 1 \tag{4}$$

Consider now the problem of finding an equation for a line when we are given the slope and the point where the line intersects the y-axis.

EXAMPLE 7 Find an equation for the line with slope 2 that intersects the y-axis at the point (0, 5).

Solution Substituting 2 for m and (0, 5) for (x_1, y_1) in Equation (3), we obtain

$$y - 5 = 2(x - 0)$$

Solving for y we have

$$y = 2x + 5$$

More generally, if (x, y) is any point except $(0, b)$ on the line that has slope m and intersects the y-axis at $(0, b)$, then from Equation (3) we obtain

$$y - b = m(x - 0)$$

Solving for y we have

$$y = mx + b$$

called the **slope–intercept form** of an equation for the line. Note that b is the y-intercept of the line.

Slope–Intercept Form The equation of the line that has slope m and intersects the y-axis at the point $(0, b)$ is

$$y = mx + b \tag{5}$$

As an illustration, the slope–intercept form of the line that has slope 3 and intersects the y-axis at $(0, -2)$ can be obtained by substituting $m = 3$ and $b = -2$ in Equation (5). This yields $y = 3x - 2$.

EXAMPLE 8 Find the slope of the line

$$y = 7x + 8$$

Solution This is the slope–intercept form of a line with $m = 7$ and $b = 8$. Thus, the line has slope $m = 7$ and y-intercept equal to 8.

Remark A nonvertical line has infinitely many different point–slope forms because the coordinates of *any* point on the line can be substituted for x_1 and y_1 in Equation (3). However, such a line has only *one* slope–intercept form. As an illustration, in Example 6, we could have substituted the coordinates of $P_2(2, 7)$ in Equation (3) rather than the coordinates of $P_1(1, 3)$, in which case we would obtain $y - 7 = 4(x - 2)$ as the point–slope form, which is different from the point–slope form $y - 3 = 4(x - 1)$ obtained in this example. However, if we solve this equation for y, we obtain $y = 4x - 1$ (verify this), which is the slope–intercept form we obtain when we solve for y [Equation (4)].

EXAMPLE 9 Find an equation for the line parallel to the x-axis and 3 units above it.

Solution As shown in Figure 9, the line intersects the y-axis at the point $(0, 3)$. Also, as in Example 2, a line parallel to the x-axis has slope $m = 0$. Thus, substituting the values $x_1 = 0$, $y_1 = 3$, and $m = 0$ in the point–slope form (3) yields

$$y - 3 = 0(x - 0)$$
$$y = 3$$

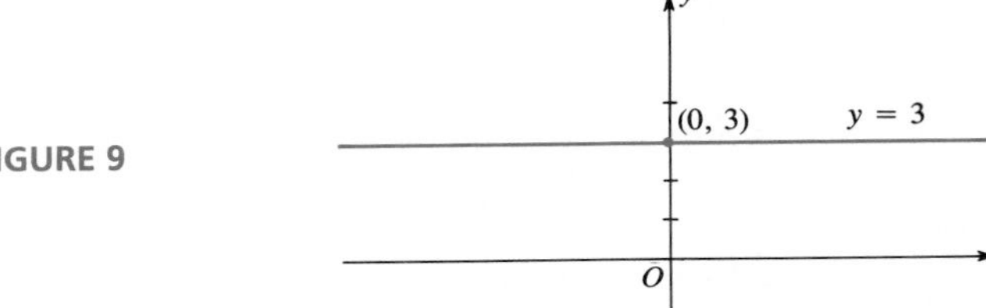

FIGURE 9

This result makes sense since the horizontal line through (0, 3) consists precisely of those points (x, y) with a y-coordinate of 3.

Example 9 is a special case of the following general result.

Horizontal Lines The horizontal line crossing the y-axis at $(0, b)$ has the equation

$$y = b$$

A corresponding result holds for vertical lines.

Vertical Lines The vertical line crossing the x-axis at $(a, 0)$ has the equation

$$x = a$$

For example, the line in Figure 10 represents the equation $x = 4$.

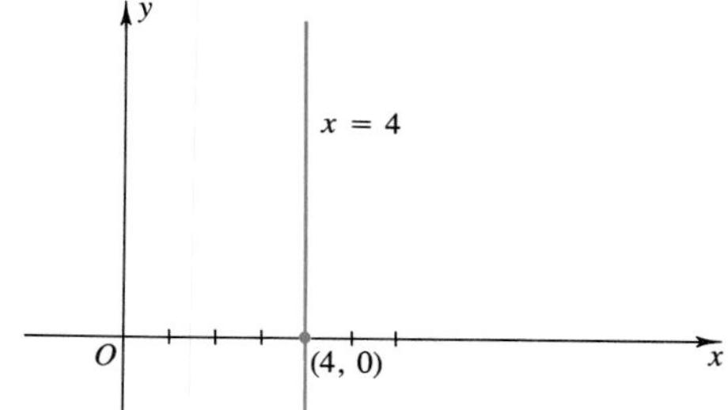

FIGURE 10

The next result shows us how to recognize when a given equation has a straight line as its graph.

Any equation that can be expressed in the form

$$Ax + By = C \tag{6}$$

where A, B, and C are constants and A and B are not both zero, has a straight line as its graph.

To see this, we distinguish between two cases, $B = 0$ and $B \neq 0$.

If $B \neq 0$, we can solve (6) for y in terms of x to obtain

$$y = -\frac{A}{B}x + \frac{C}{B}$$

But this is the slope–intercept form $y = mx + b$ of the line with

$$m = -\frac{A}{B} \quad \text{and} \quad b = \frac{C}{B}$$

If $B = 0$, then A cannot be zero; therefore, Equation (6) reduces to $Ax = C$, which can be written as

$$x = \frac{C}{A}$$

But this is the equation of a vertical line. Thus, in either case $Ax + By = C$ (where A and B are not both zero) has a straight line for its graph. For this reason, an equation that is expressible in this form is called a **linear equation** in x and y.

EXAMPLE 10 The graph of the equation

$$x = 7y + 8$$

is a line since this equation can be rewritten as $x - 7y = 8$, which is of form (6) with $A = 1$, $B = -7$, and $C = 8$. Solving the given equation for y, we obtain

$$y = \frac{1}{7}x - \frac{8}{7}$$

So the slope of the line is $\frac{1}{7}$ and its y-intercept is $-\frac{8}{7}$.

EXAMPLE 11 Sketch the graph of the equation

$$x + 4y - 8 = 0$$

and find its slope.

Solution The graph is a line since the equation can be rewritten as $x + 4y = 8$, which is of form (6). Thus, to sketch the graph we need only plot two points and draw the line through them. We will plot the intersections with the coordinate axes. The line crosses the y-axis where $x = 0$ and crosses the x-axis where $y = 0$. Substituting $x = 0$ in the given equation and solving for y yields $y = 2$, so $(0, 2)$ is the intersection with the y-axis. Substituting $y = 0$ in the given equation and solving for x yields $x = 8$, so $(8, 0)$ is the intersection with the x-axis. The graph is shown in Figure 11. Solving the given equation for y, we obtain

$$y = -\frac{1}{4}x + 2$$

so the slope of the line is $-\frac{1}{4}$.

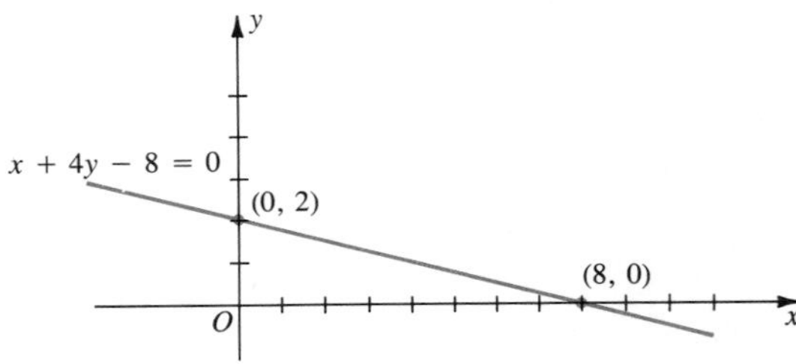

FIGURE 11

Linear equations may involve variables other than x and y. The procedure for graphing such equations is the same as in Example 11, except that the coordinate axes must be labeled appropriately.

EXAMPLE 12 Sketch the graph of

$$2r + 3s = 6$$

Solution This is of form (6), except that the variables are r and s rather than x and y. If we choose a horizontal r-axis and vertical s-axis, then the intersection of the graph with the r-axis is (3, 0) and with the s-axis is (0, 2). (Verify this.) The graph is shown in Figure 12.

FIGURE 12

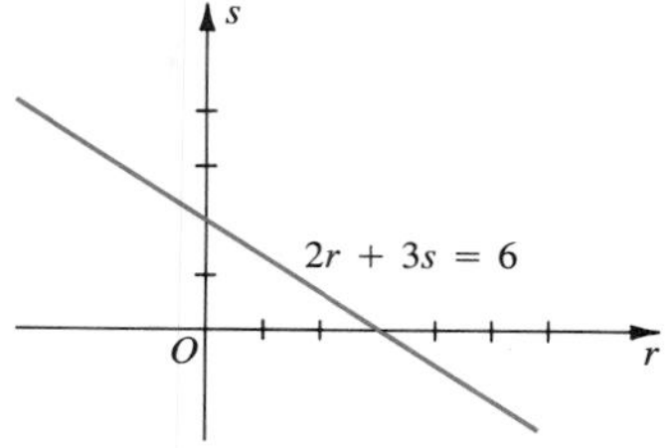

Remark In Example 12 we could have chosen a horizontal s-axis and vertical r-axis—the choice is arbitrary. However, the graph would not be the same since the r and s would be interchanged.

EXERCISE SET 1.3

1. In each part find the slope of the line determined by the points P_1 and P_2.
(a) $P_1(2, 4)$, $P_2(-3, 5)$
(b) $P_1(3, 2)$, $P_2(5, 6)$
(c) $P_1(0, 0)$, $P_2(-3, -2)$
(d) $P_1(4, 2)$, $P_2(2, 2)$
(e) $P_1(1.846, 2.772)$, $P_2(0.927, 2.253)$

2. Suppose that a certain line has slope $m = 3$.
(a) Find the run if the rise is 2 units.
(b) Find the rise if the run is 5 units.

3. Suppose that a certain line has slope $m = -2$.
(a) Find the run if the rise is -3 units.
(b) Find the rise if the run is 4 units.

4. Which of the following lines are parallel to the line $y = 5x - 3$?
(a) $y = 5x + 6$ (b) $y + 2 = 5(x - 3)$
(c) $10x - 2y + 12 = 0$ (d) $y = 5$

5. Given that the line $y = mx + b$ is parallel to the line $4x - 2y + 7 = 0$, find m.

6. Find an equation for
(a) the vertical line through (1, 3)
(b) the horizontal line through (1, 3)

7. In each part let L be the line determined by the points P_1 and P_2, and let L' be the line determined by the points P_1' and P_2'. Decide whether L and L' are parallel and sketch the lines.
(a) $P_1(4, 3)$, $P_2(1, 4)$, $P_1'(6, 5)$, $P_2'(0, 7)$
(b) $P_1(2, 3)$, $P_2(3, 2)$, $P_1'(5, 5)$, $P_2'(2, 2)$
(c) $P_1(0, 0)$, $P_2(2, 3)$, $P_1'(4, 6)$, $P_2'(0, 0)$

8. In each part sketch the line with slope m that passes through the point P.
(a) $m = 1$, $P(2, 3)$
(b) $m = -\frac{2}{3}$, $P(0, 2)$
(c) $m = -3$, $P(3, 3)$
(d) $m = 0$, $P(-2, 3)$

9. In each part use the given information to find the slope–intercept form of the line.
(a) $m = 2$; the point (0, 4) lies on the line
(b) $m = -\frac{4}{3}$; the point (0, −2) lies on the line
(c) $m = 0$; the point (2, 3) lies on the line

10. Find the slope of the given line.
(a) $y = 2x + 3$ (b) $y = -\frac{3}{2}x + 5$
(c) $y = 7$ (d) $x = 2$
(e) $2x + 3y = 2$ (f) $x = \frac{1}{2}y + 5$

11. In each part use the given information to find the point–slope form of the line.
(a) $m = 1, P(2, 3)$ (b) $m = -\frac{2}{3}, P(0, 2)$
(c) $m = -3, P(3, 3)$ (d) $m = 0, P(-2, 3)$
(e) $m = 6, P(2, -4)$ (f) $m = -4, P(\frac{1}{2}, \frac{2}{3})$

12. In each part find the slope–intercept form of the line passing through the points P_1 and P_2.
(a) $P_1(4, 2), P_2(2, 2)$
(b) $P_1(-3, -2), P_2(0, 0)$
(c) $P_1(-3, 5), P_2(2, 4)$
(d) $P_1(5, 6), P_2(3, 2)$

13. Find the point–slope form of the line passing through the points P_1 and P_2.
(a) $P_1(2, 3), P_2(-3, 4)$
(b) $P_1(\frac{3}{2}, 1), P_2(2, -\frac{2}{3})$

14. (a) Find an equation for the line parallel to the x-axis and 2 units below it.
(b) Find an equation for the line parallel to the y-axis and 3 units to the left of it.

15. Find an equation for the line in each part of Figure 13.

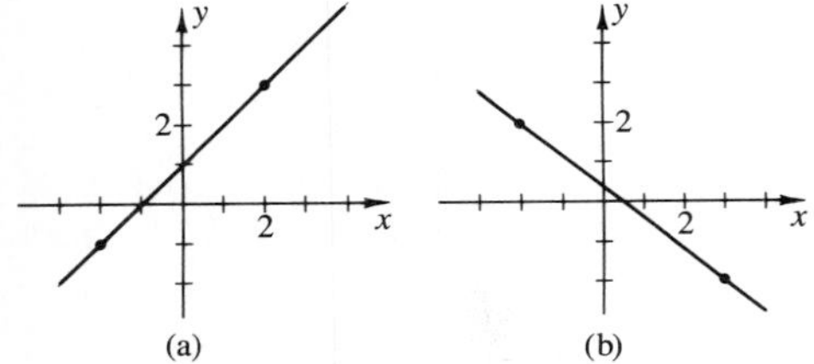

FIGURE 13

In Exercises 16–18 find an equation for the line satisfying the given conditions.

16. the line with slope 3 and x-intercept -5

17. the line with x-intercept 2 and y-intercept 4

18. the vertical line that passes through the point $(-1, 3)$

19. Express each of the following equations in the form $Ax + By = C$ and give the values of A, B, and C.
(a) $x = \frac{1}{3}y - 7$ (b) $y = 2$
(c) $y = 7 + 4(x - 3)$ (d) $x = 3(y - 2)$

20. Which of the following are linear equations?
(a) $2x + y^2 = 3$
(b) $xy = 1$
(c) $2x - 3y + 2 = 0$
(d) $2(x - 1) + 3(y - 1) = 4$

21. Graph the following equations.
(a) $3x + 2y = 6$ (b) $4x - 8y = 16$
(c) $x - 5y = 10$ (d) $6x + y = 12$

22. Graph the following equations.
(a) $y = 7x + 3$ (b) $y - 4 = 3(x + 2)$
(c) $y + 1 = 2(x - 1)$ (d) $x = 5(y - 2)$

23. Graph the following equations in an appropriate coordinate system.
(a) $3r - 2s = 12$ (Use a horizontal r-axis.)
(b) $8s + 4w = 16$ (Use a horizontal s-axis.)
(c) $v = 7t + 1$ (Use a horizontal t-axis.)

24. Find the slopes of the sides of the triangle with vertices $(2, 3)$, $(3, -4)$, and $(0, 3)$.

25. Find the equation of the line passing through the point $(3, 4)$ and parallel to the line $y = 2x + 3$.

26. In each part use slopes to determine whether the given three points lie on the same line.
(a) $(1, 5), (-2, -1), (\frac{1}{2}, 4)$
(b) $(0, 0), (3, 2), (-4, 3)$

27. Using plane geometry, show that the value of the slope m of a line L is the same no matter which two points are used in Equation (1). [*Hint:* Use similar triangles.]

28. Show that $(0, 2)$, $(2, 0)$, $(8, 5)$, and $(10, 3)$ are vertices of a parallelogram.

29. A company determines that the graph of the profit P (in dollars) versus advertising expenditure x (in dollars) is a straight line. Find an equation for this line if $P = 5000$ when $x = 0$, and $P = 7000$ when $x = 1000$.

1.4 SYSTEMS OF TWO LINEAR EQUATIONS IN TWO UNKNOWNS

In many applied problems two linear equations are given, and it is of interest to find values of the unknowns that satisfy *both* of the equations. In this section we study this problem, and in the next section we give some applications. In Chapter 3 we consider generalizations of this work.

If we are given two linear equations

$$\begin{aligned} A_1x + B_1y &= C_1 \\ A_2x + B_2y &= C_2 \end{aligned}$$

(where the A's, B's, and C's are constants), then we can seek values of x and y that satisfy both equations. The two equations are then said to form a **system** of two linear equations, the variables x and y are called the **unknowns**, and an ordered pair of real numbers (a, b) is called a **solution of the system** if both equations are satisfied when we substitute

$$x = a \quad \text{and} \quad y = b$$

EXAMPLE 1 Consider the system

$$\begin{aligned} 2x + 3y &= 8 \\ 3x - y &= 1 \end{aligned}$$

The ordered pair (1, 2) is a solution of the system since both equations are satisfied when we substitute $x = 1$ and $y = 2$. However, the ordered pair $(-2, 4)$ is not a solution since only the first equation is satisfied when we substitute $x = -2$ and $y = 4$.

Not all systems of linear equations have solutions. For example, consider the following system of equations.

$$\begin{aligned} x + y &= 2 \\ 3x + 3y &= 9 \end{aligned}$$

If we multiply both sides of the second equation in the system by $\frac{1}{3}$, it becomes evident that there are no solutions since the two equations in the resulting system

$$\begin{aligned} x + y &= 2 \\ x + y &= 3 \end{aligned}$$

contradict each other. Note that the lines are parallel.

To illustrate the possibilities that can occur in solving systems of linear equations, consider a general system of two linear equations in the unknowns x and y:

$$\begin{aligned} A_1x + B_1y &= C_1 \\ A_2x + B_2y &= C_2 \end{aligned}$$

The graphs of these equations are lines, call them l_1 and l_2. Since a point (a, b) lies on a line if and only if the numbers $x = a$, $y = b$ satisfy the equation of the line, the solutions of the system will correspond to points of intersection of l_1 and l_2. There are three possibilities (Figure 1).

FIGURE 1

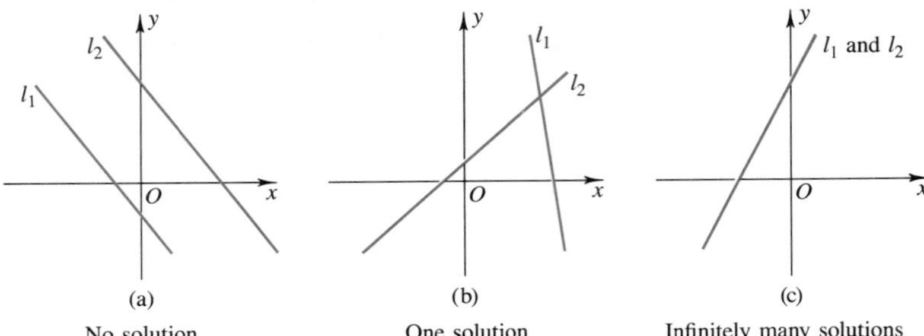

(a) No solution

(b) One solution

(c) Infinitely many solutions

(a) The lines l_1 and l_2 can be parallel, in which case there is no intersection and consequently no solution of the system.

(b) The lines l_1 and l_2 can intersect at only one point, in which case the system has exactly one solution.

(c) The lines l_1 and l_2 can coincide, in which case there are infinitely many solutions of the system.

To summarize, a system of two linear equations in x and y has either no solutions, exactly one solution, or infinitely many solutions.

There are two elementary algebraic methods for solving systems of two linear equations in two unknowns, **substitution** and **elimination**. We will only consider the method of elimination, since it will be generalized in Chapter 2 to solve systems with more equations and unknowns.

In this method we replace the given system of equations with a new system that has the same set of solutions but is easier to solve. The following example illustrates the method.

EXAMPLE 2 Solve the following system by elimination.

$$\begin{aligned} 4x + y &= 5 \\ 2x - 3y &= 13 \end{aligned} \tag{1}$$

Solution The variable y can be eliminated by multiplying the top equation in (1) by 3 and adding the result to the bottom equation:

$$\begin{array}{r} 12x + 3y = 15 \\ 2x - 3y = 13 \\ \hline 14x \qquad = 28 \end{array}$$

(That is, we have multiplied $4x + y = 5$ by 3 so that the coefficients of y in both equations will have the same numerical value but be of opposite signs.) Solving for x yields $x = 2$. To find y we substitute this value into either equation in (1), for example, the first. This yields

$$\begin{aligned} 8 + y &= 5 \\ y &= -3 \end{aligned}$$

Thus, the solution is $x = 2$, $y = -3$. To check the solution we substitute $x = 2$ and $y = -3$ into each of the original equations. The solution of system (1) can be seen geometrically by graphing the equations in the system (Figure 2). The x- and y-values of the solution are the x- and y-coordinates of the point of intersection.

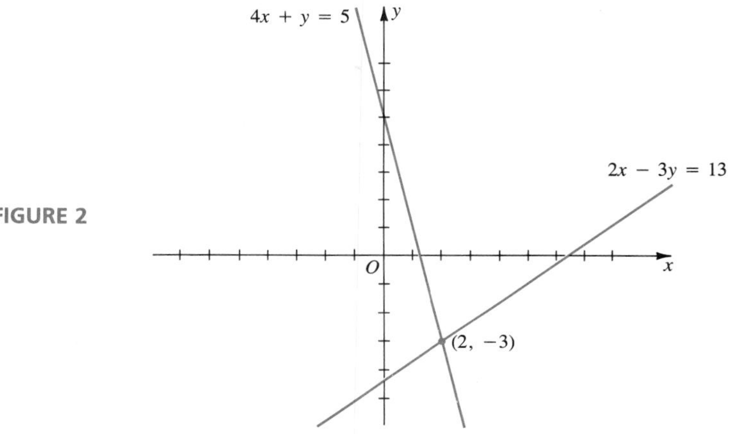

FIGURE 2

The general method of elimination can be stated as follows:

Solution by Elimination

Step 1 Choose the unknown to be eliminated.

Step 2 Multiply each equation (both sides) by an appropriate constant so that the coefficients of the unknowns to be eliminated are numerically equal but of opposite signs.

Step 3 Add the equations to eliminate this unknown.

Step 4 Solve for the remaining unknown and substitute this value into either equation to obtain the other unknown.

EXAMPLE 3 Solve the following system by elimination:

$$\begin{aligned} 3x + 2y &= -12 \\ 2x - 3y &= 5 \end{aligned} \qquad (2)$$

Solution We can eliminate the variable x if we multiply the top equation by 2, the bottom by -3, and add:

$$\begin{aligned} 6x + 4y &= -24 \\ -6x + 9y &= -15 \\ \hline 13y &= -39 \end{aligned}$$

Solving for y yields $y = -3$. To find x we substitute this value into either equation in (2), for example, the first. This yields

$$3x + 2(-3) = -12$$

or

$$3x = -6$$

or

$$x = -2$$

Thus, the solution is $x = -2$, $y = -3$.

The next two examples illustrate what occurs when the elimination method is applied to systems with no solutions or infinitely many solutions.

EXAMPLE 4 Solve the following system by elimination:

$$\begin{aligned} x - 3y &= -1 \\ 2x - 6y &= 6 \end{aligned}$$

Solution We can eliminate x by multiplying the top equation by -2 and adding:

$$\begin{array}{r} -2x + 6y = 2 \\ 2x - 6y = 6 \\ \hline 0x + 0y = 8 \end{array}$$

Since the last equation is not satisfied by any values of x and y, the given system has no solution. This indicates that the lines are parallel.

EXAMPLE 5 Solve the following system by elimination:

$$\begin{aligned} x + 3y &= -1 \\ 2x + 6y &= -2 \end{aligned} \tag{3}$$

Solution We can eliminate x by multiplying the top equation by -2 and adding:

$$\begin{array}{r} -2x - 6y = 2 \\ 2x + 6y = -2 \\ \hline 0x + 0y = 0 \end{array}$$

Note that the second equation in (3) is twice the first. It follows that any solution of the first equation in (3) will satisfy the second automatically, and so the solutions of the system consist of all values of x and y that satisfy the first equation

$$x + 3y = -1 \tag{4}$$

The system thus has infinitely many solutions. They can be obtained from (4) by assigning arbitrary values to either variable and solving for the other variable. For example, substituting $y = 3$ in (4) and solving for x yields the solution $x = -10$ and $y = 3$, and substituting $y = -2$ yields the solution $x = 5$ and $y = -2$.

Remark You should be able to see from this example that when one linear equation is a multiple of another, they have the same solutions. Geometrically, the graphs of the equations coincide.

EXAMPLE 6 A bag contains 6 coins consisting of dimes and nickels. If the total value of the coins is 50 cents, how many coins of each type are in the bag?

Solution By letting d = the number of dimes and n = the number of nickels, we obtain a linear system:

$$\begin{aligned} d + n &= 6 \qquad \text{(there are 6 coins)} \\ 10d + 5n &= 50 \qquad \text{(the total value is 50 cents)} \end{aligned} \tag{5}$$

Multiply the first equation of (5) by -10 to obtain

$$\begin{aligned} -10d - 10n &= -60 \\ 10d + 5n &= 50 \end{aligned}$$

Adding, we eliminate d and obtain

$$-5n = -10$$

or

$$n = 2$$

Substituting $n = 2$ into the first equation of (5) yields $d = 4$, so the bag contains 4 dimes and 2 nickels. We can check our answer by returning to the verbal statement of the problem and observing that the 6 coins, 4 dimes and 2 nickels, do yield a total value of 50 cents.

EXERCISE SET 1.4

1. Which of the following ordered pairs are solutions of the system

$$\begin{aligned} 3x + 2y &= 1 \\ 2x - y &= -4 \end{aligned}$$

(a) $(-1, 2)$ (b) $(-1, -2)$
(c) $(0, \frac{1}{2})$ (d) $(1, -1)$

2. Consider the system

$$\begin{aligned} x + 2y &= 5 \\ 3x - y &= 1 \end{aligned}$$

(a) Solve the system by elimination.
(b) Graph the equations and find the coordinates of the point of intersection.

3. Follow the directions of Exercise 2 for the system

$$\begin{aligned} 3x - 2y &= -10 \\ 2x - y &= -6 \end{aligned}$$

4. Solve the following systems by elimination.

(a) $\begin{aligned} x + 4y &= -8 \\ 3x - 2y &= 4 \end{aligned}$ (b) $\begin{aligned} 5x + y &= 0 \\ -3x - 2y &= -7 \end{aligned}$

(c) $\begin{aligned} 2x + 3y &= 7 \\ 3x - 2y &= -9 \end{aligned}$ (d) $\begin{aligned} 3x + 5y &= -5 \\ 2x + 4y &= -2 \end{aligned}$

5. For each system in Exercise 4, graph the equations and find the coordinates of the point of intersection.

6. Solve the following systems by elimination.

(a) $\begin{aligned} 4x + y &= -8 \\ -2x + 3y &= 4 \end{aligned}$ (b) $\begin{aligned} x + 5y &= 0 \\ 2x + 3y &= 7 \end{aligned}$

(c) $\begin{aligned} 2x - 3y &= 9 \\ 3x + 2y &= 7 \end{aligned}$ (d) $\begin{aligned} 5x + 3y &= -5 \\ 4x + 2y &= -2 \end{aligned}$

7. For each system in Exercise 6, graph the equations and find the coordinates of the point of intersection.

8. Solve the following systems.

(a) $\begin{aligned} 2x + 3y &= -5 \\ 3x + 2y &= 12 \end{aligned}$ (b) $\begin{aligned} 4x + 3y &= 6 \\ 3x - 4y &= -2 \end{aligned}$

(c) $\begin{aligned} 3x - 2y &= 2 \\ 2y &= -1 \end{aligned}$ (d) $\begin{aligned} -2x - 4y &= 3 \\ 5x - 3y &= -2 \end{aligned}$

9. Solve the following systems.

(a) $\begin{aligned} 2x + y &= -3 \\ 3x + 2y &= -3 \end{aligned}$ (b) $\begin{aligned} 2x - 7y &= 10 \\ 3x + 2y &= 15 \end{aligned}$

(c) $\begin{aligned} 3x - 4y &= -1 \\ -2x + 3y &= 2 \end{aligned}$ (d) $\begin{aligned} \tfrac{1}{4}x - \tfrac{1}{2}y &= 0 \\ \tfrac{3}{4}x + \tfrac{5}{2}y &= 8 \end{aligned}$

10. Solve each of the following systems geometrically by graphing the equations in the system and determining their point of intersection. Check your results by substituting the solution into the equations in the system.

(a) $2x + 3y = 6$, $3x - 4y = -8$ (b) $-3x + 5y = 1$, $2x + 8y = 5$

11. (a) Show that the following system has no solutions.

$$3x - y = 5$$
$$6x - 2y = 3$$

(b) What does the result in (a) tell you about the graphs of the equations in the system?
(c) Graph these equations.

12. (a) Show that the following system has infinitely many solutions.

$$3x - y = 2$$
$$6x - 2y = 4$$

(b) What does the result in (a) tell you about the graphs of the equations in the system?
(c) Graph these equations.
(d) Find a solution for which $x = 3$ and another for which $y = -2$.

13. Solve the following systems, if possible.

(a) $2x + 3y = 4$, $4x + 6y = 2$ (b) $2x - 3y = 5$, $4x - 6y = 10$
(c) $-3x + 4y = 8$, $-\frac{3}{2}x + 2y = 4$ (d) $0.2x + 0.5y = 0.8$, $0.4x + y = 1$

14. Solve the following systems.

(a) $5s + 8t = -3$, $2s - 3t = 5$ (b) $-V + 4W = -6$, $2V + 3W = 1$

15. **(Harder)** Solve the following system for x and y, where a, b, c, and d are real numbers such that $ad - bc \neq 0$.

$$ax + by = e$$
$$cx + dy = f$$

16. **(Harder)** Show that the following system has infinitely many solutions or no solutions if a, b, c, and d are real numbers such that $ad - bc = 0$.

$$ax + by = e$$
$$cx + dy = f$$

17. A bag contains 8 coins consisting of dimes and nickels. If the total value of the coins is 55 cents, how many of each type are in the bag?

18. A health food shop mixes nuts and raisins in a snack pack. How many pounds of nuts selling for \$2.00 per pound and how many pounds of raisins selling for \$1.50 per pound must be mixed to produce a 50-pound mixture selling for \$1.80 per pound?

19. **(Investment)** A fund is planning to invest \$6000 in two types of bonds, A and B. Bond A pays an annual dividend of 8% and B an annual dividend of 10%. How much should be invested in each bond to obtain an annual dividend of \$520?

1.5 APPLICATIONS OF LINEAR EQUATIONS

Simple Interest

Interest is the fee charged for borrowing money. The fee is called **simple interest** if the fee is a fixed percentage per year of the amount borrowed. For example, if one borrows money at a simple interest rate of 6% per year, then for each year the borrowed money is held, the interest fee is 6% of the amount borrowed. The sum borrowed is called the **principal**. We now obtain a formula for the amount that would be owed if a principal P is borrowed for t years at a simple interest rate r. Assume that the annual interest rate r is expressed as a decimal. For example, for 6% interest per year, the value of r would be $r = .06$. By definition, the interest due each year the money is held is

$$Pr$$

so that if the money is held for t years, the total interest due is

$$Prt$$

Consequently, if S is the amount (in dollars) owed after t years, then

$$S = P + Prt \tag{1}$$

since we must repay the borrowed money (principal) and the interest.

Formula (1) is often rewritten in factored form as

$$S = P(1 + rt)$$

EXAMPLE 1 If we borrow \$8000 at the simple interest rate of 8% per year, then from Equation (1) with $P = 8000$ and $r = .08$, the amount S owed after t years would be

$$S = 8000 + (8000)(.08)t$$

or

$$S = 8000 + 640t \qquad t \geq 0 \tag{2}$$

Thus, the amount owed after 20 years ($t = 20$) would be

$$S = 8000 + 640(20) = 20{,}800 \qquad \text{(dollars)}$$

and the amount owed after three months ($t = \frac{3}{12} = \frac{1}{4}$ years) would be

$$S = 8000 + 640\left(\frac{1}{4}\right) = 8160 \qquad \text{(dollars)}$$

Observe that Equation (2) is a linear equation, where the variables are t and S rather than x and y. See Figure 1 for the graph of Equation (2). The graph, which is a straight line, shows how the amount S changes with time. The slope of this line is 640 (dollars per year), which is the constant annual interest amount. See Chapter 9 for additional material on interest.

FIGURE 1

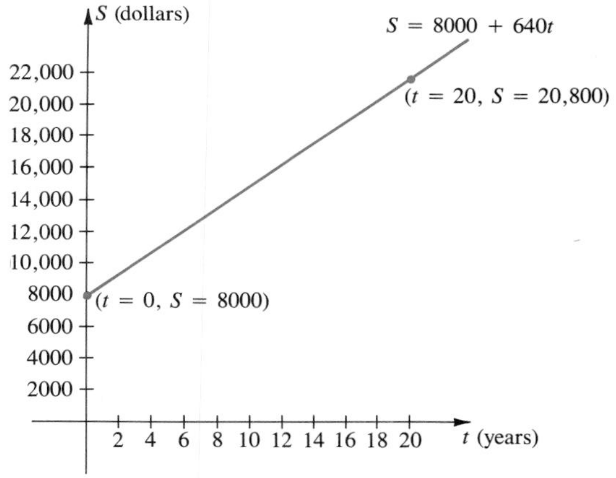

Linear Depreciation

Some property loses all or part of its value over a period of time. The decrease in value is called the **depreciation** of the property. For example, if a used car is purchased for \$2500 and 2 years later its resale value is \$1400, then the depreciation of the car after 2 years is

$$\$2500 - \$1400 = \$1100$$

The depreciation is called **linear depreciation** or **straight-line depreciation** if the loss in value is a fixed percentage per year of the original value. For example, if an item depreciates linearly at the rate of 10% per year, then for each year the item is held, its loss in value is 10% of the original value.

Assume the original cost of a certain item is C dollars and that the item depreciates linearly in value at an annual rate of r. For example, if the depreciation is 10% per year, the value of r would be $r = .10$. By definition, the value lost each year the item is owned is Cr, so if the item is owned for t years, the total loss in value is Crt. Consequently, if V is the value of the item after t years, then

$$V = C - Crt \tag{3}$$

since V is the original cost less the loss in value.

Formula (3) is often written in factored form as

$$V = C(1 - rt)$$

EXAMPLE 2 If we purchase an item for \$6000 and if the item depreciates linearly at the rate of 10% per year, then from (3) with $C = 6000$ and $r = .10$, the value V of this item after t years would be

$$V = 6000 - (6000)(.10)t$$

or

$$V = 6000 - 600t \qquad t \geq 0 \tag{4}$$

Thus, the value of the item after 3 years ($t = 3$) would be

$$V = 6000 - 600(3) = 4200 \qquad \text{(dollars)}$$

and the value of the item after 2 months ($t = \frac{2}{12} = \frac{1}{6}$ years) would be

$$V = 6000 - 600\left(\frac{1}{6}\right) = 5900 \qquad \text{(dollars)}$$

Observe that Equation (4) is a linear equation, where the variables are t and V rather than x and y. See Figure 2 for a graph of Equation (4). The graph, which is a straight line, shows how the value V changes with time. The slope of this line is -600 (dollars per year). Thus the constant yearly loss in value of the item is \$600.

FIGURE 2

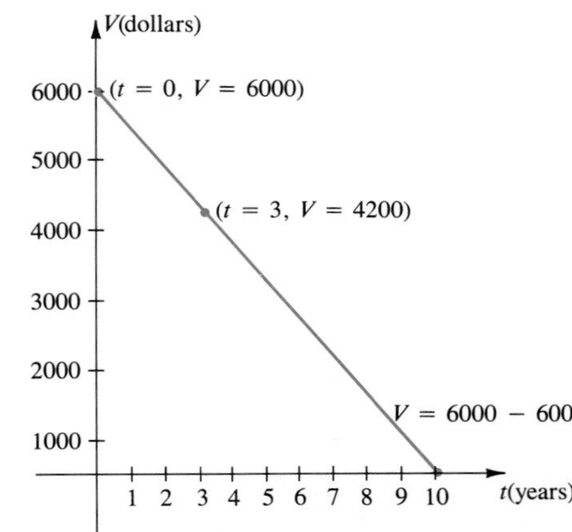

Prediction Linear equations can sometimes be used to predict future results based on past or current data. The following example illustrates this idea.

EXAMPLE 3 **WATER POLLUTION** Suppose that tests performed at the beginning of 1986 showed that each 1000 liters of water in Lake Michigan contained 6 milligrams of polluting mercury compounds, and similar tests performed at the beginning of 1988 showed that each 1000 liters of water contained 8 milligrams of polluting mercury compounds. Assuming that the quantity of polluting mercury compounds will continue to increase at a constant rate, predict the number of milligrams of polluting mercury compounds per 1000 liters of water that will be present at the beginning of 1996.

Solution Let y denote the number of milligrams of polluting mercury compounds per 1000 liters of water and let x denote time (in years). We are interested in finding the value of y when $x = 1996$. We know the values of y when $x = 1986$ and when $x = 1988$.

Since we are assuming that y will grow at a constant rate, the graph of y versus x will be a straight line. Moreover, from the given data, this line will pass through the points

$$(1986, 6) \quad \text{and} \quad (1988, 8)$$

(Figure 3). From the given data, the slope of the line is

$$m = \frac{8 - 6}{1988 - 1986} = \frac{2}{2} = 1$$

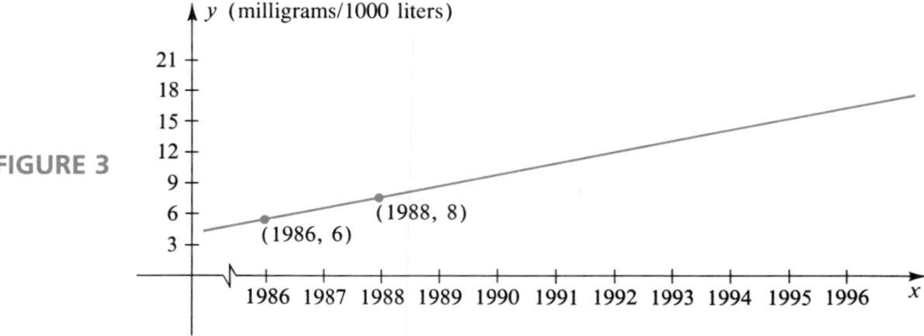

FIGURE 3

Using this value of m and the point (1986, 6) together with the point–slope form of the line [Equation (3) of Section 1.3], we obtain

$$y - 6 = 1(x - 1986)$$

or

$$y = x - 1980 \tag{5}$$

The value of y at the beginning of the year 1996 can now be obtained by substituting the value $x = 1996$ in (5). This yields

$$y = 1996 - 1980 = 16$$

Thus, we would predict that at the beginning of 1996, Lake Michigan will contain 16 milligrams of polluting mercury compounds per 1000 liters.

We emphasize that the result obtained in this example depends on the assumption that the polluting mercury compounds will continue to increase at the same rate. Should it turn out that this assumption is not correct, then the predicted results would, of course, not be accurate.

Trends in Data (Method of Least Squares)

In many physical problems we are interested in determining relationships between two variable quantities x and y. One method of obtaining such relationships consists of gathering data that show the values of y that are associated with various values of x and then plotting the points (x, y) on graph paper. We hope that after examining the resulting graph, we can describe how the variables are related. To illustrate this idea, let us consider the following example.

EXAMPLE 4 **ASTRONAUT REACTION TIME** In an experiment designed to test the effect of weightlessness on an astronaut's reaction time, an astronaut is required to respond to a flashing light by depressing a button. The time elapsed between the light flash and the depression of the button is denoted by y, and the length of time that the astronaut is weightless is denoted by x. The resulting data appears in Table 1. The points in this table are plotted in Figure 4a.

TABLE 1

x (hours)	y (seconds)	x (hours)	y (seconds)
5	0.70	10	1.42
6	1.08	11	1.47
7	1.02	12	1.85
8	1.26	13	2.02
9	1.62	14	1.93

FIGURE 4

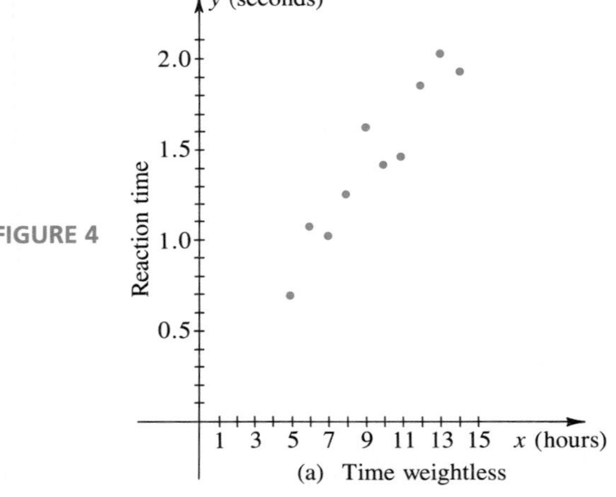

(a) Time weightless

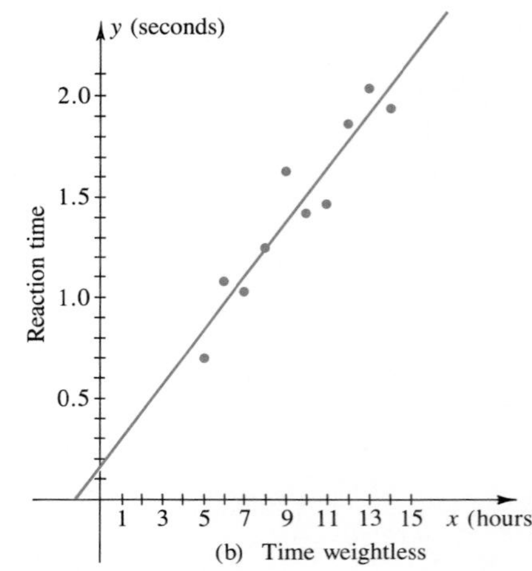

(b) Time weightless

In a problem like this we would not usually try to relate x and y by means of a curve *passing through* the plotted points since the data are really "probabilistic" in nature. That is, we would not expect the astronaut to exhibit exactly the same reaction times if the experiment were repeated. Instead, we observe that the trend of the data is upward as x increases and that this upward trend can reasonably be described by a straight line (Figure 4b). We now discuss a technique, called the **method of least squares**, for determining this "trend line." Once this line is obtained, it is useful for predicting values of y that are to be expected for values of x not appearing in the data.

To illustrate the idea of the method of least squares, let us consider the four points (x_1, y_1), (x_2, y_2), (x_3, y_3), and (x_4, y_4) in Figure 5a. Since the four points are not collinear, no line will pass through all four of them. Thus, if we want to describe the upward trend of these points by a line, we must settle for a line that in some sense comes closest to passing through all the points. Although there are various ways of measuring how close a line comes to passing through a set of noncollinear points, the most widely used criterion is the **principle of least squares**, which states:

The line should be chosen so that the sum of the squares of the "vertical distances" from the data points to the line is as small as possible.

FIGURE 5

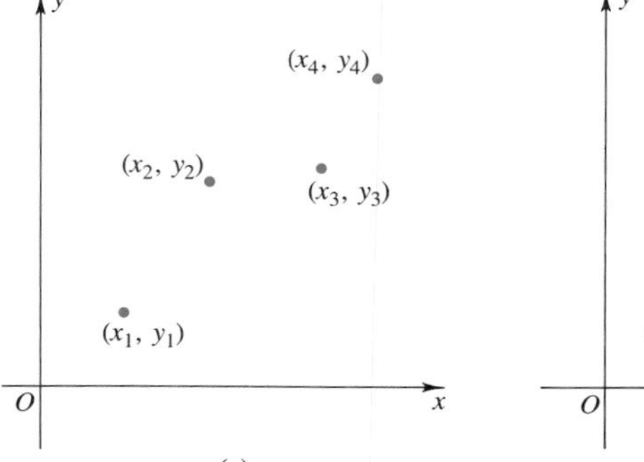

(a)

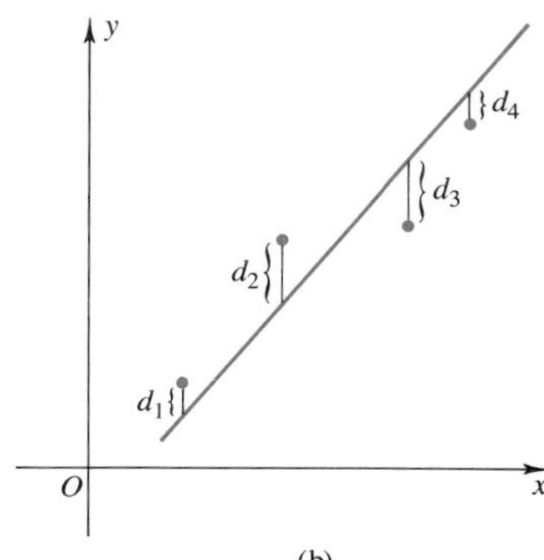

(b)

This line is called the **line of best fit**. Thus, the line of best fit for the points in Figure 5a is the one that makes the sum of

$$d_1^2 + d_2^2 + d_3^2 + d_4^2$$

as small as possible (Figure 5b).

Since the mathematical theory of least squares is most easily worked out with calculus, we simply state the main result and give some examples.

Line of Best Fit Given n data points $(x_1, y_1), (x_2, y_2), \ldots, (x_n, y_n)$, the line of best fit for the data is

$$y = mx + b \tag{6}$$

where m and b satisfy the system of equations

$$\begin{aligned} Am + nb &= B \\ Cm + Ab &= D \end{aligned} \tag{7}$$

with

$$\begin{aligned} A &= x_1 + x_2 + \cdots + x_n \\ B &= y_1 + y_2 + \cdots + y_n \\ C &= x_1^2 + x_2^2 + \cdots + x_n^2 \\ D &= x_1 y_1 + x_2 y_2 + \cdots + x_n y_n \end{aligned}$$

EXAMPLE 5 Find the line of best fit for the data (1, 3), (2, 1), (3, 4), and (4, 3).

Solution Denoting the given data by

$$(x_1, y_1) = (1, 3) \qquad (x_2, y_2) = (2, 1)$$
$$(x_3, y_3) = (3, 4) \qquad (x_4, y_4) = (4, 3)$$

we obtain

$$x_1 = 1 \qquad x_2 = 2 \qquad x_3 = 3 \qquad x_4 = 4$$
$$y_1 = 3 \qquad y_2 = 1 \qquad y_3 = 4 \qquad y_4 = 3$$

From Table 2 we obtain the values of A, B, C, and D.

TABLE 2

x_i	y_i	x_i^2	$x_i y_i$
1	3	1	3
2	1	4	2
3	4	9	12
4	3	16	12
$A = 10$	$B = 11$	$C = 30$	$D = 29$

Since $n = 4$, substituting in Equations (7) we obtain

$$\begin{aligned} 10m + 4b &= 11 \\ 30m + 10b &= 29 \end{aligned}$$

whose solution is (verify)

$$m = 0.3 \qquad b = 2$$

Thus, the line of best fit is

$$y = 0.3x + 2$$

The graph of this line and the given data are shown in Figure 6.

FIGURE 6

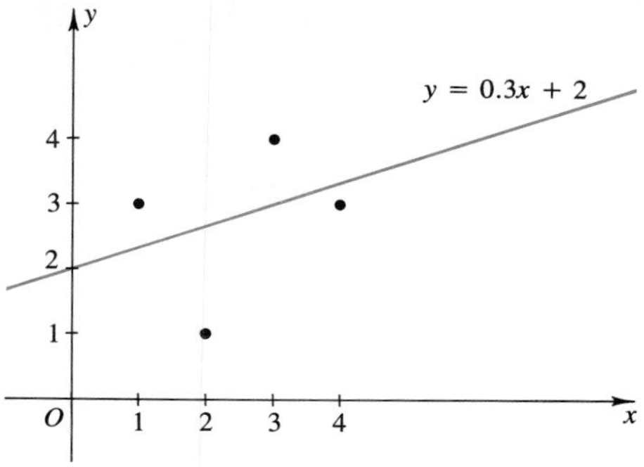

EXAMPLE 6 (a) Find the equation of the line of best fit for the data in Example 4.

(b) Use the equation obtained in part (a) to predict the astronaut's reaction time after 20 hours of weightlessness.

Solution (a) Table 3 tabulates the data needed in Equations (7) to find the line of best fit. According to the table

$$A = 95 \qquad B = 14.3 \qquad C = 985 \qquad D = 147.63$$

TABLE 3

x_i	y_i	x_i^2	$x_i y_i$
5	0.70	25	3.50
6	1.08	36	6.48
7	1.02	49	7.14
8	1.26	64	10.08
9	1.62	81	14.58
10	1.42	100	14.20
11	1.47	121	16.17
12	1.85	144	22.20
13	2.02	169	26.26
14	1.93	196	27.02
$A = 95$	$B = 14.37$	$C = 985$	$D = 147.63$

Since $n = 10$, substituting in Equations (7), we obtain

$$\begin{aligned} 95m + 10b &= 14.37 \\ 985m + 95b &= 147.63 \end{aligned}$$

whose solution is (verify)

$$m = 0.1347 \qquad b = 0.1574$$

Thus, the line of best fit (shown in Figure 4b) is

$$y = 0.1347x + 0.1574$$

Solution (b) To predict the astronaut's reaction time after 20 hours of weightlessness, we substitute $x = 20$ in the equation for the line of best fit obtained in part (a). This yields

$$y = (0.1347)(20) + 0.1574 = 2.8514$$

Thus, the astronaut's reaction time will be approximately 2.8514 seconds.

Break-even Analysis

One of the basic problems of business is to determine how many units of a product must be sold in order to make a profit. Suppose

R = the revenue received from selling x units of the product
C = the cost of manufacturing or purchasing x units of the product

If $R > C$ (revenue exceeds cost), then there is a profit on the x units, but if $R < C$ (cost exceeds revenue), then there is a loss. If $R = C$, then there is neither a profit nor a loss. The value of x for which $R = C$ is called the **break-even point**.

EXAMPLE 7 A manufacturer of cleanser determines that the total cost (in thousands of dollars) of producing x tons of cleanser is

$$C = 400 + 2x \tag{8}$$

and that the revenue (in thousands of dollars) from selling x tons of cleanser is

$$R = 4x \tag{9}$$

(a) Find the break-even point.
(b) At the break-even point, what are the cost and revenue?
(c) Graph Equations (8) and (9) and mark the break-even point.

Solution (a) The break-even point occurs when $R = C$, or from (8) and (9),

$$4x = 400 + 2x$$
$$2x = 400$$
$$x = 200$$

Thus, the break-even point is 200 tons.

Solution (b) Substituting the break-even value $x = 200$ in either Equation (8) or (9) yields

$$C = R = 800 \qquad \text{(thousands of dollars)}$$

Solution (c) Equations (8) and (9) are linear equations, and so each has a straight-line graph. Graphing these lines in the same coordinate system, we obtain the results shown in Figure 7. (The vertical axis represents both R and C.) As noted in the figure, a profit occurs when $x > 200$ because then $R > C$, and a loss occurs when $x < 200$ because then $R < C$.

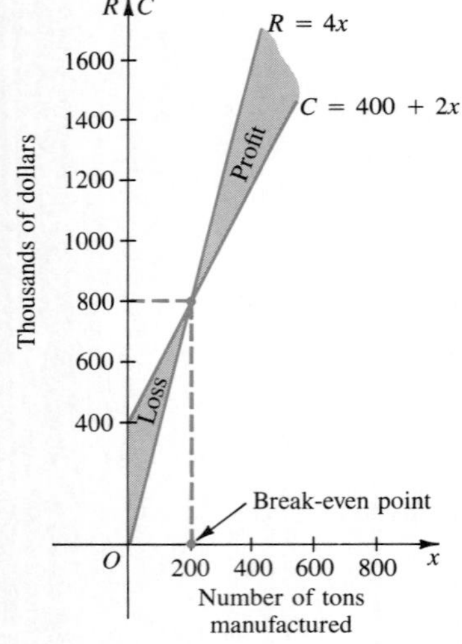

FIGURE 7

Supply and Demand

One of the basic principles of economics is the **law of supply and demand**. It states that:

1. The willingness of manufacturers to produce a product (the supply) increases as the price increases.
2. The willingness of consumers to buy a product (the demand) decreases as the price increases.

If q is the quantity of a particular product that its manufacturers are willing to supply at a price of p per unit, then the graph of q versus p is called the **supply curve**. A typical supply curve is shown in Figure 8. If q is the quantity of a particular product that its consumers are willing to buy at a price of p per unit, then the graph of q versus p is called the **demand curve**. A typical demand curve is shown in Figure 9.

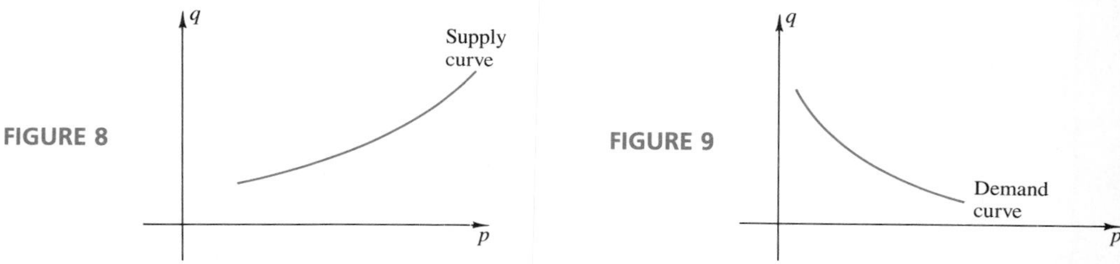

FIGURE 8

FIGURE 9

If the supply and demand curves of a product are graphed in the same coordinate system and if they intersect at a point (p_0, q_0) (Figure 10a), then p_0 is the price at which the supply and demand are equal. This is called the **equilibrium price** since every unit supplied at that price is purchased. This price for the product produces a stable market. However, if p exceeds p_0, then supply exceeds demand and a product surplus will develop. Similarly, if p is less than p_0, demand exceeds supply and a product shortage will develop (Figure 10b). The value q_0 is called the **equilibrium quantity**. It is the quantity of the product that will be purchased at the equilibrium price.

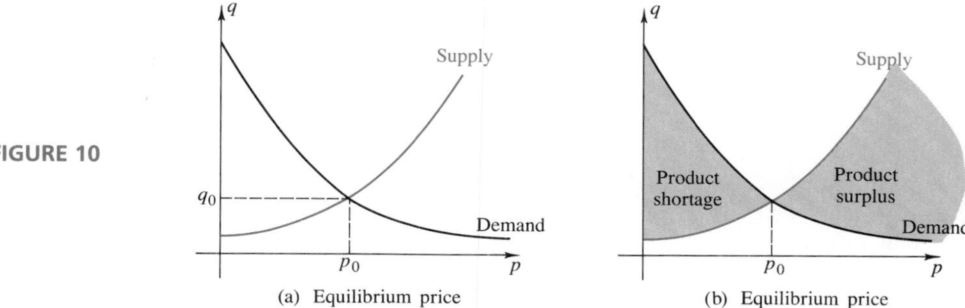

FIGURE 10

(a) Equilibrium price

(b) Equilibrium price

In this section we consider the simplified case in which the supply and demand curves are straight lines.

EXAMPLE 8 Suppose that a market research firm has obtained the following equations for the monthly supply and demand in New York City for a certain solar-powered calculator:

$$\text{supply: } q = 2p - 12$$
$$\text{demand: } q = -3p + 28$$

where p is the price (in dollars) per calculator and q is the number (in thousands) of calculators to be supplied and bought.

(a) Find the equilibrium price and determine the number of calculators that will be sold at this price.

(b) Sketch the supply and demand curves in the same coordinate system.

Solution (a) Rewrite the given equations for the supply and demand curves as

$$2p - q = 12$$
$$-3p - q = -28$$

We can eliminate q from this linear system by adding -1 times the second equation to the first equation, obtaining

$$5p = 40$$
$$p = 8$$

Substituting this value into either the equation for supply or the equation for demand, we find that $q = 4$ (thousands). Thus, the equilibrium price is \$8 per calculator, and at this price 4000 calculators will be supplied and bought.

Solution (b) See Figure 11.

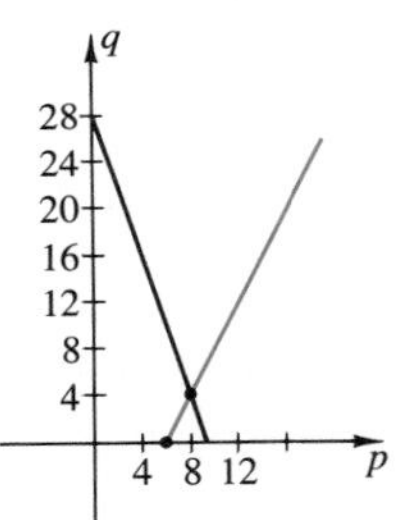

FIGURE 11

EXERCISE SET 1.5

1. **(Simple interest)** Suppose a person borrows \$12,000 at the simple interest rate of 7% per year.
   (a) Find a formula for the amount S that will be owed after t years.
   (b) Use this formula to determine how much will be owed after 5 years.
   (c) Use this formula to determine how much will be owed after 42 months.

2. **(Simple interest)** Draw a Cartesian coordinate system with vertical axis S and horizontal axis t. Sketch the graph of the equation you obtained in Exercise 1a.

3. **(Simple interest)** Suppose a certain sum of money is borrowed for 3.5 years at the simple interest rate of 12% per year. At the end of the loan period, \$8520 is repaid to the lender. How much money was borrowed?

4. **(Simple interest)** Suppose \$7000 is borrowed for a certain period of time at the simple interest rate of 10% per year. At the end of the loan period, \$9975 is repaid to the lender. What was the loan period?

5. **(Depreciation)** Suppose a piece of heavy machinery costs \$24,000 and depreciates linearly at the rate of 8% per year.
   (a) Find a formula for the value V of the item after t years.
   (b) Use this formula to determine the value V of the item after 7 years.
   (c) After how many years will the item be worth \$5000?
   (d) After how many years will the item have lost all its value?

6. **(Depreciation)** Draw a Cartesian coordinate system with vertical axis V and horizontal axis t. Sketch the graph of the equation you obtained in Exercise 5a.

7. **(Depreciation)** Suppose a business college buys a computer at a certain price and depreciates it linearly at the rate of 6% per year. If the computer is worth \$24,000 after 10 years, what was its purchase price?

8. **(Depreciation)** An audiovisual department buys a television set for \$2200 and depreciates it linearly at a certain rate per year. If the set is worth \$1320 after 4 years, what was the annual depreciation rate?

9. **(Biology)** Suppose in a biological experiment a certain type of bacteria is treated with a growth-inhibiting chemical. After 2 hours 64,000 bacteria per unit volume are found in a culture, and 128,000 are found after 5 hours. Let N be the number (in thousands) of bacteria per unit volume and let t denote time (in hours) of growth. Assuming that N continues to increase with t at a constant rate, find
   (a) an equation relating N and t
   (b) the graph of the equation in (a) in a Cartesian coordinate system with horizontal axis t and vertical axis N
   (c) the number of bacteria per unit volume that are present after 9 hours of growth

10. **(Communications)** Suppose at the present time a communications satellite is handling 1.2 million messages per year. Solve (a) and (b) below assuming that the number N (in millions) of messages will increase linearly at a constant rate of 10% per year. Let t denote time in years and take the present time as $t = 0$.
    (a) Find an equation relating N and t.
    (b) Predict the number of messages that will be handled 12 years from now.

11. **(Temperature conversion)** There are two common systems for measuring temperature, centigrade (Celsius) and Fahrenheit. Water freezes at 0° centigrade and 32° Fahrenheit; it boils at 100° centigrade and 212° Fahrenheit. Since the two are related by a linear equation, show that a temperature of $C°$ centigrade corresponds to a temperature of $F°$ Fahrenheit by means of the formula

$$F = \frac{9}{5}C + 32$$

12. Suppose a taxi fare is 40 cents plus 25 cents for each $\frac{1}{4}$ mile traveled.
    (a) Write an equation for the cost c (in dollars) for traveling x miles.
    (b) Graph the equation in (a) in a Cartesian coordinate system with vertical axis c and horizontal axis x.
    (c) What is the slope of the line in (a)?

13. Find the line of best fit for the data points (2, 4), (3, 5), (4, 7), and (5, 9).

14. Find the line of best fit for the data points (2, 1), (3, 6), (4, 9), (5, 13), (6, 15), (7, 18), and (8, 21).

15. Show that the solutions to Equation (7) are given by

$$m = \frac{AB - nD}{A^2 - nC} \qquad b = \frac{AD - BC}{A^2 - nC}$$

16. Solve Exercise 13 by using the fomulas given in Exercise 15.

17. **(Business)** A department store reports the following sales data over a 6-year period.

Year	**1985**	**1986**	**1987**	**1988**	**1989**	**1990**
Yearly sales (in millions of dollars)	5.57	6.87	7.13	8.11	8.20	8.90

Let y denote the yearly sales (in millions of dollars) and let the years 1985, 1986, . . . , 1990 be "coded," respectively as 0, 1, 2, . . . , 5. That is, $x = 0$ represents 1985, $x = 1$ represents 1986, and so on.
   (a) Find the line of best fit relating x and y.
   (b) Use the equation obtained in (a) to estimate the yearly sales in the year 1996 (i.e., when $x = 11$).
   (c) Explain why it was convenient to "code" the years in this problem.

18. **(Break-even analysis)** A manufacturer of photographic developer determines that the total cost (in dollars) of producing x liters of developer is

$$C = 550 + 0.40x$$

The manufacturer sells the developer for \$0.50 per liter.

(a) What is the total revenue received when x liters of developer are sold?
(b) Find the break-even point.
(c) At the break-even point, what are the cost and revenue?
(d) Graph the cost and revenue equations and show the break-even point.

19. **(Break-even analysis)** A school cafeteria manager finds that the weekly cost of operation is \$1375 plus \$1.25 per meal served. Assuming that each meal produces a revenue of \$2.50, find the break-even point.

20. **(Break-even analysis)** An ice cream vendor finds that her weekly operating cost is \$300 plus \$0.60 per ice cream portion served. If each ice cream portion retails for \$0.80, what is the break-even point?

21. **(Break-even analysis)** A newspaper vendor finds that his monthly operating cost is \$400 plus \$0.15 per newspaper sold. If each newspaper is sold for \$0.25, what is the break-even point?

22. **(Supply and demand)** Suppose the equations for a certain month's supply and demand of umbrellas in London are given by

$$\begin{aligned} \text{supply: } q &= 4p - 3 \\ \text{demand: } q &= -3p + 11 \end{aligned}$$

where p is the price (in pounds sterling) per umbrella and q is the number (in thousands) of umbrellas to be supplied and bought.

(a) Find the equilibrium price and the number of umbrellas to be sold at this price.
(b) Sketch the supply and demand curves in the same coordinate system.

23. **(Supply and demand)** Suppose the equations for the weekly supply and demand of compact discs in Chicago are given by

$$\begin{aligned} \text{supply: } -5p + q &= -71 \\ \text{demand: } 2p + q &= 34 \end{aligned}$$

where p is the price (in dollars) per compact disc and q is the number (in thousands) to be supplied and bought.

(a) Find the equilibrium price and the number of compact discs to be sold at this price.
(b) Sketch the supply and demand curves in the same coordinate system.

24. **(Supply and demand)** Suppose the equations for the monthly supply and demand of a certain product are given by

$$\begin{aligned} \text{supply: } q &= 3p - 8 \\ \text{demand: } q &= -5p + 24 \end{aligned}$$

(a) Find the equilibrium price.
(b) If the product is priced at \$4.00 per unit, then does the supply exceed the demand? If so, by how much?
(c) If the product is priced at \$2.50 per unit, then does the demand exceed the supply? If so, by how much?

25. **(Supply and demand)** Suppose that the equations for the weekly supply and demand of a certain product are given by

$$\begin{aligned} \text{supply: } q &= 4p - 10 \\ \text{demand: } q &= -2p + 20 \end{aligned}$$

(a) Find the equilibrium price.
(b) If the product is priced at \$6.50 per unit, then does the supply exceed the demand? If so, by how much?
(c) If the product is priced at \$4.00 per unit, then does the demand exceed the supply? If so, by how much?

1.6 LINEAR INEQUALITIES AND THEIR GRAPHS

In the last few sections we studied linear *equations* and their graphs. In this section we study linear *inequalities* and their graphs. The ideas we develop in this section are applied in Chapter 3 to a variety of important practical problems.

As illustrated in Figure 1 a line divides the plane into two parts, called **half planes**. If the line is included with the half plane, the half plane is called a **closed half plane** (Figure 1b). If the line is not included, then it is called an **open half plane** (Figure 1c). We use a dashed line to emphasize that the points on the line are not included in the half plane and a solid line to indicate that the points are included. Just as the points on the line can be described by means of an equation

A closed half plane includes the boundary line.

An open half plane does not include the boundary line.

(a) (b) (c)

FIGURE 1

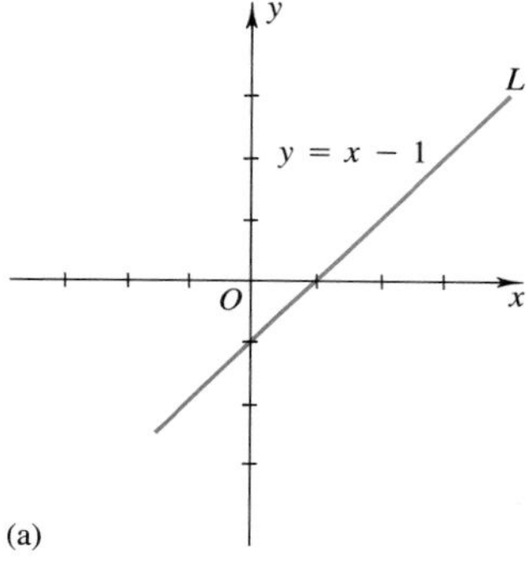

(a)

P(x, y)

(b)

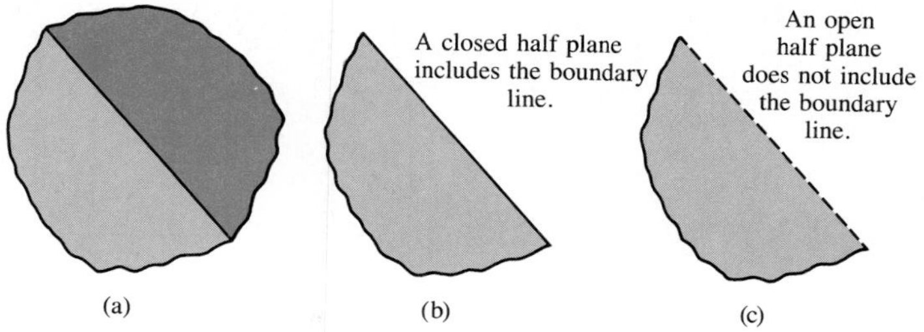

FIGURE 2

(or equality), so the points in the half planes can be described by means of inequalities. For example, consider the line

$$y = x - 1 \tag{1}$$

shown in Figure 2a. Call this line L. If $P(x, y)$ is any point on L, then x and y satisfy the equation for L; that is, $y = x - 1$. If A is any point directly above P (Figure 2b), then the y-coordinate of A is greater than the y-coordinate of P so that

$$y > x - 1 \tag{2}$$

Similarly, if B is any point directly below P, then the y-coordinate of B is less than the y-coordinate of P so that

$$y < x - 1 \tag{3}$$

Thus, the line L consists of all points (x, y) satisfying *equality* (1), while the open half plane above L consists of all points (x, y) satisfying *inequality* (2) and the open half plane below L consists of all points (x, y) satisfying *inequality* (3).

More generally we can state the following result.

The line

$$Ax + By = C$$

determines two open half planes given by the inequalities

$$Ax + By > C \quad \text{and} \quad Ax + By < C$$

and it determines two closed half planes given by the inequalities

$$Ax + By \geq C \quad \text{and} \quad Ax + By \leq C$$

The line $Ax + By = C$ is called the **boundary line** of each half plane.

EXAMPLE 1 Sketch the set of points satisfying

$$3x + 2y > 6$$

Solution The set of points we want to sketch is one of the open half planes determined by the line $3x + 2y = 6$. First, sketch the boundary line $3x + 2y = 6$

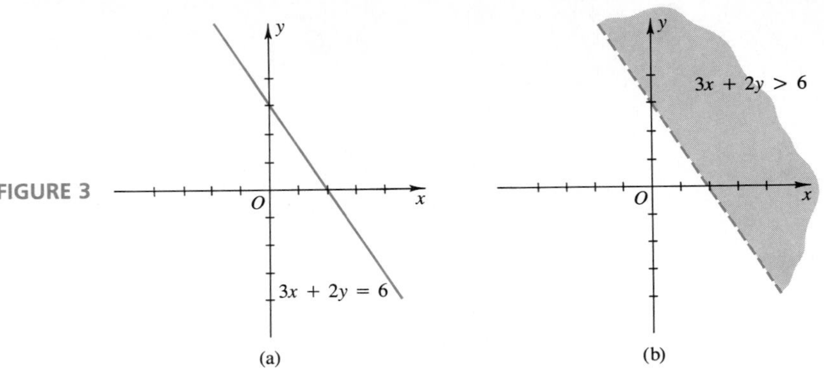

FIGURE 3

(Figure 3a). To determine which of the two open half planes is described by the inequality

$$3x + 2y > 6 \tag{4}$$

we consider a "test point": (0, 0), for example. If we substitute the coordinates $x = 0$ and $y = 0$ of the test point into (4), we obtain

$$3(0) + 2(0) > 6$$

or

$$0 > 6$$

which is false. This indicates that the test point (0, 0) does not lie in the open half plane determined by (4). Thus, the solution set is shaded in Figure 3b.

Remark In Example 1 we can use any point not on the line $3x + 2y = 6$ as a test point; we picked (0, 0) for simplicity.

EXAMPLE 2 Sketch the set of points satisfying

$$3x + 2y \geq 6 \tag{5}$$

Solution The points in the closed half plane determined by (5) are those points in the open half plane $3x + 2y > 6$ together with those points on the line $3x + 2y = 6$ (Figure 4).

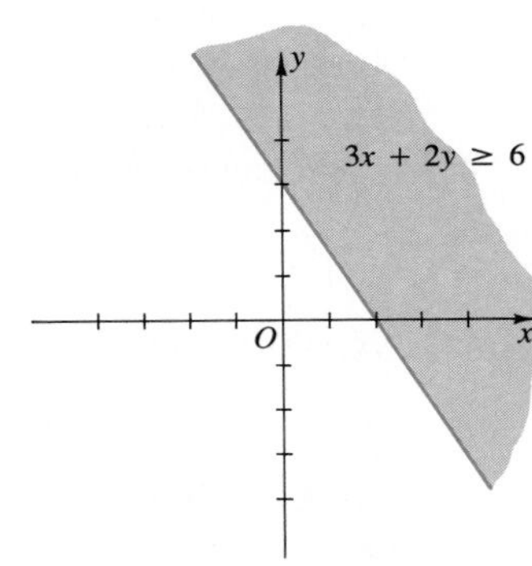

FIGURE 4

EXAMPLE 3 Sketch the set of points satisfying

$$x + y > 0 \tag{6}$$

Solution The set of points we want to sketch is one of the open half planes determined by the line $x + y = 0$. First sketch the boundary line $x + y = 0$ (see Figure 5a). Since (0, 0) lies on this line, we cannot use it as a test point. Instead, we pick any other convenient point that is not on the line, for example, (1, 1). If we substitute the coordinates $x = 1$ and $y = 1$ of the test point into (6), we obtain

$$1 + 1 > 0 \quad \text{or} \quad 2 > 0$$

which is true. This indicates that the test point lies in the open half plane determined by (6). Thus, the solution set is the open half plane shaded in Figure 5b.

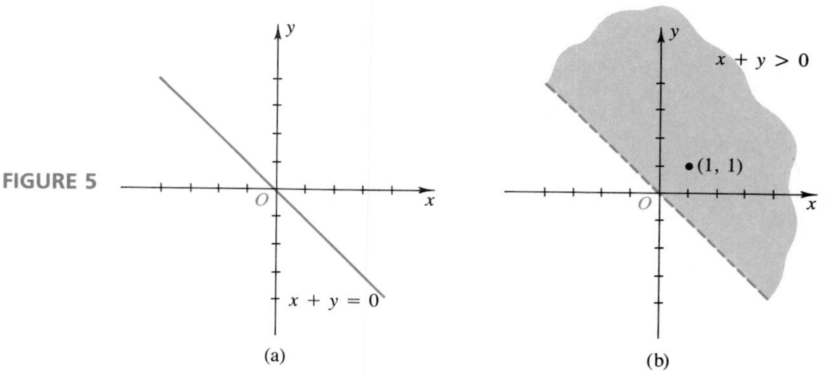

FIGURE 5

Systems of Inequalities

Just as we considered systems of equations in earlier sections, we now consider systems of inequalities.

If we are given two or more inequalities in x and y such as

$$\begin{aligned} 2x + 3y &\geq 6 \\ x + y &\geq 0 \end{aligned} \qquad \text{or} \qquad \begin{aligned} 10x + 15y &\leq 60 \\ x + y &\leq 25 \\ x &\geq 0 \\ y &\geq 0 \end{aligned}$$

we can seek values of x and y that satisfy all the inequalities. The inequalities are then said to form a **system** of inequalities. An ordered pair of real numbers (a, b) is called a **solution** of the system if all the inequalities are satisfied when we substitute $x = a$ and $y = b$.

EXAMPLE 4 Consider the system

$$\begin{aligned} 2x + 3y &\geq 6 \\ x - y &\leq 0 \end{aligned}$$

The ordered pair (1, 2) is a solution of the system since both inequalities are satisfied when we substitute $x = 1$ and $y = 2$. However, $(-3, -2)$ is not a solution since only the second inequality is satisfied when we substitute $x = -3$ and $y = -2$.

EXAMPLE 5 Sketch the set of solutions of the system

$$\begin{aligned} 2x + 3y &\leq 6 \\ 3x - y &\leq 1 \end{aligned}$$

Solution Using the method illustrated in the preceding examples, you should be able to show that the solutions of the first inequality form the set shaded in Figure 6a and the solutions of the second inequality form the set shaded in Figure 6b. Since the solutions of the system satisfy both inequalities, the set of solutions of the system is the intersection of the shaded regions in (a) and (b) of Figure 6. This set is sketched in Figure 6c.

FIGURE 6

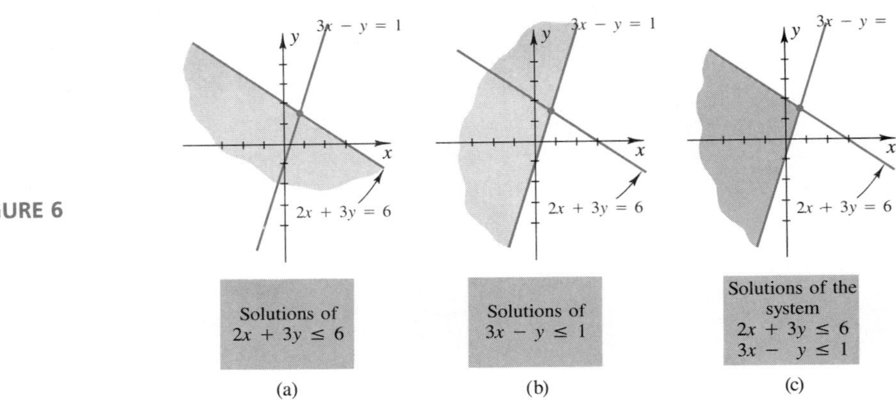

EXAMPLE 6 Sketch the set of solutions of the system

$$\begin{aligned} x + y &\leq 2 \\ 2x + 2y &\geq 7 \end{aligned}$$

Solution The solutions of the first inequality are shaded in Figure 7a, and the solutions of the second inequality are shaded in Figure 7b. Since the solutions of the system satisfy both inequalities, the set of solutions is the intersection of the shaded regions in (a) and (b) of Figure 7. As shown in Figure 7c, the intersection of these regions is empty; thus, the system has no solutions.

FIGURE 7

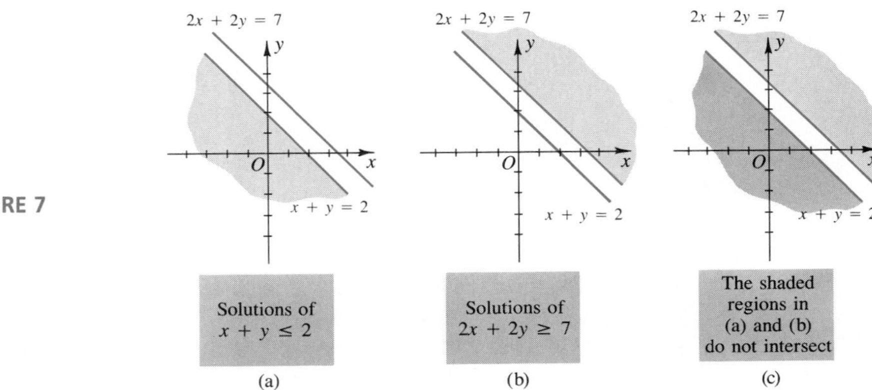

EXAMPLE 7 Sketch the set of solutions of the system

$$\begin{aligned} x + 2y &\leq 8 \\ -x + 2y &\leq 6 \\ y &\geq 0 \end{aligned}$$

Solution The solutions of each inequality in the system are shown as shaded regions in (a), (b), and (c) of Figure 8. The set of solutions of the system, shown in Figure 8d, is obtained by forming the intersection of these three regions.

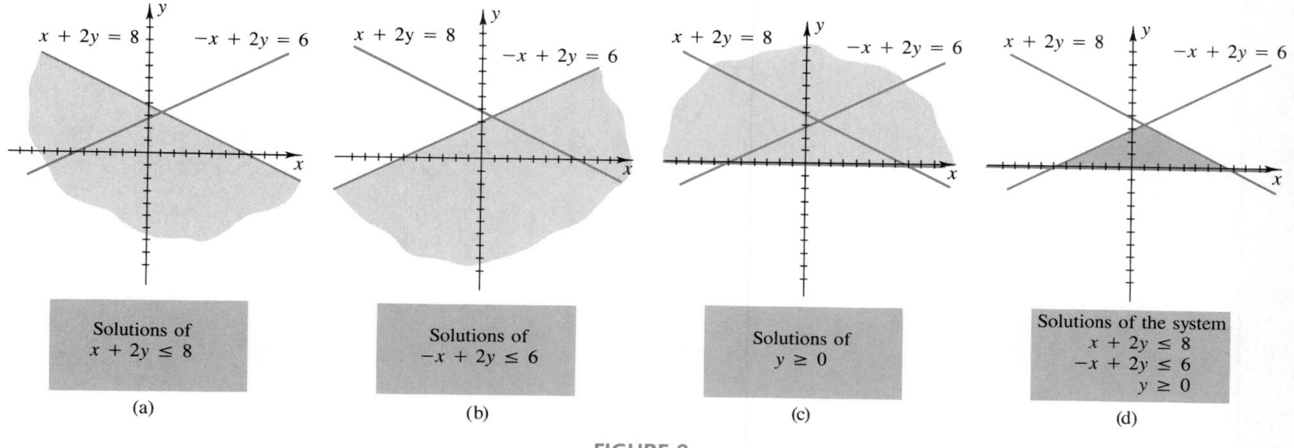

FIGURE 8

EXERCISE SET 1.6

In each part of Exercises 1–5 sketch the set of points satisfying the given inequality.

1. (a) $3x - 2y < 3$ (b) $3x - 2y \geq 3$
2. (a) $2x + y \leq 2$ (b) $2x + y \geq 2$
3. (a) $y > x$ (b) $y \leq x$
4. (a) $x \geq 2$ (b) $y < -1$
5. (a) $7x - 8y < 0$ (b) $7x - 8y > 0$

In Exercises 6–16 sketch the set of points satisfying the given system of inequalities.

6. $\begin{aligned} x + y &\leq 3 \\ x &\geq 0 \end{aligned}$

7. $\begin{aligned} 2x + 3y &< -5 \\ 3x - 2y &> 12 \end{aligned}$

8. $\begin{aligned} 2x + 3y &> 4 \\ 4x + 6y &< 4 \end{aligned}$

9. $\begin{aligned} 5x - 2y &\leq 4 \\ 3x + y &\geq 9 \end{aligned}$

10. $\begin{aligned} 7x + y &\leq 11 \\ 2x - 3y &\geq -10 \end{aligned}$

11. $\begin{aligned} 4x + 2y &\leq 5 \\ -x + y &\leq 0 \\ y &\geq 0 \end{aligned}$

12. $\begin{aligned} 2x - 7y &\geq 5 \\ 4x + 3y &\leq 3 \\ x - y &\geq 2 \end{aligned}$

13. $\begin{aligned} 2x + 3y &< -5 \\ 3x - 2y &> 12 \\ x &\geq 0 \\ y &\geq 0 \end{aligned}$

14. $\begin{aligned} x + y &\leq 1 \\ x + y &> 0 \\ 2x + 2y &> 7 \end{aligned}$

15. $\begin{aligned} 4x + 3y &\leq 17 \\ 2x - 5y &\geq -11 \\ x &\geq 0 \\ y &\geq 0 \end{aligned}$

16. $\begin{aligned} x + 4y &\leq 12 \\ 3x - 2y &\leq 8 \\ x &\geq 0 \\ y &\geq 0 \end{aligned}$

In Exercises 17–20 find an inequality whose solutions form the shaded region.

17.

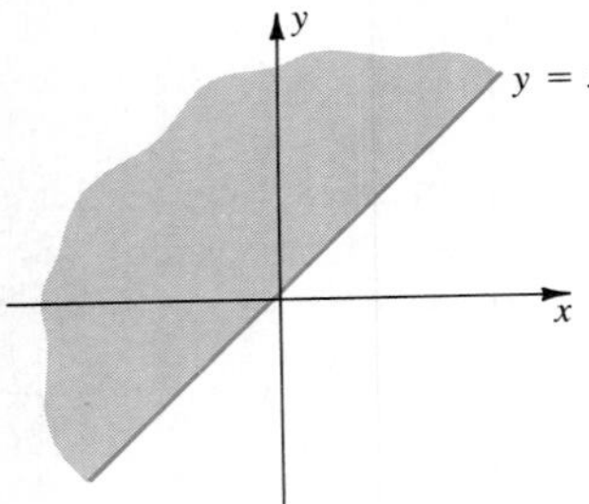

18.

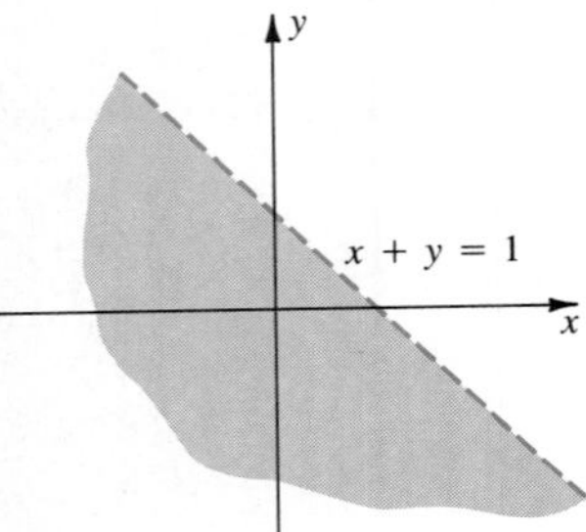

19.

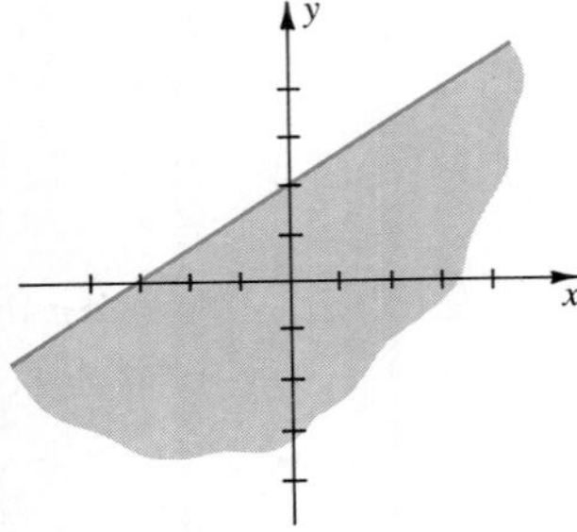

20.

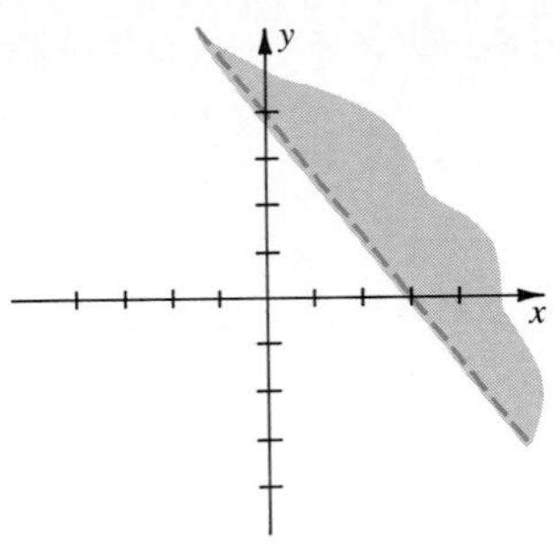

In Exercises 21 and 22 find a system of inequalities whose solutions form the shaded region.

21.

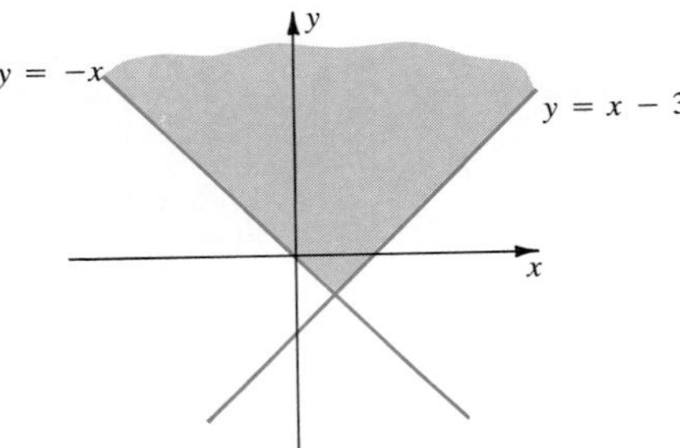

22.

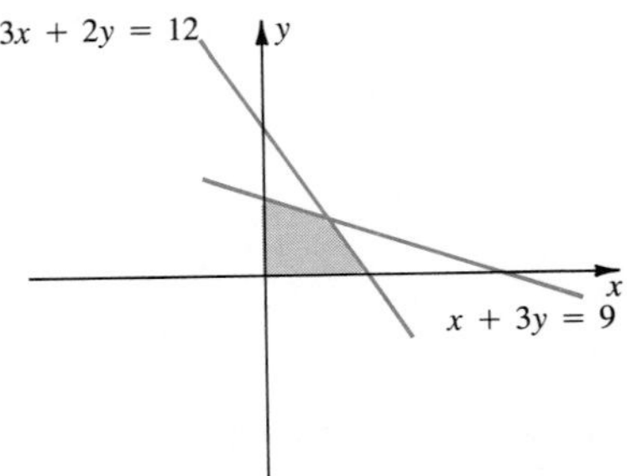

KEY IDEAS FOR REVIEW

- **Solution of an equation in x and y** Ordered pair that satisfies the equation.
- **Solution set** The set of all solutions of an equation.
- **Slope of a nonvertical line**

$$m = \frac{y_2 - y_1}{x_2 - x_1}$$

where $P_1(x_1, y_1)$ and $P_2(x_2, y_2)$ are any two distinct points on the line.

- **Vertical line** has undefined slope.
- **Horizontal line** has slope $m = 0$.
- **Parallel lines** have the same slope, and lines with the same slope are parallel or coincide.
- **Slope–intercept form of a line** $y = mx + b$
- **Point–slope form of a line** $y - y_1 = m(x - x_1)$
- **Linear equation** An equation that can be expressed in the form $Ax + By = C$, with A and B not both zero, has a line as its graph.
- **Linear systems with two unknowns** can be solved by elimination (p. 27).
- **Open and closed** half planes **The line**

$$Ax + By = C$$

determines two open half planes given by the inequalities

$$Ax + By > C \quad \text{and} \quad Ax + By < C$$

and it determines two closed half planes given by the inequalities

$$Ax + By \geq C \quad \text{and} \quad Ax + By \leq C$$

SUPPLEMENTARY EXERCISES

In Exercises 1–4 determine the x- and y-intercepts of the graphs of the given equations.

1. $y = x^2 - 9$ **2.** $x = y + 3$

3. $x^2 - y = 4$ **4.** $Ax + By = C \quad A \neq 0, B \neq 0$

In Exercises 5–12 sketch the graph of the given equation.

5. $2x = 3$ **6.** $3x - y = 4$

7. $y = x^2 + 2x + 1$ **8.** $y = \dfrac{1}{x + 1}$

9. $yx^2 = 2$ **10.** $y = (x + 1)^2$

11. $y = \sqrt{x - 1}$ **12.** $y = \sqrt{x^2 + 2}$

In Exercises 13–17 find an equation for the line satisfying the given conditions.

13. parallel to the line $y = 4x - 2$; passes through the point $(-2, 3)$

14. parallel to the line $3x + 2y = 5$; passes through the point $(4, 3)$

15. horizontal; passes through the point $(-2, 1)$

16. has x-intercept 4 and y-intercept -3

17. passes through the points $(-2, 1)$ and $(3, 5)$

18. Find the x- and y-intercepts of the line

$$3x + 5y = 5$$

19. Find the x- and y-intercepts of the line

$$\frac{x}{a} + \frac{y}{b} = 1$$

20. Two lines of slopes m_1 and m_2, respectively, are **perpendicular** if $m_1 m_2 = -1$. Find an equation for the line perpendicular to the line satisfying $y = 2x - 3$ and passing through the point $(4, -2)$.

21. Show that (1, 2), (4, 6), (−1, 8), and (−4, 4) are the vertices of a parallelogram.

22. Referring to Figure 1, show that the midpoint of the line segment whose endpoints are P_1 and P_2 is $P(x, y)$, where

$$x = \frac{x_1 + x_2}{2} \qquad y = \frac{y_1 + y_2}{2}$$

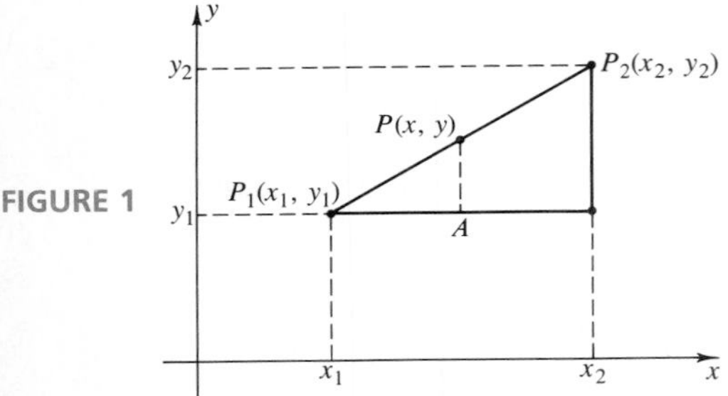

FIGURE 1

Hint: Use the fact that corresponding sides of similar triangles are proportional:

$$\frac{\text{length of } P_1P_2}{\text{length of } P_2B} = \frac{\text{length of } P_1P}{\text{length of } PA}$$

23. Use the result of Exercise 22 to find the midpoint of the line segment whose endpoints are (2, 3) and (3, 7).

24. The **perpendicular bisector** of the line segment whose endpoints are P_1 and P_2 is the straight line that is perpendicular to the line segment and passes through its midpoint. Find an equation for the perpendicular bisector of the line segment whose endpoints are given (see Exercises 20 and 22).
(a) (−2, −5) and (3, 4)
(b) (−3, 5) and (1, −3)

25. Referring to Figure 2, show that the line $y = x$ is the perpendicular bisector of the line segment whose endpoints are (a, b) and (b, a) (see Exercise 24).

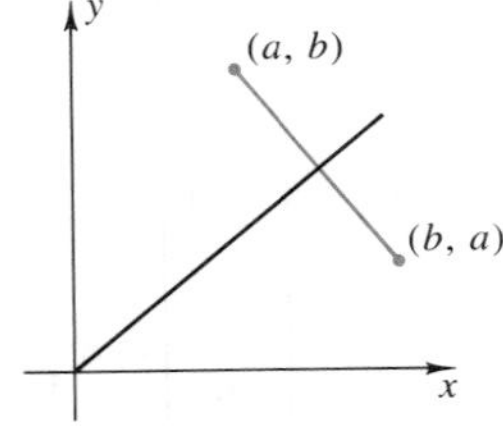

FIGURE 2

26. (a) Graph $y = 2x$ and $x = 2y$ in the same coordinate system.
(b) Note that the graphs in (a) are mirror images with respect to the line $y = x$. Use Exercise 25 to explain why this is the case.
(c) Given an equation in x and y, how would you obtain the equation for its mirror image about the line $y = x$?
(d) Use the result in (c) to find an equation for the mirror image of $y = x^2$ about the line $y = x$.

In Exercises 27–30 solve the linear system.

27. $3x + y = 3$
$2x - 5y = 19$

28. $2x + y = 0$
$3x - 4y = -22$

29. $2x - 3y = 0$
$x + 2y = 0$

30. $3x - y = 0$
$6x - 2y = 0$

31. Suppose that $8000 is borrowed for 10 years at a certain annual simple interest rate. At the end of the loan period, $14,400 is repaid. What was the interest rate?

32. Suppose that an X-ray machine is purchased for $120,000 by a hospital and is depreciated linearly at the rate of 15% per year. After how many years will the machine be worth $12,000?

33. Suppose that an automobile manufacturer that spent $3.4 million on advertising had a total revenue of $580 million, and when the advertising budget was reduced to $2.6 million, it had a total revenue of $420 million. Assuming that the total revenue R and the amount spent on advertising are related by a linear equation, what total revenue would result if $4.3 million were spent on advertising?

34. Find the line of best fit for the data points (3, 5), (4, 7), (5, 9), and (6, 12).

35. Find the line of best fit for the data points (2, 3), (3, 5), (4, 6), (5, 9), (6, 10), and (7, 12).

36. A photocopy center has determined that its weekly cost of operation is $250 plus 4 cents per page. If the center charges 9 cents per page, what is the break-even point?

37. Suppose that the equations for a certain week's supply and demand of bananas in San Francisco are given by

$$\text{supply: } q = 2p - 12$$
$$\text{demand: } q = -5p + 268$$

where p is the price (in cents) per pound of bananas and q is the number (in thousands of pounds) to be supplied and bought. Find the equilibrium price and the number of bananas to be sold at this price.

38. Sketch the set of points satisfying the following system of inequalities:

$$\begin{aligned} 3x + 4y &< 12 \\ 2x - 3y &> -6 \end{aligned}$$

39. Find a system of inequalities whose solutions form the shaded region in Figure 3.

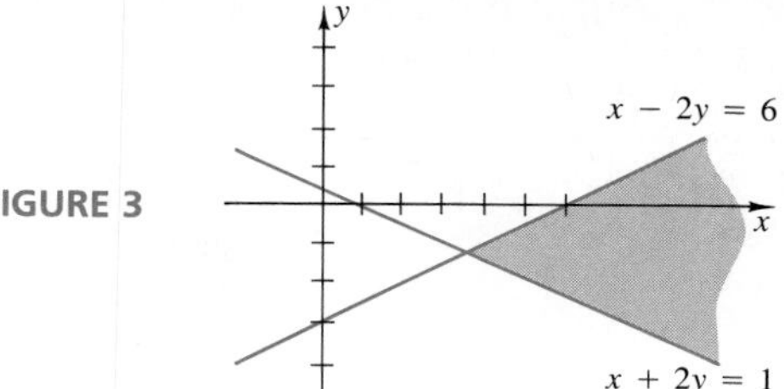

FIGURE 3

CHAPTER TEST

1. Determine the intercepts of the graph of

$$3x^2 + y = 27$$

2. Sketch the graph of the equation $y = x^2 - 2x$.

3. Find an equation for the line that is parallel to the line satisfying $3x + 2y = 5$ and passes through the point $(-1, 3)$.

4. Find an equation for the line with x-intercept 2 and y-intercept 3.

5. Solve the linear system

$$\begin{aligned} 3x + 2y &= 7 \\ x - 2y &= -4 \end{aligned}$$

6. Suppose that $15,000 is borrowed for a certain period of time at the simple interest rate of 8% per year. At the end of the loan period, $22,200 is repaid to the lender. What was the loan period in years?

7. Suppose a study of air pollution near a pesticide manufacturing plant shows that when 4 million tons of the chemical are manufactured, 30,000 pounds of pollutants are emitted, and when 6 million tons are manufactured, 40,000 pounds of the pollutant are emitted. If the number of pounds of pollutants emitted is related, by a linear equation, to the number of tons of the product manufactured, how much of the pollutant will be emitted when 8 million tons of the product are manufactured?

8. A manufacturer of ballpoint pens finds that the total cost (in dollars) of manufacturing x pens is given by $C = 400 + 0.60x$. If each pen sells for 80 cents, find the break-even point. At the break-even point, what are the cost and revenue?

9. Sketch the set of points satisfying the following system of inequalities:

$$\begin{aligned} x + 2y &\geq 6 \\ 3x - 4y &\leq 12 \\ y &\geq 0 \end{aligned}$$

2

Matrices and Linear Systems

Section 1.4 describes a method for solving systems of two linear equations in two unknowns. In this chapter we examine systems that may involve more than two equations or more than two unknowns. We also develop a new concept (a matrix), which has important applications in business and economics as well as in the behavioral and social sciences. We discuss some of these applications here and in Chapters 4 and 10.

2.1 LINEAR SYSTEMS

Recall from Section 1.3 that an equation of the form

$$ax + by = c$$

represents a straight line in the xy-plane. Thus, such an equation is called a **linear equation** in the variables x and y. Sometimes it is more convenient to use letters with subscripts to denote constants and variables in a linear equation. Thus, a linear equation in the variables x_1 and x_2 might be written

$$a_1x_1 + a_2x_2 = c$$

where a_1 and a_2 are constants. In general, an equation expressible in the form

$$a_1x_1 + a_2x_2 + \cdots + a_nx_n = c$$

where $a_1, a_2, \ldots, a_n$ and c are constants, is called a **linear equation** in the variables (or unknowns) $x_1, x_2, \ldots, x_n$.

EXAMPLE 1 The following are linear equations:

$$\begin{aligned} 2x + 3y &= 4 \\ x &= 6 \\ -2x + y + z &= 6 \\ 2x_1 - \frac{3}{2}x_2 + 5x_3 + x_4 &= 8 \end{aligned}$$

Observe that a linear equation does not contain any products or quotients of variables and that all variables occur only to the first power. The following are not linear equations:

$$\begin{aligned} 2xy + 3z &= 5 \\ 2x^2 + 3y &= 4 \\ \sqrt{x_1} + 2x_2 - x_3 &= 7 \\ \frac{1}{x} - 2y + 3z &= 9 \end{aligned}$$

Application to Production Problems

EXAMPLE 2 Each day, a pharmaceutical firm produces 100 ounces of a perishable ingredient, called alpha, which is used in the manufacture of drugs A, B, C, and D. These drugs require the following amounts of alpha in their manufacture:

Each ounce of A requires 0.1 ounce of ingredient alpha.
Each ounce of B requires 0.3 ounce of ingredient alpha.
Each ounce of C requires 0.5 ounce of ingredient alpha.
Each ounce of D requires 0.2 ounce of ingredient alpha.

To prevent any waste of the perishable ingredient alpha, the firm wants to produce x_1 ounces of A, x_2 ounces of B, x_3 ounces of C, and x_4 ounces of D each day so that the entire 100 ounces of ingredient alpha is used. For this to happen, x_1, x_2, x_3, and x_4 must satisfy the condition

$$0.1x_1 + 0.3x_2 + 0.5x_3 + 0.2x_4 = 100 \tag{1}$$

which is a linear equation in x_1, x_2, x_3, and x_4.

A set of linear equations in the variables $x_1, x_2, \ldots, x_n$ is called a **system of linear equations**, or more briefly a **linear system**, in the unknowns $x_1, x_2, \ldots, x_n$. A **solution** to a linear system in $x_1, x_2, \ldots, x_n$ is a sequence of n numbers $s_1, s_2, \ldots, s_n$ such that each equation is satisfied when we substitute

$$x_1 = s_1, x_2 = s_2, \ldots, x_n = s_n$$

For a system with $n = 2$ variables, we sometimes write the variables as x and y rather than x_1 and x_2; similarly, for a system with $n = 3$ variables, we sometimes write the variables as x, y, and z rather than x_1, x_2, and x_3.

EXAMPLE 3 The linear system

$$\begin{aligned} 2x + 3y + z &= 2 \\ -3x + 2y - 5z &= -7 \\ 5x \qquad + 2z &= -4 \end{aligned} \tag{2}$$

is a system of three equations in the unknowns x, y, and z. The sequence

$$x = -2 \qquad y = 1 \qquad z = 3$$

is a solution to (2), as can be verified by substituting these values into every equation in (2). On the other hand,

$$x = 2 \qquad y = -1 \qquad z = 1$$

is not a solution to (2) since these values do not satisfy the second and third equations.

EXAMPLE 4 The linear system

$$\begin{aligned} 2x_1 + 3x_2 - x_3 &= 1 \\ 3x_1 + 2x_2 + x_3 &= -1 \end{aligned} \tag{3}$$

is a system of two equations in the unknowns x_1, x_2, and x_3. The sequence

$$x_1 = -2 \qquad x_2 = 2 \qquad x_3 = 1$$

is a solution to the system, and

$$x_1 = -3 \qquad x_2 = 3 \qquad x_3 = 2$$

is another solution. On the other hand,

$$x_1 = -2 \qquad x_2 = 3 \qquad x_3 = 4$$

is not a solution to (3) since these values do not satisfy the second equation.

EXAMPLE 5 **APPLICATION TO PRODUCTION PROBLEMS** Let us expand on the production problem discussed in Example 2. In addition to the ingredient alpha discussed in that example, suppose the drugs A, B, C, and D also require the following amounts of a second ingredient, beta, in their manufacture:

Each ounce of A requires 0.4 ounce of ingredient beta.
Each ounce of B requires 0.2 ounce of ingredient beta.
Each ounce of C requires 0.3 ounce of ingredient beta.
Each ounce of D requires 0.8 ounce of ingredient beta.

If the firm makes 300 ounces of ingredient beta each day and if it is desired to produce x_1 ounces of A, x_2 ounces of B, x_3 ounces of C, and x_4 ounces of D each day so that the entire 300 ounces of ingredient beta is used, then x_1, x_2, x_3, and x_4 must satisfy the condition

$$0.4x_1 + 0.2x_2 + 0.3x_3 + 0.8x_4 = 300 \tag{4}$$

which is a linear equation in x_1, x_2, x_3, and x_4. Thus, if ingredients alpha and beta are both to be used, conditions (1) and (4) must both be satisfied. In other words, x_1, x_2, x_3, x_4 must be a solution of the linear system

$$\begin{aligned} 0.1x_1 + 0.3x_2 + 0.5x_3 + 0.2x_4 &= 100 \\ 0.4x_1 + 0.2x_2 + 0.3x_3 + 0.8x_4 &= 300 \end{aligned}$$

There is more than one solution to this system. For example, the firm could produce

$$x_1 = 305 \qquad x_2 = 15 \qquad x_3 = 50 \qquad x_4 = 200$$

ounces of A, B, C, and D, respectively, or it could produce

$$x_1 = 104 \qquad x_2 = 32 \qquad x_3 = 40 \qquad x_4 = 300$$

ounces of A, B, C, and D, respectively (verify this).

Solving Systems of Linear Equations

Not every linear system has a solution. For example, if we multiply the first equation of the system

$$\begin{aligned} 2x + 3y - z &= 4 \\ 4x + 6y - 2z &= 7 \end{aligned} \tag{5}$$

by 2, system (5) becomes

$$\begin{aligned} 4x + 6y - 2z &= 8 \\ 4x + 6y - 2z &= 7 \end{aligned}$$

It is now evident that there is no solution since these equations cannot be true simultaneously.

A linear system that has no solutions is said to be **inconsistent**; if it has at least one solution, it is called **consistent**.

In Section 1.4 we show that a system of two linear equations

$$\begin{aligned} a_1x + b_1y &= c_1 \\ a_2x + b_2y &= c_2 \end{aligned}$$

has either no solutions, exactly one solution, or infinitely many solutions (Figure 1 of Section 1.4). Although we omit the proof, it can be shown that this same result holds for any linear system. In other words, either a linear system is inconsistent (has no solutions) or it is consistent, in which case it has one solution or infinitely many solutions.

Given a system of linear equations, we want to decide whether the system is consistent; if it is consistent, we want to find its solutions. Section 1.4 discusses a method for solving systems of two linear equations in two unknowns. We now develop a method that can be used to solve any linear system. The method we use is based on replacing the given system by a new system that has the same solutions but is easier to solve. Two linear systems are said to be **equivalent** if they have the same solutions. An equivalent system is obtained in a series of steps by applying the following operations.

Operations Used to Solve a Linear System

1. Interchange two equations.
2. Multiply an equation by a nonzero constant. (6)
3. Add a multiple of one equation to another equation.

The following example illustrates how these operations can be used to solve a linear system. In reading this example, you should not worry about how the steps were selected (we discuss this later). Instead, just concentrate on following the computations.

EXAMPLE 6 Consider the linear system

$$\begin{aligned} x + y - 2z &= -3 \\ 2x + 3y + z &= 10 \\ 3x - y + 2z &= 11 \end{aligned} \tag{7}$$

Solution

First, add -2 times the first equation of (7) to the second equation of (7):

$$\begin{array}{r} -2x - 2y + 4z = 6 \\ 2x + 3y + z = 10 \\ \hline y + 5z = 16 \end{array}$$

Then replace the second equation of (7) by $y + 5z = 16$ and obtain the equivalent system

$$\begin{aligned} x + y - 2z &= -3 \\ y + 5z &= 16 \\ 3x - y + 2z &= 11 \end{aligned} \tag{8}$$

Next, add -3 times the first equation of (8) to the third equation of (8):

$$\begin{array}{r} -3x - 3y + 6z = 9 \\ 3x - y + 2z = 11 \\ \hline -4y + 8z = 20 \end{array}$$

Then replace the third equation of (8) by $-4y + 8z = 20$ and obtain another equivalent system. This yields the new system

$$\begin{aligned} x + y - 2z &= -3 \\ y + 5z &= 16 \\ -4y + 8z &= 20 \end{aligned} \tag{9}$$

Now, add 4 times the second equation of (9) to the third equation of (9):

$$\begin{array}{r} 4y + 20z = 64 \\ -4y + 8z = 20 \\ \hline 28z = 84 \end{array}$$

Then replace the third equation of (9) by $28z = 84$ and obtain another equivalent system:

$$\begin{aligned} x + y - 2z &= -3 \\ y + 5z &= 16 \\ 28z &= 84 \end{aligned} \tag{10}$$

Multiply the third equation of (10) by $\frac{1}{28}$ to obtain the equivalent system

$$\begin{aligned} x + y - 2z &= -3 \\ y + 5z &= 16 \\ z &= 3 \end{aligned} \tag{11}$$

Add -5 times the third equation of (11) to the second equation of (11):

$$\begin{array}{r} -5z = -15 \\ \underline{y + 5z = 16} \\ y = 1 \end{array}$$

and 2 times the third equation of (11) to the first equation of (11):

$$\begin{array}{r} 2z = 6 \\ \underline{x + y - 2z = -3} \\ x + y = 3 \end{array}$$

Then replace the second equation of (11) by $y = 1$ and its first equation by $x + y = 3$, obtaining the equivalent, simpler system

$$\begin{aligned} x + y &= 3 \\ y &= 1 \\ z &= 3 \end{aligned} \tag{12}$$

Add -1 times the second equation of (12) to the first equation of (12):

$$\begin{array}{r} -y = -1 \\ \underline{x + y = 3} \\ x = 2 \end{array}$$

Finally replace the first equation of (12) by $x = 2$, obtaining

$$\begin{aligned} x &= 2 \\ y &= 1 \\ z &= 3 \end{aligned}$$

Thus, $x = 2$, $y = 1$, $z = 3$ is the solution to system (7).

Since in the above steps we work with the coefficients of x, y, and z, if we mentally keep track of the location of the unknowns and the equal signs, we can abbreviate a system of linear equations by writing only a rectangular array of numbers. For example, the system

$$\begin{aligned} x + y - 2z &= -3 \\ 2x + 3y + z &= 10 \\ 3x - y + 2z &= 11 \end{aligned}$$

can be abbreviated by writing

$$\left[\begin{array}{rrr|r} 1 & 1 & -2 & -3 \\ 2 & 3 & 1 & 10 \\ 3 & -1 & 2 & 11 \end{array}\right]$$

This array is called the **augmented matrix** for the system. When forming the augmented matrix, the system should be written so that like unknowns are in the same columns and only the constants appear to the right of the equal signs. The dashed vertical line helps us to remember that the entries to the right of it are the numbers appearing on the right sides of the equations.

EXAMPLE 7 The augmented matrix

$$\left[\begin{array}{cccc|c} 3 & -1 & 5 & 2 & 3 \\ -2 & 3 & 4 & 1 & -5 \\ 3 & 2 & -1 & 4 & 6 \end{array}\right]$$

corresponds to the linear system

$$\begin{aligned} 3x_1 - x_2 + 5x_3 + 2x_4 &= 3 \\ -2x_1 + 3x_2 + 4x_3 + x_4 &= -5 \\ 3x_1 + 2x_2 - x_3 + 4x_4 &= 6 \end{aligned}$$

where x_1, x_2, x_3, x_4 represent four unknowns.

Since the rows (horizontal lines) of an augmented matrix correspond to the equations in the associated linear system, the three operations listed in (6) correspond to the following operations on the rows of the augmented matrix.

Row Operations

1. Interchange two rows.
2. Multiply a row by a nonzero constant. (13)
3. Add a multiple of one row to another row.

The next example illustrates how these row operations can be applied to an augmented matrix to solve a linear system.

EXAMPLE 8 In this example we show how to solve the system

$$\begin{aligned} 3x - 4y &= 15 \\ x + 5y &= -14 \end{aligned}$$

by applying the row operations in (13) to the augmented matrix

$$\left[\begin{array}{cc|c} 3 & -4 & 15 \\ 1 & 5 & -14 \end{array}\right]$$

For illustrative purposes, we also solve the system by operating on the equations in the system, so you can see the two methods side by side.

System	Augmented Matrix
$\begin{aligned} 3x - 4y &= 15 \\ x + 5y &= -14 \end{aligned}$	$\left[\begin{array}{cc\|c} 3 & -4 & 15 \\ 1 & 5 & -14 \end{array}\right]$
Interchange the first and second equations to obtain	Interchange the first and second rows to obtain
$\begin{aligned} x + 5y &= -14 \\ 3x - 4y &= 15 \end{aligned}$	$\left[\begin{array}{cc\|c} 1 & 5 & -14 \\ 3 & -4 & 15 \end{array}\right]$

$$\begin{array}{r} -3x - 15y = 42 \\ 3x - 4y = 15 \\ \hline -19y = 57 \end{array}$$

Add -3 times the first equation to the second to obtain

$$\begin{aligned} x + 5y &= -14 \\ -19y &= 57 \end{aligned}$$

Add -3 times the first row to the second to obtain

$$\left[\begin{array}{rr|r} 1 & 5 & -14 \\ 0 & -19 & 57 \end{array}\right]$$

$$\begin{array}{rrr} -3 & -15 & 42 \\ 3 & -4 & 15 \\ \hline 0 & -19 & 57 \end{array}$$

Multiply the second equation by $-\frac{1}{19}$ to obtain

$$\begin{aligned} x + 5y &= -14 \\ y &= -3 \end{aligned}$$

Multiply the second row by $-\frac{1}{19}$ to obtain

$$\left[\begin{array}{rr|r} 1 & 5 & -14 \\ 0 & 1 & -3 \end{array}\right]$$

$$\begin{array}{r} x + 5y = -14 \\ -5y = 15 \\ \hline x \quad = 1 \end{array}$$

Add -5 times the second equation to the first to obtain

$$\begin{aligned} x &= 1 \\ y &= -3 \end{aligned}$$

Add -5 times the second row to the first to obtain

$$\left[\begin{array}{rr|r} 1 & 0 & 1 \\ 0 & 1 & -3 \end{array}\right]$$

$$\begin{array}{rrr} 1 & 5 & -14 \\ 0 & -5 & 15 \\ \hline 1 & 0 & 1 \end{array}$$

The solution is therefore $x = 1$, $y = -3$.

EXERCISE SET 2.1

1. Which of the following are linear equations?
(a) $\frac{5}{6}x + 3y - z = 6$
(b) $0.8x^2 + 2y = 5$
(c) $\sqrt{x_1} + \frac{3}{2}x_2 + 3x_3 - x_4 = 8$
(d) $-3x + 2y + 5z = 16$

2. Which of the following are linear equations?
(a) $2x_1 + 3x_2 - \frac{7}{8}x_3 + x_4 + \frac{3}{2}x_5 = 2$
(b) $2x + xy + z = -8$
(c) $\frac{2}{7}x = 12$
(d) $2x^2 + y - 3z = 7$

3. Which of the following are linear systems?

(a)
$$\begin{aligned} 2x + 3y &= 8 \\ \tfrac{3}{2}x - 3y^2 &= -8 \end{aligned}$$

(b)
$$\begin{aligned} 4x - 5y + 4z &= 18 \\ \tfrac{3}{2}x + 7y + 2z &= 6 \end{aligned}$$

(c)
$$\begin{aligned} \tfrac{2}{5}x + \tfrac{4}{7}y + z &= -4 \\ 2x - y + z &= \tfrac{8}{7} \\ x + 3\sqrt[3]{y} + 2z &= 6 \end{aligned}$$

(d)
$$\begin{aligned} 5x_1 + 3x_2 + 2x_3 &= 6 \\ -2x_1 + 2x_2 + \tfrac{7}{3}x_3 &= 6 \\ 3x_1 + 2x_2 \qquad &= 4 \\ \tfrac{4}{5}x_1 \qquad\qquad &= 15 \end{aligned}$$

4. Consider the linear system

$$\begin{aligned} 3x + y + z &= -4 \\ 3x + 2y - z &= 4 \end{aligned}$$

Which of the following are solutions?
(a) $x = -5,\ y = 10,\ z = 1$
(b) $x = 1,\ y = 1,\ z = -8$
(c) $x = -2,\ y = 4,\ z = -2$

5. Consider the linear system

$$\begin{aligned} 3x_1 + 4x_2 + x_3 - 17x_4 &= 15 \\ 3x_1 + 5x_2 + x_3 - 20x_4 &= 20 \\ 2x_1 + 3x_2 + x_3 - 12x_4 &= 13 \end{aligned}$$

Which of the following are solutions?
(a) $x_1 = -2,\ x_2 = 6,\ x_3 = 5,\ x_4 = 1$
(b) $x_1 = -1,\ x_2 = 8,\ x_3 = 3,\ x_4 = 1$
(c) $x_1 = -3,\ x_2 = 5,\ x_3 = 4,\ x_4 = 0$

6. Write the augmented matrix of each of the following systems.

(a)
$$\begin{aligned} 2x_1 + 3x_2 - x_3 \qquad &= 8 \\ 3x_1 + x_2 \qquad + x_4 &= 6 \end{aligned}$$

(b)
$$\begin{aligned} x + 2y - 3z &= 6 \\ -2x \qquad + z &= -6 \\ 3x + 4y + z &= 8 \end{aligned}$$

(c)
$$\begin{aligned} 4x_1 + 2x_2 + 3x_3 &= 7 \\ 5x_1 + 6x_2 + 7x_3 &= 8 \end{aligned}$$

(d)
$$\begin{aligned} x_1 \qquad &= 3 \\ x_2 &= -5 \end{aligned}$$

7. Write a linear system corresponding to each augmented matrix.

(a) $\left[\begin{array}{rrr|r} 4 & -2 & 0 & 7 \\ -2 & 3 & 5 & -4 \\ 3 & 2 & -5 & 3 \\ 2 & 1 & 3 & 4 \end{array}\right]$ (b) $\left[\begin{array}{rr|r} 2 & 3 & 5 \\ 4 & 2 & 1 \end{array}\right]$

(c) $\left[\begin{array}{rrrr|r} 3 & 2 & 4 & 2 & 5 \\ 2 & 3 & 1 & 3 & 1 \end{array}\right]$ (d) $\left[\begin{array}{rrr|r} 1 & 0 & 0 & 1 \\ 0 & 1 & 0 & 2 \\ 0 & 0 & 1 & 3 \end{array}\right]$

8. For what value(s) of the constant k is the following linear system consistent? Inconsistent?

$$\begin{aligned} 3x - 3y &= k \\ x - y &= 2 \end{aligned}$$

9. Carry out the computations in Example 6 using augmented matrices.

10. Consider the linear system

$$\begin{aligned} ax + by &= k \\ cx + dy &= l \\ ex + fy &= m \end{aligned}$$

Each of the equations in this system represents a line. Sketch the possible positions of these lines when

(a) the system has no solutions
(b) the system has one solution
(c) the system has infinitely many solutions

11. **(Nutrition)** A meal containing two foods is proposed for astronauts by a dietitian. Each ounce of food I provides 2 units of riboflavin, 3 units of iron, and 2 units of carbohydrates. Each ounce of food II provides 2 units of riboflavin, 1 unit of iron, and 4 units of carbohydrates. Let x_1 and x_2 be the number of ounces of foods I and II contained in the meal. Write a system of equations that x_1 and x_2 must satisfy if the meal must provide 12 units of riboflavin, 16 units of iron, and 14 units of carbohydrates.

12. **(Business)** A shipping company has three categories of vessels, A, B, and C, which carry containerized cargos of types I, II, and III. The load capacities of the vessels are given by the rectangular array

Vessel	Types I	II	III
A	4	3	2
B	5	2	3
C	2	2	3

Let x_1, x_2, and x_3 be the number of vessels in each of the categories A, B, and C. Write a system of equations that x_1, x_2, and x_3 must satisfy if the company must ship 42 containers of type I, 27 containers of type II, and 33 containers of type III.

13. Using Figure 1, determine if the system

$$\begin{aligned} a_1x + b_1y &= c_1 \\ a_2x + b_2y &= c_2 \\ a_3x + b_3y &= c_3 \end{aligned}$$

is consistent. Justify your answer.

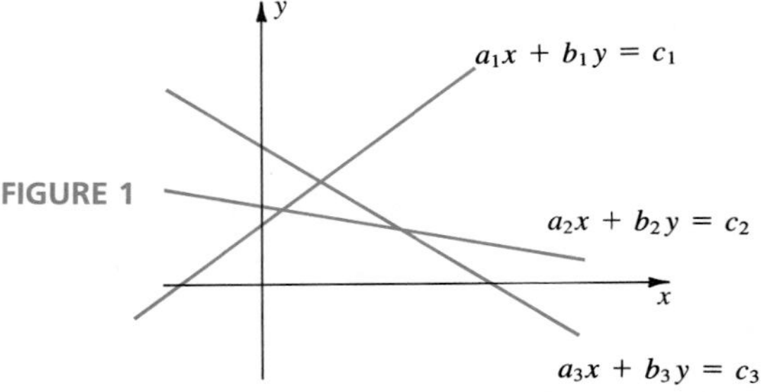

FIGURE 1

14. The graph of a linear equation in three unknowns is a plane in three-dimensional space. A system of three linear equations corresponds to three planes that may not intersect, or may intersect in a line, a plane, or a point. In each part of Figure 2, determine whether the corresponding linear system has one solution, infinitely many solutions, or no solution.

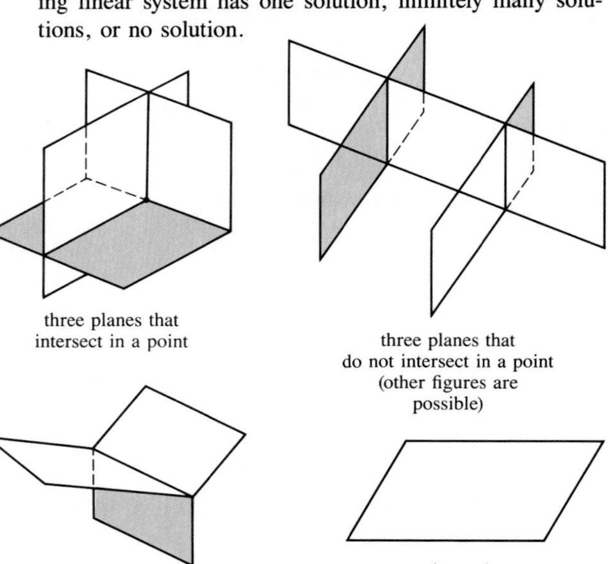

three planes that intersect in a point
three planes that do not intersect in a point (other figures are possible)
three planes that intersect in a line
three planes that intersect in a plane

FIGURE 2

2.2 GAUSS–JORDAN ELIMINATION

In this section we develop a systematic procedure that can be used to solve any system of linear equations. In this method, we reduce the augmented matrix of the given system to a simple form, from which the solutions of the system can be obtained by inspection.

Consider the augmented matrix

$$\left[\begin{array}{ccc|c} 1 & 0 & 0 & 2 \\ 0 & 1 & 0 & 1 \\ 0 & 0 & 1 & 3 \\ 0 & 0 & 0 & 0 \end{array}\right] \tag{1}$$

This is an example of an augmented matrix that is in **reduced row echelon form**. This means the augmented matrix has the following four properties:

Properties of an Augmented Matrix in Reduced Row Echelon Form

1. If a row is not made up entirely of zeros, then the leftmost nonzero number in the row is a 1.
2. If there are any rows consisting entirely of zeros, then they are all together at the bottom of the matrix.
3. In two successive rows not consisting entirely of zeros, the first nonzero number in the lower row is to the right of the first nonzero number in the upper row.
4. Each column that contains the first nonzero number of some row has zeros everywhere else.

EXAMPLE 1 The following augmented matrices are in reduced row echelon form:

$$\left[\begin{array}{cc|c} 1 & 0 & 0 \\ 0 & 1 & 0 \\ 0 & 0 & 1 \end{array}\right] \quad \left[\begin{array}{ccc|c} 1 & 0 & 0 & 1 \\ 0 & 1 & 0 & 5 \\ 0 & 0 & 1 & -2 \\ 0 & 0 & 0 & 0 \end{array}\right] \quad \left[\begin{array}{ccc|c} 1 & 7 & 0 & 0 \\ 0 & 0 & 1 & 0 \\ 0 & 0 & 0 & 0 \end{array}\right]$$

You should check to see that each of the above augmented matrices satisfies all the necessary requirements.

EXAMPLE 2 The following augmented matrices are not in reduced row echelon form:

$$\left[\begin{array}{ccc|c} 1 & 0 & 0 & 4 \\ 0 & 3 & 0 & 1 \\ 0 & 0 & 1 & 2 \\ 0 & 0 & 0 & 0 \end{array}\right] \quad \left[\begin{array}{cc|c} 0 & 0 & 0 \\ 0 & 1 & 0 \\ 0 & 0 & 1 \end{array}\right] \quad \left[\begin{array}{cc|c} 0 & 0 & 1 \\ 0 & 1 & 0 \\ 0 & 0 & 0 \end{array}\right] \quad \left[\begin{array}{cc|c} 1 & 2 & 0 \\ 0 & 1 & 0 \\ 0 & 0 & 1 \end{array}\right]$$

Violates property 1 (The number in color should be 1.)	Violates property 2 (The numbers in color should be at the bottom.)	Violates property 3 (The number in color in the first row should be to the left of the number in color in the second row.)	Violates property 4 (The number in color should be zero.)

The strategy for solving a linear system is to perform a sequence of row operations on the augmented matrix for the system until we obtain a matrix in reduced row echelon form. (We next show how to do this.) Once the augmented matrix has been transformed to reduced row echelon form, the solutions of the system are easy to obtain. The following example illustrates this point.

EXAMPLE 3 In each part, suppose the augmented matrix for a linear system has been reduced by row operations to the given augmented matrix (which is in reduced row echelon form). Solve the system.

(a) $\left[\begin{array}{ccc|c} 1 & 0 & 0 & -2 \\ 0 & 1 & 0 & 4 \\ 0 & 0 & 1 & 3 \end{array}\right]$ (b) $\left[\begin{array}{ccc|c} 1 & 0 & 0 & 0 \\ 0 & 1 & 0 & 0 \\ 0 & 0 & 0 & 1 \end{array}\right]$

(c) $\left[\begin{array}{cccc|c} 1 & 0 & 0 & 2 & 3 \\ 0 & 1 & 0 & 3 & 4 \\ 0 & 0 & 1 & -1 & 1 \\ 0 & 0 & 0 & 0 & 0 \end{array}\right]$ (b) $\left[\begin{array}{ccccc|c} 1 & 2 & 0 & 0 & 3 & 5 \\ 0 & 0 & 1 & 0 & 3 & -2 \\ 0 & 0 & 0 & 1 & -4 & 4 \\ 0 & 0 & 0 & 0 & 0 & 0 \end{array}\right]$

Solution (a) The linear system with unknowns x, y, and z corresponding to the augmented matrix is

$$\begin{aligned} x \qquad &= -2 \\ y \quad &= 4 \\ z &= 3 \end{aligned}$$

Thus, the solution is $x = -2$, $y = 4$, $z = 3$.

Solution (b) The linear system with unknowns x, y, and z corresponding to the augmented matrix is

$$\begin{aligned} x \qquad\qquad &= 0 \\ y \qquad &= 0 \\ 0x + 0y + 0z &= 1 \end{aligned}$$

The third equation in this system is not satisfied by any values of x, y, and z, so the given system is inconsistent (i.e., has no solutions).

Solution (c) The linear system with unknowns x_1, x_2, x_3, and x_4 corresponding to the augmented matrix is

$$\begin{aligned} x_1 \qquad + 2x_4 &= 3 \\ x_2 \quad + 3x_4 &= 4 \\ x_3 - x_4 &= 1 \end{aligned}$$

Since x_1, x_2, x_3 each "begins" one of these equations, we call them **beginning variables**. If we solve for the beginning variables in terms of the remaining variables, we obtain

$$\begin{aligned} x_1 &= 3 - 2x_4 \\ x_2 &= 4 - 3x_4 \\ x_3 &= 1 + x_4 \end{aligned}$$

We can now assign x_4 an arbitrary value t to obtain infinitely many solutions. These solutions are given by the formulas

$$x_1 = 3 - 2t \qquad x_2 = 4 - 3t \qquad x_3 = 1 + t \qquad x_4 = t$$

Particular numerical solutions can be obtained from these formulas by substituting values for t. For example, letting $t = 0$ yields the solution

$$x_1 = 3 \qquad x_2 = 4 \qquad x_3 = 1 \qquad x_4 = 0$$

and letting $t = -1$ yields the solution

$$x_1 = 5 \qquad x_2 = 7 \qquad x_3 = 0 \qquad x_4 = -1$$

Solution (d) The linear system corresponding to the augmented matrix is

$$\begin{aligned} x_1 + 2x_2 \qquad\quad + 3x_5 &= 5 \\ x_3 \quad + 3x_5 &= -2 \\ x_4 - 4x_5 &= 4 \end{aligned}$$

Here the beginning variables are x_1, x_3, and x_4. If we solve for these beginning variables in terms of the remaining variables x_2 and x_5, we obtain

$$\begin{aligned} x_1 &= 5 - 2x_2 - 3x_5 \\ x_3 &= -2 \qquad\quad - 3x_5 \\ x_4 &= 4 \qquad\quad\;\; + 4x_5 \end{aligned}$$

We can now assign x_2 and x_5 arbitrary values s and t, respectively, to obtain infinitely many solutions. These solutions are given by the formulas

$$x_1 = 5 - 2s - 3t \qquad x_2 = s \qquad x_3 = -2 - 3t \qquad x_4 = 4 + 4t \qquad x_5 = t$$

These examples show how to solve a linear system once its augmented matrix is in reduced row echelon form. We now give a procedure, called **Gauss–Jordan* elimination**, for transforming an augmented matrix to reduced row echelon form. We illustrate the method using the following augmented matrix:

$$\left[\begin{array}{rrrrr|r} 0 & 0 & -2 & 0 & 7 & 6 \\ 2 & 4 & 0 & 6 & 12 & 8 \\ 2 & 4 & 5 & 6 & -5 & -2 \end{array}\right]$$

---

* *Wilhelm Jordan* (1842–1899) was a German professor of geodesy who made significant contributions to this field. The Gauss–Jordan elimination method has been incorrectly attributed to Camille Jordan, a celebrated French mathematician.

Carl Friedrich Gauss (1777–1855), sometimes called the "prince of mathematicians," made profound contributions to number theory, theory of functions, probability, and statistics. He discovered a way to calculate the orbits of asteroids, made basic discoveries in electromagnetic theory, and invented a telegraph.

Step 1 Find the leftmost column (vertical line) that does not consist entirely of zeros.

$$\left[\begin{array}{ccccc:c} 0 & 0 & -2 & 0 & 7 & 6 \\ 2 & 4 & 0 & 6 & 12 & 8 \\ 2 & 4 & 5 & 6 & -5 & -2 \end{array}\right]$$

↑ Leftmost nonzero column

Step 2 Interchange the top row with a row below, if necessary, so that the number at the top of the column found in Step 1 is nonzero.

$$\left[\begin{array}{ccccc:c} 2 & 4 & 0 & 6 & 12 & 8 \\ 0 & 0 & -2 & 0 & 7 & 6 \\ 2 & 4 & 5 & 6 & -5 & -2 \end{array}\right]$$

The first and second rows in the preceding augmented matrix were interchanged.

Step 3 If the number that is now at the top of the column found in Step 1 is a, multiply the first row by $1/a$ so that the top entry in the column becomes a 1.

$$\left[\begin{array}{ccccc:c} 1 & 2 & 0 & 3 & 6 & 4 \\ 0 & 0 & -2 & 0 & 7 & 6 \\ 2 & 4 & 5 & 6 & -5 & -2 \end{array}\right]$$

The top row of the preceding augmented matrix was multiplied by $\frac{1}{2}$.

Step 4 Add suitable multiples of the top row to all rows below so that in the column located in Step 3, all numbers below the top row become zeros.

$$\begin{array}{ccccc:cl} -2 & -4 & 0 & -6 & -12 & -8 & (-2 \times \text{row } 1) \\ 2 & 4 & 5 & 6 & -5 & -2 & (\text{row } 3) \\ \hline 0 & 0 & 5 & 0 & -17 & -10 & \end{array}$$

$$\left[\begin{array}{ccccc:c} 1 & 2 & 0 & 3 & 6 & 4 \\ 0 & 0 & -2 & 0 & 7 & 6 \\ 0 & 0 & 5 & 0 & -17 & -10 \end{array}\right]$$

-2 times the first row of the preceding augmented matrix was added to the third row.

Step 5 Ignore (shade) the top row of the last matrix and begin again with Step 1 applied to the *sub*matrix that remains. Continue in this way until it is impossible to follow the steps any further.

$$\left[\begin{array}{ccccc:c} 1 & 2 & 0 & 3 & 6 & 4 \\ 0 & 0 & -2 & 0 & 7 & 6 \\ 0 & 0 & 5 & 0 & -17 & -10 \end{array}\right]$$

↑ Leftmost nonzero column in the submatrix obtained by covering the first row.

$$\left[\begin{array}{ccccc:c} 1 & 2 & 0 & 3 & 6 & 4 \\ 0 & 0 & 1 & 0 & -\frac{7}{2} & -3 \\ 0 & 0 & 5 & 0 & -17 & -10 \end{array}\right]$$

The first row of the preceding submatrix was multiplied by $-\frac{1}{2}$.

$$\begin{array}{ccccc:cl} 0 & 0 & -5 & 0 & \frac{35}{2} & 15 & (-5 \times \text{row } 2) \\ 0 & 0 & 5 & 0 & -17 & -10 & (\text{row } 3) \\ \hline 0 & 0 & 0 & 0 & \frac{1}{2} & 5 & \end{array}$$

$$\left[\begin{array}{ccccc:c} 1 & 2 & 0 & 3 & 6 & 4 \\ 0 & 0 & 1 & 0 & -\frac{7}{2} & -3 \\ 0 & 0 & 0 & 0 & \frac{1}{2} & 5 \end{array}\right]$$

-5 times the first row of the preceding submatrix was added to the second row of the submatrix.

$$\left[\begin{array}{ccccc|c} 1 & 2 & 0 & 3 & 6 & 4 \\ 0 & 0 & 1 & 0 & -\frac{7}{2} & -3 \\ 0 & 0 & 0 & 0 & \frac{1}{2} & 5 \end{array}\right]$$

The top row of the submatrix was ignored and we returned to Step 1 again.

Leftmost nonzero column in the new submatrix.

$$\left[\begin{array}{ccccc|c} 1 & 2 & 0 & 3 & 6 & 4 \\ 0 & 0 & 1 & 0 & -\frac{7}{2} & -3 \\ 0 & 0 & 0 & 0 & 1 & 10 \end{array}\right]$$

The top (and only) row of the new submatrix was multiplied by 2.

Step 6 Consider this last matrix without any shading. Beginning with the last nonzero row and working upward, add suitable multiples of each row to the rows above so that the fourth property in the definition of reduced row echelon form is satisfied.

$$\left[\begin{array}{ccccc|c} 1 & 2 & 0 & 3 & 6 & 4 \\ 0 & 0 & 1 & 0 & -\frac{7}{2} & -3 \\ 0 & 0 & 0 & 0 & 1 & 10 \end{array}\right]$$

$$\begin{array}{ccccc|cl} 0 & 0 & 0 & 0 & \frac{7}{2} & 35 & (\frac{7}{2} \times \text{row } 3) \\ 0 & 0 & 1 & 0 & -\frac{7}{2} & -3 & (\text{row } 2) \\ \hline 0 & 0 & 1 & 0 & 0 & 32 & \end{array}$$

$$\begin{array}{ccccc|cl} 0 & 0 & 0 & 0 & -6 & -60 & (-6 \times \text{row } 3) \\ 1 & 2 & 0 & 3 & 6 & 4 & (\text{row } 1) \\ \hline 1 & 2 & 0 & 3 & 0 & -56 & \end{array}$$

$\frac{7}{2}$ times the third row of the preceding matrix was added to the second row and -6 times the third row was added to the first row.

$$\left[\begin{array}{ccccc|c} 1 & 2 & 0 & 3 & 0 & -56 \\ 0 & 0 & 1 & 0 & 0 & 32 \\ 0 & 0 & 0 & 0 & 1 & 10 \end{array}\right]$$

The final matrix shown is in reduced row echelon form.

We now illustrate how Gauss–Jordan elimination can be used to solve linear systems.

EXAMPLE 4 Use Gauss–Jordan elimination to solve

$$\begin{aligned} 4x + 4y + 8z &= 4 \\ 3x + 4y + 6z &= 3 \\ -2x + 3y - 3z &= -1 \end{aligned}$$

Solution The augmented matrix for this system is

$$\left[\begin{array}{ccc|c} 4 & 4 & 8 & 4 \\ 3 & 4 & 6 & 3 \\ -2 & 3 & -3 & -1 \end{array}\right]$$

Multiply the first row by $\frac{1}{4}$ to obtain

$$\left[\begin{array}{ccc|c} 1 & 1 & 2 & 1 \\ 3 & 4 & 6 & 3 \\ -2 & 3 & -3 & 1 \end{array}\right]$$

Add -3 times the first row to the second and add 2 times the first row to the third to obtain

$$\begin{array}{rrr|r} -3 & -3 & -6 & -3 \\ 3 & 4 & 6 & 3 \\ \hline 0 & 1 & 0 & 0 \end{array} \qquad \begin{array}{rrr|r} 2 & 2 & 4 & 2 \\ -2 & 3 & -3 & -1 \\ \hline 0 & 5 & 1 & 1 \end{array}$$

$$\left[\begin{array}{ccc|c} 1 & 1 & 2 & 1 \\ 0 & 1 & 0 & 0 \\ 0 & 5 & 1 & 1 \end{array}\right]$$

Add -5 times the second row to the third to obtain

$$\begin{array}{rrr|r} 0 & -5 & 0 & 0 \\ 0 & 5 & 1 & 1 \\ \hline 0 & 0 & 1 & 1 \end{array}$$

$$\left[\begin{array}{ccc|c} 1 & 1 & 2 & 1 \\ 0 & 1 & 0 & 0 \\ 0 & 0 & 1 & 1 \end{array}\right]$$

Add -2 times the third row to the first to obtain

$$\begin{array}{rrr|r} 1 & 1 & 2 & 1 \\ 0 & 0 & -2 & -2 \\ \hline 1 & 1 & 0 & -1 \end{array}$$

$$\left[\begin{array}{ccc|c} 1 & 1 & 0 & -1 \\ 0 & 1 & 0 & 0 \\ 0 & 0 & 1 & 1 \end{array}\right]$$

Add -1 times the second row to the first to obtain

$$\begin{array}{rrr|r} 1 & 1 & 0 & -1 \\ 0 & -1 & 0 & 0 \\ \hline 1 & 0 & 0 & -1 \end{array}$$

$$\left[\begin{array}{ccc|c} 1 & 0 & 0 & -1 \\ 0 & 1 & 0 & 0 \\ 0 & 0 & 1 & 1 \end{array}\right]$$

which is in reduced row echelon form. The linear system corresponding to the last augmented matrix is

$$\begin{aligned} x \quad\quad &= -1 \\ y \quad &= 0 \\ z &= 1 \end{aligned}$$

so the solution to the system is $x = -1$, $y = 0$, $z = 1$. Check your answer.

EXAMPLE 5 Use Gauss–Jordan elimination to solve

$$\begin{aligned} 2x_1 + 3x_2 - 3x_3 + 7x_4 + x_5 + 4x_6 &= -2 \\ x_2 - x_3 + 3x_4 - x_5 + 2x_6 &= 2 \\ 2x_1 + 2x_2 - x_3 + 6x_4 + 2x_5 - x_6 &= -7 \\ 2x_1 + x_2 - x_3 + x_4 + 3x_5 \quad &= -6 \end{aligned} \tag{2}$$

Solution The augmented matrix of this linear system is

$$\left[\begin{array}{rrrrrr|r} 2 & 3 & -3 & 7 & 1 & 4 & -2 \\ 0 & 1 & -1 & 3 & -1 & 2 & 2 \\ 2 & 2 & -1 & 6 & 2 & -1 & -7 \\ 2 & 1 & -1 & 1 & 3 & 0 & -6 \end{array}\right]$$

Multiply the first row by $\frac{1}{2}$ to obtain

$$\left[\begin{array}{cccccc:c} 1 & \frac{3}{2} & -\frac{3}{2} & \frac{7}{2} & \frac{1}{2} & 2 & -1 \\ 0 & 1 & -1 & 3 & -1 & 2 & 2 \\ 2 & 2 & -1 & 6 & 2 & -1 & -7 \\ 2 & 1 & -1 & 1 & 3 & 0 & -6 \end{array}\right]$$

Add -2 times the first row to both the third and fourth rows to obtain

$$\begin{array}{cccccc:c} -2 & -3 & 3 & -7 & -1 & -4 & 2 \\ 2 & 2 & -1 & 6 & 2 & -1 & -7 \\ \hline 0 & -1 & 2 & -1 & 1 & -5 & -5 \\ \\ -2 & -3 & 3 & -7 & -1 & -4 & 2 \\ 2 & 1 & -1 & 1 & 3 & 0 & -6 \\ \hline 0 & -2 & 2 & -6 & 2 & -4 & -4 \end{array}$$

$$\left[\begin{array}{cccccc:c} 1 & \frac{3}{2} & -\frac{3}{2} & \frac{7}{2} & \frac{1}{2} & 2 & -1 \\ 0 & 1 & -1 & 3 & -1 & 2 & 2 \\ 0 & -1 & 2 & -1 & 1 & -5 & -5 \\ 0 & -2 & 2 & -6 & 2 & -4 & -4 \end{array}\right]$$

Add the second row to the third row and 2 times the second row to the fourth row to obtain

$$\begin{array}{cccccc:c} 0 & 1 & -1 & 3 & -1 & 2 & 2 \\ 0 & -1 & 2 & -1 & 1 & -5 & -5 \\ \hline 0 & 0 & 1 & 2 & 0 & -3 & -3 \\ \\ 0 & 2 & -2 & 6 & -2 & 4 & 4 \\ 0 & -2 & 2 & -6 & 2 & -4 & -4 \\ \hline 0 & 0 & 0 & 0 & 0 & 0 & 0 \end{array}$$

$$\left[\begin{array}{cccccc:c} 1 & \frac{3}{2} & -\frac{3}{2} & \frac{7}{2} & \frac{1}{2} & 2 & -1 \\ 0 & 1 & -1 & 3 & -1 & 2 & 2 \\ 0 & 0 & 1 & 2 & 0 & -3 & -3 \\ 0 & 0 & 0 & 0 & 0 & 0 & 0 \end{array}\right]$$

Add the third row to the second row to obtain

$$\begin{array}{cccccc:c} 0 & 0 & 1 & 2 & 0 & -3 & -3 \\ 0 & 1 & -1 & 3 & -1 & 2 & 2 \\ \hline 0 & 1 & 0 & 5 & -1 & -1 & -1 \end{array}$$

$$\left[\begin{array}{cccccc:c} 1 & \frac{3}{2} & -\frac{3}{2} & \frac{7}{2} & \frac{1}{2} & 2 & -1 \\ 0 & 1 & 0 & 5 & -1 & -1 & -1 \\ 0 & 0 & 1 & 2 & 0 & -3 & -3 \\ 0 & 0 & 0 & 0 & 0 & 0 & 0 \end{array}\right]$$

Add $\frac{3}{2}$ times the third row to the first row to obtain

$$\begin{array}{cccccc:c} 0 & 0 & \frac{3}{2} & 3 & 0 & -\frac{9}{2} & -\frac{9}{2} \\ 1 & \frac{3}{2} & -\frac{3}{2} & \frac{7}{2} & \frac{1}{2} & 2 & -1 \\ \hline 1 & \frac{3}{2} & 0 & \frac{13}{2} & \frac{1}{2} & -\frac{5}{2} & -\frac{11}{2} \end{array}$$

$$\left[\begin{array}{cccccc:c} 1 & \frac{3}{2} & 0 & \frac{13}{2} & \frac{1}{2} & -\frac{5}{2} & -\frac{11}{2} \\ 0 & 1 & 0 & 5 & -1 & -1 & -1 \\ 0 & 0 & 1 & 2 & 0 & -3 & -3 \\ 0 & 0 & 0 & 0 & 0 & 0 & 0 \end{array}\right]$$

Add $-\frac{3}{2}$ times the second row to the first row to obtain

$$\begin{array}{cccccc:c} 0 & -\frac{3}{2} & 0 & -\frac{15}{2} & \frac{3}{2} & -\frac{3}{2} & \frac{3}{2} \\ 1 & \frac{3}{2} & 0 & \frac{13}{2} & \frac{1}{2} & -\frac{5}{2} & -\frac{11}{2} \\ \hline 1 & 0 & 0 & -1 & 2 & -1 & -4 \end{array}$$

$$\left[\begin{array}{cccccc:c} 1 & 0 & 0 & -1 & 2 & -1 & -4 \\ 0 & 1 & 0 & 5 & -1 & -1 & -1 \\ 0 & 0 & 1 & 2 & 0 & -3 & -3 \\ 0 & 0 & 0 & 0 & 0 & 0 & 0 \end{array}\right] \tag{3}$$

which is in reduced row echelon form. The linear system corresponding to (3) is

$$\begin{aligned} x_1 \qquad\quad - x_4 + 2x_5 - x_6 &= -4 \\ x_2 \quad + 5x_4 - x_5 - x_6 &= -1 \\ x_3 + 2x_4 \qquad\quad - 3x_6 &= -3 \end{aligned}$$

(We have dropped the last equation, $0x_1 + 0x_2 + 0x_3 + 0x_4 + 0x_5 + 0x_6 = 0$, because it will automatically be satisfied.) Solving for the beginning variables, x_1, x_2, and x_3, we obtain

$$\begin{aligned} x_1 &= -4 + x_4 - 2x_5 + x_6 \\ x_2 &= -1 - 5x_4 + x_5 + x_6 \\ x_3 &= -3 - 2x_4 \qquad\quad + 3x_6 \end{aligned}$$

If we assign x_4, x_5, and x_6 the arbitrary values r, s, and t, respectively, we obtain infinitely many solutions. They are given by the formulas

$$x_1 = -4 + r - 2s + t \qquad x_2 = -1 - 5r + s + t$$
$$x_3 = -3 - 2r + 3t \qquad x_4 = r \qquad x_5 = s \qquad x_6 = t$$

EXAMPLE 6 **APPLICATION TO PRODUCTION PROBLEMS** In Example 5 of Section 2.1, we show that if a pharmaceutical manufacturer wants to produce x_1 ounces of A, x_2 ounces of B, x_3 ounces of C, and x_4 ounces of D each day so that the entire daily stock of 100 ounces of alpha and 300 ounces of beta is used, then x_1, x_2, x_3, and x_4 must satisfy

$$\begin{aligned} 0.1x_1 + 0.3x_2 + 0.5x_3 + 0.2x_4 &= 100 \\ 0.4x_1 + 0.2x_2 + 0.3x_3 + 0.8x_4 &= 300 \end{aligned}$$

We solve this system by Gauss–Jordan elimination. The augmented matrix for the system is

$$\left[\begin{array}{cccc:c} 0.1 & 0.3 & 0.5 & 0.2 & 100 \\ 0.4 & 0.2 & 0.3 & 0.8 & 300 \end{array}\right]$$

To transform this augmented matrix to reduced row echelon form, we first multiply the top row by 10 to obtain

$$\left[\begin{array}{cccc:c} 1 & 3 & 5 & 2 & 1000 \\ 0.4 & 0.2 & 0.3 & 0.8 & 300 \end{array}\right]$$

Next add -0.4 times the top row to the bottom to obtain

$$\left[\begin{array}{cccc:c} 1 & 3 & 5 & 2 & 1000 \\ 0 & -1 & -1.7 & 0 & -100 \end{array}\right]$$

Now multiply the second row by -1. This gives

$$\left[\begin{array}{cccc:c} 1 & 3 & 5 & 2 & 1000 \\ 0 & 1 & 1.7 & 0 & 100 \end{array}\right]$$

Finally, add -3 times the bottom row to the top row to obtain the reduced row echelon form

$$\left[\begin{array}{cccc:c} 1 & 0 & -0.1 & 2 & 700 \\ 0 & 1 & 1.7 & 0 & 100 \end{array}\right]$$

The corresponding system of equations is

$$\begin{aligned} x_1 \qquad - 0.1x_3 + 2x_4 &= 700 \\ x_2 + 1.7x_3 \qquad\quad &= 100 \end{aligned}$$

Solving for the beginning variables, we obtain

$$x_1 = 700 + 0.1x_3 - 2x_4$$
$$x_2 = 100 - 1.7x_3$$

If we assign x_3 and x_4 arbitrary values s and t, respectively, we obtain

$$x_1 = 700 + 0.1s - 2t \qquad x_2 = 100 - 1.7s \qquad x_3 = s \qquad x_4 = t$$

Thus, the problem has infinitely many solutions. Some of the solutions and the values of s and t that generate them are shown in Table 1.

TABLE 1

Ounces of A, x_1	**Ounces of B,** x_2	**Ounces of C,** x_3	**Ounces of D,** x_4	**Values of** s **and** t
305	15	50	200	$s = 50,\ t = 200$
104	32	40	300	$s = 40,\ t = 300$
503	49	30	100	$s = 30,\ t = 100$
670.5	91.5	5	15	$s = 5,\ t = 15$
700	100	0	0	$s = 0,\ t = 0$

In a practical situation, the choice of the values for s and t would be dictated by such considerations as demand for the products, cost of manufacture, available materials and labor, and so on. Also, to obtain a physically sensible solution, the values of s and t would have to be chosen so that $x_1 \geq 0$, $x_2 \geq 0$, $x_3 \geq 0$, and $x_4 \geq 0$.

EXERCISE SET 2.2

1. Which of the following augmented matrices are in reduced row echelon form?

(a) $\left[\begin{array}{cccc|c} 0 & 1 & 0 & 0 & 0 \\ 0 & 0 & 1 & 0 & 1 \\ 0 & 0 & 0 & 0 & 0 \end{array}\right]$

(b) $\left[\begin{array}{cccc|c} 0 & 1 & 0 & 0 & 2 \\ 0 & 0 & 1 & 0 & -1 \\ 0 & 0 & 0 & 0 & 1 \end{array}\right]$

(c) $\left[\begin{array}{ccccc|c} 0 & 0 & 1 & 0 & 0 & 1 \\ 0 & 0 & 0 & 1 & 0 & -3 \\ 0 & 0 & 0 & 0 & 1 & 4 \end{array}\right]$

(d) $\left[\begin{array}{cc|c} 1 & 0 & 2 \\ 0 & 0 & 1 \end{array}\right]$

2. Which of the following augmented matrices are in reduced row echelon form?

(a) $\left[\begin{array}{cccc|c} 1 & 0 & 0 & 0 & 0 \\ 0 & 0 & 1 & 0 & 2 \\ 0 & 0 & 0 & 1 & -3 \\ 0 & 0 & 0 & 0 & 0 \end{array}\right]$

(b) $\left[\begin{array}{ccc|c} 0 & 1 & 0 & 0 \\ 0 & 0 & -1 & 2 \\ 0 & 0 & 0 & 0 \end{array}\right]$

(c) $\left[\begin{array}{cccc|c} 1 & 0 & 0 & 0 & 1 \\ 0 & 0 & 1 & 2 & 3 \\ 0 & 0 & 1 & 0 & 0 \end{array}\right]$

(d) $\left[\begin{array}{cccc|c} 0 & 1 & 0 & 0 & -8 \\ 0 & 0 & 1 & 0 & 6 \\ 0 & 0 & 0 & 1 & -2 \end{array}\right]$

3. In each of the following, the augmented matrix for a linear system has been transformed by row operations to the given reduced row echelon form. Solve the linear system.

(a) $\left[\begin{array}{ccc|c} 1 & 0 & 0 & 3 \\ 0 & 1 & 0 & -2 \\ 0 & 0 & 1 & 4 \end{array}\right]$

(b) $\left[\begin{array}{cccc|c} 1 & 0 & 0 & 2 & 3 \\ 0 & 1 & 0 & 3 & 6 \\ 0 & 0 & 1 & 4 & -2 \end{array}\right]$

(c) $\left[\begin{array}{ccccc|c} 1 & 2 & 0 & 0 & 5 & 1 \\ 0 & 0 & 1 & 0 & 4 & -2 \\ 0 & 0 & 0 & 1 & 2 & 3 \\ 0 & 0 & 0 & 0 & 0 & 0 \end{array}\right]$

(d) $\left[\begin{array}{ccc|c} 1 & 2 & 0 & 0 \\ 0 & 0 & 1 & 0 \\ 0 & 0 & 0 & 1 \end{array}\right]$

4. In each of the following, the augmented matrix for a linear system has been transformed by row operations to the given reduced row echelon form. Solve the linear system.

(a) $\left[\begin{array}{ccc|c} 1 & 0 & 5 & 2 \\ 0 & 1 & 3 & 2 \\ 0 & 0 & 0 & 0 \end{array}\right]$ (b) $\left[\begin{array}{ccc|c} 1 & 4 & 0 & 0 \\ 0 & 0 & 1 & 0 \\ 0 & 0 & 0 & 1 \end{array}\right]$

(c) $\left[\begin{array}{ccccc|c} 1 & 0 & 0 & 0 & 3 & 4 \\ 0 & 1 & 0 & 0 & 2 & 0 \\ 0 & 0 & 0 & 1 & 2 & -1 \end{array}\right]$

(d) $\left[\begin{array}{cccc|c} 1 & 0 & 3 & 0 & 4 \\ 0 & 1 & 2 & 0 & 2 \\ 0 & 0 & 0 & 1 & -5 \\ 0 & 0 & 0 & 0 & 0 \end{array}\right]$

5. Transform each of the following augmented matrices to reduced row echelon form.

(a) $\left[\begin{array}{cccc|c} 2 & -4 & -2 & 0 & 4 \\ 2 & -2 & -4 & 6 & 10 \\ -2 & 4 & 2 & 0 & 1 \end{array}\right]$

(b) $\left[\begin{array}{cccc|c} 1 & 1 & 2 & 1 & 5 \\ 1 & 2 & 4 & 2 & 8 \\ 1 & -1 & -2 & -1 & -1 \end{array}\right]$

(c) $\left[\begin{array}{cccc|c} 1 & 0 & 1 & 5 & -1 \\ 2 & 0 & -1 & 1 & 7 \\ 4 & 0 & 1 & 11 & 5 \end{array}\right]$

6. Transform each of the following augmented matrices to reduced row echelon form.

(a) $\left[\begin{array}{ccc|c} 2 & 4 & 4 & 6 \\ 2 & 2 & -2 & -4 \\ 3 & 3 & 12 & 15 \end{array}\right]$

(b) $\left[\begin{array}{ccc|c} 1 & 1 & 1 & 6 \\ 1 & 5 & 1 & -2 \\ 1 & 1 & 1 & 3 \\ -1 & 0 & 0 & -3 \end{array}\right]$

(c) $\left[\begin{array}{cccc|c} 1 & 1 & -1 & 0 & 3 \\ 3 & 5 & 1 & 2 & 11 \\ 1 & 2 & 1 & 1 & 4 \end{array}\right]$

7. Solve each of the following linear systems by Gauss–Jordan elimination.

(a) $$\begin{aligned} 2x - 2y + 4z &= 6 \\ 2x - y + 5z &= 15 \\ -x + y + 3z &= 7 \end{aligned}$$

(b) $$\begin{aligned} x_1 + 2x_2 + x_3 &= 0 \\ -2x_1 + x_2 + 3x_3 &= 0 \\ 2x_1 + 3x_2 + 5x_3 &= 0 \end{aligned}$$

(c) $$\begin{aligned} x_1 - x_2 + 3x_3 + 2x_4 - 3x_5 &= 2 \\ 2x_1 + x_2 + x_4 + 3x_5 &= 1 \\ 3x_1 + x_2 - 3x_3 + 2x_4 - x_5 &= 10 \end{aligned}$$

(d) $$\begin{aligned} -x_1 + 3x_2 - x_3 + x_4 &= 3 \\ x_1 - 2x_2 - x_3 + x_4 &= 1 \\ 3x_1 - 8x_2 + 3x_3 - x_4 &= -5 \end{aligned}$$

8. Solve each of the following linear systems by Gauss–Jordan elimination.

(a) $$\begin{aligned} 2x + 3y &= 2 \\ -5x + 12y &= -5 \\ -3x + 2y &= -3 \end{aligned}$$

(b) $$\begin{aligned} x + 2y - z &= 0 \\ 5x + 2y + z &= 0 \\ 3x + 2y + 3z &= 0 \end{aligned}$$

(c) $$\begin{aligned} 2x_1 + 3x_2 + x_3 + x_4 &= 0 \\ 3x_1 - x_2 + x_3 - x_4 &= 0 \end{aligned}$$

(d) $$\begin{aligned} x_1 + 2x_2 &= 3 \\ 2x_1 + x_2 + 3x_3 &= 5 \\ x_2 + x_3 &= -3 \end{aligned}$$

9. Solve each of the following linear systems by Gauss–Jordan elimination.

(a) $$\begin{aligned} x_1 + 2x_2 + 3x_3 + x_4 &= 3 \\ 2x_1 + 3x_2 + 3x_3 + 2x_4 &= -5 \\ -5x_2 - x_3 + 28x_4 &= -13 \\ x_1 + x_2 + 2x_3 - 3x_4 &= -2 \end{aligned}$$

(b) $$\begin{aligned} x_1 + 2x_2 &= 3 \\ 2x_1 - 3x_2 &= -1 \end{aligned}$$

(c) $$\begin{aligned} x_1 + 2x_2 - x_3 &= 1 \\ -x_1 + x_2 + x_3 &= 5 \\ -2x_1 - 3x_2 + 4x_3 &= 4 \end{aligned}$$

(d) $$\begin{aligned} 2x_1 + x_2 - x_3 &= 3 \\ -2x_1 - x_2 + 3x_3 &= 5 \end{aligned}$$

10. Show that if $ad \neq 1$, then the linear system

$$\begin{aligned} ax + y &= 3 \\ x + dy &= 5 \end{aligned}$$

has exactly one solution.

11. For which values of a will the following linear system have no solution? Exactly one solution? Infinitely many solutions?

$$\begin{aligned} x + y &= 2 \\ x + (a^2 - 3)y &= a \end{aligned}$$

12. Show that $x = 0$, $y = 0$ is a solution to the system

$$\begin{aligned} ax + by &= 0 \\ cx + dy &= 0 \end{aligned}$$

13. Show that if $ad - bc = 0$, then the system

$$\begin{aligned} ax + by &= 0 \\ cx + dy &= 0 \end{aligned}$$

has infinitely many solutions.

14. Verify Exercise 13 for the system

$$\begin{aligned} 2x + 3y &= 0 \\ 4x + 6y &= 0 \end{aligned}$$

by giving two different solutions.

15. **(Harder)** Show that if

$$\begin{aligned} x &= x_0 \qquad y = y_0 \\ x &= x_1 \qquad y = y_1 \end{aligned}$$

are solutions to the system

$$\begin{aligned} ax + by &= 0 \\ cx + dy &= 0 \end{aligned}$$

then so is $x = x_0 + rx_1$, $y = y_0 + ry_1$, where r is an arbitrary real number.

16. In Exercise 11 of Exercise Set 2.1, find the number of ounces of foods I and II that should be included in the astronauts' meal.

17. Solve the problem in Exercise 12 of Exercise Set 2.1.

18. A furniture manufacturer makes chairs, coffee tables, and dining-room tables. Each chair requires 2 minutes of sanding, 2 minutes of staining, and 4 minutes of varnishing. Each coffee table requires 5 minutes of sanding, 4 minutes of staining, and 3 minutes of varnishing. Each dining-room table requires 5 minutes of sanding, 4 minutes of staining, and 6 minutes of varnishing. The sanding and varnishing benches are each available 6 hours per day, and the staining bench is available 5 hours per day. How many of each type of furniture can be made daily?

19. A manufacturer produces 12-, 16-, and 19-inch television sets that require assembly, testing, and packing. The 12-inch sets each require 45 minutes to assemble, 30 minutes to test, and 10 minutes to package. The 16-inch sets each require 1 hour to assemble, 45 minutes to test, and 15 minutes to package. The 19-inch sets each require $1\frac{1}{2}$ hours to assemble, 1 hour to test, and 15 minutes to package. If the assembly line operates for $17\frac{3}{4}$ hours per day, the test facility is available for $12\frac{1}{2}$ hours per day, and the packing equipment is available for $3\frac{3}{4}$ hours per day, how many sets of each type can be produced daily?

2.3 MATRICES; MATRIX ADDITION AND SCALAR MULTIPLICATION

We have already seen how rectangular arrays of numbers are used in solving systems of linear equations. Such arrays arise in other contexts as well. They occur in problems of business, science, engineering, economics, and the behavioral sciences. In this section we study rectangular arrays in detail and indicate several applications. Other, more detailed applications appear in Chapter 10.

A rectangular array of numbers is called a **matrix**. The numbers in the array are called the **entries** of the matrix.

EXAMPLE 1 The following are matrices:

$$\begin{bmatrix} 1 & 2 \\ 3 & -4 \end{bmatrix} \quad [\tfrac{2}{5} \;\; 3] \quad \begin{bmatrix} 0 \\ 1 \\ \sqrt{2} \end{bmatrix} \quad [7] \quad \begin{bmatrix} -1 & 1 & 2 \\ 3 & 8 & -5 \end{bmatrix}$$

A matrix is said to have **size** $m \times n$ (read "m by n") if it has m rows (horizontal lines) and n columns (vertical lines). If $m = n$, then the matrix is called a **square matrix.**

EXAMPLE 2 The matrices in Example 1 are of size

$$2 \times 2 \qquad 1 \times 2 \qquad 3 \times 1 \qquad 1 \times 1 \qquad 2 \times 3$$

respectively. The first and fourth matrices are square.

We denote matrices by uppercase letters, such as A, B, and C. In discussions involving matrices, it is common to call real numbers **scalars.** We use lowercase letters, such as a, b, and c to denote scalars. For example, we might write

$$A = \begin{bmatrix} a & b \\ c & d \end{bmatrix} \quad \text{or} \quad B = \begin{bmatrix} 1 & 3 & 5 \\ 7 & 2 & 1 \end{bmatrix}$$

If A is a matrix, we often denote the entry in row i and column j of A by a_{ij}. For example, a_{13} denotes the entry in row 1 and column 3 of matrix A. Thus, a general 2×3 matrix might be written

$$A = \begin{bmatrix} a_{11} & a_{12} & a_{13} \\ a_{21} & a_{22} & a_{23} \end{bmatrix}$$

Similarly, in a matrix B, we denote the entry in row i and column j by b_{ij}.

EXAMPLE 3 If

$$B = \begin{bmatrix} 2 & -4 & 3 & 4 \\ 6 & 0 & 2 & 3 \\ 4 & 2 & -5 & 0 \end{bmatrix}$$

then B is a 3×4 matrix and

$$b_{12} = -4 \qquad b_{14} = 4 \qquad b_{21} = 6 \qquad b_{23} = 2 \qquad b_{31} = 4 \qquad b_{33} = -5$$

EXAMPLE 4 A manufacturer has three machines that produce goods. These machines are operated by two shifts, shift 1 and shift 2. If a_{ij} denotes the output on machine j by shift i, then the daily production can be described by a 2×3 matrix A such as

$$A = \begin{matrix} & \text{Machines} & & \\ & 1 & 2 & 3 \\ \text{Shift 1} & 300 & 217 & 64 \\ \text{Shift 2} & 271 & 201 & 83 \end{matrix}$$

From this matrix we see that the output on machine 2 by shift 1 is $a_{12} = 217$ units.

EXAMPLE 5 **ECONOMICS** Consider a simplified economic society consisting of a farming sector (1), a housing sector (2), and a clothing sector (3). In this economy, just as in our own, each sector uses the products of the other sectors. For example, farmers need food, houses, and clothing and so do the people who produce clothes or housing. If c_{ij} denotes the fraction of the output of sector j which is consumed by sector i, then the economic relationships between the three sectors can be described by a 3×3 matrix C such as

$$C = \begin{array}{r} \text{Farming (1)} \\ \text{Housing (2)} \\ \text{Clothing (3)} \end{array} \overset{\begin{array}{ccc} \text{Farming} & \text{Housing} & \text{Clothing} \\ (1) & (2) & (3) \end{array}}{\begin{bmatrix} \frac{3}{16} & \frac{4}{7} & \frac{7}{16} \\ \frac{5}{16} & \frac{2}{7} & \frac{5}{16} \\ \frac{1}{2} & \frac{1}{7} & \frac{1}{4} \end{bmatrix}}$$

For example, entry $c_{23} = \frac{5}{16}$, so sector (2) (housing) consumes $\frac{5}{16}$ of the output of sector (3) (clothing).

Equality of Matrices Two matrices A and B are said to be **equal**, written $A = B$, if they have the same size and their corresponding entries are equal.

EXAMPLE 6 In order for the matrices

$$A = \begin{bmatrix} 2 & 3 \\ -1 & 5 \end{bmatrix} \quad \text{and} \quad B = \begin{bmatrix} 2 & a \\ b & 5 \end{bmatrix}$$

to be equal, we must have $a = 3$ and $b = -1$.

EXAMPLE 7 Consider the matrices

$$A = \begin{bmatrix} 4 & 2 \\ -1 & 2 \end{bmatrix} \qquad B = \begin{bmatrix} 4 & 2 & 0 \\ -1 & 2 & 0 \end{bmatrix} \qquad C = \begin{bmatrix} 4 & 2 \\ 3 & 2 \end{bmatrix}$$

Here, $A \neq B$ since the matrices A and B have different sizes. For the same reason, $B \neq C$. Further, $A \neq C$ since not all the corresponding entries are equal.

Matrix Addition If A and B are two matrices with the same size, then the **sum** $A + B$ is the matrix obtained by adding the corresponding entries in A and B. We do not define addition for matrices with different sizes.

EXAMPLE 8 Let

$$A = \begin{bmatrix} 3 & 4 & -1 \\ 2 & 0 & 1 \\ 3 & 1 & 2 \end{bmatrix} \qquad B = \begin{bmatrix} 2 & -1 & 3 \\ -2 & 1 & 2 \\ 6 & 4 & -3 \end{bmatrix} \qquad C = \begin{bmatrix} 1 & 2 & -3 & 4 \\ 2 & 1 & 5 & 0 \\ 6 & 2 & 7 & 1 \end{bmatrix}$$

Then

$$A + B = \begin{bmatrix} 3+2 & 4+(-1) & -1+3 \\ 2+(-2) & 0+1 & 1+2 \\ 3+6 & 1+4 & 2+(-3) \end{bmatrix} = \begin{bmatrix} 5 & 3 & 2 \\ 0 & 1 & 3 \\ 9 & 5 & -1 \end{bmatrix}$$

The sum $A + C$ is undefined since A and C have different sizes. Similarly, $B + C$ is undefined.

EXAMPLE 9 Let

$$A = \begin{bmatrix} -4 & 3 & -2 \\ 5 & 6 & 3 \end{bmatrix} \quad \text{and} \quad B = \begin{bmatrix} 2 & -1 & 0 \\ 3 & 2 & 4 \end{bmatrix}$$

Then

$$A + B = \begin{bmatrix} -4 + 2 & 3 + (-1) & -2 + 0 \\ 5 + 3 & 6 + 2 & 3 + 4 \end{bmatrix} = \begin{bmatrix} -2 & 2 & -2 \\ 8 & 8 & 7 \end{bmatrix}$$

and

$$B + A = \begin{bmatrix} 2 + (-4) & -1 + 3 & 0 + (-2) \\ 3 + 5 & 2 + 6 & 4 + 3 \end{bmatrix} = \begin{bmatrix} -2 & 2 & -2 \\ 8 & 8 & 7 \end{bmatrix}$$

Thus, $A + B = B + A$.

It should be evident from this example that for matrices A and B with the same size, we have

$$A + B = B + A$$

which is called the **commutative law for matrix addition**.

EXAMPLE 10 An importing firm handles three items imported from Japan: item 1, item 2, and item 3. On each item the firm must pay Japanese export duties, Japanese transportation costs, U.S. import duties, and U.S. transportation costs. These expenses, in dollars per item, are contained in the following matrices:

$A =$

	Japanese transportation	Japanese export duties
Item 1	30	12
Item 2	20	7
Item 3	19	8

$B =$

	U.S. transportation	U.S. import duties
Item 1	40	15
Item 2	9	22
Item 3	27	16

The matrix representing total transportation costs and total duties is

$A + B =$

	Total transportation	Total duties
Item 1	70	27
Item 2	29	29
Item 3	46	24

If A, B, and C are matrices with the same sizes, then the following equality holds (we omit the proof):

$$A + (B + C) = (A + B) + C$$

This result is called the **associative law for matrix addition**.

EXAMPLE 11 Let

$$A = \begin{bmatrix} -4 & 3 & -2 \\ 5 & 6 & 3 \end{bmatrix} \quad B = \begin{bmatrix} 2 & -1 & 0 \\ 3 & 2 & 4 \end{bmatrix} \quad C = \begin{bmatrix} 3 & -4 & 0 \\ 4 & 2 & 5 \end{bmatrix}$$

Then

$$A + (B + C) = \begin{bmatrix} -4 & 3 & -2 \\ 5 & 6 & 3 \end{bmatrix} + \left(\begin{bmatrix} 2 & -1 & 0 \\ 3 & 2 & 4 \end{bmatrix} + \begin{bmatrix} 3 & -4 & 0 \\ 4 & 2 & 5 \end{bmatrix} \right)$$

$$= \begin{bmatrix} -4 & 3 & -2 \\ 5 & 6 & 3 \end{bmatrix} + \begin{bmatrix} 5 & -5 & 0 \\ 7 & 4 & 9 \end{bmatrix} = \begin{bmatrix} 1 & -2 & -2 \\ 12 & 10 & 12 \end{bmatrix}$$

$$(A + B) + C = \left(\begin{bmatrix} -4 & 3 & -2 \\ 5 & 6 & 3 \end{bmatrix} + \begin{bmatrix} 2 & -1 & 0 \\ 3 & 2 & 4 \end{bmatrix} \right) + \begin{bmatrix} 3 & -4 & 0 \\ 4 & 2 & 5 \end{bmatrix}$$

$$= \begin{bmatrix} -2 & 2 & -2 \\ 8 & 8 & 7 \end{bmatrix} + \begin{bmatrix} 3 & -4 & 0 \\ 4 & 2 & 5 \end{bmatrix} = \begin{bmatrix} 1 & -2 & -2 \\ 12 & 10 & 12 \end{bmatrix}$$

Thus, $A + (B + C) = (A + B) + C$.

A matrix each of whose entries is zero is called a **zero matrix**. It is denoted by O.

EXAMPLE 12 The following are zero matrices.

$$\begin{bmatrix} 0 & 0 \\ 0 & 0 \end{bmatrix} \qquad \begin{bmatrix} 0 & 0 & 0 & 0 \\ 0 & 0 & 0 & 0 \end{bmatrix} \qquad [0] \qquad \begin{bmatrix} 0 & 0 \\ 0 & 0 \\ 0 & 0 \end{bmatrix}$$

EXAMPLE 13 Let

$$A = \begin{bmatrix} 4 & -2 \\ 3 & 5 \\ -1 & 2 \end{bmatrix} \quad \text{and} \quad O = \begin{bmatrix} 0 & 0 \\ 0 & 0 \\ 0 & 0 \end{bmatrix}$$

Then

$$A + O = \begin{bmatrix} 4 & -2 \\ 3 & 5 \\ -1 & 2 \end{bmatrix} + \begin{bmatrix} 0 & 0 \\ 0 & 0 \\ 0 & 0 \end{bmatrix} = \begin{bmatrix} 4 & -2 \\ 3 & 5 \\ -1 & 2 \end{bmatrix} = A$$

It is evident from the last example that if A is any matrix and O is a zero matrix of the same size, then

$$A + O = O + A = A$$

Thus, a zero matrix behaves much like the scalar zero.

If A is any matrix, then the **negative** of A, denoted by $-A$, is the matrix whose entries are the negatives of the corresponding entries in A.

EXAMPLE 14 The negative of the matrix

$$A = \begin{bmatrix} 3 & 4 & -2 \\ -5 & 6 & 3 \end{bmatrix}$$

is

$$-A = \begin{bmatrix} -3 & -4 & 2 \\ 5 & -6 & -3 \end{bmatrix}$$

If A is any matrix and O is the zero matrix with the same size as A, then it is evident that

$$A + (-A) = O$$

We omit the formal proof.

If A and B are matrices of the same size, then the **difference** between A and B, denoted $A - B$, is defined by

$$A - B = A + (-B)$$

EXAMPLE 15 Let

$$A = \begin{bmatrix} -4 & 3 & -2 \\ 5 & 6 & 3 \end{bmatrix} \quad \text{and} \quad B = \begin{bmatrix} 2 & -1 & 0 \\ 3 & 2 & 4 \end{bmatrix}$$

Then

$$\begin{aligned} A - B = A + (-B) &= \begin{bmatrix} -4 & 3 & -2 \\ 5 & 6 & 3 \end{bmatrix} + \begin{bmatrix} -2 & 1 & 0 \\ -3 & -2 & -4 \end{bmatrix} \\ &= \begin{bmatrix} -6 & 4 & -2 \\ 2 & 4 & -1 \end{bmatrix} \end{aligned}$$

Observe that the difference $A - B$ can be obtained directly by subtracting each entry in B from the corresponding entry in A.

There are two kinds of multiplication involving matrices, multiplication of matrices by scalars, to be considered below, and multiplication of matrices by matrices, to be discussed in the next section. To motivate the definition of multiplication by scalars, consider a 2×2 matrix

$$A = \begin{bmatrix} a_{11} & a_{12} \\ a_{21} & a_{22} \end{bmatrix}$$

From the definition of matrix addition.

$$A + A = \begin{bmatrix} a_{11} & a_{12} \\ a_{21} & a_{22} \end{bmatrix} + \begin{bmatrix} a_{11} & a_{12} \\ a_{21} & a_{22} \end{bmatrix} = \begin{bmatrix} 2a_{11} & 2a_{12} \\ 2a_{21} & 2a_{22} \end{bmatrix}$$

Since it is natural to think of $A + A$ as $2A$, we obtain

$$2A = \begin{bmatrix} 2a_{11} & 2a_{12} \\ 2a_{21} & 2a_{22} \end{bmatrix}$$

Thus, to add A to itself, we multiply each entry in A by 2. This motivates the following definition.

Scalar Multiplication If A is a matrix and c is a scalar, then the **scalar product** cA is defined to be the matrix obtained by multiplying each entry in A by c.

EXAMPLE 16 Let

$$A = \begin{bmatrix} 3 & 4 \\ 6 & 2 \end{bmatrix}$$

Then

$$5A = \begin{bmatrix} 5(3) & 5(4) \\ 5(6) & 5(2) \end{bmatrix} = \begin{bmatrix} 15 & 20 \\ 30 & 10 \end{bmatrix}$$

and

$$-\tfrac{1}{2}A = \begin{bmatrix} -\frac{3}{2} & -2 \\ -3 & -1 \end{bmatrix}$$

If A and B are matrices of the same size and c and d are scalars, then the following rules of matrix arithmetic hold (we omit the proof):

$$\begin{aligned} (c + d)A &= cA + dA \\ c(A + B) &= cA + cB \\ cd(A) &= c(dA) \end{aligned}$$

EXAMPLE 17 Let $c = 3$,

$$A = \begin{bmatrix} 3 & 4 & -2 \\ 5 & 2 & 3 \end{bmatrix} \quad \text{and} \quad B = \begin{bmatrix} -2 & 3 & 4 \\ 5 & 1 & 2 \end{bmatrix}$$

Then

$$\begin{aligned} c(A + B) &= 3\left(\begin{bmatrix} 3 & 4 & -2 \\ 5 & 2 & 3 \end{bmatrix} + \begin{bmatrix} -2 & 3 & 4 \\ 5 & 1 & 2 \end{bmatrix}\right) \\ &= 3\begin{bmatrix} 1 & 7 & 2 \\ 10 & 3 & 5 \end{bmatrix} = \begin{bmatrix} 3 & 21 & 6 \\ 30 & 9 & 15 \end{bmatrix} \end{aligned}$$

Also,

$$\begin{aligned} cA + cB &= 3\begin{bmatrix} 3 & 4 & -2 \\ 5 & 2 & 3 \end{bmatrix} + 3\begin{bmatrix} -2 & 3 & 4 \\ 5 & 1 & 2 \end{bmatrix} \\ &= \begin{bmatrix} 9 & 12 & -6 \\ 15 & 6 & 9 \end{bmatrix} + \begin{bmatrix} -6 & 9 & 12 \\ 15 & 3 & 6 \end{bmatrix} = \begin{bmatrix} 3 & 21 & 6 \\ 30 & 9 & 15 \end{bmatrix} \end{aligned}$$

Thus, $c(A + B) = cA + cB$.

EXAMPLE 18 **CLASSIFICATION OF INFORMATION** Matrices are useful for recording information that involves a two-way classification. For example, suppose a retail paint store stocks three kinds of exterior paints: regular, deluxe, and commercial. Suppose also that these paints come in four colors: white, black, yellow, and red. The store's weekly sales can be recorded in matrix form as

$$S = \begin{bmatrix} s_{11} & s_{12} & s_{13} & s_{14} \\ s_{21} & s_{22} & s_{23} & s_{24} \\ s_{31} & s_{32} & s_{33} & s_{34} \end{bmatrix} \begin{matrix} \text{Regular} \\ \text{Deluxe} \\ \text{Commerical} \end{matrix}$$

(columns: White, Black, Yellow, Red)

where s_{32} denotes the number of gallons of commerical black paint that was sold during the week.

Similarly, the store's inventory can be recorded in matrix form as

$$A = \begin{bmatrix} a_{11} & a_{12} & a_{13} & a_{14} \\ a_{21} & a_{22} & a_{23} & a_{24} \\ a_{31} & a_{32} & a_{33} & a_{34} \end{bmatrix} \begin{matrix} \text{Regular} \\ \text{Deluxe} \\ \text{Commercial} \end{matrix}$$

(columns: White, Black, Yellow, Red)

If A is the inventory matrix at the beginning of the week and S is the sales matrix for the week, then the matrix

$$A - S$$

represents the inventory at the end of the week. If the store then receives a new shipment of stock represented by a matrix

$$B = \begin{bmatrix} b_{11} & b_{12} & b_{13} & b_{14} \\ b_{21} & b_{22} & b_{23} & b_{24} \\ b_{31} & b_{32} & b_{33} & b_{34} \end{bmatrix} \begin{matrix} \text{Regular} \\ \text{Deluxe} \\ \text{Commercial} \end{matrix}$$

(columns: White, Black, Yellow, Red)

its new inventory would be

$$A - S + B$$

EXERCISE SET 2.3

1. List the sizes of the following matrices:

(a) $\begin{bmatrix} 1 & 2 & -1 \\ 3 & 4 & 2 \end{bmatrix}$ (b) $\begin{bmatrix} 2 & 1 \\ 3 & 2 \\ 4 & -1 \end{bmatrix}$

(c) $[3 \quad 2]$ (d) $[-4]$

2. If

$$A = \begin{bmatrix} 3 & 2 & -4 \\ -2 & 5 & 0 \\ 2 & 1 & 5 \\ -3 & 4 & 6 \end{bmatrix}$$

give values of the following entries.
(a) a_{12} (b) a_{23} (c) a_{32}
(d) a_{41} (e) a_{43}

3. Let

$$A = \begin{bmatrix} 2 & -3 \\ u & 0 \end{bmatrix} \quad \text{and} \quad B = \begin{bmatrix} v & -3 \\ 5 & w \end{bmatrix}$$

For what values of u, v, and w are A and B equal?

4. Solve the following equation for a, b, c, and d.

$$\begin{bmatrix} a + b & b + 2c \\ 2c + d & 2a - d \end{bmatrix} = \begin{bmatrix} -1 & 4 \\ 8 & 0 \end{bmatrix}$$

5. Let A and B be 3×4 matrices and let C and D be 3×3 matrices. Which of the following matrix expressions are defined? For those that are defined, give the size of the resulting matrix.
(a) $A + B$ (b) $B + D$ (c) $3A - 2C$
(d) $7C + 2D$

6. Let

$$A = \begin{bmatrix} -1 & 2 & 3 \\ 4 & 2 & 0 \\ -3 & 2 & 5 \end{bmatrix} \quad B = \begin{bmatrix} 3 & -1 & 2 \\ -5 & 3 & 4 \\ -3 & -4 & 0 \end{bmatrix}$$

$$C = \begin{bmatrix} 2 & -3 & 6 \\ 0 & 4 & -1 \\ -5 & 1 & 3 \end{bmatrix}$$

Compute, if possible,
(a) $A + 2B$ (b) $3A - 4B$
(c) $(A + B) - A$ (d) $A + (B + C)$
(e) $(A + B) + C$

7. Let $c = 2$, and $d = -4$.

$$A = \begin{bmatrix} 2 & -3 \\ 4 & 5 \end{bmatrix} \quad B = \begin{bmatrix} 2 & 5 \\ -1 & 3 \end{bmatrix} \quad C = \begin{bmatrix} 3 & -1 \\ 0 & 4 \end{bmatrix}$$

Verify that
(a) $A + (B + C) = (A + B) + C$
(b) $(c + d)A = cA + dA$
(c) $c(A + B) = cA + cB$
(d) $cd(A) = c(dA)$

8. Let

$$A = \begin{bmatrix} 2 & -1 & 0 \\ 3 & 1 & 4 \end{bmatrix} \qquad B = \begin{bmatrix} 6 & 2 & -4 \\ 1 & 0 & 2 \end{bmatrix}$$

$$C = \begin{bmatrix} -1 & -2 & -3 \\ 1 & 4 & 0 \end{bmatrix}$$

Compute, if possible,
(a) $2A - B$ (b) $5A + 2B$
(c) $A + (B - 3C)$ (d) $A - (B - C)$
(e) $(A + C) + 2B$

9. Let $c = -3$, $d = 2$,

$$A = \begin{bmatrix} 3 & 2 & 0 \\ -1 & 4 & 1 \end{bmatrix} \qquad B = \begin{bmatrix} 2 & 2 & -3 \\ 5 & 0 & 1 \end{bmatrix}$$

$$C = \begin{bmatrix} -1 & -4 & 0 \\ 0 & 2 & 3 \end{bmatrix}$$

Verify that
(a) $A + (B - C) = (A + B) - C$
(b) $(c - d)B = cB - dB$
(c) $c(A - B) = cA - cB$
(d) $(abc)A = ab(cA)$

10. Show that if $kA = O$, then $k = 0$ or $A = O$.

11. **(Data classification)** A number of people were questioned about their income and their political affiliation. The following results were obtained:

205 were Republicans earning less than \$30,000 per year.
317 were Democrats earning less than \$30,000 per year.
96 were Independents earning less than \$30,000 per year.
192 were Republicans earning more than \$30,000 per year.
128 were Democrats earning more than \$30,000 per year.
63 were Independents earning more than \$30,000 per year.

Display this information in matrix form.

12. **(Data classification)** In the communication network illustrated in Figure 1, station 1 has two transmission channels to receiver A and three transmission channels to receiver B, while station 2 has three transmission channels to A and four transmission channels to B. Show how you might describe this information in matrix form.

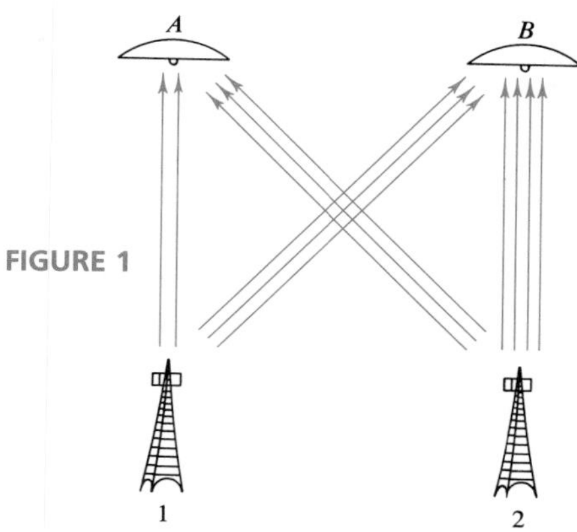

FIGURE 1

13. **(Inventory control)** A supermarket chain with three stores records the following inventory matrices at the beginning and end of June.

June 1 inventory:	Store 1	Store 2	Store 3
Produce	5	2	4
Meat	3	1	8
Canned goods	6	3	5

June 30 inventory:	Store 1	Store 2	Store 3
Produce	3	1	2
Meat	1	0	2
Canned goods	5	1	3

The matrix entries are retail inventory values in thousands of dollars. Use matrix subtraction to determine how much should be reordered to bring the inventory back up to the June 1 level.

14. To adjust for demand, the supermarket chain in Exercise 13 increases all its inventories by 2% on June 1. What matrix operation would you perform on the June 1 inventory matrix to obtain the new inventory values?

2.4 MATRIX MULTIPLICATION

In this section we discuss the multiplication of matrices and illustrate some applications. On the surface it might seem natural to multiply two matrices together by multiplying their corresponding entries. However, such a definition turns out to have almost no practical applications. The following more complicated definition has proved to be more useful.

> **Matrix Multiplication** If A is an $m \times r$ matrix and B is an $r \times n$ matrix, then the **product** AB is the $m \times n$ matrix obtained as follows. To find the entry in row i and column j of AB, multiply the entries in the ith row of A by the corresponding entries in the jth column of B and add up the resulting products.

Note that the definition of the product AB requires that the number of columns of the first factor A must be the same as the number of rows of the second factor B. If this condition is not satisfied, then the product of A and B cannot be formed.

EXAMPLE 1 Suppose A is a 3×4 matrix, B is a 4×2 matrix, and C is a 4×3 matrix. Then AB is a 3×2 matrix, AC is 3×3, and CA is a 4×4 matrix, while BC, CB, and BA are undefined.

EXAMPLE 2 Let

$$A = \begin{bmatrix} 3 & 4 & -1 & 2 \\ 0 & 1 & 2 & 3 \end{bmatrix} \quad \text{and} \quad B = \begin{bmatrix} -1 & 0 & 2 \\ -2 & 4 & 1 \\ 3 & 2 & 3 \\ 5 & -1 & 2 \end{bmatrix}$$

Since A is a 2×4 matrix and B is a 4×3 matrix, the product AB is a 2×3 matrix. Suppose we want the entry in row 2 and column 3 of AB. As illustrated below, we multiply the corresponding entries of row 2 in A and column 3 in B and add the resulting products:

$$\begin{bmatrix} 3 & 4 & -1 & 2 \\ 0 & 1 & 2 & 3 \end{bmatrix} \begin{bmatrix} -1 & 0 & 2 \\ -2 & 4 & 1 \\ 3 & 2 & 3 \\ 5 & -1 & 2 \end{bmatrix} = \begin{bmatrix} \square & \square & \square \\ \square & \square & 13 \end{bmatrix}$$

$$(0)(2) + (1)(1) + (2)(3) + (3)(2) = 13$$

The entry in row 1 and column 2 of AB is computed as follows:

$$\begin{bmatrix} 3 & 4 & -1 & 2 \\ 0 & 1 & 2 & 3 \end{bmatrix} \begin{bmatrix} -1 & 0 & 2 \\ -2 & 4 & 1 \\ 3 & 2 & 3 \\ 5 & -1 & 2 \end{bmatrix} = \begin{bmatrix} \square & 12 & \square \\ \square & \square & \square \end{bmatrix}$$

$$(3)(0) + (4)(4) + (-1)(2) + (2)(-1) = 12$$

Although the definition of matrix multiplication seems complicated, it was devised because it arises naturally in various applications. The following example illustrates one such application.

EXAMPLE 3 **ECOLOGY** A manufacturer makes two types of products, I and II, at two plants X and Y. In the manufacture of these products, the following pollutants result: sulfur dioxide, carbon monoxide, and particulate matter. At either plant, the daily pollutants resulting from the production of product I are

300 pounds of sulfur dioxide
100 pounds of carbon monoxide
200 pounds of particulate matter

and of product II are

400 pounds of sulfur dioxide
50 pounds of carbon monoxide
300 pounds of particulate matter

This information can be tabulated in matrix form as

$$A = \begin{bmatrix} 300 & 100 & 200 \\ 400 & 50 & 300 \end{bmatrix} \begin{matrix} \text{Product I} \\ \text{Product II} \end{matrix}$$

(columns: Sulfur dioxide, Carbon monoxide, Particulate matter)

To satisfy federal regulations, these pollutants must be removed. Suppose the daily cost in dollars for removing each pound of pollutant at plant X is

\$5 per pound of sulfur dioxide
\$3 per pound of carbon monoxide
\$2 per pound of particulate matter

and at plant Y is

\$8 per pound of sulfur dioxide
\$4 per pound of carbon monoxide
\$1 per pound of particulate matter

This cost information can be represented in matrix form as

$$B = \begin{bmatrix} 5 & 8 \\ 3 & 4 \\ 2 & 1 \end{bmatrix} \begin{matrix} \text{Sulfur dioxide} \\ \text{Carbon monoxide} \\ \text{Particulate matter} \end{matrix}$$

(columns: Plant X, Plant Y)

From the first row of A and the first column of B, we see that the daily cost of removing all pollutants resulting from the manufacture of product I at plant X is

$$300(5) + 100(3) + 200(2) = 2200 \qquad \text{(dollars)}$$

But this is the entry in row 1 and column 1 of

$$AB = \begin{bmatrix} 300 & 100 & 200 \\ 400 & 50 & 300 \end{bmatrix} \begin{bmatrix} 5 & 8 \\ 3 & 4 \\ 2 & 1 \end{bmatrix} = \begin{bmatrix} 2200 & 3000 \\ 2750 & 3700 \end{bmatrix}$$

Similarly, from the second row of A and the second column of B, the daily cost of removing all pollutants resulting from the manufacture of product II at plant Y is

$$400(8) + 50(4) + 300(1) = 3700$$

But this is just the entry in row 2 and column 2 of the product AB. Thus, each entry in AB has a physical interpretation as the daily cost of removing all pollutants from one of the plants in the manufacture of one of the products.

Properties of Matrix Multiplication

If a and b are real numbers, then we know that $ab = ba$. In matrix arithmetic, however, the product AB and the product BA need not be equal. For example, if A is a 2×3 matrix and B is a 3×4 matrix, then AB is a 2×4 matrix but BA is undefined (why?). However, as the next example shows, AB and BA may not be equal, even if they are both defined.

EXAMPLE 4 Let

$$A = \begin{bmatrix} 2 & 3 \\ -4 & 2 \end{bmatrix} \quad \text{and} \quad B = \begin{bmatrix} 3 & -1 \\ 5 & 2 \end{bmatrix}$$

Then

$$AB = \begin{bmatrix} 21 & 4 \\ -2 & 8 \end{bmatrix} \quad \text{and } BA = \begin{bmatrix} 10 & 7 \\ 2 & 19 \end{bmatrix}$$

Thus, $AB \neq BA$.

In ordinary arithmetic we know:

(i) If $ab = 0$, then $a = 0$ or $b = 0$.

(ii) If $ab = ac$ and $a \neq 0$, then $b = c$ (we can cancel a on both sides of the equation).

These properties do not hold for matrix multiplication, as is shown in the next example.

EXAMPLE 5 Let

$$A = \begin{bmatrix} 0 & 2 \\ 0 & 3 \end{bmatrix} \quad B = \begin{bmatrix} -1 & 4 \\ 3 & -5 \end{bmatrix} \quad C = \begin{bmatrix} -2 & 5 \\ 3 & -5 \end{bmatrix} \quad D = \begin{bmatrix} 2 & 3 \\ 0 & 0 \end{bmatrix}$$

Then

$$AB = AC = \begin{bmatrix} 6 & -10 \\ 9 & -15 \end{bmatrix}$$

Thus, $AB = AC$, but $B \neq C$. Also,

$$AD = \begin{bmatrix} 0 & 0 \\ 0 & 0 \end{bmatrix}$$

but $A \neq O$ and $D \neq O$.

As the above examples show, some properties of ordinary arithmetic do not carry over to matrix arithmetic. There are many that do, however. We now give a partial list of some of the more important of these properties. We omit the proofs.

Assuming the sizes of the matrices are such that the indicated operations can be performed, the following rules of matrix arithmetic hold:

Other Rules of Matrix Arithmetic

$A(BC) = (AB)C$ **(Associative law for matrix multiplication)**

$A(B + C) = AB + AC$
$(A + B)C = AC + BC$ **(Distributive laws)**

$A(cB) = c(AB) = (cA)B$

EXAMPLE 6 Let

$$A = \begin{bmatrix} 4 & 2 \\ 0 & -3 \\ 3 & -2 \end{bmatrix} \qquad B = \begin{bmatrix} 4 & 2 \\ -2 & 3 \end{bmatrix} \qquad C = \begin{bmatrix} 1 & 2 \\ -3 & 4 \end{bmatrix}$$

Then

$$A(B + C) = \begin{bmatrix} 4 & 2 \\ 0 & -3 \\ 3 & -2 \end{bmatrix}\left(\begin{bmatrix} 4 & 2 \\ -2 & 3 \end{bmatrix} + \begin{bmatrix} 1 & 2 \\ -3 & 4 \end{bmatrix}\right)$$

$$= \begin{bmatrix} 4 & 2 \\ 0 & -3 \\ 3 & -2 \end{bmatrix}\begin{bmatrix} 5 & 4 \\ -5 & 7 \end{bmatrix} = \begin{bmatrix} 10 & 30 \\ 15 & -21 \\ 25 & -2 \end{bmatrix}$$

On the other hand,

$$AB + AC = \begin{bmatrix} 4 & 2 \\ 0 & -3 \\ 3 & -2 \end{bmatrix}\begin{bmatrix} 4 & 2 \\ -2 & 3 \end{bmatrix} + \begin{bmatrix} 4 & 2 \\ 0 & -3 \\ 3 & -2 \end{bmatrix}\begin{bmatrix} 1 & 2 \\ -3 & 4 \end{bmatrix}$$

$$= \begin{bmatrix} 12 & 14 \\ 6 & -9 \\ 16 & 0 \end{bmatrix} + \begin{bmatrix} -2 & 16 \\ 9 & -12 \\ 9 & -2 \end{bmatrix} = \begin{bmatrix} 10 & 30 \\ 15 & -21 \\ 25 & -2 \end{bmatrix}$$

Thus,

$$A(B + C) = AB + AC$$

as guaranteed by the distributive law.

EXAMPLE 7 Let $c = 2$,

$$A = \begin{bmatrix} 4 & 2 \\ 0 & -3 \\ 3 & -2 \end{bmatrix} \quad \text{and} \quad B = \begin{bmatrix} 4 & 2 \\ -2 & 3 \end{bmatrix}$$

Then

$$cB = 2\begin{bmatrix} 4 & 2 \\ -2 & 3 \end{bmatrix} = \begin{bmatrix} 8 & 4 \\ -4 & 6 \end{bmatrix}$$

and

$$A(cB) = \begin{bmatrix} 4 & 2 \\ 0 & -3 \\ 3 & -2 \end{bmatrix}\begin{bmatrix} 8 & 4 \\ -4 & 6 \end{bmatrix} = \begin{bmatrix} 24 & 28 \\ 12 & -18 \\ 32 & 0 \end{bmatrix}$$

Also

$$c(AB) = 2\left(\begin{bmatrix} 4 & 2 \\ 0 & -3 \\ 3 & -2 \end{bmatrix}\begin{bmatrix} 4 & 2 \\ -2 & 3 \end{bmatrix}\right)$$

$$= 2\begin{bmatrix} 12 & 14 \\ 6 & -9 \\ 16 & 0 \end{bmatrix} = \begin{bmatrix} 24 & 28 \\ 12 & -18 \\ 32 & 0 \end{bmatrix}$$

so $A(cB) = c(AB)$. Moreover,

$$cA = 2\begin{bmatrix} 4 & 2 \\ 0 & -3 \\ 3 & -2 \end{bmatrix} = \begin{bmatrix} 8 & 4 \\ 0 & -6 \\ 6 & -4 \end{bmatrix}$$

and

$$(cA)B = \begin{bmatrix} 8 & 4 \\ 0 & -6 \\ 6 & -4 \end{bmatrix}\begin{bmatrix} 4 & 2 \\ -2 & 3 \end{bmatrix} = \begin{bmatrix} 24 & 28 \\ 12 & -18 \\ 32 & 0 \end{bmatrix}$$

Thus

$$A(cB) = c(AB) = (cA)B$$

If A is a square matrix, then the entries $a_{11}, a_{22}, \ldots, a_{nn}$ are said to form the **main diagonal** of A (Figure 1):

FIGURE 1

$$A = \begin{bmatrix} a_{11} & a_{12} & \cdots & a_{1n} \\ a_{21} & a_{22} & \cdots & a_{2n} \\ \vdots & \vdots & & \vdots \\ a_{n1} & a_{n2} & \cdots & a_{nn} \end{bmatrix}$$

(the main diagonal of A)

Square matrices which have 1s on the main diagonal and 0s elsewhere are of particular importance. They are called **identity matrices**. The $n \times n$ identity matrix is denoted by I_n or sometimes just I.

EXAMPLE 8 The 2×2 identity matrix is

$$I_2 = \begin{bmatrix} 1 & 0 \\ 0 & 1 \end{bmatrix}$$

and the 3×3 identity matrix is

$$I_3 = \begin{bmatrix} 1 & 0 & 0 \\ 0 & 1 & 0 \\ 0 & 0 & 1 \end{bmatrix}$$

In ordinary arithmetic, the number 1 has the properties

$$1 \cdot a = a \quad \text{and} \quad a \cdot 1 = a$$

In matrix arithmetic, we have the analogous relationships

$$I_mA = A \quad \text{and} \quad AI_n = A \tag{1}$$

for any $m \times n$ matrix A. The following examples illustrates the equations in (1).

EXAMPLE 9 Let

$$A = \begin{bmatrix} a_{11} & a_{12} & a_{13} \\ a_{21} & a_{22} & a_{23} \end{bmatrix}$$

Then

$$I_2A = \begin{bmatrix} 1 & 0 \\ 0 & 1 \end{bmatrix}\begin{bmatrix} a_{11} & a_{12} & a_{13} \\ a_{21} & a_{22} & a_{23} \end{bmatrix} = \begin{bmatrix} a_{11} & a_{12} & a_{13} \\ a_{21} & a_{22} & a_{23} \end{bmatrix} = A$$

and

$$AI_3 = \begin{bmatrix} a_{11} & a_{12} & a_{13} \\ a_{21} & a_{22} & a_{23} \end{bmatrix}\begin{bmatrix} 1 & 0 & 0 \\ 0 & 1 & 0 \\ 0 & 0 & 1 \end{bmatrix} = \begin{bmatrix} a_{11} & a_{12} & a_{13} \\ a_{21} & a_{22} & a_{23} \end{bmatrix} = A$$

Matrix multiplication is important in the further study of linear systems. Consider a general linear system of m equations in n unknowns:

$$\begin{array}{ccccccccc} a_{11}x_1 & + & a_{12}x_2 & + & \cdots & + & a_{1n}x_n & = & b_1 \\ a_{21}x_1 & + & a_{22}x_2 & + & \cdots & + & a_{2n}x_n & = & b_2 \\ \vdots & & \vdots & & \vdots & & \vdots & & \vdots \\ a_{m1}x_1 & + & a_{m2}x_2 & + & \cdots & + & a_{mn}x_n & = & b_m \end{array} \tag{2}$$

Since two matrices are equal if and only if their corresponding entries are equal, we can replace the m equations in (2) by the single matrix equation

$$\begin{bmatrix} a_{11}x_1 + a_{12}x_2 + \cdots + a_{1n}x_n \\ a_{21}x_1 + a_{22}x_2 + \cdots + a_{2n}x_n \\ \vdots \\ a_{m1}x_1 + a_{m2}x_2 + \cdots + a_{mn}x_n \end{bmatrix} = \begin{bmatrix} b_1 \\ b_2 \\ \vdots \\ b_m \end{bmatrix} \tag{3}$$

If we now define

$$A = \begin{bmatrix} a_{11} & a_{12} & \cdots & a_{1n} \\ a_{21} & a_{22} & \cdots & a_{2n} \\ \vdots & \vdots & & \vdots \\ a_{m1} & a_{m2} & \cdots & a_{mn} \end{bmatrix} \qquad X = \begin{bmatrix} x_1 \\ x_2 \\ \vdots \\ x_n \end{bmatrix} \qquad B = \begin{bmatrix} b_1 \\ b_2 \\ \vdots \\ b_m \end{bmatrix}$$

then the matrix equation (3) can be written as

$$AX = B \tag{4}$$

Thus the *entire* system of linear equations in (2) can be represented by the *single* matrix equation in (4). The matrix A in this equation is called the **coefficient matrix** of the system. In the next section we give some applications of (4).

EXAMPLE 10 Consider the linear system

$$\begin{aligned} 2x_1 + 3x_2 - 3x_3 - 5x_4 &= 8 \\ -2x_1 \qquad\quad + 5x_3 + 6x_4 &= 12 \\ 3x_1 + 7x_2 - 8x_3 + 2x_4 &= 6 \end{aligned} \tag{5}$$

Letting

$$A = \begin{bmatrix} 2 & 3 & -3 & -5 \\ -2 & 0 & 5 & 6 \\ 3 & 7 & -8 & 2 \end{bmatrix} \qquad X = \begin{bmatrix} x_1 \\ x_2 \\ x_3 \\ x_4 \end{bmatrix} \qquad B = \begin{bmatrix} 8 \\ 12 \\ 6 \end{bmatrix}$$

we can write (5) as

$$AX = B$$

EXAMPLE 11 The matrix equation

$$\begin{bmatrix} 3 & 4 & -1 & 8 \\ 2 & -3 & 5 & 2 \\ -4 & 2 & 1 & 7 \\ 1 & 1 & -5 & 5 \end{bmatrix} \begin{bmatrix} x_1 \\ x_2 \\ x_3 \\ x_4 \end{bmatrix} = \begin{bmatrix} 8 \\ 2 \\ 7 \\ 5 \end{bmatrix}$$

represents the linear system

$$\begin{aligned} 3x_1 + 4x_2 - x_3 + 8x_4 &= 8 \\ 2x_1 - 3x_2 + 5x_3 + 2x_4 &= 2 \\ -4x_1 + 2x_2 + x_3 + 7x_4 &= 7 \\ x_1 + x_2 - 5x_3 + 5x_4 &= 5 \end{aligned}$$

Transpose of a Matrix

We conclude this section with a matrix operation that is needed in later sections. If A is any $m \times n$ matrix, then the **transpose** of A, denoted by A^t, is defined to be the $n \times m$ matrix whose first column is the first row of A, whose second column is the second row of A, whose third column is the third row of A, and so on. That is, the transpose of A is the matrix obtained from A by interchanging corresponding rows and columns. For example, if

$$A = \begin{bmatrix} 1 & -3 & 2 \\ 6 & 1 & 7 \end{bmatrix} \quad \text{and} \quad B = \begin{bmatrix} b_{11} & b_{12} \\ b_{21} & b_{22} \end{bmatrix}$$

then

$$A^t = \begin{bmatrix} 1 & 6 \\ -3 & 1 \\ 2 & 7 \end{bmatrix} \quad \text{and} \quad B^t = \begin{bmatrix} b_{11} & b_{21} \\ b_{12} & b_{22} \end{bmatrix}$$

It can be shown that the transpose operation satisfies the following properties.

Properties of the Transpose

(i) $(A^t)^t = A$

(ii) $(A + B)^t = A^t + B^t$

(iii) $(cA)^t = cA^t$

(iv) $(AB)^t = B^tA^t$

(We omit the proof.)

The following example illustrates property (iv).

EXAMPLE 12 If

$$A = \begin{bmatrix} 3 & -2 & 4 \\ 5 & -3 & 2 \end{bmatrix} \quad \text{and} \quad B = \begin{bmatrix} 1 & 0 \\ 2 & 4 \\ -3 & 6 \end{bmatrix}$$

then

$$A^t = \begin{bmatrix} 3 & 5 \\ -2 & -3 \\ 4 & 2 \end{bmatrix} \quad \text{and} \quad B^t = \begin{bmatrix} 1 & 2 & -3 \\ 0 & 4 & 6 \end{bmatrix}$$

so that

$$B^tA^t = \begin{bmatrix} 1 & 2 & -3 \\ 0 & 4 & 6 \end{bmatrix} \begin{bmatrix} 3 & 5 \\ -2 & -3 \\ 4 & 2 \end{bmatrix} = \begin{bmatrix} -13 & -7 \\ 16 & 0 \end{bmatrix}$$

Also,

$$AB = \begin{bmatrix} 3 & -2 & 4 \\ 5 & -3 & 2 \end{bmatrix} \begin{bmatrix} 1 & 0 \\ 2 & 4 \\ -3 & 6 \end{bmatrix} = \begin{bmatrix} -13 & 16 \\ -7 & 0 \end{bmatrix}$$

so that

$$(AB)^t = B^tA^t$$

EXERCISE SET 2.4

1. Let A and B be 3×4 matrices and let C, D, and E be 4×2, 3×2, and 4×3 matrices, respectively. Which of the following matrix expressions are defined? For those that are defined, give the size of the resulting matrix.

(a) AC (b) AB (c) $BA + C$
(d) $EA + D$ (e) $E(B + A)$ (f) $AC + D$
(g) $(AE)D$

2. Let

$$A = \begin{bmatrix} -1 & 2 & 3 \\ 4 & 2 & 0 \\ -3 & 2 & 5 \end{bmatrix} \quad B = \begin{bmatrix} 3 & -1 & 2 \\ -5 & 3 & 4 \\ -3 & -4 & 0 \end{bmatrix}$$

$$C = \begin{bmatrix} 2 & -5 \\ 0 & 4 \\ -5 & 1 \end{bmatrix} \quad D = \begin{bmatrix} 1 & 3 & 2 \\ 4 & 5 & -2 \end{bmatrix}$$

$$E = \begin{bmatrix} 3 & 5 & -3 \\ 0 & 2 & 4 \end{bmatrix}$$

Compute the following where possible.

(a) AB (b) BA (c) AC (d) CA
(e) CD (f) DC (g) EE (h) AA

3. Using the matrices in Exercise 2, compute the following where possible.

(a) $E(A - 2B)$ (b) ABE (c) EBC

4. Let $c = -2$,

$$A = \begin{bmatrix} 2 & -1 \\ 3 & 0 \end{bmatrix} \quad B = \begin{bmatrix} 4 & 8 \\ 6 & -3 \end{bmatrix} \quad C = \begin{bmatrix} 1 & 4 \\ 5 & 6 \end{bmatrix}$$

Verify the following.
(a) $A(BC) = (AB)C$
(b) $A(B + C) = AB + AC$
(c) $(A + B)C = AC + BC$
(d) $A(cB) = (cA)B$

5. Let I_3 be the 3×3 identity matrix and let

$$A = \begin{bmatrix} -4 & 3 & 2 \\ 5 & 1 & 2 \\ 3 & -2 & 5 \end{bmatrix}$$

Verify that

$$AI_3 = I_3A = A$$

6. If

$$A = \begin{bmatrix} 1 & 2 \\ 2 & 4 \end{bmatrix} \quad \text{and} \quad B = \begin{bmatrix} 2 & 2 \\ -1 & -1 \end{bmatrix}$$

show that $AB = O$.

In Exercises 7–10 all matrices are square matrices and C^2 denotes the matrix CC.

7. If

$$A = \begin{bmatrix} 0 & 1 \\ 1 & 0 \end{bmatrix}$$

show that $A^2 = I_2$.

8. If

$$A = \begin{bmatrix} 1 & 2 \\ 3 & 4 \end{bmatrix}$$

find $A^2 - 2A + 3I_2$.

9. Show that if $AB = BA$, then

$$(A + B)^2 = A^2 + 2AB + B^2$$

10. Show that if $AB = BA$, then $(AB)^2 = A^2B^2$.

11. Find the matrices A, X, and B that result when the system

$$\begin{aligned} 3x_1 - x_2 + 2x_3 + x_4 + x_5 &= 7 \\ -2x_1 + x_2 + 4x_3 + x_4 - 2x_5 &= -8 \\ 3x_2 + 6x_3 + 3x_4 + x_5 &= 9 \end{aligned}$$

is written in the matrix form $AX = B$.

12. Write out the linear system that is represented in matrix form as

$$\begin{bmatrix} 3 & 2 & 0 \\ 2 & 4 & 5 \\ -3 & 0 & 2 \end{bmatrix} \begin{bmatrix} x_1 \\ x_2 \\ x_3 \end{bmatrix} = \begin{bmatrix} 8 \\ 0 \\ -2 \end{bmatrix}$$

13. (a) Show that if A has a row of zeros and B is a matrix for which AB is defined, then AB also has a row of zeros.
(b) Discover a similar result involving a column of zeros.

14. Let X_1 be a solution to the linear system

$$AX = B$$

and let Y be a solution to the linear system

$$AX = O$$

Show that $X_1 + Y$ is also a solution to the linear system

$$AX = B$$

15. Suppose the matrix

$$P = \begin{bmatrix} 3 & 4 \\ 2 & 1 \end{bmatrix}$$

describes 1 hour's output in a small bicycle factory according to Table 1.

TABLE 1

	Production line 1	Production line 2
3-speed bikes	3	4
10-speed bikes	2	1

Let $M = \begin{bmatrix} 7 \\ 8 \end{bmatrix}$ denote the number of hours in a day that the production lines operate; that is, line 1 operates 7 hours per day and line 2 operates 8 hours per day. Compute PM and give its physical interpretation.

16. Let $c = 3$,

$$A = \begin{bmatrix} 1 & 2 \\ -3 & 6 \\ 0 & 1 \end{bmatrix} \quad B = \begin{bmatrix} 0 & 3 \\ 5 & 7 \\ 1 & -4 \end{bmatrix}$$

(a) Compute A^t and B^t.
(b) Verify the transpose properties (i), (ii), and (iii) stated in this section.

17. Let

$$A = \begin{bmatrix} 1 & 0 & -1 \\ 2 & 0 & 6 \end{bmatrix} \quad \text{and} \quad B = \begin{bmatrix} 1 & 7 \\ -8 & 4 \\ 0 & 1 \end{bmatrix}$$

Show that $(AB)^t = B^tA^t$.

18. Show that $I_n^t = I_n$.

19. **(Manufacturing)** A manufacturer of calculators has an east coast and a west coast plant, each of which produces business calculators and standard calculators. The manufacturing time requirements (in hours per calculator) and the assembly and packing costs (in dollars per hour) are given by the following matrices:

$$A = \begin{bmatrix} 0.2 & 0.1 \\ 0.3 & 0.1 \end{bmatrix} \begin{matrix} \text{Business calculators} \\ \text{Standard calculators} \end{matrix}$$

(Hours per unit; columns: Assembly, Packaging)

$$B = \begin{bmatrix} 5 & 6 \\ 4 & 5 \end{bmatrix} \begin{matrix} \text{Assembly} \\ \text{Packaging} \end{matrix}$$

(Dollars per hour; columns: East coast, West coast)

(a) Find the total cost of manufacturing a business calculator on the east coast. You can do this by calculating the entry in the first row and first column of AB.

(b) Compute AB and state the meaning of each entry.

20. **(Nutrition)** A pet food manufacturer blends two nutrient sources in varying amounts to produce three kinds of dog biscuits. The amounts (in grams) of calcium, protein, and carbohydrate in each nutrient source are shown in matrix R, and the amount (in grams) of each nutrient source used in each biscuit is shown in matrix S.

$$R = \begin{bmatrix} 3 & 4 \\ 8 & 6 \\ 5 & 2 \end{bmatrix} \begin{matrix} \text{Calcium} \\ \text{Protein} \\ \text{Carbohydrate} \end{matrix}$$

(Columns: Source I, Source II)

$$S = \begin{bmatrix} 14 & 8 & 15 \\ 16 & 22 & 15 \end{bmatrix} \begin{matrix} \text{Source I} \\ \text{Source II} \end{matrix}$$

(Columns: Biscuit A, Biscuit B, Biscuit C)

(a) Find the total amount of calcium in biscuit B by calculating the entry in the first row and second column of RS.

(b) Compute RS and state the meaning of each entry.

2.5 INVERSES OF MATRICES

Section 2.4 shows that a system of m linear equations in n unknowns can be written as a single matrix equation

$$AX = B$$

In this section we develop a method that can be used to solve this equation when A is a special kind of square matrix.

If A is a *square* matrix and if we can find a matrix B such that

$$AB = I \quad \text{and} \quad BA = I$$

then A is called an **invertible** or **nonsingular** matrix. The matrix B is called an **inverse** of A, and A is an inverse of B.

EXAMPLE 1 The matrix

$$B = \begin{bmatrix} -3 & 2 \\ 2 & -1 \end{bmatrix}$$

in an inverse of

$$A = \begin{bmatrix} 1 & 2 \\ 2 & 3 \end{bmatrix}$$

since

$$AB = \begin{bmatrix} 1 & 2 \\ 2 & 3 \end{bmatrix}\begin{bmatrix} -3 & 2 \\ 2 & -1 \end{bmatrix} = \begin{bmatrix} 1 & 0 \\ 0 & 1 \end{bmatrix} = I$$

and

$$BA = \begin{bmatrix} -3 & 2 \\ 2 & -1 \end{bmatrix}\begin{bmatrix} 1 & 2 \\ 2 & 3 \end{bmatrix} = \begin{bmatrix} 1 & 0 \\ 0 & 1 \end{bmatrix} = I$$

The next example shows that not every matrix is invertible.

EXAMPLE 2 Consider the matrix

$$A = \begin{bmatrix} 2 & 3 \\ 4 & 6 \end{bmatrix}$$

If A were invertible, we would have a matrix

$$B = \begin{bmatrix} b_{11} & b_{12} \\ b_{21} & b_{22} \end{bmatrix}$$

such that

$$AB = I \quad \text{and} \quad BA = I$$

But the equation

$$AB = I \tag{1}$$

can be written

$$\begin{bmatrix} 2 & 3 \\ 4 & 6 \end{bmatrix} \begin{bmatrix} b_{11} & b_{12} \\ b_{21} & b_{22} \end{bmatrix} = \begin{bmatrix} 1 & 0 \\ 0 & 1 \end{bmatrix}$$

or equivalently

$$\begin{bmatrix} 2b_{11} + 3b_{21} & 2b_{12} + 3b_{22} \\ 4b_{11} + 6b_{21} & 4b_{12} + 6b_{22} \end{bmatrix} = \begin{bmatrix} 1 & 0 \\ 0 & 1 \end{bmatrix} \tag{2}$$

Since two matrices are equal if and only if their corresponding entries are equal, it follows from (2) that

$$\begin{aligned} 2b_{11} + 3b_{21} &= 1 \\ 4b_{11} + 6b_{21} &= 0 \end{aligned}$$

But this system is inconsistent (why?), so there can be no matrix B that satisfies (1). Thus, A is not invertible.

By definition, every invertible matrix has an inverse. The following result, whose proof is omitted, shows that an invertible matrix cannot have more than one inverse.

THEOREM An invertible matrix has exactly one inverse.

Since an invertible matrix A has exactly one inverse, it is now proper to talk about *the* inverse of A. We denote the inverse by A^{-1} (read "A inverse"). Thus,

$$AA^{-1} = 1 \quad \text{and} \quad A^{-1}A = I$$

We now turn to the problem of finding the inverse of an invertible matrix. Suppose we want to find the inverse of

$$A = \begin{bmatrix} 1 & 2 \\ 2 & 3 \end{bmatrix}$$

If we let

$$X = \begin{bmatrix} x_1 & x_2 \\ x_3 & x_4 \end{bmatrix}$$

denote the unknown inverse of A, then we must have

$$AX = I \tag{3}$$

and

$$XA = I \tag{4}$$

The matrix equation (3) can be written

$$\begin{bmatrix} 1 & 2 \\ 2 & 3 \end{bmatrix} \begin{bmatrix} x_1 & x_2 \\ x_3 & x_4 \end{bmatrix} = \begin{bmatrix} 1 & 0 \\ 0 & 1 \end{bmatrix}$$

or

$$\begin{bmatrix} x_1 + 2x_3 & x_2 + 2x_4 \\ 2x_1 + 3x_3 & 2x_2 + 3x_4 \end{bmatrix} = \begin{bmatrix} 1 & 0 \\ 0 & 1 \end{bmatrix}$$

Since two matrices are equal if and only if their corresponding entries are equal, we must have

$$\begin{aligned} x_1 + 2x_3 &= 1 \\ 2x_1 + 3x_3 &= 0 \end{aligned} \quad \text{and} \quad \begin{aligned} x_2 + 2x_4 &= 0 \\ 2x_2 + 3x_4 &= 1 \end{aligned} \tag{5}$$

To solve these systems we must reduce their augmented matrices

$$\left[\begin{array}{cc|c} 1 & 2 & 1 \\ 2 & 3 & 0 \end{array}\right] \quad \text{and} \quad \left[\begin{array}{cc|c} 1 & 2 & 0 \\ 2 & 3 & 1 \end{array}\right] \tag{6}$$

to reduced row echelon form. You can check that these reduced row echelon forms are

$$\left[\begin{array}{cc|c} 1 & 0 & -3 \\ 0 & 1 & 2 \end{array}\right] \quad \text{and} \quad \left[\begin{array}{cc|c} 1 & 0 & 2 \\ 0 & 1 & -1 \end{array}\right]$$

Thus,

$$\begin{aligned} x_1 &= -3 \\ x_3 &= 2 \end{aligned} \quad \text{and} \quad \begin{aligned} x_2 &= 2 \\ x_4 &= -1 \end{aligned}$$

so the matrix X satisfying (3) is

$$X = \begin{bmatrix} x_1 & x_2 \\ x_3 & x_4 \end{bmatrix} = \begin{bmatrix} -3 & 2 \\ 2 & -1 \end{bmatrix}$$

You can verify that this X satisfies (4) so that X is the inverse of A, that is,

$$A^{-1} = \begin{bmatrix} -3 & 2 \\ 2 & -1 \end{bmatrix}$$

We now show how to simplify the above computations. To obtain the inverse, we solved the two linear systems in (5) by reducing their augmented matrices (6). Since the two systems have the same coefficient matrix A, we can solve these systems *at the same time* by the following procedure: Write down the coefficient matrix A, and to the right of it, put the constants appearing on the right sides in (5). This gives

$$\left[\begin{array}{cc|cc} 1 & 2 & 1 & 0 \\ 2 & 3 & 0 & 1 \end{array}\right] \tag{7}$$

(For convenience, we separated A from the constants by a dashed line.) Observe that the matrix to the right of the dashed line is the 2×2 identity matrix I. If we now reduce (7) to reduced row echelon form, we will be reducing simultaneously the two augmented matrices in (6), and therefore we will be solving the two systems in (5) at the same time. These computations appear as follows:

$$\overset{\displaystyle A \qquad\qquad I}{\left[\begin{array}{cc|cc} 1 & 2 & 1 & 0 \\ 2 & 3 & 0 & 1 \end{array}\right]}$$

$$\left[\begin{array}{cc|cc} 1 & 2 & 1 & 0 \\ 0 & -1 & -2 & 1 \end{array}\right] \quad \text{We added } -2 \text{ times the first row to the last row.}$$

$$\left[\begin{array}{cc|cc} 1 & 2 & 1 & 0 \\ 0 & 1 & 2 & -1 \end{array}\right] \quad \text{We multiplied the last row by } -1.$$

$$\left[\begin{array}{cc|cc} 1 & 0 & -3 & 2 \\ 0 & 1 & 2 & -1 \end{array}\right] \quad \text{We added } -2 \text{ times the last row to the first.}$$

This 2×4 matrix is now in reduced row echelon form, and the inverse of A appears on the right side of the dashed line. This example suggests the following:

Procedure for Finding A^{-1} Let A be an invertible $n \times n$ matrix and let $[A \mid I]$ be the matrix obtained by adjoining the $n \times n$ identity matrix on the right side of A. Then the reduced row echelon form of $[A \mid I]$ is $[I \mid A^{-1}]$.

EXAMPLE 3 Find the inverse of

$$A = \begin{bmatrix} 1 & 0 & 1 \\ 2 & 1 & 0 \\ 0 & 1 & -1 \end{bmatrix}$$

Solution We adjoin the 3×3 identity matrix on the right side of A and transform $[A \mid I]$ to reduced row echelon form:

$$\overset{\displaystyle A \qquad\qquad\qquad I}{\left[\begin{array}{ccc|ccc} 1 & 0 & 1 & 1 & 0 & 0 \\ 2 & 1 & 0 & 0 & 1 & 0 \\ 0 & 1 & -1 & 0 & 0 & 1 \end{array}\right]}$$

$$\left[\begin{array}{ccc|ccc} 1 & 0 & 1 & 1 & 0 & 0 \\ 0 & 1 & -2 & -2 & 1 & 0 \\ 0 & 1 & -1 & 0 & 0 & 1 \end{array}\right] \quad \begin{array}{l} -2 \text{ times the first row} \\ \text{was added to the second} \\ \text{row.} \end{array}$$

$$\left[\begin{array}{ccc|ccc} 1 & 0 & 1 & 1 & 0 & 0 \\ 0 & 1 & -2 & -2 & 1 & 0 \\ 0 & 0 & 1 & 2 & -1 & 1 \end{array}\right] \quad \begin{array}{l} -1 \text{ times the second row} \\ \text{was added to the third} \\ \text{row.} \end{array}$$

$$\left[\begin{array}{ccc|ccc} 1 & 0 & 1 & 1 & 0 & 0 \\ 0 & 1 & 0 & 2 & -1 & 2 \\ 0 & 0 & 1 & 2 & -1 & 1 \end{array}\right]$$

2 times the third row was added to the second row.

$$\left[\begin{array}{ccc|ccc} 1 & 0 & 0 & -1 & 1 & -1 \\ 0 & 1 & 0 & 2 & -1 & 2 \\ 0 & 0 & 1 & 2 & -1 & 1 \end{array}\right]$$

-1 times the third row was added to the first row.

This matrix is in reduced row echelon form, so

$$A^{-1} = \begin{bmatrix} -1 & 1 & -1 \\ 2 & -1 & 2 \\ 2 & -1 & 1 \end{bmatrix}$$

Often, we do not know in advance whether a given matrix A is invertible. If the procedure in the above example is attempted on a matrix that is not invertible, then at some point in the computations a row with all zeros to the left of the dashed line will occur. When this happens, it can be shown that A is not invertible, so we stop our computations.

EXAMPLE 4 Find the inverse of

$$A = \begin{bmatrix} 1 & 3 & 4 \\ 2 & 7 & -1 \\ -1 & -4 & 5 \end{bmatrix}$$

Solution If we form the matrix $[A \mid I_3]$ and begin to transform it to reduced row echelon form, we obtain

$$\left[\begin{array}{ccc|ccc} 1 & 3 & 4 & 1 & 0 & 0 \\ 2 & 7 & -1 & 0 & 1 & 0 \\ -1 & -4 & 5 & 0 & 0 & 1 \end{array}\right]$$

$$\left[\begin{array}{ccc|ccc} 1 & 3 & 4 & 1 & 0 & 0 \\ 0 & 1 & -9 & -2 & 1 & 0 \\ 0 & -1 & 9 & 1 & 0 & 1 \end{array}\right]$$

-2 times the first row was added to the second row and the first row was added to the third row.

$$\left[\begin{array}{ccc|ccc} 1 & 3 & 4 & 1 & 0 & 0 \\ 0 & 1 & -9 & -2 & 1 & 0 \\ 0 & 0 & 0 & -1 & 1 & 1 \end{array}\right]$$

The second row was added to the third row.

Since we have a row with all zeros to the left of the dashed line, the matrix A is not invertible.

Using Inverses to Solve Linear Systems

The inverse provides a useful alternative tool for solving certain systems of n linear equations in n unknowns. To illustrate this method, consider the system

$$\begin{aligned} x + \phantom{y - {}} z &= 5 \\ 2x + y \phantom{{} - z} &= -2 \\ y - z &= 3 \end{aligned} \tag{8}$$

We can write this system in matrix form as

$$AX = B \tag{9}$$

where

$$A = \begin{bmatrix} 1 & 0 & 1 \\ 2 & 1 & 0 \\ 0 & 1 & -1 \end{bmatrix} \qquad X = \begin{bmatrix} x \\ y \\ z \end{bmatrix} \qquad B = \begin{bmatrix} 5 \\ -2 \\ 3 \end{bmatrix}$$

In Example 3 we showed that A is an invertible matrix and we computed its inverse

$$A^{-1} = \begin{bmatrix} -1 & 1 & -1 \\ 2 & -1 & 2 \\ 2 & -1 & 1 \end{bmatrix}$$

If we multiply both sides of (9) on the left by A^{-1}, we obtain

$$A^{-1}(AX) = A^{-1}B$$

or

$$(A^{-1}A)X = A^{-1}B$$

or

$$IX = A^{-1}B$$

or

$$X = A^{-1}B \tag{10}$$

Substituting the matrices X, A^{-1}, and B into (10) gives

$$\begin{bmatrix} x \\ y \\ z \end{bmatrix} = \begin{bmatrix} -1 & 1 & -1 \\ 2 & -1 & 2 \\ 2 & -1 & 1 \end{bmatrix} \begin{bmatrix} 5 \\ -2 \\ 3 \end{bmatrix} = \begin{bmatrix} -10 \\ 18 \\ 15 \end{bmatrix}$$

Thus, the solution to the given linear system (8) is

$$x = -10 \qquad y = 18 \qquad z = 15$$

This example illustrates the following result:

If $AX = B$ is a linear system of n equations in n unknowns and if the coefficient matrix A is invertible, then the system has one solution, namely, $X = A^{-1}B$.

EXAMPLE 5 A plastics firm manufactures two types of commercial Plexiglas rods, flexible and rigid. Each rod undergoes a molding process and a smoothing process. The number of hours per ton required by each process is displayed in the following production matrix:

$$A = \begin{array}{c} \\ \\ \end{array}\begin{matrix} \text{Flexible} & \text{Rigid} \\ \end{matrix}$$

$$A = \begin{bmatrix} 1 & 2 \\ 1 & 3 \end{bmatrix} \begin{matrix} \text{Molding} \\ \text{Smoothing} \end{matrix}$$

(columns: Flexible, Rigid)

If the firm manufactures x_1 tons of flexible rods and x_2 tons of rigid rods, then the matrix product

$$AX = \begin{bmatrix} 1 & 2 \\ 1 & 3 \end{bmatrix}\begin{bmatrix} x_1 \\ x_2 \end{bmatrix} = \begin{bmatrix} x_1 + 2x_2 \\ x_1 + 3x_2 \end{bmatrix}$$

tells us how many total hours will be needed for molding and smoothing; that is, to produce x_1 tons of flexible rods and x_2 tons of rigid rods, the firm will require

$$x_1 + 2x_2 \qquad \text{hours for molding}$$

and

$$x_1 + 3x_2 \qquad \text{hours for smoothing}$$

Suppose the firm's production manager wants answers to the following three questions: How many tons of rods of each type can be produced in a day if

1. the molding plant operates 16 hours and the smoothing plant operates 20 hours?
2. the molding plant operates 14 hours and the smoothing plant operates 18 hours?
3. the molding plant operates 10 hours and the smoothing plant operates 11 hours?

To answer these questions, we form the matrices

$$B_1 = \begin{bmatrix} 16 \\ 20 \end{bmatrix} \qquad B_2 = \begin{bmatrix} 14 \\ 18 \end{bmatrix} \qquad B_3 = \begin{bmatrix} 10 \\ 11 \end{bmatrix}$$

which tell us the amount of time available for each process. To answer the production manager's questions, we must solve the three systems

$$AX = B_1 \qquad AX = B_2 \qquad AX = B_3$$

You can verify that the inverse of the matrix A is

$$A^{-1} = \begin{bmatrix} 3 & -2 \\ -1 & 1 \end{bmatrix}$$

Thus the solution to $AX = B_1$ is

$$X = A^{-1}B_1 = \begin{bmatrix} 3 & -2 \\ -1 & 1 \end{bmatrix}\begin{bmatrix} 16 \\ 20 \end{bmatrix} = \begin{bmatrix} 8 \\ 4 \end{bmatrix} \tag{11}$$

the solution to $AX = B_2$ is

$$X = A^{-1}B_2 = \begin{bmatrix} 3 & -2 \\ -1 & 1 \end{bmatrix}\begin{bmatrix} 14 \\ 18 \end{bmatrix} = \begin{bmatrix} 6 \\ 4 \end{bmatrix} \tag{12}$$

and the solution to $AX = B_3$ is

$$X = A^{-1}B_3 = \begin{bmatrix} 3 & -2 \\ -1 & 1 \end{bmatrix}\begin{bmatrix} 10 \\ 11 \end{bmatrix} = \begin{bmatrix} 8 \\ 1 \end{bmatrix} \tag{13}$$

CODED MESSAGES

A	B	C	D
↕	↕	↕	↕
1	2	3	4
E	F	G	H
↕	↕	↕	↕
5	6	7	8
I	J	K	L
↕	↕	↕	↕
9	10	11	12
M	N	O	P
↕	↕	↕	↕
13	14	15	16
Q	R	S	T
↕	↕	↕	↕
17	18	19	20
U	V	W	X
↕	↕	↕	↕
21	22	23	24
Y	Z		
↕	↕		
25	26		

Cryptography is the study of methods for encoding and decoding messages. One of the very simplest techniques for doing this involves the use of the inverse of a matrix.

First, attach a different number to every letter of the alphabet. For example, we can let A be 1, B be 2, and so on, as shown in the accompanying table. Suppose that we then want to send the message

MATH WORKS

Substituting for each letter, we send the message

$$13, 1, 20, 8, 23, 15, 18, 11, 19 \tag{1}$$

Unfortunately, this simple code can be easily cracked. A better method involves the use of matrices.

Break the message (1) into three 3×1 matrices:

$$X_1 = \begin{bmatrix} 13 \\ 1 \\ 20 \end{bmatrix} \qquad X_2 = \begin{bmatrix} 8 \\ 23 \\ 15 \end{bmatrix} \qquad X_3 = \begin{bmatrix} 18 \\ 11 \\ 19 \end{bmatrix}$$

The sender and receiver jointly select an invertible 3×3 matrix such as

$$A = \begin{bmatrix} 1 & 1 & 2 \\ 1 & 1 & 1 \\ 1 & 0 & 1 \end{bmatrix}$$

The sender forms the 3×1 matrices

$$AX_1 = \begin{bmatrix} 54 \\ 34 \\ 33 \end{bmatrix} \qquad AX_2 = \begin{bmatrix} 61 \\ 46 \\ 23 \end{bmatrix} \qquad AX_3 = \begin{bmatrix} 67 \\ 48 \\ 37 \end{bmatrix}$$

and sends the message

$$54, 34, 33, 61, 46, 23, 67, 48, 37 \tag{2}$$

To decode the message, the receiver uses the inverse of matrix A:

$$A^{-1} = \begin{bmatrix} -1 & 1 & 1 \\ 0 & 1 & -1 \\ 1 & -1 & 0 \end{bmatrix}$$

and forms

$$A^{-1}\begin{bmatrix} 54 \\ 34 \\ 33 \end{bmatrix} = X_1 \qquad A^{-1}\begin{bmatrix} 61 \\ 46 \\ 23 \end{bmatrix} = X_2 \qquad A^{-1}\begin{bmatrix} 67 \\ 48 \\ 37 \end{bmatrix} = X_3$$

which, of course, is the original message (1) and which can be understood by using the accompanying table.

If the receiver sends back the message

$$44, 26, 21, 39, 30, 29, 76, 51, 39$$

what is the response?

From (11), (12), and (13) we obtain the following answers to the production manager's questions:

1. If the molding plant operates 16 hours and the smoothing plant 20 hours, then the firm can produce 8 tons of flexible rods and 4 tons of rigid rods.
2. If the molding plant operates 14 hours and the smoothing plant 18 hours, then the firm can produce 6 tons of flexible rods and 4 tons of rigid rods.
3. If the molding plant operates 10 hours and the smoothing plant 11 hours, then the firm can produce 8 tons of flexible rods and 1 ton of rigid rods.

EXERCISE SET 2.5

1. Are either of the following matrices inverses of $A = \begin{bmatrix} 1 & 2 \\ 3 & 4 \end{bmatrix}$?

(a) $\begin{bmatrix} 3 & -2 \\ 1 & 1 \end{bmatrix}$ (b) $\begin{bmatrix} -2 & 1 \\ \frac{3}{2} & -\frac{1}{2} \end{bmatrix}$

In Exercises 2–5 find the inverse, if possible.

2. (a) $\begin{bmatrix} 2 & -2 \\ 5 & 1 \end{bmatrix}$ (b) $\begin{bmatrix} 2 & 5 \\ 3 & -4 \end{bmatrix}$ (c) $\begin{bmatrix} 1 & -1 \\ 3 & 2 \end{bmatrix}$

3. (a) $\begin{bmatrix} 1 & 2 & 1 \\ 0 & 1 & 1 \\ 1 & 0 & 1 \end{bmatrix}$ (b) $\begin{bmatrix} 2 & 1 & -1 \\ 1 & 1 & 3 \\ 0 & 1 & 1 \end{bmatrix}$

(c) $\begin{bmatrix} 2 & 1 & -2 \\ 3 & 1 & -5 \\ 3 & 2 & -1 \end{bmatrix}$

4. (a) $\begin{bmatrix} -1 & 2 & 1 \\ 0 & 1 & 1 \\ 2 & 3 & 1 \end{bmatrix}$ (b) $\begin{bmatrix} 1 & -2 & 2 \\ 0 & -1 & 5 \\ 2 & 1 & 3 \end{bmatrix}$

(c) $\begin{bmatrix} 2 & 0 & 0 & 0 \\ 4 & 3 & 0 & 0 \\ -1 & 3 & -2 & 0 \\ 1 & 1 & 2 & 4 \end{bmatrix}$

5. (a) $\begin{bmatrix} -1 & -2 & 0 & 0 \\ -4 & -6 & -3 & -3 \\ 2 & 4 & 2 & 3 \\ 2 & 3 & 1 & 1 \end{bmatrix}$ (b) $\begin{bmatrix} 2 & 3 & 1 \\ -2 & 1 & 0 \\ 1 & 1 & 0 \end{bmatrix}$

(c) $\begin{bmatrix} 6 & 5 & -8 \\ 1 & 1 & -2 \\ 3 & 2 & -2 \end{bmatrix}$

In Exercises 6–9 solve the given linear system by finding the inverse of the coefficient matrix.

6. $\begin{aligned} x - 3y &= 5 \\ 4x + 6y &= 7 \end{aligned}$

7. $\begin{aligned} 2x + 3y - z &= 15 \\ x - y + z &= 1 \\ 3x - y + 2z &= 8 \end{aligned}$

8. $\begin{aligned} x_1 - 2x_2 + x_3 &= -2 \\ 2x_1 - x_2 - x_3 &= -10 \\ 3x_1 + x_2 + 2x_3 &= -2 \end{aligned}$

9. $\begin{aligned} 2x_1 + 2x_2 - x_3 &= 2 \\ x_1 + x_2 + x_3 &= 4 \\ x_1 \qquad - x_3 &= -3 \end{aligned}$

10. Solve the system

$$\begin{aligned} x + 2y &= -4 \\ 3x + 5y &= -9 \end{aligned}$$

by

(a) Gauss–Jordan elimination.
(b) finding the inverse of the coefficient matrix.

11. Follow the directions of Exercise 10 for the system

$$\begin{aligned} x + 2y - 3z &= 7 \\ 2x - y - z &= 4 \\ 3x + 4y + z &= 1 \end{aligned}$$

12. By finding the inverse of the coefficient matrix, solve the linear system

$$\begin{aligned} x + 3y &= b_1 \\ 2x - y &= b_2 \end{aligned}$$

when

(a) $b_1 = 1, b_2 = -1$ (b) $b_1 = -2, b_2 = 3$
(c) $b_1 = 3, b_2 = 2$

13. By finding the inverse of the coefficient matrix, solve the linear system

$$\begin{aligned} 2x_1 - x_2 + x_3 &= b_1 \\ 3x_1 - x_2 - 3x_3 &= b_2 \\ 2x_1 + 2x_2 - x_3 &= b_3 \end{aligned}$$

when

(a) $b_1 = 2, b_2 = 1, b_3 = 3$
(b) $b_1 = 3, b_2 = -4, b_3 = 2$
(c) $b_1 = -3, b_2 = 2, b_3 = 2$

14. Show that the matrix

$$\begin{bmatrix} a & b & c \\ 0 & 0 & 0 \\ d & e & f \end{bmatrix}$$

is not invertible. Can any square matrix with a row of zeros be invertible?

15. Show that if A is an invertible matrix, then $(A^{-1})^{-1} = A$.

16. Let

$$\begin{bmatrix} 2 & 3 \\ -1 & 2 \end{bmatrix}$$

be the inverse of a matrix A. Find A. [*Hint:* See Exercise 15.]

17. Show that if A and B are invertible matrices, then AB is invertible and $(AB)^{-1} = B^{-1}A^{-1}$.

18. Verify the result of Exercise 17 for the matrices

$$A = \begin{bmatrix} 1 & 1 \\ 2 & -1 \end{bmatrix} \quad \text{and} \quad B = \begin{bmatrix} 1 & 1 \\ 3 & 2 \end{bmatrix}$$

19. Given that

$$A^{-1} = \begin{bmatrix} 1 & -2 \\ 3 & 5 \end{bmatrix}$$

find the solution of $AX = B$ if

$$B = \begin{bmatrix} -2 \\ 4 \end{bmatrix}$$

20. Show that the matrix

$$D = \begin{bmatrix} a & 0 & 0 \\ 0 & b & 0 \\ 0 & 0 & c \end{bmatrix}$$

is invertible if a, b, and c are nonzero.

21. If

$$A^{-1} = \begin{bmatrix} 1 & 2 \\ -3 & 4 \end{bmatrix} \quad \text{and} \quad B^{-1} = \begin{bmatrix} 2 & -1 \\ 5 & 3 \end{bmatrix}$$

find $(AB)^{-1}$. [*Hint*: See Exercise 17.]

22. **(Harder)** Show that if A is an invertible matrix and c is a nonzero number, then

$$(cA)^{-1} = \frac{1}{c}A^{-1}$$

23. **(Harder)** If A is an $n \times n$ matrix, then A^3 denotes AAA and A^2 denotes AA. Verify that if A is an $n \times n$ matrix such that $A^3 = O$, then

$$(I - A)^{-1} = I + A + A^2$$

24. **(Investment)** Suppose a trust fund has \$3000 that must be invested in two different types of bonds. The first bond pays 5% interest per year and the second bond pays 7% interest per year. Determine how to divide the \$3000 among the two types of bonds if the fund must obtain an annual total interest of

(a) \$180
(b) \$200
(c) \$160
(d) Is it possible to obtain a total annual interest of \$250?

25. Find the inverse of $A = \begin{bmatrix} 3 & 1 \\ 5 & 2 \end{bmatrix}$ and use A^{-1} to calculate the points of intersection of the following pairs of lines.

(a) $\begin{aligned} 3x + y &= 1 \\ 5x + 2y &= 0 \end{aligned}$ (b) $\begin{aligned} 3x + y &= -4 \\ 5x + 2y &= 7 \end{aligned}$

KEY IDEAS FOR REVIEW

- **Linear equation** An equation of the form $a_1x_1 + a_2x_2 + \cdots + a_nx_n = c$.
- **System of linear equations (a linear system)** A set of linear equations.
- **Solution of a linear system** See page 56.
- **Inconsistent system** One that has no solutions.
- **Consistent system** One that has at least one solution.

- **Augmented matrix** See page 57.
- **Equivalent systems** Two linear systems having the same solutions.
- **Reduced row echelon form for an augmented matrix** See page 61.
- **Gauss–Jordan elimination** See page 63.
- **Matrix** A rectangular array of numbers.
- **Entries of a matrix** The numbers in the matrix.
- **Scalars** Ordinary real numbers.
- **Size of a matrix** A is $m \times n$ if it has m rows and n columns.
- **Equal matrices** Matrices of the same size whose corresponding entries are equal.
- $\mathbf{A + B}$ The matrix obtained by adding corresponding entries in A and B.
- **Zero matrix** A matrix each of whose entries is zero.
- **The negative of a matrix A** The matrix whose entries are the negatives of the corresponding entries in A.
- $\mathbf{cA}$ The matrix obtained by multiplying each entry of A by c.
- $\mathbf{AB}$ See page 80.
- **Identity matrix (I or I_n)** A square matrix with 1s on the main diagonal and 0s elsewhere.
- **Coefficient matrix** The matrix A in a linear system $AX = B$.
- $\mathbf{A^t}$ **(the transpose of A)** The matrix obtained from A by converting columns to rows.
- **Inverse of a square matrix A** A matrix, denoted by A^{-1}, such that

$$AA^{-1} = A^{-1}A = I$$

SUPPLEMENTARY EXERCISES

1. For what value(s) of k is the following system inconsistent?

$$\begin{aligned} 2x - y + 4z &= 2 \\ 4x - 2y + 8z &= k \end{aligned}$$

2. For what value(s) of a is the following system inconsistent?

$$\begin{aligned} x + \qquad\quad y &= 3 \\ x + (a^2 - 9)y &= a \end{aligned}$$

3. Transform each of the following augmented matrices to reduced row echelon form.

(a) $\left[\begin{array}{rrr|r} 3 & -1 & 5 & 2 \\ 2 & 4 & 0 & -1 \\ 2 & 1 & 1 & 3 \end{array}\right]$

(b) $\left[\begin{array}{rrrr|r} 2 & 4 & 3 & 1 & 3 \\ -1 & 1 & 2 & 0 & -2 \\ 3 & 5 & -2 & 3 & 1 \end{array}\right]$

4. Show that if A is an $n \times n$ matrix, which is in reduced row echelon form and not equal to the identity matrix, then A has a row consisting entirely of zeros.

5. Solve each of the following linear systems by Gauss–Jordan elimination.

(a) $\begin{aligned} x + y + z &= 4 \\ 2x - y - z &= 2 \\ 3x + 2y + 2z &= 6 \end{aligned}$

(b) $\begin{aligned} 2x - y + 3z &= 13 \\ -2x + 2y + z &= -6 \\ 5x \qquad - 2z &= 11 \end{aligned}$

(c) $\begin{bmatrix} 2 & 3 & 1 & -1 \\ 1 & -2 & 2 & 2 \\ -3 & 4 & -1 & 3 \end{bmatrix}\begin{bmatrix} x_1 \\ x_2 \\ x_3 \\ x_4 \end{bmatrix} = \begin{bmatrix} 2 \\ -3 \\ 4 \end{bmatrix}$

(d) $\begin{bmatrix} 2 & -1 & 3 & -3 \\ 4 & 1 & 5 & 2 \end{bmatrix}\begin{bmatrix} x_1 \\ x_2 \\ x_3 \\ x_4 \end{bmatrix} = \begin{bmatrix} 2 \\ 4 \end{bmatrix}$

6. Solve each of the following linear systems by Gauss–Jordan elimination.

(a) $$\begin{aligned} 3x + 3y - 2z &= -11 \\ -4x - 2y + 5z &= 26 \\ 3x + y + 2z &= 3 \\ 3x + y + 5z &= 18 \end{aligned}$$

(b) $$\begin{aligned} 2x + y - z &= 8 \\ 3x - 2y &= 17 \\ -2x + y - 2z &= -13 \end{aligned}$$

(c) $\begin{bmatrix} 2 & -1 & 2 & 3 & 1 \\ -1 & 3 & 4 & -2 & 3 \\ 3 & 2 & -1 & 2 & -2 \\ 3 & 1 & 8 & 4 & 5 \end{bmatrix}\begin{bmatrix} x_1 \\ x_2 \\ x_3 \\ x_4 \\ x_5 \end{bmatrix} = \begin{bmatrix} 1 \\ 2 \\ 3 \\ 4 \end{bmatrix}$

(d) $\begin{bmatrix} 2 & -1 & 3 \\ 3 & 2 & -1 \\ -4 & -5 & 5 \end{bmatrix}\begin{bmatrix} x_1 \\ x_2 \\ x_3 \end{bmatrix} = \begin{bmatrix} 0 \\ 0 \\ 0 \end{bmatrix}$

7. Show that if $ad - bc \neq 0$, then the augmented matrix

$$\left[\begin{array}{cc|c} a & b & e \\ c & d & f \end{array}\right]$$

when transformed to reduced row echelon form is

$$\left[\begin{array}{cc|c} 1 & 0 & e' \\ 0 & 1 & f' \end{array}\right]$$

where e' and f' are constants.

8. Show that the linear system

$$\begin{aligned} ax + by &= e \\ cx + dy &= f \end{aligned}$$

has exactly one solution if $ab - cd \neq 0$ [*Hint*: See Exercise 7.]

9. For which values of a will the following linear system have no solution? Exactly one solution? Infinitely many solutions?

$$\begin{aligned} x + y &= 4 \\ x + (a^2 - 15)y &= a \end{aligned}$$

In Exercises 10–12 let

$$A = \begin{bmatrix} 2 & -1 & 3 \\ 4 & 1 & 2 \end{bmatrix} \qquad B = \begin{bmatrix} 3 & -2 & 3 \\ 4 & 5 & 1 \\ 2 & -1 & 0 \end{bmatrix}$$

$$C = \begin{bmatrix} -2 & 2 \\ 3 & -1 \end{bmatrix} \qquad D = \begin{bmatrix} 2 & 3 \\ -1 & 4 \end{bmatrix}$$

$$E = \begin{bmatrix} 4 & 2 & -1 \\ 5 & -3 & 2 \end{bmatrix}$$

10. Compute, if possible,
(a) $A + 3B$ (b) $2A + 4E$
(c) $AB + 2E$ (d) $CB + A$

11. Compute, if possible,
(a) $(2D)A - EB$ (b) $CA + B$
(c) $EA + D$ (d) EBC

12. Compute, if possible,
(a) E^tD (b) $C^tA^t + D$
(c) $A^t + (EB)^t$ (d) $C^tE^t - 2A$

13. Show that the jth column of the matrix product AB is the matrix product AB_j, where B_j is the matrix consisting of the jth column of B.

14. If A and B are $n \times n$ matrices, when is

$$(A - B)(A + B) = AA - BB$$

15. If X_0 and X_1 are solutions to the linear system $AX = 0$, show that $rX_0 + sX_1$ is also a solution, where r and s are any real numbers.

In Exercises 16–19 find the inverse, if possible.

16. $\begin{bmatrix} 3 & 2 \\ -1 & 4 \end{bmatrix}$

17. $\begin{bmatrix} 3 & -1 & 2 \\ 4 & 1 & 3 \\ 2 & -3 & 1 \end{bmatrix}$

18. $\begin{bmatrix} 0 & 1 & 2 \\ 1 & -1 & 2 \\ 3 & 1 & 4 \end{bmatrix}$

19. $\begin{bmatrix} 1 & 2 & -1 \\ 3 & 2 & 4 \\ 1 & 2 & -2 \end{bmatrix}$

In Exercises 20–23 solve the given linear systems by finding the inverse of the coefficient matrix.

20. $$\begin{aligned} x + 2y &= 8 \\ 2x - 3y &= -5 \end{aligned}$$

21. $$\begin{aligned} 2x - y + z &= -1 \\ x + y - 2z &= -5 \\ 3x - y + z &= -2 \end{aligned}$$

22. $$\begin{aligned} 2x - y + z &= 4 \\ x + y - z &= 8 \\ 4x - z &= 18 \end{aligned}$$

23. $$\begin{aligned} x_1 - x_2 + x_3 &= -3 \\ 2x_1 + 2x_2 &= 14 \\ 3x_1 - x_2 + x_3 &= 3 \end{aligned}$$

24. Solve the linear system $AX = B$ if

$$A^{-1} = \begin{bmatrix} 2 & -1 \\ 3 & 4 \end{bmatrix} \quad \text{and} \quad B = \begin{bmatrix} 4 \\ 5 \end{bmatrix}$$

25. If

$$A^{-1} = \begin{bmatrix} 2 & 3 \\ 4 & -1 \end{bmatrix}$$

find A. [*Hint*: See Exercise 15 of Section 2.5.]

26. Let A, B, and C be $n \times n$ matrices, where A is invertible. Show that if $AB = AC$, then $B = C$.

27. (Nutrition) A veterinarian is preparing a special blend of cat food by mixing three different foods. Each ounce of food A contains 5 grams of fat, 6 grams of ash, and 12 grams of protein; each ounce of food B contains 10 grams of fat, 3 grams of ash, and 20 grams of protein; each ounce of food C contains 25 grams of fat, 8 grams of ash, and 15 grams of protein. How many ounces of each type of food should be used to provide a meal that contains exactly 115 grams of fat, 45 grams of ash, and 129 grams of protein?

28. Suppose the financial advisor of a university's endowment fund must invest exactly \$130,000 in three types of securities: Bond AAA, paying a dividend of 7%; stock BB, paying a dividend of 9%; and certificate of deposit CC, paying a dividend of 8%. The advisor has been told that the amount invested in bond AAA must be twice the amount invested in certificate of deposit CC. How much money should be invested in each security to have an annual income of \$10,500?

29. A pharmaceutical firm prepares three different chemotherapy compounds, Chem 1, Chem 2, and Chem 3, by blending three substances A, B, and C. Suppose each cubic centimeter (cc) of Chem 1 contains 4 units of A, 3 units of B, and 5 units of C; each cc of Chem 2 contains 5 units of A, 2 units of B, and 4 units of C; and each cc of Chem 3 contains 3 units of A, 6 units of B, and 5 units of C. How many cc's of each chemotherapy compound should be given to a patient who needs 25 units of substance A, 32 units of B, and 34 units of C?

30. A potato chip manufacturer makes two types of chips, cheese- and pizza-flavored, each of which goes through a flavoring process and a packing process. The following matrix gives the time (in seconds) required by these processes:

	Flavoring process	Packing process
Cheese flavor	20	10
Pizza flavor	30	10

$$A = \begin{bmatrix} 20 & 10 \\ 30 & 10 \end{bmatrix}$$

There are two plants, one in San Diego and the other in Orlando. The following matrix gives the hourly rates (in dollars) for each of the processes:

	San Diego	Orlando
Flavoring process	12	10
Packing process	15	12

$$B = \begin{bmatrix} 12 & 10 \\ 15 & 12 \end{bmatrix}$$

What do the entries in the matrix AB represent?

CHAPTER TEST

1. For what values of a does the following linear system have exactly one solution? No solution? Infinitely many solutions?

$$\begin{aligned} x + y &= 5 \\ x + (a^2 - 24)y &= a \end{aligned}$$

2. Solve each of the following linear systems by Gauss–Jordan elimination.

(a)
$$\begin{aligned} x_1 + 2x_2 + 3x_3 + 4x_4 - x_5 &= 2 \\ 2x_1 + 4x_2 + 5x_3 + 3x_4 + 2x_5 &= -1 \\ 3x_1 + 6x_2 + 8x_3 + 11x_4 - 3x_5 &= 7 \end{aligned}$$

(b)
$$\begin{bmatrix} 2 & -1 & 3 \\ -3 & 2 & 1 \\ 3 & 2 & -1 \\ -2 & -8 & 6 \end{bmatrix} \begin{bmatrix} x_1 \\ x_2 \\ x_3 \end{bmatrix} = \begin{bmatrix} 0 \\ -10 \\ 6 \\ -4 \end{bmatrix}$$

3. If

$$A = \begin{bmatrix} 1 & 2 \\ 3 & 2 \\ 1 & 0 \end{bmatrix} \quad B = \begin{bmatrix} 4 & -2 \\ 3 & 2 \\ -1 & 5 \end{bmatrix}$$

$$C = \begin{bmatrix} 4 & -2 & 1 \\ 5 & 3 & 2 \end{bmatrix} \quad D = \begin{bmatrix} 2 \\ -3 \end{bmatrix} \quad E = [5 \quad 3]$$

compute, if possible,

(a) $CA + B$ (b) $(2A - 3B)C'$ (c) $C'D + E$ (d) DE (e) ED

4. Find the values of x for which the following matrix equation is true.

$$[1 \quad -3 \quad 1] \begin{bmatrix} x \\ x^2 \\ 2 \end{bmatrix} = [0]$$

5. (a) Find A^{-1} if

$$A = \begin{bmatrix} 1 & 1 & 0 \\ 1 & 0 & 1 \\ 0 & 1 & 1 \end{bmatrix}$$

(b) Use A^{-1} to solve the following linear system.

$$\begin{aligned} x + y \phantom{{}+z} &= 6 \\ x \phantom{{}+y} + z &= -3 \\ y + z &= 4 \end{aligned}$$

6. A special low-calorie diet consists of dishes A, B, and C. Each unit of A has 2 grams of fat, 1 gram of carbohydrate, and 3 grams of protein. Each unit of B has 1 gram of fat, 2 grams of carbohydrate, and 1 gram of protein. Each unit of C has 1 gram of fat, 2 grams of carbohydrate, and 3 grams of protein. The diet must provide exactly 10 grams of fat, 14 grams of carbohydrate, and 18 grams of protein. How much of each dish should be used?

3

Linear Programming (A Geometric Approach)

In this chapter we introduce a relatively new area of mathematics called linear programming. The importance of this subject was recognized in the late 1940s when it was first used to help solve certain logistics problems for the U.S. government. Since that time linear programming has been applied to problems in such diverse areas as economics, engineering, biology, agriculture, business, and the social sciences.

Our approach in this introductory chapter is geometric; as a consequence, the methods we develop are only applicable to a limited variety of problems. In Chapter 4 we develop linear programming more extensively from an algebraic point of view.

3.1 WHAT IS LINEAR PROGRAMMING?

In many problems we are interested in choosing a course of action that maximizes or minimizes some quantity of importance. For example, an investor may want to select investments that maximize profits, a nutritionist may want to design a diet with a minimal number of calories, or a businessperson may want to select the sources of supply to minimize the shipping distances. It is often the case that the available courses of action are subject to certain constraints. The nutritionist, for example, may have to design a diet to meet certain daily vitamin requirements; the investor may be required to invest 20% of the available money in corporate bonds; and the businessperson may have to place an order of at least 10 truckloads in order to deal with a certain supplier. In problems like these, linear programming helps to select the optimal course of action subject to the given constraints. The following examples illustrate more precisely the kinds of problems to which linear programming applies and how these problems are formulated mathematically.

EXAMPLE 1 **PRODUCTION** A meat-packer mixes beef and pork to make two types of frankfurters, regular and deluxe. Suppose each pound of regular frankfurters contains

0.2 pound of beef
0.2 pound of pork

while each pound of deluxe frankfurters contains

0.4 pound of beef
0.2 pound of pork

Suppose also that the profits on the two types of frankfurters are

10¢ per pound on the regular
12¢ per pound on the deluxe

If the meat-packer has 30 pounds of beef and 20 pounds of pork in stock, how many pounds of each kind of frankfurter can be made to obtain the largest profit?

Solution As a first step toward finding the solution, we reformulate the problem in mathematical terms. This can be done as follows. Let x be the number of pounds of regular frankfurters to be made and let y be the number of pounds of deluxe frankfurters. Since the profit on each pound of regular is 10¢ and the profit on each pound of deluxe is 12¢, the total profit z (in cents) will be

$$z = 10x + 12y$$

Since each pound of regular frankfurters contains 0.2 pound of beef and each pound of deluxe frankfurters contains 0.4 pound of beef, the mixture will contain $0.2x + 0.4y$ pounds of beef. Similarly, the mixture will contain $0.2x + 0.2y$ pounds of pork. Since the packer can use no more than 30 pounds of beef and no more than 20 pounds of pork, we must have

$$\begin{aligned} 0.2x + 0.4y &\leq 30 \\ 0.2x + 0.2y &\leq 20 \end{aligned}$$

Since x and y cannot be negative, we must also have $x \geq 0$ and $y \geq 0$. Therefore, this problem can be reformulated mathematically as follows.

Find those values of x and y that make $z = 10x + 12y$ as large as possible, where x and y must satisfy

$$\begin{aligned} 0.2x + 0.4y &\leq 30 \\ 0.2x + 0.2y &\leq 20 \\ x &\geq 0 \\ y &\geq 0 \end{aligned}$$

In the next section we discuss techniques for solving this kind of mathematical problem. For the present, you should just concentrate on understanding how the verbal problems are translated into mathematical terms.

EXAMPLE 2 **NUTRITION** A nutritionist wants to design a breakfast menu for certain hospital patients. The menu is to include cereal and bread. Suppose each ounce of cereal provides

1 unit of iron
2 units of vitamin D
3 calories

while each ounce of bread provides

2 units of iron
2 units of vitamin D
4 calories

If the breakfast must provide at least 8 units of iron and at least 10 units of vitamin D, how many ounces of each item should be provided to meet the iron and vitamin D requirements with the smallest possible intake of calories?

Solution To formulate this problem mathematically, let x be the number of ounces of cereal to be provided and let y be the number of ounces of bread. Since there are 3 calories in 1 ounce of cereal and 4 calories in 1 ounce of bread, the number z of calories provided by the two items is

$$z = 3x + 4y$$

Since each ounce of cereal contains 1 unit of iron and each ounce of bread contains 2 units of iron, the two items provide a total of $x + 2y$ units of iron. Similarly, the two items provide $2x + 2y$ units of vitamin D.

Thus, to meet iron and vitamin D requirements we must have

$$\begin{aligned} x + 2y &\geq 8 \\ 2x + 2y &\geq 10 \end{aligned}$$

Since x and y cannot be negative, we must also have $x \geq 0$ and $y \geq 0$. Therefore, this problem can be formulated mathematically as follows.

Find those values of x and y that make $z = 3x + 4y$ as small as possible, where x and y must satisfy

$$\begin{aligned} x + 2y &\geq 8 \\ 2x + 2y &\geq 10 \\ x &\geq 0 \\ y &\geq 0 \end{aligned}$$

EXAMPLE 3 **INVESTMENT** The managers of a pension plan want to invest *up to* \$5000 in two stocks X and Y. Stock X is considered conservative, whereas stock Y is considered speculative. The managers agree that the investment in stock X should be at most \$4000 and that the investment in stock Y should be at least \$600. Suppose also that for each dollar invested,

X is expected to return \$0.08
Y is expected to return \$0.10

If the bylaws of the pension plan require that investment in the speculative stock Y can be at most one-third of the investment in the conservative stock X, how much should be invested in X and how much in Y to maximize the return on the investment?

Solution To formulate this problem mathematically, let x be the number of dollars to be invested in stock X and let y be the number of dollars to be invested in stock Y. Since stock X will return \$0.08 for each dollar invested and stock Y will return \$0.10 for each dollar invested, the number z of dollars returned by the two stocks together is

$$z = 0.08x + 0.10y$$

Since the managers can invest up to \$5000 in the stocks, we must have

$$x + y \leq 5000 \qquad \text{(total amount invested)}$$

Since the investment in stock X can be no more than \$4000 and since the investment in stock Y must be at least \$600, we obtain $x \leq 4000$ and $y \geq 600$. Since the investment in Y is required to be at most one-third of the investment in X, we also have $y \leq \frac{1}{3}x$. Finally, since x and y must be nonnegative, we have $x \geq 0$ and $y \geq 0$. Therefore, this problem can be formulated mathematically as follows.

Find those values of x and y that make $z = 0.08x + 0.10y$ as large as possible, where x and y must satisfy

$$\begin{aligned} x + y &\leq 5000 \\ x &\leq 4000 \\ y &\geq 600 \\ y &\leq \tfrac{1}{3}x \\ x &\geq 0 \\ y &\geq 0 \end{aligned}$$

EXAMPLE 4 **ALLOCATION OF RESOURCES** A fuel manufacturer makes two grades of commercial gasoline, regular (R) and unleaded (U). Each fuel goes through two processes, cracking and refining. Suppose each liquid unit of type R requires

0.2 hour for cracking
0.5 hour for refining

whereas each liquid unit of type U requires

0.4 hour for cracking
0.2 hour for refining

Suppose also that the profits on the two types of gasoline are

\$15 per liquid unit of R
\$30 per liquid unit of U

If the cracking plant can remain open for at most 8 hours a day and the refining plant can also remain open for at most 8 hours a day, how many liquid units of each type of gasoline should be manufactured each day to maximize the manufacturer's daily profit?

THE DEVELOPMENT OF LINEAR PROGRAMMING

During World War II, the U.S. military establishment embarked on an unprecedented effort to use all available scientific tools and techniques in the planning of military activities. These efforts included the deployment of personnel and equipment to the combat theaters, the solving of supply and maintenance problems, and the development and operation of training programs.

After the war, the U.S. Air Force continued to sponsor vigorously the development and application of scientific techniques to the solution of the Air Force's problems. In June 1947 a group called Project SCOOP (Scientific Computation of Optimum Programs) composed of Marshall Woods, George B. Dantzig, John Norton, and Murray Geisler began to work on the interindustry model, a way to describe the interindustry relations of an economy. The linear programming model was developed by July 1947, and by the end of the summer of 1947 Dantzig had formulated the simplex method for solving the linear programming problem (to be discussed in Chapter 4).

Mathematicians and economists quickly became interested in linear programming and its applications. One of its early applications occurred in the Berlin airlift. Economists such as T. C. Koopmans, Robert Dorfman, and Paul Samuelson started to reexamine the classical problems of economics from a new point of view. John von Neumann, A. W. Tucker, and a group of his students, including David Gale and H. W. Kuhn, worked on problems in linear inequalities, game theory, and their connections with linear programming. At the same time, the Air Force sponsored work at the U.S. Bureau of Standards to develop computer programs for solving large linear programming problems. The extraordinary growth of computing power and its decreasing cost, coupled with theoretical breakthroughs, have made linear programming one of the major tools of decision making.

Solution To formulate this problem mathematically, let x be the number of liquid units of type R to be manufactured each day and let y be the number of liquid units of type U. Since the profit on each unit of R is \$15 and the profit on each unit of U is \$30, the total daily profit z (in dollars) is

$$z = 15x + 30y$$

Since each unit of type R requires 0.2 hour for cracking and each unit of type U requires 0.4 hour for cracking, the two types of fuel will require a total of $0.2x + 0.4y$ hours each day for cracking. Similarly, the two types of fuel will require a total of $0.5x + 0.2y$ hours for refining. Thus, since the cracking and refining plants can be open at most 8 hours a day, we must have

$$\begin{aligned} 0.2x + 0.4y &\leq 8 \\ 0.5x + 0.2y &\leq 8 \end{aligned}$$

Since x and y cannot be negative, we must also have $x \geq 0$ and $y \geq 0$. Therefore, this problem can be reformulated mathematically as follows.

Find those values of x and y that make $z = 15x + 30y$ as large as possible, where x and y must satisfy

$$\begin{aligned} 0.2x + 0.4y &\leq 8 \\ 0.5x + 0.2y &\leq 8 \\ x &\geq 0 \\ y &\geq 0 \end{aligned}$$

Observe that the four examples above have much in common; in each problem we want to find nonnegative values of x and y that maximize or minimize an expression of the form

$$z = ax + by \tag{1}$$

called the **objective function**; further, x and y must also satisfy certain other inequalities called **contraints**. Mathematically, the expression $ax + by$ on the right side of (1) and the expressions involving x and y in the inequalities are called **linear functions** of x and y—hence the word *linear* in linear programming. The word *programming* is derived from the early applications of the subject to problems in the programming or allocation of supplies. The general approach followed in translating each of the above problems into a linear programming problem can be summarized as follows.

Formulating a Linear Programming Problem Mathematically

Step 1 Read the problem carefully.

Step 2 Choose variables to represent the unknown quantities.

Step 3 Set up the objective function.

Step 4 Derive the constraints (inequalities) that must be satisfied by the variables. Don't forget the nonnegative constraints. (In practical problems, quantities of a product cannot be negative.)

EXERCISE SET 3.1

1. A firm has budgeted \$1500 for display space at a toy show. Two types of display booths are available: preferred space costs \$18 per square foot, with a minimum rental of 60 square feet, and regular space costs \$12 per square foot, with a minimum rental of 30 square feet. It is estimated that there will be 120 visitors for each square foot of preferred space and 60 visitors for each square foot of regular space. Let x = the number of square feet of preferred space and y = the number of square feet of regular space.
   (a) Write an expression for the cost of the display space in terms of x and y.
   (b) Write a system of inequalities that must be satisfied by x and y.
   (c) Write an expression for the objective function giving the number of visitors in terms of x and y.
   (d) Repeat part (c) if it is estimated that there will be 110 visitors for each square foot of preferred space and 70 visitors for each square foot of regular space.

2. A coffee packer uses Jamaican and Colombian coffee to prepare a mild blend and a strong blend. Each pound of mild blend requires $\frac{1}{2}$ pound of Jamaican coffee and $\frac{1}{2}$ pound of Colombian coffee. Each pound of strong blend

requires $\frac{1}{4}$ pound of Jamaican coffee and $\frac{3}{4}$ pound of Colombian coffee. Each pound of mild blend coffee yields a profit of \$3, and each pound of strong blend coffee yields a profit of \$4. The packer has available 100 pounds of Jamaican coffee and 125 pounds of Colombian coffee. Let x = the number of pounds of mild blend coffee prepared and y = the number of pounds of strong blend coffee prepared.

(a) Write an expression for the amount of Jamaican coffee used.

(b) Write an expression for the amount of Colombian coffee used.

(c) Write a system of inequalities that must be satisfied by x and y.

(d) Write an expression for the objective function giving the total profit.

(e) Repeat part (d) if each pound of mild blend coffee yields a profit of \$4 and each pound of strong blend yields a profit of \$3.

In Exercises 3–23 formulate the given linear programming problem mathematically. Do not attempt to solve the problem.

3. A manufacturer of lawn products wants to prepare a fertilizer by mixing two ingredients, GROW and THRIVE. Each pound of GROW contains 4 ounces of nitrogen and 4 ounces of phosphate, while each pound of THRIVE contains 10 ounces of nitrogen and 2 ounces of phosphate. The final product must contain at least 100 ounces of nitrogen and at least 60 ounces of phosphate. If each pound of GROW costs \$1.00 and each pound of THRIVE costs \$1.50, how many pounds of each ingredient should be mixed if the manufacturer wants to minimize the cost?

4. John Smith installs and then demonstrates burglar alarms. There are two burglar alarms that he installs, types A and B. Type A requires 1 hour to install and $\frac{1}{4}$ hour to demonstrate. Type B requires 2 hours to install and $\frac{1}{2}$ hour to demonstrate. Union rules require Smith to work a minimum of 20 hours per week as an installer and a maximum of 20 hours as a demonstrator. If he gets paid \$5 per hour for installing and \$4 per hour for demonstrating, how many alarms of each type should Smith install and demonstrate each week to maximize his earnings?

5. A hospital wants to design a dinner menu containing two items, M and N. Each ounce of M provides 1 unit of vitamin A and 2 units of vitamin B. Each ounce of N provides 1 unit of vitamin A and 1 unit of vitamin B. The two dishes must provide at least 7 units of vitamin A and at least 10 units of vitamin B. If each ounce of M costs 8 cents and each ounce of N costs 12 cents, how many ounces of each item should the hospital serve to minimize its cost?

6. Repeat Exercise 5 if each ounce of M costs 12 cents and each ounce of N costs 8 cents.

7. A small furniture-finishing plant finishes two kinds of tables, A and B. Each table must be sanded, stained, and varnished. Table A requires 10 minutes of sanding, 6 minutes of staining, and 9 minutes of varnishing. Table B requires 5 minutes of sanding, 12 minutes of staining, and 9 minutes of varnishing. The profit is \$5 on each A table and \$3 on each B table. If the sanding and varnishing facilities are each available at most 450 minutes per day and the staining facility is available at most 480 minutes per day, how many tables of each type should be made each day to maximize the plant's profit?

8. Repeat Exercise 7 if the profit on each A table is \$3 and on each B table is \$5.

9. A shipper has trucks that carry cardboard containers. Each container from the Jones Corp. weighs 5 pounds and is 5 cubic feet in volume. Each container from the Jackson Corp. weighs 6 pounds and is 3 cubic feet in volume. For each trip, a contract requires the shipper to charge the Jones Corp. 30 cents for each container and the Jackson Corp. 40 cents for each container. If the truck cannot carry more than 12,000 pounds and cannot hold more than 9000 cubic feet of cargo, how many containers from each customer should the shipper carry to maximize the profit?

10. Repeat Exercise 9 if the shipper must carry at least 240 containers of the Jones Corp. on the trip.

11. Holiday Airline Service wants to fly 1000 members of a travel club to Rome. The airline owns two types of planes. Type A can carry 100 passengers, and type B can carry 200 passengers. Type A will cost the airline \$10,000 for the trip, and type B will cost \$12,000 for the trip. If each airplane requires 8 flight attendants and there are only 48 flight attendants available, how many planes of each type should be used to minimize the airline's cost for the trip?

12. An investment banker has funds available for investing. She can purchase a type A bond yielding a 5% return on the amount invested, and she can purchase a type C bond yielding a 10% return on the amount invested. Her client insists that she invest at least twice as much in A as in C but no more than \$6000 in A and no more than \$1500 in C. How much should be invested in each kind of bond to maximize the client's return?

13. A farmer has a 120-acre farm on which he plants two crops, alfalfa and soybeans. For each acre of alfalfa planted, his expenses are \$12; for each acre of soybeans planted, they are \$24. Each acre of alfalfa requires 32 bushels of storage and yields a profit of \$25; each acre of soybeans requires 8 bushels of storage and yields a profit of \$35. If the total amount of storage available is 160 bushels and the farmer has only \$1200 on hand, how many acres of soybeans and alfalfa should the farmer plant to maximize the profit?

14. A clothes manufacturer has 10 square yards of cotton material, 10 square yards of wool material, and 6 square yards of silk material. A pair of slacks requires 1 square yard of cotton, 2 square yards of wool, and 1 square yard of silk. A skirt requires 2 square yards of cotton, 1 square yard of wool, and 1 square yard of silk. The net profit on a pair of slacks is \$3, and the net profit on a skirt is \$4. How many skirts and how many slacks should be made to maximize the profit?

15. A factory uses two kinds of petroleum products in its manufacturing process, regular (R) and low sulfur (L). Each gallon of R used emits 0.03 pound of sulfur dioxide and 0.01 pound of lead pollutants; each gallon of L used emits 0.01 pound of sulfur dioxide and 0.01 pound of lead pollutants. Each gallon of R costs \$0.50, and each gallon of L costs \$0.60. The factory needs to use at least 100 gallons of the petroleum products each day. If federal pollution regulations allow the factory to emit no more than 6 pounds of sulfur dioxide and no more than 4 pounds of lead pollutants daily, how many gallons of each type of product should the factory use to minimize its costs?

16. A handbag manufacturer makes two types of handbags, patent leather and lizard. The sewing machine limits the total daily production to at most 600 handbags. On the other hand, the mechanical clasp fastener limits daily production to at most 400 patent leather bags and 500 lizard bags. If the net profit on each patent leather bag is 80 cents and on each lizard bag is \$1.00, how many bags of each type should the manufacturer produce daily to maximize the profit?

17. A computer user is planning to buy up to 30 minutes of computer time from the Computer Corp. High-priority time (H) provides the customer with more memory than does low-priority time (L). The user wishes to buy at least 3 minutes of time H. Computer Corp. will not sell more than 10 minutes of time H and insists on selling at least 18 minutes of time L. Time H sells for \$1000 per minute, and time L sells for \$600 per minute. How much time H and how much time L should the customer buy to minimize the cost?

18. A pharmaceutical firm develops two drugs, Curine I and Curine II. Each gram of Curine I contains 1 milligram of beneficial factor Z, and each gram of Curine II contains 2 milligrams of beneficial factor Z. Both drugs contain factors X and Y, which produce undesirable side effects. Each gram of Curine I contains 3 milligrams of X and 1 milligram of Y, while each gram of Curine II contains 1 milligram of X and 1 milligram of Y. If a patient can tolerate no more than 6 milligrams of X and no more than 4 milligrams of Y daily, how many grams of Curine I and how many grams of Curine II should be administered daily to provide the patient with the maximum amount of factor Z?

19. (a) Repeat Exercise 18 if each gram of Curine I contains 3 milligrams of Z and each gram of Curine II contains 1 milligram of Z.

(b) Repeat Exercise 18 if each gram of Curine I contains 4 milligrams of Z and each gram of Curine II contains 1 milligram of Z.

Exercises 20 and 21 illustrate an important type of linear programming problem, the ***transportation problem****. In this kind of problem there are various warehouses and various stores, and the cost of shipping an item from any warehouse to any store is known. The problem is "How many items should be shipped from each warehouse to each store so that each store's demands are met and the shipping costs are minimized?"*

20. A seafood broker has two docks, one in New Jersey, the other in Maine. Seafood is shipped to three stores: one in Ohio, one in New York, and one in Virginia. The cost (in dollars) of sending one case of seafood from each dock to each store is given in Table 1.

TABLE 1

	Ohio	New York	Virginia
New Jersey	9	7	8
Maine	10	5	12

The demand at each of the stores is Ohio: 55 cases, New York: 70 cases, and Virginia: 25 cases. The number of cases available at each warehouse is New Jersey: 60 cases and Maine: 90 cases. How many cases of seafood should be shipped from each dock to each store to minimize the total shipping cost? Formulate this transportation problem as a linear programming problem, but do not attempt to solve. [*Hint*: There are six unknowns; call them x_1, x_2, $\ldots$, x_6.]

21. Three pipeline companies—Regional Natural Gas, Union Natural Gas, and Greater Gas Incorporated—are controlled by one conglomerate. They supply natural gas to Chicago, Philadelphia, Detroit, and Huntsville. Over a certain time period, the needs of each city (in millions of cubic feet) are Chicago: 180, Philadelphia: 175, Detroit: 170, and Huntsville: 200. The available gas (in millions of cubic feet) is Regional: 150, Union: 300, and Greater: 275. The profit (in thousands of dollars) from sending 1 million cubic feet of gas from each pipeline company to each city is given in Table 2. How much gas should be sent by each company to each city to maximize the total profit for the conglomerate? Formulate this transportation problem as a linear programming problem, but do not attempt to solve. [*Hint*: There are 12 unknowns, call them x_1, x_2, $\ldots$, x_{12}.]

TABLE 2

	Chicago	Philadelphia	Detroit	Huntsville
Regional Gas	10	9	11	8
Union Gas	7	12	16	10
Greater Gas	13	10	9	15

22. An investor will invest a total of \$300,000 in gold, silver, and diamonds. Experience has shown her that she should put at least twice the amount in gold as in silver. Her broker insists that she spend at least \$200,000 on diamonds. If gold, silver, and diamonds are currently selling at \$650 per ounce, \$16 per ounce, and \$400 per carat, respectively, and the projected prices at the time of sale are estimated at \$725 per ounce, \$18 per ounce, and \$500 per carat, respectively, how should the \$300,000 be spent to maximize the profit? [*Hint*: Use one of the constraints to reduce the problem to one with two unknowns.]

23. **(Harder)** A farmer grows corn, tomatoes, and eggplant. There are 250 acres available to plant, which must all be used. Each acre of corn will produce 30 bushels. One acre of tomatoes will yield 3 tons. One acre of eggplant will yield 32 crates. Because of crop rotation, a crop cannot be planted on the same ground where it grew the year before. Last year the planting consisted of 100 acres of corn, 100 acres of tomatoes, and 50 acres of eggplants. The farmer realizes a profit of \$3 per bushel of corn, \$31 per ton of tomatoes, and \$2.50 per crate of eggplants. How many acres should be allotted to each crop to maximize the profit? [*Hint*: Use one of the constraints to reduce the problem to one with two unknowns.]

3.2 SOLVING LINEAR PROGRAMMING PROBLEMS GEOMETRICALLY

In the previous section we show how to express several linear programming problems mathematically. In this section we develop a geometric method for solving these mathematical problems.

Example 1 in the previous section reduces to the following mathematical problem.

Find those values of x and y that make the objective function

$$z = 10x + 12y \tag{1}$$

as large as possible, where x and y must satisfy the constraints

$$\begin{aligned} 0.2x + 0.4y &\leq 30 \\ 0.2x + 0.2y &\leq 20 \\ x &\geq 0 \\ y &\geq 0 \end{aligned} \tag{2}$$

Using the techniques in Section 1.6, you should have no trouble showing that the points (x, y) satisfying the four inequalities in (2) form the shaded region in

Figure 1. Since each point (x, y) in this region has a chance of being a solution to the problem, we call these points **feasible solutions**, and the set S of all feasible solutions is called the **feasible region**.

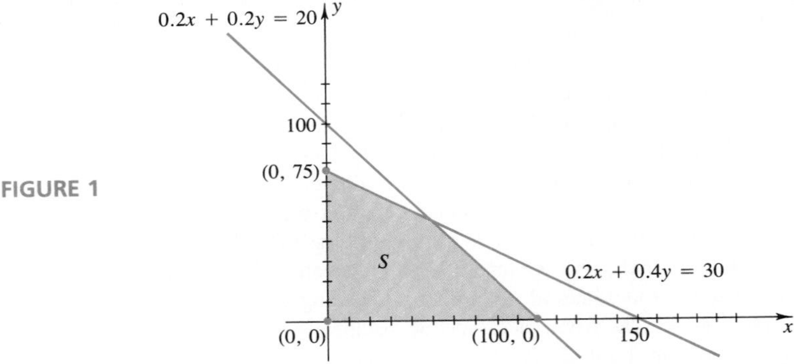

FIGURE 1

To solve the problem, we must look among the feasible solutions for one that makes the objective function (1) as large as possible. Such a solution is called an **optimal solution**. This is, however, easier said than done. To see why, consider the two feasible solutions $D(25, 50)$ and $E(60, 20)$ shown in Figure 2. At the point $D(25, 50)$ the value of the objective function is

$$z = 10x + 12y = 10(25) + 12(50) = 850$$

while at the point $E(60, 20)$ the value of the objective function is

$$z = 10x + 12y = 10(60) + 12(20) = 840$$

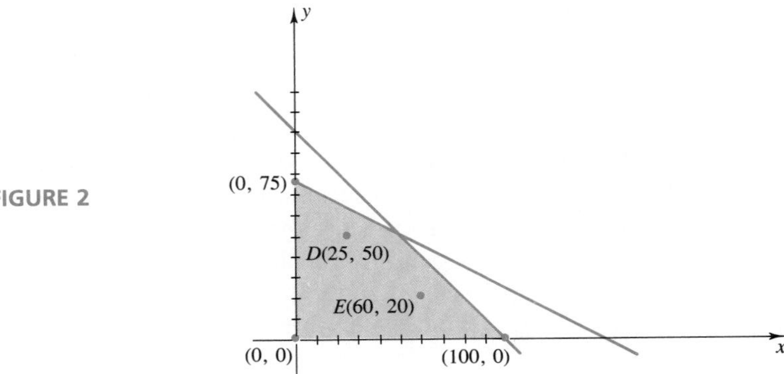

FIGURE 2

Since the value of the objective function is larger at the point $D(25, 50)$ than at the point $E(60, 20)$, E cannot be an optimal solution. However, we cannot be sure that D itself is an optimal solution since there may be other feasible solutions at which the objective function has an even larger value than at D. Since there are infinitely many feasible solutions, it is impossible to compute z at each one of them to see where it has the largest value. We therefore need some techniques to avoid this difficulty. Before we describe these techniques, we discuss some preliminary ideas.

The set of feasible solutions shaded in Figure 1 is an example of a **convex set**. This means that if we connect *any* two points in the set by a line segment, then the line segment will be completely in the set.

EXAMPLE 1 The shaded set in Figure 3a is *not* convex since the line segment connecting the points P_1 and P_2 does not lie entirely in the set. The shaded set in Figure 3b is convex since, as illustrated in the figure, a line segment connecting *any* two points of the set will lie in the set.

FIGURE 3

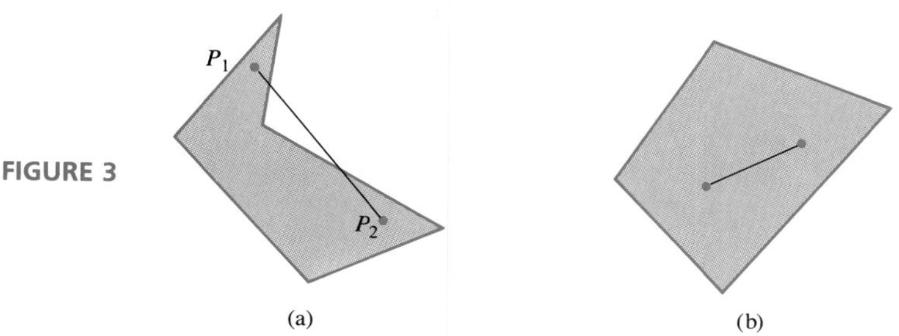

Convex sets are of two types: **bounded** and **unbounded**. A bounded set is one that can be enclosed by some suitably large circle, and an unbounded set is one that cannot be so enclosed. For example, the convex sets (a) and (b) in Figure 4 are bounded, while the convex sets (c) and (d) are unbounded.

FIGURE 4

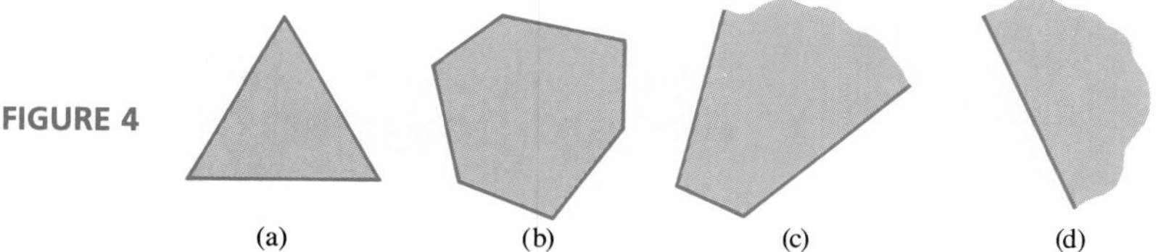

We also need the following definition as we solve linear programming problems.

> A **corner point** or **vertex** of a convex set is any point in the set that is the intersection of two boundary lines.

EXAMPLE 2 In Figure 5 we have indicated the corner points of the convex sets in parts (a), (b), and (c) of Figure 4. The convex set in part (d) of Figure 4 has no corner points.

FIGURE 5

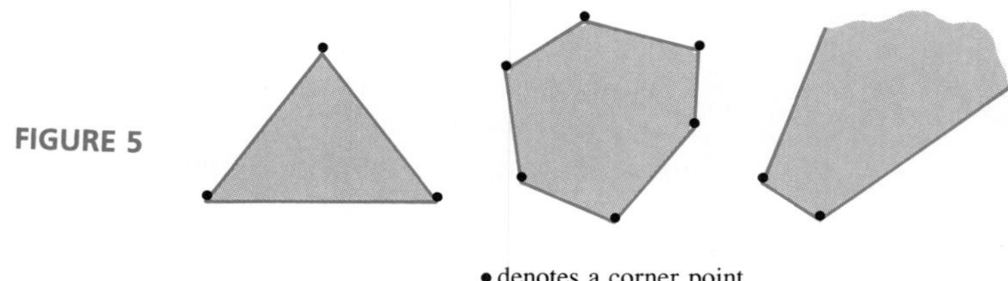

• denotes a corner point

EXAMPLE 3 The set of feasible solutions S shaded in Figure 1 has four corner points. Three of these lie on the coordinate axes: (0, 0), (0, 75), and (100, 0). The fourth corner point is the point of intersection of the lines satisfying the equations

$$0.2x + 0.2y = 20$$
$$0.2x + 0.4y = 30$$

To find this point of intersection we solve simultaneously by subtracting the first equation from the second (to eliminate x). This yields $0.2y = 10$, or $y = 50$. Substituting $y = 50$ into the first equation we obtain $x = 50$, or (50, 50) is the fourth corner point. Thus, the corner points of the set of feasible solutions are (0, 0), (0, 75), (50, 50), and (100, 0) (Figure 6).

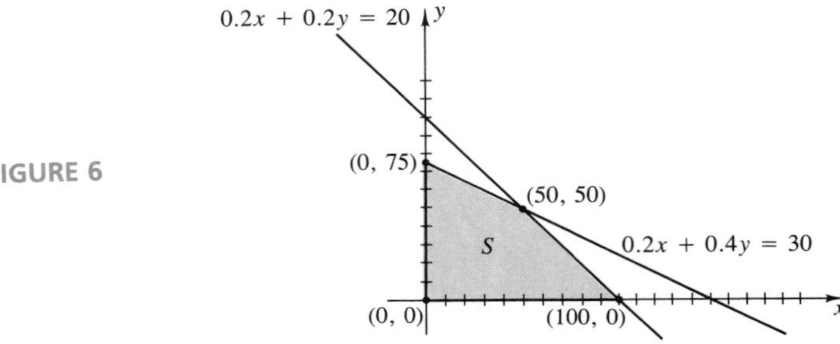

FIGURE 6

Our interest in convex sets and corner points stems from the following result, which we state without proof.

> The set S of feasible solutions to a linear programming problem is convex. Further, if S is bounded, then the objective function
>
> $$z = ax + by$$
>
> has both a largest value and a smallest value on S, and these values occur at corner points of S. If S is unbounded, there may not be a largest or smallest value on S. However, if a largest or smallest value on S exists, it must occur at a corner point.
>
> If the largest or smallest value of the objective function z occurs at two points of S, then it must occur at every point on the line segment joining the two points.

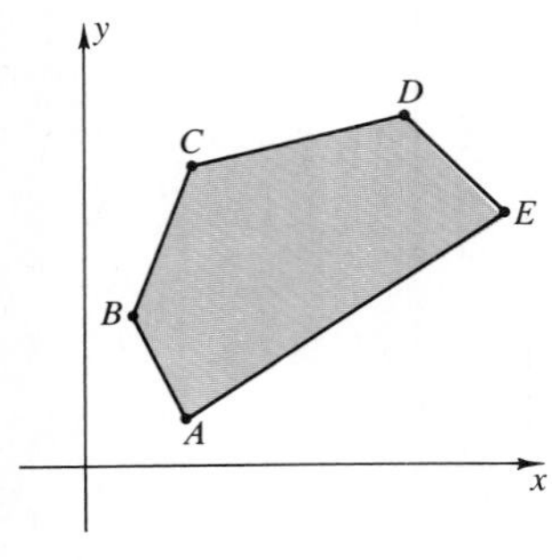

FIGURE 7

An intuitive understanding of this result can be gained by considering the example of minimizing $z = 3x + 2y$ over the bounded convex set shown in Figure 7. If we solve $z = 3x + 2y$ for y,

$$y = -\frac{3}{2}x + \frac{1}{2}z \qquad (3)$$

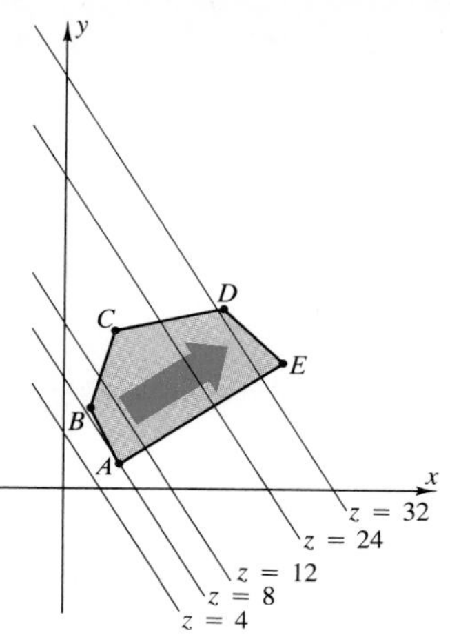

FIGURE 8

then for a particular value of z, for example, $z = 4$, Equation (3) is the equation of a straight line

$$y = -\frac{3}{2}x + 2$$

with slope equal to $-\frac{3}{2}$.

If the value of z is changed to 2, then Equation (3) becomes

$$y = -\frac{3}{2}x + 1$$

which is a straight line parallel to the first line. Similarly, as the value of z changes, we obtain a family of parallel lines (Figure 8). As z increases, the lines move in the direction of the arrow. As z decreases, the lines move in the opposite direction. (In other examples, the directions of increase and decrease may be reversed.) In all cases the optimal (maximum or minimum) value of z occurs when the line intersects the convex set only at a corner point (Figure 9a) or at its boundary (Figure 9b). The second case occurs in Example 7 of this section.

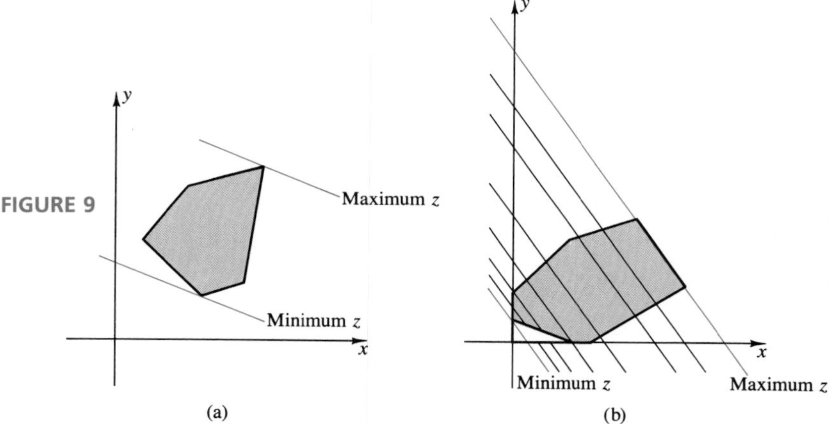

FIGURE 9

If the convex set is unbounded, then the objective function may have no maximum value (Figure 10).

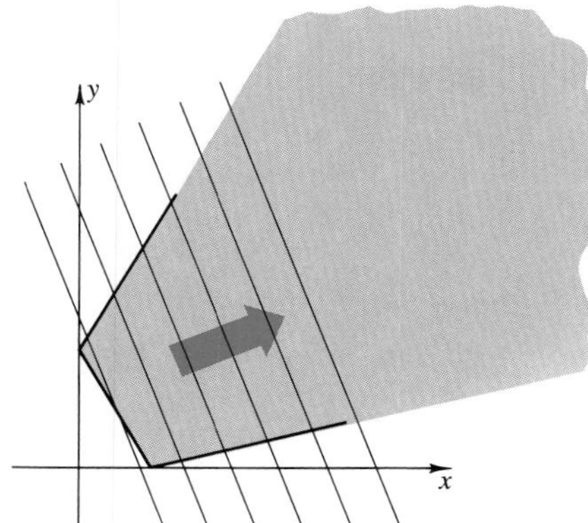

FIGURE 10

This result suggests the following procedure for solving linear programming problems when the set of feasible solutions is bounded.

Step 1 Determine the set of feasible solutions.

Step 2 Find the corner points of the set.

Step 3 Evaluate the objective function

$$z = ax + by$$

at each corner point to determine where the largest or smallest values of z occur.

Step 2 in the above procedure requires finding the point of intersection of two lines. As Example 3 illustrates, we can find this point by using the methods described in Section 1.4.

EXAMPLE 4 The problem in Example 1 of the previous section reduces to finding values of x and y that make

$$z = 10x + 12y$$

as large as possible, where x and y must satisfy

$$\begin{aligned} 0.2x + 0.4y &\leq 30 \\ 0.2x + 0.2y &\leq 20 \\ x &\geq 0 \\ y &\geq 0 \end{aligned}$$

TABLE 1

Corner Point (x, y)	Value of $z = 10x + 12y$
(0, 0)	0
(0, 75)	900
(50, 50)	1100
(100, 0)	1000

The set of feasible solutions is sketched in Figure 1 of this section, and the corner points (0, 0), (0, 75), (50, 50), and (100, 0) are determined in Example 3 (Figure 6). Thus, to find values of x and y that satisfy the constraints and make z as large as possible, we must first evaluate z at each corner point. These values are listed in Table 1. From this table, we see that the largest value of z is 1100 and this value occurs when $x = 50$ and $y = 50$. These values are shaded in the table.

We have now solved the problem in Example 1 of Section 3.1. The meatpacker should make $x = 50$ pounds of regular frankfurters and $y = 50$ pounds of deluxe frankfurters, in which case the profit $z = 1100¢ = \$11.00$ will be as large as possible.

EXAMPLE 5 Solve the following problem, which was posed in Example 3 of Section 3.1: Find those values of x and y that make

$$z = 0.08x + 0.10y \tag{4}$$

as large as possible, where x and y must satisfy

$$\begin{aligned} x + y &\leq 5000 \\ x &\leq 4000 \\ y &\geq 600 \\ y &\leq \tfrac{1}{3}x \\ x &\geq 0 \\ y &\geq 0 \end{aligned} \tag{5}$$

Solution Using the methods presented in Section 1.6, you should be able to show that the points (x, y) satisfying the six inequalities in (5) form the shaded region S in Figure 11.

FIGURE 11

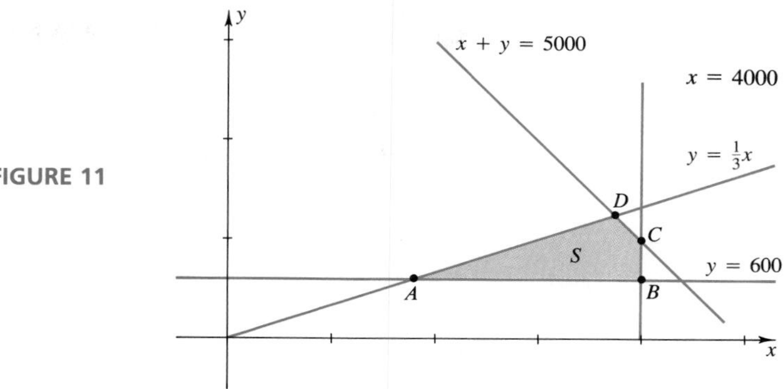

To find the coordinates of the corner points we proceed as follows: corner point A is the intersection of the lines with equations

$$y = \frac{1}{3}x \quad \text{and} \quad y = 600$$

Solving this system yields $x = 1800$ and $y = 600$, so A has coordinates (1800, 600). Corner point B is the intersection of the lines with equations

$$x = 4000 \quad \text{and} \quad y = 600$$

so B has coordinates (4000, 600). Corner point C is the intersection of the lines with equations

$$x = 4000 \quad \text{and} \quad x + y = 5000$$

Solving this system yields $x = 4000$ and $y = 1000$, so C has coordinates (4000, 1000). Corner point D is the intersection of the lines satisfying the equations

$$x + y = 5000 \quad \text{and} \quad y = \frac{1}{3}x$$

Solving this system yields $x = 3750$ and $y = 1250$, so D has coordinates (3750, 1250).

The region S is bounded so that the maximum value of the objective function (4) will occur at one of the corner points of S. The values of the objective function at the corner points are listed in Table 2. The maximum values are shaded.

TABLE 2

Corner Point (x, y)	Value of $z = 0.08x + 0.10y$
A(1800, 600)	204
B(4000, 600)	380
C(4000, 1000)	420
D(3750, 1250)	425

From Table 2, we see that the largest value of z on S is 425, and this value occurs when $x = 3750$ and $y = 1250$. Thus, the managers of the pension plan should invest \$3750 in stock X and \$1250 in stock Y, in which case their return of \$425 will be the largest possible.

Observe from Table 2 that the poorest possible investment would occur if the managers invested \$1800 in stock X and \$600 in stock Y, in which case their return of \$204 would be the smallest possible.

EXAMPLE 6 We previously stated that if the set S of feasible solutions is bounded, then the objective function in a linear programming problem has both a largest and a smallest value on S, and these values occur at corner points of S. In this example we show that if S is not bounded, then this result does not apply.

Consider the problem: Find those values of x and y that make

$$z = 3x + 4y$$

as large as possible, where x and y satisfy

$$\begin{aligned} x + 2y &\geq 8 \\ 2x + 2y &\geq 10 \\ x &\geq 0 \\ y &\geq 0 \end{aligned}$$

Using the techniques of Section 1.6, you should have no trouble obtaining the set S of feasible solutions shaded in Figure 12. Observe that S is an unbounded set. In Table 3 we have listed the values of z at the corner points of S. The largest value of z in this table is $z = 24$, which occurs at the point (8, 0). However, this is not the largest possible value of z on S. For example, $z = 3x + 4y$ has the value $z = 38$ at the point (10, 2). In fact, z does not have a largest value on S. (Can you see why?)

TABLE 3

Corner Point (x, y)	Value of $z = 3x + 4y$
(0, 5)	20
(2, 3)	18
(8, 0)	24

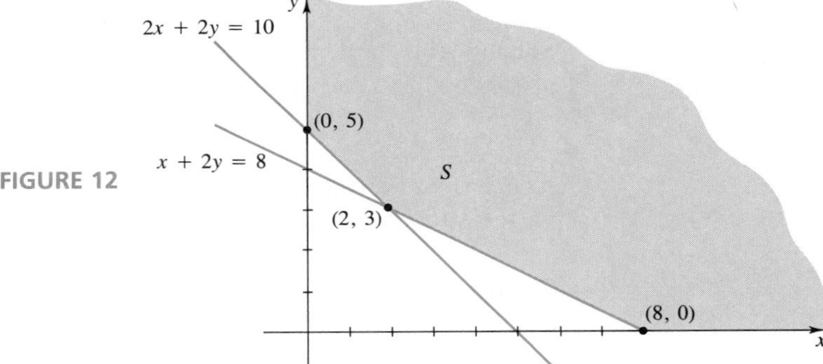

FIGURE 12

Observe that a minimum value of 18 for $z = 3x + 4y$ occurs at the corner point (2, 3). Thus, in an unbounded region the objective function may have a minimum value although it has no maximum value.

Example 6 illustrates what can happen when the set of feasible solutions is unbounded. In such cases, the objective function can fail to have a largest value, a smallest value, or both. It can be shown, however, that if there is a largest (or smallest) value for the objective function, it must occur at a corner point.

EXAMPLE 7 Solve the following problem, which was posed in Example 4 of Section 3.1: Find those values of x and y that make

$$z = 15x + 30y \tag{6}$$

as large as possible, where x and y must satisfy

$$\begin{aligned} 0.2x + 0.4y &\le 8 \\ 0.5x + 0.2y &\le 8 \\ x &\ge 0 \\ y &\ge 0 \end{aligned} \tag{7}$$

Solution Using the methods of Section 1.6, you can show that the points (x, y) satisfying the four inequalities in (7) form the shaded region S in Figure 13.

The region S is bounded, so the maximum value of the objective function (6) will occur at one of the corner points of S. The values of the objective function at the corner points are listed in Table 4. The maximum values are shaded.

FIGURE 13

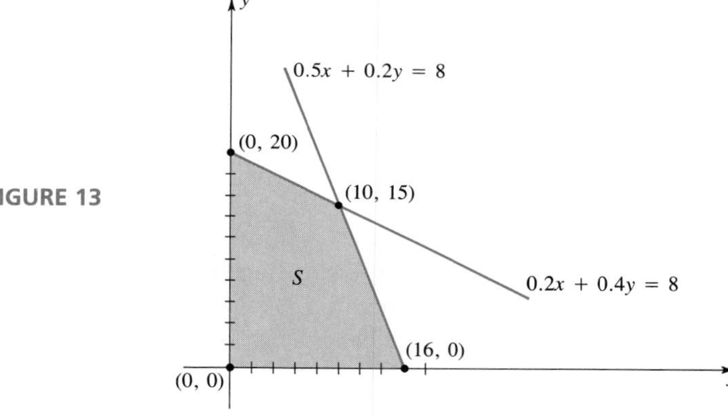

TABLE 4

Corner Point (x, y)	Value of $z = 15x + 30y$
(0, 0)	0
(16, 0)	240
(10, 15)	600
(0, 20)	600

This example shows that a linear programming problem can have more than one solution. From this table, we see that the largest value of z is 600, and this value occurs at two corner points

$$x = 10 \qquad y = 15$$

and

$$x = 0 \qquad y = 20$$

It then follows from the discussion preceding Example 4 that any point on the line segment joining the points (10, 15) and (0, 20) will also be an optimal solution. This can be seen as follows:

The equation of the line joining these two points is

$$0.2x + 0.4y = 8$$

Multiplying both sides by 75, we obtain

$$15x + 30y = 600$$

The left side of this equation is the objective function $z = 15x + 30y$. Thus, the objective function is 600 at every point on the line joining the points (10, 15) and (0, 20).

In summary, the fuel manufacturer can maximize the profit in many ways. For example,

$$x = 10 \text{ liquid units of type R} \quad \text{and} \quad y = 15 \text{ liquid units of type U}$$

or

$$x = 0 \text{ liquid units of type R} \quad \text{and} \quad y = 20 \text{ liquid units of type U}$$

or

$$x = 12 \text{ liquid units of type R} \quad \text{and} \quad y = 14 \text{ liquid units of type U}$$

and so forth. In all cases, the daily profit of $600 is the largest possible.

So far in this chapter, we have considered linear programming problems involving two unknowns. In actual practice, most linear programming problems involve more than two unknowns—some problems deal with *hundreds* or even *thousands** of unknowns. Problems in three unknowns, such as

$$\begin{aligned} &\text{Maximize} \quad z = 2x + 3y + 5t \\ &\text{subject to} \\ &\qquad 5x + 3y + 5t \leq 15 \\ &\qquad 10x + 4y + 5t \leq 20 \\ &\qquad x \geq 0 \\ &\qquad y \geq 0 \\ &\qquad t \geq 0 \end{aligned}$$

can also be solved geometrically. However, the set of feasible solutions must be drawn in three-dimensional space, a considerably more difficult problem than that for two variables, which is drawn in a plane. The set of feasible solutions for the above problem is shown in Figure 14. For most problems involving more than three unknowns, geometric methods are usually not applicable. In Chapter 4 we discuss algebraic methods used to solve problems with any number of unknowns.

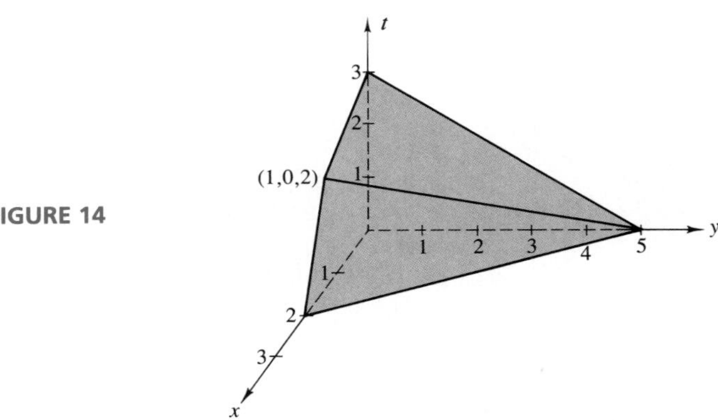

FIGURE 14

* For example, problems of this type occur in the oil industry and in econometrics.

EXERCISE SET 3.2

1. Which of the following points is a feasible solution to the given linear programming problem?

$$\text{Maximize } z = 4x + 5y$$
subject to
$$\begin{aligned} 2x + 3y &\le 4 \\ x - y &\le 3 \\ x &\ge 0 \\ y &\ge 0 \end{aligned}$$

(a) $(1, 1)$ (b) $(2, 0)$ (c) $(\frac{1}{2}, 1)$ (d) $(2, 1)$

2. Which of the following points is a feasible solution to the given linear programming problem?

$$\text{Minimize } z = 3x_1 - 2x_2$$
subject to
$$\begin{aligned} 3x_1 + x_2 &\ge 4 \\ 2x_1 - x_2 &\ge 2 \\ x_1 - 2x_2 &\ge 5 \\ x_1 &\ge 0 \\ x_2 &\ge 0 \end{aligned}$$

(a) $(1, 2)$ (b) $(8, 1)$ (c) $(3, 1)$ (d) $(5, 0)$

3. Consider a linear programming problem with the objective function $z = 4x - 2y$, which is to be maximized. Find an optimal solution given that the corner points of the feasible region are $(0, 4)$, $(5, 2)$, $(2, 6)$, and $(8, 4)$.

4. Consider a linear programming problem with the objective function $z = -3x + 2y$, which is to be minimized. Find an optimal solution given that the corner points of the feasible region are $(0, 3)$, $(3, 0)$, $(4, 4)$, $(5, 2)$, and $(6, 5)$.

5. For which values of x and y is $z = 18x + 12y$ a maximum subject to

$$\begin{aligned} x + 3y &\le 12 \\ x + 2y &\le 9 \\ x &\ge 0 \\ y &\ge 0 \end{aligned}$$

6. For which values of x and y is $z = 15x - 10y$ a minimum subject to

$$\begin{aligned} x + 3y &\le 8 \\ x + 2y &\le 7 \\ x &\ge 0 \\ y &\ge 0 \end{aligned}$$

7. For which values of x and y is $z = 2x + 3y$ a maximum subject to

$$\begin{aligned} 5x + 2y &\ge 9 \\ 6x - 2y &\le 2 \\ y &\le 6 \\ x &\ge 0 \\ y &\ge 0 \end{aligned}$$

8. For which values of x and y is $z = 2x + 3y$ a minimum subject to the constraints of Exercise 7?

9. (a) For which values of x and y is $z = 3x - y$ a maximum subject to the constraints of Exercise 7?
   (b) For which values of x and y is it a minimum?

For these exercises, solve the indicated problem from Exercises 3–23 of Exercise Set 3.1:

10. Exercise 4
11. Exercise 7
12. Exercise 8
13. Exercise 9
14. Exercise 10
15. Exercise 11
16. Exercise 12
17. Exercise 13
18. Exercise 14
19. Exercise 15
20. Exercise 16
21. Exercise 17
22. Exercise 18
23. Exercise 19

KEY IDEAS FOR REVIEW

- **Linear programming problem** Find nonnegative values of x and y that will maximize or minimize an expression of the form $z = ax + by$ subject to certain inequalities that must be satisfied by x and y.
- **Objective function** The quantity $z = ax + by$ to be maximized or minimized in a linear programming problem.

- **Constraints** Inequalities that must be satisfied by x and y in a linear programming problem.
- **Feasible solution** Values of x and y that satisfy all the constraints of a linear programming problem.
- **Optimal solution** A feasible solution that makes the objective function as large or as small as possible.
- **Convex set** A set with the property that a line segment joining any two points in the set lies entirely in the set.
- **Bounded convex set** A convex set that can be enclosed by a sufficiently large circle.
- **Unbounded convex set** A convex set that cannot be enclosed by any circle.
- **Corner point** A point in a convex set that is the intersection of two boundary lines.

SUPPLEMENTARY EXERCISES

1. Which of the following points is a feasible solution to the given linear programming problem?

$$\text{Minimize} \quad z = 3x - 2y$$
$$\text{subject to} \quad \begin{aligned} x + 2y &\le 6 \\ -x + y &\le 1 \\ 2x - y &\ge 2 \\ x &\ge 0 \\ y &\ge 0 \end{aligned}$$

(a) $(2, 1)$ (b) $(1, 1)$ (c) $(1, -\frac{1}{2})$
(d) $(3, 2)$

2. For which values of x and y is $z = 3x - 4y$ a maximum subject to

$$\begin{aligned} 3x + 4y &\le 15 \\ 5x + 4y &\le 20 \\ x &\ge 0 \\ y &\ge 0 \end{aligned}$$

3. For which values of x and y is $z = 5x + 2y$ a minimum subject to

$$\begin{aligned} 5x + 6y &\le 30 \\ x - y &\le 2 \\ x &\ge 3 \\ y &\le 3 \\ y &\ge 0 \end{aligned}$$

4. Consider a linear programming problem with the objective function $z = 5x + 4y$, which is to be maximized. Find an optimal solution given that the corner points of the feasible region are $(3, 0)$, $(5, 2)$, $(7, 5)$, $(4, 4)$, and $(1, 3)$.

5. A bakery makes yellow cake and white cake. Each pound of yellow cake requires $\frac{1}{4}$ pound of flour and $\frac{1}{4}$ pound of sugar; each pound of white cake requires $\frac{1}{3}$ pound of flour and $\frac{1}{3}$ pound of sugar. The baker finds that 100 pounds of flour and 80 pounds of sugar are available. If yellow cake sells for \$3 per pound and white cake sells for \$2.50 per pound, how many pounds of each cake should the bakery produce to maximize income, assuming that all cakes baked can be sold?

6. A shop sells a mixture of Java and Colombian coffee beans for \$4 per pound. The shopkeeper has allocated \$1000 for buying fresh beans and finds that it costs \$1.50 per pound for Java beans and \$2.00 per pound for Colombian beans. In a satisfactory mixture, the weight of Colombian beans will be at least twice and no more than four times the weight of the Java beans. How many pounds of each type of coffee bean should be ordered to maximize the profit if the entire mixture can be sold?

7. A pension fund plans to invest up to \$50,000 in U.S. Treasury bonds yielding 12% interest per year and corporate bonds yielding 15% interest per year. The fund manager is told to invest a minimum of \$25,000 in the Treasury bonds and a minimum of \$10,000 in the corporate bonds, with no more than one-fourth of the total investment to be in corporate bonds. How much should the manager invest in each type of bond to achieve the maximum annual interest? What is the maximum annual interest?

8. A farmer intends to plant crops A and B on all or part of a 100-acre field. Seed for crop A costs \$60 per acre, and labor costs \$200 per acre. For crop B, seed costs \$90 per acre, and labor costs \$150 per acre. The farmer cannot spend more than \$8100 for seed and no more than \$18,000 for labor. If the income per acre is \$1500 for crop A and \$1750 for crop B, how many acres of each crop should be planted to maximize total income?

9. The farmer in Exercise 8 finds that a worldwide surplus in crop B reduces the income to \$1400 per acre, while the income for crop A remains steady at \$1500 per acre. How many acres of each crop should be planted to maximize total income?

10. In preparing food for the college cafeteria, a dietitian will combine Volume Pack A and Volume Pack B. Each pound of Volume Pack A costs \$2.50 and contains 4 units of carbohydrate, 3 units of protein, and 5 units of fat. Each pound of Volume Pack B costs \$1.50 and contains 3 units of carbohydrate, 4 units of protein, and 1 unit of fat. If minimum monthly requirements are 60 units of carbohydrates, 52 units of protein, and 42 units of fat, how many pounds of each food pack will the dietitian use to minimize costs?

11. A lawn service uses a rider mower that cuts a 5000-square-foot area per hour and a small mower that cuts a 3000-square-foot area per hour. Each mower uses $\frac{1}{2}$ gallon of gasoline per hour. Near the end of a long summer day, the supervisor finds that both mowers are empty and that 0.6 gallon of gasoline remains in the storage cans. To satisfactorily conclude the day, at least 4000 square feet of lawn must still be mowed. If the cost of operating the riding mower is \$9 per hour and the cost of operating the smaller mower is \$5 per hour, how much of the remaining gasoline should be allocated to each mower to do the job at the least cost?

CHAPTER TEST

1. Which of the following points is a feasible solution to the given system of inequalities?

$$\begin{aligned} x - y &\le 4 \\ 2x + y &\ge 6 \end{aligned}$$

(a) (1, 1) (b) (3, 2) (c) (2, 1) (d) (4, 1)

2. Suppose a linear programming problem has the feasible region shown in Figure 1. For which values of x and y is $z = 3x - 4y$
(a) a maximum (b) a minimum

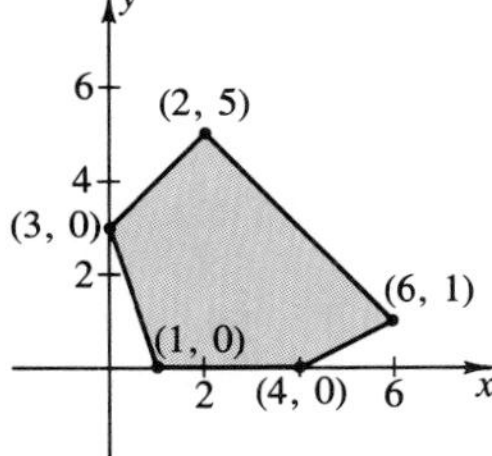

FIGURE 1

3. For which values of x and y is $z = 2x + 3y$ a minimum subject to

$$\begin{aligned} -2x + 3y &\le 6 \\ 3x + 4y &\ge 12 \\ x &\ge 0 \\ y &\ge 0 \end{aligned}$$

4. A company manufactures an 8-bit computer and a 16-bit computer. To meet existing orders, the company must schedule at least fifty 8-bit computers for the next production cycle and can produce no more than one hundred fifty 8-bit computers. The manufacturing facilities are adequate to produce no more than three hundred 16-bit computers, but the total number of these computers that can be produced cannot exceed 400. The profit is \$310 on each 8-bit computer and \$275 on each 16-bit computer. Find the number of computers of each type that should be manufactured to maximize profit.

5. A trucking firm is negotiating a contract with a spice supplier that uses two sizes of containers: large, 4-cubic-foot containers weighing 10 pounds, and small, 2-cubic-foot containers weighing 8 pounds. The trucking firm will use a vehicle that can handle a maximum load of 3280 pounds and a cargo size of up to 1000 cubic feet. The firms have agreed on a shipping rate of 50 cents for each large container and 30 cents for each small container. How many containers of each type should the trucking firm place on a truck to maximize income?

4

Linear Programming (An Algebraic Approach)

In Chapter 3 we show how linear programming problems involving two variables x and y can be solved geometrically. The geometric method we examine in that chapter can be extended to linear programming problems involving three variables. However, for problems involving more than three variables, geometric techniques must usually be replaced by algebraic ones. In this chapter we describe an important algebraic technique, called the simplex method, for solving linear programming problems.

4.1 INTRODUCTION; SLACK VARIABLES

The simplex method is applicable only when a linear programming problem satisfies certain conditions. In this section we discuss a simplified version of these conditions as well as certain other preliminaries; in the next two sections we discuss the simplex method itself.

Consider the following linear programming problems:

(a) Maximize $z = 10x + 12y$
subject to

$$\begin{aligned} 0.2x + 0.4y &\le 30 \\ 0.2x + 0.2y &\le 20 \\ x &\ge 0 \\ y &\ge 0 \end{aligned}$$

(b) Minimize $z = 10x - 12y + 2t$
subject to

$$\begin{aligned} 2x - 3y + t &\le 6 \\ x + y + 9t &\le 8 \\ x &\ge 0 \\ y &\ge 0 \\ t &\ge 0 \end{aligned}$$

(c) Maximize $z = 0.08x_1 + 0.1x_2$
subject to

$$\begin{aligned} x_1 + x_2 &\le 5000 \\ 3x_1 - x_2 &\le 17 \end{aligned}$$

(d) Maximize $z = 3x + 4y$
subject to

$$\begin{aligned} x + 2y &\le 8 \\ 2x + 2y &\ge 10 \\ x &\ge 0 \\ y &\ge 0 \end{aligned}$$

Problem (a) is an example of a **standard linear programming problem**.* This means that

(i) The objective function is to be maximized.

(ii) All variables are required to be nonnegative.

(iii) In the other constraints the expressions involving the variables are less than or equal to ($\leq$) a *nonnegative* constant.

Problem (b) is not a standard linear programming problem since the objective function is to be minimized and not maximized, as required by condition (i).

Problem (c) is not a standard linear programming problem since the variables x_1 and x_2 are not constrained to be nonnegative, as required by condition (ii).

Problem (d) is not a standard linear programming problem since the constraint

$$2x + 2y \geq 10 \tag{1}$$

violates condition (iii) (the inequality goes the wrong way).

Sometimes linear programming problems that violate condition (iii) can be rewritten in an equivalent form so that this condition is satisfied. For example, consider the problem

(e) Maximize $z = 3x_1 + 2x_2 - x_3$
subject to

$$\begin{aligned} x_1 - 2x_2 - x_3 &\geq -4 \\ x_1 &\geq 0 \\ x_2 &\geq 0 \\ x_3 &\geq 0 \end{aligned}$$

This is not a standard linear programming problem since condition (iii) is violated by the constraint

$$x_1 - 2x_2 - x_3 \geq -4 \tag{2}$$

However, if we multiply both sides of this inequality by -1, we obtain the equivalent inequality†

$$-x_1 + 2x_2 + x_3 \leq 4 \tag{3}$$

If constraint (2) is rewritten in form (3), then problem (e) becomes a standard linear programming problem. To check your understanding of the ideas presented thus far, you may wish to show that this trick cannot be applied to inequality (1) to convert problem (d) into a standard linear programming problem.

In Section 4.4 we discuss the solution of certain nonstandard linear programming problems. However, *until that time we work with standard linear programming problems only*.

We now come to the main objective of this section: to show that a standard linear programming problem can be reformulated as a problem concerned with

---

*The term *standard linear programming problem* is our own; it is not used universally.

† Recall that if both sides of an inequality are multiplied by a negative number, then the inequality is reversed. For example, if we multiply both sides of the inequality $-2 \leq 4$ by -1, the result is $2 \geq -4$ (see Section A.4 in Appendix A).

systems of equations rather than inequalities. To illustrate how this is done, consider the following problem, which we refer to as the "original problem."

Original Problem

$$\begin{aligned} \text{Maximize} \quad & z = 10x + 12y \\ \text{subject to} \quad & 0.2x + 0.4y \le 30 \qquad \text{(4a)} \\ & 0.2x + 0.2y \le 20 \qquad \text{(4b)} \\ & x \ge 0 \\ & y \ge 0 \end{aligned}$$

Since the right sides of inequalities (4a) and (4b) are at least as large as the left sides, it is possible to convert these inequalities into equalities by adding *nonnegative* quantities v and w on the left. When this is done, (4a) and (4b) become the equations

$$0.2x + 0.4y + v = 30 \qquad \text{(4a')}$$
$$0.2x + 0.2y + w = 20 \qquad \text{(4b')}$$

where

$$v \ge 0 \quad \text{and} \quad w \ge 0 \qquad (5)$$

The quantities v and w are called **slack variables** because they take up the "slack" or difference between the two sides.

If we replace inequalities (4a) and (4b) by equalities (4a′) and (4b′) and use the constraints in (5), we obtain a modification of the original problem called the "new problem with slack variables."

New Problem with Slack Variables

$$\begin{aligned} \text{Maximize} \quad & z = 10x + 12y \\ \text{subject to} \quad & 0.2x + 0.4y + v = 30 \\ & 0.2x + 0.2y + w = 20 \\ & x \ge 0 \qquad (6) \\ & y \ge 0 \\ & v \ge 0 \\ & w \ge 0 \end{aligned}$$

Athough we omit the details, it is not difficult to show that the original problem and the new problem with slack variables are equivalent in the following sense:

If $x = a$, $y = b$ is any solution of the original problem, then there are values of the slack variables, for example, $v = c$ and $w = d$, such that $x = a$, $y = b$, $v = c$, and $w = d$ is a solution of the new problem with slack variables.	If $x = 10$, $y = 20$ is a solution of (4), then $x = 10$, $y = 20$, $v = 20$, $w = 14$ is a solution of the above new problem with slack variables.
Conversely, if $x = a$, $y = b$, $v = c$, $w = d$ is any solution of the new problem with slack variables, then $x = a$, $y = b$ is a solution of the original problem.	If $x = 15$, $y = 5$, $v = 25$, $w = 16$ is a solution of the new problem with slack variables, then $x = 15$, $y = 5$ is a solution of (4).

EXAMPLE 1 In Example 7 of Section 3.2 we show that

$$x = 0 \qquad y = 20$$

is a solution of the problem

$$\begin{aligned} \text{Maximize} \quad & z = 15x + 30y \\ \text{subject to} \quad & 0.2x + 0.4y \le 8 \\ & 0.5x + 0.2y \le 8 \\ & x \ge 0 \\ & y \ge 0 \end{aligned}$$

Use this result to find a solution of the new problem with slack variables.

Solution The new problem with slack variables is

$$\begin{aligned} \text{Maximize} \quad & z = 15x + 30y \\ \text{subject to} \quad & \end{aligned}$$

$$0.2x + 0.4y + v = 8 \tag{7}$$

$$0.5x + 0.2y + w = 8 \tag{8}$$

$$x \ge 0 \quad y \ge 0 \quad v \ge 0 \quad w \ge 0$$

Substituting the values $x = 0$, $y = 20$ in (7) and (8), we obtain

$$\begin{aligned} 0.2(0) + 0.4(20) + v &= 8 \\ 0.5(0) + 0.2(20) + w &= 8 \end{aligned}$$

so $v = 0$ and $w = 4$. Since these values satisfy $v \ge 0$, $w \ge 0$, we have

$$x = 0 \qquad y = 20 \qquad v = 0 \qquad w = 4$$

as a solution of the new problem.

In Example 1 we started with a solution of the original problem and then obtained a solution of the new problem with slack variables. We are now interested in reversing the procedure. The simplex method described in later sections enables us to solve the new problem with slack variables. Using this solution, we then obtain a solution of the original problem.

We conclude this section by showing how linear programming problems can be formulated using matrix terminology. Consider the standard linear programming problem

$$\text{Maximize} \quad z = 10x + 12y \tag{9}$$

subject to

$$0.2x + 0.4y \le 30 \tag{10a}$$

$$0.2x + 0.2y \le 20 \tag{10b}$$

$$x \ge 0 \tag{11a}$$

$$y \ge 0 \tag{11b}$$

and let

$$A = \begin{bmatrix} 0.2 & 0.4 \\ 0.2 & 0.2 \end{bmatrix} \qquad B = \begin{bmatrix} 30 \\ 20 \end{bmatrix} \qquad X = \begin{bmatrix} x \\ y \end{bmatrix} \qquad C = [10 \quad 12]$$

If we agree to omit the brackets on 1×1 matrices, then the quantity z in statement (9) can be written as the matrix product

$$z = [10 \quad 12]\begin{bmatrix} x \\ y \end{bmatrix}$$

or equivalently

$$z = CX$$

Further, constraints (10a) and (10b) can be written as*

$$\begin{bmatrix} 0.2x + 0.4y \\ 0.2x + 0.2y \end{bmatrix} \le \begin{bmatrix} 30 \\ 20 \end{bmatrix}$$

or

$$\begin{bmatrix} 0.2 & 0.4 \\ 0.2 & 0.2 \end{bmatrix}\begin{bmatrix} x \\ y \end{bmatrix} \le \begin{bmatrix} 30 \\ 20 \end{bmatrix}$$

or

$$AX \le B$$

Finally, constraints (11a) and (11b) can be written as

$$\begin{bmatrix} x \\ y \end{bmatrix} \ge \begin{bmatrix} 0 \\ 0 \end{bmatrix}$$

or equivalently

$$X \ge 0$$

Combining these results, we obtain this matrix formulation of the problem:

$$\begin{aligned} &\text{Maximize} \quad z = CX \\ &\text{subject to} \\ &\qquad AX \le B \\ &\qquad X \ge 0 \end{aligned}$$

EXERCISE SET 4.1

1. Which of the following are standard linear programming problems?

(a) Maximize $z = 2x - y$
subject to
$$x + y \le 1$$
$$x \ge 0, \quad y \ge 0$$

(b) Maximize $z = 3x + y + t$
subject to
$$2x - y + t \le 3$$
$$x + t \le 5$$

(c) Minimize $z = 2x_1 - 3x_2 + 3x_3$
subject to
$$2x_1 + 3x_2 + 3x_3 \le 20$$
$$3x_1 + 2x_2 - x_3 \le 30$$
$$x_1 \ge 0, \quad x_2 \ge 0, \quad x_3 \ge 0$$

(d) Maximize $z = 3x + 2y$
subject to
$$x + 2y \le 7$$
$$5x - 8y \le -3$$
$$x \ge 0, \quad y \ge 0$$

* If E and F are matrices of the same size, we write $E \le F$ if each entry in E is less than or equal to the corresponding entry in F.

2. Rewrite the following as standard linear programming problems.

(a) Maximize $z = 4x - 2y$
subject to
$$\begin{aligned} x + 2y &\le 7 \\ 5x - 8y &\ge -3 \\ x \ge 0, \quad y &\ge 0 \end{aligned}$$

(b) Maximize $z = 2x_1 + 3x_2 + x_3$
subject to
$$\begin{aligned} 2x_1 + x_2 - x_3 &\ge -5 \\ 3x_1 + x_2 + 5x_3 &\ge -8 \\ x_1 \ge 0, \quad x_2 \ge 0, \quad x_3 &\ge 0 \end{aligned}$$

3. Can the following problem be rewritten as a standard linear programming problem? Explain your answer.

Maximize $z = 5x - 8y$
subject to
$$\begin{aligned} x - 3y &\ge -5 \\ 2x + y &\ge 4 \\ x \ge 0, \quad y &\ge 0 \end{aligned}$$

4. Rewrite the following as equivalent new problems with slack variables:

(a) Maximize $z = 6x - 9y$
subject to
$$\begin{aligned} 6x - 7y &\le 4 \\ 2x + 5y &\le 5 \\ x \ge 0, \quad y &\ge 0 \end{aligned}$$

(b) Maximize $z = 4x_1 + 3x_2 + 2x_3$
subject to
$$\begin{aligned} 2x_1 + x_2 - 3x_3 &\le 8 \\ x_1 + 2x_2 + 4x_3 &\le 14 \\ 2x_1 + 3x_2 - x_3 &\le 12 \\ x_1 \ge 0, \quad x_2 \ge 0, \quad x_3 &\ge 0 \end{aligned}$$

5. Rewrite the following as equivalent new problems with slack variables:

(a) Maximize $z = 5x + 7y$
subject to
$$\begin{aligned} 4x - 6y &\le 9 \\ 2x + 7y &\le 3 \\ 5x - 8y &\le 2 \\ x \ge 0, \quad y &\ge 0 \end{aligned}$$

(b) Maximize $z = -x_1 + x_2 - x_3 + x_4$
subject to
$$\begin{aligned} 2x_1 + 3x_3 + x_4 &\le 8 \\ x_1 - 2x_2 + x_4 &\le 6 \\ x_1 \ge 0, \quad x_2 &\ge 0 \\ x_3 \ge 0, \quad x_4 &\ge 0 \end{aligned}$$

6. The following are linear programming problems with slack variables u, v, and w. Find the original problem.

(a) Maximize $z = x - 3y$
subject to
$$\begin{aligned} 2x + 4y + u &= 5 \\ 3x - 2y + v &= 7 \\ x \ge 0, \quad y &\ge 0 \\ u \ge 0, \quad v &\ge 0 \end{aligned}$$

(b) Maximize $z = 3x_1 + 2x_2 - x_3$
subject to
$$\begin{aligned} 2x_1 + 3x_2 - x_3 + u &= 12 \\ 2x_1 + x_2 + v &= 18 \\ 3x_1 + x_2 + 4x_3 + w &= 13 \\ x_1 \ge 0, \quad x_2 \ge 0, \quad x_3 &\ge 0 \\ u \ge 0, \quad v \ge 0, \quad w &\ge 0 \end{aligned}$$

7. (a) Write the problem in Example 4 of Section 3.2 as a new problem with slack variables.

(b) Use the solution obtained in Example 4 of Section 3.2 to obtain a solution of the new problem with slack variables.

8. The following problems can be written in matrix notation as

Maximize $z = CX$
subject to
$$\begin{aligned} AX &\le B \\ X &\ge 0 \end{aligned}$$

Find the matrices A, B, C, and X.

(a) Maximize $z = 6x - 9y$
subject to
$$\begin{aligned} 6x - 7y &\le 4 \\ 2x + 5y &\le 5 \\ x \ge 0, \quad y &\ge 0 \end{aligned}$$

(b) Maximize $z = x_1 + x_2 - x_3 + x_4$
subject to
$$\begin{aligned} 2x_1 + 3x_3 + x_4 &\le 8 \\ x_1 - 2x_2 + x_4 &\le 6 \\ x_1 \ge 0, \quad x_2 &\ge 0 \\ x_3 \ge 0, \quad x_4 &\ge 0 \end{aligned}$$

9. Repeat the directions of Exercise 8 for the problems

(a) Maximize $z = 4x - 2y + 7t$
subject to
$$\begin{aligned} x - t &\le 1 \\ y - t &\le 2 \\ x - y &\le 3 \\ x \ge 0, \quad y \ge 0, \quad t &\ge 0 \end{aligned}$$

(b) Maximize $z = x_1 - x_2$
subject to
$$\begin{aligned} x_1 + 3x_2 &\le 5 \\ x_1 \ge 0, \quad x_2 &\ge 0 \end{aligned}$$

10. Let

$$A = \begin{bmatrix} 3 & 4 & -3 \\ 2 & 5 & -3 \end{bmatrix} \qquad B = \begin{bmatrix} 2 \\ 5 \end{bmatrix}$$

$$X = \begin{bmatrix} x_1 \\ x_2 \\ x_3 \end{bmatrix} \qquad C = \begin{bmatrix} 8 & 4 & 2 \end{bmatrix}$$

Rewrite the problem

$$\begin{aligned} &\text{Maximize} \quad z = CX \\ &\text{subject to} \\ &\qquad AX \le B \\ &\qquad X \ge 0 \end{aligned}$$

without matrix notation.

11. Write the following linear programming problem without matrix notation:

$$\text{Maximize} \quad z = \begin{bmatrix} 6 & 7 & 9 \end{bmatrix} \begin{bmatrix} x \\ y \\ t \end{bmatrix}$$

subject to

$$\begin{bmatrix} 2 & -4 & 5 \\ 7 & 1 & 3 \end{bmatrix} \begin{bmatrix} x \\ y \\ t \end{bmatrix} \le \begin{bmatrix} 3 \\ 9 \end{bmatrix}$$

$$\begin{bmatrix} x \\ y \\ t \end{bmatrix} \ge \begin{bmatrix} 0 \\ 0 \\ 0 \end{bmatrix}$$

12. For the problems in Exercise 8, the new problems with slack variables can be written in matrix notation as

$$\begin{aligned} &\text{Maximize} \quad z = C_s X_s \\ &\text{subject to} \\ &\qquad A_s X_s = B \\ &\qquad X_s \ge 0 \end{aligned}$$

Find A_s, B, C_s, and X_s for
(a) the problem in Exercise 8a
(b) the problem in Exercise 8b

4.2 THE KEY IDEAS OF THE SIMPLEX METHOD

We are now ready to discuss the basic ideas needed to solve linear programming problems by the simplex method. Since the technique is somewhat involved, we use this section to motivate and develop the basic concepts. In the next section we formulate a step-by-step procedure for using the ideas developed here.

Section 4.1 shows how to reformulate the standard linear programming problem

$$\text{Maximize} \quad z = CX \quad (1) \qquad \text{Maximize} \quad z = \begin{bmatrix} c_1 & c_2 & \cdots & c_m \end{bmatrix} \begin{bmatrix} x_1 \\ x_2 \\ \vdots \\ x_m \end{bmatrix} \quad (1)$$

subject to

$$AX \le B \quad (2) \qquad \begin{bmatrix} a_{11} & a_{12} & \cdots & a_{1m} \\ a_{21} & a_{22} & \cdots & a_{2m} \\ \vdots & \vdots & & \vdots \\ a_{k1} & a_{k2} & \cdots & a_{km} \end{bmatrix} \begin{bmatrix} x_1 \\ x_2 \\ \vdots \\ x_m \end{bmatrix} \le \begin{bmatrix} b_1 \\ b_2 \\ \vdots \\ b_k \end{bmatrix} \quad (2)$$

$$X \ge 0 \quad (3) \qquad \begin{bmatrix} x_1 \\ x_2 \\ \vdots \\ x_m \end{bmatrix} \ge \begin{bmatrix} 0 \\ 0 \\ \vdots \\ 0 \end{bmatrix} \quad (3)$$

as a new equivalent problem with slack variables. This is accomplished by adding slack variables to convert each of the inequalities in (2) into equalities. In our discussion of slack variables, we emphasize problems with two variables in which (2) has two constraints. However, we now want to consider the general case, where we have m variables and $AX \leq B$ has k constraints. Thus, to convert each of the k inequalities in $AX \leq B$ into equalities, we would have to introduce k slack variables. Therefore, where the original problem contains m variables, the new problem contains these m variables together with k slack variables, making a total of $m + k$ variables. A **basic feasible solution** of this new problem with slack variables is a solution of this new system of k *equations* in $m + k$ variables in which m variables have value zero and all other variables are nonnegative. At first, it might seem that we have complicated affairs by introducing these extra variables. Fortunately, this is not the case. To explain why, we need the following result, which we state without proof.

Basic Feasible Solutions If a standard linear programming problem

$$\begin{aligned} \text{Maximize} \quad & z = CX \\ \text{subject to} \quad & AX \leq B \\ & X \geq 0 \end{aligned}$$

with m unknowns has a solution, then the new problem with slack variables has a basic feasible solution which yields the same maximum value of the objective function z, and conversely.

This result shows that the solutions of a standard linear programming problem can be found by considering the basic feasible solutions of the new problem. The following example illustrates how this result can be used to solve linear programming problems.

EXAMPLE 1 Consider the problem

$$\begin{aligned} \text{Maximize} \quad & z = -x + y \\ \text{subject to} \quad & 0.2x + 0.4y \leq 30 \\ & 0.2x + 0.2y \leq 20 \\ & x \geq 0, \quad y \geq 0 \end{aligned}$$

The corresponding new problem with slack variables is

$$\begin{aligned} \text{Maximize} \quad & z = -x + y \\ \text{subject to} \quad & 0.2x + 0.4y + v = 30 && (4) \\ & 0.2x + 0.2y + w = 20 && (5) \\ & x \geq 0, \quad y \geq 0, \quad v \geq 0, \quad w \geq 0 && (6) \end{aligned}$$

According to the result we stated above, this new problem with slack variables has a solution in which at least two of the variables have value zero (since the original problem has $m = 2$ unknowns, x and y). Thus, there is a solution of the new problem in which one of the following six possibilities occurs:

$$\begin{array}{lll} x = 0 \text{ and } y = 0 & \qquad & y = 0 \text{ and } v = 0 \\ x = 0 \text{ and } v = 0 & & y = 0 \text{ and } w = 0 \\ x = 0 \text{ and } w = 0 & & v = 0 \text{ and } w = 0 \end{array}$$

In each of these cases we can use Equations (4) and (5) to find the values of the remaining two variables. For example, if

$$x = 0 \quad \text{and} \quad v = 0$$

then we can substitute these values into (4) and (5) to obtain the system of equations

$$\begin{aligned} 0.4y &= 30 \\ 0.2y + w &= 20 \end{aligned}$$

Solving for y and w we obtain

$$y = 75 \qquad w = 5$$

Thus,

$$x = 0 \qquad v = 0 \qquad y = 75 \qquad w = 5$$

is a possible solution of the new problem with slack variables.

If we follow this procedure in each of the six cases, we obtain the following list of possible solutions of the new problem with slack variables:

$$x = 0 \qquad y = 0 \qquad v = 30 \qquad w = 20 \tag{7a}$$
$$x = 0 \qquad v = 0 \qquad y = 75 \qquad w = 5 \tag{7b}$$
$$x = 0 \qquad w = 0 \qquad y = 100 \qquad v = -10 \tag{7c}$$
$$y = 0 \qquad v = 0 \qquad x = 150 \qquad w = -10 \tag{7d}$$
$$y = 0 \qquad w = 0 \qquad x = 100 \qquad v = 10 \tag{7e}$$
$$v = 0 \qquad w = 0 \qquad x = 50 \qquad y = 50 \tag{7f}$$

We know there is a solution of the new problem somewhere among these six possibilities. However, we can eliminate (7c) and (7d) immediately since they are not basic feasible solutions (v and w violate the nonnegativity constraints). This leaves (7a), (7b), (7e), and (7f) as the only candidates. To finish, we need to evaluate the objective function

$$z = -x + y$$

at each basic feasible solution to find where the maximum value occurs. Table 1 summarizes the results of these computations.

TABLE 1

Basic Feasible Solutions of the New Problem with Slack Variables				Value of $z = -x + y$	
$x = 0$	$y = 0$	$v = 30$	$w = 20$	0	
$x = 0$	$v = 0$	$y = 75$	$w = 5$	75	← maximum z
$v = 0$	$w = 0$	$x = 50$	$y = 50$	0	
$y = 0$	$w = 0$	$x = 100$	$v = 10$	−100	

From this table we see that the maximum value of z is 75, so

$$x = 0 \qquad v = 0 \qquad y = 75 \qquad w = 5 \tag{8}$$

is a solution of the new problem with slack variables. As the previous section describes, a solution of the original problem can be obtained by discarding the slack variables from a solution of the new problem. Thus, discarding $v = 0$ and $w = 5$ from (8) we obtain the solution

$$x = 0 \qquad y = 75$$

of the original problem. You may wish to check this result by solving the problem geometrically, as discussed in Chapter 3. Observe, however, that the solution in this example is completely *algebraic*.

Relationship between Algebraic and Geometric Methods

There is a noteworthy relationship between the geometric solution and the algebraic solution. In the geometric solution we look for the maximum value of z at the corner points of the convex set determined by the constraints, while in the algebraic solution we look for the maximum value of z at the basic feasible solutions of the new problem with slack variables. This suggests that there must be a relationship between the basic feasible solutions of the new problem and the corner points in the original problem. This conjecture is correct. As an illustration of the relationship, Figure 1 shows the convex set determined by the constraints in the original problem of Example 1, and Table 2 lists the corner points of this convex set and the basic feasible solutions of the new problem. (These basic feasible solutions are also shown in Table 1.)

FIGURE 1

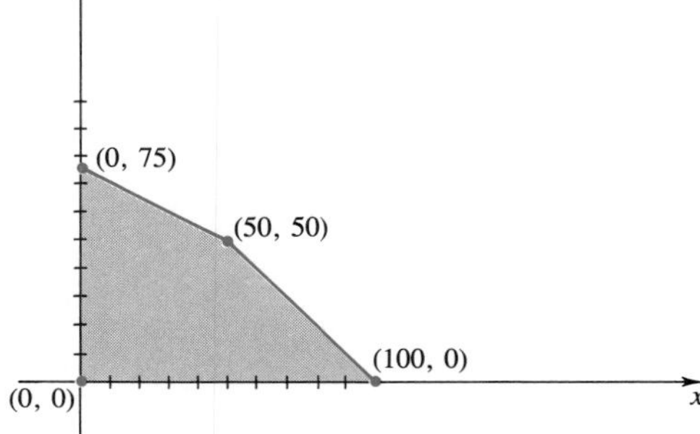

TABLE 2

Corner Points	Basic Feasible Solutions			
(0, 0)	$x = 0$	$y = 0$	$v = 30$	$w = 20$
(0, 75)	$x = 0$	$y = 75$	$v = 0$	$w = 5$
(50, 50)	$x = 50$	$y = 50$	$v = 0$	$w = 0$
(100, 0)	$x = 100$	$y = 0$	$v = 10$	$w = 0$

The relationship between corner points and basic feasible solutions is now evident from Table 2: the coordinates of each corner point in the left column are simply the x- and y-values of the corresponding basic feasible solution in the right column. For example, the coordinates of the corner point (0, 75) are simply the x- and y-values in the corresponding basic feasible solution

$$x = 0 \qquad y = 75 \qquad v = 0 \qquad w = 5$$

Since the objective function

$$z = -x + y$$

does not involve the slack variables v and w, it is clear that evaluating z at a basic feasible solution is equivalent to evaluating z at the corresponding corner point.

Although the algebraic method used in Example 1 is fine in theory, in practical problems the amount of computation involved can become prohibitively large. For example, in a problem with 20 variables and 10 constraints, there could conceivably be more than 30 million basic feasible solutions to check if we used this method! To remedy this difficulty we must develop ways of refining the technique in Example 1 to reduce the amount of computation. We proceed as follows. For the linear programming problem given by (1), (2), and (3), we have m variables and $AX \leq B$ has k constraints. Introducing the slack variables, we obtain the new problem which has $m + k$ variables and k equations. It is convenient to introduce the equation defining the objective function as an additional equation with the added unknown z so that we now have $m + k + 1$ variables with $k + 1$ equations. The simplex method is an organized way of searching for the m variables that are zero in the solution giving the maximum value of z. Before discussing the simplex method in Section 4.3, we first develop some preliminary ideas.

In Example 1 the new problem with slack variables involves nonnegativity constraints and the equations

$$\begin{aligned} z &= -x + y \\ 0.2x + 0.4y + v &= 30 \\ 0.2x + 0.2y + w &= 20 \end{aligned}$$

If we move the x and y terms in the first equation to the left side of the equals sign, we can rewrite this system with the variables aligned:

$$\begin{aligned} x - y \qquad\qquad + z &= 0 \\ 0.2x + 0.4y + v \qquad\qquad &= 30 \\ 0.2x + 0.2y \qquad + w \qquad &= 20 \end{aligned} \tag{9a}$$

Observe that the variables z, v, and w each occur in exactly one equation and where they occur they have a coefficient of $+1$. We call these the **explicit variables** or sometimes **basic variables**. More precisely, basic variables are those variables that have a coefficient $+1$ in only one equation and coefficient 0 in the remaining equations. We call the remaining variables **implicit** or **nonbasic** variables.

Equations (9a) are particularly convenient for finding the basic feasible solution in which the implicit variables x and y are zero. If we set $x = 0$, $y = 0$ in (9a), we obtain

$$\begin{aligned} z &= 0 \\ v &= 30 \\ w &= 20 \end{aligned}$$

Thus the basic feasible solution in which x and y are zero is

$$x = 0 \qquad y = 0 \qquad v = 30 \qquad w = 20$$

and the value of the objective function at this solution is $z = 0$.

The augmented matrix for (9a) is

$$\begin{bmatrix} 1 & -1 & 0 & 0 & 1 & 0 \\ 0.2 & 0.4 & 1 & 0 & 0 & 30 \\ 0.2 & 0.2 & 0 & 1 & 0 & 20 \end{bmatrix} \tag{9b}$$

It is called the **tableau with x and y implicit** or sometimes the **tableau corresponding to the basic feasible solution where $x = 0$ and $y = 0$**. For mnemonic reasons, it is helpful to indicate along the upper margin of the tableau which columns go with which variables, and to indicate along the left side of the tableau which variable is explicit in the corresponding equation. We also omit the dashed line. Thus we would write (9b) as

$$\begin{array}{c} \\ z \\ v \\ w \end{array} \begin{array}{c} \begin{array}{cccccc} x & y & v & w & z & \end{array} \\ \begin{bmatrix} 1 & -1 & 0 & 0 & 1 & 0 \\ 0.2 & 0.4 & 1 & 0 & 0 & 30 \\ 0.2 & 0.2 & 0 & 1 & 0 & 20 \end{bmatrix} \end{array} \tag{9c}$$

Observe that the entries in the last column tell us the values of z, v, and w when the implicit variables are set equal to zero.

If we were interested in finding the basic feasible solution where y and w have value zero, it would be more convenient to have Equations (9a) written so that y and w are implicit (or equivalently so that z, x, and v are explicit). Let us illustrate how to do this. Since z and v are already explicit in (9a), we need only replace the old explicit variable w in the last equation by the new explicit variable x. To do this we must get rid of the x terms in the first two equations and give x a coefficient of $+1$ in the last equation. To accomplish this we multiply the last equation

$$0.2x + 0.2y + w = 20$$

by $\frac{1}{0.2}$, obtaining

$$x + y + 5w = 100 \tag{10}$$

Replace the first equation of (9a) by itself minus Equation (10) and replace the second equation of (9a) by itself minus 0.2 times Equation (10). Upon simplifying, we obtain the following replacements for the first and second equations of (9a):

$$-2y \qquad -5w + z = -100 \tag{11}$$

$$0.2y + v - \quad w \qquad = 10 \tag{12}$$

Writing (11), (12), and (10) with the variables aligned, we obtain the equations

$$\begin{aligned} -2y \qquad - 5w + z &= -100 \\ 0.2y + v - \quad w \qquad &= 10 \\ x + \quad y \qquad + 5w \qquad &= 100 \end{aligned} \tag{13}$$

Observe that x, v, and z are now explicit since they each occur in exactly one equation and where they occur they have a coefficient of $+1$. From these equations it is easy to obtain the basic feasible solution in which the implicit variables y and w have value zero; we simply set $y = 0$, $w = 0$ in (13) to obtain

$$\begin{aligned} z &= -100 \\ v &= 10 \\ x &= 100 \end{aligned}$$

Thus the basic feasible solution in which y and w have value zero is

$$x = 100 \qquad y = 0 \qquad v = 10 \qquad w = 0$$

and the value of the objective function at this solution is

$$z = -100$$

The augmented matrix for (13) yields the tableau with x, v, and z explicit, namely,

$$\begin{array}{c} \\ z \\ v \\ x \end{array} \begin{array}{c} \begin{array}{cccccc} x & y & v & w & z & \end{array} \\ \begin{bmatrix} 0 & -2 & 0 & -5 & 1 & -100 \\ 0 & 0.2 & 1 & -1 & 0 & 10 \\ 1 & 1 & 0 & 5 & 0 & 100 \end{bmatrix} \end{array} \tag{14}$$

You may wish to check your understanding of these ideas by showing that Equations (13) can be rewritten with z, y, and x explicit as

$$\begin{aligned} 10v - 15w + z &= 0 \\ y + 5v - 5w \quad &= 50 \\ x - 5v + 10w \quad &= 50 \end{aligned} \tag{15}$$

where the corresponding tableau is

$$\begin{array}{c} \\ z \\ y \\ x \end{array} \begin{array}{c} \begin{array}{cccccc} x & y & v & w & z & \end{array} \\ \begin{bmatrix} 0 & 0 & 10 & -15 & 1 & 0 \\ 0 & 1 & 5 & -5 & 0 & 50 \\ 1 & 0 & -5 & 10 & 0 & 50 \end{bmatrix} \end{array} \tag{16}$$

You can also check that Equations (15) can be rewritten with z, y and w explicit as

$$\begin{aligned} 1.5x \quad + 2.5v \quad + z &= 75 \\ 0.5x + y + 2.5v \quad &= 75 \\ 0.1x \quad -0.5v + w \quad &= 5 \end{aligned}$$

where the corresponding tableau is

$$\begin{array}{c} \\ z \\ y \\ w \end{array} \begin{array}{c} \begin{array}{cccccc} x & y & v & w & z & \end{array} \\ \begin{bmatrix} 1.5 & 0 & 2.5 & 0 & 1 & 75 \\ 0.5 & 1 & 2.5 & 0 & 0 & 75 \\ 0.1 & 0 & -0.5 & 1 & 0 & 5 \end{bmatrix} \end{array} \tag{17}$$

EXAMPLE 2 For the problem in Example 1, use a tableau to find the value of z at the basic feasible solution where x and v equal zero.

Solution We look for the tableau in which x and v are implicit. This is tableau (17). If we set $x = 0$ and $v = 0$, then the entries in the last column of this tableau tell us that

$$\begin{aligned} z &= 75 \\ y &= 75 \\ w &= 5 \end{aligned}$$

Thus, at the basic feasible solution

$$x = 0 \qquad y = 75 \qquad v = 0 \qquad w = 5$$

the value of z is 75.

In the above discussion, each row of the tableau is labeled with the explicit variable in the corresponding equation. Although convenient, this is really unnecessary because this information can be obtained from the entries of the tableau. To illustrate, consider tableau (17). The z column of the tableau tells us that z appears in the first equation with a coefficient of $+1$ and occurs in no other equation. Thus z is explicit. Similarly, the y and w columns,

$$\begin{array}{cc} y & w \\ 0 & 0 \\ 1 & 0 \\ 0 & 1 \end{array}$$

tell us that y and w are explicit since y occurs in the second equation with coefficient $+1$ but occurs in no other equation, while w occurs in the third equation with coefficient $+1$ and occurs in no other equation.

These observations are important in the next section.

EXAMPLE 3 The explicit variables in the tableau

$$\begin{array}{cccccc} x & y & v & w & z & \\ \end{array}$$
$$\begin{bmatrix} 0 & -2 & 0 & -5 & 1 & 100 \\ 0 & 0.2 & 1 & -1 & 0 & 10 \\ 1 & 1 & 0 & 5 & 0 & -100 \end{bmatrix}$$

are x, v, and z. (Why?)

EXERCISE SET 4.2

1. Solve using the method of Example 1:

$$\text{Maximize } z = 5x + 7y$$
$$\text{subject to } \quad x + y \leq 1$$
$$x \geq 0, \quad y \geq 0$$

2. Solve the problem in Exercise 1 geometrically and match the corner points with the basic feasible solutions computed in Exercise 1 (as in Table 2).

3. Solve using the method of Example 1:

$$\text{Maximize } z = 6x + 2y$$
$$\text{subject to}$$
$$-3x + 2y \leq 6$$
$$5x + 3y \leq 15$$
$$x \geq 0, \quad y \geq 0$$

4. Solve the problem in Exercise 3 geometrically and match the corner points with the basic feasible solutions computed in Exercise 3 (as in Table 2).

5. Solve using the method of Example 1:

$$\text{Maximize } z = -2x - 5y$$
$$\text{subject to}$$
$$3x + 2y \leq 2$$
$$2x + 5y \leq 8$$
$$x \geq 0, \quad y \geq 0$$

6. In each of the following, name the explicit variables.

(a)
$$\begin{aligned} 3y \qquad -5w + z &= 105 \\ x + 2y \qquad + 8w \qquad &= 20 \\ 7y + v - 4w \qquad &= 13 \end{aligned}$$

(b)
$$\begin{aligned} -10x - 12y \qquad + z &= 0 \\ x + 2y + v \qquad &= 150 \\ x + y \qquad + w \qquad &= 40 \end{aligned}$$

(c)
$$\begin{aligned} 6v \qquad + z &= 0 \\ y + 2v + w \qquad &= 75 \\ x \qquad - 2v + 2w \qquad &= 13 \end{aligned}$$

7. Obtain tableau (16) from Equations (13).

8. Obtain tableau (17) from Equations (15).

9. Consider the problem

$$\text{Maximize } z = 2x + 5y$$
$$\text{subject to}$$
$$3x + 2y \leq 2$$
$$2x + 5y \leq 8$$
$$x \geq 0, y \geq 0$$

(a) Form the new problem with slack variables v and w.
(b) Find a tableau in which x and y are implicit.
(c) Find a tableau in which x and w are implicit.
(d) Find a tableau in which y and w are implicit.

10. Consider the problem

$$\text{Maximize } z = -3x_1 + 5x_2 + 2x_3$$
$$\text{subject to}$$
$$2x_1 + 3x_2 + 4x_3 \leq 18$$
$$3x_1 + 2x_2 + 3x_3 \leq 12$$
$$x_1 \geq 0, \quad x_2 \geq 0, \quad x_3 \geq 0$$

(a) Form the new problem with slack variables v_1 and v_2.
(b) Find a tableau in which x_1, x_2, and x_3 are implicit.
(c) Find a tableau in which x_1, x_2, and v_1 are implicit.
(d) Find a tableau in which x_1, v_1, and v_2 are implicit.

11. Name the explicit variables in the following tableaux.

(a)
$$\begin{array}{cccccc} x & y & v & w & z & \\ \end{array}$$
$$\begin{bmatrix} 0 & 2 & 0 & 4 & 1 & 18 \\ 0 & -3 & 1 & 2 & 0 & 12 \\ 1 & 5 & 0 & 1 & 0 & 22 \end{bmatrix}$$

(b)
$$\begin{array}{ccccccc} x & y & t & u & v & z & \\ \end{array}$$
$$\begin{bmatrix} 4 & 0 & 4 & 5 & 0 & 1 & -50 \\ -2 & 1 & -2 & 2 & 0 & 0 & 8 \\ 5 & 0 & 3 & 0 & 1 & 0 & 4 \end{bmatrix}$$

(c)
$$\begin{array}{ccccccc} x_1 & x_2 & x_3 & v_1 & v_2 & z & \\ \end{array}$$
$$\begin{bmatrix} 3 & -5 & 0 & 0 & 2 & 1 & 43 \\ -2 & -1 & 1 & 0 & -3 & 0 & 18 \\ -3 & 2 & 0 & 1 & 4 & 0 & 12 \end{bmatrix}$$

(d)
$$\begin{array}{ccccccc} x & y & u & v & w & z & \\ \end{array}$$
$$\begin{bmatrix} 0 & 0 & -11 & 0 & 2 & 1 & 12 \\ 1 & 0 & 6 & 0 & 7 & 0 & 13 \\ 0 & 1 & 4 & 0 & 1 & 0 & 14 \\ 0 & 0 & -3 & 1 & 6 & 0 & 2 \end{bmatrix}$$

12. For each of the tableaux in Exercise 11, find the corresponding basic feasible solution and the value of the objective function at that solution.

4.3 THE SIMPLEX METHOD

In this section we describe a step-by-step procedure for solving certain linear programming problems. This technique, called the **simplex method**, was developed by a contemporary mathematician, George B. Dantzig, professor of operations research and computer science at Stanford University, during the late 1940s.

In the simplex method, we begin with a certain easily obtained basic feasible solution, then by a definite procedure we move through a succession of basic feasible solutions in such a way that the objective function increases in value at each step. We continue this process until either (1) we reach a basic feasible solution for which the objective function is maximum or (2) we find that there is no solution. As we discuss the details of each step, we illustrate the idea using the problem

$$\begin{aligned} \text{Maximize} \quad & z = 10x + 12y \\ \text{subject to} \quad & x + 2y \le 150 \\ & x + y \le 100 \\ & x \ge 0 \\ & y \ge 0 \end{aligned}$$

The corresponding new problem with slack varibles is

$$\begin{aligned} \text{Maximize} \quad & z = 10x + 12y \\ \text{subject to} \quad & x + 2y + v = 150 \\ & x + y + w = 100 \\ & x \ge 0, \quad y \ge 0, \quad v \ge 0, \quad w \ge 0 \end{aligned} \tag{1}$$

Throughout this section we assume we have a standard linear programming problem. This being the case, we can always get a basic feasible solution by setting the original (nonslack) variables equal to zero. We call this the **initial basic feasible solution**. For example, if we set $x = 0$ and $y = 0$ in (1), we obtain the initial basic feasible solution

$$x = 0 \qquad y = 0 \qquad v = 150 \qquad w = 100$$

Moreover, at this solution the objective function $z = 10x + 12y$ has the value $z = 0$.

The important information about the initial basic feasible solution can be obtained from the last column of the tableau in which the slack variables are explicit. We call this the **initial tableau**. In our illustrative example, the initial tableau is

$$\begin{array}{c} \\ z \\ v \\ w \end{array} \begin{array}{c} \begin{array}{cccccc} x & y & v & w & z & \end{array} \\ \begin{bmatrix} -10 & -12 & 0 & 0 & 1 & 0 \\ 1 & 2 & 1 & 0 & 0 & 150 \\ 1 & 1 & 0 & 1 & 0 & 100 \end{bmatrix} \end{array} \tag{2}$$

The construction of the initial tableau is the first step in the simplex method. For reference, we formally state this fact.

Step 1 in the Simplex Method (*Forming the Initial Tableau*) Construct the tableau in which the slack variables and z are explicit.

Once any initial tableau is constructed, there are two questions to consider:

(a) Does the tableau correspond to an optimal solution?

(b) If not, how can we obtain a new tableau in which the value of z is closer to the maximum value?

Let us consider question (a) first. In our example, the first row of the initial tableau is

x	y	v	w	z		
−10	−12	0	0	1	0	(3)

so that the corresponding equation is

$$-10x - 12y + z = 0$$

or

$$z = 10x + 12y \tag{4}$$

Since x and y occur with positive coefficients $+10$ and $+12$ in this last equation, it is obvious that the value of z will increase whenever x increases in value or y increases in value. With this in mind, we can see that the value $z = 0$ that occurs at the initial basic feasible solution cannot be the maximum possible value of z since larger values of z can be obtained by increasing either x or y. The key to this argument is the fact that the expression for z in terms of the implicit variables contained at least one term with a positive coefficient. Comparing (3) and (4) above, we see that the expression for z will contain terms with positive coefficients whenever the z row of the tableau contains *negative* entries in one of the columns headed by a variable. This suggests the following test for determining from the tableau whether the maximum value of z has been obtained.

Step 2 in the Simplex Method (*Test for Maximality*) If the z row of a tableau contains no negative entries in the columns labeled with variables, we have arrived at an optimal solution.

EXAMPLE 1 In Table 1 of the preceding section we listed the basic feasible solutions of the new problem with slack variables given in Example 1. From that table it is evident that the maximum value $z = 75$ occurs at the basic feasible solution

$$x = 0 \qquad v = 0 \qquad y = 75 \qquad w = 5$$

Let us see if we can obtain this same conclusion by examining the tableaus corresponding to the basic feasible solutions. These tableaus were all computed in the last section [see (9c), (14), (16), and (17)], but for convenience we list them here.

Basic feasible solution	Corresponding tableau
(a) $x = 0, \quad y = 0$ $v = 30, \quad w = 20$	$\begin{array}{c} \\ z \\ v \\ w \end{array}\begin{array}{c} \begin{array}{cccccc} x & y & v & w & z & \end{array} \\ \left[\begin{array}{cccccc} 1 & -1 & 0 & 0 & 1 & 0 \\ 0.2 & 0.4 & 1 & 0 & 0 & 30 \\ 0.2 & 0.2 & 0 & 1 & 0 & 20 \end{array}\right] \end{array}$
(b) $y = 0, \quad w = 0$ $x = 100, \quad v = 10$	$\begin{array}{c} \\ z \\ v \\ x \end{array}\begin{array}{c} \begin{array}{cccccc} x & y & v & w & z & \end{array} \\ \left[\begin{array}{cccccc} 0 & -2 & 0 & -5 & 1 & -100 \\ 0 & 0.2 & 1 & -1 & 0 & 10 \\ 1 & 1 & 0 & 5 & 0 & 100 \end{array}\right] \end{array}$
(c) $v = 0 \quad w = 0$ $x = 50, \quad y = 50$	$\begin{array}{c} \\ z \\ y \\ x \end{array}\begin{array}{c} \begin{array}{cccccc} x & y & v & w & z & \end{array} \\ \left[\begin{array}{cccccc} 0 & 0 & 10 & -15 & 1 & 0 \\ 0 & 1 & 5 & -5 & 0 & 50 \\ 1 & 0 & -5 & 10 & 0 & 50 \end{array}\right] \end{array}$
(d) $x = 0, \quad v = 0$ $y = 75, \quad w = 5$	$\begin{array}{c} \\ z \\ y \\ w \end{array}\begin{array}{c} \begin{array}{cccccc} x & y & v & w & z & \end{array} \\ \left[\begin{array}{cccccc} 1.5 & 0 & 2.5 & 0 & 1 & 75 \\ 0.5 & 1 & 2.5 & 0 & 0 & 75 \\ 0.1 & 0 & -0.5 & 1 & 0 & 5 \end{array}\right] \end{array}$

Observe that the z rows of tableaux (a), (b), and (c) all have one or more negative entries in the columns labeled with variables. Thus the corresponding basic feasible solutions are not optimal. However, the z row of tableau (d) contains no negative entries in these columns, so the corresponding basic feasible solution

$$x = 0 \qquad v = 0 \qquad y = 75 \quad w = 5$$

is optimal. This agrees with our previous observation.

Returning to our illustrative example, we know that the initial basic feasible solution is not optimal since the z row in the initial tableau (2) has negative entries in the x and y columns. The values of x and y in this initial basic feasible solution are $x = 0$ and $y = 0$, and, as we previously pointed out, the value of z can be increased by increasing either x or y. In the simplex method we try to increase only *one* variable at a time. Thus we have two choices:

1. Look for a new tableau in which x remains at $x = 0$ and $y > 0$.

Or

2. Look for a new tableau in which y remains at $y = 0$ and $x > 0$.

To determine which of these alternatives is better, let us reexamine Equation (4). From this equation we see that by increasing y by 1 unit, we increase z by 12 units, while increasing x by 1 unit only increases z by 10 units. Thus, the best strategy is to keep x fixed at $x = 0$ and increase y, since this choice tends to bring us toward the maximum value of z more rapidly. The variable to be increased is usually called the **entering variable** (because in going to the next tableau it will leave the set of implicit variables and *enter* the set of explicit variables). For convenience we mark the entering variable with an arrow at the top of the tableau as follows:

$$\begin{array}{c} \\ z \\ v \\ w \end{array}\begin{array}{c} \text{Entering variable} \\ \downarrow \\ \begin{array}{cccccc} x & y & v & w & z & \end{array} \\ \left[\begin{array}{rrrrr|r} -10 & -12 & 0 & 0 & 1 & 0 \\ 1 & 2 & 1 & 0 & 0 & 150 \\ 1 & 1 & 0 & 1 & 0 & 100 \end{array}\right] \end{array} \tag{5}$$

As tableau (5) illustrates, the entering variable in a tableau is the variable whose entry in the z row is the smallest negative number. We can now formally state the next step in the simplex method.

Step 3 in the Simplex Method *(Selecting the Entering Variable)* If the test for maximality in Step 2 shows that an optimal solution has not been reached, then label the variable whose entry in the z row is the most negative as the entering variable.

In tableau (5) we labeled y as the entering variable. This means we want to find a new tableau for which the value of y is positive. But if y is to be positive in the new tableau, it will be explicit (the implicit variables in a tableau have value zero). Thus we must look for a new tableau in which y is explicit and some other variable replaces y as an implicit variable. The *new* implicit variable is usually called the **departing variable** (since it *departs* from the set of explicit variables and becomes implicit).

Soon we discuss criteria for choosing the departing variable. For illustrative purposes, however, let us suppose that we have somehow settled on making v the departing variable. We mark the departing variable with an arrow on the left side of the tableau as follows:

$$\begin{array}{r} \\ \\ z \\ \text{Departing variable} \rightarrow v \\ w \end{array}\begin{array}{c} \text{Entering variable} \\ \downarrow \\ \begin{array}{cccccc} x & y & v & w & z & \end{array} \\ \left[\begin{array}{rrrrr|r} -10 & -12 & 0 & 0 & 1 & 0 \\ 1 & \boxed{2} & 1 & 0 & 0 & 150 \\ 1 & 1 & 0 & 1 & 0 & 100 \end{array}\right] \end{array} \tag{6}$$

Now that we have specified the entering and departing variables, the new tableau is partially determined. The new explicit variables are z, y, and w, so that the new tableau will have the form

$$\begin{array}{c} \\ z \\ y \\ w \end{array}\begin{array}{c} \begin{array}{cccccc} x & y & v & w & z & \end{array} \\ \left[\begin{array}{ccccc|c} \cdot & 0 & \cdot & 0 & 1 & \cdot \\ \cdot & 1 & \cdot & 0 & 0 & \cdot \\ \cdot & 0 & \cdot & 1 & 0 & \cdot \end{array}\right] \end{array} \tag{7}$$

where the dots designate numbers that are yet to be determined. To see how these unknown entries can be obtained, we need only recall that tableaux (6) and (7) are augmented matrices for the *same* system of equations but written in two different ways. This suggests that tableau (7) can be obtained from tableau (6) using row operations. We now show how this can be done.

Since the w and z columns of (6) and (7) are identical, we need only find row operations that will transform the y column

$$\begin{matrix} y \\ -12 \\ 2 \\ 1 \end{matrix}$$

in (6) into the y column

$$\begin{matrix} y \\ 0 \\ 1 \\ 0 \end{matrix}$$

in (7) without disturbing the w and z columns. This can be accomplished by the following procedure.

If we extend the arrows marking the entering and departing variables in a tableau, they intersect at an entry called the **pivot entry**. To illustrate, the pivot entry in tableau (6) is the circled entry below:

$$\begin{array}{cc} & \text{Entering variable} \\ & \downarrow \\ & \begin{matrix} \;\;\; x & \;\; y & \;\; v & \;\; w & \;\; z & \end{matrix} \\ \begin{matrix} z \\ \text{Departing variable} \rightarrow v \\ w \end{matrix} & \begin{bmatrix} -10 & -12 & 0 & 0 & 1 & 0 \\ 1 & \textcircled{2} & 1 & 0 & 0 & 150 \\ 1 & 1 & 0 & 1 & 0 & 100 \end{bmatrix} \end{array} \tag{6a}$$

The row and column containing the pivot entry are called, respectively, the **pivot row** and **pivot column**. To transform tableau (6) into tableau (7) we can use the following method, called **pivotal elimination**.

Pivotal Elimination

(i) If the pivot entry is k, multiply the pivot row by $1/k$ so that the pivot entry becomes a 1.

(ii) Add suitable multiples of the pivot row to the other rows so that the remaining entries in the pivot column become zeros.

Note that only multiples of the pivot row can be used to carry out the elimination.

To apply step (i) to tableau (6a), we multiply the pivot row 2 by $\frac{1}{2}$ to obtain

$$\begin{array}{cc} & \begin{matrix} \;\;\; x & \;\; y & \;\; v & \;\; w & \;\; z & \end{matrix} \\ \begin{matrix} z \\ v \\ w \end{matrix} & \begin{bmatrix} -10 & -12 & 0 & 0 & 1 & 0 \\ \frac{1}{2} & 1 & \frac{1}{2} & 0 & 0 & 75 \\ 1 & 1 & 0 & 1 & 0 & 100 \end{bmatrix} \end{array}$$

To apply step (ii) to this matrix, we add 12 times the pivot row to the first row and −1 times the pivot row to the third row. This yields the new tableau

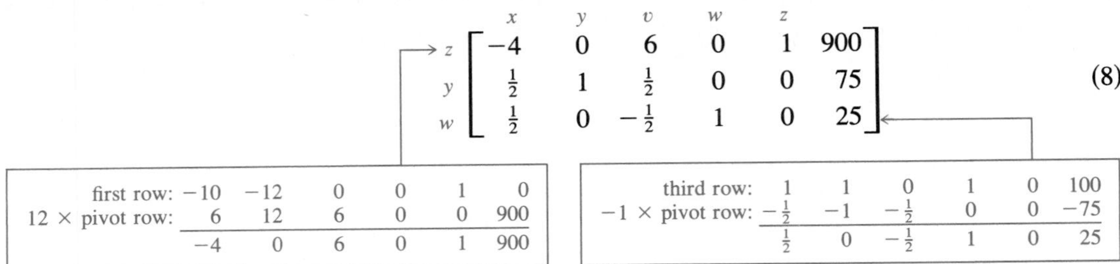

$$\begin{array}{c} \\ z \\ y \\ w \end{array}\begin{bmatrix} x & y & v & w & z & \\ -4 & 0 & 6 & 0 & 1 & 900 \\ \frac{1}{2} & 1 & \frac{1}{2} & 0 & 0 & 75 \\ \frac{1}{2} & 0 & -\frac{1}{2} & 1 & 0 & 25 \end{bmatrix} \tag{8}$$

first row:	−10	−12	0	0	1	0
12 × pivot row:	6	12	6	0	0	900
	−4	0	6	0	1	900

third row:	1	1	0	1	0	100
−1 × pivot row:	$-\frac{1}{2}$	−1	$-\frac{1}{2}$	0	0	−75
	$\frac{1}{2}$	0	$-\frac{1}{2}$	1	0	25

EXAMPLE 2 Use pivotal elimination to find the new tableau if w is taken to be the departing variable in (5).

Solution The pivot entry is the circled entry in the following tableau.

Entering variable ↓ (y)

$$\begin{array}{c} \\ z \\ v \\ \text{Departing variable} \rightarrow w \end{array}\begin{bmatrix} x & y & v & w & z & \\ -10 & -12 & 0 & 0 & 1 & 0 \\ 1 & 2 & 1 & 0 & 0 & 150 \\ 1 & \textcircled{1} & 0 & 1 & 0 & 100 \end{bmatrix}$$

Since the pivot entry is a 1, we can go immediately to step (ii) in pivotal elimination. Thus, we add 12 times the pivot row to the first row and −2 times the pivot row to the second row; this yields the new tableau

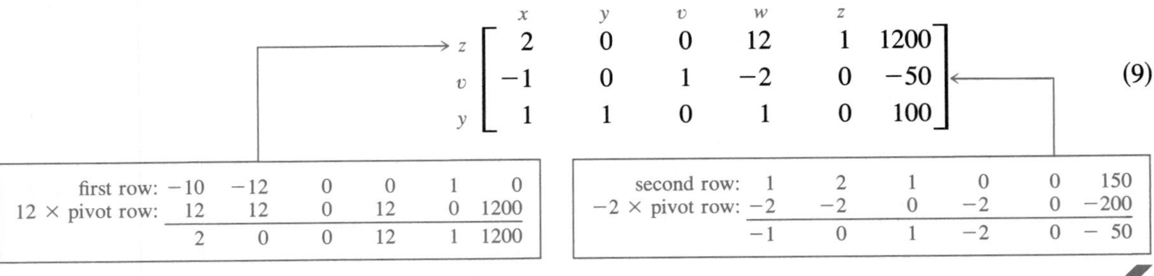

$$\begin{array}{c} \\ z \\ v \\ y \end{array}\begin{bmatrix} x & y & v & w & z & \\ 2 & 0 & 0 & 12 & 1 & 1200 \\ -1 & 0 & 1 & -2 & 0 & -50 \\ 1 & 1 & 0 & 1 & 0 & 100 \end{bmatrix} \tag{9}$$

first row:	−10	−12	0	0	1	0
12 × pivot row:	12	12	0	12	0	1200
	2	0	0	12	1	1200

second row:	1	2	1	0	0	150
−2 × pivot row:	−2	−2	0	−2	0	−200
	−1	0	1	−2	0	− 50

So far, we have shown how to select the entering variable in a tableau and how to obtain the new tableau once the entering and departing variables are known. We have not explained how to choose the departing variable.

To illustrate the pitfalls that can occur in choosing the departing variable, let us compare tableaux (8) and (9). Both were obtained from tableau (5), but using different departing variables. In tableau (8) the implicit variables are x and v; if we set these variables equal to zero, this tableau indicates that

$$x = 0 \qquad v = 0 \qquad y = 75 \qquad w = 25$$

is a basic feasible solution and the value of the objective function there is $z = 900$. On the other hand, the implicit variables in tableau (9) are x and w. If we set these variables equal to zero, the tableau tells us that

$$x = 0 \qquad w = 0 \qquad v = -50 \qquad y = 100$$

But this is not a basic feasible solution since the variable v violates the constraint $v \geq 0$. In other words, by making a wrong choice of the departing variable, we can obtain new values of the variables that violate the nonnegativity constraints. Thus, the objectives in selecting the departing variable are twofold:

1. Obtain the largest possible increase in z.
2. Avoid violating the nonnegativity constraints.

For reasons too technical to discuss here, it is difficult to determine which choice of the departing variable will meet objective 1. Nevertheless, we can give a procedure for selecting the departing variable that will meet objective 2 and still give a reasonably good increase in z. Since any increase in z will lead us closer to a solution, this is adequate for our purposes. We omit the mathematical theory underlying this procedure.

First we need some terminology. If, as illustrated in Figure 1, b is the last element in a row of a tableau and a is the element of that row in the column headed by the entering variable, then we call the ratio

$$b/a$$

the **quotient** for the row.

FIGURE 1

$$\begin{array}{ccccc} & \text{Entering variable} & & & \text{Row quotient} \\ & \downarrow & & & \downarrow \\ \cdots & a & \cdots & b & \dfrac{b}{a} \end{array}$$

EXAMPLE 3 Since y is the entering variable, the quotients for the v and w rows of tableau (5) are $150/2 = 75$ and $100/1 = 100$, respectively.

Step 4 in the Simplex Method *(Selecting the Departing Variable)* Consider the quotients for all but the z row of the tableau. Then choose as the departing variable the explicit variable in the row where the quotient is the least nonnegative number and has a positive denominator.

EXAMPLE 4 As shown in Example 3, the quotients for the v and w rows of tableau (5) are 75 and 100. Since 75 is the smaller of these quotients and is nonnegative, the departing variable is v.

We now have all the necessary ingredients to state the simplex method:

The Simplex Method

Step 1 Compute the initial tableau.

Step 2 Test for maximality. If the test shows we have obtained an optimal solution, stop; otherwise go to Step 3.

Step 3 Determine the entering variable.

Step 4 Determine the departing variable.

Step 5 Using pivotal elimination, construct the new tableau and return to Step 2.

Figure 2 presents the simplex method in diagram form.

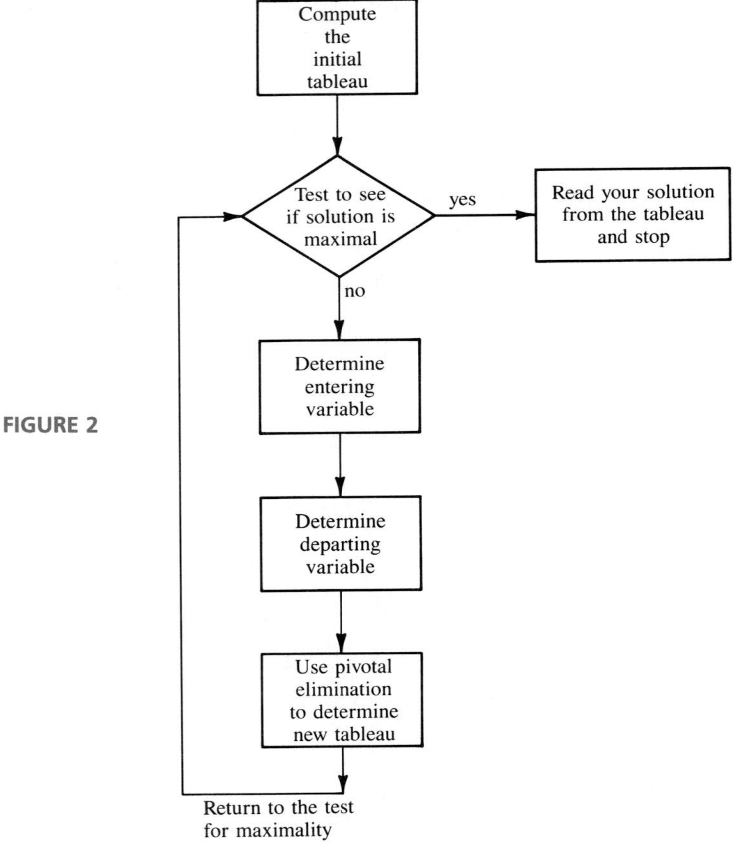

FIGURE 2

EXAMPLE 5 Solve the following problem by the simplex method:

$$\begin{aligned} \text{Maximize} \quad & z = 10x + 12y \\ \text{subject to} \quad & \\ & x + 2y \le 150 \\ & x + y \le 100 \\ & x \ge 0 \\ & y \ge 0 \end{aligned}$$

(This is the illustrative problem we have been working with in this section.)

Solution The corresponding new problem with slack variables is

$$\begin{aligned} \text{Maximize} \quad & z = 10x + 12y \\ \text{subject to} \quad & \\ & x + 2y + v \qquad = 150 \\ & x + y \qquad + w = 100 \\ & x \ge 0, \quad y \ge 0, \quad v \ge 0, \quad w \ge 0 \end{aligned}$$

As shown in (2), the initial tableau is

$$\begin{array}{c} \\ z \\ v \\ w \end{array} \begin{array}{c} \begin{array}{cccccc} x & y & v & w & z & \end{array} \\ \left[\begin{array}{cccccc} -10 & -12 & 0 & 0 & 1 & 0 \\ 1 & 2 & 1 & 0 & 0 & 150 \\ 1 & 1 & 0 & 1 & 0 & 100 \end{array}\right] \end{array}$$

Since the x and y columns of the z row have negative entries, we have not yet reached an optimal solution. Since -12 is the smallest negative entry in the z row occurring under a variable, y is the entering variable, that is,

$$\begin{array}{c} \\ \\ z \\ v \\ w \end{array} \begin{array}{c} \text{Entering variable} \\ \downarrow \\ \begin{array}{cccccc} x & y & v & w & z & \end{array} \\ \left[\begin{array}{cccccc} -10 & -12 & 0 & 0 & 1 & 0 \\ 1 & 2 & 1 & 0 & 0 & 150 \\ 1 & 1 & 0 & 1 & 0 & 100 \end{array}\right] \end{array} \begin{array}{c} \text{Row} \\ \text{quotient} \\ \\ 75 \\ 100 \end{array}$$

The quotient for the v row is $150/2 = 75$ and the w row quotient is $100/1 = 100$; thus, v is the departing variable since its row has the smallest nonnegative quotient, that is,

$$\begin{array}{r} \\ \\ z \\ \text{Departing variable} \rightarrow v \\ w \end{array} \begin{array}{c} \text{Entering variable} \\ \downarrow \\ \begin{array}{cccccc} x & y & v & w & z & \end{array} \\ \left[\begin{array}{cccccc} -10 & -12 & 0 & 0 & 1 & 0 \\ 1 & \boxed{2} & 1 & 0 & 0 & 150 \\ 1 & 1 & 0 & 1 & 0 & 100 \end{array}\right] \end{array} \qquad (6a)$$

As previously shown [see tableau (8)], pivotal elimination yields the new tableau

$$\begin{array}{c} \\ z \\ y \\ w \end{array} \begin{array}{c} \begin{array}{cccccc} x & y & v & w & z & \end{array} \\ \left[\begin{array}{cccccc} -4 & 0 & 6 & 0 & 1 & 900 \\ \frac{1}{2} & 1 & \frac{1}{2} & 0 & 0 & 75 \\ \frac{1}{2} & 0 & -\frac{1}{2} & 1 & 0 & 25 \end{array}\right] \end{array}$$

Since the z row of this tableau contains a negative entry in the x column, the test for maximality shows we have not yet obtained an optimal solution. In the z row, -4 is the smallest negative entry under a variable, so x is the new entering variable, that is,

Entering variable ↓

$$\begin{array}{c} \\ z \\ y \\ w \end{array} \begin{array}{c} \begin{array}{cccccc} x & y & v & w & z & \end{array} \\ \left[\begin{array}{cccccc} -4 & 0 & 6 & 0 & 1 & 900 \\ \frac{1}{2} & 1 & \frac{1}{2} & 0 & 0 & 75 \\ \frac{1}{2} & 0 & -\frac{1}{2} & 1 & 0 & 25 \end{array}\right] \end{array} \begin{array}{c} \text{Row quotient} \\ \\ 150 \\ 50 \end{array}$$

The quotient for the y row is $75/\frac{1}{2} = 150$ and the quotient for the w row is $25/\frac{1}{2} = 50$. Thus, w is the new departing variable since its row has the smallest nonnegative quotient, that is,

Entering variable ↓

$$\begin{array}{r} \\ z \\ y \\ \text{Departing variable} \rightarrow w \end{array} \begin{array}{c} \begin{array}{cccccc} x & y & v & w & z & \end{array} \\ \left[\begin{array}{cccccc} -4 & 0 & 6 & 0 & 1 & 900 \\ \frac{1}{2} & 1 & \frac{1}{2} & 0 & 0 & 75 \\ \boxed{\frac{1}{2}} & 0 & -\frac{1}{2} & 1 & 0 & 25 \end{array}\right] \end{array}$$

Using pivotal elimination, we first multiply the pivot row by 2. This yields

$$\begin{array}{c} \\ z \\ y \\ w \end{array} \begin{array}{c} \begin{array}{cccccc} x & y & v & w & z & \end{array} \\ \left[\begin{array}{cccccc} -4 & 0 & 6 & 0 & 1 & 900 \\ \frac{1}{2} & 1 & \frac{1}{2} & 0 & 0 & 75 \\ 1 & 0 & -1 & 2 & 0 & 50 \end{array}\right] \end{array}$$

Next we add 4 times the pivot row to the first row and $-\frac{1}{2}$ times the pivot row to the second row; this yields the new tableau

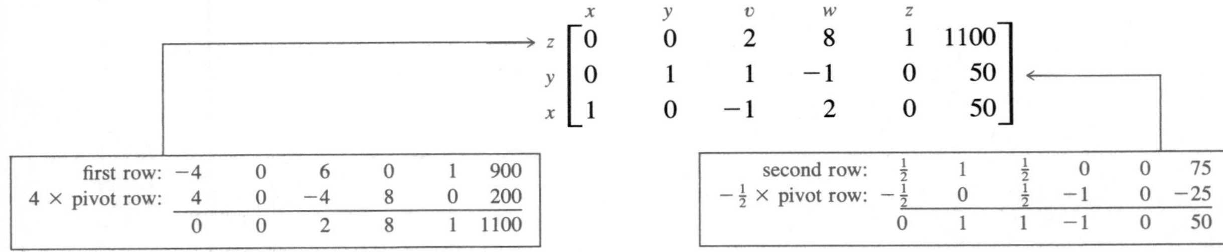

$$\begin{array}{c} \\ z \\ y \\ x \end{array} \begin{array}{c} \begin{array}{cccccc} x & y & v & w & z & \end{array} \\ \left[\begin{array}{cccccc} 0 & 0 & 2 & 8 & 1 & 1100 \\ 0 & 1 & 1 & -1 & 0 & 50 \\ 1 & 0 & -1 & 2 & 0 & 50 \end{array}\right] \end{array}$$

first row:	-4	0	6	0	1	900
$4 \times$ pivot row:	4	0	-4	8	0	200
	0	0	2	8	1	1100

second row:	$\frac{1}{2}$	1	$\frac{1}{2}$	0	0	75
$-\frac{1}{2} \times$ pivot row:	$-\frac{1}{2}$	0	$\frac{1}{2}$	-1	0	-25
	0	1	1	-1	0	50

LINEAR PROGRAMMING IN THE NEWS

Mathematical developments are always published in professional journals but only major results are reported (and only sometimes) in newspapers. Thus, it is especially significant that the front page of the *New York Times* carried the following story on November 19, 1984 (only partially reprinted here).

Breakthrough in Problem Solving By James Gleick

A 28-year-old mathematician at A.T.&T. Bell Laboratories has made a startling theoretical breakthrough in the solving of systems of equations that often grow too vast and complex for the most powerful computers.

The discovery, which is to be formally published next month, is already circulating rapidly through the mathematical world. It has also set off a deluge of inquiries from brokerage houses, oil companies and airlines, industries with millions of dollars at stake in problems known as linear programming.

These problems are fiendishly complicated systems, often with thousands of variables. They arise in a variety of commercial and government applications ranging from allocating time on a communications satellite to routing millions of telephone calls over long distances, or whenever time must be allocated most efficiently among competing users. Investment companies use them to devise portfolios with the best mix of stocks and bonds.

Faster Solutions Seen

The Bell Labs mathematician, Dr. Narendra Karmarkar, has devised a radically new procedure that may speed the routine handling of such problems by businesses and Government agencies and also make it possible to tackle problems that are now far out of reach.

"This is a path-breaking result," said Dr. Ronald L. Graham, director of mathematical sciences for Bell Labs in Murray Hill, N.J. "Science has its moments of great progress, and this may well be one of them."

Because problems in linear programming can have billions or more possible answers, even high-speed computers cannot check every one. So computers must use a special procedure, an algorithm, to examine as few answers as possible before finding the best one—typically the one that minimizes cost or maximizes efficiency.

A procedure devised in 1947, the simplex method, is now used for such problems, usually in the form of highly refined computer codes sold by the International Business Machines Corporation, among others.

The new Karmarkar approach exists so far only in rougher computer code. Its full value will be impossible to judge until it has been tested experimentally on a wide range of problems. But those who have tested the early versions at Bell Labs say that it already appears many times faster than the simplex method, and the advantage grows rapidly with more complicated problems.

"The problems that people would really like to solve are larger than can be done today," Dr. Karmarkar said. "I felt strongly that there must be a better solution."

Corporations Seek Answers

American Airlines, among others, has begun working with Dr. Karmarkar to see whether his technique will speed their handling of linear programming problems, from the scheduling of flight crews to the planning of fuel loads. Finding the best answer to the fuel problem, where each plane should refuel and how much it should carry, cuts fuel costs substantially.

"It's big dollars," said Thomas M. Cook, American's director of operations research. "We're hoping we can solve harder problems faster, and we think there's a definite potential."

The Exxon Corporation uses linear programming for a variety of applications, such as deciding how to spread its crude oil among refineries. It is one of several oil companies studying the Karmarkar algorithm.

"It promises a more rapid solution of linear programming problems," said David Smith of Exxon's communications and computer sciences department. "It's most important at times when conditions are changing rapidly, for example, the price of crude oil."

If Dr. Karmarkar's procedure performs as well as expected, it will be able to handle many linear programming problems faster than the simplex method can, saving money by using less computer time. But it may also be applied to problems that are left unsolved now because they are too big and too complex to tackle with the simplex method.

For example, A.T.&T. believes the discovery may provide a new approach to the problem of routing long-distance telephone calls through hundreds or thousands of cities with maximum efficiency. Because of the different volumes of calls between different places, the different capacities of the telephone lines and the different needs of users at different hours, the problem is extraordinarily difficult. (Reprinted with permission.)

Since the z row of this tableau has no negative entries in the columns labeled with variables, the test for maximality indicates that we have arrived at an optimal solution. From the tableau, this optimal solution is

$$v = 0 \qquad w = 0 \qquad x = 50 \qquad y = 50$$

and the maximum value of z is

$$z = 1100$$

To obtain a solution of the original problem, we discard the slack variables. Thus, in the original problem the maximum value $z = 1100$ occurs at $x = 50$, $y = 50$.

EXAMPLE 6 Use the simplex method to solve the standard linear programming problem

$$\begin{aligned} &\text{Maximize} \quad z = 6x + 9y + 6t \\ &\text{subject to} \\ &\qquad 2x + 2y + t \le 20 \\ &\qquad 2x + 3y + 3t \le 30 \\ &\qquad 2x + 5y + t \le 40 \\ &\qquad x \ge 0, \quad y \ge 0, \quad t \ge 0 \end{aligned}$$

Solution The corresponding problem with slack variables is

$$\begin{aligned} &\text{Maximize} \quad z = 6x + 9y + 6t \\ &\text{subject to} \\ &\qquad 2x + 2y + t + u = 20 \\ &\qquad 2x + 3y + 3t + v = 30 \\ &\qquad 2x + 5y + t + w = 40 \\ &\qquad x \ge 0, \quad y \ge 0, \quad t \ge 0 \\ &\qquad u \ge 0, \quad v \ge 0, \quad w \ge 0 \end{aligned}$$

The initial tableau is

	x	y	t	u	v	w	z		Row quotient
z	−6	−9	−6	0	0	0	1	0	
u	2	2	1	1	0	0	0	20	10
v	2	3	3	0	1	0	0	30	10
w	2	5	1	0	0	1	0	40	8

Since the z row of this tableau has negative entries in some of the columns labeled with variables, we have not yet obtained an optimal solution. Since −9 is the smallest negative of these entries, y is the entering variable. Consequently, the quotients for the u, v, and w rows are

$$\frac{20}{2} = 10 \qquad \frac{30}{3} = 10 \qquad \frac{40}{5} = 8$$

respectively. Thus, w is the departing variable since its row has the smallest non-negative quotient; the pivot element is now determined, that is,

Entering variable ↓ (column y); Departing variable → w

$$\begin{array}{c} \\ z \\ u \\ v \\ w \end{array}\begin{array}{c}\begin{array}{cccccccc} x & y & t & u & v & w & z & \end{array}\\ \left[\begin{array}{cccccccc} -6 & -9 & -6 & 0 & 0 & 0 & 1 & 0 \\ 2 & 2 & 1 & 1 & 0 & 0 & 0 & 20 \\ 2 & 3 & 3 & 0 & 1 & 0 & 0 & 30 \\ 2 & \boxed{5} & 1 & 0 & 0 & 1 & 0 & 40 \end{array}\right]\end{array}$$

Using pivotal elimination we first multiply the pivot row by $\frac{1}{5}$ to obtain

$$\begin{array}{c} \\ z \\ u \\ v \\ w \end{array}\begin{array}{c}\begin{array}{cccccccc} x & y & t & u & v & w & z & \end{array}\\ \left[\begin{array}{cccccccc} -6 & -9 & -6 & 0 & 0 & 0 & 1 & 0 \\ 2 & 2 & 1 & 1 & 0 & 0 & 0 & 20 \\ 2 & 3 & 3 & 0 & 1 & 0 & 0 & 30 \\ \frac{2}{5} & 1 & \frac{1}{5} & 0 & 0 & \frac{1}{5} & 0 & 8 \end{array}\right]\end{array}$$

Next we add 9 times the pivot row to the first row, add -2 times the pivot row to the second row, and add -3 times the pivot row to the third row to obtain

$$\begin{array}{c} \\ z \\ u \\ v \\ y \end{array}\begin{array}{c}\begin{array}{cccccccc} x & y & t & u & v & w & z & \end{array}\\ \left[\begin{array}{cccccccc} -\frac{12}{5} & 0 & -\frac{21}{5} & 0 & 0 & \frac{9}{5} & 1 & 72 \\ \frac{6}{5} & 0 & \frac{3}{5} & 1 & 0 & -\frac{2}{5} & 0 & 4 \\ \frac{4}{5} & 0 & \frac{12}{5} & 0 & 1 & -\frac{3}{5} & 0 & 6 \\ \frac{2}{5} & 1 & \frac{1}{5} & 0 & 0 & \frac{1}{5} & 0 & 8 \end{array}\right]\end{array}$$

This new tableau still contains negative entries in the z row under the variables, so we have not yet reached the optimal solution. We thus repeat the process of choosing entering and departing variables. We leave it to you to obtain the following result.

Entering variable ↓ (column t); Departing variable → v

$$\begin{array}{c} \\ z \\ u \\ v \\ y \end{array}\begin{array}{c}\begin{array}{cccccccc} x & y & t & u & v & w & z & \end{array}\\ \left[\begin{array}{cccccccc} -\frac{12}{5} & 0 & -\frac{21}{5} & 0 & 0 & \frac{9}{5} & 1 & 72 \\ \frac{6}{5} & 0 & \frac{3}{5} & 1 & 0 & -\frac{2}{5} & 0 & 4 \\ \frac{4}{5} & 0 & \boxed{\frac{12}{5}} & 0 & 1 & -\frac{3}{5} & 0 & 6 \\ \frac{2}{5} & 1 & \frac{1}{5} & 0 & 0 & \frac{1}{5} & 0 & 8 \end{array}\right]\end{array}$$

After pivotal elimination we obtain the new tableau (verify this)

$$\begin{array}{c} \\ z \\ u \\ t \\ y \end{array}\begin{array}{c}\begin{array}{cccccccc} x & y & t & u & v & w & z & \end{array}\\ \left[\begin{array}{cccccccc} -1 & 0 & 0 & 0 & \frac{7}{4} & \frac{3}{4} & 1 & \frac{165}{2} \\ 1 & 0 & 0 & 1 & -\frac{1}{4} & -\frac{1}{4} & 0 & \frac{5}{2} \\ \frac{1}{3} & 0 & 1 & 0 & \frac{5}{12} & -\frac{1}{4} & 0 & \frac{5}{2} \\ \frac{1}{3} & 1 & 0 & 0 & -\frac{1}{12} & \frac{1}{4} & 0 & \frac{15}{2} \end{array}\right]\end{array}$$

Once again we have a negative entry in the z row under a variable, so we do not have an optimal solution. Choosing the new entering and departing variables, we obtain

Entering variable ↓ (column x); Departing variable → (row u)

$$\begin{array}{c} \\ z \\ u \\ t \\ y \end{array}\begin{array}{c}\begin{array}{cccccccc} x & y & t & u & v & w & z & \end{array}\\ \left[\begin{array}{cccccccc} -1 & 0 & 0 & 0 & \frac{7}{4} & \frac{3}{4} & 1 & \frac{165}{2} \\ \boxed{1} & 0 & 0 & 1 & -\frac{1}{4} & -\frac{1}{4} & 0 & \frac{5}{2} \\ \frac{1}{3} & 0 & 1 & 0 & \frac{5}{12} & -\frac{1}{4} & 0 & \frac{5}{2} \\ \frac{1}{3} & 1 & 0 & 0 & -\frac{1}{12} & \frac{1}{4} & 0 & \frac{15}{2} \end{array}\right]\end{array}$$

After pivotal elimination we obtain the new tableau (verify this)

$$\begin{array}{c} \\ z \\ x \\ t \\ y \end{array}\begin{array}{c}\begin{array}{cccccccc} x & y & t & u & v & w & z & \end{array}\\ \left[\begin{array}{cccccccc} 0 & 0 & 0 & 1 & \frac{3}{2} & \frac{1}{2} & 1 & 85 \\ 1 & 0 & 0 & 1 & -\frac{1}{4} & -\frac{1}{4} & 0 & \frac{5}{2} \\ 0 & 0 & 1 & -\frac{1}{3} & \frac{1}{2} & -\frac{1}{6} & 0 & \frac{5}{3} \\ 0 & 1 & 0 & -\frac{1}{3} & 0 & \frac{1}{3} & 0 & \frac{20}{3} \end{array}\right]\end{array}$$

Since the z row in this tableau has no negative entries under the variables, we have obtained an optimal solution. From the tableau, the maximal value $z = 85$ occurs when

$$u = 0 \qquad v = 0 \qquad w = 0 \qquad x = \frac{5}{2} \qquad t = \frac{5}{3} \qquad y = \frac{20}{3}$$

To obtain a solution of the original problem we discard the slack variables. Thus, in the original problem, the maximum value $z = 85$ occurs at

$$x = \frac{5}{2} \qquad t = \frac{5}{3} \qquad y = \frac{20}{3}$$

Not every standard linear programming problem has a solution. If the simplex method is attempted on a standard linear programming problem that has no solution, then at some point in the computations we will obtain a tableau in which the quotients used to select the departing variable are all negative. Thus, the rule for choosing the departing variable will not work. We can then stop the computations and conclude that the problem has no solution (see Exercise 19).

Certain difficulties occur when the minimum nonnegative quotient, computed to determine the departing variable, occurs in two or more rows. This situation is called **degeneracy**. Students interested in how degeneracy is handled are referred to more detailed books on linear programming, such as

Garvin, W. W. *Introduction to Linear Programming* (New York: McGraw-Hill Book Company, 1960).

Gass, Saul I. *Linear Programming,* 4th ed. (New York: McGraw-Hill Book Company, 1975).

Kolman, Bernard, and Robert E. Beck. *Elementary Linear Programming with Applications* (Orlando, Fla.: Academic Press, Inc., 1980).

Strang, Gilbert. *Introduction to Applied Mathematics* (Wellesley, Mass.: Wellesley-Cambridge Press, 1985).

EXERCISE SET 4.3

In Exercises 1–3 find the initial tableau and the corresponding basic feasible solution.

1. Maximize $z = 4x + 3y$
subject to
$$\begin{aligned} 2x + 3y &\leq 12 \\ -3x + 2y &\leq 6 \\ x \geq 0, \quad y &\geq 0 \end{aligned}$$

2. Maximize $z = -3x_1 + 2x_2 + 3x_3$
subject to
$$\begin{aligned} 2x_1 + x_2 + x_3 &\leq 12 \\ 3x_1 - x_2 + 3x_3 &\leq 8 \\ x_1 \geq 0, \quad x_2 \geq 0, \quad x_3 &\geq 0 \end{aligned}$$

3. Maximize $z = 8x + 6y$
subject to
$$\begin{aligned} x + y &\leq 4 \\ x + 3y &\leq 6 \\ -x + y &\leq 1 \\ x \geq 0, \quad y &\geq 0 \end{aligned}$$

4. (a) Find the equations represented by the tableau

$$\begin{array}{c} \\ z \\ u \\ y \end{array} \begin{array}{c} \begin{array}{cccccc} x & y & u & v & z & \end{array} \\ \begin{bmatrix} 12 & 0 & 0 & 6 & 1 & 100 \\ -3 & 0 & 1 & 7 & 0 & 2 \\ 5 & 1 & 0 & -5 & 0 & 7 \end{bmatrix} \end{array}$$

(b) Find the basic feasible solution corresponding to this tableau.

(c) Find the value of z at the basic feasible solution in part (b).

5. Repeat the directions of Exercise 4 for the tableau

$$\begin{array}{c} \\ z \\ x_1 \\ x_2 \\ v \end{array} \begin{array}{c} \begin{array}{cccccccc} x_1 & x_2 & x_3 & u & v & w & z & \end{array} \\ \begin{bmatrix} 0 & 0 & 5 & 7 & 0 & 6 & 1 & 40 \\ 1 & 0 & -2 & 1 & 0 & 2 & 0 & 6 \\ 0 & 1 & 4 & 1 & 0 & -1 & 0 & 14 \\ 0 & 0 & 3 & 0 & 1 & 3 & 0 & 12 \end{bmatrix} \end{array}$$

6. Explain why none of the following tableaux correspond to optimal solutions.

(a)
$$\begin{array}{c} \begin{array}{cccccc} x & y & u & v & z & \end{array} \\ \begin{bmatrix} 4 & 0 & -6 & 0 & 1 & 9 \\ 7 & 1 & 2 & 0 & 0 & 12 \\ 1 & 0 & -3 & 1 & 0 & 4 \end{bmatrix} \end{array}$$

(b)
$$\begin{array}{c} \begin{array}{cccccc} x & y & u & v & z & \end{array} \\ \begin{bmatrix} 0 & 0 & 12 & 6 & 1 & 9 \\ 1 & 0 & -5 & 3 & 0 & -2 \\ 0 & 1 & 2 & 0 & 0 & 1 \end{bmatrix} \end{array}$$

(c)
$$\begin{array}{c} \begin{array}{ccccccc} x_1 & x_2 & v_1 & v_2 & v_3 & z & \end{array} \\ \begin{bmatrix} 7 & 0 & 0 & 0 & -3 & 1 & 100 \\ 2 & 0 & 0 & 1 & 6 & 0 & 4 \\ 4 & 1 & 0 & 0 & 7 & 0 & 3 \\ 5 & 0 & 1 & 0 & 2 & 0 & 9 \end{bmatrix} \end{array}$$

7. Which of the tableaux in Exercise 6 correspond to basic feasible solutions?

For the tableaux in Exercises 8–10 find the entering variable, the departing variable, and the pivot entry.

8.
$$\begin{array}{c} \\ z \\ u \\ y \end{array} \begin{array}{c} \begin{array}{ccccccc} x & y & t & u & v & z & \end{array} \\ \begin{bmatrix} -8 & 0 & 3 & 0 & 0 & 1 & 140 \\ 2 & 0 & 6 & 1 & 2 & 0 & 100 \\ 3 & 1 & 5 & 0 & 3 & 0 & 180 \end{bmatrix} \end{array}$$

9.
$$\begin{array}{c} \\ z \\ x \\ w \end{array} \begin{array}{c} \begin{array}{ccccccc} x & y & t & v & w & z & \end{array} \\ \begin{bmatrix} 0 & -3 & -2 & 4 & 0 & 1 & 40 \\ 1 & 4 & 1 & 3 & 0 & 0 & 10 \\ 0 & -3 & 5 & 7 & 1 & 0 & 5 \end{bmatrix} \end{array}$$

10.
$$\begin{array}{c} \\ z \\ v_1 \\ x_2 \\ x_3 \\ v_2 \end{array} \begin{array}{c} \begin{array}{cccccccccc} x_1 & x_2 & x_3 & v_1 & v_2 & v_3 & v_4 & z & \end{array} \\ \begin{bmatrix} -8 & 0 & 0 & 0 & 0 & 5 & -4 & 1 & 120 \\ 4 & 0 & 0 & 1 & 0 & 3 & 0 & 0 & 200 \\ -3 & 1 & 0 & 0 & 0 & 2 & 1 & 0 & 300 \\ 0 & 0 & 1 & 0 & 0 & -1 & 2 & 0 & 150 \\ 6 & 0 & 0 & 0 & 1 & 2 & 3 & 0 & 180 \end{bmatrix} \end{array}$$

11. Use pivotal elimination on the tableau in Exercise 8 to find the new tableau.

12. Use pivotal elimination on the tableau in Exercise 9 to find the new tableau.

13. Use pivotal elimination on the tableau in Exercise 10 to find the new tableau.

In Exercises 14–18 solve by the simplex method.

14. Maximize $z = 3x_1 - 4x_2 + x_3$
subject to
$$\begin{aligned} 2x_1 + 3x_2 - 2x_3 &\leq 8 \\ 2x_1 - 3x_2 + x_3 &\leq 12 \\ x_1 \geq 0, \quad x_2 \geq 0, \quad x_3 &\geq 0 \end{aligned}$$

15. Maximize $z = 2x + 3y - t$
subject to
$$\begin{aligned} x + 3y + 4t &\leq 12 \\ x + y + 2t &\leq 18 \\ 2x - y - t &\leq 16 \\ x \geq 0, \quad y \geq 0, \quad t &\geq 0 \end{aligned}$$

16. Maximize $z = 3x_1 - 5x_2 + 2x_3 + x_4$
subject to

$$\begin{aligned} 2x_1 + 2x_2 - x_3 \quad &\leq 14 \\ x_1 - 2x_2 + x_3 + x_4 &\leq 16 \\ 2x_1 + 4x_2 \quad + 5x_4 &\leq 25 \\ x_1 \geq 0, \quad x_2 \geq 0, \quad x_3 \geq 0, \quad & x_4 \geq 0 \end{aligned}$$

17. Maximize $z = 2x_1 - 4x_2 - x_3$
subject to

$$\begin{aligned} 5x_1 - 4x_2 - 2x_3 &\leq 6 \\ 4x_1 + 3x_2 + x_3 &\leq 5 \\ 3x_1 + 2x_2 - x_3 &\leq 4 \\ x_1 \geq 0, \quad x_2 \geq 0, \quad & x_3 \geq 0 \end{aligned}$$

18. Maximize $z = 2x - 3y$
subject to

$$\begin{aligned} 2x + 6y &\leq 9 \\ 3x - 2y &\leq 8 \\ 5x + y &\leq 5 \\ x \geq 0, \quad & y \geq 0 \end{aligned}$$

19. Use the simplex method to show that the following problem has no solution.
Maximize $z = 2x + 5y$
subject to

$$\begin{aligned} -4x + y &\leq 5 \\ -x + y &\leq 4 \\ x \geq 0, \quad & y \geq 0 \end{aligned}$$

20. An investment banker wants to invest \$18,000 or less in three types of bonds: type A bond yielding a 5% profit on the amount invested, type B bond yielding a 7% profit on the amount invested, and type C bond yielding a 10% profit on the amount invested. He wants to invest no more than \$10,000 in bonds of types A and B together, no more than \$8000 in bonds of types B and C together, and no more than \$6000 in bonds of type C. How much should he invest in each kind of bond to maximize the profit?

21. An ice cream manufacturer blends ingredients A, B, and C in varying proportions to make three types of ice cream: low calorie (L), regular (R), and extrarich (E). Each gallon of L contains 0.2 gallon of A, 0.3 gallon of B, and 0.2 gallon of C. Each gallon of R contains 0.4 gallon of A, 0.2 gallon of B, and 0.2 gallon of C; each gallon of E contains 0.5 gallon of A, 0.2 gallon of B, and 0.1 gallon of C. Suppose that the profits on the ice creams L, R, and E are 20, 40, and 20 cents per gallon, respectively. If the manufacturer has 50, 80, and 100 gallons of A, B, and C available, how many gallons of each type of ice cream should be made to maximize the profit?

22. A company has three methods for obtaining high-grade aluminum from recycled cans. With method A, each ton of cans requires 4 hours of smelting, 2 hours of centrifugal separation, and 1 hour of chemical purification; it yields 5 pounds of pure aluminum per ton of cans. With method B, each ton of cans requires 2 hours of smelting, 2 hours of centifugal separation, and 3 hours of chemical purification; it yields 3 pounds of pure aluminum per ton of cans. With method C, each ton of cans requires 4 hours of smelting, 3 hours of centrifugal separation, and 2 hours of chemical purification; it yields 4 pounds of pure aluminum per ton of cans. Assuming that the smelter is available for 80 hours per week, the centrifugal separator for 50 hours per week, and the chemical purifier for 40 hours per week, how many tons of cans should be processed by each method to maximize the weekly output of pure aluminum?

4.4 NONSTANDARD LINEAR PROGRAMMING PROBLEMS; DUALITY (OPTIONAL)

Recall that a standard linear programming problem is one that satisfies the following conditions.

(i) The objective function is to be maximized.

(ii) The variables are all constrained to be nonnegative.

(iii) In the remaining constraints, the expressions involving the variables are less than or equal to ($\leq$) a nonnegative constant.

In this section we discuss methods for solving certain nonstandard problems. Consider the problem

Minimize $z = -10x - 12y$
subject to

$$\begin{aligned} x + 2y &\leq 150 \\ x + y &\leq 100 \\ x \geq 0, \quad & y \geq 0 \end{aligned}$$

This problem is not a standard linear programming problem because it violates condition (i) above. Before we can show how to solve such a problem, we have to develop some preliminary results.

Let S be any set of real numbers and suppose S contains a *smallest* number a. If we form the set $\bar{S}$ consisting of the negatives of the numbers in S, then it is evident that $-a$ is the *largest* number in $\bar{S}$. For example, the smallest number in the set

$$S = \{3, 4, 5, 6\}$$

is 3, whereas the largest number in

$$\bar{S} = \{-3, -4, -5, -6\}$$

is -3. In summary,

$$\text{minimum value in } S = -(\text{maximum value in } \bar{S})$$

Applying this result to the set of possible z values in problem (1) above, we obtain

$$\text{minimum value of } z = -(\text{maximum value of } z') \tag{2}$$

where z' is the negative of z.

The following example shows how this result can be used to solve problem (1).

EXAMPLE 1 Solve the nonstandard linear programming problem

$$\begin{aligned} &\text{Minimize} \quad z = -10x - 12y \\ &\text{subject to} \\ &\qquad x + 2y \le 150 \\ &\qquad x + y \le 100 \\ &\qquad x \ge 0, \quad y \ge 0 \end{aligned}$$

Solution From (2) above, the minimum value of z is the negative of the maximum value of

$$z' = -z = 10x + 12y$$

We thus consider the problem

$$\begin{aligned} &\text{Maximize} \quad z' = 10y + 12y \\ &\text{subject to} \\ &\qquad x + 2y \le 150 \\ &\qquad x + y \le 100 \\ &\qquad x \ge 0, \quad y \ge 0 \end{aligned}$$

This problem is solved in Example 5 of Section 4.3, where we show that the maximum value $z' = 1100$ occurs when $x = 50$, $y = 50$. Thus, the minimum value of z is $z = -1100$, and this occurs when $x = 50$, $y = 50$.

Other Nonstandard Problems

We now turn to some other kinds of nonstandard linear programming problems. Consider the problem

$$\begin{aligned} \text{Minimize} \quad & z = 3x_1 + 4x_2 \\ \text{subject to} \quad & x_1 + 2x_2 \geq 8 \\ & 2x_1 + 2x_2 \geq 10 \\ & x_1 + 4x_2 \geq 12 \\ & x_1 \geq 0, \quad x_2 \geq 0 \end{aligned} \tag{3}$$

This is not a standard linear programming problem for two reasons. First, the objective function is to be minimized, which violates condition (i) stated at the beginning of this section. Second, the inequalities in the constraints

$$x_1 + 2x_2 \geq 8 \qquad 2x_1 + 2x_2 \geq 10 \qquad x_1 + 4x_2 \geq 12$$

point the wrong way, violating condition (iii). Before explaining how to solve problems of this type, we must develop some preliminary ideas:

In matrix notation, problem (3) has the form

$$\begin{aligned} \text{Minimize} \quad & z = CX \\ \text{subject to} \quad & AX \geq B \\ & X \geq 0 \end{aligned}$$

where

$$C = [3 \quad 4] \qquad X = \begin{bmatrix} x_1 \\ x_2 \end{bmatrix} \qquad A = \begin{bmatrix} 1 & 2 \\ 2 & 2 \\ 1 & 4 \end{bmatrix} \qquad B = \begin{bmatrix} 8 \\ 10 \\ 12 \end{bmatrix}$$

In this problem we have two unknowns x_1 and x_2 in X and three constraints in $AX \geq B$. More generally, we want to solve

$$\begin{aligned} \text{Minimize} \quad & z = CX \\ \text{subject to} \quad & AX \geq B \\ & X \geq 0 \end{aligned} \tag{4}$$

when there are n unknowns $x_1, x_2, \ldots, x_n$ in X and k constraints in $AX \geq B$. To solve problems like (4), we begin by considering the following maximization problem involving k new unknowns $y_1, y_2, \ldots, y_k$.

$$\begin{aligned} \text{Maximize} \quad & z' = B^tY \\ \text{subject to} \quad & A^tY \leq C^t \\ & Y \geq 0 \end{aligned} \tag{5}$$

This problem is called the **dual problem**, and the original problem is called the **primal problem**. The matrices A^t, B^t, and C^t in the dual problem are the transposes of the matrices A, B, and C in the primal problem. The matrix Y is the $k \times 1$ matrix of new unknowns $y_1, y_2, \ldots, y_k$, and z' is the quantity to be maximized (we cannot use z here since the letter z is already used in the primal problem).

EXAMPLE 2 Find the dual problem for (3).

Solution As previously shown, the matrices A, B, and C in problem (3) are

$$A = \begin{bmatrix} 1 & 2 \\ 2 & 2 \\ 1 & 4 \end{bmatrix} \qquad B = \begin{bmatrix} 8 \\ 10 \\ 12 \end{bmatrix} \qquad C = \begin{bmatrix} 3 & 4 \end{bmatrix}$$

Thus,

$$A^t = \begin{bmatrix} 1 & 2 & 1 \\ 2 & 2 & 4 \end{bmatrix} \qquad B^t = \begin{bmatrix} 8 & 10 & 12 \end{bmatrix} \qquad C^t = \begin{bmatrix} 3 \\ 4 \end{bmatrix}$$

Since the primal problem had $k = 3$ constraints in $AX \geq B$, the dual problem will have three unknowns, y_1, y_2, and y_3. Thus,

$$Y = \begin{bmatrix} y_1 \\ y_2 \\ y_3 \end{bmatrix}$$

Therefore, the dual problem is

$$\text{Maximize} \quad z' = \begin{bmatrix} 8 & 10 & 12 \end{bmatrix} \begin{bmatrix} y_1 \\ y_2 \\ y_3 \end{bmatrix}$$

subject to

$$\begin{bmatrix} 1 & 2 & 1 \\ 2 & 2 & 4 \end{bmatrix} \begin{bmatrix} y_1 \\ y_2 \\ y_3 \end{bmatrix} \leq \begin{bmatrix} 3 \\ 4 \end{bmatrix}$$

$$\begin{bmatrix} y_1 \\ y_2 \\ y_3 \end{bmatrix} \geq \begin{bmatrix} 0 \\ 0 \\ 0 \end{bmatrix}$$

or without matrix notation,

$$\begin{aligned} \text{Maximize} \quad & z' = 8y_1 + 10y_2 + 12y_3 \\ \text{subject to} \quad & y_1 + 2y_2 + y_3 \leq 3 \\ & 2y_1 + 2y_2 + 4y_3 \leq 4 \\ & y_1 \geq 0, \quad y_2 \geq 0, \quad y_3 \geq 0 \end{aligned} \tag{6}$$

Observe that this is a standard linear programming problem.

The dual problem is important because of the following result, which we state without proof.

> **The Duality Principle** If either the dual problem or the primal problem has a solution, then so does the other. Moreover, the minimum value of z and the maximum value of z' are equal.

The following example illustrates how this result can be used.

EXAMPLE 3 Find the minimum value of z in (3).

Solution The minimum value of z is the same as the maximum value of z' in the dual problem. We previously showed the dual problem to be (6). Since this is a standard linear programming problem, we can use the simplex method. The corresponding problem with slack variables is

$$\begin{aligned} &\text{Maximize} \quad z' = 8y_1 + 10y_2 + 12y_3 \\ &\text{subject to} \quad \begin{aligned} y_1 + 2y_2 + y_3 + v_1 \quad &= 3 \\ 2y_1 + 2y_2 + 4y_3 \quad + v_2 &= 4 \end{aligned} \\ &y_1 \geq 0, \quad y_2 \geq 0, \quad y_3 \geq 0, \quad v_1 \geq 0, \quad v_2 \geq 0 \end{aligned}$$

You can check that the initial tableau and pivot entry are as follows:

Entering variable (y_3)

$$\begin{array}{r|ccccccc} & y_1 & y_2 & y_3 & v_1 & v_2 & z' & \\ z' & -8 & -10 & -12 & 0 & 0 & 1 & 0 \\ v_1 & 1 & 2 & 1 & 1 & 0 & 0 & 3 \\ \text{Departing variable} \rightarrow v_2 & 2 & 2 & \boxed{4} & 0 & 1 & 0 & 4 \end{array}$$

After pivotal elimination we obtain

Entering variable (y_2)

$$\begin{array}{r|ccccccc} & y_1 & y_2 & y_3 & v_1 & v_2 & z' & \\ z' & -2 & -4 & 0 & 0 & 3 & 1 & 12 \\ \text{Departing variable} \rightarrow v_1 & \frac{1}{2} & \boxed{\frac{3}{2}} & 0 & 1 & -\frac{1}{4} & 0 & 2 \\ y_3 & \frac{1}{2} & \frac{1}{2} & 1 & 0 & \frac{1}{4} & 0 & 1 \end{array}$$

After pivotal elimination again, we obtain

Entering variable (y_1)

$$\begin{array}{r|ccccccc} & y_1 & y_2 & y_3 & v_1 & v_2 & z' & \\ z' & -\frac{2}{3} & 0 & 0 & \frac{8}{3} & \frac{7}{3} & 1 & \frac{52}{3} \\ y_2 & \frac{1}{3} & 1 & 0 & \frac{2}{3} & -\frac{1}{6} & 0 & \frac{4}{3} \\ \text{Departing variable} \rightarrow y_3 & \boxed{\frac{1}{3}} & 0 & 1 & -\frac{1}{3} & \frac{1}{3} & 0 & \frac{1}{3} \end{array}$$

After one more pivotal elimination, we obtain the final tableau

$$\begin{array}{r|ccccccc} & y_1 & y_2 & y_3 & v_1 & v_2 & z' & \\ z' & 0 & 0 & 2 & 2 & 3 & 1 & 18 \\ y_2 & 0 & 1 & -1 & 1 & -\frac{1}{2} & 0 & 1 \\ y_1 & 1 & 0 & 3 & -1 & 1 & 0 & 1 \end{array}$$

From this tableau we see that the maximum value of z' is 18. Therefore the minimum value of z is also 18.

The technique illustrated in Example 3 gave us the minimum value of z; it did not, however, tell us the values of x_1 and x_2 where this minimum occurs. For reasons too technical to pursue here, it can be shown that these values appear in the z' row of the final tableau for the dual problem; the value of x_1 occurs under the slack variable v_1, and the value of x_2 occurs under the slack variable v_2. Thus, the minimum value $z = 18$ in problem (3) occurs when $x_1 = 2$ and $x_2 = 3$. To summarize, we state the following result:

> The optimal solution to the primal problem occurs under the slack variables in the z' row of the final tableau for the dual problem. The value of x_1 occurs under v_1, the value of x_2 occurs under v_2, the value of x_3 occurs under v_3, and so on.

EXERCISE SET 4.4

1. Solve by the method of Example 1:

$$\begin{aligned} \text{Minimize} \quad & z = -4x + 2y \\ \text{subject to} \quad & 6x + 2y \leq 18 \\ & 3x - 2y \leq 6 \\ & x \geq 0, \quad y \geq 0 \end{aligned}$$

2. Solve by the method of Example 1:

$$\begin{aligned} \text{Minimize} \quad & z = 2x_1 - 3x_2 + x_3 \\ \text{subject to} \quad & x_1 + x_2 + 2x_3 \leq 5 \\ & x_1 - x_2 + 3x_3 \leq 4 \\ & 2x_1 + x_2 \leq 6 \\ & x_1 \geq 0, \quad x_2 \geq 0, \quad x_3 \geq 0 \end{aligned}$$

3. Solve by the method of Example 1:

$$\begin{aligned} \text{Minimize} \quad & z = 2x - 2y + t \\ \text{subject to} \quad & 3x + y + t \geq -6 \\ & x - 3y - t \geq -8 \\ & x \geq 0, \quad y \geq 0, \quad t \geq 0 \end{aligned}$$

4. Find the dual problem for

$$\begin{aligned} \text{Minimize} \quad & z = 3x_1 + 4x_2 + x_3 \\ \text{subject to} \quad & 2x_1 + 3x_2 + x_3 \geq 8 \\ & 5x_1 + 2x_2 + 2x_3 \geq 5 \\ & x_1 \geq 0, \quad x_2 \geq 0, \quad x_3 \geq 0 \end{aligned}$$

5. Solve the problem in Exercise 4.

6. Find the dual problem for

$$\begin{aligned} \text{Minimize} \quad & z = 3x_1 + 5x_2 \\ \text{subject to} \quad & 3x_1 + 2x_2 \geq 6 \\ & 4x_1 + x_2 \geq 4 \\ & 14x_1 + 6x_2 \geq 21 \\ & x_1 \geq 0, \quad x_2 \geq 0 \end{aligned}$$

7. Solve the problem in Exercise 6.

8. Find the dual problem for

$$\begin{aligned} \text{Minimize} \quad & z = 6x_1 + 5x_2 \\ \text{subject to} \quad & 2x_1 + 3x_2 \geq 6 \\ & 5x_1 + 2x_2 \geq 10 \\ & x_1 \geq 0, \quad x_2 \geq 0 \end{aligned}$$

9. Solve the problem in Exercise 8.

10. Solve the problem in Exercise 3 of Exercise Set 3.1. (Use the dual problem.)

11. Solve the problem in Exercise 5 of Exercise Set 3.1. (Use the dual problem.)

KEY IDEAS FOR REVIEW

- **Standard linear programming problem** See page 125.
- **Slack variables** New variables added to a linear programming problem to convert inequality constraints to equalities.
- **Basic feasible solution to a standard linear programming problem with m unknowns** See page 131.
- **Basic variable or explicit variable** A variable that occurs in exactly one equation and there it occurs with coefficient $+1$.
- **Entering variable** See page 141.
- **Departing variable** See page 142.
- **Primal and dual problems** See page 156.

SUPPLEMENTARY EXERCISES

1. Rewrite the following as standard linear programming problems, if possible.

(a)
$$\begin{array}{ll} \text{Maximize} & z = 3x + 5y \\ \text{subject to} & \\ & 2x - y \le 4 \\ & 3x - 2y \ge -2 \\ & x \ge 0, \quad y \ge 0 \end{array}$$

(b)
$$\begin{array}{ll} \text{Minimize} & z = 4x_1 - 3x_2 + x_3 \\ \text{subject to} & \\ & -5x_1 + 3x_2 - x_3 \ge -2 \\ & x_1 - 2x_2 \le 4 \\ & x_1 \ge 0, \quad x_2 \ge 0, \quad x_3 \ge 0 \end{array}$$

2. Write the following linear programming problem in matrix notation.

$$\begin{array}{ll} \text{Maximize} & z = 4x_1 - x_2 + x_3 \\ \text{subject to} & \\ & x_1 + 2x_2 - x_3 \le 4 \\ & -4x_1 + x_2 + 2x_3 \le 5 \\ & 3x_1 - x_3 \le 8 \\ & x_1 \ge 0, \quad x_2 \ge 0, \quad x_3 \ge 0 \end{array}$$

3. Solve the following linear programming problem geometrically and match the corner points with the feasible solutions.

$$\begin{array}{ll} \text{Maximize} & z = 3x + 4y \\ \text{subject to} & \\ & 3x - y \le 3 \\ & 2x + y \le 7 \\ & x \ge 0, \quad y \ge 0 \end{array}$$

4. Consider the linear programming problem

$$\begin{array}{ll} \text{Maximize} & z = -5x_1 - x_2 + 2x_3 \\ \text{subject to} & \\ & 3x_1 - x_2 + x_3 \le 12 \\ & 2x_1 + x_2 - 3x_3 \le 15 \\ & x_1 \ge 0, \quad x_2 \ge 0, \quad x_3 \ge 0 \end{array}$$

(a) Form the new problem with slack variables v_1 and v_2.

(b) Find a tableau in which x_1, x_2, and x_3 are implicit.

(c) Find a tableau in which x_1, x_2, and v_1 are implicit.

(d) Find a tableau in which x_2, v_1, and v_2 are implicit.

5. Consider the tableau

$$\begin{array}{cccccccc} x_1 & x_2 & x_3 & u & v & w & z & \\ \end{array}$$
$$\begin{bmatrix} 3 & 0 & -3 & 0 & 1 & 1 & 1 & 12 \\ 0 & 1 & 2 & 0 & 0 & -1 & 0 & 40 \\ 2 & 0 & 4 & 1 & 0 & 2 & 0 & 20 \end{bmatrix}$$

(a) Name the explicit variables.

(b) Find the corresponding basic feasible solution and the value of the objective function at that solution.

6. Set up the initial tableau

$$\begin{array}{ll} \text{Maximize} & z = 2x + 5y \\ \text{subject to} & \\ & 3x + 5y \le 8 \\ & 2x + 7y \le 12 \\ & x \ge 0, \quad y \ge 0 \end{array}$$

7. Consider the tableau

$$\begin{array}{c} \\ z \\ v \\ x_1 \\ x_3 \end{array}\begin{array}{c} \begin{array}{cccccccc} x_1 & x_2 & x_3 & u & v & w & z & \end{array} \\ \left[\begin{array}{cccccccc} 0 & 5 & 0 & 4 & 0 & 6 & 1 & 120 \\ 0 & 3 & 0 & -4 & 1 & 10 & 0 & 15 \\ 1 & 5 & 0 & 2 & 0 & 5 & 0 & 8 \\ 0 & 16 & 1 & 1 & 0 & -2 & 0 & 4 \end{array}\right] \end{array}$$

Find the basic feasible solution corresponding to this tableau, and the value of z at this basic feasible solution.

8. Consider the tableau

$$\begin{array}{c} \\ z \\ v_1 \\ x_1 \\ v_3 \\ x_2 \end{array}\begin{array}{c} \begin{array}{ccccccccc} x_1 & x_2 & x_3 & v_1 & v_2 & v_3 & v_4 & z & \end{array} \\ \left[\begin{array}{ccccccccc} 0 & 0 & 4 & 0 & 10 & 0 & 8 & 1 & 100 \\ 0 & 0 & 2 & 1 & \frac{5}{2} & 0 & 0 & 0 & \frac{6}{7} \\ 1 & 0 & 5 & 0 & -3 & 0 & -2 & 0 & \frac{2}{7} \\ 0 & 0 & 3 & 0 & 4 & 1 & -4 & 0 & \frac{5}{7} \\ 0 & 1 & 0 & 0 & \frac{3}{2} & 0 & 0 & 0 & \frac{1}{7} \end{array}\right] \end{array}$$

Find the departing variable if the entering variable is
(a) v_2 (b) x_3

9. Use pivotal elimination on the following tableau to find the new tableau.

$$\begin{array}{c} \\ z \\ t \\ x \\ w \end{array}\begin{array}{c} \begin{array}{cccccccc} x & y & t & u & v & w & z & \end{array} \\ \left[\begin{array}{cccccccc} 0 & 4 & 0 & -2 & -3 & 0 & 1 & 100 \\ 0 & 2 & 1 & 6 & 10 & 0 & 0 & 20 \\ 1 & -3 & 0 & -5 & 4 & 0 & 0 & 10 \\ 0 & 5 & 0 & 2 & 3 & 1 & 0 & 5 \end{array}\right] \end{array}$$

10. Solve by the simplex method.

Maximize $z = x_1 + 2x_2 + x_3 + x_4$
subject to

$$\begin{aligned} 2x_1 + x_2 + 3x_3 + x_4 &\le 8 \\ 2x_1 + 3x_2 \qquad + 4x_4 &\le 12 \\ 3x_1 + x_2 + 2x_3 \qquad &\le 18 \\ x_1 \ge 0, \quad x_2 \ge 0, \quad x_3 \ge 0, \quad x_4 &\ge 0 \end{aligned}$$

11. Solve the linear programming problem.

Maximize $z = 4x_1 + 2x_2 + 3x_3$
subject to

$$\begin{aligned} 2x_1 + 3x_2 + x_3 &\le 12 \\ x_1 + 4x_2 + 2x_3 &\le 10 \\ 3x_1 + x_2 + x_3 &\le 10 \\ x_1 \ge 0, \quad x_2 \ge 0, \quad x_3 &\ge 0 \end{aligned}$$

12. Solve the linear programming problem.

Minimize $z = 40x_1 + 24x_2 + 8x_3 + 10x_4$
subject to

$$\begin{aligned} 5x_1 + 4x_2 + x_3 + 5x_4 &\ge 20 \\ 4x_1 + 3x_2 + 2x_3 + x_4 &\ge 60 \\ x_1 \ge 0, \quad x_2 \ge 0, \quad x_3 \ge 0, \quad x_4 &\ge 0 \end{aligned}$$

13. Find the dual problem of the linear programming problem.

Minimize $z = 2x_1 + 3x_2 + x_3$
subject to

$$\begin{aligned} 2x_1 + x_2 + x_3 &\ge 8 \\ 3x_1 - x_2 + 4x_3 &\ge 15 \\ x_1 \ge 0, \quad x_2 \ge 0, \quad x_3 &\ge 0 \end{aligned}$$

14. Solve

Minimize $z = 2x_1 + 3x_2 + 8x_3$
subject to

$$\begin{aligned} x_1 + 2x_2 + 7x_3 &\ge 8 \\ x_1 + 3x_2 + 6x_3 &\ge 9 \\ 2x_1 + 4x_2 + 2x_3 &\ge 4 \\ x_1 \ge 0, \quad x_2 \ge 0, \quad x_3 &\ge 0 \end{aligned}$$

15. The R.H. Lawn Products Company has available 80 metric tons of nitrate and 50 metric tons of phosphate to use during the coming week in producing their three types of fertilizer. The mixture ratios and profit figures are given in Table 1. Determine how the current inventory should be used to maximize profit.

TABLE 1

	Nitrate	Phosphate	Profit ($/1000 bags)
Regular Lawn (metric tons/1000 bags)	4	2	300
Super Lawn (metric tons/1000 bags)	4	3	500
Garden (metric tons/1000 bags)	2	2	400

16. A health food store packages a nut sampler consisting of walnuts, pecans, and almonds. Suppose each ounce of walnuts contains 12 units of protein and 3 units of iron and costs 12 cents, each ounce of pecans contains 1 unit of protein and 3 units of iron and costs 9 cents, and each ounce of almonds contains 2 units of protein and 1 unit of iron and costs 6 cents. If each package of the nut sampler is to contain at least 24 units of protein and at least 18 units of iron, how many ounces of each type of nut should be used to minimize the cost of the sampler?

CHAPTER TEST

1. Solve the following linear programming problem geometrically and match the corner points with the basic feasible solutions.

$$\text{Maximize} \quad z = 2x - 4y$$
$$\text{subject to} \quad \begin{aligned} 2x + y &\le 8 \\ 3x - 2y &\le 5 \\ x \ge 0, \quad y &\ge 0 \end{aligned}$$

2. Solve by the simplex method.

$$\text{Maximize} \quad z = 8x_1 + 9x_2 + 5x_3$$
$$\text{subject to} \quad \begin{aligned} x_1 + x_2 + 2x_3 &\le 2 \\ 2x_1 + 3x_2 + 4x_3 &\le 3 \\ 6x_1 + 6x_2 + 2x_3 &\le 8 \\ x_1 \ge 0, \quad x_2 \ge 0, \quad x_3 &\ge 0 \end{aligned}$$

3. Determine the dual problem of the linear programming problem.

$$\text{Minimize} \quad z = 3x_1 + x_2 + 4x_3$$
$$\text{subject to} \quad \begin{aligned} 2x_1 + x_2 - x_3 &\ge 20 \\ 4x_1 - x_2 + 2x_3 &\ge 15 \\ x_1 + x_2 + 3x_3 &\ge 12 \\ x_1 \ge 0, \quad x_2 \ge 0, \quad x_3 &\ge 0 \end{aligned}$$

4. Solve.

$$\text{Minimize} \quad z = 2x_1 + x_2 + 2x_3$$
$$\text{subject to} \quad \begin{aligned} 3x_1 + x_2 + x_3 &\ge 10 \\ x_1 + 3x_2 - x_3 &\ge 20 \\ x_1 \ge 0, \quad x_2 \ge 0, \quad x_3 &\ge 0 \end{aligned}$$

5. **(Agriculture problem)** A farmer owns a farm that produces corn, soybeans, and oats. There are 12 acres of land available for cultivation. Each crop planted has certain requirements for labor and capital. These data along with the net profit figures are given in Table 2. The farmer has available \$360 for capital and knows that there are 48 hours available for working these crops. How much of each crop should the farmer plant to maximize profit?

TABLE 2

	Labor (hours)	Capital (\$)	Net Profit (\$)
Corn (per acre)	6	36	40
Soybeans (per acre)	6	24	30
Oats (per acre)	2	18	20

6. A lawn service blends two of its products, CARPET and GROW, to produce a new product. Each pound of CARPET costs \$3.00 and contains 6 units of fertilizer, 4 units of crabgrass control, and 3 units of grub control. Each pound of GROW costs \$2.50 and contains 5 units of fertilizer, 6 units of crabgrass control, and 2 units of grub control. It has been determined that the minimum requirements per application are 40 units of fertilizer, 30 units of crabgrass control, and 40 units of grub control. How many pounds of each product should be blended per application to minimize cost?

5

Set Theory

A herd of buffalo, a bunch of bananas, the collection of all positive even integers, and the collection of all stocks listed on the New York Stock Exchange have something in common: they are all examples of objects that have been grouped together and viewed as a single object. This idea of grouping objects together gives rise to the mathematical notion of a set, which we examine in this chapter. This material is used in later chapters to help solve a variety of important problems.

5.1 INTRODUCTION TO SETS

A **set** is a collection of objects; the objects are called the **elements** or **members** of the set.

One way to describe a set is to list the elements of the set between braces. Thus, the set of all positive integers that are less than 4 can be written

$$\{1, 2, 3\}$$

and the set of all U.S. Presidents whose last names begin with the letter T can be written

{Taft, Tyler, Taylor, Truman}

When set notation is used, an element in a set is listed only once. Thus, the set of letters in the word *bookkeeper* is {b, o, k, e, p, r}.

We denote sets by uppercase letters such as A, B, C, . . . and members of a set by lowercase letters such as a, b, c, ... With this notation, an arbitrary set with five members might be written $A = \{a, b, c, d, e\}$. To indicate that an element a is a member of the set A, we write $a \in A$, which is read "a is an element of A" or "a belongs to A." To indicate that the element a is *not* a member of the set A, we write $a \notin A$, which is read "a is not an element of A" or "a does not belong to A."

EXAMPLE 1 Let

$$A = \{1, 2, 3\}$$

Then $1 \in A$, $2 \in A$, $3 \in A$, but $4 \notin A$.

At times it is inconvenient or impossible to list the elements of a set. In such cases, the set can often be described by specifying a property that the elements of the set have in common. A convenient way of doing this is to use what is sometimes called **set-builder** notation. To illustrate, the set A consisting of all the positive integers that are less than 10 can be written in set-builder notation as

$$A = \{x | x \text{ is a positive integer less than } 10\}$$

which is read "A equals the set of all x such that x is a positive integer less than 10."

In this notation x denotes a typical element of the set, the vertical bar $|$ is read "such that," and following the bar are the conditions that x must satisfy to be a member of the set A.

The following examples further illustrate this notation.

EXAMPLE 2 The set of all IBM stockholders can be written in set-builder notation as

$$\{x | x \text{ is an IBM stockholder}\}$$

which is read "the set of all x such that x is an IBM stockholder." Note that it would be very inconvenient to list all the members of this set.

EXAMPLE 3 We read $\{x | x$ is a letter in the word *stock*$\}$ as "the set of all x such that x is a letter in the word *stock*." This set can also be described by listing its elements as $\{s, t, o, c, k\}$.

Two sets A and B are said to be **equal** if they have the same elements, in which case we write

$$A = B$$

EXAMPLE 4 If A is the set

$$\{1, 2, 3, 4\}$$

and B is the set

$$\{x | x \text{ is a positive integer and } x^2 \text{ is less than } 25\}$$

then $A = B$.

EXAMPLE 5 Consider the sets of stocks

$$A = \{\text{IBM, Du Pont, General Electric}\}$$

and

$$B = \{\text{Du Pont, General Electric, IBM}\}$$

Even though the members of these sets are listed in different orders, the sets A and B are equal since they have the same members.

RUSSELL'S PARADOX* AND THE BARBER'S DILEMMA

The British philosopher and logician Bertrand Russell proposed the following paradox: Let R be the set of all sets that are not members of themselves (note that a set is a subset of itself, but it is not a member of itself). Is R a member of itself?

1. If you answer yes, then you are including R in the realm of sets that are not members of themselves, so R is not a member of itself—which contradicts your answer.
2. If you answer no, then you are excluding R from the realm of sets that are not members of themselves, so R must be a member of itself—which contradicts your answer.

In either case a contradiction occurs. To avoid this paradox, it is necessary to place restrictions on the way sets are formed. Fortunately, we are not concerned with such matters here.

Closely related to the Russell paradox is the following mind-teaser: In a certain village there is a barber who shaves only the set of villagers who do not shave themselves. Does the barber shave himself?

* *Bertrand Russell* (1872–1970) was one of the outstanding figures of twentieth-century British philosophy, publishing numerous works on logic and the theory of knowledge. One of his most important works is the three-volume treatise *Principia Mathematica*, which he co-authored with Alfred North Whitehead. In this work Russell and Whitehead show how mathematics is derivable from symbolic logic and therefore an extension of logic. Russell was imprisoned for his pacifist activities in 1918 and again in 1961 for civil disobedience in the cause of nuclear disarmament. He received the Order of Merit in 1949 and the Nobel Prize for Literature in 1950.

EXAMPLE 6 In a study of the effectiveness of antipollution devices attached to the exhaust systems of 11 buses, the following percentage decreases in carbon monoxide emissions were observed.

$$12,\ 15,\ 11,\ 12,\ 14,\ 15,\ 11,\ 16,\ 15,\ 17,\ 12$$

The set of observations is

$$\{11,\ 12,\ 14,\ 15,\ 16,\ 17\}$$

There are sets that have no members. For example, the set

$$\{x \mid x \text{ is an integer and } x^2 = -1\}$$

has no members since no integer has a square that is negative.

A set with no members is called the **empty set** or sometimes the **null set**. Such a set is denoted by the symbol $\varnothing$.

Note that $\{0\}$ and $\{\varnothing\}$ are not empty sets; the first contains the element 0, the second contains the set $\varnothing$.

Consider the sets

$$B = \{a, e, i, o, u\} \quad \text{and} \quad A = \{a, o, u\}$$

Every member of the set A is also a member of the set B. This suggests the following definition.

If every member of the set A is also a member of set B, then we say **A is a subset of B** and write

$$A \subset B$$

In addition, it can be shown that the empty set $\varnothing$ is a subset of *every* set.

EXAMPLE 7 Let

$$B = \{a, b, c\}$$

All possible subsets of B are

$$\varnothing, \{a\}, \{b\}, \{c\}, \{a, b\}, \{a, c\}, \{b, c\}, \{a, b, c\}$$

EXAMPLE 8 If

$$A = \{x | x \text{ is a positive even integer}\}$$

and

$$B = \{x | x \text{ is an integer}\}$$

then $A \subset B$.

EXAMPLE 9 If S is the set of all stocks listed with the New York Stock Exchange on July 26, 1990, and T is the set of all stocks on the New York Stock Exchange that traded over 100 shares on July 26, 1990, then $T \subset S$.

EXAMPLE 10 If A is any set, then $A \subset A$.

If set A is not a subset of set B, we write

$$A \not\subset B$$

EXAMPLE 11 Let A be the set of points inside the left circle in Figure 1 and B be the set of points inside the right circle. Then

$$A \not\subset B \quad \text{and} \quad B \not\subset A$$

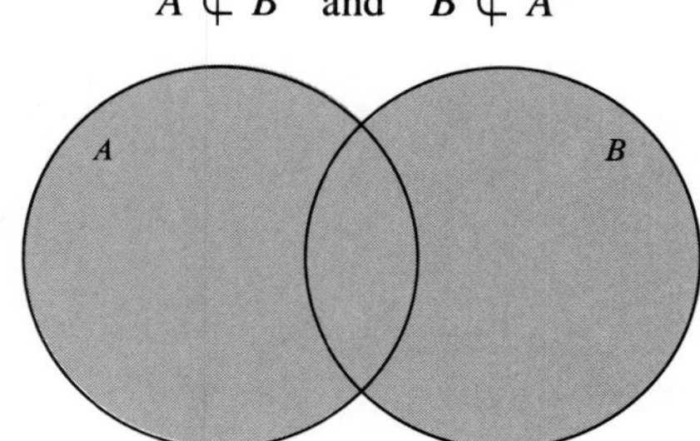

FIGURE 1

EXAMPLE 12 Let A be the set of points inside the larger circle in Figure 2 and let B be the set of points inside the smaller circle. Then

$$B \subset A \quad \text{and} \quad A \not\subset B$$

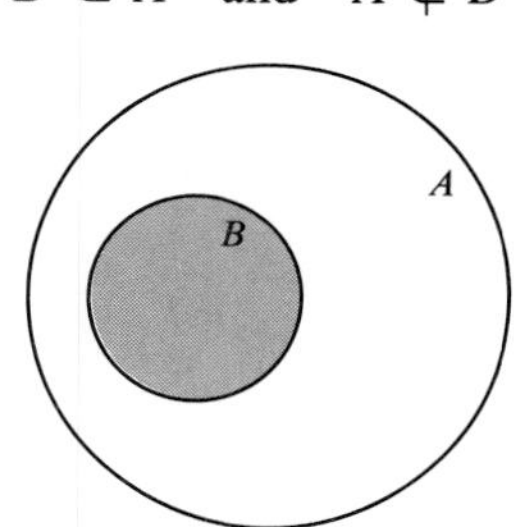

FIGURE 2

EXERCISE SET 5.1

1. Let $A = \{a, b, c, 1, 2, 5, e, f\}$. Answer the following as true or false.
   (a) $4 \in A$ (b) $4 \notin A$ (c) $a \in A$
   (d) $a \notin A$ (e) $\varnothing \notin A$ (f) $A \in \{a\}$

2. Let $A = \{x | x \text{ is an integer greater than } 5\}$. Answer the following as true or false.
   (a) $2 \in A$ (b) $5.5 \in A$ (c) $7 \notin A$
   (d) $5 \in A$ (e) $3 \notin A$ (f) $3 \in A$

3. Consider the set of water pollutants: A = {sulfur, crude oil, phosphates, mercury}. Answer the following as true or false.
   (a) sulfur $\in A$ (b) phosphate $\notin A$
   (c) arsenic $\notin A$ (d) crude oil $\in A$

4. Let $A = \{x | x \text{ is an integer and } x^2 = 9\}$. List the elements of A.

5. In each part form a set from the letters in the given words.
   (a) *aardvark* (b) *mississippi* (c) *table*

6. Write $A = \{1, 2, 3, 4, 5\}$ in set-builder notation.

7. Write the following in set-builder notation.
   (a) the set of U.S. citizens
   (b) the set of U.S. citizens over 40 years of age

8. Let $A = \{1, 2, 3, 4\}$. Which of the following sets are equal to A?
   (a) $\{3, 2, 1, 4\}$ (b) $\{1, 2, 3\}$
   (c) $\{1, 2, 3, 4, 0\}$
   (d) $\{x | x \text{ is a positive integer and } x^2 \text{ is less than } 16\}$
   (e) $\{x | x \text{ is a positive integer and } x^2 \text{ equals } 4\}$
   (f) $\{x | x \text{ is a positive integer and } x \text{ is less than } 4\}$

9. Consider the set of psychological disorders: $A = $ {schizophrenia, paranoia, depression, megalomania}. Which of the following sets are equal to A?
(a) {schizophrenia, paranoia, depression}
(b) {schizophrenia, paranoia, megalomania, depression}

10. Which of the following sets are empty?
(a) $\{x|x$ is an integer and $x^2 = 4\}$
(b) $\{x|x$ is an integer and $x^2 = -4\}$
(c) $\{x|x$ is an integer satisfying $x^2 + 1 = 0\}$
(d) $\{0\}$ (e) $\{\varnothing\}$

11. List all subsets of the set $\{2, 5\}$.

12. List all subsets of the set {Roosevelt, Truman, Kennedy}.

13. List all subsets of
(a) $\{a_1, a_2, a_3\}$ (b) $\varnothing$

14. Let $A = \{3, 5, 7, 9\}$. Answer the following as true or false.
(a) $\{5, 3\} \subset A$ (b) $A \subset A$
(c) $\{3, 5, 9, 7\} \not\subset A$ (d) $A \subset \{3, 5\}$
(e) $A \subset \{3, 5, 7, 9, 0\}$ (f) $\varnothing \subset A$
(g) $\{1\} \not\subset A$

15. Let $A = \{x|x$ is an integer and x^2 is less than $25\}$. Answer the following as true or false.
(a) $\{x|x$ is a positive integer and x^2 is less than $16\} \subset A$
(b) $\{2, -2, 6\} \subset A$
(c) $A \subset \{1, 2, 3, 4, 5\}$
(d) $\{x|x$ is an integer and x is less than $5\} \subset A$

16. In Figure 3 let $S =$ the set of points inside the square, $T =$ the set of points inside the triangle, and $C =$ the set of points inside the circle, and let x and y be the indicated points. Answer the following as true or false.
(a) $C \subset T$ (b) $C \subset S$
(c) $T \not\subset C$ (d) $x \notin C$
(e) $x \in T$ (f) $y \in C$ and $y \in T$*
(g) $x \in C$ or $x \in T$ (or x is in both)

FIGURE 3
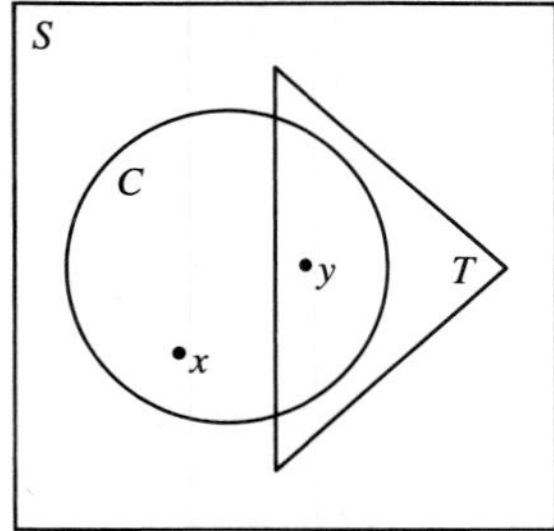

* The expression $y \in C$ and $y \in T$ means that y must be an element of both C and T.

17. In Figure 4 let $R =$ the set of points inside the rectangle, $C =$ the set of points inside the circle, and $E =$ the set of points inside the ellipse. Copy the figure and indicate points u, v, w, x, y, and z satisfying the given property.
(a) $u \in E$ (b) $v \in R$
(c) $w \in C$ and $w \notin E$ (d) $x \in E$ and $x \notin R$
(e) $y \in E$ and $y \in R$
(f) $z \in R$ and $z \in C$, but $z \notin E$

FIGURE 4
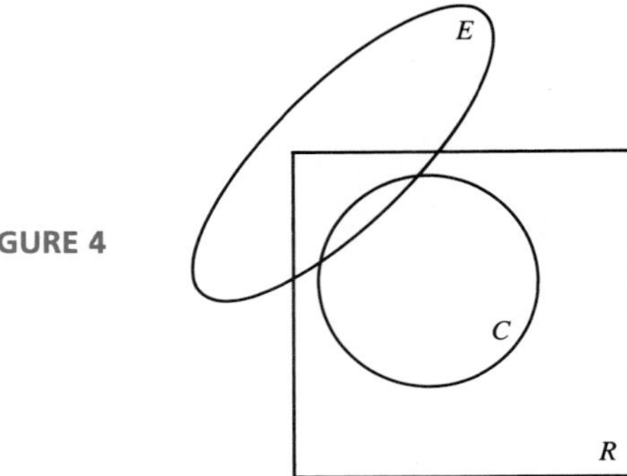

18. In Figure 5 let $C =$ the set of points inside the circle, $R =$ the set of points inside the rectangle, and $T =$ the set of points inside the triangle. In each part copy the figure, and shade sets D, E, F, G, H, and I satisfying the given properties.
(a) $D \subset C$ (b) $E \subset R$ and $E \subset T$
(c) $F \subset C$ or $E \subset R$ (d) $G \subset T$ and $G \not\subset R$
(e) $H \subset C$ and $H \subset T$, but $H \not\subset R$
(f) $I \subset C$, $I \subset R$, and $I \subset T$

FIGURE 5
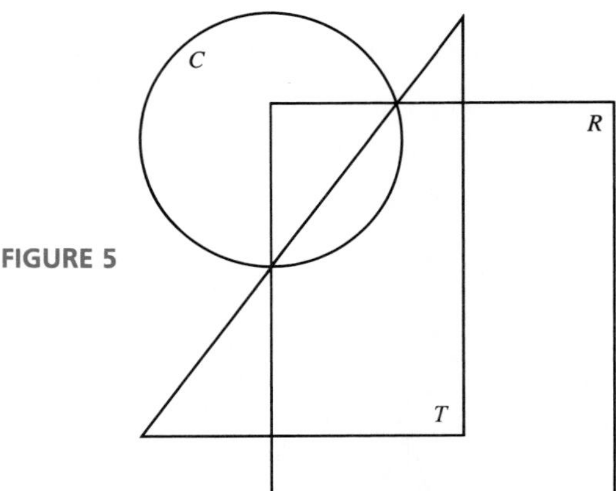

19. In each part find the "smallest" possible set that contains the given sets as subsets.
(a) $\{1, 3, 7\}$, $\{3, 5, 9, 2\}$, $\{1, 2, 3, 4, 6\}$, $\{3\}$
(b) $\{a, b, c\}$, $\varnothing$

20. In each part find a set that contains the given sets as subsets.
(a) {IBM, Du Pont, Xerox}, {Polaroid, Honeywell, Xerox, IBM, Avco}
(b) $\{1, 3, 5\}$, $\{a, b, 3\}$, $\{a\}$, $\{a, b\}$

21. Is it true that $\varnothing \in \varnothing$? Is it true that $\varnothing \subset \varnothing$?

22. Is the set of letters in the word *latter* the same as the set of letters in the word *later*?

23. How many different subsets can be formed from a set of n objects? [*Hint*: Solve the problem for $n = 1, 2, 3$, and 4, and then guess at the general result. Don't forget that the empty set is a subset of every set.]

24. Let $S = \{1, 2, 3, 4, 5, 6\}$.
(a) How many different subsets of S contain the number 6?
(b) How many different subsets of S contain at least one even integer?
[*Hint*: Use the results of Exercise 23.]

25. Let $S = \{1, 2, 3, 4, 5, 6\}$. List all the subsets of S containing the set $\{1, 2, 3\}$ as a subset.

5.2 INTERSECTION AND UNION OF SETS

We all know that the arithmetic operations of addition, subtraction, multiplication, and division can be used to solve a variety of problems. Analogously, we can introduce operations on sets to solve many important problems. In this section we discuss two such operations and in later sections illustrate their applications.

If A and B are sets, then the set of all elements that belong to both A and B is called the **intersection** of A and B. It is denoted by

$$A \cap B$$

EXAMPLE 1 If, as in Figure 1, A is the set of points inside the circle on the left and B is the set of points inside the circle on the right, then $A \cap B$ is the set of points in the shaded region of the figure.

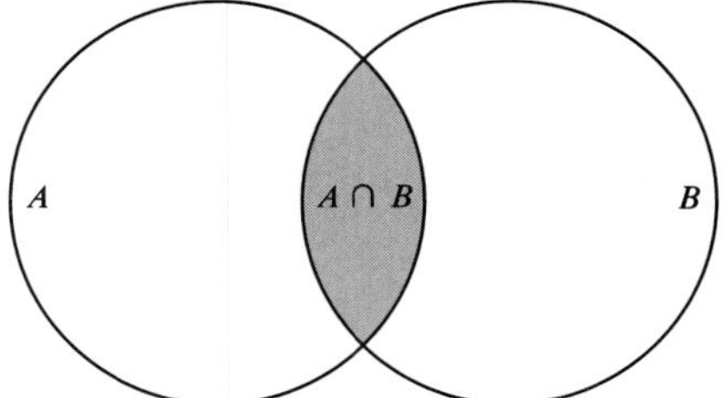

FIGURE 1

The idea of using shading and circles to illustrate relationships between sets is attributed to the British logician John Venn.* In his honor, these figures are called **Venn diagrams**. In the exercises, we have indicated some other useful set relationships that can be verified using Venn diagrams.

**John Venn* (1834–1923) was the son of a minister. He graduated from Gonville and Caius College in Cambridge, England, in 1853, after which he pursued theological interests as a curate in the parishes of London. As a result of his contact with intellectual agnostics and the works of Augustus DeMorgan, George Boole, and John Stuart Mill, Venn became interested in logic. In addition to his work in logic, he made important contributions to the mathematics of probability. He was an accomplished linguist, a botanist, and a noted mountaineer. Also see Appendix B on logic.

EXAMPLE 2 Let

$$A = \{a, b, c, d, e\} \qquad B = \{b, d, e, g\} \qquad C = \{a, h\}$$

Find $A \cap B$, $A \cap C$, and $B \cap C$.

Solution The only elements that belong to both A and B are b, d, and e. Therefore,

$$A \cap B = \{b, d, e\}$$

Similarly,

$$A \cap C = \{a\}$$

Since the sets B and C have no elements in common, their intersection is the empty set; that is,

$$B \cap C = \varnothing$$

The sets B and C in Example 2 motivate the following definition.

Two sets that have no elements in common are said to be **disjoint**.

Note that disjoint sets have an empty intersection. In other words, A and B are disjoint if $A \cap B = \varnothing$.

EXAMPLE 3 In Figure 2 let A, B, C, and D be the sets of points inside the indicated circles. Since A and B overlap,

$$A \cap B \neq \varnothing$$

that is, A and B are not disjoint. On the other hand, C and D are disjoint since

$$C \cap D = \varnothing$$

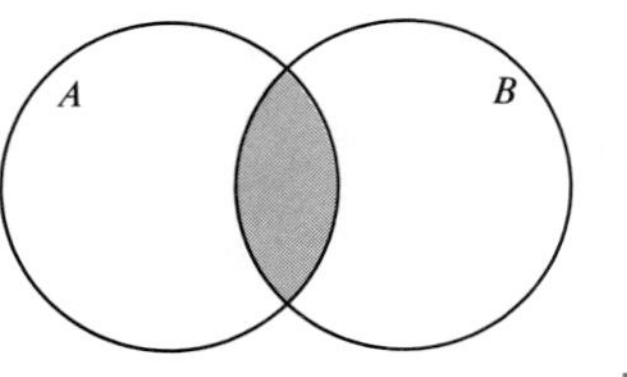

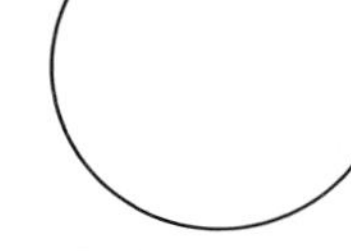

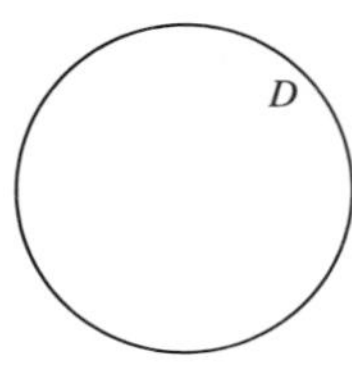

FIGURE 2

We can define intersections of more than two sets as follows:

The **intersection** of any collection of sets is the set of elements that belong to every set in the collection.

EXAMPLE 4 If A, B, and C are the sets of points inside the circles indicated in Figure 3, then the intersection of these sets, denoted by

$$A \cap B \cap C$$

is the shaded region in the figure.

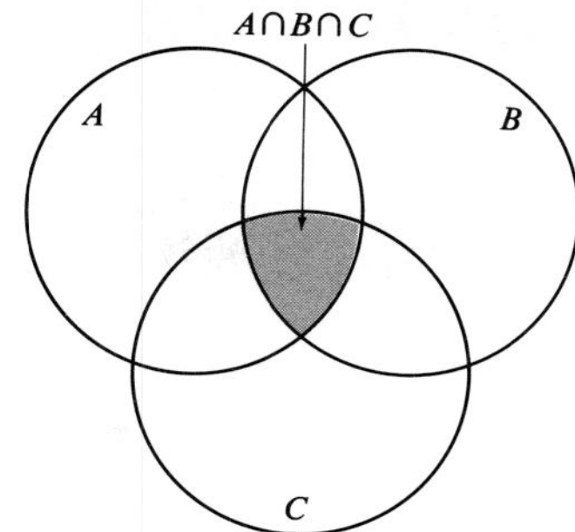

FIGURE 3

EXAMPLE 5 Let

$$A = \{1, 2, 4, 7, 9, 11\} \qquad B = \{2, 7, 9, 11, 17, 19\}$$
$$C = \{0, 2, 5, 7, 19, 24\} \qquad D = \{2, 7, 9\}$$

Then

$$A \cap B \cap C \cap D = \{2, 7\}$$

EXAMPLE 6 Let S be the set of stocks on the New York Stock Exchange that have paid a dividend for each of the past 40 years, and let R be the set of railroad stocks listed on the New York Stock Exchange. Describe the set $S \cap R$.

Solution The members of $S \cap R$ belong to both S and R, so $S \cap R$ consists of all railroad stocks on the New York Stock Exchange that have paid a dividend for each of the past 40 years.

> If A and B are sets, then the set of all elements that belong to A or B is called the **union** of A and B. It is denoted by
>
> $$A \cup B$$

The word *or** is used in the inclusive sense, so the statement "p or q" is true when at least one of those statements is true. In other words, "p or q" is true in the following three cases:

1. p is true and q is not.

2. q is true and p is not.

3. Both p and q are true.

It also follows that $x \in A \cup B$ if x belongs to at least one of the sets.

---

*The word *or* has two meanings in the English language: the inclusive sense (defined above) and the exclusive sense, such that "p or q" is true only in cases 1 and 2 (below). In the English language the meaning of *or* is determined from context. For example, the "or" in the sentence "would you like tea or coffee?" is generally interpreted as the "exclusive or." However, in mathematics the word *or* is always used in the inclusive sense, and if you intend the exclusive case, the statement must be worded as "p or q, but not both." See also Appendix B.

EXAMPLE 7 Let

$$A = \{a, b, c, d, e\} \quad \text{and} \quad B = \{b, d, e, g\}$$

Find $A \cup B$.

Solution The elements that belong to A or to B are a, b, c, d, e, and g. Therefore,

$$A \cup B = \{a, b, c, d, e, g\}$$

EXAMPLE 8 If A and B are the sets of points inside the indicated circles in Figure 4a, then $A \cup B$ is the set of points in the shaded region shown in Figure 4b.

FIGURE 4

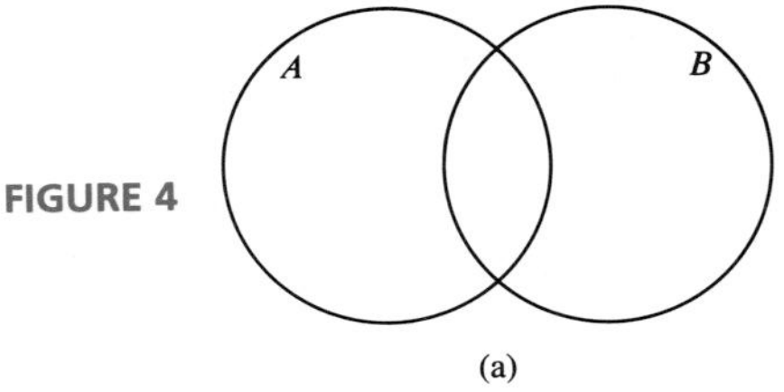

(a)

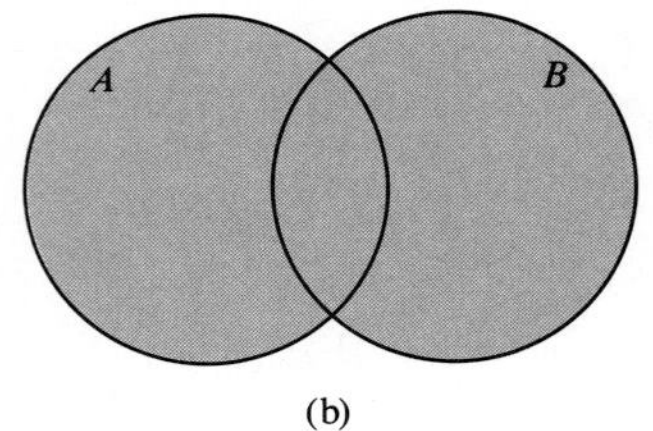

(b)

EXAMPLE 9 Let A be any set. Find $A \cup \varnothing$.

Solution The members of $A \cup \varnothing$ are those elements that lie in A or in $\varnothing$. Since $\varnothing$ has no elements, we obtain

$$A \cup \varnothing = A$$

We can define unions of more than two sets as follows:

> Given any collection of sets, their **union** is the set of elements that belong to one or more of the sets in the collection.

EXAMPLE 10 If A, B, and C are the sets of points inside the circles indicated in Figure 5a, then the union of these sets, denoted by

$$A \cup B \cup C$$

is the shaded region in Figure 5b.

FIGURE 5

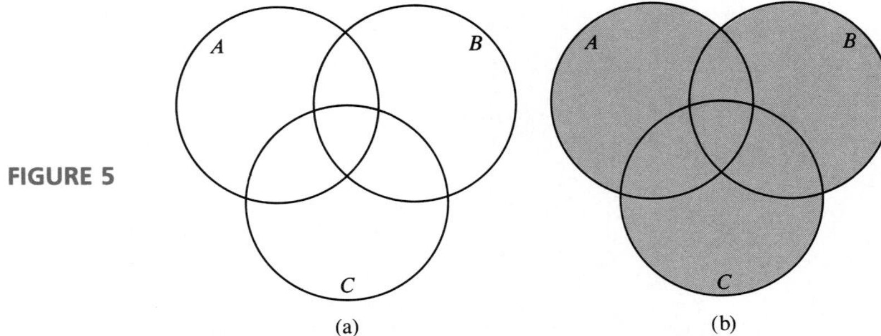

(a) (b)

EXAMPLE 11 Let

$$A = \{1, 2, 4, 7, 9, 11\} \qquad C = \{0, 11\}$$
$$B = \{2, 7, 9, 11, 17\} \qquad D = \{1, 4\}$$

Find the following sets:
(a) $A \cup B$ (b) $B \cup C$ (c) $A \cup D$ (d) $A \cup B \cup C \cup D$

Solution
(a) $A \cup B = \{1, 2, 4, 7, 9, 11, 17\}$ (b) $B \cup C = \{0, 2, 7, 9, 11, 17\}$
(c) $A \cup D = \{1, 2, 4, 7, 9, 11\}$
(d) $A \cup B \cup C \cup D = \{0, 1, 2, 4, 7, 9, 11, 17\}$

EXAMPLE 12 Let A, B, and C be the points inside the circles indicated in Figure 5a. Shade the sets

(a) $A \cup B$ (b) $C \cap (A \cup B)$ (c) $(C \cap A) \cup (C \cap B)$

Solution (a) See Figure 6.

Solution (b) To find $C \cap (A \cup B)$ we intersect C with the shaded set $A \cup B$ in Figure 6. As in arithmetic, the expression in parentheses is computed first. This yields the shaded set in Figure 7.

FIGURE 6

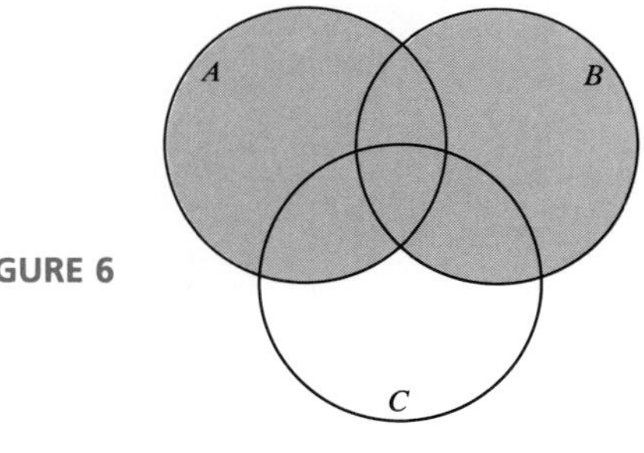

$A \cup B$

FIGURE 7

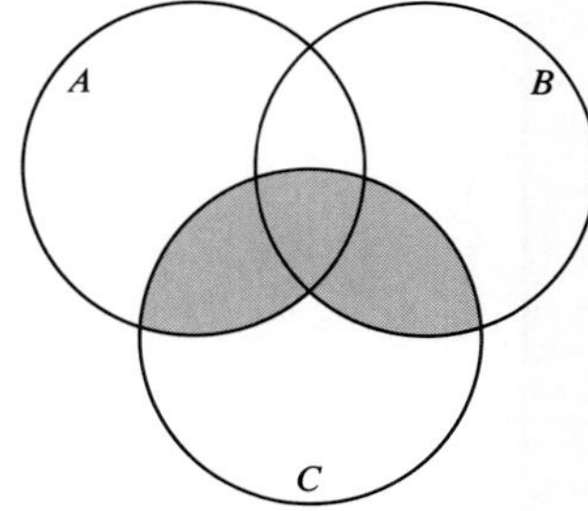

$C \cap (A \cup B)$

Solution (c) We begin by shading the sets $C \cap A$ and $C \cap B$; this gives the diagrams in Figure 8. To find $(C \cap A) \cup (C \cap B)$, we form the union of the shaded sets $C \cap A$ and $C \cap B$; this yields the shaded set in Figure 9.

FIGURE 8

FIGURE 9

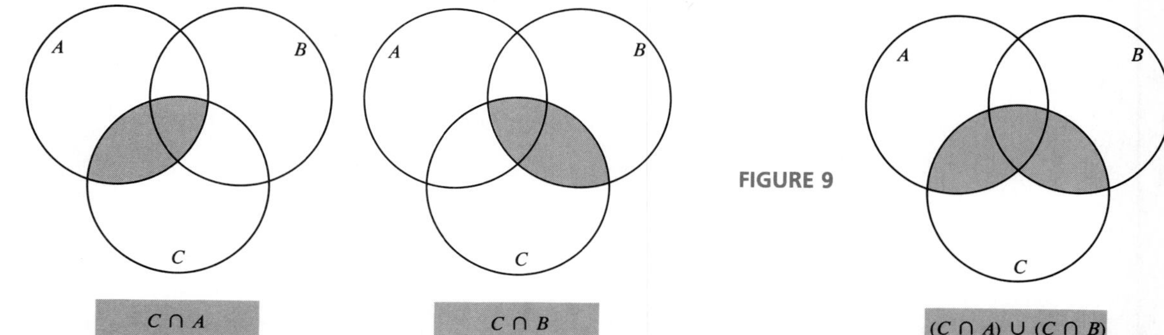

$C \cap A$ $C \cap B$ $(C \cap A) \cup (C \cap B)$

Observe that the Venn diagrams of the sets $C \cap (A \cup B)$ and $(C \cap A) \cup (C \cap B)$ obtained in parts (b) and (c) are identical. We have thus illustrated the following:

$$C \cap (A \cup B) = (C \cap A) \cup (C \cap B) \qquad \textbf{first distributive law for sets}$$

We leave it as an exercise to establish

$$C \cup (A \cap B) = (C \cup A) \cap (C \cup B) \qquad \textbf{second distributive law for sets}$$

Remark When performing operations on sets, pay careful attention to parentheses. In Figure 10 the shaded region is the set $(C \cap A) \cup B$, which is different from the set $C \cap (A \cup B)$, shown in Figure 7. In general, operations inside parentheses should be performed first.

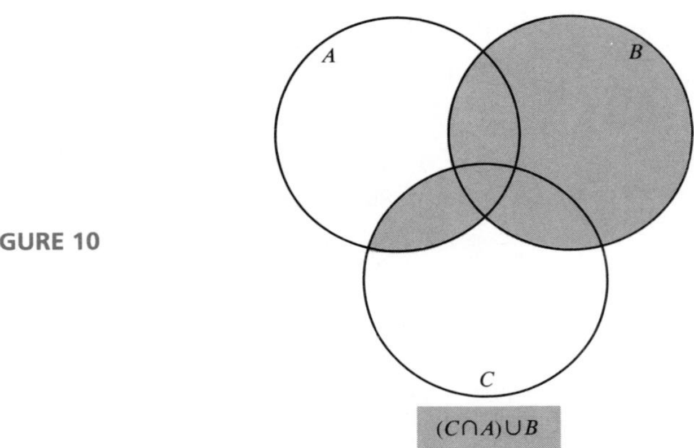

FIGURE 10

In translating a verbal problem to mathematical form, usually *and* means intersection while *or* means union.

EXAMPLE 13 In an experiment with hybrid corn, the corn plants were classified into sets as follows:

Q = quick-growing $\qquad R$ = rust-resistant
W = all white kernels $\qquad Y$ = all yellow kernels

Describe the characteristics of the plants in the following sets:

(a) $Q \cap Y$ (b) $R \cup W$
(c) $(R \cap Q) \cup (R \cap Y)$ (d) $Q \cup (W \cap Y)$

Represent the following statements by set notation:

(e) The plants that are quick-growing and rust-resistant
(f) The plants that are quick-growing, white-kerneled, and rust-resistant
(g) The plants that are quick-growing or yellow-kerneled

Solution (a) The plants in $Q \cap Y$ are in both Q and Y. Thus $Q \cap Y$ consists of quick-growing, yellow-kerneled plants.

Solution (b) The plants in $R \cup W$ are either in R or W. Thus $R \cup W$ consists of plants that are either rust-resistant or white-kerneled.

Solution (c) The plants in $R \cap Q$ are rust-resistant and quick-growing. The plants in $R \cap Y$ are rust-resistant and yellow-kerneled. Thus $(R \cap Q) \cup (R \cap Y)$ consists of plants that are either rust-resistant and quick-growing or rust-resistant and yellow-kerneled.

Solution (d) The set $W \cap Y$ is empty since the kernels cannot be both all white and all yellow. Therefore,

$$Q \cup (W \cap Y) = Q \cup \varnothing = Q$$

Thus $Q \cup (W \cap Y)$ is the set of quick-growing plants only.

Solution (e) Since these plants have both characteristics, they are in the set $Q \cap R$.

Solution (f) Since these plants have all three characteristics, they are in the set $Q \cap W \cap R$.

Solution (g) Since these plants have at least one of the characteristics, they are in the set $Q \cup Y$.

EXERCISE SET 5.2

1. Let $A = \{1, 3, 5, 7\}$, $B = \{2, 3, 4, 7\}$, $C = \{x|x$ is an integer such that x^2 is less than $9\}$, and $D = \{5, 6\}$. Compute
(a) $A \cap B$ (b) $A \cap C$ (c) $B \cap C$
(d) $B \cap D$ (e) $A \cap B \cap C$ (f) $D \cap \varnothing$

2. Let A, B, C, and D be the sets in Exercise 1. Compute
(a) $A \cup B$ (b) $A \cup C$ (c) $B \cup C$
(d) $A \cup B \cup D$ (e) $A \cup \varnothing$ (f) $D \cup D$

3. In Figure 11 let S = the set of points inside the square, T = the set of points inside the triangle, and C = the set of points inside the circle. Shade the following sets.
(a) $S \cap T$ (b) $S \cap C$
(c) $T \cap C$ (d) $S \cap T \cap C$

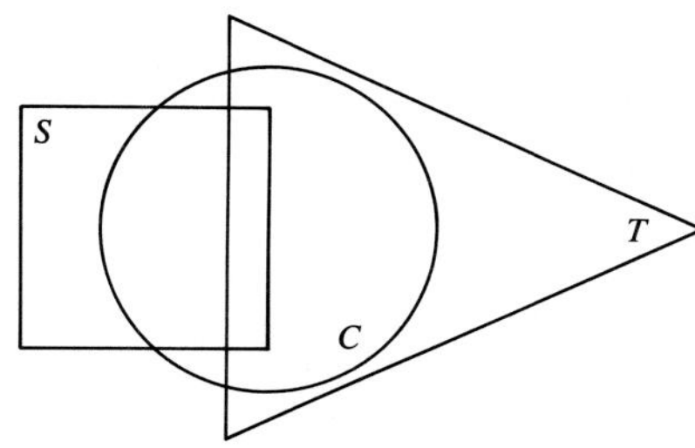

FIGURE 11

4. Let S, T, and C be the sets in Exercise 3. Shade the following sets.
(a) $S \cup T$
(b) $S \cup C$
(c) $T \cup C$
(d) $S \cup T \cup C$

5. Let S, T, and C be the sets in Exercise 3. Shade the following sets.
(a) $C \cup (S \cap T)$
(b) $C \cap (S \cup T)$
(c) $(C \cap S) \cup T$
(d) $(C \cup S) \cap (C \cap T)$

6. In each part determine whether the given sets are disjoint.
(a) $\{a, b, d\}$, $\{e, f, g\}$
(b) $\{1, 2, 3\}$, $\{3, 7, 9\}$
(c) $\varnothing$, $\{1, 2\}$
(d) {book, candle, bell}, {page, fire, ring}

For Exercises 7–9 refer to Figure 12 and let S = the set of points inside the square, T = the set of points inside the triangle, and C = the set of points inside the circle, and let v, w, x, y, z be the indicated points.

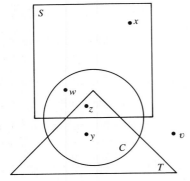

FIGURE 12

7. Answer the following as true or false.
(a) $x \notin S \cap C$ (b) $y \in C \cap T$
(c) $z \in S \cap T \cap C$ (d) $w \in S \cap C \cap T$
(e) $v \notin S \cap C$

8. Answer the following as true or false.
(a) $x \in C \cup T$ (b) $x \in C \cup S$
(c) $z \in C \cup S \cup T$ (d) $y \in C \cup C$
(e) $v \in C \cup S \cup T$

9. Answer the following as true or false.
(a) $w \in C \cup (S \cap T)$
(b) $w \in C \cap (S \cup T)$
(c) $x \in (C \cup T) \cap S$
(d) $x \in (C \cup S) \cap (C \cup T)$

10. Let $A = \{1, 2, 3, 4, 5\}$, $B = \{3, 4, 7\}$, and $C = \{1, 3, 7, 8\}$. Verify the distributive laws:
(a) $C \cap (A \cup B) = (C \cap A) \cup (C \cap B)$
(b) $C \cup (A \cap B) = (C \cup A) \cap (C \cup B)$

11. Use Venn diagrams to establish the second distributive law: $C \cup (A \cap B) = (C \cup A) \cap (C \cup B)$.

12. Let A = {ATT, IBM, GE},
B = {Du Pont, GM, GE, Kodak},
C = {Ford, Sun Oil, IBM, GE, Du Pont}.
Compute
(a) $(A \cap B) \cup C$ (b) $A \cup (B \cap C)$
(c) $A \cup B \cup C$ (d) $A \cap B \cap C$
(e) $A \cap (B \cup C)$ (f) $(A \cup B) \cap C$

13. Explain why the following are true.
(a) $A \cup B = B \cup A$ (b) $A \cap B = B \cap A$

14. Use Venn diagrams to establish the following.
(a) $(A \cap B) \cap C = A \cap (B \cap C)$
(b) $(A \cup B) \cup C = A \cup (B \cup C)$

15. Use Venn diagrams to establish the following.
(a) $A \subset (A \cup B)$ and $B \subset (A \cup B)$
(b) $(A \cap B) \subset A$ and $(A \cap B) \subset B$

16. Use Venn diagrams to establish the following.
(a) If $A \subset C$ and $B \subset C$, then $(A \cup B) \subset C$
(b) If $C \subset A$ and $C \subset B$, then $C \subset (A \cap B)$

17. Use Venn diagrams to establish the following.
(a) If $C \subset A$ and $C \subset B$, then $C \subset (A \cup B)$
(b) If $A \subset C$ and $B \subset C$, then $(A \cap B) \subset C$

18. (a) If $A \cup B = A \cup C$, must $B = C$? Justify your answer.
(b) If $A \cap B = A \cap C$, must $B = C$? Justify your answer.

19. Describe the shaded regions in Figure 13 using the unions and intersections of sets C, R, and T.

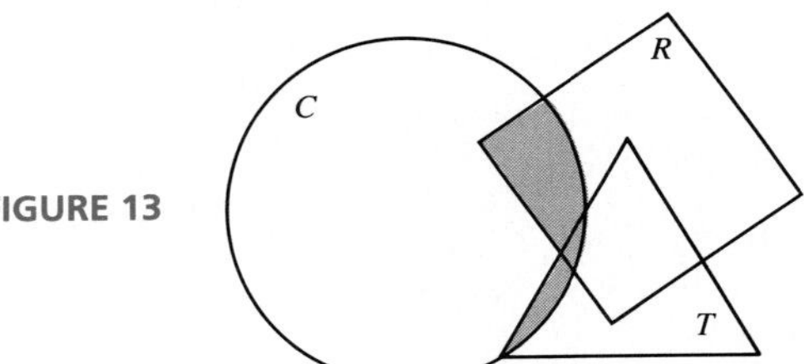

FIGURE 13

20. Describe the shaded region in Figure 14 using the unions and intersections of sets C, D, and S.

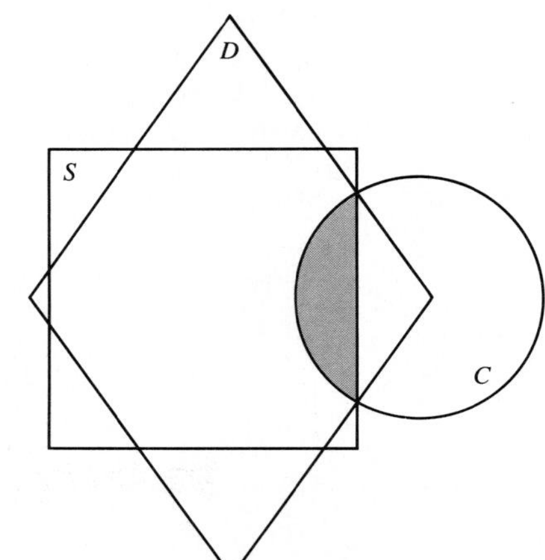

FIGURE 14

21. The personnel department of a major company classifies its employees into the following categories:

M = the set of all male employees
F = the set of all female employees
A = the set of all administrative employees
T = the set of all technical employees
S = the set of all employees working for the company at least 5 years.

Describe the members of the following sets.

(a) $M \cap A$ (b) $M \cup F$
(c) $A \cap T \cap F$ (d) $A \cup T \cup F$
(e) $M \cap A \cap S$

22. Let M, F, A, and S be the sets in Exercise 21. Let x designate a male employee who has worked for the company at least 5 years. Answer the following as true or false.

(a) $x \in M \cup F \cup A$ (b) $x \in M \cap S$
(c) $x \in F \cap S$ (d) $x \in (M \cap S) \cup F$

23. An automobile insurance company classifies its policyholders into the following categories:

A = the set of all policyholders who drive cars with engines over 200 horsepower
B = the set of all policyholders who drive cars with engines over 250 horsepower
C = the set of all policyholders who are over 25 years of age
D = the set of all policyholders who are over 20 years of age
M = the set of all male policyholders
F = the set of all female policyholders

Describe the policyholders in the following sets.

(a) $A \cap B$ (b) $A \cup B$
(c) $A \cap C \cap M$ (d) $A \cup D \cup F$
(e) $B \cap (D \cup F)$

24. Let A, B, C, D, M, and F be the sets in Exercise 23. Write the following sets using unions and intersections of A, B, C, D, M, and F.

(a) the set of all female policyholders who are over 25 years of age
(b) the set of all policyholders who are either male or drive cars with engines over 200 horsepower
(c) the set of all female policyholders over 20 years of age who drive cars with engines over 250 horsepower
(d) the set of all male policyholders who are either over 25 years of age or drive cars with engines over 200 horsepower

25. Human blood contains three possible antigens, denoted by A, B, and Rh. Depending on which antigens are present, there are eight possible blood types, denoted by A^-, A^+, B^-, B^+, AB^-, AB^+, O^+, and O^-. The antigens present in each blood type can be described by the Venn diagram in Figure 15, which shows the eight possible blood types resulting from the various combinations of antigens. In this diagram a blood type within circle A contains antigen A, one within circle B contains antigen B, and one within circle Rh contains antigen Rh. In the following table, write yes if the antigen is present and no if it is not, for each of the eight blood types.

FIGURE 15

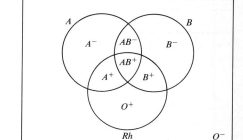

Blood Type	Antigen A	Antigen B	Antigen Rh
A^-			
A^+			
B^-			
B^+			
AB^-			
AB^+			
O^-			
O^+			

5.3 COMPLEMENT OF A SET

In this section we introduce another set operation that is useful in our later work.

Given a set A, we may want to consider those elements that are not in A. Usually, however, there are elements not in A that are extraneous to the problem being studied. For example, suppose A is the set of all positive integers and we describe "an element x that is not in A." Clearly, if x is a negative integer, then x is not in A. However, if x is a cabbage or a walrus, then x is also not in A since cabbages and walruses are not positive integers. If we are concerned with a problem about integers, then what we really mean when we describe

an element x that is not in A

is

an integer x that is not in A

—cabbages and walruses are extraneous to the problem.

One way of eliminating extraneous elements is to specify in advance that *all* elements in the problem under consideration come from some fixed set U, called the **universal set** for the problem. For example, in a problem concerned with integers, we might agree to take the universal set U to be the set of all integers. Then, if we speak of

an element x that is not in A

we must mean

an integer x that is not in A

since we have agreed that all elements come from the universal set U of integers.

In many problems, the universal set is intuitively clear from the nature of the problem and is not explicitly stated. For example, in problems concerned with letters of the English alphabet, U would be the set of *all* letters in the English alphabet. In problems concerned with sets of stocks on the New York Stock Exchange, U would be the set of *all* stocks on the New York Stock Exchange, and so on.

In Figure 1, we describe the above ideas using a Venn diagram. The points inside the rectangle form the universal set U. Since all the sets under study must have their members in U, they lie inside the rectangle. In particular, if A is the set of points inside the circle, then the shaded region in Figure 1 forms the set of all elements not in A.

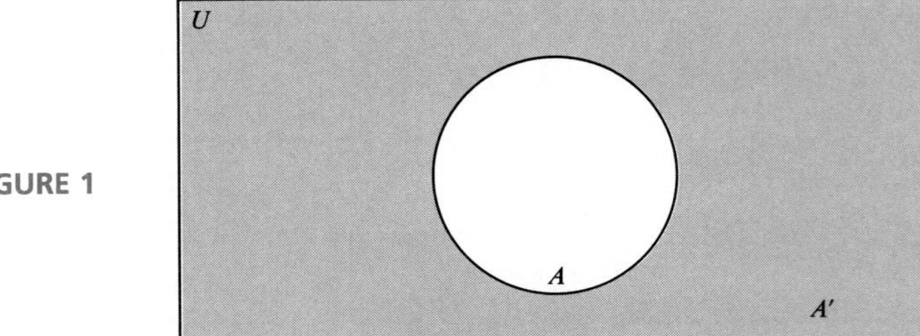

FIGURE 1

If U is the universal set and A is a subset of U, then the set of all elements in U that are not members of A is called the **complement** of A. It is denoted by the symbol

$$A'$$

read "A prime."

EXAMPLE 1 Let A be the subset of the integers consisting of all positive integers. Find A'.

Solution Since A is a subset of the set of integers, it is natural to take the universal set U to be the set of all integers. Thus A' is the set of all integers that are not positive. That is, A' is the subset of the integers consisting of the negative integers and zero.

EXAMPLE 2 Let W be the set of U.S. citizens who file tax returns listing less than \$10,000 taxable income in 1990. Find the complement of W.

Solution Since W is a subset of the set of U.S. citizens filing tax returns in 1990, it is reasonable to let the universal set U be the set of all U.S. citizens filing tax returns in 1990. Thus W' is the set of all U.S. citizens filing tax returns in 1990 listing at least \$10,000 taxable income.

EXAMPLE 3 Let

$$U = \{1, 3, 5, 7, 9, 11\} \qquad V = \{1, 5, 7\}$$

then

$$V' = \{3, 9, 11\}$$

EXAMPLE 4 Let U be the universal set for a given problem. Find (a) U' (b) $\varnothing'$

Solution (a) By definition of complement, U' consists of all elements that are not in U. But every element lies in U, so

$$U' = \varnothing$$

Solution (b) The set $\varnothing'$ consists of all elements that are not in $\varnothing$. Since $\varnothing$ has no elements, all elements are not in $\varnothing$; thus,

$$\varnothing' = U$$

EXAMPLE 5 Let A, B, and C be the sets of points inside the circles indicated in Figure 2. Shade the following sets.

(a) $A' \cap B \cap C$ (b) $A \cap B' \cap C'$

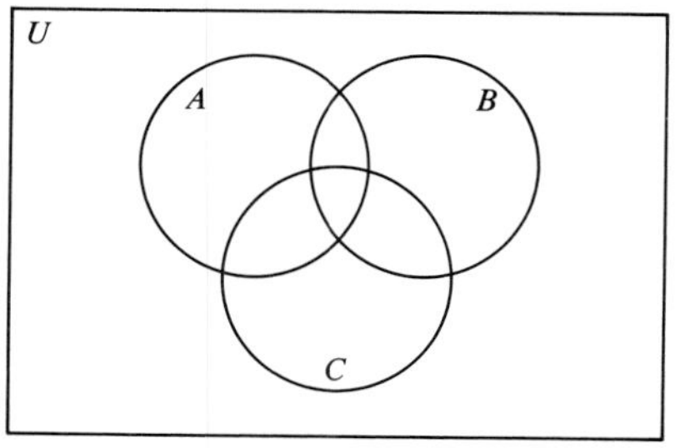

FIGURE 2

Solution (a) The members of $A' \cap B \cap C$ belong to both B and C but not A. This yields the shaded region in Figure 3.

Solution (b) The members of $A \cap B' \cap C'$ belong to A but not to B and not to C. This yields the shaded region in Figure 4.

FIGURE 3

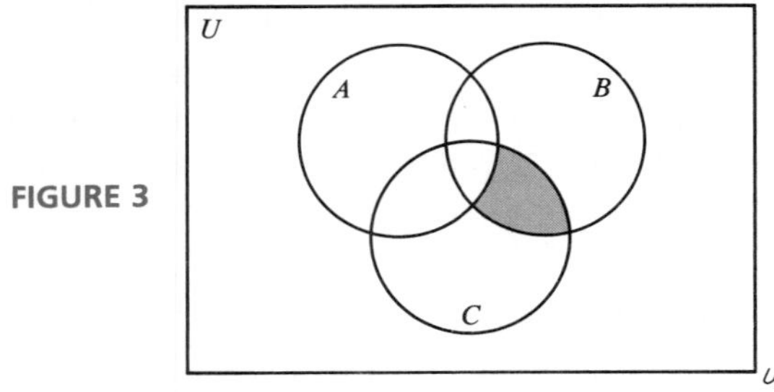

FIGURE 4

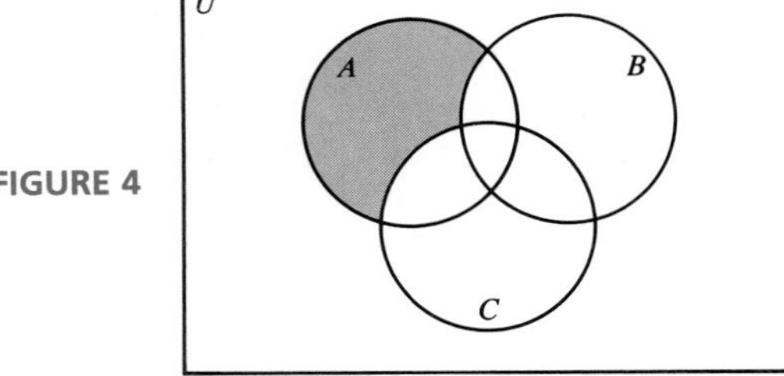

There are a number of useful properties of the complementation operation. We discuss a few of the more important ones here. The following two results are called **De Morgan's* laws**.

De Morgan's Laws If A and B are any two sets, then

1. $(A \cup B)' = A' \cap B'$

2. $(A \cap B)' = A' \cup B'$

We verify the first De Morgan law using Venn diagrams and leave the second as an exercise. The set $(A \cup B)'$ consists of all elements that are not in $A \cup B$. This is given by the shaded region in Figure 5a.

In Figures 5b and 5c the regions A' and B' are shaded. In Figure 5d their intersection yields $A' \cap B'$. Comparing Figures 5a and 5d, we see that

$$(A \cup B)' = A' \cap B'$$

which is the first law of De Morgan.

**Augustus De Morgan* (1806–1871), British mathematician and logician, the son of a British army officer, was born in Madura, India. He graduated from Trinity College in Cambridge, England, in 1827 but was denied a teaching position there for refusing to subscribe to religious tests. He was, however, appointed to a mathematics professorship at the newly opened University of London.

He was a man of firm principles who was described as "indifferent to politics and society and hostile to the animal and vegetable kingdoms." He twice resigned his teaching position on matters of principle (but later returned), refused to accept honorary degrees, and declined memberships in many learned societies.

He is best known for his work *Formal Logic*, which appeared in 1847, but he also wrote papers on the foundation of algebra, philosophy of mathematical methods, and probability as well as several successful elementary textbooks.

FIGURE 5

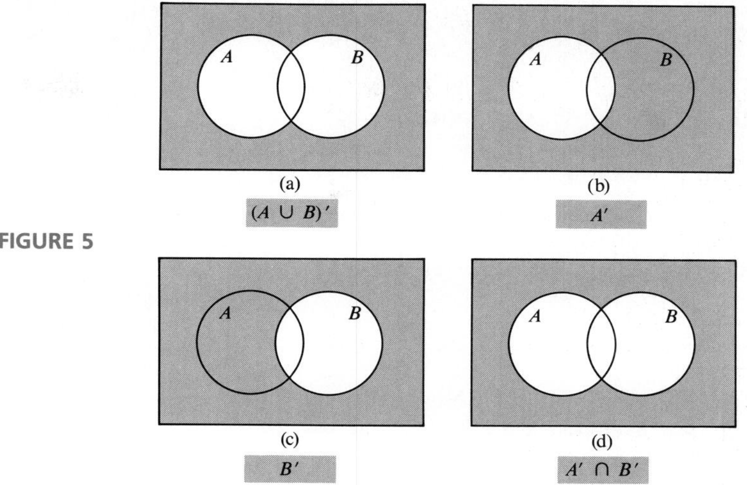

The next example illustrates the first De Morgan law.

EXAMPLE 6 Let

$$U = \{a, b, c, d, e, f, g, h\} \qquad A = \{a, b, d, e\} \qquad B = \{b, d, f, g\}$$

Then

$$A \cup B = \{a, b, d, e, f, g\}$$

and

$$(A \cup B)' = \{c, h\}$$

On the other hand,

$$A' = \{c, f, g, h\} \quad \text{and} \quad B' = \{a, c, e, h\}$$

so

$$A' \cap B' = \{c, h\}$$

Therefore,

$$(A \cup B)' = A' \cap B'$$

as guaranteed by the first De Morgan law.

The verifications of the following results are left as exercises.

Properties of the Complement If U is the universal set and A is a subset of U, then

$$(A')' = A$$
$$A \cup A' = U$$
$$A \cap A' = \varnothing$$

Remark Other commonly used notations for the complement of A are $\overline{A}$ and $U - A$.

EXERCISE SET 5.3

1. Let $U = \{a, b, c, d, e, f, g, h\}$ be the universal set and let $A = \{a, d, f, h\}$, $B = \{a, d\}$, $C = \{e, f\}$. Find
(a) A' (b) B'
(c) $(A \cup B)'$ (d) $(B \cap C)'$ (e) U'

2. Let U, A, B, and C be the sets in Exercise 1. Find
(a) $A' \cap B'$ (b) $B' \cup C'$
(c) $(A \cup A)'$ (d) C' (e) $(C \cap C)'$

3. Let
U = the set of all stocks traded on the New York Stock Exchange
A = the set of stocks traded on the New York Stock Exchange that have paid a dividend for the past 10 years without any interruption
B = the set of stocks traded on the New York Stock Exchange that have a price-to-earnings ratio of no more than 12

Describe the following sets.
(a) A' (b) B' (c) $(A \cup B)'$ (d) $(A \cap B)'$

4. Let U = the set of all integers, $A = \{x | x \text{ is a solution of } x^2 = 1\}$, and $B = \{-1, 2\}$. Find
(a) A' (b) B' (c) $(A \cup B)'$ (d) $(A \cap B)'$

5. Let $A = \{x | x \text{ is an integer greater than } 4\}$. Specify a universal set and compute A'.

6. Let D = the set of all Democratic U.S. senators. Specify a universal set and compute D'.

7. Let C be the set of all consonants in the English alphabet. Specify a universal set and compute C'.

8. Let A, B, and C be the sets of points in Figure 2 of this section. Shade the following sets.
(a) $A \cap B' \cap C$ (b) $A \cap B \cap C'$
(c) $A' \cap B \cap C'$ (d) $A' \cap B' \cap C$

9. Let A, B, and C be the sets of points in Figure 2 of this section. Shade the following sets.
(a) $A' \cap B' \cap C'$ (b) $(A \cup B \cup C)'$
(c) $(A \cap B \cap C)'$ (d) $[(A \cup B) \cap C]'$

10. Use a Venn diagram to verify that if $A \subset B$, then $B' \subset A'$.

11. Use a Venn diagram to verify the second De Morgan law: $(A \cap B)' = A' \cup B'$.

12. Use a Venn diagram to verify the following De Morgan laws for three sets A, B, and C.

$$(A \cup B \cup C)' = A' \cap B' \cap C'$$
$$(A \cap B \cap C)' = A' \cup B' \cup C'$$

13. (a) Extend De Morgan's laws for four sets A, B, C, and D.
(b) Extend De Morgan's laws for n sets

$$A_1, A_2, \ldots, A_n$$

14. Explain why the following are true.
(a) $(A')' = A$ (b) $A \cup A' = U$
(c) $A \cap A' = \varnothing$

In Exercises 15 and 16 let A, B, and C be the indicated sets and let x, y, z and w be the indicated points in Figure 6.

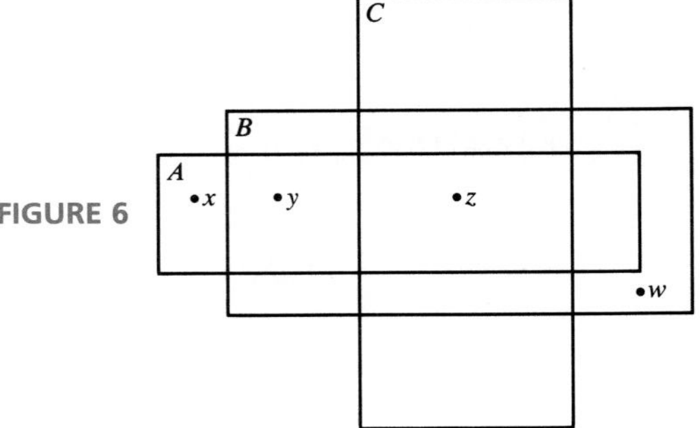

FIGURE 6

15. In each case determine which of the points x, y, z, w belong to the indicated set.
(a) $C' \cap B'$ (b) $A \cap B \cap C'$
(c) $A \cap B' \cap C$ (d) $A' \cap B' \cap C$

16. Follow the directions of Exercise 15 for the following sets.
(a) $A' \cup B$ (b) $B' \cup C'$
(c) $A' \cup B' \cup C'$ (d) $A' \cup B \cup C'$

In Exercises 17–19 copy the Venn diagram in Figure 7 and shade the indicated sets.

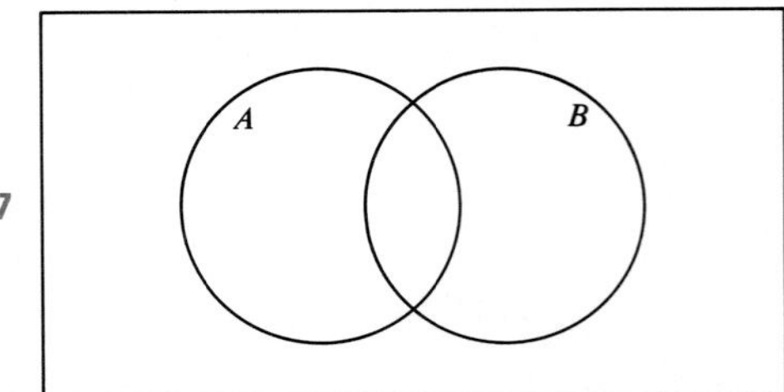

FIGURE 7

17. (a) $A' \cup B$ (b) $A' \cap B$ (c) $A' \cap B'$

18. (a) $A' \cup B'$ (b) $(A \cup B)'$ (c) $(A \cap B)'$

19. (a) $(A' \cup B)'$ (b) $(A' \cup B')'$
(c) $(A' \cap B')'$

In Exercises 20 and 21 copy the Venn diagram in Figure 8 and shade the indicated sets.

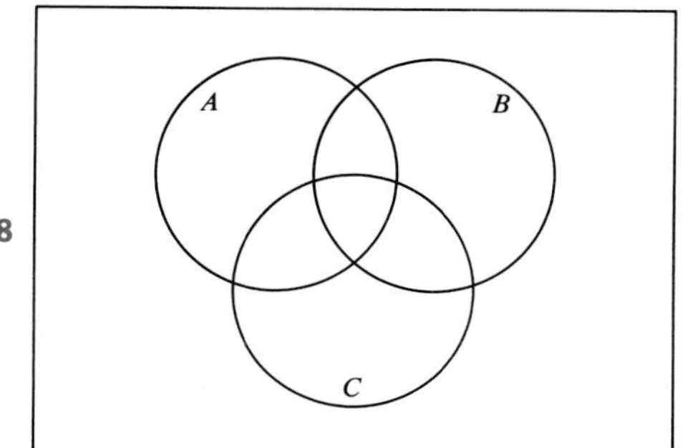

FIGURE 8

20. (a) $A \cup (B \cap C)$ (b) $A \cap (B \cup C)$
(c) $A \cap (B \cap C)$

21. (a) $A \cup (B \cap C)'$ (b) $A' \cup (B' \cap C)$
(c) $A \cup (B \cup C')'$

22. Exercise 25 of Section 5.2 describes the eight possible human blood types in terms of the antigens that they contain. Using the Venn diagram in that example, complete the table below by writing yes if the antigen is present in the given set and no if it is not.

23. When using a computer to compile a bibliography, we must use proper set operations to compile the desired bibliography. For example, let U be the data base listing all reports on pollution, and let

S = the set of all reports on pollution caused by sulfur dioxide
P = the set of all reports on pollution caused by particulate matter
A = the set of all reports on pollution in the air
R = the set of all reports on respiratory problems caused by pollution
G = the set of all reports written by S. A. Gould

Using the operations of union, intersection, and complement, obtain the set of all reports on respiratory problems due to pollution in the air caused by sulfur dioxide *or* particulate matter but excluding all reports written by S. A. Gould (this exclusion is due to the fact that you already are familiar with all of Gould's reports).

Set	Antigen A	Antigen B	Antigen Rh
$A' \cup B$			
$B' \cup Rh'$			
$Rh' \cap B'$			
$A \cap B \cap Rh'$			
$A' \cup B' \cup Rh'$			
$A \cap B' \cap Rh$			
$A' \cup B \cup Rh'$			
$A' \cap B' \cap Rh$			

KEY IDEAS FOR REVIEW

- **Set** A collection of objects.
- **$a \in A$** a is an element of the set A.
- **Equal sets** Sets with the same elements.
- **$\varnothing$ (empty set or null set)** The set with no elements.
- **$A \subset B$ (A is a subset of B)** Every element of A is also an element of B.
- **$A \cap B$ (intersection of A and B)** The set of all elements that belong to both A and B.

- **Disjoint sets** Two sets with no elements in common.
- **$A \cup B$ (union of A and B)** The set of all elements that belong to A or to B, or to both.
- **Distributive law** $C \cap (A \cup B) = (C \cap A) \cup (C \cap B)$
- **Distributive law** $C \cup (A \cap B) = (C \cup A) \cap (C \cup B)$
- **Universal set** The set of all elements under consideration in a given problem.
- **A' (complement of A)** The set of all elements in the universal set that are not in A.
- **De Morgan's laws** 1. $(A \cup B)' = A' \cap B'$ 2. $(A \cap B)' = A' \cup B'$

SUPPLEMENTARY EXERCISES

In Exercises 1 and 2 let $A = \{a, b, c, e, f, g\}$, $B = \{b, c, f, g\}$, $C = \{a, b\}$, *and* $D = \{c\}$.

1. In each part replace the symbol □ by the symbol $\in$, $\notin$, $\subset$, or $\not\subset$ to make the statement true.
(a) $a \square A$ (b) $a \square B$
(c) $a \square C$ (d) $c \square C$
(e) $D \square A$ (f) $A \square A$

2. In each part replace the symbol □ by the symbol $\in$, $\notin$, $\subset$, or $\not\subset$ to make the statement true.
(a) $A \square B$ (b) $B \square A$
(c) $D \square C$ (d) $A \square \varnothing$
(e) $a \square \varnothing$ (f) $\varnothing \square A$

In Exercises 3 and 4 let $A = \{1, 2, 3, 5, 7\}$, $B = \{2, 3, 4, 5\}$, $C = \{x|x$ *is an integer and* $2x + 1 = 5\}$, $D = \{2\}$, *and* $E = \{x|x$ *is an integer and* $1/x = 0\}$.

3. Answer the following as true or false.
(a) $A \subset A$ (b) $A \subset B$
(c) $B \subset A$ (d) $C = D$
(e) $B = \{5, 2, 3, 4\}$ (f) $\{1, 2\} \in A$

4. Answer the following as true or false.
(a) $\varnothing \in A$ (b) $D \in A$
(c) $1 \in E$ (d) $E \subset D$
(e) $\{1, 2\} \not\subset C$ (f) $\{2, 5\} \in B$

5. Answer whether the following statement is true or false. $\{x|x$ is a letter in the word *red*$\}$ = $\{x|x$ is a letter in the word *reed*$\}$.

6. List all the subsets of the set $\{a, b\}$.

7. List all subsets of $\{a, b, c, d\}$ that contain $\{a, b\}$ as a subset.

8. Show that if $A \subset B$ and $B \subset C$, then $A \subset C$.

In Exercises 9 and 10 let $A = \{1, 2, 4\}$, $B = \{x|x$ *is a positive integer and* x^2 *is less than* $25\}$, $C = \{3\}$, $D = \{2, 4, 7\}$.

9. Compute the following.
(a) $A \cap D$ (b) $A \cap B$
(c) $C \cap D$ (d) $A \cup B$
(e) $A \cup C$ (f) $B \cup C$

10. Compute the following.
(a) $B \cap C$ (b) $B \cap (C \cup D)$
(c) $(A \cap C) \cup D$ (d) $A \cap B \cap C$
(e) $A \cup D \cup \varnothing$ (f) $A \cup B \cup C$

For Exercises 11 and 12 refer to Figure 1.

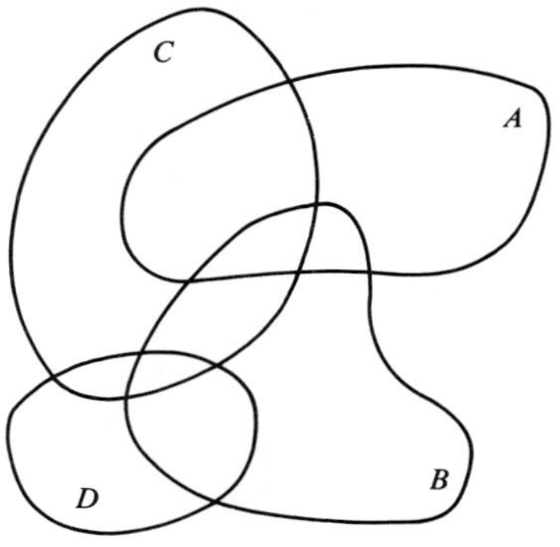

FIGURE 1

11. Shade the following sets.
(a) $A \cap C$ (b) $B \cap C \cap D$
(c) $A \cup C$ (d) $A \cup B \cup D$

12. Shade the following sets.
(a) $A \cup (B \cap D)$ (b) $(A \cap B) \cap C$
(c) $(B \cap C) \cup D$ (d) $(A \cup C) \cap B$

13. Let $A = \{2, 3, 4\}$, $B = \{1, 3, 5\}$, $C = \{x | x \text{ is an integer and } x^2 = 9\}$, and $D = \{-1, -2, -3\}$. Answer the following as true or false.
(a) $2 \in A \cap B$ (b) $3 \in A \cup B$
(c) $C \subset A \cup D$ (d) $3 \in B \cap D$
(e) $A \cap B \cap D \neq \varnothing$ (f) $3 \in (A \cup B) \cap C$

14. Use a Venn diagram to illustrate the following statements.
(a) $A \cup (A \cap B) = A$ (b) $A \cap (A \cup B) = A$

15. Use a Venn diagram to show that if $A \cap B = A$, then $A \subset B$.

16. Use a Venn diagram to show that if $A \cup B = B$, then $A \subset B$.

17. Show that $A \cap \varnothing = \varnothing$.

18. Use a Venn diagram to show that if $A \subset B$ and $A \subset C$, then $A \subset B \cup C$.

19. Use a Venn diagram to show that if $B \subset A$ and $C \subset A$, then $B \cap C \subset A$.

20. Use a Venn diagram to show that if $A \subset B$ and $C \subset D$, then $A \cup C \subset B \cup D$.

21. Express the shaded region in Figure 2 in terms of A, B, and C using set operations.

FIGURE 2

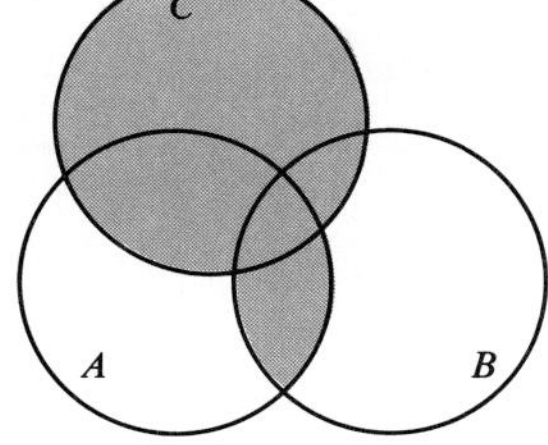

22. Express the shaded region in Figure 3 in terms of A, B, and C using set operations.

FIGURE 3

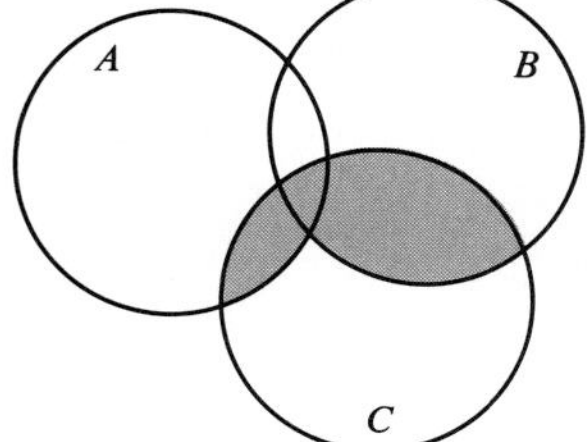

If A and B are two sets, we define the ***complement*** *of B with respect to A, written A − B, as the set of all elements that belong to A but not to B. The shaded regions in Figure 4 show the sets A − B and B − A, respectively.*

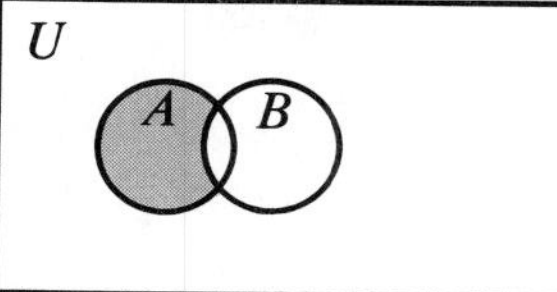

$A - B$
(a)

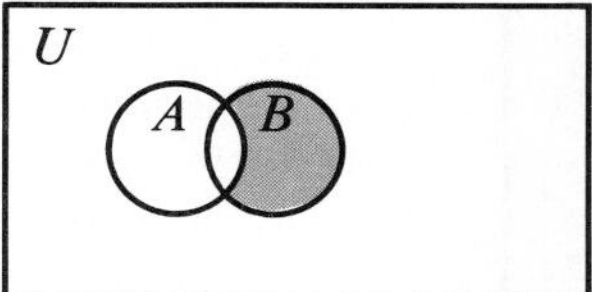

$B - A$
(b)

FIGURE 4

23. Let $A = \{1, 3, 4, 5, 7, 8\}$ and $B = \{2, 3, 5, 6, 9\}$. Compute the following.
(a) $A - B$ (b) $B - A$

24. Let $A = \{a, b, c, e, g, h\}$ and $B = \{d, f, h, i\}$. Compute the following.
(a) $A - B$ (b) $B - A$

If A and B are two sets, we define the ***symmetric difference*** *of A and B, written A ⊕ B, as the set of all elements that belong to A or to B but not to both A and B. The shaded regions in Figure 5 shows the set A ⊕ B.*

FIGURE 5

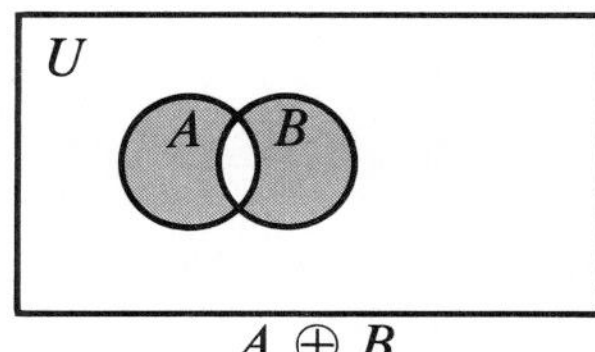

$A \oplus B$

25. Let $A = \{1, 2, 4, 5, 8\}$ and $B = \{3, 4, 5, 7\}$. Compute $A \oplus B$.

26. Let $A = \{a, b, d, e, g\}$ and $B = \{c, f, h, i\}$. Compute $A \oplus B$.

27. A computer store classifies its microcomputers into the following categories:

E = the set of all 8-byte machines
S = the set of all 16-byte machines
T = the set of all 32-byte machines
F = the set of all machines that support FORTRAN
B = the set of all machines that support BASIC
P = the set of all machines that support PASCAL

Use set notation to describe the members of the following sets.
(a) $E \cap F$ (b) $S \cap F \cap B$
(c) $(S \cup T) \cap P$ (d) $S \cap (F \cup B)$

28. A municipal bond broker in New York City classifies her offerings into the following categories:

I = the set of insured bonds
U = the set of uninsured bonds
N = the set of New York state bonds
P = the set of Pennsylvania bonds
T = the set of bonds maturing after 1992
S = the set of school bonds
H = the set of hospital bonds

Express the following sets in terms of the given sets and the operations of union, intersection, and complement.
(a) the set of all New York state insured bonds
(b) the set of all New York state school or hospital bonds
(c) the set of all New York or Pennsylvania bonds maturing after 1992
(d) the set of all insured New York state hospital bonds maturing after 1992

In Exercises 29 and 30 let $U = \{1, 2, 3, 4, 5, 6, 7\}$ be the universal set, and let $A = \{1, 3, 4\}$, $B = \{2, 3, 5\}$, $C = \{x|3x - 1 = 2\}$, and $D = \{1, 5, 6\}$.

29. Find
(a) $A' \cap C$ (b) $B' \cap C$
(c) $(A' \cup C) \cap D'$ (d) $(A' \cup B) \cap C'$

30. Find
(a) $D' \cap (A \cup D)$ (b) $(A \cup C)' \cap B$
(c) $C \cup (A \cap D)'$ (d) $(A \cup B \cup C)' \cap D$

For Exercises 31 and 32 refer to Figure 6.

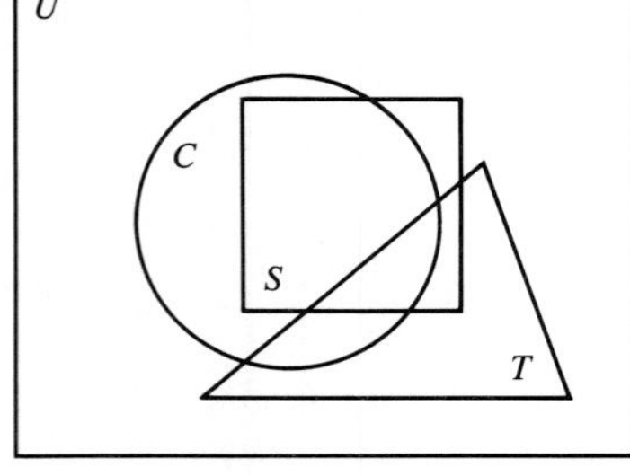

FIGURE 6

31. Shade the following sets.
(a) $S \cap C'$ (b) $S' \cup T'$
(c) $(S \cup C) \cap T'$ (d) $(C \cap T) \cup S'$

32. Shade the following sets.
(a) $(S' \cap T)' \cup C$ (b) $(C \cup T) \cap S'$
(c) $S \cup C' \cup T'$ (d) $S \cap C' \cap T'$

33. Express the shaded region in Figure 7 in terms of A, B, and C and their complements.

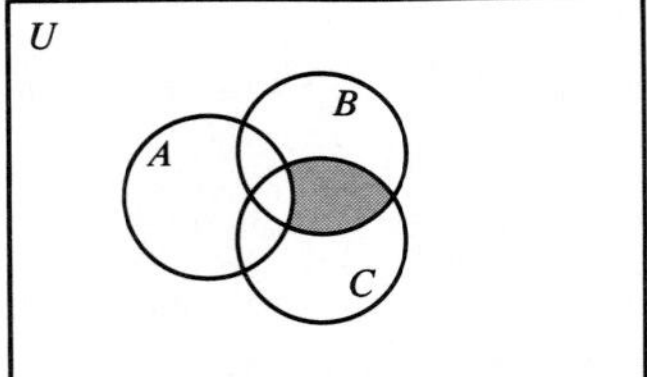

FIGURE 7

34. Express the shaded region in Figure 8 in terms of A, B, and C and their complements.

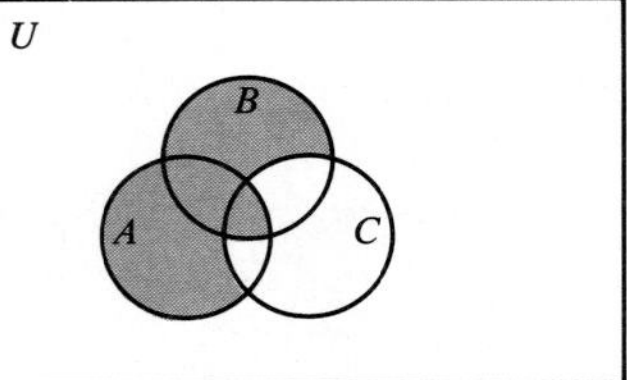

FIGURE 8

In Exercises 35–37 refer to the definition of $A - B$ that precedes Exercise 23.

35. Use a Venn diagram to show that $A - A = \varnothing$.

36. Use a Venn diagram to show that $A - B = A \cap B'$.

37. Use a Venn diagram to show that $A - (A - B) \subset B$.

38. Use a Venn diagram to show that $A \oplus B = B \oplus A$ (see the definition of $A \oplus B$ that precedes Exercise 25).

CHAPTER TEST

Let

$U = \{-4, -3, -2, -1, 0, 1, 2, 3, 4, 5, 6, 7\}$
$A = \{2, 5\}$
$B = \{x | x \text{ is an integer and } x^2 \text{ is less than } 16\}$
$C = \{3\}$
$E = \{0, 1, 2, 3, 4\}$
$F = \{-3, -2, 0, 2, 7\}$

1. Answer the following as true or false.
   (a) $3 \in C$ (b) $\{-5\} \in A \cup B$
   (c) $4 \in \{4\}$ (d) $\varnothing \in A$
2. Answer the following as true or false.
   (a) $C \subset A$ (b) $C \subset B$
   (c) $B = E$ (d) $C \subset (B \cap E)$
3. Compute $(A \cup B) \cap F$.
4. Compute $(A \cap F) \cup B$.
5. Compute $(A' \cap C) \cup E$.
6. Compute $(B' \cup F') \cap (A \cup C)$.

Refer to Figure 1 for Problems 7 and 8.

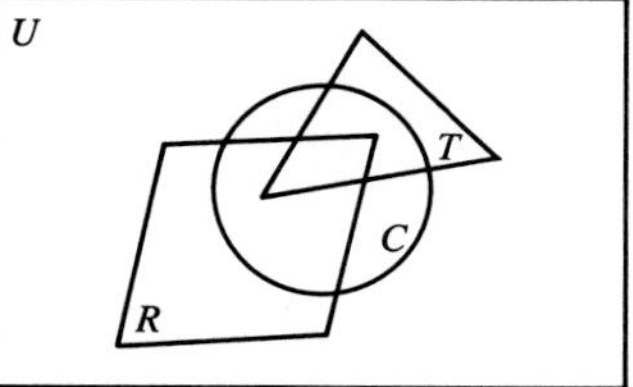

FIGURE 1

7. Shade the set $(R \cap C') \cup T$.
8. Shade the set $(R \cup C) \cap T'$.
9. Express the shaded region in Figure 2 in terms of A, B, and C and their complements.

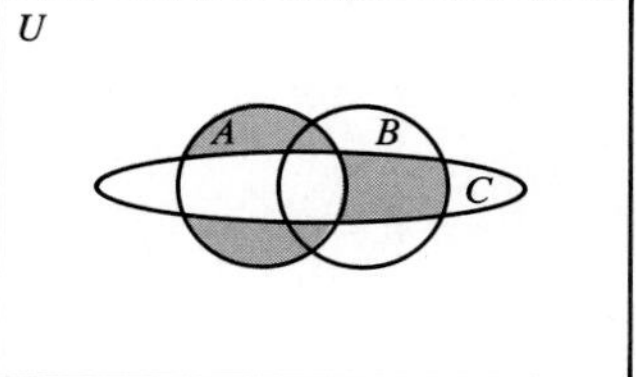

FIGURE 2

10. Let

M = the set of all men
D = the set of all people who drink alcohol
S = the set of all people who smoke
E = the set of all people who exercise
H = the set of all people who will have a heart attack before the age of 50

Express the following sets in terms of the given sets and the operations of union, intersection, and complement.

(a) The set of all men who drink alcohol or smoke and will have a heart attack before the age of 50.
(b) The set of all women who drink alcohol or do not exercise and will not have a heart attack before the age of 50.

6

Counting Techniques: Permutations and Combinations

In this chapter we discuss two kinds of counting problems—counting the number of elements in a set and counting the number of ways that the elements of a set can be arranged. The material we examine here has a variety of applications, especially in the study of probability, which we take up in the next chapter.

6.1 CARTESIAN PRODUCTS

In many problems we are interested in paired data, such as the height and weight of an individual, the wind speed and wind direction at a certain time, or the total assets and total liabilities of a business firm. Often, the order in which the information is listed is important. For example, suppose we want to describe the financial status of a business firm by listing a pair of numbers, the first number indicating its total assets in millions of dollars and the second indicating its total liabilities in millions of dollars.

If the pair of Company A is (9, 0) and the pair for Company B is (0, 9), then Company A is in an excellent financial position, whereas Company B is likely to be on the verge of bankruptcy. Thus, even though the pairs (9, 0) and (0, 9) involve the same numbers, they convey very different information because of their order. This notion of ordered data gives rise to the following concept.

An **ordered pair** (a, b) is a listing of two objects a and b in a definite order. The element a is called the **first entry** in the ordered pair, and the element b is called the **second entry** in the ordered pair.

Two ordered pairs are **equal** if they list the same objects in the same order. Thus, for (a, b) and (c, d) to be equal ordered pairs, we must have

$$a = c \quad \text{and} \quad b = d$$

We are now in a position to define an important new set operation.

If A and B are two sets, then the set of all ordered pairs (a, b), where $a \in A$ and $b \in B$, is called the **Cartesian* product** of A and B; it is denoted by the symbol

$$A \times B$$

(read "A cross B")

EXAMPLE 1 Find $A \times B$ and $B \times A$ if

$$A = \{r, s, t\} \quad \text{and} \quad B = \{1, 2, 3, 4\}$$

Solution We must form the set of all ordered pairs whose first entry belongs to A and whose second entry belongs to B; $A \times B$ is

$$\{(r, 1), (r, 2), (r, 3), (r, 4), (s, 1), (s, 2), (s, 3), (s, 4), (t, 1), (t, 2), (t, 3), (t, 4)\}$$

Note that the members of $A \times B$ can be conveniently arranged in tabular form, as in Table 1.

TABLE 1

A \ B	1	2	3	4
r	$(r, 1)$	$(r, 2)$	$(r, 3)$	$(r, 4)$
s	$(s, 1)$	$(s, 2)$	$(s, 3)$	$(s, 4)$
t	$(t, 1)$	$(t, 2)$	$(t, 3)$	$(t, 4)$

Alternatively, the ordered pairs in $A \times B$ can be constructed systematically using what is called a **tree diagram**. This is done as follows:

Step 1 The first entry in each ordered pair in $A \times B$ comes from A, so at the start there are three possible choices: r, s, or t. In Figure 1 these choices are represented by the three dots (or **nodes**) labeled r, s, and t.

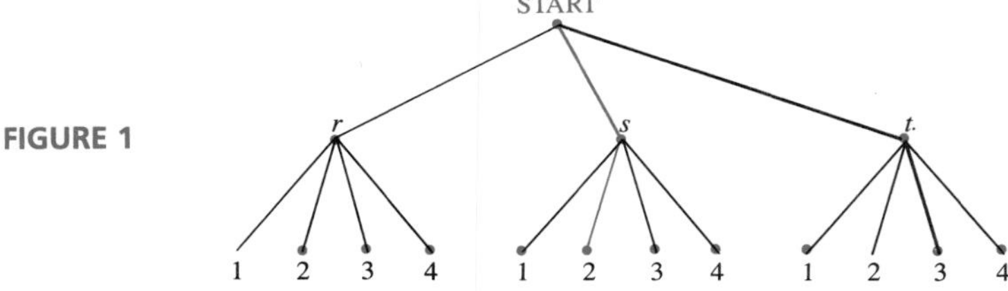

FIGURE 1

* Named after René Descartes (See page 6).

Step 2 The four branches emanating from each of these nodes correspond to the possible choices for the second entry in the ordered pair, namely, 1, 2, 3, or 4. The various ordered pairs can be listed by tracing out all possible paths or "branches" from the top of the tree to the bottom of the tree. For example, the path in color corresponds to the pair $(s, 2)$ and the heavy black path to the pair $(t, 3)$.

You will find it helpful to compute the set $A \times B$ by this method.

Similarly, $B \times A$ is

$$\{(1, r), (1, s), (1, t), (2, r), (2, s), (2, t), (3, r), (3, s), (3, t), (4, r), (4, s), (4, t)\}$$

Note that in general $A \times B \neq B \times A$.

It is possible to consider Cartesian products of more than two sets. For example, the Cartesian product $A \times B \times C$ of the three sets, A, B, and C would consist of all ordered *triples* (a, b, c), where

$$a \in A \qquad b \in B \qquad c \in C$$

EXAMPLE 2 A firm that conducts political polls classifies people for its files according to sex, income, and political registration:

***S* (sex)**	m = male	f = female	
***I* (income)**	h = high	a = average	l = low
***P* (political registration)**	d = Democrat	r = Republican	i = Independent

The Cartesian product $S \times I \times P$ contains all possible classifications. For example, a person filed under (f, h, r) would be a female with high income who is registered as a Republican.

From the tree in Figure 2 the different possible classifications are

$$\begin{array}{llllll} (m, h, d) & (m, h, r) & (m, h, i) & (f, h, d) & (f, h, r) & (f, h, i) \\ (m, a, d) & (m, a, r) & (m, a, i) & (f, a, d) & (f, a, r) & (f, a, i) \\ (m, l, d) & (m, l, r) & (m, l, i) & (f, l, d) & (f, l, r) & (f, l, i) \end{array}$$

FIGURE 2

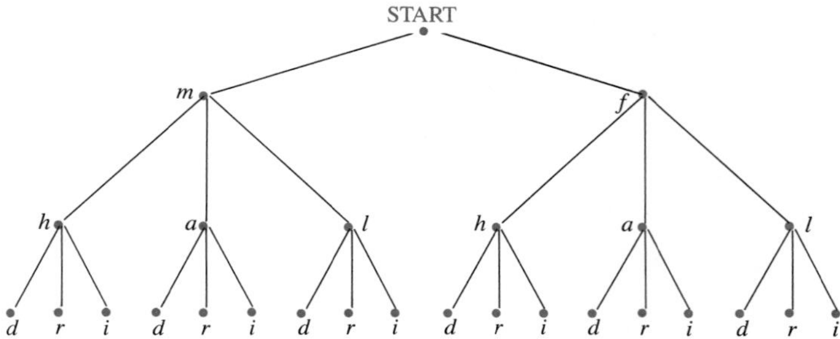

EXAMPLE 3 A license plate consists of two letters followed by two numbers. The first letter can be X or Y, the second W or Z, and the numbers can be 1, 2, or 3.

(a) Represent the set of all such license plates as a Cartesian product of four sets.

(b) Use a tree diagram to help list all the possible license plates.

Solution (a) Let

$$A = \{X, Y\} \qquad B = \{W, Z\} \qquad C = \{1, 2, 3\}$$

Then, a license plate consists of four entries; the first from A, the second from B, and the third and fourth from C. Thus, the set of all license plates can be represented as the Cartesian product

$$A \times B \times C \times C$$

Solution (b) The tree diagram for $A \times B \times C \times C$ is shown in Figure 3. Tracing all possible paths through the tree yields this list of possible license plates.

XW11	XW21	XW31
XW12	XW22	XW32
XW13	XW23	XW33
XZ11	XZ21	XZ31
XZ12	XZ22	XZ32
XZ13	XZ23	XZ33
YW11	YW21	YW31
YW12	YW22	YW32
YW13	YW23	YW33
YZ11	YZ21	YZ31
YZ12	YZ22	YZ32
YZ13	YZ23	YZ33

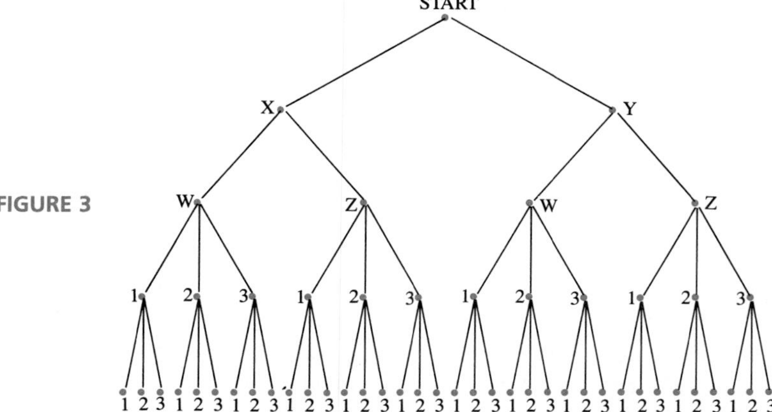

FIGURE 3

EXERCISE SET 6.1

1. In each part find the values of x and y for which the given ordered pairs of integers are equal.
(a) $(x, 7) = (3, 7)$
(b) $(2x, 3) = (6, y)$
(c) $(4, y + 7) = (2x + 2, 14)$
(d) $(x^2, 9) = (16, 9)$

2. In each part find the values of r, s, and t for which the given ordered triples of real numbers are equal.
(a) $(r, s, 4) = (-3, 2, t)$
(b) $(3r, 5s, \frac{1}{2}t) = (-2, 5, 4)$
(c) $(r + 2, s - 3, 2t + 5) = (0, 0, 0)$
(d) $(r^2 + 1, 2s + 6, 4) = (10, -2, t^2)$

In Exercises 3–6 find $A \times B$, $B \times A$, $A \times A$, and $B \times B$.

3. $A = \{u, v\}$, $B = \{q, r\}$

4. $A = \{a, b, c, d\}$, $B = \{0, 1, 2\}$

5. $A = \{-2, 1, 4\}$, $B = \{-2, 1, 4\}$

6. $A = \{m, o, u, s, e\}$, $B = \{k, e, y\}$

In Exercises 7–10 draw a tree diagram and then make a list of all elements in the indicated Cartesian product.

7. $A = \{l, p, q, t\}$, $B = \{1, 2\}$; $A \times B \times B$

8. $Q = \{\text{juice, soup, salad}\}$, $R = \{\text{beef, chicken, fish}\}$, $S = \{\text{pie, ice cream}\}$; $Q \times R \times S$

9. $B = \{\text{male, female}\}$, $B \times B \times B \times B$

10. $A = \{1, 2, 3\}$, $B = \{v, w\}$, $C = \{r, s\}$; $A \times B \times B \times C$

11. A coin is tossed 4 times and the sequence of heads (h) and tails (t) is recorded. For example, $\{h, h, t, h\}$ is one possibility.
(a) Find a set H such that $H \times H \times H \times H$ is the set of all possible sequences.
(b) Make a tree diagram and list all the possible sequences.

12. **(Traffic control)** An air traffic control station supplies the following data to airline pilots:

Traffic	crowded (c), average (a), light (l)
Visibility	poor (p), good (g)
Windspeed	negligible (n), medium (m), high (h)

Flying conditions are described by an ordered triple: for example (c, p, n) means crowded traffic, poor visibility, negligible windspeed.
(a) Find sets T, V, and W such that $T \times V \times W$ is the set of all possible flying conditions.
(b) Make a tree diagram and list all possible flying conditions.

13. **(Psychology)** In a psychology experiment, a rat is placed in a cage with three doors, a, b, and c (Figure 4). The rat leaves the cage through one of the doors. On reaching the intersection it turns either left or right, and at the next intersection it turns either left or right again. The rat then proceeds to the reward section. If the rat chose door a, went left at the first intersection and right at the second, then this path could be denoted by the triple (a, l, r).

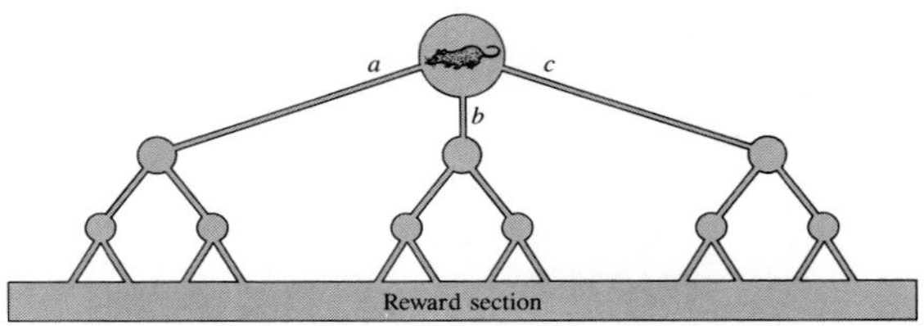

FIGURE 4

(a) Find sets E and T such that $E \times T \times T$ is the set of all possible paths.
(b) Make a tree diagram and list all possible paths.

14. Find values of x and y such that

$$(x - 2y, 3) = (-3, x + y)$$

15. Find values of r, s, and t such that

$$(2r + s + t, r - s - t, r + 2s - 2t) = (3, -6, -3)$$

16. In Example 1 we show that $A \times B$ and $B \times A$ are different sets; find two sets such that $A \times B$ and $B \times A$ are the same.

In Exercises 17 and 18 use a tree diagram to answer the given question.

17. (a) If A has 2 elements and B has 3 elements, how many elements are there in $A \times B$?
(b) If A has m elements and B has n elements, how many elements are there in $A \times B$?

18. (a) If A has 2 elements, how many elements are there in $A \times A$?
(b) How many in $A \times A \times A$?
(c) How many in $\underbrace{A \times A \times \cdots \times A}_{n \text{ factors}}$?

19. (a) If A has m elements, how many elements are there in $A \times A$?
(b) How many in $A \times A \times A$?
(c) How many in $\underbrace{A \times A \times \cdots \times A}_{n \text{ factors}}$?

20. A menu lists a choice of soup, salad, or juice for the appetizer; a choice of beef, chicken, or fish for the entree; and a choice of pie or fruit for dessert. A complete dinner consists of one choice for each course. Draw a tree diagram for the possible complete dinners.
(a) How many different complete dinners are possible?
(b) If a man refuses to eat chicken or pie, how many different complete dinners can he choose?
(c) A certain customer eats pie for dessert if he has soup as an appetizer; otherwise, he chooses fruit for dessert. How many different complete dinners are available to him?

21. A prisoner can escape from the penitentiary by digging a tunnel under the east wall, climbing over the west wall, or hiding in a laundry truck leaving the north gate. He can then proceed along the stream or along the back road to the highway intersection, where he can take the north, south, east, or west highway. Draw a tree diagram for the possible escape routes.
(a) How many escape routes are possible?
(b) If the north and west highways are blockaded, how many escape routes are possible?
(c) If it is known that the prisoner did not hide in the laundry truck and did not take the back road, how many escape routes are possible?
(d) If it is known that the prisoner took the south highway, how many escape routes are possible?

22. The street system in a certain city can be represented schematically by the diagram in Figure 5, in which A, B, C, D, E, and F denote intersections. Use a tree diagram to help determine the number of ways in which a car can travel from A to F without passing through the same intersection twice.

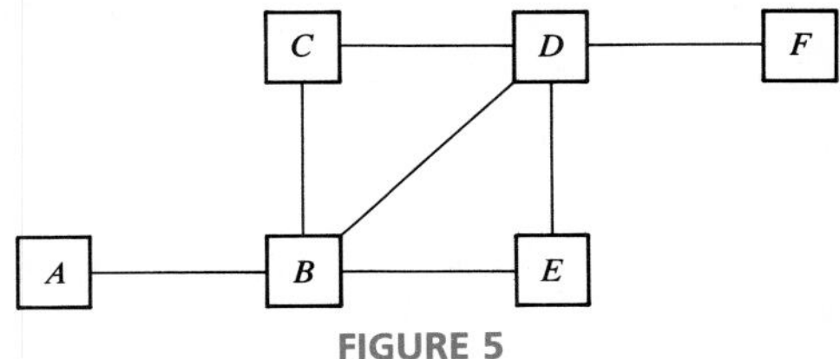

FIGURE 5

6.2 COUNTING ELEMENTS IN SETS

Suppose we know the number of elements in a set A and the number of elements in a set B. What can we say about the number of elements in $A \cup B$ and $A \times B$? In this section we investigate problems like these and illustrate some of their applications.

If S is a set with a finite number of elements, then we denote the **number of elements in S** by

$$n(S)$$

EXAMPLE 1 Consider the sets

$$A = \{a, e, g, l\}$$
$$B = \varnothing$$
$$C = \{x | x \text{ is a positive integer}\}$$

Find $n(A)$, $n(B)$, and $n(C)$.

Solution The set A has four elements, so

$$n(A) = 4$$

The set B has no elements, so

$$n(B) = 0$$

BEAN COUNTING AND THE INFINITE

Imagine you are the chieftain of a remote primitive tribe whose currency is dried beans and whose culture is so undeveloped that no one, not even you, can count. A feud develops between two tribal members, each of whom claims to have more beans than the other. To settle the argument you ask each to put his or her beans in a pot and bring them to you so you can determine who is correct. To your surprise each brings a gigantic pot with mounds of beans. How do you decide who is correct given that you can't count? The solution is simple—you take a bean from each pot and set the pair aside, then take another bean from each pot and set that pair aside. You continue in this way until one or both pots is empty. If one pot empties before the other, then it had the smaller number of beans; if both pots empty at that same time, then they had the same number. You are a wise chieftain indeed for you have shown that by matching elements from two sets it is possible to compare the sizes of the sets without counting the members.

In the late 1800s the German mathematician Georg Cantor* used the "bean counting" principle to deduce some strange results about infinite sets. By doubling each positive integer and matching the integer with its double, Cantor deduced the shocking result that there are as many even integers as even and odd integers together!

$$\begin{array}{cccccccccc} 1 & 2 & 3 & 4 & 5 & 6 & 7 & 8 & 9 & \dots \\ \updownarrow & \updownarrow & \updownarrow & \updownarrow & \updownarrow & \updownarrow & \updownarrow & \updownarrow & \updownarrow & \\ 2 & 4 & 6 & 8 & 10 & 12 & 14 & 16 & 18 & \dots \end{array}$$

This very odd result as well as others about infinite sets led mathematicians to realize that these sets behave very differently from those with finitely many members.

---

* *Georg Ferdinand Ludwig Philipp Cantor* (1845–1918) was born in St. Petersburg, Russia. His father was a prosperous merchant, born in Denmark, and his mother was of German descent. At age 11, Cantor and his family moved to Frankfurt, Germany, where Cantor's extraordinary mathematical talent was recognized at an early age. Cantor decided to study mathematics despite his father's strong opposition. After receiving his Ph.D. from the University of Berlin, he taught at the University of Halle, a rather undistinguished institution.

Cantor's theory of the infinite was so startlingly original that it aroused a great deal of controversy. The highly regarded German mathematician Leopold Kronecker led the assault not only on Cantor's work but on Cantor himself. Cantor, who was sensitive, could not withstand these attacks and in 1884 suffered the first of many mental breakdowns. He died in a mental hospital in 1918. However, before his death he was honored for his major contribution to the foundations of mathematics.

The symbol $n(S)$ is defined only if S has finitely many elements. Since C has infinitely many elements, $n(C)$ is undefined.

Suppose A and B are sets with finitely many elements. If, as illustrated in Figure 1, A and B are disjoint, then the number of elements in $A \cup B$ can be obtained by adding the number of elements in A and the number of elements in B. In other words,

> If A and B are disjoint sets,
> $$n(A \cup B) = n(A) + n(B) \tag{1}$$

FIGURE 1

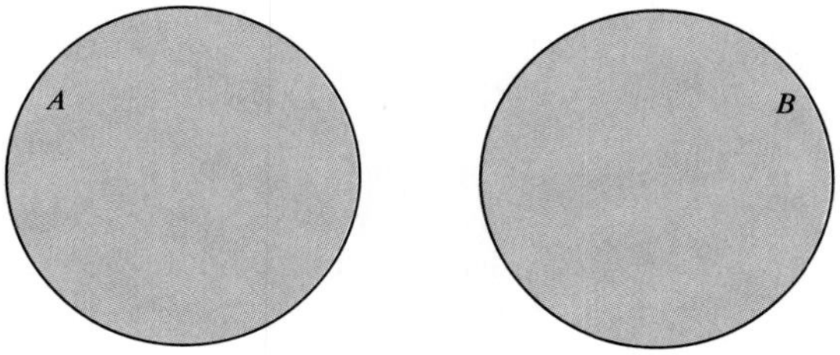

EXAMPLE 2 Consider the disjoint sets

$$A = \{1, 2, 5, 7\} \quad \text{and} \quad B = \{3, 9\}$$

Since

$$A \cup B = \{1, 2, 3, 5, 7, 9\}$$

we have

$$n(A \cup B) = 6$$

On the other hand,

$$n(A) = 4 \quad \text{and} \quad n(B) = 2$$

so

$$n(A \cup B) = n(A) + n(B)$$

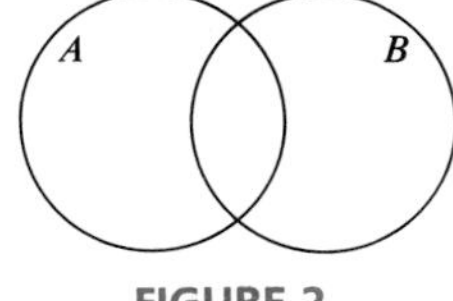

FIGURE 2

If A and B are sets with finitely many elements, but are *not disjoint* (Figure 2), then Formula (1) does not apply. The problem is that the points in $A \cap B$ are counted once in $n(A)$ and again in $n(B)$, so they are counted twice in forming the sum $n(A) + n(B)$. Since they should only be counted once when calculating $n(A \cup B)$, we can correct for this duplication by subtracting $n(A \cap B)$ from $n(A) + n(B)$. This yields the formula

$$n(A \cup B) = n(A) + n(B) - n(A \cap B) \tag{2}$$

which is valid for *any* sets A and B with finitely many elements. Note that if A and B are disjoint sets, then $A \cap B = \varnothing$, so $n(A \cap B) = 0$ and (2) reduces to (1).

EXAMPLE 3 Let

$$A = \{a, b, c, d, e\} \quad \text{and} \quad B = \{d, e, f, g, h, i\}$$

so that

$$A \cup B = \{a, b, c, d, e, f, g, h, i\}$$

and

$$A \cap B = \{d, e\}$$

Since

$$n(A) = 5 \qquad n(B) = 6 \qquad n(A \cup B) = 9 \qquad n(A \cap B) = 2$$

we have

$$n(A \cup B) = n(A) + n(B) - n(A \cap B)$$

as guaranteed by Formula (2).

EXAMPLE 4 Philadelphia has two newspapers, *The Inquirer* and *The Daily News*. Assume that a survey of 1000 people who subscribe to at least one of the papers reveals that 500 subscribe to *The Inquirer* and 100 subscribe to both. How many of the 1000 surveyed subscribe to *The Daily News*?

Solution Let A be the set of people surveyed who subscribe to *The Inquirer* and let B be the set of people surveyed who subscribe to *The Daily News* so that $A \cup B$ is the entire set of people surveyed. From the above data,

$$n(A \cup B) = 1000 \qquad n(A) = 500 \qquad n(A \cap B) = 100$$

Substituting these values in the formula

$$n(A \cup B) = n(A) + n(B) - n(A \cap B)$$

yields

$$1000 = 500 + n(B) - 100 \quad \text{so} \quad n(B) = 600$$

Therefore, 600 of the 1000 surveyed subscribed to *The Daily News*.

EXAMPLE 5 A manufacturer of magnetic tape for digital computers finds that an experimental production process can introduce two possible defects in a reel of tape: the tape can contain creases or the magnetic coating can be insufficient. In testing 100 reels, the manufacturer recorded that

15 have creased tape
12 have tape with insufficient magnetic coating
7 have both defects

(a) How many reels were defective (i.e., had at least one defect)?

(b) How many reels were nondefective?

Solution (a) Let A be the set of reels with creased tape and let B be the set of reels containing tape with insufficient coating. Then

$$A \cup B$$

is the set of defective reels, so we must find $n(A \cup B)$. From the given data,

$$n(A) = 15 \qquad n(B) = 12 \qquad n(A \cap B) = 7$$

Substituting these values into (2) yields

$$\begin{aligned} n(A \cup B) &= n(A) + n(B) - n(A \cap B) \\ &= 15 + 12 - 7 \\ &= 20 \end{aligned}$$

Therefore, 20 reels were defective.

Solution (b) From part (a), 20 of the 100 reels tested were defective. Therefore,

$$100 - 20 = 80 \qquad \text{(reels)}$$

were nondefective.

Formula (2) can be extended to the unions of three sets. More precisely,

If A, B, and C are sets with finitely many elements, then

$$\begin{aligned} n(A \cup B \cup C) = n(A) &+ n(B) + n(C) \\ &- n(A \cap B) - n(A \cap C) - n(B \cap C) \\ &+ n(A \cap B \cap C) \end{aligned} \tag{3}$$

We omit the proof.

EXAMPLE 6 Let

$$A = \{a, b, c, d\} \qquad B = \{d, e, f, g, h, i\} \qquad C = \{a, g, h, i, j, k\}$$

Then

$$\begin{aligned} A \cup B \cup C &= \{a, b, c, d, e, f, g, h, i, j, k\} \\ A \cap B &= \{d\} \\ A \cap C &= \{a\} \\ B \cap C &= \{g, h, i\} \\ A \cap B \cap C &= \varnothing \end{aligned}$$

As guaranteed by Formula (3)

$$\begin{aligned} 11 = n(A \cup B \cup C) &= n(A) + n(B) + n(C) \\ &\quad - n(A \cap B) - n(A \cap C) - n(B \cap C) \\ &\quad + n(A \cap B \cap C) \\ &= 4 + 6 + 6 - 1 - 1 - 3 + 0 \end{aligned}$$

EXAMPLE 7 A new car dealership offers three options to its customers: power steering, a high-performance engine, and air-conditioning. The dealership listed the following information in its yearly tax records:

200 cars sold
75 without any options
10 with all three options
40 included high-performance engine and air-conditioning
30 included power steering and air-conditioning
20 included power steering and high-performance engine
80 included power steering
60 included high-performance engine
70 included air-conditioning

Explain why the Bureau of Internal Revenue ordered a complete audit of the dealership's records.

Solution Let

A = the set of all cars sold with power steering
B = the set of all cars sold with a high-performance engine
C = the set of all cars sold with air-conditioning.

The $A \cup B \cup C$ is the set of all cars sold with at least one of the three options. From Formula (3),

$$\begin{aligned} n(A \cup B \cup C) &= n(A) + n(B) + n(C) \\ &\quad - n(A \cap B) - n(A \cap C) - n(B \cap C) \\ &\quad + n(A \cap B \cap C) \\ &= 80 + 60 + 70 - 20 - 30 - 40 + 10 \\ &= 130 \end{aligned}$$

Thus 130 cars were sold with at least one of the three options. Since the dealership indicated that 75 cars were sold with no options, it should have reported

$$130 + 75 = 205$$

cars sold, rather than 200, as the records showed.

EXAMPLE 8 In a pollution study of 1500 U.S. rivers the following data were reported:

520 were polluted by sulfur compounds
335 were polluted by phosphates
425 were polluted by crude oil
100 were polluted by crude oil and sulfur compounds
180 were polluted by sulfur compounds and phosphates
150 were polluted by phosphates and crude oil
28 were polluted by sulfur compounds, phosphates, and crude oil

EXAMPLE 10 A coin is tossed four times and the resulting sequence of heads and tails is recorded. Some of the possible sequences are

$$(h, h, h, h) \qquad (h, t, h, t) \quad (t, h, t, t)$$

How many different sequences are possible?

Solution The different sequences can be regarded as the members of the set

$$A \times A \times A \times A$$

where $A = \{h, t\}$. Can you see why? Thus, the number of different sequences is

$$\begin{aligned} n(A \times A \times A \times A) &= n(A) \cdot n(A) \cdot n(A) \cdot n(A) \\ &= 2 \cdot 2 \cdot 2 \cdot 2 = 2^4 = 16 \end{aligned}$$

To check your answer, you may wish to list the 16 possible sequences using a tree diagram.

EXERCISE SET 6.2

1. Find $n(A)$ if it is defined.
(a) $A = \{x|x$ is a consonant in the English alphabet$\}$
(b) $A = \{x|x$ is a solution of $x^2 = 1\}$
(c) $A = \{3, -2, 5, 9\}$
(d) $A = \{f, i, c, k, l, e\}$
(e) $A = \{x|x$ is an even integer$\}$
(f) $A = \{x|x$ is a real number satisfying $x^2 = -4\}$

2. Find $n(A)$ if it is defined.
(a) $A = \{x|x$ is an ace in a standard deck of cards$\}$
(b) $A = \{x||x| = 4\}$
(c) $A = \{-5, -4, -3, -2, -1, 0\}$
(d) $A = \{b, i, r, d, s\}$
(e) $A = \{x|x$ is an odd integer$\}$
(f) $A = \{x|\sqrt{x} = -8\}$

In Exercises 3 and 4 verify the formula

$$n(A \cup B) = n(A) + n(B)$$

for the given disjoint sets.

3. (a) $A = \{3, 7, -8, 6, 2\}$, $B = \{1, -5, 4\}$
(b) $A = \{x|x^2 = 9\}$, $B = \{x||x| = 9\}$

4. (a) $A = \{a, e, f, z\}$ and $B = \{c, h, k\}$
(b) $A = \{x|x$ is a real number satisfying $x^2 < 0\}$ and $B = \{x|x$ is a real number satisfying $x^2 = 4\}$

5. If A and B are disjoint sets such that

$$n(A \cup B) = n(A)$$

what can we say about B?

In Exercises 6–9 verify the formula

$$n(A \cup B) = n(A) + n(B) - n(A \cap B)$$

for the given sets.

6. $A = \{a, c, e, g, i, k\}$, $B = \{e, f, g, h, i, j\}$

7. $A = \{x|x$ is a positive integer $<10\}$ and $B = \{x|x$ is an integer satisfying $1 \leq x \leq 7\}$

8. $A = \{$Xerox, GE, AT&T, IBM, Polaroid$\}$ and $B = \{$Chrysler, Ford, Kodak, Xerox, GE$\}$

9. $A = \{$personal income, gross national product, unemployment$\}$, $B = \{$productivity, unemployment$\}$

10. Let A and B be sets of integers such that $A \cap B = \{-1, 1\}$. Given that $A \cup B$ has 7 elements and B has 4, find the number of elements in A.

11. Let A and B be sets of integers such that

$$A \cup B = \{1, 2, 3, 4, 5, 6, 7, 8, 9\}$$

Given that $A \cap B$ has 5 elements and A has 8, find the number of elements in B.

12. Fill in the missing numbers in each line of Table 1.

TABLE 1

	$n(A)$	$n(B)$	$n(A \cup B)$	$n(A \cap B)$
(a)	6	3	8	
(b)	8	7		4
(c)		5	8	1
(d)	1		9	0

13. Let A, B, and C be sets such that

$$n(A) = 18 \quad n(A \cap B) = 9$$
$$n(B) = 21 \quad n(A \cap C) = 7 \quad n(A \cap B \cap C) = 2$$
$$n(C) = 22 \quad n(B \cap C) = 11$$

Find
(a) $n(A \cup B \cup C)$ (b) $n(A \cup B)$
(c) $n(A \cup C)$ (d) $n(B \cup C)$

14. Solve the problem in Example 8 given that

499 rivers were polluted by sulfur compounds
314 rivers were polluted by phosphates
404 rivers were polluted by crude oil
79 rivers were polluted by crude oil and sulfur compounds
159 rivers were polluted by sulfur compounds and phosphates
129 rivers were polluted by phosphates and crude oil
7 rivers were polluted by sulfur compounds, phosphates, and crude oil

15. Show that the equation

$$n(A \cup B) = n(A) + n(B) - n(A \cap B)$$

reduces to the equation

$$n(A \cup B) = n(A) + n(B)$$

if A and B are disjoint.

16. Under what conditions is it true that

$$n(A \cup B \cup C) = n(A) + n(B) + n(C)$$

17. (a) Use Formula (5) to find $n(A \times B \times C)$ for the sets $A = \{1, 2, 3\}$, $B = \{4, 5, 6\}$, and $C = \{7, 8\}$.
(b) Check the result in (a) by listing the elements in $A \times B \times C$ and then counting them.

18. A license plate consists of 3 letters followed by 2 numbers with repetitions allowed. How many license plates are possible?

19. (a) An ordinary six-sided die is tossed 5 times and the resulting sequence of numbers tossed is recorded. How many sequences are possible?
(b) How many sequences are possible if the die is tossed n times?

20. (a) A coin is tossed 8 times and the resulting sequence of heads and tails is recorded. How many sequences are possible?
(b) How many sequences are possible if the coin is tossed n times?

21. A bank account number consists of seven of the digits $\{0, 1, 2, 3, 4, 5, 6, 7, 8, 9\}$, possibly with repetitions. How many different account numbers are possible?

22. An environmental agency needs to hire a total of 50 employees to monitor water pollution and a total of 60 employees to monitor air pollution. Of those hired, 15 will monitor both. How many employees must be hired?

23. A certain kind of item is considered defective if it has a major defect, a minor defect, or both. In a batch of 25 defective items, 20 have major defects and 14 have minor defects. How many items in the batch have both major and minor defects?

24. According to a political poll before the 1984 presidential election,

420 people surveyed liked Mondale
310 people surveyed liked Ferraro
280 people surveyed like both

If every person who liked either candidate voted the Mondale–Ferraro ticket, how many votes did that ticket get from the group of people surveyed?

25. Solve the problem in part (a) of Example 8 using Formula (3).

26. A sales firm wants to use the following system to keep records on the performance of its salespeople. One cabinet will be devoted to the eastern region, a second to the central region, and a third to the western region. Each cabinet will be divided into three sections:

I: salespeople selling between \$100,000 and \$199,999 per year
II: salespeople selling between \$200,000 and \$499,999 per year
III: salespeople selling more than \$500,000 per year

In each section there will be a folder for each letter of the alphabet to hold the salesperson's file. What is the total number of folders the sales firm will need?

27. In a genetics experiment, the members of a generation of fruit flies were classified as follows:

	Short-winged	**Medium-winged**	**Long-winged**
Male	26	17	8
Female	21	22	9

(a) Find the number of male fruit flies.
(b) Find the number of long-winged fruit flies.
(c) How many fruit flies were either male or medium-winged?
(d) What percentage of the fruit flies were long-winged males?
(e) How many fruit flies were not short-winged females?
(f) How many fruit flies were either female, long-winged, or short-winged?

28. A survey of 125 people revealed that

27 regularly smoked cigars
42 regularly smoked cigarettes
24 regularly smoked a pipe
9 regularly smoked a pipe and cigars
8 regularly smoked cigars and cigarettes
6 regularly smoked a pipe and cigarettes
2 regularly smoked all three

(a) How many nonsmokers were surveyed?
(b) How many of those surveyed regularly smoked only cigarettes?

29. A telephone company routes overseas calls in the following manner. The incoming call is routed to one of 200 transmitting stations. Each transmitting station relays the message to one of three satellites. Each satellite in turn relays the call to one of 60 receiving stations, which in turn sends the message directly to the listener. In how many different ways can a message be routed from the sender to the listener?

30. Excel University sends a student recruiting team composed of Ms. Jane Welcome, Director of Admissions; Mr. William Friendly, Dean of Students; and Mr. Robert Cash, Director of Student Aid on an extensive trip. Each member of the team will talk briefly about a fixed number of topics on a list and will then talk at greater length about one topic on the list.

Ms. Welcome will talk about

Admissions Standards
Work Load
History of Excel University
Excel University's Graduates' Achievements.

Mr. Friendly will discuss the

Sciences Program
Social Sciences Program
Pre-Professional Programs
Athletics Program
Fraternities.

Mr. Cash will cover the

Excel Fund
Federal Aid
State Aid and Minority-Program Aid
Work-Study Program.

(a) How many different presentations are possible?
(b) If Mr. Friendly's slides for the Sciences Program and Mr. Cash's handouts for the Work-Study Program are not available, how many presentations are possible?
(c) If Ms. Welcome's longer talk is on the History of Excel University and Mr. Friendly's longer talk is on the Pre-Professional Programs, how many presentations are possible?

31. One hundred people were polled on whether or not each had been to the opera, theater, or ballet last year. The responses revealed that

45 had been to the opera
40 had been to the theater
41 had been to the ballet
15 had been to the opera and ballet
11 had been to the opera and theater
9 had been to the theater and ballet
5 had been to all three.

(a) How many of the people polled had not attended any of the events?
(b) How many had been to the theater and opera but not to the ballet?
(c) How many had gone only to the theater?

32. One hundred people were polled to determine whether during the previous month they had bought brand X, Y, or Z gasoline. The survey revealed the following:

21 bought brand X
30 bought brand Y
36 bought brand Z
6 bought brands X and Y
16 bought brands Y and Z
2 bought brands X and Z

If two people bought all three brands,
(a) how many people did not buy any of the three brands?
(b) how many bought brands X and Y but not brand Z?
(c) how many bought only brand X?

33. At Superior College, the 1000-student freshman class must take at least one laboratory science course. During the past year,

200 students registered for physics
290 students registered for biology
340 students registered for chemistry
40 students registered for chemistry and physics
70 students registered for physics and biology
30 students registered for chemistry and biology
300 students did not register for any of these three courses (they opted for geology)

(a) How many students registered for chemistry, physics, and biology?
(b) How many students registered for chemistry and physics but not biology?

6.3 PERMUTATIONS AND COMBINATIONS

In this section we are concerned with counting arrangements of the elements of a set. To illustrate what we have in mind, suppose we have three cards labeled a, b, and c, respectively; consider the following problems:

Problem I If 2 cards at a time are selected from the set of 3 and placed side by side, how many different sequences of letters can result?

Answer There are six possible sequences, namely,

$$ab,\ ac,\ ba,\ bc,\ ca,\ cb$$

Problem II If 2 cards are dealt from the set of 3, how many different hands are possible?

Answer In a hand of cards, the order in which the cards are received does not matter. (For example, *ac* and *ca* represent the same 2-card hand.) Thus, three hands are possible, namely,

$$ab,\ ac,\ bc$$

The distinction between these problems is that the order of the letters is important in the first problem, but not in the second. Arrangements of objects are called **permutations** if the order is important, and **combinations** if it is not. This section is devoted to counting permutations and combinations.

There is a general counting rule, called the **multiplication principle**, which we use repeatedly. It can be stated as follows.

Multiplication Principle If there are k ways to make a decision D_1 and for each of these ways there are l ways to make a decision D_2, then there are kl ways to make the two decisions D_1 and D_2.

To see why the multiplication principle holds, let

$$A = \{a_1, a_2, \ldots, a_k\}$$

be the set of possibilities for decision D_1, and let

$$B = \{b_1, b_2, \ldots, b_l\}$$

be the set of possibilities for decision D_2. To make decision D_1 and then decision D_2 involves selecting an element from set A and then selecting an element from set B. Some possibilities are (a_1, b_1), (a_1, b_2), and (a_2, b_1). But these are just ordered pairs from the Cartesian product $A \times B$. As shown in Section 6.2 [Formula (4)], this Cartesian product has kl members; thus, there are kl ways to make the two decisions.

Extended Multiplication Principle If there are n_1 ways to make a decision D_1 and for each of these ways there are n_2 ways to make a decision D_2, and no matter what choices were made for D_1 and D_2 there are n_3 ways to make decision D_3, and so on, then there are

$$n_1 n_2 n_3 \cdots$$

ways to make all the decisions.

EXAMPLE 1 Assume that an employee identification number consists of two letters of the alphabet followed by a sequence of seven digits selected from the set

$$\{0, 1, 2, 3, 4, 5, 6, 7, 8, 9\}$$

If repetitions are allowed, how many identification numbers are possible?

Solution There are 26 possibilities for each of the two letters, and there are 10 possibilities for each of the seven digits. Thus, by the multiplication principle there are

$$26 \cdot 26 \cdot 10 \cdot 10 \cdot 10 \cdot 10 \cdot 10 \cdot 10 \cdot 10 = 26^2 \cdot 10^7 = 676 \cdot 10^7$$

different possible identification numbers.

EXAMPLE 2 A coin is tossed six times, generating a sequence of h's (heads) and t's (tails). Some of the possiblities are

hhhtht *tttht h* *ththth*

How many different sequences are possible?

Solution On each of the six tosses there are two possibilities, h or t. Thus, by the multiplication principle there are

$$2 \cdot 2 \cdot 2 \cdot 2 \cdot 2 \cdot 2 = 2^6 = 64$$

different sequences.

Given a set of distinct objects, an arrangement of these objects in a definite order without repetitions is called a **permutation** of the set.

EXAMPLE 3 There are six different permutations of the set {1, 2, 3}. They are

1 2 3 2 1 3 3 1 2
1 3 2 2 3 1 3 2 1

EXAMPLE 4 How many different permutations can be formed from the set {1, 2, 3, 4}?

Solution Each permutation consists of four digits. The first digit can be selected in any one of four ways; then, since repetitions are not allowed, the second digit can be selected in any one of three ways, the third digit in any of two ways, and the last digit in one way. Thus, by the multiplication principle, there are

$$4 \cdot 3 \cdot 2 \cdot 1 = 24$$

different permutations.

EXAMPLE 5 List the 24 permutations of the set {1, 2, 3, 4}.

Solution Consider the tree diagram shown in Figure 1.

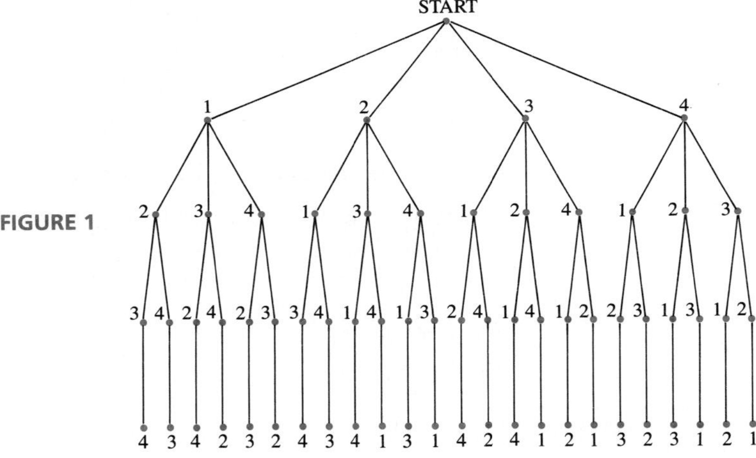

FIGURE 1

The four dots labeled 1, 2, 3, 4 in Figure 1 represent the possible choices for the first number in the permutation. The three branches emanating from each of these dots represent the possible choices for the second position in the permutation. Thus, if the permutation begins

3 _ _ _

the three possibilities for the second position are 1, 2, and 4. The two branches emanating from each dot in the second position represent the possible choices for the third position. Thus, if the permutation begins

3 2 _ _

the two possible choices for the third position are 1 and 4. Finally, the single branch emanating from each dot in the third position represents the only possible choice for the fourth position. Thus, if the permutation begins

3 2 4 _

the only possible choice for the fourth position is 1. The different permutations can now be listed by tracing all possible paths through the tree from the first position to the last position. We obtain the following list by this process:

1 2 3 4	2 1 3 4	3 1 2 4	4 1 2 3
1 2 4 3	2 1 4 3	3 1 4 2	4 1 3 2
1 3 2 4	2 3 1 4	3 2 1 4	4 2 1 3
1 3 4 2	2 3 4 1	3 2 4 1	4 2 3 1
1 4 2 3	2 4 1 3	3 4 1 2	4 3 1 2
1 4 3 2	2 4 3 1	3 4 2 1	4 3 2 1

By imitating the argument given in Example 4, you can show that the number of permutations of $\{1, 2, 3, 4, 5\}$ is

$$5 \cdot 4 \cdot 3 \cdot 2 \cdot 1 = 120$$

and the number of permutations of $\{a, b, c, d, e, f\}$ is

$$6 \cdot 5 \cdot 4 \cdot 3 \cdot 2 \cdot 1 = 720$$

The general principle is now evident:

Counting Permutations of Distinct Objects The number of permutations of a set with n distinct objects is

$$n(n-1)(n-2) \cdot \cdots \cdot 3 \cdot 2 \cdot 1$$

This formula occurs so frequently that we use the symbol $n!$ (read "n factorial") to denote the product of the integers from 1 to n; that is,

$$n! = n(n-1)(n-2) \cdot \cdots \cdot 3 \cdot 2 \cdot 1$$

Thus,

$$\begin{aligned}
1! &= 1 \\
2! &= 2 \cdot 1 = 2 \\
3! &= 3 \cdot 2 \cdot 1 = 6 \\
4! &= 4 \cdot 3 \cdot 2 \cdot 1 = 24 \\
5! &= 5 \cdot 4 \cdot 3 \cdot 2 \cdot 1 = 120 \\
6! &= 6 \cdot 5 \cdot 4 \cdot 3 \cdot 2 \cdot 1 = 720
\end{aligned}$$

Once the value of $n!$ is known, this value can be exploited in computing $(n + 1)!$ For example, using the value of 6! given above, we can compute 7! by writing

$$7! = 7 \cdot 6 \cdot 5 \cdot 4 \cdot 3 \cdot 2 \cdot 1 = 7 \cdot (6!) = 7(720) = 5040$$

Similarly,

$$8! = 8 \cdot (7!) = 8 \cdot (5040) = 40{,}320$$

In general, $(n + 1)!$ and $n!$ are related by

$$(n + 1)! = (n + 1) \cdot n! \qquad (1)$$

Since $n!$ is defined to be the product of the integers from 1 up to n, the symbol $0!$ is not yet defined. By convention

$$0! = 1$$

With this definition, Equation (1) will hold when $n = 0$.

In forming a permutation of a set, all the elements of the set are used. In some problems, however, we want to consider permutations or arrangements where not all the elements are used. For example, we might want to consider all possible permutations of two letters selected from the set $\{a, b, c\}$; they would be

$$\begin{array}{ccc} ab & ac & bc \\ ba & ca & cb \end{array}$$

A permutation of r objects selected from a set of n objects is called a **permutation of the n objects taken r at a time.**

EXAMPLE 6 List all permutations of $\{a, b, c, d\}$ taken two at a time.

Solution This problem can be solved by constructing the tree diagram shown in Figure 2. The dots labeled a, b, c, d correspond to the possible choices for the first letter in the arrangement, the dots at the bottom of the tree diagram correspond to the possible choices for the second letter. By tracing the different paths from the top of the tree diagram to the bottom, we obtain the following list of possible arrangements:

$$\begin{array}{cccc} ab & ba & ca & da \\ ac & bc & cb & db \\ ad & bd & cd & dc \end{array}$$

Thus, there are 12 different permutations of $\{a, b, c, d\}$ taken two at a time.

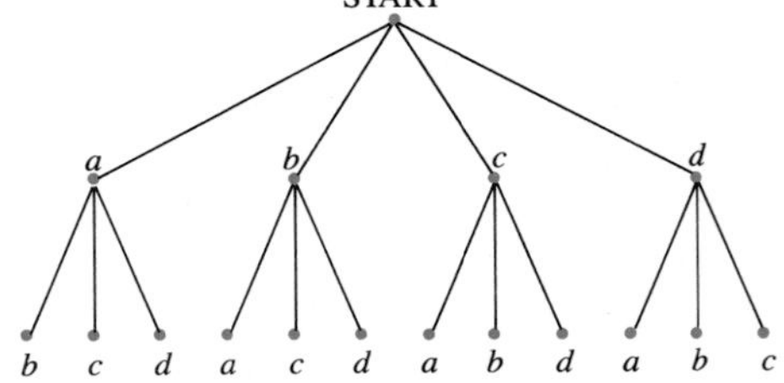

FIGURE 2

We now derive a formula for the number of permutations of n distinct objects taken r at a time. Since we have n distinct objects, there are n ways to choose the first object in the permutation. Then, since repetitions are not allowed, there are

$n - 1$ ways to choose the second object in the permutation, $n - 2$ ways to choose the next object, and so on. Since each permutation contains r objects, it follows from the extended multiplication principle that there are

$$\underbrace{n(n-1)(n-2)\cdots}_{r \text{ factors}}$$

different permutations. This expression occurs so often that it is common to denote it by the symbol $P_{n,r}$. Thus, we have the following result.

Counting Permutations of *n* Distinct Objects Taken *r* at a Time The number of permutations of n distinct objects taken r at a time is

$$P_{n,r} = \underbrace{n(n-1)(n-2)\cdots}_{r \text{ factors}} \tag{2}$$

For example,

$$P_{5,3} = 5 \cdot 4 \cdot 3 = 60 \qquad P_{9,1} = 9$$
$$P_{8,2} = 8 \cdot 7 = 56 \qquad P_{4,4} = 4 \cdot 3 \cdot 2 \cdot 1 = 24$$

EXAMPLE 7 Find the number of permutations of

$$\{a, b, c, d, e, f, g\}$$

taken four at a time.

Solution From (2), the number of permutations of seven distinct objects taken four at a time is

$$P_{7,4} = 7 \cdot 6 \cdot 5 \cdot 4 = 840$$

Formula (2) can be rewritten in another way that is sometimes useful. In (2),

the first factor is n
the second factor is $n - 1$
the third factor is $n - 2$

From the pattern emerging here, you can see that the rth, or last, factor will be

$$n - (r - 1) = n - r + 1$$

Therefore, Formula (2) can be written in the following alternate form, which explicitly shows the last factor in the product.

$$P_{n,r} = n(n-1)(n-2)\cdots(n-r+1) \tag{3}$$

If we multiply Equation (3) by

$$\frac{(n-r)!}{(n-r)!}$$

we obtain still another useful formula for $P_{n,r}$ (see Exercise 47):

$$P_{n,r} = \frac{n!}{(n-r)!} \tag{4}$$

EXAMPLE 8 Use Formulas (2), (3), and (4) to compute $P_{7,4}$ three ways.

Solution We have $n = 7$ and $r = 4$, so Formula (2) yields

$$P_{7,4} = \underbrace{7 \cdot 6 \cdot 5 \cdot 4}_{4 \text{ factors}} = 840$$

Since $n - r + 1 = 7 - 4 + 1 = 4$, Formula (3) yields

$$P_{7,4} = 7 \cdot 6 \cdot 5 \cdot 4 = 840$$

Finally, from Formula (4)

$$P_{7,4} = \frac{7!}{3!} = \frac{7 \cdot 6 \cdot 5 \cdot 4 \cdot 3 \cdot 2 \cdot 1}{3 \cdot 2 \cdot 1} = 7 \cdot 6 \cdot 5 \cdot 4 = 840$$

It should be evident from Example 8 that Formulas (2) and (3) are better for computations than is (4). However, if your calculator has a factorial key, then (4) is simpler to use. Moreover, (4) has other uses that will be important to us.

Combinations

In many counting problems we want to determine the number of different subsets with r objects that can be selected from a set with n distinct objects. For example, there are six different subsets with two objects that can be selected from the set $\{a, b, c, d\}$. They are

$$\{a, b\} \qquad \{a, c\} \qquad \{a, d\} \qquad \{b, c\} \qquad \{b, d\} \qquad \{c, d\}$$

Note, however, that there are 12 different *permutations* of $\{a, b, c, d\}$ taken two at a time (see Example 6). There are many more permutations with two objects than there are subsets with two objects because permutations distinguish among different orderings of the objects. Subsets do not. For example, *ab* and *ba* are different permutations of $\{a, b, c, d\}$ taken two at a time, whereas the sets

$$\{a, b\} \qquad \text{and} \qquad \{b, a\}$$

are not different (they have the same members).

To illustrate why we are interested in counting subsets, suppose we want to find the number of different five-card hands that can be dealt from a standard deck of 52 cards. A hand of five cards can be viewed as a set of five objects selected from a set of 52 distinct objects. Thus, the number of different five-card hands that can be dealt is the same as the number of different subsets with five objects that can be selected from a set with 52 distinct objects.

Observe that it would be *incorrect* to argue that the number of five-card hands is the number of permutations of 52 objects taken five at a time. The reason is that permutations distinguish between different orders so that

$$2H \quad 3H \quad 4H \quad 5H \quad 6H$$

($2H$ = 2 of hearts, $3H$ = 3 of hearts, and so on) and

$$3H \quad 4H \quad 6H \quad 2H \quad 5H$$

would be counted *incorrectly* as different hands, even though they involve the same cards.

For historical reasons, subsets are often called **combinations**. The following definition is more precise.

> A subset of r objects selected from a set of n objects is called a **combination of *n* objects taken *r* at a time**.

The number of different combinations of n objects taken r at a time is denoted by

$$C_{n,r} \quad \text{or} \quad \binom{n}{r}$$

(read "n choose r").

Since each of these symbols is widely used, we use both of them in this text.

To obtain a formula for $C_{n,r}$ we need only observe that a list of all *permutations* of n objects taken r at a time can be obtained by the following two-step procedure:

Step 1 List all combinations of the n objects taken r at a time.

Step 2 Take each combination and list all the permutations of its elements.

When these two steps are completed, we will have a complete list of all permutations of the n objects taken r at a time. For example, the combinations of $\{a, b, c, d\}$ taken three at a time are

$$\{a, b, c\} \quad \{a, b, d\} \quad \{a, c, d\} \quad \{b, c, d\}$$

If we now take each of these combinations and form all possible permutations of the elements, we obtain the list shown in the tinted block in Figure 3.

FIGURE 3

$\{a, b, c\} \rightarrow$	*abc*	*acb*	*bac*	*bca*	*cab*	*cba*
$\{a, b, d\} \rightarrow$	*abd*	*adb*	*bad*	*bda*	*dab*	*dba*
$\{a, c, d\} \rightarrow$	*acd*	*adc*	*cad*	*cda*	*dac*	*dca*
$\{b, c, d\} \rightarrow$	*bcd*	*bdc*	*cbd*	*cdb*	*dbc*	*dcb*

If we are given a set with n distinct objects, there are $C_{n,r}$ different combinations of these objects taken r at a time. Since each of these combinations has r elements, the elements can be permuted in $r!$ different ways. Thus, by the multiplication principle, there are

$$C_{n,r} \cdot r!$$

different ways of forming a combination and permuting its elements. Since this process yields all permutations of the n objects taken r at a time, we obtain

$$P_{n,r} = C_{n,r} \cdot r! \qquad \text{or} \qquad C_{n,r} = \frac{P_{n,r}}{r!}$$

$$C_{n,r} = \frac{n!}{r!(n-r)!} \qquad \text{(from Formula (4))}$$

Thus, we have the following result:

Counting Combinations of *n* Distinct Objects Taken *r* at a Time The number of combinations of n distinct objects taken r at a time is

$$C_{n,r} = \frac{n!}{r!(n-r)!} \tag{5}$$

EXAMPLE 9 Find

(a) the number of combinations of 30 objects taken 8 at a time

(b) the number of combinations of 30 objects taken 22 at a time

Solution (a) Applying Formula (5) with $n = 30$ and $r = 8$ yields

$$C_{30,8} = \frac{30!}{8!\,22!} \tag{6}$$

To evaluate this expression, the best strategy is to rewrite the numerator in a form that will enable us to cancel the larger factor in the denominator, that is, 22! We do this by writing

$$C_{30,8} = \frac{30!}{8!\,22!} = \frac{(30 \cdot 29 \cdot 28 \cdot 27 \cdot 26 \cdot 25 \cdot 24 \cdot 23) \cdot \cancel{22!}}{8! \cdot \cancel{22!}}$$

$$= \frac{30 \cdot 29 \cdot 28 \cdot 27 \cdot 26 \cdot 25 \cdot 24 \cdot 23}{8 \cdot 7 \cdot 6 \cdot 5 \cdot 4 \cdot 3 \cdot 2 \cdot 1}$$

$$= 5{,}852{,}925$$

where the final computation is done with a calculator after canceling common factors in the numerator and denominator of the preceding expression.

Solution (b) Applying Formula (5) with $n = 30$ and $r = 22$ yields

$$C_{30,22} = \frac{30!}{22!\,8!}$$

This expression is the same as (6) except that the order of the factors in the denominator is reversed. Thus,

$$C_{30,22} = C_{30,8} = 5{,}852{,}925$$

The result in part (b) of Example 9 is a special case of the following general result (see Exercise 48).

$$C_{n,r} = C_{n,n-r} \tag{7}$$

EXAMPLE 10 The following are applications of Formula (7):

$$C_{12,3} = C_{12,9} \qquad C_{46,2} = C_{46,44} \qquad C_{100,1} = C_{100,99}$$

Sometimes you must judge whether to count permutations or combinations. Always keep the following rules in mind:

> If order matters, use permutations.
> If order does not matter, use combinations.

EXAMPLE 11 The board of directors authorizes a bank to promote three of its top 10 executives to the position of vice president. In how many different ways can the promoted persons be selected?

Solution We must first decide whether order matters. To do this, label the executives 1, 2, 3, . . . , 10. Should we, for example, regard

$$1, 5, 7 \quad \text{and} \quad 7, 1, 5$$

as different selections, in which case order matters, or should we regard them as the same selection, in which case order does not matter?

Clearly, the two selections should be regarded as the same since the same people become vice president in both cases. Thus, order does not matter, and the number of ways in which the individuals can be selected is the number of combinations of 10 objects taken three at a time. Thus, there are

$$C_{10,3} = \frac{10!}{3!\,7!} = \frac{(10 \cdot 9 \cdot 8) \cdot \cancel{7!}}{3!\,\cancel{7!}} = \frac{10 \cdot 9 \cdot 8}{3 \cdot 2 \cdot 1} = 120$$

different possible selections.

EXAMPLE 12 The board of directors authorizes a bank to promote three of its top 10 executives, one to the rank of first vice president, one to the rank of second vice president, and one to the rank of third vice president. In how many ways can the individuals be selected for promotion?

Solution As in Example 11, label the executives 1, 2, 3, . . . , 10 and suppose that we first choose the first vice president, then the second vice president, and then the third vice president. In this case

$$1, 5, 7 \quad \text{and} \quad 7, 1, 5$$

are different selections, so order matters. Thus, there are

$$P_{10,3} = 10 \cdot 9 \cdot 8 = 720$$

possible selections.

EXAMPLE 13 How many different five-card hands can be dealt from a deck of 52 cards?

Solution A hand of cards is determined only by the cards received and *not* by the order in whch they are dealt. Thus, the number of five-card hands is the num ber of combinations of 52 objects taken five at a time:

$$C_{52,5} = \frac{52!}{5!\,47!} = \frac{(52 \cdot 51 \cdot 50 \cdot 49 \cdot 48) \cdot \cancel{47!}}{5 \cdot 4 \cdot 3 \cdot 2 \cdot 1 \cdot \cancel{47!}} = 2{,}598{,}960$$

EXERCISE SET 6.3

In Exercises 1–10 evaluate the expressions.

1. (a) $6!$ (b) $7!$ (c) $8!$ (d) $\frac{33!}{31!}$ (e) $\frac{67!}{66!}$ (f) $\frac{14!}{8!}$ (g) $\frac{0!}{2!}$

2. (a) $9!$ (b) $10!$ (c) $11!$ (d) $\frac{22!}{19!}$ (e) $\frac{100!}{98!}$ (f) $\frac{12!}{7!}$ (g) $\frac{9!}{0!}$

3. (a) $\frac{8!}{4!\,2!}$ (b) $\frac{27!}{14!\,13!}$ (c) $\frac{0!}{3!\,0!}$ (d) $\frac{5!}{3! + 2!}$ (e) $\frac{5!}{(3 + 2)!}$

4. (a) $\frac{12!}{3!\,5!}$ (b) $\frac{5!}{12!\,3!}$ (c) $\frac{9!}{0!\,0!}$ (d) $\frac{(4 + 2)!}{6!}$ (e) $\frac{4! + 2!}{6!}$

5. (a) $P_{7,2}$ (b) $P_{12,4}$ (c) $P_{5,5}$ (d) $P_{n,1}$

6. (a) $P_{32,4}$ (b) $P_{6,5}$ (c) $P_{17,6}$ (d) $P_{n,n}$

7. (a) $C_{7,2}$ (b) $C_{12,4}$ (c) $C_{5,5}$ (d) $C_{40,6}$

8. (a) $C_{9,3}$ (b) $C_{56,4}$ (c) $C_{56,52}$ (d) $C_{20,12}$

9. (a) $\frac{4! + 5!}{4! - 6!}$ (b) $\frac{5!\,7!}{9! - 7!}$ (c) $\frac{5! + 4!}{(5 + 4)!}$

10. (a) $\frac{17!\,19!}{18!\,17!}$ (b) $\frac{8! - 5!}{(8 - 5)!}$ (c) $\frac{4!\,4!}{4! + 4!}$

11. Find the value of s that creates an equality.
(a) $C_{10,2} = C_{10,s}$ (b) $C_{9,3} = C_{s,6}$
(c) $C_{5,s} = 1$

12. Simplify:
(a) $\frac{(n + 1)!}{n!}$ (b) $\frac{n!}{(n - 2)!}$ (c) $\frac{(n!)(n + 1)}{(n + 1)!}$
(d) $\frac{n!\,n!}{2n!}$

13. A communication signal can be transmitted from point a to one of seven possible satellites, after which it is relayed to one of 50 possible ground stations and sent to point b. Use the multiplication principle to compute the number of different ways to send a signal from a to b.

14. A license plate consists of two letters of the alphabet followed by three of the integers from 0 to 9. Use the multiplication principle to determine how many different license plates are possible.

15. A medical researcher classifies humans according to one of three skin colors, one of eight blood types, and one of two sexes. Use the multiplication principle to determine how many classes are possible.

16. A coin is tossed 10 times and the sequence of heads and tails is recorded. Use the multiplication principle to determine how many sequences are possible. How many sequences are possible if the coin is tossed n times?

17. In a VAX 11/785 digital computer, a *bit* is one of the integers 0 or 1; a *word* is a sequence of 32 bits. Use the multiplication principle to determine how many different words are possible.

18. A psychological test consists of 10 questions, each of which can be answered *always, never,* or *sometimes*. Use the multiplication principle to determine the number of ways the 10-question test can be answered.

19. A die is tossed fives times and the sequence of numbers is recorded. Use the multiplication principle to determine how many sequences are possible. How many sequences are possible if the die is tossed n times?

20. How many different permutations can be formed from the following sets?

(a) $\{1, 2\}$ (b) $\{a, b, c, d, e\}$ (c) $\{1, 2, \ldots, 10\}$

21. Use a tree diagram to list all permutations of

$$\{a, b, c, d\}$$

22. Use a tree diagram to list all permutations of

$$\{a, b, c, d, e\}$$

23. (a) How many permutations can be formed from $\{a, b, c, d, e\}$ taken two at a time?
(b) Use a tree diagram to list them.

24. (a) How many permutations can be formed from $\{a, b, c, d, e\}$ taken three at a time?
(b) Use a tree diagram to list them.

25. In how many ways can a jury of 12 people be selected from an available pool of 25 people?

26. Repeat Exercise 25 if one person is to be designated as foreman.

27. In how many ways can a jury consisting of 8 women and 4 men be chosen from an available pool of 15 women and 10 men?

28. A video shop is displaying seven new videocassettes on a shelf. Suppose that four are sport games and three are science fiction games.

(a) In how many ways can these be arranged on the shelf?
(b) In how many ways can these be arranged on the shelf if the sport games and the science fiction games are to be kept together?

29. A university has to recruit 5 new freshmen for its football team. In how many ways can these be chosen from a group of 20 eligible applicants?

30. In a certain township each police car is staffed by 2 police officers. If seven cars have to be staffed from a pool of 14 male officers and 14 female officers, how many different teams can be formed?

31. In the runoff of a state lottery, there are 10 ticket holders in a drawing for three prizes, a \$50,000 prize, a \$25,000 prize, and a \$10,000 prize. If no person can win more than one prize, how many outcomes are possible?

32. How many three-letter words (real or fictitious) can be formed from the first 10 letters of the alphabet if

(a) repetitions are not allowed?
(b) repetitions are allowed?

33. A combination lock is opened by making three turns to the left, stopping at digit a, then two turns to the right, stopping at digit b, and finally one turn to the left, stopping at digit c. How many lock combinations are possible if a, b, and c are selected from the digits 0 to 9 inclusive?

34. A testing program to determine the pollutants present in river water involves selecting one of three filtering processes, then one of five precipitation procedures, and then one of two evaporation methods. How many testing programs are possible?

35. In a customer preference survey, a person is asked to test 3 of 10 breakfast cereals and rate them 1, 2, and 3 in order of preference. How many ratings are possible?

36. List all combinations of $\{a, b, c, d, e\}$ taken three at a time.

37. Use the results of Exercise 36 and the two-step procedure illustrated in Figure 3 of this section to obtain a list of all permutations of $\{a, b, c, d, e\}$ taken three at a time. Compare your results to those obtained in Exercise 24.

38. How many different seven-card hands can be dealt from a deck of 52 cards?

39. How many different seven-card hands with all black cards can be dealt from a deck of 52 cards?

40. (a) How many different 11-card hands can be dealt from a deck of 52 cards?
(b) How many different 11-card hands with 4 aces can be dealt from a deck of 52 cards?

41. In a state election, there are 9 candidates running for 3 positions entitled State Judge. How many different election outcomes are possible?

42. At a national presidential convention, party rules require that the Credentials Committee contain 3 men and 2 women. If there are 9 men and 6 women to choose from, in how many different ways can the committee be formed?

43. A soils engineer divides a proposed building site into 25 equal square plots and randomly chooses four plots for testing. In how many different ways can the four plots be selected for testing, if they are all tested in the same manner?

44. A coin is tossed seven times and the resulting sequence of heads and tails is recorded.
(a) How many sequences with exactly 5 heads are possible?
(b) How many sequences with exactly 2 heads are possible?
(c) How many sequences with at least 5 heads are possible?

45. A die is tossed seven times in succession and the resulting sequence of numbers is observed.
(a) How many sequences are possible?
(b) How many sequences with exactly one 6 are possible?
(c) How many sequences with exactly two 6's are possible?
(d) How many sequences are possible which contain at most two 6's?

46. A psychologist has a maze with five compartments. For a certain experiment, each compartment is to be painted with one color: red, black, or orange.
(a) In how many different ways can the maze be painted?
(b) In how many different ways can the maze be painted if no more than two colors can be used?

47. Show that

$$n(n-1)(n-2)\cdots(n-r+1) = \frac{n!}{(n-r)!}$$

48. Show that $C_{n,r} = C_{n,n-r}$.

49. **(Social security)** A U.S. social security number consists of 9 digits with repetitions allowed. What is the maximum U.S. population that can be supported with this system if every individual must have a different social security number?

6.4 MORE ON COUNTING PROBLEMS

In this section we consider counting problems that require careful analysis and the right strategy to solve. Although these problems are primarily concerned with flipping coins, dealing cards, and such, the techniques involved are directly applicable to real-world problems.

EXAMPLE 1 How many 5-card hands with 3 red cards and 2 black cards can be dealt from a deck of 52 cards?

Solution We need the multiplication principle to solve this problem. Imagine a hand with 3 red cards and 2 black cards to be formed in two steps:

Step 1 Deal 3 cards from the 26 red cards in the deck.

Step 2 Deal 2 cards from the 26 black cards in the deck.

The first step can be performed in $C_{26,3}$ ways and the second in $C_{26,2}$ ways. Therefore, by the multiplication principle, the two steps can be performed in $C_{26,3} \cdot C_{26,2}$ ways. So the number of hands with 3 red cards and 2 black cards is

$$C_{26,3} \cdot C_{26,2} = \frac{26 \cdot 25 \cdot 24}{3 \cdot 2 \cdot 1} \cdot \frac{26 \cdot 25}{2 \cdot 1} = 2600 \cdot 325 = 845{,}000$$

EXAMPLE 2 A hand of 7 cards is dealt from a deck of 52 cards.

(a) How many different hands are possible?

(b) How many hands with 3 spades are possible?

(c) How many hands with 3 spades and 2 diamonds are possible?

Solution (a) This problem is similar to Example 13 of Section 6.3. The number of hands is

$$C_{52,7} = \frac{52!}{7!\,45!} = \frac{52 \cdot 51 \cdot 50 \cdot 49 \cdot 48 \cdot 47 \cdot 46}{7 \cdot 6 \cdot 5 \cdot 4 \cdot 3 \cdot 2 \cdot 1} = 133{,}784{,}560$$

Solution (b) This problem is similar to Example 1 above: imagine dealing 3 spades from the 13 spades in the deck, then dealing 4 cards from the remaining 39 cards (nonspades) in the deck. The number of possible hands is

$$C_{13,3} \cdot C_{39,4} = \frac{13!}{3!\,10!} \cdot \frac{39!}{4!\,35!} = 23{,}523{,}786$$

Solution (c) Imagine dealing 3 spades from the 13 spades in the deck, then 2 diamonds from the 13 diamonds in the deck, and then 2 cards from the remaining 26 cards. The number of possible hands is

$$C_{13,3} \cdot C_{13,2} \cdot C_{26,2} = \frac{13!}{3!\,10!} \cdot \frac{13!}{2!\,11!} \cdot \frac{26!}{2!\,24!} = 7{,}250{,}100$$

The following two examples show how you can sometimes solve counting problems by drawing analogies to card-dealing problems.

EXAMPLE 3 Four balls are selected from an urn that contains 7 red balls and 3 black balls. How many different selections containing 3 red balls and 1 black ball are possible?

Solution This is equivalent to a card-dealing problem. Imagine a deck of 10 cards marked $R_1, R_2, \ldots, R_7$ (corresponding to the 7 red balls) and B_1, B_2, B_3 (corresponding to the 3 black balls). The selection of 4 balls is equivalent to dealing a hand of 4 cards from this deck. Thus, the number of selections with 3 red balls and 1 black ball is equal to the number of hands that have 3 cards with an R and 1 card with a B. The number of such selections is

$$C_{7,3} \cdot C_{3,1} = \frac{7!}{3!\,4!} \cdot \frac{3!}{1!\,2!} = 105$$

EXAMPLE 4 A sample of 8 items is selected from a lot of 100 items known to contain 10 defectives and 90 nondefectives. How many different samples with 3 defectives and 5 nondefectives are possible?

Solution This can also be viewed as a card-dealing problem. We can imagine the 100 items to be a deck of 100 cards. In this deck, 10 cards are marked $D_1, D_2, \ldots, D_{10}$ (corresponding to the 10 defectives) and 90 of the cards are marked N_1,

$N_2, \ldots, N_{90}$ (corresponding to the 90 nondefectives). The sample of 8 items corresponds to a hand of 8 cards dealt from this deck, and a sample with 3 defectives and 5 nondefectives corresponds to a hand with 3 D's and 5 N's. Thus, the number of such samples is

$$C_{10,3} \cdot C_{90,5} = \frac{10!}{3!\,7!} \; \frac{90!}{5!\,85!} = 5{,}273{,}912{,}160$$

Permutations of Objects with Repetitions

So far we have considered only permutations of distinct objects. We now consider some problems that lead to counting permutations of objects with repetitions.

EXAMPLE 5 A coin is tossed 6 times, and the resulting sequence of heads and tails is recorded. Example 2 of Section 6.3 shows that there are $2^6 = 64$ different sequences possible. How many of these sequences contain exactly 2 heads?

Solution A sequence with 2 heads (h's) will necessarily have 4 tails (t's). Some possibilities are

$$hhtttt \quad \text{and} \quad tththt \tag{1}$$

It is evident that any sequence with 2 h's and 4 t's is completely determined once we specify the two tosses on which the heads occur or the four tosses on which the tails occur. For example, the first sequence in (1) is completely determined by stating that the 2 h's occur on tosses 1 and 2; it is also completely determined by stating that the 4 t's occur on tosses 3, 4, 5, and 6. Thus, the number of sequences with 2 h's and 4 t's can be obtained in two ways:

1. Find the number of ways of selecting a set of two toss numbers for the h's from the set of possible toss numbers $\{1, 2, 3, 4, 5, 6\}$.
2. Find the number of ways of selecting a set of four toss numbers for the t's from the set of possible toss numbers $\{1, 2, 3, 4, 5, 6\}$.

Method 1 yields

$$C_{6,2} = \frac{6!}{2!\,4!} = \frac{6 \cdot 5}{2 \cdot 1} = 15$$

and method 2 yields

$$C_{6,4} = \frac{6!}{4!\,2!} = \frac{6 \cdot 5}{2 \cdot 1} = 15$$

The last example can be viewed as a problem of counting arrangements (permutations) of objects with repetitions—we were interested in counting the number of different permutations of 2 h's and 4 t's. The following result is useful for solving such problems.

If n objects can be divided into two categories such that the objects within each category are identical but those in different categories are not, then the number of distinct permutations of the n objects is

$$\frac{n!}{n_1!\,n_2!} \tag{2}$$

where n_1 is the number of objects in the first category and n_2 is the number in the second category.

This is a special case of the following more general result, whose proof we omit.

If n objects can be divided into k categories such that the objects within each category are identical but those in different categories are not, then the number of distinct permutations of the n objects is

$$\frac{n!}{n_1!\,n_2!\cdots n_k!} \tag{3}$$

where n_1 is the number of objects in the first category, n_2 the number in the second category, and so on.

The next example illustrates the result above.

EXAMPLE 6 In how many ways can we permute the letters of the word *MISSISSIPPI*?

Solution We have

$$\begin{aligned} n &= 11 \text{ total letters} \\ n_1 &= 1\ M \\ n_2 &= 4\ I\text{'s} \\ n_3 &= 4\ S\text{'s} \\ n_4 &= 2\ P\text{'s} \end{aligned}$$

Thus, from Equation (3) the number of different permutations is

$$\frac{n!}{n_1!\,n_2!\,n_3!\,n_4!} = \frac{11!}{1!\,4!\,4!\,2!} = 34{,}650$$

The phrases *exactly*, *at most*, *at least*, *less than*, and *more than* are common in counting problems. The following examples illustrate typical counting problems involving these phrases.

EXAMPLE 7 A coin is tossed 5 times, and the resulting sequence of heads and tails is recorded.

(a) How many sequences have exactly 3 heads?

(b) How many sequences have at most 3 heads?

(c) How many sequences have at least 3 heads?

(d) How many sequences have more than 3 heads?

Solution (a) If there are exactly 3 heads, then there are exactly 2 tails, so the number of such sequences is the number of permutations of 3 heads and 2 tails. From (2), that number is

$$\frac{5!}{3!\,2!} = 10 \text{ sequences}$$

Solution (b) To say that there are at most 3 heads means that one of the following is true:

there are no heads
there is exactly 1 head
there are exactly 2 heads
there are exactly 3 heads

Thus, we calculate the number of sequences satisfying each condition and add the results. We obtain

$$\frac{5!}{0!\,5!} + \frac{5!}{1!\,4!} + \frac{5!}{2!\,3!} + \frac{5!}{3!\,2!} = 1 + 5 + 10 + 10 = 26 \text{ sequences}$$

Solution (c) To say that a sequence has at least 3 heads means that one of the following is true:

there are exactly 3 heads
there are exactly 4 heads
there are exactly 5 heads

Proceeding as in part (b), we obtain

$$\frac{5!}{3!\,2!} + \frac{5!}{4!\,1!} + \frac{5!}{5!\,0!} = 10 + 5 + 1 = 16 \text{ sequences}$$

Solution (d) To say that the sequence has more than 3 heads means that one of the following is true:

there are exactly 4 heads
there are exactly 5 heads

Proceeding as in part (b), we obtain

$$\frac{5!}{4!\,1!} + \frac{5!}{5!\,0!} = 5 + 1 = 6 \text{ sequences}$$

Counting problems involving the phrases *at least* or *at most* can sometimes be recast in an alternate form, saving time in computation.

EXAMPLE 8 A coin is tossed 100 times, and the sequence of heads and tails is recorded. How many sequences have at least 2 heads?

Solution A direct solution of this problem would involve a sum of 99 terms: the number of sequences with exactly 2 heads plus the number with exactly 3 heads, and so on. A better approach is to count the number of sequences that do *not* have at least 2 heads and subtract this from the total number of sequences. This will leave us with the number of sequences having at least 2 heads.

Using the method of Example 10 in Section 6.2, the total number of sequences is

$$\underbrace{2 \cdot 2 \cdot 2 \cdot \cdots \cdot 2}_{100 \text{ factors}} = 2^{100}$$

Those that do *not* have at least 2 heads are those with no heads or exactly 1 head. The number of these is

$$\frac{100!}{0!\ 100!} + \frac{100!}{1!\ 99!} = 1 + 100 = 101$$

Thus, the number of sequences with at least 2 heads is

$$2^{100} - 101$$

EXERCISE SET 6.4

1. How many seven-card hands with 5 red cards and 2 black cards can be dealt from a deck of 52 cards?
2. How many seven-card hands with all black cards can be dealt from a deck of 52 cards?
3. A hand of six cards is dealt from a deck of 52 cards.
   (a) How many hands are possible?
   (b) How many hands that contain exactly 2 diamonds are possible?
   (c) How many hands with 3 hearts and 3 clubs are possible?
   (d) How many hands with 3 hearts and 3 black cards are possible?
4. A hand of eight cards is dealt from a deck of 52 cards.
   (a) How many hands are possible?
   (b) How many hands with no hearts are possible?
   (c) How many hands with 4 aces are possible?
   (d) How many hands with 3 diamonds and 5 black cards are possible?
5. An urn contains 9 balls; 6 are green and 3 are red. Six balls are selected from the urn.
   (a) How many different selections are possible?
   (b) How many selections with 4 green balls and 2 red balls are possible?
   (c) How many selections with all green balls are possible?
6. An urn contains 12 balls; 6 are blue and 6 are yellow. Six balls are selected from the urn.
   (a) How many different selections are possible?
   (b) How many selections with half blue balls and half yellow balls are possible?
   (c) How many selections in which all balls have the same color are possible?
7. A sample of 6 items is selected from a lot of 50 items known to contain 10 defectives and 40 nondefectives.
   (a) How many different samples are possible?
   (b) How many different samples with 2 defectives and 4 nondefectives are possible?
   (c) How many different samples with no defectives are possible?

8. A sample of 3 light bulbs is selected from a lot of 25 light bulbs known to contain 3 defectives and 22 nondefectives.
   (a) How many different samples are possible?
   (b) How many different samples with 2 defective bulbs and 1 nondefective bulb are possible?
   (c) How many different samples are possible in which all the bulbs are defective or all are nondefective?

9. A coin is tossed 9 times and the sequence of heads and tails is recorded.
   (a) How many different sequences are possible?
   (b) How many sequences with 5 heads and 4 tails are possible?
   (c) How many sequences with exactly 2 heads or exactly 2 tails are possible?

10. A coin is tossed 12 times and the sequence of heads and tails is recorded.
   (a) How many different sequences are possible?
   (b) How many different sequences with 4 heads and 8 tails are possible?
   (c) How many sequences with half heads and half tails are possible?

11. In each part determine how many different permutations of the given letters are possible.
   (a) BBBBBBGGGGG (b) AARDVARK
   (c) MISSOURI

12. In each part determine how many different permutations of the given letters are possible.
   (a) QQQTTVVVVH (b) CADILLAC
   (c) BOOKKEEPER

13. A hand of nine cards is dealt from a deck of 52 cards.
   (a) How many hands with at least 1 spade are possible?
   (b) How many hands with at most 7 spades are possible?

14. A hand of seven cards is dealt from a deck of 52 cards.
   (a) How many hands with at most 2 black cards are possible?
   (b) How many hands with at least 2 black cards are possible?

15. A coin is tossed 50 times and the sequence of heads and tails is recorded.
   (a) How many sequences have at least 3 heads?
   (b) How many sequences have at most 48 tails?

16. A sample of 50 items is selected from a lot of 100 items containing 50 defectives and 50 nondefectives.
   (a) How many samples have at least 1 defective?
   (b) How many samples have at most 48 nondefectives?

17. **(Computer science)** In some computers, a *word* consists of a sequence of 16 *bits*, each of which can be a 0 or a 1.
   (a) How many words with at least twelve 0s are possible?
   (b) How many words with at most two 1s are possible?

18. **(Personnel selection)** A set of 5 people is selected from a personnel pool that consists of 12 men and 12 women. Half the men in the pool have B.A. degrees and half have M.A. degrees. The same is true for the women in the pool.
   (a) How many different selections are possible?
   (b) How many different selections with 2 B.A. degrees and 3 M.A. degrees are possible?
   (c) How many different selections with 3 women having M.A. degrees and 2 men are possible?

19. **(Medicine)** A blood bank has the following inventory:

   10 units of type A^-
   5 units of type A^+
   6 units of type AB^-
   4 units of type AB^+
   2 units of type O^+

   The blood is shipped in cartons of 8 units each to hospitals.
   (a) How many cartons with 8 units of type A^- are possible?
   (b) How many different cartons with 3 units of type A^-, 2 units of type AB^+, and 1 unit of type O^+ are possible?
   (c) How many different cartons are possible?

20. **(Harder)** A department store receives 8 dresses: 4 green and 4 yellow. In how many different ways can these be distributed between two show windows if each window must display at least 3 dresses and at least 1 dress of each color? (Assume that dresses of the same color are of different types.)

21. **(Harder)** A modeling agency agrees to provide a fashion house with 3 models each week for 52 weeks. The contract requires that in no two weeks will exactly the same 3 models be sent. What is the minimum number of models needed?

22. The mathematics curriculum committee is composed of 12 faculty members. They intend to form 3 subcommittees: 1 consisting of 4 members, 1 of 2 members, and 1 of 6 members. How many different subcommittees can be formed?

6.5 THE BINOMIAL FORMULA

In this section we discuss the binomial formula and its relationship to counting problems.

By direct calculation it is easy to obtain the following powers of $a + b$:

$$
\begin{aligned}
(a+b)^0 &= 1\\
(a+b)^1 &= a+b\\
(a+b)^2 &= a^2+2ab+b^2\\
(a+b)^3 &= a^3+3a^2b+3ab^2+b^3\\
(a+b)^4 &= a^4+4a^3b+6a^2b^2+4ab^3+b^4\\
(a+b)^5 &= a^5+5a^4b+10a^3b^2+10a^2b^3+5ab^4+b^5
\end{aligned}
$$

The expressions on the right side of the equal signs are called the **expansions** of the powers on the left side. For centuries, mathematicians have been fascinated with the patterns in the expansions of $(a + b)^n$. For example, the six expansions above suggest the following general results:

Pattern in the expansion of $(a + b)^n$	**Example**
There are $n + 1$ terms.	$(a + b)^4$ has 5 terms.
The first term is a^n and the last term is b^n.	The first term of $(a + b)^4$ is a^4 and the last term is b^4.
In each term the sum of the exponents is n.	In each term of $(a + b)^4$ the sum of the exponents is 4.
Moving left to right, the exponents of a decrease by 1.	In the terms of $(a + b)^4$, the successive powers of a are a^4, a^3, a^2, a^1, a^0 (equals 1).
Moving left to right, the exponents of b increase by 1.	In the terms of $(a + b)^4$, the successive powers of b are b^0 (equals 1), b^1, b^2, b^3, b^4.

The coefficients in the successive expansions of $(a + b)^n$ also exhibit important patterns. For example, if we arrange the coefficients of $(a + b)^n$ for $n = 1$ to 5 in a triangular array, we obtain

$$
\begin{array}{lccccccccccc}
(a+b)^0\text{-------} & & & & & & 1 & & & & & \\
(a+b)^1\text{-------} & & & & & 1 & & 1 & & & & \\
(a+b)^2\text{-------} & & & & 1 & & 2 & & 1 & & & \\
(a+b)^3\text{-------} & & & 1 & & 3 & & 3 & & 1 & & \\
(a+b)^4\text{-------} & & 1 & & 4 & & 6 & & 4 & & 1 & \\
(a+b)^5\text{-------} & 1 & & 5 & & 10 & & 10 & & 5 & & 1
\end{array}
$$

This array, called **Pascal's* triangle**, has the property that *each row begins and ends with a 1, and each intermediate number in a row (if any) is the sum of the two diagonal entries immediately above it.*

**Blaise Pascal* (1623–1662), born in France, was the son of a government official. Pascal was in poor health throughout his lifetime. Although he was an extremely gifted child who showed great interest in mathematics at an early age, his father tried to keep him away from mathematics so that he would not strain himself. However, Pascal persisted in learning geometry by himself at the age of 12. When he was 19 years old he invented the first computing machine to help his father in his duties as a tax assessor. He made important contributions to the foundations of calculus and to projective geometry. Together with Fermat, Pascal was one of the founders of the field of probability. He was also the author of several literary classics. From the age of 24 until his death, religious thought and practice were his most important concerns.

EXAMPLE 1 Find the expansion of $(a + b)^6$.

Solution The row in the Pascal triangle corresponding to $(a + b)^6$ can be generated from the preceding row as follows:

$$\begin{array}{lccccccccccccc}
(a+b)^5\text{-------------} & & 1 & & 5 & & 10 & & 10 & & 5 & & 1 & \\
& & & \searrow & \swarrow\ \searrow & & \swarrow\ \searrow & & \swarrow\ \searrow & & \swarrow\ \searrow & \swarrow & & \\
(a+b)^6\text{----------} & 1 & & 6 & & 15 & & 20 & & 15 & & 6 & & 1
\end{array}$$

From these coefficients and the patterns above we obtain

$$(a + b)^6 = a^6 + 6a^5b + 15a^4b^2 + 20a^3b^3 + 15a^2b^4 + 6ab^5 + b^6$$

Because $(a + b)^n$ is a power of the binomial $a + b$, the coefficients in the expansion of $(a + b)^n$ are called **binomial coefficients**. We leave it as an exercise to show that starting with the second row in Pascal's triangle the binomial coefficients can be written in terms of the combination symbol

$$C_{n,r} = \binom{n}{r}$$

as follows:

$$\begin{array}{ll}
(a+b)^0\text{--------------------------------} & \binom{0}{0} = 1 \\
(a+b)^1\text{-----------------------------} & \binom{1}{0}\quad\binom{1}{1} \\
(a+b)^2\text{------------------------} & \binom{2}{0}\quad\binom{2}{1}\quad\binom{2}{2} \\
(a+b)^3\text{---------------------} & \binom{3}{0}\quad\binom{3}{1}\quad\binom{3}{2}\quad\binom{3}{3} \\
(a+b)^4\text{-----------------} & \binom{4}{0}\quad\binom{4}{1}\quad\binom{4}{2}\quad\binom{4}{3}\quad\binom{4}{4} \\
(a+b)^5\text{-------------} & \binom{5}{0}\quad\binom{5}{1}\quad\binom{5}{2}\quad\binom{5}{3}\quad\binom{5}{4}\quad\binom{5}{5}
\end{array}$$

This result suggests that in general the binomial coefficients of $(a + b)^n$ are

$$\binom{n}{0}, \binom{n}{1}, \binom{n}{2}, \ldots, \binom{n}{n-1}, \binom{n}{n}$$

From this fact and the patterns above, we are led to the following general result, called the **binomial formula**.

Binomial Formula

$$(a+b)^n = \binom{n}{0}a^n + \binom{n}{1}a^{n-1}b + \binom{n}{2}a^{n-2}b^2 + \cdots + \binom{n}{n-1}ab^{n-1} + \binom{n}{n}b^n \qquad (1)$$

EXAMPLE 2 Use the binomial formula to find the expansion of $(a + b)^6$.

Solution

$$(a+b)^6 = \binom{6}{0}a^6 + \binom{6}{1}a^5b + \binom{6}{2}a^4b^2 + \binom{6}{3}a^3b^3 + \binom{6}{4}a^2b^4 + \binom{6}{5}ab^5 + \binom{6}{6}b^6$$

Evaluating the binomial coefficients yields

$$(a+b)^6 = a^6 + 6a^5b + 15a^4b^2 + 20a^3b^3 + 15a^2b^4 + 6ab^5 + b^6$$

(verify this), which agrees with the result in Example 1.

To understand why the combinatorial symbol enters into the binomial formula observe that

$$(a+b)^n = \underbrace{(a+b)(a+b)\cdots(a+b)}_{n \text{ factors}} \qquad (2)$$

To expand the expression on the right side of the equals sign in (2), we must form all possible products, where one term is selected from each of the n factors, and then combine like terms. For example,

$$\begin{aligned}(a+b)(a+b)(a+b) &= aaa + aab + aba + abb + baa + bab + bba + bbb \\ &= a^3 + a^2b + a^2b + ab^2 + a^2b + ab^2 + ab^2 + b^3 \\ &= a^3 + 3a^2b + 3ab^2 + b^3\end{aligned}$$

In this calculation, the first term is obtained by selecting factor a from each of the three factors on the left side of the equals sign. The second term is obtained by selecting factor a from the first two factors on the left and factor b from the third factor on the left. Remaining terms are obtained in the same manner.

In general, if we expand

$$\underbrace{(a+b)(a+b)\cdots(a+b)}_{n \text{ factors}}$$

a term of the form a^rb^{n-r} will result whenever we select an a from each of the r factors and a b from each of the remaining $n - r$ factors. To determine how many such selections are possible, we can either count the number of factors from which to obtain the a's or the number of factors from which to obtain the b's. Taking the first approach, we see that the number of terms of the form $a^r b^{n-r}$ is

$$\binom{n}{r}$$

(r factors chosen from the n available factors). Thus, in the expansion of $(a + b)^n$, the term $a^r b^{n-r}$ has coefficient $\binom{n}{r}$.

Counting Subsets

The following counting problem arises in various applications:

Problem How many different subsets can be formed from a set of n elements?

To solve this problem we count the number of subsets of every possible size. Since there are $C_{n,r} = \binom{n}{r}$ subsets with r elements, it follows that there are

$\binom{n}{0}$ sets with 0 elements

$\binom{n}{1}$ sets with 1 element

$\binom{n}{2}$ sets with 2 elements

$\vdots$

$\binom{n}{n}$ sets with n elements

Thus, the total number of possible subsets is

$$\binom{n}{0} + \binom{n}{1} + \binom{n}{2} + \cdots + \binom{n}{n}$$

A simpler version of this formula can be obtained by substituting $a = 1$ and $b = 1$ into the binomial formula, (1), and using the fact that all powers of 1 are equal to 1. This yields

$$2^n = \binom{n}{0} + \binom{n}{1} + \binom{n}{2} + \cdots + \binom{n}{n}$$

Thus, we have the following result.

Counting Subsets There are 2^n different subsets that can be formed from a set of n objects.

EXAMPLE 3 There are $2^4 = 16$ different subsets that can be formed from the set $\{a, b, c, d\}$. They are

$$\varnothing \quad \{a\} \quad \{b\} \quad \{c\} \quad \{d\}$$
$$\{a, b\} \quad \{a, c\} \quad \{a, d\} \quad \{b, c\} \quad \{b, d\} \quad \{c, d\}$$
$$\{a, b, c\} \quad \{a, b, d\} \quad \{a, c, d\} \quad \{b, c, d\}$$
$$\{a, b, c, d\}$$

EXAMPLE 4 An ice cream parlor makes its sundaes using 1 scoop of vanilla ice cream to which any number of 5 possible toppings are added. How many different sundaes are possible?

Solution Making a sundae requires selecting a set of toppings from the 5 possible toppings. (The empty set of toppings corresponds to plain ice cream.) Since the number of subsets of the 5 toppings is 2^5, there are $2^5 = 32$ different sundaes possible.

EXERCISE SET 6.5

1. Find the binomial expansion of
(a) $(a + b)^7$ (b) $(x + y)^9$ (c) $(r + s)^{11}$

2. Find the binomial expansion of
(a) $(a + b)^8$ (b) $(w + z)^{10}$ (c) $(c + d)^{12}$

3. (a) Find the binomial expansions of $(a - b)^2$, $(a - b)^3$, $(a - b)^4$, and $(a - b)^5$ by substituting $-b$ for b in the binomial expansions of $(a + b)^2$, $(a + b)^3$, $(a + b)^4$, and $(a + b)^5$ and then simplifying.
(b) Based on your results in (a), make a general statement about the signs of the terms in the binomial expansion of $(a - b)^n$.

4. Use the result in Exercise 3b to find the binomial expansion of
(a) $(a - b)^7$ (b) $(x - y)^9$ (c) $(r - s)^{11}$

5. Use the result in Exercise 3b to find the binomial expansion of
(a) $(a - b)^8$ (b) $(w - z)^{10}$ (c) $(c - d)^{12}$

In Exercises 6–11 find the binomial expansion of the expression and multiply where possible.

6. $(a + 2b)^5$ 7. $(1 + x)^7$ 8. $(3a + 2b)^6$

9. $(a - 3b)^3$ 10. $(y - 1)^8$ 11. $(2r - 2s)^9$

12. Find the first five terms in the binomial expansion of
(a) $(x + y)^{15}$ (b) $(x - y)^{15}$

13. Find the first five terms in the binomial expansion of
(a) $(1 + p)^{20}$ (b) $(1 - p)^{20}$

14. Find the term involving b^3 in the binomial expansion of $(a + b)^{17}$.

15. Find the term involving y^7 in the binomial expansion of $(x - y)^{25}$.

16. (a) How many subsets can be formed from the set $\{a, b, c, d, e\}$?
(b) List the subsets.

17. How many subsets can be formed from a set with six elements?

18. A section from a pizzeria menu is shown. How many different pizza combinations does the restaurant offer?

ANTHONY'S PIZZERIA

Plain Cheese **\$6.00**

Add as many of the following delicious toppings as you like (**\$.75 each**):
anchovies, peppers, onions, mushrooms, sausage, black olives

19. A salad bar serves a plate of lettuce to which the patrons may add as many items as they like (see Figure 1). When the patrons return to their table, the waitress offers 3 dressings from which the patron can choose at most 1. How many salad combinations are possible?

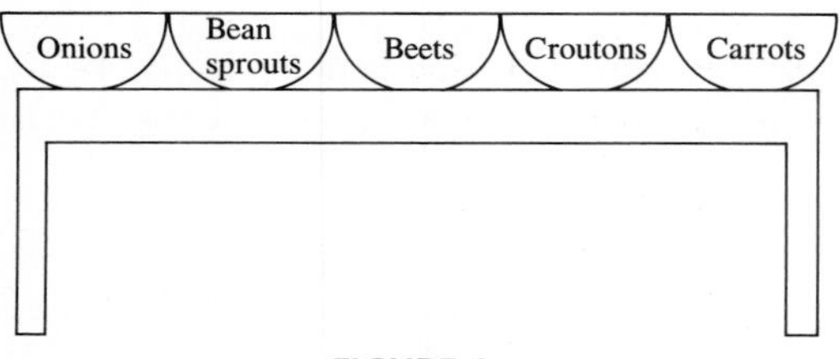

FIGURE 1

20. A section from an ice cream parlor menu is shown. How many different kinds of "olde-fashioned sundaes" does the parlor offer?

> OLDE-FASHIONED SUNDAES
>
> One or two scoops of our homemade vanilla ice cream ($.90/scoop).
> Choose as many of the following as you like ($.30 each): walnuts, chocolate syrup, peanuts, cherries, crushed pineapple, marshmallow, sprinkles

KEY IDEAS FOR REVIEW

- **Ordered pair** A listing of two objects in a definite order.
- ***A* × *B* (Cartesian product)** The set of all ordered pairs (a, b) with $a \in A$ and $b \in B$.
- **$n(S)$** The number of elements in S.
- **Number of elements in a union of sets**

$$n(A \cup B) = n(A) + n(B) - n(A \cap B)$$
$$n(A \cup B \cup C) = n(A) + n(B) + n(C) - n(A \cap B) - n(A \cap C) - n(B \cap C) + n(A \cap B \cap C)$$

- **Number of elements in a Cartesian product**

$$n(A \times B) = n(A) \cdot n(B)$$

- **Multiplication principle** See page 204.
- **Permutation** An arrangement of objects in a definite order without repetitions.
- **Permutation of *n* objects taken *r* at a time** A permutation of r objects selected from a set of n objects.
- **Number of permutations of *n* objects taken *r* at a time**

$$P_{n,r} = \underbrace{n(n-1)(n-2)\cdots}_{r \text{ factors}} = \frac{n!}{(n-r)!}$$

- **Combination** A subset of r objects selected from a set of n objects.

Number of combinations of n objects taken r at a time

$$C_{n,r} = \binom{n}{r} = \frac{n!}{r!(n-r)!}$$

$C_{n,r} = C_{n,n-r}$

Pascal's triangle See page 223.

Binomial coefficients The coefficients in the expansion of $(a + b)^n$.

Binomial formula

$$(a + b)^n = \binom{n}{0}a^n + \binom{n}{1}a^{n-1}b + \binom{n}{2}a^{n-2}b^2 + \cdots + \binom{n}{n-1}ab^{n-1} + \binom{n}{n}b^n$$

SUPPLEMENTARY EXERCISES

1. (a) Find $A \times B$ if $A = \{a, b, c\}$ and $B = \{1, 2\}$.
   (b) Determine $n(A \times B)$.

2. Freshman students at Learned University must select one course in English, one in mathematics, one in a foreign language, and one in science. If the courses open to freshmen are

   English 10 and English 50

   Mathematics 55, Mathematics 75, and Mathematics 85

   Spanish 51, French 51, German 51, and Japanese 51

   Chemistry 71, Chemistry 81, and Biology 81

   how many different schedules are available?

3. A survey of 100 freshmen at Learned University indicates that 70 are enrolled in English 50 and 60 are enrolled in Mathematics 55. If five of the freshmen are not enrolled in either class, how many are enrolled in both classes?

4. License plates in New Jersey consist of 3 letters followed by 3 digits or 3 digits followed by 3 letters. If repetition of digits and letters is allowed, how many different plates are possible?

5. Telephone numbers consist of 7 digits (repetition allowed). If the first digit cannot be 0 or 1 and the second and third digit cannot be 0, what is the maximum number of different telephone numbers within one area code?

6. Evaluate (a) $P_{9,3}$, (b) $C_{8,3}$, and (c) $C_{n,0}$.

Use the following information for Exercises 7–9: A committee of 4 must be selected from a group of 8 men and 6 women.

7. If the committee must consist of 2 men and 2 women, how many different committees can be selected?

8. If the committee must consist of at least 2 men, how many different committees can be selected?

9. If John Smith, a member of the group, must be on this committee, in how many ways can a committee consisting of 2 men and 2 women be selected?

10. There are 20 applicants for a job at Company ABC. The personnel manager can only interview 5 applicants each day. In how many ways can 5 people be chosen for Monday interviews?

11. Once the 5 applicants of Exercise 10 are chosen for Monday interviews, in how many ways can they be scheduled during the day if interviews are scheduled at 9 A.M., 10 A.M., 1 P.M., 2 P.M., and 3 P.M.?

12. In how many ways can 4 persons be selected to be president, vice president, secretary, and treasurer from a group of 20 people?

Use the following information for Exercises 13–16: A die is tossed five times in succession and the resulting sequence of outcomes is noted. (A toss of a die results in a 1, 2, 3, 4, 5, or 6.)

13. How many different sequences of outcomes can occur?

14. How many different sequences containing exactly three 6s can occur?

15. How many different sequences of outcomes containing at least three 6s can occur?

16. How many different sequences of outcomes with three 6s and two 4s can occur?

Use the following information for Exercises 17–20: Of 100 students who went sightseeing,

40 visited the art museum
30 visited the planetarium
25 visited the Liberty Bell
15 visited the art museum and the planetarium
10 visited the planetarium and the Liberty Bell
15 visited the art museum and the Liberty Bell
5 visited all three of these sights

17. How many students visited none of these sights?

18. How many students visited only the art museum?

19. How many students visited the art museum *or* the planetarium?

20. If a student visited all three of these sights, in how many different ways could the visits be made?

21. A multiple choice test consists of 10 questions, each of which has 5 possible answers. In how many ways can this 10-question test be answered?

Use the following information for Exercises 22–24: A farmer wants to plant 10 fruit trees in a row. He has 6 apple trees and 4 pear trees. (Assume that all trees are distinguishable from one another.)

22. In how many different arrangements can these trees be planted?

23. How many different arrangements are there if the first three trees are to be apple trees?

24. How many different arrangements are there if the apple trees are to be planted next to one another?

25. Write the binomial expansion for $(1 + x)^4$.

26. Write the binomial expansion for $(1 + 2x)^4$.

27. In the Super 6 lottery, a person must select six different numbers from 1, 2, 3, 4, . . . , 40. If the order of the numbers selected is not counted, how many different sets of six such numbers are possible?

28. How many 13-card hands with all red cards can be dealt from a deck of 52 cards?

29. How many 13-card hands with all cards of one suit can be dealt from a deck of 52 cards?

30. How many 13-card hands with 7 spades and 6 hearts can be dealt from a deck of 52 cards?

CHAPTER TEST

1. Find $n(A \times B)$ if

$$A = \{M, F\} \quad \text{and} \quad B = \{21, 22, 23\}$$

2. A survey of 1000 shoppers at the SHOPMUCH mall indicated that 700 shopped at Highway Department Store, 600 shopped at Midway Department Store, and 550 shopped at both stores. How many of the shoppers did not go to either store?

3. Four girls and three boys are to be seated in a row.
(a) In how many different ways can they be seated?
(b) In how many different ways can they be seated if all the girls sit together and all the boys sit together?
(c) In how many different ways can they be seated if all the girls sit together?
(d) In how many different ways can they be seated if Mary and Joan have seats at the ends of the row?
(e) In how many different ways can they be seated if girls and boys alternate?

4. Evaluate (a) $C_{10,6}$ and (b) $P_{10,6}$.

5. How many different arrangements are there of the letters of the word EXCEEDED?

6. (a) How many different four-card hands can be dealt from a deck of 52 cards?
(b) How many different four-card hands consisting of 4 hearts can be dealt from a deck of 52 cards?

7. A survey of 200 children indicated that

100 liked gum
90 liked mints
80 liked gumdrops
60 liked gum and mints
50 liked mints and gumdrops
40 liked gum and gumdrops
30 liked all three types of candy

(a) How many of these children did not like any of the three types of candy?

(b) How many of these children only liked gumdrops?

8. A 5-question multiple choice test is given. There are 3 possible answers—(i), (ii), and (iii)—to each question. Only 1 answer is correct.

(a) How many different sets of 5 answers are possible?

(b) How many different sets of answers will contain 3 (i) answers?

(c) How many different sets of answers will contain *at least* 3 (i) answers?

9. Write the binomial expansion of $(1 - x)^5$.

10. Jane has 7 different books. She wants to give 3 to Tom and 4 to Mary. In how many different ways can she give away the books?

7 Probability

There is an element of unpredictability in many physical phenomena. Even under apparently identical conditions, many observable quantities vary in an uncertain way. For example, the total number of phone calls arriving at a switchboard varies unpredictably from day to day, the total annual rainfall in Chicago varies unpredictably from year to year, the Dow Jones industrial average varies unpredictably from hour to hour, and the number of heads obtained in five tosses of a coin varies unpredictably with each group of tosses. In this chapter we study probability theory, which is the branch of mathematics concerned with making rational statements about phenomena that are subject to an element of uncertainty.

7.1 INTRODUCTION; SAMPLE SPACE AND EVENTS

The meaning that we give to the term *probability* can best be described by considering what happens when we toss a coin. If we toss a coin once, it is impossible to predict in advance (with certainty) whether the outcome will be a head or a tail. However, if we toss the coin again and again, the proportions of heads and tails tend toward certain fixed values. To illustrate, consider Table 1, where we have recorded results relating to 20,000 tosses of an ordinary coin. In the early stages of tossing, the proportion of heads varied considerably. As the coin tossing continued, however, the proportion of heads began to settle down so that during the last 5000 tosses the proportion of heads changed by only .008. These data suggest that if we continued the coin tossing beyond 20,000 tosses, the proportion of heads would approach some fixed constant value close to $\frac{1}{2}$. We think of this value as the *probability of heads*. Intuition suggests that if the coin were perfectly balanced, then the probability of heads would be exactly $\frac{1}{2}$, or .5. However, since no physical coin is perfectly balanced, the probability of heads for the coin used to obtain Table 1 would presumably be close to but not exactly equal to $\frac{1}{2}$.

TABLE 1 Summary of 20,000 coin tosses

Number of tosses (n)	Number of heads (h)	Proportion of tosses that were heads, $\left(\frac{h}{n}\right)$
10	8	.8000
100	62	.6200
1,000	473	.4730
5,000	2,550	.5100
10,000	5,098	.5098
15,000	7,649	.5099
20,000	10,038	.5019

To generalize the above idea, assume we have an experiment that can be repeated indefinitely under fixed conditions, and suppose that during n repetitions of this experiment, a certain event occurs f times. We call the ratio

$$\frac{f}{n}$$

the **relative frequency** of the event after n repetitions. If this relative frequency approaches a number p as n becomes larger and larger, then p is called the **probability** of the event. Thus as n (the number of repetitions) becomes larger and larger, the approximation

$$p \cong \frac{f}{n}$$

becomes better and better.

This definition of probability is somewhat unsatisfactory since the meaning of the phrase "this relative frequency *approaches* a number p" has not been precisely explained. Nevertheless, your intuitive feeling for the meaning of this statement should be perfectly adequate for most purposes.

Informally, the probability of an event is its **long-term relative frequency**, that is, the proportion of the time that the event would occur if the experiment were repeated indefinitely under fixed conditions.

In some situations, probabilities can be obtained using logical reasoning and intuition; for example, if we have an ordinary six-sided die that we assume to be perfectly balanced and perfectly symmetrical, then intuition tells us that the probability of tossing a 2 with this die is $\frac{1}{6}$. We obtain this conclusion by arguing that each of the six possible outcomes has an equal chance of occurring, so that over the long term, the number 2 will appear one-sixth of the time. Some probabilities cannot be obtained using intuition; they can only be estimated from experimental data. For example, suppose we have a certain production process for manufacturing photographic flashbulbs, and we are interested in the probability that the process will produce a defective bulb. Intuition will not tell us what this probability is. However, if we test 10,000 bulbs and find that 3 bulbs are defective, we can approximate the probability of a defective bulb by the relative frequency

$$\frac{3}{10{,}000} = .0003$$

The long-term relative frequency interpretation of probability is appropriate for experiments that can be repeated over and over under fixed conditions. For experiments that cannot be repeated under fixed conditions, the **subjective interpretation of probabilities** is more appropriate. With this approach, the probability of an event is viewed as a measure (on a scale of 0 to 1) of one's **strength of belief** that the event will occur when the experiment is performed:

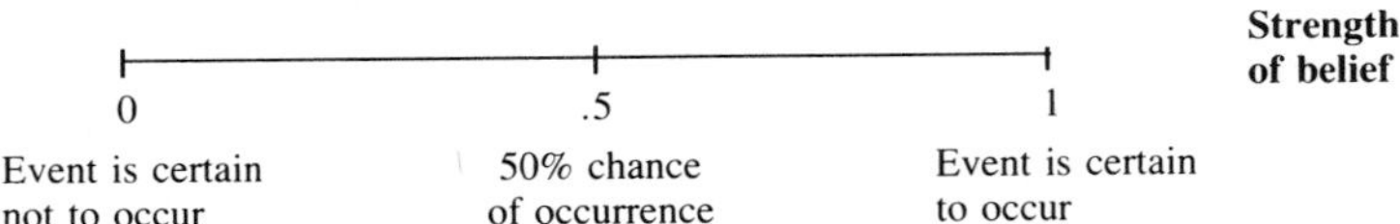

For example, before the first soft landing on Mars, experts estimated a 40% chance (probability .4) of finding life on Mars. This was a subjective assignment of probability based on expert opinion.

Throughout this chapter we are concerned with (1) experiments whose outcomes are not predictable with certainty in advance and (2) the outcomes themselves. The term **experiment** is used in a broad sense to mean the observation of any physical occurrence. Since we are interested in probabilities of events associated with these experiments, it is necessary that the term *event* be given a precise mathematical meaning. This is our objective for the remainder of this section.

We begin by discussing experiments and their outcomes since most probability investigations start from this point.

> The set of all possible outcomes of an experiment is called the **sample space** for the experiment. The outcomes in the sample space are called the **sample points**.

The sample points and sample space of an experiment depend on what the experimenter chooses to observe. For example, if the experiment consists of tossing a coin twice, the experimenter can record the sequence of heads (h) and tails (t) that result, in which case the sample space is

$$S = \{hh, ht, th, tt\}$$

(A tree diagram illustrates this in Figure 1.) On the other hand, the experimenter could record the total number of tails observed, in which case the sample space S is

$$S = \{0, 1, 2\}$$

or whether the two tosses match (m) or do not (d), in which case the sample space is

$$S = \{m, d\}$$

FIGURE 1

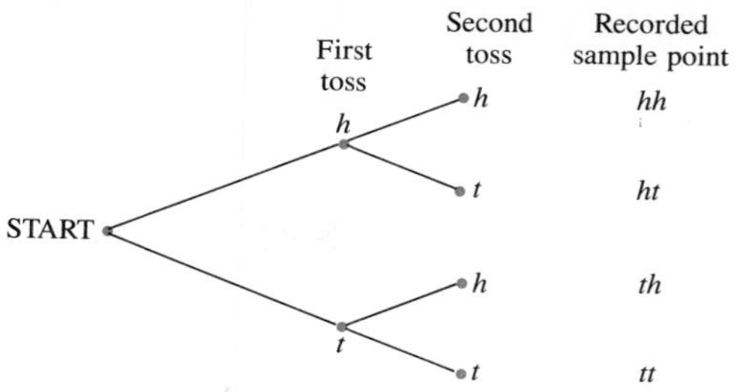

First toss | Second toss | Recorded sample point

START

h — h — hh

h — t — ht

t — h — th

t — t — tt

EXAMPLE 1 Consider the experiment of tossing a die and recording the number on the top face. There are six possible outcomes or sample points for this experiment. Thus, the sample space S is the set

$$S = \{1, 2, 3, 4, 5, 6\}$$

EXAMPLE 2 A light bulb manufacturer tests a bulb by letting it burn until it burns out. The manufacturer records the total time t in hours that the bulb stays lit. Since any nonnegative real number may be recorded for t, the sample space S consists of all nonnegative real numbers; that is,

$$S = \{t \mid t \geq 0\}$$

EXAMPLE 3 A traffic engineer records the number of cars entering a tunnel between 9:00 A.M. and 10:00 A.M. The sample space S consists of the nonnegative integers; that is,

$$S = \{0, 1, 2, \ldots\}$$

EXAMPLE 4 A stock market analyst observes General Electric common stock for one market day and records whether the stock increases in value, decreases in value, or undergoes no change in value. The sample space S is

$$S = \{i, d, n\}$$

where i, d, and n denote increase, decrease, and no change, respectively.

EXAMPLE 5 A geneticist records the sex of successive children in three-child families. If m denotes a male and f a female, then the sample space is

$$S = \{mmm, mmf, mfm, mff, fmm, fmf, ffm, fff\}$$

(See the tree diagram in Figure 2.)

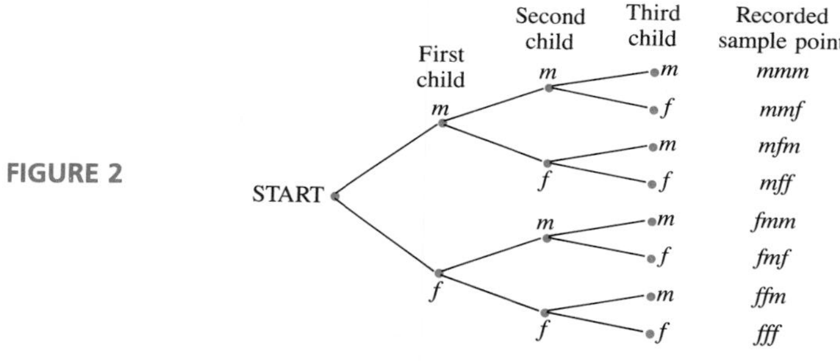

FIGURE 2

EXAMPLE 6 Horses a, b, and c are running in a race and the winning horse is recorded. Assuming there are no ties, the sample space of possible winners is

$$S = \{a, b, c\}$$

We are interested in making probability statements about *events*. Intuitively, we all know examples of events: the event that an odd number is tossed with a die, the event that more than 10 cars enter a certain tunnel between 9:00 A.M. and 10:00 A.M., the event that a bulb burns out in fewer than 10 hours, and so on. We now extend this intuitive notion of an event to a precise definition.

For motivation, consider the die-tossing experiment in Example 1, and let E denote the event that an odd number comes up when the die is tossed. If, when the experiment is performed, one of the sample points 1, 3, or 5 results, then the event E occurs. Thus, to say that the event E occurs is equivalent to saying that the experiment results in an outcome in the set $\{1, 3, 5\}$. This suggests the following definition.

An **event** is a subset of the sample space of an experiment. An event E is said to **occur** if the outcome of the experiment is an element of E.

Since events are defined as subsets of the sample space, we can use the concepts developed in Chapter 5 throughout this chapter.

EXAMPLE 7 Consider the experiment of tossing a die and recording the number on the top face. If E is the event that an even number occurs, then

$$E = \{2, 4, 6\}$$

If F is the event that a number larger than 4 occurs, then

$$F = \{5, 6\}$$

EXAMPLE 8 In a germination experiment, 10 corn seeds are planted and the number of seeds that germinate within 30 days is recorded. Let E be the event that more than half the seeds germinate.

The sample space is

$$S = \{0, 1, 2, 3, 4, 5, 6, 7, 8, 9, 10\}$$

and the event E is

$$E = \{6, 7, 8, 9, 10\}$$

EXAMPLE 9 Consider the light bulb testing experiment in Example 2. If E is the event that the life of the bulb is between 10 and 20 hours, inclusive, then

$$E = \{t \mid 10 \leq t \leq 20\}$$

If S is the sample space of an experiment, then S and $\varnothing$ are events since they are subsets of S. In any performance of the experiment, the event S must occur and the event $\varnothing$ cannot occur. (Why?) For this reason, S is called a **certain event** and $\varnothing$ an **impossible event**.

EXAMPLE 10 Consider the experiment of tossing a coin twice and recording the ordered pair

$$(n_h, n_t)$$

where n_h is the total number of heads tossed and n_t is the total number of tails tossed. Then the sample space S is the set of ordered pairs

$$S = \{(2, 0), (1, 1), (0, 2)\}$$

If E is the event that $n_h + n_t$ is even and F is the event that $n_h + n_t$ is less than 1, then

$$E = \{(2, 0), (1, 1), (0, 2)\} = S \quad \text{and} \quad F = \varnothing$$

so E is a certain event and F is an impossible event.

Since events are sets, we can apply the set operations (union, intersection, and complementation) to them. Thus, given events E and F in a sample space S, we can form the new events

$$E \cap F, \quad E \cup F, \quad \text{and} \quad E'$$

The meanings of these events follow from the definitions given in Chapter 5 of intersection, union, and complement of sets (Figure 3). The event $E \cup F$ consists of those sample points that belong to E or to F. The word *or* is used in the *inclusive* sense as defined in Section 5.2. The event $E \cap F$ consists of those sample points common to both E and F. The event E' consists of the sample points in S but not in E.

FIGURE 3

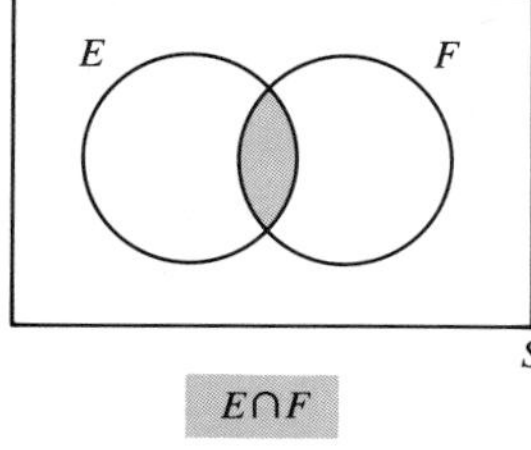

$E \cap F$

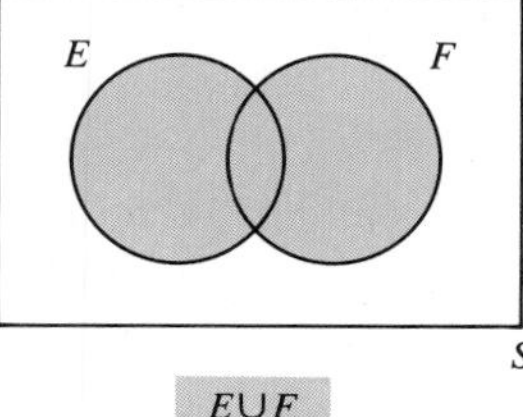

$E \cup F$

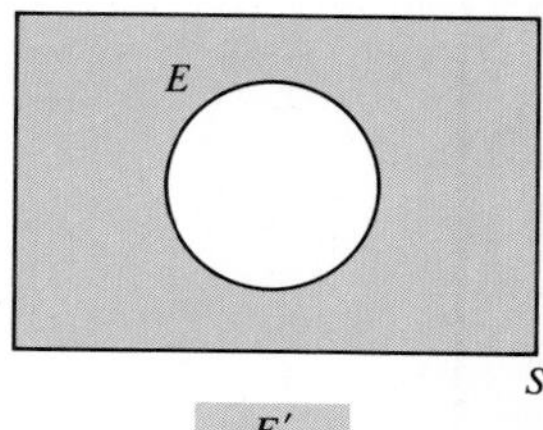

E'

Therefore, we have the following:

> Stating that $E \cup F$ occurs is equivalent to stating that E or F occurs.
> Stating that $E \cap F$ occurs is equivalent to stating that E and F both occur.
> Stating that E' occurs is equivalent to stating that E does not occur.

EXAMPLE 11 An experimenter tosses a die and records the number on the top face. Let E be the event that the number is divisible by 3 and F the event that the number is odd. Thus,

$$E = \{3, 6\} \quad \text{and} \quad F = \{1, 3, 5\}$$

The event that the number is divisible by 3 *and* odd is

$$E \cap F = \{3\}$$

The event that the number is divisible by 3 *or* odd is

$$E \cup F = \{1, 3, 5, 6\}$$

The event that the number is *not* divisible by 3 is

$$E' = \{1, 2, 4, 5\}$$

Table 2 summarizes these results:

TABLE 2

Verbal expression	Set interpretation
E or F occurs	$E \cup F$
E and F occur	$E \cap F$
E does *not* occur	E'

If two events E and F have no common sample points, then the two events cannot both occur on any single performance of the experiment; such events are said to be **mutually exclusive** or **disjoint**.

Symbolically, E and F are mutually exclusive or disjoint if $E \cap F = \varnothing$.

EXAMPLE 12 An experimenter tosses a die and records the number on the top face. Let E be the event that an even number is tossed and let F be the event that an odd number is tossed. Then

$$E = \{2, 4, 6\} \quad \text{and} \quad F = \{1, 3, 5\}$$

Since $E \cap F = \varnothing$, these events are mutually exclusive.

The notion of mutually exclusive can be extended to more than two events. We call n events $E_1, E_2, \ldots, E_n$ **mutually exclusive** if no two of them have any sample points in common (Figure 4). That is, to say n events are mutually exclusive means no two of them can occur on any single performance of the experiment.

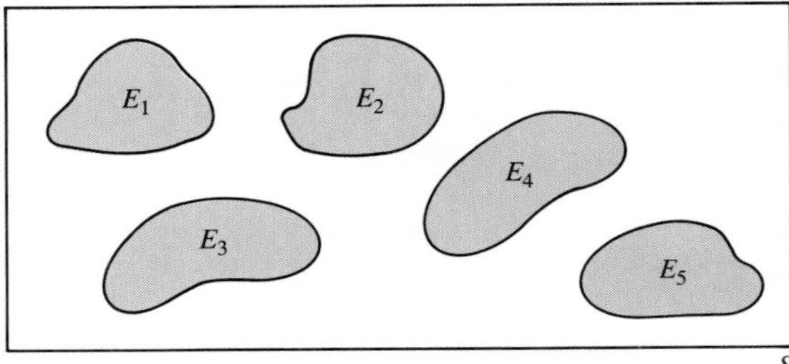

FIGURE 4

EXERCISE SET 7.1

1. For each of the following experiments, describe the sample space S in set notation:
   (a) Ten people are asked if they smoke regularly, and the number of people answering affirmatively is recorded.
   (b) A traffic engineer records the number of cars crossing the George Washington Bridge in a 24-hour period.
   (c) A chemist measures and records the percentage of ethylene glycol in a solution. (Assume the percentage can be measured with perfect accuracy.)
   (d) A transistor is operated until it fails and its life in hours is recorded. (Assume the length of life can be measured with perfect accuracy.)
   (e) A coin is tossed three times and the sequence of h's (heads) and t's (tails) is recorded.

2. A person's weight is measured and recorded. Write the sample space S in set notation assuming:
   (a) The weight is recorded with perfect accuracy.
   (b) The weight is recorded to the nearest half pound (e.g., 161, 161.5, 162).

3. The sample space of a certain experiment is $S = \{a, b, c\}$. List all possible events.

4. A die is tossed and the number on the top face is recorded. Write the following events in set notation:
   (a) E: a number greater than 3 is tossed.
   (b) F: a number other than 2 is tossed.
   (c) G: the number tossed is either odd or less than 3.

5. A die is tossed twice and the resulting ordered pair of numbers is recorded. Describe the following events *in words*.
   (a) $\{(1, 1), (2, 2), (3, 3), (4, 4), (5, 5), (6, 6)\}$
   (b) $\{(3, 1), (3, 2), (3, 3), (3, 4), (3, 5), (3, 6)\}$
   (c) $\{(4, 4)\}$

6. Items coming off a production line are checked to see whether they are defective (d) or nondefective (n). The checker continues to check until two consecutive defectives are found or four items have been checked, whichever occurs first; the checker then records the resulting sequence of d's and n's.
   (a) Write the sample S in set notation.
   (b) Let E be the event that the checker stops at the third item. Write E in set notation.
   (c) Let F be the event that the checker stops at the fourth item. Write F in set notation.

7. A firm that conducts political polls classifies people according to three characteristics:

   Sex: male (m), female (f)
   Income: high (h), average (a), low (l)
   Political registration: Democrat (d), Republican (r), Independent (i)

   Thus a person classified (f, h, r) would be a female with high income who is a registered Republican.
   (a) A person is randomly selected and classified according to the above system. Express the sample space for this experiment in set notation.
   (b) Let E be the event that the person selected is a Republican; write E in set notation.
   (c) Let F be the event that the person selected is a female registered as an Independent; write F in set notation.
   (d) Let G be the event that the person selected is either a male or a Democrat. Write G in set notation.

8. A geneticist successively selects 3 newly hatched fruit flies and observes whether each is long-winged (l) or short-winged (s). The resulting sequence of l's and s's is then recorded. Write the following events in set notation:
   (a) Exactly one long-winged fly is observed.
   (b) At least one long-winged fly is observed.
   (c) No more than two short-winged flies are observed.

9. A die is tossed and the number on the top face is recorded. Let E, F, and G be the events:

   E: the number tossed is even
   F: the number tossed is divisible by 3
   G: the number tossed is less than 5

   Describe the following events *in words*:
   (a) $E \cup G$ (b) $F \cap G$ (c) $F \cup G'$
   (d) $F' \cap G$ (e) $E' \cap F' \cap G'$

10. List the sample points for each event in Exercise 9.

11. Consider the experiment described in Example 2 of the text, and let E and F be the events

    E = the bulb lasts 10 or more hours
    F = the bulb lasts fewer than 15 hours

    Express the following events using set-builder notation:
    (a) E (b) F (c) E'
    (d) F' (e) $E \cap F'$ (f) $E \cup F'$

12. Are the events E and F in Exercise 11 mutually exclusive? Explain your answer.

13. A medical file is selected from the records of a hospital and information about the person's weight, age, and marital status is noted. Let E, F, G, and H be the events:

E: the person is overweight
F: the person is underweight
G: the person is more than 50 years of age
H: the person is married

(a) Are E and F mutually exclusive? Explain.
(b) Are E and H mutually exclusive? Explain.
(c) Is $E \cup F$ a certain event? Explain.
(d) Is $E \cup E'$ a certain event? Explain.

14. Let E, F, and G be three events associated with an experiment. Express the following events in terms of E, F, and G using the operations $\cup$, $\cap$, and $'$:
(a) E occurs or F occurs.
(b) E occurs or G does not occur.
(c) F occurs and G occurs.
(d) E occurs and G does not occur.
(e) At least one of the three events occurs.
(f) Exactly one of the three events occurs.
(g) None of the events occurs.

15. A communications transmission line can be open (o) or busy (b). Suppose the line is monitored three times and the resulting sequence of o's and b's is recorded. Determine whether the events E, F, and G are mutually exclusive:

$$E = \{ooo, oob, boo\}$$
$$F = \{bob, obo\}$$
$$G = \{bbb, obb\}$$

Explain your answer.

16. A geneticist records the sex of successive children in a four-child family. If m denotes a male and f a female, list the points in the sample space using the method of Example 5.

7.2 PROBABILITY MODELS FOR FINITE SAMPLE SPACES

In this section we discuss procedures for assigning probabilities to the events in a sample space. We limit the discussion to experiments with finitely many outcomes; such experiments are said to have a **finite sample space**.

If $S = \{s_1, s_2, \ldots, s_k\}$ is the sample space of an experiment, then the events

$$\{s_1\}, \{s_2\}, \ldots, \{s_k\}$$

that consist of exactly one sample point are called **elementary** or **simple events**.

EXAMPLE 1 If we toss a coin, then the sample space S is

$$S = \{h, t\}$$

where h = heads and t = tails. The elementary events are

$$\{h\} \quad \text{and} \quad \{t\}$$

EXAMPLE 2 If we toss a die and observe the number on the top face, then the sample space S is

$$S = \{1, 2, 3, 4, 5, 6\}$$

and the elementary events are

$$\{1\} \quad \{2\} \quad \{3\} \quad \{4\} \quad \{5\} \quad \text{and} \quad \{6\}$$

EXAMPLE 3 If we pick a name at random from a list of 8000 different names, the sample space has 8000 sample points; consequently, there are 8000 different elementary events.

Probability is used to measure the likelihood that an event will occur. To motivate our development of the concept of probability, consider the experiment of tossing one die and noting the number on its face. The sample space is

$$S = \{1, 2, 3, 4, 5, 6\}$$

The die is tossed 100 times and the number of times, f_i, that each elementary event occurred is noted in Table 1. The relative frequency $f_i/100$ is denoted by $\tilde{p}_i$ (read "p tilde sub i").

TABLE 1

Elementary event	f_i = number of times the elementary event occurred in 100 tosses of the die	$\tilde{p}_i = \frac{f_i}{100}$
{1}	20	$\frac{20}{100}$
{2}	15	$\frac{15}{100}$
{3}	16	$\frac{16}{100}$
{4}	19	$\frac{19}{100}$
{5}	18	$\frac{18}{100}$
{6}	12	$\frac{12}{100}$
Total	100	$\frac{100}{100} = 1$

The relative frequencies $\tilde{p}_i$ of the elementary events {1}, {2}, {3}, {4}, {5}, {6} are noted in the third column of Table 1. Observe also that each of the relative frequencies satisfies the inequality

$$0 \le \tilde{p}_i = \frac{f_i}{100} \le 1$$

Now consider the relative frequency $\tilde{p}_S$ of the event $S = \{1, 2, 3, 4, 5, 6\}$. Since S occurs 100 times out of 100 tosses,

$$\begin{aligned}
\tilde{p}_S = 1 &= \frac{100}{100} \\
&= \frac{20 + 15 + 16 + 19 + 18 + 12}{100} \\
&= \frac{20}{100} + \frac{15}{100} + \frac{16}{100} + \frac{19}{100} + \frac{18}{100} + \frac{12}{100} \\
&= \tilde{p}_1 + \tilde{p}_2 + \tilde{p}_3 + \tilde{p}_4 + \tilde{p}_5 + \tilde{p}_6
\end{aligned}$$

Note that the relative frequencies of the elementary events are nonnegative fractions that add up to 1.

Now consider the events

$$E = \{1, 3, 5\} \qquad F = \{1, 2\} \qquad G = \{3, 6\} \qquad H = \{7\}$$

The number of times that each of these events occurs in 100 tosses of the die is recorded in the second column of Table 2.

TABLE 2

Event A	f_A = number of times the event occurred in 100 tosses of the die	$\tilde{p}_A = \frac{f_A}{100}$ = the relative frequency of the occurrence of the event
$E = \{1, 3, 5\}$	$54 = 20 + 16 + 18$	$\frac{54}{100}$
$F = \{1, 2\}$	$35 = 20 + 15$	$\frac{35}{100}$
$G = \{3, 6\}$	$28 = 16 + 12$	$\frac{28}{100}$
$H = \{7\}$	0	$\frac{0}{100} = 0$
$S = \{1, 2, 3, 4, 5, 6\}$	$100 = 20 + 15 + 16 + 19 + 18 + 12$	$\frac{100}{100} = 1$

The relative frequencies of the above events are noted in the third column of Table 2. Observe that each of the relative frequencies of these events satisfies the inequalities

$$0 \leq \tilde{p}_A \leq 1$$

and $\tilde{p}_H = 0$ (where H is the empty set for this experiment) and $\tilde{p}_S = 1$.

Now consider the events

$$E \cup G = \{1, 3, 5, 6\} \quad \text{and} \quad F \cup G = \{1, 2, 3, 6\}$$

Table 3 indicates the number of times $E \cup G$ and $F \cup G$ occur, respectively, and their relative frequencies. Observe that the relative frequency of $F \cup G$ ($\tilde{p}_{F \cup G}$) equals the sum of the relative frequencies $\tilde{p}_F$ and $\tilde{p}_G$ of F and G, respectively.

$$\tilde{p}_{F \cup G} = \frac{63}{100} = \frac{35}{100} + \frac{28}{100} = \tilde{p}_F + \tilde{p}_G \qquad \text{(from Table 2)}$$

TABLE 3

Event A	f_A = number of times the event occurred in 100 tosses of the die	$\tilde{p}_A$ = the relative frequency of the event
$E \cup G = \{1, 3, 5, 6\}$	$66 = 20 + 16 + 18 + 12$	$\frac{66}{100}$
$F \cup G = \{1, 2, 3, 6\}$	$63 = 20 + 15 + 16 + 12$	$\frac{63}{100}$

However, the relative frequency $\tilde{p}_{E \cup G}$ of $E \cup G$ is not equal to the sum of the relative frequencies $\tilde{p}_E$ and $\tilde{p}_G$ of E and G:

$$\tilde{p}_{E \cup G} = \frac{66}{100} \neq \frac{54}{100} + \frac{28}{100} = \frac{82}{100} = \tilde{p}_E + \tilde{p}_G \qquad \text{(from Table 2)}$$

Since F and G are mutually exclusive events, the number of times that $F \cup G$ occurs is the sum of the number of times (35) that F occurs and the number of times (28) that G occurs (F and G cannot occur simultaneously), so

$$\tilde{p}_{F \cup G} = \tilde{p}_F + \tilde{p}_G$$

However, in the case of $E \cup G$,

$$\tilde{p}_{E \cup G} \neq \tilde{p}_E + \tilde{p}_G$$

because the outcome 3 occurs in both E and G. In other words, E and G are not mutually exclusive events, and the number of times that 3 occurs is counted in both f_E and f_G.

If we look again at the relative frequencies of the elementary events, we again observe that they are nonnegative fractions that add up to 1. If the experiment is repeated indefinitely, then we would expect these relative frequencies to come closer and closer to the exact probabilities of the elementary events. This suggests that the probabilities of the elementary events are also nonnegative fractions that add up to 1. Similarly, we would expect the probability of an event E to be a nonnegative number less than or equal to 1.

In general, since probability is used to measure the likelihood of the occurrence of events, probabilities should behave like relative frequencies. To ensure that they do, probabilities assigned to events must satisfy the following fundamental properties [$P(E)$, read "P of E," represents the probability of an event E]:

Fundamental Properties of Probability for Finite Sample Spaces

1. For an event E, $0 \leq P(E) \leq 1$.

2. $P(S) = 1$, where S is the sample space.

3a. $P(E \cup F) = P(E) + P(F)$ when E and F are mutually exclusive events.

3b. $P(E_1 \cup E_2 \cup \cdots \cup E_k) = P(E_1) + P(E_2) + \cdots + P(E_k)$ when $E_1, E_2, \ldots, E_k$ are mutually exclusive events.

(In the general definition of probability for nonfinite spaces, property 3b holds for the union of an infinite number of mutually exclusive events.)

In most probability problems, we need to determine $P(E)$ for an event E. The first step in this process is to assign a probability to each elementary event; these probabilities are obtained in one of two ways:

(a) by using experimental data

(b) by logical reasoning and intuition

In either case we need to make some assumptions about the problem; it is these assumptions that are based on intuition or past experience. The validity of these assumptions, and hence the usefulness of the value assigned $P(E)$, depends on how well this work fits with future experimental data. Regardless of the method that is used, the probabilities assigned to the elementary events must satisfy the two conditions that follow from the fundamental properties of probability for finite sample spaces.

> **1.** Each probability must be a number between 0 and 1, inclusive.
> **2.** The probabilities assigned to the elementary events must add up to 1.

Thus, if the sample space is

$$S = \{s_1, s_2, \ldots, s_k\}$$

and if we denote the probability assigned to the elementary event $\{s_i\}$ by $P(\{s_i\})$, then we must have

1. $0 \le P(\{s_i\}) \le 1$

2. $P(\{s_1\}) + P(\{s_2\}) + \cdots + P(\{s_k\}) = 1$

That is, the probabilities of the elementary events are nonnegative numbers that add up to 1.

When probabilities are assigned to the elementary events of an experiment so that these two fundamental properties of probabilities are satisfied, we call that assignment a **probability model** for the experiment. To assign probabilities to the elementary events, we must assess the relative frequencies of these events.

EXAMPLE 4 A six-sided die is tossed and the number on the top face is recorded. Thus, the sample space is $S = \{1, 2, 3, 4, 5, 6\}$.

If the die is symmetric and perfectly balanced, then the probability model for the experiment would be based on

$$P(\{1\}) = P(\{2\}) = P(\{3\}) = P(\{4\}) = P(\{5\}) = P(\{6\}) = x$$

Since the probabilities of the elementary events must add up to 1,

$$x + x + x + x + x + x = 6x = 1$$
$$x = \frac{1}{6}$$

and

$$P(\{1\}) = P(\{2\}) = P(\{3\}) = P(\{4\}) = P(\{5\}) = P(\{6\}) = \frac{1}{6}$$

If the die is "loaded" so that the 1, 3, and 5 have the same chance of occurring, whereas each of 2, 4, and 6 is twice as likely to occur as the 1, then the probability model can be obtained by assuming

$$P(\{1\}) = P(\{3\}) = P(\{5\}) = x \quad \text{and} \quad P(\{2\}) = P(\{4\}) = P(\{6\}) = 2x$$

Again, since the probabilities of the elementary events must add up to 1,

$$\begin{aligned} x + x + x + 2x + 2x + 2x &= 1 \\ 9x &= 1 \\ x &= \frac{1}{9} \end{aligned}$$

and

$$P(\{1\}) = P(\{3\}) = P(\{5\}) = \frac{1}{9} \qquad \text{while} \qquad P(\{2\}) = P(\{4\}) = P(\{6\}) = \frac{2}{9}$$

Note that in both models the probabilities assigned to the elementary events are nonnegative numbers adding up to 1, as required, and both are valid probability models. But each is based on a different assessment of the relative frequencies of the occurrences of the outcomes.

EXAMPLE 5 Horses a, b, c, and d are in a race. Assuming no ties, the set of possible winners is

$$S = \{a, b, c, d\}$$

John Jones studies the past performance records of these horses and decides that a is twice as likely to win as b, b is three times as likely to win as c, and c has exactly the same chance to win as d. Based on John's assessment (handicapping),

$$P(\{d\}) = x \qquad P(\{c\}) = x \qquad P(\{b\}) = 3x \qquad P(\{a\}) = 2(3x) = 6x$$

Since the sum of the probabilities of the elementary events must add up to 1,

$$\begin{aligned} x + x + 3x + 6x &= 1 \\ 11x &= 1 \\ x &= \frac{1}{11} \end{aligned}$$

Therefore, $P(\{a\}) = \frac{6}{11}$, $P(\{b\}) = \frac{3}{11}$, $P(\{c\}) = \frac{1}{11}$, and $P(\{d\}) = \frac{1}{11}$.

Janet Jones studies the past performance records of the same horses and decides that a and b have equal chances of winning, b is twice as likely to win as c, and c is twice as likely to win as d. Based on Janet's handicapping,

$$P(\{d\}) = x \qquad P(\{c\}) = 2x \qquad P(\{b\}) = 2(2x) = 4x \qquad P(\{a\}) = 4x$$

Since the probabilities of the elementary events must add up to 1,

$$\begin{aligned} x + 2x + 4x + 4x &= 1 \\ 11x &= 1 \\ x &= \frac{1}{11} \end{aligned}$$

Therefore, $P(\{a\}) = \frac{4}{11}$, $P(\{b\}) = \frac{4}{11}$, $P(\{c\}) = \frac{2}{11}$, and $P(\{d\}) = \frac{1}{11}$.

Both Janet and John have constructed valid probability models. These models are not the same because in the case of horse racing, assessments are subjective, and so the probabilities assigned to the elementary events depend on the handicapper.

Once the probabilities of the elementary events are determined, the probabilities of all other events can be obtained using the following addition principle, which follows from fundamental property 3.

Addition Principle If an experiment has finitely many points in its sample space, then the probability $P(E)$ of an event

$$E = \{s_1, s_2, \ldots, s_m\}$$

is

$$P(E) = P(\{s_1\}) + P(\{s_2\}) + \cdots + P(\{s_m\})$$

where $P(\{s_1\}), P(\{s_2\}), \ldots, P(\{s_m\})$ are the probabilities of the elementary events $\{s_1\}, \{s_2\}, \ldots, \{s_m\}$.

The probability of an elementary event $\{s_i\}$ is also called the **probability of the sample point** s_i. With this terminology the addition principle states:

The probability of an event is the sum of the probabilities of its sample points.

Since the empty set has no sample points, the addition principle cannot be used to obtain its probability. We agree to assign the impossible event $\varnothing$ a probability of zero. This is reasonable since this event never occurs. Actually, the result that $P(\varnothing) = 0$ follows from the three fundamental properties of probability of finite sample spaces (see Theorem 1, Section 7.3).

EXAMPLE 6 A six-sided die is loaded so that the numbers from 1 to 6 occur with the following probabilities when the die is tossed,

$$P(\{1\}) = \frac{2}{12} \qquad P(\{2\}) = \frac{1}{12} \qquad P(\{3\}) = \frac{4}{12}$$

$$P(\{4\}) = \frac{1}{12} \qquad P(\{5\}) = \frac{2}{12} \qquad P(\{6\}) = \frac{2}{12}$$

Find the probability of

(a) the event E that an even number is tossed

(b) the event G that a number divisible by 3 is tossed

(c) the event H that a number greater than 7 is tossed

Solution The events E, G, and H are

$$E = \{2, 4, 6\}$$
$$G = \{3, 6\}$$
$$H = \varnothing$$

Thus

$$P(E) = P(\{2\}) + P(\{4\}) + P(\{6\}) = \frac{1}{12} + \frac{1}{12} + \frac{2}{12} = \frac{1}{3}$$

$$P(G) = P(\{3\}) + P(\{6\}) = \frac{4}{12} + \frac{2}{12} = \frac{1}{2}$$

$$P(H) = 0$$

We noted earlier that probabilities may be assigned to the elementary events either by logical reasoning or by using experimental data. Probability models obtained by logical reasoning are called **assumed** or **a priori models**, and those obtained from experimental data are called **empirical** or **a posteriori models**. The simplest a priori models occur when the experiment has finitely many outcomes, all of which have equal chances of occurring. We now study such models in detail.

Experiments with Equally Likely Outcomes

In many experiments there is a certain symmetry or uniformity among the sample points that suggests that all sample points have the same chance of occurring. Such sample points are called **equally likely**, and the outcomes are said to occur **at random**, or randomly.

EXAMPLE 7 A fair coin is one that is perfectly balanced, with a head on one side and a tail on the other. If we toss a fair coin, it is reasonable to assume that heads and tails are equally likely.

EXAMPLE 8 A fair die is a perfectly symmetric and perfectly balanced cube with faces labeled 1, 2, 3, 4, 5, and 6. If we toss a fair die, it is reasonable to assume that the six possible outcomes are equally likely.

EXAMPLE 9 If we pick a name at *random* from a list of 8000 different names, we mean each of the names has the same chance of being selected as any other name. By definition, the randomness of the 8000 outcomes indicates that they are equally likely.

If an experiment has k *equally likely* sample points, then we assign each sample point the same probability. Moreover, since the probabilities of the sample points must add up to 1, the probability of each sample point must be $1/k$. Thus, we have the following result:

> If the sample space for an experiment has k equally likely sample points, then each elementary event should be assigned probability $1/k$. This assignment of probabilities is called the **uniform probability model** for the experiment.

EXAMPLE 10 If we toss a *fair* coin, then heads h and tails t are equally likely, so

$$P(\{h\}) = \frac{1}{2} \quad \text{and} \quad P(\{t\}) = \frac{1}{2}$$

Thus, we expect each outcome to occur half the time over the long run.

EXAMPLE 11 If we toss a *fair* die, then the six possible numbers on the top face are equally likely, so

$$P(\{1\}) = P(\{2\}) = P(\{3\}) = P(\{4\}) = P(\{5\}) = P(\{6\}) = \frac{1}{6}$$

Thus, each number occurs one-sixth of the time over the long run.

EXAMPLE 12 The probability model for the die-tossing experiment in Example 6 is not a uniform probability model since the elementary events are not equally likely.

EXAMPLE 13 If a card is picked at random from a standard deck of 52 cards, what is the probability that the card selected is the ace of spades?

Solution Since the card is picked at random, all possible selections are considered to be equally likely. Since there are 52 different outomes for this experiment, each has probability $\frac{1}{52}$. Thus, the probability of picking the ace of spades is $\frac{1}{52}$.

If an experiment has a uniform probability model, then the probability of any event can be obtained by counting sample points. To be specific, suppose the sample space has k sample points and E is an event with m sample points, for example,

$$E = \{s_1, s_2, \ldots, s_m\}$$

Since the sample space contains k equally likely outcomes, each outcome has probability $1/k$; thus, by the addition principle,

$$\begin{aligned} P(E) &= P(\{s_1\}) + P(\{s_2\}) + \cdots + P(\{s_m\}) \\ &= \underbrace{\frac{1}{k} + \frac{1}{k} + \cdots + \frac{1}{k}}_{m \text{ terms}} = \frac{m}{k} \end{aligned}$$

Therefore, we have the following result:

THEOREM 1 If an experiment can result in any one of k equally likely outcomes and if an event E contains m sample points, then the probability of the event E is

$$P(E) = \frac{m}{k} = \frac{\text{number of sample points in } E}{\text{number of sample points in the sample space}} = \frac{n(E)}{n(S)}$$

(From Chapter 6, $n(E)$ = the number of elements in E.)

EXAMPLE 14 Consider the experiment of tossing a *fair* die and observing the number on the top face. Find the probability of

(a) the event E that an even number is tossed

(b) the event G that a number divisible by 3 is tossed

Solution The die is fair, so we assign the uniform probability model. The events are

$$E = \{2, 4, 6\} \qquad \text{and} \qquad G = \{3, 6\}$$

Since E has three sample points, G has two sample points, and the entire sample space has six sample points, it follows from Theorem 1 that

$$P(E) = \frac{3}{6} = \frac{1}{2} \qquad \text{and} \qquad P(G) = \frac{2}{6} = \frac{1}{3}$$

EXAMPLE 15 A batch of 7 resistors contains 2 defectives. If a resistor is selected at random from the batch, what is the probability that it is defective?

Solution Since the resistor is selected at random, the 7 possible choices are equally likely. Since 2 of the 7 choices are defective, it follows from Theorem 1 that the probability of choosing a defective is

$$P(D) = \frac{2}{7}$$

where D = the subset of defective resistors.

EXAMPLE 16 If a pair of fair dice is tosed, what is the probability that the sum of the numbers tossed is 7?

Solution Assume that one die is red and the other green. Label the dice r and g.* Each die can show one of the integers from 1 to 6, so the two dice can yield any one of the 36 pairs in Table 4. Since the dice are fair, we assume that each of these 36 possible outcomes is equally likely. The event E that the sum of the numbers is 7 contains the six shaded sample points in Table 4. Therefore, the probability that the sum of the numbers is 7 is $\frac{6}{36} = \frac{1}{6}$.

TABLE 4

r \ g	1	2	3	4	5	6
1	(1, 1)	(1, 2)	(1, 3)	(1, 4)	(1, 5)	(1, 6)
2	(2, 1)	(2, 2)	(2, 3)	(2, 4)	(2, 5)	(2, 6)
3	(3, 1)	(3, 2)	(3, 3)	(3, 4)	(3, 5)	(3, 6)
4	(4, 1)	(4, 2)	(4, 3)	(4, 4)	(4, 5)	(4, 6)
5	(5, 1)	(5, 2)	(5, 3)	(5, 4)	(5, 5)	(5, 6)
6	(6, 1)	(6, 2)	(6, 3)	(6, 4)	(6, 5)	(6, 6)

*Considering one die to be red and the other green does not affect the result that we record, so it does not change the experiment. It does make it easier to determine the proper equally likely sample space.

EXAMPLE 17 In an experiment 2 fair coins are tossed. Find the probability of obtaining exactly 1 head on the 2 tosses.

Solution Figure 1 suggests there are four equally likely outcomes, *hh, ht, th,* and *tt*. If E is the event that exactly one head occurs, then

$$E = \{ht, th\}$$

FIGURE 1

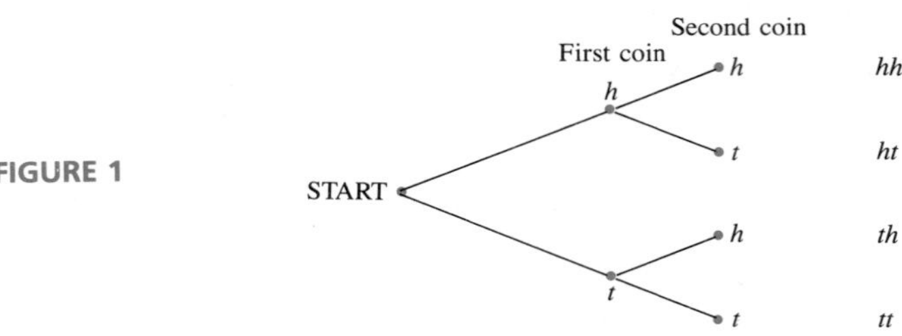

If we assume that the four possible outcomes are equally likely, then

$$P(E) = \frac{2}{4} = \frac{1}{2}$$

Sometimes it is difficult to tell whether the outcomes of an experiment are equally likely. As an illustration, suppose in the last example we had argued as follows: Since the number of heads on the two tosses is zero, one, or two, the sample space is $S = \{0, 1, 2\}$, and the event E that exactly one head is tossed is $E = \{1\}$. If we assume that the three sample points are equally likely, then $P(E) = \frac{1}{3}$.

Obviously, this result disagrees with that of Example 17. But which is correct? Actually, both are correct in the sense that each of them arises from a valid probability model for the experiment, one based on three equally likely outcomes and the other on four equally likely outcomes. It is more meaningful to ask which theory (model) will most accurately describe a physical coin-tossing experiment. You may find it interesting to solve Exercise 15 and show experimentally that the solution of Exercise 15 most accurately describes the physical situation. (If you further consider the problem, you may be able to convince yourself that the three sample points in the second model are not really equally likely.)

Empirical Probability Models

In problems in which it is difficult or impossible to obtain probabilities for the experimental outcomes by logical reasoning, we can sometimes obtain approximate values for these probabilities by performing the experiment repetitively (the more times the better); we can then use the relative frequencies of the sample points as *approximate* probabilities. Probabilities obtained in this way are called **empirical probabilities.**

EXAMPLE 18 A thumbtack, tossed 2000 times, lands point up 910 times and point down 1090 times (Figure 2). Find the empirical probability of each outcome.

FIGURE 2

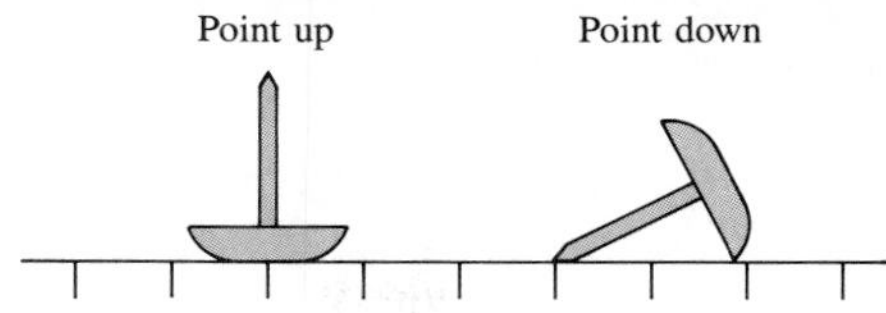

Solution

$$P(\{\text{point up}\}) \cong \frac{910}{2000} = .455$$

$$P(\{\text{point down})\} \cong \frac{1090}{2000} = .545$$

EXAMPLE 19 A loaded six-sided die is tossed 1000 times with the results shown in Table 5.

(a) Find the empirical probability of each outcome.
(b) Find the empirical probability of tossing an odd number with this die.
(c) Find the empirical probability of tossing a number greater than 4 with this die.

TABLE 5

Outcome	Number of occurrences
1	102
2	156
3	28
4	214
5	113
6	387

Solution (a)

$$P(\{1\}) \cong \frac{102}{1000} = .102 \qquad P(\{2\}) \cong \frac{156}{1000} = .156$$

$$P(\{3\}) \cong \frac{28}{1000} = .028 \qquad P(\{4\}) \cong \frac{214}{1000} = .214$$

$$P(\{5\}) \cong \frac{113}{1000} = .113 \qquad P(\{6\}) \cong \frac{387}{1000} = .387$$

Solution (b) From the addition principle, the probability of tossing an odd number is

$$P(\text{odd number}) = P(\{1\}) + P(\{3\}) + P(\{5\}) = .102 + .028 + .113 = .243$$

Solution (c) From the addition principle, the probability of tossing a number greater than 4 is

$$P(\text{number greater than 4}) = P(\{5\}) + P(\{6\}) = .113 + .387 = .500$$

EXERCISE SET 7.2

1. List the elementary events for each of the following experiments:
   (a) A coin is tossed two times and the resulting sequence of heads and tails is observed.
   (b) A stock on the New York Stock Exchange is observed for one market day, and it is recorded whether the stock increases in value (i), decreases in value (d), or undergoes no change in value (n).
   (c) The stock in part (b) is observed for 2 market days.
   (d) A letter is picked at random from the set $\{a, b, c, d, e\}$.
   (e) A die is tossed and it is observed whether the outcome is even (e) or odd (o).
   (f) A card is picked at random from an ordinary deck of 52 cards and the suit—clubs (c), diamonds (d), hearts (h), or spades (s)—is observed.
   (g) A lot of 10 items is tested and the number of defectives recorded.

2. State the number of elementary events for each of the following experiments:
   (a) A botanist successively observes 4 corn plants to decide whether the plants have rust (r) or are free from rust (f). The botanist then records the resulting sequence of observations.
   (b) A card is picked at random from a standard deck of 52 cards and the card picked is recorded.
   (c) A set of 2 different letters is picked from the set $\{a, b, c, d, e\}$.
   (d) In the investigation of human births, the month of birth and the sex of a child are recorded as an ordered pair, for example, (February, male).

3. Use the probability model in Example 6 and the addition principle to compute the probabilities of the following events:
   (a) The event A that a number less than 5 is tossed.
   (b) The event B that an odd number is tossed.
   (c) The event C that a number less than 1 is tossed.

4. Compute the probabilities of the events A, B, and C in Exercise 3 assuming that the die is fair.

5. Let $S = \{s_1, s_2, s_3, s_4, s_5\}$ be the sample space for an experiment, and assume the experiment is assigned the probability model

$$P(\{s_1\}) = \frac{4}{11} \qquad P(\{s_2\}) = \frac{1}{11} \qquad P(\{s_3\}) = \frac{1}{11}$$

$$P(\{s_4\}) = \frac{2}{11} \qquad P(\{s_5\}) = \frac{3}{11}$$

   (a) Find the probability of the event $A = \{s_1, s_3\}$.
   (b) Find the probability of the event

$$B = \{s_2, s_3, s_5\}$$

   (c) Find the probability of the event S.

6. Using the probability model and the events A and B of Exercise 5, compute
   (a) $P(A \cap B)$ (b) $P(A \cup B)$
   (c) $P(A \cap B')$ (d) $P(A' \cap B')$
   (e) $P(A \cap A')$

7. Which of the following are valid probability models for an experiment with sample space

$$S = \{s_1, s_2, s_3, s_4\}$$

   (a) $P(\{s_1\}) = .3$, $P(\{s_2\}) = .6$, $P(\{s_3\}) = .1$, $P(\{s_4\}) = .2$
   (b) $P(\{s_1\}) = -.2$, $P(\{s_2\}) = .8$, $P(\{s_3\}) = .2$, $P(\{s_4\}) = .2$
   (c) $P(\{s_1\}) = 0$, $P(\{s_2\}) = .7$, $P(\{s_3\}) = .2$, $P(\{s_4\}) = .1$
   (d) $P(\{s_1\}) = .75$, $P(\{s_2\}) = .05$, $P(\{s_3\}) = .15$, $P(\{s_4\}) = .05$

8. Let $S = \{s_1, s_2, s_3\}$ be the sample space of an experiment, and suppose $P(\{s_1\}) = .3$ and $P(\{s_2\}) = .2$. Find $P(\{s_3\})$.

9. In each of the following experiments, list the elementary events and decide whether the outcomes are equally likely:
   (a) A die is tossed and it is recorded whether the top face is even or odd.
   (b) A die is tossed and it is recorded whether the top face is a 3 or not.
   (c) A card is picked at random from a standard deck of 52 cards and it is recorded whether the card is an ace or not.
   (d) A card is picked at random from a deck of 52 cards and it is recorded whether the card is black or red.
   (e) A set of 2 letters is chosen at random from the set $\{a, b, c, d\}$.

10. Assume a uniform probability model is assigned to each of the following experiments. In each case state the probability assigned to each outcome.
   (a) A card is drawn from a standard deck of 52 cards and the card picked is recorded.
   (b) A coin is tossed 5 times and the resulting sequence of outcomes is recorded.

(c) In a three-child family, suppose the sexes of the children are listed from oldest to youngest as an ordered triple. For example, (m, m, f) means the oldest and middle children are males and the youngest is a female.

11. Assume an experiment with sample space

$$S = \{s_1, s_2, s_3\}$$

is assigned the uniform probability model. List all possible events and their probabilities.

12. A survey shows that 20% of all cars entering a certain intersection turn left, 25% of all cars turn right, and 55% of all cars proceed straight ahead. A car entering the intersection is observed and it is recorded whether the car turns left (l), turns right (r), or goes straight (s). Assign a probability model to this experiment and use the model to compute the probability that the car makes a turn.

13. Assume that in a three-child family, the 8 possible sex distributions of the children, namely, *mmm*, *mmf*, . . . , *fff*, are equally likely. Compute the probabilities of the following events:
(a) Exactly one child is female.
(b) At least one child is female.
(c) There are more female children than male children.
(d) The family has at most one male child.

14. Compute the probabilities of the following events for the experiment in Example 16:
(a) At least one of the dice shows a 3.
(b) The sum of the numbers on the dice is 6.
(c) Neither a 2 nor a 5 appears.

15. Toss a pair of coins 300 times and record the number of times that the pair shows zero heads, exactly one head, and exactly two heads. On the basis of your data, do you think that the three outcomes are equally likely?

16. A six-sided die is loaded so that 1, 2, and 3 occur equally often, and 4, 5, and 6 occur equally often, but 4, 5, and 6 occur three times as often as 1, 2, and 3. Find a probability model for the experiment of tossing the die and recording the number on the top face.

17. The popular game of Dungeons and Dragons uses a symmetric 12-sided die (a dodecahedron), with sides numbered from 1 to 12, and a symmetric 20-sided die (an icosahedron), whose sides are numbered from 0 to 9 with each number appearing on two faces. Solve the following problems assuming that the dice are fair (perfectly balanced):
(a) Find a probability model for the experiment of tossing the 12-sided die and recording the number on the top face.
(b) Repeat (a) for the 20-sided die.
(c) If both dice are tossed, what is the probability that the numbers on the top face add up to 7? [*Hint*: See Example 16.]

18. A fair die is tossed and the number on the top face is recorded. Find the probabilities of the events

$$E = \{1, 3, 5, 6\} \qquad F = \{1, 6\} \qquad G = \{1, 2, 3, 4, 5\}$$

19. A pair of fair dice is tossed and the sum of the numbers on the faces is recorded. Use Table 4 and the method of Example 16 to construct a list of probabilities for all the sums from 2 to 12.

20. In a certain experiment with finite sample space S, two events E and F have the properties

$$E \cup F = S \quad \text{and} \quad E \cap F = \varnothing$$

Given that $P(E) = \frac{1}{4}$, find $P(F)$.

21. An urn contains 8 black balls and 4 red balls. Find a probability model for the experiment of drawing a ball at random and recording its color.

22. Each of 15 cards is labeled with a digit from 1 to 5 as follows:

three cards are labeled with a 1
two cards are labeled with a 2
one card is labeled with a 3
seven cards are labeled with a 4
two cards are labeled with a 5

(a) Find a probability model for the experiment of drawing a card at random and observing its digit.
(b) Find the probability of the event E that the digit is even.
(c) Find the probability of the event F that the digit is greater than 2.

23. In his lifetime, Babe Ruth batted 129 times in World Series play and got 42 hits, 15 of which were home runs.
(a) Find the empirical probability of the Babe getting a hit in any single time at bat during a World Series.
(b) Find the empirical probability of the Babe hitting a home run in any single time at bat during a World Series.

24. A sampling of 20,000 flowering plants results in the findings that

8250 have yellow flowers
3172 have blue flowers
8578 have green flowers

If a plant is chosen at random, find the empirical probability that
(a) it has blue flowers.
(b) it has blue or green flowers.
(c) it does not have green flowers.

25. In a market survey, the data in Table 6 relating family income and telephone service are recorded. Find the empirical probability that
(a) a household has two telephones.
(b) a household with income in the \$13,000–18,000 bracket has more than one telephone.
(c) a household with income of \$8000 or more has no telephone.

TABLE 6

	Number of telephones in the household				
Annual household income	**0**	**1**	**2**	**3**	**4 (or more)**
< \$8,000	19	48	3	0	0
\$8000–13,000	15	40	10	3	2
\$13,000–18,000	9	38	14	7	2
\$18,000–23,000	2	35	21	9	3
> \$23,000	0	22	32	11	5

7.3 BASIC THEOREMS OF PROBABILITY

In this section we investigate how the probabilities of various events are related. For example, how is the probability that an event occurs related to the probability that it does not occur, and how are the probabilities of $E \cup F$ and $E \cap F$ related to the probabilities of E and F? The results obtained here are useful in solving complicated probability problems. In this section we only consider experiments involving finite sample spaces.

We must use the three fundamental properties of probability for a finite sample space to obtain these results. The properties are repeated here:

1. For an event E, $0 \leq P(E) \leq 1$.
2. $P(S) = 1$, where S is the sample space.
3. $P(E_1 \cup E_2 \cup \cdots \cup E_k) = P(E_1) + P(E_2) + \cdots + P(E_k)$, where E_1, E_2, . . . , E_k are mutually exclusive events.

In Section 7.1 we interpret some expressions about events in terms of sets. Using that information, you can obtain the results in Table 1.

TABLE 1

Verbal expression	Set interpretation
The probability that E does *not* occur	$P(E')$
The probability that E *and* F occur	$P(E \cap F)$
The probability that E *or* F occurs	$P(E \cup F)$

Note that fundamental property 3 states:

For two mutually exclusive events E and F,

$$P(E \cup F) = P(E) + P(F) \tag{1}$$

See Figure 1.

FIGURE 1

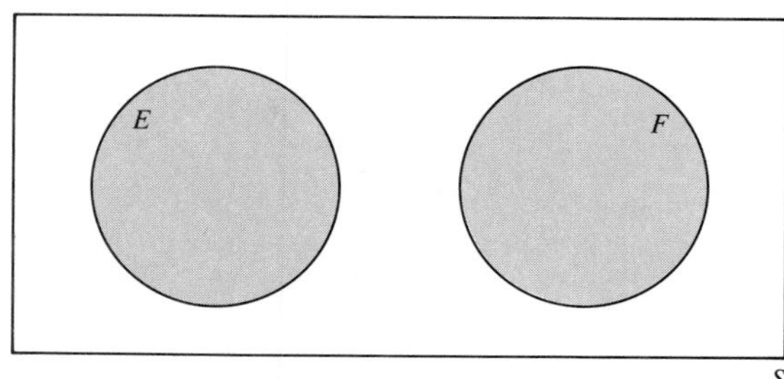

In the previous section, we state that $P(\varnothing) = 0$; we now prove this statement.

THEOREM 1 $P(\varnothing) = 0$ where $\varnothing$ is the empty set

Proof For any event E, $E \cup \varnothing = E$ and $E \cap \varnothing = \varnothing$. ($E$ and $\varnothing$ are mutually exclusive events.) Therefore, by fundamental property 3,

$$P(E) = P(E \cup \varnothing) = P(E) + P(\varnothing)$$

So $P(\varnothing) = 0$.

EXAMPLE 1 If a fair die is tossed, what is the probability that the number on the top face is even or a 5? What is the probability that the number on the top face is a 7?

Solution We assign the uniform probability model to this experiment. The event G that the number tossed is either even or a 5 is

$$G = \{2, 4, 5, 6\}$$

The event H that the number showing is a 7 is the empty set: $H = \varnothing$. Thus,

$$P(G) = \frac{4}{6} = \frac{2}{3} \qquad \text{and} \qquad P(H) = P(\varnothing) = 0$$

Note that we can obtain $P(G)$ in another way. We have

$$G = E \cup F$$

where E is the event that an even number is tossed and F is the event that a 5 is tossed. The events E and F are

$$E = \{2, 4, 6\} \qquad \text{and} \qquad F = \{5\}$$

Since $E \cap F = \varnothing$, the events E and F are mutually exclusive. Thus, Equation (1) tells us that

$$P(E \cup F) = P(E) + P(F) = \frac{3}{6} + \frac{1}{6} = \frac{2}{3}$$

Therefore,

$$P(G) = P(E \cup F) = \frac{2}{3}$$

which agrees with the result obtained above.

Warning Equation (1) does not apply if E and F are not mutually exclusive. The next theorem applies to any two events, whether or not they are mutually exclusive.

THEOREM 2 If E and F are events in a finite sample space, then

$$P(E \cup F) = P(E) + P(F) - P(E \cap F) \tag{2}$$

Proof As shown in Figure 2b,

$$E \cup F = \underbrace{(E \cap F') \cup (E \cap F)}_{E} \cup (E' \cap F)$$

(with $(E \cap F) \cup (E' \cap F)$ marked F)

where the three events on the right side of the equal signs are mutually exclusive. Therefore,

$$P(E \cup F) = \underbrace{P(E \cap F') + P(E \cap F)}_{P(E)} + P(E' \cap F)$$

(with $P(E \cap F) + P(E' \cap F)$ marked $P(F)$)

FIGURE 2

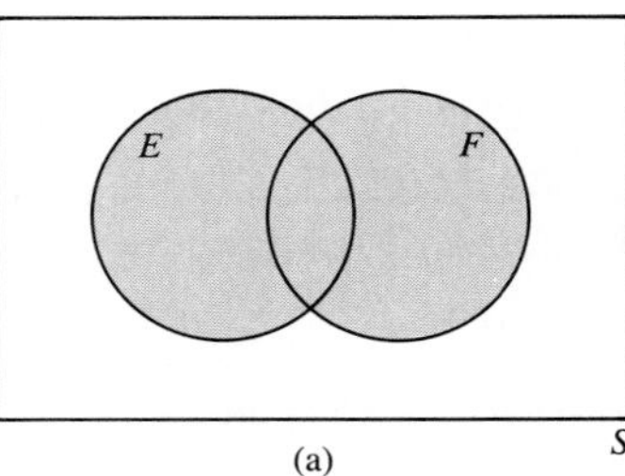

(a)

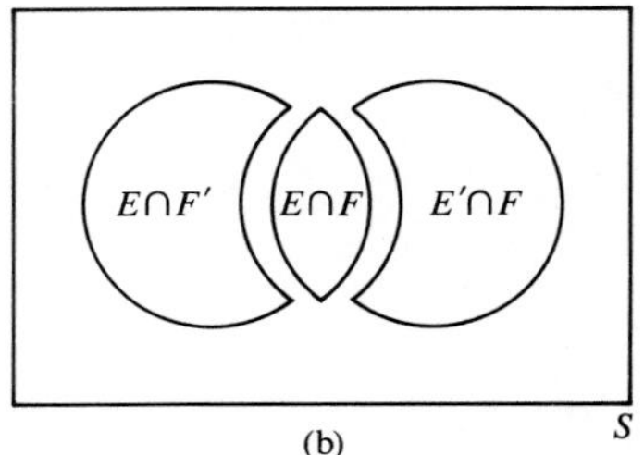

(b)

Since $P(E \cap F)$ is included twice in $P(E) + P(F)$,

$$P(E \cup F) = P(E) + P(F) - P(E \cap F).$$

EXAMPLE 2 A card is picked at random from an ordinary deck of 52 cards. Find:

(a) the probability that the card is a club.

(b) the probability that the card is an ace.

(c) the probability that the card is either an ace or a club.

Solution (a) Since the card is picked at random, we assign the uniform probability model; thus, each sample point is assigned probability $\frac{1}{52}$. Let C be the event that the card is a club. Since there are 13 clubs in the deck,

$$P(C) = \frac{13}{52}$$

Solution (b) Let A be the event that the card is an ace. Since there are 4 aces in the deck,

$$P(A) = \frac{4}{52}$$

Solution (c) The probability that the card is either an ace or a club is

$$P(C \cup A)$$

From Theorem 2,

$$P(C \cup A) = P(C) + P(A) - P(C \cap A) \tag{3}$$

To belong to $C \cap A$, a sample point must be both a club and an ace. There is one such sample point, the ace of clubs. Thus,

$$P(C \cap A) = \frac{1}{52}$$

Substituting this value and the values obtained in (a) and (b) into Equation (3) gives

$$P(C \cup A) = \frac{13}{52} + \frac{4}{52} - \frac{1}{52} = \frac{16}{52} = \frac{4}{13}$$

EXAMPLE 3 Assume that a reel of magnetic tape for a digital computer is classified defective if it has either of the following imperfections: (1) the magnetic coating is improperly applied or (2) the tape has a fold or tear; otherwise, it is classified as nondefective.

A survey of defective reels returned to the manufacturer shows

0.5% of all reels produced have improperly applied coatings
0.3% of all reels produced have tape folds or tears
0.1% of all reels produced have both imperfections

Find the probability that a purchaser will receive a defective reel of tape.

Solution Let E be the event that the purchased reel has an improperly applied coating and let F be the event that the tape has a fold or tear. Thus,

$$P(E) = .005 \qquad P(F) = .003 \qquad P(E \cap F) = .001$$

The event that the purchased reel is defective is $E \cup F$, and the probability of this event is

$$\begin{aligned} P(E \cup F) &= P(E) + P(F) - P(E \cap F) \\ &= .005 + .003 - .001 \\ &= .007 \end{aligned}$$

In other words, 0.7% of all reels produced are defective.

Remark To determine $P(E \cup F)$, proceed as follows:

1. If $E \cup F = \{s_1, s_2, \ldots, s_k\}$, then
$$P(E \cup F) = P(\{s_1\}) + P(\{s_2\}) + \cdots + P(\{s_k\})$$
2. If E and F are mutually exclusive events ($E \cap F = \varnothing$), then
$$P(E \cup F) = P(E) + P(F)$$
3. In general, $P(E \cup F) = P(E) + P(F) - P(E \cap F)$

We conclude this section with a theorem relating the probability that an event occurs to the probability that it does not occur.

THEOREM 3 If E is an event in a finite sample space, then
$$P(E') = 1 - P(E)$$

Proof Since $E \cup E' = S$ and $E \cap E' = \varnothing$ (Figure 3), we can apply fundamental properties 2 and 3 to obtain
$$1 = P(S) = P(E \cup E') = P(E) + P(E')$$
Thus,
$$P(E') = 1 - P(E)$$

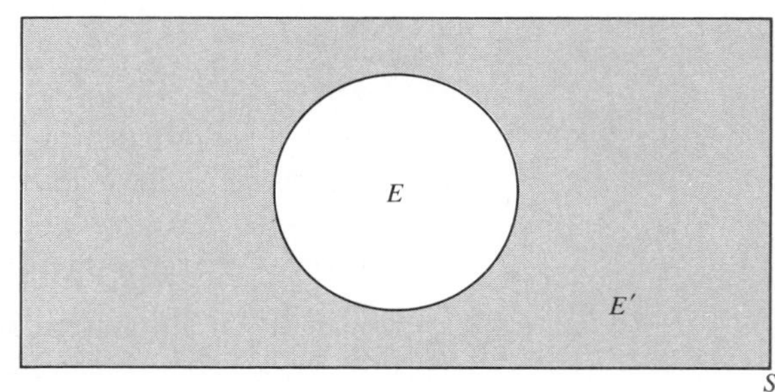

FIGURE 3

EXAMPLE 4 In Example 3, we show that the probability of a purchaser receiving a defective reel of tape is .007. Thus, the probability of the purchaser receiving a nondefective reel is
$$1 - .007 = .993$$

Probabilities Expressed as Odds

In betting circles, the likelihood of an event is often expressed in terms of **odds**,* either **odds in favor** of the event or **odds against** the event. These terms are defined as follows: If
$$\frac{P(E)}{P(E')} = \frac{b}{w}$$

* In betting circles, the odds in favor of E = bet:net win. For example, odds of 1:2 in favor of E means that if you bet \$1 that E will occur, you will win \$2 (and get your bet back) if E does occur and you will lose \$1 if E does not occur.

then we say the odds in favor of E are b to w (also denoted by $b:w$), and we say the odds against E are w to b, where

$$\frac{P(E')}{P(E)} = \frac{w}{b}$$

Therefore, if E is the event that a 2 is rolled with a fair die, then $P(E) = \frac{1}{6}$ and $P(E') = \frac{5}{6}$. Thus,

$$\frac{P(E)}{P(E')} = \frac{1/6}{5/6} = \frac{1}{5}$$

The odds in favor of rolling a 2 are 1 to 5, and the odds against rolling a 2 are 5 to 1. These statements can be interpreted to mean that a number other than 2 will be rolled five times more frequently than a 2 over the long run.

Note that the odds in favor of an event and the odds against that event are reciprocals of each other.

Since the odds in favor of E are b to w,

$$\frac{b}{w} = \frac{P(E)}{P(E')} = \frac{P(E)}{1 - P(E)} \tag{4}$$

Equation (4) can be solved for $P(E)$ to obtain

$$P(E) = \frac{b}{b + w} \tag{5}$$

(The proof is left for Exercise 20.)

EXAMPLE 5 A card is drawn at random from a standard deck of 52 cards.

(a) Find the odds in favor of drawing an ace.

(b) Find the odds against drawing an ace.

Solution (a) Let E be the event that an ace is drawn. Since there are 4 aces in the deck, it follows that

$$P(E) = \frac{4}{52} = \frac{1}{13} \quad \text{and} \quad P(E') = 1 - P(E) = \frac{12}{13}$$

Thus,

$$\frac{P(E)}{P(E')} = \frac{1/13}{12/13} = \frac{1}{12}$$

so the odds are 1 to 12 in favor of drawing an ace.

Solution (b) Since

$$\frac{P(E')}{P(E)} = \frac{12/13}{1/13} = \frac{12}{1}$$

the odds are 12 to 1 against drawing an ace.

EXAMPLE 6 In a championship boxing match the oddsmakers state that the champion is a 3 to 2 favorite over the challenger (i.e., the odds are 3 to 2 in favor of the champion winning the match). According to the oddsmakers, what is the probability that the champion will win?

Solution If we let E be the event that the champion will win, then by Equation (5)

$$P(E) = \frac{b}{b + w} = \frac{3}{3 + 2} = \frac{3}{5}$$

EXERCISE SET 7.3

1. Suppose E and F are mutually exclusive events such that $P(E) = .2$ and $P(F) = .5$.
   (a) Find the probability that E or F occurs.
   (b) Find the probability that E and F occur.
   (c) Find the probability that E does not occur.

2. Suppose E_1, E_2, and E_3 are mutually exclusive events such that $P(E_1) = .2$, $P(E_2) = .3$, and $P(E_3) = .4$.
   (a) Find the probability that at least one of the events occurs.
   (b) Find the probability that E_2 or E_3 occurs.
   (c) Find the probability that all three events occur.

3. Let E and F be events such that $P(E) = .3$, $P(F) = .1$, and $P(E \cap F) = .05$. Find
   (a) $P(E')$ (b) $P(F')$ (c) $P(E \cup F)$

4. Let E and F be events such that $P(E) = .3$, $P(F) = .7$, and $P(E \cup F) = .8$. Find $P(E \cap F)$.

5. Let E and F be events such that $P(E') = .3$, $P(F) = .2$, and $P(E \cup F) = .8$. Find $P(E \cap F)$.

6. Let E and F be the events described in Exercise 3. Use De Morgan's laws to find $P(E' \cup F')$.

7. Let E and F be mutually exclusive events such that $P(E) = .4$ and $P(F) = .3$. Find
   (a) $P(E \cup F)$ (b) $P(E \cap F)$ (c) $P(E')$
   (d) $P(F')$ (e) $P(E' \cap F')$
   [*Hint*: Use De Morgan's laws in part (e).]

8. The drug alphacidin can produce two unwanted side effects, drowsiness and headache. Drowsiness occurs in 2% of all users, headaches in 3%, and both occur in 1%. What is the probability that
   (a) a user will experience drowsiness or headache?
   (b) a user will not experience a headache?

9. In Exercise 8 what is the probability that the user will experience neither side effect? [*Hint*: Use De Morgan's laws.]

10. An item produced by a manufacturing process is either nondefective, has defect a, has defect b, or both defect a and defect b. In each part, *verbally* describe the complementary event E'.
    (a) E: the item has at least one defect.
    (b) E: the item has at most one defect.
    (c) E: the item has defect a.
    (d) E: the item has two defects.

11. Let $S = \{s_1, s_2, s_3, s_4, s_5\}$ be the sample space of an experiment and let

    $$P(\{s_1\}) = .2 \qquad P(\{s_2\}) = .1 \qquad P(\{s_3\}) = .3$$
    $$P(\{s_4\}) = .1 \qquad P(\{s_5\}) = .3$$

    If

    $$E = \{s_1, s_3\} \qquad F = \{s_1, s_4, s_5\} \qquad G = \{s_2, s_4, s_5\}$$

    find
    (a) $P(E')$ (b) $P(E \cup F)$ (c) $P(G \cap F)$
    (d) $P(E \cap G)$ (e) $P(E \text{ and } F)$ (f) $P(F \text{ or } G)$

12. Using the probability model and events given in Exercise 11, verify that the following equations are true by computing both sides.
    (a) $P(G \cup F) = P(G) + P(F) - P(G \cap F)$
    (b) $P(G') = 1 - P(G)$
    (c) $P(E \text{ or } F) = P(E) + P(F) - P(E \text{ and } F)$
    (d) $P(E \cup G) = P(E) + P(G)$

13. For a bill to come before the President of the United States for signing, it must be passed by the House of Representatives and the Senate. A lobbyist estimates the probability of her bill passing the House to be .5 and the prob-

ability of passing the Senate to be .7. She also estimates that the probability the bill will be passed either by the House or by the Senate to be .8. What is the probability the bill will come before the President?

14. River water entering a filtration plant is classified as polluted if it contains (1) an intolerable percentage of dangerous organic materials or (2) an intolerable percentage of inorganic materials; otherwise, it is classified as unpolluted. A survey of the plant's records show that

27% of all water entering the plant is polluted because of organic materials.

46% of all water entering the plant is polluted because of inorganic materials.

23% of all water entering the plant is polluted because of both organic and inorganic materials.

Find the probability that a quantity of water in the plant is unpolluted.

15. A study of air traffic patterns at a major metropolitan airport yields the following probabilities for the number of aircraft waiting to land on one of the runways:

Number of aircraft waiting to land	0	1	2	3	More than 3
Probability	.1	.2	.4	.2	.1

Find the probability that
(a) at least 3 aircraft are waiting to land.
(b) at most 2 aircraft are waiting to land.
(c) more than 2 aircraft are waiting to land.

16. Find the odds in favor of and against
(a) rolling a number less than 3 with a fair die.
(b) drawing a club from a standard deck of 52 cards.
(c) rolling a total of 7 with two fair dice.
(d) tossing a head with a fair coin.

17. Find the odds in favor of and against
(a) obtaining 2 tails by tossing a pair of fair coins.
(b) drawing a face card (jack, queen, or king) from a standard deck of 52 cards.
(c) rolling "snake eyes" (a total of 2) or "boxcars" (a total of 12) with a pair of fair dice.
(d) obtaining a consonant by choosing a letter at random from the alphabet.

18. A roulette wheel is pictured in Figure 4. The wheel contains the numbers from 1 to 36, half on a colored background and half on a black background; the two numbers 0 and 00 are on a white background.*
(a) What is the probability of the number 23 occurring?
(b) What is the probability that a number on a colored background will occur?
(c) What are the odds against winning if you bet that the ball will land on number 23?
(d) What are the odds against winning if you bet that the ball will land on a colored position?

FIGURE 4

19. (a) Suppose the odds in favor of an event E are 7 to 1. What is the probability of E?
(b) Suppose the odds against an event F are 8 to 3. What is the probability of F?

20. Show that if the odds in favor of an event E are b to w, then the probability of E is

$$P(E) = \frac{b}{b + w}$$

*In an actual roulette wheel, half the numbers from 1 to 36 are on a red background, half are on a black background, and 0 and 00 are on a green background.

21. In Exercise 31 of Section 6.2, suppose that a person from the group is chosen at random.
(a) What is the probability that he or she had been only to the opera?
(b) What is the probability that he or she had not attended any of the events?

22. In Exercise 32 of Section 6.2, suppose that a person from the group is chosen at random.
(a) What is the probability that he or she did not buy any of the brands X, Y, or Z?
(b) What is the probability that he or she bought brands X and Y but not brand Z?

7.4 PERMUTATIONS, COMBINATIONS, AND PROBABILITY

In this section we show how permutations and combinations can be used to solve certain probability problems involving equally likely outcomes. Counting techniques from Chapter 6 are used.

EXAMPLE 1 In a psychological test for extrasensory perception (ESP), 3 colored cards are used: 1 red, 1 yellow, and 1 blue. The cards are placed in some order and a blindfolded subject is asked to guess the order of the cards. Assuming that the subject has no ESP, what is the probability that he or she will guess the correct order by chance?

Solution There are $3! = 6$ different ways to order the cards, one of which is the correct order. In the absence of ESP, it is reasonable to assume that the subject is selecting an order at random. Thus, each of the six possible orders is equally likely to be selected. The probability the correct order will be selected by chance is thus $\frac{1}{6}$.

EXAMPLE 2 A fair coin is tossed 6 times. What is the probability that exactly 2 of the tosses are heads?

Solution If we denote heads by h and tails by t, then six tosses of a coin result in a sequence of h's and t's. As shown in Example 2 of Section 6.3, there are $2^6 = 64$ different sequences possible. Since the coin is fair, it is reasonable to assume that these 64 sequences are equally likely. Thus, by Theorem 1 in Section 7.2,

$$P(\text{exactly 2 } h\text{'s}) = \frac{\text{number of sequences with exactly 2 } h\text{'s}}{\text{number of all possible sequences}}$$

Now a sequence with 2 h's will necessarily have 4 t's. It is evident that any sequence with 2 h's and 4 t's is completely determined once we specify the two tosses on which the heads occur. Thus, the number of different sequences with 2 h's and 4 t's is simply the number of ways of selecting a set of two toss numbers for the h's from

$$\{1, 2, 3, 4, 5, 6\}$$

This number is

$$C_{6,2} = \frac{6!}{2!\,4!} = \frac{6 \cdot 5}{2 \cdot 1} = 15$$

Therefore,

$$P(\text{exactly 2 } h\text{'s}) = \frac{C_{6,2}}{2^6} = \frac{15}{64}$$

EXAMPLE 3 A hand of 7 cards is dealt from a well-shuffled deck of 52 cards. Find the probability that

(a) the hand has 7 spades.

(b) the hand has 2 spades and 5 red cards.

Solution (a) Since the deck is well-shuffled, it is reasonable to assume that all possible 7-card hands are equally likely. Thus, by Theorem 1 of Section 7.2 the probability that the hand has 7 spades is

$$P(7 \text{ spades}) = \frac{\text{number of hands with 7 spades}}{\text{total number of 7-card hands}}$$

Since there are 52 cards in the deck, there are $C_{52,7}$ different 7-card hands, and since there are 13 spades in the deck, there are $C_{13,7}$ different 7-card hands consisting entirely of spades. Thus,

$$P(7 \text{ spades}) = \frac{C_{13,7}}{C_{52,7}} = \frac{1716}{133{,}784{,}560} \cong .000013$$

Solution (b) The number of different 7-card hands is $C_{52,7}$. To find the number of 7-card hands with 2 spades and 5 red cards, we can argue as follows. Since the deck has 13 spades, there are $C_{13,2}$ possible ways to choose a set of 2 spades from the 13. Moreover, the deck has 26 red cards, so there are $C_{26,5}$ ways to choose 5 red cards from the 26 cards. Thus, by the multiplication principle, there are

$$C_{13,2} \cdot C_{26,5}$$

different ways to choose 2 spades and 5 red cards to form a hand. Therefore,

$$P(2 \text{ spades and 5 red cards}) = \frac{C_{13,2} \cdot C_{26,5}}{C_{52,7}} = \frac{(78)(65{,}780)}{133{,}784{,}560} \cong .038$$

EXAMPLE 4 An urn contains 7 red balls and 3 black balls. If 4 balls are selected at random without replacement, what is the probability that all 4 are red?

Solution This is equivalent to a card-dealing problem. Imagine a deck of 10 cards marked $R_1, R_2, \ldots, R_7$ (corresponding to the 7 red balls) and B_1, B_2, B_3 (corresponding to the 3 black balls). The selection of 4 balls is equivalent to dealing a hand of 4 cards from this deck. We want to find the probability that the hand contains all red cards. Since there are 10 cards in the deck, the number of possible 4-card hands is $C_{10,4}$, and since there are 7 red cards, the number of 4-card hands with all red cards is $C_{7,4}$. Thus,

$$P(\text{all 4 balls are red}) = \frac{C_{7,4}}{C_{10,4}} = \frac{35}{210} = \frac{1}{6}$$

EXAMPLE 5 A lot of 100 items from a manufacturing process is known to contain 10 defectives and 90 nondefectives. If a sample of 8 items is selected at random, what is the probability that

(a) the sample has 3 defectives and 5 nondefectives?
(b) the sample has at least 1 defective?
(c) the sample has more than 6 defectives?

Solution (a) This can also be viewed as a card-dealing problem. Imagine the 100 items to be a deck of 100 cards. In this deck, 10 cards are marked D_1, $D_2, \ldots, D_{10}$ (corresponding to the 10 defectives) and 90 of the cards are marked $N_1, N_2, \ldots, N_{90}$ (corresponding to the 90 nondefectives). The sample of 8 items corresponds to a hand of 8 cards dealt from this deck. Thus the probability that the sample has 3 defectives and 5 nondefectives is just the probability that the hand of 8 cards contains 3 D's and 5 N's. By imitating the argument in Example 3(b), you can show that this probability is

$$P(3 \text{ defectives and } 5 \text{ nondefectives}) = \frac{C_{10,3} \cdot C_{90,5}}{C_{100,8}} \cong .0283$$

Solution (b) By Theorem 3 in Section 7.3, the probability that the sample has at least one defective is 1 minus the probability that the sample does *not* have at least one defective, or equivalently,

$$P(\text{at least 1 defective}) = 1 - P(\text{no defectives}) \qquad (1)$$

If, as in part (a), we view this as a card-dealing problem, then the probability of no defectives is simply the probability of dealing a hand with 8 N's from the deck with 90 N's and 10 D's described in part (a). Since there are $C_{90,8}$ different ways to select 8 N's from the 90 N's and since there are $C_{100,8}$ different 8-card hands possible, the probability of no defectives is

$$P(\text{no defectives}) = \frac{C_{90,8}}{C_{100,8}} \cong .42$$

Substituting this value in Equation (1) we obtain

$$P(\text{at least 1 defective}) = 1 - \frac{C_{90,8}}{C_{100,8}} \cong 1 - .42 = .58$$

Solution (c) The probability that the sample has more than 6 defectives is just the probability that the sample has exactly 7 defectives or exactly 8 defectives. Thus, if we let E and F denote the events

E: the sample has exactly 7 defectives
F: the sample has exactly 8 defectives

then

$$P(\text{more than 6 defectives}) = P(E \cup F)$$

PLAYING THE LOTTERY

In the Transylvania Lottery a player selects a set of six distinct integers from 1 to 40. If this set matches the set of six integers drawn by the lottery commission (order does not count), the player wins the grand prize.

If Hugo Frankenstein picks *two such sets,* what is the probability of winning the grand prize?

Solution Since there are

$$C_{40,6} = \frac{40!}{6!\ 34!} = 3{,}838{,}380$$

ways of selecting 6 integers out of 40 and Hugo has two chances to win, the probability that Hugo wins is

$$\frac{2}{C_{40,6}} = \frac{2}{3{,}838{,}380} = \frac{1}{1{,}919{,}190} \cong .00000052$$

The odds against Hugo winning are 1,919,189 to 1.

But E and F are mutually exclusive. (Why?) Thus, by Equation (1) of Section 7.3,

$$P(\text{more than 6 defectives}) = P(E) + P(F) \tag{2}$$

By viewing this as a card-dealing problem, you can now show that

$$P(E) = P(\text{exactly 7 defectives}) = \frac{C_{10,7}C_{90,1}}{C_{100,8}} \cong .0000000580$$

and

$$P(F) = P(\text{exactly 8 defectives}) = \frac{C_{10,8}}{C_{100,8}} \cong .000000000242$$

Substituting these values in (2) yields

$$P(\text{more than 6 defectives}) = \frac{C_{10,7}C_{90,1}}{C_{100,8}} + \frac{C_{10,8}}{C_{100,8}} \cong .0000000583$$

EXERCISE SET 7.4

1. If the letters t, c, a are typed in a random order, what is the probability that the word *cat* is typed?

2. In a 5-horse race, each horse has an equal chance of winning. Assuming there are no ties, find the probability that horse number 1 wins.

3. Cards numbered from 1 to 6 are well-shuffled and then placed side by side in a line.
   (a) What is the probability that the cards are in numerical order reading left to right?
   (b) What is the probability that card number 6 is at the far left?
   (c) What is the probability that card number 6 is at the far left and card number 1 is at the far right?

4. In Exercise 3, what is the probability that an even-numbered card is on the far left?

5. A fair coin is tossed 7 times. What is the probability that exactly 5 of the tosses are heads?

6. A fair coin is tossed 6 times. Find the probability that
   (a) at least 5 tosses are heads.
   (b) at most 4 tosses are heads. [*Hint*: Use the result in (a).]

7. A fair coin is tossed 8 times. Find the probability that at least 2 of the tosses are heads. [*Hint*: Use Theorem 3 of Section 7.3.]

8. A hand of 7 cards is dealt from a well-shuffled deck of 52 cards. Find the probability that
   (a) the hand has 4 clubs.
   (b) the hand has 3 clubs and 4 red cards.

9. A hand of 5 cards is dealt from a well-shuffled deck of 52 cards.
   (a) Find the probability that all 5 cards are spades.
   (b) Find the probability that all 5 cards are of the same suit.

10. A hand of 8 cards is dealt from a standard deck of 52 cards. What is the probability that half are red and half black?

11. An urn contains 3 white balls and 6 red balls. If 2 balls are selected at random, what is the probability that
   (a) both are red?
   (b) both are white?
   (c) one is red and one is white?

12. An urn contains 3 green balls and 3 yellow balls. If 3 balls are selected at random, what is the probability that
   (a) at least 2 are yellow?
   (b) at most 1 is yellow?

13. A lot of 50 items from a manufacturing process is known to contain 20 defectives and 30 nondefectives. If a sample of 10 items is selected at random, what is the probability that
   (a) the sample has 6 defectives and 4 nondefectives?
   (b) the sample has no defectives?
   (c) the sample has at least 1 defective?

14. A box containing 5 light bulbs is checked by picking 2 bulbs at random and testing them. If both bulbs light, the box passes inspection, otherwise it fails.
   (a) What is the probability that a box with exactly 1 bad bulb will pass inspection?
   (b) What is the probability that a box with exactly 3 bad bulbs will pass inspection?

15. Suppose 2 cards are dealt from a standard deck of 52 cards.
   (a) What are the odds against both cards being black?
   (b) What are the odds against both being aces?

16. In Exercise 15, what are the odds in favor of obtaining at least one red card?

17. **(Harder)** In draw poker each player is dealt a hand of five cards from a standard deck of 52 cards. According to *Hoyles' Rules of Games* (Albert H. Morehead and Geoffrey Mott-Smith [New York: Signet, 1963]), the probabilities of certain hands are as shown in Table 1. Verify these results:

TABLE 1

Hand	Probability
Exactly one pair	$\frac{1,098,240}{2,598,960} \cong .42$
Three of a kind	$\frac{54,912}{2,598,960} \cong .021$
Four of a kind	$\frac{624}{2,598,960} \cong .00024$
Full house (3 of one kind, 2 of another kind)	$\frac{3,744}{2,598,960} \cong .0014$

18. **(Birthday problem)** A group of n people is gathered together. What is the probability that at least two of them have the same birthday (i.e., the same day and month of birth, although not necessarily the same year)? [*Hint*: Let E be the event that at least two of the people have the same birthday. Instead of finding $P(E)$ directly, it is easier to use the formula $P(E) = 1 - P(E')$, where E' is the event that no two people in the group have the same birthday.] To simplify the problem, we ignore leap years and assume that each year has 365 days. To finish the problem, you should show that

$$P(E') = \frac{P_{365,n}}{365^n}$$

$$= \frac{365 \cdot 364 \cdot 363 \cdot \cdots \cdot (365 - n + 1)}{365^n}$$

Remark It is tedious to evaluate $P(E')$ for a specific value of n. However, with the aid of a calculator, we obtain the following rather surprising table of probabilities, where n = the number of people and $P(E)$ = the probability that at least two have the same birthday.

n	5	10	20	22	23	30	50	60
$P(E)$	.027	.117	.411	.476	.507	.706	.970	.994

This table shows that when 23 people are gathered together, the probability is greater than $\frac{1}{2}$ that at least two will have the same birthday. Further, when 60 people are gathered together, it is almost certain that at least two will have the same birthday!

7.5 CONDITIONAL PROBABILITY; INDEPENDENCE

From our previous work, we know that the probability of tossing a 1 with a fair die is $\frac{1}{6}$. Suppose, however, that a fair die is tossed and we are told that the number on the top face is odd. What is the probability that the number tossed is a 1? Since the number tossed must be odd, we have three equally likely outcomes for the experiment, 1, 3, or 5. In other words, there is a new or **reduced sample space**, $S = \{1, 3, 5\}$. Thus, the probability that the number tossed is a 1, given that the number is odd, is $\frac{1}{3}$. This simple illustration shows that probabilities can be affected by given information about the outcome of the experiment. In this section we develop systematic methods for computing probabilities when additional information is known about the outcome of an experiment.

In the die-tossing experiment above, let E and F be the events

E: a 1 is tossed.
F: an odd number is tossed

We can summarize our above discussion by stating that $P(E) = \frac{1}{6}$, while

$$P(E \text{ occurs given that } F \text{ occurs}) = \frac{1}{3} \tag{1}$$

In probability problems, it is usual to abbreviate the phrase

E occurs given that F occurs

by writing

$$E \mid F$$

With this notation, Equation (1) can be written as

$$P(E \mid F) = \frac{1}{3}$$

The probability $P(E \mid F)$ is called the **conditional probability of E given F.**

Suppose now that A and B are two events associated with an experiment having a finite sample space. Our next objective is to formulate a relationship between the conditional probability $P(A \mid B)$ and the probabilities $P(A)$ and $P(B)$. To motivate the ideas, we consider the simplest situation, an experiment with n equally likely outcomes. Suppose, as illustrated in Figure 1, that A and B are two events in the sample space S for this experiment. If we assume that the event B contains l sample points and the event $A \cap B$ contains m sample points, then

$$P(B) = \frac{l}{n} \quad \text{and} \quad P(A \cap B) = \frac{m}{n} \tag{2}$$

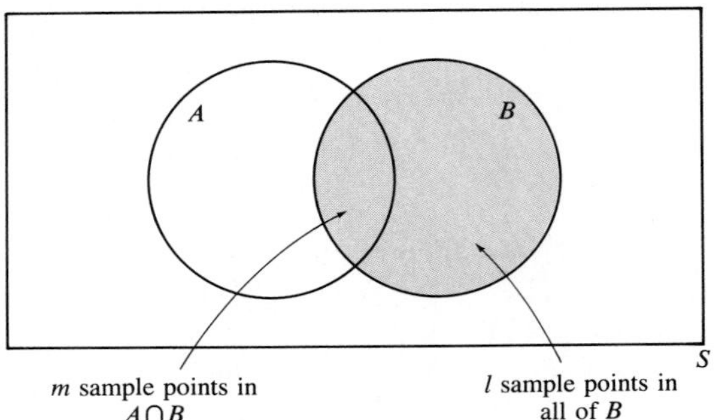

FIGURE 1

We are interested in finding $P(A \mid B)$, that is, the probability that A occurs given that B occurs.

Since we are given that the event B occurs, the outcome of the experiment must be one of the l sample points in B. In other words, B is the new sample space. Of these l sample points, there are m sample points for which A occurs; these are the m sample points in $A \cap B$. Thus, since all outcomes are equally likely,

$$P(A \mid B) = \frac{n(A \cap B)}{n(B)} = \frac{m}{l} \tag{3}$$

If we divide the numerator and denominator of Equation (3) by n, we obtain

$$P(A \mid B) = \frac{m/n}{l/n}$$

which can be rewritten, using Equation (2), as

$$P(A \mid B) = \frac{P(A \cap B)}{P(B)} \tag{4}$$

Although Formula (4) was derived under the assumption that the outcomes in S are equally likely, it serves to motivate the following definition that applies to any sample space.

If A and B are events and $P(B) \neq 0$, then we define the **conditional probability of A given B** by

$$P(A \mid B) = \frac{P(A \cap B)}{P(B)} \tag{5}$$

EXAMPLE 1 If two fair dice are tossed, what is the probability that the sum is 7 if we know that at least one face shows a 4?

Solution Let A and B be the events

A: sum is 7
B: at least one face shows a 4

Method 1: We want to compute $P(A \mid B)$. From Table 4 in Section 7.2 we obtain

$$A = \{(6, 1), (5, 2), (4, 3), (3, 4), (2, 5), (1, 6)\}$$
$$B = \{(4, 1), (4, 2), (4, 3), (4, 4), (4, 5), (4, 6), (1, 4), (2, 4), (3, 4), (5, 4), (6, 4)\}$$

so that

$$A \cap B = \{(4, 3), (3, 4)\}$$

Thus,

$$P(A \cap B) = \frac{2}{36} \quad \text{and} \quad P(B) = \frac{11}{36}$$

so that, from Formula (5),

$$P(A \mid B) = \frac{2/36}{11/36} = \frac{2}{11}$$

Method 2: From Formula (3)

$$P(A \mid B) = \frac{n(A \cap B)}{n(B)} = \frac{2}{11} \qquad \text{(since } B \text{ is the reduced or new sample space)}$$

EXAMPLE 2 Table 1 shows the results of a survey of 1100 traffic accidents. The data is tabulated according to the cause of the accident and the sex of the driver at fault.

(a) Given that an accident was caused by mechanical failure, what is the probability the driver was a male?

(b) Given that the driver was a male, what is the probability that the accident was caused by mechanical failure?

TABLE 1

	Mechanical failure	Intoxication	Poor judgment	Total
Male	310	102	208	620
Female	280	45	155	480
Total	590	147	363	1100

Solution (a) Method 1: Let A and B be the events

A: the driver was a male
B: the accident was caused by mechanical failure

We want to find $P(A \mid B)$. Of the 1100 traffic accidents, it follows from Table 1 that 310 + 280 accidents were caused by mechanical failure and 310 were caused by mechanical failure with a male driver. Thus,

$$P(B) = \frac{310 + 280}{1100} = \frac{590}{1100}$$

and

$$P(A \cap B) = \frac{310}{1100}$$

Thus, from Formula (5),

$$P(A \mid B) = \frac{P(A \cap B)}{P(B)} = \frac{310/1100}{590/1100} = \frac{31}{59}$$

Method 2: From Formula (3)

$$P(A \mid B) = \frac{n(A \cap B)}{n(B)} \qquad \text{(since } B \text{ is the reduced sample space)}$$

$$= \frac{310}{590} = \frac{31}{59} \qquad \text{(from Table 1)}$$

Solution (b) Method 1: In this part we want to find $P(B \mid A)$. From Formula (5) with the roles of A and B interchanged, we have

$$P(B \mid A) = \frac{P(B \cap A)}{P(A)}$$

From Table 1,

$$P(A) = \frac{310 + 102 + 208}{1100} = \frac{620}{1100}$$

and

$$P(B \cap A) = \frac{310}{1100}$$

so that

$$P(B \mid A) = \frac{P(B \cap A)}{P(A)} = \frac{310/1100}{620/1100} = \frac{1}{2}$$

Method 2: From Formula (3)

$$P(B \mid A) = \frac{n(B \cap A)}{n(A)} \qquad \text{(since } A \text{ is the reduced sample space)}$$

$$= \frac{310}{620} = \frac{1}{2} \qquad \text{(from Table 1)}$$

Remark When information is given in tabular form, as in Example 2, it is generally simpler to use the reduced sample space to compute $P(A \mid B)$ and $P(B \mid A)$ as in Method 2 of Example 2. In this case, A is represented by the first row of the table and B is represented by the first column of the table. Thus, $n(A \cap B)$ is the number in the intersection of the first row and first column.

The equation

$$P(A \mid B) = \frac{P(A \cap B)}{P(B)}$$

gives a way to obtain $P(A \cap B)$. Multiply both sides by $P(B)$ to obtain the following.

The Product Rule for Probabilities

$$P(A \cap B) = P(B)P(A \mid B)$$

In problems where the value of $P(A \mid B)$ is known, this formula can be used to compute $P(A \cap B)$. The next example illustrates this idea.

EXAMPLE 3 In 95% of all manned lunar flights, a midcourse trajectory correction is required. This is done by sending a fire signal from ground control to ignite small correction thrusters. For technical reasons, this fire signal is sometimes not executed by the thrusters. Tests show that the probability is .0001 that a fire signal will not be executed when it is required. If the correction is required and not executed, the rocket will plunge into the sun's gravitational field. What is the probability that this will happen?

Solution Let A and B be the events

A: correction is not executed
B: correction is required

We want to find $P(A \cap B)$. From the data in the problem we have

$$P(B) = .95$$
$$P(A \mid B) = P(\text{correction is not executed} \mid \text{correction is required}) = .0001$$

Thus,

$$P(A \cap B) = P(A \mid B)P(B) = (.0001)(.95) = .000095$$

Many practical problems involve a sequence of experiments in which the possible outcomes and probabilities associated with any one experiment depend on the outcomes of the preceding experiments. Such a sequence of experiments is called a **stochastic process**. As a simple example, consider the following.

EXAMPLE 4 Suppose we have two identical jars, I and II. Jar I contains 4 red balls and 3 blue balls. Jar II contains 5 red balls and 1 blue ball. Select a jar at random and then select 1 ball from this jar. A typical question we might ask about this process is: What is the probability that the ball drawn is blue?

Solution The analysis can be simplified by considering the tree diagram in Figure 2a. There are two stages to this diagram, the first representing the outcomes of experiment 1 (select a jar at random) and the second, the outcome of experiment 2 (select a ball at random). Since the jar selection is random, the probability of selecting jar I is $\frac{1}{2}$, as is the probability of selecting jar II. These probabilities are shown at the appropriate branches of the first stage of the tree in Figure 2b (note branches in color). The chances of obtaining a blue ball now depend on which jar was selected at stage 1. If jar I was initially selected, $P(\text{blue} \mid \text{jar I}) = \frac{3}{7}$, whereas if jar II was initially selected, $P(\text{blue} \mid \text{jar II}) = \frac{1}{6}$. These probabilities are noted on the appropriate branches at the second stage of the tree, as are $P(\text{red} \mid \text{jar I}) = \frac{4}{7}$ and $P(\text{red} \mid \text{jar II}) = \frac{5}{6}$. (See the paths in color in Figure 2c.)

To find the probability of drawing a blue ball, we must take all the possibilities into account:

we can select jar I *and* then a blue ball
or we can select jar II *and* then a blue ball

(The two paths on the tree are indicated in color in Figure 3.)

FIGURE 2

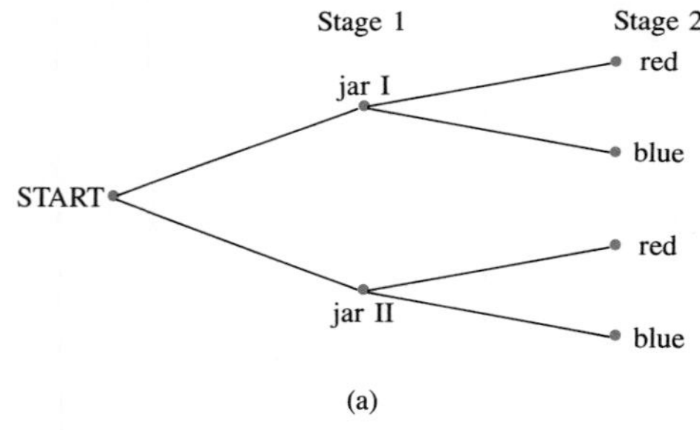

(a)

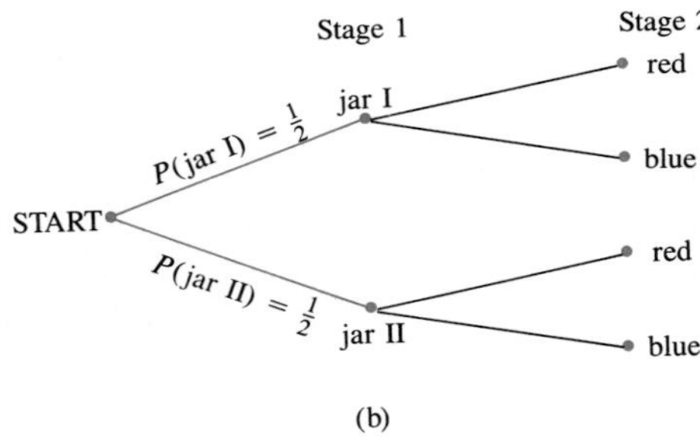

(b)

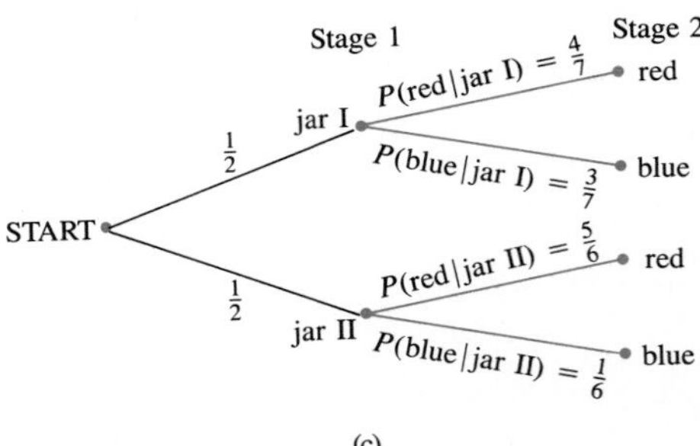

(c)

FIGURE 3

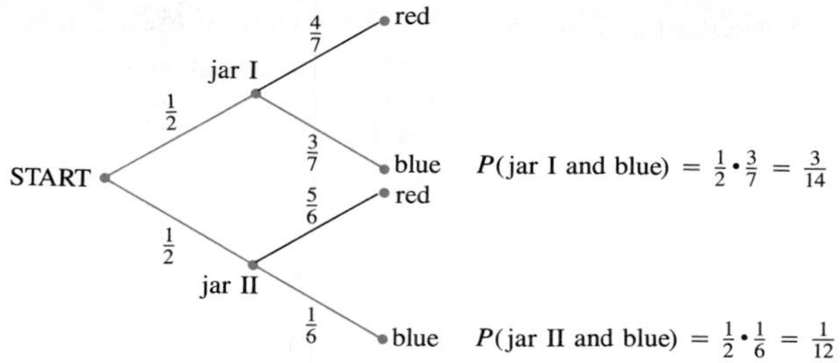

By the product rule for probabilities,

$$\begin{aligned} P(\text{jar I and blue}) &= P(\text{jar I}) \cdot P(\text{blue} \mid \text{jar I}) \\ &= \frac{1}{2} \cdot \frac{3}{7} = \frac{3}{14} \quad \text{(product along upper colored path of Figure 3)} \\ P(\text{jar II and blue}) &= P(\text{jar II}) \cdot P(\text{blue} \mid \text{jar II}) \\ &= \frac{1}{2} \cdot \frac{1}{6} = \frac{1}{12} \quad \text{(product along lower colored path of Figure 3)} \end{aligned}$$

Since the cases are mutually exclusive,

$$\begin{aligned} P(\text{blue}) &= P(\text{jar I and blue}) + P(\text{jar II and blue}) \\ &= \frac{3}{14} + \frac{1}{12} = \frac{25}{84} \end{aligned}$$

Remark Tree diagrams are often helpful in problems of this type. Note that in Figure 4a the branch from A to B is labeled with $P(B \mid A)$.

FIGURE 4

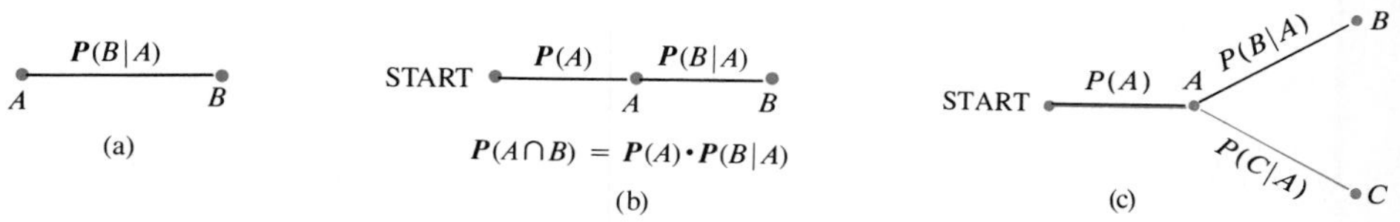

1. Read "and" along the branches of Figure 4b. This path represents the occurrence of A and B, A occurring first. In this case $P(A \cap B) = P(A)P(B \mid A)$; $P(A \cap B)$ is called the **path probability**.
2. Read "or" for alternate paths as in Figure 4c. One of these paths is shown in color and the other in black. These paths represent the occurrence of "B or C":

$$\begin{aligned} P(B \text{ or } C) &= P[(A \cap B) \text{ or } (A \cap C)] \\ &= P(A \cap B) + P(A \cap C) \\ &= P(A)P(B \mid A) + P(A)P(C \mid A) \end{aligned}$$

3. If $P(B \mid A) = 0$, then the branch from A to B is omitted.

EXAMPLE 5 Consider the setting of Example 4. But now, without replacing the ball that was selected, select a second ball.

(a) What is the probability of selecting two red balls from jar I?

(b) What is the probability that the second bail selected is red?

Solution Extend the tree diagram to a third stage, that of selecting a second ball (Figure 5). Note that the colored path in Figure 5 represents the case where the second ball selected is red given that jar I was selected and the first ball selected was red. In this case, of the six balls left in the jar, only three are red; therefore,

$$P(\text{second ball is red} \mid \text{jar I and first ball is red}) = \frac{3}{6} = \frac{1}{2}$$

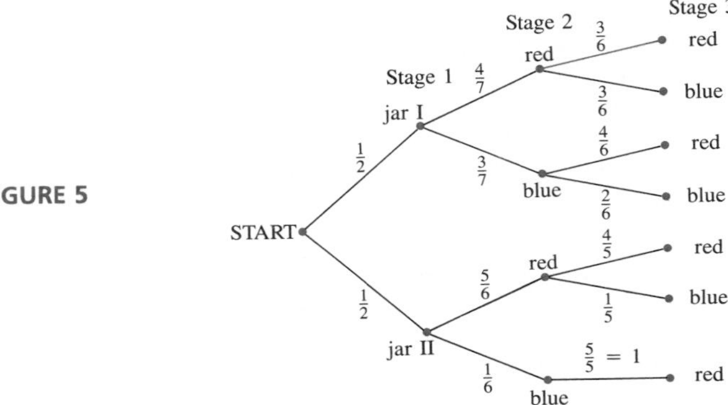

FIGURE 5

(a) $P(\text{selecting two red balls from jar I}) = P(\text{jar I and red and red}) = \frac{1}{2} \cdot \frac{4}{7} \cdot \frac{3}{6} = \frac{1}{7}$

Note: $P(\text{second ball is blue} \mid \text{jar II and first ball is blue}) = 0$ since there is only 1 blue ball in this jar.

(b) $P(\text{second ball selected is red}) = P(\text{jar I and red and red}) + P(\text{jar I and blue and red}) + P(\text{jar II and red and red}) + P(\text{jar II and blue and red})$

$$= \frac{1}{2} \cdot \frac{4}{7} \cdot \frac{3}{6} + \frac{1}{2} \cdot \frac{3}{7} \cdot \frac{4}{6} + \frac{1}{2} \cdot \frac{5}{6} \cdot \frac{4}{5} + \frac{1}{2} \cdot \frac{1}{6} \cdot 1$$

$$= \frac{1}{7} + \frac{1}{7} + \frac{1}{3} + \frac{1}{12} = \frac{59}{84}$$

EXAMPLE 6 Three manufacturing plants A, B, and C supply 20, 30, and 50%, respectively, of all shock absorbers used by a certain automobile manufacturer. Records show that the percentage of defective items produced by A, B, and C is 3, 2, and 1%, respectively. What is the probability that a randomly chosen shock absorber installed by the manufacturer will be defective?

Solution The possibilities are shown in the two-stage tree diagram of Figure 6. At the first stage of the tree, the probabilities .2, .3, and .5 represent the chances of the selected item being produced by A, B, or C. The probabilities at the second stage represent the chances of the item being defective or not.

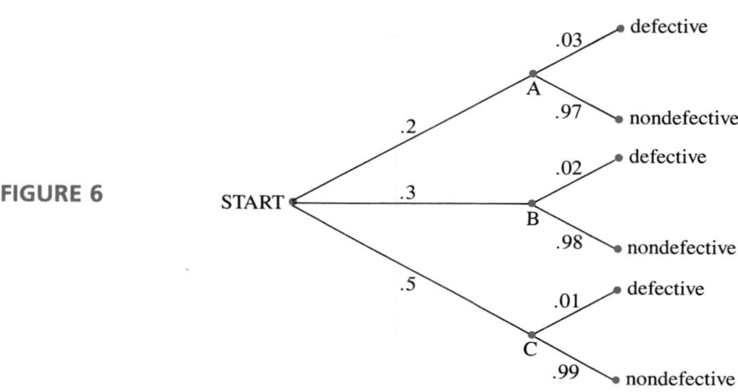

FIGURE 6

There are three paths through the tree that result in a defective shock absorber. Multiplying the probabilities along each such path and adding the resulting products yields

$$\begin{aligned} P(\text{defective is selected}) &= (.2)(.03) + (.3)(.02) + (.5)(.01) \\ &= .006 + .006 + .005 \\ &= .017 \end{aligned}$$

In other words, we can expect that 1.7% of all shock absorbers installed will be defective.

Independent Events

At times, $P(A \mid B) = P(A)$. For example:

EXAMPLE 7 A fair coin is tossed twice. Find the probability that

(a) the second toss is a head.

(b) the second toss is a head given that the first toss is a head.

Solution If A and B are the events

A: the second toss is a head
B: the first toss is a head

then the sample space S for this experiment is

$$S = \{hh, ht, th, tt\}$$

The events A and B are

$$A = \{hh, th\} \quad \text{and} \quad B = \{hh, ht\}$$

and the event $A \cap B$ is

$$A \cap B = \{hh\}$$

Since the coin is fair, we assume that the four sample points in S are equally likely, so

$$P(A) = \frac{2}{4} = \frac{1}{2} \qquad P(B) = \frac{2}{4} = \frac{1}{2} \qquad P(A \cap B) = \frac{1}{4}$$

Thus, the probability that the second toss is a head is

$$P(A) = \frac{1}{2} \tag{6}$$

and the probability that the second toss is a head given that the first toss is a head is

$$P(A \mid B) = \frac{P(A \cap B)}{P(B)} = \frac{1/4}{1/2} = \frac{1}{2} \tag{7}$$

Comparing Equations (6) and (7), we see that

$$P(A \mid B) = P(A)$$

In other words, the probability of A is unaffected by the additional knowledge that B occurs. Intuitively, this tells us that the two events are independent, that is, they do not influence the probabilities of one another. This result is not unexpected. It states that the coin has no memory. When the second toss is made, the coin does not "remember" what happened on the first toss. Thus, the chances of a head on the second toss will be $\frac{1}{2}$ regardless of the outcome of the first toss.

Let us pursue these ideas in more detail. Suppose A and B are two events such that

$$P(A \mid B) = P(A)$$

or equivalently

$$\frac{P(A \cap B)}{P(B)} = P(A)$$

If we multiply both sides of this equation by $P(B)$, we obtain

$$P(A \cap B) = P(A)P(B)$$

This suggests the following definition:

Two events A and B are called **independent** if

$$P(A \cap B) = P(A)P(B) \tag{8}$$

We leave it for the exercises at the end of this section to show the following result.

THEOREM If A and B are independent events with nonzero probabilities, then

$$P(A|B) = P(A) \quad \text{and} \quad P(B|A) = P(B) \tag{9}$$

Conversely, if either of the equations in (9) holds, then A and B are independent.

Thus, when A and B are independent events, knowledge that B occurs does not affect the probability that A occurs and knowledge that A occurs does not affect the probability that B occurs.

EXAMPLE 8 The events A and B in Example 7 are independent because

$$P(A) = \frac{1}{2} \qquad P(B) = \frac{1}{2} \qquad P(A \cap B) = \frac{1}{4}$$

so that

$$P(A \cap B) = P(A)P(B)$$

EXAMPLE 9 A card is drawn from a standard deck of 52 cards and then a second card is drawn without replacing the first card. Let A be the event that the second card is a spade and let B be the event that the first card is a spade. Determine whether A and B are independent.

Solution Intuitively we should suspect that the events are *not* independent since the chances of getting a spade on the second draw are better if a spade is not removed from the deck on the first draw than if one is removed. Let us confirm our suspicions mathematically. To show that A and B are not independent, we can proceed in one of three ways. We can show that

$$P(A \cap B) \neq P(A)P(B)$$

or that

$$P(B|A) \neq P(B)$$

or that

$$P(A|B) \neq P(A)$$

We use the last approach in this example.

The conditional probability $P(A|B)$ is the probability of getting a spade on the second draw given that a spade was obtained on the first draw. In this case, a spade is removed from the deck on the first draw, so 12 of the 51 cards available for the second draw are spades. Thus,

$$P(A|B) = \frac{12}{51} = \frac{4}{17}$$

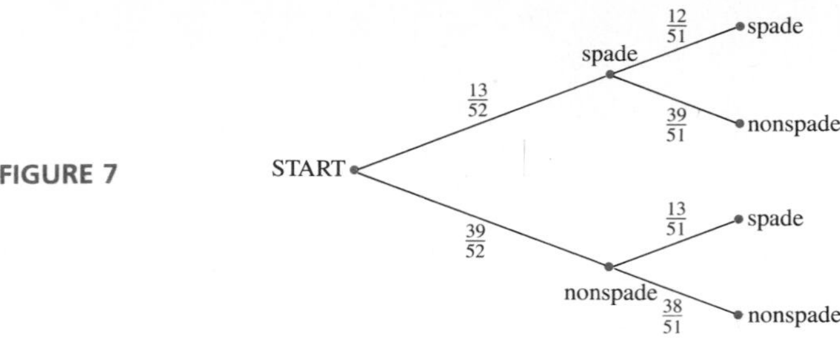

FIGURE 7

$P(A)$ can be determined from the tree diagram shown in Figure 7. Thus,

$$P(A) = \frac{13}{52} \cdot \frac{12}{51} + \frac{39}{52} \cdot \frac{13}{51} = \frac{1}{4}$$

and therefore

$$P(A \mid B) \neq P(A)$$

so A and B are not independent events.

In many practical problems, it is difficult or even impossible to determine with certainty whether two events are independent. In such situations, the experimenter often relies on intuition to decide whether to assume the events are independent or not. For example, suppose we know from past records that the probability is .9 that a certain sharpshooter will hit the bull's-eye on a target. Suppose also we want to find the probability that this sharpshooter will fire two bull's-eyes on two successive shots. We might begin by introducing the events

A: the sharpshooter gets a bull's-eye on the first shot
B: the sharpshooter gets a bull's-eye on the second shot

Thus, $A \cap B$ corresponds to getting bull's-eyes on both shots. If we *assume* that A and B are independent events, then

$$P(A \cap B) = P(A)P(B) = (.9)(.9) = .81$$

Remark Be careful! Many times in practice, events are considered to be independent and really are not. Since this computation is erroneous if A and B are not independent, we should carefully examine this independence assumption. For A and B to be independent, the chances of getting a bull's-eye on the second shot should not be affected by the result of the first shot. It is possible, however, that a miss on the first shot might disturb the sharpshooter's concentration and reduce the chances of a bull's-eye on the second shot, or it might increase the desire for a bull's-eye and thus improve the shooter's concentration, thereby bettering the chances. In this situation the independence assumption would not be warranted. On the other hand, if the experimenter does not believe this psychological factor is significant, he or she may choose to accept the independence assumption. Although there are statistical tests that can sometimes be applied to test for independence experimentally, they are beyond the scope of this text. In any event, these tests cannot always be applied, so the applicability of the independence assumption often depends on the judgment of the experimenter.

EXAMPLE 10 A certain manufacturing process produces defective items 1% of the time. What is the probability that two items selected at random will both be defective? Assume the selections are independent.

Solution Let A and B be the events

A: the first item is defective
B: the second item is defective

We want to find $P(A \cap B)$. Assuming A and B are independent, we obtain

$$P(A \cap B) = P(A)P(B) = (.01)(.01) = .0001$$

The notion of independence can be extended to more than two events, as follows.

> A set of events is called **independent** if for each finite subset $\{A_1, A_2, \ldots, A_n\}$ we have
>
> $$P(A_1 \cap A_2 \cap \cdots \cap A_n) = P(A_1)P(A_2) \cdots P(A_n)$$

EXAMPLE 11 For a set of three events $\{A, B, C\}$ to be independent, we must have

$$P(A \cap B) = P(A)P(B)$$
$$P(A \cap C) = P(A)P(C)$$
$$P(B \cap C) = P(B)P(C)$$

and

$$P(A \cap B \cap C) = P(A)P(B)P(C)$$

EXAMPLE 12 In the manufacture of light bulbs, filaments, glass casings, and bases are manufactured separately and then assembled into the final product. Assume that past records show the following:

2% of all filaments are defective
3% of all glass casings are defective
1% of all bases are defective

What is the probability that one of these bulbs will have no defects?

Solution Let A, B, and C be the events

A: the bulb has no defect in the filament
B: the bulb has no defect in the glass casing
C: the bulb has no defect in the base

From the given data,

$$P(A) = .98 \qquad P(B) = .97 \qquad P(C) = .99$$

We are interested in finding

$$P(A \cap B \cap C)$$

Since filaments, glass casings, and bases are manufactured separately, it seems reasonable to assume A, B, and C are independent. Thus,

$$P(A \cap B \cap C) = P(A)P(B)P(C) = (.98)(.97)(.99) \cong .94$$

Remark To determine $P(E \cap F)$, proceed as follows:

1. If $E \cap F = \{s_1, s_2, \ldots, s_k\}$, then

$$P(E \cap F) = P(\{s_1\}) + P(\{s_2\}) + \cdots + P(\{s_k\})$$

2. If E and F are mutually exclusive events, then

$$E \cap F = \varnothing \quad \text{and} \quad P(E \cap F) = 0$$

3. If E and F are independent events, then $P(E \cap F) = P(E)P(F)$.

4. If $P(E \cup F)$, $P(E)$, and $P(F)$ are known, use

$$P(E \cup F) = P(E) + P(F) - P(E \cap F)$$

to solve for $P(E \cap F)$.

5. In general, $P(E \cap F) = P(E)P(F \mid E) = P(F)P(E \mid F)$.

EXERCISE SET 7.5

1. Let A and B be events such that $P(A) = .7$, $P(B) = .4$, and $P(A \cap B) = .2$.
   (a) Determine $P(A \mid B)$. (b) Determine $P(B \mid A)$.
   (c) Are A and B independent? Explain your answer.
   (d) Are A and B mutually exclusive? Explain your answer.

2. Let A and B be the events in Exercise 1. Find
   (a) $P(A \mid B')$ (b) $P(B \mid A')$

3. Let A and B be events such that $P(A \mid B) = .5$ and $P(B) = .2$. Find $P(A \cap B)$.

4. A certain missile guidance device contains 100 components. Some of the components are resistors (r) and some of the components are transistors (t). Some of the components are made in the United States (u), whereas others are made in Japan (j). The number of components in each category is listed in Table 2. Assume each component is equally likely to malfunction during a lunar flight.
   (a) If a malfunction occurs in a transistor, what is the probability that it was made in the United States?
   (b) If a malfunction occurs in a component made in Japan, what is the probability that it is a resistor?

TABLE 2

	r	t
j	20	32
u	18	30

5. A lot contains 25 defective items and 75 nondefective items. If we choose two items in succession without replacement, what is the probability that both items are defective?

6. A telephone repairperson knows that two circuits are working properly and two circuits are working improp-

erly. The circuits are tested, one by one, until both defective circuits are located.

(a) What is the probability that the two defective circuits will be located on the first two tests?

(b) What is the probability that the last defective circuit is located on the third test?

7. In a certain state, 45% of all major manufacturers violate some federal pollution standard and 30% violate both a state and federal pollution standard. Given that a major manufacturer violates a federai standard, what is the probability the manufacturer violates a state pollution standard?

8. A card is drawn from a standard deck of 52 cards. What is the probability that it is an ace given that it is not a king?

9. A student must answer a multiple-choice question with one of five possible answers.

(a) What is the probability the student will get the correct answer by guessing?

(b) What is the probability the student will get the correct answer if the first two possibilities can be ruled out?

10. Identical twins come from a single egg and must therefore be of the same sex. Fraternal twins, on the other hand, come from different eggs and can be of opposite sex. Assuming the probability that twins are fraternal is $\frac{2}{3}$ and the probability that a fraternal twin is a female is $\frac{1}{2}$, find the probability that twins have the same sex. [*Hint*: Let S: twins have the same sex, I: twins are identical, and F: twins are fraternal. Then, $S = (S \cap I) \cup (S \cap F)$.]

11. In a medical study of the common cold, 100 cold sufferers exhibited the following symptoms:

18 people had fevers
32 people had coughs
50 people had stuffy noses
4 people had coughs and stuffy noses
5 people had fevers and stuffy noses
7 people had fevers and coughs
2 people had fevers, coughs, and stuffy noses

If a person selected at random from the group has a stuffy nose, what is the probability that the person also has a cough?

12. Let A and B be independent events such that $P(A) = \frac{1}{5}$ and $P(B) = \frac{1}{6}$. Find $P(A \cap B)$.

13. Let $S = \{s_1, s_2, s_3, s_4, s_5\}$ be the sample space for an experiment and assume that $P(\{s_1\}) = P(\{s_2\}) = P(\{s_3\}) = \frac{1}{4}$ and $P(\{s_4\}) = P(\{s_5\}) = \frac{1}{8}$.

(a) Show that $A = \{s_1, s_2\}$ and $B = \{s_2, s_3\}$ are independent events.

(b) Show that $C = \{s_3, s_4\}$ and $D = \{s_4, s_5\}$ are not independent events.

14. Let $P(E) = \frac{1}{2}$ and $P(E \cup F) = \frac{3}{4}$.

(a) If E and F are independent events, compute $P(F)$.

(b) If E and F are mutually exclusive events, compute $P(F)$.

15. In a federal study of unemployment, 1000 people were classified as employed (E), unemployed (E'), having a high school diploma (H), or not having a high school diploma (H'); the data obtained are shown in Table 3.

(a) Find $P(E \mid H)$ and $P(E)$.

(b) Are E and H independent events?

TABLE 3

	H	H'
E	917	34
E'	9	40

16. A submarine detection system consists of three units, a sonar device, a magnetic detector, and a visual spotter. To enter a certain area without being detected, a submarine must escape detection by all three units. The probability of escaping the sonar is .5, the probability of escaping the magnetic detector is .4, and the probability of escaping the visual spotter is .8. If we assume that the units act independently, what is the probability that a submarine can enter the area undetected?

17. Prove statement (9).

18. In each part, decide if E and F can reasonably be assumed to be independent events.

(a) A person is picked at random;

E: the person is more than 6 feet tall
F: the person weighs more than 150 pounds

(b) The weather is observed for two consecutive days;

E: rain occurs on the first day
F: rain occurs on the second day

(c) A person is picked at random;

E: the person has blue eyes
F: the person is wearing a ring

19. Let A and B be two events associated with an experiment, and suppose $P(A) = .5$ and $P(A \cup B) = .8$. Let $P(B) = p$. For which values of p are

(a) A and B mutually exclusive?

(b) A and B independent?

20. Assume we have two identical urns. Urn A contains 3 black balls and 1 white ball, while urn B contains 2 black balls and 4 white balls. An urn is chosen at random and then a ball is selected at random from this urn. What is the probability that the ball is white?

21. Choose a card at random from a standard deck of 52 cards. Let E and F be the following events.

E: the chosen card is a heart
F: the chosen card is a picture card

(a) Show that E and F are independent events.
(b) Show that E and F' are independent events.

22. Choose a card at random from a standard deck of 52 cards. Let

E be the event that the card is a heart
F be the event that the card is an ace

(a) Are E and F independent events?
(b) Are E and F' independent events?

23. Roll a fair die.

E is the event that an odd number shows on the top face.
F is the event that a 3, 4, 5, or 6 shows on the top face.

(a) Show that E and F are independent events.
(b) Show that E' and F are independent events.

24. Toss two fair dice.

E is the event that the sum of the numbers showing on the top faces is 6.
F is the event that the number showing on the top face of the first die is a 2.

(a) Are E and F independent events?
(b) Are E and F' independent events?

25. Toss two fair dice.

E is the event that the sum of the numbers showing on the top faces is 7
F is the event that the number showing on the top face of the second die is a 3.

(a) Show that E and F are independent events.
(b) Show that E' and F' are independent events.

26. Suppose that you are given two dice that are not fair.

A red die has two faces showing a 1, one face showing a 2, one face showing a 3, one face showing a 4, and one face showing a 5.

A green die has one face showing a 2, one face showing a 3, one face showing a 4, one face showing a 5, and two faces showing a 6.

(a) Show that the events E and F defined in Exercise 24 are not independent.
(b) Show that the events E and F defined in Exercise 25 are not independent.

27. A box contains 10 light bulbs, 2 of which are defective. Suppose 3 bulbs are selected in succession without replacing those already selected.
(a) What is the probability that all 3 bulbs selected are good?
(b) What is the probability that the third bulb selected is good?
(c) What is the probability that 2 are good and 1 is defective?
(d) What is the probability that at least 2 are good?

28. A fair coin is tossed four times in succession. Use a tree diagram to find
(a) the probability of tossing 4 heads
(b) the probability of tossing 3 heads and 1 tail

29. In a manufacturing plant, a certain part is made by both the day shift and the night shift. Of the parts made by the day shift, 80% are good and 20% defective. The night shift produces good parts only 75% of the time. Suppose one of the shifts is chosen at random and two pieces of its output are selected independently for inspection.
(a) What is the probability that both pieces are good?
(b) What is the probability that the pieces are good, given that they came from the day shift?
(c) What is the probability that the pieces came from the night shift, given that both are defective?

30. A mouse is trapped in the wall of a house. There are 2 holes by which the mouse can escape. The family cat is always guarding one of the two holes but spends the same amount of time at each hole. However, 3 out of the 5 times when the cat guards a hole he is asleep. The other times there is only a 50–50 chance that the cat will catch the mouse should the mouse leave by the hole the cat is guarding. The mouse chooses a hole at random and runs out. What is the probability that it is not caught by the cat?

31. A certain computer has a probability of 10^{-10} of transmitting a character incorrectly. If the sentence *Don't give up the ship* is to be transmitted, what is the probability that there will be no errors? (Spaces and apostrophes count as characters.)

7.6 BAYES' FORMULA (OPTIONAL)

In many practical problems an experimenter observes the outcome of an experiment and then asks for the probability that the outcome was caused by one of several possible factors.

EXAMPLE 1 Jar I contains 4 red balls and jar II contains 3 red and 2 blue balls. The experiment is to choose a jar at random and from this jar select a ball and note the color of the ball. The tree diagram with branch probabilities for this experiment is shown in Figure 1.

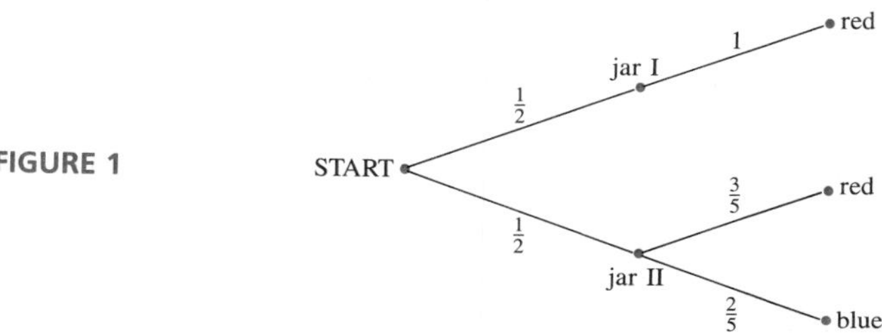

FIGURE 1

In Section 7.5 we address the question: What is the probability that a blue ball is drawn given that it is drawn from jar II? This is a conditional probability that equals $\frac{2}{5}$ since jar II contains 2 blue balls out of 5.

Now suppose the experiment is completed and we note that the selected ball is blue. We now ask: What is the probability that the selected ball came from jar II given that the ball is blue? The answer must be 1 since there are no blue balls in jar I.

Starting over again, suppose the selected ball is red and consider the probability that the selected ball is from jar I, given that the ball is red. The answer is not obvious, but we can determine it from the definition of conditional probability

$$P(\text{jar I} \mid \text{red}) = \frac{P(\text{jar I and red})}{P(\text{red})}$$

$$= \frac{1/2 \cdot 1}{1/2 \cdot 1 + 1/2 \cdot 3/5} = \frac{1/2}{8/10} = \frac{5}{8}$$

where these probabilities are obtained by the tree diagram methods of Section 7.5 applied to Figure 1. In fact, to answer the given question, we do not need the whole tree, only the paths that give the outcome red (Figure 2).

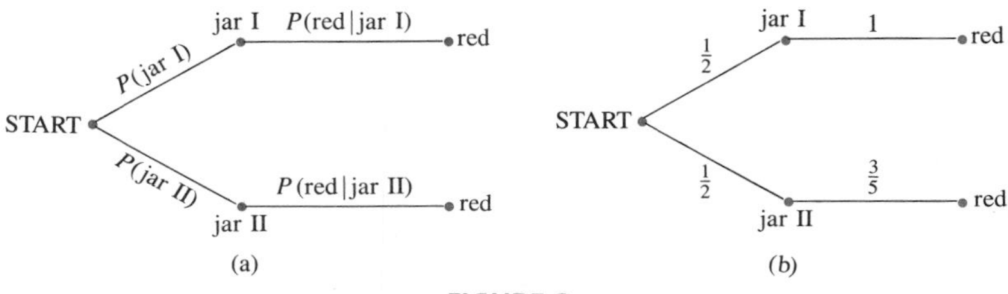

FIGURE 2

EXAMPLE 2 A card is drawn from a standard deck of 52 cards and then a second card is drawn without replacing the first card. If we know that the second card is a spade, what is the probability that the first card is a spade?

Solution Let A and B be the events

A: the second card is a spade
B: the first card is a spade

We want to find $P(B \mid A)$. From the tree diagram of Figure 3,

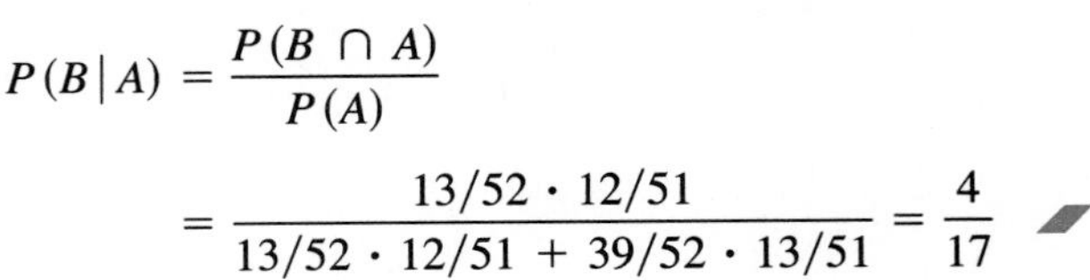

$$P(B \mid A) = \frac{P(B \cap A)}{P(A)}$$

$$= \frac{13/52 \cdot 12/51}{13/52 \cdot 12/51 + 39/52 \cdot 13/51} = \frac{4}{17}$$

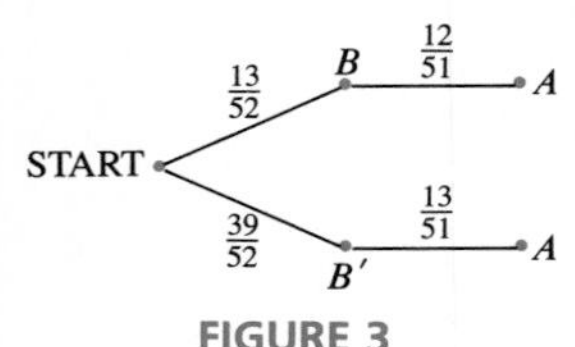

FIGURE 3

From the tree diagram of Figure 4, we see that $P(A) = P(B)P(A \mid B) + P(B')P(A \mid B')$. This leads to the following result known as **Bayes' theorem** or **Bayes' formula** for the case when events B and B' are each followed by event A.

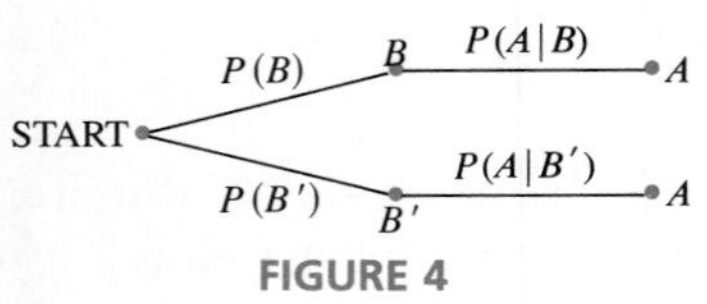

FIGURE 4

$$P(B \mid A) = \frac{P(B \cap A)}{P(A)} = \frac{P(B)P(A \mid B)}{P(B)P(A \mid B) + P(B')P(A \mid B')} \tag{1}$$

Verify that Equation (1) gives the same result as the tree method used in Examples 1 and 2.

EXAMPLE 3 A certain item is manufactured by three factories, I, II, and III. Suppose we have a stockpile of these items in which

30% of the items were made in factory I
20% of the items were made in factory II
50% of the items were made in factory III

Moreover, suppose

2% of all items produced by I are defective
3% of all items produced by II are defective
4% of all items produced by III are defective

Assume that an item selected at random from the stockpile is observed to be defective.

(a) What is the probability that the item came from factory I?

(b) From factory II? (c) From factory III?

Solution Let us introduce the events

A : the item is defective
B_1: the item came from factory I
B_2: the item came from factory II
B_3: the item came from factory III

In parts (a), (b), and (c) we want to find

$$P(B_1 | A) \qquad P(B_2 | A) \qquad P(B_3 | A)$$

respectively. Before attempting to find these values, let us see what information is immediately available from the data. Since 30% of all items in the stockpile comes from factory I, we have

$$P(B_1) = .3$$

Similarly,

$$P(B_2) = .2 \quad \text{and} \quad P(B_3) = .5$$

Since 2% of all items from factory I are defective, we have

$$P(A | B_1) = .02$$

Similarly,

$$P(A | B_2) = .03 \quad \text{and} \quad P(A | B_3) = .04$$

The tree diagram for this experiment is shown symbolically in Figure 5a and numerically in Figure 5b.

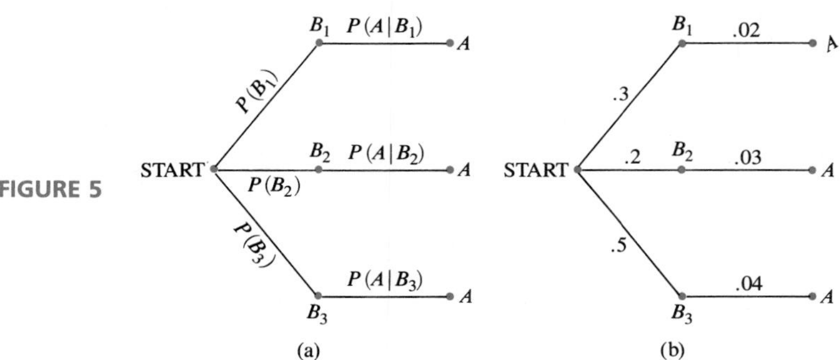

FIGURE 5

Solution (a) We now have

$$P(B_1 | A) = \frac{P(B_1 \cap A)}{P(A)} = \frac{P(B_1)P(A | B_1)}{P(A)} = \frac{.3(.02)}{.3(.02) + .2(.03) + .5(.04)} = \frac{3}{16}$$

Solution (b) We obtain the following from the tree diagram of Figure 5b:

$$P(B_2 | A) = \frac{P(B_2 \cap A)}{P(A)} = \frac{.2(.03)}{.3(.02) + .2(.03) + .5(.04)} = \frac{3}{16}$$

Solution (c)

$$P(B_3 | A) = \frac{P(B_3 \cap A)}{P(A)} = \frac{.5(.04)}{.3(.02) + .2(.03) + .5(.04)} = \frac{5}{8}$$

In general, when we are given events $B_1, B_2, B_3, \ldots, B_k$ each followed by the event A with probabilities

$$P(B_1), P(B_2), P(B_3), \ldots, P(B_k) \quad \text{and} \quad P(A | B_1), P(A | B_2), P(A | B_3), \ldots, P(A | B_k)$$

and want to find $P(B_1|A)$, $P(B_2|A)$, . . . , $P(B_k|A)$, we have the following statement of **Bayes' theorem** or **Bayes' formula**.

BAYES' THEOREM

$$P(B_i|A) = \frac{P(A|B_i)P(B_i)}{P(A|B_1)P(B_1) + P(A|B_2)P(B_2) + \cdots + P(A|B_k)P(B_k)}$$

Bayes' theorem is named after the Reverend Thomas Bayes* who published this result in a paper in 1763 entitled "An Essay toward Solving a Problem in the Doctrine of Chances."

It is not necessary to memorize Bayes' formula since these problems can be solved by using tree diagrams. To find $P(B_1|A)$, $P(B_2|A)$, . . . , $P(B_k|A)$ we take the following steps:

Step 1 From a common initial point draw a separate branch to represent each of the events B_1, B_2, B_3, . . . , B_k and label these branches with the probabilities $P(B_1)$, $P(B_2)$, $P(B_3)$, . . . , $P(B_k)$. (See Figure 6 for the case when $k = 3$.)

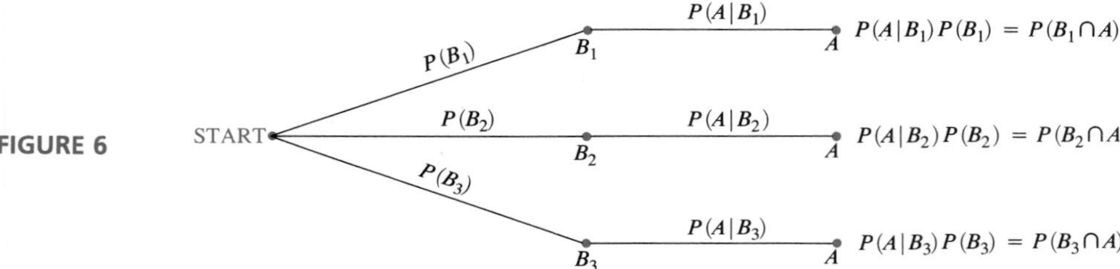

FIGURE 6

Step 2 From the end of each of these branches, draw a single branch to represent the event A and label these new branches with the conditional probabilities $P(A|B_1)$, $P(A|B_2)$, $P(A|B_3)$, . . . , $P(A|B_k)$.

Step 3 Each possible path through the tree ends at a different point. Label each endpoint with the product of the probabilities on the path leading to it. We call these the **path probabilities**.

A tree diagram constructed in this way is called a **Bayes tree**.

Observe that the sum of all the path probabilities is precisely $P(A)$, the denominator in Bayes' formula,

$$P(A) = P(B_1 \cap A) + P(B_2 \cap A) + P(B_3 \cap A)$$

where

$$P(B_i \cap A) = P(B_i)P(A|B_i)$$

**Thomas Bayes* (1702–1763) was the son of a Presbyterian minister. He began his own ministry by assisting his father, and he continued in the ministry until his death. Bayes published several theological papers, but he is most famous for his paper on probability, which was published after his death by a friend who found it in his effects. The work is noteworthy because it is the first discussion of inductive inference in precise quantitative form.

in Bayes' formula is precisely the path probability for the path through events A and B_i. Thus, the probability

$$P(B_i \mid A) = \frac{P(A \cap B_i)}{P(A)}$$

can be obtained from the tree diagram as follows:

Obtaining Conditional Probabilities from a Bayes Tree

$$P(B_i \mid A) = \frac{\text{path probability for the path through } A \text{ and } B_i}{\text{sum of all path probabilities}}$$

EXAMPLE 4 A box contains 4 coins. One of the coins has 2 heads; the other 3 coins are fair. One of the coins is picked at random and tossed. If we know the outcome is heads, what is the probability that the coin is fair?

Solution Let F represent the event that the coin is fair; then F' is the event that the coin is not fair. Let $H =$ the event that the coin turns up heads. The tree diagram is shown in Figure 7.

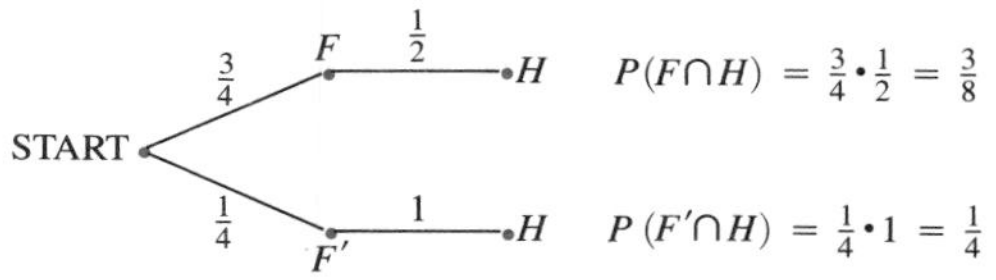

FIGURE 7

START → F ($\frac{3}{4}$) → H ($\frac{1}{2}$): $P(F \cap H) = \frac{3}{4} \cdot \frac{1}{2} = \frac{3}{8}$

START → F' ($\frac{1}{4}$) → H (1): $P(F' \cap H) = \frac{1}{4} \cdot 1 = \frac{1}{4}$

Since there are only 3 fair coins out of 4 in the box,

$$P(F) = \frac{3}{4} \qquad P(F') = \frac{1}{4}$$

and

$P(H \mid F) = \frac{1}{2}$ (since heads and tails are equally likely when a fair coin is tossed)

$P(H \mid F') = 1$ (since this coin has two heads)

Therefore,

$$P(F \mid H) = \frac{P(F \cap H)}{P(H)} = \frac{3/4 \cdot 1/2}{3/4 \cdot 1/2 + 1/4 \cdot 1} = \frac{3/8}{5/8} = \frac{3}{5}$$

EXAMPLE 5 **OIL EXPLORATION** Records show that when drilling for oil, the probability of a successful strike is .1. However, it has been observed that if there is oil, then the probability is .6 that there is permeable, porous, sedimentary rock present. Records also show that when there is no oil, the probability is .3 that such rock formations are present. What is the probability of oil beneath permeable, porous, sedimentary rock?

Solution We can solve this problem using a Bayes tree (Figure 8), where

O: oil
O': no oil
R: permeable, porous, sedimentary rock

Then,

$$P(O|R) = \frac{P(O \cap R)}{P(R)} = \frac{(.1)(.6)}{(.1)(.6) + (.9)(.3)} = \frac{.06}{.33} = \frac{2}{11}$$

FIGURE 8

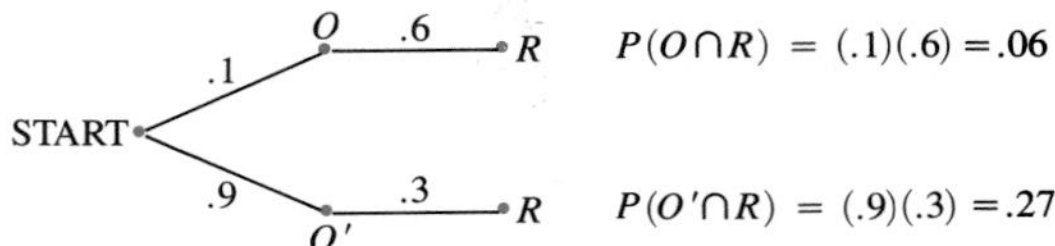

Applications of Bayes' Theorem

Bayes' theorem has many varied and interesting applications. One of the most fascinating is given by F. Mosteller and D. L. Wallace in a book called *Inference and Disputed Authorship: The Federalist* (Reading, Mass.: Addison-Wesley, 1964). In that book Mosteller and Wallace use Bayes' theorem to resolve a long-standing historical question: Between 1787 and 1788 a series of 77 essays, called the *Federalist Papers,* appeared in New York newspapers under the pen name Publius. It is known that James Madison wrote 14 of the essays, John Jay wrote 5, Alexander Hamilton wrote 43, and 3 were jointly authored. The authorship of the remaining 12 essays has been in dispute for years. By analyzing the writing styles and applying Bayes' theorem, Mosteller and Wallace provide a convincing argument to show that Madison was the most probable author of the disputed papers. You will find a more extensive discussion of the Mosteller–Wallace method in an excellent elementary textbook called *Sets, Functions and Probability* by Johnston, Price, and Van Vleck (Reading, Mass.: Addison-Wesley, 1968).

EXERCISE SET 7.6

1. In a tire factory, assembly lines A, B, and C account for 60, 30, and 10% of the total production. If .3% of the tires from line A are defective, .6% of the tires from B are defective, and .8% of the tires from line C are defective, what is the probability that:
(a) a defective tire comes from line A?
(b) a defective tire comes from line B?
(c) a defective tire comes from line C?

2. Company insurance records show that a new driver who has finished a driver training program has a probability of .9 of completing the first year of driving without an accident, whereas a new driver who has not finished a driver training program has a probability of only .7. If 60% of all new drivers have completed a driver training program, what is the probability that a person who is involved in an accident during his or her first year of driving finished a driver training program?

3. Records show that 20% of all students taking calculus at Rigor University fail the course. To reduce the failure rate, the university designs a screening test to determine beforehand which students are likely to fail. These students are to be given remedial coursework. An experiment shows that the probability is .9 that a student who passes calculus can pass the screening test, and the probability is .3 that a student who fails calculus can pass the screening test. A student passes the screening test. What is the probability that he or she will pass calculus?

4. A pediatrician knows from past experience that when a parent calls and states that a child has a low-grade fever, the probabilities for various causes are shown in Table 1. The pediatrician also knows that when the parent is advised to administer aspirin and give lots of liquid, the probability that the fever will be gone (G) within 24 hours is

$$P(G \mid A) = .6 \qquad P(G \mid B) = .5$$
$$P(G \mid C) = 0 \qquad P(G \mid D) = .7$$

(a) Given that a parent has called and has been advised to administer aspirin and lots of liquid, what is the probability that the child's fever will be gone within 24 hours?

(b) Given that the child's fever is not gone within 24 hours after the advice to administer aspirin and lots of liquid, what is the probability that the child has the flu?

TABLE 1

Cause	Probability
(A) Flu	.3
(B) Strep infection	.2
(C) Inaccurate thermometer	.1
(D) Other causes	.4

5. Teleview and Radion Corporations are competing for a contract to build television equipment for a Mars probe; they both have equal chances of getting the contract. If Teleview gets the contract, then the probability of a total success is .6, the probability of a partial success is .1, and the probability of failure is .3. If Radion gets the contract, then the probability of a total success is .3, the probability of a partial success is .5, and the probability of failure is .2. If the contract is awarded and the result is a total success, what is the probability that Teleview got the contract?

6. A shipper must send fresh produce to a distribution center. The produce can be sent by one of four different routes, and the probabilities of the produce reaching the center within 24 hours on the four routes are .3, .2, .5, and .1, respectively. The shipper picks a route at random.

(a) If the produce arrives within 24 hours, what is the probability that the first route was picked?

(b) If the produce does not arrive within 24 hours, what is the probability that the third route was picked?

7. Theoretical genetic considerations suggest that three-fourths of all marigold seeds produced by a certain crop of marigolds should yield pure yellow offspring; the rest should yield multicolored offspring. A seed packager has a test to separate the seeds that produce pure yellow marigolds from the seeds that produce multicolored marigolds. However, the test is not perfect, and 5% of the pure yellow will be classified as multicolored and 10% of the multicolored will be classified as pure yellow. Of 1000 seeds packaged as pure yellow, how many will actually be pure yellow?

8. A test is designed to detect cancer. If a person has cancer, then the probability that the test will detect it is .95; if the person does not have cancer, then the probability that the test will erroneously indicate that he or she does have cancer is .1. If 3% of the population who take the test have cancer, what is the probability that a person described by the test as having cancer does not really have it?

9. A patient has one of two diseases, D_1 or D_2. Each disease can produce one of the following sets of symptoms:

$$S_1 = \{\text{fever, aching}\} \quad \text{or} \quad S_2 = \{\text{vomiting, fatigue}\}$$

It is known that

$$P(S_1 \mid D_1) = .4 \qquad P(S_2 \mid D_1) = .6$$
$$P(S_1 \mid D_2) = .8 \qquad P(S_2 \mid D_2) = .2$$

Moreover, D_1 and D_2 occur equally often and never occur together. If a patient has fever and aching, what is the probability that the disease is D_1?

10. A box contains two coins, an ordinary fair dime and a dime with two heads. One of the coins is picked at random and tossed twice. If both tosses are heads, what is the probability the coin has two heads?

11. In the sheik's palace are three vaults; each vault contains 2 chests and each chest contains 1 gem. In one vault each chest contains a ruby; in the second vault each chest contains a diamond; and in the last vault one chest contains a ruby and the other contains a diamond. A thief randomly chooses a vault and steals one chest. When the chest is opened, the thief finds a ruby. What is the probability that the other chest of the same vault contains a ruby?

KEY IDEAS FOR REVIEW

- **Relative frequency of an event** The ratio m/n, where m is the number of occurrences of the event in n repetitions of the experiment.
- **Probability of an event** See pages 232–234.
- **Sample space** The set of all possible outcomes of an experiment.
- **Sample point** An element of the sample space.
- **Event** A subset of the sample space.
- **Certain event** An event that occurs every time the experiment is performed.
- **Impossible event** An event that never occurs when the experiment is performed.
- **Mutually exclusive events (two)** Events that have no outcomes in common.
- **Elementary event** An event consisting of one sample point.
- **Probability model** An assignment of probabilities to the outcomes of an experiment.
- **Probability of an event E** $P(E)$ denotes the probability of the event E.
- **Addition principle** See page 246.
- **Assumed (a priori) probability model** A probability model obtained by logical reasoning.
- **Empirical (a posteriori) probability model** A probability model obtained from experimental relative frequencies.
- **Equally likely sample points** Sample points with equal chances of occurring.
- **Uniform probability model** A model in which each sample point is assigned the same probability.
- **Odds in favor of E** $\dfrac{P(E)}{P(E')}$
- **Odds against E** $\dfrac{P(E')}{P(E)}$
- **Conditional probability of A, given B** $P(A \mid B) = \dfrac{P(A \cap B)}{P(B)}$
- **Product principle for probabilities** $P(A \cap B) = P(A \mid B)P(B)$
- **Independent events A and B** Events with the property $P(A \cap B) = P(A)P(B)$.
- **Bayes' theorem** See page 286.

SUPPLEMENTARY EXERCISES

Use the following information for Exercises 1–4. Let $S = \{s_1, s_2, s_3, s_4, s_5\}$ *be the sample space of an experiment and let*

$$P(\{s_1\}) = .1 \quad P(\{s_2\}) = .2 \quad P(\{s_3\}) = .3$$
$$P(\{s_4\}) = .3 \quad E = \{s_1, s_2, s_3\} \quad F = \{s_2, s_3, s_4\}$$

1. Find $P(\{s_5\})$.

2. Find $P(E \cup F)$.

3. Find $P(E \cap F)$.

4. Find $P(E' \cap F)$.

Use the following information for Exercise 5–7:

$$P(A \cup B) = \tfrac{3}{4} \quad P(A) = \tfrac{1}{2} \quad P(A \cap B) = \tfrac{1}{8}$$

5. Find $P(B)$.

6. Find $P(A \mid B)$.

7. (a) Are A and B independent events? Why?
(b) Are A and B mutually exclusive events? Why?

8. Let $P(A) = \frac{1}{2}$ and $P(A \cap B) = \frac{1}{8}$. If A and B are independent events, what is $P(B)$?

9. If $P(A) = \frac{3}{4}$ and $P(B) = \frac{3}{4}$, is it possible for A and B to be mutually exclusive events? Why?

10. A six-sided die is symmetrical and balanced so that each face is as likely to occur as any of the others when the die is tossed. However, the numbers on the faces are

1, 1, 2, 3, 4, and 5 (note there is no 6)

If we toss the die, what is
(a) $P(\{1\})$ (b) $P(\{4\})$ (c) $P(\{6\})$

11. Consider two dice, each as in Exercise 10. If two such dice are tossed, what is the probability of obtaining
(a) a sum of 7? (b) a sum of 5?
(c) What are the odds in favor of obtaining a sum of 7?

12. A box of 10 batteries contains 2 defectives.
(a) If a battery is selected at random, what is the probability that it is defective?
(b) What are the odds against it being defective?
(c) If 2 batteries are selected at random, what is the probability that they are both defective?
(d) If 2 batteries are selected at random, what is the probability that at least 1 is defective?

13. Toss 2 fair dice.
(a) What are the odds in favor of rolling a sum of 7?
(b) If you bet \$1, what would you expect your net win to be?
(c) At a casino in Atlantic City, if you bet \$1 on 7 on the roll of two dice, you will win \$4 (net) if a 7 occurs and lose \$1 if a 7 does not occur. Do you like this bet? Why?

14. Consider the game of roulette (see Exercise 18 of Section 7.3).
(a) What is the probability that black will occur?
(b) What is the probability that black will occur on 4 consecutive spins of the wheel?
(c) What is the probability of obtaining *red, black, black, red* on the next 4 spins of the wheel?
(d) What are the odds against obtaining *red, black, black, red* on the next 4 spins of the wheel?

15. Five girls and three boys are to be seated in a row. If seats are taken randomly, what is the probability that the girls will be seated together?

16. Five girls and three boys are to be seated in a row. If seats are taken randomly, what is the probability that Mary and John will sit next to one another?

17. In the Big A Daily Lottery, the object is to pick a 3-digit number that matches the 3-digit number drawn by the lottery commission.
(a) What is the probability that the number 000 will be drawn?
(b) What is the probability that the number 126 will be drawn?
(c) What is the probability that the number 126 will be drawn on 2 consecutive days?
(d) What is the probability that the same number will be drawn on 2 consecutive days?
(e) Why are the answers to (c) and (d) different?

18. An urn contains 10 red balls and 5 blue balls.
(a) Select 3 balls at random. What is the probability that 2 of the balls selected will be blue and 1 will be red?
(b) What is the probability that all 3 of the balls selected will be red?

19. Assume that a nationwide survey of hospital records produced the data in Table 1 on 6000 people who had *exactly one* of the diseases d_1: heart condition or d_2: pneumonia. Suppose each of the diseases can produce at least one of the following observable symptoms: loss of appetite, chest pain, shortness of breath

TABLE 1

	Number of people with disease d_i	Number of people with disease d_i who also had at least one of the symptoms
Disease d_1	3750	3000
Disease d_2	2250	2050
Total	6000	5050

If

D_1: event that a person has disease d_1

D_2: event that a person has disease d_2

A: event that a person has at least one of the above symptoms

compute:

(a) $P(D_1)$ (b) $P(D_2)$ (c) $P(A)$
(d) $P(D_1 | A)$ (e) $P(A | D_1)$

Use the following information for Exercises 20–22: One hundred students went sightseeing. Of these,

40 visited the art museum
30 visited the planetarium
25 visited the Liberty Bell
15 visited the art museum and the planetarium
10 visited the planetarium and the Liberty Bell
15 visited the art museum and the Liberty Bell
5 visited all three of these sights

20. If a student is selected at random from the group, what is the probability that he or she visited none of these sights?

21. If we know that a student has visited the art museum, what is the probability that he or she visited all three sights?

22. What is the probability that a student selected at random visited only the art museum?

Use the following information for Exercises 23 and 24: A box contains 5 coins: 3 are fair and 2 are two-headed. A coin is selected at random and tossed twice.

23. What is the probability that both tosses result in heads?

24. **(Optional)** If the coin is tossed and both tosses result in heads, what is the probability that a fair coin was selected?

Use the following information for Exercises 25–28: Because of inexperienced personnel, 36 men's shirts were not sorted correctly by size. Box I contains 8 shirts sized small and 4 shirts sized medium. Box II contains 3 shirts sized small, 4 shirts sized medium, and 5 shirts sized large. Box III contains 2 shirts sized small, 9 shirts sized medium, and 1 shirt sized large. A box is chosen at random and 3 shirts are selected randomly from the box.

25. What is the probability that exactly 1 of the shirts selected is sized large?

26. What is the probability that at least 1 of the shirts is sized large?

27. **(Optional)** If exactly one of the selected shirts is sized large, what is the probability that box I was selected?

28. **(Optional)** If exactly one of the selected shirts is sized large, what is the probability that box II was selected?

Use the following information for Exercises 29 and 30: Two dice are rolled repeatedly until a 6 or a 7 occurs:

29. What is the probability that a 6 will occur before a 7?

30. What are the odds in favor of a 6 occurring before a 7?

CHAPTER TEST

1. Which of the following represent probability models for $S = \{s_1, s_2, s_3, s_4\}$? If not, why not?

(a) $P(\{s_1\}) = \frac{1}{4}$, $P(\{s_2\}) = \frac{1}{3}$, $P(\{s_3\}) = \frac{5}{12}$, $P(\{s_4\}) = 0$

(b) $P(\{s_1\}) = \frac{1}{4}$, $P(\{s_2\}) = \frac{1}{3}$, $P(\{s_3\}) = \frac{7}{12}$, $P(\{s_4\}) = -\frac{1}{6}$

(c) $P(\{s_1\}) = \frac{1}{4}$, $P(\{s_2\}) = \frac{1}{3}$, $P(\{s_3\}) = \frac{1}{6}$, $P(\{s_4\}) = \frac{1}{6}$

2. Let A and B be events with $P(A) = \frac{3}{4}$, $P(B) = \frac{1}{2}$, $P(A \cap B) = \frac{1}{4}$. Compute

(a) $P(A \cup B)$ (b) $P(A | B)$
(c) $P(B | A)$ (d) $P(A')$
(e) Are A and B independent events? Why?
(f) Are A and B mutually exclusive events? Why?
(g) What are the odds in favor of A?
(h) What are the odds against A?

3. Let A and B be events with $P(A) = \frac{1}{4}$ and $P(B) = \frac{1}{2}$.
   (a) Find $P(A \cup B)$ if A and B are independent events.
   (b) Find $P(A \cup B)$ if A and B are mutually exclusive events.

4. Two people are to be chosen from a group of 10 to be president and vice president of the group. If the selection is random, what is the probability that Alice will be chosen as president and Bob as vice president?

5. Suppose each of the diseases d_1, d_2, and d_3 can produce one or more of the observable symptoms: fainting spells or high blood pressure. Assume also that a nationwide survey of hospital records produced the following data on 50,000 people who had *exactly one* of the diseases d_1, d_2, or d_3:

	Number of people with disease d_i	**Number of people with disease d_i who also had one or more of the symptoms**
Disease d_1	15,000	9,000
Disease d_2	25,000	20,000
Disease d_3	10,000	4,000
Total	50,000	33,000

If

A: event that a person has at least one of the above symptoms
D_1: event that a person has disease d_1
D_2: event that a person has disease d_2
D_3: event that a person has disease d_3

Compute
(a) $P(D_1)$ (b) $P(D_2)$ (c) $P(D_3)$
(d) $P(A)$ (e) $P(D_2 \mid A)$ (f) $P(A \mid D_2)$
(g) If a patient has at least one of the symptoms, for which disease should he or she be treated?

6. Jack has 3 dimes and 5 pennies in his left pocket and 2 dimes and 6 pennies in his right pocket. He chooses a pocket at random and randomly selects 2 coins from that pocket.
   (a) Construct a tree diagram labeling all branch probabilities.
   (b) What is the probability that Jack has chosen at least 11 cents?
   (c) Jack looks in his hand and notes that he actually selected 2 pennies. What is the probability that these coins were from his left pocket?

8

Statistics and Probability

Statistics, the science of collecting and analyzing data, had its origins in early history when numerical data such as population and death counts were recorded and organized for use by heads of state for various purposes, such as taxation. Today, the majority of statisticians are concerned with the problem of statistical inference, that is, using numerical data to make logical decisions and inferences. It is assumed that the data generated satisfies some unknown probability distribution—the statistician is concerned with inferring the characteristics of this distribution.

So we see that the fields of probability and statistics are related and yet different. To illustrate the difference, consider two groups of eligible voters: group A contains 2 Republicans and 3 Democrats; the political preference of the five members of group B is not known.

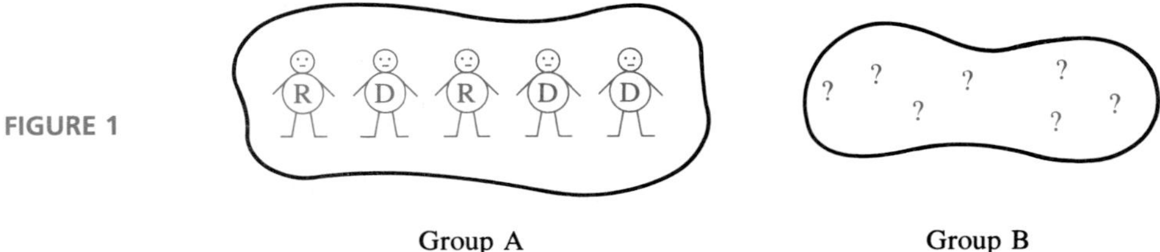

FIGURE 1

From a probability point of view, since the distribution of Republicans and Democrats is known, the answer to the question If a person is selected at random from group A, what is the probability that he or she is Republican? is easily determined as $\frac{2}{5} = 0.4$.

From a statistical point of view, the composition of group B (**population**) is not known. A subset, or **sample**, of persons is chosen randomly from group B, and the distribution of Republicans and Democrats in this sample is noted. Statisticians then try to make inferences about the distribution of the total population of group B based on the information obtained from the sample.

In this chapter, we first take another look at probability distributions and see the kinds of characteristics that are important to consider. Chapter 8 describes how some probability models can be determined by mathematical principles, whereas others can be obtained only from experimental data. This chapter describes how statisticians organize this data and determine estimations of the desired characteristics. Finally, we touch on one of the main areas of statistical inference called **hypothesis testing**. This area is concerned with procedures for determining whether the results predicted by a probability model are in reasonable agreement with physical observations.

8.1 INTRODUCTION; RANDOM VARIABLES

Many times the outcomes of an experiment can be described numerically. As an example, consider the experiment of tossing a fair coin three times and recording the resulting sequence of heads and tails. The sample space is

$$S = \{hhh, hht, hth, thh, htt, tht, tth, ttt\}$$

For this experiment we have outcomes that are sequences of three letters. Suppose, however, we are interested in making probability statements about the number of heads that occur in the three tosses. For example, what is the probability of obtaining two heads in the three tosses? In this case we are not primarily concerned with the particular sequence of h's and t's in the outcome, but rather we are interested in the number of h's that occur. Thus, associated with each outcome of this experiment, we have a number that is of importance to us, the number of h's in the sequence. This idea of associating a number with an outcome of an experiment occurs so frequently that statisticians have found it convenient to introduce the following terminology.

A **random variable** is a rule that assigns a numerical value to each outcome of an experiment.

We denote random variables by uppercase letters such as X, Y, and Z.

EXAMPLE 1 Suppose we toss a coin three times and introduce the random variable X to denote the number of heads that occur in the three tosses. In Table 1 we have listed the sample points for the experiment, and next to each sample point we have given the associated value of the random variable X.

TABLE 1

Sample point	Value of X
hhh	3
hht	2
hth	2
thh	2
htt	1
tht	1
tth	1
ttt	0

We see that the random variable X takes on the values 0, 1, 2, and 3, and we can consider the sample space $S_h = \{0, 1, 2, 3\}$, where the outcomes of the tosses are represented by the number of heads that occur. (Note that these outcomes are not equally likely.)

EXAMPLE 2 Suppose we toss a coin repeatedly until a head occurs and then we stop. Let the random variable Y denote the number of tosses performed in the experiment. In Table 2 we have listed some of the sample points and the associated values of the random variable.

TABLE 2

Sample point	Value of Y
h	1
th	2
tth	3
ttth	4
tttth	5
ttttth	6
⋮	⋮

EXAMPLE 3 A traffic engineer records the times at which two consecutive buses arrive at a checkpoint. Let the random variable Z denote the time elapsed (in minutes) between the arrivals of the buses. Thus, if the arrival times are

9:07 A.M. and 9:36 A.M.

then the value of Z associated with this experimental outcome is

$$Z = 29 \text{ minutes}$$

EXAMPLE 4 A geneticist records the length of life (in hours) of a fruit fly. Let the random variable W denote the number recorded. If we assume for simplicity that the time can be recorded with perfect accuracy, then the value of W can be any nonnegative real number.

A random variable is often described according to the number of values it can take on. A random variable is **finite discrete** if it can take on only finitely many possible values. For example, the random variable X in Example 1 is finite discrete since it can take on only the values

$$X = 0, 1, 2, \text{ or } 3$$

A random variable is **infinite discrete** if it can take on infinitely many values that can be arranged in a sequence. For example, the random variable Y in Example 2 is infinite discrete since its possible values can be arranged in the sequence

$$Y = 1, 2, 3, 4, \ldots$$

Finally, a random variable is **continuous** if its possible values form an entire interval of numbers. For example, the random variable W in Example 4 can take on any nonnegative value. The random variable W is thus continuous since its pos-

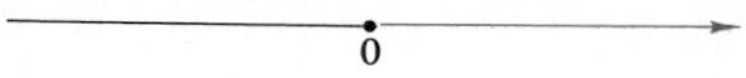

0

FIGURE 1

sible values form the interval shown in color in Figure 1. *For the remainder of this section we restrict our discussion to finite discrete random variables.*

Since the value that a random variable takes on is determined by the outcome of an experiment, and since the outcomes of an experiment occur with various probabilities, it follows that the values of the random variable also occur with various probabilities. To illustrate, consider the following.

EXAMPLE 5 We return to the coin-tossing experiment in Example 1. In the left column of Table 1 we listed the eight possible outcomes for this experiment. If we assume the coin is fair, then each of these eight outcomes will be equally likely so that each outcome will have probability $\frac{1}{8}$ of occurring. From the right column of Table 1, we see that X takes on the value 2 for three of these sample points; thus the probability that $X = 2$ will be $\frac{3}{8}$. We denote this by

$$P(X = 2) = \frac{3}{8}$$

(read "the probability that X equals 2 is $\frac{3}{8}$"). Similarly, Table 1 shows that

$$P(X = 0) = \frac{1}{8} \qquad P(X = 1) = \frac{3}{8} \qquad P(X = 3) = \frac{1}{8}$$

We can also determine the probability that X is less than or equal to 2:

$$P(X \leq 2) = P(X = 0) + P(X = 1) + P(X = 2) = \frac{1}{8} + \frac{3}{8} + \frac{3}{8} = \frac{7}{8}$$

Note that

$$P(X > 2) = 1 - P(X \leq 2) = 1 - \frac{7}{8} = \frac{1}{8}$$

Also,

$$P(X > 2) = P(X = 3) = \frac{1}{8}$$

EXAMPLE 6 Let X denote the sum of the numbers tossed with two fair dice. Find the probabilities of the various possible X values, and determine $P(X \leq 5)$ and $P(1 < X \leq 3)$.

Solution This experiment is discussed in Example 16 of Section 7.2 and the possible outcomes are shown in Table 1 of that section. That example shows that $P(X = 7) = \frac{6}{36}$. Similarly,

$$P(X = 2) = \frac{1}{36} \qquad P(X = 3) = \frac{2}{36} \qquad P(X = 4) = \frac{3}{36}$$

$$P(X = 5) = \frac{4}{36} \qquad P(X = 6) = \frac{5}{36} \qquad P(X = 7) = \frac{6}{36}$$

$$P(X = 8) = \frac{5}{36} \qquad P(X = 9) = \frac{4}{36} \qquad P(X = 10) = \frac{3}{36}$$

$$P(X = 11) = \frac{2}{36} \qquad P(X = 12) = \frac{1}{36}$$

Also,

$$P(X \leq 5) = P(X = 2) + P(X = 3) + P(X = 4) + P(X = 5)$$

$$= \frac{1}{36} + \frac{2}{36} + \frac{3}{36} + \frac{4}{36} = \frac{10}{36} = \frac{5}{18}$$

$$P(1 < X \leq 3) = P(X = 2) + P(X = 3)$$

$$= \frac{1}{36} + \frac{2}{36} = \frac{1}{12}$$

EXAMPLE 7 Company records show that 5% of all resistors produced at a certain electronics plant are defective. Suppose two resistors are picked independently from the plant's production line and assume the random variable X denotes the number of defectives in the sample.

(a) Find $P(X = 0)$, $P(X = 1)$, $P(X = 2)$, $P(X < 2)$, assuming that resistors are picked independently.

(b) Find the probability that at least one of the resistors is defective.

(c) Find the probability that at most one of the resistors is defective.

Solution If we let n denote a nondefective resistor and d a defective one, then the tree diagram of Figure 2 indicates the possible selections and the branch prob-

FIGURE 2

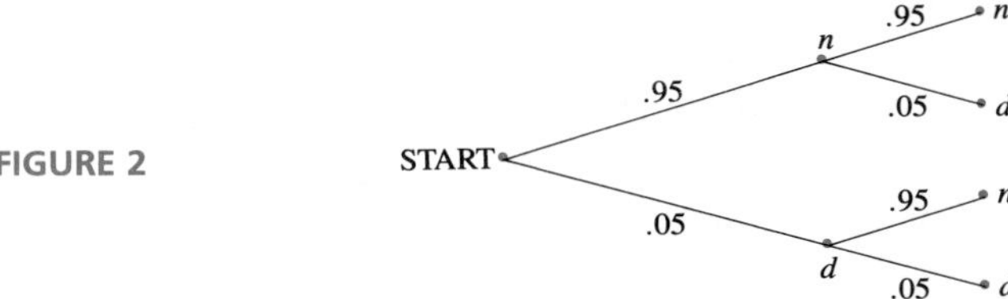

abilities. The outcomes may be viewed as

$$(n, n) \qquad (n, d) \qquad (d, n) \qquad (d, d)$$

where, for example, (n, d) means that the first resistor picked was nondefective and the second was defective. Since 5% of all resistors produced are defective, the probability of picking a defective resistor is .05 and the probability of picking a nondefective resistor is .95. Moreover, since we assume that the resistors are selected independently, the probability of selecting a nondefective and then a defective one will be the product (.95)(.05); that is, the probability of the outcome (n, d) is

$$(.95)(.05) = .0475$$

The probabilities of all the outcomes and the associated values of X are listed in Table 3.

TABLE 3

Outcome	Probability	Value of X associated with the outcome
(n, d)	$(.95)(.05) = .0475$	$X = 1$
(d, n)	$(.05)(.95) = .0475$	$X = 1$
(n, n)	$(.95)(.95) = .9025$	$X = 0$
(d, d)	$(.05)(.05) = .0025$	$X = 2$

Solution (a) From Table 3 we obtain

$$\begin{aligned} P(X = 0) &= .9025 \\ P(X = 1) &= .0475 + .0475 = .0950 \\ P(X = 2) &= .0025 \\ P(X < 2) &= .9025 + .0950 = .9975 \end{aligned} \tag{1}$$

Solution (b) The probability that at least one of the resistors is defective is

$$P(X \geq 1) = 1 - P(X = 0) = 1 - .9025 = .0975$$

Solution (c) The probability that at most one of the resistors is defective is

$$P(X \leq 1) = 1 - P(X = 2) = 1 - .0025 = .9975$$

Probability Functions: Graphs and Histograms

Given a random variable X, we are interested in the probability that X takes on a particular real value x. In Example 5, we see that X takes on the values 0, 1, 2, and 3. We now represent the probability that X takes on a real value x by

$$p(x) = P(X = x)$$

where $p(x)$ (read "p of x") is called the **probability function of the random variable X**. The notation $p(x)$ means that for each value of x, a corresponding value of p is determined. Thus, the first three answers to part (a) of Example 7 may also be written as

$$\begin{aligned} p(0) &= .9025 \\ p(1) &= .0950 \\ p(2) &= .0025 \end{aligned}$$

We can now describe the various probabilities of X geometrically. To illustrate, consider again the experiment of tossing three fair coins (Example 5), where the random variable X denotes the total number of heads tossed. Table 4 summarizes the results already obtained in terms of the values x of the random variable X.

TABLE 4

x	0	1	2	3
$p(x)$	$\frac{1}{8}$	$\frac{3}{8}$	$\frac{3}{8}$	$\frac{1}{8}$

Using this table, we can pair up each value of x with $p(x)$ to obtain the four ordered pairs

$$\left(0, \frac{1}{8}\right) \quad \left(1, \frac{3}{8}\right) \quad \left(2, \frac{3}{8}\right) \quad \left(3, \frac{1}{8}\right)$$

These ordered pairs can now be plotted in a Cartesian coordinate system, as shown in Figure 3a. Since the first coordinate of each point is a value of x and the second is the probability $p(x)$, we have labeled the coordinate axes x and p. The configuration of points obtained by plotting these ordered pairs is called the **graph** of the probability function (Figure 3a).

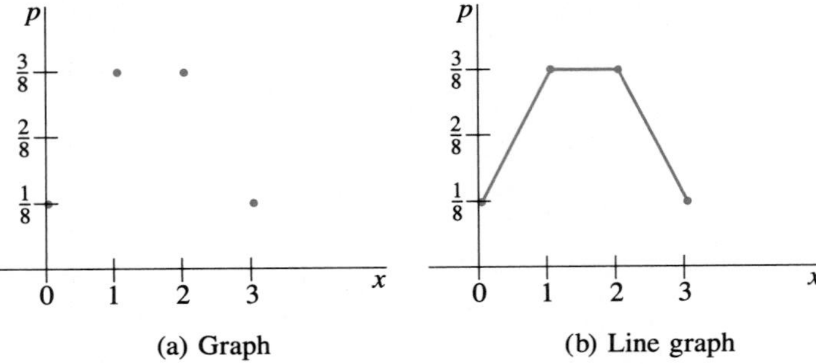

(a) Graph (b) Line graph

FIGURE 3

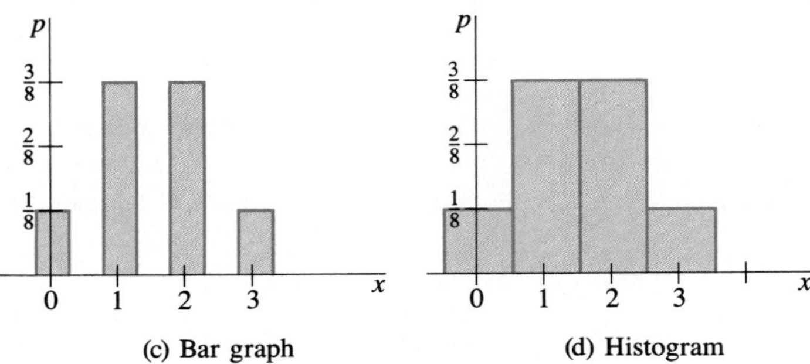

(c) Bar graph (d) Histogram

For visual emphasis, other types of graphs can be used:

1. A **line graph** is obtained by connecting the points on the graph by straight-line segments (Figure 3b).
2. A **bar graph** is obtained by constructing rectangles with base on the x-axis, centered about each value of x, and heights equal to $p(x)$ (Figure 3c).
3. A **histogram** is a bar graph with touching bars having equal bases (Figure 3d). In particular, when a histogram is used to represent probability data where the random variable X takes on integer values, each bar has its base centered at $x = k$ with width equal to 1 and height equal to $p(k) = P(X = k)$. Note that in this case $P(X = k)$ equals the area of this rectangle (Figure 3d). This property is extremely useful in Section 8.6.

EXAMPLE 8 Graph the probability function of the random variable X in Example 7 using the four types of graphs mentioned above.

Solution From the calculations in Example 7, the values of x and their probabilities are summarized by Table 5. The graph thus consists of the ordered pairs

$$(0, .9025) \quad (1, .0950) \quad (2, .0025)$$

TABLE 5

x	0	1	2
$p(x)$	.9025	.0950	.0025

The four types of graphs are shown in Figure 4.

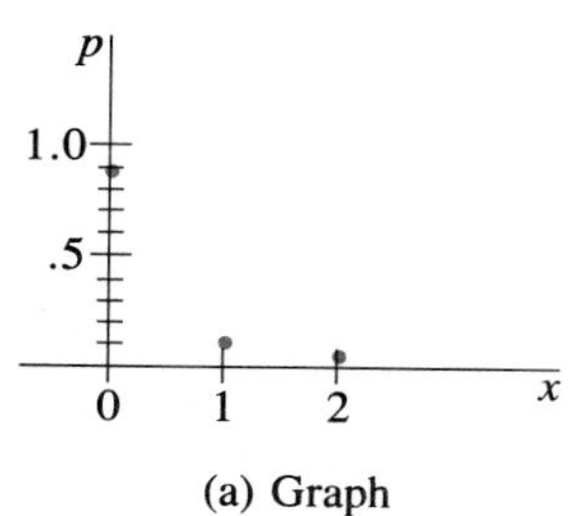

(a) Graph

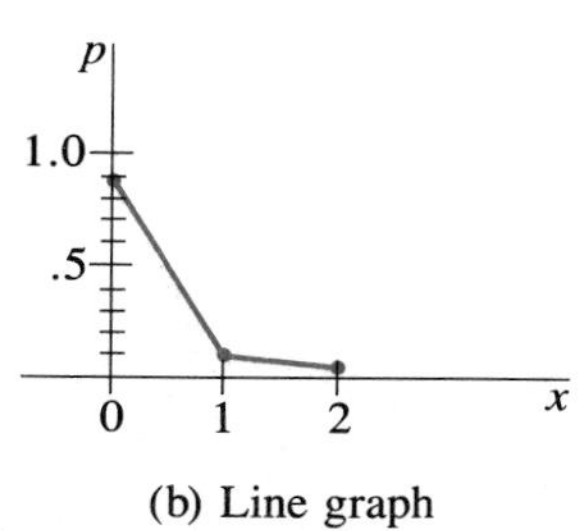

(b) Line graph

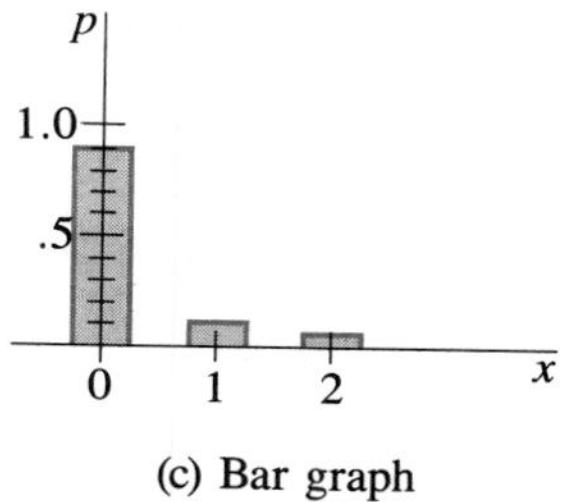

(c) Bar graph

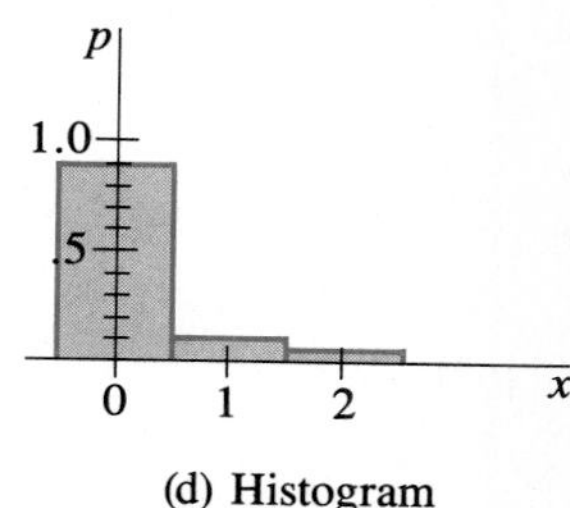

(d) Histogram

FIGURE 4

EXERCISE SET 8.1

In Exercises 1–5 list the values of X and
(a) their probabilities of occurrence
(b) the probability that X is at least 1
(c) the probability that X is at most 1

1. A fair die is tossed once. The random variable X is the number showing on the top face.

2. A fair coin is tossed four times. The random variable X is the number of heads obtained on the four tosses.

3. Let X be the number of male children in a family with two children. (Assume the probability of a male birth is $\frac{1}{2}$.)

4. A coin is weighted so that on each toss the probability of a head is $\frac{2}{3}$ and the probability of a tail is $\frac{1}{3}$. The coin is tossed three times and X is the total number of tails obtained.

5. A student takes a three-question true–false examination and guesses at every answer. Let X be the number right minus the number wrong.

6–10. Graph the probability functions for the random variables in Exercises 1–5 as
(a) line graphs
(b) histograms

11. Classify the following random variables as finite discrete, infinite discrete, or continuous:
(a) X is the number of calls entering a switchboard in a 24-hour period.
(b) X is the number of defectives in a lot of 1000 items.
(c) X is the length of time a person must wait in line to check out at a supermarket.

12. Let X be a random variable whose probability function is graphed in Figure 5. Find:
(a) $P(X = 3)$
(b) $P(X \leq 3)$
(c) $P(X = \frac{1}{2})$
(d) the probability that X is at least 3
(e) the probability that X is at most 3

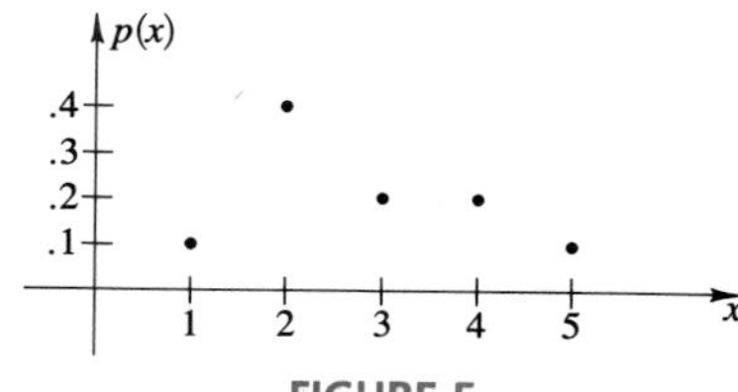

FIGURE 5

8.2 EXPECTED VALUE OF A RANDOM VARIABLE

We are all familiar with the idea of using an average value to summarize a collection of numerical data. The use of averages has been so ingrained in most of us by our technological environment that we are constantly comparing aspects of ourselves against the average. Are we overweight or underweight compared to the average? Is our temperature too high or too low compared to the average? Is our IQ high or low compared to the average?

In the process of melting down data into an average, we lose certain information, and thus averages can be misleading. For example, knowing that the average weight of men inducted into the Army since 1950 is 161.5 pounds does not enable us to determine the highest and lowest weights recorded. Since these numbers could have been obtained from the original data, we have lost this information in the averaging process. However, since the original data would take volumes to list individually, the conciseness of the average as a description of the data may more than compensate for this loss of information.

In this section we develop a notion called the expected value of a random variable. The expected value is akin to the idea of an average in the sense that it is a single number that is meant to summarize certain data, namely, the values of a random variable and their probabilities of occurrence. We also establish an important mathematical relationship between expected values and averages.

Recall that the *arithmetic average* or *mean* of a finite set of numbers is computed by adding the numbers in the set and then dividing by the number of terms.

The **arithmetic mean** or **average** of $x_1, x_2, \ldots, x_n$ is

$$\overline{x} = \frac{x_1 + x_2 + \cdots + x_n}{n} \tag{1}$$

For example, consider the set of test scores for 10 students in a finite mathematics class:

$$65, 90, 70, 65, 70, 90, 80, 65, 90, 90$$

The average grade ($\overline{x}$) of the class is

$$\overline{x} = \frac{65 + 90 + 70 + 65 + 70 + 90 + 80 + 65 + 90 + 90}{10} = 77.5 \tag{2}$$

(Note that the average value does not have to equal any of the given values.) Since there are repetitions in this set of scores (65 appears three times, 70 appears twice, 80 appears once, and 90 appears four times), Equation (2) can be written as

$$\overline{x} = \frac{3 \cdot 65 + 2 \cdot 70 + 1 \cdot 80 + 4 \cdot 90}{10} = 77.5 \tag{3}$$

We can also write (3) in the alternative form

$$\overline{x} = \frac{3}{10} \cdot 65 + \frac{2}{10} \cdot 70 + \frac{1}{10} \cdot 80 + \frac{4}{10} \cdot 90 = 77.5 \tag{4}$$

Note that

3/10 is the relative frequency of the occurrence of 65
2/10 is the relative frequency of the occurrence of 70
1/10 is the relative frequency of the occurrence of 80
4/10 is the relative frequency of the occurrence of 90

To formalize this idea, suppose we want to compute the arithmetic average of n pieces of data with repetitions. Specifically, suppose that the distinct numbers occurring in the data are $x_1, x_2, \ldots, x_t$ and

x_1 occurs f_1 times
x_2 occurs f_2 times
x_3 occurs f_3 times
⋮
x_t occurs f_t times

The following result is suggested by Equations (3) and (4).

Arithmetic Mean or Average The arithmetic mean of the above data with repetitions is given by the formulas

$$\bar{x} = \frac{f_1x_1 + f_2x_2 + f_3x_3 + \cdots + f_tx_t}{n} \tag{5}$$

or

$$\bar{x} = \frac{f_1}{n}x_1 + \frac{f_2}{n}x_2 + \frac{f_3}{n}x_3 + \cdots + \frac{f_t}{n}x_t \tag{6}$$

where n is the total number of values being averaged.

Note that $f_1 + f_2 + \cdots + f_t = n$, where $f_1, f_2, \ldots, f_t$ are called the **frequencies** of the values $x_1, x_2, \ldots, x_t$; the ratios $f_1/n, f_2/n, \ldots, f_t/n$ are called the **relative frequencies** of the numbers $x_1, x_2, \ldots, x_t$. In fact, it is often convenient to use a table called a **frequency table**, which indicates the frequencies of the values $x_1, x_2, \ldots, x_t$. Table 1 is a frequency table for the test scores in the above example.

TABLE 1

Test scores	Frequency
65	3
70	2
80	1
90	4
	Total = 10

EXAMPLE 1 Suppose the Bureau of the Census surveys a group of married couples and records the data shown in Table 2 concerning the number of children born to them. Find the average number of children born to the couples surveyed.

TABLE 2

Number of couples	Number of children born to the couples
50	0
107	1
201	2
102	3
25	4
9	5
3	6
2	7
1	9
Total = 500	

Solution The numbers denoted by $x_1, x_2, \ldots$ in Formula (4) are the numbers in the right column of Table 2; their frequencies $f_1, f_2, \ldots$ are the numbers in the left column of the table. As indicated in Formula (6), the total number n of values being averaged is the total number of couples and is obtained by adding the frequencies; thus,

$$n = 500$$

On substituting the data into Formula (6), we obtain

$$\bar{x} = \frac{50}{500} \cdot 0 + \frac{107}{500} \cdot 1 + \frac{201}{500} \cdot 2 + \frac{102}{500} \cdot 3 + \frac{25}{500} \cdot 4 + \frac{9}{500} \cdot 5$$

$$+ \frac{3}{500} \cdot 6 + \frac{2}{500} \cdot 7 + \frac{1}{500} \cdot 9 = \frac{1001}{500} = 2.002$$

Thus, the average number of children born to each couple is 2.002.

Suppose now that X is a discrete random variable associated with a certain experiment. If we perform the experiment repeatedly n times, we will observe n values of X. We might ask if it is possible to estimate the arithmetic mean of these n values. Obviously, it is impossible to determine the mean *exactly* since the values of X obtained in one group of n repetitions need not be the same as the values obtained in another group of n repetitions. For example, if we toss a die three times and obtain the numbers 6, 6, and 3 whose arithmetic mean is

$$\frac{6 + 6 + 3}{3} = 5$$

it is conceivable that on the next three tosses we might obtain the numbers 2, 1, and 3 whose arithmetic mean is

MISLED BY THE AVERAGE

Many states have periodic lotteries in which the jackpot continues growing until there is a winner. Some of these jackpots can exceed $25 million.

Consider Robert Smith, a 45-year-old factory worker, who in 1986 won a $25 million jackpot. Mr. Smith's winnings will be paid out over a 20-year period. Thus, he wins $1,250,000 per year. The following lists Mr. Smith's weekly income for the 10 years before and including his winning the lottery:

	1977	$ 255	
	1978	$ 274	
	1979	$ 296	
	1980	$ 336	
	1981	$ 364	
	1982	$ 408	
	1983	$ 452	
	1984	$ 492	
	1985	$ 546	
wins the lottery	1986	$24,038	(= 1,250,000/52)
		$27,461	(total)

The average weekly income for the 10-year period is $2746 per week. Of course, this is a rather misleading number since the salaries during the first 9 years were so much less than the income during the tenth year.

$$\frac{2 + 1 + 3}{3} = 2$$

Nevertheless, it is possible to make a reasonable guess about the mean of n observed values of X when the number n is large. To see why, consider a finite discrete random variable whose only possible values are x_1, x_2, and x_3. Assume also that these values occur with known probabilities p_1, p_2, and p_3. Suppose that after n repetitions of the experiment,

x_1 occurs f_1 times
x_2 occurs f_2 times
x_3 occurs f_3 times

Applying Formula (6),

$$\overline{x} = \frac{f_1}{n}x_1 + \frac{f_2}{n}x_2 + \frac{f_3}{n}x_3 \tag{7}$$

Since

$$\frac{f_1}{n} \quad \frac{f_2}{n} \quad \frac{f_3}{n}$$

represent the proportion of x_1's, x_2's, and x_3's in the n repetitions, it follows that as n becomes larger, these proportions usually approach the probabilities p_1, p_2, and p_3. That is, when n is large, the approximations

$$\frac{f_1}{n} \cong p_1 \qquad \frac{f_2}{n} \cong p_2 \qquad \frac{f_3}{n} \cong p_3$$

should be good. Thus, from (7) we obtain the following approximation to $\bar{x}$ which is likely to be good when n is large:

$$\bar{x} \cong p_1x_1 + p_2x_2 + p_3x_3$$

More generally, if X is a finite discrete random variable whose possible values are $x_1, x_2, \ldots, x_k$, and if these values occur with probabilities $p_1, p_2, \ldots, p_k$, then when n is large, the arithmetic mean of n observed values of X will be approximately

$$\bar{x} \cong p_1x_1 + p_2x_2 + \cdots + p_kx_k \tag{8}$$

EXAMPLE 2 An experiment consists of tossing a fair die and observing the number X showing on the top face. Use the approximation in (8) to estimate the arithmetic mean of the observed values if the experiment is repeated many times.

Solution The possible values of X are

$$x_1 = 1 \qquad x_2 = 2 \qquad x_3 = 3 \qquad x_4 = 4 \qquad x_5 = 5 \qquad x_6 = 6$$

and since the die is fair, each of these values has a probability $\frac{1}{6}$ of occurring; that is,

$$p_1 = \frac{1}{6} \qquad p_2 = \frac{1}{6} \qquad p_3 = \frac{1}{6} \qquad p_4 = \frac{1}{6} \qquad p_5 = \frac{1}{6} \qquad p_6 = \frac{1}{6}$$

Substituting these values in (7) gives

$$\bar{x} \cong \frac{1}{6} \cdot 1 + \frac{1}{6} \cdot 2 + \frac{1}{6} \cdot 3 + \frac{1}{6} \cdot 4 + \frac{1}{6} \cdot 5 + \frac{1}{6} \cdot 6$$

or

$$\bar{x} \cong \frac{21}{6} = 3.5$$

The quantity on the right side of the equal sign in (8) is of such importance that it has its own notation and name:

Expected Value If the values

$$x_1, x_2, \ldots, x_k$$

of a finite discrete random variable X occur with probabilities

$$p_1, p_2, \ldots, p_k$$

then the **expected value** of X, denoted by $E(X)$, is defined by

$$E(X) = p_1x_1 + p_2x_2 + \cdots + p_kx_k \tag{9}$$

Remark In other words, *the expected value of X is the sum of the possible values of X times their probabilities of occurrence*. The expected value of X is also called the **mean of X** or the **expectation of X**. It is also denoted by μ_X or sometimes just μ when the random variable involved is evident (μ is the Greek letter *mu*, pronounced "mew").

EXAMPLE 3 If X is the number showing on the top face when a fair die is tossed, then, as shown in Example 2,

$$E(X) = \mu_X = 3.5$$

This example shows that the expected value $E(X)$ need not be a possible value of X; the expected value represents the long-term average value of X.

EXAMPLE 4 In Example 7 of Section 8.1 we consider a random variable X denoting the number of defective resistors in a sample of two resistors from a production process in which 5% of all resistors produced are defective. Find the expected value of X.

Solution Since the expected value of a finite discrete random variable is the sum of the values of the random variable times their probabilities of occurrence, we obtain

$$\begin{aligned} E(X) &= (.9025)(0) + (.0950)(1) + (.0025)(2) \\ &= .1 \end{aligned}$$

[see Equation (1) in Example 7 of Section 8.1]. This result can also be written as

$$\mu_X = .1 \quad \text{or} \quad \mu = .1$$

To summarize, the expected value $E(X)$ of a finite discrete random variable is the long-term average value of X when the experiment is repeated many times. However, even when an experiment is performed once, the expected value can be of importance, especially if a decision has to be made involving two or more choices of action. For example, consider the following situation:

EXAMPLE 5 A businesswoman has two investment possibilities: project A and project B. If she invests in project A, there is a 20% chance that she will lose \$10,000, a 70% chance that she will break even, and a 10% chance that she will make \$200,000. If she invests in project B, then there is a 15% chance that she will lose \$10,000, a 25% chance that she will break even, and a 60% chance that she will make \$50,000. In which project should she invest?

Solution One method of comparison involves comparing the expected values for the two projects (tree diagrams are shown in Figure 1):

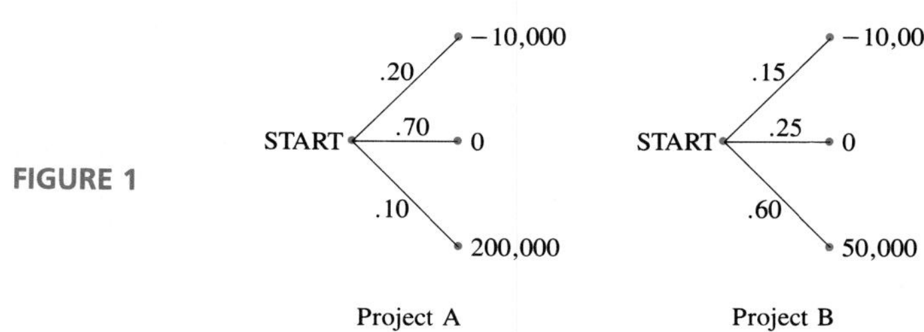

FIGURE 1

$$\text{Expected value of project A} = .20(-10{,}000) + .70(0) + .10(200{,}000) = \$18{,}000$$
$$\text{Expected value of project B} = .15(-10{,}000) + .25(0) + .6(50{,}000) = \$28{,}500$$

If her decision is based on expected-value comparisons, she will choose project B even though the possible profit of \$50,000 from this investment is significantly lower than the possible profit of \$200,000 if she invests in project A.

Gamblers use expected value to determine the expected win or loss if a game is played repeatedly. In this type of problem, the random variable X takes on the possible dollar outcomes. We can then compare the expected values per dollar bet, $E(X)/\text{bet}$, to choose which game to play. The house percentage (the average win for the gambling house or casino) is

$$HP = -100\frac{E(X)}{\text{bet}}\%$$

Note that the expected value per dollar bet and the house percentage are independent of the amount of the bet. The house percentage indicates the long-run percentage of profit for the house from this type of bet.

EXAMPLE 6 Consider the game of roulette with the numbers 0 and 00 (green) and 1, 2, 3, 4, . . . , 36 (half are red, half are black) on the wheel (see Exercise 18, Section 7.3). You bet \$1 on the number 4. If 4 occurs, your net win is \$35 (the payoff is 35 to 1). If 4 does not occur, then your loss is \$1. What is the expected win for this game? What is your expected win per dollar bet? What is the house percentage?

Solution The random variable X takes on the values $+35$ and -1 with probabilities

$$P(X = 35) = \frac{1}{38} \quad \text{(only one position on the wheel, 4, gives a win of \$35)}$$

$$P(X = -1) = \frac{37}{38} \quad \text{(the other 37 positions on the wheel give a loss of \$1)}$$

Therefore, the expected win is

$$E(X) = \frac{1}{38}(35) + \frac{37}{38}(-1) = -\frac{2}{38} = -\$.0526$$

(An expected win of $-\$.0526$ is an expected loss of \$.0526.) The expected value per dollar bet is

$$\frac{E(X)}{\text{bet}} = -\frac{.0526}{1} = -.0526$$

The house percentage is

$$HP = -100\frac{E(X)}{\text{bet}}\% = -100(-.0526)\% = 5.26\%$$

(This bet favors the house.)

EXAMPLE 7 Consider the game of roulette of Example 6. You bet \$2 on red. If red occurs, then the net win is \$2; if black occurs, then your loss is \$2; if green occurs, then your loss is \$1. What is your expected value? What is the expected value per dollar bet? What is the house percentage?

Solution The random variable $X = +2, -2, -1$ with

$$P(X = 2) = \frac{18}{38} \qquad P(X = -2) = \frac{18}{38} \qquad \text{and} \qquad P(X = -1) = \frac{2}{38}$$

Therefore, your expected win is

$$E(X) = \frac{18}{38}(2) + \frac{18}{38}(-2) + \frac{2}{38}(-1) = -\frac{2}{38} = -.0526$$

The expected value per dollar bet is

$$\frac{E(X)}{\text{bet}} = -\frac{.0526}{2} = -.0263$$

The house percentage is

$$-100\frac{E(X)}{\text{bet}}\% = -100(-.0263)\% = 2.63\%$$

EXAMPLE 8 Which of the two bets would you rather make, the one in Example 6 or the one in Example 7?

Solution Since the expected loss per dollar bet is less when you play \$2 on red than it is when you play \$1 on 4, betting on red is a better bet. (Note, however, that it is possible to lose heavily no matter which game you play and that in the long run you should expect to lose in either case.)

EXERCISE SET 8.2

1. Find the arithmetic mean of 3.2, 1.7, 6.3, .2, and 5.6.

2. Find the mean of the following numbers in two ways, using the methods of (1) and (5).

   15, 14, 14, 16, 13, 15, 15, 14, 17, 16

3. Table 3 is a record of visits for dental care made by 250 clinic patients during a 1-year period. Find the mean number of visits per patient.

TABLE 3

Number of visits	Frequency	Number of visits	Frequency
0	41	6	11
1	25	7	0
2	33	8	6
3	49	9	0
4	60	10	1
5	24		

4. Table 4 is a record of the yearly income for a group of people eligible for Medicare benefits. Find the mean yearly income for the group.

TABLE 4

Yearly income ($)	Frequency	Yearly income ($)	Frequency
1000	225	6,000	315
2000	342	7,000	263
3000	516	8,000	101
4000	822	9,000	94
5000	491	10,000	16

5. Find the expected value of the random variable X in Exercise 2 of Section 8.1.

6. Find the expected value of the random variable X in Exercise 3 of Section 8.1.

7. Find the expected value of the random variable X in Exercise 4 of Section 8.1.

8. A random variable Y has the probability distribution shown in Table 5. Find the expectation of Y.

TABLE 5

y	−2	−1	0	1	7	12	18
$p(y)$	.01	.09	.2	.1	.4	.005	.195

9. (a) Find the expectation of the random variable X whose probability function is graphed in Figure 2.

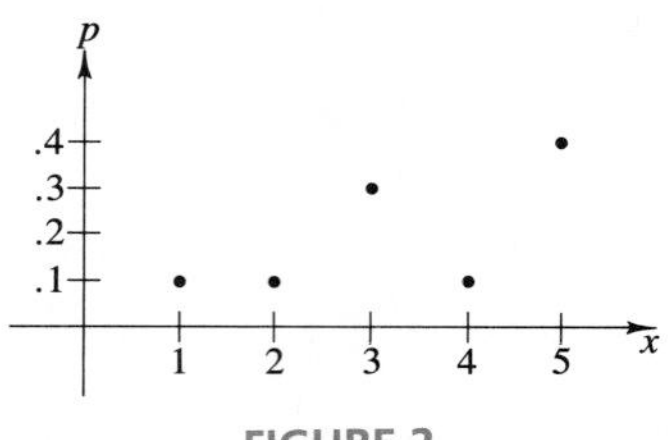

FIGURE 2

(b) Find the expectation of the random variable X whose probability function is graphed in Figure 3.

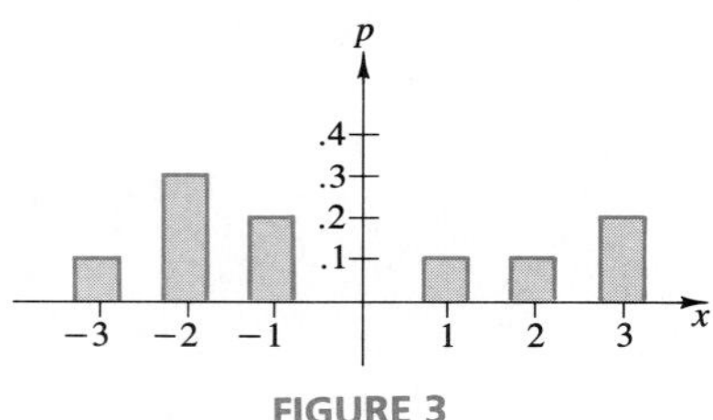

FIGURE 3

10. The number of accidents on the New Jersey Turnpike during the Friday rush hour is 0, 1, 2, or 3 with probabilities .93, .02, .03, and .02, respectively.
    (a) Find the expected number of accidents during the Friday rush hour.
    (b) How many accidents can be expected to occur during Friday rush hours over a 1-year period (52 weeks)?

11. A retailer purchases men's shirts for $3 each and sells them for $5 each. Based on past sales records, he estimates the following probabilities for his weekly sales:

Number of shirts sold	0	1	2	3	4
Probability	.1	.1	.3	.4	.1

If X is the random variable representing the retailer's profit, compute $E(X)$.

12. For each part of Figure 4, determine $E(X)$ *by inspection* (i.e., no calculations) from the given graph.

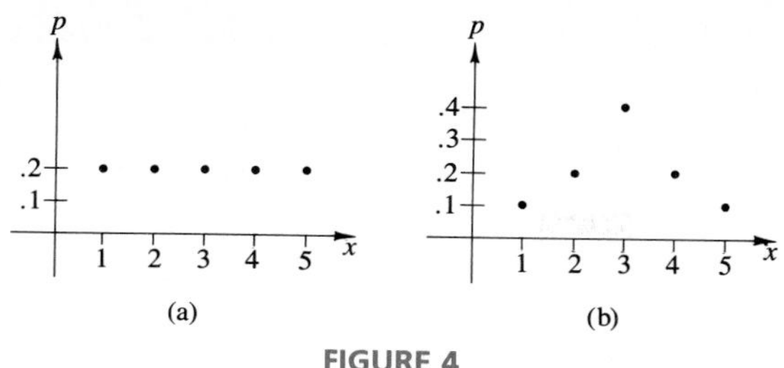

FIGURE 4

13. A U.S. lumber firm is allowed to bid for rights to Soviet forest reserves in one of two regions of Siberia, region A or region B. If the firm gets rights to region A, its estimated profit is $3,000,000; if the firm gets rights to region B, its estimated profit is $5,000,000. The cost of submitting bids for the two regions is $100,000 for A and $300,000 for B. Due to a difference in the number of competitors, the firm estimates its probability of getting rights to A as .7 and its probability of getting rights to B as .5. Should the firm bid for region A or region B? [*Hint*: Compare the expected profits for the bids.]

14. A certain insurance company charges a 45-year-old man $150 for a 1-year term life insurance policy that will pay the beneficiary $10,000 if the man dies within the year. Assuming that the probability is .005 that a 45-year-old man will die during the next year, and that the $150 premium will be worth $165 to the company at the end of the year (because of investment), determine the expected profit at the end of the year, per policy, if the insurance company sells many such policies.

15. Pilferage is a serious problem in business. Suppose the manager of a store finds from the analysis of the records that the store loses, on the average, $500 worth of material each week from pilfering. Further, suppose that for $300 a week, a security program can be implemented. Table 6 shows the results found by other managers of comparable stores where such a program has been used.

TABLE 6

Pilferage total per week in dollars	Probability
500	.05
400	.10
300	.15
200	.60
100	.10

(a) If the program is implemented, how much of a loss per week from pilferage should be expected over the long term?
(b) Should the security program be undertaken?

16. The Hollingsworth Candy Store had the following gross receipts for 6 business days:

Monday	$1215
Tuesday	$1329
Wednesday	$1290
Thursday	$1510
Friday	$1025
Saturday	$1483

The manager computed the average daily receipts to be $1336 for the first 4 days and $1254 for the last 2 days. The manager concluded that the daily average for the 6 days was

$$\frac{1336 + 1254}{2} = 1295$$

The manager was fired. Explain why.

17. (a) Suppose you are at a casino in Atlantic City. On the roll of two dice you bet $1 on the number 7 (see Example 16 of Section 7.2). If a 7 occurs, then you win $4 and get your bet back. If a 7 does not occur, then you lose $1. What is the expected value of this game and what is the house percentage?
(b) On the roll of two dice you bet $2 on the number 2. If a 2 occurs on this roll of the dice, then you win $60 and get your bet back. If a 2 does not occur, then you lose $2. What is the expected value of this game per dollar bet and what is the house percentage?
(c) Would you rather play the game of (a) or the game of (b)? Why?

8.3 VARIANCE OF A RANDOM VARIABLE

The expected value of a finite discrete random variable X gives us information about the long-term average value of X when the experiment is repeated over and over. For many purposes, however, as in the case of Mr. Smith (the lucky winner of the state lottery discussed in Section 8.2), an average does not supply sufficient information. As another example, suppose a pharmaceutical manufacturer wants to buy automatic machinery designed for counting and then packaging 10,000 aspirin tablets in bulk containers. Since no production process is perfectly accurate, sometimes the device may package too many aspirin and sometimes not enough. What criteria might the pharmaceutical firm use to judge the quality of the automatic machinery it intends to buy? One criterion is immediately obvious. Since the machine is designed to package 10,000 aspirin, it seems reasonable to require that over a long period of time the average number of aspirin packaged should be 10,000. More formally, if X denotes the number of aspirin packaged in a container, then we would want the expected value of X to be 10,000. This criterion alone, however, does not adequately describe the quality of the machinery. To see why, suppose machine A packages aspirin according to the pattern

$$10{,}001,\ 9999,\ 10{,}001,\ 9999,\ \ldots$$

and machine B packages aspirin according to the pattern

$$0,\ 20{,}000,\ 0,\ 20{,}000,\ \ldots$$

In both cases, the long-term average number of aspirin packaged is 10,000. However, machine A is obviously preferable to B since it is less variable in its performance. Thus, not only should the pharmaceutical firm look for a machine that packages a long-term average of 10,000 aspirin, but it should also seek a machine for which most of the output is close to the average. In terms of X, the machine should have a high probability of producing values of X close to the expected value. In this section we introduce a way to measure the variability of a finite discrete random variable.

Let X and Y be the random variables whose probability functions are graphed in Figure 1. Since the origin is a center of symmetry for both probability functions, the two random variables have 0 as their expected values (why?); that is,

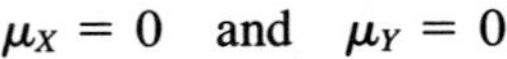

$$\mu_X = 0 \quad \text{and} \quad \mu_Y = 0$$

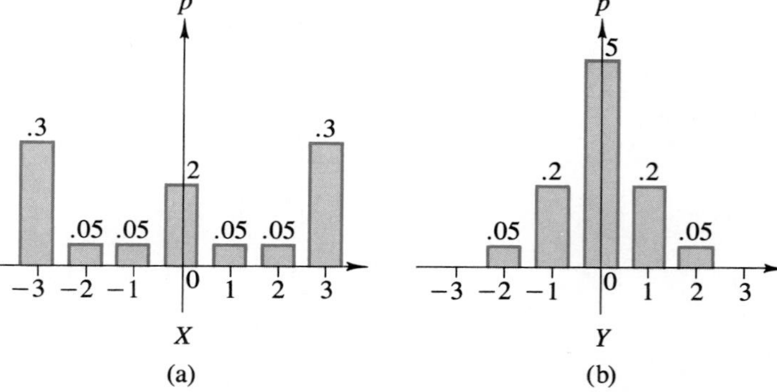

FIGURE 1

Observe, however, that it is more likely for the random variable Y to take on values close to the mean than it is for the random variable X. For example, the random variable Y takes on one of the values -1, 0, or 1 with probability

$$.2 + .5 + .2 = .9$$

whereas the random variable X takes on one of the values -1, 0, or 1 with probability

$$.05 + .2 + .05 = .3$$

Thus, Y takes on values within 1 unit (on either side) of the mean 90% of the time, whereas X takes on values within 1 unit of the mean only 30% of the time. We can summarize this idea loosely by stating that X is more variable about its mean than is Y. Our next objective is to give a numerical way of describing the variability of a finite discrete random variable about its mean.

Let X be a finite discrete random variable with expected value μ. If X takes on the value x when the experiment is performed, then the difference

$$x - \mu$$

called the **deviation of x from μ**, tells us how far the observed value x is from μ. For example, a deviation of $+3$ means x is 3 units to the right of μ and a deviation of -3 means x is 3 units to the left of μ. For most purposes, it is unnecessary and bothersome to worry about the side of μ on which x falls. For this reason as well as others, statisticians prefer to work with the square of the deviation:

$$(x - \mu)^2$$

For example, a squared deviation of 4 tells us that x is 2 units from μ but does not tell whether x is 2 units to the left of μ or 2 units to the right of μ.

Suppose now that X is a finite discrete random variable and assume that we perform the associated experiment repeatedly n times. We observe n values of X, and for each of these n values of X we can compute the squared deviation from the mean μ. If we now average these squared deviations, we can produce a single number that gives us some indication, although exaggerated, of how the data were spread out. The following example illustrates this idea.

EXAMPLE 1 An experiment consists of tossing a fair die and observing the number X showing on the top face. Suppose, after four repetitions, we obtain the values

$$4 \quad 3 \quad 2 \quad 3 \tag{1}$$

Since $\mu_X = 3.5$ (see Example 3 of Section 8.2), the squared deviations from the mean are

$$(4 - 3.5)^2 \qquad (3 - 3.5)^2 \qquad (2 - 3.5)^2 \qquad (3 - 3.5)^2$$

and the average of these squared deviations is

$$\frac{(4 - 3.5)^2 + (3 - 3.5)^2 + (2 - 3.5)^2 + (3 - 3.5)^2}{4}$$

$$= \frac{0.25 + 0.25 + 2.25 + 0.25}{4} = .75$$

On the other hand, if we had obtained the values

$$1 \quad 2 \quad 6 \quad 5 \tag{2}$$

then the average of the squared deviations would be

$$\frac{(1-3.5)^2+(2-3.5)^2+(6-3.5)^2+(5-3.5)^2}{4}$$

$$= \frac{6.25 + 2.25 + 6.25 + 2.25}{4}$$

$$= 4.25$$

Since data set (2) has a larger average squared deviation than data set (1), we describe data set (2) as more *spread out* or more *variable* about the mean than data set (1).

Suppose now that X is a finite discrete random variable associated with a certain experiment. We might ask if it is possible to determine what the average of the squared deviations is after n repetitions of the experiment. Obviously, we cannot determine this value exactly since the values of X obtained in one group of n repetitions need not be the same as the values obtained in another group of n repetitions. Nevertheless, when n is large, it is possible to make a reasonable guess about the average of the squared deviations. To see why, consider a finite discrete random variable whose only possible values are

$$x_1 \quad x_2 \quad x_3$$

Assume also that these values occur with known probabilities

$$p_1 \quad p_2 \quad p_3$$

Suppose that in n repetitions of the experiment,

x_1 occurs f_1 times
x_2 occurs f_2 times
x_3 occurs f_3 times

On computing the squared deviations of these values, we see that

$(x_1-\mu)^2$ occurs f_1 times
$(x_2-\mu)^2$ occurs f_2 times
$(x_3-\mu)^2$ occurs f_3 times

Thus, the average of the squared deviations is

$$\text{average squared deviation} = \frac{f_1(x_1-\mu)^2 + f_2(x_2-\mu)^2 + f_3(x_3-\mu)^2}{n}$$

which can be rewritten as

$$\text{average squared deviation} = \frac{f_1}{n}(x_1-\mu)^2 + \frac{f_2}{n}(x_2-\mu)^2 + \frac{f_3}{n}(x_3-\mu)^2 \tag{3}$$

Since

$$\frac{f_1}{n} \qquad \frac{f_2}{n} \qquad \frac{f_3}{n}$$

represent the proportion of x_1, x_2, and x_3 values in the n repetitions, it follows that when n is large, these proportions are likely to be close to the probabilities

$$p_1 \qquad p_2 \qquad p_3$$

that is, when n is large, the approximations

$$\frac{f_1}{n} \cong p_1 \qquad \frac{f_2}{n} \cong p_2 \qquad \frac{f_3}{n} \cong p_3$$

should be good. Thus, from (3) we obtain the following approximation to the average squared deviation, good when n is large:

$$\text{average squared deviation} \cong p_1(x_1 - \mu)^2 + p_2(x_2 - \mu)^2 + p_3(x_3 - \mu)^2$$

More generally, if X takes on values

$$x_1, x_2, \ldots, x_k$$

with probabilities

$$p_1, p_2, \ldots, p_k$$

then for large n,

$$\text{average squared deviation} \cong p_1(x_1 - \mu)^2 + p_2(x_2 - \mu)^2 + \cdots + p_k(x_k - \mu)^2 \tag{4}$$

EXAMPLE 2 An experiment consists of tossing a fair die and observing the number X showing on the top face. Use the approximation in (4) to estimate the average squared deviation of the observed values if the experiment is repeated many times.

Solution The possible values of X are

$$x_1 = 1 \qquad x_2 = 2 \qquad x_3 = 3 \qquad x_4 = 4 \qquad x_5 = 5 \qquad x_6 = 6$$

and since the die is fair, each of these values has probability $\frac{1}{6}$ of occurring; that is,

$$p_1 = \frac{1}{6} \qquad p_2 = \frac{1}{6} \qquad p_3 = \frac{1}{6} \qquad p_4 = \frac{1}{6} \qquad p_5 = \frac{1}{6} \qquad p_6 = \frac{1}{6}$$

Moreover, from Example 3 of Section 8.2, we have $\mu = 3.5$. Substituting these values in (4) we obtain the approximation

$$\begin{aligned}\text{average squared deviation} &\cong \frac{1}{6}(1 - 3.5)^2 + \frac{1}{6}(2 - 3.5)^2 + \frac{1}{6}(3 - 3.5)^2 \\ &\quad + \frac{1}{6}(4 - 3.5)^2 + \frac{1}{6}(5 - 3.5)^2 + \frac{1}{6}(6 - 3.5)^2 \\ &= \frac{17.5}{6} \cong 2.92\end{aligned}$$

The quantity on the right side of (4) is of such importance that it has its own notation and name:

Variance If the values

$$x_1, x_2, \ldots, x_k$$

of a finite discrete random variable X occur with probabilities

$$p_1, p_2, \ldots, p_k$$

then the **variance of X**, denoted by $\text{Var}(X)$, is defined by

$$\text{Var}(X) = p_1(x_1 - \mu)^2 + p_2(x_2 - \mu)^2 + \cdots + p_k(x_k - \mu)^2 \tag{5}$$

where μ is the expected value of X.

In other words, *the variance of X is the sum of the squared deviations of the possible values of X times the probabilities of their occurrence.*

EXAMPLE 3 If X is the random variable in the die-tossing experiment of Example 2, then the computations in that example show that

$$\text{Var}(X) = \frac{17.5}{6} \cong 2.92$$

Since the variance formula involves the squares of the deviations, the units of $\text{Var}(X)$ are the squares of the units of X. For example, if the values of X represent feet, then the units of $\text{Var}(X)$ would be square feet. Many people prefer to have the variability of X about its mean described by a number having the same units as X. This can be accomplished by taking the positive square root of $\text{Var}(X)$. The resulting number, called the **standard deviation of X**, is denoted either by σ_X or just σ when the random variable is evident (σ is the lowercase Greek letter *sigma*, pronounced "sigma").

Standard Deviation If $\text{Var}(X)$ is denoted by σ_X^2, then the standard deviation of X is

$$\sigma_X = \sqrt{\text{Var}(X)}$$

EXAMPLE 4 The standard deviation of the random variable X in Example 2 is

$$\sigma_X = \sqrt{\text{Var}(X)} = \sqrt{\frac{17.5}{6}} \cong 1.71$$

The computation of $\text{Var}(X)$ can be simplified by rewriting Equation (5) in an alternative form. To see how, consider a random variable X that takes on values

$$x_1 \quad x_2 \quad x_3$$

with probabilities

$$p_1 \quad p_2 \quad p_3$$

Equation (5) states

$$\text{Var}(X) = p_1(x_1 - \mu)^2 + p_2(x_2 - \mu)^2 + p_3(x_3 - \mu)^2 \tag{6}$$

If we expand the squares, the terms on the right side of this equation become

$$\begin{aligned} p_1(x_1 - \mu)^2 &= p_1x_1^2 - 2p_1x_1\mu + p_1\mu^2 \\ p_2(x_2 - \mu)^2 &= p_2x_2^2 - 2p_2x_2\mu + p_2\mu^2 \\ p_3(x_3 - \mu)^2 &= p_3x_3^2 - 2p_3x_3\mu + p_3\mu^2 \end{aligned}$$

Summing up both aides of these equations and using (6), we obtain

$$\begin{aligned} \text{Var}(X) = (p_1x_1^2 + p_2x_2^2 + p_3x_3^2) - 2\mu(p_1x_1 + p_2x_2 + p_3x_3) \\ + \mu^2(p_1 + p_2 + p_3) \end{aligned} \tag{7}$$

But

$$p_1x_1 + p_2x_2 + p_3x_3 = \mu$$

and

$$p_1 + p_2 + p_3 = 1 \qquad \text{(why?)}$$

Substituting these relations in (7) yields

$$\text{Var}(X) = (p_1x_1^2 + p_2x_2^2 + p_3x_3^2) - 2\mu^2 + \mu^2$$

or

$$\text{Var}(X) = (p_1x_1^2 + p_2x_2^2 + p_3x_3^2) - \mu^2$$

More generally, if X takes on values

$$x_1, x_2, \ldots, x_k$$

with probabilities

$$p_1, p_2, \ldots, p_k$$

then the formula for variance can be written in the following form.

Alternate Formula for Variance

$$\text{Var}(X) = (p_1x_1^2 + p_2x_2^2 + \cdots + p_kx_k^2) - \mu^2 \tag{8}$$

Note that this formula involves only one subtraction, whereas Formula (5) involves k subtractions. This is one reason why Formula (8) is preferable for numerical computations.

EXAMPLE 5 Use Formula (8) to compute the variance of the random variable X in the die-tossing experiment of Example 2.

Solution In solving this problem, we arrange our computations in a format you may wish to follow whenever a variance must be computed. See Table 1. To construct this table, we have listed the values of X in the first column and the probabilities of these values in the second column. We have multiplied each x_i in column 1 by the corresponding p_i in column 2 to obtain the third column. Finally, we multiplied each $p_i x_i$ in column 3 by the corresponding x_i from column 1 to obtain $p_i x_i^2$ in column 4. Observe that the sum of the entries in column 3 is

$$\mu = p_1 x_1 + p_2 x_2 + p_3 x_3 + p_4 x_4 + p_5 x_5 + p_6 x_6 = \frac{21}{6}$$

TABLE 1

x_i	p_i	$p_i x_i$	$p_i x_i^2$
1	$\frac{1}{6}$	$\frac{1}{6}$	$\frac{1}{6}$
2	$\frac{1}{6}$	$\frac{2}{6}$	$\frac{4}{6}$
3	$\frac{1}{6}$	$\frac{3}{6}$	$\frac{9}{6}$
4	$\frac{1}{6}$	$\frac{4}{6}$	$\frac{16}{6}$
5	$\frac{1}{6}$	$\frac{5}{6}$	$\frac{25}{6}$
6	$\frac{1}{6}$	$\frac{6}{6}$	$\frac{36}{6}$
		Sum $= \frac{21}{6}$	Sum $= \frac{91}{6}$

[See Formula (9) in Section 8.2.] The sum of the entries in column 4 is the first term on the right side of Formula (8) of this section. Thus, we obtain

$$\text{Var}(X) = \frac{91}{6} - \left(\frac{21}{6}\right)^2 = \frac{105}{36} \cong 2.92$$

which agrees with the result obtained in Example 3.

EXAMPLE 6 Let X be the random variable in the resistor sampling problem of Example 7 of Section 8.1. Find the variance and standard deviation of X.

Solution As shown in Example 7 of Section 8.1, X takes on values

$$0 \qquad 1 \qquad 2$$

with probabilities

$$.9025 \qquad .0950 \qquad .0025$$

so its variance can be computed using Table 2 and Formula (8). The variance is

$$\begin{aligned} \sigma_X^2 = \text{Var}(X) &= (0.1050) - (0.1000)^2 \\ &= 0.1050 - 0.0100 \\ &= 0.0950 \end{aligned}$$

and the standard deviation is

$$\sigma_X = \sqrt{0.0950} \cong 0.3082$$

TABLE 2

x_i	p_i	$p_i x_i$	$p_i x_i^2$
0	.9025	0.0000	0.0000
1	.0950	0.0950	0.0950
2	.0025	0.0050	0.0100
		Sum = 0.1000	Sum = 0.1050

EXERCISE SET 8.3

1. Let X be the random variable whose probability function is described in the table. Compute the mean and variance of X.

Value of X	1	2	3	4
Probability	.2	.4	.3	.1

In Exercises 2–4 compute the mean, variance, and standard deviation of the random variable whose probability function is described in the given table. (A table of square roots appears as Appendix Table C.4.)

2.

Value of X	−2	−3	2	3
Probability	.3	.2	.3	.2

3.

Value of X	0.01	0.25	0.20	0.90	0.40
Probability	.4	.2	.2	.1	.1

4.

Value of X	4000	2500	3000	9100	2600
Probability	.8	.01	.01	.02	.16

5. Find the mean and variance of the random variable X in Exercise 2 of Section 8.1.

6. Find the mean and variance of the random variable X in Exercise 3 of Section 8.1.

7. Find the mean and variance of the random variable X in Exercise 4 of Section 8.1.

8. Compute the mean and variance of the random variable whose probability function is graphed in Figure 2.

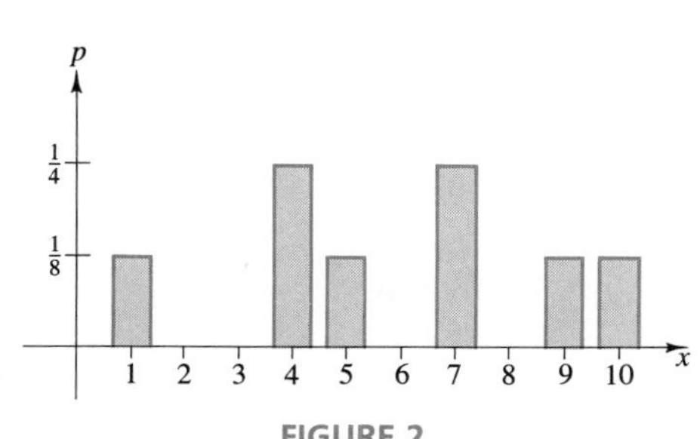

FIGURE 2

9. In Figure 3 we have graphed probability functions for random variables X and Y. By *inspection* (no computations), determine which variable has the larger variance.

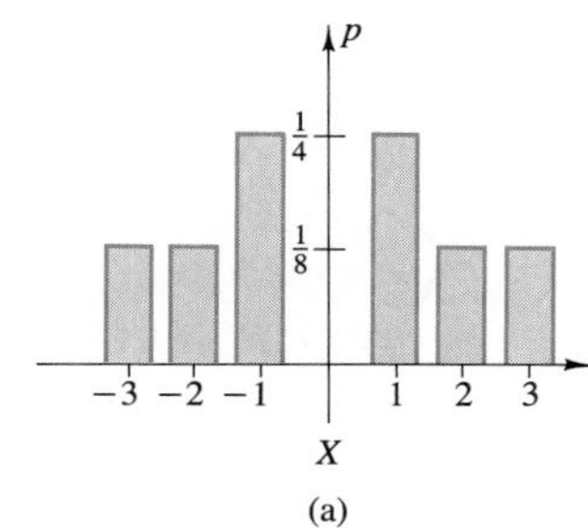

FIGURE 3

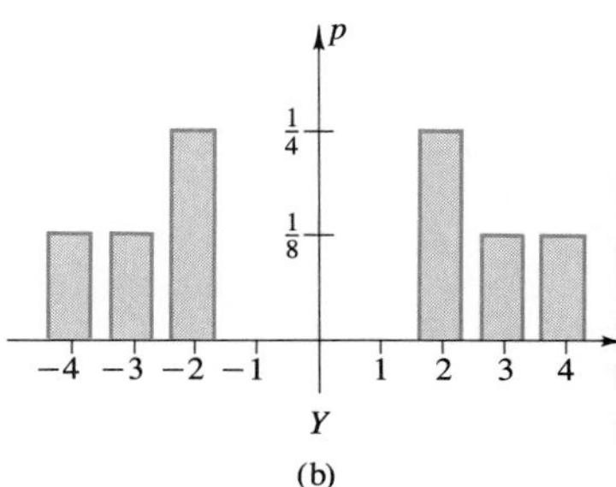

10. In Figure 4 we have graphed probability functions for random variables X and Y. By *inspection* (no computations), determine which variable has the larger variance.

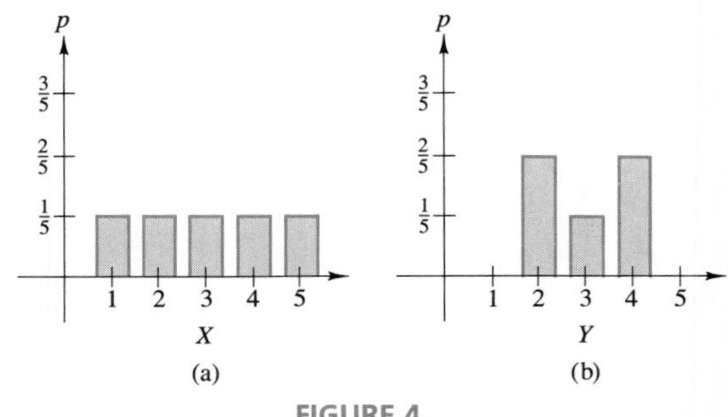

FIGURE 4

11. Company A is accused by the Department of Labor of not paying its employees a fair wage. The president of the company notes that the average salary paid to its employees is greater than the average salary paid to the employees of company B, which the Department of Labor said is paying a fair wage to its employees. Should the Department of Labor accept the argument as proof that company A is treating its employees fairly?

8.4 CHEBYSHEV'S INEQUALITY: APPLICATIONS OF MEAN AND VARIANCE (OPTIONAL)

In the previous section we discuss criteria that a pharmaceutical firm might consider when planning to purchase automatic machinery for counting and packaging 10,000 aspirin tablets in bulk containers. If X denotes the number of tablets packaged in a container, we observe that the firm would want

1. the mean of X to be 10,000

2. the probability of obtaining values of X close to the mean to be high

In this section we develop some mathematical techniques for investigating the second of these criteria.

Every time the automatic machinery produces a value of X greater than the mean of 10,000, the pharmaceutical firm loses money since it is giving the customer more aspirin than the customer paid for. By the same token, every time the machinery produces a value of X less than 10,000, the firm risks losing the customer since it has not given the customer all the aspirin expected. Thus a value of X too far from the mean, on either side, is undesirable. Recognizing that no machinery is perfect, the pharmaceutical firm may settle on a certain error that it feels is tolerable. For example, the firm's marketing staff might decide that an error of 10 aspirins in either direction would not be too detrimental; that is, overpackaging by 10 aspirin would not be too costly and underpackaging by 10 aspirin would still be good enough to satisfy the customer. Thus, the pharmaceutical firm will be content if the machinery it buys always produces values of X that lie within 10 units of the mean, that is, values of X satisfying

$$9990 \le X \le 10{,}010 \tag{1}$$

Unfortunately, even this relaxed requirement may be too stringent. Occasionally, for example, aspirin may chip and disturb the counting mechanism or still other unforeseen problems may occur. Recognizing this fact, the firm may relax its requirements still further and require only that the conditions of (1) hold most of the time, for example, at least 95% of the time. This requirement can be formulated mathematically as follows. If X denotes the number of aspirin packed in a container and if

$$P(9990 \le X \le 10{,}010)$$

denotes the probability that X is between 9990 and 10,010, inclusive, then this probability should be at least .95; that is,

$$P(9990 \le X \le 10{,}010) \ge .95 \tag{2}$$

We next show that once the mean μ and variance σ^2 of the random variable X are known,* it is possible to devise a mathematical procedure for determining if a criterion like (2) is satisfied. But first we must develop some preliminary ideas.

Let X be a random variable with mean μ and standard deviation σ. Statisticians often find it convenient to measure distance from the mean in terms of **standard deviation units** or σ-units. For example, the point A in Figure 1a can be described as 2 σ-units to the right of the mean μ and the point B as 3.5 σ-units to the left of the mean. In general, to say that a point is h σ-units from μ (Figure 1b) means that its distance from μ is $h\sigma$.

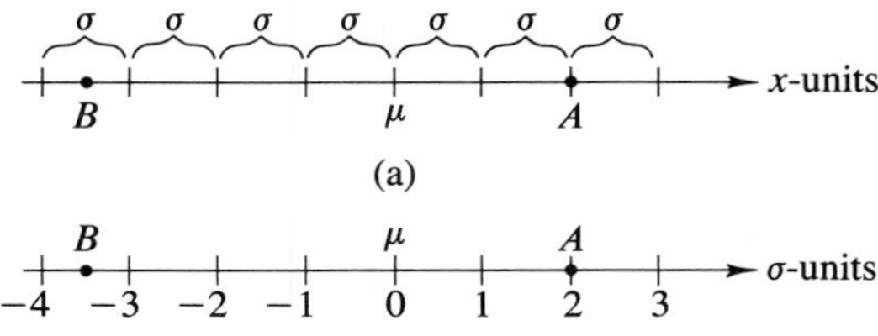

FIGURE 1

EXAMPLE 1 Let X be a random variable with mean $\mu = 3$ and variance $\sigma^2 = 25$. In terms of σ-units, how far to the right of μ is the point $x = 9$?

Solution The distance between x and μ is

$$x - \mu = 9 - 3 = 6$$

Since $\sigma = 5$, this distance in terms of σ-units is

$$\frac{x - \mu}{\sigma} = \frac{6}{5} = 1.2 \ \sigma\text{-units}$$

The following result, named in honor of P. L. Chebyshev,† shows that once the mean and variance of a random variable X are known, it is possible to estimate the chances that a value of X will fall within a specified distance of the mean.

> **Chebyshev's Theorem** For *any* random variable X having mean μ and standard deviation σ, the probability that this random variable will take on a value within h σ-units of the mean μ is at least $1 - (1/h^2)$.

* Methods of estimating μ and σ statistically are discussed in Section 8.7.

† *Pafnuti Lvovich Chebyshev* (1821–1894) was one of the most distinguished Russian mathematicians. He served most of his life as a professor of mathematics at the University of St. Petersburg. In addition to his work in probability and statistics, he made important contributions to the field of number theory, solved basic problems in the construction of geographical maps, and devised a way for mechanically linking bars to obtain rectilinear motion.

Note that this result depends only on the mean and standard deviation of X, not on the specific distribution of X.

Symbolically, Chebyshev's theorem states that for *any* random variable X, we have

$$P(\mu - h\sigma \leq X \leq \mu + h\sigma) \geq 1 - \frac{1}{h^2}$$

This is called **Chebyshev's inequality**.

We omit the proof of this important result and concentrate, instead, on some of its applications.

EXAMPLE 2 From Chebyshev's theorem, the probability that a random variable takes on a value within 2 σ-units of the mean (i.e., $h = 2$) is at least

$$1 - \frac{1}{h^2} = 1 - \frac{1}{4} = .75$$

This result has the following geometric interpretation. Let X be *any* random variable with mean μ and standard deviation σ. Suppose, as shown in Figure 2, we construct an interval extending a distance 2σ on each side of μ. If we observe many values of X, then we expect at least 75% of these values to fall in the constructed interval and consequently no more than 25% of the values to fall outside the interval. For example, if X is a random variable with mean 10 and standard deviation .1, then the probability is at least .75 that X takes on a value between 9.8 and 10.2. Thus, if we observe a large number of X values, 75% of these values can be expected to fall between 9.8 and 10.2.

FIGURE 2

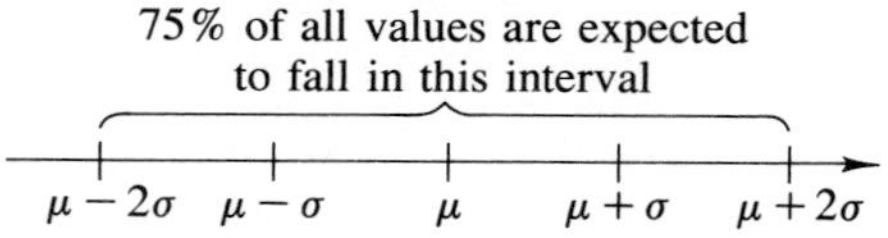

75% of all values are expected to fall in this interval

$\mu - 2\sigma$ $\mu - \sigma$ μ $\mu + \sigma$ $\mu + 2\sigma$

EXAMPLE 3 Let X be a random variable with mean $\mu = 100$ and variance $\sigma^2 = 4$. If we are going to observe many X values and if we want to state that at least 96% of the observed values are likely to fall between $100 - k$ and $100 + k$, what value should we take for k? Use Chebyshev's theorem to obtain the answer.

Solution To ask that at least 96% of the observed values fall between $100 - k$ and $100 + k$ is equivalent to asking that each observed value fall between $100 - k$ and $100 + k$ with probability at least .96.

If we choose h in Chebyshev's theorem so that

$$1 - \frac{1}{h^2} = .96 \tag{3}$$

then, since $\mu = 100$, X will fall between $100 - h\sigma$ and $100 + h\sigma$ with probability of at least .96. To solve (3) for h^2 we multiply both sides by h^2 to obtain

$$h^2 - 1 = .96h^2$$

or

$$.04h^2 = 1$$

or

$$h^2 = 25$$

Taking positive square roots gives $h = 5$. Thus, the probability is at least .96 that X falls between

$$100 - 5\sigma \quad \text{and} \quad 100 + 5\sigma$$

Since $\sigma = 2$, we obtain

$$k = 5\sigma = 10$$

Therefore, 96% of the values fall between 90 and 110.

Again, observe that Chebyshev's theorem applies to *all* random variables for which the mean and variance are known. Since the theorem has such broad applicability, it is natural to expect that for some random variables the estimate $1 - (1/h^2)$ may be rather poor. It is in fact the case that for many random variables, the probability is much greater than $1 - (1/h^2)$ of obtaining a value within h σ-units of the mean. For this reason, statisticians usually regard the estimate $1 - (1/h^2)$ as very conservative.

EXAMPLE 4 **LIGHT BULB LIFE** A certain high-quality commercial light bulb has an expected lifetime of $\mu = 750$ hours with a variance $\sigma^2 = 100$.

(a) Estimate the probability that one of these light bulbs will live between 735 and 765 hours.

(b) If 20,000 of these bulbs are installed in city streetlights, how many bulbs do you estimate would need replacement between 735 and 765 hours of use?

Solution (a) As shown in Figure 3, the time interval from 735 to 765 hours extends 15 units to the left of the mean $\mu = 750$ and 15 units to the right of the mean. Thus, we are interested in finding the probability that the lifetime is within 15 units of the mean. To apply Chebyshev's theorem, we must express 15 in terms of σ-units. Since $\mu^2 = 100$, we have $\sigma = \sqrt{100} = 10$, so 15 units is

FIGURE 3

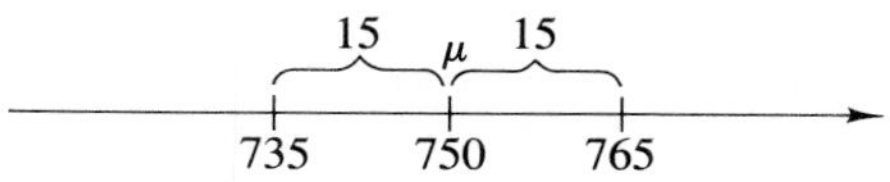

$$h = \frac{15}{\sigma} = \frac{15}{10} = \frac{3}{2} \ \sigma\text{-units}$$

By Chebyshev's theorem the probability that the lifetime is within $h = \frac{3}{2}$ σ-units of the mean is at least

$$1 - \frac{1}{h^2} = 1 - \frac{1}{(3/2)^2} = \frac{5}{9}$$

Solution (b) From (a) the probability is at least $\frac{5}{9}$ that one of the bulbs will burn out between 735 and 765 hours; therefore, we would expect at least

$$\left(\frac{5}{9}\right)(20{,}000) \cong 11{,}111 \text{ bulbs}$$

to need replacement during the given time interval.

EXERCISE SET 8.4

1. Let X be a random variable with mean $\mu_X = 0$ and variance $\sigma_X^2 = 16$. In each part express the distance between a and μ_X in terms of σ-units.
(a) $a = 4$ (b) $a = 8$ (c) $a = 6$
(d) $a = -11$

2. Let X be a random variable with mean $\mu_X = 5$ and variance $\sigma_X^2 = 4$. In each part express the distance between a and μ in terms of σ-units.
(a) $a = 7$ (b) $a = 11$ (c) $a = 2$
(d) $a = 0$

3. Let X be a random variable with mean $\mu_X = 10$ and variance $\sigma_X^2 = 9$. In each part the distance between a point a and the mean μ_X is expressed in terms of σ-units. Find the value of a, where:
(a) a is 2 σ-units to the right of μ_X
(b) a is 3.5 σ-units to the left of μ_X
(c) a is 4.7 σ-units to the right of μ_X
(d) a is 0.3 σ-units to the left of μ_X

4. Let X be a random variable with mean $\mu_X = 0$ and variance $\sigma_X^2 = 4$. Use Chebyshev's theorem to estimate the probability that X takes on a value within:
(a) 1 unit of the mean
(b) 2 units of the mean
(c) 3 units of the mean

5. Let X be a random variable with mean $\mu_X = 12$ and variance $\sigma_X^2 = 9$. Use Chebyshev's theorem to estimate the probability that:
(a) X takes on a value between 9 and 15
(b) X takes on a value between 11 and 13
(c) X takes on a value between 0 and 24

6. Let X be a random variable with mean $\mu_X = 10$ and variance $\sigma_X^2 = 4$. If we are going to observe many X values and if we want to state that at least 90% of the observed values are likely to fall between $10 - k$ and $10 + k$, what value should we take for k? Use Chebyshev's theorem to obtain the answer.

7. A camera used in a weather satellite has an expected lifetime of $\mu = 900$ days with a variance of $\sigma^2 = 81$. Use Chebyshev's theorem to estimate the probability that this camera will live between 870 and 930 days.

8. Heavy-duty resistors have an expected lifetime of 5 years with a variance of 0.16.
(a) If 3000 such resistors are installed, how many do you estimate will last between 3 and 7 years?
(b) If 3000 such resistors are installed, how many do you estimate will last between 2 and 8 years?

9. When a raw, 10-carat diamond is cut and polished, the finished diamond has an expected weight of 6.3 carats with a variance of .04. Estimate the probability that a raw, 10-carat diamond will yield a finished diamond with a weight between 5.5 and 7.1 carats.

10. A manufacturer of an automatic device for filling bottles with 16 ounces of soda claims that its machine will fill bottles to a mean of 16 ounces of soda with a variance of 0.04. If a bottling plant wants a device that will fill bottles to between 15.5 and 16.5 ounces at least 70% of the time, will this device meet its needs? Assume the manufacturer's claims are correct and base your conclusion on Chebyshev's inequality.

11. Consider the automatic device described in Exercise 10. For what value of k can the manufacturer claim that his machine will fill bottles to between $16 - k$ and $16 + k$ ounces at least 90% of the time?

12. Let X be a random variable whose probability function is described by

Value of X	1	2	3	4	5
Probability	$\frac{1}{5}$	$\frac{1}{5}$	$\frac{1}{5}$	$\frac{1}{5}$	$\frac{1}{5}$

(a) Find μ_X.
(b) Find σ_X.
(c) Use Chebyshev's inequality to estimate the probability $P(2 \leq X \leq 4)$.
(d) Find the exact value of $P(2 \leq X \leq 4)$.
(e) Are the results obtained in (c) and (d) contradictory?

13. A manufacturer of printed circuits produces an average of 1 defective circuit out of 300. It promises a customer to take back a shipment of 3000 boards and to pay a penalty of $700 if there are more than x defective boards in the lot. What should x be so that the manufacturer can be 90% sure that it will not have to take back the shipment and pay the penalty? Assume $\mu = 10$ and $\sigma^2 = 10$.

8.5 BINOMIAL RANDOM VARIABLES

In the previous section we saw how Chebyshev's theorem could be used to *estimate* certain probabilities when the mean and variance of a random variable were known. In cases where it is possible to find the probability function for the random variable, probabilities can, in theory, be computed exactly so that approximations often become unnecessary. In this section we discuss an important class of problems where probability functions are readily obtainable.

In many experiments we are primarily interested in whether a certain result does or does not occur. For example, does a tossed coin show a head or not, does an individual innoculated with a flu vaccine contract flu or not, is an item selected from the output of a production process defective or not, is a newborn child a girl or not.

Experiments for which there are only two possible outcomes are called **Bernoulli* experiments** or sometimes **Bernoulli trials.** It is traditional in statistical work to arbitrarily label one of the outcomes *success* and the other *failure*. For example, if we toss a coin we might call a head a success and a tail a failure. The selection of the outcome to be labeled success is completely arbitrary and need not connote a successful outcome in everyday life. For example, an investigator studying the effect of a new drug for the prevention of malignant tumors in rats may possibly choose to call the appearance of a tumor a "success" even though one would not ordinarily think of this outcome as successful.

Experiments with more than two outcomes can, if desired, be viewed as Bernoulli experiments by grouping the outcomes into two categories. For example, if we toss a die, there are six possible outcomes:

$$1 \quad 2 \quad 3 \quad 4 \quad 5 \quad 6$$

However, if we call the outcome 3 a success and call the outcomes 1, 2, 4, 5, and 6 failures, then we can view this as a Bernoulli experiment with only two outcomes, success and failure.

In a Bernoulli experiment, we use $p = P(s)$ for the probability of a success and $q = P(f)$ for the probability of a failure. Since success and failure are the

---

**Jakob Bernoulli* (1654–1705) was one of a family of distinguished Swiss mathematicians. He was born in Basel, Switzerland, and studied theology at the insistence of his father. Later, he refused a church appointment and began lecturing on experimental physics at the University of Basel. His work included studies in astronomy, the motion of comets, and applications of (the newly invented) calculus.

only two outcomes, we must have

$$p + q = 1$$

or equivalently

$$q = 1 - p$$

EXAMPLE 1 Toss a fair die and let success $= 3$ and failure $= 1, 2, 4, 5,$ or 6. Then,

$$p = \frac{1}{6} \quad \text{and} \quad q = \frac{5}{6}$$

Many experiments consist of a sequence of Bernoulli trials. For example, a single toss of a coin is a Bernoulli trial and an experiment consisting of five tosses of a coin is a sequence of five Bernoulli trials. Examine the following problem:

Problem If a Bernoulli experiment is repeated n times, what is the probability of obtaining exactly x successes in the n repetitions?

This problem is most easily solved when we assume that the trials are *independent*. Physically, this means we are assuming that the occurrence of a success s or failure f in one trial has no effect on the chances of success or failure in any other trial. Mathematically, independence allows us to multiply probabilities. For example, if we have three independent repetitions of a Bernoulli experiment, where the probability of success on each trial is p, then the probability of the outcome

$$ssf$$

denoted $P(ssf)$, would be

$$P(ssf) = P(s)P(s)P(f) = ppq = p^2q$$

and the probability of the outcome fff would be

$$P(fff) = P(f)P(f)P(f) = qqq = q^3$$

Suppose now, we want the probability of obtaining exactly two successes in the three trials (see the tree diagram in Figure 1; the paths indicating these cases are in color). Let the random variable X equal the number of successes in three trials.

FIGURE 1

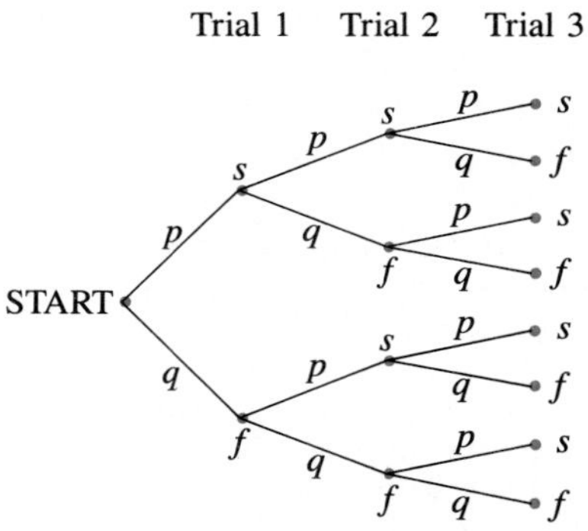

From Figure 1,

$$P(X = 0) = q^3 \qquad P(X = 2) = 3p^2q$$
$$P(X = 1) = 3pq^2 \qquad P(X = 3) = p^3 \tag{1}$$

Therefore, the probability of exactly two successes in three trials is

$$P(X = 2) = 3p^2q$$

The following examples further illustrate this idea.

EXAMPLE 2 A fair die is tossed three times. What is the probability of obtaining exactly two 4s in the three tosses?

Solution On each toss of the die, let success indicate that a 4 is tossed and failure that a 4 is not tossed. Thus,

$$p = \text{probability of success} = \frac{1}{6}$$

$$q = \text{probability of failure} = \frac{5}{6}$$

Since the tosses are independent, we may use the results of (1) where the random variable X is the number of successes in three trials. Thus,

$$P(\text{exactly two 4s}) = P(X = 2) = 3p^2q = 3\left(\frac{1}{6}\right)^2\frac{5}{6} = \frac{15}{216} = \frac{5}{72}$$

EXAMPLE 3 In the die-tossing experiment of the previous example, find the probability of obtaining no 4s, obtaining exactly one 4, and obtaining three 4s.

Solution Assuming independent tosses and again letting the random variable X = the number of successes in three trials,

$$P(\text{no 4s}) = P(X = 0) = q^3 = \left(\frac{5}{6}\right)^3 = \frac{125}{216}$$

$$P(\text{exactly one 4}) = P(X = 1) = 3pq^2 = 3\left(\frac{1}{6}\right)\left(\frac{5}{6}\right)^2 = \frac{75}{216} = \frac{25}{72}$$

$$P(\text{three 4s}) = P(X = 3) = p^3 = \left(\frac{1}{6}\right)^3 = \frac{1}{216}$$

The problem described in Example 3 is easily solved with the aid of the tree diagram of Figure 1, and (1) can be used to determine the probability of X successes in three independent trials for any problem with three independent Bernoulli trials. If there are more than three trials, then the tree diagram can be extended. However, the tree becomes unwieldy, so we must explore another method of obtaining the above results.

Note that to obtain exactly two successes in three trials we have the cases *ssf*, *sfs*, and *fss*. In other words, there are three cases with exactly two successes out of the three trials, or three different ways of choosing two places out of three for the successes to occur. For each arrangement, the probability of its occurrence is p^2q. By the counting methods of Chapter 6, there are $C_{3,2}$ ways of making such a choice. Consequently,

$$P(X = 2) = C_{3,2}p^2q = \frac{3!}{2!1!}p^2q = 3p^2q$$

Similarly,

$$P(X = 0) = C_{3,0}q^3 = \frac{3!}{0!3!}q^3 = 1q^3 = q^3$$

$$P(X = 1) = C_{3,1}pq^2 = \frac{3!}{1!2!}pq^2 = 3pq^2$$

$$P(X = 3) = C_{3,3}p^3 = \frac{3!}{3!0!}p^3 = 1p^3 = p^3$$

This method generalizes to cases when the number of trials is larger.

EXAMPLE 4 Assuming the probability of success in a single trial of a Bernoulli experiment is p, find the probability of exactly four successes in 10 independent repetitions of this experiment.

Solution One way of obtaining exactly four successes is

$$ssssffffff$$

Another way is

$$ssffsfsfff$$

and still another way is

$$fffssfffss$$

Observe that each of these results is an arrangement of 4 s's and 6 f's. Obviously, the number of different ways to obtain four successes is just the number of different ways to arrange 4 s's and 6 f's. It is evident that any such arrangement is completely determined once we specify the four positions where the s's occur. Thus the number of arrangements with 4 s's and 6 f's is just the number of ways of selecting four positions for the s's from 10 possibilities. Therefore, the number of such arrangements is

$$C_{10,4} = \frac{10!}{4!6!} = 210$$

Consequently,

$$P(\text{exactly 4 successes}) = \underbrace{P(ssssffffff) + P(ssffsfsfff) + \cdots}_{210\text{ terms}}$$

Since each probability on the right side has value p^4q^6 (why?), we obtain

$$P(\text{exactly 4 successes}) = 210p^4q^6$$

We are now in a position to solve the following general problem.

Problem If the probability of success in a certain Bernoulli experiment is p, what is the probability of obtaining exactly x successes in n independent repetitions of the experiment?

One way of obtaining exactly x successes in the n repetitions is to have the x successes immediately followed by $(n - x)$ failures, that is,

$$\underbrace{sss \cdots s}_{x \text{ successes}}\underbrace{fff \cdots f}_{(n-x) \text{ failures}} \tag{2}$$

The other ways of obtaining x successes can be obtained by rearranging these x s's and $(n - x)f$'s. As illustrated in Example 4 there are $C_{n,x}$ different arrangements of these letters. Since each of these arrangements has x s's and $(n - x)f$'s, the probability of each arrangement is

$$p^xq^{n-x}$$

for example,

$$\begin{aligned} P(\underbrace{sss \cdots s}_{x \text{ successes}}\underbrace{fff \cdots f}_{(n-x) \text{ failures}}) &= \underbrace{P(s)P(s)P(s) \cdots P(s)}_{x \text{ factors}}\underbrace{P(f)P(f)P(f) \cdots P(f)}_{(n-x) \text{ factors}} \\ &= \underbrace{ppp \cdots p}_{x \text{ factors}}\underbrace{qqq \cdots q}_{(n-x) \text{ factors}} \\ &= p^xq^{n-x} \end{aligned}$$

Thus:

> The probability of exactly x successes in n independent repetitions of a Bernoulli experiment with success probability p is given by
>
> $$P(\text{exactly } x \text{ successes}) = C_{n,x}p^xq^{n-x} \tag{3}$$

EXAMPLE 5 Find the probability of obtaining exactly four successes in 10 independent repetitions of a Bernoulli experiment if the probability of success on each trial is .6.

Solution Applying Formula (3) with

$$n = 10 \qquad p = .6 \qquad q = .4 \qquad x = 4$$

we obtain

$$\begin{aligned} P(\text{exactly 4 successes}) &= C_{10,4}(.6)^4(.4)^6 \\ &= \frac{10!}{4!\,6!}(.6)^4(.4)^6 \\ &\cong .111 \end{aligned}$$

An experiment that consists of n independent repetitions of a Bernoulli experiment is called a **binomial experiment**. If the random variable X denotes the number of successes in a binomial experiment, then X is called a **binomial random variable**.

As evidenced by Example 5, it can be a tedious job to compute the probabilities of a binomial random variable. In applications where many binomial probabilities must be computed, computers are usually used. However, in problems where only a few binomial probabilities are needed, the cost of using a computer is usually unwarranted, so statisticians often resort to ready-made tables.

In Table C.2 in Appendix C we list the binomial probabilities, accurate to three decimal places, for certain values of n and p. These tables are somewhat incomplete, so in practical problems you may have to consult more extensive tables.* To illustrate how these tables are used, we use them to solve the problem in Example 6.

EXAMPLE 6 Use Appendix Table C.2 to find the probability of obtaining four successes in 10 independent repetitions of a Bernoulli experiment if the probability of success on each trial is .6.

Solution The answer can be found as follows:

Step 1 Find the section of the table for which $n = 10$ (see Table 1).

TABLE 1

		p													
n	x	.01	.05	.10	.20	.30	.40	.50	.60	.70	.80	.90	.95	.99	x
10	0	.904	.599	.349	.107	.028	.006	.001	.000	.000	.000	.000	.000	.000	0
	1	.091	.315	.387	.268	.121	.040	.010	.002	.000	.000	.000	.000	.000	1
	2	.004	.075	.194	.302	.233	.121	.044	.011	.001	.000	.000	.000	.000	2
	3	.000	.010	.057	.201	.267	.215	.117	.042	.009	.001	.000	.000	.000	3
	4	.000	.001	.011	.088	.200	.251	.205	.111	.037	.006	.000	.000	.000	4
	5	.000	.000	.001	.026	.103	.201	.246	.201	.103	.026	.001	.000	.000	5

Step 2 Locate the column labeled $p = .60$.

Step 3 Look down the column to the block in the row labeled $x = 4$. We obtain the number .111, which is the probability we want.

EXAMPLE 7 Toss a fair coin 12 times. Find the probability of obtaining at least 9 heads in the 12 tosses.

Solution If X denotes the number of heads in the 12 tosses, then we are interested in the probability that $X \geq 9$ or equivalently the probability that X has one

* National Bureau of Standards, *Tables of the Binomial Probability Distribution*, Applied Mathematics Series, **6** (1950) ($n = 1$ to $n = 49$). Harry G. Romig. *Binomial Tables* (New York: Wiley, 1953) ($n = 50$ to $n = 100$).

of the values 9, 10, 11, or 12. The probability of a head on each toss is .5. From Appendix Table C.2 we obtain

$$\begin{aligned} P(X \geq 9) &= P(X = 9) + P(X = 10) + P(X = 11) + P(X = 12) \\ &= .054 + .016 + .003 + .000 \\ &= .073 \end{aligned}$$

The following are a number of examples that illustrate how binomial random variables may arise in practical problems. In each of these examples, we assume that the trials are independent.

EXAMPLE 8 **RADIOACTIVE DECAY** Physical evidence shows that the probability a given type of atom will split in a 24-hour period is $1/10^{24}$. In a sample containing 10^{30} of these atoms, the number X of atoms that will split in a 24-hour period is a binomial random variable.

To see this, imagine that each atom in turn is observed for a 24-hour period and let success indicate that the atom splits. Thus, X represents the number of successes in 10^{30} repetitions of a Bernoulli experiment, where the probability of success on each trial is $1/10^{24}$.

EXAMPLE 9 **GENETIC MUTATIONS** A generation of 2500 fruit fly larvae is subjected to radiation to induce mutations. Genetic evidence suggests that the probability of a mutation under the given conditions is .001. The number X of mutations is a binomial random variable.

To see this, imagine that each larva in turn is observed for evidence of mutation and let success denote that a mutation occurs. Thus, X is the number of successes in 2500 repetitions of a Bernoulli experiment, where the probability of success on each trial is .001.

EXAMPLE 10 **DIVORCE PROCEEDINGS** In a certain state, "mental cruelty" is the legal reason given in 4 of 5 divorce proceedings. If 25 divorce cases are selected at random, then the number X of cases based on mental cruelty is a binomial random variable.

To see this, imagine that each case in turn is observed and let success indicate that the legal reason for the divorce is mental cruelty. Thus, X is the number of successes in 25 repetitions of a Bernoulli experiment, where the probability of success on each trial is $4/5 = .8$.

These examples, diverse as they are, all follow the same basic pattern. In each situation we can interpret the problem as a repetition of a simple Bernoulli experiment. It is important to keep in mind, however, that in order to have a binomial experiment the trials must be *independent*. In each of the examples above, we assumed the trials were independent. In many physical problems, however, the assumption of independence is unwarranted. For example, suppose medical records show that 40% of all U.S. citizens will contract German measles by age 25. What is the probability that in a group of 6 people more than 25 years old, exactly 2 have contracted German measles at some time before their twenty-fifth birthday? If we let success indicate that an individual has contracted German measles and let X be the number of successes in the group of 6, then we might argue that we have six repetitions of a Bernoulli experiment, where the probability of success on each trial is .4. If we assume that the trials are independent, then X is a binomial random variable, and from Appendix Table C.2, the probability of exactly 2 people of the 6 having contracted German measles is

$$P(X = 2) \cong .311$$

However, let us examine the independence assumption. Suppose the 6 people selected are brothers who live in the same house. Since German measles is contagious, it is evident that if one brother contracts the disease, then the chances that some of the other brothers will also contract it are greatly increased over the national probability of .4. Thus success or failure on one trial affects the chances for success on the other trials, so the assumption of independence is unwarranted. If, on the other hand, the 6 people selected are not from the same family, then the independence assumption would be warranted if we knew that the 6 people had no direct contact with one another and had no direct contact through mutual acquaintances who might have carried the disease from one individual to the other.

If X is a binomial random variable, then X can be interpreted as the number of successes in n independent repetitions of a Bernoulli experiment with success probability p. Since X is a random variable, we can ask for its expected value μ and its variance σ^2. Although we omit the mathematical details, the following result can be proved.

Expected Value and Variance of Binomial Random Variables If a binomial random variable X represents the number of successes in n independent repetitions of a Bernoulli experiment with success probability p, then the expected value and variance of X are given by

$$\mu = E(X) = np \tag{4}$$
$$\sigma^2 = \text{Var}(X) = npq \tag{5}$$

where $q = 1 - p$.

EXAMPLE 11 If X is a binomial random variable with $n = 48$ and $p = \frac{1}{4}$, then

$$\mu = E(X) = np = 48\left(\frac{1}{4}\right) = 12$$

$$\sigma^2 = \text{Var}(X) = npq = 48\left(\frac{1}{4}\right)\left(\frac{3}{4}\right) = 9$$

EXAMPLE 12 **EXAMPLE 4 OF SECTION 8.2 AND EXAMPLE 6 OF SECTION 8.3 REVISITED** Let the random variable X denote the number of defective resistors in a sample of two resistors selected independently from a production process in which 5% of all resistors produced are defective. Since the resistors are selected independently, we can interpret X as the number of successes (defectives) in 2 independent repetitions of a Bernoulli experiment with success probability $p = .05$. By Equation (4),

$$\mu = E(X) = np = 2(0.05) = 0.10$$

which agrees with the solution to Example 4 in Section 8.2. The variance of X, by Equation (5), is

$$\begin{aligned} \text{Var}(X) &= npq \\ &= 2(0.05)(0.95) \\ &= 0.095 \end{aligned}$$

which agrees with the result obtained in the solution of Example 6 in Section 8.3.

EXERCISE SET 8.5

1. Assume there are four independent repetitions of a Bernoulli experiment with success probability $p = \frac{1}{5}$. Use Formula (3) to find the probability of
   (a) no successes in the four repetitions
   (b) exactly one success
   (c) exactly two successes
   (d) exactly three successes
   (e) exactly four successes
   (f) at most three successes
   (g) at least two successes

2. Use Appendix Table C.2 to check your work in Exercise 1.

3. Let X be the number of successes in 12 independent repetitions of a Bernoulli experiment with success probability $p = .6$. Find
   (a) $P(X = 4)$ (b) $P(X = 11)$
   (c) $P(X = 0)$ (d) $P(X \geq 10)$
   (e) $P(X \leq 5)$ (f) $P(X > 11)$

4. In each part, assume X is a binomial random variable with the given values of n and p. Use Appendix Table C.2 to find the requested probability.
   (a) $n = 3$, $p = .5$; find $P(X = 2)$
   (b) $n = 8$, $p = .8$; find $P(X = 6)$
   (c) $n = 11$, $p = .2$; find $P(X \leq 5)$
   (d) $n = 9$, $p = .4$; find $P(X > 6)$

5. Let X be a binomial random variable with $n = 7$ and $p = .4$. Graph the probability function for X.

6. A certain system for betting on horses produces winners 40% of the time. What is the probability the system produces exactly 3 winners out of 8 on a race day?

7. On a large boulevard, 10 traffic lights operate independently. Each light is red for 1 minute and green for 4 minutes.
   (a) What is the probability a car traveling on the boulevard will not meet any red lights?
   (b) What is the probability that the car will meet at least 1 red light?
   (c) What is the probability the car will meet exactly 3 red lights?

8. Assuming the probability of a male birth is $\frac{1}{2}$, find the probability that among 12 births at Columbia Presbyterian Hospital, more than half will be females.

9. For the Bernoulli experiment in Exercise 1, what is the mean and variance?

10. Find the mean and variance for a Bernoulli experiment with $n = 3$ and $p = .8$.

11. A package of crimson giant radish seeds states that the probability of germination for each seed is .9. Suppose 10 such seeds are planted.
    (a) What is the probability that all 10 seeds germinate?
    (b) What is the probability that 8 or more seeds germinate?
    (c) What is the probability that 5 or fewer germinate?
    (d) Find the mean and variance.

12. A fair die is tossed four times.
    (a) What is the probability of obtaining exactly two 5s in the four tosses?
    (b) What is the mean and variance?

13. Suppose that in a lunar rocket flight, each midcourse-correction thruster has a probability of .7 of working and

that the rocket has 4 thrusters that work independently. If the rocket can carry out its correction when at least 2 of the thrusters are working, what is the probability that the correction can be made? What is the probability if the rocket has 6 thrusters?

14. Assume 70% of the voting population supports the President's foreign policy. If 10 people are selected at random and asked for an opinion, what is the probability that the majority will indicate that they oppose the President's foreign policy?

15. Which has a better chance of occurring: tossing 5 heads in a row with a fair coin or obtaining 8 heads out of 10 tosses with a fair coin?

16. Suppose that 5% of all items coming off a production line are defective. Assume the manufacturer packages the items in boxes of 6 and guarantees "double your money back" if more than 2 items in a box are defective. On what percentage of the boxes will the manufacturer have to pay double money back?

17. A manufacturer of polygraphs (lie detectors) claims that the machine can correctly distinguish between the truth and a lie 90% of the time. The machine is to be tested on a series of 10 questions. If it correctly identifies at least 8 of the 10 questions as a truth or a lie, the machine will be purchased by the Rambler City Police Department, otherwise it will be rejected.

(a) What is the probability that the machine will be purchased if it is absolutely worthless (i.e., its probability of producing a correct result is $\frac{1}{2}$)?

(b) What is the probability that the machine will be rejected even if it can perform as the manufacturer claims?

18. Let X be a binomial random variable with $n = 2$ and success probability p. Determine the mean and variance of X.

8.6 THE NORMAL APPROXIMATION TO THE BINOMIAL

In Example 7 of Section 8.5 we consider the binomial experiment of tossing a fair coin 12 times. We find that if the random variable X equals the number of heads in the 12 tosses of the coin, then

$$\begin{aligned} P(X = 9) &= C_{12,9}\left(\frac{1}{2}\right)^9\left(\frac{1}{2}\right)^3 \\ &\cong .054 \qquad \text{(by Appendix Table C.2)} \end{aligned}$$

If, however, we consider tossing the coin 1000 times and are interested in the probability of getting exactly 499 heads in the 1000 tosses, then

$$P(X = 499) = C_{1000,499}\left(\frac{1}{2}\right)^{499}\left(\frac{1}{2}\right)^{501}$$

In this case, the table would not help since n is too large. To handle such problems, statisticians have found methods for approximating binomial probabilities. In this section, we discuss one such method.

To motivate our discussion, we first consider some examples.

EXAMPLE 1 Consider an experiment that consists of three independent repetitions of a Bernoulli experiment with success probability $p = \frac{1}{5} = .2$. Substituting $n = 3$ and $p = \frac{1}{5}$ into Formula (3) of Section 8.5, we see that the probability of getting x successes in the three repetitions is given by

$$P(X = x) = C_{3,x}\left(\frac{1}{5}\right)^x\left(\frac{4}{5}\right)^{3-x}$$

From this formula or from Appendix Table C.2, we obtain

$$P(X = 0) = \frac{64}{125} = .512 \qquad P(X = 1) = \frac{48}{125} = .384$$

$$P(X = 2) = \frac{12}{125} = .096 \qquad P(X = 3) = \frac{1}{225} = .008$$

As discussed in Section 8.1, these results can be described geometrically by considering the histogram of the random variable X (see Figure 1).

FIGURE 1

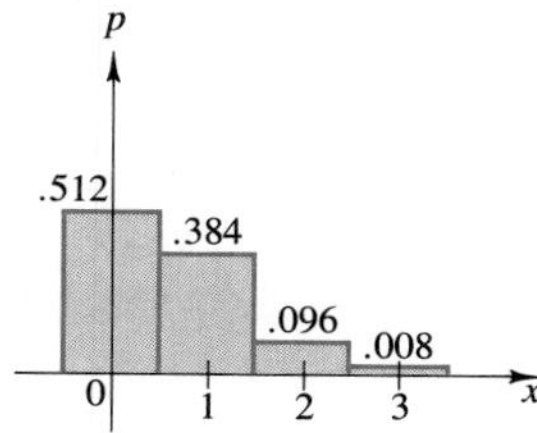

Note that $P(X = 0)$ is the area of the rectangular shaded region centered about $x = 0$. Thus,

$$P(X = 0) = 1\left(\frac{64}{125}\right) = \frac{64}{125} = .512$$

Similarly, $P(X = 1)$, $P(X = 2)$, and $P(X = 3)$ are equal, respectively, to the areas of the rectangular shaded regions centered about $x = 1$, $x = 2$, and $x = 3$. Also, the total area of the histogram rectangles is

$$\begin{aligned} &P(X = 0) + P(X = 1) + P(X = 2) + P(X = 3) \\ &= 1\left(\frac{64}{125}\right) + 1\left(\frac{48}{125}\right) + 1\left(\frac{12}{125}\right) + 1\left(\frac{1}{125}\right) = 1 \end{aligned}$$

and $P(X \leq 2)$ is the sum of the areas of the rectangles centered about $x = 0$, $x = 1$, and $x = 2$. Histograms are extremely important because they can be used to describe probabilities geometrically as areas.

EXAMPLE 2 Let us now see what happens to the histogram of a binomial random variable as n, the number of repetitions, becomes larger and larger. To illustrate, we have constructed in Figure 2 three histograms for binomial random variables. In these histograms the number of repetitions increases from $n = 5$ to $n = 25$ with the probability of success in each case fixed at $p = \frac{1}{5} = .2$.

FIGURE 2

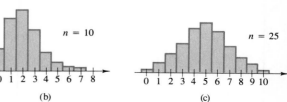

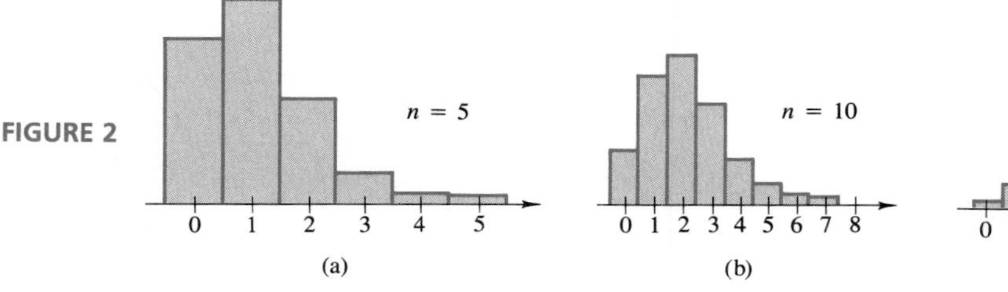

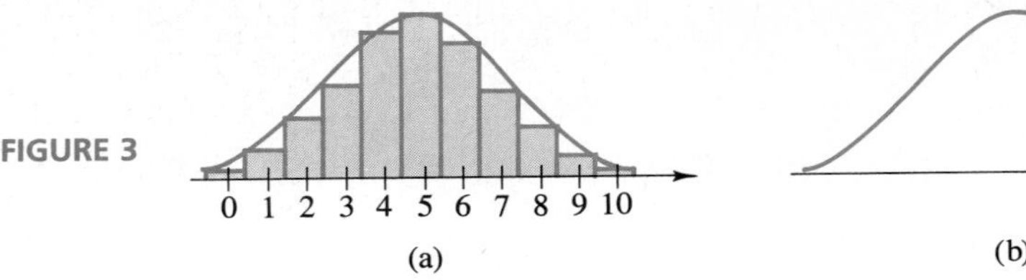

FIGURE 3

The histograms in Figure 2 suggest that as n increases, the area occupied by the rectangles in the histogram (Figure 3a) begins to look more and more like the area under a bell-shaped curve (Figure 3b).

Curves of this shape are called **normal** or **Gaussian** curves. We now discuss these curves and their properties and see how we can use them to approximate binomial probabilities.

Normal Curves and the Normal Distribution

The normal curve represents a probability distribution called the **normal, Gaussian**, or, more familiarly, the **bell-shaped** distribution. This distribution is important not only because it represents many important applications, but also because it can be used to approximate many important discrete distributions, such as the binomial distribution (for large values of n). Individual weight variations, for example, tend to be approximately normally distributed, as are SAT scores for high school students. College students often want to know whether or not grades in a class are "curved." In other words, does the professor assume that the grades are normally distributed?

Some important properties of the normal curve include:

1. The equation of the bell-shaped curve is

$$y = \frac{1}{\sqrt{2\pi}\,\sigma} e^{-\frac{1}{2}\left(\frac{x-\mu}{\sigma}\right)^2}$$

where μ is the mean, σ^2 is the variance, and σ is the standard deviation of this distribution. Here π and e are constants with $\pi \cong 3.14159$ and $e \cong 2.71828$. (See Appendix A.5 for a practical way to compute y for given values of x.)

2. A normal curve never touches the x-axis, although it comes very close to it; the curve is symmetrical about a vertical line through its highest point. This vertical line has equation $x = \mu$, where μ is the mean of the normal distribution (Figure 4).

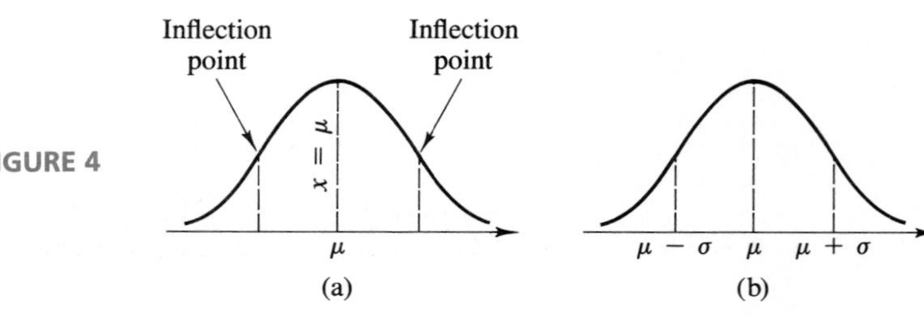

FIGURE 4

3. If we travel along a normal curve away from the center, we pass through points where the curvature changes from the downward direction to the upward direction; these are called **inflection points** for the curve (Figure 4a). Note that between the inflection points, the curve looks like an upside-down bowl; to the left or right of the inflection points, the curve looks like part of a right-side-up bowl. The inflection points occur at the points $x = \mu - \sigma$ and $x = \mu + \sigma$, where σ is the standard deviation of the normal distribution (Figure 4b).
4. The area between the normal curve and the x-axis equals 1.
5. If X is a normal random variable, then
   (a) $P(X \leq a)$ is the area under the normal curve to the left of $x = a$ (Figure 5a). Also,
   $$P(X < a) = P(X \leq a)$$
   since for continuous distributions $P(X = a) = 0$.
   (b) $P(x_1 \leq X \leq x_2) = P(x_1 < X \leq x_2)$ is the area under the normal curve between $x = x_1$ and $x = x_2$ (Figure 5b).

FIGURE 5

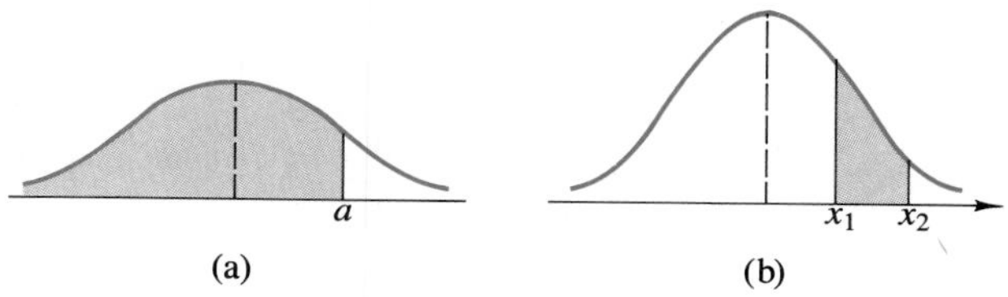

a

(a)

x_1 x_2

(b)

We now consider a normal random variable X with mean μ and standard deviation σ. We want to determine $P(x_1 \leq X \leq x_2)$, which is the area under the normal curve between $x = x_1$ and $x = x_2$ (Figure 5b).

To determine this area, table values, obtained by methods of calculus, must be used (see Appendix Table C.1). Notice that the shape of a normal curve is determined by the values of μ and σ. The value of μ determines the center, and the value of σ determines how flat the curve is. To illustrate, we superimposed three normal curves in Figure 6, each with $\mu = 0$. As shown in this figure, the larger the value of σ, the flatter the curve. The normal curve with $\mu = 0$ and $\sigma = 1$ is of particular importance; it is called the **standard normal curve**.

FIGURE 6

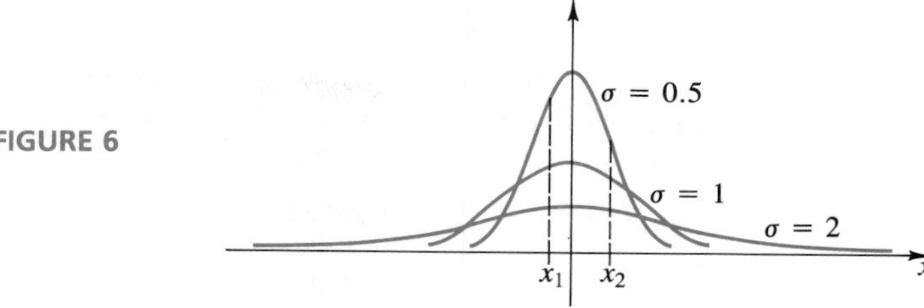

Notice, too, that the areas under these curves between x_1 and x_2 are not equal (Figure 6). Therefore, it seems that we need different tables of areas for different μ and σ. Fortunately, this is not the case. By using a technique that we now describe, you can always work with tables of areas under the standard normal curve. This technique is based on the following result, which we state without proof.

Areas under Normal Curves For a normal curve with center of symmetry μ and inflection points at $\mu + \sigma$ and $\mu - \sigma$, the area under the curve to the left of a point x is the same as the area to the left of

$$z = \frac{x - \mu}{\sigma}$$

under the standard normal curve (Figure 7).

FIGURE 7

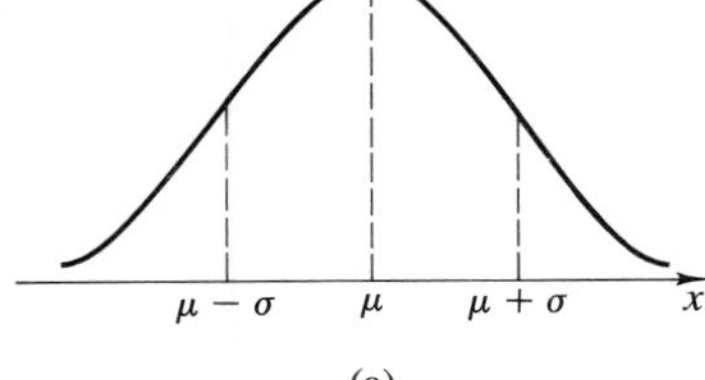

(a)

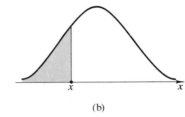

(b)

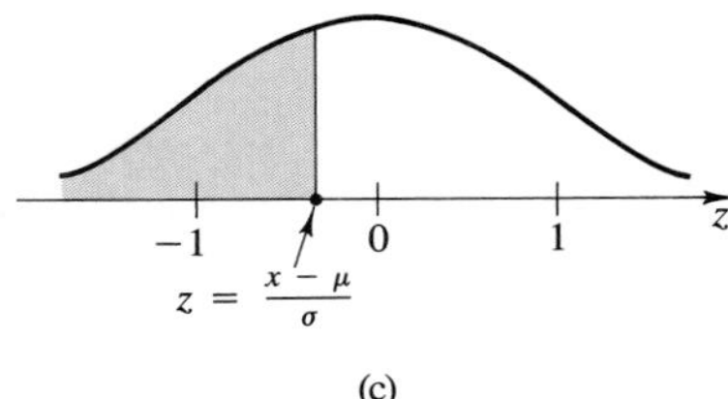

(c)

In other words, Z is the standard normal random variable that corresponds to the normal random variable X, and

$$Z = \frac{X - \mu}{\sigma}$$

If $z = (x - \mu)/\sigma$, then $P(X \leq x) = P(Z \leq z)$. Therefore, if we are given a normal curve with $\mu = 3200$ and $\sigma = 40$ and if we want to determine $P(X \leq 3240)$, we let $x = 3240$:

$$z = \frac{x - \mu}{\sigma} = \frac{3240 - 3200}{40} = 1$$

so $P(X \leq 3240) = P(Z \leq 1)$.

We illustrate our method of determining $P(X \leq x)$ by considering the following example.

EXAMPLE 3 For a normal curve with $\mu = 3200$ and $\sigma = 40$, find $P(X \leq 3286)$, which is the area under the normal curve to the left of $X = 3286$.

Solution

Method to Determine $P(X \leq x)$

1. Transform x to z:

$$z = \frac{x - \mu}{\sigma}$$

1. $x = 3286$

$$z = \frac{3286 - 3200}{40} = \frac{86}{40} = 2.15$$

In Appendix Table C.1, the z values (to one decimal place) are listed in the left-most column. The numbers in the second decimal position of the z values are listed in the top row. $P(Z \leq z)$ is in the corresponding row and column intersection.

2. Find the z value in Appendix Table C.1.

2.

z	. . .	5
⋮		⋮
2.1	. . .	.9842

3. $P(X \leq x) = P(Z \leq z)$ = value in this table

3. $P(X \leq 3286) = P(Z \leq 2.15) \cong .9842$

EXAMPLE 4 For a normal curve with $\mu = 2500$ and $\sigma = 25$, find
(a) $P(X \leq 2560)$ (b) $P(X > 2560)$

Solution (a) The z value corresponding to $x = 2560$ is

$$z = \frac{2560 - \mu}{\sigma} = \frac{2560 - 2500}{25} = \frac{60}{25} = 2.40$$

From Appendix Table C.1,

$$P(Z \leq 2.40) \cong .9918$$

Therefore, $P(X \leq 2560) \cong .9918$.

Solution (b) Since $P(X > 2560) = 1 - P(X \leq 2560)$,

$$P(X > 2560) \cong 1 - .9918 = .0082$$

EXAMPLE 5 (a) For the normal curve with $\mu = 3200$ and $\sigma = 40$, find the area between the points 3260 and 3270.
(b) For this normal distribution determine $P(3260 \leq X \leq 3270)$.

Solution (a) The area between $x_1 = 3260$ and $x_2 = 3270$ can be obtained by subtracting the area to the left of x_1 from the area to the left of x_2 (Figure 8). The area to the left of x_1 is the area under the standard normal curve to the left of

$$z_1 = \frac{x_1 - \mu}{\sigma} = \frac{3260 - 3200}{40} = 1.5$$

which from Appendix Table C.1 is approximately .9332. Similarly, the area to the left of x_2 is the area under the standard normal curve to the left of

$$z_2 = \frac{x_2 - \mu}{\sigma} = \frac{3270 - 3200}{40} = 1.75$$

which from Appendix Table C.1 is approximately .9599.

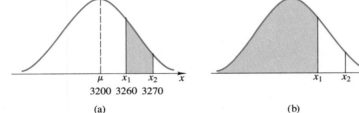

FIGURE 8

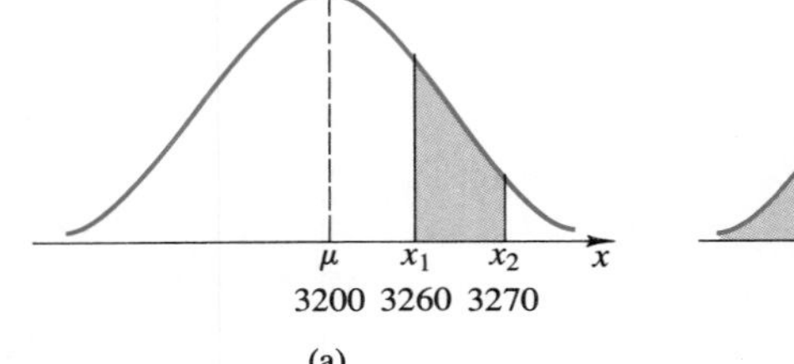

(a)

x1 x2 x
(b)

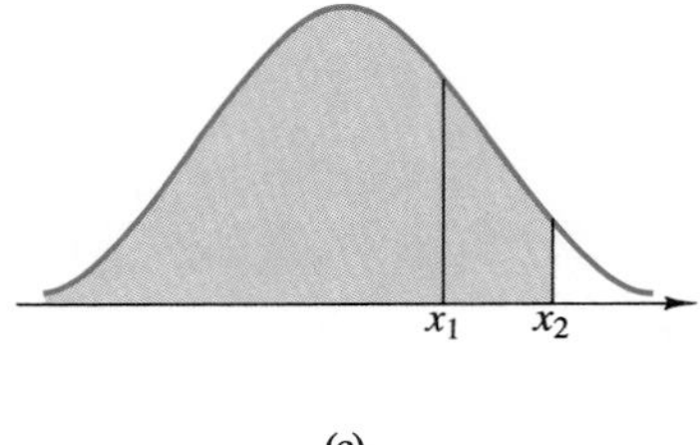

(c)

Thus, the area between x_1 and x_2 is approximately $.9599 - .9332 \cong .0267$.

Solution (b) $P(3260 \leq X \leq 3270) \cong .0267$ from part (a).

EXAMPLE 6 (a) For the normal curve with $\mu = 3200$ and $\sigma = 40$, find the area to the left of the point 3120 (Figure 9).

FIGURE 9

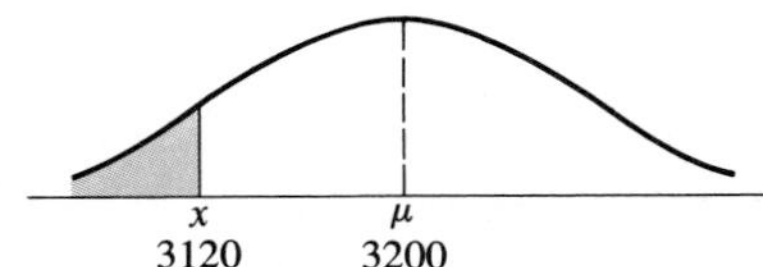

(b) For this normal distribution determine $P(X < 3120)$.

Solution (a) The area to the left of $x = 3120$ is the area to the left of

$$z = \frac{x - \mu}{\sigma} = \frac{3120 - 3200}{40} = -2$$

under the standard normal curve. From Appendix Table C.1, the area to the left of $z = -2$ is approximately .0028.

Solution (b) $P(X < 3120) = P(Z < -2) \cong .0028$ from part (a).

Remark For the standard normal curve note that the z values of Appendix Table C.1 are between -3.49 and 3.49. For z values greater than 3.49, the area to the left of these values is so close to 1 that it is assumed equal to 1. Therefore, $P(Z \le 3.5) \cong 1$, $P(Z \le 4.3) \cong 1$; and for z values less than -3.49, the area to the left of these values is so close to 0 that it is assumed equal to 0. Thus,

$$P(Z \le -3.5) \cong 0 \qquad \text{and} \qquad P(Z \le -5.5) \cong 0$$

EXAMPLE 7 Given the normal distribution with $\mu = 3200$ and $\sigma = 40$, determine

(a) $P(X \le 3300)$ (b) $P(3000 \le X \le 3300)$

Solution (a) Since $P(X \le 3300) =$ area to the left of $x = 3300$,

(i) convert x to z by

$$z = \frac{x - \mu}{\sigma} = \frac{3300 - 3200}{40} = \frac{100}{40} = 2.50$$

(ii) $P(X \le 3300) = P(Z \le 2.50) =$ area to the left of $z = 2.50$
$\cong .9938$ (Appendix Table C.1 value at $z = 2.5$)

Solution (b) Since $P(3250 \le X \le 3300) =$ area between $x_1 = 3250$ and $x_2 = 3300$,

(i) convert x_1 to z_1 by

$$z_1 = \frac{x_1 - \mu}{\sigma} = \frac{3250 - 3200}{40} = \frac{50}{40} = 1.25$$

(ii) convert x_2 to z_2 by

$$z_2 = \frac{x_2 - \mu}{\sigma} = \frac{3300 - 3200}{40} = \frac{100}{40} = 2.50$$

(iii) $P(3250 \le X \le 3300) = P(1.25 \le Z \le 2.50)$
= Appendix Table C.1 value at $z = 2.50$
− Appendix Table C.1 value at $z = 1.25$
$\cong .9938 - .8944 = .0994$

In the above examples the values of μ and σ are given. In general, these values must be determined. How they are determined depends on the use of the normal distribution curve. If it is used to approximate another distribution, as, for example, the binomial distribution (see Example 8 below), then μ and σ are the values of the mean and standard deviation of the particular binomial distribution.* In other cases σ and μ might be estimated statistically. We explore these cases further in Section 8.7.

*There are many common distributions other than the binomial and normal that also have a mean and variance. A reference is *Applied Statistics* by Neber, Wasserman, and Whitmore (Boston: Allyn and Bacon, 1978).

Approximating Binomial Probabilities

The next two examples illustrate the basic computational techniques for approximating binomial probabilities by areas under normal curves. The method we use is as follows:

Method of Approximating Binomial Probabilities by Normal Probabilities
We refer to the binomial random variable as Y, where Y denotes the number of successes in n independent Bernoulli trials. Our objective is to determine $P(y_1 \leq Y \leq y_2)$. Note that Y is nonnegative.

We refer to the normal random variable, which we use to approximate the binomial random variable, as X. The random variable X has mean μ and standard deviation σ equal to the mean and standard deviation of the binomial random variable it is approximating. From Equations (4) and (5) of Section 8.5,

$$\mu = np \quad \text{and} \quad \sigma = \sqrt{npq}$$

The probability $P(y_1 \leq Y \leq y_2)$ is the area of the histogram of Y between $y_1 - \frac{1}{2}$ and $y_2 + \frac{1}{2}$, since the histogram rectangles are centered about integer values (Figure 10). Therefore, we approximate the area of the histogram between $x_1 = y_1 - \frac{1}{2}$ and $x_2 = y_2 + \frac{1}{2}$ by the area under the normal curve between $x_1 = y_1 - \frac{1}{2}$ and $x_2 = y_2 + \frac{1}{2}$. Thus,

$$\underset{\text{(binomial)}}{P(y_1 \leq Y \leq y_2)} \cong \underset{\text{(normal approximation)}}{P(x_1 \leq X \leq x_2)}$$

Step 1 Convert the normal random variable X to the standard normal random variable Z:

$$Z = \frac{X - \mu}{\sigma} = \frac{X - np}{\sqrt{npq}}$$

Step 2 Proceed, as in the previous examples, to determine

$$P(z_1 \leq Z \leq z_2) = P(x_1 \leq X \leq x_2) \cong P(y_1 \leq Y \leq y_2)$$

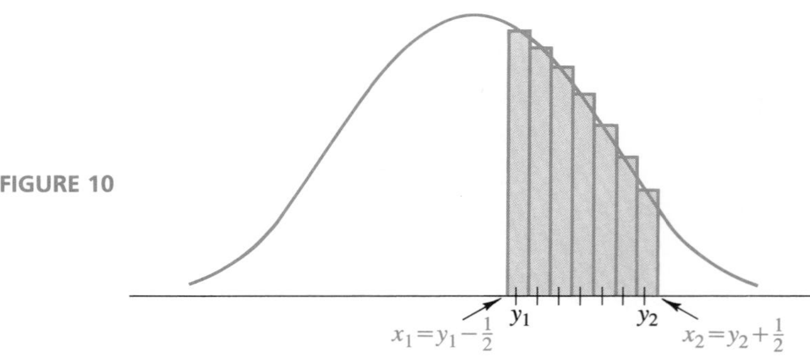

FIGURE 10

EXAMPLE 8 Use a normal curve to estimate the probability that in 324 tosses of a fair coin, the number of heads is between 171 and 180, inclusive.

Solution If Y denotes the number of heads in 324 tosses, then Y is a binomial random variable with $n = 324$ and $p = \frac{1}{2}$. We want to estimate

$$P(171 \leq Y \leq 180)$$

This probability is the area between 170.5 and 180.5 in the histogram of Y (Figure 11).

FIGURE 11

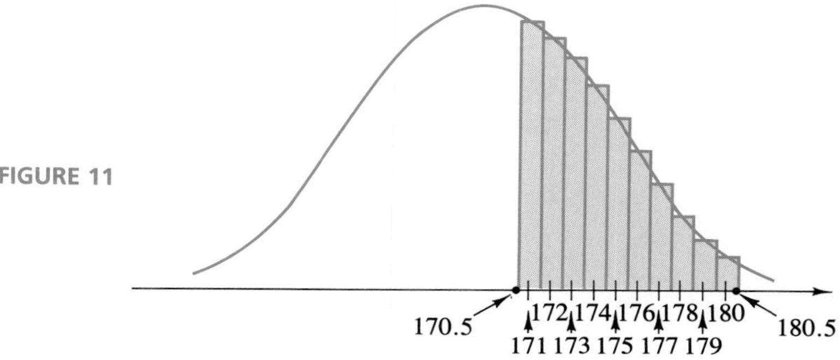

This is approximately the area between

$$x_1 = 170.5 \quad \text{and} \quad x_2 = 180.5$$

under the normal curve with μ and σ determined by Equations (4) and (5) of Section 8.5. We are using the normal distribution to approximate a binomial distribution with

$$\mu = np = (324)\left(\frac{1}{2}\right) = 162$$

and

$$\sigma = \sqrt{npq} = \sqrt{(324)\left(\frac{1}{2}\right)\left(\frac{1}{2}\right)} = 9$$

Proceeding as in Example 7, $P(170.5 \leq X \leq 180.5)$ is the area between $x_1 = 170.5$ and $x_2 = 180.5$ under the normal curve with $\mu = 162$ and $\sigma = 9$. Converting to standard normal z values,

$$z_1 = \frac{x_1 - \mu}{\sigma} = \frac{170.5 - 162}{9} \cong .94$$

and

$$z_2 = \frac{x_2 - \mu}{\sigma} = \frac{180.5 - 162}{9} \cong 2.06$$

From Appendix Table C.1, the area between z_1 and z_2 is

$$P(.94 \leq Z \leq 2.06) \cong .9803 - .8264 = .1539$$

Thus,

$$P(171 \leq Y \leq 180) \cong P(.94 \leq Z \leq 2.06) \cong .1539$$

(where Y is the binomial random variable defined above).

EXAMPLE 9 Use a normal curve to estimate the probability that in 150 independent Bernoulli trials with success probability .4, we observe 70 or fewer successes.

Solution If Y is the random variable representing the number of successes, then Y is a binomial random variable with $n = 150$ and $p = .4$. We want to estimate

$$P(Y \leq 70) = P(0 \leq Y \leq 70)$$

This probability is the area between -0.5 and 70.5 in the histogram of Y, which is approximately the area between $x_1 = -0.5$ and $x_2 = 70.5$ under the normal curve with μ and σ determined from Equations (4) and (5) of Section 8.5. Because the normal curve is being used to approximate a binomial distribution with $n = 150$ and $p = .4$,

$$\mu = np = 150(.4) = 60$$

and

$$\sigma = \sqrt{npq} = \sqrt{150(.4)(.6)} = 6$$

In terms of the standard normal curve, the desired area is the area between

$$z_1 = \frac{x_1 - np}{\sqrt{npq}} = \frac{-0.5 - 60}{6} = -10.08$$

and

$$z_2 = \frac{x_2 - np}{\sqrt{npq}} = \frac{70.5 - 60}{6} = 1.75$$

As in Example 4 the area between z_1 and z_2 is the area to the left of z_2 minus the area to the left of z_1. But the area to the left of -10.08 is approximately zero (see the remark following Example 6). Therefore,

$$\begin{aligned} P(Y \leq 70) &\cong P(-0.5 \leq X \leq 70.5) \\ &= P(-10.08 \leq Z \leq 1.75) \\ &\cong .9599 - 0 \\ &= .9599 \end{aligned}$$

The idea of using a normal curve to estimate probabilities of binomial random variables is based on the assumption that n is large enough so that the areas under the histogram are approximately the same as the corresponding areas under the normal curve. But how large should n be to use this method of approximation? The answer to this question depends to some degree on the nature of the problem. However, as a rule of thumb, most statisticians only use normal curve approximations when $npq \geq 3$. Thus, if $p = \frac{1}{2}$, then n can be as small as 12. However, if p

is near 0 or 1, larger values of n are required; for example, if $p = \frac{9}{10}$, we would need

$$n\left(\frac{9}{10}\right)\left(\frac{1}{10}\right) \geq 3 \quad \text{or} \quad n(.09) \geq 3$$

Thus n would have to be at least as large as 34.

EXERCISE SET 8.6

1. In each part, draw the histogram for the binomial random variable with the given values of n and p.
(a) $n = 3, p = \frac{1}{2}$ (b) $n = 5, p = .2$
(c) $n = 5, p = .8$ (d) $n = 10, p = .6$

2. Let X be the random variable whose histogram is shown in Figure 12. Determine the following:
(a) $P(X = 2)$ (b) $P(1 \leq X \leq 4)$
(c) $P(X \geq 3)$ (d) $P(X < 1)$

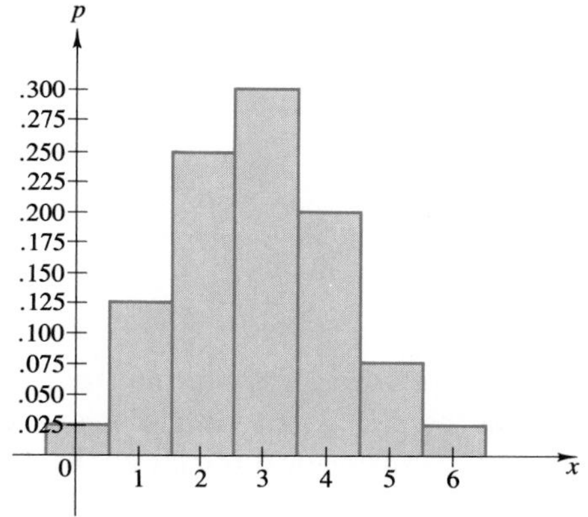

FIGURE 12

3. In each part, find the mean and variance of the binomial random variable with the given values of n and p.
(a) $n = 3, p = \frac{1}{2}$ (b) $n = 5, p = .2$
(c) $n = 1000, p = \frac{1}{4}$ (d) $n = 2500, p = .9$

4. In each part, let X be the binomial random variable with the given values of n and p. Find the center of symmetry and the inflection points of the normal curve that approximates the histogram for X.
(a) $n = 6400, p = \frac{1}{2}$ (b) $n = 2500, p = \frac{4}{5}$
(c) $n = 1000, p = \frac{1}{4}$ (d) $n = 500, p = .6$

5. Find the following areas under the standard normal curve by using Appendix Table C.1:
(a) the area to the left of 1.25
(b) the area to the right of 0.84
(c) the area between 0.37 and 1.84

6. Find the following areas under the standard normal curve:
(a) the area to the left of -0.72
(b) the area to the right of -1.80
(c) the area between -1.31 and -0.38
(d) the area between -2.81 and 1.47

7. In each part, find the value of c that satisfies the given conditions:
(a) The area to the left of c under the standard normal curve is .8749.
(b) The area to the right of c under the standard normal curve is .1423.
(c) The area to the right of c under the standard normal curve is .9726.
(d) The area to the left of c under the standard normal curve is .0078.

8. Find the following areas under the normal curve with $\mu = 10$ and $\sigma = 2$:
(a) the area to the left of 13
(b) the area to the right of 8
(c) the area between 11.6 and 13.8
(d) the area between 5.2 and 15.4

9. Use a normal curve to estimate the following binomial probabilities:
(a) $P(Y \leq 10)$ when Y is a binomial random variable with $n = 25$ and $p = \frac{1}{2}$.
(b) $P(25 \leq Y \leq 35)$ when Y is a binomial random variable with $n = 80$ and $p = .4$.
(c) $P(Y \geq 155)$ when Y is binomial random variable with $n = 200$ and $p = .75$.

10. Use a normal curve to estimate the following binomial probabilities:
   (a) $P(Y \leq 6)$ when Y is a binomial random variable with $n = 36$ and $p = \frac{1}{6}$
   (b) $P(10 \leq Y \leq 20)$ when Y is a binomial random variable with $n = 60$ and $p = .2$
   (c) $P(Y \geq 90)$ when Y is a binomial random variable with $n = 121$ and $p = .8$

11. Find the following areas under the normal curve with $\mu = 1500$ and $\sigma^2 = 3600$:
   (a) the area to the right of 1400
   (b) the area to the left of 1375
   (c) the area between 1420 and 1580

12. Find the following areas under the normal curve with $\mu = 2000$ and $\sigma^2 = 6400$:
   (a) the area to the right of 1600
   (b) the area to the left of 1800
   (c) the area between 1800 and 2180

13. Given a normal distribution with $\mu = 10$ and $\sigma = 2$, evaluate:
   (a) $P(X \leq 11)$ (b) $P(10.5 \leq X \leq 11)$
   (c) $P(X \leq 9)$ (d) $P(9 \leq X \leq 11.5)$

14. Given a normal distribution with $\mu = 1500$ and $\sigma^2 = 3600$, evaluate
   (a) $P(X \leq 1600)$
   (b) $P(1400 \leq X \leq 1600)$

Solve Exercises 15–19 using a normal approximation.

15. Assuming that a penicillin injection causes an allergic reaction in 1 of 10 people, estimate the probability that fewer than 470 reactions will occur among 5000 people injected with penicillin.

16. A bag of 400 dimes is dumped onto a table. Estimate the probability that the number of heads is between 175 and 225, inclusive.

17. Assume 10% of the items coming off a production line are defective; estimate the probability that more than 14% of the items in a sample of size n will be defective if:
   (a) $n = 100$ (b) $n = 500$ (c) $n = 1000$

18. The death rate for yellow fever is five deaths per 1000 cases. If 700 individuals contract yellow fever, what is the probability of:
   (a) exactly 3 deaths (b) more than 3 deaths

19. Genetic theory predicts that when the right two varieties of peas are crossed, the probability of obtaining a smooth seed is .5. How many seeds must be produced to assure with probability .9 that at least 100 smooth seeds are obtained?

8.7 DESCRIPTIVE STATISTICS

So far in this chapter the examples we have considered fall into two categories. The first involves an experiment, such as tossing two dice, for which the probabilities of the outcomes are determined using mathematical principles. In this type of problem, the mean and variance of the distribution are determined by Equation (9) of Section 8.2 and Equation (5) of Section 8.3. The second type involves an experiment for which the probabilities of the outcomes must in some way be determined empirically, and the mean and variance must somehow be estimated. Consider the following examples:

EXAMPLE 1 A certain high-quality commercial light bulb has an expected lifetime of $\mu = 750$ hours with a variance of $\sigma^2 = 100$.

EXAMPLE 2 In the city of Metropolis (with 100,000 eligible voters) the probability that an eligible voter will vote for a Republican for mayor is .65.

EXAMPLE 3 The number of accidents on the New Jersey Turnpike during the Friday rush hour is 0, 1, 2, or 3 with corresponding probabilities .85, .10, .03, and .02.

The mean and variance in Example 1 and the probabilities given in Examples 2 and 3 are estimated empirically by the methods of descriptive statistics, a branch of mathematics that includes the collection, organization, and description of numerical information called **data**. In this section, we formally consider some methods of descriptive statistics.

First we must define some terms used in statistics. A **population** is the set of objects or individuals whose properties are to be considered. The second column of Table 1 shows the population sets of Examples 1, 2, and 3. An n-element subset of the population is called a **sample of size n** (see column 3 of Table 1).*

TABLE 1

Example	Population	An example of a sample of size 4	Data set for this sample
1	Set of all light bulbs in a batch: $\{B_1, B_2, \ldots, B_n\}$	Sample of 4 light bulbs: $\{B_3, B_8, B_9, B_{20}\}$	751, 725, 735, 752 hours
2	Set of all eligible voters: $\{V_1, V_2, \ldots, V_{100,000}\}$	Sample of 4 eligible voters: $\{V_{10}, V_{18}, V_{19}, V_{1000}\}$	(R represents a vote for a Republican, D represents a vote for a Democrat) R, R, D, R
3	Set of all Fridays in a 100-week period: $\{F_1, F_2, \ldots, F_{100}\}$	Sample of four Fridays: $\{F_1, F_{55}, F_{80}, F_{90}\}$	Number of accidents: 0, 3, 1, 1

In general, statisticians are interested in certain properties of a population, as, for example, the average lifetime of the light bulbs in Example 1, the preference of voters in Example 2, and the average number of accidents in Example 3. A **data set** is the set of values that are observed for each element of the population or sample. The data (plural) are the values. In Example 1, a data set is the set of lifetimes in hours for the bulbs in a sample. In Example 2, the data set could be a set of R's and D's corresponding to members of the sample, where R represents a vote for a Republican and D represents a vote against. In Example 3, the data set could be the set of the numbers of accidents observed on the Friday mornings of the sample (Table 1).

Remark Since a data set consists of observations, there can be repeated values in it. Since set notation implies no repetition of elements, we will not use set notation when data are presented.

At times we are interested in a particular numerical characteristic (an average, for example) of a whole population; at other times we are interested in the corresponding characteristic of a sample. A **parameter** is a value representing a characteristic of the entire population. In Example 1, a parameter could be the average lifetime of the light bulbs. Although the parameter is a constant, it usually is

*Methods of general sampling are not discussed here. For further information, see *Elementary Survey Sampling* by R. Schaeffer, W. Mendenhall, and L. Ott (Boston: Duxbury Press, 1986).

not known. The corresponding value representing the characteristic of a sample is called a **statistic**.

The statistic representing the average lifetime of the bulbs in the sample of Table 1 is

$$\frac{751 + 725 + 735 + 752}{4} = 740.75$$

where the average is the arithmetic average defined in Equation (1) of Section 8.2. Many times the statistic of a random sample is used to estimate the parameter of the population.

EXAMPLE 4 Lee walks to school five days a week. Each day last week, she recorded the duration of her walk:

Day	Data (time in minutes)
Monday	4.9
Tuesday	4.5
Wednesday	5.0
Thursday	5.1
Friday	4.5

What is the sample and what are the data?

Solution The sample is

$$\{M, T, W, Th, F\}$$

The corresponding data are

$$4.9, 4.5, 5.0, 5.1, 4.5$$

Frequency Tables and Histograms

Once the data are collected for a sample, the next step usually is to organize these values into a frequency table. In this table, the distinct values of the data are listed with corresponding frequencies of occurrence. Numerical data are listed in increasing order of size with corresponding frequencies of occurrence.

EXAMPLE 5 The test grades for 30 students in a finite mathematics class are

85	75	90	64	38	100
70	68	85	80	72	70
10	70	75	80	69	72
85	85	85	70	75	90
85	45	68	72	70	90

The **frequency** of a value of the data is the number of times f that it appears in the set of data. A frequency table for this set of data appears in Table 2; x is the test grade and f is the number of times this grade appears in the data.

TABLE 2

x	f
10	1
38	1
45	1
64	1
68	2
69	1
70	5
72	3
75	3
80	2
85	6
90	3
100	1
Total =	30

There are times when we may want to group the data in ranges of scores as 1–10, 11–20, 21–30, . . . , 91–100. This is done in Table 3.

TABLE 3

Group	Frequency
1–10	1
11–20	0
21–30	0
31–40	1
41–50	1
51–60	0
61–70	9
71–80	8
81–90	9
91–100	1
Total =	30

Histograms (similar to those discussed in Section 8.1) can also be used to represent a frequency distribution. In this type of histogram, the height of each bar indicates the frequency. A histogram corresponding to the data of Example 5 and Table 2 is shown in Figure 1a (zero values are indicated by a heavy line along the x-axis). Figure 1b shows a frequency histogram for the grouped data in Table 3.

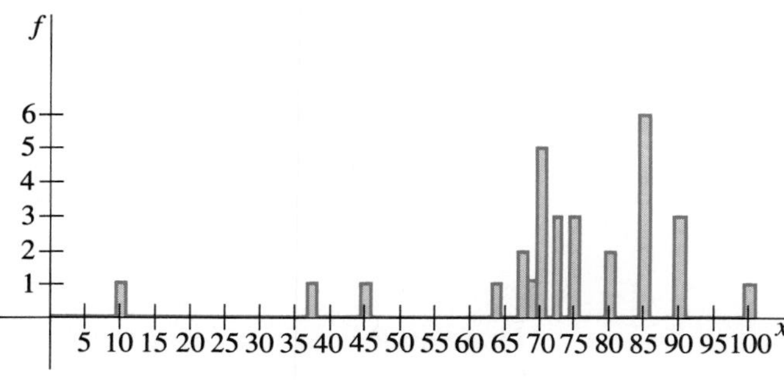

(a)

FIGURE 1

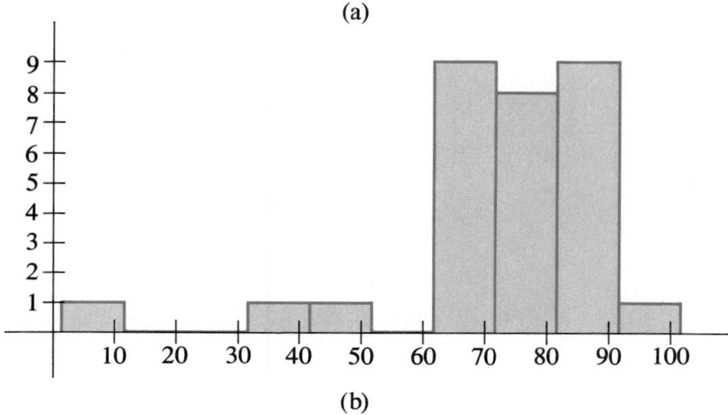

(b)

A relative frequency histogram is formed by changing the vertical scale to a relative frequency f/n, where f is the frequency of the outcome and n is the total number of data values. The relative frequency is used to represent the empirical probability $\hat{p}$ (read "p hat") of the corresponding outcome:

$$\hat{p} = \frac{f}{n}$$

EXAMPLE 6 Records are kept over a 100-week period of the number of accidents occurring during the Friday rush hour on the New Jersey Turnpike (see Example 3). The following data were recorded.

0	1	3	0	0	2	0	0	0	1
0	1	0	0	0	0	0	0	0	0
0	0	0	0	1	0	0	0	1	0
0	0	0	1	0	0	0	0	0	0
0	0	0	0	0	0	0	0	0	0
0	0	0	0	0	0	0	0	0	0
2	0	0	0	0	0	0	0	0	0
0	0	1	0	0	0	1	0	0	0
0	0	0	0	1	0	0	0	1	0
0	3	0	2	0	0	0	0	0	0

(a) Construct a frequency table.

(b) Construct a frequency histogram.

(c) Construct a relative frequency histogram.

(d) What are the empirical probabilities of having 0 accidents, 1 accident, 2 accidents, and 3 accidents during Friday rush hours?

(e) What is the empirical probability of having at least one accident on the New Jersey Turnpike on a Friday?

Solution (a) Table 4 gives the frequency and relative frequency for 0, 1, 2, and 3 accidents. The number of accidents is represented by x, the total number of observations by n.

TABLE 4

x	f	$\frac{f}{n} = \hat{p}$
0	85	$\frac{85}{100} = .85$
1	10	$\frac{10}{100} = .10$
2	3	$\frac{3}{100} = .03$
3	2	$\frac{2}{100} = .02$

Solution (b) See Figure 2a.

Solution (c) See Figure 2b (note the shapes are identical).

FIGURE 2

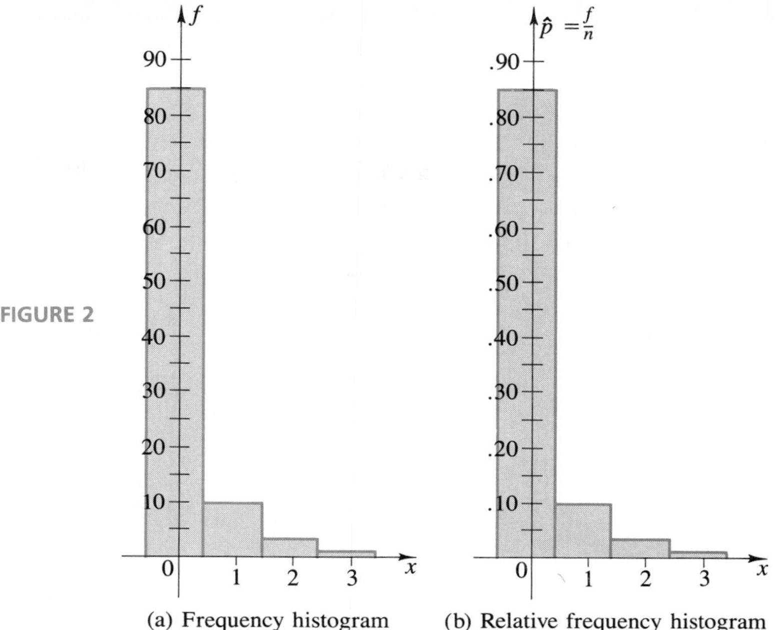

(a) Frequency histogram (b) Relative frequency histogram

Solution (d) The relative frequencies give the empirical probabilities of 0, 1, 2, and 3 accidents occurring over this 100-week period. If the random variable X denotes the number of accidents, then the empirical probabilities are

$$P(X = 0) = .85 \qquad P(X = 1) = .10 \qquad P(X = 2) = .03 \qquad P(X = 3) = .02$$

Solution (e) $P(X \geq 1) = 1 - P(X = 0) = 1 - .85 = .15$

In Example 2 the population of eligible voters is 100,000. In this case it is impractical to poll every voter. To determine the probability that a person will vote Republican, a random sample of the population is chosen and polled. The true probability p can be estimated by

$$\hat{p} = \frac{\text{number of } R\text{'s in the sample}}{\text{number in the sample}}$$

and we say that $\hat{p}$ is a **point estimate** of the parameter p. For the particular sample in Table 1,

$$\hat{p} = \frac{3}{4} = .75$$

There are methods of determining minimum values of n (the sample size) so that $\hat{p}$ (i.e., the estimated value of p) is likely to be close to the true value. These methods, which are beyond the level of this book, involve techniques of statistical inference. For further information, see *Elementary Statistics*, 4th edition, by Robert Johnson (Boston: Duxbury Press, 1984).

Measures of Central Tendency: Mean, Median, and Mode

We now look at some other characteristic properties of samples that can be obtained from the set of data. For example, in a testing situation, students are usually interested in where they stand relative to the rest of the class and what the average test score is. We now examine three averages, or measures of central tendency—the sample mean, median, and mode. Each has its own use.

The **sample mean** ($\overline{x}$), or **mean,** is the arithmetic average given in Equation (1) of Section 8.2. To restate this definition:

If $x_1, x_2, \ldots, x_n$ are the n pieces of data from a sample, then the sample mean is

$$\overline{x} = \frac{x_1 + x_2 + \cdots + x_n}{n} \tag{1}$$

As we see in Equation (5) of Section 8.2,

If $x_1, x_2, \ldots, x_k$ are distinct data points with corresponding frequencies $f_1, f_2, \ldots, f_k$, then the sample mean is

$$\overline{x} = \frac{f_1x_1 + f_2x_2 + \cdots + f_kx_k}{n} \tag{2}$$

EXAMPLE 7 What is the mean test score for the data of Example 5?

Solution Using the frequencies in Table 2 and Equation (2),

$$\overline{x} = \frac{1\cdot 10 + 1\cdot 38 + 1\cdot 45 + 1\cdot 64 + 2\cdot 68 + 1\cdot 69 + 5\cdot 70 + 3\cdot 72 + 3\cdot 75 + 2\cdot 80 + 6\cdot 85 + 3\cdot 90 + 1\cdot 100}{30}$$

$$= \frac{2193}{30} = 73.1$$

Thus, the mean test score is 73.1.

In many cases it is not practical to find the mean μ of a population, so the sample mean $\overline{x}$ of a random sample is used to estimate the population mean; $\overline{x}$ is then a point estimate for μ.

The second measure of central tendency is the **median**, which is the middle value of the sample data when the data are ranked according to size (including repetitions). If there are an even number of data values, then the median is the average of the two middle values.

EXAMPLE 8 Determine the median grade of the data set of Example 5.

Solution Method 1: List all the grades in increasing order:

10, 38, 45, 64, 68, 68, 69, 70, 70, 70, 70, 70, 72, 72, 72, 75, 75, 75, 80, 80, 85, 85, 85, 85, 85, 85, 90, 90, 90, 100

Since there are 30 grades, the median is obtained by averaging the middle two (circled in color):

$$\text{median} = \frac{72 + 75}{2} = 73.5$$

Method 2: Using the frequencies listed in Table 2, we note that since there are 30 data points, the median is the average of the 15th and 16th data points. The frequency table indicates that the 15th value is 72 and the 16th is 75; therefore, the median is 73.5.

Remark Half the data points are greater than or equal to the median and half are less than or equal to it. We sometimes refer to the median as the **50th percentile point**.

The third measure of central tendency is the **mode**, which is the data value that occurs most frequently. Constructing a frequency table or histogram is generally the easiest way to determine the mode of a data set. Sometimes there is more than one mode. If there are two, then the distribution is said to be **bimodal**; there is not a unique mode in this case.

EXAMPLE 9 What is the mode of the set of grades of Example 5?

Solution We note that in Table 2 the grade of 85 occurs with frequency 6—the highest frequency shown in the table. Therefore, the mode is 85.

EXAMPLE 10 Find the (a) sample mean, (b) median, and (c) mode of the data of Example 6.

Solution (a) Using Equation (2) and Table 4, the sample mean is

$$\bar{x} = \frac{85(0) + 10(1) + 3(2) + 2(3)}{100} = \frac{22}{100} = .22$$

Solution (b) To determine the median, we could list the data values in increasing order and average the two middle values (since there are an even number of data values). However, we can obtain the information we want from Table 4. Since the median in this case is the average of the 50th and 51st data values and since these values must be 0 because the frequency of 0 is 85, the median is

$$\frac{0 + 0}{2} = 0$$

Solution (c) Again referring to Table 4, we see that the mode is 0 since 0 occurs most often (with a frequency of 85) among the values 0, 1, 2, 3.

EXAMPLE 11 Talltree College has 9 students eligible to play basketball. The heights of these players are

6′1″, 6′3″, 6′4″, 5′11″, 6′3″, 6′0″, 6′1″, 6′1″, 6′0″

(a) What is the mean height of these players?

(b) What is the median height of these players?

(c) What is the mode of the heights of these players?

TABLE 5

Height	Frequency f
5′11″	1
6′0″	2
6′1″	3
6′3″	2
6′4″	1
	Total = 9

Solution (a) Table 5 is the frequency table for the data. Then

$$\bar{x} = \frac{1(5'11'') + 2(6'0'') + 3(6'1'') + 2(6'3'') + 1(6'4'')}{9}$$

$$= \frac{55'}{9} = \left(6\frac{1}{9}\right)' = 6'\frac{4''}{3}$$

Solution (b) The median is the middle or fifth value, which is 6′1″.

Solution (c) The mode is 6′1″ because this height occurs most often (three times) in the frequency table.

There are some instances when we are interested in one of these measures of central tendency and at other times in all of them.

EXAMPLE 12 Consider the annual salaries of the employees of Union Inc.:

President:	\$60,000
Salespersons:	\$10,000
	\$10,000
	\$10,000
	\$ 7,000

The mean salary is

$$\frac{1(60{,}000) + 3(10{,}000) + 1(7000)}{5} = \frac{97{,}000}{5} = \$19{,}400$$

The median salary is \$10,000. The mode salary is \$10,000. In this example the mean salary is in no way representative of the salary of the average worker. The extreme value of \$60,000 (president) is out of line with the other salaries, so the median or mode might be considered to be the "average" salary.

EXAMPLE 13 A shoe manufacturer is coming out with a new line of shoes. Which category of shoe size would be most useful to the manufacturer: the mean, median, or mode?

Solution The mode would be most helpful because it is the size worn most often. The mean size would be of no value, neither would the median.

EXAMPLE 14 SAT examinations are given. Excel College wants to accept students with at least 50th-percentile scores. All students with scores above the median grade will be accepted into the college.

Variance and Standard Deviation

Other measures of variability correspond to the variance σ^2 and standard deviation σ defined for probability distributions in Section 8.3. Usually it is impossible or impractical to determine μ and σ^2; therefore, a sample mean $\overline{x}$ [see Equation (1)] and a sample variance s^2 are determined from a sample of size n and are used as point estimates for μ and σ^2, respectively.

Sample Variance Given data $x_1, x_2, \ldots, x_n$ for a sample with sample mean $\overline{x}$, then the **sample variance** is

$$s^2 = \frac{(x_1 - \overline{x})^2 + (x_2 - \overline{x})^2 + \cdots + (x_n - \overline{x})^2}{n - 1}$$

Sample Standard Deviation

$$s = \sqrt{s^2} = \sqrt{\frac{(x_1 - \overline{x})^2 + (x_2 - \overline{x})^2 + \cdots + (x_n - \overline{x})^2}{n - 1}}$$

The sample variance s^2 is usually used to estimate the variance of the population (denoted by σ^2). Intuitively it seems that the sum of the squares should be divided by n rather than $n - 1$. However, it can be shown that the above definition of s^2 gives a better approximation to σ^2, the population variance. We omit the proof.

EXAMPLE 15 Consider Example 1. Since it would be impossible to measure the lifetime of all bulbs in a manufacturing lot, μ and σ are estimated by taking a random sample and computing $\overline{x}$ and s^2.

Solution If we use the sample of size 4 in Table 1, the data (in hours) are

751 725 735 752

Then

$$\overline{x} = \frac{751 + 725 + 735 + 752}{4} = 740.75$$

$$s^2 = \frac{(751 - 740.75)^2 + (725 - 740.75)^2 + (735 - 740.75)^2 + (752 - 740.75)^2}{3}$$

$$= 170.92$$

EXAMPLE 16 Consider the data from the sample of Example 4. Find the sample mean $\overline{x}$ and the sample variance s^2.

Solution

$$\overline{x} = \frac{4.9 + 4.5 + 5.0 + 5.1 + 4.5}{5} = 4.8$$

$$s^2 = \frac{(4.9 - 4.8)^2 + (4.5 - 4.8)^2 + (5.0 - 4.8)^2 + (5.1 - 4.8)^2 + (4.5 - 4.8)^2}{4}$$

$$= \frac{.32}{4} = .08$$

Now consider the data from Monday, Tuesday, Wednesday, and Thursday as a *sample* of size 4. We find $\overline{x}$ and s^2:

$$\overline{x} = \frac{4.9 + 4.5 + 5.0 + 5.1}{4} = \frac{19.5}{4} = 4.875$$

$$s^2 = \frac{(4.9 - 4.875)^2 + (4.5 - 4.875)^2 + (5.0 - 4.875)^2 + (5.1 - 4.875)^2}{3}$$

$$= .069$$

In summary, descriptive statistics gives us methods for estimating probabilities and for estimating the mean μ and variance σ^2 for a large population. For further information see *Elementary Statistics,* 4th edition, by Robert Johnson (Boston: Duxbury Press, 1984).

EXERCISE SET 8.7

1. Consider the sample data 3, 4, 8, 4, 3, 3, 7.
(a) Construct a frequency table and histogram.
(b) Find the mean.
(c) Find the median.
(d) Find the mode.
(e) Find the sample variance and standard deviation.

2. Consider the sample data 2, 5, 5, 3, 5, 4, 5, 2.
(a) Construct a frequency table and histogram.
(b) Find the mean.
(c) Find the median.
(d) Find the mode.
(e) Find the sample variance and standard deviation.

3. The rainfall per day over the first 10 days of April in Oshkosh was recorded as shown in Table 6.
(a) Construct a frequency table and histogram.
(b) Find the mean.
(c) Find the median.

TABLE 6

April	Inches of rainfall	April	Inches of rainfall
1	0	6	0
2	0.5	7	0.8
3	1.5	8	1.2
4	2.3	9	0.3
5	0	10	0

(d) Find the mode.
(e) Find the sample variance and standard deviation.

4. The heights of 12 of the players on Tall Gals basketball team were recorded as follows:

5′10″, 6′0″, 6′1″, 5′10″, 5′11″, 5′10″, 6′0″, 6′1″, 5′11″, 6′0″, 6′0″, 6′0″

(a) What are the sample mean, median, mode, sample variance, and sample standard deviation?

(b) Consider a sample of size 4: 5′10″, 6′0″, 6′1″, 5′11″. What are the sample mean $\overline{x}$, sample variance s^2, and sample standard deviations?

5. Toss a coin 50 times and record the sequence of heads and tails. Construct a frequency table.
   (a) From your data determine the empirical probability $\hat{p}$ of heads.
   (b) Determine the mean number of heads of this sample.
   (c) Compute the sample variance of your data.

6. Toss a die 50 times and record the sequence of face values. Construct a frequency table.
   (a) From your data determine the empirical probability $\hat{p}$ of tossing a 4.
   (b) Determine the mean face value for this sample.
   (c) Compute the sample variance of your data.

7. The traffic light at Broad and Market Streets is adjusted so that the red light is on for 60 seconds and the green is on for 60 seconds. A survey is taken to determine the average number of cars stopped at a red light on Broad Street and the average number of cars stopped at a red light on Market Street during the rush hour on a randomly selected weekday (there are 30 observations per street). The results are shown in Table 7.

TABLE 7

Market Street				
20	18	16	12	20
19	20	18	17	16
20	25	19	20	18
19	15	20	18	20
16	18	18	19	21
20	19	20	17	16

Broad Street				
10	7	8	12	9
5	10	6	9	10
8	7	9	10	5
7	9	10	12	8
10	7	8	6	5
6	10	7	5	9

   (a) Construct a frequency table for each of the sets of data.
   (b) What is the mean number of cars stopped at a red light on Market Street during rush hour?
   (c) What is the mean number of cars stopped at a red light on Broad Street during rush hour?
   (d) Based on the information obtained from this survey, do you think the timing of the traffic lights should be changed?

8. Rapid Foods stocks Brand A canned corn and Brand B canned corn. Tests indicate that the mean weight of 1000 cans of Brand A is 10 ounces with a standard deviation of 1 ounce. The mean weight of Brand B is 10 ounces with a standard deviation of 0.5 ounces. What brand should you choose and why?

9. Maxtime Corporation has two locations in mind for relocation of the main plant. A survey was made to determine traveling time for each employee to each location. The data collected (time in minutes) are shown in Table 8.

TABLE 8

Location A				
35	40	80	65	30
45	80	40	35	40
50	35	30	35	50
15	50	35	40	35
25	75	50	40	40
15	60	45	35	40
70	60	25	40	20
70	80	25	30	25

Location B				
95	35	40	65	45
80	25	15	30	15
70	60	15	20	10
80	10	20	30	75
20	10	60	10	40
20	25	25	60	55
15	30	50	30	80
60	10	5	10	95

   (a) What is the mean traveling time for each location?
   (b) What is the sample standard deviation s for each location?
   (c) What is the median time for each location?
   (d) What is the mode for each location?
   (e) If the location is to be chosen based on traveling time, which location would you choose? Explain your answer.

8.8 HYPOTHESIS TESTING; THE CHI-SQUARE TEST (OPTIONAL)

Suppose we observe a finite number of values of a random variable X and we hypothesize that this set satisfies a particular probability model. We want to decide if our hypothesis regarding this model is correct. In this section we discuss a statistical procedure, called the chi-square test of goodness of fit, which can be used to test the validity of our hypothesis.

To motivate some basic ideas, consider the following situation. A good friend, Mr. X, produces a coin that he insists is fair. We watch him flip this coin three times, and each time the coin turns up heads. If the coin really is fair, as Mr. X claims, the probability of tossing three heads in a row is

$$\frac{1}{2^3} = \frac{1}{8}$$

Although this probability is small, Mr. X insists that the coin is fair, so we have no strong reason to doubt his claim. Mr. X continues to flip the coin and still after 10 tosses every single flip has turned up heads. If the coin is fair, as Mr. X claims, the probability of tossing 10 heads in a row is

$$\frac{1}{2^{10}} = \frac{1}{1024}$$

Since this probability is so small, we are beginning to have doubts that the coin is fair. But Mr. X insists so vigorously that the coin is fair that we are willing to give him the benefit of the doubt. We view his feat of tossing 10 heads in a row as a freak occurrence. Mr. X continues to flip the coin and still after 30 tosses every flip has turned up heads. If the coin is fair, as Mr. X claims, the probability of tossing 30 heads in a row is

$$\frac{1}{2^{30}} \cong .000000001$$

Thus, the chances of flipping 30 heads in a row with a fair coin are less than 1 in a billion. The odds against tossing 30 heads in a row with a fair coin are now so great that we are forced to reject Mr. X's claim and conclude that the coin is unfair.

This simple example illustrates an important principle about probability models. In any probabilistic experiment we expect some discrepancy between observed frequencies and frequencies suggested by the model. If the discrepancy is reasonably small, we are inclined to accept the validity of the model and attribute the discrepancy to chance. If, as in the case of tossing 30 heads in a row, there is a big difference between observed frequencies and the frequencies suggested by the model, we are inclined to reject the validity of the model.

Before we try to make these ideas mathematically more precise, let us consider some examples that illustrate the types of problems we want to consider.

EXAMPLE 1 A die is tossed 60 times. If the die is fair, we expect each sample point in the sample space

$$S = \{1, 2, 3, 4, 5, 6\}$$

to occur 10 times in the 60 tosses. These theoretical frequencies together with the frequencies actually observed are shown in Table 1. We must decide whether the observed data in the problem are consistent with the assumption that the die is fair. In other words, is the difference between the observed and theoretical frequencies due to chance or is the die loaded?

TABLE 1

	1	2	3	4	5	6
Observed frequencies	7	12	8	11	9	13
Theoretical frequencies	10	10	10	10	10	10

EXAMPLE 2 A coin is tossed 1000 times. If the coin is fair we expect each point in the sample space

$$S = \{h, t\}$$

to occur 500 times in the 1000 tosses. These theoretical frequencies together with the frequencies actually observed are shown in Table 2. We must decide whether the difference between the observed and theoretical frequencies is due to chance or whether the coin is unfair. In other words, should we accept or reject the uniform probability model for this experiment?

TABLE 2

	h	*t*
Observed frequencies	450	550
Theoretical frequencies	500	500

EXAMPLE 3 **GENETICS** The Mendelian* inheritance theory in genetics states that in crossing two kinds of peas, four types of seeds will result:

$$A \quad B \quad C \quad D$$

According to the theory, the different types occur with probabilities

$$P(A) = \frac{9}{16} \qquad P(B) = \frac{3}{16} \qquad P(C) = \frac{3}{16} \qquad P(D) = \frac{1}{16}$$

***Gregor Johann Mendel* (1822–1884) was an Austrian Augustinian monk. Working in a small monastery garden, he discovered the first laws of heredity and thereby laid the foundation for the science of genetics. For many years Mendel taught science in the technical high school at Brünn, Austria (later called Brno, Czechoslovakia) without a teacher's license. The reason—he failed the *biology* portion of the license examination!

If 320 seeds are obtained from crossing the two kinds of peas, then Mendelian theory predicts that the number of type A seeds will be

$$\frac{9}{16}(320) = 180$$

Similarly, the number of seeds of the remaining types should theoretically be

$$\frac{3}{16}(320) = 60 \qquad \text{type } B \text{ seeds}$$

$$\frac{3}{16}(320) = 60 \qquad \text{type } C \text{ seeds}$$

$$\frac{1}{16}(320) = 20 \qquad \text{type } D \text{ seeds}$$

These theoretical frequencies together with the number of seeds of each type actually observed are shown in Table 3. We must decide whether the difference between the observed and theoretical values is due to chance or whether the data in this experiment conflict with the Mendelian theory.

TABLE 3

	A	*B*	*C*	*D*
Observed frequencies	168	65	68	19
Theoretical frequencies	180	60	60	20

Each of the above examples is a special case of the general problem we now describe. Suppose an experiment with a finite sample space

$$S = \{s_1, s_2, \ldots, s_k\}$$

has been assigned a probability model. Specifically, suppose the elementary events have probabilities

$$P(\{s_1\}) = p_1 \qquad P(\{s_2\}) = p_2 \qquad \cdots \qquad P(\{s_k\}) = p_k$$

If the experiment is repeated n times, then the sample points should theoretically occur with frequencies

$$f_1 = np_1 \qquad f_2 = np_2 \qquad \cdots \qquad f_k = np_k \tag{1}$$

in the n repetitions. For example, if $p_1 = P(\{s_1\}) = .3$, then in $n = 50$ repetitions of the experiment we would expect sample point s_1 to occur $np_1 = 50(.3) = 15$ times. We call the frequencies $f_1, f_2, \ldots, f_k$ in (1) the **expected frequencies** or **theoretical frequencies**. When the experiment is performed physically n times, we observe certain frequencies

$$o_1, o_2, \ldots, o_k \tag{2}$$

for the sample points $s_1, s_2, \ldots, s_k$. We call the numbers in (2) the **observed frequencies**.

Usually the observed frequencies and the theoretical frequencies are different, and the problem is to decide whether the difference can be attributed to chance or whether the difference results from an incorrect probability model. To make this decision, we need a way of measuring how much the observed frequencies

$$o_1, o_2, \ldots, o_k$$

differ from the theoretical frequencies

$$f_1, f_2, \ldots, f_k$$

Although there are many possible ways of measuring this difference, statisticians have settled on the following quantity called **chi-square** (written χ^2):

$$\chi^2 = \frac{(o_1 - f_1)^2}{f_1} + \frac{(o_2 - f_2)^2}{f_2} + \cdots + \frac{(o_k - f_k)^2}{f_k} \tag{3}$$

[*Chi* (χ) is a letter in the Greek alphabet; it is pronounced "kai."]

EXAMPLE 4 Find the value of χ^2 for the data in Example 3.

Solution From Table 3, the observed frequencies are

$$o_1 = 168 \qquad o_2 = 65 \qquad o_3 = 68 \qquad o_4 = 19$$

and the theoretical frequencies are

$$f_1 = 180 \qquad f_2 = 60 \qquad f_3 = 60 \qquad f_4 = 20$$

Substituting these values in Formula (3) yields

$$\chi^2 = \frac{(168 - 180)^2}{180} + \frac{(65 - 60)^2}{60} + \frac{(68 - 60)^2}{60} + \frac{(19 - 20)^2}{20}$$

$$= \frac{420}{180} \cong 2.33$$

If the observed frequencies are the same as the theoretical frequencies, then

$$o_1 - f_1 = 0 \qquad o_2 - f_2 = 0 \qquad \ldots \qquad o_k - f_k = 0$$

so that from Formula (3), $\chi^2 = 0$. As the differences

$$o_1 - f_1 \qquad o_2 - f_2 \qquad \ldots \qquad o_k - f_k$$

become larger, the value of χ^2 increases. Thus, χ^2 is a measure of how closely the observed frequencies agree with the theoretical frequencies predicted by the probability model; a small value of χ^2 means good agreement and a large value of χ^2 means poor agreement.

This suggests the following procedure for using experimental data to decide whether to accept or reject the correctness of a probability model:

Step 1 Pick a cutoff point that divides "large values" of χ^2 from "small values."

Step 2 If the value of χ^2 obtained from the data is "small," then accept the correctness of the model; if the value of χ^2 is "large," then reject the correctness of the model (Figure 1).

FIGURE 1

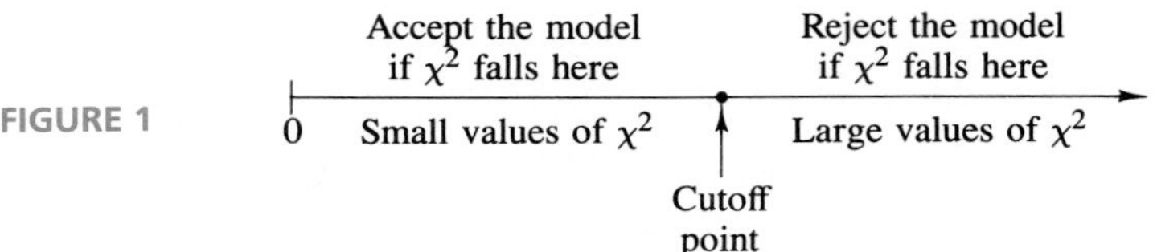

Before we can show how to select the cutoff point, we need some preliminary ideas. If a probability model is assigned to an experiment with a finite sample space, then the quantity χ^2 becomes a random variable whose value is determined by the frequencies observed in n repetitions of the experiment. Moreover, since the original experiment has a finite sample space, χ^2 will be a finite discrete random variable. (We omit the details.) Thus the values of χ^2 have various probabilities of occurring, and these probabilities can be represented as areas in a histogram. Just as we were able to approximate areas in a binomial histogram by areas under a normal curve, so can we, in certain cases, approximate the areas in a chi-square histogram by areas under a curve called a **chi-square curve**. The exact shape of the chi-square curve depends on the value of k in Formula (3). Some typical chi-square curves are shown in Figure 2. As was the case for normal curves, the area under every chi-square curve is 1.

FIGURE 2

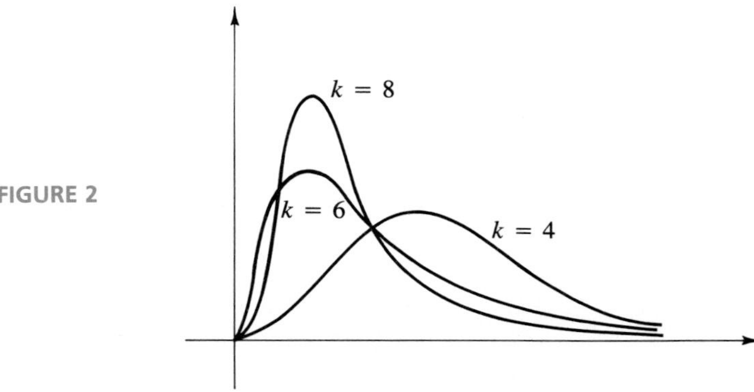

In selecting the cutoff point c that divides "large" values of χ^2 from "small" values, we must recognize that no matter how the value of c is chosen, we will make an error a certain percentage of the time. Even if the probability model is correct, there is an outside chance that the value of χ^2 will be greater than c, in which case we would *incorrectly* reject the model. This is called a type I error. It is common in statistical work to select the cutoff point c so that this kind of error occurs only 5% of the time. In this case, c is called the 5% **critical level**. To illustrate how the 5% critical level can be obtained, let us assume that the areas of the rectangles in the chi-square histogram can be approximated by the corresponding areas under a chi-square curve. Since the total area under a chi-square curve is 1, there is a point c on the x-axis that divides the area under the curve into two parts: an area of .95 to the left of c and an area of .05 to the right of c (Figure 3). Since areas under the chi-square curve correspond to probabilities for the χ^2 random variable, the probability of obtaining a value of χ^2 less than c is .95 and the probability of obtaining a value of χ^2 greater than c is .05 when the probability model being tested is correct. Thus, if we use c as the cutoff point between large and small values of χ^2, the probability of rejecting a correct model will be .05.

FIGURE 3

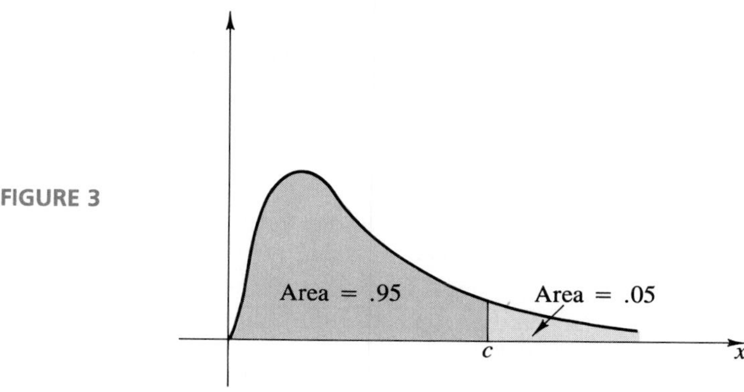

Appendix Table C.3 lists the 5% critical levels for various chi-square curves. Observe that the first row in Table C.3 is labeled ν (Greek letter *nu*). The quantity ν is called the **number of degrees of freedom** for the chi-square curve, and its value is

$$\nu = k - 1 \tag{4}$$

where k is the number of terms in the formula for χ^2 [Formula (3)]. To use the table, determine the value of ν for the problem and read off the critical level c next to the value of ν just determined. To illustrate, consider the die-tossing problem in Example 1. Since the formula for χ^2 contains $k = 6$ terms, the approximating chi-square curve has $\nu = k - 1 = 5$ degrees of freedom, so from Table C.3 the 5% critical level is $c = 11.1$.

To summarize, we have the following procedure for using experimental data to decide whether to accept or reject the validity of a hypothesis of a probability model for a finite-sample-space experiment:

Step 1 Compute the value of χ^2 from the experimental data.

Step 2 Compute the number of degrees of freedom from Formula (4).

Step 3 Determine the 5% critical level c from Appendix Table C.3.

Step 4 If the value of χ^2 is greater than c, reject the validity of the probability model; if the value of χ^2 is less than c, accept the validity of the model.

This four-step procedure is called the **chi-square test of goodness of fit at the 5% significance level**. The phrase "5% significance level" reminds us that this test is designed with the understanding that there is a probability of .05, or 5%, that a correct model will be rejected.

EXAMPLE 5 Use the chi-square test at the 5% significance level to determine whether the data in Table 1 support the assumption that the die is fair.

Solution From the data in Table 1 and from Formula (3), we obtain

$$\chi^2 = \frac{(7-10)^2}{10} + \frac{(12-10)^2}{10} + \frac{(8-10)^2}{10} + \frac{(11-10)^2}{10} + \frac{(9-10)^2}{10} + \frac{(13-10)^2}{10}$$
$$= 2.8$$

Since the above formula for χ^2 contains $k = 6$ terms, the number of degrees of freedom is $\nu = k - 1 = 5$, so from Appendix Table C.3 the cutoff point is $c = 11.1$. Since the value of χ^2 is less than 11.1, we accept the probability model and conclude that the observed data are consistent with the assumption that the die is fair.

EXAMPLE 6 Use the chi-square test at the 5% significance level to determine whether the data in Table 2 support the assumption that the coin is fair.

Solution From the data in Table 2 and from Formula (3), we obtain

$$\chi^2 = \frac{(450-500)^2}{500} + \frac{(550-500)^2}{500} = 10$$

Since this formula for χ^2 contains $k = 2$ terms, the number of degrees of freedom is $\nu = k - 1 = 1$, so from Appendix Table C.3 the 5% critical level is $c = 3.84$. The value of χ^2 is greater than c, so we conclude that the coin is not fair.

In this section we restrict our discussion to chi-square tests at the 5% significance level, which means that a correct probability model will be erroneously rejected 5% of the time (a type I error). In problems where severe consequences result from rejecting a correct model, it may be desirable to reduce this error to less than 5%. In still other kinds of problems an error of more than 5% may be tolerable. For such problems, there are chi-square tests at other significance levels. These tests are performed exactly as those at the 5% level with the exception that different tables are used to obtain the cutoff points.

EXERCISE SET 8.8

1. Find the value of χ^2 for the data in the following tables.

(a)

Observed frequencies	41	59
Theoretical frequencies	50	50

(b)

Observed frequencies	147	316	537
Theoretical frequencies	200	300	500

(c)

Observed frequencies	27	33	31	49
Theoretical frequencies	20	30	40	50

2. Find the 5% critical level for the chi-square curve with
(a) $\nu = 4$ degrees of freedom
(b) $\nu = 10$ degrees of freedom

3. In New York City, 210 automobile accidents involving an intoxicated driver were classified according to the day of the week on which the accident occurred; the following data were obtained:

Day	Mon.	Tues.	Wed.	Thurs.	Fri.	Sat.	Sun.
Number of accidents	27	18	28	20	31	46	40

Use a chi-square test at the 5% significance level to determine if these data support the assumption that an accident involving an intoxicated driver is equally likely to occur on any day of the week.

4. In an experiment designed to control the sex of baby rabbits by chemical injections to the mother, a sample of 50 baby rabbits yielded 30 males and 20 females. Use a chi-square test at the 5% significance level to determine if these data support the hypothesis that male births and female births are equally likely.

5. In a survey of consumer opinion, 500 people were asked to use toothpastes A, B, and C, each for 1 week. At the end of 3 weeks, the individuals were asked to name the product that they felt was most effective. The results were

Toothpaste	A	B	C
Number of people naming it most effective	176	159	165

Use a chi-square test at the 5% significance level to determine whether these data support the conclusion that the consumers regard the three products to be equally effective.

6. According to genetic theory, the offspring of a certain cross of bearded iris plants should be colored pink, blue, or red with probabilities $\frac{9}{16}$, $\frac{3}{16}$, and $\frac{4}{16}$, respectively. If an experiment yields 68, 34, and 42 in each category, does the experiment support the theory? Use a chi-square test at the 5% significance level.

7. In a certain manufacturing plant, 800 of the employees are men and 200 are women. Company records show that during the previous year 33 job-related accidents involved men and 7 job-related accidents involved women. Use a chi-square test at the 5% significance level to decide whether these data support the contention that men and women are equally likely to be involved in job-related accidents.

8. Workers on an assembly line install a particular part on an engine block. Fifty engines are assembled per hour, and the probability that the part will be improperly installed varies with the time of day. A consulting firm constructs the following model to describe the probability of an improper installation:

Time interval	8 to 10 A.M.	10 A.M. to 12 P.M.	1 to 3 P.M.	3 to 5 P.M.
Probability of improper installation	.01	.05	.02	.06

Suppose a count of actual improper installations yields the following results:

Time interval	8 to 10 A.M.	10 A.M. to 12 P.M.	1 to 3 P.M.	3 to 5 P.M.
Number of improper installations	1	1	5	9

Is the model valid at the 5% significance level?

9. Suppose the following revision of the model in Exercise 8 is made:

Time interval	8 to 10 A.M.	10 A.M. to 12 P.M.	1 to 3 P.M.	3 to 5 P.M.
Probability of improper installation	.01	.02	.04	.07

Is the new model valid at the 5% significance level?

KEY IDEAS FOR REVIEW

- **Random variable** A rule that assigns a numerical value to each outcome of an experiment.
- **Finite discrete random variable** A random variable that can take on only finitely many values.
- **Infinite discrete random variable** A random variable that can take on infinitely many values that can be arranged in a sequence.
- **Continuous random variable** A random variable whose values form an interval.
- **Probability function** The set of ordered pairs relating the values of a discrete random variable with their probabilies.
- **Mean (or arithmetic average) of numbers $x_1, x_2, \ldots, x_n$**

$$\overline{x} = \frac{x_1 + x_2 + \cdots + x_n}{n}$$

- **Expected value (mean, expectation) of X**

$$\mu = E(X) = p_1x_1 + p_2x_2 + \cdots + p_kx_k$$

- **Deviation of x from μ** $x - \mu$
- **Variance of X**

$$\sigma^2 = \text{Var}(X) = p_1(x_1 - \mu)^2 + p_2(x_2 - \mu)^2 + \cdots + p_k(x_k - \mu)^2$$
$$\sigma^2 = \text{Var}(X) = (p_1x_1^2 + p_2x_2^2 + \cdots + p_kx_k^2) - \mu^2$$

- **Standard deviation of X** $\sigma_X = \sqrt{\text{Var}(X)}$
- **Chebyshev's inequality (theorem)** See pages 321–322.
- **Bernoulli experiment (trial)** An experiment with two possible outcomes.
- **Binomial experiment** n independent repetitions of a Bernoulli experiment.
- **Binomial random variable** The number of successes in a binomial experiment.

Mean and variance of a binomial random variable

$$\mu = np \qquad \sigma^2 = npq$$

Normal (or Gaussian) curve See page 336.

Standard normal curve The normal curve with $\mu = 0$ and $\sigma = 1$.

Population The set of objects whose properties are to be considered.

Sample of size n An n-element subset of the population.

Parameter A numerical characteristic of the population.

Statistic A numerical characteristic of the sample.

Data The values that are observed for each element of the population or sample.

Sample mean $\bar{x} = \dfrac{x_1 + x_2 + \cdots + x_n}{n}$

where n is the number of elements in the sample.

Sample variance $s^2 = \dfrac{(x_1 - \bar{x})^2 + (x_2 - \bar{x})^2 + \cdots + (x_n - \bar{x})^2}{n - 1}$

where n is the number of elements in the sample.

Point estimate The sample statistic.

Chi-square (χ^2)

$$\chi^2 = \frac{(o_1 - f_1)^2}{f_1} + \frac{(o_2 - f_2)^2}{f_2} + \cdots + \frac{(o_k - f_k)^2}{f_k}$$

Critical level (in chi-square test) See page 363.

Degrees of freedom (in chi-square test) See page 363.

SUPPLEMENTARY EXERCISES

Use the following information for Exercises 1–4. A fair die is tossed three times. The random variable X represents the number of 6s appearing in the three tosses.

1. Graph the probability function for this random variable as a line graph and as a histogram.
2. Determine $P(X = 3)$, $P(X < 3)$, and the probability that at least one 6 will occur on the three tosses.
3. Determine $E(X)$.
4. Determine $\text{Var}(X)$.

Use the following information for Exercises 5–8. You have a die that has the numbers 2, 3, 4, 5, 6, 6 on its faces. (Note there is no number 1.) Toss this die four times. The random variable X represents the number of 6s appearing in the four tosses.

5. Graph the probability function for this random variable as a line graph and as a histogram.
6. Determine $P(X = 3)$, $P(X < 3)$, and the probability that at most one 6 will occur on the four tosses.
7. Determine $E(X)$.
8. Determine $\text{Var}(X)$.
9. You are at a casino in Atlantic City playing roulette and are placing a \$5 bet on the two numbers 35 and 36. If neither number comes up, you lose your bet. If one of

these numbers comes up, you win \$85 (and get your bet back). What is the expected value per dollar bet, and what is the house percentage?

10. You are playing roulette and you place a \$10 bet on the numbers 1–12. If one of these numbers comes up, you win \$20 (and get your bet back). If any other number comes up, you lose your bet. What is the expected value of this game per dollar bet, and what is the house percentage?

11. You are playing roulette and place a \$5 bet on the numbers 0, 00, 1, 2, and 3. If one of these numbers comes up, you win \$30 (and get your bet back). If any other number comes up, you lose your bet. What is the expected value of this game per dollar bet, and what is the house percentage?

12. You are at the roulette table and want to place a \$20 bet. If your decision is to be based on expected values per dollar bet, which of the plays of Exercises 10 or 11 would you choose?

13. A businessman has two investment opportunities: project A and project B. If he invests in project A, there is a 25% chance that he will lose \$50,000, a 60% chance that he will lose \$10,000, and a 15% chance that he will make \$300,000. If he invests in project B, there is a 15% chance that he will lose \$70,000, a 25% chance that he will lose \$10,000, and a 60% chance that he will make \$80,000. If his decision is to be based on expected value comparisons, which project will he choose?

Use the following information for Exercises 14–16. There are six independent repetitions of a Bernoulli experiment with success probability $= \frac{1}{4}$.

14. Determine:
(a) the probability of no successes
(b) the probability of one success
(c) the probability of at least one success

15. Determine the mean.

16. Determine the variance.

17. A fair die is tossed 5 times.
(a) What is the probability of obtaining exactly two 3s in the 5 tosses?
(b) What are the mean and variance?

18. Given the normal curve with $\mu = 1500$ and $\sigma = 30$, for this normal distribution find:
(a) $P(X \leq 1540)$
(b) $P(1500 \leq X \leq 1540)$
(c) $P(X > 1540)$

19. Given a normal curve with $\mu = 80$ and $\sigma = 4$, for this normal distribution find:
(a) $P(X \leq 70)$
(b) $P(70 \leq X \leq 84)$
(c) $P(X > 84)$

20. Let Y be the binomial random variable with $n = 1800$ and $p = \frac{1}{3}$. If the binomial distribution is approximated by the normal distribution, determine:
(a) μ (b) σ (c) $P(Y \leq 600)$
(d) $P(590 \leq Y \leq 650)$

21. Let Y be a binomial random variable with $n = 1200$ and $p = \frac{1}{4}$. If the binomial distribution is approximated by the normal distribution, determine:
(a) μ (b) σ (c) $P(Y \geq 350)$
(d) $P(250 \leq Y \leq 320)$

22. A survey is given to determine consumer preference of brand B to brand X. Of 300 people selected at random, 200 preferred brand B, 50 preferred brand X, and 50 had no opinion.
(a) On the basis of this survey, what is the estimated probability that a consumer will prefer brand B?
(b) What is the probability that of two consumers selected at random, both will prefer brand B? Assume independence.

23. Mr. White wants to determine the number of words he can type in 1 minute. He records the outcomes from 20 different trials:

45	48	43	40	38
48	42	39	39	48
42	43	40	43	48
49	45	42	46	40

(a) Construct a frequency table.
(b) What is the mean?
(c) What is the sample variance?
(d) What is the median?
(e) What is the mode?

24. A survey is made of a random sample of 20 women buying a new book that has just been published. The question of interest is the mean age of purchasers of this book. The ages of the 20 women are

25	20	40	35	25
20	18	25	35	22
35	22	20	22	25
25	30	25	19	23

(a) What is the mean age?
(b) What is the median age?

(c) What is the mode for the data?
(d) What is the sample variance?

25. To determine the proportion of television viewers watching channels 1, 2, and 3 at 9 P.M. Sunday, 50 viewers were randomly selected and polled. The results were

1	1	1	2	3	1	2	2	1	1
2	2	1	1	2	1	3	3	3	3
1	3	1	1	2	2	1	2	2	3
1	1	1	1	2	3	1	3	3	1
3	1	3	2	3	1	1	1	1	2

(a) What is the point estimate for the proportion of viewers watching channel 1?
(b) According to this survey, which channel had the most viewers at 9 P.M. Sunday?

26. Toss two dice 50 times and record the sum of each toss. Construct a frequency table.

(a) What is the empirical probability of tossing a sum of 7?
(b) What is the mean sum?
(c) Compute the sample variance of the data.
(d) Compute the sample standard deviation of the data.

27. **(Optional)** In a survey of viewer opinion, 500 persons were asked which of channels 1, 2, or 3 each watched the most during primetime Sunday night. The results were

Channel	1	2	3
Number of viewers naming it the preferred channel	175	160	165

Use a chi-square 5% significance level to determine if these data support the conclusion that the viewers regard the three channels to be equally good.

CHAPTER TEST

1. In a batch of light bulbs, there is a 1% chance that a light bulb is defective. Select three bulbs at random and let the random variable X represent the number of defective bulbs in this sample.
(a) Assuming independence, determine

$P(X = 0)$ $\quad$ $P(X = 1)$ $\quad$ $P(X = 2)$ $\quad$ $P(X = 3)$

(b) Determine the probability that at most one bulb is defective.
(c) Construct a histogram for this distribution.
(d) Determine $E(X)$ and $\text{Var}(X)$.

2. Let X be the random variable whose probability function is described in the table. Compute the mean and variance of X.

x	1	2	3	4	5
$p(x)$	.1	.1	.3	.4	.1

3. There are five independent repetitions of a Bernoulli experiment with success probability of $\frac{1}{3}$. Determine:
(a) the probability of no successes
(b) the probability of one success
(c) the probability of at least one success
(d) the mean (e) the variance

4. We are given the normal curve with $\mu = 1800$ and $\sigma = 30$. For this normal distribution, find:

(a) $P(X \leq 1700)$ $\quad$ (b) $P(1810 \leq X \leq 1850)$
(c) $P(X > 1700)$

5. Let Y be the binomial random variable with $n = 1500$ and $p = \frac{1}{5}$. Using the normal approximation for the binomial distribution, determine
(a) $P(Y \leq 300)$ $\quad$ (b) $P(300 \leq Y \leq 350)$

6. Of a sample of 500 eligible voters, a survey indicates that 150 prefer the Republican candidate, 250 prefer the Democratic candidate, and 100 are undecided.
(a) On the basis of this survey, what is the estimated probability that a voter selected at random will vote for the Democratic candidate?
(b) If 3 voters are selected at random, what is the estimated probability that at least 2 will vote for the Democratic candidate? (Assume independence.)

7. A survey is made of a random sample of 24 people entering a supermarket to determine the average salary of shoppers at this market. The data collected (in thousands of dollars) are

21	15	30	40	45	30	90	28
28	25	28	35	35	40	50	32
30	40	40	38	18	20	25	35

(a) What is the mean salary?
(b) What is the sample variance?
(c) What is the median salary?

9

Mathematics of Finance

In this chapter we discuss mathematical methods and formulas that are useful in business and personal finance. Tables are included in Appendix C to simplify the computations; however, many of the computations can also be performed on microcomputers, standard hand-held calculators, or special-purpose business calculators that are in current use. Regardless of whether tables or calculators are used, it is essential to understand the basic principles developed here.

One of the fundamental concepts in the mathematics of finance is the "time value of money"—the value of a particular amount of money at various points in time. For example, suppose you have \$100 today, what will it be worth at the end of 1 year?* You might lend the \$100 to a friend who promises to pay back \$110 at the end of the year, in which case your \$100 will have a value of \$110 at the end of the year, or you might buy shares of a stock that decreases in value to \$90 at the end of the year. In this case, if you sell the stock, then your \$100 of today will have a value of only \$90 at the end of the year.

Every financial transaction can be considered as one between two parties: the lender and the borrower. For example, if you buy a U.S. Series EE savings bond, you are lending the government an amount of money, and the government will return the money with interest at the time you cash in the bond. When you deposit money in a savings account, you are lending the bank this money, and in return the bank pays you interest.

The initial investment is called the **principal** or **present value** of a transaction and is denoted by P. The amount that we will have after t years is called the **future value** or **amount** of P and is denoted by S. The interest I is the increase in the value of P after t years:

$$I = S - P$$

---

* In this chapter the effect of inflation is ignored.

You can think of interest as the rent a borrower pays for the use of money lent to him or her. In this chapter we consider three types of interest: simple interest, compound interest, and simple bank discount, as well as related applications.

If there is a decrease in the value of P after t years, this decrease is referred to as **depreciation**. For example, if a new car costs \$10,000 and its resale value at the end of 3 years is \$6000, then we say that the value of the car has depreciated by \$4000. See also Section 1.5.

9.1 SIMPLE INTEREST

In Section 1.4 we discuss simple interest as an application of linear equations. Let us review the basic ideas. Simple interest involves three considerations.

P = **principal** (amount borrowed)
r = **interest rate** (a percentage per year that must be converted to decimal form for computation)
t = **time** (in years that the principal is held)

By definition, simple interest I is given by

$$I = Prt \qquad (r \text{ is the decimal form of the interest rate})^*$$

and the amount S after t years is

$$\begin{aligned} S &= P + I \\ &= P + Prt \\ &= P(1 + rt) \end{aligned} \tag{1}$$

From (1), solving for P,

$$P = \frac{S}{1 + rt} \tag{2}$$

Therefore, at simple interest the principal P today will have the value S in t years; and S in t years has the value P today.

A time diagram helps visualize a transaction (Figure 1). Note that we represent "now" by $t = 0$.

FIGURE 1

If P is given, then S is the **future value** or **amount** of P and $S = P(1 + rt)$ (Figure 1a).

If S is given, then P is the **present** or **past value** of S and $P = S/(1 + rt)$ (Figure 1b).

*See Appendix Section A.1.

Note that the head of the arrow in the time diagram points toward the time at which we want to determine the value of the given sum of money. The value of money can be thought of as moving either forward or backward in time. If it is moved forward, use the formula for amount:

$$S = P(1 + rt)$$

If it is moved backward, use the formula for present value:

$$P = \frac{S}{1 + rt}$$

EXAMPLE 1 John borrows \$1000 on January 1, 1990; the simple interest rate is 6%.

(a) What amount must be paid back on January 1, 1992?

(b) What is the interest on the loan?

(c) What is the value of the loan on January 1, 1992?

Solution (a) Note that 6% must be converted to $\frac{6}{100} = .06$ for computation purposes. Let $P = 1000$, $r = .06$, and $t = 2$. Since we want the future value of the \$1000, we use

$$S = P(1 + rt) = 1000[1 + (.06)2] = \$1120$$

The time diagram is shown in Figure 2. (The arrow indicates that $P = \$1000$ is given and S is determined 2 years later.)

FIGURE 2

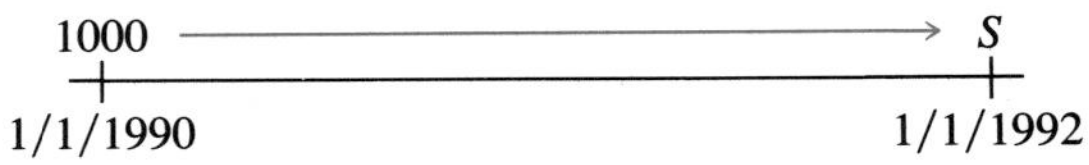

Solution (b) $I = S - P = 1120 - 1000 = \$120$

Solution (c) The value of the loan on January 1, 1992, is $S = \$1120$.

EXAMPLE 2 (a) How much should an investor deposit in the bank now in order to have \$1000 in the bank $\frac{1}{2}$ year from now? The simple interest rate is 8%.

(b) What is the interest received?

(c) What is the present value of \$1000 due in $\frac{1}{2}$ year at the simple interest rate of 8%?

Solution (a) The time line is shown in Figure 3. Let $S = 1000$, $r = .08$, and $t = \frac{1}{2}$. (Note the arrow in Figure 3 goes from $S = \$1000$ to P because S is given and P is to be determined.) Then,

$$P = \frac{S}{1 + rt} = \frac{1000}{1 + .08\left(\frac{1}{2}\right)} = \frac{1000}{1.04} = \$961.54$$

FIGURE 3

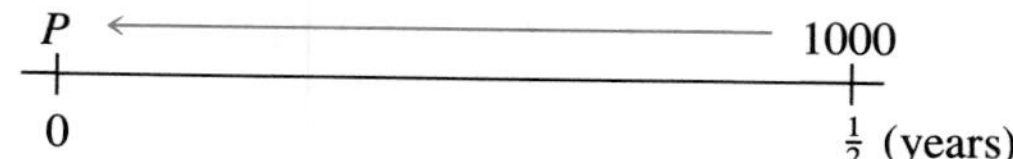

Thus, the investor should deposit $961.54 now.

Solution (b) $I = S - P = 1000 - 961.54 = \$38.46$

Solution (c) P is the present value (at $t = 0$). Therefore, the answer is the same as in (a).

EXAMPLE 3 A widow deposits $100 on February 1, 1990, into an account paying 7% simple interest. One-half year later, she deposits $300 into this account. What will her bank balance S be on February 1, 1992? Assume no other deposits and no withdrawals during this time period.

Solution The time line is shown in Figure 4. Let

$$\begin{aligned} S &= \text{sum of the (future) values of the two deposits on February 1, 1992} \\ &= S_1 + S_2 \end{aligned}$$

where

$$S_1 = P_1(1 + rt_1) = 100[1 + .07(2)] = 114.00$$

$$S_2 = P_2(1 + rt_2) = 300[1 + .07\left(\frac{3}{2}\right)] = 331.50$$

Therefore,

$$S = 114.00 + 331.50 = \$445.50$$

Thus, the widow's bank balance will be $445.50 on February 1, 1992.

FIGURE 4

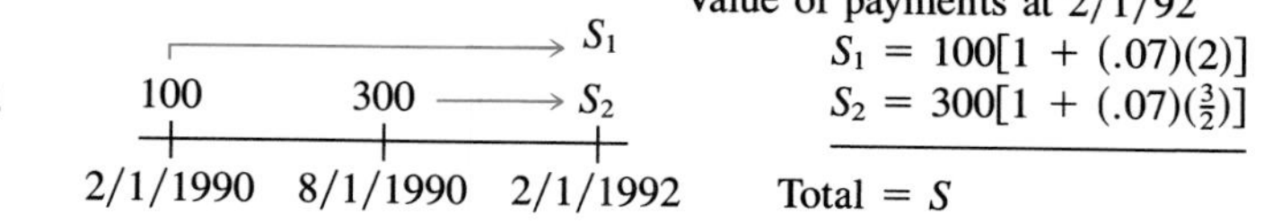

For many transactions, the time may be given in months, weeks, or days. However, in the simple interest formula, t must be in years, and so a conversion must be made:

$$k \text{ months} = \frac{k}{12} \text{ years} \qquad 3 \text{ months} = \frac{3}{12} = \frac{1}{4} \text{ years}$$

$$n \text{ weeks} = \frac{n}{52} \text{ years} \qquad 32 \text{ weeks} = \frac{32}{52} = \frac{8}{13} \text{ years}$$

If the time is given in days, historically more than one method has been used to determine the time in years (mainly to facilitate computations before the time of calculators). The exact number of days between two dates is referred to as **exact**

time. (An approximate method, based on the assumption that each month has 30 days, is occasionally used. We will use this method only when we assume that k months $= k/12$ years.)

EXAMPLE 4 The exact time from March 3, 1993, to June 6, 1993, is obtained by determining the number of days between the two dates:

Number of remaining days in March	$31 - 3 = 28$
Number of days in April	30
Number of days in May	31
Number of days in June	6
	Total = 95 days

Therefore, the exact time is 95 days.

EXAMPLE 5 The exact time from February 1, 1988, to March 10, 1992, is found by determining the number of days between the two dates. Note that 1992 is a leap year.

Number of remaining days in February	$29 - 1 = 28$
Number of days in March	10
	Total = 38 days

Therefore, the exact time is 38 days.

Given an annual simple interest rate with the time d in days, there are two methods used to convert days into years:

1. If t (in years) $= \dfrac{d \text{ (in days)}}{365}$, then the interest is said to be **exact**.

2. If t (in years) $= \dfrac{d \text{ (in days)}}{360}$, then the interest is said to be **ordinary**.

EXAMPLE 6 Jackson borrows \$1000 on June 1, 1991, for 60 days. The simple interest rate is $7\frac{1}{2}\%$.

(a) Compute the exact simple interest.

(b) Compute the ordinary simple interest.

Solution (a) We have $P = 1000$ and $r = .075$. For exact interest, $t = 60/365$ and

$$I = Prt = 1000(.075)\left(\frac{60}{365}\right) = \$12.33$$

Solution (b) For ordinary interest, $t = 60/360 = \frac{1}{6}$ and

$$I = Prt = 1000(.075)\left(\frac{1}{6}\right) = \$12.50$$

Equations of Value and Time Value of Money

Consider now the case where Jack borrows \$100 from Jill at 6% simple interest and agrees to pay \$50 on the loan in 6 months. What payment 1 year from now will settle the debt?

Set up the information on a time diagram (Figure 5) and let x represent the payment in 12 months. In this case, since there is more than one transaction, write the money borrowed above the line and the payments made below the line.

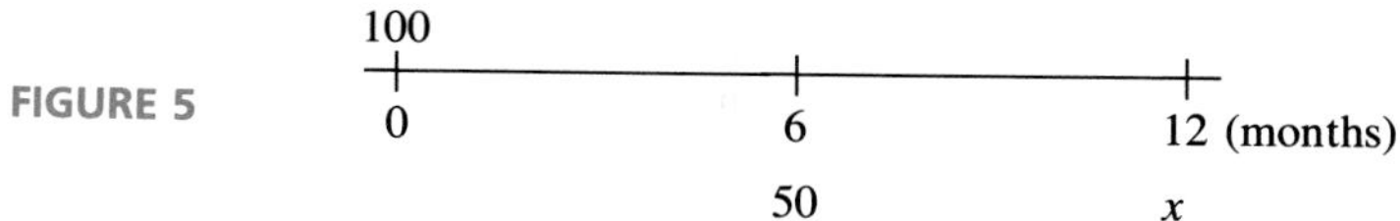

FIGURE 5

To handle this type of problem in which payments are made at different dates, we need another fundamental principle of the mathematics of finance:

> Amounts of money payable at different times can only be compared at a particular date, which is referred to as the **focal date**.

The focal date is decided on by the lender and the borrower. Once the date is selected, the value of each amount of money in the transaction is determined at the focal date and an equation of value is set up:

> **Equation of Value**
>
> value of loan at focal date = value of payments at focal date

In this problem if the focal date is selected to be 1 year from now, then the value of each sum of money must be determined at the focal date (Figure 6). The value of the \$100 loan at the focal date is $100[1 + .06(1)] = \$106$; we use $S = P(1 + rt)$ because we want the future value of \$100. The value of the \$50 payment at the focal date is $50[1 + .06(\frac{1}{2})] = \$51.50$; we use $S = P(1 + rt)$ because we want the future value of \$50. The value of the payment x at the focal date is x. (There is no shift in time.) Therefore, the equation of value is

$$\text{value of loan at focal date} = \text{value of payments at focal date}$$

$$\underbrace{\text{value of \$100 at focal date}}_{106.00} = \underbrace{\text{value of \$50 at focal date}}_{51.50} + \underbrace{\text{value of } x \text{ at focal date}}_{x}$$

Solve for x,

$$\begin{aligned} x &= 106 - 51.50 \\ &= \$54.50 \end{aligned}$$

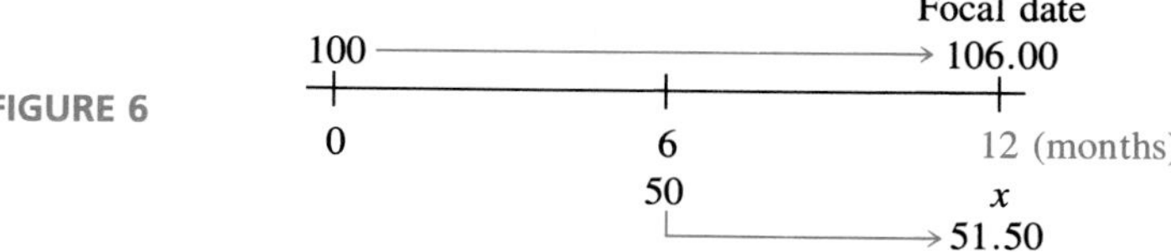

FIGURE 6

If, in this problem, the focal date is taken to be 6 months from now, then the time diagram is as shown in Figure 7. The value of the \$100 loan at the focal date is $100[1 + .06(\frac{1}{2})] = \$103$; we use $S = P(1 + rt)$ since we want the future value of \$100. The value of the \$50 payment at the focal date is \$50 (there is no shift in time). The value of the payment x at the focal date is

$$\frac{x}{1 + .06\left(\frac{1}{2}\right)} = \frac{x}{1.03}$$

We use $P = S/(1 + rt)$ since we want the past value of x.

FIGURE 7

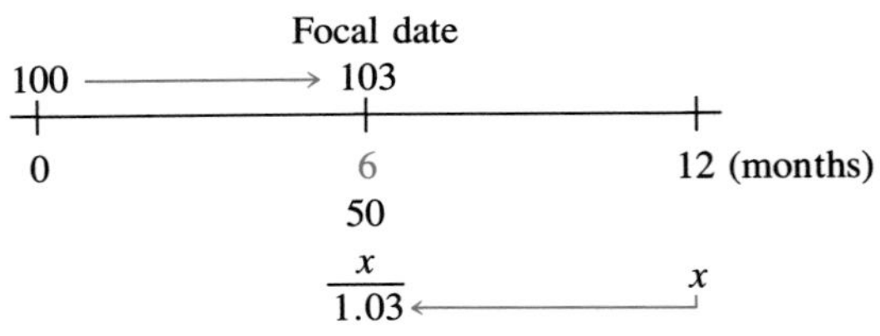

Focal date

100 ⟶ 103

0 — 6 — 12 (months)

50

$\frac{x}{1.03}$ ⟵ x

Therefore the equation of value is

$$\text{value of loan at focal date} = \text{value of payments at focal date}$$

$$\underbrace{\text{value of 100 at focal date}}_{103} = \underbrace{\text{value of 50 at focal date}}_{50} + \underbrace{\text{value of } x \text{ at focal date}}_{\frac{x}{1.03}}$$

Solving for x,

$$\frac{x}{1.03} = 103 - 50 = 53$$

$$x = 53(1.03)$$

$$= \$54.59$$

Note that different focal dates give different values of x. This difference will occur in simple interest transactions and so it is important for the parties involved to agree on the focal date.

EXAMPLE 7 Johanna borrows \$2000 now and agrees to pay \$500 in 2 months and \$700 in 6 months. What final payment should she make 18 months from now to settle this debt if the simple interest rate is 12% and the focal date is now?

Solution Construct a time diagram (Figure 8) and let x represent the final payment. The values of the loan and payments at the focal date are given in Table 1.

FIGURE 8

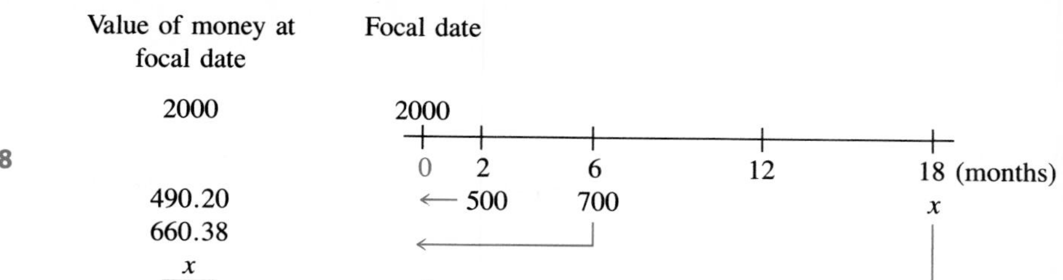

TABLE 1

Money ($)	Formula to Use	Value at Focal Date
2000	No shift in time	2000
500	$P = \dfrac{S}{1 + rt}$	$\dfrac{500}{1 + .12(2/12)} = \dfrac{500}{1.02} = 490.20$
700	$P = \dfrac{S}{1 + rt}$	$\dfrac{700}{1 + .12(6/12)} = \dfrac{700}{1.06} = 660.38$
x	$P = \dfrac{S}{1 + rt}$	$\dfrac{x}{1 + .12(18/12)} = \dfrac{x}{1.18}$

Therefore the equation of value at the focal date is

$$2000 = 490.20 + 660.38 + \frac{x}{1.18}$$

Solving for x,

$$\begin{aligned}\frac{x}{1.18} &= 2000 - 490.20 - 660.38\\ &= 849.42\\ x &= (849.42)(1.18)\\ &= \$1002.32\end{aligned}$$

EXERCISE SET 9.1

1. A principal of $3000 is borrowed at 12% simple interest. Find the future value after:
   (a) 1 year (b) 3 months (c) 3 years (d) 13 weeks

2. An investor lends $10,000 at 15% simple interest. Determine how much the investor will be owed after:
   (a) 2 years (b) 5 years (c) 18 months (d) 8 weeks

3. A certain principal is to be invested at 10% simple interest so that the future value of the investment after 1 year will be $5000. How much should be invested?

4. How much should you deposit in a Christmas Club account paying $5\frac{1}{2}$% simple interest if you want to use the money to buy a $700 television set 6 months later?

5. A depositor buys a certificate of deposit for $1000. In 90 days she will cash it in. How much will she collect if the simple interest rate is $9\frac{1}{2}$% at:
   (a) exact interest (b) ordinary interest

6. An investment club buys a certificate of deposit for $1500 and will cash it in 180 days. How much will the club collect if the simple interest rate is $8\frac{3}{4}$% at:
   (a) exact interest (b) ordinary interest

7. How much should an investor deposit in the bank now if he needs $2000 in 3 months? The simple interest rate is 9%.

8. How much should a depositor invest now at 8% simple interest so that she will have $3000 in 6 months?

9. Robert Silver deposits $100 into an account paying 8% simple interest. He makes two more deposits of $200 each: the first in 3 months and the second in 6 months. How much will be in the account at the end of the year if he makes no other deposits and no withdrawals during this time?

10. Barbara Gold deposits $500 into an account paying 7% simple interest. She makes three more deposits of $200 each at 2-month intervals. How much is in the account at the end of 1 year? Assume no other deposits and no withdrawals during this time.

11. A student borrows $1500 and agrees to pay $500 in 8 months, $500 in 1 year, and to settle the debt in 18 months. If the simple interest rate is 8% and the focal date is in 18 months, what must his final payment be?

12. An ice cream vendor borrows $1000 and agrees to make two equal payments, one in 6 months and one in 12 months to settle the debt. If the simple interest rate is 9%, what should each payment be? The focal date is in 12 months. [*Hint*: Let x be the value of each payment.]

13. Show that the simple interest formula can be rewritten as:

(a) $r = \dfrac{S - P}{Pt}$ (b) $t = \dfrac{S - P}{Pr}$

In Exercises 14–16 use the formulas obtained in Exercise 13.

14. An individual deposits $5000 in an account that pays simple interest. If 2 years later the account contains $5500, what is the interest rate?

15. An individual wants to deposit $1000 in an account that pays 12% simple interest. How long must the money be left on deposit if he wants to withdraw $1300?

16. How long will it take for an investment at 10% simple interest to double in value? [*Hint*: The value will have doubled when $S = 2P$.]

9.2 COMPOUND INTEREST

For many transactions, interest is added to the principal at regular time intervals so that the interest itself earns interest. This is called **compounding** of interest. The time interval between successive additions of interest is called the **conversion period**. Typical conversion periods are given in Table 1.

TABLE 1

Conversion Period	Compounded
1 year	Annually
6 months	Semiannually
3 months	Quarterly
1 month	Monthly
1 week	Weekly
1 day	Daily

An interest rate per period i is needed. Usually an annual rate, called the **nominal rate** j, is quoted together with the frequency of conversion. The interest rate per period i is determined from

$$i = \frac{j}{m} \quad \text{where } m = \text{number of conversion periods per year}$$

In all formulas given in this chapter, i and j are decimal forms of the corresponding interest rates. Thus, 10% converted semiannually implies that $i = \frac{10}{2}\% = 5\%$ is the interest rate for a 6-month period since there are two 6-month periods per year, so $m = 2$, and for computational purposes $i = .05$.

EXAMPLE 1 The rate per period equivalent to a nominal rate of 12% compounded (or **converted**) for various periods is given in Table 2.

TABLE 2

12% Interest Compounded	m	$i = \frac{j}{m}(\%)$	Decimal Value of i
Annually	1	12	.12
Semiannually	2	6	.06
Quarterly	4	3	.03
Monthly	12	1	.01
Weekly	52	$\frac{12}{52} \cong .23$	.0023
Daily	365	$\frac{12}{365} \cong .033$	.00033

To illustrate the mechanics of compound interest, consider the following example.

EXAMPLE 2 Find the interest on \$1000 for 1 year at:

(a) 8% simple interest

(b) 8% compounded semiannually

(c) 8% compounded quarterly

Solution (a) The simple interest is

$$\begin{aligned} I &= Prt \\ &= (1000)(.08)(1) \\ &= \$80 \end{aligned}$$

Solution (b) Since the nominal interest rate is 8%, the semiannual interest rate is $\frac{1}{2}(8\%) = 4\%$. Thus, at the end of the first 6 months the interest earned is

$$(\$1000)(.04) = \$40$$

which, when added to the principal, yields \$1040. For the second 6 months this new principal earns interest of

$$(\$1040)(.04) = \$41.60$$

Combining the interest for the two 6-month periods we obtain a total interest of

$$\$40.00 + \$41.60 = \$81.60$$

Solution (c) Since the annual interest rate is 8%, the quarterly interest rate is $\frac{1}{4}(8\%) = 2\%$. Thus,

$$\begin{aligned}
\text{first-quarter interest} &= (\$1000)(.02) = \$20.00 \\
\text{second-quarter interest} &= (\$1020)(.02) = \$20.40 \\
\text{third-quarter interest} &= (\$1040.40)(.02) = \$20.81 \\
\text{fourth-quarter interest} &= (\$1061.21)(.02) = \underline{\$21.22} \\
\text{Total} &= \$82.43
\end{aligned}$$

This example illustrates an important point: the more frequent the compounding, the greater the total interest.

Our next objective is to obtain the following result.

Compound Interest Formulas If P dollars is invested at $j\%$ compounded m times a year, then after n conversion periods the investment will have grown to an amount S given by

$$S = P(1 + i)^n \tag{1}$$

where $i = j/m$. And to have S dollars at the end of n periods, an initial investment of

$$P = S(1 + i)^{-n} \tag{2}$$

is needed.

To demonstrate this result let V_{beg} denote the value of the investment at the beginning of any conversion period. This period will have a duration of $t = 1/m$ years (since there are m conversion periods per year). Since the nominal interest rate is j, we let

$$rt = j \cdot \frac{1}{m} = i \quad \text{(interest rate per period)}$$

Therefore, the amount V_{end} at the end of this period is obtained from Equation (1) of Section 9.1:

$$V_{\text{end}} = V_{\text{beg}}(1 + rt) = V_{\text{beg}}(1 + i) \tag{3}$$

If the initial investment is P, Table 3 shows the amount at the end of n periods obtained by the repeated use of Equation (3).

TABLE 3

Number of Period n	Value at Beginning of Period	Value at End of Period
1	P	$P(1 + i)$
2	$P(1 + i)$	$[P(1 + i)](1 + i) = P(1 + i)^2$
3	$P(1 + i)^2$	$[P(1 + i)^2](1 + i) = P(1 + i)^3$
4	$P(1 + i)^3$	$[P(1 + i)^3](1 + i) = P(1 + i)^4$
⋮	⋮	⋮

Thus, we see that if P dollars is invested for n periods at $j\%$ compounded m times a year, then the amount at the end of the n periods is

$$S = P(1 + i)^n$$

Solving for P,

$$P = \frac{S}{(1 + i)^n} = S(1 + i)^{-n}$$

Appendix Table C.7 lists values of $(1 + i)^n$ in column 1 and values of $(1 + i)^{-n}$ in column 2. Of course, you can also use your calculator to evaluate $(1 + i)^n$ and $(1 + i)^{-n}$.

EXAMPLE 3 Evaluate $(1 + .04)^5$ and $(1 + .04)^{-5}$.

Solution Method 1: In Appendix Table C.7 find the page for $i = 4\%$, partially reproduced in Table 4. According to these values,

TABLE 4

	$i = 4\%$	
n	$(1 + i)^n$	$(1 + i)^{-n}$
1	⋮	⋮
2		
3		
4	1.16985856	0.85480419
5	1.21665290	0.82192711

$$(1 + .04)^5 = 1.21665290 \qquad \text{and} \qquad (1 + .04)^{-5} = 0.82192711$$

Method 2: Using the "y^x" key on your calculator, enter the following series of keystrokes*:

[1.04] [y^x] [5] [=]

The result on a calculator displaying eight digits is 1.2166529.

Similarly, to obtain $(1.04)^{-5}$, enter the following series of keystrokes:

[1.04] [y^x] [−5] [=]

The result is 0.8219271.

*Some calculators may require a different sequence of keystrokes. Check the instructions provided with your calculator.

The terms **principal**, **amount**, **future value**, **present value**, and **past value** have the same meaning in compound interest problems that they have in simple interest problems with compound interest formulas used in place of simple interest formulas. Time-line diagrams are drawn as in simple interest problems.

The compound interest CI over n periods is

$$CI = S - P$$

EXAMPLE 4 What is the compound amount of \$1000 invested for 1 year at:

(a) 8% compounded annually

(b) 8% compounded semiannually

(c) 8% compounded quarterly

Solution (a) Since $P = 1000$, and

$$i = \frac{.08}{1} = .08 \qquad n = 1 \text{ (year)}$$
$$S = P(1 + i)^1 = 1000(1 + .08) = \$1080$$

Solution (b)
$$i = \frac{.08}{2} = .04 \qquad n = 2 \text{ (half years)}$$
$$S = P(1 + i)^2 = 1000(1 + .04)^2 = \$1081.60$$

Solution (c)
$$i = \frac{.08}{4} = .02 \qquad n = 4 \text{ (quarters)}$$
$$S = P(1 + i)^4 = 1000(1 + .02)^4 = \$1082.43$$

(Compare these results with those of Example 2.)

EXAMPLE 5 (a) What is the compound amount of \$1000 invested for 3 years at 12% compounded quarterly?

(b) What is the compound interest over this time?

Solution (a) Construct a time line (Figure 1). S is now determined from Equation (1):

$$P = 1000 \qquad i = \frac{.12}{4} = .03 \qquad n = 3 \cdot 4 = 12 \text{ quarters}$$

Therefore,

$$S = 1000(1 + .03)^{12} = \$1425.76$$

FIGURE 1

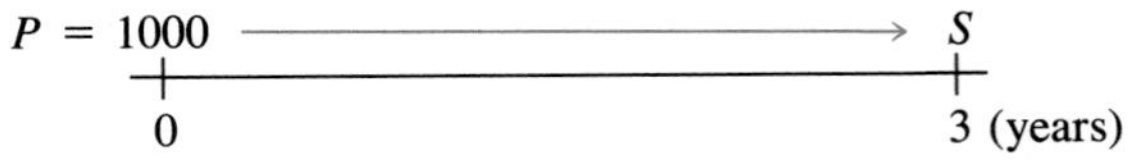

$P = 1000$ ⟶ S

0 ——— 3 (years)

Solution (b) $CI = S - P = 1425.76 - 1000 = \$425.76$

EXAMPLE 6 What is the present value of \$1000 due in 3 years if the interest rate is 12% compounded monthly?

Solution Construct a time line (Figure 2). Here

$$S = 1000 \qquad n = 3 \cdot 12 = 36 \text{ months} \qquad i = \frac{.12}{12} = .01$$

From Equation (2),

$$P = S(1 + i)^{-n} = 1000(1 + .01)^{-36} = \$698.92$$

FIGURE 2

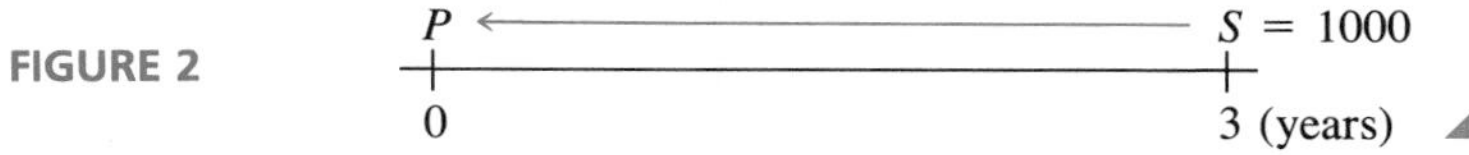

Remark In compound interest problems, to compute the future value of money, use the formula for amount: $S = P(1 + i)^n$; to compute the past value of money, use the formula for present value: $P = S(1 + i)^{-n}$.

EXAMPLE 7 One thousand dollars is deposited into an account paying 8% compounded quarterly on September, 1991; \$500 is deposited 3 months later and \$400 is deposited 6 months after that. What amount is in the bank on September 1, 1993? Assume no other deposits and no withdrawals during this time interval.

Solution The time line is shown in Figure 3.

FIGURE 3

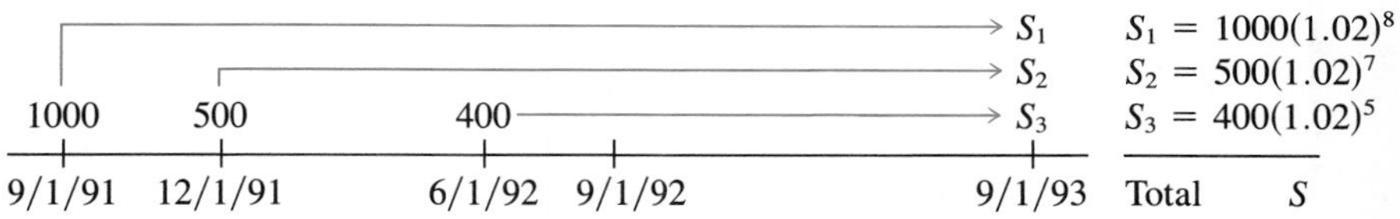

1000 (9/1/91) → S_1; 500 (12/1/91) → S_2; 400 (6/1/92) → S_3; 9/1/92; 9/1/93

$S_1 = 1000(1.02)^8$
$S_2 = 500(1.02)^7$
$S_3 = 400(1.02)^5$
Total S

Here

$$i = \frac{.08}{4} = .02$$

Let $S =$ the amount in the account on September 1, 1993. Then,

$$\begin{aligned} S_1 &= \text{the value of the \$1000 deposit on September 1, 1993} \\ &= 1000(1 + .02)^8 = 1000(1.17165938) = \$1171.66 \end{aligned}$$

$$\begin{aligned} S_2 &= \text{the value of the \$500 deposit on September 1, 1993} \\ &= 500(1 + .02)^7 = 500(1.14868567) = \$574.34 \end{aligned}$$

$$\begin{aligned} S_3 &= \text{the value of the \$400 deposit on September 1, 1993} \\ &= 400(1 + .02)^5 = 400(1.10408080) = \$441.63 \end{aligned}$$

Therefore,

$$\begin{aligned} S &= S_1 + S_2 + S_3 \\ &= 1171.66 + 574.34 + 441.63 \\ &= \$2187.63 \end{aligned}$$

EXAMPLE 8 What is the present value of \$1000 due in 6 months and \$2000 due in 18 months if the interest rate is 8% compounded quarterly?

Solution Draw a time line (Figure 4) with $i = .08/4 = .02$. The present value of these two amounts will equal the sum of the present values of the two amounts. Thus,

$$\begin{aligned} P_1 &= \text{present value of \$1000} = 1000(1 + .02)^{-2} \\ &= 1000(0.96116878) = \$961.17 \end{aligned}$$

$$\begin{aligned} P_2 &= \text{present value of \$2000} = 2000(1 + .02)^{-6} \\ &= 2000(0.88797138) = \$1775.94 \end{aligned}$$

$$\begin{aligned} P = P_1 + P_2 &= \$961.17 + \$1775.94 \\ &= \$2737.11 \end{aligned}$$

FIGURE 4

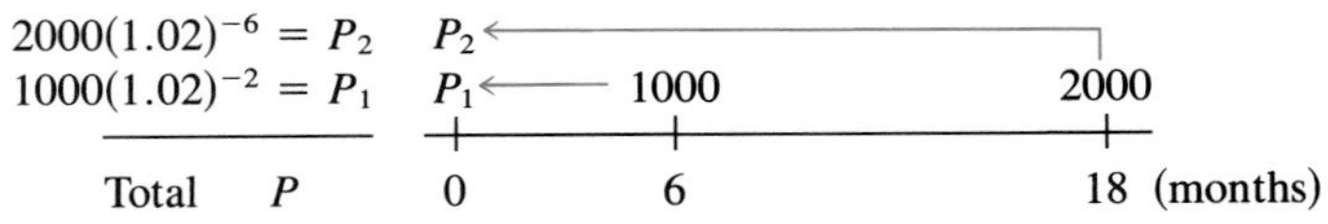

$2000(1.02)^{-6} = P_2$ P_2 ← 2000

$1000(1.02)^{-2} = P_1$ P_1 ← 1000

Total P 0 6 18 (months)

Equations of Value and Compound Interest

We again consider transactions in which one or more debts are repaid with one or more payments due at different points in time, as illustrated in Example 7 of Section 9.1.

EXAMPLE 9 Johanna borrows \$2000 and agrees to pay \$500 in 2 months and \$700 in 6 months. What final payment should she make 18 months from now to settle her debt if interest is 12% compounded monthly?

The only difference between Example 7 of Section 9.1 and this example is the type of interest involved. In Section 9.1 simple interest is used; in this example, compound interest is used. Otherwise the method of solution is essentially the same. Again, a focal date is selected and an equation of value is set up.

Equation of Value

value of loan at focal date = value of payments at focal date

When compound interest is involved, however, the focal date may be any date at which interest is compounded, and the resulting equations of value will give the same result for the quantity to be determined.

Solution to Example 9 We solve the problem first by selecting the focal date to be 18 months from now. Let x be the amount of the final payment; set up a time diagram (Figure 5).

Focal date | Value of money at focal date

2000 → $2000(1.01)^{18}$

0 2 6 18 (months)

500 700 x

x

$700(1.01)^{12}$

$500(1.01)^{16}$

FIGURE 5

Table 5 indicates the values of each amount of money at the focal date. The interest is 12% compounded monthly, so $i = 1\%$.

TABLE 5

Amount ($)	Formula used	n	Value at focal date
2000	$S = P(1 + i)^n$	18	$2000(1 + .01)^{18}$
500	$S = P(1 + i)^n$	16	$500(1 + .01)^{16}$
700	$S = P(1 + i)^n$	12	$700(1 + .01)^{12}$
x	No shift in time	—	x

Therefore the equation of value at the focal date is

value of loan at focal date = value of payments at focal date

$$2000(1 + .01)^{18} = 500(1 + .01)^{16} + 700(1 + .01)^{12} + x \qquad (4)$$

$$2000(1.19614748) = 500(1.17257864) + 700(1.12682503) + x$$

$$2392.29 = 586.29 + 788.78 + x$$

$$x = 2392.29 - 586.29 - 788.78$$

$$= \$1017.22 \qquad (5)$$

We now solve the same problem by selecting the focal date to be now. The time line is shown in Figure 6. Table 6 indicates the values of each amount of money at time $t = 0$.

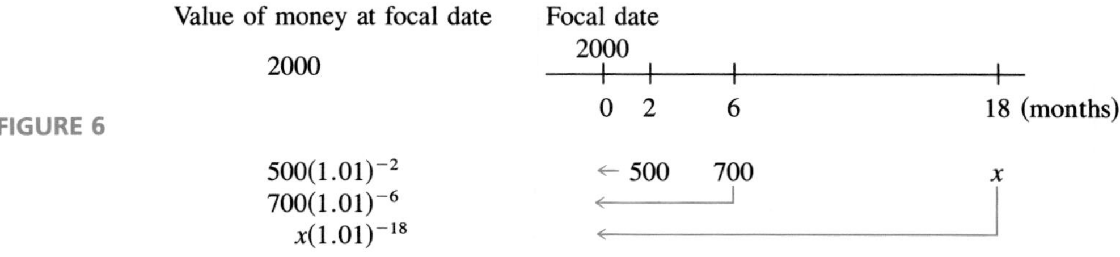

FIGURE 6

TABLE 6

Amount ($)	Formula Used	n	Value at Focal Date
2000	No shift in time	—	2000
500	$P = S(1 + i)^{-n}$	2	$500(1 + .01)^{-2}$
700	$P = S(1 + i)^{-n}$	6	$700(1 + .01)^{-6}$
x	$P = S(1 + i)^{-n}$	18	$x(1 + .01)^{-18}$

Therefore the equation of value is

$$\text{value of loan at focal date} = \text{value of payments at focal date}$$

$$2000 = 500(1 + .01)^{-2} + 700(1 + .01)^{-6} + x(1 + .01)^{-18} \quad (6)$$

$$2000 = 500(0.98029605) + 700(0.94204524) + x(0.83601731)$$

$$= 490.15 + 659.43 + 0.83601731x$$

$$0.83601731x = 2000 - 490.15 - 659.43 = 850.42$$

$$x = \frac{850.42}{0.83601731} = 1017.23$$

[The 0.01 difference between this value of x and the one in (5) is due to rounding error.] Note that if we multiply both sides of Equation (6) by $(1 + .01)^{18}$ we obtain Equation (4). This indicates that the two equations (for the two different focal dates) are algebraically equivalent and therefore will give the same value for x (except possibly for rounding error).

EXAMPLE 10 Janet deposits $1000 into an account paying interest at a rate of 8% compounded semiannually. Six months later she deposits $500 into the account. She withdraws $400 six months after her second deposit. She wishes to withdraw all the money in the account the year after the $400 withdrawal. How much will she withdraw at that time?

Solution Let the final amount in the account be x. Set up the time diagram (Figure 7). Let the focal date be 2 years from now. Here

$$i = \frac{.08}{2} = .04$$

FIGURE 7

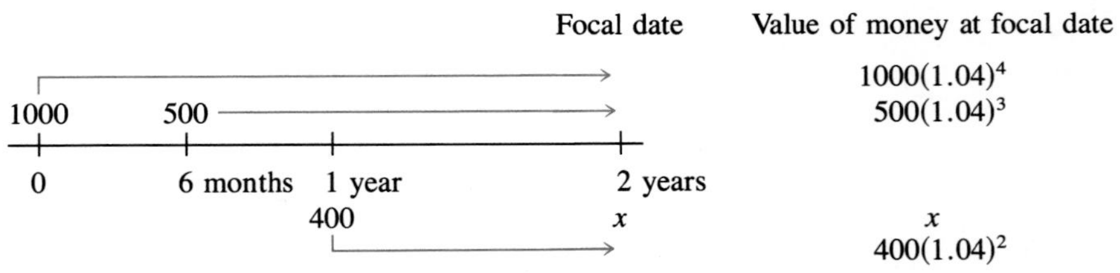

Table 7 indicates the values of the deposits and withdrawals at the focal date.

TABLE 7

Amount ($)	Formula Used	n	Value at Focal Date
1000	$S = P(1 + i)^n$	4	$1000(1 + .04)^4$
500	$S = P(1 + i)^n$	3	$500(1 + .04)^3$
400	$S = P(1 + i)^n$	2	$400(1 + .04)^2$
x	No shift in time	—	x

The equation of value is

$$\begin{aligned} \text{value of deposits at focal date} &= \text{value of withdrawals at focal date} \\ 1000(1 + .04)^4 + 500(1 + .04)^3 &= 400(1 + .04)^2 + x \\ 1000(1.16985856) + 500(1.12486400) &= 400(1.08160000) + x \\ 1169.86 + 562.43 &= 432.64 + x \\ x &= 1169.86 + 562.43 - 432.64 \\ &= \$1299.65 \end{aligned}$$

EXAMPLE 11 A woman needs $4000 at the end of 4 years to help pay for her son's college tuition. She will make 4 equal deposits into an account paying 8% compounded annually: one deposit at the end of each year for the 4 years. How much should each deposit be?

Solution Let x be the amount of each deposit, and let the focal date be the date of the last deposit. Construct a time line (Figure 8), and set up the equation of value. Here

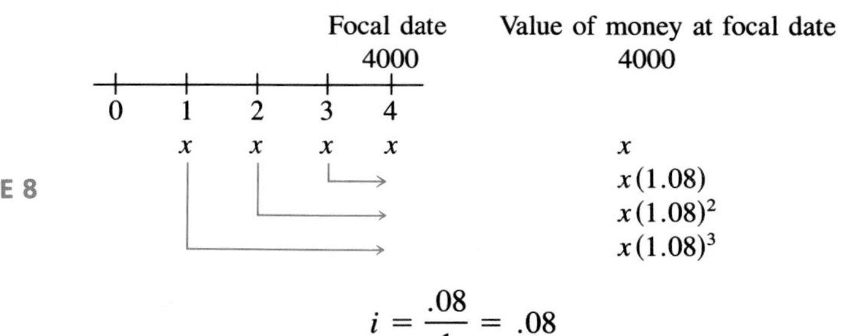

FIGURE 8

$$i = \frac{.08}{1} = .08$$

The equation of value is

$$\begin{aligned} \text{value of final amount at focal date} &= \text{value of deposits at focal date} \\ 4000 &= x(1 + .08)^3 + x(1 + .08)^2 + x(1 + .08) + x \\ &= x[(1.08)^3 + (1.08)^2 + (1.08) + 1] \\ &= x(1.259712 + 1.166400 + 1.08 + 1) \\ &= 4.506112x \\ x &= \frac{4000}{4.506112} = \$887.68 \end{aligned}$$

(Suppose there were 20 equal yearly payments instead of 4. The method would be the same but the computation horrendous. We examine a shorter method of solving this type of problem in Section 9.4.)

Continuous Compounding of Interest

For a fixed principal, time period, and annual interest rate, the more frequent the compounding, the greater is the return on the investment. You might ask if the return can be increased without bound by increasing the frequency of compounding. The answer is no; there is a theoretical upper limit on the return that can be achieved in this way. If we imagine the number of annual conversions to increase indefinitely, we approach a situation in which interest is compounded "continuously." That is, at each instant of time the investment grows in proportion to its current value. This is called **continuous compounding**. In Exercise 13 we help you develop the following formula for continuous compound interest.

Continuous Compound Interest Formula If P dollars is invested at $j\%$ compounded continuously, then after t years the investment will have grown to an amount S given by

$$S = Pe^{jt} \qquad (j \text{ in decimal form}) \tag{7}$$

and to have S dollars at the end of t years, an initial investment of

$$P = Se^{-jt} \tag{8}$$

is needed. The letter e represents an irrational number whose value is approximately

$$e \cong 2.7182818$$

Remark A table of values for e^x and e^{-x} can be found in Appendix Table C.5. For example, to determine $e^{.03}$ and $e^{-.03}$, we use Appendix Table C.5 (see also Table 8).

TABLE 8

x	e^x	e^{-x}
0.00	1.0000	1.0000
0.01	1.0101	0.9900
0.02	1.0202	0.9802
0.03	1.0305	0.9704
0.04	1.0408	0.9608
0.05	1.0513	0.9512
0.06	1.0618	0.9418

Therefore $e^{.03} \cong 1.0305$ and $e^{-.03} \cong 0.9704$. The result for $e^{.03}$ on a calculator displaying eight digits is 1.0304545.

EXAMPLE 12 Suppose \$10,000 is invested at 8% compounded continuously. What is the value of the investment after 2 years?

Solution Substituting the principal $P = \$10{,}000$ and the annual interest rate $j = .08$ in Formula (7) yields

$$S = 10{,}000e^{.08t}$$

After $t = 2$ years have elapsed, the value S of the investment will be

$$\begin{aligned} S &= 10{,}000e^{.08(2)} \\ &= 10{,}000e^{.16} \\ &= 10{,}000(1.1735) \qquad \text{(from Appendix Table C.5)} \end{aligned}$$

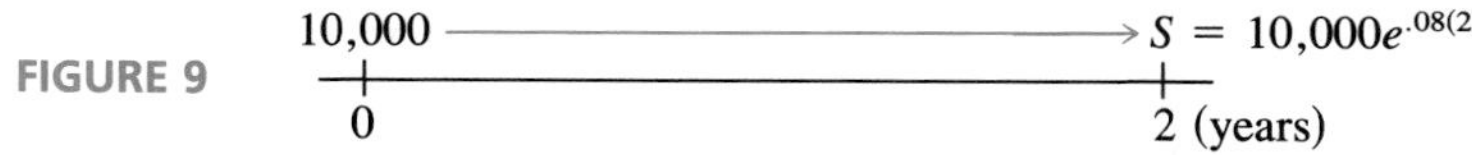

FIGURE 9

Thus, the value will be \$11,735. The time line for this example is shown in Figure 9.

EXAMPLE 13 John needs \$5000 in $2\frac{1}{2}$ years. How much should he invest now at 10% compounded continuously so that he will have \$5000 in $2\frac{1}{2}$ years? (Figure 10 shows the time diagram.)

$5000e^{-.10(2.5)} = P$ ← 5000

0 — $2\frac{1}{2}$ (years)

FIGURE 10

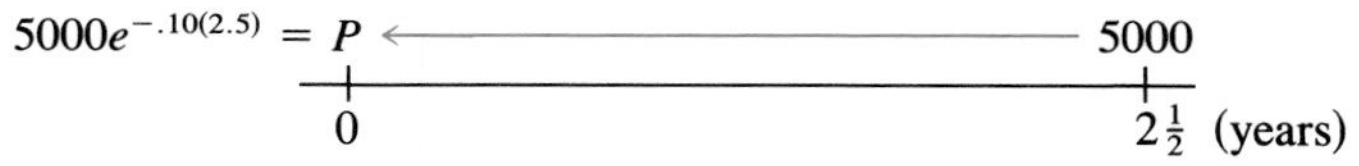

Solution Here $S = \$5000$, $t = 2.5$ years, and $j = .10$:

$$\begin{aligned} P &= Se^{-jt} \qquad \text{[by Formula (8)]} \\ &= 5000e^{-.10(2.5)} = 5000e^{-0.25} \\ &= 5000(0.7788) \qquad \text{(from Appendix Table C.5)} \\ &= \$3894.00 \end{aligned}$$

Remark When interest is compounded continuously, the value of money is moved forward in time by $S = Pe^{jt}$ and backward in time by $P = Se^{-jt}$.

EXAMPLE 14 Suppose \$1000 is deposited into an account. Determine the amount at the end of 1 year if the interest rate is

(a) 12% compounded annually (b) 12% compounded semiannually
(c) 12% compounded monthly (d) 12% compounded weekly
(e) 12% compounded daily (f) 12% compounded continuously

Solution See Table 9.

TABLE 9

12% Interest Compounded	Formula	Amount ($)
Annually	$S = 1000(1 + .12)^1$	1120.00
Semiannually	$S = 1000(1 + .06)^2$	1123.60
Monthly	$S = 1000(1 + .01)^{12}$	1126.83
Weekly	$S = 1000(1 + .12/52)^{52}$	1127.34
Daily	$S = 1000(1 + .12/365)^{365}$	1127.47
Continuously	$S = 1000e^{.12(1)}$	1127.50

EXAMPLE 15 Johanna borrows \$2000 and agrees to pay \$500 in 2 months and \$700 in 6 months. What final payment should she make 18 months from now to settle her debt if interest is 12% compounded continuously?

Solution This problem is solved, as is Example 7 of Section 9.1, by constructing a time line, specifying a focal date, and setting up the equation of value. The only difference is that interest is now compounded continuously. As in compound interest, it can be shown that any choice of the focal date will yield the same result. Let x be the final payment.

The time line is shown in Figure 11. Set the focal date at 18 months. The equation of value is

$$\begin{aligned}
\text{value of loan at focal date} &= \text{value of payments at focal date} \\
2000e^{.12(3/2)} &= 500e^{.12(16/12)} + 700e^{.12(12/12)} + x \\
2000e^{.18} &= 500e^{.16} + 700e^{.12} + x \\
2000(1.1972) &= 500(1.1735) + 700(1.1275) + x \\
2394.40 &= 586.75 + 789.25 + x \\
x &= 2394.40 - 586.75 - 789.25 \\
&= 1018.40
\end{aligned}$$

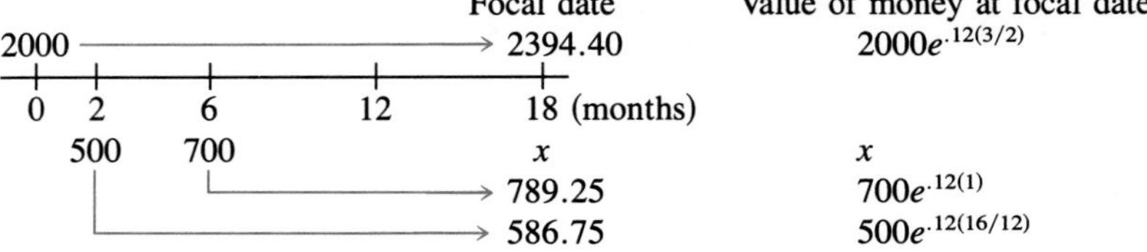

FIGURE 11

Thus, the final payment should be \$1018.40.

EXERCISE SET 9.2

1. Find the compound amount of \$5000 deposited for 2 years at:
   (a) 6% compounded annually
   (b) 6% compounded monthly
   (c) 6% compounded quarterly
   (d) 6% compounded semiannually
   (e) 6% compounded continuously

2. What is the compound amount of \$2000 deposited for 2 years at:
   (a) 12% compounded annually
   (b) 12% compounded semiannually
   (c) 12% compounded monthly
   (d) 12% compounded continuously

3. Find the present value of \$2000 due in 3 years at:
   (a) 12% compounded monthly
   (b) 12% compounded quarterly
   (c) 12% compounded continuously

4. Find the present value of \$3000 due in 18 months at:
   (a) 8% compounded monthly
   (b) 8% compounded quarterly
   (c) 8% compounded continuously

5. If \$10,000 is invested at an annual interest rate of 9% compounded quarterly, find the value of the investment after:
   (a) 1 year (b) 6 months (c) 2 years

6. If \$1500 is invested at an annual interest rate of 15% compounded monthly, what is the value of the investment after 30 months?

7. A principal P is to be invested to yield \$5000 in 6 years. Find the principal if the interest is
   (a) 8% compounded annually
   (b) 8% compounded quarterly
   (c) 8% compounded semiannually
   (d) 8% compounded continuously

8. A principal P is to be invested at 6% per year compounded semiannually to yield \$20,000 for a child's college education 18 years later. How much should the principal be?

9. Bob borrows \$3000 and agrees to make a payment of \$1000 in 6 months and \$1000 18 months from now. Determine how much he must pay 2 years from now to settle the debt if interest is
   (a) 12% compounded semiannually
   (b) 12% compounded monthly
   (c) 12% compounded continuously

10. Dick deposits \$500 on January 1, 1990, \$1000 on January 1, 1991, and \$2000 on January 1, 1992. If interest is 10% compounded semiannually, how much is in the account on January 1, 1995?

11. Carol borrows \$5000 and agrees to make three equal payments, the first in 6 months, the second in 12 months, and the third in 18 months to settle the debt. If the interest is 9% compounded semiannually, how much is each payment?

12. Alice makes three equal annual deposits into a savings account. The first deposit is made now, the second 1 year from now, and the third 2 years from now. If interest is 10% compounded annually and she wants to have \$4000 in her account 3 years from now, how much should each deposit be?

13. Using calculus, it can be shown that the value of the quantity

$$\left(1 + \frac{1}{n}\right)^n$$

approaches the number e as n becomes larger. In this exercise we help you derive the continuous compound interest formula from this result.

   (a) Suppose a principal P is invested at an annual interest rate of j compounded m times a year. Show that the amount of the investment after t years is

$$S = P\left(1 + \frac{j}{m}\right)^{tm}$$

   (b) Let $n = m/j$ and show that the result in (a) can be written

$$S = P\left(1 + \frac{1}{n}\right)^{tjn}$$

   (c) Explain why the result in (b) can be written

$$S = P\left[\left(1 + \frac{1}{n}\right)^n\right]^{jt}$$

   (d) Since $n = m/j$, it follows that n increases as m (the number of annual conversion periods) increases. Use this observation to deduce the continuous compound interest formula from the result in part (c).

9.3 EFFECTIVE RATE OF INTEREST; BANK DISCOUNT

Because compound interest is affected by both the nominal interest rate and the frequency of compounding, it is sometimes hard to tell offhand which of two compound interest procedures is more advantageous. For example, is it better to invest at 4% compounded monthly or $4\frac{1}{4}$% compounded semiannually? To make such comparisons, it is common to use the notion of **effective rate of interest** or **annual yield**. By definition this is the annual interest rate that produces the same yearly return as the compound interest procedure.

A formula for effective rate of interest can be derived as follows. Suppose a principal P is invested at a nominal interest rate of j% compounded m times a year, and let i_{eff} be the effective rate of interest. It follows from Formula (1) of Section 9.2 that in one year the compound interest procedure yields an amount

$$S = P\left(1 + \frac{j}{m}\right)^m = P(1 + i)^m$$

(m = number of conversion periods in 1 year), whereas the amount resulting in 1 year from the annual interest rate i_{eff} is

$$S = P(1 + i_{eff})$$

Since the two amounts are the same,

$$P(1 + i_{eff}) = P(1 + i)^m$$

dividing both sides by P yields

$$1 + i_{eff} = (1 + i)^m$$
$$i_{eff} = (1 + i)^m - 1$$

Effective Rate of Interest Formula for Compound Interest

$$i_{eff} = (1 + i)^m - 1 \qquad (1)$$

where

i_{eff} = effective rate of interest
j = nominal interest rate
m = number of conversion periods per year
$i = j/m$ = interest rate per period

The Truth in Lending Law enacted in 1969 requires the effective interest rate or annual yield to appear on all contracts.

EXAMPLE 1 Find the effective rate of interest equivalent to 8% compounded quarterly.

Solution We have $j = .08$, $m = 4$ (conversion periods per year), and $i = .02$. Substituting in Equation (1),

$$\begin{aligned} i_{\text{eff}} &= (1 + .02)^4 - 1 \\ &= 1.08243216 - 1 \qquad \text{(from Appendix Table C.7, } i = 2\%,\ n = 4) \\ &= .08243216 \end{aligned}$$

which is an effective interest rate of 8.243216%.

Remark To compare investments with different compound interest rates, we need only compare the effective rates. The larger effective rate benefits the lender, the smaller benefits the borrower.

EXAMPLE 2 Which is a better investment, 12% compounded monthly or 12.5% compounded annually?

Solution We must compute the effective rate of interest for each investment. The effective rate of interest for the second investment is the annual rate of 12.5%. (Why?) For the first investment we have

$$j = .12 \qquad m = 12 \text{ conversion periods per year}$$

$$i = \frac{j}{m} = \frac{.12}{12} = .01$$

Thus,

$$\begin{aligned} i_{\text{eff}} &= (1 + .01)^{12} - 1 \\ &= 1.12682503 - 1 \qquad \text{(from Appendix Table C.7, } i = 1\%,\ n = 12) \\ &= .12682503 \end{aligned}$$

which is an effective interest rate of 12.68%. Therefore, 12% compounded monthly is more favorable to the investor than the annual rate of 12.5%.

Effective Rate for Continuous Compounding

A formula for the effective interest rate equivalent to $j\%$ compounded continuously is found in a manner similar to the derivation of Equation (1).

Suppose a principal P is invested at $j\%$ compounded continuously for 1 year. Then in 1 year, Equation (7) of Section 9.2 gives an amount S, where

$$S = Pe^{j(1)} = Pe^j \qquad (t = 1)$$

Let i_{eff} be the effective rate of interest. Therefore,

$$S = P[1 + i_{\text{eff}}]$$

since the two amounts are the same,

$$P(1 + i_{\text{eff}}) = Pe^j$$

dividing both sides by P gives

$$\begin{aligned} 1 + i_{\text{eff}} &= e^j \\ i_{\text{eff}} &= e^j - 1 \end{aligned}$$

Effective Rate of Interest for Continuous Compounding

$$i_{\text{eff}} = e^{j} - 1 \tag{2}$$

where j is the nominal interest rate compounded continuously.

EXAMPLE 3 Find the effective rate of interest if money is invested at 8% compounded continuously.

Solution Substituting the interest rate $j = .08$ in Equation (2) yields

$$\begin{aligned} i_{\text{eff}} &= e^{.08} - 1 \\ &= 1.0833 - 1 \qquad \text{(from Appendix Table C.5)} \\ &= .0833 \end{aligned}$$

Thus the effective rate is 8.33%.

Simple Bank Discount and the Effective Simple Rate

Another common business procedure is for a lender to deduct the interest due in advance. For example, if you make a bank loan of \$300, the bank will compute the interest due and.deduct it in advance, giving you \$300 minus the interest. At the end of the lending period you then pay the bank \$300. The money deducted in advance is called the **simple bank discount**, and the money received by the borrower is called the **proceeds**.

Simple bank discount or simple discount or bank discount is computed in much the same way as simple interest. However it is based on the amount rather than the principal. More precisely, let

P = proceeds (amount received by borrower)
d = discount rate (a percentage per year)
t = time (in years that the proceeds will be held)
S = amount (to be paid back by borrower)

By definition, **simple discount** is computed as

$$D = Sdt$$

so that the proceeds received by the borrower is the amount S minus the discount. Thus,

$$\begin{aligned} P &= S - D \\ &= S - Sdt \\ &= S(1 - dt) \end{aligned}$$

Simple Discount Formula

$$P = S(1 - dt) \tag{3}$$

or

$$S = \frac{P}{1 - dt} \tag{4}$$

EXAMPLE 4 Jones borrows \$600 from his bank for 2 years at an 8% simple bank discount rate. Thus, the proceeds are

$$P = S(1 - dt) = 600[1 - (.08)(2)] = 600(.84) = \$504$$

Therefore, Jones will receive \$504 now and pay \$600 at the end of 2 years.

EXAMPLE 5 If Jones needs \$500 now, how much should he borrow from his bank for 2 years at 8% bank discount?

Solution Let $P = 500$, $t = 2$, and $d = .08$. From Equation (4),

$$\begin{aligned} S &= \frac{P}{1 - dt} \\ &= \frac{500}{1 - (.08)2} = \frac{500}{1 - .16} = \frac{500}{.84} \\ &= \$595.24 \end{aligned}$$

Therefore, Jones should borrow \$595.24.

Remark Compound discount can be defined in a manner analogous to compound interest. However, we will not discuss it in this text. (See Stephen Kellerson, *The Theory of Interest*, (Illinois: Irwin, 1970.) Just as simple interest is used for short-term transactions, so is simple discount.

Just as we computed the effective annual interest rate equivalent to a given compound rate, we can compute an effective simple interest rate equivalent to a simple discount rate d. We are looking for an annual simple interest rate so that $I = D$. Let

$$\begin{aligned} r_{\text{eff}} &= \text{simple interest rate} \\ t &= \text{time of the loan in years} \\ d &= \text{simple discount rate} \\ P &= \text{principal or proceeds} \\ S &= \text{amount} \end{aligned}$$

Then

$$I = D$$

so

$$\begin{aligned} Pr_{\text{eff}}t &= Sdt \\ &= \frac{P}{1 - dt}\,dt \qquad \text{[by Equation (4)]} \end{aligned}$$

We cancel P and t to obtain

$$r_{\text{eff}} = \frac{d}{1 - dt} \tag{5}$$

(Note that r_{eff} depends on t.)

INVESTORS: BEWARE OF THE PERCENTAGES

BUSINESS NEWS

July

Treasury Department sells \$8 billion in 3-month Treasury bills at a simple discount rate of 6.99%.

A recent news item reported that the Treasury Department sold \$8 billion in 3-month Treasury bills at a discount rate of 6.99%. An advertisement on the same page indicated that Bank XYZ was offering 3-month certificates of deposit at a simple interest rate of 7%.

Bob and Mary Jones, recent retirees, bought the certificate of deposit, thinking that a 7% return was better than 6.99%. Were they right?

BANK XYZ
ANNOUNCES
3-Month Certificates of Deposit paying 7% simple interest.

Solution Treasury bill rates are simple discount rates, whereas certificate of deposit rates are simple interest rates. To compare these rates, the investor must first compute the effective simple rate equivalent to the simple discount rate of 6.99%:

$$r_{\text{eff}} = \frac{d}{1 - dt} = \frac{.0699}{1 - (.0699)(.25)} = .0711$$

The effective rate is 7.11%, which is more favorable to the investor than the 7% offered by Bank XYZ. The treasury bill is the better investment.

EXAMPLE 6 From Jones' point of view in Example 4, Jones paid interest of

$$\$600 - \$504 = \$96$$

for the use of \$504 for 2 years. His simple interest rate can be obtained from Formula (5) when $d = .08$ and $t = 2$:

$$r_{\text{eff}} = \frac{d}{1 - dt} = \frac{.08}{1 - (.08)2} = \frac{.08}{.84} = .09524$$
$$= 9.524\%$$

EXERCISE SET 9.3

1. Find the effective annual rate of interest i_{eff} on each of these investments:
   (a) 6% compounded monthly
   (b) 6% compounded quarterly
   (c) 6% compounded semiannually
   (d) 6% compounded continuously

2. Find the effective annual rate of interest i_{eff} on each of these investments:
   (a) 12% compounded monthly
   (b) 9% compounded quarterly
   (c) 7% compounded semiannually
   (d) 7% compounded continuously

3. Which is better for the investor, an investment paying 9% compounded quarterly or 9.1% compounded annually?

4. Find the effective annual rate of interest i_{eff} on each of these investments:
   (a) 6% compounded continuously
   (b) 10% compounded continuously

5. Find the effective annual rate of interest i_{eff} on each of these investments:
   (a) 12% compounded continuously
   (b) 15% compounded continuously

6. Which is better for the investor, an investment paying 8% compounded semiannually or 7.9% compounded continuously?

7. With semiannual compounding, what nominal interest rate is equivalent to $3\frac{1}{2}$% compounded annually?

8. From a practical viewpoint there is not much difference between daily compounding (use a 365-day year) and continuous compounding. For example, with an annual interest rate of 10%, the effective rate of interest with daily compounding is

$$i_{\text{eff}} = \left(1 + \frac{.1}{365}\right)^{365} - 1 \cong .10516 = 10.516\%$$

Calculate the effective annual interest rate with continuous compounding.

9. If you borrow $400 for 6 months from a bank that charges 8% simple discount:
   (a) How much will you receive from the bank?
   (b) What is the discount?
   (c) How much will you repay at the end of 6 months?
   (d) What is the effective simple interest rate r_{eff}?
   (e) From your point of view, what is the simple interest rate you are paying for the loan?

10. (a) Find the bank discount on $5000 for 2 years at 9%.
    (b) Find the proceeds.

11. How much should be borrowed at 12% bank discount for 5 months to obtain proceeds of $1000?

12. The proceeds of a $2000 loan for 1 year at simple discount were $1800. What was the simple discount rate?

13. (a) If a woman wants to borrow money for 6 months at 8% simple discount rate, how much should she borrow if she needs $500 now?
    (b) What is the effective simple interest rate r_{eff}?

14. (a) If John Smith needs $1000, how much should he borrow at a simple discount rate of 10% if he will pay back the loan in 8 months?
    (b) What is the effective simple interest rate r_{eff}?

9.4 ANNUITIES

An **annuity** is a sequence of payments paid or received at equal time intervals. Some examples are:

1. A sequence of equal monthly investments (the annuity) is made with the objective of accumulating a certain lump sum at a specified future time (e.g., monthly deposits in a savings account).
2. A sequence of equal monthly payments (the annuity) to pay off an interest-bearing debt (e.g., a home mortgage).
3. A lump sum to be invested is placed with an investment company and paid back with interest in 10 equal annual installments (the annuity) on retirement.

The time period between successive annuity payments is called the **payment period** or **payment interval** for the annuity. When the payments are due at the end of the payment periods, the annuity is called an **ordinary annuity**.* The

*There are other types of annuities which we do not consider here. For example, if payments are made at the beginning of each period, the annuity is referred to as an **annuity due**. A reference for these other types is *Mathematics of Finance*, by Robert Cissell, Helen Cissell, David C. Flaspohler, 7th edition (Boston: Houghton Mifflin, 1986).

time from the beginning of the first payment period to the end of the last period is called the **term** of the annuity (Figure 1 shows an ordinary annuity with a term of 5 periods). In the following we only consider annuities with equal payments.

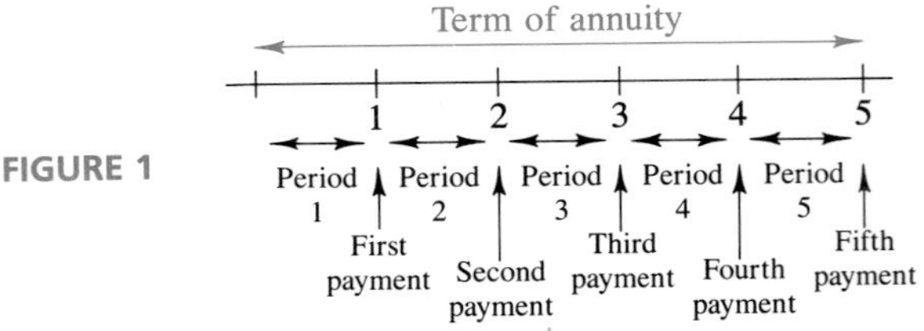

FIGURE 1

Annuity problems involve the determination of the value of the annuity at various points in time. For example, in the first example we might want to know the amount in the savings account at the date of the last deposit. In the second example we might ask the amount of the monthly payment required to pay off a $50,000 mortgage on a house. These problems are compound interest problems and can be handled by the methods of Section 9.2. However, in annuity problems with equal payments made at equal intervals of time, the equations can be simplified algebraically so that simple formulas can be derived to facilitate the computation.

Amount of an Annuity; Sinking Funds

The first type of problem that we consider is the determination of the amount of an ordinary annuity at the date of the last payment.

EXAMPLE 1 At the end of each month $100 is invested in a savings account paying 6% compounded monthly. What is the value S_5 of this ordinary annuity just after the fifth payment?

Solution This problem is a compound interest problem of a type that we consider in Section 9.2. It is identified as an annuity problem because there are equal payments made at equal intervals of time. Our method of solution follows that of Example 7 in Section 9.2.

Let S_5 be the value of these payments at the date of the last payment. The time line shown in Figure 2 indicates the payments. We now determine the value of these payments immediately after the fifth payment is made. We have $i = .06/12 = .005$. Figure 3 indicates the values of each of the five payments at the end of the fifth month ($t = 5$).

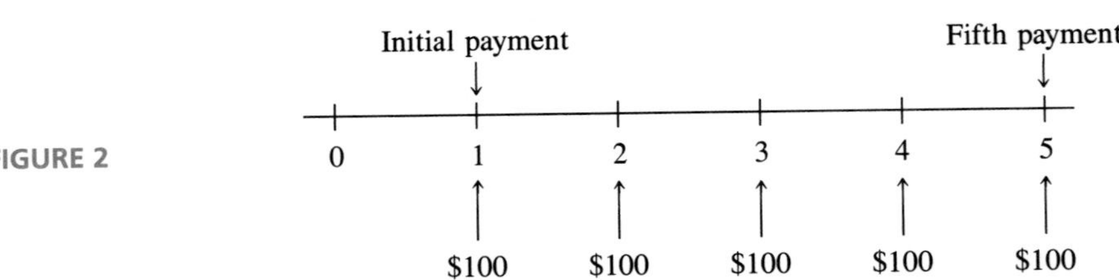

FIGURE 2

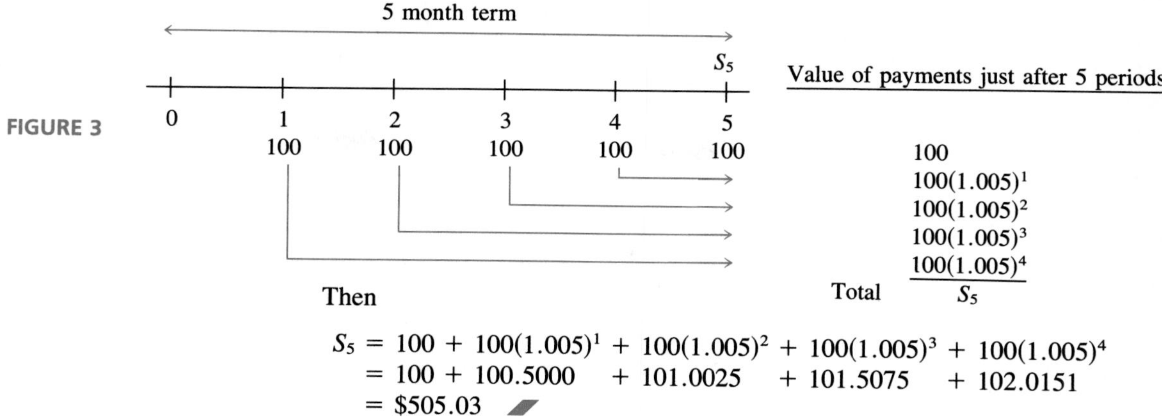

FIGURE 3

Then

$$
\begin{aligned}
S_5 &= 100 + 100(1.005)^1 + 100(1.005)^2 + 100(1.005)^3 + 100(1.005)^4 \\
&= 100 + 100.5000 \quad + 101.0025 \quad + 101.5075 \quad + 102.0151 \\
&= \$505.03
\end{aligned}
$$

If in Example 1 there were 50 monthly payments rather than 5, the method used would involve many more computations. We now see how this kind of calculation can be simplified. Consider a more general problem. Suppose we are interested in the value S_n of an ordinary annuity at the end of n payment periods, where

$$
\begin{aligned}
\text{each payment} &= R \text{ dollars} \\
i &= \text{the interest rate per period}
\end{aligned}
$$

and payments are made at the end of each period. Then, as suggested by Figure 4, the value S_n of the annuity will be

$$
\begin{aligned}
S_n &= R + R(1 + i) + R(1 + i)^2 + \cdots + R(1 + i)^{n-2} + R(1 + i)^{n-1} \\
&= R[1 + (1 + i) + (1 + i)^2 + \cdots + (1 + i)^{n-2} + (1 + i)^{n-1}] \qquad (1)
\end{aligned}
$$

Equation (1) can be simplified algebraically to obtain

$$
S_n = R\frac{(1 + i)^n - 1}{i} \qquad (2)
$$

(For a derivation, see Exercise 10 at the end of this section.)

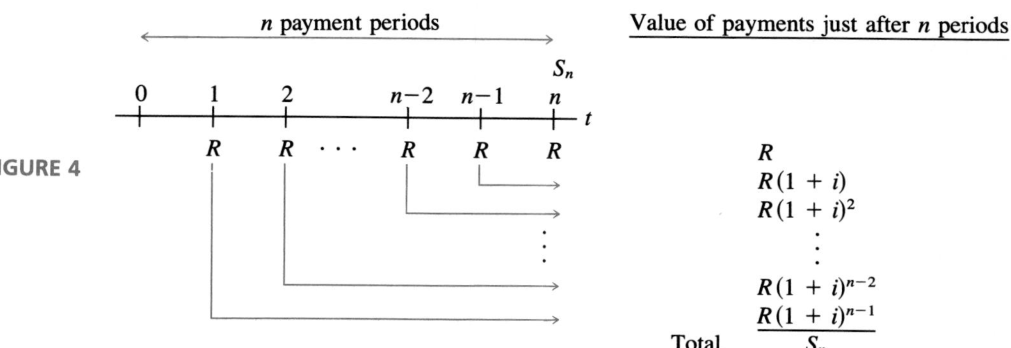

FIGURE 4

In Formulas (1) and (2) the quantity S_n is called the **amount** of the ordinary annuity of n payments and the quantity R is called the **payment** or **rent**. The expression $[(1 + i)^n - 1]/i$ is commonly denoted by $s_{\overline{n}|i}$ (read "s sub n angle i").

With this notation Formulas (1) and (2) become the following:

Amount of an Ordinary Annuity

$$S_n = R\frac{(1+i)^n - 1}{i} = Rs_{\overline{n}|i} \tag{3}$$

where

S_n = amount at the time of the nth payment i = interest rate per period
R = periodic payment n = number of periods

We can compute S_n with a calculator. However, for convenience, values of $s_{\overline{n}|i}$ and $1/s_{\overline{n}|i}$ are given in Appendix Table C.7.

EXAMPLE 2 Solve the problem of Example 1 using Formula (3).

Solution

$$\begin{aligned} S_5 &= Rs_{\overline{n}|i} \\ &= 100s_{\overline{5}|.005} \\ &= 100(5.05025063) \qquad \text{(from Appendix Table C.7, } i = \tfrac{1}{2}\%,\ n = 5) \\ &= \$505.03 \end{aligned}$$

which agrees with the result in Example 1.

Note that by using Appendix Table C.7 it would take the same number of computations to find the amount of the annuity with 50 monthly payments of $100 each. In this case,

$$\begin{aligned} S_{50} &= Rs_{\overline{n}|i} \\ &= 100s_{\overline{50}|.005} \\ &= 100(56.64516) \qquad \text{(from Appendix Table C.7, } i = \tfrac{1}{2}\%,\ n = 50) \\ &= \$5664.52 \end{aligned}$$

EXAMPLE 3 To save for a child's education, a family decides to invest $300 at the end of each 6-month period in a fund paying 8% compounded semiannually. Find the amount of the investment at the end of 18 years.

Solution From the given information

$$i = \frac{.08}{2} = .04$$

$$n = 18 \cdot 2 = 36 \qquad \text{(18 years, 2 payments per year)}$$

$$R = 300$$

Thus, at the end of 36 payment periods, the amount S_{36} of the investment will be

$$\begin{aligned} S_{36} &= Rs_{\overline{n}|i} \\ &= 300s_{\overline{36}|.04} \\ &= 300(77.59831385) \qquad \text{(from Appendix Table C.7, } i = 4\%,\ n = 36) \\ &= \$23{,}279.49 \end{aligned}$$

In some annuity problems, we are given the amount S_n of the annuity at the nth payment date and want to determine the periodic payment R that will give this amount.

EXAMPLE 4 Debbie needs to have $4000 in 18 months to pay part of her college tuition. How much should she deposit into her savings account at the end of each month if interest is 9% compounded monthly?

Solution Since Debbie is making equal monthly deposits at the end of each month, her payments form an ordinary annuity where $S_{18} = \$4000$. We have $i = .09/12 = .0075$. Therefore,

$$S_n = Rs_{\overline{n}|i} \quad \text{with} \quad n = 18$$

and

$$\begin{aligned} R &= \frac{S_n}{s_{\overline{n}|i}} \\ &= \frac{S_{18}}{s_{\overline{18}|.0075}} = \frac{4000}{19.19471849} \qquad \text{(from Appendix Table C.7, } i = \tfrac{3}{4}\%,\ n = 18) \\ &= \$208.39 \end{aligned}$$

A **sinking fund** is a fund that is accumulated from a series of equal payments made at equal intervals of time (the annuity) for the purpose of paying off a financial obligation at some future date. We are given the amount S_n and want to find the periodic payment R.

EXAMPLE 5 A firm anticipates a capital expenditure of $10,000 for new equipment needed in 5 years. How much should be deposited quarterly in a sinking fund earning 10% compounded quarterly to provide for the purchase?

Solution The future value of the investment after 5 years must be $S_{20} = \$10{,}000$. We must determine the value R of each payment.

From Formula (3) for the amount of an ordinary annuity at the date of the last payment, we obtain

$$S_n = Rs_{\overline{n}|i} \quad \text{or} \quad R = \frac{S_n}{s_{\overline{n}|i}} = S_n \cdot \frac{1}{s_{\overline{n}|i}} \tag{4}$$

Substituting the given information

$$i = \frac{.10}{4} = .025 \quad n = 5 \cdot 4 = 20 \text{ periods}$$

into Formula (4) yields

$$\begin{aligned} R &= \frac{10{,}000}{s_{\overline{20}|.025}} = 10{,}000 \frac{1}{s_{\overline{20}|.025}} \\ &= 10{,}000(.03914713) \qquad \text{(from Appendix Table C.7, } i = 2\tfrac{1}{2}\%,\ n = 20) \\ &= \$391.47 \end{aligned}$$

Thus, $391.47 should be invested quarterly in the sinking fund.

The following formula summarizes the discussion in these examples.

Periodic Payment of an Ordinary Annuity

$$R = \frac{S_n}{s_{\overline{n}|i}} \tag{5}$$

where

i = the interest rate per period
S_n = desired amount
n = number of periods

Present Value of an Annuity; Mortgages

So far we have considered annuity problems in which payments are made with the objective of accumulating a certain lump sum at a future date. We now consider the reverse problem. A lump sum is invested at compound interest with the objective of obtaining a series of payments (an annuity) over some future period of time. The lump-sum investment is called the **present value of the ordinary annuity** (Figure 5).

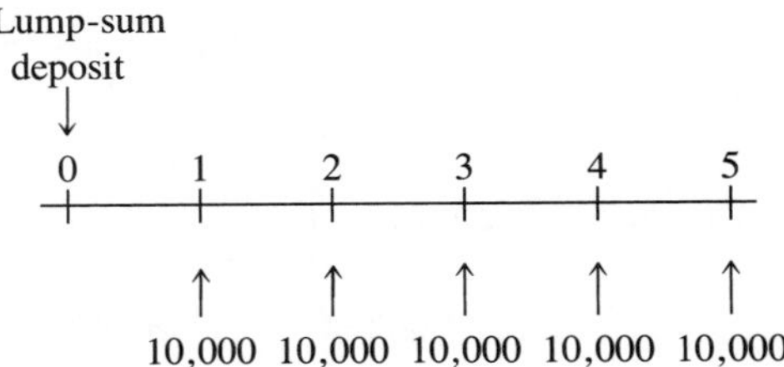

FIGURE 5

EXAMPLE 6 On retirement, a couple wants to make a lump-sum investment paying 8% compounded annually so that they can receive annuity payments of \$10,000 at the end of each year for the following 5 years. How much must they invest?

Solution In this problem we must determine the value of this annuity at the beginning of the first payment period ($t = 0$). This value A_5 is just the sum of present values of the five payments at $t = 0$ (Figure 6). We have

$$i = .08/1 = .08.$$

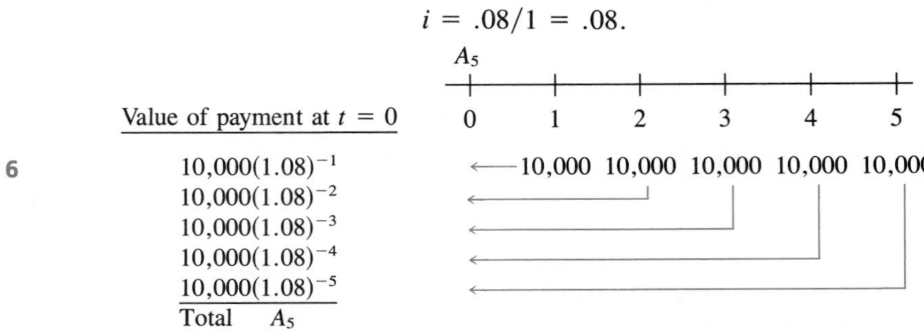

FIGURE 6

Then

$$\begin{aligned} A_5 &= 10{,}000(1.08)^{-1} + 10{,}000(1.08)^{-2} + 10{,}000(1.08)^{-3} + 10{,}000(1.08)^{-4} \\ &\quad + 10{,}000(1.08)^{-5} \\ &= 9259.2593 + 8573.3882 + 7938.3224 + 7350.2985 + 6805.8320 \\ &= \$39{,}927.10 \end{aligned}$$

Thus, the couple should invest $39,927.10.

To consider a more general problem, suppose we are interested in finding the sum A_n that must be invested at an interest rate of i per conversion period. Our goal is to obtain an ordinary annuity of n payments of R dollars each with payments beginning one conversion period after the initial investment A_n.

The quantity A_n is called the **present value** of the ordinary annuity of n payments of R dollars each; A_n is the sum of the present values of the n payments (Figure 7):

FIGURE 7

Value of payment at $t = 0$
$R(1+i)^{-1}$
$R(1+i)^{-2}$
$R(1+i)^{-3}$
⋮
$R(1+i)^{-n}$
Total A_n

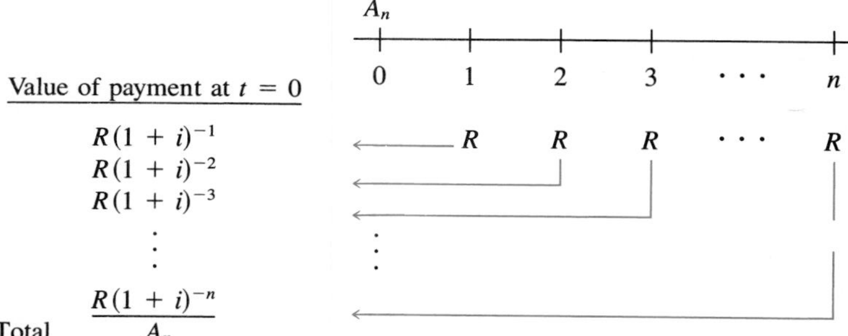

$$\begin{aligned} A_n &= R(1+i)^{-1} + R(1+i)^{-2} + R(1+i)^{-3} + \cdots + R(1+i)^{-n} \\ &= R[(1+i)^{-1} + (1+i)^{-2} + (1+i)^{-3} + \cdots + (1+i)^{-n}] \end{aligned} \tag{6}$$

Equation (6) can be simplified algebraically as

$$A_n = R\left[\frac{1-(1+i)^{-n}}{i}\right] \tag{7}$$

(See Exercise 19 at the end of this section.) The quantity A_n is the **present value** of the annuity. The expression in the square brackets of Formula (7) is commonly denoted by the symbol

$$a_{\overline{n}|i}$$

(read "a sub n angle i"). With this notation, Formulas (6) and (7) become the following:

Present Value of an Ordinary Annuity

$$A_n = Ra_{\overline{n}|i} \tag{8}$$

where

A_n = present value R = periodic payment
i = interest rate per period n = number of periods

You can use a calculator to compute A_n. For convenience, however, values of $a_{\overline{n}|i}$ and $1/a_{\overline{n}|i}$ are given in Appendix Table C.7.

EXAMPLE 7 What is the present value of an ordinary annuity with 20 quarterly payments of \$100 each if the interest is 8% compounded quarterly?

Solution We have $R = 100$, $n = 20$, and $i = .08/4 = .02$. By Equation (8),

$$\begin{aligned} A_{20} &= Ra_{\overline{20}|.02} \\ &= 100(16.35143334) \qquad \text{(from Appendix Table C.7, } i = 2\%, n = 20) \\ &= \$1635.14 \end{aligned}$$

Thus, the present vaue is \$1635.14.

An important example involving the present value of an ordinary annuity involves home mortgages. In this type of problem the value of the mortgage (or loan) A_n is given, and we are interested in determining the periodic payment R so that the set of payments have A_n as their present value.

An interest-bearing debt is said to be **amortized** if the principal and interest are paid by a sequence of equal payments made over equal time periods. Most home mortgages, for example, are paid in this way.

In other words, an amortized mortgage is a lump sum lent by the bank to the borrower, and in return the bank will receive fixed periodic payments consisting of principal and interest.

EXAMPLE 8 A developer purchases a parcel of land for \$10,000 to be amortized over 2 years at 9% compounded monthly. How much will each monthly payment be?

Solution Let R be the unknown amount of each monthly payment and consider the problem from the lender's viewpoint. The lender is investing a sum of \$10,000 (present value) at 9% compounded monthly to receive an annuity of R dollars per month for 2 years. Thus, from the formula for the present value of an annuity, Formula (8), we obtain

$$A_n = Ra_{\overline{n}|i}$$

or on solving for R,

$$R = \frac{A_n}{a_{\overline{n}|i}} = A_n \cdot \frac{1}{a_{\overline{n}|i}}$$

THE LOTTERY PROBLEM (REVISITED)

Robert Smith, the 45-year-old factory worker, who in 1986 won a $25 million jackpot (see Section 8.2) receives $1,250,000 at the end of each year for 20 years. How much money must the lottery commission invest now at 8% compounded annually to meet these 20 payments?

Solution The amount to be invested is the present value of these payments:

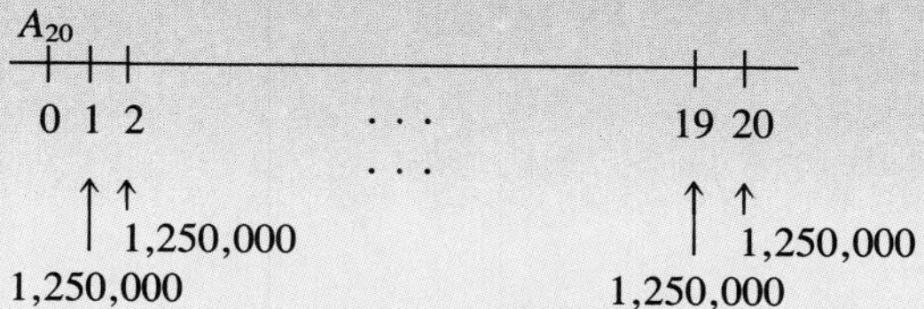

Using Formula (7) we have

$$\begin{aligned} A_{20} &= 1{,}250{,}000 a_{\overline{20}|.08} \\ &= 1{,}250{,}000(9.81814741) \\ &= 12{,}272{,}684 \end{aligned}$$

Thus, the lottery commission only has to invest $12,272,684 now.

Substituting $A = 10{,}000$, $n = 24$, and $i = .09/12 = .0075$ in that formula yields

$$\begin{aligned} R &= \frac{10{,}000}{a_{\overline{24}|.0075}} \\ &= (10{,}000)\left(\frac{1}{a_{\overline{24}|.0075}}\right) \\ &= (10{,}000)(0.04568474) \qquad \left(\text{from Appendix Table C.7, } i = \frac{3}{4}\%,\ n = 24\right) \\ &= 456.8474 \end{aligned}$$

Thus, each monthly payment will be $456.85.

The following formula summarizes the discussion in this example.

Periodic Payment on an Amortized Loan

$$R = \frac{A_n}{a_{\overline{n}|i}} \tag{9}$$

where

R = periodic payment $\quad$ A_n = amount of the loan

i = interest rate per period $\quad$ n = number of periods

EXERCISE SET 9.4

1. At the end of each month \$200 is invested in bonds paying 6% compounded monthly. What is the value of this ordinary annuity after:
   (a) the third payment (b) the tenth payment

2. Find the future value of the following ordinary annuities:
   (a) \$1000 a year for 5 years at 7% compounded annually
   (b) \$500 per quarter for 10 years at 8% compounded quarterly
   (c) \$4000 every 6 months for 15 years at 5% compounded semiannually
   (d) \$40 every month for 5 years at 6% compounded monthly

3. A newly married couple plans to save for a house by depositing \$300 a month in a money market fund paying 12% compounded monthly. How much will the couple have available for their house after 5 years in this savings program?

4. A self-employed photographer deposits \$7500 a year in a Keogh retirement account paying 8% compounded annually. How much will the photographer have in this account after 20 years?

5. An employer deposits \$9000 quarterly into a worker's pension fund invested at 12% compounded quarterly. How much will be in the fund after 25 years?

6. A firm anticipates a capital expenditure of \$50,000 for new equipment needed in 10 years. How much should be deposited annually in a sinking fund earning 8% compounded annually to provide for the purchase?

7. A municipality issues \$25,000,000 worth of revenue bonds that are due in 30 years. To pay off the debt, a sinking fund earning 4% compounded semiannually will be established. What semiannual deposit is needed?

8. The XYZ Corporation must retire (buy back) \$200,000 worth of bonds in 20 years. If they will receive 6% compounded semiannually, how much must they deposit every 6 months to retire the bonds?

9. (a) To save for a child's education a family invests \$250 at the end of each 3-month period in a fund paying 12% compounded quarterly. How much money will be in the fund after 10 years?
   (b) Suppose the family wants \$60,000 in the fund after 10 years. How much will they have to invest every 3 months to achieve this goal?

10. In the study of geometric progressions in algebra, it is shown that if b is any number other than 1, then

$$S = 1 + b + b^2 + \cdots + b^m = \frac{1 - b^{m+1}}{1 - b} = \frac{b^{m+1} - 1}{b - 1}$$

   (a) Establish this result as follows:
      1. Let $S = 1 + b + b^2 + \cdots + b^m$.
      2. Compute bS.
      3. Compute $S - bS = S(1 - b)$.
      4. Solve for S.
   (b) Use this result with $b = 1 + i$ and $m = n - 1$ to show that

$$s_{\overline{n}|i} = 1 + (1 + i) + (1 + i)^2 + \cdots + (1 + i)^{n-1} = \frac{(1 + i)^n - 1}{i}$$

11. In each part, use a calculator and the formula in Exercise 10(b) to evaluate $s_{\overline{n}|i}$. Check your results against those in Appendix Table C.7.
   (a) $n = 5,\ i = .01$
   (b) $n = 10,\ i = .02$
   (c) $n = 20,\ i = .045$

12. Use a calculator and Formula (2) of this section to find the future value of an ordinary annuity after 15 semiannual payments of \$100 at 20% compounded semiannually.

13. Find the present value of the following ordinary annuities:
   (a) \$1000 per year for 5 years at 7% compounded annually
   (b) \$500 per quarter for 10 years at 8% compounded quarterly
   (c) \$4000 every 6 months for 15 years at 5% compounded semiannually
   (d) \$40 per month for 5 years at 6% compounded monthly

14. Find the present value of the following ordinary annuities:
   (a) \$1000 per month for 2 years at 12% compounded monthly
   (b) \$15,000 per year for 10 years at 7% compounded annually

15. How much should you deposit now in an account paying 14% compounded semiannually to receive semiannual payments of \$2000 for the next 8 years?

16. The stipulations of a will call for establishing a trust that will yield the beneficiary \$500 a month for 5 years. How much should be deposited in the trust account if the interest rate is 8% compounded monthly?

17. A television set is financed at \$75 per month for 12 months at an interest rate of 12% compounded monthly on the unpaid balance. How much of the total amount paid goes for interest and how much for the television itself? [*Hint*: $nR - A_n$ = interest paid.]

18. A boat is financed for \$200 per month for 24 months at an interest rate of 15% compounded monthly on the unpaid balance. How much of the total amount paid goes for interest and how much for the boat itself? [*Hint*: $nR - A_n$ = interest paid.]

19. In Exercise 10 we state the progression formula

$$1 + b + b^2 + \cdots + b^m = \frac{1 - b^{m+1}}{1 - b}$$

Multiplying this formula through by c yields

$$c + cb + cb^2 + \cdots + cb^m = \frac{c - cb^{m+1}}{1 - b}$$

Use this result with $c = (1 + i)^{-1}$, $b = (1 + i)^{-1}$, and $m = n - 1$ to show that

$$a_{\overline{n}|i} = (1 + i)^{-1} + (1 + i)^{-2} + \cdots + (1 + i)^{-n} = \frac{1 - (1 + i)^{-n}}{i}$$

20. In each part use a calculator and the third formula in Exercise 19 to evaluate $a_{\overline{n}|i}$. Check your results against those in Appendix Table C.7.
(a) $n = 5$, $i = .01$
(b) $n = 10$, $i = .02$
(c) $n = 20$, $i = .045$

21. Find the monthly payment if:
(a) \$5000 is amortized for 5 years at 12% compounded monthly
(b) \$10,000 is amortized for 2 years at 12% compounded monthly

22. Find the periodic payment if:
(a) \$8000 is amortized for 10 years at 14% compounded semiannually
(b) \$1000 is amortized for 3 years at 10% compounded quarterly

23. A real estate speculator purchases a lot for \$25,000 to be amortized over 3 years at 8% compounded semiannually. How much will each semiannual payment be?

24. An automobile costs \$6000 and can be financed at 9% compounded monthly. If a down payment of 10% is made, what is the amount of the monthly payment if:
(a) the automobile is financed for 3 years
(b) the automobile is financed for 4 years

KEY IDEAS FOR REVIEW

- **Simple interest formula** $S = P(1 + rt)$
- **Compound interest formula** $S = P(1 + \frac{j}{m})^n = P(1 + i)^n$
- **Present value formula for compound interest** $P = S(1 + i)^{-n}$
- **Continuous compound interest formula** $S = Pe^{jt}$
- **Present value for continuous compound interest** $P = Se^{-jt}$
- **Effective rate of interest formula** $i_{\text{eff}} = (1 + \frac{j}{m})^m - 1 = (1 + i)^m - 1$
- **Effective rate of interest formula for continuous compounding** $i_{\text{eff}} = e^j - 1$
- **Bank discount formula** $P = S(1 - dt)$
- **Effective simple interest rate formula for bank discount** $r_{\text{eff}} = \dfrac{d}{1 - dt}$
- **Annuity** A sequence of payments paid or received at equal time intervals.
- **Ordinary annuity** An annuity in which the payments are due at the end of the conversion period.

- **Amount formula for an ordinary annuity** $S_n = Rs_{\overline{n}|i}$
- **Periodic payment for an ordinary annuity** $R = \dfrac{S_n}{s_{\overline{n}|i}}$
- **Sinking fund** An ordinary annuity in which R dollars is deposited at the end of each period to accumulate a given amount S_n at the time of the nth payment.
- **Present value formula for an ordinary annuity** $A_n = Ra_{\overline{n}|i}$
- **Periodic payment for an ordinary annuity with present value A_n**

$$R = \frac{A_n}{a_{\overline{n}|i}}$$

- **Mortgage** An ordinary annuity in which R dollars is paid at the end of each period to pay off a loan of a given amount A_n.
- **Periodic payment on an amortized loan** $R = \dfrac{A_n}{a_{\overline{n}|i}}$

SUPPLEMENTARY EXERCISES

1. (a) A bank certificate of deposit can be bought for $10,000. At 8% simple interest, what will be the value of this certificate in 3 months?
   (b) What is the interest accrued over this 3-month period?

2. A 3-month treasury bill sells for $9832.30 and will pay $10,000 in 3 months.
   (a) What is the simple discount rate?
   (b) What is the effective simple interest rate r_{eff}?

3. John can invest money at 9% compounded continuously or at 9.1% compounded semiannually. Which investment should he choose?

4. Jill deposits $1000 into an account paying 6% interest compounded semiannually. Two years later the interest rate increases to 8% compounded quarterly. What is the amount in the account 5 years after the $1000 deposit? Assume there are no other deposits and no withdrawals during this time.

5. Mary deposits $500 into an account. She deposits $100 1 month later and withdraws $300 5 months after this last deposit. If interest in 12% compounded monthly, how much is in the account $1\frac{1}{2}$ years after she opened the account?

6. If Mike needs $1000 in 6 months and $2000 in 18 months, how much should he deposit now into a savings account paying 8% compounded quarterly?

7. Mr. Jones needs to borrow $10,000 now. He will make a payment of $3000 in 6 months and $4000 in 1 year. What final amount should he pay in 2 years to settle this debt if interest is 12% compounded semiannually?

8. Ms. Jones deposits $100 at the end of each month into an account paying 12% compounded monthly. How much will be in the account at the time of the twentieth deposit?

9. The Cox family buys a home for $100,000. They pay $20,000 down and take out a 30-year mortgage on the balance. If payments are made at the end of each year and interest is 8% compounded annually, how much is each payment?

10. Margaret takes out a loan to purchase a car. She will pay $200 at the end of each month for 3 years. If the interest rate is 9% compounded monthly, what is the amount of the loan?

11. **(Harder)** What is the outstanding balance of the loan of Exercise 10 at the end of 1 year? [*Hint*: The outstanding balance is the present value (at the time of the last payment) of the 24 payments which have not yet been made.]

12. A sinking fund is to be established to accumulate $20,000 at the end of 10 years. If payments are to be made at the end of every 3 months and interest is 8% compounded quarterly, how much should each payment be?

CHAPTER TEST

1. What is $1000 invested now worth in 18 months if interest is
   (a) 10% simple interest
   (b) 10% compounded quarterly
   (c) 10% compounded continuously

2. What is the present value of $2000 due in 3 years if interest is
   (a) 6% simple interest
   (b) 6% compounded semiannually
   (c) 6% compounded continuously

3. What is the effective rate i_{eff} of
   (a) 10% compounded quarterly
   (b) 10% compounded continuously

4. Mark borrows $2000 at a simple discount rate of 7% for 6 months.
   (a) How much will he receive now for this loan?
   (b) What is the effective simple interest rate r_{eff}?

5. Merle wants to borrow $4000 now. She will pay off the loan in two equal payments: one in 1 year and one in 18 months. How much is each payment if interest is 9% compounded semiannually?

6. Mr. and Mrs. Smith wish to accumulate $10,000 in 6 years to pay part of their son's college tuition. How much should they deposit at the end of every 6 months if interest is 10% compounded semiannually?

7. If Mr. and Mrs. Smith deposit $900 at the end of every 6 months into an account paying 10% interest compounded semiannually, how much will be in the account at the end of 6 years?

8. If Mr. and Mrs. Smith of Exercise 7 leave the money in the account for another 2 years and make no more deposits, how much will be in the account at the end of that time?

9. Stephen buys a television for $500. He pays $50 down and pays the balance in 12 equal monthly payments (at the end of each month). If interest is 12% compounded monthly, how much is each payment?

10. Rachelle can afford to make 36 monthly payments of $150 each (at the end of each month) to pay off a car loan. What price car can she afford if interest is 12% compounded monthly?

10

Applications

10.1 APPLICATIONS OF PROBABILITY TO LIFE INSURANCE AND MORTALITY

Probability and statistics play an important role in the insurance industry. Companies that want to insure individuals against fire, death, or automobile accidents use probability and statistics to determine their **premiums**, that is, what they will charge for the insurance. To determine the premiums, the company must be able to estimate the probability that it will have to pay off on each policy it sells. In this section we illustrate how this is done in the field of life insurance. (The mathematics of insurance is referred to as **actuarial mathematics**. An actuary is an applied mathematician who uses mathematics to analyze and solve business problems related to the field of insurance.)

Life insurance companies have compiled statistics, called **mortality tables**, that they use to determine the probability that a policyholder will die while the insurance policy is in effect. Mortality tables are constructed by selecting a large sample of individuals and recording how many deaths occur among individuals in the sample at each age. Mortality tables are constantly revised to reflect changes in medical technology, health standards, and environmental conditions. There are many different mortality tables available, and each insurance company tries to pick a mortality table based on individuals whose characteristics closely match those of its policyholders. Tables 1a and 1b are typical mortality tables: Table 1a is the 1980 CSO ALB Table for males, Table 1b is the 1980 CSO ALB Table for females.*

In these tables l_x represents the number of individuals of the original population of $l_0 = 10{,}000{,}000$ individuals, who are expected to be alive at (survive to) age x; d_x is the number of individuals who are expected to die between the ages of x and $x + 1$. Therefore,

$$d_x = l_x - l_{x+1} \tag{1}$$

*CSO stands for Commissioners Standard Ordinary, and ALB stands for age at last birthday. These tables are published in the *Transactions of the Society of Actuaries*, **33**(1981):673.

TABLE 1a

1980 CSO Mortality Table ALB—Males

Age Last Birthday x	Number Living l_x	Number Dying d_x	Age Last Birthday x	Number Living l_x	Number Dying d_x
0	10,000,000	26,300	50	8,955,550	62,689
1	9,973,700	10,273	51	8,892,861	67,853
2	9,963,427	9,864	52	8,825,008	73,512
3	9,953,563	9,655	53	8,751,496	79,901
4	9,943,908	9,248	54	8,671,595	86,803
5	9,934,660	8,743	55	8,584,792	94,089
6	9,925,917	8,239	56	8,490,703	101,634
7	9,917,678	7,736	57	8,389,069	109,393
8	9,909,942	7,432	58	8,279,676	117,406
9	9,902,510	7,328	59	8,162,270	125,862
10	9,895,182	7,421	60	8,036,408	135,012
11	9,887,761	8,009	61	7,901,396	145,070
12	9,879,752	9,089	62	7,756,326	156,057
13	9,870,663	10,562	63	7,600,269	167,890
14	9,860,101	12,227	64	7,432,379	180,384
15	9,847,874	13,984	65	7,251,995	193,048
16	9,833,890	15,636	66	7,058,947	205,627
17	9,818,254	16,887	67	6,853,320	217,867
18	9,801,367	17,838	68	6,635,453	229,918
19	9,783,529	18,393	69	6,405,535	242,193
20	9,765,136	18,554	70	6,163,342	254,977
21	9,746,582	18,519	71	5,908,365	268,417
22	9,728,063	18,289	72	5,639,948	282,449
23	9,709,774	17,866	73	5,357,499	296,484
24	9,691,908	17,445	74	5,061,015	309,228
25	9,674,463	16,930	75	4,751,787	319,558
26	9,657,533	16,611	76	4,432,229	326,655
27	9,640,922	16,486	77	4,105,574	329,965
28	9,624,436	16,362	78	3,775,609	329,686
29	9,608,074	16,526	79	3,445,923	326,536
30	9,591,548	16,785	80	3,119,387	321,110
31	9,574,763	17,235	81	2,798,277	313,659
32	9,557,528	17,873	82	2,484,618	304,142
33	9,539,655	18,602	83	2,180,476	291,835
34	9,521,053	19,518	84	1,888,641	275,968
35	9,501,535	20,618	85	1,612,673	256,383
36	9,480,917	21,996	86	1,356,290	233,567
37	9,458,921	23,553	87	1,122,723	208,523
38	9,435,368	25,287	88	914,200	182,410
39	9,410,081	27,289	89	731,790	156,376
40	9,382,792	29,556	90	575,414	131,442
41	9,353,236	31,988	91	443,972	108,378
42	9,321,248	34,582	92	335,594	87,734
43	9,286,666	37,425	93	247,860	69,929
44	9,249,241	40,419	94	177,931	55,153
45	9,208,822	43,558	95	122,778	43,201
46	9,165,264	46,926	96	79,577	33,501
47	9,118,338	50,424	97	46,076	24,927
48	9,067,914	54,135	98	21,149	15,759
49	9,013,779	58,229	99	5,390	5,390

*Adapted from tables in *Transactions of the Society of Actuaries*, **33** (1981). Used with permission of the Society of Actuaries.

TABLE 1b

1980 CSO Mortality Table ALB—Females

Age Last Birthday x	Number Living l_x	Number Dying d_x	Age Last Birthday x	Number Living l_x	Number Dying d_x
0	10,000,000	18,800	50	9,210,440	47,250
1	9,981,200	8,384	51	9,163,190	50,398
2	9,972,816	7,978	52	9,112,792	53,948
3	9,964,838	7,773	53	9,058,844	57,795
4	9,957,065	7,667	54	9,001,049	61,657
5	9,949,398	7,462	55	8,939,392	65,526
6	9,941,936	7,258	56	8,873,866	69,216
7	9,934,678	7,054	57	8,804,650	72,638
8	9,927,624	6,949	58	8,732,012	75,969
9	9,920,675	6,845	59	8,656,043	79,636
10	9,913,830	6,741	60	8,576,407	84,049
11	9,907,089	6,935	61	8,492,358	89,509
12	9,900,154	7,227	62	8,402,849	96,549
13	9,892,927	7,618	63	8,306,300	104,909
14	9,885,309	8,106	64	8,201,391	114,163
15	9,877,203	8,593	65	8,087,228	123,654
16	9,868,610	9,079	66	7,963,574	133,071
17	9,859,531	9,465	67	7,830,503	141,967
18	9,850,066	9,850	68	7,688,536	150,618
19	9,840,216	10,135	69	7,537,918	160,030
20	9,830,081	10,420	70	7,377,888	170,872
21	9,819,661	10,605	71	7,207,016	183,995
22	9,809,056	10,790	72	7,023,021	199,945
23	9,798,266	10,974	73	6,823,076	218,270
24	9,787,292	11,255	74	6,604,806	238,103
25	9,776,037	11,438	75	6,366,703	258,233
26	9,764,599	11,718	76	6,108,470	277,630
27	9,752,881	12,094	77	5,830,840	295,507
28	9,740,787	12,468	78	5,535,333	311,750
29	9,728,319	12,841	79	5,223,583	326,840
30	9,715,478	13,310	80	4,896,743	341,156
31	9,702,168	13,777	81	4,555,587	354,561
32	9,688,391	14,242	82	4,201,026	366,540
33	9,674,149	14,898	83	3,834,486	375,396
34	9,659,251	15,551	84	3,459,090	379,185
35	9,643,700	16,394	85	3,079,905	376,642
36	9,627,306	17,522	86	2,703,263	367,157
37	9,609,784	18,835	87	2,336,106	350,836
38	9,590,949	20,429	88	1,985,270	328,324
39	9,570,520	22,204	89	1,656,946	300,802
40	9,548,316	24,157	90	1,356,144	269,669
41	9,524,159	26,191	91	1,086,475	236,504
42	9,497,968	28,304	92	849,971	202,880
43	9,469,664	30,303	93	647,091	170,450
44	9,439,361	32,471	94	476,641	140,719
45	9,406,890	34,617	95	335,922	114,556
46	9,372,273	36,739	96	221,366	91,619
47	9,335,534	39,116	97	129,747	69,705
48	9,296,418	41,648	98	60,042	44,669
49	9,254,770	44,330	99	15,373	15,373

*Adapted from tables in *Transactions of the Society of Actuaries,* **33** (1981). Used with permission of the Society of Actuaries.

The number of individuals who are expected to die between the ages of x and y is

$$l_x - l_y$$

Table 1a gives values for l_x and d_x for males, Table 1b for females.

EXAMPLE 1 (a) The number of males who are expected to live until age 22 is obtained from Table 1a:

$$l_{22} \text{ (male)} = 9{,}728{,}063$$

(b) The number of males who are expected to die between the ages of 22 and 23 is obtained from Table 1a:

$$d_{22} \text{ (male)} = 18{,}289$$

(c) The number of males who are expected to die between the ages of 22 and 28 is:

$$l_{22} - l_{28} \text{ (male)} = 9{,}728{,}063 - 9{,}624{,}436 = 103{,}627$$

(d) The number of females who are expected to live until age 22 is obtained from Table 1b:

$$l_{22} \text{ (female)} = 9{,}809{,}056$$

(e) The number of females who are expected to die between the ages of 22 and 23 is obtained from Table 1b:

$$d_{22} \text{ (female)} = 10{,}790$$

Estimated probabilities of survival and death are obtained from the mortality tables in the following manner:

The probability that an individual will survive to age x is

$$\frac{l_x}{l_0} \tag{2}$$

The probability that an individual will die between the ages of y and z is

$$\frac{l_y - l_z}{l_0} \tag{3}$$

Note that in (2) and (3) the sample space consists of $l_0 = 10{,}000{,}000$ individuals of age 0, the original population.

The probability that an individual of age x will survive to age y is

$$\frac{l_y}{l_x} \tag{4}$$

The probability that an individual of age x will die between the ages of y and z is

$$\frac{l_y - l_z}{l_x} \tag{5}$$

Equations (4) and (5) are conditional probabilities since we are given that the individual has survived to age x. Therefore, the denominator is l_x, which is the number of individuals in the reduced sample space, the number of people who have survived to age x.

EXAMPLE 2 (a) The probability that a male will survive to age 50 is

$$\frac{l_{50}}{l_0}\text{ (male)} = \frac{8{,}955{,}550}{10{,}000{,}000}\text{ (from Table 1a)} = .8956$$

(b) The probability that a female will die between the ages of 50 and 51 is

$$\frac{l_{50} - l_{51}}{l_0} = \frac{d_{50}}{l_0}\text{ (female)} = \frac{47{,}250}{10{,}000{,}000}\text{ (from Table 1b)} = .004725$$

(c) The probability that a 20-year-old female will survive to age 50 is

$$\frac{l_{50}}{l_{20}}\text{ (female)} = \frac{9{,}210{,}440}{9{,}830{,}081}\text{ (from Table 1b)} = .9370$$

(d) The probability that a 20-year-old male will die between the ages of 50 and 55 is

$$\frac{l_{50} - l_{55}}{l_{20}}\text{ (male)} = \frac{8{,}955{,}550 - 8{,}584{,}792}{9{,}765{,}136}\text{ (from Table 1a)} = .03797$$

The following sections are for readers who are familiar with Chapter 9.

Premiums for Insurance Policies

A **premium** is the amount a person pays to purchase an insurance policy. In return, the company pays out **benefits** according to the terms of the policy. Since the premiums are invested by the company, interest rates are taken into account when determining their value.

The **net single premium** for a policy is the present value of the future benefits that the company *expects* to pay out on that type of policy, where *present value* is as defined in Section 9.2. In general, the true premium is composed of the net single premium plus a **loading factor**. The loading factor involves the insurance company's expenses and profit.

In this section we compute the net single premium for two types of policies: n-year pure endowment and n-year term insurance. Our method can be used to determine net single premiums for other types of policies as well.* Keep in mind that the true premium will be higher, in general, than the net single premium. We will assume that interest is compounded annually and that benefits are paid at the end of the year in which they are due. In the following discussion, x represents the age of the person buying the policy at the time of purchase.

Pure Endowment Policy An n-year pure endowment with face value F is a policy that guarantees the purchaser will receive one payment of F dollars at the end of n years, provided that he or she is alive at that time. We denote the net single premium for this policy by P. (In actuarial mathematics, the net single premium for an n-year pure endowment with face value of \$1 is denoted by ${}_nE_x$. In this notation, $P = F{}_nE_x$.)

EXAMPLE 3 A 27-year-old man buys a 10-year pure endowment policy with a face value of \$1000. What is the net single premium for this policy? Assume an interest rate of 8% compounded annually.

Solution Let $x = 27$ (male), $n = 10$, $F = 1000$, and $i = .08$. We now show how probability is involved in this computation. We assume that all males of age 27 buy this policy (there are l_{27} such males). If the net single premium for this policy is denoted by P, then the company will receive $P \cdot l_{27} = Pl_{27}$ dollars at the time when $x = 27$. In 10 years the company will have to pay \$1000 to all males who have survived to age 37 (there are l_{37} such males). Therefore, the company expects to pay out $1000l_{37}$ in 10 years. (See the time line in Figure 1.)

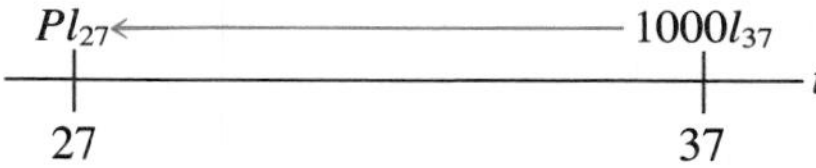

Then

$$
\begin{aligned}
Pl_{27} &= \text{present value of } \$1000l_{37} \\
&= 1000l_{37}(1 + i)^{-10} \\
&= 1000l_{37}(1 + .08)^{-10} \\
P &= 1000\frac{l_{37}}{l_{27}}(1.08)^{-10} \\
&= 1000\left(\frac{9,458,921}{9,640,922}\right)(1.08)^{-10} \qquad \text{(from Table 1a)} \\
&= \$454.45
\end{aligned}
$$

This means that the net single premium is the present value of the proportion of \$1000 that on the average the company expects to pay out in 10 years. This

*See Robert Cissel, Helen Cissel, and David C. Flaspohler, *Mathematics of Finance*, 7th ed. (Boston: Houghton Mifflin, 1986).

proportion is, by (4),

$$\frac{l_{37}}{l_{27}} = \text{probability that a 27-year-old male will survive to age 37}$$

In general the net single premium for an n-year pure endowment policy with face value F with interest rate i per year bought by a person age x is

$$P = F\frac{l_{x+n}}{l_x}(1 + i)^{-n} \qquad (6)$$

The proof follows the method of Example 3 and is left for Exercise 13.

EXAMPLE 4 What is the net single premium for a 20-year pure endowment with a face value of \$15,000 issued to a woman age 25? The interest rate is 8% compounded annually.

Solution Here $F = 15{,}000$, $x = 25$ (female), $n = 20$, and $i = .08$. Therefore, by (6),

$$P = 15{,}000\frac{l_{25+20}}{l_{25}}(1 + .08)^{-20}$$

$$= 15{,}000\frac{l_{45}}{l_{25}}(1.08)^{-20}$$

$$= 15{,}000\left(\frac{9{,}406{,}890}{9{,}776{,}037}\right)(.214548) = \$3096.70 \qquad \text{(by Table 1b)}$$

n-year Term Insurance

The insurance policy known as ***n*-year term insurance** guarantees that if the policyholder dies within n years of the purchase of this policy, his or her beneficiary will collect the face value of the policy at the time of the insured's death (as mentioned earlier, we assume that the payment is made at the end of the year of death). If the insured survives the n years, the policy is terminated. We denote the net single premium for this policy by Q. (In actuarial mathematics, the net single premium for this type of policy with a face value of \$1 is denoted by $A^{1}_{x:\overline{n|}}$. In this notation $Q = FA^{1}_{x:\overline{n|}}$.)

EXAMPLE 5 What is the net single premium for 2-year term insurance with a face value of \$1000 issued to a 25-year-old woman? The interest rate is 6% compounded annually.

Solution Again, we let F = the face value of the policy: $F = 1000$, $x = 25$ (female), $n = 2$, and $i = .06$. We assume that all females of age 25 buy this policy (there are l_{25} such females). The net single premium is Q, so the company will receive Ql_{25} dollars at the time when $x = 25$. (See the time line of Figure 2.)

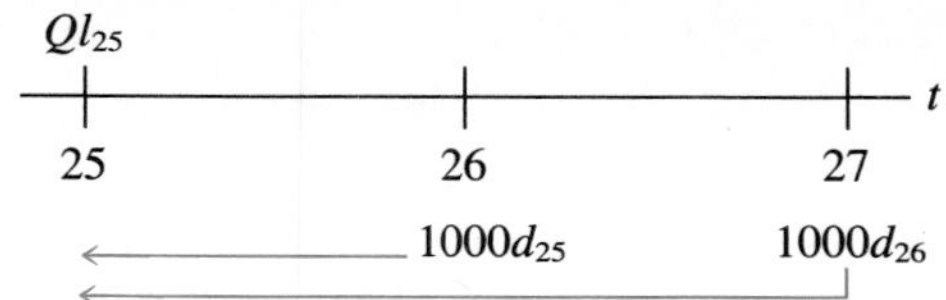

At the end of the first year, the company expects to pay $1000 to each of the beneficiaries of the females who are expected to die between the ages of 25 and 26 (there are d_{25} such females). That is, at the end of the first year the company expects to pay out $\$1000d_{25}$ (see Figure 2).

At the end of the second year, the company expects to pay $1000 to each of the beneficiaries of the females who are expected to die during that year (there are d_{26} such females). Therefore at the end of the second year, the company expects to pay out $\$1000d_{26}$ (see Figure 2). There are no further payments since the policy is in effect only 2 years. Therefore, the total net single premium is

$$\begin{aligned} Ql_{25} &= \text{present value of future benefits} \\ &= 1000d_{25}(1 + .06)^{-1} + 1000d_{26}(1 + .06)^{-2} \end{aligned}$$

and

$$\begin{aligned} Q &= 1000\frac{d_{25}}{l_{25}}(1 + .06)^{-1} + 1000\frac{d_{26}}{l_{25}}(1 + .06)^{-2} \\ &= 1000\left[\frac{d_{25}}{l_{25}}(1.06)^{-1} + \frac{d_{26}}{l_{25}}(1.06)^{-2}\right] \\ &= 1000\left[\frac{11{,}438}{9{,}776{,}037}(1.06)^{-1} + \frac{11{,}718}{9{,}776{,}037}(1.06)^{-2}\right] \qquad \text{(from Table 1b)} \\ &= \$2.17 \end{aligned}$$

Note that

$$\frac{d_{25}}{l_{25}} = \frac{l_{25} - l_{26}}{l_{25}} \qquad \text{[from Equation (1)]}$$

This is the probability that a 25-year-old woman will die before age 26. Similarly, d_{26}/l_{25} is the probability that a 25-year-old woman will die between the ages of 26 and 27. Therefore, the net single premium is the present value of the proportions of $1000 that the company expects to pay out each of the 2 years that the policy is in force.

In general, the net single premium for n-year term insurance with face value F and interest rate i per year, bought by a person age x, is

$$Q = F\left[\frac{d_x}{l_x}(1 + i)^{-1} + \frac{d_{x+1}}{l_x}(1 + i)^{-2} + \cdots + \frac{d_{x+n-1}}{l_x}(1 + i)^{-n}\right] \qquad (7)$$

The proof follows the method of Example 5 and is left for Exercise 14.

EXAMPLE 6 What is the net single premium for a 3-year term insurance policy with face value \$5000 bought by a 65-year-old man? Interest is 7% compounded annually.

Solution Here $F = 5000$, $x = 65$ (male), $n = 3$, and $i = .07$. By Equation (7), and Table 1a,

$$Q = 5000\left[\frac{d_{65}}{l_{65}}(1 + .07)^{-1} + \frac{d_{66}}{l_{65}}(1 + .07)^{-2} + \frac{d_{67}}{l_{65}}(1 + .07)^{-3}\right]$$

$$= 5000\left[\frac{193{,}048}{7{,}251{,}995}(1.07)^{-1} + \frac{205{,}627}{7{,}251{,}995}(1.07)^{-2} + \frac{217{,}867}{7{,}251{,}995}(1.07)^{-3}\right]$$

$$= \$370.84$$

The methods used in this section are applicable for computing net single premiums for other types of life insurance. For further reading on the subject see the following books:

1. Robert Cissell, Helen Cissell, and David C. Flaspohler, *Mathematics of Finance,* 7th ed. (Boston: Houghton Mifflin, 1986).
2. Lewis C. Workman, *Mathematical Foundations of Life Insurance* (Atlanta: FLMI Insurance Education Program, Life Management Institute LOMA, 1982).

EXERCISE SET 10.1

1. Use Table 1a or 1b to determine:
(a) the number of males alive at age 60.
(b) the number of females who are expected to die between the ages of 30 and 40.

2. Use Table 1a or 1b to determine:
(a) the number of females who are expected to survive to age 90.
(b) the number of males who are expected to die between the ages of 70 and 90.

3. What is the probability that
(a) a male age 40 will survive to age 60?
(b) a female age 40 will survive to age 60? Why does this answer differ from that of (a)?
(c) a female will survive to age 60?
(d) a male will survive to age 60? Why does this answer differ from that of (c)?
(e) a female age 70 will die between the ages of 80 and 85?
(f) a male age 70 will die between the ages of 80 and 85?

4. What is the probability that
(a) a male age 20 will survive to age 60?
(b) a female age 20 will survive to age 60? Why does this answer differ from that of (a)?
(c) a female will survive to age 90?
(d) a male will survive to age 90? Why does this answer differ from that of (c)?
(e) a female age 60 will die between the ages of 70 and 80?
(f) a male age 60 will die between the ages of 70 and 80?

5. When John is 20 years old, he buys a 25-year pure endowment policy with a face value of \$10,000. What is the net single premium for this policy? Interest is 7% compounded annually.

6. Jack, age 65, buys a 5-year pure endowment policy with a face value of \$5000. What is the net single premium for this policy if interest is 6% compounded annually?

7. Jennifer buys a 25-year pure endowment policy with a face value of \$10,000 when she is 20 years old. What is the net single premium for this policy? Interest is 7% compounded annually. Why does this answer differ from that of Exercise 5?

8. Jill, age 65, buys a 5-year pure endowment policy with a face value of $5000. What is the net single premium for this policy if interest is 6% compounded annually? Why does this answer differ from that of Exercise 6?

9. Mr. Smith, age 65, purchases 4-year term insurance with a face value of $10,000. What is the net single premium for this policy if interest is 7% compounded annually?

10. Mr. Jones, age 40, purchases 5-year term insurance with a face value of $12,000. What is the net single premium for this policy if interest is 6% compounded annually?

11. Mrs. Smith, age 65, purchases 4-year term insurance with a face value of $10,000. What is the net single premium for this policy if interest is 7% compounded annually? Why does this answer differ from that of Exercise 9?

12. Mrs. Jones, age 40, purchases 5-year term insurance with a face value of $12,000. What is the net single premium for this policy if interest is 6% compounded annually? Why does this answer differ from that of Exercise 10?

13. Prove Equation (6). [*Hint*: Assume that the l_x individuals who are alive at age x buy this policy. How many will be alive at age $x + n$ to collect F dollars? Set up a time diagram.]

14. Prove Equation (7). [*Hint*: Assume that the l_x individuals who are alive at age x buy this policy. How many beneficiaries are expected to collect F dollars at time $x + 1, x + 2, \ldots, x + n$? Set up a time diagram.]

10.2 INTRODUCTION TO GAME THEORY AND APPLICATIONS

Many activities in business, economics, politics, and the sciences call for decisions in competitive situations. As a result of the pioneering efforts by John von Neumann,* Émile Borel,† and Oskar Morgenstern,‡ beginning in the 1920s, it is now recognized that many of these competitive situations are analogous to games played according to formal rules. The players are groups or individuals pursuing their own objectives in direct conflict with the other players. In this section and the next we consider the simplest kinds of games; we show how matrices and probability theory can be used to determine optimal strategies for the players, and we discuss several applications.

We begin with some examples that illustrate the kinds of problems to which game theory applies.

**John von Neumann* (1903–1957) was born in Budapest, Hungary, and was a child prodigy in mathematics and general science. In 1923 he received a doctorate in mathematics from the University of Budapest and a degree in chemical engineering from the Federal Institute of Technology at Zurich, Switzerland. After lecturing in Europe he came to Princeton University, where he stayed until his death. Von Neumann was probably the greatest mathematical genius of his century. He contributed to quantum mechanics, economics, computers, many areas of pure mathematics, and he developed a technique that accelerated the production of the first atomic bomb. His genius was recognized with many awards and honors.

†*Émile Borel* (1871–1956) was a French mathematician who made important contributions to several branches of mathematics. His pioneering work helped launch the field of measure theory on which the modern advanced notions of length, area, volume, and probability rest. In addition to his work in mathematics, he sat in the Chamber of Deputies for a period of time and was Minister of the Navy.

‡*Oskar Morgenstern* (1902–1977) was a German-born American economist at Princeton University. His book, *Theory of Games and Economic Behavior*, coauthored with John von Neumann, is a landmark in economics and game theory.

EXAMPLE 1 **FRANCHISE LOCATION: A BUSINESS GAME** A new regional shopping center, Oxford Mall, plans to have only two restaurants, one to be operated by McDonald's and the other by Gino's. As illustrated in Figure 1, McDonald's can locate at position 1 (an end) or at position 2 (near the center), and Gino's can locate at position 3 (near the center) or at position 4 (an end). If both franchises choose central locations or both choose end locations, they will each get 50% of the business. If McDonald's chooses the center and Gino's the end, then McDonald's will capture 75% of the business (and Gino's 25%), while if McDonald's chooses an end and Gino's the center, then McDonald's will get 30% of the business (and Gino's 70%).

FIGURE 1

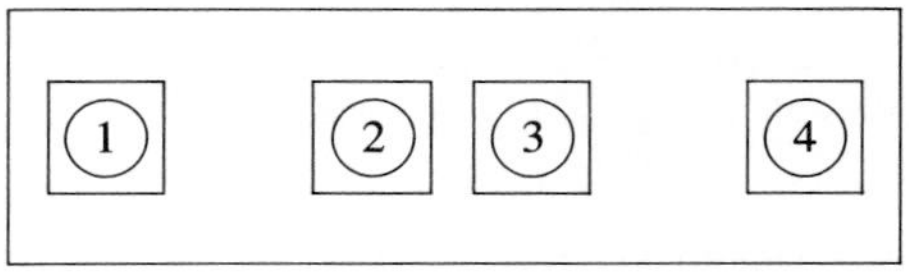

1 2 3 4

Oxford Mall

This situation is analogous to a game; each player (franchise) can control where it locates, but it cannot control where its opponent locates. The objective of each player is to develop an "optimal strategy" that will ensure the maximum possible business.

EXAMPLE 2 **PENNY MATCHING** Consider a game involving two players, I and II, in which each player selects one side of a penny without knowing the opponent's choice. If the choices match, then player I pays player II $1; otherwise, player II pays player I $1. As in the previous example, each player can control his or her own move but cannot control the opponent's move. The objective of each player is to develop an optimal strategy that will ensure the best possible result.

EXAMPLE 3 **A MATRIX GAME** Consider the 3×3 matrix

$$\begin{bmatrix} 1 & 6 & -1 \\ 3 & -2 & -3 \\ 4 & 5 & 3 \end{bmatrix}$$

We can associate a game with this matrix as follows. Consider two players, Richard and Carol, who will be referred to as R and C. Player R picks a row without telling C his choice, and player C picks a column without telling R. After the players make their choices, money or some other item of value changes hands as specified by the entry in the chosen row and column of the matrix. Positive entries in the matrix represent amounts player C pays to player R, and negative signs precede amounts player R pays to player C. For example, if player R picks row 2 and player C picks column 1, then player C pays 3 units to player R; and if player R picks row 1 and player C picks column 3, then player R pays 1 unit to player C.

Games of the type described in Example 3 are called **matrix games**, and the matrix is called the **payoff** matrix.

The games in Examples 1, 2, and 3 are **two-person** games, which means that they are played by two opposing players. These games are also examples of **constant-sum** games, which means that the sum of the payoffs realized by the players is a constant that does not depend on which moves are made. To illustrate, in Example 1 McDonald's percentage of the business plus Gino's percentage of the business is always 100%, regardless of their moves. In Example 2, player I's payoff plus player II's payoff is always zero, regardless of their moves, since the amount lost by one player is won by the other player. Constant-sum games in which the constant is zero are called **zero-sum** games. Examples 2 and 3 are zero-sum games. Example 3 is of particular importance because any two-person constant-sum game in which each player has only a finite number of possible moves can be viewed as a matrix game. The following example illustrates this point.

EXAMPLE 4 Consider the business game described in Example 1. The players, McDonald's and Gino's, each have two possible moves: locate in the center or locate at an end. The possible outcomes resulting from the moves can be tabulated in matrix form as

$$\begin{array}{ccc} & & \text{Gino's moves} \\ & & \begin{array}{cc} \text{Center} & \text{End} \end{array} \\ \text{McDonald's moves} & \begin{array}{c} \text{Center} \\ \text{End} \end{array} & \begin{bmatrix} 50 & 75 \\ 30 & 50 \end{bmatrix} \end{array} \tag{1}$$

where the numbers in this matrix indicate the percentage of shoppers that McDonald's will service as a result of the players' moves. Thus, if McDonald's locates at an end and Gino's locates in the center, McDonald's will get 30% of the business.

Using (1) as the payoff matrix, Example 1 can be viewed as a matrix game, where McDonald's picks a row and Gino's picks a column; the entry in the chosen row and column then indicates what percentage of the shoppers Gino's "pays" or concedes to McDonald's.

EXAMPLE 5 Consider the penny-matching game described in Example 2. The players, I and II, each have two possible moves: show a head or show a tail. The possible outcomes resulting from the moves can be tabulated in matrix form as

$$\begin{array}{ccc} & & \text{Player II} \\ & & \begin{array}{cc} \text{Heads} & \text{Tails} \end{array} \\ \text{Player I} & \begin{array}{c} \text{Heads} \\ \text{Tails} \end{array} & \begin{bmatrix} -1 & 1 \\ 1 & -1 \end{bmatrix} \end{array}$$

Using this as the payoff matrix, Example 2 can be viewed as a matrix game, where player I picks a row and player II picks a column. The entry in the chosen row and column then indicates the amount that player II pays player I. For example, if player I picks row 2 (tails) and player II picks column 1 (heads), then II pays I \$1. Similarly, if player I picks row 2 and player II picks column 2, then I pays II \$1.

We now describe how the players of a matrix game can determine their optimal moves. In our analysis, we make the following assumptions:

1. Each player must choose his or her move without knowing what move the opponent has made or is planning to make.
2. Each player chooses his or her move with the assumption that the player faces an intelligent opponent who will make the best possible move.

The next example illustrates the reasoning process the players of a matrix game might use to determine their optimal moves.

EXAMPLE 6 Consider a game with payoff matrix

$$\begin{bmatrix} -2 & 5 & -1 \\ 4 & -4 & 0 \\ 3 & 4 & 1 \\ 3 & 3 & 2 \end{bmatrix}$$

Recall that positive entries in the matrix are amounts won by the row player R and negative signs precede amounts won by the column player C.

In this game, player R would like to win the 5 units appearing in row 1. If he plays row 1, however, then he must assume that his opponent will make the best possible move and play column 1. Thus, instead of winning 5 units he would *lose* 2 units. With this in mind, it is clear that row 3, for example, would be a better move than row 1 since player R is assured of winning at least 1 unit, even against player C's best countermove (column 3). By the same token, row 4 would be an even better move than row 3 since player R could then assure himself of winning at least 2 units against C's best countermove.

The analysis in the last example suggests the following method for player R to determine his best move:

Strategy for Player R

Step 1 For each row, player R should find the smallest element in the row (called the **row minimum**); this number represents the payoff to R when R chooses that row and C makes the best possible countermove. In other words, the row minimum is the worst possible payoff to R for that row.

Step 2 Then, to obtain the best possible move, player R should choose a row that yields the largest value among the row minima. This assures that player R will get the largest possible payoff against player C's best countermoves.

We now use these two steps to determine player R's best move in the game from Example 6. See Figure 2.

FIGURE 2

$$\begin{bmatrix} -2 & 5 & -1 \\ 4 & -4 & 0 \\ 3 & 4 & 1 \\ 3 & 3 & 2 \end{bmatrix} \quad \begin{matrix} \text{Row minima} \\ -2 \\ -4 \\ 1 \\ 2 \end{matrix}$$

Step 1
Compute the row minima.

$$\begin{bmatrix} -2 & 5 & -1 \\ 4 & -4 & 0 \\ 3 & 4 & 1 \\ 3 & 3 & 2 \end{bmatrix} \quad \begin{matrix} \text{Row minima} \\ -2 \\ -4 \\ 1 \\ 2 \leftarrow \text{Largest row minimum} \end{matrix}$$

Step 2
Row 4 is player R's best move.

Player C will use a similar analysis to determine her best move.

Strategy for Player C

Step 1 For each column, player C should find the largest element in the column (called the **column maximum**); this number represents the payoff to R when C plays that column and R makes the best possible countermove.

Step 2 To obtain the best possible move, player C should choose a column that yields the smallest value among the column maxima. This assures that player C will give the smallest payoff against player R's best countermoves.

In Figure 3 we have used these steps to determine player C's best move in the game from Example 6.

FIGURE 3

$$\begin{matrix} & \begin{bmatrix} -2 & 5 & -1 \\ 4 & -4 & 0 \\ 3 & 4 & 1 \\ 3 & 3 & 2 \end{bmatrix} \\ \text{Column maxima} & \begin{matrix} 4 & 5 & 2 \end{matrix} \end{matrix}$$

Step 1
Compute the column maxima.

$$\begin{matrix} & \begin{bmatrix} -2 & 5 & -1 \\ 4 & -4 & 0 \\ 3 & 4 & 1 \\ 3 & 3 & 2 \end{bmatrix} \\ \text{Column maxima} & \begin{matrix} 4 & 5 & 2 \\ & & \uparrow \end{matrix} \end{matrix}$$

Smallest column maximum

Step 2
Column 3 is player C's best move.

EXAMPLE 7 Find the best moves for Gino's and McDonald's in the business game of Example 1.

Solution As shown in Example 4, the payoff matrix for this game is

		Gino's Center	Gino's End
McDonald's	Center	50	75
	End	30	50

Computing the row minima and column maxima we obtain

$$\begin{array}{rcc} & & \text{Row minima} \\ & \begin{bmatrix} 50 & 75 \\ 30 & 50 \end{bmatrix} & \begin{array}{c} 50 \\ 30 \end{array} \\ \text{Column maxima} & \begin{array}{cc} 50 & 75 \end{array} & \end{array}$$

Since the largest row minimum is 50, McDonald's should play row 1, and since the smallest column maximum is 50, Gino's should play column 1. Thus, both McDonald's and Gino's should choose central locations, in which case they will each get 50% of the business.

The following notions are useful in the analysis of matrix games.

Definition A matrix game is **strictly determined** if there is an entry in the payoff matrix that is both the smallest element in its row and the largest element in its column. Such an entry is called a **saddle point** for the game.

To illustrate, the game in Example 6 is strictly determined because the entry 2, which appears in row 4 and column 3, is a saddle point.

As the next example shows, there are matrix games that are not strictly determined; that is, they have no saddle points.

EXAMPLE 8 Consider the game with payoff matrix

$$\begin{array}{rcc} & & \text{Row minima} \\ & \begin{bmatrix} 2 & -1 & 3 \\ 0 & 1 & -2 \\ -3 & 2 & 1 \end{bmatrix} & \begin{array}{c} -1 \\ -2 \\ -3 \end{array} \\ \text{Column maxima} & \begin{array}{ccc} 2 & 2 & 3 \end{array} & \end{array}$$

It is clear from the indicated row minima and column maxima that no element is simultaneously a minimum for its row and a maximum for its column; thus, the game has no saddle points.

The next example shows that a game can have several saddle points.

EXAMPLE 9 Consider the game with payoff matrix

$$\begin{bmatrix} 2 & 3 & -2 & -1 \\ 6 & 7 & 8 & 6 \\ 0 & 1 & -2 & 3 \\ 6 & 8 & 9 & 6 \end{bmatrix}$$

Each of the entries marked in color is a minimum for its row and maximum for its column; this game has four saddle points.

When a game has more than one saddle point, player R or player C or both have more than one optimal move. Thus, player R has two best moves, row 2 and row 4; player C also has two best moves, column 1 and column 4.

It is not accidental that the same number appears at each saddle point in Example 9. It can be proved, although we omit the proof, that the same number *must* occur at every saddle point. This number is called the **value** of the game. *Thus, in a strictly determined game, the value of the game is the payoff that results when both players make their best moves.*

It is interesting to note that in strictly determined games, neither player can benefit from discovering which move the opponent will make. In other words, *it does not do any good to spy on your opponent when playing a strictly determined game*. To see why, consider the game from Example 6. Even if player C announced in advance that she would play column 3 (Figures 2 and 3), it would still be best for player R to play row 4, which was the choice he made with no knowledge of player C's intentions.

On the other hand, consider the penny-matching game from Example 2, where the payoff matrix is

		Player II Heads	Player II Tails
Player I	Heads	-1	1
	Tails	1	-1

This game has no saddle point and consequently is not strictly determined. Since the minimum in each row is -1, both rows are equally good moves for player I. Similarly, since both column maxima are 1, both columns are equally good moves for player II. Suppose, however, that player I discovers in advance that player II will play the first column. Player I would then definitely play row 2 and win \$1. Thus, for games that are not strictly determined, discovery of the opponent's intended move affects the play of the game.

We conclude this section with a game theory analysis of an important battle that occurred in the Pacific Theater during World War II.*

EXAMPLE 10 **A WAR GAME: THE BATTLE OF RABAUL–LAE IN WORLD WAR II** In 1943 General Kenney was commander of the Allied Air Forces in the Southwest Pacific Area. In the critical stages of the struggle for New Guinea, intelligence reports indicated that a Japanese troop and supply convoy was going to be moved from the port of Rabaul in New Britain to Lae in New Guinea (Figure 4). The Japanese commander had two possible courses of action: he could either travel north of New Britain or south of New Britain. Either trip would take 3 days. Rain and poor visibility were predicted for the northern route; clear weather was predicted for the southern route. General Kenney could concentrate most of his reconnaissance aircraft on either the northern or southern route. Once the convoy was sighted, it would be bombed until its arrival in Lae.

*This example was presented by O. G. Haywood, Jr., in the *Journal of the Operations Research Society of America,* **2**(1954): 365–385. It is also discussed in reference 2 at the end of Section 10.3.

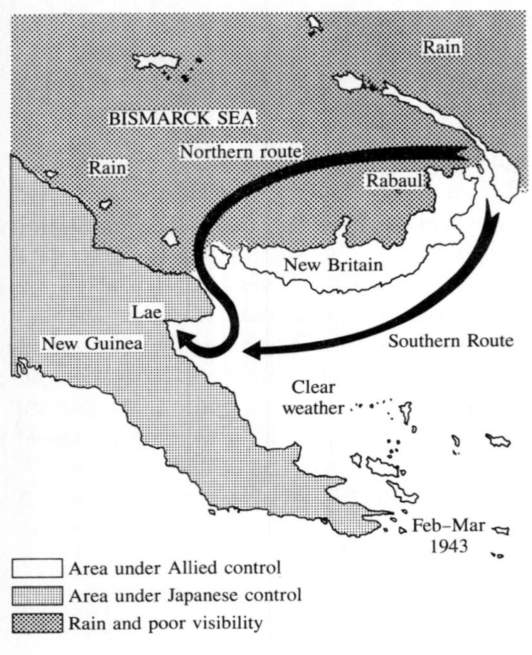

FIGURE 4

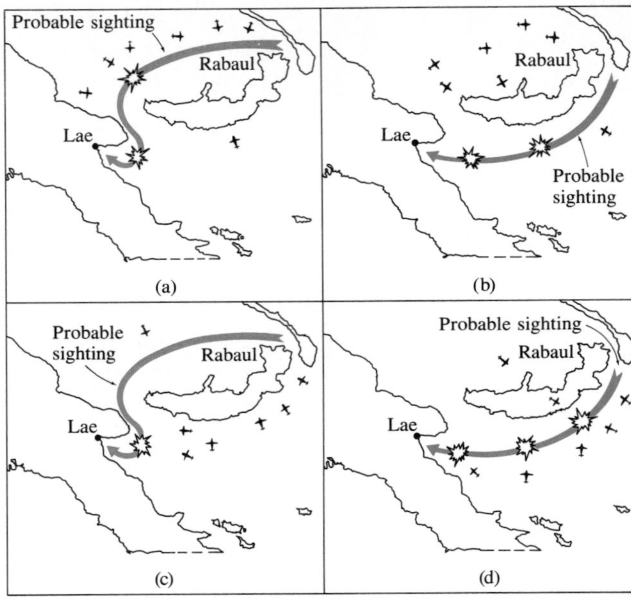

FIGURE 5

General Kenney's staff analyzed the possible alternatives as follows:

1. If Kenney concentrated on the northern route and the Japanese also went the northern route then, due to bad weather, the convoy would be discovered on the second day, allowing 2 days of bombing (Figure 5a).
2. If Kenney concentrated on the northern route and the Japanese went the southern way, the convoy would be missed on the first day (most of the aircraft would be on the northern route), allowing 2 days of bombing (Figure 5b).
3. If Kenney concentrated on the southern route and the Japanese went the northern way, then, due to poor visibility and limited aircraft on the northern route, the convoy would not be spotted for the first 2 days, allowing 1 day of bombing (Figure 5c).
4. If Kenney concentrated on the southern route and the Japanese went the southern way, then the convoy would be spotted at once, allowing 3 days of bombing (Figure 5d).

This military problem can be viewed as a two-person game; the Japanese commander wanted to minimize the number of days he would be bombed, while General Kenney wanted to maximize the number of days he could bomb the Japanese convoy. The number of bombing days in each of the four possible situations is given by the following payoff matrix:

		Japanese commander Northern route	Japanese commander Southern route
Kenney	Northern route	2 days	2 days
	Southern route	1 day	3 days

Computing the row minima and column maxima for this game we obtain

$$\begin{bmatrix} 2 & 2 \\ 1 & 3 \end{bmatrix} \quad \begin{matrix} \text{Row minima} \\ 2 \\ 1 \end{matrix}$$

$$\text{Column maxima} \quad 2 \quad 3$$

Since the largest row minimum is 2, Kenney's best move is row 1; since the smallest column maximum is 2, the best move for the Japanese commander is column 1. Thus, the best course of action for each side is to pick the northern route.

As a historical note, these were precisely the choices made by the opposing sides. The convoy sailed along the northern route; it was sighted 1 day after it sailed and was bombed for 2 days, resulting in a disastrous defeat for the Japanese. As our game-theory analysis shows, however, the Japanese commander made the best possible decision in a hopeless situation.

It is interesting to observe that the Rabaul–Lae game is strictly determined because the entry in row 1 and column 1 is a saddle point. (It is the minimum for its row and a maximum for its column.) Thus, neither commander would have benefited if an intelligence operation had uncovered his opponent's decision in advance.

EXERCISE SET 10.2

1. Which of the following matrix games are strictly determined?

(a) $\begin{bmatrix} -1 & 2 \\ 3 & -2 \end{bmatrix}$ (b) $\begin{bmatrix} -1 & -2 \\ 3 & 2 \end{bmatrix}$

(c) $\begin{bmatrix} -1 & 1 & 0 \\ 4 & 5 & 3 \\ -1 & 2 & 0 \end{bmatrix}$ (d) $\begin{bmatrix} 1 & 0 & 2 \\ 3 & 2 & -1 \\ -1 & 2 & 4 \end{bmatrix}$

(e) $\begin{bmatrix} 1 & 0 & 0 & -1 \\ 2 & 3 & 1 & 4 \\ 2 & -1 & 3 & 2 \\ 3 & -2 & -2 & 3 \end{bmatrix}$ (f) $\begin{bmatrix} 1 & -2 & 0 & 4 \\ 1 & 1 & 1 & 3 \\ 1 & 1 & 1 & 3 \\ 0 & 0 & 0 & 3 \end{bmatrix}$

2. Find all saddle points for the following two-person constant-sum matrix games:

(a) $\begin{bmatrix} 4 & 6 \\ 2 & -7 \end{bmatrix}$ (b) $\begin{bmatrix} 7 & 3 \\ 3 & -4 \end{bmatrix}$

(c) $\begin{bmatrix} -4 & 1 & 1 \\ -3 & 4 & 7 \\ -10 & 3 & 5 \end{bmatrix}$ (d) $\begin{bmatrix} 0 & 5 & 2 \\ 3 & 3 & 7 \\ 5 & 7 & -1 \end{bmatrix}$

(e) $\begin{bmatrix} 1 & -2 & 0 & 4 \\ 1 & 1 & 1 & 3 \\ 1 & 1 & 1 & 3 \\ 0 & 0 & 0 & 3 \end{bmatrix}$

3. Find the best moves for players R and C in the following strictly determined games, and give the value of the game:

(a) $\begin{bmatrix} 3 & -5 \\ 2 & -6 \end{bmatrix}$ (b) $\begin{bmatrix} 1 & 1 \\ 2 & 0 \end{bmatrix}$

(c) $\begin{bmatrix} 1 & -4 & 7 \\ 6 & -5 & 9 \end{bmatrix}$ (d) $\begin{bmatrix} 7 & 4 & 6 \\ -2 & 3 & 1 \\ 1 & 2 & 3 \end{bmatrix}$

4. Follow the directions of Exercise 3 for the following games:

(a) $\begin{bmatrix} 0 & 2 \\ -1 & 3 \end{bmatrix}$ (b) $\begin{bmatrix} 4 & 2 & 8 & 3 \\ -6 & 0 & 7 & -1 \\ 3 & 2 & 4 & 5 \end{bmatrix}$

(c) $\begin{bmatrix} 1 & 2 & 1 \\ 1 & 3 & 1 \\ 1 & -8 & 1 \end{bmatrix}$

In Exercises 5–8 write the payoff matrix for each game.

5. Each of two players has two cards, a 3 and a 4. Each player selects one of his cards, and then the players simultaneously show their cards. If the sum of the numbers is even, player II pays player I an amount equal to the sum; if the sum is odd, player I pays player II an amount equal to the sum.

6. Each of two players shows one or two fingers. Player R pays player C a number equal to the total number of fingers shown.

7. **(The children's game "stone, scissors, paper")** Each of two players writes down one of the words *stone, scissors,* or *paper*. Stone beats scissors, and scissors beats paper, and paper beats stone. The winner receives \$1 from the loser, and the payoff is \$0 in case of a tie.

8. **(Morra)** Each of two players shows one or two fingers and at the same time each guesses at the total number of fingers that will be shown. If one player guesses the correct number, he or she wins that amount from the opponent. If both or neither guesses correctly, no money changes hands. [*Hint*: A move is one of the ordered pairs (1, 2), (1, 3), (2, 3), or (2, 4), where, for example, (2, 4) means 2 fingers shown and number 4 called.]

9. Show that the game in Exercise 6 is strictly determined and find the best move for each player.

10. A region is served by two shopping centers, center A and center B. Center A serves 60% of the region, and center B serves 40% of the region. Valley Bank and Colonial Bank are each planning to open a branch in one of the centers. If both open at the same center, they will split the business in that center, and if they open in different centers, each will get all the business in its center.
    (a) Show that the percentage of *regional* shoppers serviced by Valley Bank, in each case, can be expressed in matrix form as

		Colonial Bank A	Colonial Bank B
Valley Bank	A	30%	60%
	B	40%	20%

    (b) Show that this game is not a constant-sum game.

11. Show that the matrix game

$$\begin{bmatrix} a & -2 \\ 7 & -1 \end{bmatrix}$$

is strictly determined, regardless of the value of a.

12. **(The bootlegger's game)*** A bootlegger can cross a border by 2 different routes: over a highway or through the mountains. The border patrol has 3 possible plans: guard only the highway heavily, guard only the mountain road heavily, or guard both the highway and mountain routes lightly. If the bootlegger can travel the highway undetected, he can smuggle in a fully loaded truck and make a profit of \$500. If he travels the highway and there is a light patrol on the highway, he can avoid arrest but will have to abandon his load, losing \$200. If he travels the highway and there is a heavy patrol, he will be arrested, losing his load (\$200), and fined \$300. The mountain road is much narrower and can only allow a smaller truck. If the road is unguarded, he will have no problem in getting through and will make a profit of \$200. If it is lightly or heavily guarded, he can still get through but will have to bribe the locals to get him past the border patrol. In either of these cases his profit will be \$100. What is the best course of action for the border patrol and for the bootlegger?

* Exercises 12 and 13 are variations of problems discussed in an article by Martin Shubik, "Game Theory and Management Science," *Management Science,* **2**(1955): 40–54.

13. **(Media selection in advertising)** Firms A and B, competing for the same customers, each have one million dollars for an advertising campaign. Each firm must choose whether to use no advertising (NA) or to use exactly one of the media: radio (R), television (TV), or printed matter (PM). A neutral consulting firm has developed the following payoff matrix, where each entry is the amount of extra revenue (in millions of dollars) that will be earned by firm A, in each case:

		Firm B			
		R	TV	PM	NA
Firm A	R	0	−.5	0	2.5
	TV	2	0	1.5	5
	PM	1	−.5	0	3.5
	NA	−2	−4	−3	0

How should each firm best spend its advertising money?

14. Show that the game with the following payoff matrix is strictly determined regardless of the value of x:

$$\begin{bmatrix} 3 & 4 \\ 2 & x \end{bmatrix}$$

(a) What is the optimal strategy for each player?
(b) What is the payoff?

15. Find a value of x that will make the matrix games below strictly determined.

(a) $\begin{bmatrix} -2 & -3 \\ -1 & x \end{bmatrix}$ (b) $\begin{bmatrix} 1 & -2 \\ x & 10 \end{bmatrix}$

16. A submarine wolf pack is planning to attack a convoy of supply ships protected by destroyers. The attack can be made from deep water or shallow water, each of which has advantages and disadvantages. The destroyers can search for the wolf pack in deep or shallow water. The following damage estimates are known to both sides:

1. If the wolf pack attacks from deep water and the destroyers are searching in deep water, assign the destroyers a 6-point advantage.
2. If the wolf pack attacks from shallow water and the destroyers are searching in shallow water, assign the destroyers a 1-point disadvantage.
3. If the wolf pack attacks from deep water and the destroyers are searching in shallow water, assign the destroyers a 4-point disadvantage.
4. If the wolf pack attacks from shallow water and the destroyers are searching in deep water, assign the destroyers a 7-point disadvantage.

(a) Construct a payoff matrix for this war game.
(b) What are optimal strategies for each side?
(c) Which side does the game favor?

10.3 GAMES WITH MIXED STRATEGIES

In the preceding section we observe that the players of a strictly determined game cannot improve on their best moves, even if they discover their opponent's move in advance. However, for games that are not strictly determined, a player can benefit by uncovering the opponent's intended move. In this section we examine games that are *not* strictly determined and that the players intend to play over and over again. In this situation a player must avoid making the same move in every game; otherwise, the opponent will detect the pattern and use the information to his own advantage. Our objective in this section is to explain how the players should mix their moves from game to game to obtain the most favorable results.

Consider a two-person matrix game with payoff matrix

$$\begin{bmatrix} 3 & -4 \\ 0 & 5 \end{bmatrix}$$

As usual, positive entries in the matrix represent the amounts that the column player (player C) pays to the row player (player R) after each move, and negative signs precede amounts that player R pays to player C. Since no entry is simultaneously the smallest element in its row and the largest element in its column, this game has no saddle point and consequently is not strictly determined.

The row minima and column maxima are

$$\begin{array}{r} \\ \\ \text{Column maxima} \end{array} \begin{array}{c} \begin{bmatrix} 3 & -4 \\ 0 & 5 \end{bmatrix} \\ \;3 \quad\; 5 \end{array} \quad \begin{array}{c} \text{Row minima} \\ -4 \\ 0 \end{array}$$

Since row 2 contains the largest row minimum and column 1 contains the smallest column maximum, player R might initially play row 2 and player C might initially play column 1, resulting in a payoff of zero. If, however, player C continued playing the first column game after game, player R would eventually detect the pattern and switch to row 1, thereby winning 3 units. By the same token, if player R continues to play row 1 game after game, then player C would detect the pattern and choose column 2, resulting in a loss of 4 units for player R.

It is evident from this analysis that each player should avoid using the same move in every game; rather, the players should mix their moves in an unpredictable fashion. Such mixtures of moves are called **mixed strategies** to distinguish them from **pure strategies**, where the player uses the same move in every game. Each player can mix moves by using a chance device such as cards, dice, or tables of random numbers to determine which move to make in each game. The question that remains, however, is how often each move should be used. For example, should player R in the game above choose each row 50% of the time, or should he mix his moves in some other proportion, say the first row 80% of the time and the second row 20% of the time? To answer this question, we need some way to evaluate the effectiveness of various mixtures of moves. One way of doing this is to compare the **expected winnings** that result from different mixtures of moves. The following example illustrates this idea.

EXAMPLE 1 Suppose that the matrix game whose payoff matrix is

$$\begin{bmatrix} 3 & -4 \\ 0 & 5 \end{bmatrix} \tag{1}$$

will be played over and over, and assume that player C will play each column 50% of the time. Determine which is the better strategy for player R:

Strategy (a) To play the first row 80% of the time and the second row 20% of the time.

Strategy (b) To play each row 50% of the time.

Solution We can regard each play of the game as a probability experiment in which the four combinations of moves

row 1 and column 1	row 1 and column 2
row 2 and column 1	row 2 and column 2

are the possible outcomes. To illustrate how the probabilities of these outcomes can be obtained, consider strategy (a), where player R plays the first row 80% of the times and the second row 20% of the time. Since neither player knows in advance what the other's choice will be, it is reasonable to assume that the row and column are chosen *independently*. Thus we obtain the probabilities

$$\begin{aligned} P(\text{row 1 and column 1}) &= p(\text{row 1})p(\text{column 1}) \\ &= (.8)(.5) \\ &= .4 \end{aligned}$$

$$\begin{aligned} P(\text{row 2 and column 1}) &= p(\text{row 2})p(\text{column 1}) \\ &= (.2)(.5) \\ &= .1 \end{aligned}$$

and so forth. In Table 1 we have listed the probabilities of the outcomes for strategy (a) and strategy (b). Also, we have listed the payoff that results to player R in each case.

TABLE 1

Outcome	Payoff to R	Probability of Outcome
Strategy (a):		
Row 1 and column 1	3	(.8)(.5) = .4
Row 2 and column 1	0	(.2)(.5) = .1
Row 1 and column 2	−4	(.8)(.5) = .4
Row 2 and column 2	5	(.2)(.5) = .1
Strategy (b):		
Row 1 and column 1	3	(.5)(.5) = .25
Row 2 and column 1	0	(.5)(.5) = .25
Row 1 and column 2	−4	(.5)(.5) = .25
Row 2 and column 2	5	(.5)(.5) = .25

If we denote the payoff to R by X, then X is a random variable whose expected value represents the long-term average payoff to R when the game is played over and over. Thus, to compare the relative merits of strategy (a) and strategy (b), player R can examine the expected payoff $E(X)$ in each case to determine which strategy will produce the larger expected payoff.

Recall from Section 8.2 that the expected value of a random variable X is the sum of the possible values of X times their probabilities of occurrence. Thus from Table 1 the expected payoff to R for strategy (a) is

$$\begin{aligned} E(X) &= 3(.4) + 0(.1) + (-4)(.4) + 5(.1) \\ &= 1.2 - 1.6 + .5 \\ &= .1 \end{aligned}$$

and the expected payoff to R for strategy (b) is

$$\begin{aligned} E(X) &= 3(.25) + 0(.25) + (-4)(.25) + 5(.25) \\ &= .75 - 1.00 + 1.25 \\ &= 1.00 \end{aligned}$$

Thus strategy (b) is better than strategy (a) from player R's viewpoint *if C plays each column 50% of the time*. Of course, if C uses some other mixed strategy, then it is possible that strategy (a) might be better than strategy (b) for player R.

Let us examine each player's objectives in a non-strictly determined game.

Objectives of Player R Player R wants to find a mixed strategy that will give him the largest expected payoff. Whatever mixed strategy he uses, however, player R must assume that player C will oppose him with her best counterstrategy. Thus, from among all possible mixed strategies, player R wants to pick one that will give him the largest possible expected payoff against player C's best counterstrategy. This is called an **optimal strategy for player R**.

Objectives of Player C Player C wants to find a mixed strategy that will give player R the smallest possible expected payoff. Whatever mixed strategy she uses, however, player C must assume that player R will oppose her with his best counterstrategy. Thus, from among all possible mixed strategies, player C wants to pick one that will give player R the smallest expected payoff against player R's best counterstrategy. This is called an **optimal strategy for player C**.

Before we discuss methods for finding optimal strategies in certain kinds of games, we need some preliminary ideas. Consider a two-person constant-sum game with payoff matrix

$$A = \begin{bmatrix} a_{11} & a_{12} \\ a_{21} & a_{22} \end{bmatrix}$$

Suppose player R chooses row 1 with probability p_1 and row 2 with probability p_2 (where $p_1 + p_2 = 1$), and suppose, independently, player C chooses column 1 with probability q_1 and column 2 with probability q_2 (where $q_1 + q_2 = 1$). Each play of the game results in one of four possible outcomes:

row 1 and column 1 row 1 and column 2
row 2 and column 1 row 2 and column 2

The probabilities of these outcomes can be obtained using an argument like that in Example 1. These probabilities together with the payoffs resulting from the outcomes are shown in Table 2.

TABLE 2

Outcome	Payoff	Probability of Outcome
Row 1 and column 1	a_{11}	p_1q_1
Row 2 and column 1	a_{21}	p_2q_1
Row 1 and column 2	a_{12}	p_1q_2
Row 2 and column 2	a_{22}	p_2q_2

From Table 2, we obtain

$$\text{expected payoff to player R} = p_1q_1a_{11} + p_2q_1a_{21} + p_1q_2a_{12} + p_2q_2a_{22} \qquad (2)$$

The result in (2) can also be obtained using matrices as follows. Use the probabilities in player R's mixed strategy to form a 1×2 matrix:

$$P = \begin{bmatrix} p_1 & p_2 \end{bmatrix}$$

and use the probabilities in player C's mixed strategy to form a 2×1 matrix:

$$Q = \begin{bmatrix} q_1 \\ q_2 \end{bmatrix}$$

If we form the product PAQ, we obtain

$$\begin{aligned} PAQ &= \begin{bmatrix} p_1 & p_2 \end{bmatrix} \begin{bmatrix} a_{11} & a_{12} \\ a_{21} & a_{22} \end{bmatrix} \begin{bmatrix} q_1 \\ q_2 \end{bmatrix} \\ &= p_1q_1a_{11} + p_2q_1a_{21} + p_1q_2a_{12} + p_2q_2a_{22} \end{aligned}$$

which, as shown above in Equation (2), is the expected payoff to player R.* This is a special case of the following general result.

Expected Payoff to Player R If a matrix game has an $m \times n$ payoff matrix A and if the matrices

$$P = \begin{bmatrix} p_1 & p_2 & \cdots & p_m \end{bmatrix} \quad \text{and} \quad Q = \begin{bmatrix} q_1 \\ q_2 \\ \vdots \\ q_n \end{bmatrix}$$

list the probabilities of the moves for players R and C, respectively, then the expected payoff E to player R is

$$E = PAQ \qquad (3)$$

EXAMPLE 2 Consider the matrix game whose payoff matrix is

$$A = \begin{bmatrix} 5 & -9 \\ -4 & 8 \end{bmatrix}$$

Use matrices to find the expected payoff to player R if player R plays row 1 with probability $\frac{5}{6}$ and row 2 with probability $\frac{1}{6}$, while player C plays column 1 with probability $\frac{1}{3}$ and column 2 with probability $\frac{2}{3}$.

Solution The strategies for the players can be written in matrix form as

$$P = \begin{bmatrix} \frac{5}{6} & \frac{1}{6} \end{bmatrix} \quad \text{and} \quad Q = \begin{bmatrix} \frac{1}{3} \\ \frac{2}{3} \end{bmatrix}$$

---

*Strictly speaking, PAQ is the 1×1 matrix $[p_1q_1a_{11} + p_2q_1a_{21} + p_1q_2a_{12} + p_2q_2a_{22}]$; however, we have followed the standard practice of omitting the brackets on 1×1 matrices.

Thus, from Equation (3), the expected payoff to player R is

$$E = PAQ = \begin{bmatrix} \frac{5}{6} & \frac{1}{6} \end{bmatrix} \begin{bmatrix} 5 & -9 \\ -4 & 8 \end{bmatrix} \begin{bmatrix} \frac{1}{3} \\ \frac{2}{3} \end{bmatrix}$$

$$= \begin{bmatrix} \frac{21}{6} & -\frac{37}{6} \end{bmatrix} \begin{bmatrix} \frac{1}{3} \\ \frac{2}{3} \end{bmatrix}$$

$$= -\frac{53}{18}$$

Therefore, the long-term average payoff will be $\frac{53}{18}$ units to player C.

EXAMPLE 3 Consider a two-person zero-sum matrix game with payoff matrix

$$A = \begin{bmatrix} 3 & -2 & 3 \\ -1 & 2 & 3 \end{bmatrix}$$

If player R plays row 1 with probability $\frac{1}{4}$ and row 2 with probability $\frac{3}{4}$, while player C plays column 1 with probability $\frac{1}{6}$, column 2 with probability $\frac{1}{6}$, and column 3 with probability $\frac{2}{3}$, then the expected payoff to player R is

$$E = PAQ = \begin{bmatrix} \frac{1}{4} & \frac{3}{4} \end{bmatrix} \begin{bmatrix} 3 & -2 & 3 \\ -1 & 2 & 3 \end{bmatrix} \begin{bmatrix} \frac{1}{6} \\ \frac{1}{6} \\ \frac{2}{3} \end{bmatrix}$$

$$= \begin{bmatrix} 0 & 1 & 3 \end{bmatrix} \begin{bmatrix} \frac{1}{6} \\ \frac{1}{6} \\ \frac{2}{3} \end{bmatrix}$$

$$= \frac{13}{6}$$

Let us now consider a two-person matrix game with a 2×2 payoff matrix

$$A = \begin{bmatrix} a & b \\ c & d \end{bmatrix} \tag{4}$$

It can be shown that if the game is not strictly determined, then

$$a + d - b - c \neq 0$$

and the optimal strategies for the players are as follows:

Optimal Strategy for Player R The optimal strategy for player R is $P = [p_1 \quad p_2]$, where

$$p_1 = \frac{d - c}{a + d - b - c} \quad \text{and} \quad p_2 = \frac{a - b}{a + d - b - c} \tag{5}$$

Optimal Strategy for Player C The optimal strategy for player C is

$$Q = \begin{bmatrix} q_1 \\ q_2 \end{bmatrix}$$

where

$$q_1 = \frac{d - b}{a + d - b - c} \quad \text{and} \quad q_2 = \frac{a - c}{a + d - b - c} \tag{6}$$

Moreover,

Expected Optimal Payoff If both players use their optimal strategies, then the expected payoff is

$$E = PAQ = \frac{ad - bc}{a + d - b - c} \tag{7}$$

EXAMPLE 4 In the previous section we observe that the penny-matching game with payoff matrix

$$A = \begin{bmatrix} -1 & 1 \\ 1 & -1 \end{bmatrix}$$

is not strictly determined. Let us now try to find the optimal strategies for the players.

Solution Comparing A to matrix (4) above, we obtain $a = -1$, $b = 1$, $c = 1$, and $d = -1$. Substituting these values in (5) we obtain

$$p_1 = \frac{d - c}{a + d - b - c} = \frac{-1 - 1}{-1 - 1 - 1 - 1} = \frac{1}{2}$$

$$p_2 = \frac{a - b}{a + d - b - c} = \frac{-1 - 1}{-1 - 1 - 1 - 1} = \frac{1}{2}$$

Similarly, from (6) we obtain

$$q_1 = \frac{d - b}{a + d - b - c} = \frac{-1 - 1}{-1 - 1 - 1 - 1} = \frac{1}{2}$$

$$q_2 = \frac{a - c}{a + d - b - c} = \frac{-1 - 1}{-1 - 1 - 1 - 1} = \frac{1}{2}$$

Thus player I should choose each row 50% of the time and player II should choose each column 50% of the time. If each player uses his or her optimal strategy, then from (7) the expected payoff will be

$$E = \frac{ad - bc}{a + d - b - c} = \frac{(-1)(-1) - (1)(1)}{-1 - 1 - 1 - 1} = 0$$

EXAMPLE 5 Consider the game discussed in Example 1; the payoff matrix is

$$A = \begin{bmatrix} 3 & -4 \\ 0 & 5 \end{bmatrix}$$

Comparing A to matrix (4) above, we obtain $a = 3$, $b = -4$, $c = 0$, and $d = 5$. Substituting these values in (5), we obtain the optimal strategy for player R, which is

$$p_1 = \frac{d - c}{a + d - b - c} = \frac{5 - 0}{3 + 5 - (-4) - 0} = \frac{5}{12}$$

$$p_2 = \frac{a - b}{a + d - b - c} = \frac{3 - (-4)}{3 + 5 - (-4) - 0} = \frac{7}{12}$$

Similarly, from (6) we obtain the optimal strategy for player C, which is

$$q_1 = \frac{d - b}{a + d - b - c} = \frac{5 - (-4)}{3 + 5 - (-4) - 0} = \frac{9}{12}$$

$$q_2 = \frac{a - c}{a + d - b - c} = \frac{3 - 0}{3 + 5 - (-4) - 0} = \frac{3}{12}$$

If each player uses his or her optimal strategy, then from (7) the expected payoff will be

$$E = \frac{ad - bc}{a + d - b - c} = \frac{3(5) - (-4)(0)}{3 + 5 - (-4) - 0} = \frac{15}{12} = \frac{5}{4}$$

Thus, in the long run, the game is favorable to R since he will obtain an expected payoff of $\frac{5}{4}$ units.

Although the optimal strategies given by Formulas (5) and (6) apply only to games with a 2×2 payoff matrix, it is sometimes possible, with a little ingenuity, to use these formulas in other cases. The following example illustrates this point.

EXAMPLE 6 Consider the two-person constant-sum game with payoff matrix

$$\begin{bmatrix} 3 & -4 & -2 \\ 0 & 5 & 5 \\ -2 & 1 & 3 \end{bmatrix}$$

You can check that this game has no saddle point and consequently is not strictly determined; thus, the players should use mixed strategies. Since the payoff matrix is not 2×2, the optimal strategies cannot be determined from Formulas (5) and (6). If, however, we compare the second and third rows of the payoff matrix, we see that each element in the third row is less than the corresponding element in the second row. Thus, a third-row move is always worse than a second-row move for player R, regardless of what player C decides to do. It follows that player R will

THE PRISONER'S DILEMMA: TALK OR BE SILENT

Bonnie and Clyde have been arrested for robbing a bank. They are questioned separately by the police. If they both keep quiet, each will go to jail for 5 years. If one talks and the other keeps quiet, then the one who talked will go free and the other will go to jail for 12 years. If they both talk, each will go to jail for 8 years. By assigning the sentences each receives as payoff values in the different situations, we obtain the following payoff matrices for Bonnie and Clyde:

Bonnie's Payoff Matrix

Bonnie \ Clyde	Be quiet	Talk
Be quiet	5	12
Talk	0	8

Clyde's Payoff Matrix

Bonnie \ Clyde	Be quiet	Talk
Be quiet	5	0
Talk	12	8

Observe that the prisoner's dilemma is not a zero-sum game. We now see that if they cooperate with each other by keeping quiet, they will each do best. Let us now examine how Bonnie might arrive at a decision of what is best for *her*. If Clyde keeps quiet, she is better off talking, whereas if he talks, she is also better off talking. Thus, *her* interests are best served by talking. Similarly, Clyde discovers that no matter what Bonnie does, he is best off talking. Thus, they each talk and each goes to jail for 8 years! However, if they both keep quiet, then each goes to jail for 5 years.

If this game were played repeatedly, then each player would do best by cooperating with the other player every time. However, in any one play of the game each player is tempted to sacrifice the other by acting solely in his or her—seemingly—best interest.

The prisoner's dilemma has become extremely interesting in recent years to psychologists, social scientists, biologists, and economists who are studying how living organisms cooperate and compete.

never use the third row as a move. Thus, we can cross out the third row from the payoff matrix to obtain

$$\begin{bmatrix} 3 & -4 & -2 \\ 0 & 5 & 5 \\ -2 & 1 & 3 \end{bmatrix}$$

In the remaining 2×3 matrix we see that each element in the third column is at least as large as the corresponding element in the second column. Thus, a second-column move for player C is always at least as good as a third-column move, regardless of what player R decides to do. It follows that player C will *never* use the third column as a move. Thus, we can cross out the third column from the payoff matrix to obtain

$$\begin{bmatrix} 3 & -4 & -2 \\ 0 & 5 & 5 \\ -2 & 1 & 3 \end{bmatrix}$$

The remaining 2 × 2 matrix is the one we consider in Example 5. In that example we use Formulas (5) and (6) to obtain the optimal strategy

$$p_1 = \frac{5}{12} \qquad p_2 = \frac{7}{12}$$

and

$$q_1 = \frac{9}{12} \qquad q_2 = \frac{3}{12}$$

Thus, in this example, the optimal strategies are

$$p_1 = \frac{5}{12} \qquad p_2 = \frac{7}{12} \qquad p_3 = 0$$

and

$$q_1 = \frac{9}{12} \qquad q_2 = \frac{3}{12} \qquad q_3 = 0$$

EXAMPLE 7 **COMPETITIVE PRICING OF PRODUCTS*** Assume that a certain region has only two gasoline stations, R and C, and suppose the stations are faced with the following pricing problem. Each station can sell its gasoline for either \$1.35 or \$1.40 a gallon, and each dealer's total percentage of the regional business will depend on the prices charged. Assume also that a market survey produces the payoff matrix

		Station C	
		\$1.35	\$1.40
Station R	\$1.35	60%	50%
	\$1.40	40%	70%

where the entries represent the percentage of the regional business that will be captured by station R. How should each station price its gasoline?

Solution No entry in the matrix is both the minimum in its row and the maximum in its column, so this game is not strictly determined. Thus, the stations should use mixed strategies. Comparing the given payoff matrix to the matrix in (4), we obtain $a = 60$, $b = 50$, $c = 40$, and $d = 70$. Substituting these values in (5), we obtain the optimal strategy for station R, which is

$$p_1 = \frac{d - c}{a + d - b - c} = \frac{70 - 40}{60 + 70 - 50 - 40} = \frac{3}{4}$$

*This example is a variation of one discussed in an excellent book by Andrei Rogers, *Matrix Methods in Urban and Regional Analysis* (San Francisco: Holden-Day, 1971).

$$p_2 = \frac{a-b}{a+d-b-c} = \frac{60-50}{60+70-50-40} = \frac{1}{4}$$

Similarly, from (6) we obtain the optimal strategy for station C, which is

$$q_1 = \frac{d-b}{a+d-b-c} = \frac{70-50}{60+70-50-40} = \frac{1}{2}$$

$$q_2 = \frac{a-c}{a+d-b-c} = \frac{60-40}{60+70-50-40} = \frac{1}{2}$$

Thus, the optimal strategy for station R is to charge $1.35 a gallon three-fourths of the time and $1.40 a gallon the rest of the time; the optimal strategy for station C is to charge $1.35 half the time and $1.40 half the time.

If both stations use their optimal strategies, then from (7), the expected payoff to station R is

$$E = \frac{ad-bc}{a+d-b-c} = \frac{(60)(70)-(50)(40)}{60+70-50-40} = 55$$

Thus, over the long term, station R will have slight competitive advantage, capturing 55% of the regional business.

EXAMPLE 8 **COLUMBUS' DISCOVERY OF AMERICA AS A GAME-THEORY PROBLEM*** History tells us that Christopher Columbus, faced with a mutiny by his crew, had to decide between two alternatives: turn back to Spain as his crew demanded or continue his voyage until land was sighted. Columbus must have analyzed the possible consequences of his decision in the way we have indicated in Table 3.

TABLE 3

		Unknown Location of Land	
		Land Near	***No Land Near***
Columbus' Decision	***Turn back***	Bitter disappointment in later life	Satisfaction for saving lives
	Keep going	Personal glory	Death for himself and his men

To analyze Columbus' plight as a game-theory problem, we must assign numerical values to each of the outcomes, indicating their relative importance. It is clear that personal glory and satisfaction for saving lives are favorable outcomes for Columbus and should be assigned positive values, whereas disappointment and death are unfavorable outcomes and should be assigned negative values. Beyond this, it is difficult to pick appropriate numerical values. For example, which is more favorable, satisfaction for saving lives or personal glory? And even if we could agree that personal glory is more favorable, we must still decide how much more favorable it is. Is glory fives times as favorable as satisfaction? Ten times? A

*This problem is based on an article by Leonid Hurwicz in the book, *Mathematical Thinking in Behavioral Sciences* (San Francisco: Freeman, 1968).

hundred times? Obviously these are difficult *psychological* questions that everyone would answer differently. This difficulty is often the stumbling block in applying game theory to behavioral problems. Nevertheless, once the values are agreed on and the payoff matrix constructed, we have a matrix game that can be analyzed mathematically. For example, suppose Columbus had appraised the relative values of the outcomes in Table 3 as in Table 4. Since this game has no saddle point, Columbus should have used a mixed strategy. Comparing the payoff matrix in Table 4 to matrix (4), we obtain $a = -10$, $b = 1$, $c = 5$, and $d = -50$. Substituting these values in (5) gives Columbus' optimal strategy, which is

$$p_1 = \frac{d - c}{a + d - b - c} = \frac{-50 - 5}{-10 - 50 - 1 - 5} = \frac{-55}{-66} \cong .83$$

$$p_2 = \frac{a - b}{a + d - b - c} = \frac{-10 - 1}{-10 - 50 - 1 - 5} = \frac{-11}{-66} \cong .17$$

TABLE 4

		Unknown Location of Land	
		Land Near	*No Land Near*
Columbus' Decision	*Turn back*	−10	1
	Keep going	5	−50

Thus with probability .83, Columbus should have turned around and never discovered America!

It is important to keep in mind that the conclusion reached in this last example depends on accepting the validity of the values selected for the payoff matrix.

There are many topics in game theory we have not discussed here. If you are interested in studying game theory in more detail, consult the following references:

1. Ira Buchler and Hugo Nutini, *Game Theory in the Behavioral Sciences* (Pittsburgh, Pa.: Univ. of Pittsburgh Press, 1969).

2. R. Duncan Luce and Howard Raiffa, *Games and Decisions* (New York: Wiley, 1957).

3. Daniel P. Maki and Maynard Thompson, *Mathematical Models and Applications* (Englewood Cliffs, N.J.: Prentice-Hall, 1973).

4. J. C. C. McKinsey, *Introduction to the Theory of Games* (New York: McGraw-Hill, 1952).

5. G. Owen, *Game Theory* (Philadelphia: Saunders, 1968).

EXERCISE SET 10.3

1. Consider a game with payoff matrix

$$\begin{bmatrix} -2 & 3 \\ 3 & 1 \end{bmatrix}$$

In each part, use Formula (3) to find the expected payoff that results from the given strategies.

(a) $p_1 = \frac{1}{3}$, $p_2 = \frac{2}{3}$ and $q_1 = \frac{3}{4}$, $q_2 = \frac{1}{4}$
(b) $p_1 = \frac{2}{5}$, $p_2 = \frac{3}{5}$ and $q_1 = \frac{3}{5}$, $q_2 = \frac{2}{5}$
(c) $p_1 = 0$, $p_2 = 1$ and $q_1 = \frac{1}{6}$, $q_2 = \frac{5}{6}$

2. Consider a game with payoff matrix

$$\begin{bmatrix} -3 & 2 & 1 \\ -2 & 1 & 2 \end{bmatrix}$$

In each part, use Formula (3) to find the expected payoff that results from the given strategies.

(a) $p_1 = 0$, $p_2 = 1$ and $q_1 = \frac{1}{3}$, $q_2 = \frac{1}{3}$, $q_3 = \frac{1}{3}$
(b) $p_1 = \frac{1}{2}$, $p_2 = \frac{1}{2}$ and $q_1 = \frac{5}{12}$, $q_2 = \frac{1}{12}$, $q_3 = \frac{1}{2}$
(c) $p_1 = \frac{1}{4}$, $p_2 = \frac{3}{4}$ and $q_1 = 0$, $q_2 = 0$, $q_3 = 1$

In Exercises 3–8 find the optimal strategies and the expected payoff.

3. $\begin{bmatrix} 2 & 4 \\ 3 & 1 \end{bmatrix}$ **4.** $\begin{bmatrix} -1 & 3 \\ 5 & 0 \end{bmatrix}$

5. $\begin{bmatrix} 7 & 1 \\ -6 & 2 \end{bmatrix}$ **6.** $\begin{bmatrix} 4 & 2 \\ -8 & 3 \end{bmatrix}$

7. $\begin{bmatrix} 2 & -1 \\ -1 & 2 \end{bmatrix}$ **8.** $\begin{bmatrix} 0 & -2 \\ 1 & 1 \end{bmatrix}$

In Exercises 9–12 use the method of Example 6 to find the optimal strategies and the expected payoff.

9. $\begin{bmatrix} 0 & 2 & 4 \\ 1 & -1 & 3 \\ -3 & -4 & 0 \end{bmatrix}$ **10.** $\begin{bmatrix} 2 & 6 & 7 \\ 8 & 3 & 9 \\ 0 & 2 & 5 \end{bmatrix}$

11. $\begin{bmatrix} 1 & -9 & -2 \\ 6 & 5 & -1 \\ 3 & -8 & 0 \end{bmatrix}$ **12.** $\begin{bmatrix} -1 & -2 & 0 \\ 6 & 3 & 9 \\ 3 & 6 & 7 \end{bmatrix}$

13. In the problem of Example 8, find Christopher Columbus' optimal strategy if the payoff matrix is

$$\begin{bmatrix} -100 & 70 \\ 1 & -100 \end{bmatrix}$$

14. **(Marketing problem)** An automobile manufacturer can stress either physical appearance or sound engineering in its magazine advertising. Market studies show that ads based on physical appearance are effective on 70% of the male readers and effective on 20% of the female readers, while ads based on sound engineering are effective on 40% of the male readers and 80% of the female readers. If the ratio of male to female readers is unknown, in what proportions should the two kinds of ads be mixed to be most effective? This problem can be viewed as a game between the manufacturer and the marketplace with payoff matrix.

		Marketplace	
		Female	Male
Manufacturer	Appearance	.2	.7
	Engineering	.8	.4

15. In Exercise 14, what proportion of male and female readers would be least favorable to the advertiser?

10.4 AN APPLICATION OF PROBABILITY TO GENETICS; THE HARDY-WEINBERG STABILITY PRINCIPLE

Probability theory plays a major part in the study of heredity. In this section we briefly discuss some of the genetic principles formulated by Gregor Mendel (see Chapter 9), and then we use probability theory to derive one of the fundamental results in genetics.

One aspect of genetics is concerned with *heredity*; that is, how traits such as eye color and hair texture in humans or petal color and disease resistance in plants are passed on from parents to offspring. In the following discussion we describe the theory of heredity as originally formulated by Mendel. Mendel's theory has undergone considerable refinement in the evolution of present-day genetics; however, our simplified discussion illustrates the role that probability plays in the study of heredity.

We can illustrate Mendel's observations by considering a certain plant that can have red flowers (R), pink flowers (P), or white flowers (W). Mendel observed that when two of the red-flowered plants are crossed, the offspring always have red flowers. Also, when two of the white-flowered plants are crossed, the offspring always have white flowers, and when a red-flowered plant is crossed with a white-flowered plant, the offspring always have pink flowers; schematically, we describe this by writing

$$\mathrm{R} \times \mathrm{R} \rightarrow \mathrm{R} \tag{1a}$$

$$\mathrm{W} \times \mathrm{W} \rightarrow \mathrm{W} \tag{1b}$$

$$\mathrm{R} \times \mathrm{W} \rightarrow \mathrm{P} \tag{1c}$$

Mendel observed that the remaining crosses, R × P, W × P, and P × P, yield variable results. For example, R × P yields about 50% red offspring and 50% pink offspring. Schematically, the results of these three crosses are

$$\mathrm{P} \times \mathrm{P} \rightarrow 25\%\ \mathrm{R} \qquad 50\%\ \mathrm{P} \qquad 25\%\ \mathrm{W} \tag{1d}$$

$$\mathrm{R} \times \mathrm{P} \rightarrow 50\%\ \mathrm{R} \qquad 50\%\ \mathrm{P} \tag{1e}$$

$$\mathrm{W} \times \mathrm{P} \rightarrow 50\%\ \mathrm{P} \qquad 50\%\ \mathrm{W} \tag{1f}$$

To explain these results, Mendel postulated the existence of certain structures, called *genes,* which are passed on from parents to offspring. In the simplest case, a single physical characteristic, such as flower color, is determined by one pair of genes in the offspring. One member of the pair is inherited from the male parent and one from the female parent. Each gene in the pair can assume one of two forms, called *alleles*; it is common to denote the two possible alleles by A and a. Thus, there are three possible forms for a pair of genes:

$$AA \qquad Aa \qquad aa \tag{2}$$

(Biologically, there is no difference between Aa and aA.) Each of the combinations in (2) is called a *genotype*. In the flower example, each genotype corresponds to a different flower color: a plant of genotype AA will have red flowers, a plant of genotype Aa will have pink flowers, and a plant of genotype aa will have white flowers.

According to Mendelian theory, a parent of genotype AA can only transmit an A gene to its offspring, a parent of genotype aa can only transmit an a gene to its offspring, and a parent of genotype Aa can transmit either an A gene or an a gene, each with probability $\frac{1}{2}$. It follows that if both parents have genotype AA, then every offspring will also be of genotype AA since each parent must transmit an A gene. Schematically,

$$AA \times AA \rightarrow AA \tag{3a}$$

Similarly,

$$aa \times aa \rightarrow aa \tag{3b}$$

In the flower example, genotype AA corresponds to red flowers (R) and genotype aa corresponds to white flowers (W), so (3a) and (3b) explain observations (1a) and (1b). The following examples help explain some of the other results.

EXAMPLE 1 Assume an individual of genotype Aa mates with another individual of genotype Aa. Find the probability that the offspring is

(a) of genotype AA (b) of genotype aa (c) of genotype Aa

Solution (a) To be of genotype AA, the offspring must inherit an A gene from each parent. The probability of getting an A gene from the first parent is $\frac{1}{2}$ and the probability of getting an A gene from the second parent is $\frac{1}{2}$. Thus, assuming that the genes are inherited *independently* from each parent, the probability of genotype AA is

$$\frac{1}{2} \cdot \frac{1}{2} = \frac{1}{4}$$

Solution (b) The same argument used in (a) will show that the probability of genotype aa is $\frac{1}{4}$.

Solution (c) To be of genotype Aa, the offspring can inherit an A gene from the male parent and an a gene from the female parent (call this event E) *or* it can inherit an a gene from the male parent and an A gene from the female parent (call this event F). Thus,

$$P(\text{genotype } Aa) = P(E \cup F) = P(E) + P(F) \tag{4}$$

The probability of receiving an A gene from the father is $\frac{1}{2}$, and the probability of receiving an a gene from the mother is $\frac{1}{2}$; thus, assuming independence,

$$P(E) = \frac{1}{2} \cdot \frac{1}{2} = \frac{1}{4}$$

Similarly,

$$P(F) = \frac{1}{2} \cdot \frac{1}{2} = \frac{1}{4}$$

so that from (4)

$$P(\text{genotype } Aa) = \frac{1}{4} + \frac{1}{4} = \frac{1}{2}$$

Schematically, the results in Example 1 can be written

$$Aa \times Aa \rightarrow 25\%\ AA \qquad 50\%\ Aa \qquad 25\%\ aa \tag{5}$$

In the flower example, genotype AA corresponds to red flowers (R), aa to white flowers (W), and Aa to pink flowers (P), so (5) explains the result (1d) where two pink-flowered plants were crossed.

EXAMPLE 2 Assume an individual of genotype AA mates with an individual of genotype Aa ($AA \times Aa$). Find the probability that the offspring is

(a) of genotype AA (b) of genotype aa (c) of genotype Aa

Solution (a) To be of genotype AA, the offspring must inherit an A gene from the first parent and an A gene from the second parent. Since the first parent has genotype AA, the probability of inheriting an A gene is 1. Since the second parent has genotype Aa, the probability of inheriting an A gene is $\frac{1}{2}$. Thus, by independence,

$$P(\text{genotype } AA) = 1 \cdot \frac{1}{2} = \frac{1}{2}$$

Solution (b) To be of genotype aa, the offspring must inherit an a gene from each parent. Since the AA parent cannot transmit an a gene, we obtain

$$P(\text{genotype } aa) = 0$$

Solution (c) The only way the offspring can be of genotype Aa is to receive an A gene from the AA parent and an a gene from the Aa parent. The probability of getting an A gene from the AA parent is 1, and the probability of getting an a gene from the Aa parent is $\frac{1}{2}$. Therefore,

$$P(\text{genotype } Aa) = 1 \cdot \frac{1}{2} = \frac{1}{2}$$

Schematically, the results in this example can be written

$$AA \times Aa \rightarrow 50\%\ AA \qquad 50\%\ Aa$$

which explains observation (1e).

Table 1 is a complete list of genotype probabilities for the offspring that result from various crosses. You may wish to check your understanding of the previous discussion by verifying the results listed in this table.

TABLE 1

Mating (Male × Female)	Genotype Probabilities for the Offspring		
	AA	Aa	aa
$AA \times AA$	1	0	0
$Aa \times AA$	$\frac{1}{2}$	$\frac{1}{2}$	0
$AA \times Aa$	$\frac{1}{2}$	$\frac{1}{2}$	0
$Aa \times Aa$	$\frac{1}{4}$	$\frac{1}{2}$	$\frac{1}{4}$
$aa \times Aa$	0	$\frac{1}{2}$	$\frac{1}{2}$
$Aa \times aa$	0	$\frac{1}{2}$	$\frac{1}{2}$
$aa \times AA$	0	1	0
$AA \times aa$	0	1	0
$aa \times aa$	0	0	1

Dominant and Recessive Traits

Some physical traits in humans exhibit only two possible physical forms. For example, a person either does or doesn't have blue eyes. In such cases the genotypes AA and Aa produce one of the forms, and the genotype aa produces the second form. The trait produced by genotypes AA and Aa is called the **dominant trait**, and the trait produced by genotype aa is called the **recessive trait**. In humans, for example, having blue eyes is a recessive trait.

Table 1 suggests that overall there is a much better chance for the offspring to have the dominant trait than the recessive one. Based on this observation, many scientists in the late 1800s and early 1900s argued that for large populations, Mendel's theory implied that over successive generations the number of offspring with the recessive trait should diminish and eventually disappear. Since this conclusion was at odds with physical evidence (blue eyes in humans have not died out), they rejected Mendel's theory. In 1908 G. H. Hardy* and W. Weinberg† independently proved that Mendel's theory did not imply the disappearance of recessive traits. As a result, the work of Mendel eventually became the cornerstone of modern genetic theory. We conclude this section with a discussion of the arguments given by Hardy and Weinberg.

Let us consider a large population of individuals (plants, people, or animals) and fix our attention on a characteristic determined by a single pair of genes. We investigate the percentage of offspring of each genotype in successive generations under the assumption that matings occur randomly in the population. Suppose that in the initial parental population, the fraction of the population of each genotype is

$$\begin{aligned} \text{fraction with genotype } AA &= r \\ \text{fraction with genotype } Aa &= s \\ \text{fraction with genotype } aa &= t \end{aligned}$$

Further, assume that these fractions are the same among the males and females. It follows that if an individual is picked at random, the probability that it gives its offspring an A gene is

$$p = r + \frac{s}{2} \tag{6}$$

(Why?) Similarly, the probability that a random individual gives its offspring an a gene is

$$q = t + \frac{s}{2} \tag{7}$$

(Why?) Thus, the probability that the offspring is of genotype AA is

$$pp = p^2$$

and the probability that the offspring is of genotype aa is

$$qq = q^2$$

Finally, there are two different ways that the offspring can be of genotype Aa; the male parent can provide the A gene and the female parent the a gene, or vice versa. Each of these possibilities occurs with probability pq; therefore, the probability that the offspring is of genotype Aa is

$$pq + pq = 2pq$$

**Godfrey Harold Hardy* (1877–1947) was a world-renowned English mathematician. He taught at Cambridge and Oxford and was a prolific researcher who produced more than 300 research papers. He received numerous medals and honorary degrees for his accomplishments.

† *Wilhelm Weinberg* was a German physician who did research in blood types.

Table 2 summarizes the results obtained so far for genotype probabilities for parents and offspring.

TABLE 2

	Probability of Type		
	AA	*Aa*	*aa*
Parents	r	s	t
Offspring (First Generation)	p^2	$2pq$	q^2
	where $p = r + \frac{1}{2}s$ and $q = t + \frac{1}{2}s$		

Let us now investigate the genotype probabilities for the offspring of the offspring, that is, the second generation. Again, we assume that mating is random and the genotype probabilities are the same for both males and females. First, we need a preliminary observation. Since r, s, and t denote the fraction of parents of each genotype in the original populations, we have

$$r + s + t = 1$$

Thus,

$$\begin{aligned} p + q &= \left(r + \frac{s}{2}\right) + \left(t + \frac{s}{2}\right) \\ &= r + s + t \\ &= 1 \end{aligned}$$

Now we are in a position to compute the genotype probabilities for the second generation. If an individual is picked at random from the first generation of offspring, then from Table 2 the probability that it gives its offspring an A gene is

$$p^2 + \frac{1}{2}(2pq)$$

(Why?) But this expression can be simplified by writing

$$\begin{aligned} p^2 + \frac{1}{2}(2pq) &= p^2 + pq \\ &= (q + p)p \\ &= 1p \\ &= p \end{aligned} \tag{8}$$

Similarly, the probability that it gives its offspring an a gene is

$$\begin{aligned} q^2 + \frac{1}{2}(2pq) &= q^2 + pq \\ &= (q + p)q \\ &= 1q \\ &= q \end{aligned} \tag{9}$$

From (8) and (9) and computations like those used for the first generation, we see that for the second generation of offspring the probability of genotype AA is

$$pp = p^2$$

the probability of genotype aa is

$$qq = q^2$$

and the probability of genotype Aa is

$$pq + pq = 2pq$$

Table 3 summarizes the results we have obtained so far. From this table we see that the genotype probabilities are the same for the first and second generations of offspring. Moreover, since the genotype probabilities of any generation depend only on the genotype probabilities of the previous generation, it follows that the third generation and in fact all future generations of offspring will have the same probabilities as the first and second generations. This result is the **Hardy–Weinberg stability principle**. It is often stated in the following form.

Hardy–Weinberg Stability Principle In a population with random mating, the genotype probabilities stabilize after one generation.

TABLE 3

	Probability of Type		
	AA	***Aa***	***aa***
Parents	r	s	t
Offspring (First Generation)	p^2	$2pq$	q^2
Offspring (Second Generation)	p^2	$2pq$	q^2
	where $p = r + \frac{1}{2}s$ and $q = t + \frac{1}{2}s$		

The Hardy–Weinberg principle shows that, according to Mendelian theory, recessive traits should not die out. Rather, the proportion of the population exhibiting the recessive traits should stabilize over a period of time. We note, however, that the Hardy–Weinberg principle involves certain assumptions, such as random mating, which are not always satisfied. In such situations the genetic composition of the population can slowly change from generation to generation.

EXAMPLE 3 Assume that in an initial population of plants, 50% of the plants have red flowers (R), 40% have pink flowers (P), and 10% have white flowers (W). Assuming that the conditions of the Hardy–Weinberg principle are satisfied, find the percentage of each kind of flower that will occur in the first generation of offspring.

Solution Since red flowers correspond to genotype AA, pink flowers to genotype Aa, and white flowers to genotype aa, the fraction of each genotype in the original population is

$$\begin{aligned} r &= .5 && \text{for genotype } AA \\ s &= .4 && \text{for genotype } Aa \\ t &= .1 && \text{for genotype } aa \end{aligned}$$

From Formulas (6) and (7) we obtain

$$p = r + \frac{s}{2} = .7$$

and

$$q = t + \frac{s}{2} = .3$$

Thus, from Table 3 the probabilities for each genotype in the first generation are

$$\begin{aligned} p^2 &= (.7)^2 = .49 && \text{for genotype } AA \\ 2pq &= 2(.7)(.3) = .42 && \text{for genotype } Aa \\ q^2 &= (.3)^2 = .09 && \text{for genotype } aa \end{aligned}$$

which in turn are the new values of r, s, and t, respectively. Therefore 49% of the first generation will have red flowers, 42% pink flowers, and 9% white flowers.

EXERCISE SET 10.4

1. Compute the genotype probabilities that result from the mating $aa \times Aa$, and check your results against those given in Table 1.

2. Use your results from Exericse 1 to explain observation (1f) in the text.

In Exercises 3–5 assume a certain dominant trait is produced by genotypes AA and Aa and the recessive trait by genotype aa.

3. For the mating $Aa \times Aa$, what percentage of the offspring will have the dominant trait?

4. For the mating $Aa \times aa$, what percentage of the offspring will have the dominant trait?

5. If a female of genotype Aa is mated with a male whose genotype is unknown, what is the probability that an offspring will have the dominant trait? Assume that the possible male genotypes are equally likely.

6. Suppose that in an initial population, the genotypes AA, Aa, and aa occur in equal numbers. Assuming that the conditions of the Hardy–Weinberg principle are satisfied, find the percentage of each genotype that will occur in the first generation of offspring.

7. For the problem in Exercise 6, find the percentage of each genotype that will occur in the second generation. Explain your answer.

8. Suppose that in an initial population of plants, 50% have red flowers, 20% have pink flowers, and 30% have white flowers. Assuming that the conditions of the Hardy–Weinberg principle are satisfied, find the percentage of each color that will occur in the first generation of offspring.

9. For the problem in Exercise 8, find the percentage of each color that will occur in the second generation. Explain your answer.

10. Consider families with two children in which the male parent has genotype AA and the female parent has genotype Aa. Show that $\frac{1}{4}$ of the families have two AA children, $\frac{1}{2}$ of the families have one AA child and one Aa child, and $\frac{1}{4}$ of the families have two Aa children.

11. Consider families with three children in which the male parent has genotype AA and the female parent has genotype Aa. Show that $\frac{1}{8}$ of the families have three AA children, $\frac{3}{8}$ have two AA and one Aa, $\frac{3}{8}$ have one AA and two Aa, and $\frac{1}{8}$ have three Aa children.

12. A number of human diseases, such as amaurotic idiocy, are due to inherited recessive traits that are fatal to infants. Some of these diseases are apparently determined by a single pair of genes. Individuals of genotype *AA* and *Aa* are healthy, whereas individuals of genotype *aa* are affected and die in infancy. Because of the fatal nature of the *a* gene, it is commonly called a *lethal gene*, and a parent of genotype *Aa* is called a *carrier* of the lethal gene.

(a) If an adult had a brother or sister affected by the disease, then both parents of the adult must be carriers. Explain why.

(b) Show that if an adult had a brother or sister affected by the disease, then his or her genotype probabilities are as shown in the table.

Genotype of Adult	*AA*	*Aa*
Probability	$\frac{1}{3}$	$\frac{2}{3}$

[*Hint*: This is a conditional probability problem. We are given that the individual has survived to adulthood, so he or she is not of genotype *aa*.]

10.5 MARKOV CHAINS AND APPLICATIONS

In Section 8.5 we explore binomial random variables by examining independent sequences of Bernoulli experiments. In this section we study other kinds of sequences in which the experiments need not be independent. Broadly speaking, we are interested in studying physical **systems** and their **states**. To explain what this means, we can imagine a pond, dotted with lily pads, and a lazy frog who suns himself by randomly leaping about from pad to pad whenever he pleases. The pond, the frog, and the lily pads constitute a *physical system*. When the frog leaps from one pad to another, the physical system looks different, and we say that the system has changed its *state*.

To pursue the frog discussion further, we assume that the pond contains only two lily pads, pad 1 and pad 2, and we assume that we record the state of the system at regular time intervals. A typical record of 11 observations might look like

$$1, 1, 2, 1, 2, 2, 2, 2, 1, 2, ,2 \qquad (1)$$

where a 1 indicates that the frog was observed on pad 1 and a 2 indicates that the frog was observed on pad 2 (Figure 1).

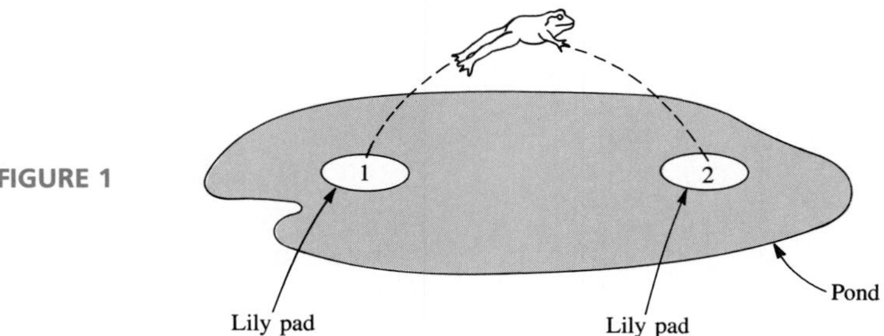

FIGURE 1

In many problems we know the present state of a physical system and are interested in predicting the state of the system at the next observation or even at some more remote future observation. In general, it is impossible to predict a future state with *certainty*; however, it is often possible to predict the *chances* that the system will be in a particular state at a later time if its present state is known. To illustrate, consider the following questions about the jumping frog problem:

(a) If the system is presently in state 1, what is the probability that it will still be in state 1 at the next observation?

(b) If the system is presently in state 1, what is the probability that it will be in state 2 at the next observation?

(c) If the system is presently in state 2, what is the probability that it will be in state 1 at the next observation?

(d) If the system is presently in state 2, what is the probability that it will still be in state 2 at the next observation?

From the recorded data in (1), we see that the system was in state 1 four times. In three of these cases the system changed to state 2 at the next observation, and in one case the system was still in state 1 at the next observation. Thus, based on the data, the probabilities requested in (a) and (b) are

$$P(\text{system remains in state 1 given that it is presently in state 1}) = \frac{1}{4} \tag{2a}$$

$$P(\text{system changes to state 2 given that it is presently in state 1}) = \frac{3}{4} \tag{2b}$$

Similarly, using the *first ten* observations in (1), the system was in state 2 six times. In two of these cases the system changed to state 1 at the next observation, and in four cases the system was still in state 2 at the next observation. Thus, based on the data, the probabilities requested in (c) and (d) are

$$P(\text{system changes to state 1 given that it is presently in state 2}) = \frac{2}{6} = \frac{1}{3} \tag{2c}$$

$$P(\text{system remains in state 2 given that it is presently in state 2}) = \frac{4}{6} = \frac{2}{3} \tag{2d}$$

The information in (2a), (2b), (2c), and (2d) can be organized conveniently in matrix form as follows:

$$\begin{array}{cc} & \text{Next observation} \\ & \begin{array}{cc}\text{State 1} & \text{State 2}\end{array} \\ \text{Now} \begin{array}{c}\text{State 1} \\ \text{State 2}\end{array} & \begin{bmatrix} \frac{1}{4} & \frac{3}{4} \\ \frac{1}{3} & \frac{2}{3} \end{bmatrix} \end{array} \tag{3}$$

Matrix (3) is called the **transition matrix** for the system; each entry is the probability of moving from a given state to another state at the next observation. For example, in matrix (3) the probability $\frac{1}{3}$ of moving from state 2 to state 1 appears in row 2 and column 1, and the probability $\frac{1}{4}$ of moving from state 1 to state 1 appears in row 1 and column 1. We note that the transition matrix (3) is based on the rather small data sample (1). More extensive data would yield a more accurate transition matrix.

Although the illustration of the frog in the pond is more descriptive than practical, it contains all the ingredients of the following more realistic examples.

EXAMPLE 1 **STOCK MARKET FLUCTUATIONS** In a stable stock market, many stocks show a tendency to cancel out one day's price change with a change in the opposite direction on the next day. Thus an increase one day tends to be followed by a decrease the next day, and similarly a decrease tends to be followed by an increase.

We can view a stock as a physical system with two possible states, increase or decrease. Suppose the probability of a price increase, given that the previous change was a decrease, is .7. (Thus, the probability of a price decrease, given that the previous price change was a decrease, is $1 - .7 = .3$.) Suppose also that the probability of a price decrease, given that the previous change was an increase, is .8. (Thus, the probability of a price increase, given that the previous change was an increase, is .2.) Then the transition matrix for the system is

$$\begin{array}{ll} & \text{Tomorrow's change} \\ & \begin{array}{cc}\text{Increase} & \text{Decrease}\end{array} \\ \begin{array}{l}\text{Today's} \\ \text{change}\end{array} \begin{array}{l}\text{Increase} \\ \text{Decrease}\end{array} & \begin{bmatrix} .2 & .8 \\ .7 & .3 \end{bmatrix} \end{array} \tag{4}$$

EXAMPLE 2 **ADJUSTMENT OF EQUIPMENT IN SPACE EXPLORATION** Much of the instrumentation used in lunar exploration or deep-space probes is "self-adjusting" in the sense that the equipment is designed to correct alignment errors automatically when they occur. Often the adjustment takes place over a period of time so that when a monitoring device finds the instrument out of alignment at a certain time, there is a fairly good chance it will still be out of alignment a short time later. Likewise, if the equipment is correctly aligned, it is likely to be aligned a short time later.

We can view a self-adjusting instrument as a physical system with two possible states, aligned or unaligned. A typical transition matrix for such a system is

$$\begin{array}{ll} & \text{State of the instrument} \\ & \text{1 second later} \\ & \begin{array}{cc}\text{Aligned} & \text{Unaligned}\end{array} \\ \begin{array}{l}\text{Present state of} \\ \text{the instrument}\end{array} \begin{array}{l}\text{Aligned} \\ \text{Unaligned}\end{array} & \begin{bmatrix} .9 & .1 \\ .2 & .8 \end{bmatrix} \end{array} \tag{5}$$

From this matrix we see that the probability that the instrument is unaligned 1 second from now, given that it is presently unaligned, is .8, and the probability that the instrument is unaligned 1 second from now, given that it is presently aligned, is .1.

EXAMPLE 3 **A TRANSPORTATION PROBLEM** The Borough of Manhattan in New York City can be divided into three sectors: uptown (sector 1), midtown (sector 2), and lower Manhattan (sector 3). Taxicabs that operate only in Manhattan can pick up a passenger in any sector and drop him or her off in any sector. We can view a taxicab and the Borough of Manhattan as a physical system. If we observe the system when the cab is picking up or discharging a passenger, the system can be in one of three states: the cab can be in sector 1, in sector 2, or in sector 3. A typical transition matrix for such a system is

$$\begin{array}{cc} & \text{Discharge sector} \\ & \begin{array}{ccc} 1 & 2 & 3 \end{array} \\ \text{Pickup sector} \begin{array}{c} 1 \\ 2 \\ 3 \end{array} & \begin{bmatrix} .5 & .4 & .1 \\ .3 & .6 & .1 \\ .2 & .1 & .7 \end{bmatrix} \end{array}$$

Thus, given that a passenger is picked up in sector 3, the probability is .2 that he or she will be discharged in sector 1, and given that a passenger is picked up in sector 1, the probability is .5 that he or she will be discharged in sector 1.

EXAMPLE 4 **A GENETICS PROBLEM FOR STUDENTS FAMILIAR WITH SECTION 10.4** Suppose a parent of unknown genotype is crossed with an individual of genotype *Aa*; then an offspring of unknown genotype is selected from this mating and crossed with another individual of genotype *Aa*; then an offspring of unknown genotype is selected from this mating and crossed with still another individual of genotype *Aa*. If we continue this process, we obtain a succession of matings

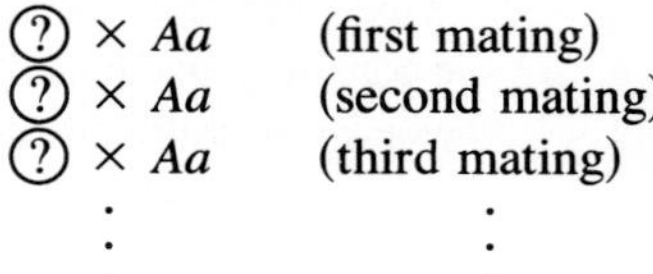

in which one parent always has an unknown genotype and the other has genotype *Aa*. We can think of the two parents as a physical system. The system can be in one of three states, *AA*, *Aa*, or *aa*, corresponding to the genotype of parent ⓘ?.

To obtain the transition matrix for the system, suppose first that the system is in state *AA*. In other words, we suppose that parent (?) has genotype *AA*. In this case the mating

$$(?) \times Aa$$

is

$$AA \times Aa$$

and from Table 1 in Section 10.4, the genotype probabilities for the offspring are

$$\begin{array}{ccc} AA & Aa & aa \\ \frac{1}{2} & \frac{1}{2} & 0 \end{array}$$

Since the new parent of unknown genotype is selected from these offspring, these probabilities form the first row of the transition matrix

$$\text{Unknown genotype of parent}\quad \begin{array}{c} \\ AA \\ Aa \\ aa \end{array} \overset{\displaystyle\text{Genotype of next parent}}{\begin{array}{ccc} AA & Aa & aa \\ \end{array}} \begin{bmatrix} \frac{1}{2} & \frac{1}{2} & 0 \\ \frac{1}{4} & \frac{1}{2} & \frac{1}{4} \\ 0 & \frac{1}{2} & \frac{1}{2} \end{bmatrix} \tag{6}$$

The second and third rows in this transition matrix are obtained using lines four and five of Table 1 in Section 10.4.

If we observe a physical system for a period of time, the system will progress from its **initial state** (its state at the first observation) through a succession of possibly different states, thereby generating a sequence or chain of observations. [See, for example, the chain of observations recorded in (1).] If a future state of the system cannot be predicted with certainty, that is, if it is only possible to give the probability of the occurrence of the future state, then the evolution of the system is an example of a **stochastic process**. To illustrate, consider the space instrumentation problem in Example 2. If we know that the system is currently aligned, we cannot tell with certainty whether it will be aligned at the next observation. At best, we can assert that it will be aligned with probability .9 and unaligned with probability .1. Thus, Example 2 illustrates a stochastic process. Similarly, the other examples in this section are stochastic processes. The stochastic processes in all of our examples have an additional important property. In each case, once the present state of the system is known, the state probabilities for the next observation can be determined without any other information. The following examples illustrate this point.

EXAMPLE 5 Consider the stock fluctuation problem of Example 1. If the present state of the system (today's change) is known to be an increase, then from the transition matrix (4) the state probabilities for tomorrow are

Increase	Decrease
.2	.8

No information, besides the present state, is needed to obtain these probabilities. In particular, it is not necessary to know the past history of the stock that led up to today's state.

EXAMPLE 6 **FOR STUDENTS FAMILIAR WITH SECTION 10.4** Consider the genetics problem of Example 4. If the present state of the system is Aa (i.e., the parent ⑦ has genotype Aa), then from transition matrix (6) the state probabilities for the next observation are

AA	Aa	aa
$\frac{1}{4}$	$\frac{1}{2}$	$\frac{1}{4}$

No information, besides the present state, is needed to obtain these probabilities.

Stochastic processes in which the probabilities for the next state are completely determined by the present state are called **Markov* processes** or sometimes **Markov chains**.

In many applications of Markov processes, we know the initial state of the system and are interested in finding the state probabilities not only for the next observation but also for more remote future observations. The next example illustrates one way of obtaining such probabilities.

EXAMPLE 7 Consider the space equipment problem of Example 2. We assume that at our initial observation the instrument is in the aligned state, and we try to find the state probabilities 2 seconds later. The analysis is easy to picture using the tree diagram in Figure 2.

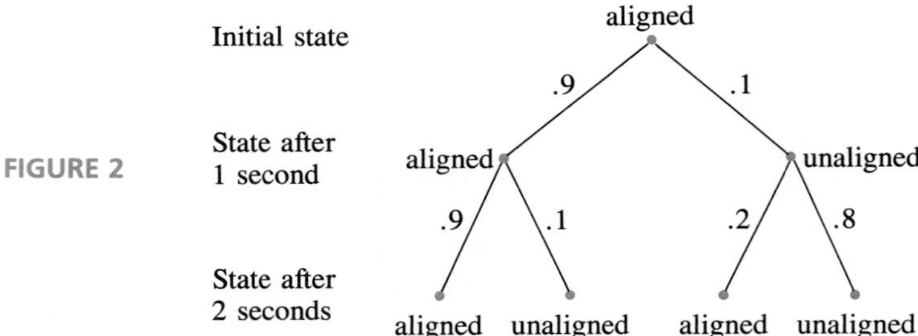

FIGURE 2

The system is assumed initially to be in the aligned state. This initial state is represented by the dot at the top of the tree. From the transition matrix (5), we see that after 1 second, the system will move into the aligned state with probability .9 and into the unaligned state with probability .1. This is indicated on the tree by the two branches emanating from the top dot. If, after 1 second, the system is in the unaligned state, the transition matrix tells us that during the next second the system will move into the aligned state with probability .2 and into the unaligned state with probability .8. This is represented on the tree by the two branches emanating from the right dot in the middle of the tree. Similarly, if the system is in the aligned state after 1 second, then (as shown on the tree) the probability is .9 that the system will move into the aligned state during the next second and .1 that it will move into the unaligned state.

We use this tree to help find the state probabilities of the system after 2 seconds. By tracing the paths from the top of the tree to the bottom, we see that there are two ways the system can move from its initial aligned state to an aligned state 2 seconds later. We can have

$$\text{aligned} \rightarrow \text{aligned} \rightarrow \text{aligned} \tag{7}$$

or

$$\text{aligned} \rightarrow \text{unaligned} \rightarrow \text{aligned} \tag{8}$$

**Andrei Andreyevich Markov* (1856–1922) was an outstanding Russian mathematician. His work was the starting point for the modern theory of stochastic processes. Markov studied and taught at the University of St. Petersburg and was a member of the Soviet Academy of Sciences.

Using the probabilities marked on the branches of the tree, we see that the probability that (7) occurs is (.9)(.9) = .81 and the probability that (8) occurs is (.1)(.2) = .02. Thus the probability that the system is aligned after 2 seconds is .81 + .02 = .83.

Similarly, there are two ways the system can move from its initial aligned state to an unaligned state 2 seconds later. We can have

$$\text{aligned} \rightarrow \text{aligned} \rightarrow \text{unaligned} \tag{9}$$

or

$$\text{aligned} \rightarrow \text{unaligned} \rightarrow \text{unaligned} \tag{10}$$

Using the probabilities marked on the branches of the tree, we see that the probability of (9) is (.9)(.1) = .09 and the probability of (10) is (.1)(.8) = .08. Thus the probability that the system is unaligned after 2 seconds is .09 + .08 = .17.

EXAMPLE 8 For the equipment problem of Example 2, find the state probabilities after 2 seconds, assuming that the system is initially unaligned.

Solution Using the transition matrix (5), we obtain the tree diagram in Figure 3.

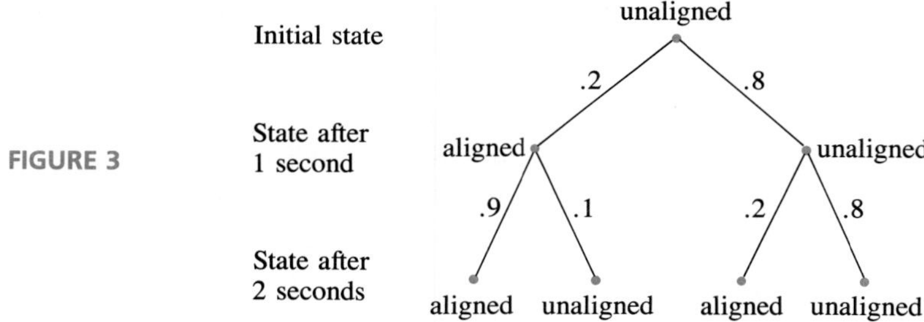

FIGURE 3

The system can move from its initial unaligned state to an aligned state 2 seconds later in two ways:

$$\text{unaligned} \rightarrow \text{aligned} \rightarrow \text{aligned} \tag{11}$$
$$\text{unaligned} \rightarrow \text{unaligned} \rightarrow \text{aligned} \tag{12}$$

From the tree diagram, (11) occurs with probability (.2)(.9) = .18 and (12) occurs with probability (.8)(.2) = .16. Thus, the probability that the system is aligned after 2 seconds is .18 + .16 = .34. Similarly, the probability that the system is unaligned after 2 seconds is (.2)(.1) + (.8)(.8) = .66.

If, in the last two examples, we had been interested in the state probabilities after 5 seconds rather than 2 seconds, the tree diagram would have contained so many branches that it would have been too cumbersome to use. To remedy this problem, we now show how matrices can be used to compute state probabilities.

Consider a physical system that can be in any one of k states. If, for a certain observation, the system is in

state 1 with probability p_1
state 2 with probability p_2
⋮
state k with probability p_k

then the $1 \times k$ matrix of probabilities

$$[p_1 \quad p_2 \quad \ldots \quad p_k]$$

is called the **state matrix** or sometimes the **state vector** for the system.

EXAMPLE 9 Consider the space equipment problem discussed in Example 7. At each observation, the system is in one of two states:

state 1 (aligned)
state 2 (unaligned)

In Example 7, the system is assumed to be aligned initially, so at the initial observation, the probability is $p_1 = 1$ that the system is aligned and $p_2 = 0$ that the system is unaligned. Thus, the initial state matrix for the system is

$$[p_1 \quad p_2] = [1 \quad 0] \tag{13}$$

As shown in Figure 2, the probability that the system is aligned after 1 second is $p_1 = .9$ and the probability that it is unaligned after 1 second is $p_2 = .1$. Thus the state matrix for the system after 1 second is

$$[p_1 \quad p_2] = [.9 \quad .1] \tag{14}$$

In Example 7 we show that after 2 seconds the probability is $p_1 = .83$ that the system is aligned and $p_2 = .17$ that the system is unaligned; thus, the state matrix after 2 seconds is

$$[p_1 \quad p_2] = [.83 \quad .17] \tag{15}$$

The following result provides a quick way to find the state matrix for any observation from the state matrix for the preceding observation.

> If P is the transition matrix for a Markov process and if X is the state matrix for any observation, then
>
> $$XP \tag{16}$$
>
> is the state matrix for the next observation.

EXAMPLE 10 For the problem in Example 9, the initial state matrix is

$$X = [1 \quad 0]$$

[see Equation (13)], and the transition matrix (5) is

$$P = \begin{bmatrix} .9 & .1 \\ .2 & .8 \end{bmatrix}$$

Thus, from result (16) the state matrix after 1 second is

$$XP = [1 \quad 0]\begin{bmatrix} .9 & .1 \\ .2 & .8 \end{bmatrix} = [.9 \quad .1]$$

which agrees with Equation (14). Similarly, the state matrix after 2 seconds is

$$[.9 \quad .1]\begin{bmatrix} .9 & .1 \\ .2 & .8 \end{bmatrix} = [.83 \quad .17]$$

which agrees with (15).

EXAMPLE 11 **APPLICATION OF MARKOV PROCESSES IN MARKETING*** A shopper's weekly purchases of a laundry cleaner (soap or detergent) can be assumed to constitute a two-state Markov process with transition matrix

		Purchases the next week: Purchases a cleaner	Purchases the next week: Does not purchase a cleaner
Purchases for 1 week	Purchases a cleaner	.78	.22
	Does not purchase a cleaner	.43	.57

Assuming that a shopper initially purchases a cleaner, find the probability that the shopper will purchase a cleaner 1 week later, 2 weeks later, and 3 weeks later.

Solution Since we are assuming that the shopper initially purchases a cleaner, the initial probability is $p_1 = 1$ that a cleaner is purchased and $p_2 = 0$ that a cleaner is not purchased. Thus, the initial state matrix is

$$[1 \quad 0]$$

From (16) the state matrix after 1 week will be

$$[1 \quad 0]\begin{bmatrix} .78 & .22 \\ .43 & .57 \end{bmatrix} = [.78 \quad .22]$$

The state matrix after 2 weeks will be

$$[.78 \quad .22]\begin{bmatrix} .78 & .22 \\ .43 & .57 \end{bmatrix} = [.703 \quad .297]$$

*The data in this problem is based on a 26-week study by G. P. Styon and H. Smith, which appeared in the paper, "Markov Chains Applied to Marketing," *Journal of Marketing Research,* **1**(1964): 50–55.

and the state matrix after 3 weeks will be

$$[.703 \quad .297]\begin{bmatrix} .78 & .22 \\ .43 & .57 \end{bmatrix} = [.676 \quad .324]$$

Thus, the probability the shopper purchases a cleaner 1 week later is .78, 2 weeks later is .703, and 3 weeks later is .676.

In many applications of Markov processes, one is interested in the "long-run" behavior of the physical system. The following example illustrates what we have in mind.

EXAMPLE 12 Consider the stock market fluctuation problem of Example 1. To illustrate how this system behaves over a period of time, we computed the state matrices for the system for a period of 10 market days, assuming that the initial state of the system was an increase. We obtained (to two decimal places)

Initial state matrix

$$[1 \quad 0]\begin{bmatrix} .2 & .8 \\ .7 & .3 \end{bmatrix} = [.20 \quad .80] \qquad \text{(after 1 day)}$$

$$[.20 \quad .80]\begin{bmatrix} .2 & .8 \\ .7 & .3 \end{bmatrix} = [.60 \quad .40] \qquad \text{(after 2 days)}$$

$$[.60 \quad .40]\begin{bmatrix} .2 & .8 \\ .7 & .3 \end{bmatrix} = [.40 \quad .60] \qquad \text{(after 3 days)}$$

$$[.40 \quad .60]\begin{bmatrix} .2 & .8 \\ .7 & .3 \end{bmatrix} = [.50 \quad .50] \qquad \text{(after 4 days)}$$

$$[.50 \quad .50]\begin{bmatrix} .2 & .8 \\ .7 & .3 \end{bmatrix} = [.45 \quad .55] \qquad \text{(after 5 days)}$$

$$[.45 \quad .55]\begin{bmatrix} .2 & .8 \\ .7 & .3 \end{bmatrix} = [.48 \quad .53] \qquad \text{(after 6 days)}$$

$$[.48 \quad .53]\begin{bmatrix} .2 & .8 \\ .7 & .3 \end{bmatrix} = [.46 \quad .54] \qquad \text{(after 7 days)}$$

$$[.46 \quad .54]\begin{bmatrix} .2 & .8 \\ .7 & .3 \end{bmatrix} = [.47 \quad .53] \qquad \text{(after 8 days)}$$

$$[.47 \quad .53]\begin{bmatrix} .2 & .8 \\ .7 & .3 \end{bmatrix} = [.47 \quad .53] \qquad \text{(after 9 days)}$$

$$[.47 \quad .53]\begin{bmatrix} .2 & .8 \\ .7 & .3 \end{bmatrix} = [.47 \quad .53] \qquad \text{(after 10 days)}$$

Thus, after 8 market days, the probability of an increase stabilizes at .47 and the probability of decrease stabilizes at .53. Similarly, if the initial state of the system had been a decrease, we would have obtained

$$[0 \quad 1]\begin{bmatrix} .2 & .8 \\ .7 & .3 \end{bmatrix} = [.70 \quad .30] \qquad \text{(after 1 day)}$$

$$[.70 \quad .30]\begin{bmatrix} .2 & .8 \\ .7 & .3 \end{bmatrix} = [.35 \quad .65] \qquad \text{(after 2 days)}$$

$$[.35 \quad .65]\begin{bmatrix} .2 & .8 \\ .7 & .3 \end{bmatrix} = [.53 \quad .48] \qquad \text{(after 3 days)}$$

$$[.53 \quad .48]\begin{bmatrix} .2 & .8 \\ .7 & .3 \end{bmatrix} = [.44 \quad .56] \qquad \text{(after 4 days)}$$

$$[.44 \quad .56]\begin{bmatrix} .2 & .8 \\ .7 & .3 \end{bmatrix} = [.48 \quad .52] \qquad \text{(after 5 days)}$$

$$[.48 \quad .52]\begin{bmatrix} .2 & .8 \\ .7 & .3 \end{bmatrix} = [.46 \quad .54] \qquad \text{(after 6 days)}$$

$$[.46 \quad .54]\begin{bmatrix} .2 & .8 \\ .7 & .3 \end{bmatrix} = [.47 \quad .53] \qquad \text{(after 7 days)}$$

$$[.47 \quad .53]\begin{bmatrix} .2 & .8 \\ .7 & .3 \end{bmatrix} = [.47 \quad .53] \qquad \text{(after 8 days)}$$

$$[.47 \quad .53]\begin{bmatrix} .2 & .8 \\ .7 & .3 \end{bmatrix} = [.47 \quad .53] \qquad \text{(after 9 days)}$$

$$[.47 \quad .53]\begin{bmatrix} .2 & .8 \\ .7 & .3 \end{bmatrix} = [.47 \quad .53] \qquad \text{(after 10 days)}$$

Thus, once again the probability of an increase stabilizes at .47, and the probability of a decrease stabilizes at .53 after a period of time.

The results observed in the last example are not accidental. As the following theorem shows, this stabilization phenomenon occurs in many Markov processes.

THEOREM Let P be the transition matrix for a Markov process. If some power of P has all positive entries, then

1. regardless of the initial state of the system, the successive state matrices will approach the same fixed state matrix Q.
2. the matrix Q satisfies the equation

$$QP = Q \tag{17}$$

and is called the **steady-state matrix** for the system.

We omit the proof. A transition matrix P is called **regular** if some power of P has all positive entries. Thus Equation (17) applies only when the transition matrix P is regular.

EXAMPLE 13 The calculations in Example 12 show that the steady-state matrix for the stock market system is (to two decimal places)

$$Q \cong [.47 \quad .53]$$

This same result can be obtained as follows by using Equation (17). Let

$$Q = [q_1 \quad q_2]$$

be the unknown steady-state matrix, where q_1 is the steady-state probability of an increase and q_2 is the steady-state probability of a decrease. (Note that $q_1 + q_2 = 1$.) Substituting Q and the transition matrix (4) into Equation (17), we obtain

$$[q_1 \quad q_2]\begin{bmatrix} .2 & .8 \\ .7 & .3 \end{bmatrix} = [q_1 \quad q_2]$$

or

$$[.2q_1 + .7q_2 \quad .8q_1 + .3q_2] = [q_1 \quad q_2] \tag{18}$$

For the matrices in (18) to be equal, their corresponding entries must be equal; that is,

$$\begin{aligned} .2q_1 + .7q_2 &= q_1 \\ .8q_1 + .3q_2 &= q_2 \end{aligned}$$

or equivalently,

$$\begin{aligned} .8q_1 - .7q_2 &= 0 \\ -.8q_1 + .7q_2 &= 0 \end{aligned}$$

On solving this system, you will see that there are infinitely many solutions, and they are given by the formulas

$$q_1 = \frac{7}{8}t \qquad q_2 = t \tag{19}$$

where t is arbitrary. However, $Q = [q_1 \quad q_2]$ is a state matrix, so, as observed above, $q_1 + q_2 = 1$. Thus from (19) we obtain

$$1 = q_1 + q_2 = \frac{7}{8}t + t = \frac{15}{8}t$$

or $t = \frac{8}{15}$. Substituting this value into (19), we obtain the steady-state matrix

$$Q = [q_1 \quad q_2] = \begin{bmatrix} \frac{7}{15} & \frac{8}{15} \end{bmatrix}$$

which agrees with the results obtained to two decimal places in Example 12.

The following examples illustrates how Markov processes can enter into the long-range planning of urban transit facilities.

EXAMPLE 14 **APPLICATION OF MARKOV PROCESSES TO PROBLEMS IN MASS TRANSIT** Commuters who work in the Delaware Valley can commute to and from their jobs either by public transportation or by automobile. After the installation of a new public transit system, the Planning Commission predicts that each year 30% of those using public transportation will change to automobile, while 70% will continue to use public transportation. The Commission also predicts that each year 60% of those using automobiles will change to public transportation, while 40% will continue to use their automobiles. This information is contained in the following transition matrix

		Mode of transportation the next year	
		Public transportation	Automobile
Mode of transportation 1 year	Public transportation	.7	.3
	Automobile	.6	.4

$$\begin{bmatrix} .7 & .3 \\ .6 & .4 \end{bmatrix} \qquad (20)$$

(In this matrix we have recorded the percentage changes as probabilities.)

Initially, assume that 20% of the commuters use public transportation and 80% use automobiles. Assuming that the population of the Delaware Valley remains constant, find:

(a) the percentage of commuters using each mode of transportation after 2 years

(b) the percentage of commuters using each mode of transportation after a "long period of time"

Solution (a) Since 20% initially use public transportation and 80% initially use their automobiles, the initial state matrix is

Public transportation	Automobile
.2	.8

$$[\,.2 \quad .8\,]$$

Thus, after 1 year the new state matrix will be

$$[.2 \quad .8]\begin{bmatrix} .7 & .3 \\ .6 & .4 \end{bmatrix} = [.62 \quad .38]$$

and after 2 years the new state matrix will be

$$[.62 \quad .38]\begin{bmatrix} .7 & .3 \\ .6 & .4 \end{bmatrix} = [.662 \quad .338]$$

Thus, after 2 years, 66.2% of the commuters will be using public transportation and 33.8% will be using automobiles.

Solution (b) To find the percentage of commuters using each mode after a long period of time, we must find the steady-state matrix for the system. Using (17) and (20), we see that the steady-state matrix

$$Q = [q_1 \quad q_2]$$

must satisfy

$$[q_1 \quad q_2]\begin{bmatrix} .7 & .3 \\ .6 & .4 \end{bmatrix} = [q_1 \quad q_2]$$

or

$$[.7q_1 + .6q_2 \quad .3q_1 + .4q_2] = [q_1 \quad q_2]$$

or

$$\begin{aligned} .7q_1 + .6q_2 &= q_1 \\ .3q_1 + .4q_2 &= q_2 \end{aligned}$$

or

$$\begin{aligned} .3q_1 - .6q_2 &= 0 \\ -.3q_1 + .6q_2 &= 0 \end{aligned}$$

On solving this system, you will see that there are infinitely many solutions, and they are given by the formulas

$$q_1 = 2t \qquad q_2 = t \tag{21}$$

where t is arbitrary. Since $Q = [q_1 \quad q_2]$ is a state matrix, we must have $q_1 + q_2 = 1$. Thus from (21) we obtain

$$1 = q_1 + q_2 = 2t + t = 3t$$

so that $t = \frac{1}{3}$. Substituting this value in (21) yields the steady-state matrix

$$Q = [q_1 \quad q_2] = \left[\tfrac{2}{3} \quad \tfrac{1}{3}\right]$$

Thus, after a long period of time, the proportion of riders using public transportation will stabilize at $\frac{2}{3}$ and the proportion of riders using their automobiles will stabilize at $\frac{1}{3}$.

EXAMPLE 15 **APPLICATION OF MARKOV PROCESSES TO GENETICS FOR STUDENTS FAMILIAR WITH SECTION 10.4** In this example we determine the long-term genotype probabilities for the offspring that result from the mating pattern described in Example 4 of this section. Let P denote the transition matrix (6). To apply result (17), some power of P must have all positive entries. Although P itself does not have all positive entries, the matrix

$$P^2 = \begin{bmatrix} \frac{3}{8} & \frac{1}{2} & \frac{1}{8} \\ \frac{1}{4} & \frac{1}{2} & \frac{1}{4} \\ \frac{1}{8} & \frac{1}{2} & \frac{3}{8} \end{bmatrix}$$

does. Thus Equation (17) can be used to compute the steady-state matrix

$$Q = [q_1 \quad q_2 \quad q_3]$$

where q_1, q_2, and q_3 are the long-term probabilities of genotypes AA, Aa, and aa, respectively. Substituting Q and P into Equation (17) gives

$$[q_1 \quad q_2 \quad q_3]\begin{bmatrix} \frac{1}{2} & \frac{1}{2} & 0 \\ \frac{1}{4} & \frac{1}{2} & \frac{1}{4} \\ 0 & \frac{1}{2} & \frac{1}{2} \end{bmatrix} = [q_1 \quad q_2 \quad q_3]$$

or equivalently

$$\begin{aligned} \tfrac{1}{2}q_1 + \tfrac{1}{4}q_2 \qquad\quad &= q_1 \\ \tfrac{1}{2}q_1 + \tfrac{1}{2}q_2 + \tfrac{1}{2}q_3 &= q_2 \\ \tfrac{1}{4}q_2 + \tfrac{1}{2}q_3 &= q_3 \end{aligned}$$

or equivalently

$$\begin{aligned} \tfrac{1}{2}q_1 - \tfrac{1}{4}q_2 \qquad\quad &= 0 \\ -\tfrac{1}{2}q_1 + \tfrac{1}{2}q_2 - \tfrac{1}{2}q_3 &= 0 \\ -\tfrac{1}{4}q_2 + \tfrac{1}{2}q_3 &= 0 \end{aligned}$$

On solving this system, you will see that there are infinitely many solutions, and they are given by the formulas

$$q_1 = t \qquad q_2 = 2t \qquad q_3 = t \tag{22}$$

Since $Q = [q_1 \quad q_2 \quad q_3]$ is a state matrix, we must have $q_1 + q_2 + q_3 = 1$. Thus from (22) we obtain

$$1 = q_1 + q_2 + q_3 = t + 2t + t = 4t$$

so that $t = \frac{1}{4}$. Substituting this value in (22) yields the steady-state matrix

$$Q = [q_1 \quad q_2 \quad q_3] = \left[\tfrac{1}{4} \quad \tfrac{1}{2} \quad \tfrac{1}{4}\right]$$

Thus, if the mating pattern is carried out repeatedly, the genotype probabilities of the offspring will stabilize at

$$\frac{1}{4} \text{ for } AA \qquad \frac{1}{2} \text{ for } Aa \qquad \frac{1}{4} \text{ for } aa$$

regardless of the original unknown genotype.

In this section we have just touched the surface of Markov processes and their applications. If you are interested in studying this subject in more detail, consult the following references:

1. J. Kemeny, H. Mirkil, J. Snell, and G. Thompson, *Finite Mathematical Structures* (Englewood Cliffs, N.J.: Prentice-Hall, 1959).

2. J. Kemeny, A. Schleifer, J. Snell, and G. Thompson, *Finite Mathematics with Business Applications* (Englewood Cliffs, N.J.: Prentice-Hall, 1962).

3. D. P. Maki and M. Thompson, *Mathematical Models and Applications* (Englewood Cliffs, N.J.: Prentice-Hall, 1973).

EXERCISE SET 10.5

1. For the frog jumping problem described in this section, replace the data in (1) by

1, 2, 1, 2, 2, 2, 1, 1, 1, 2, 1

Based on these data, find the transition matrix for the system.

2. Explain why

$$\begin{bmatrix} \frac{1}{2} & \frac{1}{2} & \frac{1}{2} \\ 0 & 1 & 0 \\ \frac{1}{4} & \frac{1}{4} & \frac{1}{8} \end{bmatrix}$$

cannot be a transition matrix for any physical system.

3. Consider a Markov process with transition matrix

$$\begin{array}{c c} & \begin{array}{cc} \text{State 1} & \text{State 2} \end{array} \\ \begin{array}{c} \text{State 1} \\ \text{State 2} \end{array} & \begin{bmatrix} \frac{1}{4} & \frac{3}{4} \\ \frac{2}{5} & \frac{3}{5} \end{bmatrix} \end{array}$$

(a) What does the entry $\frac{3}{4}$ in this matrix represent?

(b) Assuming that the system is initially in state 1, find the state matrix one observation later.

(c) Assuming that the system is initially in state 2, find the state matrix one observation later.

4. Consider a Markov process with transition matrix

	State 1	State 2
State 1	.6	.4
State 2	.9	.1

If the state matrix at a certain observation is [.75 .25], what is the state matrix at the next observation?

5. Consider the Markov process in Exercise 3.
(a) Assume the system is initially in state 1 and use a tree diagram to find the state matrix two observations later.
(b) Solve the problem in part (a) using result (16) in the text.

6. Consider a Markov process with transition matrix

	State 1	State 2	State 3
State 1	.4	.1	.5
State 2	.3	.3	.4
State 3	.6	.2	.2

(a) Assume the system is initially in state 2 and use a tree diagram to find the state matrix three observations later.
(b) Solve the problem in part (a) using result (16) in the text.

In Exercises 7–10 find the steady-state matrix for the Markov process whose transition matrix is given.

7. $\begin{bmatrix} .75 & .25 \\ .65 & .35 \end{bmatrix}$ **8.** $\begin{bmatrix} .5 & .5 \\ .8 & .2 \end{bmatrix}$

9. $\begin{bmatrix} .1 & .2 & .7 \\ .3 & .1 & .6 \\ .5 & .4 & .1 \end{bmatrix}$ **10.** $\begin{bmatrix} .4 & .4 & .2 \\ .7 & .1 & .2 \\ .5 & .3 & .2 \end{bmatrix}$

11. (a) Consider a Markov process with transition matrix

$$P = \begin{bmatrix} .7 & .3 & 0 \\ .5 & .5 & 0 \\ .6 & .4 & 0 \end{bmatrix}$$

Explain why result (17) does not apply.
(b) Does result (17) apply to a Markov process with transition matrix

$$P = \begin{bmatrix} .6 & .2 & .2 \\ .7 & .3 & 0 \\ .5 & 0 & .5 \end{bmatrix}$$

Explain your answer.

12. Recently, some resorts have been offering insurance to pay vacationers' hotel bills when it rains. Suppose that an insurer of a Florida resort has found that, after a clear day, there is a probability of .2 for rain the next day, while after a rainy day, there is a probability of .4 for rain the next day.
(a) Write down an appropriate transition matrix.
(b) Over the long term, what fraction of the time will the company have to pay off?

13. Suppose that children whose parents are in a middle-to-high-income bracket will, as adults, be in the middle-to-high-income bracket with probability .9 and in a low-income bracket with probability .1. Suppose also that children whose parents are in a low-income bracket will, as adults, be in a middle-to-high-income bracket with probability .5 and in a low-income bracket with probability .5.
(a) What is the probability that a grandchild of a low-income family will be in the low-income bracket as an adult?
(b) Over the long-term, what fraction of the adult population will be in the low-income bracket?

14. A regional planning commission estimates that each year 10% of the people in the eastern part of the region will move to the western part and 3% of the people in the western part will move to the eastern part. Assuming that the population in the region remains constant and, initially, 40% of the population lives in the eastern region and 60% in the western region, find the percentage of population in each region:
(a) after 1 year
(b) after 3 years
(c) over the long term

15. The New York metropolitan area is serviced by three major airports: J. F. Kennedy, LaGuardia, and Newark. Suppose that a car rental agency with a fleet of 500 cars has rental and storage facilities at each of the three airports. Customers rent and then return their cars to the various airports with the probabilities given in the transition matrix

		Returned to		
		Kennedy	LaGuardia	Newark
Rented from	Kennedy	.8	.1	.1
	LaGuardia	.3	.2	.5
	Newark	.2	.6	.2

How many parking spaces should the company allocate at each airport? [*Hint*: Use the steady-state matrix.]

16. A college professor has three sets of questions, A, B, and C, from which to make up exams. If she uses set A for one exam, then for the next exam she tosses a die and uses questions from set A again if 1 occurs, from set B if a 2, 3, or 4 occurs, and from set C if a 5 or 6 occurs. If she uses set B, she tosses a coin and uses set A for the next exam if it comes up heads and set C if it comes up tails. If she uses set C, then she uses set A the next time.

(a) Find the transition matrix for this process.

(b) If the chances of using sets A, B, and C are equally likely for the first exam, find the probabilities that the second exam will come from the sets A, B, and C, respectively.

(c) In the long run, which set of questions is most often used?

Exercises 17–19 refer to the following discussion: Often, the transition probabilities of a Markov process are not known and must be estimated by observation. The following is an illustration of how such transition probabilities can be estimated. Consider a system that can exist in one of three states 1, 2, and 3. Let p_{ij} denote the probability that the system will not move into state j from state i in one step. For example, if $i = 2$ and $j = 3$, then p_{23} is the probability that the system will go into state 3 from state 2 in one step. We want to estimate the values of p_{11}, p_{12}, p_{13}, p_{21}, p_{22}, p_{23}, p_{31}, p_{32}, and p_{33}. To do this, we observe the system for a number of steps (the more the better) and at each step record the state of the system. Suppose our system is in state 3 initially and we record the following list of its next 24 consecutive states:

3 1 1 1 2 2 3 1 3 3 1 1 1 2 2 1 2 1 3 2 1 3 2 1 2

From this data we obtain Table 1.

Now note that of the 11 times the system left state 1, it

entered state 1 4 times
entered state 2 4 times
entered state 3 3 times

Hence we estimate p_{11}, p_{12}, and p_{13} to be $p_{11} = \frac{4}{11}$, $p_{12} = \frac{4}{11}$, and $p_{13} = \frac{3}{11}$. Of the 7 times the system left state 2, it

entered state 1 4 times
entered state 2 2 times
entered state 3 1 time

Hence we estimate p_{21}, p_{22}, and p_{23} to be $p_{21} = \frac{4}{7}$, $p_{22} = \frac{2}{7}$, and $p_{23} = \frac{1}{7}$. Of the 6 times the system left state 3, it

entered state 1 3 times
entered state 2 2 times
entered state 3 1 time

TABLE 1

Transition From	Number of Times
State 1	
1 to 1	4
1 to 2	4
1 to 3	3
	Total = 11
State 2	
2 to 1	4
2 to 2	2
2 to 3	1
	Total = 7
State 3	
3 to 1	3
3 to 2	2
3 to 3	1
	Total = 6

Hence we estimate p_{31}, p_{32}, and p_{33} to be $p_{31} = \frac{3}{6}$, $p_{32} = \frac{2}{6}$, and $p_{33} = \frac{1}{6}$. Thus, the transition matrix for this system is given by

$$P = \begin{bmatrix} p_{11} & p_{12} & p_{13} \\ p_{21} & p_{22} & p_{23} \\ p_{31} & p_{32} & p_{33} \end{bmatrix} = \begin{bmatrix} \frac{4}{11} & \frac{4}{11} & \frac{3}{11} \\ \frac{4}{7} & \frac{2}{7} & \frac{1}{7} \\ \frac{3}{6} & \frac{2}{6} & \frac{1}{6} \end{bmatrix}$$

17. Assume the following sequence of states is recorded for a three-state system:

1 2 1 3 3 2 2 3 2 1 1 1 3 3 3 2 1 1 2 3 2 1 3 2 1 3

(a) Find the 3 × 3 transition matrix.

(b) Is the matrix a regular transition matrix? Explain.

18. Assume the following sequence of states is recorded for a two-state system:

1 1 1 1 2 2 1 2 2 2 1 1 2 1 2 2 2 2 1 1 1 2 1 2

(a) Find the 2 × 2 transition matrix.

(b) Is the matrix a regular transition matrix? Explain.

19. A manufacturer has one full-time mechanic who repairs machines as they break down. The manager of the plant is considering hiring an additional mechanic and orders a cost analysis to determine the need. It is found that hiring an additional mechanic would be worthwhile if at least 35% of the time that a machine is being repaired there is another one waiting to be serviced. Observations of repairs are made over several days. If, during the repair of a machine, there is another machine waiting to be repaired, then a 1 is recorded, otherwise 0 is recorded. The following sequence was generated:

1 0 0 0 1 1 0 0 0 1 1 1 0 1 0 0 0 1 1 0 1 0 1 1 0 0 0 1 1

Consider this situation as a two-state Markov chain.

(a) Fill in Table 2.

TABLE 2

Transition from State 0	Number of Times	Transition from State 1	Number of Times
0 to 0		1 to 0	
0 to 1	___	1 to 1	___
Total =		Total =	

(b) Find the 2×2 transition matrix

$$P = \begin{bmatrix} p_{11} & p_{12} \\ p_{21} & p_{22} \end{bmatrix}$$

where

p_{11} is the probability of going from 0 to 0
p_{12} is the probability of going from 0 to 1
p_{21} is the probability of going from 1 to 0
p_{22} is the probability of going from 1 to 1

(c) Find the long-term probabilities of being in state 0 and state 1.

(d) Should an additional mechanic be hired? Why?

KEY IDEAS FOR REVIEW

Life Insurance and Mortality

- **Mortality table** Statistics compiled by insurance companies, which are used to determine the probabilities that a person will be dead or alive at a given age.
- **Premium** The amount a person pays to purchase insurance.
- **Net single premium** The present value of the future benefits that a company expects to pay out on a type of policy.
- ***n*-year pure endowment** A policy that guarantees the purchaser will collect the value of the policy at the end of n years provided that the insured survives the n years.
- ***n*-year term insurance** A policy that guarantees the beneficiary of the insured will collect the value of the policy at the time of the insured's death if the insured dies within n years.

Game Theory and Applications

- **Payoff matrix** A matrix whose entries are the amounts won by the row player R from the column player C.
- **Constant-sum game** A game in which the sum of the payoffs is a constant that does not depend on which moves are made.
- **Zero-sum game** A constant-sum game in which the constant is zero.
- **Row minimum** The smallest element in the row of a payoff matrix.
- **Column maximum** The largest element in the column of a payoff matrix.
- **Saddle point** An entry in a payoff matrix that is both a column maximum and a row minimum.
- **Strictly determined game** A game whose payoff matrix contains a saddle point.

Value of a game The numerical value of any saddle point in a strictly determined game.

Pure strategy A strategy in which the player uses the same move in every game.

Mixed strategy A strategy in which the player mixes the moves from game to game according to certain probabilities.

Expected winnings The average winnings resulting from a strategy.

Optimal strategies See pages 432–438.

Markov Chains and Applications

Transition matrix The matrix whose entries are the probabilities of moving from a given state to another state at the next observation.

Initial state The state of a system at the first observation.

Markov process or Markov chain A process in which the probabilities for the next state are completely determined by the present state.

State matrix or state vector A $1 \times k$ matrix whose entries are the probabilities of the various states for a given observation.

Steady-state matrix See page 459.

Regular matrix A transition matrix, some power of which has all positive entries.

SUPPLEMENTARY EXERCISES

1. What is the probability that:
(a) John Smith, age 50, will survive to age 80
(b) Janet Smith, age 50, will survive to age 80

2. John Smith, age 50, wants to buy a 10-year pure endowment policy with a face value of \$10,000. If interest is 7% compounded annually, what is the net single premium for this policy?

3. Janet Smith, age 50, wants to buy a 3-year term insurance policy. If interest is 7% compounded annually, what is the net single premium for such a policy with a face value of \$10,000?

4. John Smith, now age 60, has survived to cash in his \$10,000 pure endowment policy (see Exercise 2). He wants to use this money to buy a 3-year term insurance policy. What is the face value of this policy if interest is 7% compounded annually?

5. Which of the following games are strictly determined?

(a) $\begin{bmatrix} 3 & 6 \\ 7 & 2 \end{bmatrix}$ (b) $\begin{bmatrix} 4 & 3 \\ 8 & 7 \end{bmatrix}$

(c) $\begin{bmatrix} 1 & 3 & 2 \\ 6 & 7 & 5 \\ 1 & 4 & 2 \end{bmatrix}$ (d) $\begin{bmatrix} 4 & 0 & 5 \\ 6 & 5 & 2 \\ 2 & 5 & 7 \end{bmatrix}$

6. Find all saddle points in the following two-person constant-sum matrix game:

$$\begin{bmatrix} 5 & 6 & 7 \\ 0 & 7 & 3 \\ 1 & 2 & 3 \\ 5 & 6 & 7 \end{bmatrix}$$

7. John Earnest and Jane Smart are running for mayor. Each will advertise in newspapers (N), radio (R), or television (TV). The payoff matrix below shows the number of additional votes to be received by John Earnest when each candidate chooses a particular medium for his or her advertising. What is each candidate's best advertising strategy?

$$\begin{array}{ccc} & & \text{Jane Smart} \\ & & \begin{array}{ccc} \text{N} & \text{R} & \text{TV} \end{array} \\ \text{John Earnest} & \begin{array}{c} \text{N} \\ \text{R} \\ \text{TV} \end{array} & \begin{bmatrix} 50{,}000 & -10{,}000 & 80{,}000 \\ -30{,}000 & 40{,}000 & 60{,}000 \\ 40{,}000 & 60{,}000 & 70{,}000 \end{bmatrix} \end{array}$$

8. Consider the following game: player I has three cards: 4, 5, and 6. Player II has three cards: 2, 3, and 4. Each player shows one card. If the sum of the cards shown is even, then player I receives that amount from player II. If the sum of the cards shown is odd, then player II receives that amount from player I.
(a) Write the payoff matrix for this game.
(b) Is this game strictly determined?
(c) If it is, determine the optimal moves for each player.

9. Assume that a certain dominant trait is produced by genotypes AA and Aa and the recessive trait by genotype aa. For the mating $AA \times Aa$, what percentage of the offspring will have the recessive trait?

10. Suppose that in an initial population of plants, 45% have red flowers, 35% have pink flowers, and 20% have white flowers. Assuming that the conditions of the Hardy–Weinberg principle are satisfied:
(a) Find the percentage of each color that will occur in the first generation.
(b) Find the percentage of each color that will occur in the second generation.

11. If a female of unknown genotype is mated with a male of genotype AA, what is the probability that an offspring will have the dominant trait?

12. Consider a Markov process with transition matrix

$$\begin{array}{cc} & \begin{array}{cc} \text{State 1} & \text{State 2} \end{array} \\ \begin{array}{c} \text{State 1} \\ \text{State 2} \end{array} & \begin{bmatrix} .7 & .3 \\ .8 & .2 \end{bmatrix} \end{array}$$

(a) If the state matrix at a certain observation is $[.6 \quad .4]$, what is the state matrix at the next observation?
(b) If the system is initially in state 1, use a tree diagram to find the state matrix three observations later.

13. Find the steady-state matrix for the Markov process whose transition matrix is

$$\begin{bmatrix} .55 & .45 \\ .65 & .35 \end{bmatrix}$$

14. A certain town has three supermarkets: A, B, and C. A survey of shoppers has determined the probabilities that a person shopping in one market this week will either shop at the same market or will switch to another next week. This information is given by the transition matrix

$$\begin{array}{ccc} & & \text{Next week} \\ & & \begin{array}{ccc} \text{A} & \text{B} & \text{C} \end{array} \\ \text{This week} & \begin{array}{c} \text{A} \\ \text{B} \\ \text{C} \end{array} & \begin{bmatrix} .3 & .4 & .3 \\ .6 & .3 & .1 \\ .4 & .3 & .3 \end{bmatrix} \end{array}$$

Assume that at the present time, 50% of the residents shop at A, 30% at B, and 20% at C.
(a) Find the percentage of the residents shopping at each market after 2 weeks.
(b) Over the long term, what is the percentage of shoppers at each market?

CHAPTER TEST

1. (a) What is the probability that Ruth Smart, age 40, will die before attaining the age of 60?
(b) What is the probability that Bob Smart, age 40, will die before attaining the age of 60?
(c) Why are the results of (a) and (b) different?

2. Ruth and Bob Smart are both age 40. Each buys a 4-year term insurance policy with a face value of \$10,000. If interest is 6% compounded annually, what is the net single premium each must pay? Why are the two premiums different in value?

3. Find the best moves for players R and C in the following strictly determined game and give the value of the game:

$$\begin{bmatrix} 5 & 2 & 3 \\ 0 & 1 & -1 \\ -1 & 0 & 1 \end{bmatrix}$$

4. Find the optimal strategies and the expected payoff of the matrix game whose payoff matrix is

$$\begin{bmatrix} 1 & 6 \\ 7 & 8 \\ 8 & 5 \end{bmatrix}$$

5. Consider the following game. Player I has three cards: 5 of hearts, 10 of clubs, and 6 of spades. Player II has three cards: 4 of diamonds, 3 of hearts, and 7 of spades. Each player shows one card. If the cards are of the same color, player I pays player II an amount equal to the sum of the cards. If the cards are of different colors, player II pays player I an amount equal to the sum of the cards.

(a) Write the payoff matrix for this game.

(b) Is this game strictly determined?

6. Suppose that in an initial population of plants, 60% have red flowers, 30% have pink flowers, and 10% have white flowers. Assuming that the conditions of the Hardy–Weinberg principle are satisfied, find the percentage of each color that will occur in the first-generation offspring.

7. Consider a Markov process with transition matrix

$$\begin{array}{c} \\ \text{State 1} \\ \text{State 2} \end{array} \begin{array}{c} \begin{array}{cc} \text{State 1} & \text{State 2} \end{array} \\ \begin{bmatrix} .25 & .75 \\ .40 & .60 \end{bmatrix} \end{array}$$

If the system is initially in state 2, use a tree diagram to find the state matrix 4 observations later.

11

Functions

11.1 INTRODUCTION TO FUNCTIONS

In this chapter we consider one of the more fundamental concepts in mathematics, the notion of a function. Loosely stated, a function is a rule for describing the way in which one quantity depends on another. For example, the area A of a circle depends on its radius r according to the formula

$$A = \pi r^2$$

and we say that this formula describes "A as a function of r." The quantities A and r in this equation are called **variables**. If we specify a value for the variable r, then this equation determines a unique value for the variable A. Accordingly, we say that the variable A "depends" on the variable r, and we call r the **independent variable** and A the **dependent variable**. The quantity $\pi = 3.14159\ldots$ has a fixed value that does not depend on A or r; thus, we call π a **constant**.

It is often convenient to imagine a function as an input/output relation. The independent variable is the input and the dependent variable is the output. If we use the formula $A = \pi r^2$ to calculate the area of a circle from its radius, then to determine the area (output $= A$) we must "put in" the value of its radius (input $= r$) (Figure 1). For example, when the input is 2, the output is 4π; when the input is 3, the output is 9π.

FIGURE 1

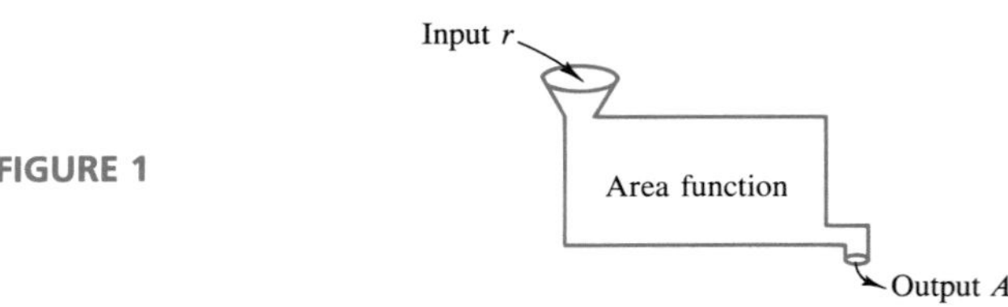

Sometimes we want to indicate that a dependent variable y is a function of an independent variable x without actually specifying a formula relating x and y. We do this by writing

$$y = f(x)$$

(read, "y equals f of x"). In this equation $f(x)$ does not mean f multiplied by x.

Definition y is a function of x,

$$y = f(x)$$

if f is a rule that associates one and only one value of y with each value of x

The symbol $f(x)$ denotes the value of y that the function f associates with x. When the input is x, the output of the function f is $f(x)$ (Figure 2).

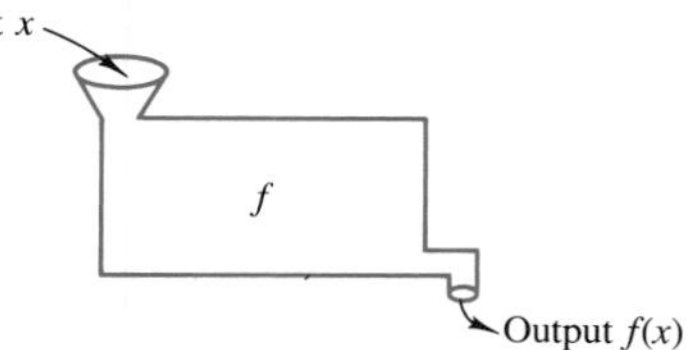

FIGURE 2

EXAMPLE 1 If the function f is given by $f(x) = 3x + 2$, then f associates the number $3x + 2$ with the number x. That is, for each input x, the output is $3x + 2$. Thus,

$$\begin{aligned} f(1) &= 3 \cdot 1 + 2 = 5 \\ f(2) &= 3 \cdot 2 + 2 = 8 \\ f(-1) &= 3 \cdot (-1) + 2 = -1 \\ f\left(\frac{1}{2}\right) &= 3 \cdot \left(\frac{1}{2}\right) + 2 = \frac{7}{2} \end{aligned}$$

If we introduce the dependent variable $y = f(x) = 3x + 2$, then it follows from these computations that

$$\begin{aligned} y &= 5 && \text{if } x = 1 \\ y &= 8 && \text{if } x = 2 \\ y &= -1 && \text{if } x = -1 \\ y &= \frac{7}{2} && \text{if } x = \frac{1}{2} \end{aligned}$$

There is nothing special about the letter f; any letter can be used to name a function. Thus, $y = f(x)$, $y = g(x)$, $y = F(x)$, and $y = \Theta_1(x)$ are all possible ways to indicate that y is a function of x.

EXAMPLE 2 The A & B Corporation finds that its profits depend on the amount of money it spends on promotion. When A & B spends x (in dollars) on promotion, then the profit (in dollars) is given by

$$P(x) = 4x^2 - 2x + 700$$

Find the profit if the amount spent on promotion is

(a) \$0 (b) \$100

Solution (a) In this case $x = 0$, so the profit is

$$P(0) = 4 \cdot 0^2 - 2 \cdot 0 + 700 = 700 \qquad \text{(dollars)}$$

Solution (b) Here $x = 100$, so the profit is

$$P(100) = 4 \cdot 100^2 - 2 \cdot 100 + 700 = 40{,}500 \qquad \text{(dollars)}$$

EXAMPLE 3 Let the function g be defined by

$$g(x) = \frac{1}{x^2 - 4}$$

Compute $g(3)$, $g(0)$, and $g(2)$.

Solution Since $g(3)$ is the value of $g(x)$ when $x = 3$, we substitute $x = 3$ into the formula for g. This yields

$$g(3) = \frac{1}{3^2 - 4} = \frac{1}{5}$$

To obtain $g(0)$ we substitute $x = 0$ into the formula for g, which yields

$$g(0) = \frac{1}{0^2 - 4} = -\frac{1}{4}$$

However, $g(2)$ is not defined since the substitution of $x = 2$ leads to a division by zero, which is not allowed:

$$g(2) = \frac{1}{2^2 - 4}$$

is not defined. Thus, 2 is not a permissible value of x for this function.

The Domain of a Function

The set of all permissible values of the independent variable of a function is called the **domain** of that function. In Example 3 the numbers 3 and 0 belong to the domain of g, but the number 2 does not.

EXAMPLE 4 The function g defined by

$$g(x) = \frac{1}{x^2 - 4}$$

has a real value for every x, except $x = 2$ and $x = -2$ since these values (and no others) result in division by zero. Thus, the domain of g consists of all real numbers except 2 and -2.

It will be understood in this book that the output value of a function must always be a real number. Thus, for the function

$$h(x) = \sqrt{x}$$

the domain of h includes only nonnegative numbers (because the square root of a negative number is not a real number).

EXAMPLE 5 The function f defined by

$$f(x) = \sqrt{x - 5}$$

does not yield a real value for $x < 5$, since $x - 5$ is negative when $x < 5$, and negative numbers do not have real square roots. The domain of f consists of all $x \geq 5$.

In applied problems sometimes practical considerations impose restrictions on the domain of a function. In such cases the nature of the application determines which input values are permissible. Thus, in Example 2 the domain of $P(x)$ consists of nonnegative real numbers because x, which represents the amount spent on promotion, cannot be negative. Similarly, for the area function $A = \pi r^2$, the domain must consist of all $r \geq 0$ since it is impossible to have a circle of negative radius.

EXAMPLE 6 If the profit $P(x)$ resulting from the sale of x units of a product is

$$P(x) = 2x + 1$$

then the domain of this function must be restricted so that x is nonnegative—because it is impossible to sell a negative number of items. To denote this restriction, we write

$$P(x) = 2x + 1 \qquad x \geq 0$$

or

$$P(x) = 2x + 1 \qquad \text{if } x \geq 0$$

Moreover, if the product can only be sold in integer units (e.g., chairs or radios), then x must be a nonnegative integer, in which case we write

$$P(x) = 2x + 1 \qquad x = 0, 1, 2, 3, \ldots$$

We can "picture" a function geometrically by means of a graph, which we define as follows.

The Graph of a Function

The Graph of a Function The graph of a function f is defined to be the graph of the equation $y = f(x)$.

EXAMPLE 7 Sketch the graphs of the functions

(a) $f(x) = x + 2$ (b) $g(x) = x^2 + 1$ (c) $h(x) = 2$

Solution (a) By definition, the graph of f is the graph of the equation $y = x + 2$. As discussed in Chapter 1, this is the equation of a straight line, so the graph can be obtained by plotting any two points on the graph and drawing a line through them. Setting $x = 0$ yields $y = 2$; therefore, $(0, 2)$ is on the graph. Setting $y = 0$ yields $x = -2$; therefore, $(-2, 0)$ is on the graph. Thus, the graph of f is the line in Figure 3a.

Solution (b) By definition, the graph of the function g is the graph of the equation $y = x^2 + 1$. As shown in Example 3 of Section 1.2, the graph is obtained by plotting points and is shown in Figure 3b.

FIGURE 3

$y = x + 2$ (a)

$y = x^2 + 1$ (b)

$y = 2$ (c)

Solution (c) By definition, the graph of h is the graph of the equation $y = 2$. As discussed in Chapter 1, this graph is the horizontal line shown in Figure 3c.

EXAMPLE 8 Sometimes the formula for a function is given in parts, each part applying to a different set of x values. For example, suppose we write

$$f(x) = \begin{cases} 2x + 1 & \text{if } x \le 1 \\ 3x & \text{if } x > 1 \end{cases}$$

Then, to calculate $f(0)$ we would use the portion of the formula that applies when $x \le 1$. And to calculate $f(5)$ we would use the portion that applies when $x > 1$. We obtain

$$f(0) = 2 \cdot 0 + 1 = 1$$
$$f(5) = 3 \cdot 5 = 15$$

To graph this function, we graph the equation $y = 2x + 1$ for $x \le 1$ and $y = 3x$ for $x > 1$. The combined graphs of these equations form the graph of $y = f(x)$ (Figure 4).

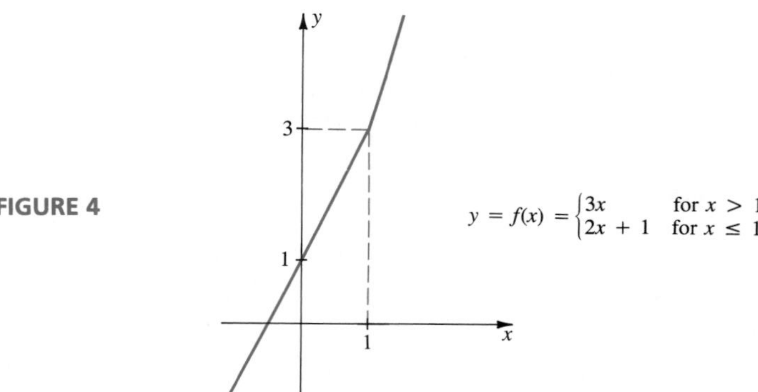

FIGURE 4

A function that is defined in "pieces," as in Example 8, is said to be defined **piecewise**. We now consider one of the most important functions of this type.

Absolute Value

The absolute value of a number is its distance from the origin when the number is plotted on a coordinate line. The absolute value of a number x is written $|x|$.

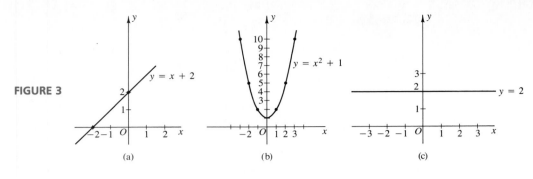

EXAMPLE 9 As shown in Figure 5, $|3| = 3$, $|-2| = 2$, and $|0| = 0$. In general,

$$|x| = \begin{cases} x & \text{if } x \geq 0 \\ -x & \text{if } x < 0 \end{cases}$$

FIGURE 5

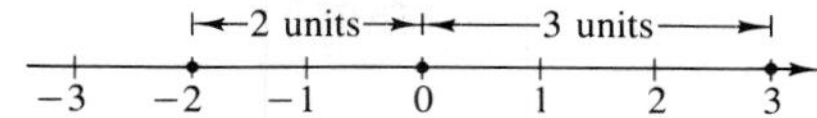

That is, the absolute value of x is equal to x if $x \geq 0$, and it is equal to $-x$ if $x < 0$. Note that this function is defined piecewise.

EXAMPLE 10 Sketch the graph of the function $f(x) = |x|$.

Solution By definition, the graph of f is the graph of the equation $y = |x|$, or equivalently,

$$y = \begin{cases} x & \text{if } x \geq 0 \\ -x & \text{if } x < 0 \end{cases}$$

Since this function is defined piecewise, we graph each piece separately. If $x \geq 0$, then the equation is $y = x$, which represents a line of slope 1 and y-intercept 0 (this is shown in the right side of Figure 6). If $x < 0$, then the equation is $y = -x$, which represents a line of slope -1 and y-intercept 0 (this is shown in the left side of Figure 6).

FIGURE 6

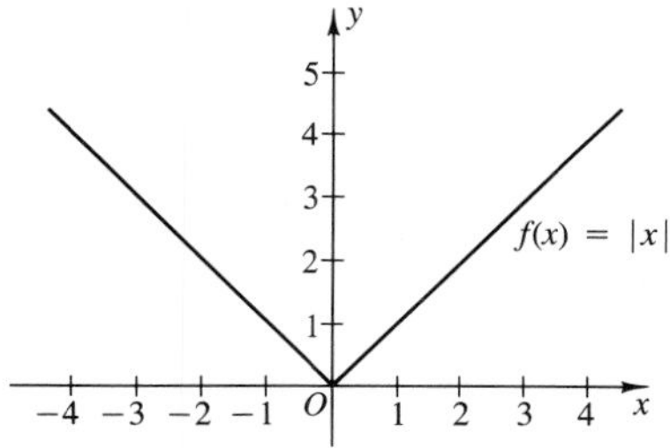

By convention, $\sqrt{a}$ denotes the nonnegative square root of a. For example, 16 has two square roots, $+4$ and -4. However, the symbol $\sqrt{16}$ denotes the nonnegative root, $+4$; that is, $\sqrt{16} = +4 = 4$. Therefore,

$$\sqrt{a^2} = |a|$$

For example, $\sqrt{(-4)^2} = \sqrt{16} = 4$.

The Range of a Function

For some purposes, it is important to know the set of all values of $f(x)$ as x varies over the domain of the function f. This set is called the **range** of f. We can picture a function as a rule f that assigns to each element x in one set (the domain) one and only one element $f(x)$ in a second set (the range) (Figure 7).

FIGURE 7

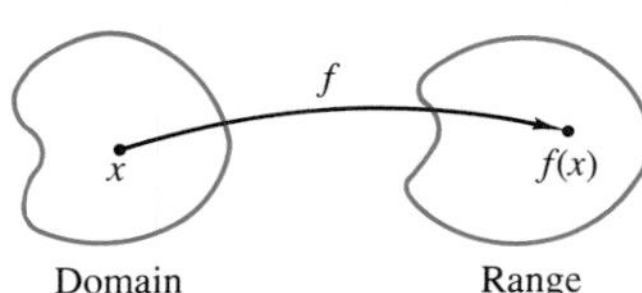

TAXES AND PIECEWISE FUNCTIONS

The following schedule of taxes for single taxpayers claiming only the taxpayer as one dependent is taken from the 1988 tax forms. Note that there are five rules for computing the tax due.

Schedule X
Single Taxpayers
Use this Schedule if you checked **Filing Status Box 1** on Form 1040—

If the Amount on Form 1040 Line 37 Is Over—	But Not Over—	Enter on Form 1040, Line 38	Of the Amount Over—
$ 0	$ 17,850	15%	$ 0
17,850	43,150	2,667.50 + 28%	17,850
43,150	89,560	9,761.50 + 33%	43,150
89,560	100,480	25,076.80 + 33%	89,560
100,480		28,680.40 + 28%	100,480

If we let $T(x)$ be the tax due on an income of x dollars, then from the above table, we see that $T(x)$ is described by the following function which is defined by the piecewise function with graph shown in the figure.

$$T(x) = \begin{cases} .15x & \text{if } 0 \le x \le 17{,}850 \\ .28(x - 17{,}850) + 2667.50 & \text{if } 17{,}850 < x \le 43{,}150 \\ .33(x - 43{,}150) + 9761.50 & \text{if } 43{,}150 < x \le 89{,}560 \\ .33(x - 89{,}560) + 25{,}076.80 & \text{if } 89{,}560 < x \le 100{,}480 \\ .28(x - 100{,}480) + 28{,}680.40 & x > 100{,}480 \end{cases}$$

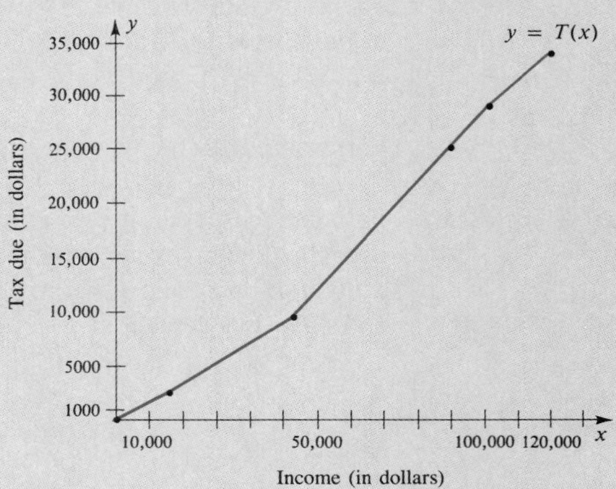

There is an important relationship between the graph of a function f and its domain and range. If we let $y = f(x)$, then the domain of f, which is the set of allowable x-values, is obtained by projecting the graph onto the x-axis; and the range of f, which is the set of y-values that correspond to x-values in the domain, is obtained by projecting the graph onto the y-axis (see Figure 8).

FIGURE 8

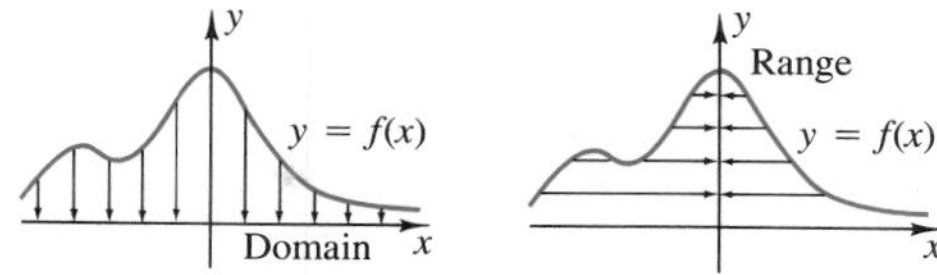

EXAMPLE 11 Find the range of

(a) $f(x) = x + 2$ (b) $g(x) = x^2 + 1$ (c) $h(x) = 2$

Solution By projecting each graph in Figure 3 onto the y-axis, we see that the

range of f = {all real numbers}
range of g = $\{y|y \geq 1\}$
range of h = {2}

Vertical Line Test Since every function corresponds to a curve (its graph) in the xy-plane, it is natural to ask whether every curve in the xy-plane is the graph of some function. The answer is no. For example, suppose the curve in Figure 9 is the graph of $y = f(x)$ for some function f. The vertical line $x = a$, shown in the figure, cuts the curve at two points, for example, (a, b) and (a, c). If (a, b) and (a, c) lie on the curve $y = f(x)$, then the coordinates of these points satisfy this equation. Thus,

$$b = f(a) \quad \text{and} \quad c = f(a)$$

indicates that input a produces *two* outputs, b and c. But this is impossible because a function cannot associate two different outputs with the same input. Thus, the curve in Figure 9 is not the graph of any function of x.

FIGURE 9

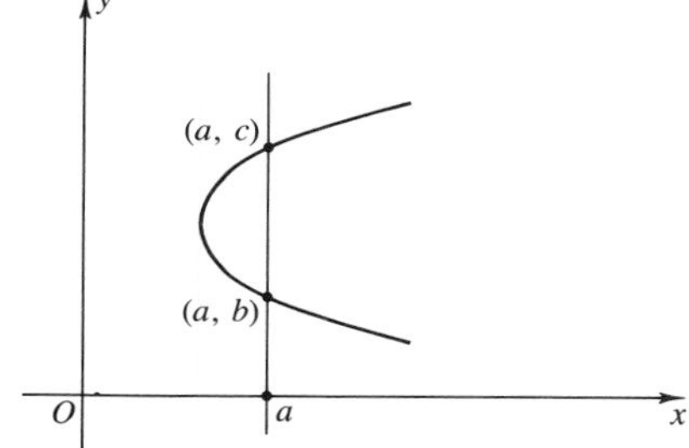

This example illustrates the following test.

The Vertical Line Test If some vertical line cuts a curve in more than one point, then the curve is not the graph of any function of x.

EXERCISE SET 11.1

1. Find the value of
(a) $|-3|$ (b) $|7|$ (c) $|2 - 8|$
(d) $\sqrt{9}$ (e) $\sqrt{(-8)^2}$

2. Find all values of x that satisfy the condition
(a) $|x| = 8$ (b) $|x - 2| = 3$
(c) $|x| = -3$ (d) $\sqrt{x^2} = 4$

3. Consider the function f defined by $f(x) = x^2 - 4$. Find:
(a) $f(3)$ (b) $f(2)$ (c) $f(-4)$
(d) $f(0)$ (e) $f(a)$ (f) $f(x - 2)$
(g) $f(x + h)$

4. Consider the function f defined by $f(x) = 3x^2 + 2x - 5$. Find:
(a) $f(2)$ (b) $f(-3)$ (c) $f(5)$
(d) $f(0)$ (e) $f(-5)$ (f) $f(a - 3)$
(g) $f(2 + h)$

5. Consider the function g defined by

$$g(x) = \frac{3}{x - 2}$$

Find:
(a) $g(5)$ (b) $g(-3)$ (c) $g(0)$
(d) $g(b)$ (e) $g(b + 7)$ (f) $g(b + h)$
(g) $g(-x)$ (h) $g(2)$

6. Consider the function F defined by

$$F(x) = \frac{3x + 2}{2x^2 - 5x + 4}$$

Find:
(a) $F(4)$ (b) $F(1)$ (c) $F(-2)$
(d) $F(r)$ (e) $F(r - 1)$ (f) $F(r + h)$

7. Let

$$f(x) = \begin{cases} x^2 + 1 & \text{if } x \geq 0 \\ -x^2 & \text{if } x < 0 \end{cases}$$

Find:
(a) $f(2)$ (b) $f(-2)$ (c) $f(0)$ (d) $f(x^2)$
(e) $f(|x|)$

8. Let

$$g(x) = \begin{cases} 3x + 7 & \text{if } x > 4 \\ -x + 1 & \text{if } x \leq 4 \end{cases}$$

Find:
(a) $g(0)$ (b) $g(3)$ (c) $g(5)$
(d) $g(6)$ (e) $g(4)$ (f) $g(-4)$

In Exercises 9–14 specify the domain of the given function.

9. $f(x) = x + 2$

10. $f(x) = \sqrt{x - 1}$

11. $f(x) = \dfrac{1}{x - 3}$

12. $f(x) = \dfrac{1}{(x - 2)(x - 1)}$

13. $f(x) = \sqrt{2 - x}$

14. $f(x) = \dfrac{x - 2}{(x - 1)^2}$

In Exercises 15–24 sketch the graph of the given function and determine the range.

15. $f(x) = 2x - 3$

16. $f(x) = x^2 - 1$

17. $f(x) = 4 - x^2$

18. $f(x) = |x - 1|$

19. $f(x) = \dfrac{1}{x}$

20. $f(x) = \sqrt{x}$

21. $f(x) = x + 1$ if $x \geq 0$

22. $f(x) = x^2 - 4,\ x \leq 4$

23. $f(x) = \begin{cases} 1 & \text{if } x \geq 0 \\ -1 & \text{if } x < 0 \end{cases}$

24. $f(x) = \begin{cases} x & \text{if } x \leq 1 \\ -x + 2 & \text{if } x > 1 \end{cases}$

In Exercises 25–30 determine which of the indicated curves are graphs of functions.

25.

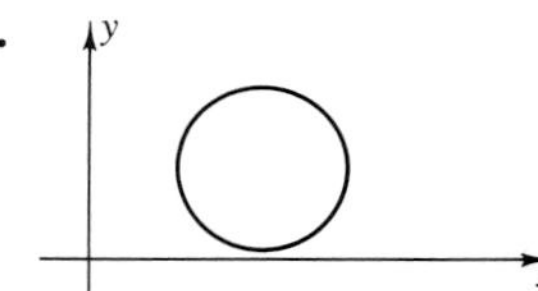

26.

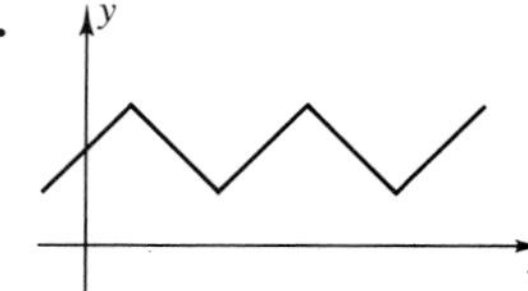

27.

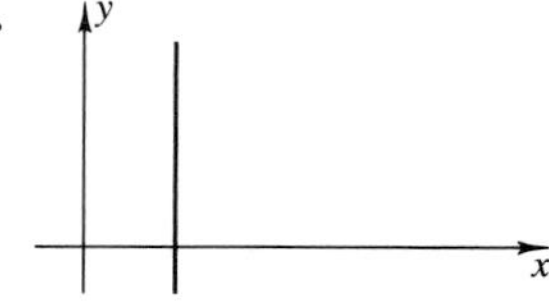

28. 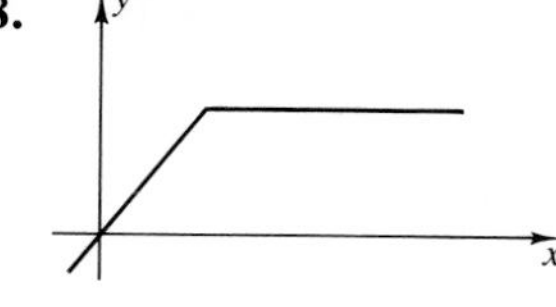

29. 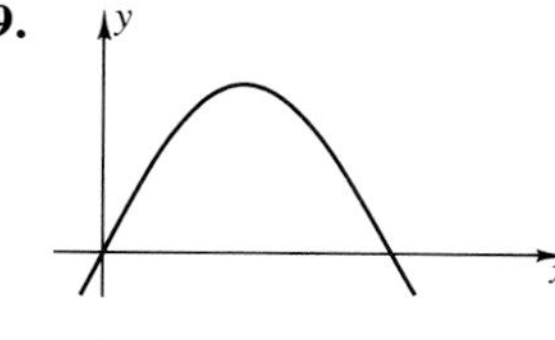

30. 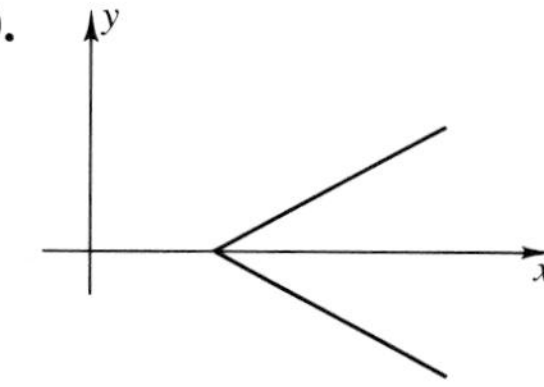

31. **(Business)** A Chicago mail-order health food store, which sells its snack pack only in 1-pound units, finds that the cost C of shipping an order to the east coast is 50 cents for the first pound and 30 cents for each additional pound.
(a) Write C as a function of the weight w (in pounds) for $0 < w \leq 10$.
(b) Sketch the graph of the function found in (a).
(c) What is the cost of shipping an order weighing 9 pounds?

32. **(Business)** A car-rental firm charges a flat fee of \$20 plus 12 cents per mile.
(a) Express the cost C of renting a car as a function of the number of miles m traveled.
(b) What is the domain of this function?
(c) How much would it cost to rent a car for a 200-mile trip?

33. **(Business)** A video club offers the following introductory plan: If you buy 4 videocassettes at the regular price of \$30 each, you may buy up to 6 more videocassettes at half price.
(a) Express the total cost C of buying videocassettes as a function of the number r of videocassettes purchased at half price.
(b) What is the domain of this function?
(c) How much will it cost to buy 10 videocassettes?

34. **(Ecology)** Suppose the amount A (in milligrams) of arsenic found in a stream near an electronics plant is given by

$$A(x) = 0.003 + 0.012x^2$$

where x is the number of units manufactured weekly by the plant. Find the amount of arsenic in the stream if the number of units manufactured by the plant is
(a) 20 (b) 60 (c) 100

11.2 POLYNOMIAL FUNCTIONS

In this section we examine the simplest and most important class of functions, the **polynomial functions**. These are functions whose formulas involve only the operations of addition, subtraction, and multiplication of variables and constants. Some examples are

$$x^2 + 2x + 1 \qquad x^5 + 2x^3 - 3 \qquad 5x \qquad \frac{1}{3}x^3 - 9 \qquad 7$$

In general, we have the following definition:

A **polynomial in *x*** is a function of the form

$$p(x) = a_n x^n + a_{n-1}x^{n-1} + \cdots + a_1 x + a_0 \tag{1}$$

where $a_0, a_1, \ldots, a_n$ are constant real numbers and n is a nonnegative integer.

The constants $a_0, a_1, a_2, \ldots, a_n$ are called the **coefficients** of the polynomial, and the expressions $a_0, a_1x, \ldots, a_{n-1}x^{n-1}, a_nx^n$ are called the **terms**. Note that a polynomial does not involve division by variables, variables raised to fractional or negative exponents, or variables inside radicals. The following are *not* polynomials in x:

$$\frac{2}{x} \qquad x^{1/3} + 1 \qquad 5x^{-2} + 3x + 2 \qquad \sqrt{x^2} + 2x - 3$$

Note also that constants are polynomials; this is the special case in which the coefficients $a_n, a_{n-1}, \ldots, a_1$ are all zero in Equation (1).

By definition, the **degree** of a polynomial is the highest power of x with a nonzero coefficient.* Polynomials are classified according to their degree. For example,

$x^2 + 2x + 1$ has degree 2
$3x^5 - 7x^2 - 8$ has degree 5
$x - 3$ has degree 1 (Recall $x^1 = x$, $x^0 = 1$)
$7 = 7x^0$ has degree 0

A polynomial of the first degree is called a **linear function**, and a polynomial of the second degree is called a **quadratic function**. For simplicity these functions are often written in the notation

$$p(x) = ax + b \qquad \text{(linear function)}$$
$$p(x) = ax^2 + bx + c \qquad \text{(quadratic function)}$$

Graphing Polynomials

The graph of a constant polynomial (polynomial of degree zero)

$$p(x) = c$$

is the graph of the equation

$$y = c$$

which is the horizontal line shown in Figure 1.

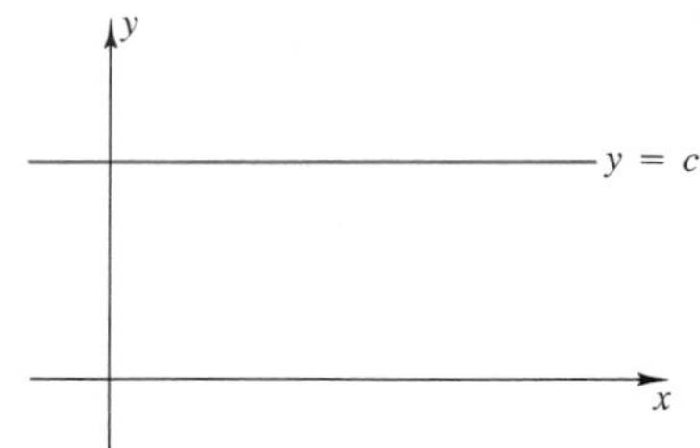

FIGURE 1

The graph of a first-degree polynomial

$$p(x) = ax + b \qquad (a \neq 0)$$

is the graph of the equation

$$y = ax + b$$

* For reasons that do not concern us in this book, it is the custom not to assign any degree to the constant zero.

which we recognize to be the slope–intercept form of a line with slope $m = a$ and y-intercept b (Figure 2). This result explains why first-degree polynomials are called **linear functions**.

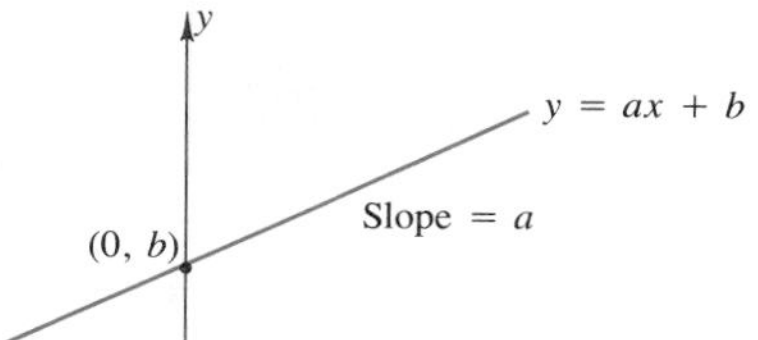

FIGURE 2

EXAMPLE 1 Sketch the graph of

(a) $p(x) = 3$ (b) $p(x) = -\frac{3}{4}x + 3$

Solution (a) The graph of $p(x) = 3$ is the graph of the equation $y = 3$, which is a horizontal line passing through the point (0, 3) (Figure 3a).

Solution (b) The graph of $p(x) = -\frac{3}{4}x + 3$, which is the graph of the equation $y = -\frac{3}{4}x + 3$, is a straight line of slope $m = -\frac{3}{4}$ and y-intercept 3 (Figure 3b).

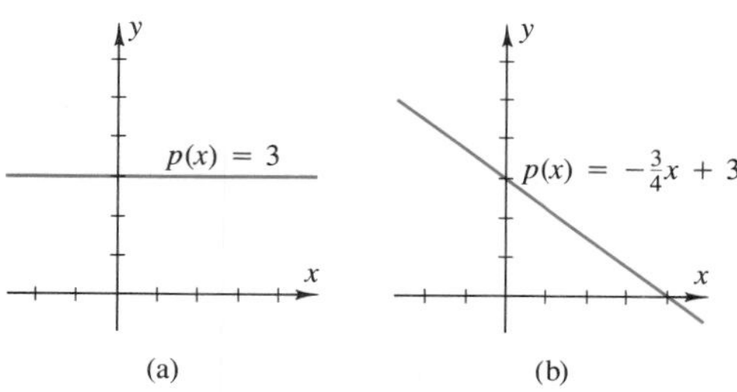

FIGURE 3

The graph of a second-degree polynomial

$$p(x) = ax^2 + bx + c \qquad (a \neq 0)$$

is the graph of the equation

$$y = ax^2 + bx + c$$

Since these graphs play an important role in a variety of applied problems, we describe some of the key facts about them.

The graph of a quadratic function is a curve called a **parabola**. If $a > 0$, then the parabola opens upward (Figure 4a), and if $a < 0$, then the parabola opens downward (Figure 4b). As indicated in Figure 4, the high or low point of the parabola is called the **vertex**. The parabola

$$y = ax^2 + bx + c$$

is symmetric about the line through the vertex parallel to the y-axis. This line is called the **axis** of the parabola (Figure 4); it cuts the parabola into two sections, each of which is the mirror image of the other with respect to the axis. It can be shown that the vertex occurs at the point

$$x = -\frac{b}{2a} \tag{2}$$

(See Supplementary Exercise 20.)

FIGURE 4

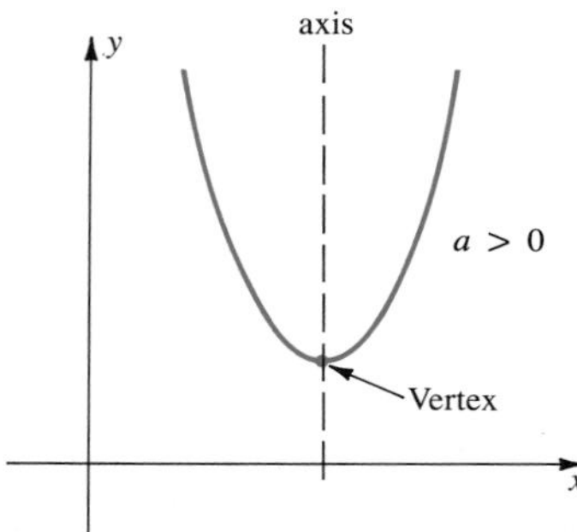

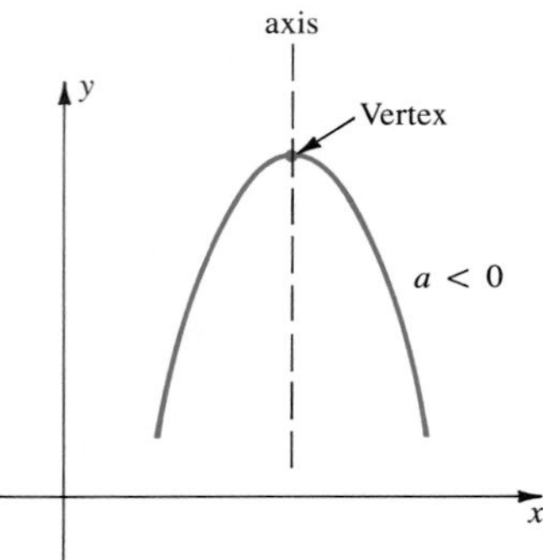

To graph $y = ax^2 + bx + c$, it is helpful to plot first the vertex and then a sufficient number of points on either side of the vertex to obtain sufficient accuracy.

EXAMPLE 2 Sketch the graph of

$$f(x) = 14x - 7x^2$$

Solution Sketch the graph of the equation $y = 14x - 7x^2$. Using Equation (2), the vertex occurs at

$$x = -\frac{b}{2a} = -\frac{14}{2(-7)} = 1$$

The y-value for $x = 1$ and some y-values for points on either side of $x = 1$ are given in Table 1. The graph is shown in Figure 5.

TABLE 1

x	$y = 14x - 7x^2$
−1	−21
0	0
1	7 (vertex)
2	0
3	−21

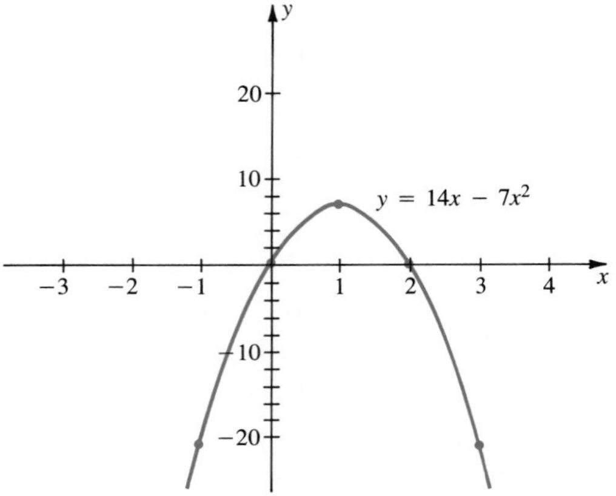

FIGURE 5

Observe that in Example 2 the parabola opens downward. This could have been predicted in advance from the fact that $a = -7$ is negative (Figure 4).

In many applications, it is important to find the x-intercepts, that is, the solutions of the equation $ax^2 + bx + c = 0$. These can be found by factoring or by using the **quadratic formula** (see Sections A.6 and A.8).

Quadratic Formula The solutions of $ax^2 + bx + c = 0$ are

$$x = \frac{-b \pm \sqrt{b^2 - 4ac}}{2a} \tag{3}$$

As illustrated in Figure 6, a quadratic function can have zero, one, or two x-intercepts, depending on the sign of the expression $b^2 - 4ac$ that appears inside the radical in Formula (3). This expression is called the **discriminant** of the polynomial.

$a > 0$

Two x - intercepts

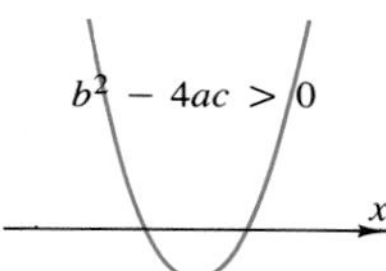

One x - intercept

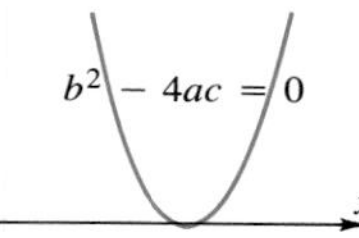

No x - intercepts

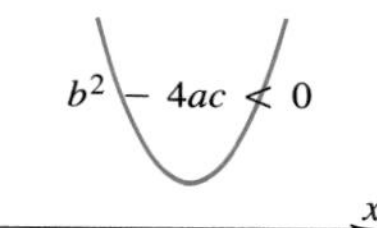

FIGURE 6

$a < 0$

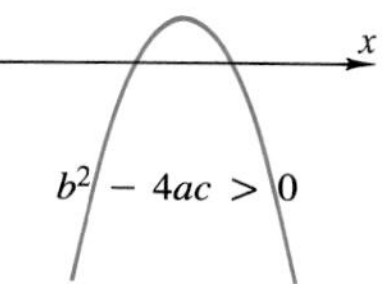

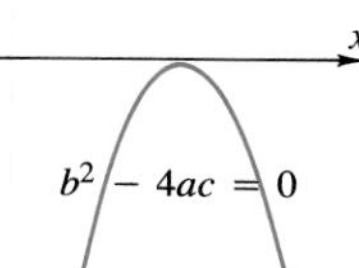

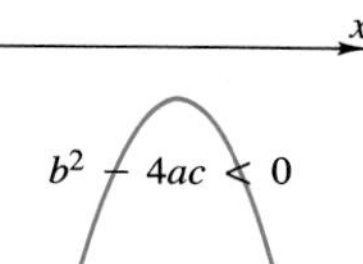

EXAMPLE 3 Use the quadratic formula to find the x-intercepts, if any, of

(a) $p(x) = 8x^2 - 6x + 1$ (b) $p(x) = 9x^2 + 12x + 4$
(c) $p(x) = x^2 - x + 1$

Solution (a) From (3) with $a = 8$, $b = -6$, and $c = 1$, we obtain

$$x = \frac{-b \pm \sqrt{b^2 - 4ac}}{2a} = \frac{6 \pm \sqrt{4}}{16} = \frac{6 \pm 2}{16}$$

Therefore,

$$x = \frac{6 + 2}{16} = \frac{1}{2}$$

and

$$x = \frac{6 - 2}{16} = \frac{1}{4}$$

the intercepts are $x = \frac{1}{2}$ and $x = \frac{1}{4}$.

Solution (b) From (3) with $a = 9$, $b = 12$, and $c = 4$, we obtain

$$x = \frac{-b \pm \sqrt{b^2 - 4ac}}{2a} = \frac{-12 \pm \sqrt{144 - 144}}{18} = \frac{-12 \pm 0}{18}$$

Thus, we only obtain one intercept, $x = -\frac{2}{3}$.

Solution (c) From (3) with $a = 1$, $b = -1$, and $c = 1$, we obtain

$$x = \frac{-b \pm \sqrt{b^2 - 4ac}}{2a} = \frac{1 \pm \sqrt{1 - 4}}{2} = \frac{1 \pm \sqrt{-3}}{2}$$

Since $\sqrt{-3}$ is not a real number, there are no x-intercepts.

EXAMPLE 4 Sketch the graph of $p(x) = x^2 - 4x + 2$.

Solution Sketch the graph of the equation $y = x^2 - 4x + 2$. From (2), the vertex occurs at

$$x = -\frac{b}{2a} = -\frac{(-4)}{2(1)} = 2$$

Table 2 includes the y-value for $x = 2$ and some y-values for some integer points on either side of $x = 2$. In addition, the x-intercepts shown in the table were obtained from the quadratic formula as follows:

$$x = \frac{-b \pm \sqrt{b^2 - 4ac}}{2a} = \frac{4 \pm \sqrt{16 - 8}}{2} = \frac{4 \pm \sqrt{8}}{2} = \frac{4 \pm 2\sqrt{2}}{2} = 2 \pm \sqrt{2}$$

Using a hand calculator, it follows that the x-intercepts are approximately

$$x = 2 + \sqrt{2} \cong 3.4 \qquad x = 2 - \sqrt{2} \cong 0.6$$

The graph is shown in Figure 7.

TABLE 2

x	$y = x^2 - 4x + 2$
0	2
1	−1
2	−2
3	−1
4	2
$2 + \sqrt{2}$ ($\cong 3.4$)	0
$2 - \sqrt{2}$ ($\cong 0.6$)	0

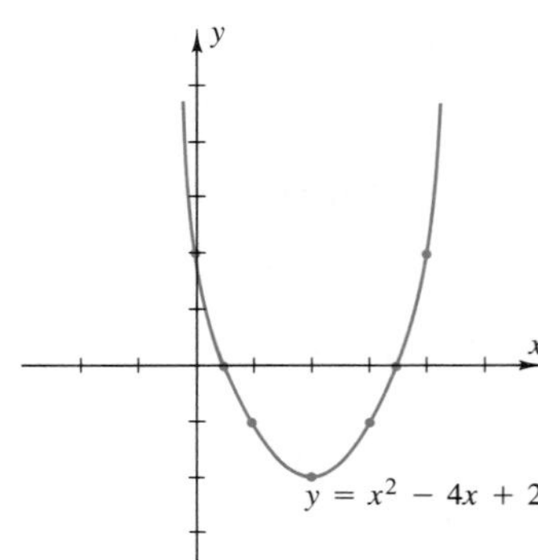

FIGURE 7

Higher-Degree Polynomials

In general, the best methods for graphing higher-degree polynomials are based on calculus. However, reasonably accurate graphs can sometimes be obtained by plotting a sufficient number of points to clarify the shape of the graph.

EXAMPLE 5 Sketch the graph of the function

$$f(x) = x^3 - 3x^2 + 3x - 1$$

Solution From Table 3 we obtain the graph in Figure 8.

TABLE 3

x	$y = x^3 - 3x^2 + 3x - 1$
-1	-8
0	-1
1	0
2	1
3	8

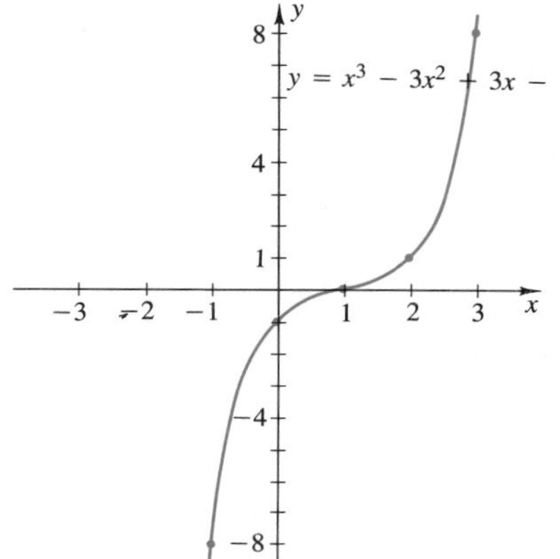

FIGURE 8

EXAMPLE 6 Sketch the graph of the function

$$f(x) = x^3 - x^2 - 2x$$

Solution From Table 4 we obtain the graph in Figure 9. Note that the point corresponding to $x = -\frac{1}{2}$ is needed to clarify the shape of the graph between $x = 0$ and $x = -1$.

TABLE 4

x	$y = x^3 - x^2 - 2x$
-2	-8
-1	0
$-\frac{1}{2}$	$\frac{5}{8} \cong 0.6$
0	0
1	-2
2	0
3	12

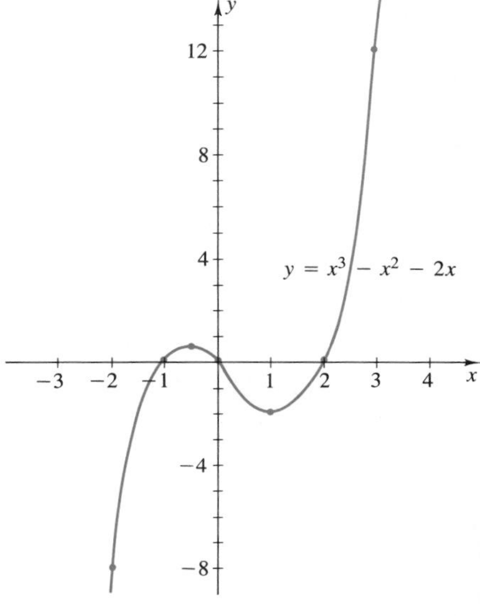

FIGURE 9

Applications

We now consider some examples of how polynomial functions arise in applications.

EXAMPLE 7 A 3- by 7-foot doorway is to be framed on three sides by molding x feet wide (Figure 10). Find the area of the molding as a function of x.

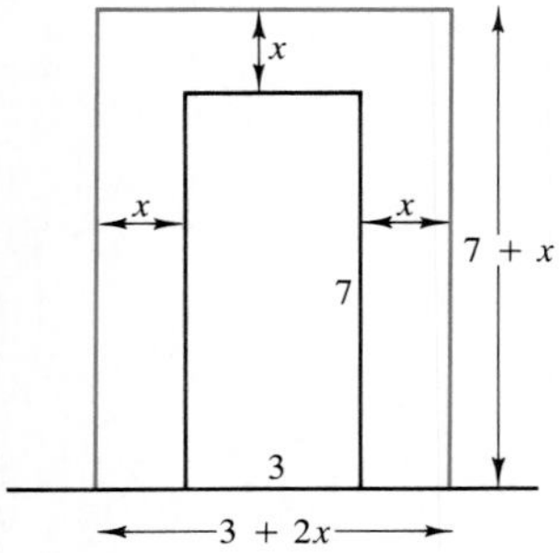

FIGURE 10

Solution The area of the doorway is $A_1 = 3 \cdot 7 = 21$ square feet. The outside dimensions of the framed door are $7 + x$ by $3 + 2x$. Thus, the area of the doorway plus molding is

$$\begin{aligned} A_2 &= (7 + x)(3 + 2x) \\ &= 21 + 17x + 2x^2 \end{aligned}$$

Therefore, the area of the molding is

$$\begin{aligned} A &= A_2 - A_1 \\ &= (21 + 17x + 2x^2) - 21 \end{aligned}$$

or, after simplifying, $A = 2x^2 + 17x$. Thus, the area of the molding is a quadratic function of x.

EXAMPLE 8 A rectangular box open at the top is to be made from a 10- by 12-inch piece of tin by cutting a small x- by x-inch square from each corner and turning up the sides (Figure 11). Find the volume of the constructed box as a function of x.

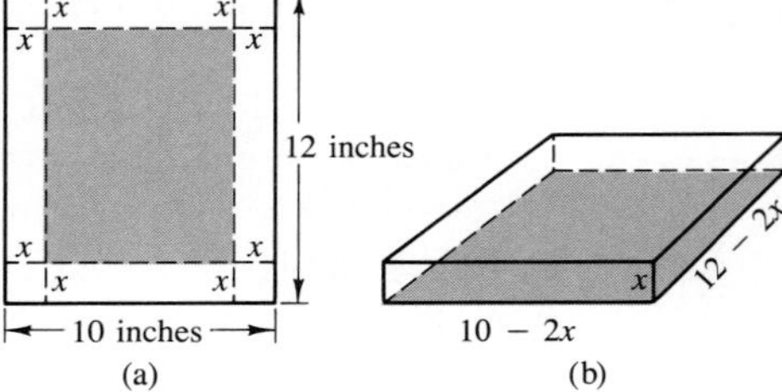

FIGURE 11

Solution The volume of a rectangular box is the product of its length, width, and height. As shown in Figure 11b, length $l = 12 - 2x$, width $w = 10 - 2x$, and height $h = x$. Therefore, the volume V is

$$\begin{aligned} V &= (12 - 2x)(10 - 2x)x \\ &= (120 - 44x + 4x^2)x \\ &= 4x^3 - 44x^2 + 120x \end{aligned}$$

which is a third-degree polynomial in x.

EXAMPLE 9 **PROFIT VERSUS REVENUE** Suppose that a candy store obtains certain candy bars at a wholesale cost of 20 cents each. If the store sells the candy bar at x cents each, then on the average it will sell $70 - x$ bars daily, where $0 \le x \le 70$. Let

$C(x)$ = cost (in cents) of buying the day's stock of candy bars
$R(x)$ = revenue (in cents) received from selling the day's stock of candy bars
$P(x)$ = profit (in cents) from selling the day's stock of candy bars

(a) Find formulas for $C(x)$, $R(x)$, and $P(x)$.

(b) Find the daily profit when $x = 40$ cents.

(c) Find the most profitable selling price.

Solution (a) Since the store will need $70 - x$ candy bars at 20 cents each,

$$C(x) = 20(70 - x) \qquad \text{(cents)}$$

Since the store will sell $70 - x$ candy bars at x cents each,

$$R(x) = x(70 - x) \qquad \text{(cents)}$$

Therefore, since profit is revenue minus cost,

$$\begin{aligned} P(x) &= x(70 - x) - 20(70 - x) \\ &= -x^2 + 90x - 1400 \qquad \text{(cents)} \end{aligned}$$

Solution (b) $P(40) = -(40)^2 + 90 \cdot 40 - 1400 = 600 \text{ cents} = \$6.00$

Solution (c) The profit function $P(x)$ is a quadratic function whose graph is a parabola opening downward (since $a = -1$ is negative). The vertex is at the point where

$$x = -\frac{b}{2a} = -\frac{90}{2(-1)} = 45$$

This is the x-coordinate of the highest point on the parabola (Figure 12). Thus, when $x = 45$, $P(x)$ is greatest. Therefore, 45 cents is the most profitable selling price for the candy bars under the conditions prescribed by the problem. In fact, when $x = 45$, the daily profit on the bars is

$$P(45) = -(45)^2 + 90 \cdot 45 - 1400 = 625 \text{ cents} = \$6.25$$

Any price higher (or lower) than 45 cents per candy bar will yield a lower daily profit.

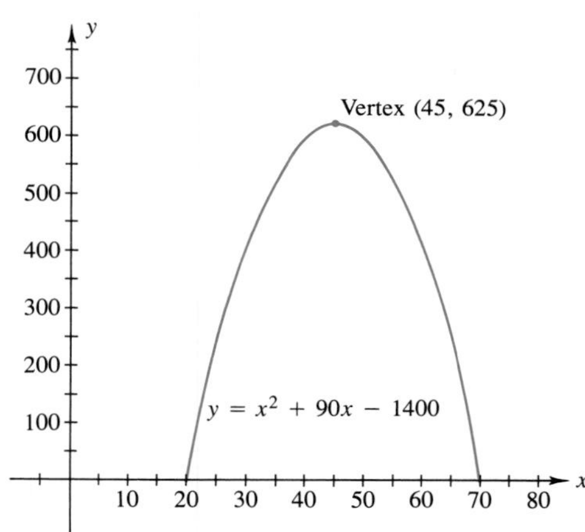

FIGURE 12

EXERCISE SET 11.2

In Exercises 1–8 sketch the graph of the given quadratic function. In each case find the coordinates of the vertex.

1. $f(x) = x^2 - 2x$ **2.** $f(x) = 3x - x^2$

3. $f(x) = x^2 + 4x + 3$ **4.** $f(x) = x^2 + x - 2$

5. $f(x) = -x^2 + x + 6$ **6.** $f(x) = 3 + 2x - x^2$

7. $f(x) = 2x^2 - 4x + 5$ **8.** $f(x) = -2x^2 + 5x + 3$

In Exercises 9–12 find the x-intercepts, if any, of the given quadratic function $p(x)$.

9. $p(x) = 2x^2 + 3x - 2$ **10.** $p(x) = x^2 + 2x + 3$

11. $p(x) = 4x^2 - 12x + 9$ **12.** $p(x) = x^2 + 3x$

In Exercises 13–18 use the discriminant to determine the number of x-intercepts in the associated graph.

13. $f(x) = 2x^2 - 12x + 18$ **14.** $f(x) = x^2 - 10$

15. $f(x) = 8 + x^2$ **16.** $f(x) = x^2 + 4x + 4$

17. $f(x) = x^2 + x - 1$ **18.** $f(x) = x^2 + x + 1$

In Exercises 19–24 graph the indicated function.

19. $f(x) = (x + 2)^3$ **20.** $f(x) = (x - 3)^3$

21. $f(x) = x^3 - 2x^2 - 3x$ **22.** $f(x) = x^3 + 2x^2$

23. $f(x) = x^3 - 2x^2 + x$ **24.** $f(x) = x^3 - x^2 - 2x$

25. Profits for the Jones Corporation depend on the amount of money it spends on advertising. If x dollars is spent on advertising, then the profit (in dollars) is given by

$$P(x) = 2x^2 - x + 1000$$

Determine the profit if the amount spent on advertising is (a) \$0 (b) \$10 (c) \$50 (d) \$1000

26. In preparing a mat for use as a border in framing a picture, the center of an 8- by 10-inch rectangular mat is removed, leaving a border x inches wide (Figure 13). Find the area of the piece that is removed as a function of x.

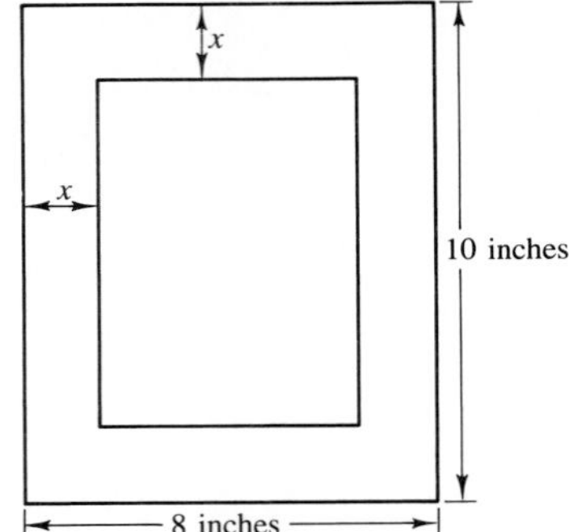

FIGURE 13

27. A homeowner must purchase paint to cover the walls and ceiling of a rectangular room. The walls are 8 feet high, and the floor is a rectangle whose width W is 1 foot shorter than twice its length L. Express the combined area of the walls and ceiling as a function of W.

28. A rectangular box has a base whose length is 5 inches longer than its width. The height of the box is twice its width. Express the volume of the box as a function of its width.

29. To meet shipping specifications, a manufacturer must use small rectangular cartons with square cross sections such that the length plus girth (distance around a cross section) is 60 inches. Find a formula for the volume of such a carton as a function of x (Figure 14).

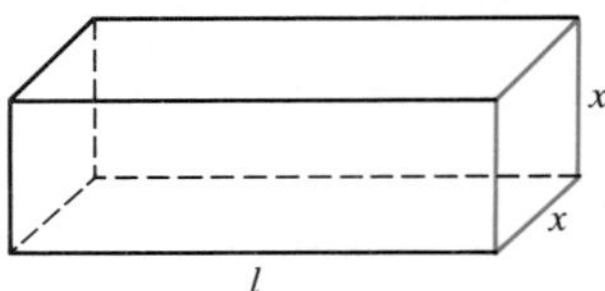

FIGURE 14

30. Jack's restaurant makes hamburgers at a cost of 50 cents each. It is estimated that if the restaurant sells the burgers at x cents each, then the restaurant will sell $600 - 5x$ hamburgers daily.

(a) Express Jack's daily profit from hamburgers as a function of x.

(b) Find the daily profit when $x = 70$ cents.

(c) Find the most profitable selling price x.

31. Organizers of a benefit concert determine that if they sell all 1250 seats for the concert at \$7 each, they will break even (i.e., there will be neither a profit nor a loss). They estimate that for each \$1 increase in the price of a ticket, 50 fewer people will attend the performance. Let x = number of dollars above \$7 charged for each ticket. Find:

(a) the total cost of the concert
(b) $R(x)$ = anticipated revenue as a function of x
(c) $P(x)$ = anticipated profit as a function of x
(d) the most profitable price per ticket. What is the anticipated attendance at that price? The profit?

32. A camera store normally sells 120 of its special cameras per month at a profit of \$16 per camera. The owner estimates that for each \$1 reduction in the price of a camera, the store will sell 10 more of the special cameras per month. Suppose the store lowers the price of each camera by x dollars. Find:

(a) $P(x)$, the monthly profit, as a function of x
(b) $P(0)$, the normal monthly profit
(c) the monthly profit if the store lowers the profit to \$13 per camera
(d) the most profitable value of x, the total monthly profit at this level, and the profit per camera

KEY IDEAS FOR REVIEW

- **Function** A rule assigning to each value of x in the domain one and only one value in the range.
- **Domain of a function** Those values of the independent variable where the function is defined and yields a real value.
- **Graph of a function** The graph of the equation $y = f(x)$.
- **Range of a function** The set of all values of $f(x)$ as x varies over the domain of the function f.
- **Vertical line test** If a vertical line cuts a curve in more than one point, then the curve is not the graph of any function of x.
- **Polynomial in x** A function of the form

$$p(x) = a_n x^n + a_{n-1}x^{n-1} + \cdots + a_1 x + a_0$$

where $a_0, a_1, \ldots, a_n$ are constants and n is a nonnegative integer.

- **Linear function** $p(x) = ax + b$
- **Quadratic function** $p(x) = ax^2 + bx + c$
- **Quadratic formula** The solutions of the equation $ax^2 + bx + c = 0$ are

$$x = \frac{-b + \sqrt{b^2 - 4ac}}{2a} \quad \text{and} \quad x = \frac{-b - \sqrt{b^2 - 4ac}}{2a}$$

SUPPLEMENTARY EXERCISES

1. On the real number line sketch all values of x satisfying the given equation.
(a) $|x| = 4$ (b) $|2x + 1| = 2$
(c) $\sqrt{(x - 2)^2} = 3$

2. Consider the function f defined by

$$f(x) = \frac{x - 1}{x^2 + 2}$$

Find:
(a) $f(3)$ (b) $f(-2)$ (c) $f(-x)$
(d) $f(x^2)$ (e) $f\left(\frac{1}{x}\right)$

3. Let

$$f(x) = \begin{cases} 2x + 1 & \text{if } x \leq 2 \\ x^2 + 1 & \text{if } x > 2 \end{cases}$$

Find:
(a) $f(-1)$ (b) $f(3)$ (c) $f(2)$ (d) $f(2x)$

In Exercises 4–7 find the domain and range of the given function and sketch the graph.

4. $f(x) = 1 - x^2$

5. $f(x) = \sqrt{x - 1}$

6. $f(x) = \begin{cases} x^2 & \text{if } x \leq 2 \\ 2x & \text{if } x > 2 \end{cases}$

7. $f(x) = |3 - 2x|$

In Exercises 8 and 9 determine whether the given curve is the graph of a function.

8.

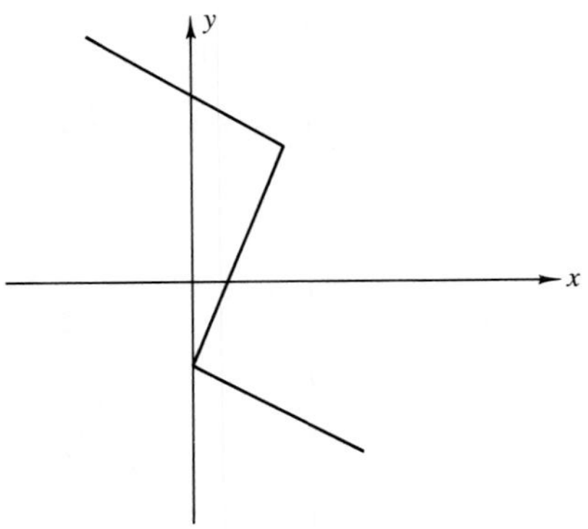

9.

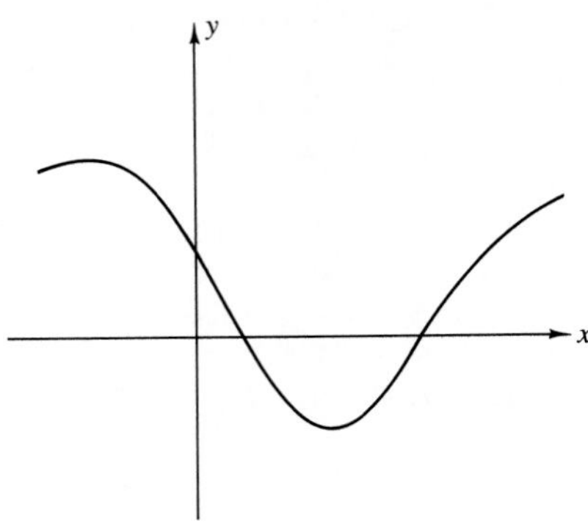

In Exercises 10 and 11 find the range of the function whose graph is pictured.

10.

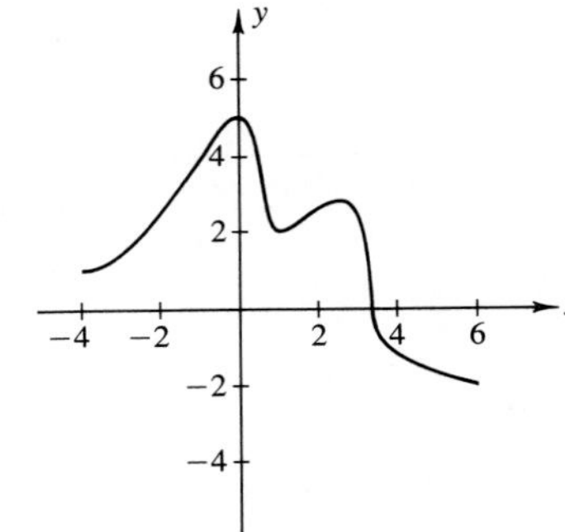

11.

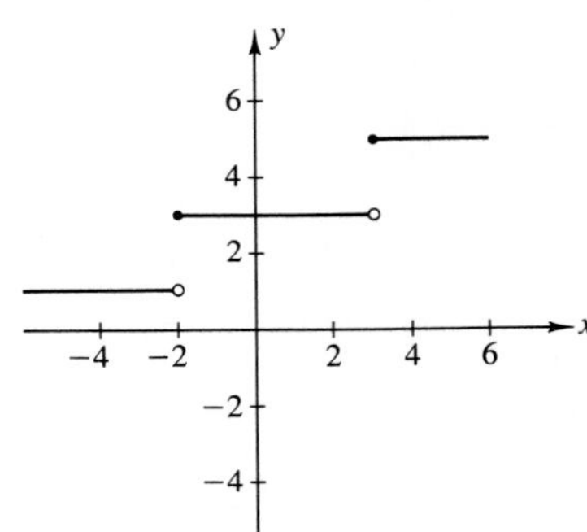

In Exercises 12 and 13 find the intercepts, if any, of the given quadratic function.

12. $p(x) = 3x^2 - 2x + 1$

13. $p(x) = x^2 + 5x$

In Exercises 14 and 15 sketch the graph of the given parabola and find the coordinates of the vertex.

14. $p(x) = 2x^2 + x - 11$

15. $p(x) = -3x^2 + 2x$

In Exercises 16 and 17 find the values of a, b, and c so that the graph of the equation satisfies the indicated conditions.

16. The graph of $p(x) = ax^2 + bx + 1$ passes through the points (1, 3) and (2, 9).

17. The graph of $p(x) = 2x^2 + bx + c$ passes through the points (0, 0) and (1, 6).

18. A telephone answering service charges a fee of \$10 per month for the first 100 message units and an additional fee of 8 cents for each of the next 100 message units. A reduced rate of 5 cents is charged for each message unit after the first 200 units.
(a) Express the monthly charge C as a function of the number of message units u.
(b) What is the monthly charge for 280 message units?
(c) Sketch the graph of $C(u)$.

19. A manufacturer of 35-millimeter film finds that the daily demand for its film in Boston is given by the equation $q = 200 - 10p$, where q is the number of units and p is the price (in dollars) per unit.
(a) Find the revenue received by the manufacturer from the sale of q units of film at p dollars per unit.
(b) Find the revenue when $p = \$8$.
(c) Find the revenue when $p = \$13$.
(d) What value of p produces the largest revenue?

20. (a) For the parabola $y = ax^2 + bx + c$ graphed in Figure 1, use the quadratic formula to find values for x_1 and x_2 in terms of a, b, and c.
(b) The x-coordinate of the vertex is the point midway between x_1 and x_2. Show that the x-coordinate of this point is $x = -b/2a$. (See Supplementary Exercise 22 of Chapter 1.)

FIGURE 1

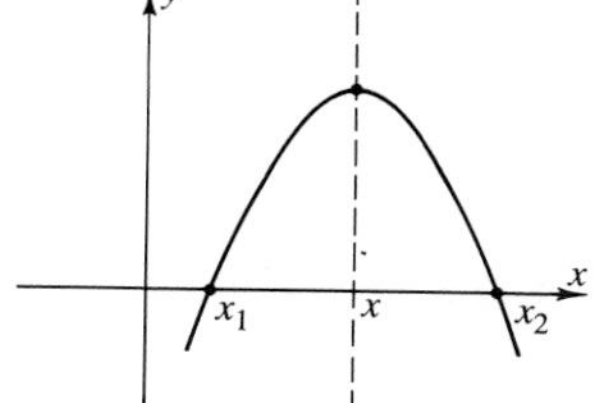

CHAPTER TEST

1. Let

$$f(x) = \begin{cases} 1 - x^2 & \text{if } x \leq 1 \\ -2x - 1 & \text{if } x > 1 \end{cases}$$

(a) Find $f(-2)$ and $f(3)$.
(b) Find the domain and range of f.
(c) Sketch the graph of f.

2. Find the intercepts, if any, of the given function.
(a) $p(x) = 2x^2 + 3x - 2$.
(b) $p(x) = 2x^2 + 3x + 1$.

3. Sketch the graph of $f(x) = 2 - |x|$. Specify the domain and range.

4. Sketch the graph of $f(x) = -x^2 + 2x - 2$.

5. Solve $x^2 - 6x - 16 = 0$.

6. Determine the number of x-intercepts of the graph of $f(x) = x^2 + x + 1$.

7. The commission earned by a door-to-door cosmetics salesperson is determined as shown in Table 1.
(a) Express the commission C as a function of sales s.
(b) Find the commission if the weekly sales are \$425.
(c) Sketch the graph of the function.

TABLE 1

Weekly Sales	Commission
Less than \$300	20% of sales
\$300 or more but less than \$400	\$60 + 35% of sales over \$300
\$400 or more	\$95 + 60% of sales over \$400

12

Limits, Rates of Change, and the Derivative

12.1 INTRODUCTION

The development of **calculus** in the seventeenth century by Isaac Newton* and Gottfried Wilhelm Leibniz† is one of the greatest achievements in the history of science and mathematics. Calculus is concerned with the mathematical study of change. Since we live in a universe of constant change, it is not surprising that calculus has been used as a fundamental tool in the physical sciences, engineering, economics, business, and the social sciences.

There are two basic geometric problems that call for the use of calculus:

1. finding a tangent to a curve
2. finding the area under a curve

Moreover, these geometric problems are related to a variety of problems in the above-mentioned areas.

---

* *Isaac Newton* (1642–1727) was born in Woolsthorpe, Lincolnshire, England. He is considered one of the greatest names in the history of human thought because of his fundamental contributions in mathematics, physics, and astronomy. In a period of 18 months (1665–1667) he laid the foundations for his later work in the development of calculus, the theory of gravity, and the theory of light and color (for example, showing that white light is made up of many colors). His work in astronomy provided an explanation of the motion of planets in their orbits around the sun. He constructed a new type of telescope, using a reflecting mirror.

† *Gottfried Wilhelm von Leibniz* (1646–1716) was a German philosopher, mathematician, physicist, and historian. He was a universal genius who made major contributions to mathematics, logic, philosophy, mechanics, geology, law and theology, all while pursuing a career as a civil servant. He was named a baron in 1700. Leibniz invented differential and integral calculus later than, but independently of, Isaac Newton, and in his later years was embroiled in a bitter dispute with friends of Newton over who invented calculus.

In this chapter we examine the close relationship between the tangent problem and the problem of determining the rate at which a variable quantity is changing in value. The portion of calculus concerned with this problem is commonly called **differential calculus**. In Chapter 15 we also discuss how the area problem is related to the problem of finding a variable quantity whose rate of change is known. The portion of calculus concerned with these ideas is commonly called **integral calculus**.

It is not our objective to make you an expert in the theoretical aspects of calculus. Rather, we aim to give you an intuitive presentation of the basic ideas and to illustrate the kinds of applied problems that can be solved by using calculus.

Appendix A, which can be consulted as needed, contains an algebra review that presents the basic algebraic skills needed in calculus.

12.2 RATE OF CHANGE OF A FUNCTION

With this topic we approach the first of two fundamental ideas of calculus: the derivative of a function. We consider this concept by studying the rate of change of a function. The prominence of calculus in the history of science and mathematics is largely due to its treatment and fruitful exploitation of this notion. Most physical quantities are in a state of constant change—the velocity of a rocket changes with time, the cost of an object changes with the available supply, the profits of a manufacturer change with sales, the size of a tumor varies with the quantity of radiation to which it is subjected, and so on. Here we focus not so much on the actual change in the quantity but rather on its *rate* of change relative to some other quantity.

Average Rate of Change

The rate at which one quantity changes relative to another can be illustrated with a familiar example, the *velocity* of a moving object. Velocity is the rate at which distance traveled changes with time. There are two very different ways to describe the velocity of a moving object: *average velocity* and *instantaneous velocity*. For example, suppose a car moves along a straight road on which we have introduced a coordinate line (Figure 1), and suppose the data in Table 1 are recorded at three checkpoints A, B, and C. To find the *average velocity* of the car between two checkpoints, we divide the distance traveled between the checkpoints by the times elapsed, that is,

$$\text{average velocity} = \frac{\text{distance traveled}}{\text{time elapsed}}$$

FIGURE 1

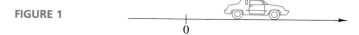

TABLE 1

Checkpoint	Distance d from the origin to the checkpoint (miles)	Time t elapsed from start of trip (hours)
A	4	$\frac{1}{4}$
B	10	$\frac{3}{4}$
C	70	$1\frac{3}{4}$

For example, between checkpoints B and C the average velocity is

$$\text{average velocity} = \frac{70 - 10}{1\frac{3}{4} - \frac{3}{4}} = \frac{60}{1} = 60 \text{ miles/hour}$$

while the average velocity between checkpoints A and B is

$$\text{average velocity} = \frac{10 - 4}{\frac{3}{4} - \frac{1}{4}} = \frac{6}{\frac{1}{2}} = 12 \text{ miles/hour}$$

More generally, if the car is d_1 miles from the origin after t_1 hours and is d_2 miles from the origin after t_2 hours, then its average velocity over this period is

$$\text{average velocity} = \frac{d_2 - d_1}{t_2 - t_1} \text{ miles/hour}$$

The notion of an average rate of change is applicable to problems other than velocity. We make the following definition.

Definition If a quantity y is a function of a quantity x and if y changes from y_1 to y_2 as x changes from x_1 to x_2, then the **average rate of change** of y with respect to x between x_1 and x_2 is

$$\frac{y_2 - y_1}{x_2 - x_1} \tag{1}$$

In this definition, if y and x are related by the equation $y = f(x)$, then

$$y_1 = f(x_1) \quad \text{and} \quad y_2 = f(x_2)$$

so (1) can be expressed in the following alternative form:

average rate of change of f with respect to x between x_1 and x_2 is

$$\frac{f(x_2) - f(x_1)}{x_2 - x_1} \tag{2}$$

EXAMPLE 1 Let

$$y = 2x + 1$$

Find the average rate of change of y with respect to x between $x = 2$ and $x = 5$.

Solution Let $x_1 = 2$ and $x_2 = 5$. Then

$$y_1 = (2 \cdot 2) + 1 = 5 \quad \text{and} \quad y_2 = (2 \cdot 5) + 1 = 11$$

Hence, the average rate of change is

$$\frac{y_2 - y_1}{x_2 - x_1} = \frac{11 - 5}{5 - 2} = \frac{6}{3} = 2$$

EXAMPLE 2 A manufacturer of an industrial liquid determines that the cost (in dollars) of manufacturing x gallons of liquid is given by the formula

$$C(x) = x^2 - 2x + 5$$

The cost of manufacturing 5 gallons is $C(5) = 20$ dollars and the cost of manufacturing 10 gallons is $C(10) = 85$ dollars, so by increasing production from 5 to 10 gallons the cost increases as follows:

$$\text{average rate of increase in cost/gallon} = \frac{C(10) - C(5)}{10 - 5} = \frac{85 - 20}{5}$$

$$= \frac{65}{5} = 13 \text{ dollars/gallon}$$

In Figure 2 we have graphed the cost function $C(x)$. Now we make a very significant observation. The average rate of increase in cost (just calculated) is the slope of the line joining the points (5, 20) and (10, 85) on the graph of $C(x)$.

FIGURE 2

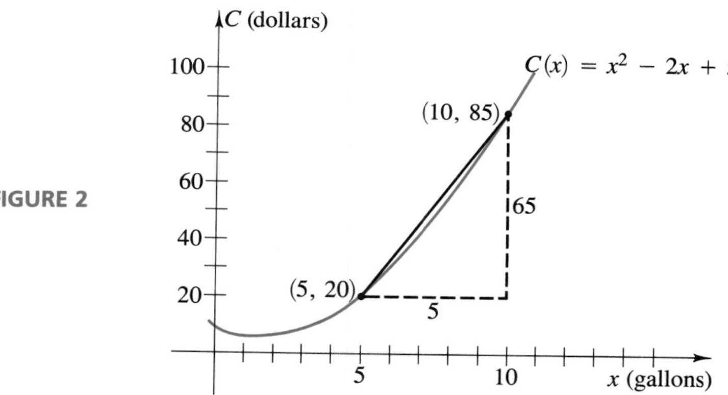

Consider the general problem of illustrating graphically the average rate of change of a function $f(x)$ as x changes from x_1 to x_2. For convenience, in Figure 3 we have drawn $x_1 < x_2$.

FIGURE 3

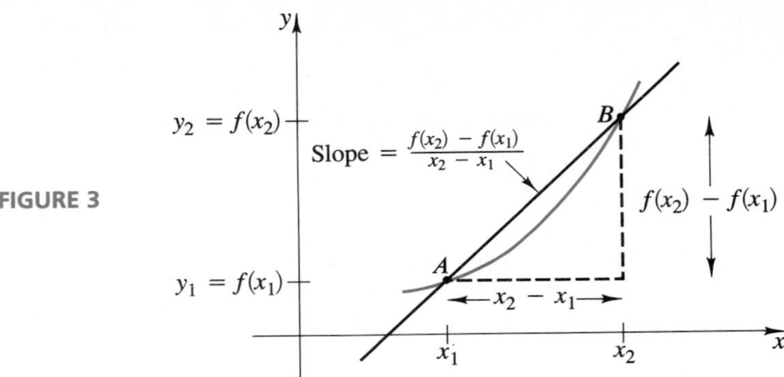

By our definition, from (2), the average rate of change is the same as the slope of the line joining the points $A(x_1, f(x_1))$ and $B(x_2, f(x_2))$. We call this line the **secant line** joining A and B.

EXAMPLE 3 After the administration of an experimental drug, a cancerous tumor undergoes a weight change that seems to follow the formula

$$W(t) = -\tfrac{1}{3}t^2 + 3$$

where $W(t)$ is the weight in grams and t is the time elapsed in months. At the end of 1 month ($t = 1$) the weight of the tumor is $W(1) = \frac{8}{3}$ grams, and at the end of 2 months ($t = 2$), the weight is $W(2) = \frac{5}{3}$ grams. Thus the average rate of change in weight over this time period is

$$\frac{W(2) - W(1)}{2 - 1} = \frac{\frac{5}{3} - \frac{8}{3}}{1} = -1 \text{ gram/month}$$

FIGURE 4

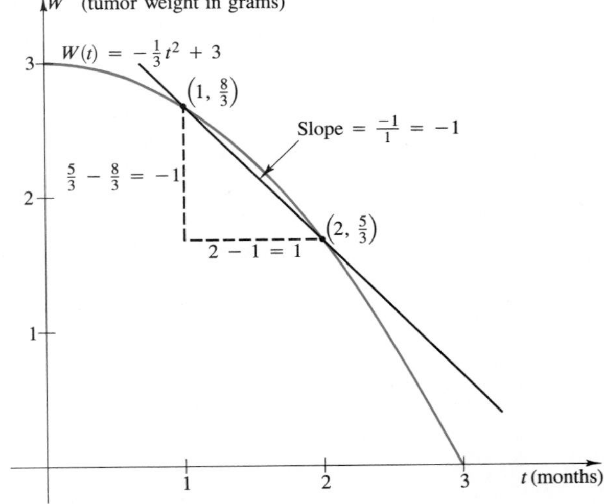

The negative sign tells us that the weight has *decreased* over the time period. Geometrically, this average decrease is represented by the slope of the secant line joining the points $(1, \frac{8}{3})$ and $(2, \frac{5}{3})$ on the graph of $W(t)$ (Figure 4).

Instantaneous Rates of Change

While the average rate of change is useful in many problems, there are times when we must use another concept, called **instantaneous rate of change**. For example, if a moving car strikes a tree, the damage sustained is not determined by the average velocity during the trip but rather by the instantaneous velocity at the precise moment of impact. The instantaneous velocity is referred to as **velocity**. This suggests that we try to define the rate at which a quantity is changing at a particular *point* as opposed to its average rate of change over an *interval*.

To motivate this definition we consider an example from the physical sciences.

EXAMPLE 4 It is known that a rock dropped from a height falls toward the Earth so that the distance fallen (in feet) after t seconds is approximately

$$f(t) = 16t^2$$

(See Figure 5.)

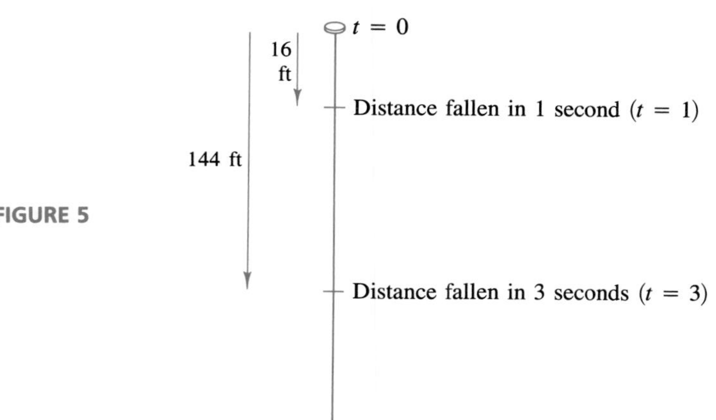

FIGURE 5

Let us try to calculate the velocity of the rock precisely 3 seconds after it has been released. Since we are not calculating an average velocity, we cannot simply divide distance traveled by the time elapsed. However, we can argue as follows. Over a very short interval of time, for example, $\frac{1}{10}$ second, the velocity of the rock cannot vary much. Thus the error will be small if we *approximate* the velocity when $t = 3$ by the average velocity over the interval from $t = 3$ to $t = 3.1$. The calculation is as follows:

$$\begin{aligned}\text{velocity when } t = 3 &\cong \text{average velocity between } t = 3 \text{ and } t = 3.1\\ &= \frac{f(3.1) - f(3)}{3.1 - 3} = \frac{16(3.1)^2 - 16(3)^2}{0.1}\\ &= 97.6 \text{ feet/second}\end{aligned}$$

To improve this approximation we might calculate the average velocity over an even smaller interval, for example, $t = 3$ to $t = 3.01$ or $t = 3$ to $t = 3.001$. The smaller the interval, the better the approximation. In Table 2 we have summarized the results of several such calculations. Since the average velocities tend toward 96 feet per second as the interval size decreases, we can reasonably conclude that 96 feet per second is the velocity when $t = 3$.

TABLE 2

Time Interval (seconds)	Average Velocity (feet/second)
$t = 3$ to $t = 3.1$	97.6
$t = 3$ to $t = 3.01$	96.16
$t = 3$ to $t = 3.001$	96.016
$t = 3$ to $t = 3.0001$	96.0016
$t = 3$ to $t = 3.00001$	96.00016

EXAMPLE 4 **ALTERNATE APPROACH** To help prepare the way for the more general mathematical approach, we take another look at Example 4. We use Table 2′, an alternate form of Table 2. Again let $f(t) = 16t^2$ and consider the time interval from $t = 3$ to $t = 3 + h$ (in seconds). Then the average velocity over this time interval is given by

$$\text{average velocity} = \frac{f(3 + h) - f(3)}{h}$$

Table 2′ gives the average velocity for selected values of h.

When we claim that the instantaneous rate of change of the function $f(t) = 16t^2$ at $t = 3$ is 96, we are saying that as h approaches 0, $\dfrac{f(3 + h) - f(3)}{h}$ approaches 96.

TABLE 2′

h	Average Velocity
0.1	97.6
0.01	96.16
0.001	96.016
0.0001	96.0016
0.00001	96.00016

Another way of saying this is

$$\frac{f(3 + h) - f(3)}{h}$$

has limit 96 as h approaches 0. The standard mathematical symbolism for this statement is

$$\lim_{h \to 0} \frac{f(3+h) - f(3)}{h} = 96$$

[read "the limit as h approaches 0 of $\dfrac{f(3+h) - f(3)}{h}$ is 96"].

To see what this limit statement means, notice that over the general time interval from $t = 3$ to $t = 3 + h$,

$$\begin{aligned}
\text{average velocity} &= \frac{f(3+h) - f(3)}{(3+h) - 3} = \frac{16(3+h)^2 - 16(3)^2}{h} \\
&= \frac{16(9 + 6h + h^2) - 16(9)}{h} \\
&= \frac{144 + 96h + 16h^2 - 144}{h} \\
&= \frac{96h + 16h^2}{h} \\
&= 96 + 16h
\end{aligned}$$

We now observe what happens to this average velocity as the interval size is decreased, that is, as $h \to 0$. We obtain

$$\lim_{h \to 0} (96 + 16h) = 96$$

This agrees with the result conjectured from Table 2.

It is important to see that we are taking a *limit* as h approaches 0 and that we are not actually letting $h = 0$. It makes no sense to merely substitute $h = 0$ into the expression $\dfrac{f(3+h) - f(3)}{h}$ since that would involve division by zero. We are thus led to a *new* concept—that of limit. This concept is the subject of the next section.

Motivated by Example 4, we present the following definition.

Definition If a quantity y is a function of a quantity x, $y = f(x)$, then the **instantaneous rate of change** of y with respect to x at $x = a$ is

$$\lim_{h \to 0} \frac{f(a+h) - f(a)}{h} \tag{3}$$

Remember that

$$\frac{f(a+h) - f(a)}{h}$$

is just the average rate of change of y with respect to x over the interval from $x = a$ to $x = a + h$ (Figure 6). Thus we are defining the instantaneous rate of change as a limit of average rates of change over smaller and smaller intervals.

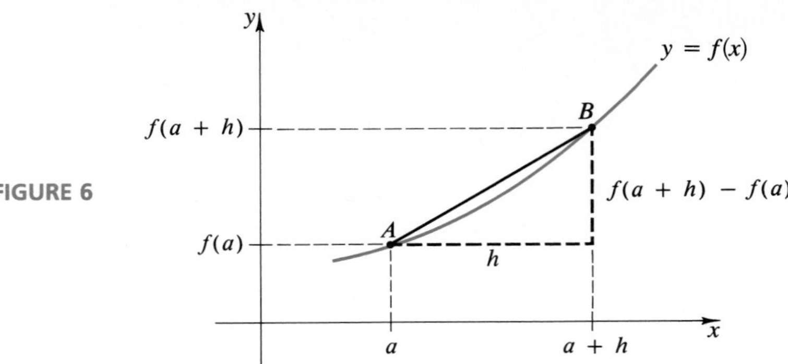

FIGURE 6

Tangent Line to a Curve; Slope of a Curve

Instantaneous rate of change has an important geometric interpretation. Since

$$\frac{f(a + h) - f(a)}{h}$$

represents the slope of the secant line joining $A(a, f(a))$ and $B(a + h, f(a + h))$ (Figure 7), the instantaneous rate of change at a,

$$\lim_{h \to 0} \frac{f(a + h) - f(a)}{h} \tag{3}$$

can be interpreted as the limit of these slopes as $h \to 0$. However, as $h \to 0$, the point B moves along the graph of f toward A, and the secant lines from A to B tend toward a line called the **tangent** line to the graph of f at A (Figure 7). Thus we can interpret (3) as the slope of the line tangent to the graph of f at the point $(a, f(a))$. We call the slope of this tangent, the **slope of the curve** $y = f(x)$ at $(a, f(a))$, or sometimes simply the **slope** of $y = f(x)$ at a.

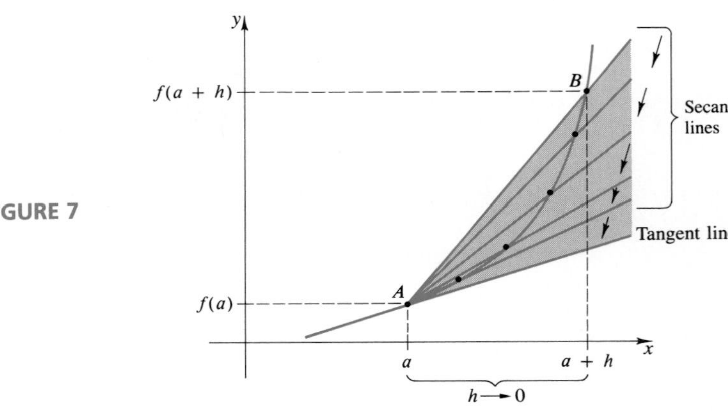

FIGURE 7

EXAMPLE 5 Let $y = f(x) = 2x + 1$. Find the instantaneous rate of change of y with respect to x at $a = 3$.

Solution From (3) the instantaneous rate of change of y with respect to x at $a = 3$ is

$$\lim_{h\to 0} \frac{f(a+h) - f(a)}{h} = \lim_{h\to 0} \frac{[2(3+h)+1] - [(2\cdot 3)+1]}{h}$$

$$= \lim_{h\to 0} \frac{6 + 2h + 1 - 6 - 1}{h}$$

$$= \lim_{h\to 0} \frac{2h}{h} = \lim_{h\to 0} 2 = 2$$

This solution is consistent with the observation that the graph itself has slope 2 at every point.

EXAMPLE 6 Let us consider again the manufacturer's problem discussed in Example 2. For each x gallons of liquid produced, the manufacturer's cost is

$$C(x) = x^2 - 2x + 5$$

As the manufacturer increases the production, the costs change. Let us try to determine the rate at which the cost will be changing at the instant a production level of $a = 5$ gallons is reached.

Solution From (3) the instantaneous rate of change in cost at a production level of x_0 is

$$\lim_{h\to 0} \frac{C(a+h) - C(a)}{h}$$

$$= \lim_{h\to 0} \frac{[(a+h)^2 - 2(a+h) + 5] - [a^2 - 2a + 5]}{h}$$

$$= \lim_{h\to 0} \frac{(a^2 + 2ah + h^2 - 2a - 2h + 5) - (a^2 - 2a + 5)}{h}$$

$$= \lim_{h\to 0} \frac{2ah + h^2 - 2h}{h}$$

$$= \lim_{h\to 0} (2a + h - 2)$$

$$= 2a - 2$$

In particular, at a production level of $a = 5$ the cost is increasing at a rate of

$$2a - 2 = 2(5) - 2 = 8 \text{ dollars/gallon}$$

Geometrically, the graph of the cost function $C(x) = x^2 - 2x + 5$ has a tangent line of slope 8 at the point where $a = 5$ (Figure 8).

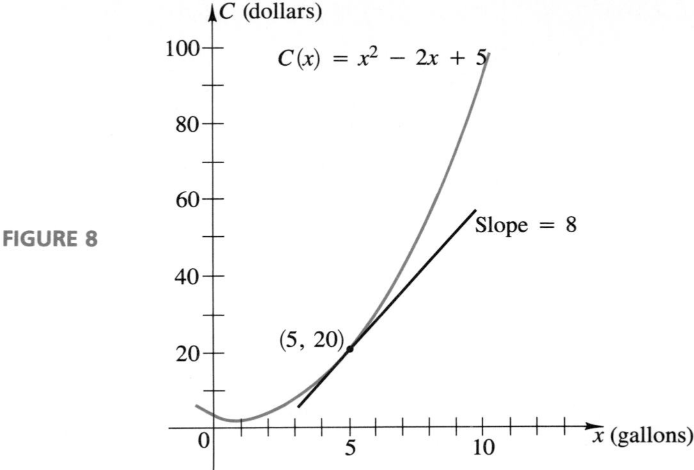

FIGURE 8

We now present the following five-step procedure for finding the instantaneous rate of change of $f(x)$ at x.

Five-Step Procedure for Finding the Instantaneous Rate of Change of $f(x)$ at x

Step	Example: $f(x) = x^2 - 3$
1. Write the formula for $f(x)$.	$f(x) = x^2 - 3$
2. Write the formula for $f(x + h)$.	$f(x+h) = (x+h)^2 - 3$ $= x^2 + 2xh + h^2 - 3$
3. Subtract $f(x)$ from $f(x + h)$.	$f(x+h) - f(x)$ $= (x^2 + 2xh + h^2 - 3) - (x^2 - 3)$ $= 2xh + h^2$ $= h(2x + h)$
4. Divide by h.	$\dfrac{f(x+h) - f(x)}{h} = \dfrac{h(2x+h)}{h}$ $= 2x + h$ (The cancellation is possible since $h \neq 0$.)
5. Take the limit of the expression obtained in step 4 as h approaches 0.	$\lim_{h \to 0} \dfrac{f(x+h) - f(x)}{h}$ $= \lim_{h \to 0} (2x + h) = 2x$

We therefore conclude that the instantaneous rate of change of the function $f(x) = x^2 - 3$ at x is $2x$.

EXERCISE SET 12.2

1. Suppose $y = 3x + 4$.
   (a) Find the average rate of change of y with respect to x between $x = 0$ and $x = 1$.
   (b) Find the average rate of change of y with respect to x between $x = -3$ and $x = 4$.
   (c) Find the average rate of change of y with respect to x between any two values $x = a$ and $x = b$.
   (d) Explain the result in (c) by graphing $y = 3x + 4$.

2. Suppose $y = \frac{1}{2}x^2$.
   (a) Find the average rate of change of y with respect to x between $x = 1$ and $x = 4$.
   (b) Sketch the graph of the curve and draw a secant line whose slope is the average rate of change obtained in (a).

In Exercises 3–9 find the average rate of change of y with respect to x between x_1 and x_2.

3. $y = 1 - 4x$; $x_1 = 4$, $x_2 = 5$

4. $y = 2x^2 + 3$; $x_1 = -2$, $x_2 = 6$

5. $y = 3$; $x_1 = 2$, $x_2 = 4$

6. $y = 1/x$; $x_1 = -3$, $x_2 = -1$

7. $y = x^3$; $x_1 = 2$, $x_2 = 3$

8. $y = |x|$; $x_1 = -1$, $x_2 = 1$

9. $y = \dfrac{3x^2 - 2}{x - 1}$; $x_1 = 2$, $x_2 = 4$

10. For the function graphed in Figure 9, find the average rate of change of y with respect to x between:
   (a) $x_1 = 1$ and $x_2 = 2$
   (b) $x_1 = 1$ and $x_2 = 3$
   (c) $x_1 = 2$ and $x_2 = 5$
   (d) $x_1 = 3$ and $x_2 = 8$
   (e) $x_1 = 2$ and $x_2 = 10$
   (f) $x_1 = 11$ and $x_2 = 12$

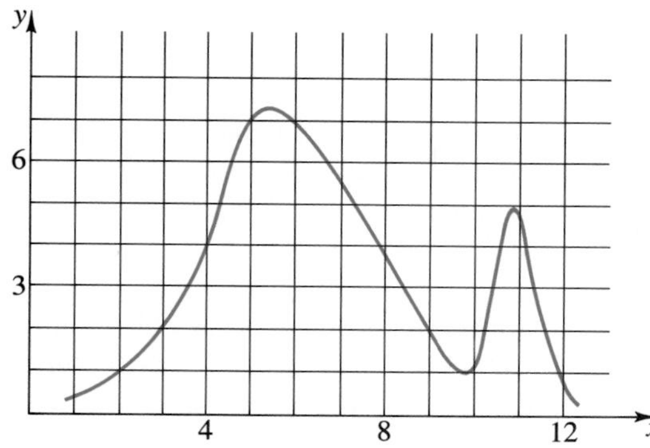

FIGURE 9

11. As discussed in Example 4, a rock dropped from a height falls to Earth so that the distance fallen (in feet) after t seconds is approximately $f(t) = 16t^2$.
   (a) Find the distance traveled by the rock between the third and sixth seconds after release (i.e., between $t = 3$ and $t = 6$).
   (b) Find the average velocity during this time interval.
   (c) Find the velocity at $t = 3$.
   (d) Find the velocity at $t = 4$.

In Exercises 12–15 find the instantaneous rate of change of y with respect to x at the given value of a. Use the five-step procedure.

12. $y = 2x$; $a = 2$

13. $y = x^2$; $a = 3$

14. $y = x^2 + 1$; $a = 0$

15. $y = 3x^2 - 2$; $a = 4$

16. **(Manufacturing)** Consider the manufacturing problem in Example 6.
   (a) Find the average rate of change of cost when the production level increases from 5 to 10 gallons.
   (b) On the graph of $C(x)$, draw a secant line whose slope is the average rate of change found in part (a).
   (c) Find the instantaneous rate of change in cost at a production level of $a = 8$ gallons.
   (d) On the graph of $C(x)$, draw a tangent line whose slope is the instantaneous rate of change found in part (b).

17. **(Population growth)** The following table lists population statistics for the city of San Francisco, California (based on data from the *World Almanac*):

Year	Population
1900	342,782
1950	775,357
1960	740,316
1970	715,674
1980	678,974

   (a) What is the average rate of change of population with respect to time between the years 1900 and 1980?

(b) What is the average rate of change of population with respect to time between the years 1950 and 1980?

(c) On the average, was the population declining more rapidly over the 10-year span between 1960 and 1970 or over the 20-year span between 1960 and 1980?

18. **(Epidemics)** In an outbreak of σ-type malaria, the total number of cases reported over a 10-day period is indicated in the following table:

Day	Total Number of Cases Reported to Date
1	20
2	80
3	140
4	220
5	230
6	245
7	247
8	250
9	250
10	250

(a) Make a graph to describe the data.

(b) What is the average rate of change in the total number of cases reported with respect to time between day 2 and day 4, inclusive?

(c) Over which two consecutive days does the largest average rate of increase occur in the number of cases reported?

(d) Over the last two days, what is the average rate of change in total number of cases reported?

(e) What does the result in (d) tell you?

19. **(Advertising)** The graph in Figure 10 illustrates a response curve for a college fund-raising campaign. With an amount x spent on advertising, the college receives donations amounting to $D(x)$.

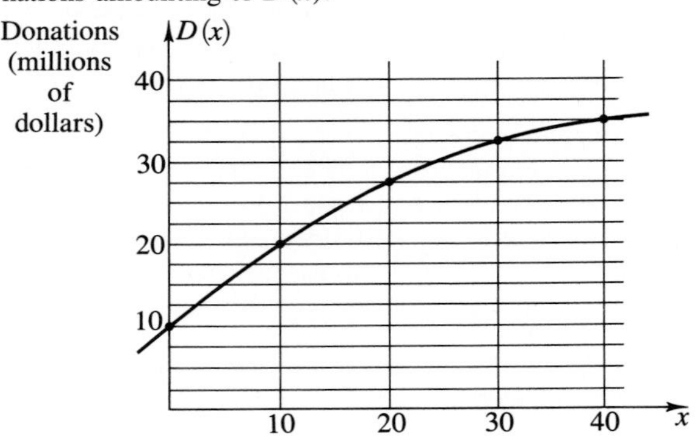

(a) Find the average rate of change of D as x changes from 0 to 10, 20 to 30, 10 to 20, and 30 to 40.

(b) Can you explain, from a practical point of view, why you should anticipate these rates of change to be decreasing?

12.3 LIMITS

As we suggest in the last section, the study of calculus requires familiarity with the mathematical concept of limit. In this section we investigate the meaning of the statement

$$\lim_{x \to a} f(x) = L$$

and present some techniques necessary for the practical computation of limits. The theory of limits forms the theoretical cornerstone of calculus; we draw on it repeatedly as the subject unfolds. We now introduce some helpful terminology and notation.

Intervals

By an **interval** we mean a set of real numbers that forms a line segment on a coordinate axis. The interval may include or exclude one or both of the endpoints. If a and b are real numbers such that $a < b$, then the **closed interval** from a to b, denoted $[a, b]$, is the set of points

$$[a, b] = \{x \mid a \leq x \leq b\}$$

and the **open interval** from a to b, denoted (a, b), is the set

$$(a, b) = \{x \mid a < x < b\}$$

These sets are shown in Figure 1.

FIGURE 1

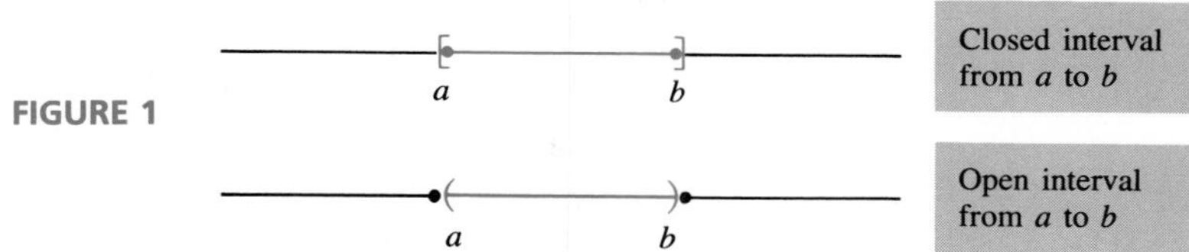

We use a square bracket [or] to indicate an endpoint that is included in the interval and a parenthesis (or) to indicate an endpoint that is excluded. Thus,

$$[a, b) = \{x \mid a \leq x < b\}$$

is the interval

which includes a but excludes b.

We also need intervals that extend indefinitely in one or both directions. Some intervals of this type and their notation are

$$(a, +\infty) = \{x \mid x > a\}$$
$$(-\infty, b] = \{x \mid x \leq b\}$$
$$(-\infty, +\infty) = \text{the entire line}$$

In this notation, the symbol $+\infty$, or ∞, (read "plus infinity") indicates that the interval extends indefinitely in the positive direction, and $-\infty$ (read "minus infinity") indicates that the interval extends indefinitely in the negative direction. The symbols ∞ and $-\infty$ do not represent actual numbers on the real number line.

EXAMPLE 1

Interval

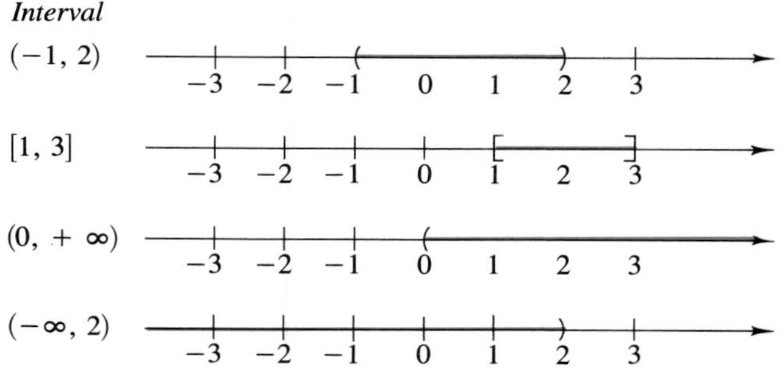

Limits In many problems we are concerned with describing the behavior of a function $f(x)$ that is defined in an open interval containing the point $x = a$, except possibly at $x = a$. For example, $f(x) = 1/(x - 2)$ is defined in the open interval (0, 3) except at $x = 2$. There are usually two questions of importance:

1. What is the value of f when $x = a$?
2. How do the values of f behave when x is near but different from a?

For example, if x represents the distance between a lunar lander and the moon's surface and $f(x)$ represents the speed of the lander at distance x, then it is important to know $f(0)$, the speed of the lander at impact. But it is just as important to know how the speed $f(x)$ varies when x is near but different from 0. For it is this information that enables the control center to make corrections to bring about a soft landing.

The following examples help distinguish between the value of a function f at a point $x = a$ and the value of f for x near but different from a.

EXAMPLE 2 Let f be the function defined by $f(x) = 3x + 1$, and let us investigate the behavior of f at the point $x = 2$ and near the point $x = 2$.

The first part is easy; at $x = 2$ the value of the function is

$$f(2) = 3(2) + 1 = 7$$

To investigate the behavior near $x = 2$, we have evaluated the function f at a succession of x values closer and closer to (but different from) $x = 2$, as shown in Figure 3.

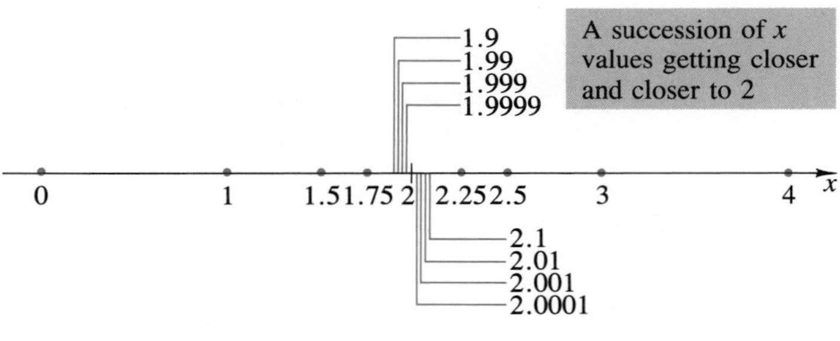

FIGURE 3

The results are shown in Figure 4 and the accompanying table.

It seems intuitively clear from these computations that the value of $f(x) = 3x + 1$ gets closer and closer to 7 as x becomes closer and closer to 2 from either side. The number 7 is called the **limit** of $f(x) = 3x + 1$ as x approaches 2, and we write

$$\lim_{x \to 2} (3x + 1) = 7$$

In this expression $x \to 2$ indicates that x is approaching (but is different from) 2.

x	$f(x) = 3x + 1$
1.75	6.25
1.9	6.70
1.99	6.97
⋮	⋮
2	7
⋮	⋮
2.01	7.03
2.1	7.30

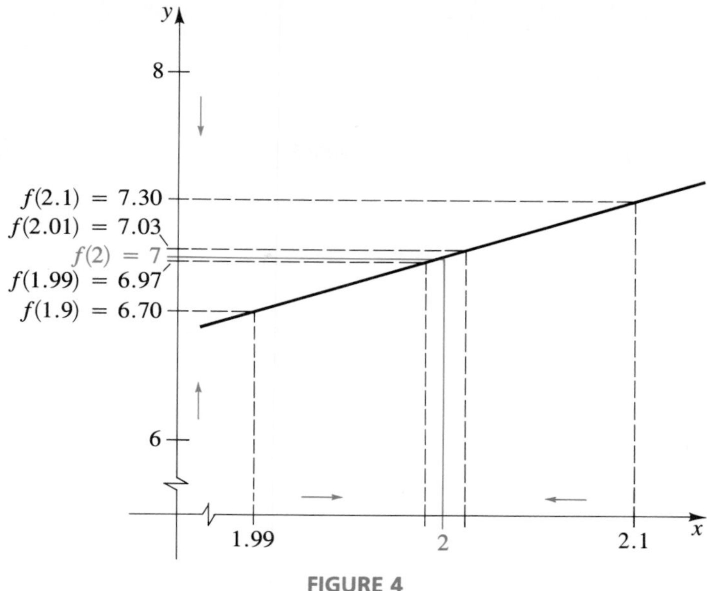

FIGURE 4

It is important to keep in mind that the limit in Equation (1) is only an educated guess based on our computations in Figure 3. It is possible to imagine that if we enlarged our table of computations to include values of x still closer to 2, the values for $f(x)$ might change their pattern and approach some number different from 7 or perhaps approach no limit at all. To be absolutely certain (1) is correct, we need mathematical proof, and to give a mathematical proof we need a precise mathematical definition of a limit. In this textbook we are not concerned with such proofs and formal definitions. However, you will have a satisfactory grasp of the limit concept if you view it as follows.

The (Intuitive) Notion of a Limit Given a function f, we interpret the statement

$$\lim_{x \to a} f(x) = L$$

to mean that the values of $f(x)$ get closer to the number L as x approaches (but remains different from) a. Stated another way, we can make the value of $f(x)$ as close as we like to L by making x sufficiently close to (but different from) a.

For the function $f(x) = 3x + 1$ in Example 2, it turns out that

$$f(2) = 7 \quad \text{and} \quad \lim_{x \to 2} f(x) = 7$$

so that the limit as x *approaches* 2 and the value *at* 2 are the same. As we see in subsequent examples, this is not always the case. However, in this case by graphing the function $f(x) = 3x + 1$, it is easy to visualize why $f(2)$ and $\lim_{x \to 2} f(x)$ have the same value (Figure 4).

EXAMPLE 3 Let us obtain the limit of the function

$$f(x) = \frac{x^2 - 9}{x - 3}$$

as x approaches 3. First, observe that $f(3)$ is undefined since substituting $x = 3$ in the formula for f leads to a division by zero. However, as Table 1 suggests, there is a value for the limit of $f(x)$ as x approaches 3, namely,

$$\lim_{x \to 3} \frac{x^2 - 9}{x - 3} = 6 \tag{2}$$

To gain some geometric insight into this limit, let us graph the function f. If $x \neq 3$, then we can write

$$f(x) = \frac{x^2 - 9}{x - 3} = \frac{(x + 3)(x - 3)}{x - 3} = x + 3$$

TABLE 1

	$x \to 3$ from the right side						
x	5	4	3.5	3.1	3.01	3.001	3.0001
$f(x) = \frac{x^2 - 9}{x - 3}$	8	7	6.5	6.1	6.01	6.001	6.0001

	$x \to 3$ from the left side						
x	1	2	2.5	2.9	2.99	2.999	2.9999
$f(x) = \frac{x^2 - 9}{x - 3}$	4	5	5.5	5.9	5.99	5.999	5.9999

Thus for $x \neq 3$, the graph of $y = f(x)$ coincides with the graph of the line $y = x + 3$; but for $x = 3$, there is no point on the graph since $f(3)$ is undefined. As shown in Figure 5, the graph of f is a straight line with a hole in it at $x = 3$.

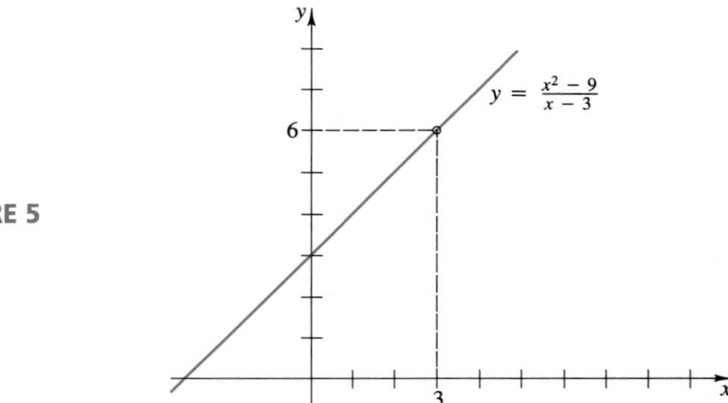

FIGURE 5

The limit statement in (2) should be intuitively evident from this graph; when x gets closer and closer to 3, $f(x)$ approaches 6.

EXAMPLE 4 Consider the function f defined by

$$f(x) = \frac{1}{x - 2}$$

Observe first that $f(2)$ is undefined since substituting $x = 2$ results in a division by zero. Next let us investigate

$$\lim_{x \to 2} f(x) = \lim_{x \to 2} \frac{1}{x - 2}$$

TABLE 2

				$x \to 2$ **from the right side**				
x	3	2.5	2.2	2.1	2.05	2.01	2.001	2.0001
$f(x) = \frac{1}{x - 2}$	1	2	5	10	20	100	1000	10,000
				$x \to 2$ **from the left side**				
x	1	1.5	1.8	1.9	1.95	1.99	1.999	1.9999
$f(x) = \frac{1}{x - 2}$	−1	−2	−5	−10	−20	−100	−1000	−10,000

As Table 2 shows, the values of $f(x)$ increase as $x \to 2$ from the right side and decrease as $x \to 2$ from the left side. Thus, $f(x)$ does not approach any fixed number as x approaches 2. The graph of f is shown in Figure 6.

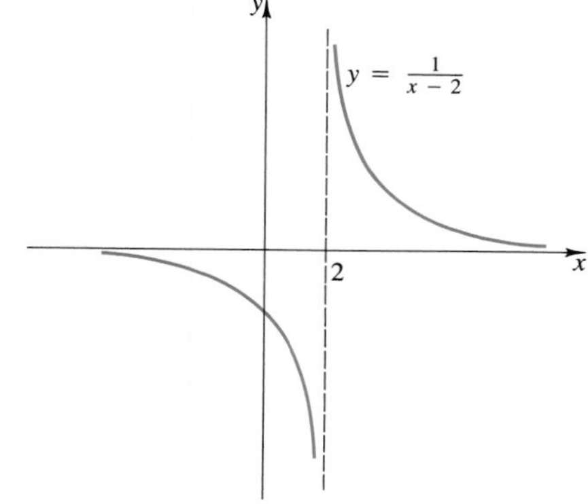

FIGURE 6

If, as in the last example, the quantity $f(x)$ approaches no single finite value as x approaches a point a, then we say

$$\lim_{x \to a} f(x) \text{ does not exist}$$

EXAMPLE 5 Let f be given by the formula

$$f(x) = \frac{|x|}{x}$$

This function is a simple example of one with a piecewise definition:

$$f(x) = \begin{cases} 1 & \text{if } x > 0 \\ \text{undefined} & \text{if } x = 0 \\ -1 & \text{if } x < 0 \end{cases}$$

The graph of f is shown in Figure 7. For this function, $f(0)$ does not exist. Moreover, $\lim_{x \to 0} f(x)$ does not exist. To see this we need only observe that as $x \to 0$ from the right side the value of $f(x)$ is constantly $+1$, while as $x \to 0$ from the left side the value of $f(x)$ is constantly -1. Thus $f(x)$ does not approach a *single* finite value as $x \to 0$. Therefore $\lim_{x \to 0} f(x)$ does not exist.

FIGURE 7

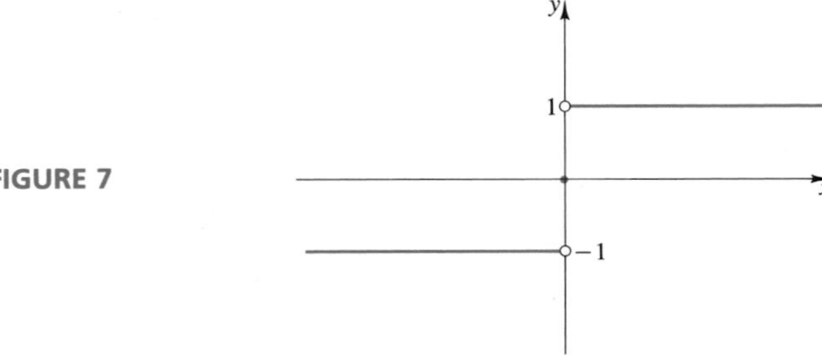

One-Sided Limits

In each of the last three examples limits are calculated by considering first $x \to a$ from one side of a and then $x \to a$ from the other side of a. When the resulting "one-sided limits" agree, as in Example 3, we say that the **limit exists**. When the one-sided limits either do not exist, as in Example 4, or exist but do not agree, as in Example 5, we say that the **limit** itself **does not exist**. One-sided limits occur naturally. For example, $f(x) = \sqrt{x}$ is defined only for $x \geq 0$, so we would only be interested in the limit of $f(x)$ as x approaches 0 from the right side.

Each real number a divides the real number line in half. Those numbers lying to the right of a are said to be on the "plus side" of a; those to the left are on the "minus side" of a (Figure 8). It is then easy to understand the following notation:

$$\lim_{x \to a^+} f(x) = \text{limit of } f(x) \text{ as } x \text{ approaches } a \text{ from the right}$$

$$\lim_{x \to a^-} f(x) = \text{limit of } f(x) \text{ as } x \text{ approaches } a \text{ from the left}$$

FIGURE 8

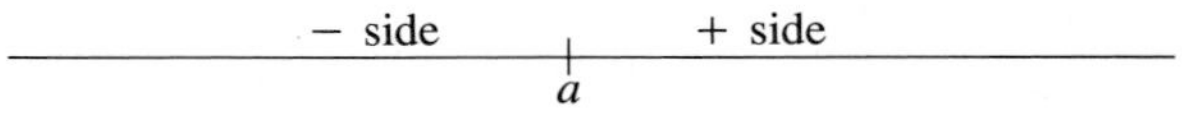

− side + side

a

The relationship between the one-sided limits and the ordinary limit is expressed by the following theorem.

THEOREM $\lim_{x\to a} f(x) = L$ if and only if both one-sided limits $\lim_{x\to a^+} f(x)$ and $\lim_{x\to a^-} f(x)$ exist and are equal to L.

EXAMPLE 6 Consider the function f defined by

$$f(x) = \begin{cases} -\frac{2}{3}x + 6 & \text{if } x \geq 3 \\ x + 1 & \text{if } x < 3 \end{cases}$$

The graph of f is shown in Figure 9. To the left of 3 the formula for f is $f(x) = x + 1$, so $f(x) = x + 1$ approaches 4 as x approaches 3 from the left. That is, $\lim_{x\to 3^-} f(x) = 4$.

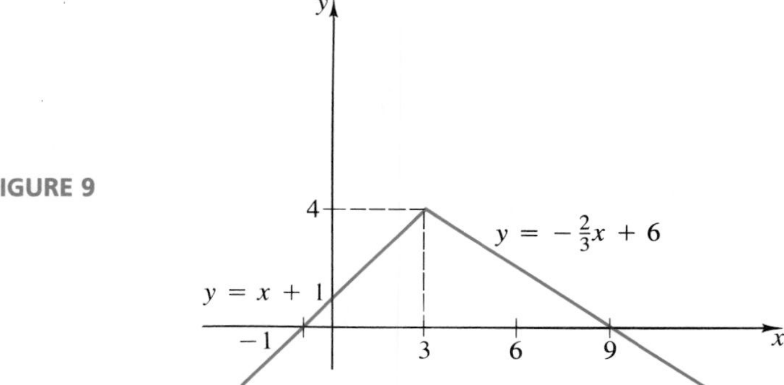

FIGURE 9

To the right of 3 the formula for f is $f(x) = -\frac{2}{3}x + 6$, so $f(x) = -\frac{2}{3}x + 6$ approaches 4 as x approaches 3 from the right. That is, $\lim_{x\to 3^+} f(x) = 4$.

Since both one-sided limits exist and equal 4, we conclude that

$$\lim_{x\to 3} f(x) = 4$$

EXAMPLE 7 **THE LIGHT BULB FUNCTION** Suppose a light bulb that has been off for a long time is suddenly turned on at time $t = a$, then off again at time $t = b$, and remains off for a long time. Let $f(t)$ = current (in amperes) flowing through the light bulb at time t. Figure 10 shows the graph of f, where c is the constant current flowing when the bulb is on. Thus $\lim_{t\to a^-} f(t) = 0$, while $\lim_{t\to a^+} f(t) = c$. Since these one-sided limits exist but are unequal, $\lim_{t\to a} f(t)$ does not exist.

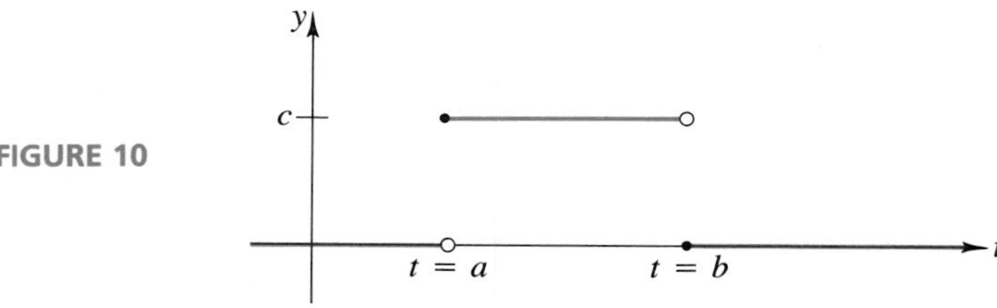

FIGURE 10

Algebra of Limits

We now consider several important algebraic properties (laws) that always hold for limits. These properties greatly facilitate the calculation of limits, and they should be regarded as useful principles rather than mere abstract formulas.

Each of these properties is stated for two-sided limits but is also true for one-sided limits.

Property 1 The limit of a constant function $f(x) = k$ is the constant k; that is,

$$\lim_{x \to a} k = k$$

EXAMPLE 8

$$\lim_{x \to 3} 5 = 5 \qquad \lim_{x \to 0} 5 = 5 \qquad \lim_{x \to -8} 5 = 5$$

Property 2

$$\lim_{x \to a} x = a$$

EXAMPLE 9

$$\lim_{x \to 0} x = 0 \qquad \lim_{x \to -2} x = -2 \qquad \lim_{x \to 6} x = 6$$

Property 3 The limit of a sum is the sum of the limits; the limit of a difference is the difference of the limits; and the limit of a product is the product of the limits; that is,

$$\lim_{x \to a}[f(x) + g(x)] = \lim_{x \to a} f(x) + \lim_{x \to a} g(x)$$
$$\lim_{x \to a}[f(x) - g(x)] = \lim_{x \to a} f(x) - \lim_{x \to a} g(x)$$
$$\lim_{x \to a}[f(x)g(x)] = \lim_{x \to a} f(x) \cdot \lim_{x \to a} g(x)$$

provided each limit on the right side exists.

EXAMPLE 10

$$\lim_{x \to 4}(x + 5) = \lim_{x \to 4} x + \lim_{x \to 4} 5 = 4 + 5 = 9$$
$$\lim_{x \to 4}(x - 5) = \lim_{x \to 4} x - \lim_{x \to 4} 5 = 4 - 5 = -1$$
$$\lim_{x \to 4} x(x + 5) = \lim_{x \to 4} x \cdot \lim_{x \to 4}(x + 5) = 4 \cdot 9 = 36$$
$$\lim_{x \to 2} x^2 = \lim_{x \to 2} x \cdot \lim_{x \to 2} x = 2 \cdot 2 = 4$$
$$\lim_{x \to 3} 7x = \lim_{x \to 3} 7 \cdot \lim_{x \to 3} x = 7 \cdot 3 = 21$$

EXAMPLE 11 If k is any constant, then using Properties 1 and 3 we can write

$$\lim_{x \to a} kf(x) = \lim_{x \to a} k \cdot \lim_{x \to a} f(x)$$

or

$$\lim_{x \to a} kf(x) = k \lim_{x \to a} f(x)$$

EXAMPLE 12

$$\begin{aligned}\lim_{x\to 3}(2x^2 - 4x + 5) &= \lim_{x\to 3} 2x^2 - \lim_{x\to 3} 4x + \lim_{x\to 3} 5\\ &= 2\lim_{x\to 3} x^2 - 4\lim_{x\to 3} x + \lim_{x\to 3} 5\\ &= 2(\lim_{x\to 3} x \cdot \lim_{x\to 3} x) - 4\lim_{x\to 3} x + \lim_{x\to 3} 5\\ &= 2(9) - 4(3) + 5\\ &= 11\end{aligned}$$

EXAMPLE 13 If

$$p(x) = a_n x^n + a_{n-1}x^{n-1} + \cdots + a_1 x + a_0$$

is a polynomial function, then

$$\begin{aligned}\lim_{x\to c} p(x) &= \lim_{x\to c}(a_n x^n + a_{n-1}x^{n-1} + \cdots + a_1 x + a_0)\\ &= \lim_{x\to c}(a_n x^n) + \lim_{x\to c}(a_{n-1}x^{n-1}) + \cdots + \lim_{x\to c}(a_1 x) + \lim_{x\to c} a_0\\ &= a_n(\lim_{x\to c} x)^n + a_{n-1}(\lim_{x\to c} x)^{n-1} + \cdots + a_1(\lim_{x\to c} x) + \lim_{x\to c} a_0\\ &= a_n c^n + a_{n-1}c^{n-1} + \cdots + a_1 c + a_0\\ &= p(c)\end{aligned}$$

Thus, $\lim_{x\to c} p(x) = p(c)$.

Property 4 The limit of a ratio is the ratio of the limits, provided the limit of the denominator is not zero; that is,

$$\lim_{x\to a}\frac{f(x)}{g(x)} = \frac{\lim_{x\to a} f(x)}{\lim_{x\to a} g(x)} \qquad \text{if } \lim_{x\to a} g(x) \neq 0$$

provided each limit on the right side exists.

EXAMPLE 14 Using Property 4,

$$\lim_{x\to 4}\frac{3x^2 - 5}{x - 2} = \frac{\lim_{x\to 4}(3x^2 - 5)}{\lim_{x\to 4}(x - 2)} = \frac{43}{2}$$

EXAMPLE 15 In Example 3 we show that

$$\lim_{x\to 3}\frac{x^2 - 9}{x - 3} = 6$$

This result cannot be obtained using Property 4 since $\lim_{x\to 3}(x - 3) = 0$. However, we can write

$$\lim_{x\to 3}\frac{x^2 - 9}{x - 3} = \lim_{x\to 3}\frac{(x - 3)(x + 3)}{x - 3} = \lim_{x\to 3}(x + 3) = 6$$

We are allowed to "cancel" the $x - 3$ common to the numerator and denominator because, as we let $x \to 3$, x is never 3, so $x - 3$ is never 0.

EXERCISE SET 12.3

1. Sketch the following intervals:
(a) $(1, 3)$ (b) $[1, 3]$ (c) $[1, 3)$
(d) $(1, 3]$ (e) $(-\infty, 1)$ (f) $(-\infty, 1]$
(g) $(3, +\infty)$ (h) $[3, +\infty)$

2. Sketch the following intervals:
(a) $(-3, -1)$ (b) $(-\infty, -4]$
(c) $[-5, +\infty)$ (d) $(0, +\infty)$
(e) $(-\infty, 0]$ (f) $[0, 3]$
(g) $[-2, 2]$ (h) $(-4, 5)$

3. (a) Complete the table of values for $f(x) = 2x - 3$:

x	2.6	2.7	2.8	2.9	2.99	2.999
$f(x) = 2x - 3$						
x	3.4	3.3	3.2	3.1	3.01	3.001
$f(x) = 2x - 3$						

(b) Use the table to help find $\lim_{x\to 3^-} (2x - 3)$, $\lim_{x\to 3^+} (2x - 3)$, and $\lim_{x\to 3} (2x - 3)$.

4. (a) Complete the table of values for $g(x) = x^2 + 2$ (a calculator may help):

x	−1.3	−1.2	−1.1	−1.01	−1.001	−1.0001
$g(x) = x^2 + 2$						
x	−0.7	−0.8	−0.9	−0.99	−0.999	−0.9999
$g(x) = x^2 + 2$						

(b) Use the table to help find $\lim_{x\to -1^-} (x^2 + 2)$, $\lim_{x\to -1^+} (x^2 + 2)$, and $\lim_{x\to -1} (x^2 + 2)$.

5. (a) Complete the table of values for $f(x) = |x|/2x$:

x	−1	−0.1	−0.01	−0.001	−0.0001
$f(x) = \frac{\lvert x\rvert}{2x}$					
x	0.0001	0.001	0.01	0.1	1
$f(x) = \frac{\lvert x\rvert}{2x}$					

(b) Use the completed table to evaluate $\lim_{x\to 0^-} |x|/2x$, $\lim_{x\to 0^+} |x|/2x$, and $\lim_{x\to 0} |x|/2x$.

6. (a) Complete the table for $f(x) = |3x| - 1$:

x		−1	−0.1	−0.001	−0.0001
$f(x) = \lvert 3x\rvert - 1$					
x	0.0001	0.001	0.01	0.1	1
$f(x) = \lvert 3x\rvert - 1$					

(b) Use the completed table to evaluate $\lim_{x\to 0^-} |3x| - 1$, $\lim_{x\to 0^+} |3x| - 1$, and $\lim_{x\to 0} |3x| - 1$.

In Exercises 7–16 evaluate the indicated limit.

7. $\lim_{x\to 3} 2x$

8. $\lim_{x\to 3} (2x + 4)$

9. $\lim_{x\to 4} (3x^2 + 2x - 5)$

10. $\lim_{x\to 3} \frac{2x + 5}{x^2 - 4}$

11. $\lim_{x\to 0} \frac{2}{x}$

12. $\lim_{x\to 0} \frac{x^2 - 2x}{x}$

13. $\lim_{x\to -1} (1 - x^2)$

14. $\lim_{x\to 1} \frac{x^2 - 36}{x - 6}$

15. $\lim_{x\to -1} \frac{x^2 + 2x + 1}{x + 1}$

16. $\lim_{x\to 4} \frac{x + 1}{x^2 - 3x - 4}$

17. Sketch the graph of the function

$$f(x) = \begin{cases} x + 3 & \text{if } x \geq 1 \\ 7x - 3 & \text{if } x < 1 \end{cases}$$

and determine $\lim_{x\to 1^-} f(x)$, $\lim_{x\to 1^+} f(x)$, and $\lim_{x\to 1} f(x)$.

18. Sketch the graph of the function

$$g(x) = \begin{cases} x + 1 & \text{if } x \geq 0 \\ 3x + 2 & \text{if } x < 0 \end{cases}$$

and determine $\lim_{x\to 0^-} g(x)$, $\lim_{x\to 0^+} g(x)$, and $\lim_{x\to 0} g(x)$.

19. For the light bulb function $f(t)$ of Example 7, determine the one-sided limits at $t = b$ and the limit there.

In Exercises 20–24 graph the function and find the indicated limit, if it exists.

20. $\lim_{x\to 4} \frac{x^2 - 16}{x - 4}$; $f(x) = \frac{x^2 - 16}{x - 4}$

21. $\lim_{x\to -5} \frac{x^2 - 25}{x + 5}$; $f(x) = \frac{x^2 - 25}{x + 5}$

22. $\lim_{x\to 1} \frac{x^3 + x^2 - 2x}{x(x + 2)}$; $f(x) = \frac{x^3 + x^2 - 2x}{x(x + 2)}$

23. $\lim_{x \to 3} \frac{|x - 3|}{x - 3}$; $f(x) = \frac{|x - 3|}{x - 3}$

24. $\lim_{x \to 5} \frac{1}{x - 5}$; $f(x) = \frac{1}{x - 5}$

In Exercises 25–28 use the graph to find the indicated limit, if it exists.

25. 

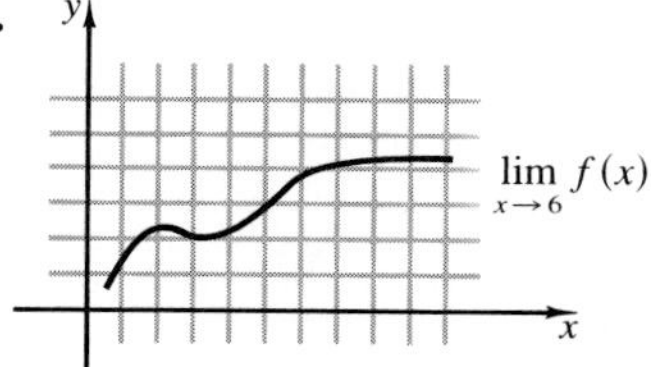

26.

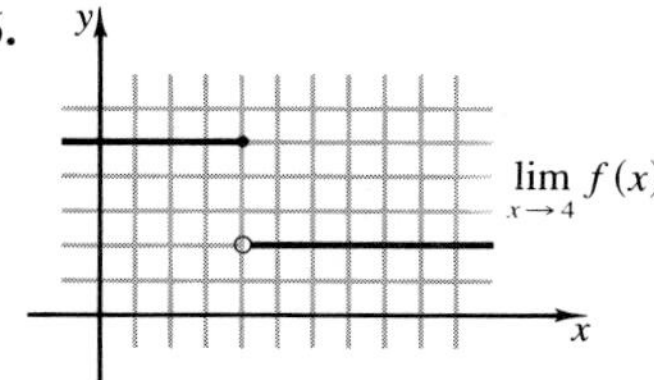

27.

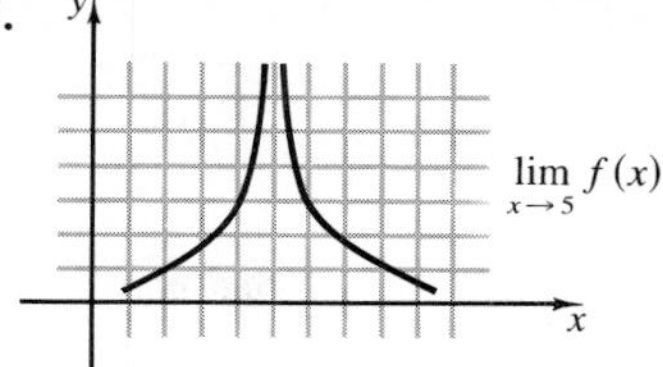

28.

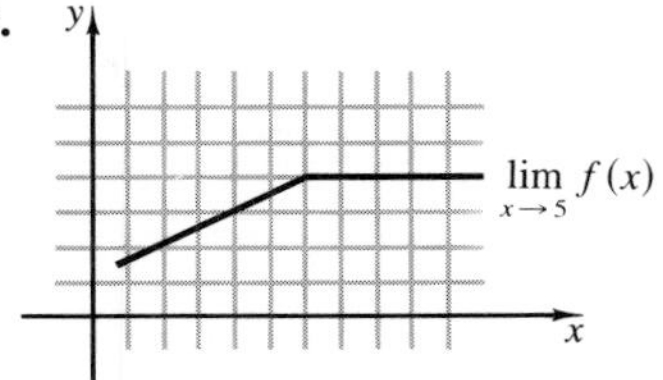

29. **(Business)** Suppose the profit received from the sale of x tons of a product is given (in terms of thousands of dollars) by

$$P(x) = \begin{cases} 3x^2 - 18 & \text{for } 0 \le x < 3 \\ 3x & \text{for } x > 3 \end{cases}$$

Find $\lim_{x \to 3} P(x)$.

12.4 INFINITE LIMITS; LIMITS AT INFINITY

The real number system has no largest element. For every real number x there is always a larger real number, for example, $x + 1$. Geometrically, the real number line extends infinitely to the right without an endpoint. Similarly, the real number system has no smallest element; the number line extends infinitely to the left without an endpoint.

Infinite Limits

In conventional mathematical language, we say that the real number system is unbounded. It has no upper bound (on the right) and no lower bound (on the left) as shown in Figure 1. We use the two symbols ∞ and $-\infty$ to denote this unboundness. They do not represent real numbers but instead indicate unboundedness and direction.

FIGURE 1

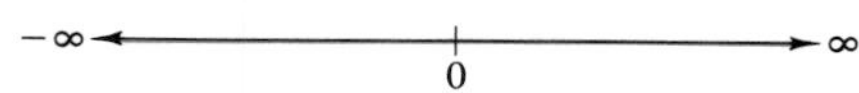

These symbols for infinity arise quite naturally when we attempt to evaluate the limit of a fraction

$$f(x) = \frac{p(x)}{q(x)}$$

as $x \to a$ (possibly one-sided) where $q(x)$ approaches 0 but $p(x)$ does not. The infinity symbols arise here because if the denominator of a fraction approaches 0 and the numerator does not, the fraction grows larger and larger (positively or negatively) without bound.

EXAMPLE 1 Let $f(x) = 1/x$. As Table 1 shows, as $x \to 0$ from the right, $f(x)$ is unbounded positively. Thus it makes sense to write

$$\lim_{x \to 0^+} \frac{1}{x} = \infty$$

TABLE 1

			$x \to 0$ from the right			
x	2	1	0.1	0.01	0.001	0.0001
$f(x) = 1/x$	$\frac{1}{2}$	1	10	100	1000	10,000
			$x \to 0$ from the left			
x	−2	−1	−0.1	−0.01	−0.001	−0.0001
$f(x) = 1/x$	$-\frac{1}{2}$	−1	−10	−100	−1000	−10,000

As $x \to 0$ from the left, $f(x)$ is unbounded negatively. Thus it makes sense to write

$$\lim_{x \to 0^-} \frac{1}{x} = -\infty$$

The graph of f is shown in Figure 2.

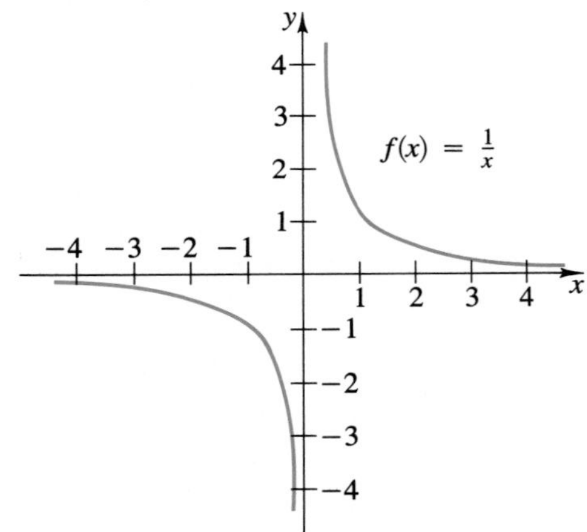

FIGURE 2

Our informal definition is this:

Infinite Limits If as $x \to a^+$, $f(x)$ increases without bound, then we write

$$\lim_{x \to a^+} f(x) = \infty$$

If as $x \to a^+$, $f(x)$ decreases without bound, then we write

$$\lim_{x \to a^+} f(x) = -\infty$$

We extend this definition analogously to cover the cases $\lim_{x \to a^-} f(x) = \infty$ and $\lim_{x \to a^-} f(x) = -\infty$.

Two precautionary notes: First, when a limit (or one-sided limit) is ∞ or $-\infty$, we do not say that the limit (or one-sided limit) exists. That is because ∞ and $-\infty$ do not exist as real numbers. Second, to claim that $\lim_{x \to a} f(x) = \infty$ (or $-\infty$), both one-sided limits must also be ∞ (or $-\infty$).

EXAMPLE 2 Let $g(x) = 1/(1 - x)$.

When $x > 1$, $1 - x < 0$, so $g(x)$ is negative. Thus as $x \to 1^+$, $1 - x \to 0^-$, and $g(x)$ is unbounded negatively. Therefore,

$$\lim_{x \to 1^+} g(x) = -\infty$$

When $x < 1$, $1 - x > 0$, so $g(x)$ is positive. Thus as $x \to 1^-$, $1 - x \to 0^+$, and $g(x)$ is unbounded positively. Therefore,

$$\lim_{x \to 1^-} g(x) = \infty$$

In this case, $\lim_{x \to 1} g(x)$ does not exist and is neither ∞ nor $-\infty$. The graph of g is shown in Figure 3.

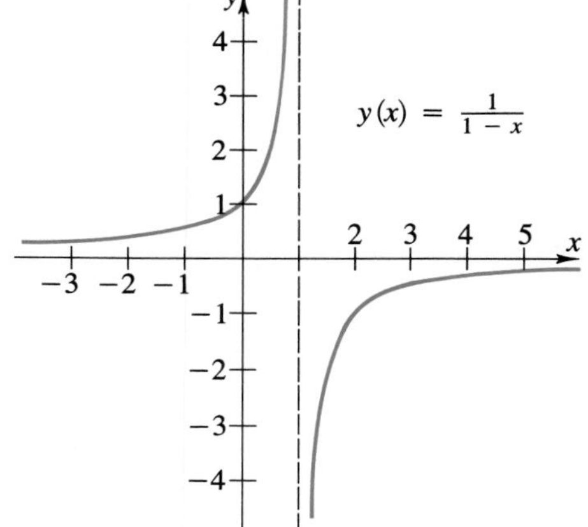

FIGURE 3

EXAMPLE 3 Let $h(x) = 1/x^2$. Because $h(x) > 0$, regardless of whether x is positive or negative,

$$\lim_{x \to 0^-} h(x) = \infty \quad \text{and} \quad \lim_{x \to 0^+} h(x) = \infty$$

Since the one-sided limits agree in this case, we can say

$$\lim_{x \to 0} h(x) = \infty$$

The graph of h is shown in Figure 4.

FIGURE 4

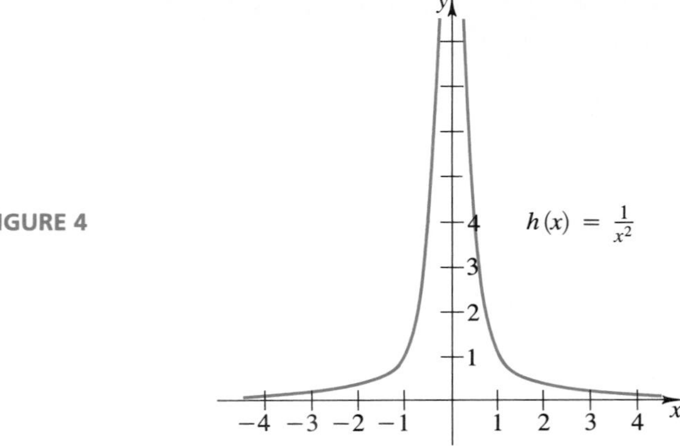

HOW MASS CHANGES WITH VELOCITY

In 1686 Isaac Newton provided an explanation for the motion of a particle. Newtonian mechanics has adequately described the motion of reasonable-size objects moving at reasonable speed. However, Newtonian mechanics does not adequately describe the motion of very small objects, such as atomic particles, or of very rapidly moving objects, such as cosmic-ray particles. A suitable explanation of the motion of these particles is provided by Einstein's special theory of relativity.

Suppose that a particle has mass m_0 when it is at rest on the Earth. If the particle moves away from the Earth at a velocity v, then to *an observer on the Earth* the mass m of the particle will be given by

$$m = \frac{m_0}{\sqrt{1 - (v^2/c^2)}}$$

where c is the velocity of light (186,000 miles per second). As v becomes larger and larger, approaching c, the mass of the particle increases without bound:

$$\lim_{v \to c} \frac{m_0}{\sqrt{1 - (v^2/c^2)}} = \frac{m_0}{\lim_{v \to c} \sqrt{1 - (v^2/c^2)}} = \infty$$

In the table below we show how the weight of an astronaut weighing 150 pounds on Earth would appear to *an observer on the Earth* as the astronaut travels away from the Earth in a spaceship traveling at an extremely high speed.

Velocity of spaceship (miles/second)	0	20,000	50,000	100,000	150,000	180,000	185,900	185,999
Mass of astronaut (pounds)	150	151	156	178	254	595	4,575	45,744

Vertical Asymptotes If $\lim_{x \to a} f(x) = \infty$ (possibly one-sided) or if $\lim_{x \to a} f(x) = -\infty$ (possibly one-sided), then the vertical line $x = a$ is said to be a **vertical asymptote** for the graph of the curve $y = f(x)$. As $x \to a$ (possibly one-sided), the curve approaches the line $x = a$. Loosely speaking, the line $x = a$ is tangent "at infinity" to the curve $y = f(x)$. In Figures 2 and 4 the y-axis is a vertical asymptote, whereas in Figure 3 the line $x = 1$ is a vertical asymptote.

Limits at Infinity It is often of interest to know how a function $f(x)$ behaves as x grows larger and larger without bound, either in the positive direction or in the negative direction. It may happen that $f(x)$ approaches a limit.

EXAMPLE 4 Consider the function

$$f(x) = 2 - \frac{1}{x} \qquad \text{for } x \geq 1$$

To see how $f(x)$ behaves as x increases without bound, we have listed in Table 2 some sample values of $y = f(x)$ for increasingly larger values of x.

TABLE 2

x	1	2	4	10	100	1000	10,000
$y = 2 - 1/x$	1	1.5	1.75	1.9	1.99	1.999	1.9999

As the table suggests, the values of $y = f(x)$ approach the number 2 as x increases positively without bound. We express this fact by writing

$$\lim_{x \to \infty} \left(2 - \frac{1}{x}\right) = 2$$

The symbol $\lim_{x \to \infty}$ is read "the limit as x approaches infinity"; it conveys the idea that x is allowed to increase positively without bound.

We now consider the function $f(x)$ for all values of x (except $x = 0$). To see how $f(x)$ behaves as x increases negatively without bound, we have listed in Table 3 some sample values of $y = f(x)$ for increasingly smaller values of x.

TABLE 3

x	−1	−2	−4	−10	−100	−1000	−10,000
$y = 2 - 1/x$	3	2.5	2.25	2.1	2.01	2.001	2.0001

As the table suggests, the values of $y = f(x)$ approach the number 2 as x decreases without bound. We express this fact by writing

$$\lim_{x \to -\infty} \left(2 - \frac{1}{x}\right) = 2$$

The symbol $\lim_{x \to -\infty}$ is read "the limit as x approaches negative infinity"; it conveys the idea that x is allowed to decrease without bound.

Example 4 suggests the following general definitions:

Limit of Infinity We write

$$\lim_{x \to \infty} f(x) = L$$

if the values of $f(x)$ approach the number L as x increases positively without bound.

Limit of Negative Infinity We write

$$\lim_{x \to -\infty} f(x) = L$$

if the values of $f(x)$ approach the number L as x decreases without bound.

Horizontal Asymptotes

Geometrically, if $\lim_{x \to \infty} f(x) = L$ [or $\lim_{x \to \infty} f(x) = L$], then the graph of $y = f(x)$ approaches the horizontal line $y = L$ as we progress in the positive x direction (or in the negative x direction). In such a case the line $y = L$ is called a **horizontal asymptote** to the curve $y = f(x)$.

EXAMPLE 5 In Example 4 we show that

$$\lim_{x \to \infty} \left(2 - \frac{1}{x}\right) = 2$$

Thus, the graph of $y = 2 - 1/x$ approaches the horizontal line $y = 2$ as x gets larger and larger (Figure 5). The line $y = 2$ is a horizontal asymptote of the curve.

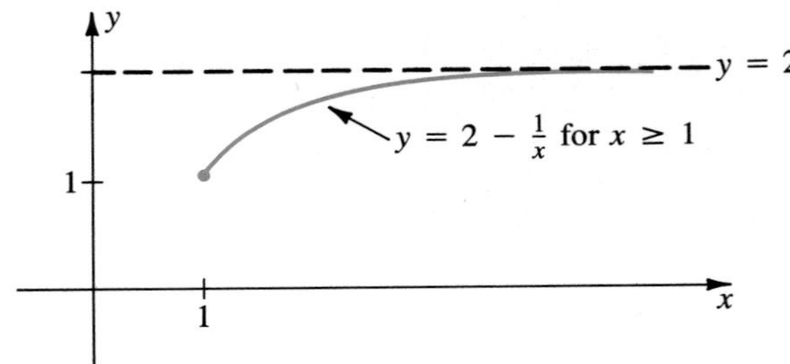

FIGURE 5

EXAMPLE 6 Find

$$\lim_{x \to \infty} \frac{5}{x^2}$$

Solution As x gets larger and larger, so does x^2, and thus the fraction $5/x^2$ approaches zero. Therefore,

$$\lim_{x \to \infty} \frac{5}{x^2} = 0$$

Similarly, $\lim_{x \to -\infty} 5/x^2 = 0$. Thus the x-axis is a horizontal asymptote in both positive and negative directions. The graph is shown in Figure 6.

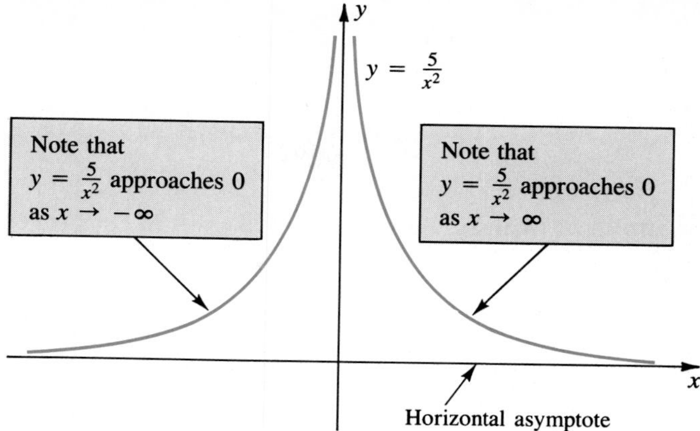

FIGURE 6

EXAMPLE 7 Find

$$\lim_{x\to\infty} 4x \quad \text{and} \quad \lim_{x\to-\infty} 4x$$

Solution As x increases positively without bound, so does $4x$. Thus, $\lim_{x\to\infty} 4x = \infty$. As x increases negatively without bound, so does $4x$. Thus, $\lim_{x\to-\infty} 4x = -\infty$. The graph is shown in Figure 7.

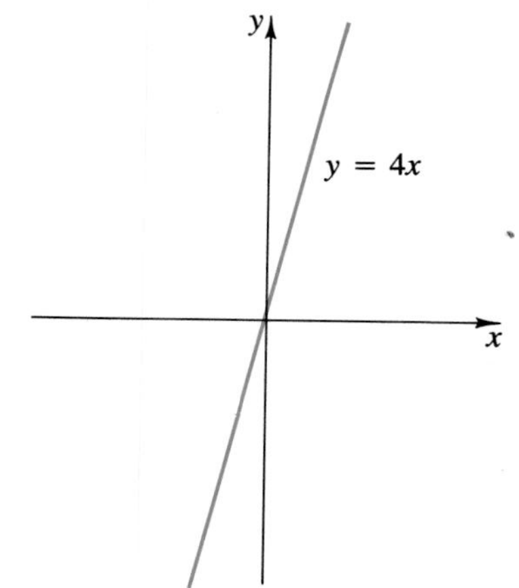

FIGURE 7

EXAMPLE 8 Find

$$\lim_{x\to\infty} \frac{4}{x^3 - 2} \quad \text{and} \quad \lim_{x\to-\infty} = \frac{4}{x^3 - 2}$$

Solution As x increases without bound, so does x^3, and so does $x^3 - 2$. Thus the entire fraction approaches zero as x approaches infinity; that is,

$$\lim_{x\to\infty} \frac{4}{x^3 - 2} = 0$$

Similarly,

$$\lim_{x \to -\infty} \frac{4}{x^3 - 2} = 0$$

Remark In Example 8, $y = 0$ is a horizontal asymptote of the curve.

Algebra of Limits at Infinity

An argument like that in Example 6 can be used to obtain the following more general result:

> If k is any constant and n a positive constant, then
>
> $$\lim_{x \to \infty} \frac{k}{x^n} = 0 \tag{1}$$
>
> and
>
> $$\lim_{x \to -\infty} \frac{k}{x^n} = 0 \tag{2}$$

For example, it follows from Equations (1) and (2) that

$$\lim_{x \to \infty} \frac{2}{x^7} = 0 \qquad \lim_{x \to \infty} \frac{5}{\sqrt{x}} = 0 \qquad \lim_{x \to -\infty} \frac{-7}{x^3} = 0$$

$$(k = 2,\ n = 7) \qquad (k = 5,\ n = \tfrac{1}{2}) \qquad (k = -7,\ n = 3)$$

Properties 1 through 4 of Section 12.3 together with Formulas (1) and (2) are extremely useful in calculating limits. These properties remain valid when applied to limits as $x \to \infty$ or $x \to -\infty$. The following two examples demonstrate their use.

EXAMPLE 9 Find

$$\lim_{x \to \infty} \frac{6x + 2}{x}$$

Solution In order to make use of Formulas (1) amd (2), we first divide by x.

$$\begin{aligned} \lim_{x \to \infty} \frac{6x + 2}{x} &= \lim_{x \to \infty} \left(6 + \frac{2}{x}\right) \\ &= \lim_{x \to \infty} 6 + \lim_{x \to \infty} \frac{2}{x} \\ &= 6 + 0 = 6 \end{aligned}$$

The limit of a quotient of two polynomials as x approaches infinity can sometimes be obtained by using Equation (1) or (2) after dividing the numerator and denominator by the highest power of x that appears in the denominator. The following example illustrates this technique.

EXAMPLE 10 Find

$$\lim_{x \to \infty} \frac{3x^2 + x}{6x^2 - 2}$$

Solution Since the highest power of x in the denominator is x^2, we divide the numerator and denominator by this expression and apply Equation (1) to obtain

$$\lim_{x\to\infty}\frac{3x^2+x}{6x^2-2}=\lim_{x\to\infty}\frac{3+1/x}{6-2/x^2}$$

$$=\frac{\lim\limits_{x\to\infty}3+\lim\limits_{x\to\infty}1/x}{\lim\limits_{x\to\infty}6-\lim\limits_{x\to\infty}2/x^2}$$

$$=\frac{3+0}{6-0}=\frac{1}{2}$$

EXAMPLE 11 Find

$$\lim_{x\to\infty}\frac{x^2-3x}{6x}\quad\text{and}\quad\lim_{x\to-\infty}\frac{x^2-3x}{6x}$$

Solution Following the method of Example 9,

$$\lim_{x\to\infty}\frac{x^2-3x}{6x}=\lim_{x\to\infty}\left(\frac{x}{6}-\frac{1}{2}\right)$$

Now $\frac{x}{6}-\frac{1}{2}$, which grows without bound as x becomes larger and larger, is positive as $x\to\infty$ but negative as $x\to-\infty$. Thus,

$$\lim_{x\to\infty}\frac{x^2-3x}{6x}=\lim_{x\to\infty}\left(\frac{x}{6}-\frac{1}{2}\right)=\infty$$

and

$$\lim_{x\to-\infty}\frac{x^2-3x}{6x}=\lim_{x\to-\infty}\left(\frac{x}{6}-\frac{1}{2}\right)=-\infty$$

EXERCISE SET 12.4

1. Consider the function $f(x)=1/(x-2)$ (see Example 4 of Section 12.3). Use Table 2 and Figure 6 in Section 12.3 to determine $\lim\limits_{x\to2^-}f(x)$, $\lim\limits_{x\to2^+}f(x)$, and $\lim\limits_{x\to2}f(x)$.
2. Consider the function $f(x)=|x|/x$ (see Example 5 in Section 12.3). Use Figure 7 of Section 12.3 to evaluate $\lim\limits_{x\to0^-}f(x)$ and $\lim\limits_{x\to0^+}f(x)$.

In Exercises 3–10 evaluate $\lim\limits_{x\to a^-}f(x)$, $\lim\limits_{x\to a^+}f(x)$, *and* $\lim\limits_{x\to a}f(x)$ *for the given function f and the specified value of a.*

3. $f(x)=\dfrac{2}{x+1}$, $a=-1$
4. $f(x)=\dfrac{1}{3-x}$, $a=3$
5. $f(x)=\dfrac{1}{1-x^2}$, $a=1$
6. $f(x)=\dfrac{1}{1+x^2}$, $a=-1$
7. $f(x)=\dfrac{x}{x-2}$, $a=2$
8. $f(x)=\dfrac{x}{x+3}$, $a=-3$
9. $f(x)=\dfrac{x-1}{x^2-1}$, $a=1$
10. $f(x)=\dfrac{x+2}{x^2-4}$, $a=2$

In Exercises 11–34 find the limit.

11. $\lim_{x \to \infty} \frac{2}{x^5}$

12. $\lim_{x \to \infty} \frac{3}{\sqrt{x}}$

13. $\lim_{x \to -\infty} \left(5 + \frac{7}{x}\right)$

14. $\lim_{x \to -\infty} \left(6 - \frac{5}{x^2}\right)$

15. $\lim_{x \to \infty} \left(-3 + \frac{6}{x^3} - \frac{2}{x}\right)$

16. $\lim_{x \to \infty} \left(\frac{4}{x^{1/3}} - \frac{5}{x^8} + 8\right)$

17. $\lim_{x \to -\infty} 6x$

18. $\lim_{x \to -\infty} (2x + 3)$

19. $\lim_{x \to \infty} (5x^2 - 2)$

20. $\lim_{x \to \infty} (x^2 + x)$

21. $\lim_{x \to -\infty} \frac{5}{x^2 + 3}$

22. $\lim_{x \to -\infty} \frac{6}{x^3 - 1}$

23. $\lim_{x \to \infty} \frac{8x + 5}{x}$

24. $\lim_{x \to \infty} \frac{9x - 2}{x}$

25. $\lim_{x \to \infty} \frac{4x - 3x^2}{x^2}$

26. $\lim_{x \to \infty} \frac{5x^2 + x^3}{x^2}$

27. $\lim_{x \to -\infty} \frac{x^2 - 5x}{8x}$

28. $\lim_{x \to -\infty} \frac{8x^3 - 3x^2}{x^3}$

29. $\lim_{x \to \infty} \frac{8x^2 + 1}{4x^2 + 2}$

30. $\lim_{x \to \infty} \frac{3x^3 + x}{9x^3 - x}$

31. $\lim_{x \to -\infty} \frac{5x^4 + 2x^2}{5x^4 - x}$

32. $\lim_{x \to -\infty} \frac{1 - x^2}{3 + x^2}$

33. $\lim_{x \to \infty} \frac{4x^3 + 2x^2 + x}{8x^3 + 1}$

34. $\lim_{x \to \infty} \frac{1 - 3x^2 + 4}{5 - x + 2x^4}$

In Exercises 35 and 36 use the given graph of $y = f(x)$ to find $\lim_{x \to \infty} f(x)$.

35.

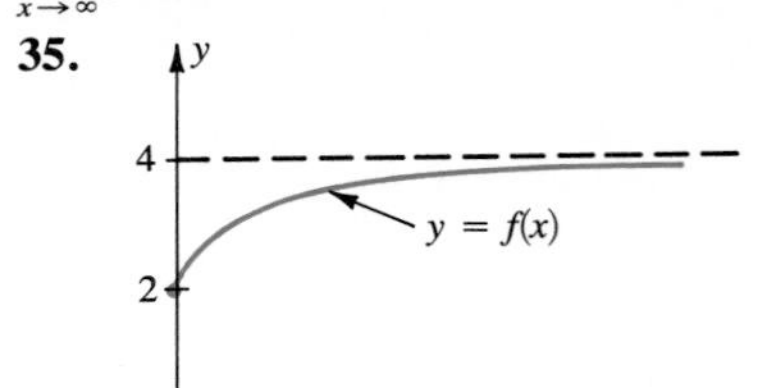

36.

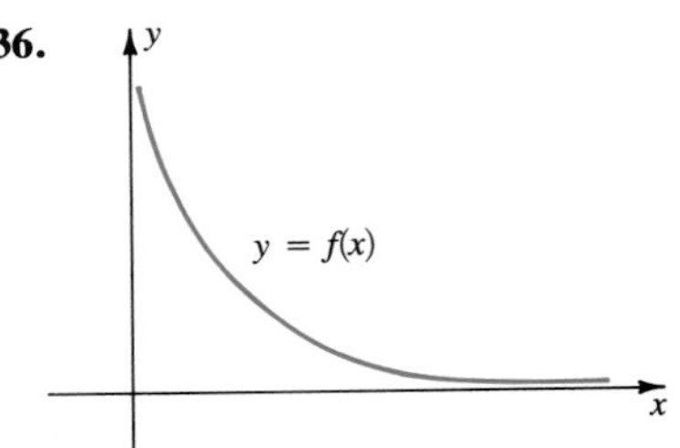

In Exercises 37–42 find all horizontal and vertical asymptotes for the given curve.

37. $y = \frac{x + 1}{x + 2}$

38. $y = \frac{3x + 5}{2x - 4}$

39. $y = \frac{x}{x^2 - x - 2}$

40. $y = \frac{2x^2}{x^2 - 9}$

41. $y = \frac{x + 1}{x^2 + 5x + 4}$

42. $y = 7 - \frac{3}{x}$

12.5 CONTINUITY

In this section we investigate conditions under which the graph of a function f is assured of forming a continuous, unbroken curve.

Loosely speaking, we use the term *continuous* to mean "without gaps or jumps." Thus we might conceive of a continuous curve as one that can be drawn without having to lift the pencil from the paper (Figure 1).

To make this intuitive notion of continuity precise, let us consider some of the ways in which a curve can *fail* to be continuous. In Figure 2 we have sketched some curves that, because of their behavior at $x = a$, are not continuous.

The curve in Figure 2a has a hole in it at the point $x = a$, indicating that the function f is undefined there. For the curves in Figure 2b and c, the function f is defined at $x = a$, but $\lim_{x \to a} f(x)$ does not exist, thereby causing a break in the

FIGURE 1

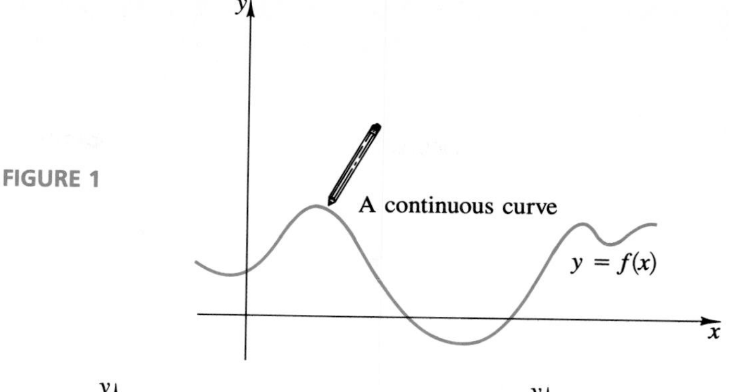

FIGURE 2

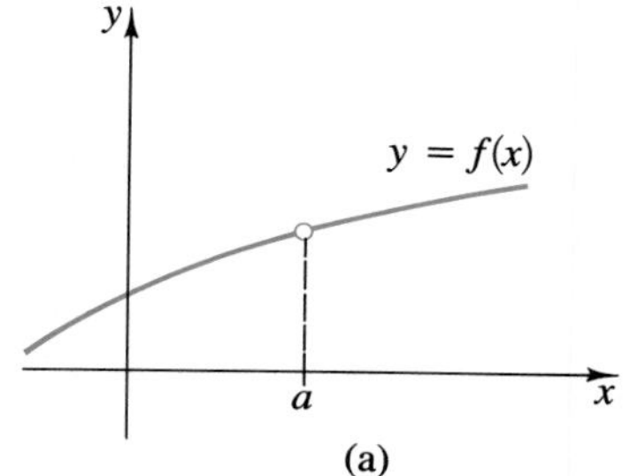

(a)

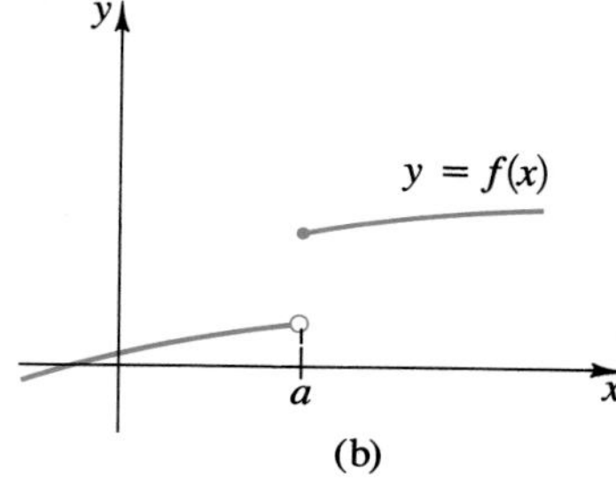

(b)

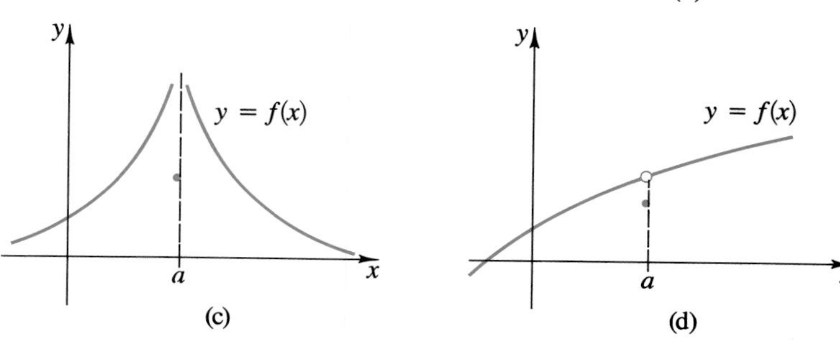

(c) (d)

graph. For the curve in (d), f is defined at $x = a$ and $\lim_{x \to a} f(x)$ exists, yet the graph has a break at the point $x = a$ because

$$\lim_{x \to a} f(x) \neq f(a)$$

As a result of this discussion, it should be evident that a function must satisfy the following three conditions so that its graph does not have a hole or break at the point $x = a$:

Definition A function f is **continuous** at $x = a$ if

1. $f(a)$ is defined.
2. $\lim_{x \to a} f(x)$ exists.
3. $\lim_{x \to a} f(x) = f(a)$.

If f is not continuous at $x = a$, we say that f is **discontinuous** at $x = a$ or that a is a **point of discontinuity** for f. A function that is continuous at every point on the real number line is called **continuous everywhere** or merely **continuous**; if there is at least one point of discontinuity, then the function is said to be **discontinuous** at each such point.

EXAMPLE 1 The function $f(x) = x$ is continuous because at any point a,

1. f is defined; the value of f at a is $f(a) = a$.
2. $\lim_{x \to a} f(x)$ exists since $\lim_{x \to a} x = a$.
3. $\lim_{x \to a} f(x) = f(a) = a$.

EXAMPLE 2 Every constant function is continuous; for if $f(x) = c$ is constant, then at any point a

$$f(a) = c \quad \text{and} \quad \lim_{x \to a} f(x) = \lim_{x \to a} c = c = f(a)$$

Algebra of Continuous Functions

It is clear from our definition that continuity is intimately connected with limit properties. In fact, a continuous function is one whose limits are easiest to calculate.

> A function $f(x)$ is continuous at $x = a$ if its limit as $x \to a$ can be calculated by merely substituting $x = a$.

From Example 13 in Section 12.3 we obtain the following result:

> 1. Every polynomial is continuous everywhere.

Properties 3 and 4 of limits in Section 12.3 allow us to conclude the following analogous properties:

> 2. The sum of two functions both continuous at $x = a$ is continuous at $x = a$.
> 3. The difference of two functions both continuous at $x = a$ is continuous at $x = a$.
> 4. A constant times a function continuous at $x = a$ is continuous at $x = a$.
> 5. The product of two functions both continuous at $x = a$ is continuous at $x = a$.
> 6. The quotient $p(x)/q(x)$ of two functions $p(x)$ and $q(x)$ both continuous at $x = a$ is continuous at $x = a$ unless $q(a) = 0$.

Although we have not yet defined "rational" functions, you have already seen many of them. Briefly, a **rational function**

$$r(x) = \frac{p(x)}{q(x)}$$

is any quotient of polynomials.

Property 6 assures us that:

> **7.** Every rational function
> $$r(x) = \frac{p(x)}{q(x)}$$
> is continuous everywhere except where the denominator $q(x)$ is 0.

EXAMPLE 3 The functions

$$f(x) = 2 + x$$
$$f(x) = x^2 - 2x + 3$$
$$f(x) = 4 + 5x - 6x^2 + 7x^3$$

are continuous everywhere since they are polynomials.

EXAMPLE 4 The rational function

$$f(x) = \frac{1}{x - 3}$$

is continuous everywhere except at $x = 3$ since the denominator is zero only at this point. Note the break in the graph at $x = 3$ (Figure 3).

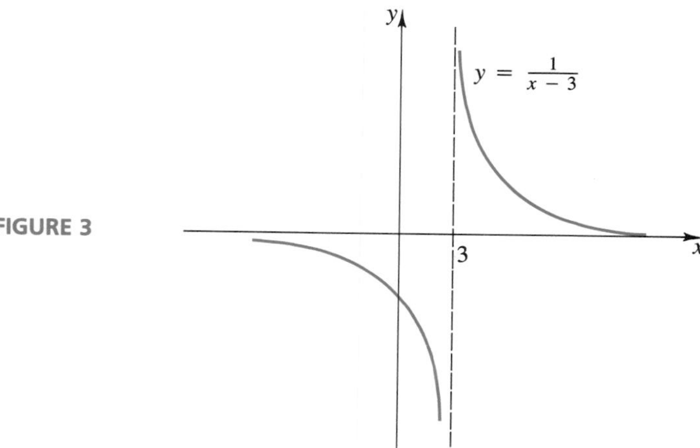

FIGURE 3

Application to Curve Sketching

When you are required to graph a function f, it will help you to know in advance whether the graph will be "all of one piece," or must be drawn in several disconnected pieces. Loosely speaking, the curve splits into separate (disconnected) pieces at each point of discontinuity. A glance back at Figure 2 will make this clear. In Example 4 the function has one point of discontinuity, $x = 3$, and thus we are not surprised that the curve separates into two pieces there. Since polynomials are continuous everywhere, all polynomial curves consist of a single connected piece.

EXAMPLE 5 **THE POSTAGE FUNCTION** The post office charges the following rates for first-class mail weighing no more than 11 ounces: 29 cents for the first ounce or fraction thereof, and for each ounce after the first ounce, 23 cents per ounce or fraction thereof. Figure 4 gives a graph of this postage function. Note that this function is discontinuous at each integer value of $x \geq 0$.

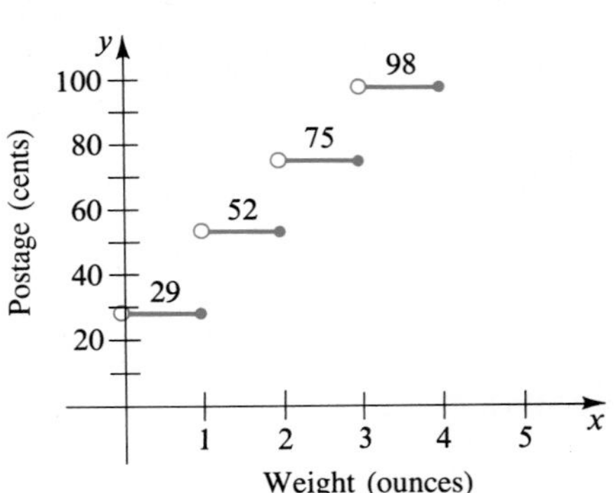

FIGURE 4

EXERCISE SET 12.5

1. In each part determine whether the function whose graph is shown is continuous everywhere. If not, locate the points of discontinuity.

(a)

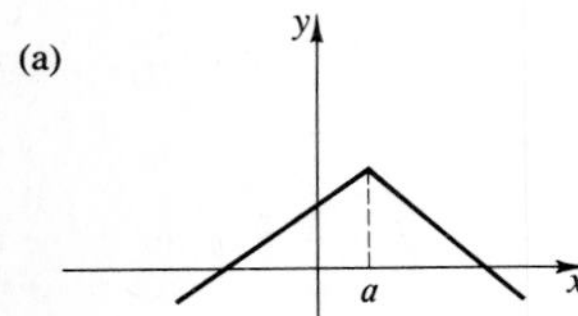

(b)

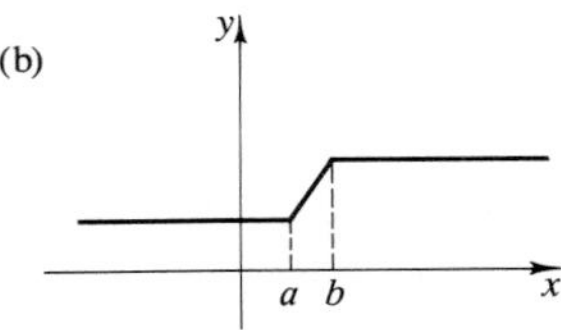

(c)

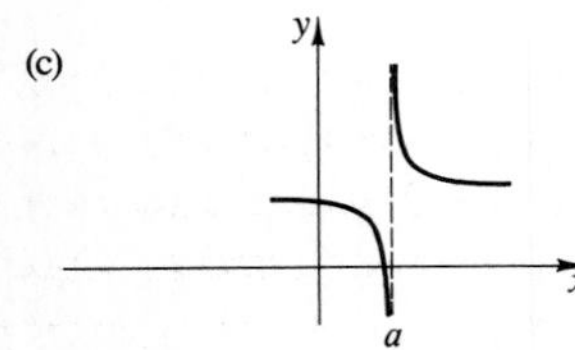

(d)

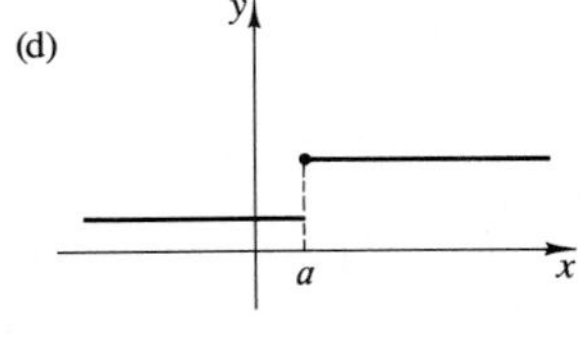

(e)

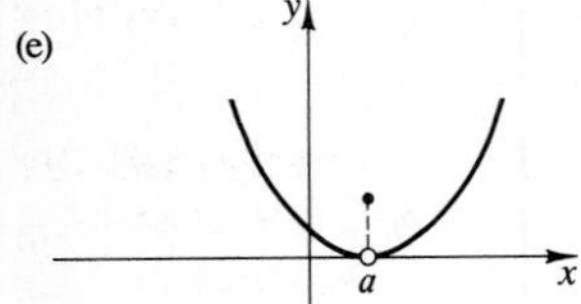

(f)

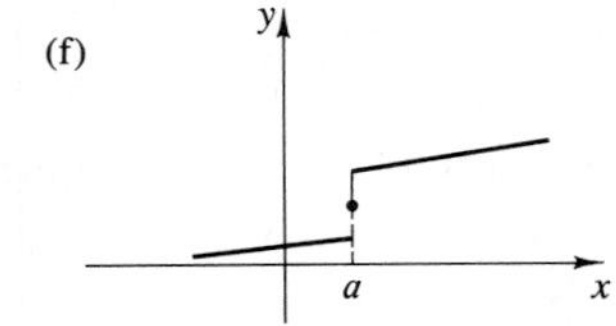

(g)

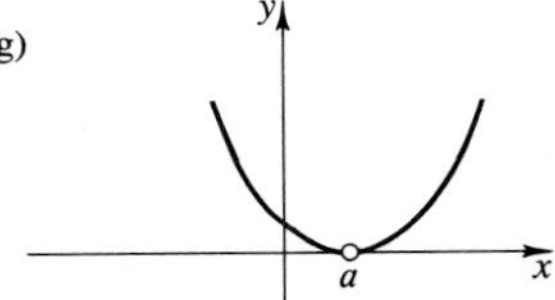

(h)

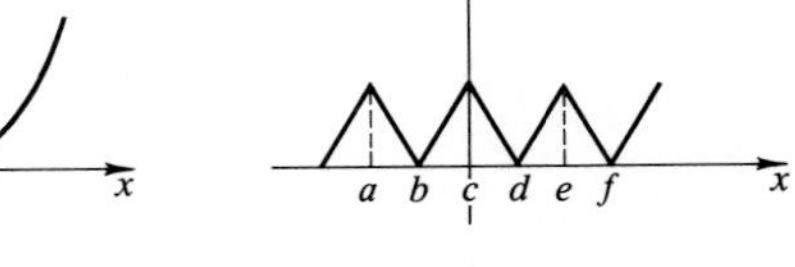

2. In each part, the given function is discontinuous at the point $x = 6$. In each case, state which of the three requirements for continuity is (are) violated at $x = 6$.

(a) $f(x) = \dfrac{x^2 - 36}{x - 6}$

(b) $f(x) = \begin{cases} x + 6 & \text{if } x \neq 6 \\ 0 & \text{if } x = 6 \end{cases}$

(c) $f(x) = \begin{cases} x + 1 & \text{if } x \geq 6 \\ x - 1 & \text{if } x < 6 \end{cases}$

3. Consider the function f defined by

$$f(x) = \begin{cases} \dfrac{x^2 - 36}{x - 6} & \text{if } x \neq 6 \\ 12 & \text{if } x = 6 \end{cases}$$

Is f continuous at $x = 6$? Justify your answer.

4. Consider the function f defined by

$$f(x) = \begin{cases} \dfrac{x^2 - 4}{x - 2} & \text{if } x \neq 2 \\ 8 & \text{if } x = 2 \end{cases}$$

Is f continuous at $x = 2$? Justify your answers.

In Exercises 5–11 determine whether the function is continuous at the indicated point, $x = a$.

5. $f(x) = 2x^2 - 5x + 1;\ a = 4$

6. $f(x) = \dfrac{1}{2x - 1};\ a = \dfrac{1}{2}$

7. $f(x) = \dfrac{(x - 1)(x - 2)}{(x - 2)};\ a = 2$

8. $f(x) = \begin{cases} 2x & \text{if } x \leq 2 \\ 4 & \text{if } x > 2 \end{cases};\ a = 2$

9. $f(x) = \begin{cases} 2x & \text{if } x \neq 2 \\ 5 & \text{if } x = 2 \end{cases};\ a = 2$

10. $f(x) = \begin{cases} 2x & \text{if } x \leq 2 \\ 3x + 1 & \text{if } x > 2 \end{cases};\ a = 2$

11. $f(x) = |x - 3|;\ a = 3$

In Exercises 12–20 determine all points of discontinuity, if any, of the given function.

12. $f(x) = 4x^2 - 2x + 5$

13. $f(x) = \dfrac{x^2 - 9}{x - 3}$

14. $f(x) = \dfrac{3}{x - 3}$

15. $f(x) = |x + 1|$

16. $f(x) = \dfrac{3}{(x - 2)(x + 3)}$

17. $f(x) = \begin{cases} 3x & \text{if } x \leq 1 \\ 2x + 1 & \text{if } x > 1 \end{cases}$

18. $f(x) = \dfrac{x}{x^2 - 5x + 6}$

19. $f(x) = \dfrac{2x}{x}$

20. $f(x) = \dfrac{(x - 1)(x + 3)}{4}$

21. A coffee merchant reorders as soon as the stock falls below 100 pounds. When this occurs, the merchant reorders enough coffee to bring the stock up to 500 pounds. Let $f(x)$ denote the stock on hand at time x. Is f a continuous function? Explain your answer.

22. **(Business)** A manufacturer of film developer finds that due to volume discounting the cost of producing x tons of the product is given (in tens of thousands of dollars) by

$$C(x) = \begin{cases} 3x + 5 & \text{if } 0 \leq x \leq 4 \\ 2x + 6 & \text{if } 4 < x \leq 8 \\ \frac{1}{2}x + 8 & \text{if } 8 < x \leq 16 \end{cases}$$

(a) Sketch the graph of $C(x)$.
(b) Discuss the continuity of the cost function.

12.6 THE DERIVATIVE OF A FUNCTION

In this section we take up the most important single idea in the entire subject of calculus: the derivative of a function. It is an extremely powerful idea that pervades the remainder of the course. You have been introduced informally to the derivative of a function earlier in this chapter. We now turn to a formal definition, and then we develop procedures and rules for computing derivatives.

Definition 1 For a given function f, the **derivative of f at a**, denoted by $f'(a)$, is the limit

$$f'(a) = \lim_{h \to 0} \frac{f(a + h) - f(a)}{h} \tag{1}$$

provided this limit exists.

You will notice that the derivative is defined "locally," meaning that a function f may have a derivative at some points but not necessarily at all points. If a function f has a derivative at a, we say that f is **differentiable** at a. The process of finding a derivative is called **differentiation**.

Recalling our discussion of rates of change and slopes of curves in Section 12.2, we make the following observations:

> If a function f has a derivative $f'(a)$ at a, then:
>
> (a) the instantaneous rate of change of f relative to x at a is $f'(a)$
>
> (b) the slope of the curve (or the slope of the tangent line to the curve) $y = f(x)$ at $(a, f(a))$ is $f'(a)$.

By Definition 1, the derivative of f is a number $f'(a)$ dependent on the number a. Thus, if we allow x to vary and consider $f'(x)$, we may regard the derivative as a function of x.

> **Definition 2** For a given function f, the **derivative of f** is the function $f'(x)$ given by
>
> $$f'(x) = \lim_{h \to 0} \frac{f(x+h) - f(x)}{h} \tag{2}$$
>
> when this limit exists. The domain of f' is the set of all points x where f is differentiable.

Thus, we may think of the derivative $f'(x)$ as a function whose output value at a is the slope of the curve $y = f(x)$ at $(a, f(a))$ or the instantaneous rate of change in $f(x)$ with respect to x at $x = a$.

EXAMPLE 1 Let f be defined by $f(x) = x^2$. Then the derivative is

$$\begin{aligned} f'(x) = \lim_{h \to 0} \frac{f(x+h) - f(x)}{h} &= \lim_{h \to 0} \frac{(x+h)^2 - x^2}{h} \\ &= \lim_{h \to 0} \frac{x^2 + 2xh + h^2 - x^2}{h} \\ &= \lim_{h \to 0} (2x + h) \\ &= 2x \end{aligned}$$

By substituting various specific values for x in the derivative

$$f'(x) = 2x$$

we can obtain the slope of the tangent to the graph of $f(x) = x^2$ at various points. For example, the slopes of the tangent lines to $f(x) = x^2$ at $x = 2$, $x = 0$, and $x = -2$ are

$$\begin{aligned} &\text{slope at } x = 2: && f'(2) = 2(2) = 4 \\ &\text{slope at } x = 0: && f'(0) = 2(0) = 0 \\ &\text{slope at } x = -2: && f'(-2) = 2(-2) = -4 \end{aligned}$$

(See Figure 1.)

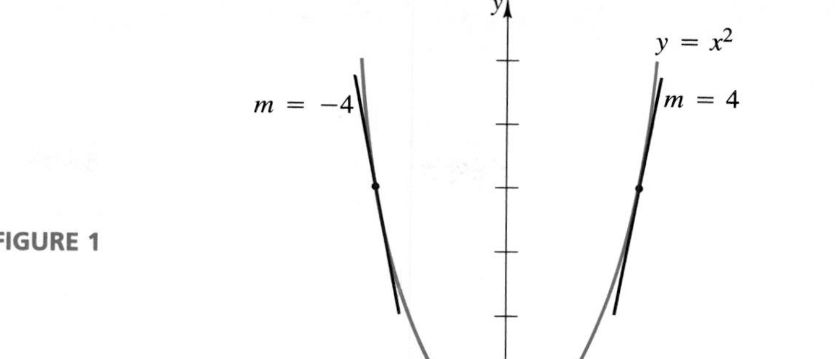

FIGURE 1

EXAMPLE 2 It is geometrically obvious that at any point a tangent to a line $y = mx + b$ will coincide with the line itself and will therefore have slope m (Figure 2). As a result, we should suspect that the derivative of $f(x) = mx + b$ will be constant with value m at each point. Let us see if this is in fact the case:

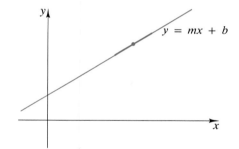

FIGURE 2

$$\begin{aligned} f'(x) &= \lim_{h\to 0} \frac{f(x+h) - f(x)}{h} \\ &= \lim_{h\to 0} \frac{[m(x+h) + b] - [mx + b]}{h} \\ &= \lim_{h\to 0} \frac{mx + mh + b - mx - b}{h} \\ &= \lim_{h\to 0} \frac{mh}{h} \\ &= \lim_{h\to 0} m \\ &= m \qquad \text{(the limit of a constant)} \end{aligned}$$

Thus $f'(x) = m$, as we suspected.

Before proceeding with more examples, let us summarize the procedure for calculating a derivative of $f(x)$.

Procedure for Calculating Derivatives

Step 1 Calculate

$$f(x + h)$$

Step 2 Form the ratio

$$\frac{f(x + h) - f(x)}{h}$$

(sometimes called the **difference quotient**).

Step 3 Find

$$\lim_{h \to 0} \frac{f(x + h) - f(x)}{h}$$

if it exists.

This is an abbreviation of the five-step procedure presented in Section 12.2. You may prefer to continue using the three-step procedure, but we will proceed directly to step 3.

EXAMPLE 3 (a) Find the derivative of the function $f(x) = x^3$.

(b) Find the equation of the line tangent to the curve $y = x^3$ at (2, 8).

Solution (a) The derivative of $f(x) = x^3$ is

$$f'(x) = \lim_{h \to 0} \frac{(x + h)^3 - x^3}{h} \qquad \text{[from Equation (2)]}$$

$$= \lim_{h \to 0} \frac{(x^3 + 3x^2h + 3xh^2 + h^3) - x^3}{h}$$

$$= \lim_{h \to 0} \frac{h(3x^2 + 3xh + h^2)}{h}$$

$$= \lim_{h \to 0} (3x^2 + 3xh + h^2)$$

$$= 3x^2$$

Thus, $f'(x) = 3x^2$.

Solution (b) The tangent line is the straight line passing through the point of tangency (2, 8) and having slope $m = f'(2) = 3(2)^2 = 12$. We use the point–slope form of the equation of the line

$$y - y_1 = m(x - x_1)$$

to get the equation

$$y - 8 = 12(x - 2) \quad \text{or} \quad y = 12x - 16$$

which is the equation of the desired tangent line.

Equation of Tangent Line If f is differentiable at x_1, then the curve $y = f(x)$ has a unique tangent line at $(x_1, f(x_1))$. The equation of this tangent line is

$$y - f(x_1) = m(x - x_1) \tag{3}$$

where $m = f'(x_1)$.

EXAMPLE 4 (a) Find the derivative of the function $f(x) = 1/x$.

(b) Find the equation of the line tangent to the curve $y = 1/x$ at the point $(2, \frac{1}{2})$.

Solution (a) The derivative of $f(x) = 1/x$ is

$$\begin{aligned} f'(x) &= \lim_{h\to 0} \frac{1/(x + h) - 1/x}{h} \\ &= \lim_{h\to 0} \frac{x - (x + h)}{hx(x + h)} \\ &= \lim_{h\to 0} \frac{-h}{hx(x + h)} \\ &= \lim_{h\to 0} \frac{-1}{x(x + h)} \\ &= -\frac{1}{x^2} \end{aligned}$$

Solution (b) The slope of the tangent line is $m = f'(2) = -1/2^2 = -\frac{1}{4}$. Using Equation (3), we find that the equation of the tangent line is

$$y - y_1 = m(x - x_1)$$

where $m = f'(2) = -\frac{1}{4}$, $x_1 = 2$, and $y_1 = \frac{1}{2}$:

$$\begin{aligned} y - \frac{1}{2} &= -\frac{1}{4}(x - 2) \\ 4y - 2 &= -(x - 2) \\ 4y &= -x + 4 \quad \text{or} \quad y = -\frac{1}{4}x + 1 \end{aligned}$$

The graph is shown in Figure 3.

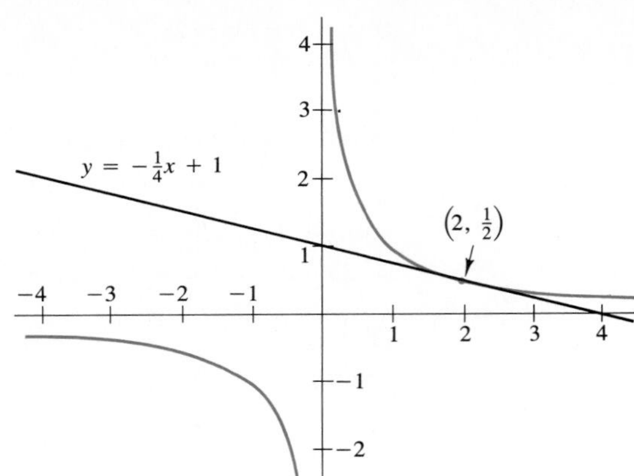

FIGURE 3

EXAMPLE 5 (a) Find the derivative of the function $f(x) = \sqrt{x}$.

(b) Find the instantaneous rate of change of $f(x) = \sqrt{x}$ with respect to x at $x = 9$.

Solution (a) The derivative of $f(x) = \sqrt{x}$ is

$$f'(x) = \lim_{h\to 0} \frac{\sqrt{x+h} - \sqrt{x}}{h}$$

$$= \lim_{h\to 0} \left(\frac{\sqrt{x+h} - \sqrt{x}}{h} \cdot \frac{\sqrt{x+h} + \sqrt{x}}{\sqrt{x+h} + \sqrt{x}} \right)$$

(Both numerator and denominator are multiplied by $\sqrt{x+h} + \sqrt{x}$ to obtain a simpler form whose limit we can compute.)

$$= \lim_{h\to 0} \frac{(x+h) - x}{h(\sqrt{x+h} + \sqrt{x})}$$

$$= \lim_{h\to 0} \frac{h}{h(\sqrt{x+h} + \sqrt{x})}$$

$$= \lim_{h\to 0} \frac{1}{\sqrt{x+h} + \sqrt{x}}$$

$$= \frac{1}{\sqrt{x} + \sqrt{x}} = \frac{1}{2\sqrt{x}}$$

Thus, $f'(x) = \dfrac{1}{2\sqrt{x}}$

Solution (b) Using (a), the instantaneous rate of change of the function $f(x) = \sqrt{x}$ at $x = 9$ is

$$f'(9) = \frac{1}{2\sqrt{9}} = \frac{1}{6}$$

Alternative Notation for Derivatives

Unfortunately, there is no general agreement on the best notation to use in representing derivatives. To enable you to read other references on the subject, we provide a list of some alternative symbols (Table 1).

The symbols D_x, D, and d/dx are operator symbols; they represent the operation of differentiation. The symbols $D_x(\ )$, $D(\ )$, and $\frac{d}{dx}[\ ]$ or $\frac{d}{dx}(\ )$ indicate that whatever function appears inside the parentheses or brackets is to be differentiated.

The delta notation is a quite common device for denoting the change in a variable or function. Using this notation with a given function $y = f(x)$,

$$\begin{aligned} \Delta x &= \text{a change in } x \qquad (\text{from } x \text{ to } x + \Delta x) \\ \Delta y &= \text{the corresponding change in } f(x) \qquad [\text{from } f(x) \text{ to } f(x + \Delta x)] \\ &= f(x + \Delta x) - f(x) \end{aligned}$$

TABLE 1

Alternative Symbol	Example of Use
$f'(x)$	If $f(x) = x^3$, then $f'(x) = 3x^2$
y'	If $y = x^3$, then $y' = 3x^2$
D_x	$D_x x^3 = 3x^2$ If $y = x^3$, then $D_x y = 3x^2$
D	$D(x^3) = 3x^2$
$\frac{dy}{dx}$	If $y = x^3$, then $\frac{dy}{dx} = 3x^2$
$\frac{d}{dx}$	$\frac{d}{dx}[x^3] = 3x^2$

With this notation,

$$f'(x) = \lim_{\Delta x \to 0} \frac{f(x + \Delta x) - f(x)}{\Delta x}$$

It is more compact to write

$$f'(x) = \lim_{\Delta x \to 0} \frac{\Delta y}{\Delta x}$$

Since the derivative is the limit of a quotient, early developers of the calculus thought of the derivative itself as a quotient. Hence they used the symbolism dy/dx to represent the derivative.

If $y = f(x)$, where f is differentiable, then

$$\frac{dy}{dx} = f'(x)$$

This is still probably the most universally accepted notation for derivatives. It is especially useful when variables other than x and y are used. For example, if $u = f(t)$, then the derivative of f with respect to t might be denoted

$$\frac{du}{dt}$$

and if $A = h(r)$, then the derivative of h with respect to r might be denoted

$$\frac{dA}{dr}$$

Thus, if $u = t^2$, then

$$\frac{du}{dt} = 2t$$

Differentiability and Continuity

It is possible that the limit in the definition of derivative (1) may not exist for certain values of a, in which case we say that f is **not differentiable** at a. Some examples of such points a (Figure 4) are:

(a) points where f has a vertical tangent (so that the tangent has infinite slope)
(b) points where f is discontinuous
(c) points where the graph of f has a sharp corner

FIGURE 4

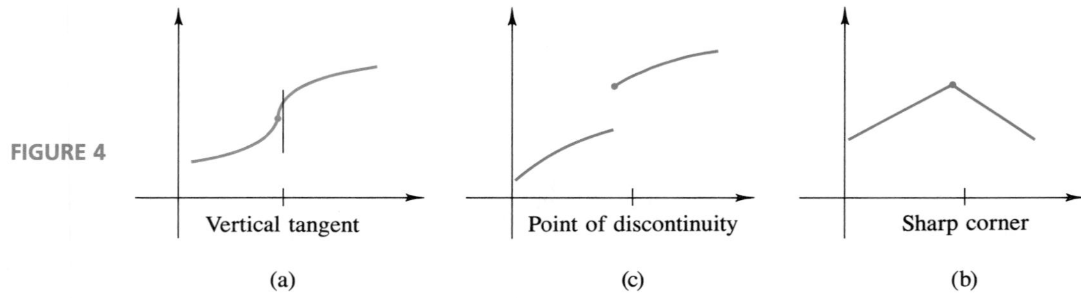

The following theorem tells us that it is impossible for a function to have a derivative at a point where it is not continuous.

THEOREM If f is differentiable at a, then f is continuous at a.

The converse of this theorem is false. That is, a function f could be continuous at a but not differentiable at a. The following example illustrates this observation.

EXAMPLE 6 The function $f(x) = |x|$ is *not* differentiable at 0. To show this we resort to the definition of the derivative as a limit and take the limit from both sides separately:

$$\lim_{h\to 0^+} \frac{f(0+h)-f(0)}{h} = \lim_{h\to 0^+} \frac{|h|-0}{h}$$

$$= \lim_{h\to 0^+} \frac{|h|}{h} = \lim_{h\to 0^+} \frac{h}{h} = 1$$

while

$$\lim_{h\to 0^-} \frac{f(0+h)-f(0)}{h} = \lim_{h\to 0^-} \frac{|h|-0}{h}$$

$$= \lim_{h\to 0^-} \frac{|h|}{h} = \lim_{h\to 0^-} \frac{-h}{h} = -1$$

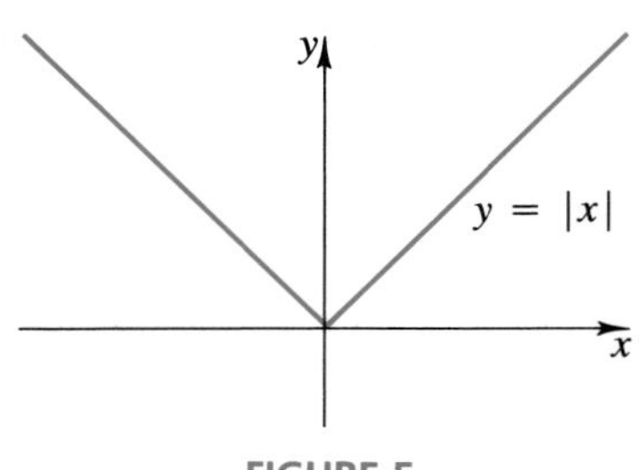

FIGURE 5

Since the above limits are different, the limit defining $f'(0)$ does not exist, and therefore the absolute value function is not differentiable at 0. Intuitively, of course, the problem is that the absolute value function has a sharp corner in its graph at $x = 0$ (Figure 5).

EXERCISE SET 12.6

1. In each part use the graph to estimate $f'(a)$.

(a)
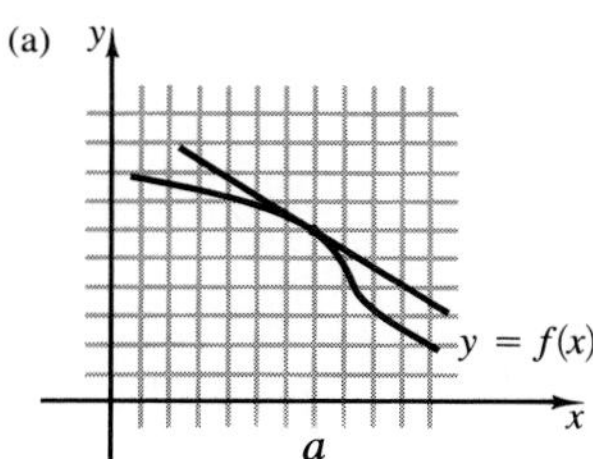

(b)
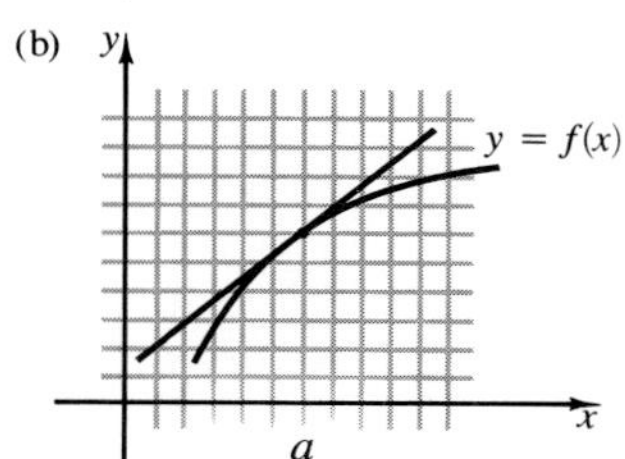

(c)
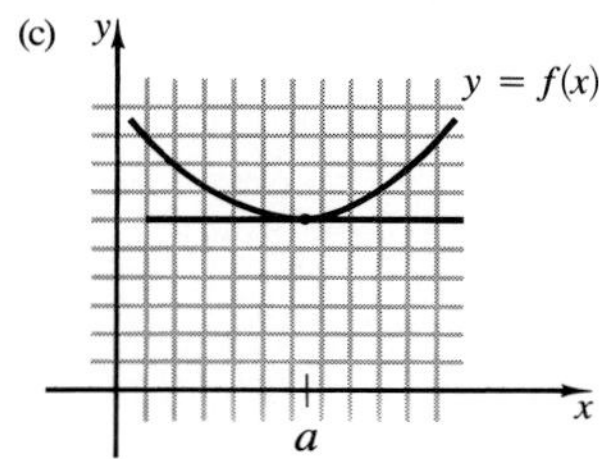

(d)
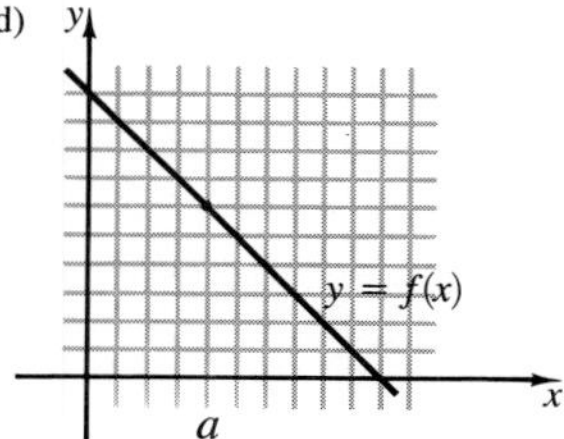

2. For the function $f(x) = \frac{1}{2}x^2$ find:
   (a) $\dfrac{f(x+h)-f(x)}{h}$
   (b) $f'(x)$
   (c) $f'(-2), f'(0), f'(4)$
   (d) Using graph paper, carefully plot the graph of f and draw approximate tangent lines at the points where $x = -2$, $x = 0$, and $x = 4$.
   (e) Are the slopes of the tangent lines in reasonable agreement with the calculations in part (c)?

In Exercises 3–14 use Definition 1 to obtain $f'(a)$ for the given $f(x)$ and a.

3. $f(x) = 4x - 5$, $a = 1$
4. $f(x) = 3x + 7$, $a = -2$
5. $f(x) = \frac{2}{3}x + 2$, $a = 3$

6. $f(x) = \frac{3}{4}x - 2$, $a = 4$

7. $f(x) = 3x^2$, $a = -1$

8. $f(x) = 2x^2 + 1$, $a = 1$

9. $f(x) = x^2 - x$, $a = 3$

10. $f(x) = x^2 - 4x$, $a = 3$

11. $f(x) = 2x^3$, $a = -2$

12. $f(x) = x^3 - 1$, $a = 2$

13. $f(x) = 1/(x + 2)$, $a = -1$

14. $f(x) = 1/(3 - x)$, $a = 0$

In Exercises 15–20 use Definition 2 to obtain $f'(x)$ from the given $f(x)$.

15. $f(x) = 5x^2$

16. $f(x) = 4x^2 - 5$

17. $f(x) = x^2 + 4x$

18. $f(x) = 2x^2 - 3x + 7$

19. $f(x) = 4/x$

20. $f(x) = x^4$

21. Find the slope and the equation of the tangent line to the graph of $f(x) = x^2 + x$ at the point where:
(a) $x = 0$ (b) $x = 1$
(c) $x = -1$ (d) $x = 2$

22. Find the slope and the equation of the tangent line to the graph of $f(x) = 1/x$ at the point where:
(a) $x = -1$ (b) $x = 1$
(c) $x = 3$ (d) $x = -3$

23. Suppose f is defined by $f(x) = 2x^2 - 3x + 4$. Find the point on the graph of f where the slope of the tangent line is 9. What is the equation of this tangent line?

24. Suppose f is defined by $f(x) = 2x^3 + 3x^2 - 12x + 5$. Find the points on the graph of f where the tangent line is horizontal.

In Exercises 25–30 rewrite the given statement using each of the other five alternative notations given in Table 1 of this section.

25. If $f(x) = 5x + 11$, then $f'(x) = 5$.

26. $D_x(x^2 + 3x) = 2x + 3$

27. If $y = 3/x$, then $dy/dx = -3/x^2$.

28. If $y = x^5$, then $y' = 5x^4$.

29. $D(3x - x^3) = 3 - 3x^2$

30. $\frac{d}{dx}(4x^2) = 8x$

31. List the points where the function graphed in Figure 6 is not differentiable.

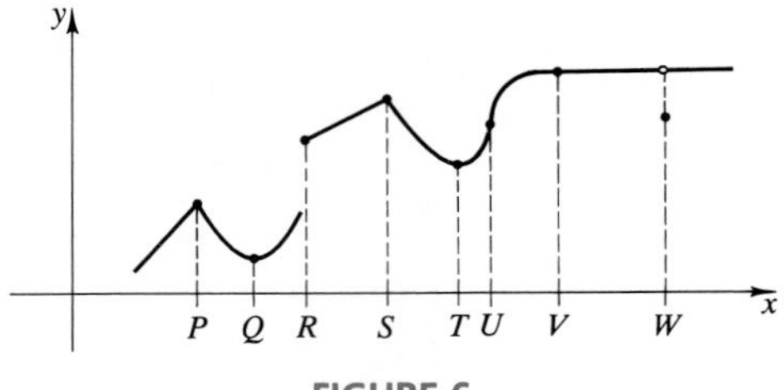

FIGURE 6

In Exercises 32–40 determine whether the given function $f(x)$ is differentiable at the given point. Is it continuous there?

32. The light bulb function

$$f(x) = \begin{cases} 0 & \text{if } x < 2 \\ 1 & \text{if } 2 \le x \le 5 \\ 0 & \text{if } x > 5 \end{cases}$$

at the point $a = 2$.

33. The light bulb function given in Exercise 32 at $a = 3$.

34. $f(x) = |x|$ at $a = 2$

35. $f(x) = |x - 1|$ at $a = 1$

36. $f(x) = 1/x$ at $a = 0$

37. $f(x) = \dfrac{x + 1}{x - 2}$ at $a = 2$

38. $f(x) = \begin{cases} x + 3 & \text{if } x \ge 1 \\ 4x & \text{if } x < 1 \end{cases}$ at $a = 1$

39. $f(x) = \begin{cases} 0 & \text{if } x < 0 \\ x & \text{if } x \ge 0 \end{cases}$ at $a = 0$

40. $f(x) = \begin{cases} 0 & \text{if } x < 0 \\ x & \text{if } x \ge 0 \end{cases}$ at $a = 1$

41. For transporting a load of goods, a trucking firm charges a service fee of \$100 and a mileage fee of \$3 per mile for the first 100 miles or less and \$2 for each additional mile.
(a) Graph the cost $C(x)$ of shipping a load of goods x miles.
(b) Is $C(x)$ a continuous function?
(c) Is $C(x)$ a differentiable function?

12.7 RULES FOR CALCULATING DERIVATIVES

In this section we use the definition of the derivative to develop some formulas for easily differentiating certain functions. We present them as a sequence of seven rules. *They must be memorized!* You will use them regularly throughout the remainder of the course. Only the first six rules are given in this section; the seventh is given in Section 12.8.

Rule 1 (Constant Rule) The derivative of a constant function is zero.

Proof Suppose $f(x) = c$. Then for all x,

$$\begin{aligned} f'(x) &= \lim_{h\to 0} \frac{f(x+h) - f(x)}{h} \\ &= \lim_{h\to 0} \frac{c - c}{h} \\ &= \lim_{h\to 0} 0 \\ &= 0 \end{aligned}$$

Rule 2 (Constant Multiplier Rule) If f is differentiable at x, then so is any constant multiple of f; moreover,

$$\frac{d}{dx}[cf(x)] = c\frac{d}{dx}[f(x)]$$

Proof

$$\begin{aligned} \frac{d}{dx}[cf(x)] &= \lim_{h\to 0} \frac{cf(x+h) - cf(x)}{h} \\ &= \lim_{h\to 0} \frac{c[f(x+h) - f(x)]}{h} \\ &= c \lim_{h\to 0} \frac{f(x+h) - f(x)}{h} \\ &= c\frac{d}{dx}[f(x)] \end{aligned}$$

Rule 3 (Power Rule) For any positive integer n, the function $f(x) = x^n$ is differentiable everywhere. Moreover,

$$\frac{d}{dx}(x^n) = nx^{n-1}$$

Proof (1) For $n = 1$,

$$\frac{d}{dx}(x^1) = \lim_{h\to 0} \frac{(x+h)^1 - x^1}{h}$$

$$= \lim_{h\to 0} \frac{h}{h}$$

$$= 1$$

$$= 1x^0 = 1x^{1-1}$$

(2) For $n > 1$,

$$\frac{d}{dx}(x^n) = \lim_{h\to 0} \frac{(x+h)^n - x^n}{h}$$

Recall from elementary algebra that $A^n - B^n$ is always factorable:

$$A^n - B^n = (A - B)(A^{n-1} + A^{n-2}B + A^{n-3}B^2 + \cdots + AB^{n-2} + B^{n-1})$$

For example, $A^5 - B^5 = (A - B)(A^4 + A^3B + A^2B^2 + AB^3 + B^4)$. Thus,

$$\frac{d}{dx}(x^n) = \lim_{h\to 0} \frac{[(x+h) - x][(x+h)^{n-1} + (x+h)^{n-2}x + \cdots + (x+h)x^{n-2} + x^{n-1}]}{h}$$

$$= \lim_{h\to 0} \frac{h[(x+h)^{n-1} + (x+h)^{n-2}x + \cdots + (x+h)x^{n-2} + x^{n-1}]}{h}$$

$$= x^{n-1} + x^{n-1} + \cdots + x^{n-1} \qquad (n \text{ terms, all} = x^{n-1})$$

$$= nx^{n-1}$$

EXAMPLE 1 Using Rules 2 and 3 together, we see that

(a) $\dfrac{d}{dx}(x^7) = 7x^6$

(b) $\dfrac{d}{dx}(3x^8) = 3\dfrac{d}{dx}(x^8) = 3(8x^7) = 24x^7$

(c) $\dfrac{d}{dx}\left(\dfrac{x^4}{5}\right) = \dfrac{1}{5}\dfrac{d}{dx}(x^4) = \dfrac{4}{5}x^3$

Rule 3a (Extended Power Rule) For any real number r, the function $f(x) = x^r$ is differentiable everywhere and

$$\frac{d}{dx}(x^r) = rx^{r-1}$$

Although this rule may appear to be the same as Rule 3, it applies to exponents that are negative or irrational. Thus its proof is quite beyond that of Rule 3 and is not presented here.

EXAMPLE 2 Using Rule 3a, we can obtain the following:

(a) $\frac{d}{dx}(\sqrt{x}) = \frac{d}{dx}(x^{1/2}) = \frac{1}{2}x^{-1/2} = \frac{1}{2\sqrt{x}}$ (See Example 5, Section 12.6)

(b) $$\frac{d}{dx}\left(\frac{6}{x}\right) = \frac{d}{dx}(6x^{-1}) = 6\frac{d}{dx}(x^{-1}) = 6(-1)x^{-1-1}$$
$$= -6x^{-2}$$
$$= -\frac{6}{x^2}$$

(c) $\frac{d}{dx}\left(\frac{-5}{x^3}\right) = -5\frac{d}{dx}(x^{-3}) = -5(-3x^{-4}) = \frac{15}{x^4}$

Rule 4 (Sum and Difference Rule) If f and g are both differentiable at x, then so are $f + g$ and $f - g$. Moreover,

$$\frac{d}{dx}[f(x) + g(x)] = \frac{d}{dx}[f(x)] + \frac{d}{dx}[g(x)]$$

$$\frac{d}{dx}[f(x) - g(x)] = \frac{d}{dx}[f(x)] - \frac{d}{dx}[g(x)]$$

In other words, *the derivative of a sum is the sum of the derivatives and the derivative of a difference is the difference of the derivatives.*

Application to Polynomials Polynomial functions are differentiable everywhere; their derivatives may be obtained by using Rules 1 through 4. [See (a), (b), and (c) of Example 3.]

EXAMPLE 3

(a) $\frac{d}{dx}(x^5 + x^4) = \frac{d}{dx}(x^5) + \frac{d}{dx}(x^4) = 5x^4 + 4x^3$

(b) $\frac{d}{dx}(1 - x^3) = \frac{d}{dx}(1) - \frac{d}{dx}(x^3) = 0 - 3x^2 = -3x^2$

(c) $$\frac{d}{dx}(3x^4 - 2x^3 + 4x^2 - 3x + 2) = 3 \cdot 4x^3 - 2 \cdot 3x^2 + 4 \cdot 2x - 3 + 0$$
$$= 12x^3 - 6x^2 + 8x - 3$$

(d) $$\frac{d}{dx}\left(\sqrt{x} + \frac{1}{x}\right) = \frac{d}{dx}(x^{1/2} + x^{-1})$$
$$= \frac{1}{2}x^{-1/2} + (-1)x^{-2}$$
$$= \frac{1}{2\sqrt{x}} - \frac{1}{x^2}$$

In light of the sum and difference rules, we might conjecture that the derivative of a product of two functions is the product of their derivatives. However, this is false in general. For example, we have $x^6 = x^4 \cdot x^2$, yet

$$\frac{d}{dx}(x^6) \neq \frac{d}{dx}(x^4) \cdot \frac{d}{dx}(x^2)$$

since

$$\frac{d}{dx}(x^6) = 6x^5$$

while

$$\frac{d}{dx}(x^4) \cdot \frac{d}{dx}(x^2) = 4x^3 \cdot 2x = 8x^4$$

The correct rule is more complicated.

Rule 5 (The Product Rule) If f and g are both differentiable at x, then so is their product $f(x)g(x)$. Moreover,

$$\frac{d}{dx}[f(x) \cdot g(x)] = f(x) \cdot \frac{d}{dx}[g(x)] + g(x) \cdot \frac{d}{dx}[f(x)]$$

In other words, *the derivative of a product of two functions is the first function times the derivative of the second plus the second function times the derivative of the first.*

Proof (Optional) Suppose f and g are both differentiable at x. Then

$$\frac{d}{dx}[f(x)g(x)] = \lim_{h\to 0}\frac{f(x+h)g(x+h) - f(x)g(x)}{h}$$

The trick here is to add and subtract the same quantity $f(x+h)g(x)$ in the numerator of this fraction, which allows us to separate the fraction into the sum of two fractions. We have

$$\begin{aligned}
\frac{d}{dx}[f(x)g(x)] &= \lim_{h\to 0}\frac{f(x+h)g(x+h) - f(x+h)g(x) + f(x+h)g(x) - f(x)g(x)}{h} \\
&= \lim_{h\to 0}\left[f(x+h)\frac{g(x+h) - g(x)}{h} + g(x)\frac{f(x+h) - f(x)}{h}\right] \\
&= \lim_{h\to 0} f(x+h) \lim_{h\to 0}\frac{g(x+h) - g(x)}{h} + g(x) \lim_{h\to 0}\frac{f(x+h) - f(x)}{h} \\
&= \lim_{h\to 0}[f(x+h)]g'(x) + g(x)f'(x)
\end{aligned}$$

Since f is differentiable at x, it is continuous there, so

$$\lim_{h\to 0} f(x+h) = f(x)$$

Therefore,

$$\frac{d}{dx}[f(x)g(x)] = f(x)g'(x) + g(x)f'(x)$$

EXAMPLE 4

$$\begin{aligned}\frac{d}{dx}(x^3\sqrt{x}) &= x^3\frac{d}{dx}(\sqrt{x}) + \sqrt{x}\frac{d}{dx}(x^3) \qquad \text{(Rule 5)}\\ &= x^3\frac{1}{2\sqrt{x}} + \sqrt{x}\,3x^2 \qquad \text{(Rule 3 and Example 2a)}\\ &= \frac{1}{2}x^2\sqrt{x} + 3x^2\sqrt{x}\\ &= \frac{7}{2}x^{5/2}\end{aligned}$$

We could have done the problem differently:

$$\frac{d}{dx}(x^3\sqrt{x}) = \frac{d}{dx}(x^{7/2}) = \frac{7}{2}x^{5/2}$$

The second approach is simpler than the first one.

EXAMPLE 5 (a) $\frac{d}{dx}[x^5 + 2x^3 - 5)(3x^4 - 2x^2 + 8)]$

$$\begin{aligned}&= (x^5 + 2x^3 - 5)\frac{d}{dx}(3x^4 - 2x^2 + 8) + (3x^4 - 2x^2 + 8)\frac{d}{dx}(x^5 + 2x^3 - 5)\\ &= (x^5 + 2x^3 - 5)(12x^3 - 4x) + (3x^4 - 2x^2 + 8)(5x^4 + 6x^2)\end{aligned}$$

If desired, we could rewrite this derivative by multiplying out and collecting terms. Also, we could have obtained the derivative by multiplying out the factors $x^5 + 2x^3 - 5$ and $3x^4 - 2x^2 + 8$ before differentiating. However, this is unnecessary extra work.

(b)
$$\begin{aligned}\frac{d}{dx}\left[(1 + x^{1/3})\left(\frac{1}{2} + x\right)\right] &= (1 + x^{1/3})\frac{d}{dx}\left(\frac{1}{2} + x\right) + \left(\frac{1}{2} + x\right)\frac{d}{dx}(1 + x^{1/3})\\ &= (1 + x^{1/3})(1) + \left(\frac{1}{2} + x\right)\left(\frac{1}{3}x^{-2/3}\right)\\ &= 1 + x^{1/3} + \frac{1}{6}x^{-2/3} + \frac{1}{3}x^{1/3}\end{aligned}$$

Just as the derivative of a product is not the product of the derivatives, so the derivative of a quotient is not the quotient of the derivatives. The correct rule is as follows. We omit the proof.

Rule 6 (The Quotient Rule) If f and g are differentiable at x, then so is their quotient f/g, provided $g(x) \neq 0$. Moreover,

$$\frac{d}{dx}\left[\frac{f(x)}{g(x)}\right] = \frac{g(x) \cdot \frac{d}{dx}[f(x)] - f(x) \cdot \frac{d}{dx}[g(x)]}{[g(x)]^2}$$

In other words, *the derivative of a quotient is the denominator times the derivative of the numerator minus the numerator times the derivative of the denominator, all divided by the square of the denominator.*

EXAMPLE 6 Since $x^4 = x^6/x^2$, we should obtain the same result whether we differentiate x^4 by the power rule or differentiate x^6/x^2 by the quotient rule. This is the case since

$$\begin{aligned} \frac{d}{dx}\left(\frac{x^6}{x^2}\right) &= \frac{x^2 \frac{d}{dx}(x^6) - x^6 \frac{d}{dx}(x^2)}{(x^2)^2} \\ &= \frac{x^2(6x^5) - x^6(2x)}{x^4} = \frac{6x^7 - 2x^7}{x^4} \\ &= \frac{4x^7}{x^4} = 4x^3 = \frac{d}{dx}(x^4) \end{aligned}$$

Application to Rational Functions A rational function, a quotient $p(x)/q(x)$ of polynomials, is differentiable everywhere except where its denominator is 0. The derivative may be found using the quotient rule.

EXAMPLE 7 If

$$y = \frac{3x^2 - 1}{x + 2}$$

then by the quotient rule

$$\begin{aligned} \frac{dy}{dx} &= \frac{(x + 2) \cdot \frac{d}{dx}(3x^2 - 1) - (3x^2 - 1) \cdot \frac{d}{dx}(x + 2)}{(x + 2)^2} \\ &= \frac{(x + 2)(6x) - (3x^2 - 1)(1)}{(x + 2)^2} \end{aligned}$$

or if desired we could multiply out and write

$$\frac{dy}{dx} = \frac{3x^2 + 12x + 1}{x^2 + 4x + 4}$$

EXAMPLE 8 Let

$$f(x) = \frac{1 + \sqrt{x}}{1 - \sqrt{x}}$$

Find $f'(x)$.

Solution By the quotient rule

$$f'(x) = \frac{(1 - \sqrt{x}) \cdot \frac{d}{dx}(1 + \sqrt{x}) - (1 + \sqrt{x}) \cdot \frac{d}{dx}(1 - \sqrt{x})}{(1 - \sqrt{x})^2}$$

$$= \frac{(1 - \sqrt{x}) \cdot \left(\frac{1}{2\sqrt{x}}\right) - (1 + \sqrt{x})\left(-\frac{1}{2\sqrt{x}}\right)}{(1 - \sqrt{x})^2}$$

$$= \frac{\frac{1}{\sqrt{x}}}{(1 - \sqrt{x})^2}$$

$$= \frac{1}{\sqrt{x}(1 - \sqrt{x})^2}$$

We conclude this section by giving a special case of Rule 6. If we apply the rule with numerator equal to 1 we obtain:

Rule 6a If f is a differentiable function, then so is its reciprocal $1/f(x)$ whenever $f(x) \neq 0$. Moreover,

$$\frac{d}{dx}\left[\frac{1}{f(x)}\right] = -\frac{f'(x)}{[f(x)]^2}$$

EXERCISE SET 12.7

In Exercises 1–34 use the rules of differentiation discussed in this section to find $f'(x)$ for the given function.

1. $f(x) = x^6$

2. $f(x) = 2x + 3$

3. $f(x) = 3x^4$

4. $f(x) = -2$

5. $f(x) = 3x^3 - 2x^2 + x - 8$

6. $f(x) = 5x^{11} - 7x^8 + 4x^4 - 25$

7. $f(x) = 5\sqrt{x}$

8. $f(x) = 8\sqrt[3]{x}$

9. $f(x) = 16/x$

10. $f(x) = 10/x^3$

11. $f(x) = 4x^{0.03}$

12. $f(x) = 0.4x^{0.5} - 28$

13. $f(x) = 2/x^2 + 3x^3$

14. $f(x) = x^{1/3} - x^{-1/2}$

15. $f(x) = 24/\sqrt{x}$

16. $f(x) = 3/\sqrt[4]{x}$

17. $f(x) = x^2 \cdot x^9$ (two ways)

18. $f(x) = (x - 2)(x + 1)$ (two ways)

19. $f(x) = (3x^2 - 2x)(2x^3 + 1)$

20. $f(x) = (x^2 + 5)(x^2 - 2x + 3)$

21. $f(x) = (2x + \sqrt{x})(x^3 + 1)$

22. $f(x) = (x^{1/3} + 7x)(x^{10} - 9x^8 + 1)$

23. $f(x) = x^{11}/x^5$ (two ways)

24. $f(x) = 1/x$ (two ways)

25. $f(x) = \dfrac{x}{100 - x}$

26. $f(x) = \dfrac{x^2 + 1}{x^2 - 1}$

27. $f(x) = \dfrac{1 + \sqrt{x}}{2x - 3}$

28. $f(x) = \dfrac{x^3 + 4x}{5x^2 + 3}$

29. $f(x) = \dfrac{3x^2 - 2x + 1}{2x^3 + 1}$

30. $f(x) = \dfrac{\sqrt{x} + x}{2 + 3\sqrt{x}}$

31. $f(x) = \dfrac{1}{x^2 + 2}$

32. $f(x) = \dfrac{1}{x^3 + 2x}$

33. $f(x) = \dfrac{1}{x + \sqrt{x}}$

34. $f(x) = x^{-1/2}$ (three ways)

35. Suppose $f(x) = 3x^2 - 2x + 1$. Compute $f'(2)$, $f'(0)$, $f'(-2)$.

36. Suppose

$$f(x) = \frac{3x - 1}{2x + 5}$$

Compute $f'(0), f'(1), f'(-2)$.

37. Find the slope of the tangent to the graph of $f(x) = x^2 - x$ at the point where:
(a) $x = 0$ (b) $x = 1$ (c) $x = -1$
(d) $x = 2$

38. Find the slope of the tangent to the graph of $f(x) = 1/x$ at the point where:
(a) $x = -2$ (b) $x = 5$ (c) $x = 3$
(d) $x = -4$

39. Suppose f is defined by $f(x) = 2x^2 - 3x + 4$. Find the point on the graph of f where the slope of the tangent line is 9.

40. Suppose f is defined by $f(x) = 3x^2 - 2x - 5$. Find the point on the graph of f where the tangent line is parallel to the line $y = x + 2$.

41. Suppose f is defined by $f(x) = 2x^3 - 3x^2 - 12x + 5$. Find the points on the graph of f where the tangent line is horizontal.

42. (a) Suppose $A = 3r^2$, find dA/dr.
(b) Find $\dfrac{d}{dt}(3t^2)$.
(c) Find y' if $y = 3x^2$.
(d) Find $\dfrac{d}{ds}(3s^2)$.

43. (a) Find du/dt if $u = 3t^2$.
(b) Find dC/dx if $C(x) = x^2$.
(c) Find $D_x(x^7 + 1)$.
(d) Suppose $V = \frac{4}{3}\pi r^3$. Find dV/dr.

44. Find dy/dx if
(a) $y = 3x^3 + x^2 - 2x + 4$
(b) $y = (2x^2 - 1)(3x^3 + 2x)$

45. Find dy/dx if
(a) $y = 3\sqrt{x} + 2x^3$ (b) $y = 3/x^2 + 2x$

12.8 THE CHAIN RULE

We have been investigating arithmetic combinations (addition, subtraction, multiplication, and division) of two or more differentiable functions with known derivatives. Rules 1 through 6 provide all the rules necessary for obtaining the derivative of any such combination. There is an entirely different way of combining two functions that is of utmost importance. It is called **composition of functions**.

EXAMPLE 1 Suppose f and g are functions, such as

$$f(x) = 3x^2 - 1 \quad \text{and} \quad g(x) = \frac{1}{x+5}$$

Then we can form the **composite of f with g**:

$$\begin{aligned}(f \circ g)(x) &= f(g(x)) \\ &= f\left(\frac{1}{x+5}\right) = 3\left(\frac{1}{x+5}\right)^2 - 1 \\ &= \frac{3}{(x+5)^2} - 1\end{aligned}$$

and the **composite of g with f**:

$$\begin{aligned}(g \circ f)(x) &= g(f(x)) \\ &= g(3x^2 - 1) = \frac{1}{(3x^2 - 1) + 5} \\ &= \frac{1}{3x^2 + 4}\end{aligned}$$

The functions $f \circ g$ (read "f of g") and $g \circ f$ (read "g of f") are entirely different. In general,

$$f \circ g \neq g \circ f$$

because, as the above example shows,

$$(f \circ g)(x) \neq (g \circ f)(x)$$

except in unusual circumstances.

In this particular example both f and g are differentiable (if $x \neq -5$), and it also happens that both $f \circ g$ and $g \circ f$ are differentiable. It is not clear at this point, however, how the derivatives of $f \circ g$ and $g \circ f$ are related to the derivatives of f and g. It is the purpose of the chain rule to establish this relationship.

Before stating the chain rule we make a few comments on notation. The relation

$$\begin{aligned}y &= (f \circ g)(x) \\ &= f(g(x))\end{aligned}$$

can be rewritten as a "chain" relation

$$y = f(u) \qquad \text{where } u = g(x)$$

For example, the equation

$$y = (2x^3 - 5x^2)^{78}$$

can be rewritten as

$$y = u^{78} \qquad \text{where } u = 2x^3 - 5x^2$$

Each of the derivatives $dy/du = 78u^{77}$ and $du/dx = 6x^2 - 10x$ is easily obtained, but the derivative dy/dx is much more difficult without the chain rule.

Rule 7 (The Chain Rule) If $y = f(u)$ is a differentiable function of u and $u = g(x)$ is a differentiable function of x, then the composite function $y = f(g(x))$ is a differentiable function of x. Moreover,

$$\frac{dy}{dx} = \frac{dy}{du} \cdot \frac{du}{dx} \tag{1}$$

In this textbook we do not attempt a proof of the chain rule. It may be found in more mathematically rigorous treatments of the calculus. Example 7 below provides an intuitive discussion of the underlying principle. However, Equation (1) is very easy to remember because of the suggestive nature of the fractional notation. While we emphasize that derivatives are *not* fractions, the *du's* here cancel as if we were merely simplifying fractions.

EXAMPLE 2 Find $\frac{d}{dx}(2x^3 - 5x^2)^{78}$.

Solution We think of the function $y = (2x^3 - 5x^2)^{78}$ as a composite function,

$$y = u^{78} \qquad \text{where} \quad u = 2x^3 - 5x^2$$

We know that

$$\begin{aligned}\frac{dy}{du} &= 78u^{77} \\ &= 78(2x^3 - 5x^2)^{77} \qquad (\text{since } u = 2x^3 - 5x^2)\end{aligned}$$

and $\frac{du}{dx} = 6x^2 - 10x$. Applying the chain rule, we obtain

$$\begin{aligned}\frac{dy}{dx} &= \frac{dy}{du} \cdot \frac{du}{dx} \\ &= 78u^{77}(6x^2 - 10x) \\ &= 78(2x^3 - 5x^2)^{77}(6x^2 - 10x)\end{aligned}$$

The basic strategy in using the chain rule is to "decompose" a function that we cannot differentiate into a composite (or "chain") of functions, each of which we can differentiate directly. Let us summarize the steps involved in using the chain rule to differentiate $y = F(x)$.

Differentiation of $y = F(x)$ by the Chain Rule

Step	Example: $y = \sqrt{x^2 + 3}$
1. Introduce a new variable $u = g(x)$ and a function $f(u)$ so that $$Y = F(x) = f(g(x))$$ Where $y = f(u)$ and $u = g(x)$	Let $u = x^2 + 3$ and $f(u) = \sqrt{u}$ ($y = \sqrt{u}$ and $u = x^2 + 3$)
2. Calculate dy/du and express as a function of x.	$\frac{dy}{du} = \frac{1}{2\sqrt{u}}$ (from Example 2, Section 12.7) $= \frac{1}{2\sqrt{x^2 + 3}}$
3. Calculate du/dx.	$\frac{du}{dx} = 2x$
4. By the chain rule, $$\frac{dy}{dx} = \frac{dy}{du} \cdot \frac{du}{dx}$$	$\frac{dy}{dx} = \frac{1}{2\sqrt{x^2 + 3}} \cdot 2x = \frac{x}{\sqrt{x^2 + 3}}$

EXAMPLE 3 Find $h'(x)$ if

$$h(x) = \left(\frac{x}{1 + x}\right)^{1/3}$$

Solution Let $y = h(x)$ and define

$$u = \frac{x}{1 + x}$$

so that $y = u^{1/3}$. Thus

$$\frac{du}{dx} = \frac{d}{dx}\left(\frac{x}{1 + x}\right) = \frac{(1 + x)(1) - x(1)}{(1 + x)^2} \qquad \text{(quotient rule)}$$

$$= \frac{1}{(1 + x)^2}$$

and

$$\frac{dy}{du} = \frac{d}{du}(u^{1/3}) = \frac{1}{3}u^{-2/3} = \frac{1}{3}\left(\frac{x}{1 + x}\right)^{-2/3}$$

Therefore, from the chain rule,

$$h'(x) = \frac{dy}{dx} = \frac{dy}{du} \cdot \frac{du}{dx} = \frac{1}{3}\left(\frac{x}{1 + x}\right)^{-2/3} \cdot \frac{1}{(1 + x)^2}$$

Differentiating a Power of a Function

In the preceding examples we use the chain rule to differentiate powers of functions, specifically,

$$(2x^3 - 5x^2)^{78} \qquad (x^2 + 3)^{1/2} \qquad \left(\frac{x}{1 + x}\right)^{1/3}$$

Rather than apply the chain rule every time we want to differentiate a power of a function, it is more efficient to obtain, once and for all, a formula for differentiating functions of the form

$$[g(x)]^k$$

If we let $y = [g(x)]^k$ and define $u = g(x)$, then we obtain

$$y = u^k$$

Thus, by the chain rule

$$\frac{d}{dx}[g(x)]^k = \frac{dy}{dx} = \frac{ay}{du} \cdot \frac{du}{dx} \tag{2}$$

But

$$\frac{dy}{du} = \frac{d}{du}(u^k) = ku^{k-1} = k[g(x)]^{k-1}$$

and

$$\frac{du}{dx} = g'(x)$$

so that (2) becomes:

Rule 7a (General Power Rule) If g is a differentiable function of x, then for every power k,

$$\frac{d}{dx}[g(x)]^k = k[g(x)]^{k-1} \cdot g'(x) \tag{3}$$

In other words, *the derivative of a function to a power is the power times the function to the power less 1, times the derivative of the function.*

EXAMPLE 4 Use the general power rule to differentiate the function $(3 + x - x^3)^{17}$.

Solution We apply (3) with $k = 17$ and $g(x) = 3 + x - x^3$ to obtain

$$\begin{aligned}\frac{d}{dx}(3 + x - x^3)^{17} &= 17(3 + x - x^3)^{16} \cdot \frac{d}{dx}(3 + x - x^3)\\ &= 17(3 + x - x^3)^{16} \cdot (1 - 3x^2)\end{aligned}$$

If variables other than y, u, and x are used, then the chain rule formula must be altered appropriately. For example, if w is a function of v and v is a function of t, then w is a function of t and the chain rule becomes

$$\frac{dw}{dt} = \frac{dw}{dv} \cdot \frac{dv}{dt}$$

EXAMPLE 5 Suppose an automobile manufacturer finds that the annual profit is given (in millions of dollars) by

$$P(x) = 30 + \frac{3}{2}\left(x^2 - 2x\right)^{5/3}$$

where x is the amount (in millions of dollars) that is spent on advertising. Find the instantaneous rate of change of $P(x)$ with respect to x when $x = 4$.

Solution We have

$$\frac{dP}{dx} = \frac{d}{dx}\left[30 + \frac{3}{2}(x^2 - 2x)^{5/3}\right]$$

$$= 0 + \frac{3}{2} \cdot \frac{5}{3}(x^2 - 2x)^{2/3} \cdot \frac{d}{dx}[x^2 - 2x] \qquad \text{[see Equation (3)]}$$

$$= \frac{5}{2}(x^2 - 2x)^{2/3}(2x - 2)$$

When $x = 4$,

$$\frac{dP}{dx} = \frac{5}{2}(4^2 - 2 \cdot 4)^{2/3}(2 \cdot 4 - 2) = 60$$

Thus, when the advertising expenditure reaches 4 million dollars ($x = 4$), the profit will be increasing at the rate of 60 million dollars per million dollars expended on advertising.

EXAMPLE 6 Suppose a manufacturer finds that the cost (in dollars) of making x units of a product is given by

$$C(x) = (5\sqrt{x} + 10)^2 \tag{4}$$

If the level of production is increasing at a constant rate of 3 units per day, find the rate at which the cost is changing with time when the level of production is $x = 36$ units.

Solution Since C is a function of x and x is a function of time t, it follows that C is a function of t. Thus, using the chain rule, we have

$$\frac{dC}{dt} = \frac{dC}{dx} \cdot \frac{dx}{dt} \tag{5}$$

We are given that the rate of change in x (level of production) is constant, with

$$\frac{dx}{dt} = 3 \tag{6}$$

We can calculate dC/dx from Equation (4):

$$\frac{dC}{dx} = \frac{d}{dx}[(5\sqrt{x} + 10)^2]$$

$$= 2(5\sqrt{x} + 10)^1 \cdot \frac{d}{dx}(5\sqrt{x} + 10)$$

$$= 2(5\sqrt{x} + 10) \cdot \frac{5}{2\sqrt{x}} \qquad \left(\text{since } \frac{d}{dx}\sqrt{x} = \frac{1}{2\sqrt{x}}\right)$$

$$= \frac{25\sqrt{x} + 50}{\sqrt{x}} \tag{7}$$

Putting together Equations (5), (6), and (7), we have

$$\frac{dC}{dt} = \frac{dC}{dx} \cdot \frac{dx}{dt} = \frac{25\sqrt{x} + 50}{\sqrt{x}} \cdot 3 \tag{8}$$

Finally, when $x = 36$, we obtain from Equation (8)

$$\frac{dC}{dt} = \frac{25\sqrt{36} + 50}{\sqrt{36}} \cdot 3 = 100 \qquad \text{(dollars per day)}$$

Additional applications of the chain rule are given in Section 13.3.

Intuitive Basis of the Chain Rule

While the chain rule is quite easy to use and remember, it may still seem somewhat mysterious. From our discussion it is not at all apparent *why* it works. We now give a simple illustration to help you see the underlying principle.

EXAMPLE 7 Imagine a race between a tortoise, a hare, and a motorcycle. Suppose the hare goes 12 times as fast as the tortoise, while the motorcycle goes 5 times as fast as the hare. How much faster does the motorcycle go than the tortoise?

Solution We can obtain the solution by observing that, in general, if A is 12 times as fast as B and B is 5 times as fast as C, then A is $12 \cdot 5 = 60$ times as fast as C. Therefore, the motorcycle goes 60 times as fast as the tortoise.

As an alternative method, we can approach this problem using the chain rule as follows. After the start of the race, let us measure how far the three racers have gone. Let s be the distance the tortoise has gone, h the distance the hare has gone, and m the distance the motorcycle has gone. Since a derivative is a rate of change in one variable relative to another, we know that

$$h = 12s \qquad m = 5h$$

$$\frac{dh}{ds} = 12 \quad \text{and} \quad \frac{dm}{dh} = 5$$

Next we must find dm/ds. By the chain rule,

$$\begin{aligned} \frac{dm}{ds} &= \frac{dm}{dh} \cdot \frac{dh}{ds} \\ &= 5 \cdot 12 \\ &= 60 \end{aligned}$$

Thus, the chain rule tells us that the motorcycle goes 60 times as fast as the tortoise. The chain rule leads to the conclusion we had already anticipated.

EXERCISE SET 12.8

1. Suppose r is a function of s and s is a function of t. Use the chain rule to relate dr/ds, dr/dt, and ds/dt.

2. Suppose C is a function of z and z is a function of w. Use the chain rule to relate dz/dw, dC/dz, and dC/dw.

3. Consider the functions $f(x) = 3x^2 - 1$ and $g(x) = 1/(x + 5)$ given in Example 1.

   (a) Calculate $\frac{d}{dx}[f(g(x))]$ using the chain rule.

(b) Calculate $\frac{d}{dx}[f(g(x))]$ *without* using the chain rule [i.e., express $f(g(x))$ as a function of x alone, and differentiate].

(c) Reconcile your answers to (a) and (b).

4. Repeat Exercise 3 with the functions $f(x) = 2x + 5$ and $g(x) = x^2 - 10$. For part (b) you will need to use Rule 7a.

5. Repeat Exercise 3, with $g(f(x))$ in place of $f(g(x))$.

6. Repeat Exercise 4, with $g(f(x))$ in place of $f(g(x))$.

In Exercises 7–32 find dy/dx.

7. $y = (9x^2 + 2x)^{18}$

8. $y = (2 - 5x^2)^{25}$

9. $y = \sqrt{2 - x}$

10. $y = (6x - x^2)^{1/3}$

11. $y = \sqrt[3]{x^2 - x}$

12. $y = (x^2 - x^3)^{-5}$

13. $y = \dfrac{1}{(2x^2 - 3x)^6}$

14. $y = \sqrt{x^3 + 2x + 1}$

15. $y = (7x^3 - 6x^2 + 2)^{2/3}$

16. $y = x\sqrt{x + 2}$

17. $y = x^2\sqrt{3x + 4}$

18. $y = \left(\dfrac{x}{1 - x^2}\right)^3$

19. $y = (3x^4 + 2)^{10}(2x^3 - 3)^{20}$

20. $y = (x^2 + 4)^7\sqrt{x}$

21. $\sqrt{\dfrac{x + 4}{x + 11}}$

22. $y = \sqrt{\dfrac{x^3}{1 + 2x}}$

23. $y = \dfrac{\sqrt{x}}{x^2 - 2}$

24. $y = x^2(x^4 + 1)^8$

25. $y = \dfrac{x^3 - 2x + 5}{x^2 + x + 1}$

26. $y = (x^2 + 3x + 16)^{-4}$

27. $y = \sqrt[3]{x^2 + 2x + 1000}$

28. $y = (x^{-3} + x^{-2})^2$

29. $y = \dfrac{1}{\sqrt{x^2 - 9}}$

30. $y = (3x^{50} + 10x + 100)^{-100}$

31. $y = \sqrt[4]{\dfrac{x}{1 + x^2}}$

32. $y = \sqrt{(x^2 + 1)^{99}x^{20}}$

33. **(Crop yield)** Assume that soybean yield y (pounds per square yard) is related to the amount x of high-phosphorus fertilizer applied (pounds per square yard) by the equation

$$y = 50[1 - 10(x^2 + 2)^{-3}]$$

Find the instantaneous rate of change of y with respect to x when $x = 1$.

34. If $y = (x^2 + 4)^{3/2}$, find all values of x at which $dy/dx = 0$.

35. If $y = \frac{1}{3}x^3 - 2x + 1$, find all points where the tangent line has a slope of 2.

36. Find all points on the curve $y = \frac{1}{2}x^2 + 5x - 2$ where the tangent line is parallel to $y = 3x - 7$.

37. **(Manufacturing)** Suppose a manufacturer finds that the cost (in dollars) of manufacturing x units of a product is given by

$$C(x) = (100 + 3\sqrt{x})^3$$

If the rate of production is maintained at a constant 50 units per day, find the rate at which cost is changing with time when the level of production is 81 units.

KEY IDEAS FOR REVIEW

Average rate of change of y with respect to x betwen x_1 and x_2

$$\frac{y_2 - y_1}{x_2 - x_1}$$

Secant line A line joining two points on the graph of $y = f(x)$.

Tangent line The tangent line to $y = f(x)$ at the point $(x_1, f(x_1))$ is the line through this point with slope

$$\lim_{h \to 0} \frac{f(x_1 + h) - f(x_1)}{h}$$

Equation of a tangent line

$$y - y_1 = m(x - x_1)$$

where (x_1, y_1) is the point of tangency and $m = f'(x_1)$

Instantaneous rate of change of $y = f(x)$ at a

$$\lim_{h \to 0} \frac{f(a + h) - f(a)}{h}$$

Slope of $y = f(x)$ at a (or equivalently, the slope of the line tangent to the graph of f at $(a, f(a))$)

$$\text{slope} = \lim_{h \to 0} \frac{f(a + h) - f(a)}{h}$$

$\lim_{x \to a} f(x) = L$ The values of $f(x)$ approach the number L as x approaches (but remains different from) a.

$\lim_{x \to a^+} f(x) = L$ The values of $f(x)$ approach the number L as x approaches a from the right.

$\lim_{x \to a^-} f(x) = L$ The values of $f(x)$ approach the number L as x approaches a from the left.

Properties of limits

$$\lim_{x \to a} k = k$$

$$\lim_{x \to a} x = a$$

$$\lim_{x \to a} [f(x) + g(x)] = \lim_{x \to a} f(x) + \lim_{x \to a} g(x)$$

$$\lim_{x \to a} [f(x) - g(x)] = \lim_{x \to a} f(x) - \lim_{x \to a} g(x)$$

$$\lim_{x \to a} [f(x)g(x)] = \lim_{x \to a} f(x) \cdot \lim_{x \to a} g(x)$$

$$\lim_{x \to a} [kf(x)] = k \lim_{x \to a} f(x)$$

$$\lim_{x \to a} \frac{f(x)}{g(x)} = \frac{\lim_{x \to a} f(x)}{\lim_{x \to a} g(x)} \qquad \left[\text{if } \lim_{x \to a} g(x) \neq 0\right]$$

$\lim_{x \to a^+} f(x) = \infty$ As x approaches a from the right, $f(x)$ increases without bound.

$\lim_{x \to a^+} f(x) = -\infty$ As x approaches a from the right, $f(x)$ decreases without bound.

$\lim_{x \to \infty} f(x) = L$ The values of $f(x)$ approach the number L as x increases without bound.

▰ $\lim_{x \to \infty} \frac{k}{x^n} = 0$ where k is any constant and n a positive constant.

▰ **Vertical asymptote** $f(x)$ has a vertical asymptote $x = a$ if $\lim_{x \to a^+} f(x) = \pm\infty$ or $\lim_{x \to a^-} f(x) = \pm\infty$.

▰ **Horizontal asymptote** $f(x)$ has a horizontal asymptote $y = b$ if $\lim_{x \to \infty} f(x) = b$ or $\lim_{x \to -\infty} f(x) = b$.

▰ **Continuity of *f* at *a*** f is continuous at a if:

(i) $f(a)$ is defined

(ii) $\lim_{x \to a} f(x)$ exists

(iii) $\lim_{x \to a} f(x) = f(a)$

▰ **Continuity of *f* and *g* at *a*** If f and g are continuous at a, then $f + g$, $f - g$, and fg are continuous at a. Moreover, f/g is continuous at a if $g(a) \neq 0$.

▰ **Derivative of *f*** $f'(x) = \lim_{h \to 0} \frac{f(x + h) - f(x)}{h}$

▰ **Derivative of a constant** $\frac{d}{dx}[k] = 0$

▰ **Power rule** $\frac{d}{dx}[x^r] = rx^{r-1}$ (for any real number r)

▰ **Constant times a function rule** $\frac{d}{dx}[k\, g(x)] = k \frac{d}{dx}[g(x)]$

▰ **Sum and difference rules**

$$\frac{d}{dx}[g(x) + h(x)] = \frac{d}{dx}[g(x)] + \frac{d}{dx}[h(x)]$$

$$\frac{d}{dx}[g(x) - h(x)] = \frac{d}{dx}[g(x)] - \frac{d}{dx}[h(x)]$$

▰ **Product rule** $\frac{d}{dx}[g(x) \cdot h(x)] = g(x) \cdot \frac{d}{dx}[h(x)] + h(x) \cdot \frac{d}{dx}[g(x)]$

▰ **Quotient rule** $\frac{d}{dx}\left[\frac{g(x)}{h(x)}\right] = \frac{h(x) \cdot \frac{d}{dx}[g(x)] - g(x) \cdot \frac{d}{dx}[h(x)]}{[h(x)]^2}$

▰ **Chain rule** $\frac{dy}{dx} = \frac{dy}{du} \cdot \frac{du}{dx}$

▰ **Derivative of a power of a function** $\frac{d}{dx}[g(x)]^k = k[g(x)]^{k-1} \cdot g'(x)$

SUPPLEMENTARY EXERCISES

1. Let $y = x^2 - x$.
(a) Find the average rate of change of y with respect to x between $x_1 = -2$ and $x_2 = 3$.
(b) Find the average rate of change of y with respect to x between $x_1 = a$ and $x_2 = b$.
(c) Find the slope of the secant line joining the points $(1, 0)$ and $(2, 2)$.

2. Let $f(x) = \dfrac{x^2 - 1}{x + 2}$.
(a) Find the average rate of change of f with respect to x between $x_1 = 1$ and $x_2 = -2$.
(b) Find the slope of the secant line joining the points $(2, \frac{3}{4})$ and $(-2, -3)$.

3. If $y = mx + b$, find the average rate of change of y with respect to x between $x_1 = 2$ and $x_2 = 2 + h$.

4. For the function graphed, find the average rate of change of y with respect to x between:

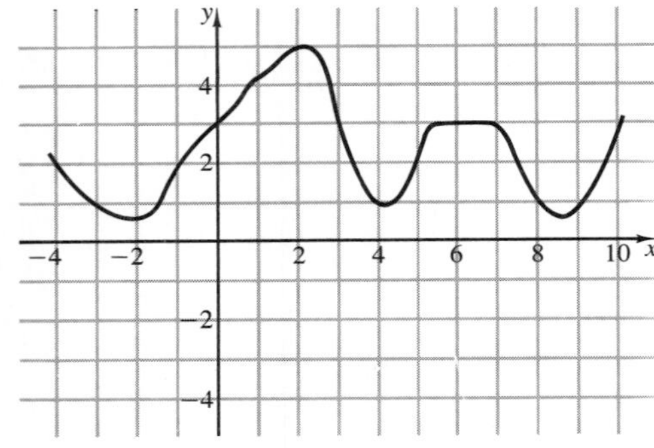

(a) $x_1 = 2$ and $x_2 = 5$
(b) $x_1 = -3$ and $x_2 = 4$

5. **(Motion)** Suppose that a car starts from rest and travels a distance d (in miles) in t hours according to the formula $d = 2t + 6t^2$.
(a) Find the average velocity between $t_1 = 0$ and $t_2 = 4$.
(b) Find the average velocity between $t_1 = 2$ and $t_2 = 5$.
(c) Find the velocity when $t = 3$.

6. **(Medicine)** A hospital patient's chart shows the following data:

Time	6 A.M.	Noon	3 P.M.	9 P.M.
Temperature (in °F)	99.2	100.4	101.3	100.3

(a) What is the average rate of change of temperature with respect to time between 6 A.M. and noon?
(b) What is the average rate of change of temperature with respect to time between 3 P.M. and 9 P.M.?
(c) Explain why the instantaneous rate of change of temperature at 1:15 P.M. cannot be determined.

7. **(Ecology)** The cooling system of a nuclear power plant is leaking waste into a nearby stream. The amount A (in gallons) of nuclear waste that flowed into a stream during the first t days after the leak occurred is given by the formula $A = 3t^2 + 2t$. What is the instantaneous rate of change of the waste flowing into the stream when $t = 3$?

8. **(Economics)** The wholesale price index for commodities from 1970 to 1976 is approximately given in the following table:

Year	1970	1971	1972	1973	1974	1975	1976
Index	110	114	119	135	160	175	183

(a) Find the average rate of change in the wholesale price index between 1970 and 1975.
(b) Find the average rate of change in the wholesale price index between 1973 and 1975.

9. **(Business)** A manufacturer of X-ray developers finds that the profit received from the sale of x million liters of the product is given (in millions of dollars) by $P(x) = 20x^2 + 12x$.
(a) Find the average rate of change of profit when the level of production changes from $x_1 = 20$ to $x_2 = 40$.
(b) Find the instantaneous rate of change of profit at $x = 15$.

10. Let $f(x) = \dfrac{|x - 1|}{x - 1}$. Use tables to compute $\lim_{x\to 1^+} f(x)$, $\lim_{x\to 5} f(x)$, and $\lim_{x\to 1} f(x)$.

In Exercises 11–24 find the indicated limit. Justify each step.

11. $\lim_{x\to 2} (x^2 - 2x + 1)$

12. $\lim_{x\to 2} \dfrac{2x^2 - x}{x^2 + 1}$

13. $\lim_{x\to 4^+} \frac{1}{x-4}$

14. $\lim_{x\to -1} \frac{x}{x+1}$

15. $\lim_{x\to 1} \frac{1}{(x-1)^2}$

16. $\lim_{x\to 2} \frac{x^2-16}{x-4}$

17. $\lim_{x\to 3} \frac{x-2}{x^2-5x+6}$

18. $\lim_{x\to 1} |x-1|$

19. $\lim_{x\to\infty} \frac{3x}{x^2-2x}$

20. $\lim_{x\to\infty} \frac{x^2-2x}{3x}$

21. $\lim_{x\to\infty} (2x^2-1)$

22. $\lim_{x\to\infty} \frac{2}{\sqrt{x-1}}$

23. $\lim_{x\to\infty} \left(3+\frac{2}{x^2}\right)$

24. $\lim_{x\to\infty} \left(\frac{3}{\sqrt{x}}-\frac{4}{x^2}+x\right)$

25. Let f be defined by

$$f(x) = \begin{cases} 2x+1 & \text{if } x \le 2 \\ -\frac{5}{2}x+10 & \text{if } x > 2 \end{cases}$$

(a) Sketch the graph of f.
(b) Evaluate $\lim_{x\to 2^-} f(x)$, $\lim_{x\to 2^+} f(x)$, and $\lim_{x\to 2} f(x)$.

26. Use the graph of f to find $\lim_{x\to 3} f(x)$, if it exists.

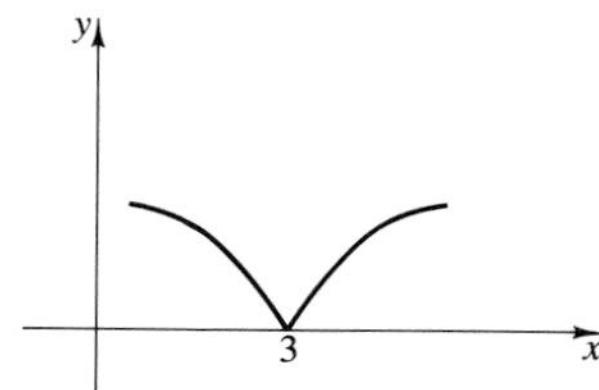

27. Use the graph of $f(x)$ to determine $\lim_{x\to 0^-} f(x)$ and $\lim_{x\to 1^+} f(x)$.

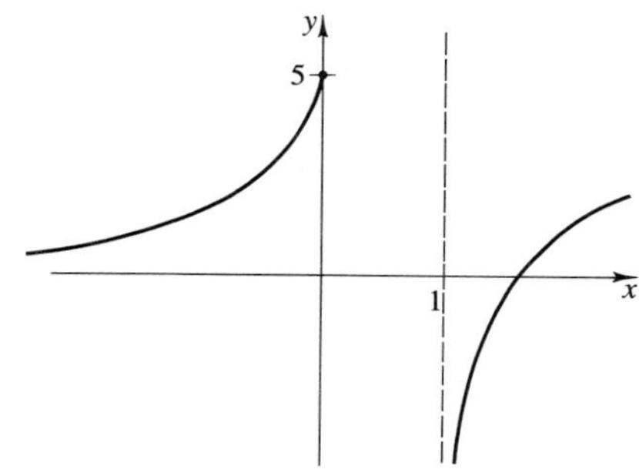

28. Discuss the continuity of the function

$$f(x) = \begin{cases} \frac{x^2-4}{x-2} & \text{if } x \ne 2 \\ 4 & \text{if } x = 2 \end{cases}$$

29. Discuss the continuity of the function graphed below at $x = a$, b, and c.

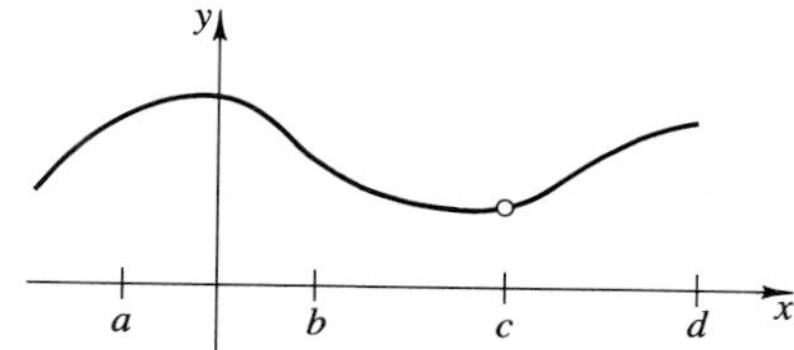

30. In each part, determine whether the given function is continuous at $x = 6$. If it is not continuous there, state which of the three continuity conditions is (are) violated. Graph each function.

(a) $f(x) = \frac{x^2-36}{x-6}$

(b) $f(x) = \begin{cases} \frac{x^2-36}{x-6} & \text{if } x \ne 6 \\ 12 & \text{if } x = 6 \end{cases}$

(c) $f(x) = \begin{cases} \frac{x^2-36}{x-6} & \text{if } x \ne 6 \\ 8 & \text{if } x = 6 \end{cases}$

31. Let

$$f(x) = \begin{cases} 2x & \text{if } x \le 2 \\ 4 & \text{if } x > 2 \end{cases}$$

Is f continuous at $x = 2$?

32. **(Medicine)** A medical research experiment established that the mass $M(t)$ of a tumor as a function of the length of time t that a patient is exposed to radiation during treatment is given by

$$M(t) = \frac{-t^2+7t-12}{t-3}$$

where $M(t)$ is in milligrams and t is in seconds. Because of a malfunction in the equipment being used, it is impossible to expose the patient for exactly 3 seconds of radiation therapy. What value should $M(3)$ be assigned so that $M(t)$ is a continuous function?

33. **(Business)** A heating oil distributor has set the following price schedule:

$1.30 per gallon for 100 or fewer gallons

$1.15 per gallon for more than 100 but no more than 150 gallons

$1.00 per gallon for more than 150 but fewer than 200 gallons

$0.85 per gallon for 200 or more gallons

(a) Determine the function $C(x)$ that gives the cost (in dollars) of purchasing x gallons of heating oil.
(b) Sketch the graph of $C(x)$.
(c) Find all points, if any, where $C(x)$ is discontinuous.

34. **(The intermediate-value property)** Explain why the following property of continuous functions holds. Suppose the function f is continuous throughout the closed interval $[a, b]$ and that $f(a)$ and $f(b)$ have opposite signs. There is at least one value of x, $x = c$, $a < c < b$ for which $f(c) = 0$. Sketch a graph of a function that satisfies these conditions.

35. In each part of the figure, calculate $f'(a)$ from the graph.

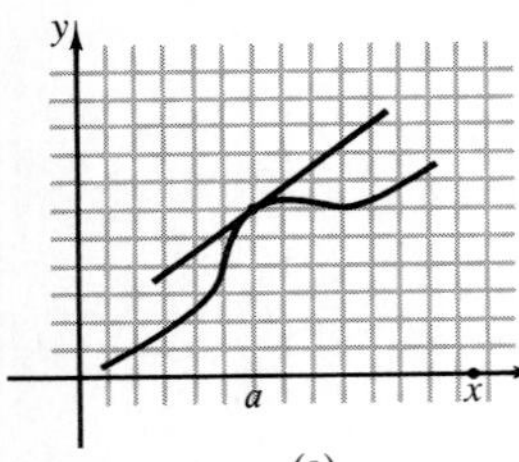

(a)

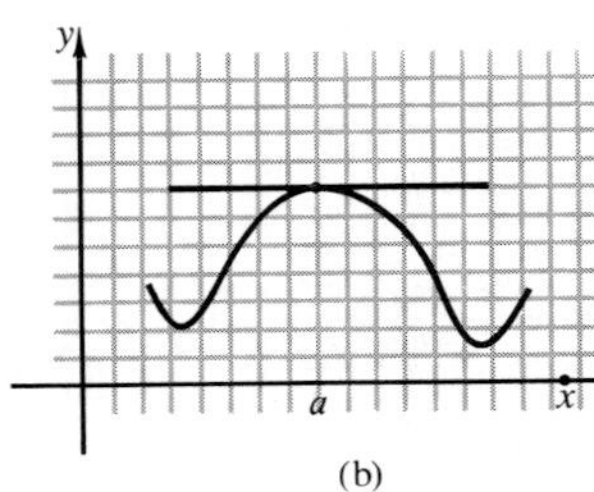

(b)

In Exercises 36–41 use Definition 1 to compute $f'(a)$ for the given function f at the indicated value a.

36. $f(x) = x^2 - x,\ a = -2$

37. $f(x) = 1 - 2x,\ a = 3$

38. $f(x) = \sqrt{x + 1},\ a = 3$

39. $f(x) = \dfrac{1}{x + 2},\ a = 2$

40. $f(x) = x\sqrt{x - 1},\ a = 5$

41. $f(x) = \dfrac{x - 1}{x + 2},\ a = 3$

In Exercises 42–53 use the rules of differentiation to find $f'(x)$.

42. $f(x) = 2x^4 - 5\sqrt{x + 3}$

43. $f(x) = 3x^{-2/3} + x^{1/3}$

44. $f(x) = \dfrac{x^2 - 4}{x^2 + 1}$

45. $f(x) = \dfrac{2x^{1/3} + 5x^2 - 2}{x^{2/3}}$ (two ways)

46. $f(x) = \dfrac{5}{\sqrt[3]{x^2}} - 2x^5 + 4x$

47. $f(x) = \dfrac{x + \sqrt{x}}{x - \sqrt{x}}$

48. $f(x) = (2x^3 - x^2 + 1)^{15}$

49. $f(x) = x^3\sqrt{x^2 - 1}$

50. $f(x) = \left(\dfrac{x^2}{x^2 - x}\right)^4$

51. $f(x) = (3x^3 + x^2 - 1)^5(x^3 + x^2 - 2x)^2$

52. $f(x) = \sqrt{\dfrac{x^2 + 1}{x + 1}}$

53. $f(x) = \dfrac{1}{(3x^3 - x^2 + 1)^4}$

54. If $y = (2x^3 - x^2 + 1)^4$, find the instantaneous rate of change of y with respect to x at $x = 2$.

55. Find the slope of the tangent line to the graph of $f(x) = x - 2x^2$ where (a) $x = 2$ and (b) $x = -3$.

56. If $f(x) = x^4\sqrt[3]{x^2 - x}$, find the instantaneous rate of change of f with respect to x at $x = 3$.

57. Find the points on the curve $y = (2x^2 - 2)^2$ where the tangent line is horizontal.

58. Find the points on the graph of $f(x) = x - x^2 + 1$ where the slope of the tangent line is 2. Write the equation of each such tangent line.

59. Find the points on the curve $y = 3x^2 - 2x + 1$ where the tangent line is parallel to the line $y = 2x - 1$.

CHAPTER TEST

1. Let $f(x) = 2x^2 - 5x$.
   (a) Find the slope of the secant line joining the points $(-1, 7)$ and $(2, -2)$ on the graph of f.
   (b) Find the slope of the tangent line to the graph of f at $(1, -3)$.

2. **(Medicine)** Suppose the growth of a human fetus is given by the function $W(t) = 3t^2$, $0 \le t \le 39$, where the weight of W is in grams and the time t is in weeks.
   (a) Find the average rate of change in the weight of the fetus between $t_1 = 4$ and $t_2 = 10$.
   (b) Find the instantaneous rate of change of the weight of the fetus at $t = 20$.

3. Find $\lim_{h \to 0} \dfrac{(x + h)^2 - x^2}{h}$

4. Find $\lim_{x \to 4} \left(x^2 + \dfrac{2}{x} - \sqrt{x}\right)$

5. Find

   (a) $\lim_{x \to 2^-} \dfrac{1}{x - 2}$ (b) $\lim_{x \to 2^+} \dfrac{1}{x - 2}$

6. Use the graph to find $\lim_{x \to \infty} f(x)$.

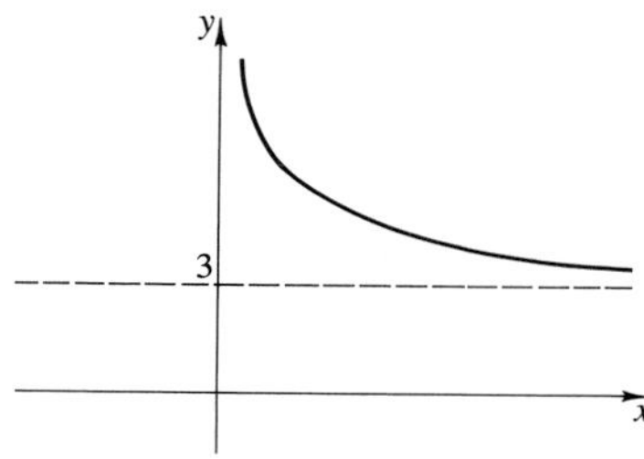

7. Let $f(x) = \dfrac{4}{4 - x}$. Find $\lim_{x \to 4^-} f(x)$, $\lim_{x \to 4^+} f(x)$, and $\lim_{x \to 4} f(x)$.

8. Discuss the continuity of the function f defined by $f(x) = (x - 1)/x$.

9. In each part find $f'(x)$:

   (a) $f(x) = \left(\dfrac{x^3 - 2x^2}{x^2 + x + 1}\right)^6$ (b) $x^4\sqrt[3]{x^2 - 1}$

10. **(Business)** A computer manufacturer finds that the profit (in dollars) received from making and selling x computers is given by $P(x) = (x^2 + x)^2$. If the rate of production is kept at 5 units per month, what is the rate of change in profit when 40 units have been made?

13

Applications of Differentiation

The derivative is useful in a wide variety of applications. In this chapter we focus on a few of the fundamental ways in which the derivative is inherently useful, especially to students of the management, life, and social sciences.

13.1 ELEMENTARY APPLICATIONS

Equations of Tangent Lines

EXAMPLE 1 Let f be defined by

$$f(x) = x^2 + 3x - 4$$

Find the equation of the tangent to the curve $y = f(x)$ at the point:

(a) $(-2, -6)$ (b) at which $x = 2$

Solution (a) As discussed earlier, the slope of the tangent to $y = f(x)$ at the point (x_1, y_1) is $f'(x_1)$. In this case,

$$f'(x) = 2x + 3$$

so that

$$f'(x_1) = 2x_1 + 3 \tag{1}$$

Then the point–slope form of the tangent line that passes through the point (x_1, y_1), see Section 1.3, is

$$y - y_1 = m(x - x_1)$$

where $m = f'(x_1)$, or

$$y = m(x - x_1) + y_1 \tag{2}$$

Substituting $x_1 = -2$ in Equation (1) we have

$$m = 2(-2) + 3 = -1$$

Then substituting $x_1 = -2$, $y_1 = -6$, and $m = -1$ in Equation (2) we obtain

$$\begin{aligned} y &= -1[x - (-2)] - 6 \\ &= -x - 8 \end{aligned}$$

Solution (b) When $x = 2$, the y-coordinate of the corresponding point on the curve is

$$y = f(2) = 2^2 + 3(2) - 4 = 6$$

From Equation (1) we have $m = 2(2) + 3 = 7$, and from (2) we obtain

$$\begin{aligned} y &= 7(x - 2) + 6 \\ &= 7x - 8 \end{aligned}$$

Rates of Change

EXAMPLE 2 **ECOLOGY** A manufacturer proposes to begin dumping a biodegradable liquid waste product into a nearby lake. The production will be such that the amount of waste dumped by time t will be

$$A = 3t^{3/2}$$

where A is in gallons and t is in weeks elapsed after the start of dumping. If the liquid decomposes at a constant rate of 27 gallons per week, how long will it take until the manufacturer is dumping liquid more rapidly than it is decomposing?

Solution The rate at which the liquid is being dumped is

$$\frac{dA}{dt} = \frac{d}{dt}(3t^{3/2}) = 3\left(\frac{3}{2}\right)t^{1/2} = \frac{9}{2}\sqrt{t} \qquad \text{gallons/week}$$

Thus the manufacturer will be dumping at the *same* rate that the material is decomposing when

$$\begin{aligned} \frac{9}{2}\sqrt{t} &= 27 \\ \sqrt{t} &= \frac{2}{9}(27) = 6 \\ t &= 36 \end{aligned}$$

Therefore, after 36 weeks the manufacturer will be dumping liquid more rapidly than the material is decomposing (Figure 1).

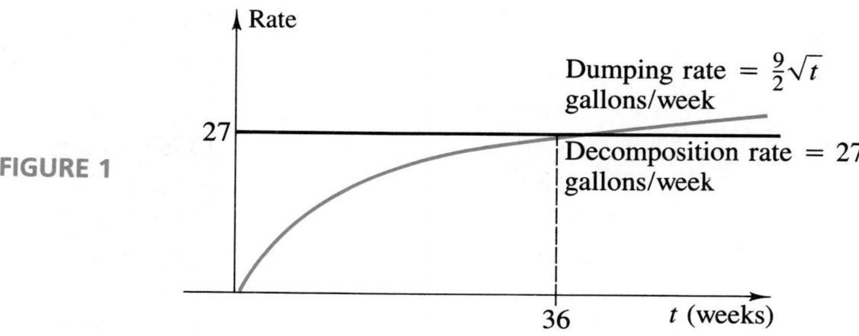

FIGURE 1

Increasing and Decreasing Functions

Consider a curve $y = f(x)$. We are often interested in knowing the regions on the real number line where the curve is rising and where it is falling.

At points where $f'(x) > 0$ the curve $y = f(x)$ has a tangent line with positive slope, and at points where $f'(x) < 0$ the curve has a tangent line with negative slope. Thus it follows that the curve $y = f(x)$ will be *increasing* (rising) on any interval where $f'(x) > 0$ for all x in the interval (Figure 2) and *decreasing* (falling) on any interval where $f'(x) < 0$ for all x in the interval (Figure 3).

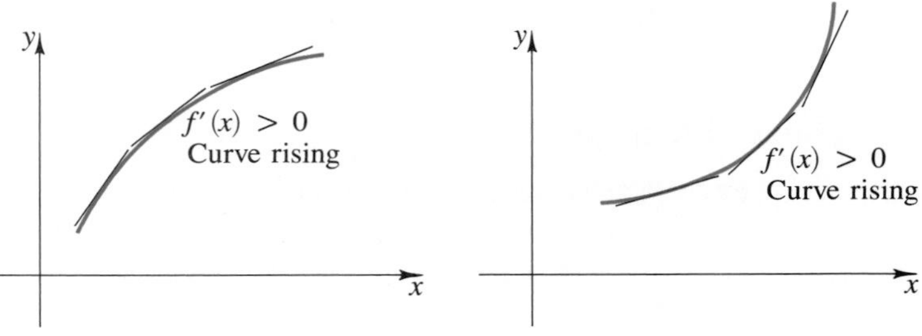

FIGURE 2

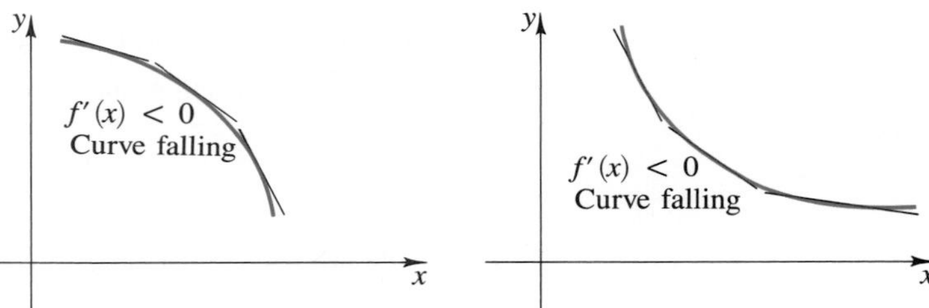

FIGURE 3

EXAMPLE 3 Let

$$f(x) = x^2 - 2x + 2$$

Where is $y = f(x)$ increasing? Decreasing?

Solution Since we obtain the information that we need from the derivative, we first find

$$f'(x) = 2x - 2$$

Since we can write

$$f'(x) = 2x - 2 = 2(x - 1)$$

we see that $f'(x) > 0$ if $x > 1$ and $f'(x) < 0$ if $x < 1$. Thus, $y = f(x)$ is increasing on $(1, +\infty)$ and decreasing on $(-\infty, 1)$. The graph of f is shown in Figure 4.

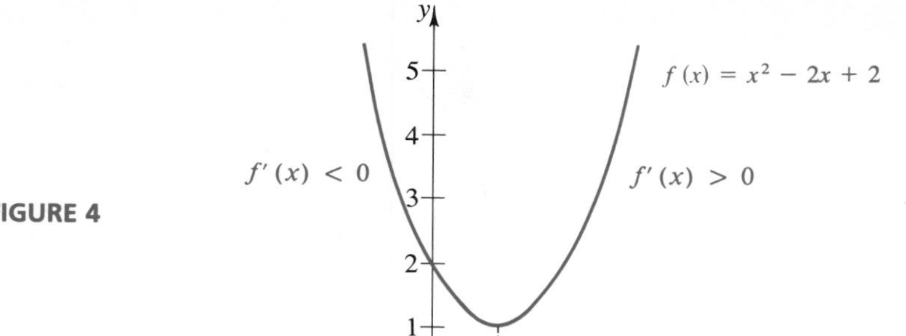

FIGURE 4

EXAMPLE 4 Let

$$f(x) = 2x^3 - 3x^2 - 12x + 2$$

Where is $y = f(x)$ increasing? Decreasing?

Solution Since the derivative of $f(x)$ gives the information we need, we must first find $f'(x)$:

$$f'(x) = 6x^2 - 6x - 12 = 6(x^2 - x - 2)$$

or

$$f'(x) = 6(x - 2)(x + 1)$$

We must determine the intervals over which $f'(x)$ is positive and those over which it is negative.

Observe that we have $f'(x) > 0$ when the factors $x - 2$ and $x + 1$ have the same sign, and we have $f'(x) < 0$ when the signs are opposite. From the analysis in Figure 5 we see that $y = f(x)$ is increasing on the intervals $(-\infty, -1)$ and $(2, +\infty)$ and decreasing on the interval $(-1, 2)$.

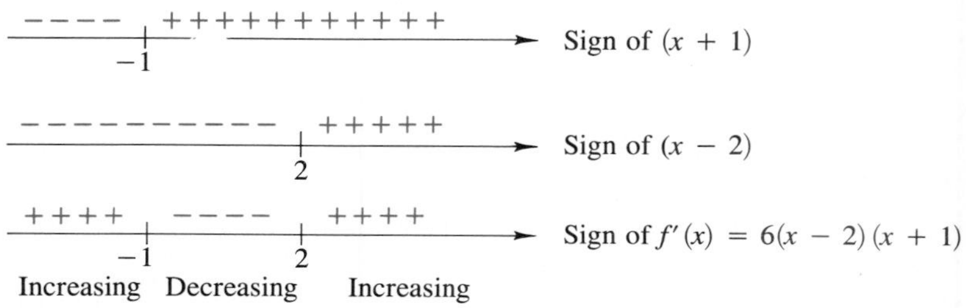

− − − − + + + + + + + + + + + (at −1) Sign of $(x + 1)$

− − − − − − − − − − − + + + + + (at 2) Sign of $(x - 2)$

+ + + + − − − − + + + + (at −1 and 2) Sign of $f'(x) = 6(x - 2)(x + 1)$

Increasing Decreasing Increasing

FIGURE 5

The graph of f is shown in Figure 6.

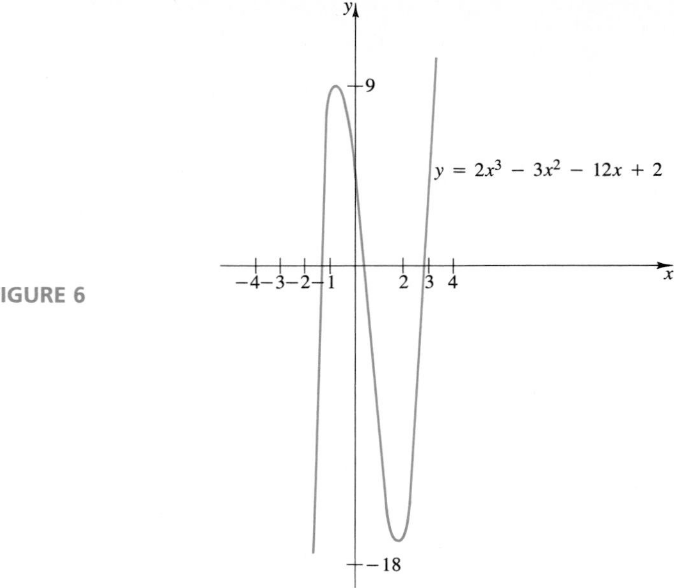

FIGURE 6

Velocity and Acceleration

Recalling our discussion of velocity in Section 12.2, we see that if an object moves along a straight line with position $x = s(t)$ at time t, then at any time t the velocity of the object is

$$v(t) = s'(t) = \frac{ds}{dt} \tag{3}$$

The *sign of the derivative ds/dt* indicates the *direction* of travel of the moving object. It can be shown that a positive derivative indicates that the object is moving "forward," in agreement with the positive direction on the line. Similarly, a negative derivative indicates that the object is moving "backward," in opposition to the positive direction on the line. A zero derivative indicates that the instantaneous velocity is 0 and that the object is stationary at that instant.

One of Newton's fundamental laws of physics states that an object in motion stays in motion, with constant velocity, unless acted on by an outside force. When the object experiences a change in velocity, it is said to undergo an acceleration (or deceleration). (Think of what happens when you press the gas pedal on a car.)

Definition For an object moving along a straight line with velocity $v(t)$ at time t, the instantaneous rate of change in $v(t)$ at time t is called the **acceleration** $a(t)$ of the object at time t. That is,

$$a(t) = v'(t) = \frac{dv}{dt} \tag{4}$$

The *sign of the acceleration* is also significant. A positive acceleration indicates that the velocity is increasing. A negative acceleration indicates that the velocity is decreasing. Note that if time is measured in seconds and distance is measured in feet, the units of acceleration are feet per second per second, or feet per second squared.

EXAMPLE 5 An object moves along an x-axis with known position $x = s(t) = 15t - 3t^2$ inches after t seconds.

(a) Where is the object at time $t = 2$?

(b) What is the object's velocity at time $t = 2$?

(c) What is the object's acceleration time at $t = 2$?

Solution (a) Since $s(2) = 15(2) - 3(2)^2 = 30 - 12 = 18$, after 2 seconds the object is 18 inches to the right of the origin.

Solution (b) Since $v(t) = \dfrac{d}{dt}(15t - 3t^2) = 15 - 6t$, $v(2) = 15 - 6(2) = 3$. At time $t = 2$, the velocity is 3 inches per second.

Solution (c) Since $a(t) = \dfrac{d}{dt}(15 - 6t) = -6$, at time $t = 2$ the acceleration is $a(2) = -6$ inches per second per second.

EXAMPLE 6 **BUSINESS** The profit derived from the sale of x hundreds of gallons of film developer is given (in dollars) by

$$P(x) = 2x^3 - 39x^2 + 240x - 50 \qquad x > 0$$

For what values of x is the profit increasing? Decreasing?

Solution The derivative of the profit function is given by

$$\begin{aligned} P'(x) &= 6x^2 - 78x + 240 \\ &= 6(x^2 - 13x + 40) \\ &= 6(x - 5)(x - 8) \end{aligned}$$

Observe that $P'(x) > 0$ when both factors $x - 5$ and $x - 8$ have the same sign and $P'(x) < 0$ when the signs are opposite. From the analysis in Figure 7 we see that $P(x)$ is increasing when $0 < x < 5$ or when $x > 8$. The profit is decreasing when $5 < x < 8$.

FIGURE 7

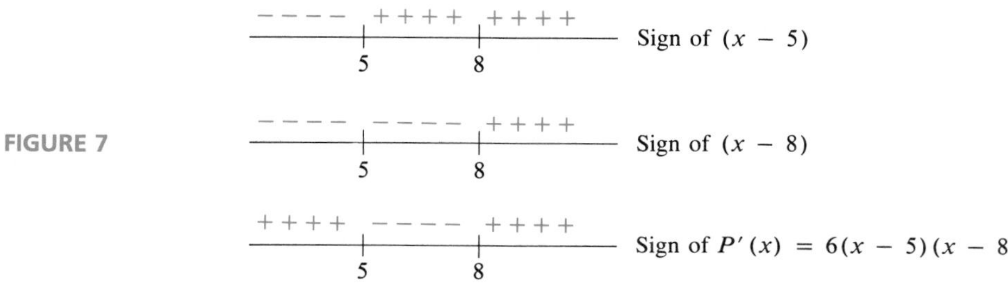

EXERCISE SET 13.1

In Exercises 1–8 find the equation of the line tangent to the graph of the given function at the given point. Sketch the graph of both the function and the tangent line.

1. $f(x) = -2x + 4$ at $(1, 2)$

2. $f(x) = x$ at $(2, 2)$

3. $f(x) = x^2$ at $(2, 4)$

4. $f(x) = 1 - x^2$ at $(1, 0)$

5. $f(x) = 1/x$ at $(1, 1)$

6. $f(x) = 1/x^2$ at $(1, 1)$

7. $f(x) = x^3$ at $(-1, -1)$

8. $f(x) = 1/(1 - x)$ at $(0, 1)$

9. Find the equation of the tangent to

$$y = \frac{x - 1}{x + 2}$$

at the point where $x = 3$.

10. Find the equation of the tangent to

$$y = (x^2 + 2x)^{18}$$

at the point where $x = -1$.

11. The volume V of a spherical malignant tumor is given by

$$V = \tfrac{4}{3}\pi r^3$$

(a) Find the rate of change of the volume with respect to radius.
(b) At what rate is V changing relative to r when $r = 0.5$ centimeter?

12. A bacteria colony has an initial population of 20,000. After t hours the colony increases to a population $P(t)$ given by

$$P(t) = 20{,}000(1 + 0.5t + t^2)$$

(a) Find the rate of change of the population P with respect to time (this is called the **growth rate**).
(b) Find the number of bacteria present after $t = 10$ hours and find the growth rate at that moment.

13. **(Advertising)** A manufacturer estimates that the firm can sell Q units of its product with an advertising expenditure of x thousand dollars where

$$Q(x) = -x^2 + 500x + 8$$

(a) What is the rate of change of the number of units sold with respect to the amount spent on advertising?
(b) What is this rate of change when $x = 20$?

14. **(Ecology)** A manufacturer proposes to begin dumping a biodegradable liquid waste product into a nearby stream. The production is such that the amount A (in gallons) of waste dumped at time t (in weeks elapsed after the start of dumping) will be

$$A = (t + 1)^{5/2}$$

If the liquid decomposes at a constant rate of 20 gallons per week, how long will it take until the manufacturer is dumping liquid more rapidly than it is decomposing?

In Exercises 15–26 determine the intervals over which the given function is increasing, and those over which it is decreasing.

15. $f(x) = 3 - x^2$

16. $f(x) = x^2 - 2x$

17. $f(x) = x^2 + 4x + 2$

18. $f(x) = |x|$

19. $f(x) = x^3 - 5$

20. $f(x) = 1 - x^3$

21. $f(x) = 1/x$

22. $f(x) = 1/(1 - x)$

23. $f(x) = |x - 2|$

24. $f(x) = x^2/(1 - x)$

25. $f(x) = 1/x^2$

26. $f(x) = x/(x^2 + 1)$

27. An object moving along the x-axis has position

$$x = s(t) = t^3 - 12t$$

at time t, where x is measured in feet and t in seconds.
(a) At time $t = 1$:
  (i) Where is the object?
  (ii) How fast is it moving?
  (iii) Is it moving toward or away from the origin?
  (iv) What is its acceleration?
(b) Answer the same questions for time $t = 3$.

28. In the wake of a successful advertising campaign, a local television retailer finds that sales increase noticeably and then gradually decrease. The number of daily sales t days after the end of the campaign is given by

$$S(t) = 20 + 8t - t^2$$

Assume this is to be true for at least a week.
(a) Is the number of sales of television sets increasing or decreasing at the end of the first day? At what rate?

(b) Answer the same questions in (a) but for sales after 3 days.
(c) Is the number of sales accelerating or decelerating? At what rate? Use the derivative of $S'(t)$.
(d) After how many days will the number of sales no longer be increasing?

In Exercises 29–32 assume that an object is moving along the x-axis with known position

$$x = s(t) \text{ inches}$$

after t seconds. Find the position, velocity, and acceleration at the specified time t.

29. $s(t) = 3t^3 - 10t^2 + 7t;\ t = 1$

30. $s(t) = 3t - \sqrt{t};\ t = 4$

31. $s(t) = 1/t;\ t = 2$

32. $s(t) = 1/(t - 1);\ t = 5$

33. A stone is thrown straight up into the air, and after t seconds its height is

$$h(t) = 48t - 16t^2 \quad \text{feet}$$

(a) What is its velocity after t seconds?
(b) What is its initial velocity (as it leaves the thrower's hand)?
(c) When will it hit the ground?
(d) How fast will it be going as it hits the ground?
(e) Find the acceleration after t seconds. Interpret your answer.
(f) As the stone ascends, its velocity is positive, and as it descends, its velocity is negative. After how many seconds does the stone reach its maximum height?
(g) What is the maximum height reached by the stone? [*Hint:* What is the velocity at the maximum height of the stone?]

13.2 MARGINAL ANALYSIS IN BUSINESS AND ECONOMICS

The concept of the derivative has significant applications in economic analysis. Economists and business managers are frequently interested in how changes in such variables as inventory, production level, supply, investment in advertising, and selling price affect other variables such as profit, demand, cost, employment, and inflation. Such relationships are often studied mathematically using **marginal analysis**. Management often makes significant policy decisions based on such analysis.

The word *marginal* is the economist's term for rate of change, or derivative. Marginal analysis is thus concerned with the rate at which one economic variable changes relative to another.

Consider a manufacturer or producer engaged in producing a product on a regular basis. We consider three important functions. If x (called the **level of production**) represents the number of units of the product, let

$C(x)$ = total cost of producing x units of the product
$R(x)$ = total revenue received from selling x units of the product
$P(x)$ = total profit obtained from selling x units of the product*

These are called, respectively, the **cost function, revenue function,** and **profit function**. If all units produced are sold, then these are related by

$$P(x) = R(x) - C(x) \tag{1}$$

$$\text{profit} = \text{revenue} - \text{cost}$$

* In Example 9 of Section 11.2 we also describe three function $C(x)$, $R(x)$, and $P(x)$. But there the cost, revenue, and profit are functions of the *selling price* x.

The per-unit average cost, revenue, and profit can be calculated easily:

$$\overline{C}(x) = \frac{C(x)}{x} = \text{average cost per unit} \tag{2}$$

$$\overline{R}(x) = \frac{R(x)}{x} = \text{average revenue per unit} \tag{3}$$

$$\overline{P}(x) = \frac{P(x)}{x} = \text{average price per unit} \tag{4}$$

Suppose our company is considering changing its level of production from x to $x + h$ units. It will expect its costs to increase from $C(x)$ to $C(x + h)$. When this increase in cost is averaged over the h additional units produced, the company can expect the **average cost of each additional unit** produced to be given by the ratio

$$\frac{C(x + h) - C(x)}{h} \tag{5}$$

Similarly, the **average revenue from each additional unit** produced above x units, up to $x + h$ units, is

$$\frac{R(x + h) - R(x)}{h} \tag{6}$$

and the **average profit earned from each additional unit** produced from x to $x + h$ units is

$$\frac{P(x + h) - P(x)}{h} \tag{7}$$

EXAMPLE 1 Ace Audio produces x amplifiers weekly at a cost of \$200 each, less $\frac{1}{5}x$ dollars each due to quantity discounts available on supplies. In addition, Ace has fixed weekly costs of \$3000. It sells the amplifiers for \$350 each, less $\frac{1}{20}x$ dollars because of quantity sales incentives. Thus the weekly cost, revenue, and profit functions for Ace Audio are

$$\begin{aligned} C(x) &= \text{fixed costs} + \text{variable costs} \\ &= 3000 \qquad + x\left(200 - \frac{x}{5}\right) \\ &= 3000 + 200x - \frac{x^2}{5} \qquad \text{(dollars)} \end{aligned} \tag{8}$$

$$\begin{aligned} R(x) &= x\left(350 - \frac{x}{20}\right) \\ &= 350x - \frac{x^2}{20} \qquad \text{(dollars)} \end{aligned} \tag{9}$$

$$P(x) = R(x) - C(x)$$

$$= \left(350x - \frac{x^2}{20}\right) - \left(3000 + 200x - \frac{x^2}{5}\right)$$

$$= \frac{3x^2}{20} + 150x - 3000 \qquad \text{(dollars)} \tag{10}$$

(a) Suppose Ace is presently producing and selling 50 amplifiers per week. Then its weekly cost, revenue, and profit are

$$C(50) = 3000 + 200(50) - \frac{50^2}{5} = \$12{,}500$$

$$R(50) = 350(50) - \frac{50^2}{20} = \$17{,}375$$

$$P(50) = 17{,}375 - 12{,}500 = \$4{,}875$$

Over these 50 units, the average per unit cost, revenue, and profit are

$$\overline{C}(50) = \frac{C(50)}{50} = \frac{12{,}500}{50} = \$250.00 \text{ per unit}$$

$$\overline{R}(50) = \frac{R(50)}{50} = \frac{17{,}375}{50} = \$347.50 \text{ per unit}$$

$$\overline{P}(50) = \frac{P(50)}{50} = \frac{4875}{50} = \$97.50 \text{ per unit}$$

(b) Imagine that Ace is contemplating raising its production (and sales) level to 60 amplifiers per week. Its new weekly cost, revenue, and profit can be calculated from Equations (8), (9), and (10):

$$C(60) = 3000 + 200(60) - \frac{60^2}{5} = \$14{,}280 \qquad \text{from (8)}$$

$$R(60) = 350(60) - \frac{60^2}{20} = \$20{,}820 \qquad \text{from (9)}$$

$$P(60) = 20{,}820 - 14{,}280 = \$6{,}540 \qquad \text{from (10)}$$

In raising its production level from 50 to 60 units per week, the per-unit average cost, revenue, and profit will be

$$\frac{C(60) - C(50)}{60 - 50} = \frac{14{,}280 - 12{,}500}{10} = \frac{1780}{10}$$
$$= \$178 \text{ per unit average cost}$$

$$\frac{R(60) - R(50)}{60 - 50} = \frac{20{,}820 - 17{,}375}{10} = \frac{3445}{10}$$
$$= \$344.50 \text{ per unit average revenue}$$

$$\frac{P(60) - P(50)}{60 - 50} = \frac{6540 - 4875}{10} = \frac{1665}{10}$$
$$= \$166.50 \text{ per unit average profit}$$

It is interesting to compare these averages with those for the first 50 units. These calculations show that although the average revenue per unit decreases, the average cost decreases so much more that the average profit is significantly improved.

Marginal Cost, Marginal Revenue, and Marginal Profit

The per-unit averages we have been calculating involve tedious arithmetic calculations and do not provide us with sufficient mathematical power. Economic analysts have found it far more productive to introduce the derivative into the following context.

The derivatives of $C(x)$, $R(x)$, and $P(x)$ are called, respectively, the **marginal cost** (MC), the **marginal revenue** (MR), and the **marginal profit** (MP):

$$\begin{aligned} C'(x) &= \text{marginal cost} \\ &= \text{rate of change of cost with respect to the number } x \text{ of units produced} \\ R'(x) &= \text{marginal revenue} \\ &= \text{rate of change of revenue with respect to the number } x \text{ of units sold} \\ P'(x) &= \text{marginal profit} \\ &= \text{rate of change of profit with respect to the number } x \text{ of units sold} \end{aligned}$$

These functions are expressed as dollars per unit. If all units are sold, then the relationship between marginal profit, marginal cost, and marginal revenue is obtained by differentiating both sides of Equation (1):

$$P(x) = R(x) - C(x) \tag{1}$$

which yields

$$\underset{\text{marginal profit}}{P'(x)} = \underset{\text{marginal revenue}}{R'(x)} - \underset{\text{marginal cost}}{C'(x)} \tag{1a}$$

or

$$\text{MP} = \text{MR} - \text{MC}$$

Common Interpretation

$$\begin{aligned} C'(x) &\cong \text{cost of producing the } (x+1)\text{st unit} \\ R'(x) &\cong \text{revenue from the sale of the } (x+1)\text{st unit} \\ P'(x) &\cong \text{profit from the manufacture and sale of the } (x+1)\text{st unit} \end{aligned}$$

EXAMPLE 2 For Ace Audio, as described in Example 1, the cost, revenue, and profit functions are

$$C(x) = 3000 + 200x - \frac{x^2}{5}$$

$$R(x) = 350x - \frac{x^2}{20}$$

$$P(x) = \frac{3x^2}{20} + 150x - 3000$$

Thus the marginal functions are

$$\text{MC} = C'(x) = 200 - \frac{2x}{5} \qquad \text{(dollars per unit)}$$

$$\text{MR} = R'(x) = 350 - \frac{x}{10} \qquad \text{(dollars per unit)}$$

$$\text{MP} = P'(x) = \frac{3x}{10} + 150 \qquad \text{(dollars per unit)}$$

(a) When $x = 50$, the marginal functions are

$$\text{MC} = 200 - \frac{100}{5} = \$180 \text{ per unit}$$

$$\text{MR} = 350 - \frac{50}{10} = \$345 \text{ per unit}$$

$$\text{MP} = \frac{150}{10} + 150 = \$165 \text{ per unit}$$

Notice how close these figures are to the per-unit averages calculated in Example 1(b).

(b) By common interpretation these marginal function values are approximately the cost, revenue, and profit of the 51st unit. Let us check out that claim:

$$\text{actual cost of producing 51st unit} = \frac{C(51) - C(50)}{51 - 50} = \frac{12{,}679.80 - 12{,}500.000}{1} = \$179.80$$

$$\text{actual revenue from 51st unit} = \frac{R(51) - R(50)}{51 - 50} = \frac{17{,}719.95 - 17{,}375.00}{1} = \$344.95$$

$$\text{actual profit from 51st unit} = \frac{P(51) - P(50)}{51 - 50} = \frac{5040.15 - 4875.00}{1} = \$165.15$$

The marginal functions are so close to the actual values that they are often used instead.

EXAMPLE 3 Consider Ace Audio once again. Suppose it is currently producing only 20 units per week because of an industrywide slump. Then,

$$C(20) = 3000 + 200(20) - \frac{(20)^2}{5} = \$6920$$

$$R(20) = 350(20) - \frac{(20)^2}{20} = \$6980$$

$$P(20) = \$60$$

The average profit per unit is $P(20)/20 = \$3$. Management is dissatisfied with this profit and wants to know the effects on profit if production increases just slightly. Since the marginal profit is

$$P'(x) = \frac{3x}{10} + 150$$

we have

$$P'(20) = \frac{3(20)}{10} + 150 = \$156 \text{ per unit}$$

We interpret this to mean that the company can expect the 21st unit to realize approximately \$156 profit. Not bad at all, compared to the previous average of \$3 per unit!

To confirm the validity of using the marginal function, we calculate the actual profit from the 21st unit:

$$\begin{aligned} P(21) - P(20) &= \frac{3(21)^2}{20} + 150(21) - 3000 \\ &= 216.15 - 60 \\ &= \$156.15 \end{aligned}$$

The MP was within 15¢ of the actual profit, a discrepancy of less than 0.1%.

Remarks Manufacturing costs can be classified as **fixed costs** or **variable costs**. Fixed costs include such items as rent, insurance, and administrative wages. They are incurred even if no items are produced. Variable costs include such items as cost of materials, cost of transporting manufactured goods, and production wages. These costs vary with the number x of items manufactured. For example, for the cost function $C(x) = x^2 + 3x + 1000$ (dollars), the fixed and variable costs break down as follows:

$$C(x) = \underbrace{x^2 + 3x}_{\text{variable costs}} + \underbrace{1000}_{\text{fixed costs}}$$

More generally, any cost function can be written in the form

$$C(x) = \underbrace{f(x)}_{\text{variable costs}} + \underbrace{k}_{\text{fixed costs}}$$

Because the derivative of a constant is zero, it follows that the marginal cost is

$$C'(x) = f'(x) + 0 = f'(x)$$

which demonstrates that *fixed costs do not affect the marginal cost*. This is a basic principle of economics.

Demand versus Revenue

It is well known that the price of a product affects the amount of the product the public is willing to purchase. Market analysts for a particular product conduct research to determine the **demand equation** for the product:

$$x = f(p) \tag{11}$$

expressing the **demand** for x units of the product if it is offered for sale at **price** p. Note that x is the desired production level for the price p. Once the demand equation is known, the revenue function

$$R(x) = xp \tag{12}$$

can be determined by solving Equation (11) for p and substituting into Equation (12).

EXAMPLE 4 Wonder Watch Company currently sells 800 of its Model W monthly at $75 each. Its market research consultant determines that for every additional dollar the company charges, the public will buy 10 fewer watches monthly. Find the demand equation and revenue function.

Solution The problem tells us that the demand x is 800 units less 10 units for each $1 by which p is greater than $75; that is,

$$x = 800 - 10(p - 75)$$

This is the demand equation. To get the revenue function we must first solve for p:

$$\begin{aligned} x &= 800 - 10p + 750 = 1550 - 10p \\ 10p &= 1550 - x \\ p &= 155 - \frac{x}{10} \end{aligned}$$

Therefore, the revenue function is

$$\begin{aligned} R(x) &= xp \\ &= x\left(155 - \frac{x}{10}\right) \\ &= -\frac{x^2}{10} + 155x \end{aligned}$$

EXERCISE SET 13.2

1. A manufacturer finds that the cost (in thousands of dollars) of producing x units monthly is

$$C(x) = 120 - 5x + 0.2x^2$$

(a) Find $\overline{C}(x)$, the average cost per unit, when $x = 20$.
(b) Find the marginal cost when $x = 20$.
(c) Is the average cost $\overline{C}(x)$ increasing or decreasing when $x = 20$? When $x = 10$? (*Hint:* Compute $\dfrac{d}{dx}[\overline{C}(x)]$).

2. The A Company finds that the revenue function for production level x is

$$R(x) = x^2 + \frac{1}{x+1} - 1$$

(a) Find $\overline{R}(x)$, the average revenue per unit. Find $\overline{R}(100)$.
(b) Find the marginal revenue when $x = 100$.
(c) Is the average revenue increasing or decreasing when $x = 50$?

3. A manufacturer's cost and revenue functions (in dollars) are

$$C(x) = 1000 + 50x + \frac{x^2}{10}$$

$$R(x) = 400 + \frac{x}{20}$$

(a) Find the marginal cost, marginal revenue, and marginal profit.
(b) At what rate is cost changing when the production level is $x = 25$ units?
(c) At what rate is revenue changing when the production level is $x = 25$ units?
(d) At what rate is the profit changing when $x = 25$ units?
(e) Estimate the cost of producing the 26th unit.

4. Repeat the directions of Exercise 3 given that

$$C(x) = \tfrac{1}{2}x^2 + 8x + 24$$
$$R(x) = -0.01x^2 + 25x - 50$$

5. A manufacturer determines that to sell x units, the price p in dollars must be

$$p(x) = 200 - \tfrac{1}{2}x$$

The cost in dollars of producing x units is

$$C(x) = 3000 + \tfrac{1}{4}x^2$$

(a) Find the revenue and profit functions.
(b) Find the marginal revenue, marginal cost, and marginal profit functions.
(c) Estimate the added revenue obtained by selling the 26th unit.
(d) Estimate the added profit (loss) incurred by selling the 26th unit.

6. Suppose the demand equation for a product at selling price p is

$$x = 4000 - 80p$$

Find the revenue function $R(x)$ when the production level is x.

7. Suppose the demand equation is

$$x = 10{,}000 - 50p$$

when the selling price is p. Find the revenue function $R(x)$ when production level is x.

8. Corner Shoe Store normally sells 50 pairs of special running shoes each month at $60 a pair. Their market research consultant reports that for every additional dollar they charge they will sell 5 fewer pairs per month.
(a) Find the demand equation, $x = f(p)$. [*Hint:* Let y = the number of increases. Write x and p in terms of y, then eliminate y.]
(b) Find the monthly revenue function when Corner Shoe Store sells all the shoes demanded.
(c) Suppose Corner Shoe Store buys these shoes wholesale for $40 each. Find the monthly cost function $C(x)$.
(d) Find the monthly profit function.
(e) Calculate the marginal cost, marginal revenue, and marginal profit functions.
(f) Find the appropriate profit on the sale of the 61st pair of shoes.

13.3 RELATED RATES PROBLEMS

In applying mathematics to solve real-world problems, equations are used to relate the variables involved in the problems. In typical applications of calculus these variables are changing. We are often interested in how their rates of change are related so that from information specifying one rate of change we can find another. The key to solving such problems is usually the chain rule (see Section 12.8).

In a typical problem, one variable may be a known function of another, for example,

$$y = f(x)$$

where both x and y are functions of a third variable, for example, t. Although x and y may not be given explicitly in terms of t, we may know how fast x is changing with respect to t and wish to find how fast y is changing with respect to t. We use the chain rule in the obvious way:

$$\frac{dy}{dt} = \frac{dy}{dx} \cdot \frac{dx}{dt}$$

Remark If a quantity y is increasing with respect to time, then dy/dt is positive; if y is decreasing with respect to time, then dy/dt is negative.

EXAMPLE 1 A stone dropped in a lake produces a circular ripple whose radius increases at a constant rate of 2 feet per second. How rapidly is the area of the ripple increasing when the radius is 10 feet?

Solution Let

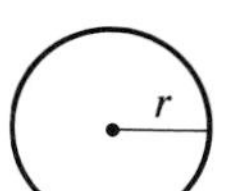

$$\begin{aligned} r &= \text{the radius of the ripple in feet} \\ A &= \text{the area of the ripple in square feet} \\ t &= \text{time in seconds} \end{aligned}$$

Note that although we are not given the explicit relationship between r and t, we are given the rate of change of r with respect to t:

$$\frac{dr}{dt} = 2 \text{ feet/second} \tag{1}$$

But we want to find another rate of change,

$$\frac{dA}{dt} \qquad \text{when } r = 10$$

The area formula for a circle gives the relationship $A = \pi r^2$, and from the chain rule,

$$\begin{aligned} \frac{dA}{dt} &= \frac{dA}{dr} \cdot \frac{dr}{dt} \\ &= 2\pi r \cdot \frac{dr}{dt} \\ &= 2\pi r(2) = 4\pi r \qquad \text{[from Equation (1)]} \end{aligned}$$

Thus when $r = 10$,

$$\frac{dA}{dt} = 40\pi \cong 126$$

Since dA/dt is positive, the area is increasing. The approximate rate of increase is 126 square feet per second when the radius is 10 feet.

EXAMPLE 2 The wholesale price of bananas in a local market is related to the supply (in crates) by

$$P = 3 + \frac{30}{S}$$

where P = price per pound (in dollars) and S = supply (in crates). At what rate will the price be changing if 100 crates are available, but the number of crates is decreasing at a rate of 10 crates per day?

Solution Let t = time in days. Since S is *decreasing* at a rate of 10 crates per day, we have

$$\frac{dS}{dt} = -10 \tag{2}$$

We thus know dS/dt and we must find dP/dt when $S = 100$. By the chain rule,

$$\frac{dP}{dt} = \frac{dP}{dS} \cdot \frac{dS}{dt} \tag{3}$$

But $P = 3 + 30S^{-1}$, so

$$\frac{dP}{dS} = \frac{d}{dS}(3 + 30S^{-1}) = -30S^{-2} = \frac{-30}{S^2} \tag{4}$$

Substituting the results of (2) and (4) into Equation (3), we have

$$\frac{dP}{dt} = \left(-\frac{30}{S^2}\right)(-10) = \frac{300}{S^2}$$

When $S = 100$ we have

$$\frac{dP}{dt} = \frac{300}{(100)^2} = 0.03 \qquad \text{dollars per day}$$

Thus the price per pound is increasing at a rate of 3 cents per day.

EXAMPLE 3 The bottom of a 20-foot ladder leaning against a wall is being pulled away from the base of the wall at the constant rate of 2 feet per second. How fast is the top of the ladder falling when the base of the ladder is 12 feet from the wall?

Solution Figure 1 depicts the problem. The bottom of the ladder is x feet from the base of the wall, and the top of the ladder is y feet from the ground. We are given that $dx/dt = 2$. We must use this information to obtain dy/dt when $x = 12$. First, we relate the variables x and y using the Pythagorean theorem about right triangles. Thus,

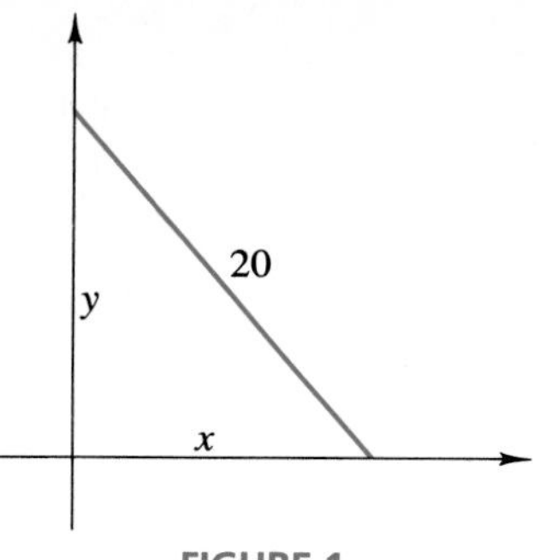

FIGURE 1

$$x^2 + y^2 = 20^2$$
$$y^2 = 400 - x^2$$
$$y = \sqrt{400 - x^2}$$

Having related the variables we now relate their derivatives using the chain rule:

$$\begin{aligned}\frac{dy}{dt} &= \frac{dy}{dx} \cdot \frac{dx}{dt} \\ &= \frac{1}{2}(400 - x^2)^{-1/2}(-2x)\frac{dx}{dt} \\ &= \frac{-x}{\sqrt{400 - x^2}} \cdot 2 \qquad \left(\text{since } \frac{dx}{dt} = 2\right) \\ &= \frac{-2x}{\sqrt{400 - x^2}}\end{aligned}$$

Now, when the base of the ladder is 12 feet from the wall ($x = 12$), we have

$$\frac{dy}{dt} = \frac{-2(12)}{\sqrt{400 - 144}} = \frac{-24}{\sqrt{256}} = \frac{-24}{16} = \frac{-3}{2} = -1.5$$

Notice that dy/dt is negative because y is decreasing. We conclude that when the base of the ladder is 12 feet from the wall and moving away from the wall at the rate of 2 feet per second, the top of the ladder is falling at the rate of 1.5 feet per second.

A Procedure for Solving Related Rates Problems

Step 1 Identify all relevant variables.

Step 2 Indicate the variables whose rates are given and the one (or more) whose rate you are asked to find.

Step 3 If the problem is a geometric one, sketch a figure.

Step 4 Construct an equation relating the relevant variables.

Step 5 Find the desired derivative from this equation (the chain rule will be used).

Step 6 Substitute the given constant values of the variables in the appropriate places.

It is important to realize that Step 6 must be done last. When variables are "frozen" at constant values, then their derivatives become zero. The variables must remain free to vary until after Step 5 has been completed. In Example 1 the last step is to substitute $r = 10$. In Example 2 the last step is to set $S = 100$, and in Example 3 it is to set $x = 12$.

EXERCISE SET 13.3

1. A stone is dropped in a lake, producing a circular ripple whose radius increases at the constant rate of 3 feet per second. How fast is the area of the circular ripple changing when the radius is 12 feet?

2. In the context of Exercise 1, how fast is the circumference of the ripple changing when the radius is 12 feet?

3. The volume of a sphere of radius r is $V = \frac{4}{3}\pi r^3$. If the radius of a balloon is increasing at the constant rate of 0.5 centimeter per second, how fast is the volume increasing when the radius is 6 centimeters?

4. A spherical balloon is inflated by blowing in air at the rate of 48 cubic inches per second. How fast is the radius increasing when it is 6 inches?

5. **(Harder)** Water is being pumped into a vertical cylindrical storage tank with a radius of 6 feet at the constant rate of 24 cubic feet per minute. How fast is the water level rising when the water is 12 feet deep?

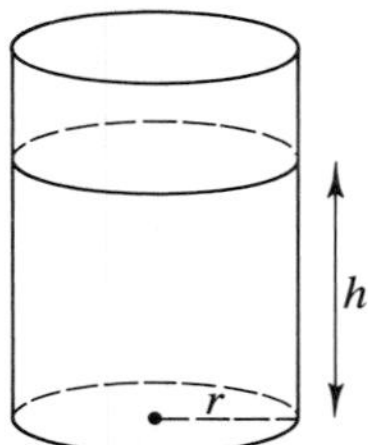

$V = \pi r^2 h$

6. **(Harder)** Sand is being poured on a conical pile at the rate of 12 cubic feet per minute. The height of this pile is always exactly twice the radius of the base. How fast is the height growing when the pile is exactly 3 feet high?

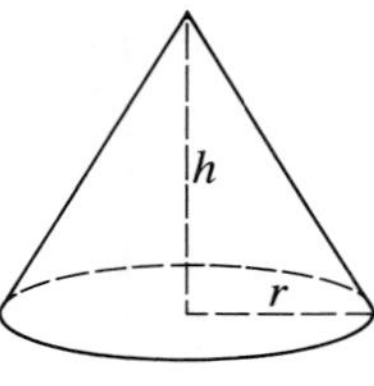

$V = \frac{1}{3}\pi r^2 h$

7. A cubical block of ice melts so that each side decreases at a rate of 1 inch per hour. How fast is the volume of the block decreasing when the sides are 6 inches?

8. The area of a circle is increasing at the rate of 6 square inches per minute. When the area of the circle reaches 25 square inches, how fast is the circumference increasing?

9. **(Manufacturing)** The total cost in dollars of refining x units of raw sugar is given by $C(x) = x^2 - x$. If sugar is refined at a constant rate of 10 units per hour, determine the rate at which the cost is changing when 40 units have been produced.

10. **(Ecology)** In a wildlife reserve, the population P of gray hawks depends on the population x of its basic food supply, rabbits. Research suggests the relationship is approximately
$$P = 0.002x + 0.0005x^2$$
If the number of rabbits is allowed to increase at a constant controlled rate of 500 per year, how rapidly will the hawk population be increasing at a time when the reserve contains 4000 rabbits?

11. Suppose the cost of producing x units monthly is
$$C(x) = 18{,}000 - 400x + 30x^2$$
If the production level is increasing steadily at the rate of 4 units per month, how fast are production costs rising when the production level reaches 30 units per month?

12. Suppose the revenue from producing x units per month is
$$R(x) = 2x + x^3 - 100$$
If the production level is increasing steadily at the rate of 5 units per month, how fast is the revenue rising when the production level reaches 50 units per month?

13. Suppose the demand equation for a product at selling price p is
$$x = 4500 - 80p$$
When the price is \$4 and decreasing at the rate of 25 cents per month, what is the instantaneous rate of change in demand?

14. Suppose the demand equation for a product at selling price p is
$$p = 500 - 12x + \tfrac{1}{8}x^2$$
When the monthly demand is at 80 units and the price is decreasing at the rate of \$24 per month, what is the instantaneous rate of change in demand?

15. For the demand equation

$$x = 1000 - 70p$$

when the production level is 150 units per month and increasing at the rate of 13 units per month, what is the instantaneous rate of change in revenue per month?

16. For the demand equation

$$p = 2500 - 5x - \tfrac{1}{18}x^2$$

find the instantaneous rate of change in the revenue per month when the production level is 90 units per month and increasing at the rate of 15 units per month. What is the actual revenue per month when the monthly production level is at 90?

13.4 APPLICATIONS OF DIFFERENTIATION TO OPTIMIZATION

For many practical problems it is important to maximize or minimize a specific function. For example, if we know the profit function $P(x)$ corresponding to production level x, then it is certainly of interest to determine what production level x will produce the *greatest* profit $P(x)$. Everyday examples abound: business managers act to maximize profits, living organisms act to minimize loss of energy, airlines attempt to minimize the flying time for a scheduled trip, leaves of plants attempt to maximize their exposure to sunlight. In this section we show how differentiation can be used to solve certain kinds of maximization and minimization problems. We begin with an example.

EXAMPLE 1 A farmer wants to fence in a rectangular field bordering on a straight stream (Figure 1) but will not fence the side along the stream. If there are 1000 feet of fence to work with, what is the maximum area that can be enclosed? What dimensions maximize this area?

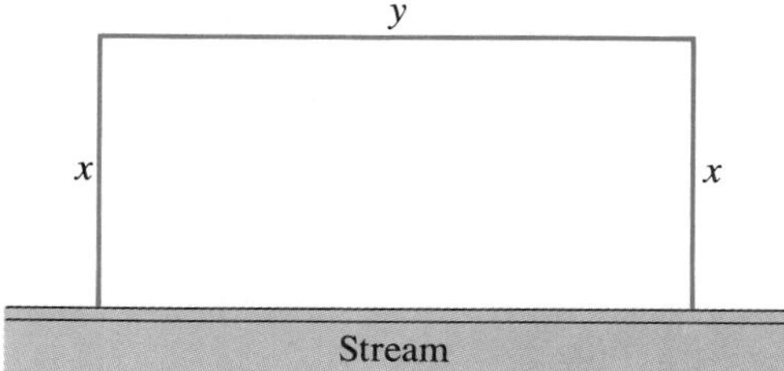

FIGURE 1

Solution To attack this problem, let us denote the lengths of the sides by x and y (as shown in Figure 1) so that the area A to be maximized is

$$A = xy \tag{1}$$

Since the farmer has 1000 feet of fence, we have the following relationship between x and y:

$$1000 = 2x + y$$

or

$$y = 1000 - 2x \tag{2}$$

Thus we can rewrite (1) as

$$A = x(1000 - 2x) = 1000x - 2x^2 \tag{3}$$

In this formula the variable x is subject to certain physical restrictions. For example, since x represents a length, we must have $x \geq 0$. Moreover, there is only 1000 feet of fence available, so we cannot use more than this amount on the two sides of length x; thus $2x \leq 1000$ or $x \leq 500$.

In light of these physical restrictions and Equation (3), we can formulate our problem as follows.

Problem Find the maximum value of

$$A = 1000x - 2x^2 \tag{4}$$

where x satisfies $0 \leq x \leq 500$.

One way to attack this problem is to graph Equation (4) and try to determine the maximum value of A from the graph. In Figure 2 we have tabulated and plotted some points on this graph. In light of this figure it appears that the maximum area is

$$A = 125{,}000 \text{ square feet}$$

FIGURE 2

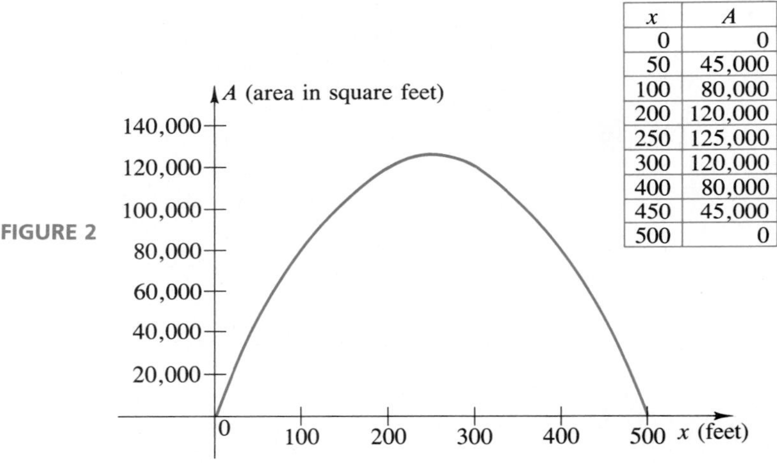

x	A
0	0
50	45,000
100	80,000
200	120,000
250	125,000
300	120,000
400	80,000
450	45,000
500	0

which occurs when

$$x = 250 \text{ feet}$$

and from (2),

$$y = 1000 - 2x = 1000 - 2(250) = 500 \text{ feet}$$

Although this graphical solution is simple, it suffers from a major defect. It is possible that some value of x different from any of those tabulated might produce a value for the area surpassing any that we have recorded and plotted. Moreover, it will do no good to plot more points since we might still omit the very value of x that produces the largest area. What we need is a definitive way to locate the highest point on the graph of Equation (4). This is where the derivative comes into play.

Observe that in this example, the highest point on the graph divides the region where the curve is rising from the region where it is falling (Figure 3). At this point the tangent to the curve is horizontal, that is, has zero slope. Thus, to determine where the curve has its peak we need only determine where the tangent has zero slope. The computations are as follows. At each point of the curve

$$A = 1000x - 2x^2$$

the slope of the tangent is

$$\frac{dA}{dx} = 1000 - 4x$$

Thus the tangent has zero slope if

$$\begin{aligned} 1000 - 4x &= 0 \\ x &= 250 \end{aligned}$$

which demonstrates definitively that the maximum area does indeed occur when $x = 250$ feet.

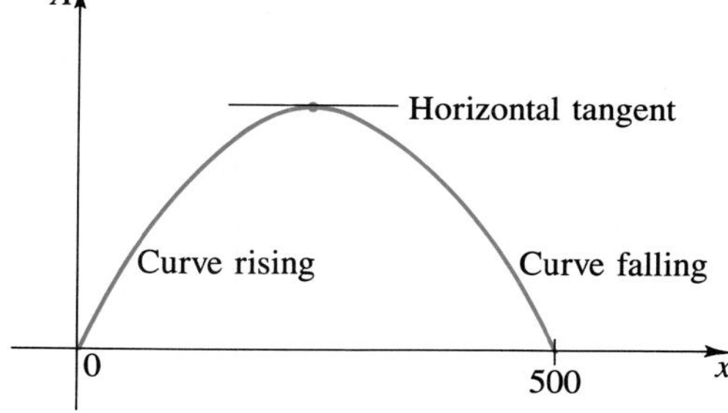

FIGURE 3

In Example 1, the physical problem of fencing a field with maximum area reduces to the mathematical problem of maximizing

$$A = 1000x - 2x^2$$

where x is required by physical considerations to lie in the closed interval [0, 500] [see the paragraph under Equation (3)]. This is typical of many optimization problems; the objective is to maximize (or minimize) some function

$$f(x)$$

where x is required to lie in some specified interval. There is some terminology associated with such problems:

Definition On a specified interval, a function f is said to have an **absolute maximum** at $x = c$ if $f(c) \geq f(x)$ for all x in the interval. Similarly, f has an **absolute minimum** at $x = c$ if $f(c) \leq f(x)$ for all x in the interval.

EXAMPLE 2 In Figure 4a, the function f has an absolute maximum at c_1 and an absolute minimum at c_2. In Figure 4b, f has an absolute maximum at the endpoint a and an absolute minimum at the endpoint b.

FIGURE 4

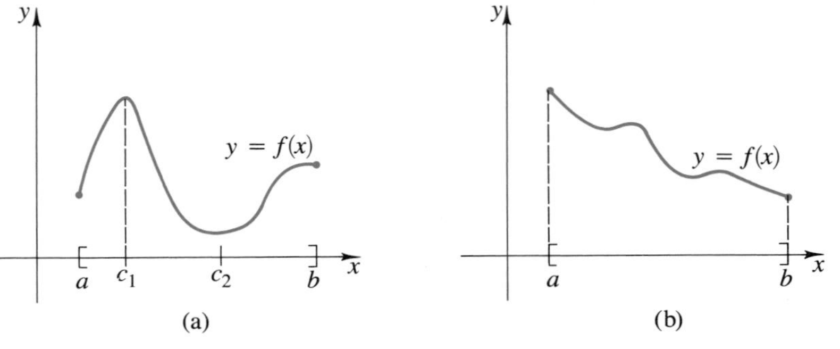

If we are given a closed interval $[a, b]$ and if we draw the graph of a continuous function by starting above the left endpoint a and not removing our pencil from the paper until we are above the right endpoint b, then it is intuitively clear that our pencil will reach a highest point and a lowest point somewhere along the curve (Figure 5). This suggests the following result, which we state without proof.

FIGURE 5

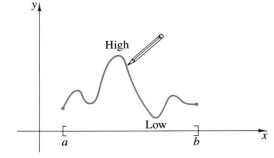

EXTREME-VALUE THEOREM If a function f is continuous at each point of a closed interval $[a, b]$, then f has an absolute maximum and an absolute minimum on $[a, b]$.

In this theorem the requirements that f be continuous and that the interval be closed are essential. For example, the discontinuous function in Figure 6a has no absolute maximum on $[a, b]$, and the continuous function in Figure 6b has no absolute minimum on the open interval (a, b).

FIGURE 6

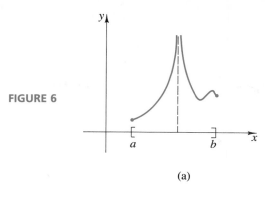

(a) (b)

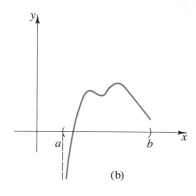

If f is differentiable at each point of a closed interval $[a, b]$, then there are two possibilities for the location of an absolute maximum; it can occur at one of the endpoints a or b (Figure 7a), or it can occur at a point where the derivative of f is zero (Figure 7b), that is, at a stationary point. The same possibilities hold for an absolute minimum. As shown in Figure 7c, a function can have an absolute maximum or minimum at a point where it is *not* differentiable.

FIGURE 7

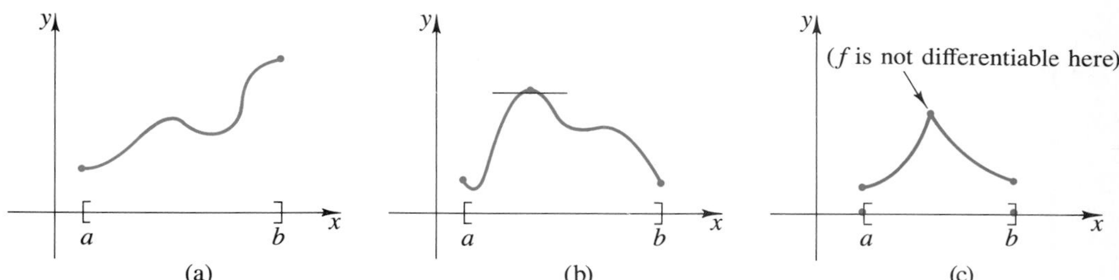

There is some standard terminology that helps to fix these ideas. A number c in the domain of f is called a **critical point** of f if either $f'(c) = 0$ or $f'(c)$ does not exist. Using this terminology we can summarize the results above as follows:

> **THEOREM** If f is defined at each point of a given interval and if f has an absolute maximum on the interval, then this maximum must occur at an endpoint (if there is one) or at a critical point. Similarly for an absolute minimum.

In the case in which we are maximizing or minimizing a continuous function over a closed interval $[a, b]$, we need not worry about the existence of this maximum or minimum; the existence is assured by the extreme value theorem.

Procedure for Finding an Absolute Maximum or Minimum of a Function Defined on a Closed Interval $[a, b]$

Step 1 Find the critical points of f on $[a, b]$.

Step 2 Evaluate f at the critical points and at endpoints a and b.

Step 3 The largest of the values obtained in Step 2 is the absolute maximum for f, and the smallest is the absolute minimum.

When we are concerned with maximum or minimum values of f over other kinds of intervals, we must worry about the possibility that the maximum or minimum may not exist. We consider such problems in the next section.

EXAMPLE 3 Find the absolute minimum and maximum of the function defined by

$$f(x) = 2x^3 - 3x^2 - 36x + 5$$

on the interval $[-5, 5]$.

Solution

Step 1 Since f is differentiable everywhere, we can find the critical values of f on $[-5, 5]$ by setting $f'(x) = 0$. We obtain

$$\begin{aligned} f'(x) = 6x^2 - 6x - 36 &= 0 \\ 6(x^2 - x - 6) &= 0 \\ 6(x - 3)(x + 2) &= 0 \end{aligned}$$

so the critical points of f are $x = -2$ and $x = 3$.

Step 2 We evaluate f at the critical points and at the endpoints. This yields

$$\begin{aligned} f(-2) &= 2(-2)^3 - 3(-2)^2 - 36(-2) + 5 = 49 \\ f(3) &= 2(3)^3 - 3(3)^2 - 36(3) + 5 = -76 \\ f(-5) &= 2(-5)^3 - 3(-5)^2 - 36(-5) + 5 = -140 \\ f(5) &= 2(5)^3 - 3(5)^2 - 36(5) + 5 = 0 \end{aligned}$$

Step 3 Hence, the absolute minimum value of f is -140, and it occurs at the endpoint $x = -5$; the absolute maximum value of f is 49, and it occurs at the critical point $x = -2$.

EXAMPLE 4 A thin rectangular sheet of cardboard 16×30 centimeters will be used to make a box by cutting a square from each corner and folding up the sides (Figure 8).

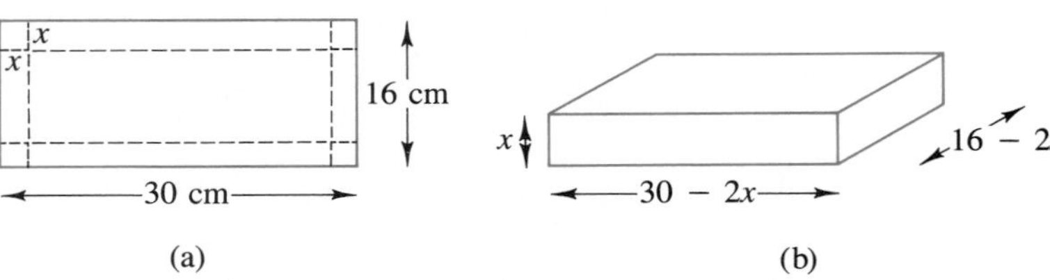

FIGURE 8

What size square should be cut from each corner to yield a box of maximum possible volume? What is the maximum volume?

Solution The volume V of the box is the product of its length, width, and height, so

$$V = (30 - 2x)(16 - 2x)x = 4x^3 - 92x^2 + 480x \tag{5}$$

(Figure 8b). The variable x in this formula satisfies certain physical restrictions. We must have $x \geq 0$ since x is a length; and since the minimum dimension of the cardboard is 16 centimeters, we must have $2x \leq 16$, or $x \leq 8$ (Figure 8a). Thus, $0 \leq x \leq 8$. Therefore, our problem reduces to maximizing the function V over the closed interval $[0, 8]$. We begin by locating the critical points of V that lie in this interval. Since

$$\frac{dV}{dx} = 12x^2 - 184x + 480$$

the critical values of V occur when $dV/dx = 0$ or

$$12x^2 - 184x + 480 = 0$$

or on dividing by 4

$$3x^2 - 46x + 120 = 0$$

Using the quadratic formula (see the algebra review in Appendix A) to solve this equation, we obtain

$$x = \frac{-b \pm \sqrt{b^2 - 4ac}}{2a} = \frac{46 \pm \sqrt{(-46)^2 - 4(3)(120)}}{2(3)} = \frac{46 \pm 26}{6}$$

which yields the critical values $x = \frac{10}{3}$ and $x = 12$. Since the value $x = 12$ is outside the interval $[0, 8]$, we need only check the value of V at the endpoints of $[0, 8]$ and at $x = \frac{10}{3}$. We obtain:

x	0	10/3	8
$V = (30 - 2x)(16 - 2x)x$	0	19,600/27	0

Thus a maximum volume (in cubic centimeters) of

$$V = \frac{19{,}600}{27} \cong 726$$

occurs when $x = \frac{10}{3}$ centimeters.

EXAMPLE 5 **PROFIT ANALYSIS** Suppose a manufacturer is limited by the firm's production facilities to a daily output of at most 80 units and suppose that the daily cost and revenue functions are

$$C(x) = x^2 + 4x + 200 \qquad \text{(dollars)}$$
$$R(x) = 108x - x^2 \qquad \text{(dollars)}$$

(see Marginal Analysis, Section 13.2). Assuming that all units produced are sold, how many units should be manufactured daily to maximize the profit?

Solution We first obtain the profit function:

$$P(x) = R(x) - C(x) = (108x - x^2) - (x^2 + 4x + 200)$$
$$= -2x^2 + 104x - 200$$

We want to determine a value of x in the interval $[0, 80]$ at which $P(x)$ has its maximum value. Since

$$P'(x) = -4x + 104$$

the only critical point of $P(x)$ occurs when

$$-4x + 104 = 0$$
$$x = 26$$

Evaluating $P(x)$ at the endpoints of $[0, 80]$ and at this critical point we obtain:

x	0	26	80
$P(x) = -2x^2 + 104x - 200$	-200	1152	-4680

Thus a maximum daily profit of \$1152 is achieved by producing and selling 26 units per day.

Most applied optimization problems are stated in words, and their solution requires a translation from the given verbal statement of the problem to a mathematical formulation. Here are some suggestions for carrying out this translation.

Solving Verbal Optimization Problems

Step 1 Read the problem carefully.

Step 2 Draw a figure if necessary to describe the problem and label the parts of the figure that enter into the problem with properly chosen variables.

Step 3 Determine the variable to be maximized or minimized and write an equation expressing this variable as a function of all other variables of the problem.

Step 4 Using the information given by the problem, express the variable to be optimized as a function of only one other variable.

Step 5 Find the domain of the function in Step 4. The domain is often a closed interval, so the extreme value theorem is applicable if the function is continuous. In Section 13.6 we discuss the situation in which the domain of the function is not a closed interval.

Step 6 Find the maximum and minimum values of the function obtained in Step 4.

Inventory Control

One of the important problems in any retail business is **inventory control**. On the one hand, there must be enough inventory to meet demand, and on the other hand, the business must avoid excess inventory since this results in unnecessary storage, insurance, and management costs. Moreover, money not committed to inventory can be earning interest elsewhere. As an example, suppose a retailer of automobile tires expects to sell 8000 tires during the year with sales occurring at a fairly constant rate. Although the retailer could order the 8000 tires all at once, this would result in high **holding costs** (insurance, storage rental, security, and so forth.) To reduce the holding costs the retailer might, instead, make many smaller orders during the year. However, this results in high **reorder costs** (delivery charges, paperwork, loading and unloading, and so forth). Thus we are led to the problem of determining an ordering strategy that will strike an optimal balance between holding costs and reorder costs. More precisely, the problem is to minimize

$$\underset{\text{inventory cost}}{\underset{\text{total annual}}{K(x)}} = \underset{\text{holding cost}}{\underset{\text{annual}}{H(x)}} + \underset{\text{reorder cost}}{\underset{\text{annual}}{O(x)}} \tag{6}$$

by choosing an appropriate **lot size** (amount to be reordered each time). The lot size that minimizes the total annual inventory cost is called the **economic ordering quantity** (EOQ).

EXAMPLE 6 A tire dealer expects to sell 8000 tires during the year with sales occurring at a relatively constant rate. The annual holding cost is \$8.00 per tire, and the contract with the wholesaler calls for a flat fee of \$80 per reorder, regardless of size. How many times per year and in what lot size should the dealer reorder to minimize the total annual inventory cost? We assume that the lot size is the same for each order and that each order is received just as the inventory on hand falls to zero.

Solution Let x denote the lot size. With the given assumptions, the largest number of tires on hand at any one time is x, and since sales are assumed to occur at a constant rate, it is reasonable that the **average inventory** during the year will be $x/2$ (Figure 9). Thus the annual holding cost $H(x)$ in dollars will be

FIGURE 9

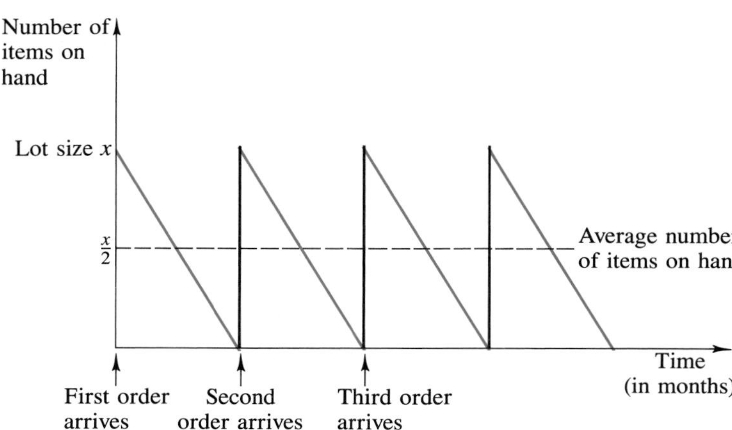

$$H(x) = \begin{pmatrix}\text{annual holding}\\ \text{cost per tire}\end{pmatrix} \cdot \begin{pmatrix}\text{average number}\\ \text{of tires}\end{pmatrix} = 8\left(\frac{x}{2}\right) = 4x \tag{7}$$

To determine the annual reorder cost we argue as follows. The number of orders is

$$\text{number of orders} = \frac{\text{annual sales}}{\text{lot size}} = \frac{8000}{x}$$

so the annual reorder cost $O(x)$ in dollars is

$$O(x) = \begin{pmatrix}\text{cost of}\\ \text{each order}\end{pmatrix} \cdot \begin{pmatrix}\text{number of}\\ \text{orders}\end{pmatrix} = 80\left(\frac{8000}{x}\right) = \frac{640{,}000}{x} \tag{8}$$

From Equation (6) the total annual inventory cost $K(x)$ is found by adding (7) and (8):

$$K(x) = H(x) + O(x) = 4x + \frac{640{,}000}{x} \tag{9}$$

Because the total annual sales are 8000 tires, the lot size x can be as small as $x = 1$ and as large as $x = 8000$. Thus the problem reduces to locating the minimum value of $K(x)$ for x in the interval $[1, 8000]$. We proceed as follows:

$$K'(x) = 4 - \frac{640{,}000}{x^2}$$

Setting $K'(x) = 0$ to obtain the critical points, we have

$$4 - \frac{640{,}000}{x^2} = 0$$

or

$$4x^2 = 640{,}000$$
$$x^2 = 160{,}000$$

Thus, $x = 400$ or $x = -400$, so $x = 400$ is the only critical point in the interval $[1, 8000]$. Evaluating $K(x)$ at the endpoints and critical point, we obtain:

x	1	400	8000
$K(x) = 4x + (640{,}000/x)$	640,004	3200	32,080

Thus a minimum total annual inventory cost of \$3200 is achieved by a lot size (EOQ) of 400 tires, which means that the dealer should meet the annual demand of 8000 tires with

$$\frac{8000}{400} = 20 \text{ orders}$$

of 400 tires each.

OPTIMIZATION IN NATURE

Many natural phenomena occur in a manner that maximizes or minimizes certain quantities. As an example, Aristotle (384–322 B.C.) observed that light travels from one point to another by following the path of minimum length. If, as shown in Figure A, the light beam is at source S and the observer is at point O (with no obstructions in between), then the path of the light beam will be a straight line, the shortest distance between the two points.

FIGURE A S O

Now suppose that the source S and the observer O are on the same side of a plane mirror (see Figure B).

FIGURE B O S Mirror

If the light beam must be aimed at the mirror and then be reflected to the observer who is located at point O, at what point M on the mirror should the light beam be focused so the light path will be of minimum length? Using Aristotle's observation, Hero of Alexandria (first century A.D.) proved that M is the point at which the angle i of incidence of the light beam equals the angle r of reflection (see Figure C).

FIGURE C

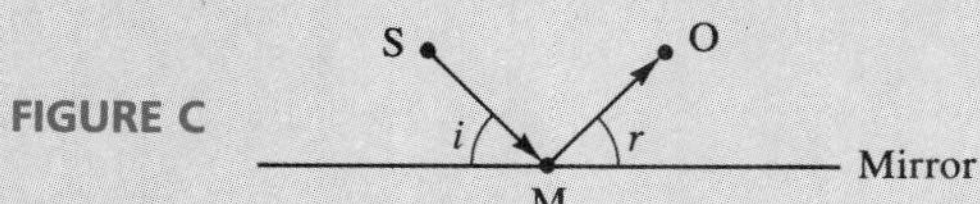

EXERCISE SET 13.4

In Exercises 1–12 find the absolute maximum and absolute minimum values for $f(x)$ on the given interval, and specify the values of x where the maximum and minimum occur.

1. $f(x) = 2x + 6$ on $[-1, 6]$
2. $f(x) = 5x - 3$ on $[1, 4]$
3. $f(x) = x^2 + 2x - 1$ on $[-2, 3]$
4. $f(x) = 1 - x^2$ on $[-1, 1]$
5. $f(x) = x - x^2$ on $[0, 2]$
6. $f(x) = x^3 - x^2$ on $[0, 1]$
7. $f(x) = x^3 - x^2$ on $[-1, 4]$
8. $f(x) = x^3 - x^2$ on $[0, \frac{1}{2}]$
9. $f(x) = x^4 - 2x^3$ on $[0, 2]$
10. $f(x) = x^4 - 2x^3$ on $[-1, 3]$
11. $f(x) = 1/x$ on $[\frac{1}{2}, 5]$
12. $f(x) = 2/(1 + x^2)$ on $[-2, 2]$
13. Find two positive numbers whose sum is 100 and whose product is maximum.

14. Find two numbers each greater than or equal to 1 whose product is 100 and whose sum is maximum.

15. Suppose that 240 feet of fencing will be used to enclose a rectangular field. Find the dimensions of the rectangle that will yield the largest possible area.

16. A garden store wants to build a rectangular enclosure to display its shrubs. Three sides will be built from chain-link fence costing \$20 per running foot, and the remaining side will be built from cedar fence costing \$10 per running foot (Figure 10). Find the dimensions of the enclosure of largest possible area that can be built with \$2000 worth of fence.

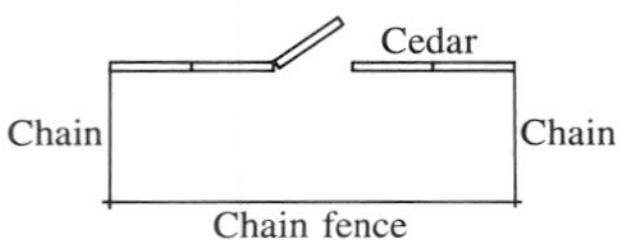

FIGURE 10

17. An agriculturalist wants to enclose two equal rectangular areas for experimentation as shown in Figure 11. If 360 yards of fencing are available, what is the largest total area that can be enclosed? [*Hint:* See the dimensions indicated in the figure.]

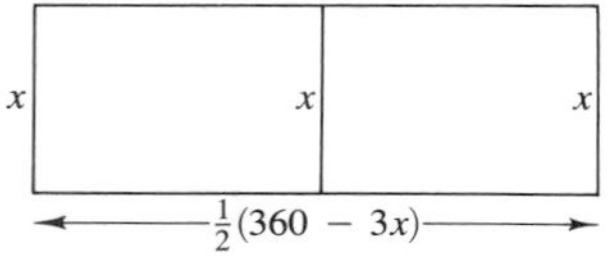

FIGURE 11

18. A square piece of cardboard 12 inches by 12 inches will be used to make an open box by cutting a square from each corner and folding up the sides (see Example 4).
(a) What size square should be cut from each corner to yield a box of maximum volume?
(b) What is the maximum volume?

19. The U.S. Postal Service has the following restriction on mailing a fourth-class parcel in the form of a rectangular box. The perimeter of one end plus the length of the box must be no more than 108 inches. What is the largest volume of a permissible rectangular parcel whose ends are squares?

20. A manufacturer of electronic components needs two wire elements: one is a circle, the other a square. These are made by cutting a piece of wire 20 centimeters long into two pieces and bending one piece into a square and the other into a circle. How long should each piece of wire be to minimize the sum of areas? [*Hint:* See Figure 12.]

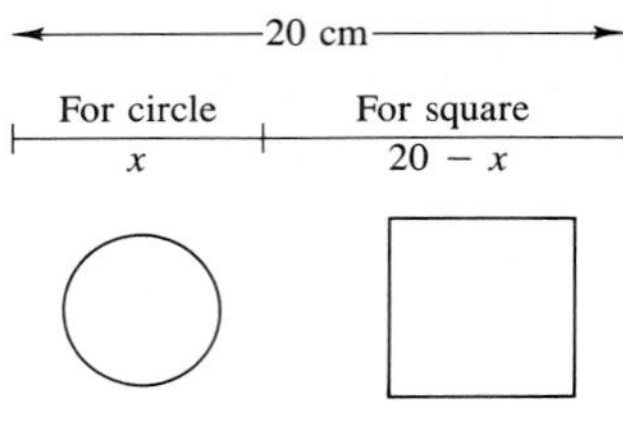

FIGURE 12

21. **(Advertising)** A manufacturer believes that the firm's yearly profit $P(x)$ (in thousands of dollars) is related to its yearly advertising expenditure x (in thousands of dollars) by $P(x) = 10 + 46x - \frac{1}{2}x^2$. Assuming the firm can afford at most \$50,000 a year for advertising, how much should be spent to maximize the profit?

22. **(Profit analysis)** A manufacturer is limited by the firm's production facilities to an output of at most 125 units per day. The daily cost and revenue functions for producing x units are

$$C(x) = 3x^2 - 750x + 100$$
$$R(x) = 50x - x^2$$

Assuming that all units produced are sold, how many units should be manufactured daily to maximize the profit?

23. **(Profit analysis)** A manufacturer whose production cannot exceed 8 units per day has cost and revenue functions given by

$$C(x) = 0.02x^3 + 0.01x^2 - x + 2$$
$$R(x) = 0.01x^2 + 0.5x$$

Assuming all units produced are sold, what production level will maximize the profit?

24. **(Pricing analysis)** A vendor of souvenir T-shirts can order them for \$4 each and determines that if she charges a selling price of s dollars each, then she will sell $x = 80 - 10s$ thousands of the shirts. How many should she order and what price should she charge to maximize the profit? Determine the maximum profit.

25. In designing a department store it is estimated that each sales counter will produce an average daily profit of \$1000 provided the number of counters is between 0 and 20. If the number of counters is above 20, then the average profit on every counter will be reduced by \$50 for each counter above 20. How many counters should be in-

stalled for maximum average daily profit if the store has room for at most 40 counters?

26. A truck traveling at x miles per hour consumes gasoline at the rate of

$$G(x) = \frac{1}{1000}\left(\frac{2000}{x} + x\right)$$

gallons per mile.

(a) If gasoline costs $1.20 per gallon, how much will it cost for gasoline (in dollars) to travel 100 miles at x miles per hour?

(b) If a 100-mile stretch of open road has a minimum speed limit of 30 miles per hour and a maximum limit of 60 miles per hour, what is the most economical speed for the truck?

(c) What is the most economical speed if the driver is paid $25 per hour for the 100-mile trip?

27. **(Inventory control)** A camera wholesaler expects to sell 1000 cameras during the year with sales occurring at a relatively constant rate. The annual holding cost is $7 per camera, and the contract with the wholesaler calls for a flat fee of $35 for each order, regardless of size. How many times per year and in what lot size should the dealer reorder to minimize the total annual inventory cost?

28. **(Inventory control)** A department store sells 2500 top-of-the-line refrigerators per year with sales occurring at a relatively constant rate. The annual holding cost per refrigerator is $10. To reorder there is a basic $20 service fee per reorder plus a $9 insurance charge for each refrigerator ordered. How many times per year and in what lot size should the store reorder to minimize the total annual inventory cost?

29. **(Cancer experimentation)** In an experiment to determine the carcinogenic effects of a questionable substance, an amount of this substance is added to the daily diet of laboratory animals so that each pound of food contains x ounces of the substance. Unfortunately, as more of the substance is added, the daily consumption y of the food decreases. In fact,

$$y = 0.3 - 0.2x \qquad \text{(in pounds)}$$

The daily consumption of the questionable substance is xy. If the researchers want the daily consumption of the questionable substance to be as large as possible, with $x \leq 1$ ounce, determine the value of x that will achieve this goal. What then will be the daily consumption of the substance?

13.5 USING FIRST AND SECOND DERIVATIVES IN SKETCHING CURVES

We have seen that by using the first derivative of a function f we can determine where it is increasing and where it is decreasing. In Section 13.1 we find the intervals where f is increasing by solving the inequality $f'(x) > 0$. Recall that the points x where $f'(x) = 0$ or where f is not differentiable are called **critical points** of f.

Geometrically, the critical points of f are the values of x at which the graph of f has a horizontal tangent line, a vertical tangent line, or no tangent line. For continuous curves, two types of critical points are of special interest because of the characteristic shapes that occur on the graph of f: "peaks" and "troughs" (Figure 1).

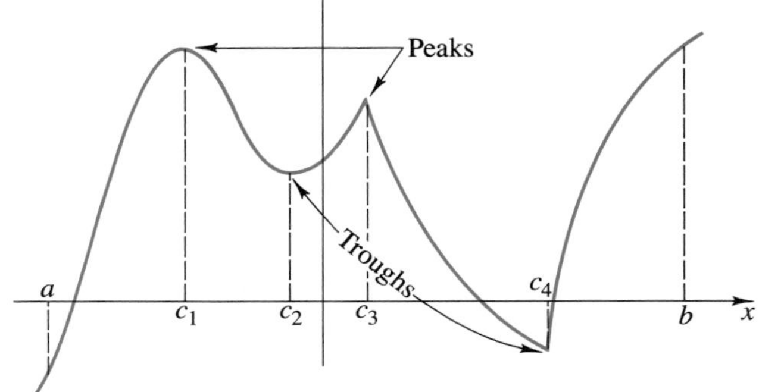

FIGURE 1

If a peak occurs at a critical point c of f, then f is said to have **relative maximum** (or **local maximum**) at the point $x = c$, and the value $f(c)$ is called a **relative maximum** of f. If a trough occurs at a critical point c, then f is said to have a **relative minimum** (or **local minimum**) at the point $x = c$, and the value $f(c)$ is called a **relative minimum** of f. If f has a relative maximum or a relative minimum at c, we say that f has a **relative extremum** at c (the plural of extremum is *extrema*).

If a function f is differentiable, then the only critical values to be considered are the values c such that $f'(c) = 0$.

EXAMPLE 1

(a) The function $f(x) = 1 - x^2$ (Figure 2a) has a relative maximum at $x = 0$ and no relative minima.

(b) The function $f(x) = |1 - x|$ (Figure 2b) has a relative minimum at $x = 1$ and no relative maxima.

(c) The function $f(x) = x^3$ (Figure 2c) has a critical point at $x = 0$, but $f(0) = 0$ is neither a relative maximum (peak) nor a relative minimum (trough).

FIGURE 2

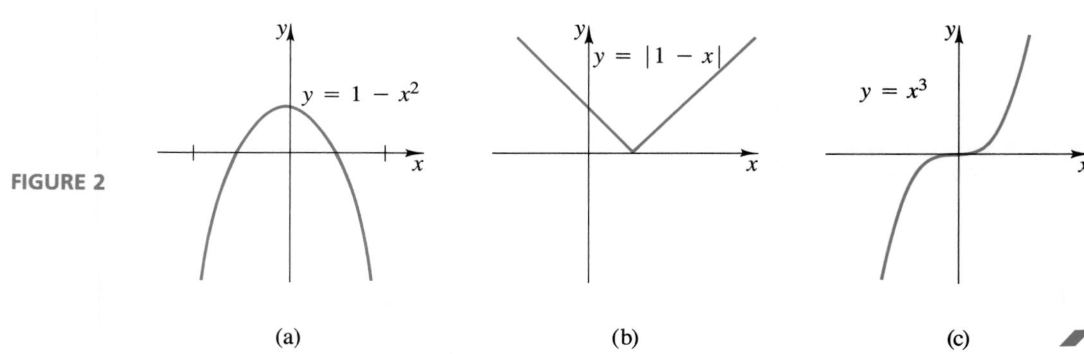

A function f can have several relative maxima and several relative minima. In Figure 1, f has relative maxima at c_1 and c_3 and relative minima at c_2 and c_4.

We now develop tests for determining the relative extrema of a function. As indicated in Figure 3 the relative maxima and minima of a function can be identified by the behavior of the derivative in the vicinity of the point. At a relative *maximum* the graph of f is increasing [$f'(x) > 0$, from Section 13.1] on an interval extending left from the point and decreasing [$f'(x) < 0$, from Section 13.1] on an interval extending right from the point; at a relative *minimum,* the graph is decreasing on the left and increasing on the right.

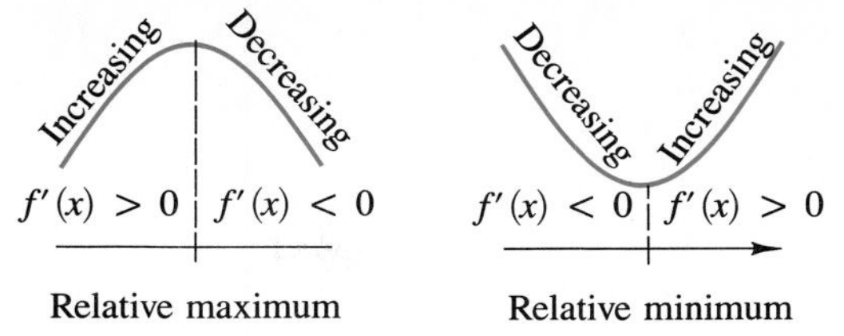

Increasing
Decreasing
f′(x) > 0
f′(x) < 0
Relative maximum
Decreasing
Increasing
f′(x) < 0
f′(x) > 0
Relative minimum

FIGURE 3

In summary, we have the following important result:

The First Derivative Test If c is a critical point of a function f, then:

(a) f has a relative maximum at c if $f'(x) > 0$ on an interval extending left from c and if $f'(x) < 0$ on an interval extending right from c.

(b) f has a relative minimum at c if $f'(x) < 0$ on an interval extending left from c and if $f'(x) > 0$ on an interval extending right from c.

(c) f has neither a relative maximum nor a relative minimum at c if the derivative $f'(x)$ has the same sign on intervals extending on both sides of c.

To paraphrase this result, a relative maximum occurs if the sign of f' changes from + to − at c, a relative minimum occurs if the sign of f' changes from − to + at c, and neither occurs if f' does not change sign at c.

EXAMPLE 2 Classify the critical points of $f(x) = \frac{1}{3}x^3 - \frac{1}{2}x^2$.

Solution Differentiating yields

$$f'(x) = x^2 - x = x(x - 1)$$

and on setting $x(x - 1) = 0$ we see that $x = 0$ and $x = 1$ are the critical points.

From the analysis in Figure 4, the sign of f' changes from + to − at 0, so a relative maximum occurs at $x = 0$; also the sign of f' changes from − to + at 1, so a relative minimum occurs at $x = 1$. The graph of f appears in Figure 5.

FIGURE 4

− − − − − − − + + + + + + + + + +
Sign of x
0
− − − − − − − − − − − − − + + + + + +
Sign of (x − 1)
1
+ + + + + + + − − − − + + + + + +
Sign of f′(x) = x(x − 1)
0
1
Increasing
Decreasing
Increasing

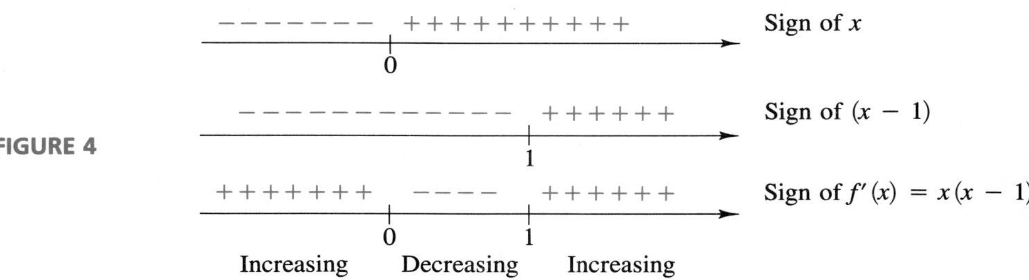

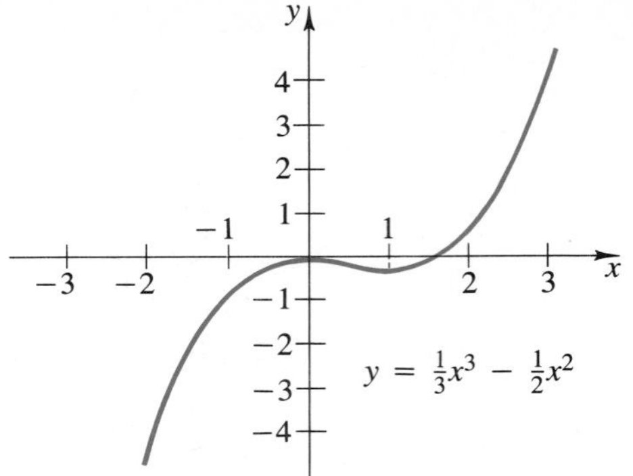

FIGURE 5

x	$y = \frac{1}{3}x^3 - \frac{1}{2}x^2$
0	0
$\frac{1}{2}$	$-\frac{1}{12}$
1	$-\frac{1}{6}$
2	$\frac{2}{3}$
3	$\frac{9}{2}$
−1	$-\frac{5}{6}$
−2	$-\frac{14}{3}$

Relative maximum: $f(0) = 0$
Relative minimum: $f(1) = -\frac{5}{6}$
Neither is an absolute maximum or minimum.

Using the Second Derivative

There is another test for relative maxima and minima that is often easier to apply than the first derivative test. This test uses the "second derivative" of the function, that is, the derivative of the derivative. More precisely, the **second derivative** of a function f is

$$\frac{d}{dx}[f'(x)]$$

and is denoted by $f''(x)$. For example, if $f(x) = x^3$, then the first derivative of f is

$$f'(x) = \frac{d}{dx}[x^3] = 3x^2$$

and the second derivative of f is

$$f''(x) = \frac{d}{dx}[3x^2] = 6x$$

If $y = f(x)$, then the second derivative is also denoted by

$$y'' \quad \text{or} \quad \frac{d^2y}{dx^2} \quad \text{or} \quad \frac{d^2}{dx^2}[f(x)]$$

Recall from Section 13.1 that if $s(t)$ denotes the distance traveled by a particle from a fixed point as a function of time, then the velocity is ds/dt and the acceleration is

$$\frac{dv}{dt} = \frac{d}{dt}\left[\frac{ds}{dt}\right] = \frac{d^2s}{dt^2}$$

To see the significance of the second derivative, suppose $f''(x) > 0$ at each point x in some interval (a, b). This would indicate that the function $f'(x)$ is increasing on this interval. Geometrically, this means that slopes of the tangents to $y = f(x)$ are increasing as we travel left to right, indicating that the tangent line is turning counterclockwise, or that the curve $y = f(x)$ bends upward over the interval (a, b) (Figure 6a). Such a curve is said to be **concave up** over this interval.

Similarly, if $f''(x) < 0$ for each x in (a, b), then the slopes of the tangents, given by $f'(x)$, will be decreasing as we travel left to right, indicating a downward bend in the curve $y = f(x)$ over (a, b) (Figure 6b). Such a curve is said to be **concave down** over the interval.

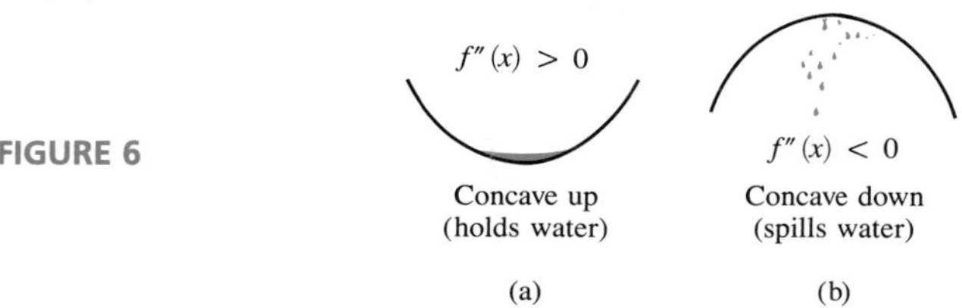

FIGURE 6

The following theorem summarizes these results.

THEOREM (CONCAVITY TEST) On a given interval, if f is twice differentiable then the curve $y = f(x)$ is:

(a) concave up if $f''(x) > 0$ at each point in the interval

(b) concave down if $f''(x) < 0$ at each point in the interval

You may think of a curve that is concave up as one that "holds" water and a curve that is concave down as one that "spills" water. With this in mind, Figure 7 may help you remember the above theorem.

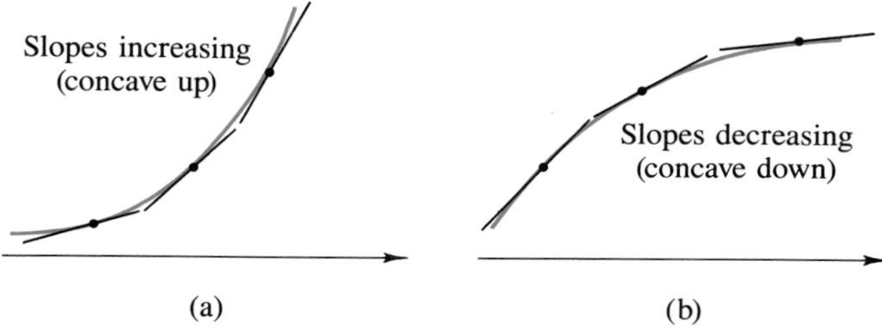

FIGURE 7

EXAMPLE 3 Let

$$f(x) = 2x^3 - 3x^2 - 12x + 2$$

Where is $y = f(x)$ concave up? Concave down?

Solution We have

$$f'(x) = 6x^2 - 6x - 12$$
$$f''(x) = 12x - 6 = 6(2x - 1)$$

Thus $f''(x) > 0$ if

$$2x - 1 > 0 \quad \text{or} \quad 2x > 1 \quad \text{or} \quad x > \frac{1}{2}$$

and $f''(x) < 0$ if

$$2x - 1 < 0 \quad \text{or} \quad 2x < 1 \quad \text{or} \quad x < \frac{1}{2}$$

Therefore, the graph is concave up if $x > \frac{1}{2}$ and concave down if $x < \frac{1}{2}$ (Figure 8).

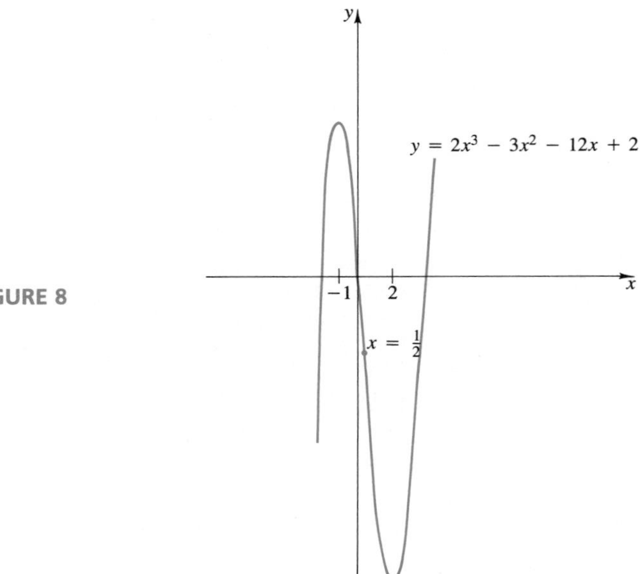

FIGURE 8

In the previous example, the curve changes from concave down to concave up at $x = \frac{1}{2}$. A point $(c, f(c))$ on the graph of f at which the curve changes from concave down to concave up, or vice versa, is called an **inflection point** of the curve.

EXAMPLE 4 **ANALYSIS OF UNEMPLOYMENT** In times of economic recession or depression, unemployment tends to grow at an ever-increasing rate until it is brought under control by appropriate economic policy and pressures from the marketplace. Figure 9 shows a typical unemployment curve. The inflection point in July indicates that unemployment is coming under control. After this point is reached, the curve is concave down, indicating a declining growth rate in unemployment. We can then anticipate that unemployment will soon "peak out" and begin to decline. In everyday language we might say that the economy "turned the corner" on unemployment in July.

FIGURE 9

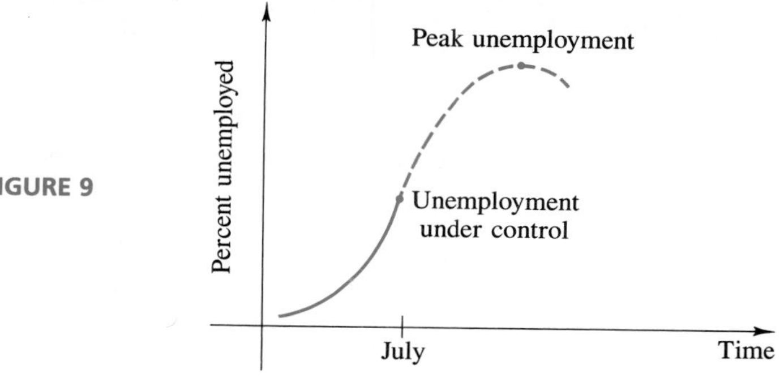

If we are given a function that has a second derivative, we can use the second derivative to "test" critical values to determine whether they correspond to relative maxima or to relative minima. Assume that c is a critical point of f such that $f'(c) = 0$. Then the graph of $y = f(x)$ has a horizontal tangent line at $x = c$. If $f''(c) > 0$, then the curve $y = f(x)$ is concave up about c (Figure 7a), indicating a relative minimum at c. On the other hand, if $f''(c) < 0$, then $y = f(x)$ is concave down about c (Figure 7b), indicating a relative maximum at c. This suggests the following result.

The Second Derivative Test

(a) If $f'(c) = 0$ and $f''(c) > 0$, then f has a relative minimum at c.

(b) If $f'(c) = 0$ and $f''(c) < 0$, then f has a relative maximum at c.

(c) If $f'(c) = 0$ and $f''(c) = 0$, then the test fails and the first derivative test must be used.

EXAMPLE 5 Use the second derivative test to locate the relative maxima and relative minima of $f(x) = \frac{1}{3}x^3 - \frac{1}{2}x^2$. Also, locate all inflection points.

Solution We have

$$f'(x) = x^2 - x$$
$$f''(x) = 2x - 1$$

On setting $f'(x) = x^2 - x = x(x - 1) = 0$, we obtain the critical points $x = 0$ and $x = 1$. We apply the second derivative test, first at $x = 0$:

$$f''(0) = 2(0) - 1 = -1 < 0$$

indicating that there is a relative maximum at $x = 0$. Applying the second derivative test at $x = 1$,

$$f''(1) = 2(1) - 1 = 1 > 0$$

we see that there is a relative minimum at $x = 1$. This agrees with the results obtained in Example 2, using the first derivative test.

Since $f''(x) > 0$ when $2x - 1 > 0$ (i.e., when $x > \frac{1}{2}$), we conclude that the curve is concave up on the interval $(\frac{1}{2}, \infty)$ and concave down on $(-\infty, \frac{1}{2})$. Thus the curve has an inflection point when $x = \frac{1}{2}$; in other words, the point $(\frac{1}{2}, -\frac{1}{12})$ is an inflection point (see Figure 5).

Caution The second derivative test cannot be applied to a critical point $x = c$ if $f''(c) = 0$. The second derivative test provides no conclusion in such a case. To determine what happens at such a critical point $x = c$, we must rely on the first derivative test. Thus, while the second derivative test is often easier to use than the first derivative test, it sometimes is not conclusive.

EXAMPLE 6 We consider three cases of a function with a critical point at $x = 0$. In all three cases we have $f''(0) = 0$; yet in one case we have a relative minimum at $x = 0$, in another we have a relative maximum at $x = 0$, and in the third we have neither.

(a) The function $f(x) = x^4$ has derivative $f'(x) = 4x^3$, therefore $x = 0$ is a critical point. Moreover, $f''(x) = 12x^2$ so $f''(0) = 0$. This function has a relative minimum at $x = 0$, as shown in Figure 10a.

(b) The function $g(x) = -x^4$ has derivative $g'(x) = -4x^3$, therefore $x = 0$ is a critical point. Moreover, $g''(x) = -12x^2$ so $g''(0) = 0$. This function has a relative maximum at $x = 0$, as shown in Figure 10b.

(c) The function $h(x) = x^3$ has derivative $h'(x) = 3x^2$, therefore $x = 0$ is a critical point. Moreover, $h''(x) = 6x$ so $h''(0) = 0$. This function has neither a relative maximum nor a relative minimum at $x = 0$, as shown in Figure 10c. In fact, h has an inflection point at $x = 0$ since its graph changes concavity from down to up as it passes through $x = 0$.

FIGURE 10

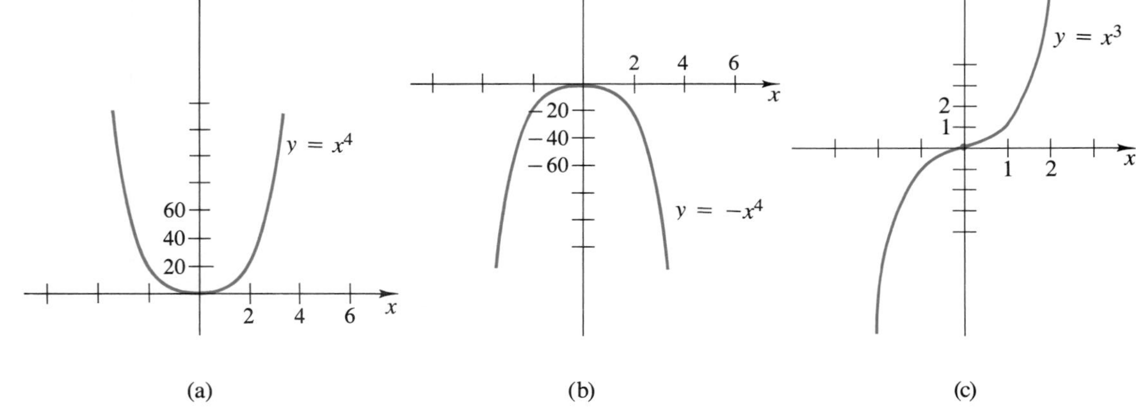

We now apply these ideas to the general problem of sketching graphs of algebraic functions. We pay particular attention to polynomial functions. Note that the polynomial functions are all twice differentiable.

How to Sketch the Graph of the Polynomial Function $f(x)$

Step 1 Obtain $f'(x)$ and $f''(x)$.

Step 2 Using $f'(x)$, find the critical points of $f(x)$ and the intervals over which f is increasing and those over which it is decreasing. Using the first derivative test or the second derivative test, examine each critical point to determine the relative minima and the relative maxima.

Step 3 Using $f''(x)$, find the intervals over which f is concave up and those over which it is concave down. Also, find the inflection points of f.

Step 4 Plot several points, if necessary, and use the above information to sketch the graph of f.

EXAMPLE 7 Sketch the graph of $f(x) = x^3 - 3x^2 + 6$.

Solution

Step 1 We have

$$f'(x) = 3x^2 - 6x = 3x(x - 2)$$

$$f''(x) = 6x - 6 = 6(x - 1)$$

Step 2 The critical points of f are

$$x = 0 \quad \text{and} \quad x = 2$$

We now determine the intervals over which f is increasing and those over which it is decreasing by examining the sign of the first derivative (Figure 11).

Thus, f is increasing on the intervals $(-\infty, 0)$ and $(2, +\infty)$ and decreasing on $(0, 2)$. By the first derivative test, we see from the sign of $f'(x)$ that f has a relative maximum at $x = 0$ and a relative minimum at $x = 2$. Moreover, $f(0) = 6$ and $f(2) = 2$.

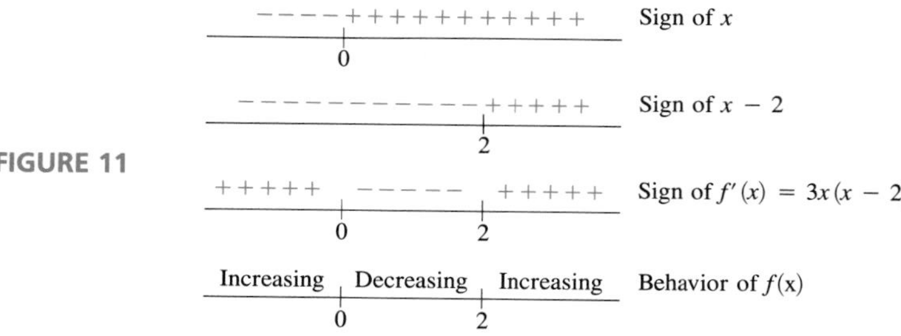

FIGURE 11

Step 3 We have

$$f''(x) > 0 \quad \text{for } x > 1$$

$$f''(x) < 0 \quad \text{for } x < 1$$

Hence, f is concave up over the interval $(1, +\infty)$ and concave down over the interval $(-\infty, 1)$. The curve has a point of inflection at $x = 1$. Moreover, $f(1) = 4$.

Step 4 Using the above information we can sketch the graph of f, as shown in Figure 12. As an aid in sketching we have plotted two additional points, $(3, f(3)) = (3, 6)$ and $(-1, f(-1)) = (-1, 2)$.

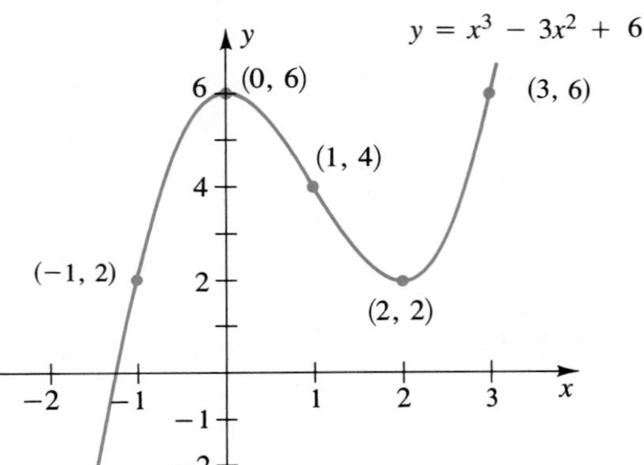

FIGURE 12

EXAMPLE 8 Sketch the graph of $f(x) = x^4 + 2x^3 - 2$.

Solution

Step 1 We have

$$f'(x) = 4x^3 + 6x^2 = 2x^2(2x + 3)$$

and

$$f''(x) = 12x^2 + 12x = 12x(x + 1)$$

Step 2 The critical points of f are

$$x = 0 \quad \text{and} \quad x = -\frac{3}{2}$$

We now determine the intervals over which f is increasing and those over which it is decreasing (Figure 13).

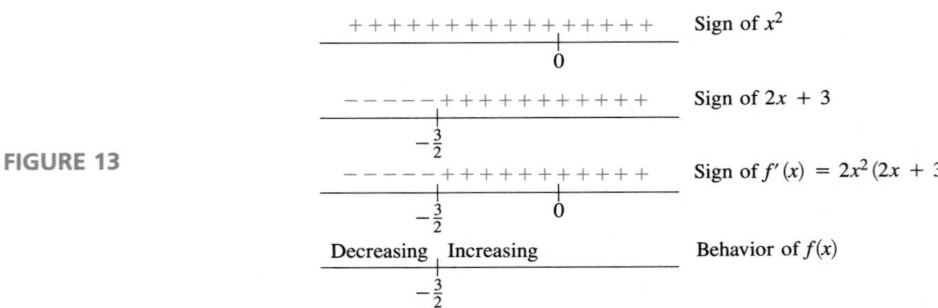

FIGURE 13

Thus, f is decreasing on the interval $(-\infty, -\frac{3}{2})$ and increasing on $(-\frac{3}{2}, +\infty)$. By the first derivative test, we see from the sign of $f'(x)$ that f has a relative minimum at $x = -\frac{3}{2}$. Moreover, $f(-\frac{3}{2}) = -\frac{59}{16}$.

Step 3 We determine the sign of $f''(x)$ (Figure 14).

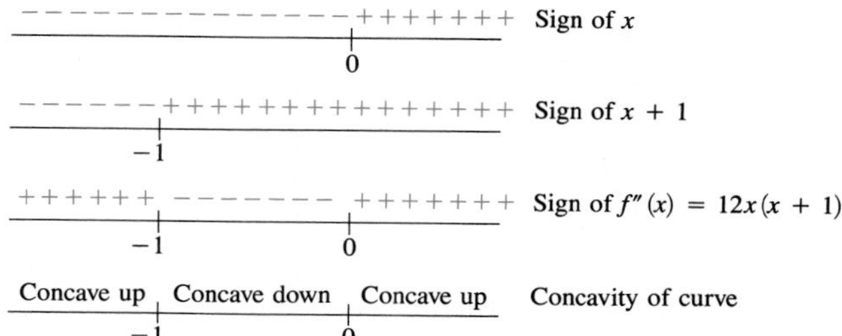

FIGURE 14

Thus, f is concave up over the intervals $(-\infty, -1)$ and $(0, +\infty)$ and concave down over $(-1, 0)$. The curve has inflection points at $x = -1$ and $x = 0$. Moreover, $f(0) = -2$ and $f(-1) = -3$.

Step 4 Using the above information, we can sketch the graph of f as shown in Figure 15. As an aid in sketching we have plotted two additional points, $(-2, f(-2)) = (-2, -2)$ and $(1, f(1)) = (1, 1)$.

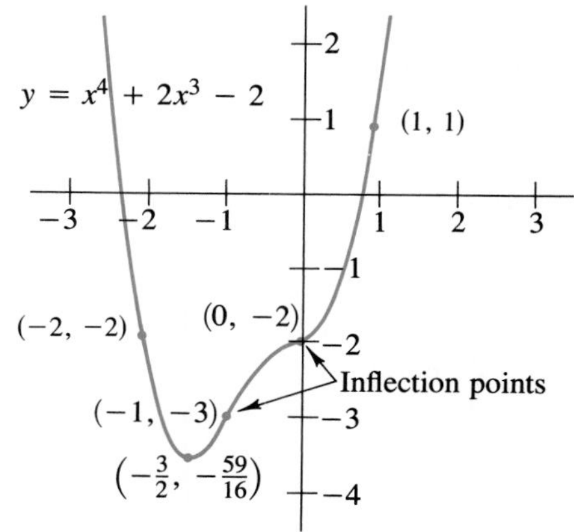

FIGURE 15

EXERCISE SET 13.5

Exercises 1–3 refer to Figure 16.

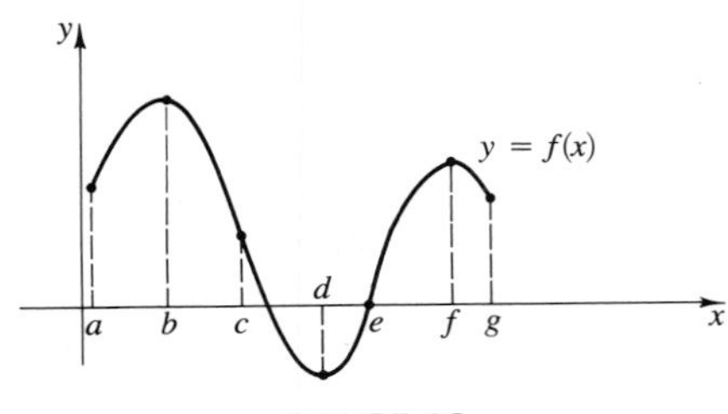

FIGURE 16

1. Over what open intervals is $y = f(x)$
   (a) increasing (b) decreasing
   (c) concave up (d) concave down

2. (a) Where is $f'(x) = 0$?
   (b) Over what open intervals is $f'(x) > 0$?
   (c) Over what open intervals is $f'(x) < 0$?
   (d) Over what open intervals is $f''(x) > 0$?
   (e) Over what open intervals is $f''(x) < 0$?

3. At what points does f have
   (a) a relative maximum
   (b) a relative minimum
   (c) an inflection point
   (d) a critical point

Exercises 4–6 refer to Figure 17.

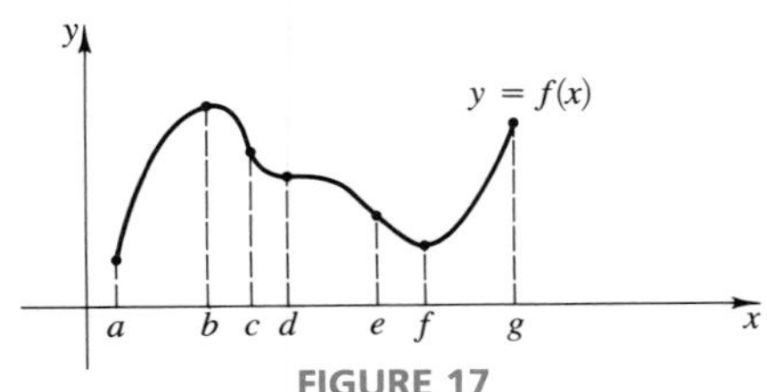

FIGURE 17

4. Over what open intervals is $y = f(x)$
   (a) increasing (b) decreasing
   (c) concave up (d) concave down

5. (a) Where is $f'(x) = 0$?
   (b) Over what open intervals is $f'(x) > 0$?
   (c) Over what open intervals is $f'(x) < 0$?
   (d) Over what open intervals is $f''(x) > 0$?
   (e) Over what open intervals is $f''(x) < 0$?

6. At what values does f have
   (a) a relative maximum
   (b) a relative minimum
   (c) an inflection point
   (d) a critical point

In Exercises 7–16 where is $y = f(x)$ increasing? Decreasing? Locate and classify the critical points as relative maxima, relative minima, or neither.

7. $f(x) = x^2 - 6x + 3$
8. $f(x) = 2x^2 + 12x - 3$
9. $f(x) = -2x + 1$
10. $f(x) = 4x - 2$
11. $f(x) = 4x^2 + 4$
12. $f(x) = \frac{5}{2}x^2 - \frac{10}{3}x^3$
13. $f(x) = x^3 + 3x^2 - 9x + 2$
14. $f(x) = 2x^3 + 3x^2 - 12x + 8$
15. $f(x) = \sqrt{x}$
16. $f(x) = x^2\sqrt{x + 2}$

In Exercises 17–22 find $f''(x)$.

17. $f(x) = 3x^2 + 2x + 1$
18. $f(x) = 4x^5 - 2x^4$
19. $f(x) = 3x + 5$
20. $f(x) = 1/x$
21. $f(x) = x/(x^2 + 1)$
22. $f(x) = x/(x - 1)$

In Exercises 23–28 where is the curve concave up? Concave down? Locate all inflection points.

23. $f(x) = x^2 + 3x - 10$
24. $f(x) = 2x^3 + 3x^2 - 36x + 5$
25. $f(x) = x^3 - 6x^2 - 36x + 15$
26. $f(x) = x^4 - 4x^3 + 8$
27. $f(x) = (x - 3)^5$
28. $f(x) = 4x^5 + 5x^4$

29. Let $f(x) = x^3 - 3x + 9$. Locate the relative maxima and minima using
    (a) the first derivative test
    (b) the second derivative test

30. Let $f(x) = 2x^3 - 3x^2 - 12x + 18$. Locate the relative maxima and minima using
    (a) the first derivative test
    (b) the second derivative test

In Exercises 31–36 use any method to locate the relative maxima and relative minima, if any.

31. $f(x) = x^4 - 2x^3$
32. $f(x) = x^4$
33. $f(x) = x^3$
34. $f(x) = \frac{1}{4}x^4 - \frac{1}{2}x^2 + 2$

35. $f(x) = x/(x + 1)$ **36.** $f(x) = 1/(x^2 + 1)$

In Exercises 37–48 sketch the graph of the given function.

37. $f(x) = x^2 + 2x + 4$ **38.** $f(x) = -2x^2 - 4x + 3$

39. $f(x) = x^3 - 3x + 3$ **40.** $f(x) = x^3 - 3x^2 + 4$

41. $f(x) = x^3 + 2x^2 + x + 2$

42. $f(x) = x^3 - x^2 - x$

43. $f(x) = x^3 + 3x^2 + 3x + 1$

44. $f(x) = \frac{1}{3}x^3 + \frac{1}{2}x^2 - 2x - 1$

45. $f(x) = 2x^3 - 3x + 1$

46. $f(x) = x^4 - 8x^2 + 3$

47. $f(x) = 3x^4 + 4x^3 + 1$

48. $f(x) = \frac{1}{4}x^4 + \frac{1}{2}x^2$

49. **(Learning)** Learning of most skills starts at a rapid rate and then slows down. A psychologist measures the learning performance of a laboratory rat by a numerical score on a standardized test. Assume the rat's score $P(t)$ after t weeks of learning is

$$P(t) = 15t^2 - t^3$$

At what point in time does the rat's *rate* of learning begin to decline?

50. **(Marginal analysis)** Suppose a manufacturer's profit function $P(x)$ has a critical value at a production level of c units. Show that the marginal cost and marginal revenue at this production level are equal.

51. Show that the quadratic polynomial $f(x) = ax^2 + bx + c$ has a relative maximum at the point $x = -b/2a$ if $a < 0$ and a relative minimum at that point if $a > 0$.

13.6 MORE OPTIMIZATION APPLICATIONS

In this section we consider some maximization and minimization problems to which the extreme value theorem of Section 13.4 does not apply. We begin with an example.

EXAMPLE 1 A manufacturer wants to design a closed can in the shape of a right circular cylinder having a storage capacity (volume) of 100 cubic inches. What dimensions should be chosen to minimize the amount of metal needed for its manufacture?

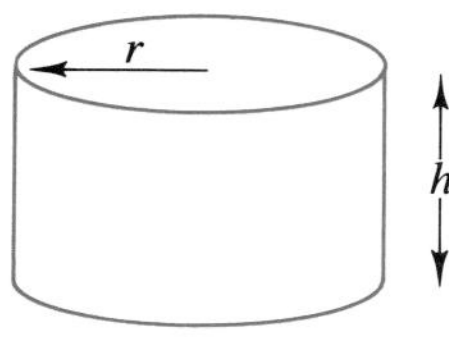

FIGURE 1

Solution The dimensions of the can may be specified by its radius r and height h (Figure 1). From geometry we know that the volume V of such a cylinder is

$$V = \pi r^2 h$$

Since the volume is required to be 100 cubic inches, r and h must satisfy $100 = \pi r^2 h$ or

$$h = \frac{100}{\pi r^2} \tag{1}$$

Thus, once the optimal value of r is obtained, the optimal value of h is automatically determined by Equation (1). If we assume the can is made from sheet metal of uniform thickness, then the amount of metal required is determined by the surface area of the can. The area is made up of a circular top and bottom, and a side is made by bending and welding a rectangular sheet (Figure 2). As indicated in the figure, the height of the rectangle will be the height of the can, and the length of the rectangle will be the circumference of the can (which is $2\pi r$, as you may recall from geometry).

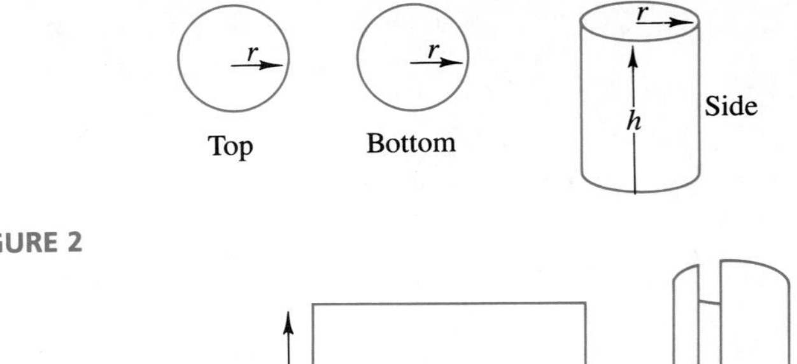

FIGURE 2

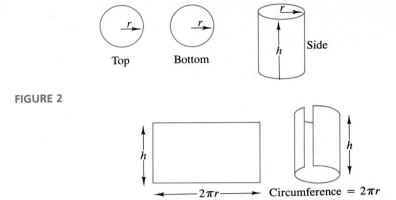

Thus the surface area S will be

$$S = \underbrace{\pi r^2}_{\substack{\text{area}\\ \text{of}\\ \text{top}}} + \underbrace{\pi r^2}_{\substack{\text{area}\\ \text{of}\\ \text{bottom}}} + \underbrace{2\pi rh}_{\substack{\text{area}\\ \text{of}\\ \text{side}}} = 2\pi r^2 + 2\pi rh$$

Since the height of the can must satisfy Equation (1), we can rewrite the surface area S in terms of r alone; we obtain

$$S = 2\pi r^2 + 2\pi r\left(\frac{100}{\pi r^2}\right) = 2\pi r^2 + \frac{200}{r} \tag{2}$$

What are the physical restrictions on r? Because a can cannot have a negative radius and because we cannot build a can of zero radius, we must have $r > 0$; that is, r must lie in the interval $(0, +\infty)$. Since this is not a closed interval $[a, b]$, the extreme value theorem does not apply; thus we have no assurance that S actually has a minimum value for $r > 0$. However, these doubts can be removed by graphing the equation

$$S = 2\pi r^2 + \frac{200}{r}$$

(See Figure 3.)

r	S
0.5	401.6
1	206.3
1.5	147.5
2	125.1
2.5	119.3
3	123.2
3.5	134.1
4	150.5
4.5	171.7
5	197.1

FIGURE 3

As before, we can find the low point on this graph by determining where the tangent is horizontal. The calculations are as follows:

$$\begin{aligned}\frac{dS}{dr} &= \frac{d}{dr}\left[2\pi r^2 + \frac{200}{r}\right]\\ &= 2\pi(2r) + 200\left(-\frac{1}{r^2}\right)\\ &= 4\pi r - \frac{200}{r^2}\end{aligned}$$

Setting $dS/dr = 0$, we obtain

$$4\pi r - \frac{200}{r^2} = 0$$

or on multiplying through by r^2

$$4\pi r^3 - 200 = 0$$

or

$$r = \sqrt[3]{\frac{200}{4\pi}} = \sqrt[3]{\frac{50}{\pi}}$$

Thus the minimum amount of metal will be used when the radius is

$$r = \sqrt[3]{\frac{50}{\pi}} \cong 2.52 \text{ inches}$$

and the height [see Equation (1)] is

$$h = \frac{100}{\pi r^2} = \frac{100}{\pi(\sqrt[3]{50/\pi})^2} \cong 5.03 \text{ inches}$$

(The approximate values for r and h were obtained using a calculator. The approximate value for r is consistent with Figure 3.)

It is interesting to note that the surface area S does not have a *maximum* value (Figure 3). Thus, had we been interested in choosing r and h to maximize S, our problem would have had no solution.

The following result is useful in optimization problems where there is only one critical point (as in the preceding example).

THEOREM Suppose a differentiable function f has *only one* critical point c in a certain interval.

(a) If f has a relative maximum at c, the value $f(c)$ is the (absolute) maximum value of f on the interval (Figure 4a).

(b) If f has a relative minimum at c, the value $f(c)$ is the (absolute) minimum value of f on the interval (Figure 4b).

FIGURE 4

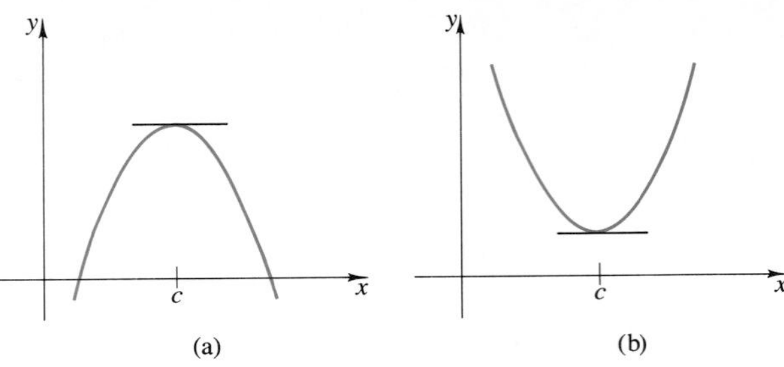

EXAMPLE 2 In Example 1 we graph the surface area function to deduce that the surface area can be minimized. This tedious graphing can be avoided by using the preceding theorem and the second derivative test. We see in Example 1 that

$$r = \sqrt[3]{\frac{50}{\pi}}$$

is the only critical point in the interval $(0, +\infty)$ for the surface area function S. Since the second derivative

$$\frac{d^2S}{dr^2} = 4\pi + \frac{400}{r^3}$$

evaluated at the critical point is positive (verify this), S has a relative minimum and thus an absolute minimum at this point.

The theorem used in the last example can also be used in maximizing or minimizing a function over a closed interval $[a, b]$. Thus, suppose f is continuous on $[a, b]$ and c is the only critical point of f in $[a, b]$. If f has a relative maximum at c, then $f(c)$ is the absolute maximum value of f on the interval; if f has a relative minimum at c, then $f(c)$ is the absolute minimum value of f on the interval. This approach is used in the next example.

EXAMPLE 3 **OIL-PRICING POLICIES** An oil-producing country is selling 1,000,000 barrels per day at \$20 per barrel. A price increase is contemplated, but it is estimated that each \$1 per barrel increase in price will result in 20,000 fewer barrels sold per day. How much of an increase, if any, should be made to maximize the daily revenue?

Solution If we denote the price increase (in dollars) by x, then $20{,}000x$ represents the decrease in the number of barrels sold per day. Thus with a price increase of x dollars per barrel, the number of barrels sold per day will be

$$1{,}000{,}000 - 20{,}000x$$

and the daily revenue $R(x)$ in dollars at the new selling price of $20 + x$ dollars will be

$$R(x) = \begin{pmatrix}\text{number of}\\ \text{barrels sold}\end{pmatrix} \cdot \begin{pmatrix}\text{price per}\\ \text{barrel}\end{pmatrix} = (1{,}000{,}000 - 20{,}000x)(20 + x)$$

or

$$R(x) = 20{,}000{,}000 + 600{,}000x - 20{,}000x^2$$

Since each \$1 increase in price reduces the sales by 20,000 barrels and since there are only 1,000,000 barrels currently produced, the maximum possible price increase is $x = 50$ dollars (why?). Thus, the problem reduces to maximizing $R(x)$ over the interval [0, 50]. We begin by finding the critical points:

$$R'(x) = 600{,}000 - 40{,}000x = 0$$

therefore $x = 15$ is the only critical point. Since

$$R''(x) = -40{,}000$$

we have

$$R''(15) = -40{,}000 < 0$$

therefore $R(x)$ has a relative maximum. By the theorem above, $R(x)$ has an absolute maximum for the interval [0, 20] at $x = \$15$. Thus the oil producer should raise the price by \$15 to \$35 per barrel to maximize the revenue.

EXAMPLE 4 A utility company is planning to run a power line from a generator located at point A on a straight shoreline to an offshore oil rig located at a point B, which is 2 miles from the closest shore point C (Figure 5). The point C is 8 miles down the coast from A. The line is to run from A along the shore to a point D between A and C and then straight to B. If it costs \$1000 per mile to run line underwater and \$800 per mile underground, where should the point D be located to minimize the total cost?

FIGURE 5

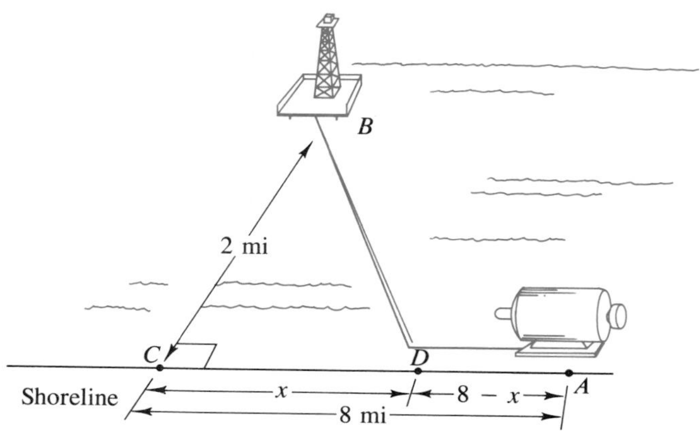

Solution Let x denote the length of CD so that x satisfies $0 \leq x \leq 8$. The length of the power line running underwater is the length of the line segment BD, which is, by the Pythagorean theorem, $\sqrt{x^2 + 2^2} = \sqrt{x^2 + 4}$. Thus, the cost of laying the line underwater is

$$C_1(x) = 1000\sqrt{x^2 + 4}$$

The length of the power line under ground is the length of AD, which is $8 - x$. Thus, the cost of laying the line underground from A to D is

$$C_2(x) = 800(8 - x)$$

The total cost (in dollars) is given by

$$\begin{aligned} C(x) &= C_1(x) + C_2(x) \\ &= 800(8 - x) + 1000\sqrt{x^2 + 4} \end{aligned}$$

To find the minimum value of $C(x)$, we find its critical points:

$$\begin{aligned} C'(x) = -800 + 1000 \cdot \frac{1}{2} \cdot \frac{2x}{\sqrt{x^2 + 4}} &= 0 \\ -800 + \frac{1000x}{\sqrt{x^2 + 4}} &= 0 \\ -800\sqrt{x^2 + 4} + 1000x &= 0 \\ 4\sqrt{x^2 + 4} &= 5x \end{aligned}$$

Squaring both sides and simplifying, we obtain

$$\begin{aligned} 16(x^2 + 4) &= 25x^2 \\ 9x^2 &= 64 \end{aligned}$$

which yields the two values,

$$x = \frac{8}{3} \quad \text{and} \quad x = -\frac{8}{3}$$

Thus $x = \frac{8}{3}$ is the only critical point in the interval $[0, 8]$. We can determine whether a minimum occurs at $x = \frac{8}{3}$ by using the second derivative test, or since $[0, 8]$ is a closed interval, by evaluating $C(x)$ at the endpoints and at the critical value.

In this case it is easier to use the latter approach:

x	0	$\frac{8}{3}$	8
$C(x) = 800(8 - x) + 1000\sqrt{x^2 + 4}$	\$8400	\$7600	$1000\sqrt{68} \cong$ \$8246.21

Thus a minimum cost of \$7600 is achieved by choosing $D = \frac{8}{3}$ miles from C.

EXERCISE SET 13.6

In Exercises 1–4 find the maximum and minimum values of f on the stated interval. If a maximum or minimum does not exist, say so.

1. $f(x) = x^2 - 3x + 2$ on $(-\infty, +\infty)$
2. $f(x) = 3 + 5x - x^2$ on $(-\infty, +\infty)$
3. $f(x) = 1 + 4x - \frac{1}{3}x^3$ on $(0, +\infty)$
4. $f(x) = x^4 - 3x^3 + x^2 - 1$ on $[0, +\infty)$
5. $f(x) = 4x^2 + 1/x$ on $(0, 1)$
6. $f(x) = 1/(1 - x^2)$ on $(-1, 1)$
7. $f(x) = x + 1/x$ on $(0, +\infty)$
8. $f(x) = x/(x^2 + 1)$ on $(0, +\infty)$

9. Suppose that x hours after a drug is administered to a patient, the drug's concentration in the body is given by

$$k(x) = \frac{2x}{x^2 + 4}$$

How long after the drug has been administered is the concentration at its maximum value?

10. **(Epidemic)** Suppose that the number of people in a certain city who are affected by an epidemic is given by

$$P(t) = -t^2 + 80t + 10$$

where t is the number of days after the disease has been detected. On what day will the maximum number of people be affected, and how many people will be affected on this day?

11. Find the minimum value of $x^2 + 432/x$ for x in the interval $(0, +\infty)$.

12. It costs a pharmaceutical firm $x^3 + 200x + 1000$ dollars to manufacture x gallons of penicillin that it sells for \$500 a gallon.
(a) What size batch should it produce to achieve the maximum profit on the batch?
(b) What is the maximum profit?

13. You are assigned the job of designing a rectangular box with a square bottom and open top. If the box is required to have a volume of 62.5 cubic inches, what dimensions would you use to obtain a box with minimum surface area? What is the minimum surface area?

14. A shipper needs a closed rectangular container with a volume of 96 cubic feet and a square bottom. The heavy-duty plastic needed for the top and bottom costs \$3 per square foot, and the standard plastic for the sides costs \$2 per square foot. What dimensions yield a container of minimum cost? What is the minimum cost?

15. If the marginal revenue and marginal cost for producing x units are $\text{MR} = 100 - \frac{1}{20}x^2$ and $\text{MC} = 10 + \frac{1}{20}x^2$, how many items should be produced to maximize the profit?

16. An airline finds that when its cargo plane flies a full load at x miles per hour, it consumes fuel at the rate of

$$F(x) = \frac{1}{500}\left(\frac{10{,}000}{x} + x\right)$$

gallons per mile. If the cost of jet fuel is \$2 per gallon, what is the most economical velocity for the plane to make a 1000-mile trip? What is the minimum fuel cost for the trip?

17. It cost $5 + x/100$ dollars per mile to drive a truck x miles per hour. If the driver receives \$16 per hour, how fast should the truck be driven to minimize the cost per mile?

18. A manufacturer of bicycles finds that when x bicycles per day are produced, the following costs are incurred: a fixed cost of \$1000, labor cost of \$10 per bicycle, and a cost of $\$25{,}000/x$ for advertising. How many bicycles should be produced daily to minimize the total cost?

19. A manufacturer wants to design an open can (no top) in the shape of a right circular cylinder having a storage capacity of 1000 cubic inches. What dimensions should be used to minimize the amount of metal needed for the can?

20. A tour operator charters an airplane with a 200-passenger capacity. If 200 passengers agree to go, the price will be \$400 per passenger. For each passenger under 200, the price will rise \$20 per passenger. How many passengers must go to give the operator the greatest revenue? In this case what will each passenger pay for the fare? [*Hint:* Let x = the number of passengers under 200.]

21. A motel owner finds that if a rent of \$20 per room is charged, then 100 rooms per night can be rented. For each \$1 increase in the rate per night, 2 fewer rooms are rented. How much of an increase, if any, should be made to maximize the daily revenue?

22. An orange grower finds that the average yield is 120 bushels per tree when 50 trees are planted in the grove. For each additional tree planted, the average yield per tree decreases by 2 bushels. How many trees should be planted to maximize the yield?

23. A utility company wants to run a power line from a generator located at a point A on a straight shoreline to an offshore oil rig located at a point B, which is 6 miles from the closest shorepoint C. The point C is 7 miles down the coast from A. The line is to run from A along the shore to a point D between A and C and then straight to B. If it costs \$5000 per mile to run line underwater and \$4000 per mile over land, where should the point D be located to minimize the total cost? [*Hint:* See Example 4.]

13.7 ELASTICITY IN BUSINESS AND ECONOMICS (OPTIONAL)

The absolute amount by which a variable changes is often of far less significance than the *relative* amount of change. For example, knowing merely that the value of an investment has grown by \$100 does not tell us whether to consider that gain a large one or a small one. An investment that grows by \$100 from \$400 to \$500 experiences a 25% gain, whereas an investment that grows by \$100 from \$2000 to \$2100 experiences only a 5% gain. In the absolute sense both gains are the same (\$100), but in the relative sense the first gain (25%) is 5 times as large as the second (5%).

A common way to compare the relative values of two quantities is divide one by the other.

Definition 1 For a function $y = f(x)$, when x undergoes a change Δx from x to $x + \Delta x$:

(a) The **relative change in *x*** is

$$\frac{\Delta x}{x}$$

(b) The **absolute change in *y*** is

$$\Delta y = f(x + \Delta x) - f(x)$$

(c) The **relative change in *y*** is

$$\frac{\Delta y}{y}$$

where $y = f(x)$

EXAMPLE 1 Suppose for some function $y = f(x)$ that x changes by an amount $\Delta x = 2$ from $x = 16$ to $x = 18$. Then the relative change in x is $\Delta x/x = \frac{2}{16} = \frac{1}{8} = 12.5\%$.

(a) If the corresponding change in y is from 72 to 99, then:
 (i) The absolute change in y is $\Delta y = 99 - 72 = 27$.
 (ii) The relative change in y is

$$\frac{\Delta y}{y} = \frac{27}{72} = \frac{3}{8} = 37.5\%$$

In this case the relative change in y (37.5%) is 3 times the relative change in x (12.5%).

(b) If, instead, the change in y is from 100 to 105, then:
 (i) The absolute change in y is $\Delta y = 105 - 100 = 5$.
 (ii) The relative change in y is

$$\frac{\Delta y}{y} = \frac{5}{100} = \frac{1}{20} = 5\%$$

This time the relative change in y (5%) is only $\frac{2}{5}$ of the relative change in x because $(1/20)/(1/8) = \frac{8}{20} = \frac{2}{5}$.

In Example 1(a) a relative change of $\frac{1}{8}$ in x brings about a relative change in y that is 3 times as large. In Example 1(b) the relative change in y is only $\frac{2}{5}$ as large as the relative change in x. Thus in (a), the function is more responsive to a change in x than is the case in (b). The concept of **elasticity** is introduced as a mathematical measure of this type of sensitivity of a function to changes in its input.

Definition 2 For a function $y = f(x)$, as x undergoes a change Δx from x to $x + \Delta x$, the **average elasticity** of f relative to x is

$$\frac{\Delta y/y}{\Delta x/x} = \frac{\text{relative change in } y}{\text{relative change in } x} \tag{1}$$

$$= \frac{x}{y}\frac{\Delta y}{\Delta x}$$

$$= \frac{x}{y}\frac{f(x + \Delta x) - f(x)}{\Delta x} \tag{2}$$

EXAMPLE 2 Consider the function f given in Example 1, where x changes by $\Delta x = 2$ from $x = 16$ to $x = 18$.

(a) If the corresponding change in y is from 72 to 99, then the average elasticity in y relative to x is

$$\frac{\Delta y/y}{\Delta x/x} = \frac{3/8}{1/8} = 3$$

(b) If, instead, the corresponding change in y is from 100 to 105, then the average elasticity in y relative to x is

$$\frac{\Delta y/y}{\Delta x/x} = \frac{1/20}{1/8} = \frac{8}{20} = \frac{2}{5}$$

We have already seen the significance of the numbers 3 and $\frac{2}{5}$ in our discussion of Example 1.

Definition 3 For a function $y = f(x)$, the **point elasticity** at x is

$$\lim_{\Delta x \to 0} \frac{\Delta y/y}{\Delta x/x} = \frac{x}{y} \lim_{\Delta x \to 0} \frac{\Delta y}{\Delta x} \tag{3}$$

$$= \frac{x}{y} f'(x) \tag{4}$$

provided this limit exists.

Thus the concept of point elasticity applies only where the function is differentiable. We write the formula for **point elasticity of y relative to x** as:

$$E_x(y) = \frac{dy/y}{dx/x} = \frac{x}{y}\frac{dy}{dx} = \frac{x}{y}f'(x) \tag{5}$$

Thus the point elasticity of a function is the limit of the average elasticity, as $\Delta x \to 0$.

If $|E_x(y)| > 1$, we say that y is *elastic* relative to x.
If $|E_x(y)| < 1$, we say that y is *inelastic* relative to x.
If $|E_x(y)| = 1$, we say that y has *unit elasticity* relative to x.

Where a function is elastic, a small change in the input will produce a relative change in the output that is *greater than* the relative change in input. Where the function is inelastic, the relative change in output will be *less than* the relative change in input. Where a function has unit elasticity, the relative change in output equals the relative change in input.

EXAMPLE 3 Consider the function $f(x) = 5x^2 - 3$ at $x = 2$.

(a) If x changes from $x = 2$ to $x = 2.03$, then:

(i) The relative change in x is

$$\frac{\Delta x}{x} = \frac{0.03}{2} = 0.015$$

(ii) The absolute change in y is

$$\begin{aligned}\Delta y &= f(2.03) - f(2)\\ &= [5(2.03)^2 - 3] - [5(2)^2 - 3]\\ &\cong 17.6045 - 17\\ &= 0.6045\end{aligned}$$

(iii) The relative change in y is

$$\frac{\Delta y}{y} = \frac{0.6045}{17} \cong 0.0356$$

(iv) The average elasticity of f relative to x is

$$\frac{\Delta y/y}{\Delta x/x} \cong \frac{0.0356}{0.015} \cong 2.3733$$

(b) The point elasticity of f is

$$E_x(y) = \frac{x}{y}f'(x) = \frac{x}{5x^2 - 3} \cdot 10x$$

At $x = 2$, the point of elasticity of f is

$$E_x(y) = \frac{2}{17}(20)$$
$$\cong 2.3529$$

The function f is elastic at $x = 2$ because its point elasticity there is greater than 1.

Elasticity of Demand

Suppose the demand function for a product is $x = f(p)$, where consumers demand x units of the product when the price is p dollars per unit. Therefore,

$$E_p(x) = \frac{p}{x}\frac{dx}{dp} \tag{6}$$

is called the **point elasticity of demand**.

We usually assume that an increase in price will result in a decrease in demand. That is, demand is a decreasing function of price. In this case the demand function $x = f(p)$ will have a negative derivative:

$$\frac{dx}{dp} < 0 \tag{7}$$

Since both price p and demand x are assumed to be positive, p/x is positive. Combining this with (7) and (6), we see that *elasticity of demand is negative in this case:*

$$E_p(x) < 0 \quad \text{when demand is a decreasing function of price} \tag{8}$$

EXAMPLE 4 Suppose the demand function for a certain product is

$$x = 200 - 4p$$

where p is the price (in dollars) per unit and x is the number of units that can be sold monthly when the price is $\$p$.

(a) Find the point elasticity of demand as a function of p.

(b) Find the point elasticity of demand when $p = 10$. Interpret your answer.

(c) Find the point elasticity of demand when $p = 30$. Interpret your answer.

(d) At what price will the demand have unit elasticity?

Solution (a) From Equation (6),

$$E_p(x) = \frac{p}{200 - 4p} \cdot (-4)$$
$$= \frac{-p}{50 - p}$$

Solution (b) When $p = 10$, $E_p(x) = -10/(50 - 10) = -\frac{1}{4}$. This means that when the price is \$10, a slight increase in price will induce a relative decrease in demand that is approximately $\frac{1}{4}$ as large as the relative increase in price.

Since $|E_p(x)| = \frac{1}{4} < 1$, the demand is inelastic at this price level.

Solution (c) When $p = 30$,

$$E_p(x) = -30/(50 - 30) = -30/20 = -\tfrac{3}{2} = -1.5$$

Thus when $p = 30$, a slight increase in price will induce a relative decrease in demand that is approximately $1\frac{1}{2}$ times as large as the relative increase in price.

Since $|E_p(x)| = 1.5 > 1$, the demand is elastic at this price level.

Solution (d) To have unit elasticity we must have $|E_p(x)| = 1$. That is,

$$\begin{aligned} \frac{p}{50 - p} &= 1 \\ p &= 50 - p \\ 2p &= 50 \\ p &= 25 \end{aligned}$$

Thus, when the price is \$25, demand has unit elasticity. At this price level, a slight increase in price will induce a relative decrease in demand that is nearly equal to the relative increase in price.

EXAMPLE 5 Suppose the demand function for a certain commodity is given in the form

$$p = 4000 - 5x^2 \tag{9}$$

This equation tells us the price that must be charged per unit to sustain a demand of x units.

(a) Find the point elasticity of demand as a function of p.

(b) Find and interpret the point elasticity of demand when $p = \$2000$.

(c) Find and interpret the point elasticity of demand when $p = \$3000$.

Solution (a) Solving Equation (9) for x and noting that x is positive, we obtain

$$x = \sqrt{800 - \frac{p}{5}} \tag{10}$$

Thus,

$$\begin{aligned} E_p(x) &= \frac{p}{x}\frac{dx}{dp} = \frac{p}{\sqrt{800 - p/5}}\left(-\frac{1}{10\sqrt{800 - p/5}}\right) \\ &= -\frac{p}{10(800 - p/5)} = -\frac{p}{8000 - 2p} \end{aligned}$$

or

$$E_p(x) = -\frac{p}{2(4000 - p)} \tag{11}$$

Solution (b) When $p = 2000$,

$$E_p(x) = \frac{-2000}{2(2000)} = -\frac{1}{2}$$

so $|E_p(x)| < 1$ and the demand is inelastic. At this price level, a slight increase in price will induce a relative decrease in demand that is approximately $\frac{1}{2}$ the relative increase in price.

Solution (c) When $p = 3000$,

$$E_p(x) = \frac{-3000}{2(1000)} = -\frac{3}{2}$$

so $|E_p(x)| > 1$ and the demand is elastic. At this price level, a slight increase in price will induce a relative decrease in demand that is approximately $1\frac{1}{2}$ times the relative increase in price.

The concept of point elasticity of demand is helpful in analyzing the responsiveness of revenue R to changes in price. If we know the demand $x = f(p)$, where x is the number of units consumers will demand if the price is $\$p$, then

$$R = x \cdot p$$

Then, using the product rule,

$$\begin{aligned} \frac{dR}{dp} &= x\frac{dp}{dp} + p\frac{dx}{dp} \\ &= x\left(1 + \frac{p}{x}\frac{dx}{dp}\right) \\ &= x[1 + E_p(x)] \end{aligned} \tag{12}$$

[from Equation (6)].

Case 1 If demand is elastic at price p, then $|E_p(x)| > 1$. That means $E_p(x) < -1$ because elasticity is always negative. Thus, $1 + E_p(x) < 0$. But $x > 0$ because demand is always positive. Thus,

$$x[1 + E_p(x)] = \text{positive} \cdot \text{negative}$$

Therefore, from Equation (12), $dR/dp < 0$; that is, revenue decreases when price increases.

Case 2 If the demand is inelastic, then $|E_p(x)| < 1$. That means $-1 < E_p(x) < 0$, and so $1 + E_p(x) > 0$. As before, $x > 0$, which implies

$$x[1 + E_p(x)] = \text{positive} \cdot \text{positive}$$

Therefore, Equation (12) tells us that $dR/dp > 0$; that is, revenue increases when price increases. In summary:

When demand is elastic, the change in revenue is in the opposite direction from the change in price. When demand is inelastic, the change in revenue is in the same direction as the change in price.

EXAMPLE 6 Miracle Manufacturing Company is currently charging \$80 each for its Miracle #5. A marketing consultant has determined that at price \$$p$ the consumer demand for Miracle #5 is $x = 1000 - 5p$. Miracle is considering raising the price by a small amount. How will that affect its revenue? Would you draw the same conclusion if the current price were \$120?

Solution We first calculate the elasticity of demand, using Equation (6):

$$E_p(x) = \frac{p}{x}\frac{dx}{dp} = \frac{p}{1000 - 5p} \cdot (-5) = \frac{-p}{200 - p}$$

When $p = \$80$,

$$|E_p(x)| = \left|\frac{-80}{200 - 80}\right| = \frac{80}{120} = \frac{2}{3} < 1$$

Thus, demand is *inelastic* at this price level. That means that a small increase in price will be accompanied by an *increase* in revenue. Thus, when the price is \$80, it is safe to recommend a price increase.

What if the current price is \$120? At this price level the elasticity of demand is

$$E_p(x) = \frac{-120}{(200 - 120)} = \frac{-120}{80} = \frac{-3}{2}$$

Since $|E_p(x)| > 1$, demand is *elastic;* therefore, a small increase in price will be accompanied by a *decrease* in revenue. Thus, when the price is \$120, it is unwise to consider a price increase. In fact, *lowering* the price will increase the revenue!

Point Elasticity of Cost

In Section 13.2 we introduce the cost function $C(x)$ to express the cost of producing x units of a product. By Definition 3, the **point elasticity of cost** relative to x is

$$E_x(C) = \frac{x}{C(x)}C'(x) \tag{13}$$

using Equation (5). Alternatively,

$$E_x(C) = \frac{C'(x)}{C(x)/x}$$

Thus

$$\text{point elasticity of cost} = \frac{\text{marginal cost}}{\text{average cost}} \tag{14}$$

We assume that cost is an increasing function of x, since producing usually costs more. With this assumption, both the marginal cost and the average cost are positive. Therefore,

$$E_x(C) > 0 \qquad \text{when } C(x) \text{ is increasing} \tag{15}$$

For a cost function $C(x)$, we say that:

Cost is *elastic* if $E_x(C) > 1$.

Cost is *inelastic* if $E_x(C) < 1$.

Cost has *unit elasticity* if $E_x(C) = 1$.

Recall from Section 13.2 that the marginal cost $C'(x)$ may be thought of as the approximate cost of producing the $(x + 1)$st unit. When this result is combined with Equation (14), we have the following application:

When the cost is *elastic,* the cost of producing the next unit will be greater than the average cost of the units already produced. When the cost is *inelastic,* the cost of producing the next unit will be less than the average cost of the units already produced.

EXAMPLE 7 Suppose the cost of producing x tires per day is

$$C(x) = 800 + 5x - \frac{x^2}{40}$$

Find the point elasticity of cost when $x = 10$, and interpret your answer.

Solution Using Equation (13) we calculate the point elasticity of cost:

$$E_x(C) = \frac{x}{800 + 5x - x^2/40}\left(5 - \frac{x}{20}\right)$$

When $x = 10$,

$$\begin{aligned} E_x(C) &= \frac{10}{800 + 50 - 100/40}\left(5 - \frac{1}{2}\right) \\ &= \frac{10}{850 - 5/2}\left(\frac{9}{2}\right) = \frac{90}{1700 - 5} = \frac{90}{1695} \\ &\cong 0.053 \end{aligned}$$

Since $E_x(c) < 1$, the cost is inelastic at this production level. Thus, the cost of producing the eleventh tire per day will be less than the average cost of producing the first 10 tires.

EXERCISE SET 13.7

1. Suppose that for some function $y = f(x)$, as x changes from 25 to 29, y changes from 150 to 160. Find:
   (a) the absolute change in x
   (b) the relative change in x
   (c) the absolute change in y
   (d) the relative change in y

2. Suppose that for the function $y = f(x)$, as x changes from 59 to 63, y changes from 95 to 120. Find the quantities (a) through (d) specified in Exercise 1.

3. Consider the function $y = x^2$. As x changes from 3 to 5, calculate:
   (a) the relative change in x
   (b) the relative change in y
   (c) the average elasticity of y relative to x

4. Consider the function $y = 1/x$ as x changes from 2 to 6. Calculate the quantities (a) through (c) specified in Exercise 3.

In Exercises 5–12 find the point elasticity of the given function $y = f(x)$ at the given point. Also, tell whether the function is elastic, inelastic, or has unit elasticity at the given point.

5. $y = x^2$ at $x = 3$
6. $y = x^3$ at $x = 5$
7. $y = 1/x$ at $x = 2$
8. $y = 1/x^2$ at $x = 4$
9. $y = 3x^2 + x$ at $x = 3$
10. $y = x^3 + 2$ at $x = 2$
11. $y = 1/(x + 3)$ at $x = 1$
12. $y = \sqrt{2x + 3}$ at $x = 3$

13. Suppose we know that $y = 2x^2 - 4x$. Find the point elasticity of y relative to x when $x = 3$, using Equation (5).

14. For the relation $y = 1/(x^2 + 4x)$, find the point elasticity of y relative to x when $x = 1$.

15. Suppose the demand function for a commodity is

$$x = \frac{500 - p^2}{10}$$

where x is the number of units that can be sold weekly when the price is p (in hundreds of dollars) per unit.
   (a) Find the point elasticity of demand as a function of p.
   (b) Find the point elasticity of demand when $p = 10$. Interpret your answer.
   (c) Find the point elasticity of demand when $p = 20$. Interpret your answer.
   (d) At what price will the demand have unit elasticity?

16. Suppose that the number of units that can be sold weekly when the price p (in hundreds of dollars) per unit is

$$x = \frac{800}{p + 50}$$

   (a) Find the point elasticity of demand when the price is \$500 per unit.
   (b) Show that this demand function is inelastic at every price p.

17. Suppose that to sustain a demand of x units monthly the price charged per unit must be

$$p = 1000 - 8x^2$$

dollars per unit.
   (a) Find the point elasticity of demand as a function of p.
   (b) Find the point elasticity of demand when $p = \$200$. Interpret your answer.
   (c) Find the point elasticity of demand when $p = \$800$. Interpret your answer.
   (d) At what price will the demand have unit elasticity?

18. To sell x units weekly, a company finds that it must set the selling price at

$$p = (60 - 12x)^2$$

dollars per unit.
   (a) Find the point elasticity of demand as a function of p.
   (b) Find and interpret the point elasticity of demand when $p = \$900$.
   (c) Find and interpret the point elasticity of demand when $p = \$1600$.

19. For the demand function of Exercise 15:
   (a) If the price is currently \$10 per unit, would a price increase bring about an increase or a decrease in revenue? Explain your answer.
   (b) For the current price of \$20, answer the question posed in (a).

20. For the demand function of Exercise 18, if the price is currently \$2500 per unit, would a price increase bring about an increase or decrease in revenue? Explain your answer.

21. (Elasticity of cost) Suppose that the cost of producing x units weekly is

$$C(x) = \tfrac{1}{2}x^2 + 8x + 100$$

Find the point elasticity of cost when $x = 10$. Is the cost elastic or inelastic at this level of production? Interpret your answer. Investigate the same questions when $x = 20$.

22. (Elasticity of cost) Suppose that the cost of producing x units daily is

$$C(x) = \sqrt{3x + 12} + 80$$

Find the point elasticity of cost when $x = 8$. Is the cost elastic or inelastic at this level of production? Interpret your answer.

23. (Elasticity of cost) Suppose that the cost of producing x units weekly is

$$C(x) = \tfrac{1}{3}x^3 - 2x^2 + 8x + 50$$

Find the point elasticity of cost when $x = 6$. Is the cost elastic or inelastic at this level of production? Interpret your answer.

KEY IDEAS FOR REVIEW

- **Equation of tangent line** A line tangent to $y = f(x)$ through (x_1, y_1):

$$y = m(x - x_1) + y_1 \qquad \text{where } m = f'(x_1)$$

- **Increasing and decreasing** $y = f(x)$, a differentiable function, is increasing on an interval where $f'(x) > 0$ for all x in the interval; it is decreasing on an interval where $f'(x) < 0$ for all x in the interval.

- **Velocity**

$$\frac{ds}{dt}$$

- **Per-unit average cost**

$$\overline{C}(x) = \frac{C(x)}{x}$$

  Per-unit average revenue

$$\overline{R}(x) = \frac{R(x)}{x}$$

  Per-unit average profit

$$\overline{P}(x) = \frac{P(x)}{x}$$

- **Marginal cost** MC $= C'(x)$
  Marginal revenue MR $= R'(x)$
  Marginal profit MP $= P'(x)$

- **Extreme-value theorem** If a function f is continuous at each point of a closed interval $[a, b]$, then f has an absolute maximum and an absolute minimum on $[a, b]$.

- **Critical point** A number c in the domain of f such that either $f'(c) = 0$ or $f'(c)$ does not exist.
- **Location of absolute maxima and minima** If f is differentiable at each point of a given interval and if f has an absolute maximum on the interval, then this maximum must occur at an endpoint (if there is one) or at a critical point. Similarly for an absolute minimum.
- **Procedure for finding an absolute maximum or minimum** See page 584.
- **First derivative test** See page 593.
- **Concavity test** $y = f(x)$ is concave up on an interval if $f''(x) > 0$ on the interval and concave down if $f''(x) < 0$.
- **Second derivative test** If $f'(c) = 0$, then f has a relative minimum at c if $f''(c) > 0$ and a relative maximum at c if $f''(c) < 0$. If $f'(c) = 0$ and $f''(c) = 0$, then the test fails and the first derivative test must be used.
- **Sketching the graph of a polynomial function** See page 599.
- **Relative change in x** $\dfrac{\Delta x}{x}$

  Absolute change in y $\Delta y = f(x + \Delta x) - f(x)$

  Relative change in y $\dfrac{\Delta y}{y}$
- **Average elasticity of f relative to x**

$$\frac{x}{y}\left(\frac{f(x + \Delta x) - f(x)}{\Delta x}\right)$$

- **Point elasticity of y relative to x** $E_x(y) = \dfrac{x}{y}\dfrac{dy}{dx} = \dfrac{x}{y}f'(x)$
- **y is elastic relative to x** $|E_x(y)| > 1$
  y is inelastic relative to x $|E_x(y)| < 1$
  y has unit elasticity relative to x $|E_x(y)| = 1$
- **Point elasticity of demand** $E_p(x) = \dfrac{p}{x}\dfrac{dx}{dp}$
- **Point elasticity of cost** $E_x(C) = \dfrac{C'(x)}{C(x)/x}$
- **Cost is elastic** $E_x(C) > 1$
  Cost is inelastic $E_x(C) < 1$
  Cost has unit elasticity $E_x(C) = 1$

SUPPLEMENTARY EXERCISES

In Exercises 1–4 determine the slope of the curve at the point (x_1, y_1) and find the equation of the tangent line at this point.

1. $f(x) = x^2 - x$, $x_1 = -2$
2. $f(x) = \sqrt{x+1}$, $x_1 = 3$
3. $f(x) = x\sqrt{x-1}$, $x_1 = 5$
4. $f(x) = 1/(x+2)$, $x_1 = 2$
5. Find the points on the curve $y = (2x^2 - 2)^2$ at which the tangent line is horizontal.
6. Find the points on the curve $y = -x^2 + x + 1$ at which the slope of the tangent line is 2.

In Exercises 7–10 find the critical points of $y = f(x)$, the interval(s) on which $f(x)$ is increasing, and the interval(s) on which it is decreasing.

7. $f(x) = x^2 - x$
8. $f(x) = 1 - |x|$
9. $f(x) = 3x^4 - 4x^3$
10. $f(x) = 1 + 1/x$

In Exercises 11–14 find the maximum and minimum values of $f(x)$ on the given interval.

11. $f(x) = x^2 - x$ on $[-1, 4]$
12. $f(x) = 1 - |x|$ on $[-1, 2]$
13. $f(x) = 1 + 1/x$ on $[1, 3]$
14. $f(x) = \sqrt{x+1}$ on $[3, 8]$

In Exercises 15 and 16 find relative maxima, relative minima, and inflection points, and sketch the curve.

15. $y = 3x^4 - 4x^3$
16. $y = x^4 - x^2$
17. **(Medicine)** Suppose that the growth of a human fetus is given by the function

$$W(t) = 3t^2 \qquad 0 \le t \le 39$$

where the weight W is in grams and the time t is in weeks. Find the instantaneous rate of change of the weight of the fetus at $t = 20$.

18. **(Medicine)** A medical research experiment established that the mass $M(t)$ of a tumor, as a function of the length of time t that a patient is exposed to radiation during treatment, is given by

$$M(t) = -\frac{3C}{10}\left(\frac{t^2 - 5t + 10}{t - 3}\right)$$

where $M(t)$ is in milligrams, C is the original mass of the tumor, and t is in seconds. Because of a malfunction in the equipment used, the patient can be exposed to at most 2.9 seconds of radiation therapy.
(a) What is the rate of change of the mass of the tumor when $t = \frac{1}{2}$ second?
(b) For what time t is the size of the tumor decreasing?

19. **(Business)** A computer manufacturer finds that the profit (in dollars) received from making and selling x computers is given by $P(x) = (x^2 + x)^2$. If the rate of production is kept at 5 units per month, what is the rate of change of profit with respect to time when 40 units have been made?

20. Given the graph of $y = f(x)$ (Figure 1), indicate the possible signs of the indicated quantity.
(a) $f(d)$ (b) $f'(a)$ (c) $f''(b)$
(d) $f(b)$ (e) $f'(b)$ (f) $f''(c)$

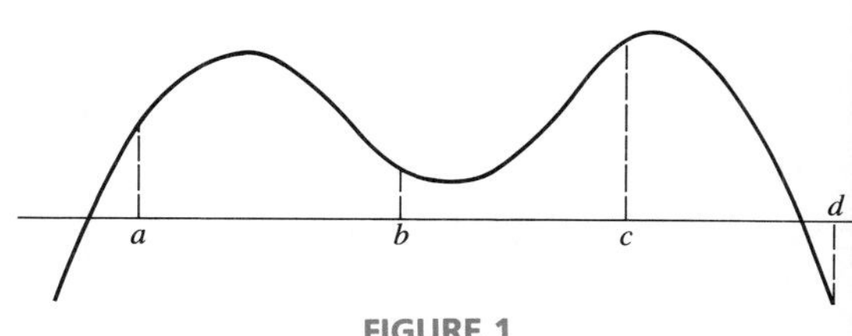

FIGURE 1

21. Farmer Jones must fence in a rectangular region with an area of 100 square feet. What is the minimum amount of fencing that he can use?

22. Paint is dripping through a hole in the floor, forming a circular puddle. If the area of the puddle is increasing at the rate of 2 square inches per minute, what is the rate of change of the radius when the radius is 3 inches?

23. A ball is thrown up into the air. At time t its distance from the ground is given by $s = -16t^2 + 64t$.
(a) What is the velocity at time $t = 2$?
(b) What is the acceleration at time $t = 2$?
(c) What is the maximum height reached by the ball?

24. Find two positive numbers whose sum is 64 and whose product is a maximum. What is the maximum product?

25. Suppose that the cost per hour of manufacturing x sweatshirts of Excel University is $C(x) = 5x + 500/x + 20$. How many shirts should be made to minimize the cost? What is the cost per shirt?

26. The profit made by a stereo dealer in selling x compact disc players per week is $P(x) = 525 + 100x - x^2$. How many players should be sold each week to maximize the profit? What is the maximum profit?

27. A movie theatre owner finds that when she charges $5 per ticket she can fill 400 seats per night. For each 50-cent increase, 20 fewer tickets are sold. How much of an increase, if any, should be made to maximize the daily revenue? What is the maximum revenue?

28. **(Optional)** Suppose that $y = x + 1/x$. Find the point elasticity of y relative to x when $x = 2$.

29. **(Optional)** Suppose the demand function for a commodity is

$$x = \frac{500}{p^2 + 25}$$

where x is the number of units that can be sold weekly when the price is p (in dollars) per unit.
(a) Find the point elasticity of demand as a function of p.
(b) Find the point elasticity of demand when $p = 10$.
(c) At what price will the demand have unit elasticity?

30. **(Optional)** Suppose that the cost of producing x units monthly is

$$C(x) = \tfrac{1}{4}x^4 + x^3 + 25$$

Find the point elasticity of cost when $x = 4$. Is the cost elastic or inelastic at this level of production? Interpret your answer.

CHAPTER TEST

1. What is the equation of the tangent line to $y = (x^2 - x)^2$ at the point with x-coordinate 2?

2. Find the critical points, and identify the relative maxima, relative minima, and inflection points of $y = (x^2 - x)^2$; sketch the graph.

3. An object is moving along the x-axis according to the equation $x = t^2 - t$.
(a) For what values of t is the object moving toward the right, and for what values is it moving toward the left?
(b) At what time is the object farthest to the left, and what is its position at that time?

4. What are the maximum and minimum values of $f(x) = x^3 - 9x + 1$ on the interval $[0, 3.5]$?

5. The profit function for Good Buy Used Cars, Inc., is $P(x) = -x^2 + 200x$, where x is the number of cars sold per week. How many cars should be sold to maximize the company's profit? What is this maximum profit?

6. A farmer wants to fence in a rectangular region and to subdivide it into 3 adjacent rectangular pens with fencing (see Figure 2). If the farmer has 1000 feet of fencing available, what are the dimensions that will yield a rectangular region of maximum area?

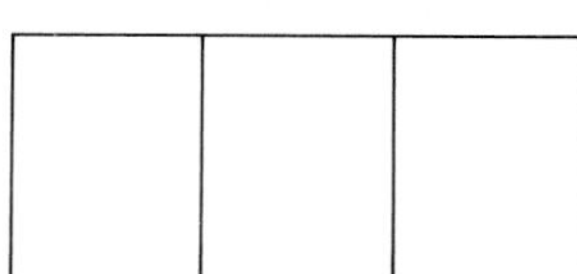

FIGURE 2

7. Suppose that the function estimating the cost of manufacturing x items of a new product is given by

$$C(x) = x^2 + \frac{250{,}000}{x} + 1000$$

How many items should the company manufacture to minimize the cost?

8. A 25-foot ladder is leaning against a wall. If the bottom of the ladder slides along the floor at a rate of 1 foot per second, at what rate is the top of the ladder sliding down at the time that the bottom of the ladder is 5 feet from the wall?

14

Exponential and Logarithmic Functions

In the preceding chapters we consider polynomial, rational, and irrational functions. There are two other important types of functions that arise often in applied mathematics: the exponential and logarithmic functions. In this chapter we define these functions, determine their derivatives, and examine their basic properties. We then investigate some practical applications in which these types of functions naturally occur.

14.1 EXPONENTIAL FUNCTIONS

For a polynomial function

$$p(x) = a_n x^n + a_{n-1} x^{n-1} + \cdots + a_1 x + a_0$$

the exponents on the variable x are all *constants*. In this section, we study functions that have *variables* as exponents. Functions of this type occur naturally in problems arising in science as well as in the management, life, and social sciences.

The simplest functions with variable exponents are called **exponential functions**. These have the form

$$f(x) = a^x \qquad a > 0,\ a \neq 1 \tag{1}$$

for all real numbers x.

Note that a is a *positive constant*, which is called the **base** of the exponential function. We have already encountered in algebra a^r where r is a rational number: for example, a^4, a^{-5}, a^0, and $a^{2/3}$. We now extend this definition to allow x to be any real number. Examples are

$$2^x \qquad 3.4^x \qquad \left(\frac{1}{2}\right)^x \qquad 79^x$$

For the sake of simplicity we often approximate the real number x by a rational number. For example, to compute $1024^{\sqrt{2}}$ we approximate $\sqrt{2}$ by $1.4 = \frac{14}{10}$ to obtain

$$1024^{\sqrt{2}} \cong (1024)^{14/10} = (\sqrt[10]{1024})^{14} = 2^{14} = 16{,}384$$

We can obtain a better answer by approximating $\sqrt{2}$ by 1.41, 1.414, 1.41421, . . .

EXAMPLE 1 Graph the following exponential functions:
(a) $f(x) = 2^x$ (b) $g(x) = (\frac{1}{2})^x$

Solution (a) Since we do not yet know any of the properties of this function, we must plot some points with integer values of x and then connect these points with a smooth curve. Note that if x is a positive integer such as $x = 3$, then $2^x = 2^3 = 8$. If x is a negative integer such as $x = -3$, then

$$2^x = 2^{-3} = \frac{1}{2^3} = \frac{1}{8} \qquad \text{(See rules of exponents in Appendix A.5.)}$$

Some values of 2^x are given in Table 1. The graph is shown in Figure 1a. Note that the graph indicates that as x increases, y increases. We prove in Section 14.4 that this is indeed so.

TABLE 1

x	-4	-3	-2	-1	0	1	2	3	4	5
$y = 2^x$	$\frac{1}{16} \cong 0.06$	$\frac{1}{8} \cong 0.13$	$\frac{1}{4} = 0.25$	$\frac{1}{2} = 0.5$	1	2	4	8	16	32

FIGURE 1

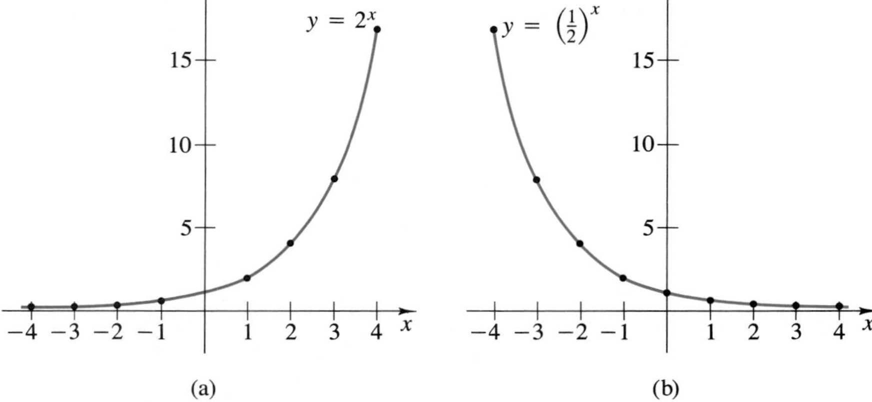

Solution (b) Again, we plot some points with integer values of x and connect these points with a smooth curve. Note that if $x = 3$, then

$$\left(\frac{1}{2}\right)^x = \left(\frac{1}{2}\right)^3 = \frac{1}{2^3} = \frac{1}{8} \qquad \text{(See rules of exponents in Appendix A.5.)}$$

and if $x = -3$, then $(\frac{1}{2})^{-3} = 2^3 = 8$. Some values of $(\frac{1}{2})^x$ are given in Table 2 and the graph is shown in Figure 1b. Note that $(\frac{1}{2})^x$ appears to be a decreasing function. In Section 14.4 we show that this is so.

TABLE 2

x	-4	-3	-2	-1	0	1	2	3	4	5
$y = (\frac{1}{2})^x$	16	8	4	2	1	$\frac{1}{2} = 0.5$	$\frac{1}{4} = 0.25$	$\frac{1}{8} \cong 0.13$	$\frac{1}{16} \cong 0.06$	$\frac{1}{32} \cong 0.03$

When we connect these plotted points with a smooth curve, we are assuming that $f(x) = 2^x$ and $g(x) = (\frac{1}{2})^x$ are continuous functions and are therefore defined for real values of the exponent. In Section 14.4 we see that this assumption is valid. In the meantime we will assume that the rules of exponents discussed in Appendix A.5 of the algebra review are now generalized to include real numbers occurring as exponents. These rules are summarized in Table 3.

TABLE 3

***a* and *b* are positive constants; *y* and *z* are any real numbers**
1. $a^y a^z = a^{y+z}$
2. $\dfrac{a^y}{a^z} = a^{y-z}$
3. $a^{-y} = \dfrac{1}{a^y}$
4. $a^0 = 1$
5. $(a^y)^z = a^{yz}$
6. $(ab)^y = a^y b^y$
7. $\left(\dfrac{a}{b}\right)^y = \dfrac{a^y}{b^y}$

The graphs in Figure 1 are typical of all functions a^x. That is, if $a > 1$, then a^x has the general shape shown in Figure 2a, and if $0 < a < 1$, then a^x has the general shape shown in Figure 2b. Note that if $0 < a < 1$, as for example $a = \frac{1}{2}$, then we can write

FIGURE 2

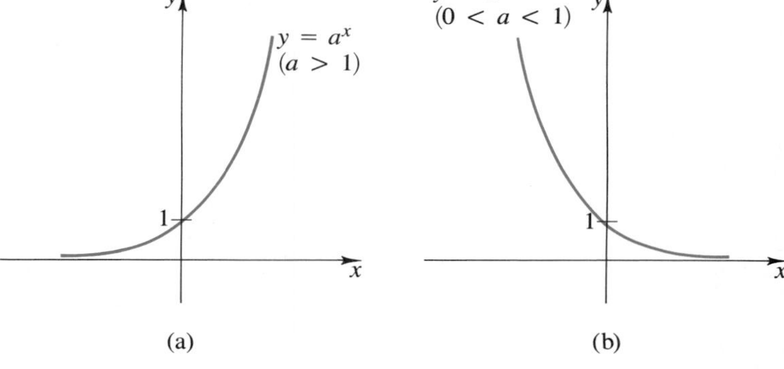

$$\left(\frac{1}{2}\right)^x = 2^{-x} \qquad \text{(by the laws of exponents)}$$

Therefore, all exponential functions can be written in the form a^x or a^{-x} where $a > 1$. The following basic facts about the exponential function a^x are summarized here:

1. For every base $a > 0$, the graph of $y = a^x$ crosses the y-axis at $y = 1$. This is because $a^0 = 1$.
2. If $0 < a < 1$, then the graph of $y = a^x$ gets closer and closer to the positive x-axis as x increases but never actually touches the x-axis (Figure 2b).
3. If $a > 1$, then the graph of $y = a^x$ gets closer and closer to the negative x-axis as x decreases but never actually touches the x-axis (Figure 2a).
4. For every base $a > 0$, $a \neq 1$, the exponential function $f(x) = a^x$ is continuous at every real number x.

EXAMPLE 2 Sketch the graphs of

$$y = 2^x \quad \text{and} \quad y = 3^x$$

on the same coordinate system.

Solution From Table 1, obtained previously, and Table 4 we obtain the graphs in Figure 3.

TABLE 4

x	-4	-3	-2	-1	0	1	2	3	4
$y = 3^x$	$\frac{1}{81} \cong 0.01$	$\frac{1}{27} \cong 0.04$	$\frac{1}{9} \cong 0.11$	$\frac{1}{3} \cong 0.33$	1	3	9	27	81

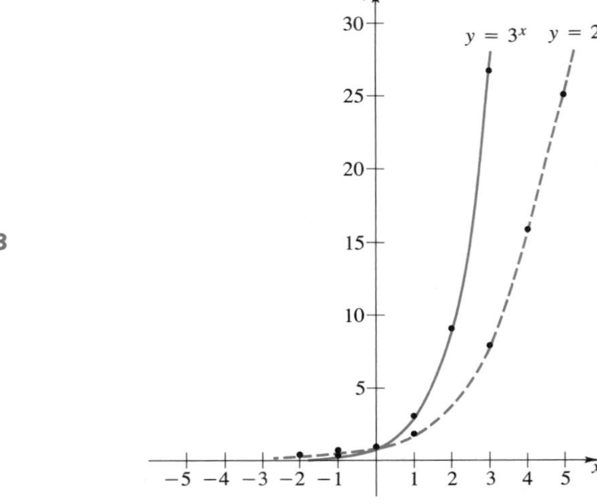

FIGURE 3

Note from Figure 3 that $2^x < 3^x$ for positive values of x, but $3^x < 2^x$ for negative values of x. In general, if $1 < a < b$, then $a^x < b^x$ for positive values of x, but $b^x < a^x$ for negative values of x. We leave it as an exercise for you to determine the relative sizes of a^x and b^x if $0 < a < b < 1$.

Frequently, we encounter functions such as $f(x) = 3(4^{2x})$, or more generally:

$$f(x) = ka^{cx} \qquad a > 0,\ a \neq 1 \tag{2}$$

where k and c are constants.* These are also called **exponential functions**.

EXAMPLE 3 **BACTERIA REPRODUCTION** In this example we show how the growth of a bacteria colony can be described by an exponential function. It is a biological fact that a bacterium reproduces by splitting into two identical bacteria at regular time intervals. The time interval is the same for all members of the species, and so a bacteria colony of one species doubles in size at uniform time intervals.

Consider a bacteria colony with an initial size of n (at time $t = 0$) that doubles in size every 30 minutes. The number of bacteria in the colony after t minutes, $f(t)$, is shown in the following table for various values of t.

t	0	30	60	90	120	150	180	210	240
$f(t)$	n	$2n$	$4n$	$8n$	$16n$	$32n$	$64n$	$128n$	$256n$

We rewrite this table in the alternate form, t = time in minutes, $f(t)$ = number of bacteria present at time t.

t	0	30	60	90	120	150	180	210	240
$f(t)$	n	$2n$	2^2n	2^3n	2^4n	2^5n	2^6n	2^7n	2^8n

Then we see that there is a relationship between t and the exponent on the factor 2 in $f(t)$: the exponent is always $t/30$. Thus, the formula for $f(t)$ is

$$f(t) = 2^{t/30} \cdot n = n2^{t/30} \qquad (t \text{ in minutes}) \tag{3}$$

(Note that the exponent applies only to 2 and not to n.) Except for the fact that the variable is t rather than x, this is an exponential function of form (2) with $k = n$ and $c = \frac{1}{30}$.

EXAMPLE 4 A certain bacteria colony is known to double in size every 30 minutes. If there are 1000 bacteria present initially, how many bacteria will be present 2 hours later?

Solution From Formula (3), the number of bacteria present t minutes later is

$$f(t) = 1000(2^{t/30})$$

---

* Note that the exponent cx belongs only to the base a and not to the constant k.

Since 2 hours = 120 minutes, it follows from this formula that 2 hours later the number of bacteria is

$$f(120) = 1000(2^{120/30}) = 1000(2^4) = 16{,}000$$

Base e The exponential base that occurs most frequently in applications is the irrational number e, which is approximately 2.71828 . . . . In Section 14.3 we see that e is defined as

$$e = \lim_{t \to \infty}\left(1 + \frac{1}{t}\right)^t \cong 2.71828$$

Computing e^x with a Calculator On a calculator such as the one shown in Figure 4, the following sequence of keystrokes is used to evaluate e^x*:

[x] [2nd] [ln x]

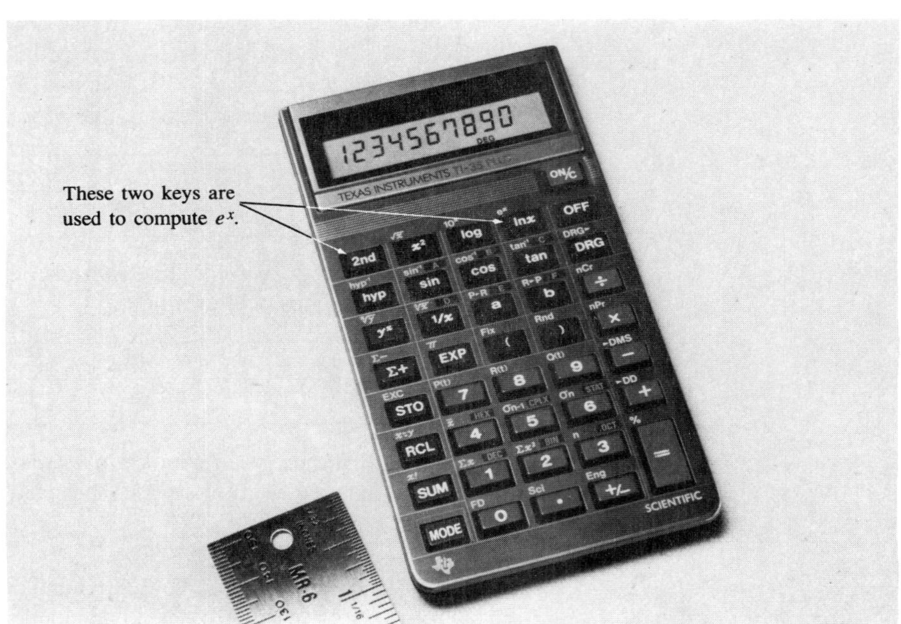
These two keys are used to compute e^x.

FIGURE 4

For example,

Keystrokes				Result	
[.7]	[2nd]	[ln x]	=	[2.013752707]	$= e^{0.7}$
[−2.3]	[2nd]	[ln x]	=	[0.1002588437]	$= e^{-2.3}$

*Other sequences of keystrokes might have to be used depending on the calculator. Consult the instructions accompanying your calculator.

Computing a^x with a Calculator

On a calculator with a y^x key, the following sequence of operations can be used to evaluate a^x:

$$\boxed{a}\ \boxed{y^x}\ \boxed{x}$$

For example,

Keystrokes			Result	
2	y^x	.7	1.624504793	$= 2^{0.7}$
9	y^x	−1.5	0.037037037	$= 9^{-1.5}$
1.8	y^x	2.6	4.610080545	$= (1.8)^{2.6}$

The actual number of digits displayed will depend on the calculator. If a calculator is not available, use the table of values of e^x and e^{-x} in Appendix Table C.5 (note that $e^{-x} = 1/e^x$).

EXAMPLE 5 Sketch the graph of $y = 4e^{0.5x}$.

Solution With the help of a calculator, a microcomputer, or Appendix Table C.5, you can construct Table 5, from which the graph in Figure 5 follows.

TABLE 5

x	$e^{0.5x}$	$y = 4e^{0.5x}$
−3	$e^{-1.5}$	$4e^{-1.5} \cong 0.89$
−2	e^{-1}	$4e^{-1} \cong 1.47$
−1	$e^{-0.5}$	$4e^{-0.5} \cong 2.43$
0	e^{0}	$4e^{0} \cong 4$
1	$e^{0.5}$	$4e^{0.5} \cong 6.59$
2	e^{1}	$4e^{1} \cong 10.87$
3	$e^{1.5}$	$4e^{1.5} \cong 17.93$

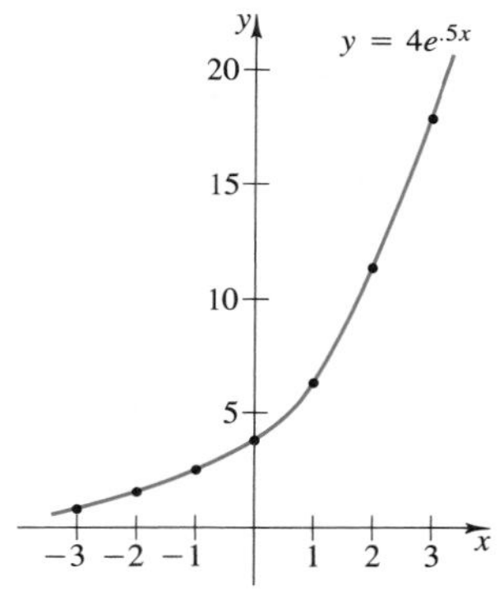

FIGURE 5

EXAMPLE 6 From Table C.5 in Appendix C, it should be evident that

$$\lim_{x\to\infty} e^x = \infty \qquad \lim_{x\to-\infty} e^x = 0 \tag{4}$$

$$\lim_{x\to\infty} e^{-x} = 0 \qquad \lim_{x\to-\infty} e^{-x} = \infty \tag{5}$$

These facts are also evident from the graphs of $y = e^x$ and $y = e^{-x}$, which are shown in Figure 6.

TALE OF THE SERVANT AND THE WHEAT

The tale is told that long ago in India, a servant was offered a chest of rubies by his master for many years of faithful service. The servant replied, "You are too generous, master, and I am not deserving of such a valuable reward." "What can I offer you then?" the master asked. "All I ask, dear master, is that tomorrow you place a single grain of wheat on the first square of my chessboard and on the following day twice that amount on the second square. Then, on each successive day, you bring me twice the grain of the previous day until I have received a supply of grain for each of the 64 squares on my board." Said the master, "Such a humble gift is hardly worth your many years of service, but I shall give you what you request." The wise men tell us that the gift made the humble servant the richest man in ancient India. Here is why: Suppose that

512 grains of wheat = 1 shaft

512 shafts of wheat = 1 bushel

512 bushels of wheat = 1 wagon load

512 wagon loads of wheat = 1 field

512 fields of wheat = 1 principality

512 principalities of wheat = all the wheat in India

On the first day the servant received one grain, on the second day 2 grains, on the third day 2^2 (= 4) grains, and by the tenth day 2^9 (= 512) grains, or 1 shaft. On the eleventh day he received 2 shafts, on the twelfth day 2^2 (= 4) shafts, and by the twentieth day 2^9 (= 512) shafts, or one bushel. Can you see how the servant eventually obtained all the wheat in India? (See Exercise 18.)

FIGURE 6

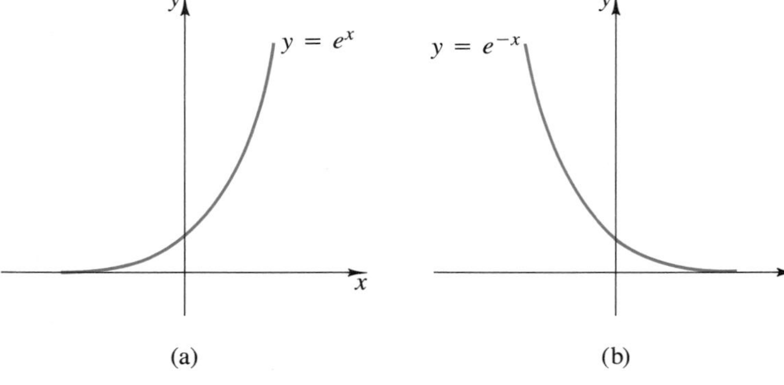

EXAMPLE 7 **MARKETING** Sales of new products often grow rapidly at first and then level off as the market becomes saturated. Suppose the daily sales receipts S (in dollars) of a new product vary with time according to the formula

$$S = 5000 - 4000e^{-t} \qquad (6)$$

where t is the time elapsed (in months) after the introduction of the product. What can be said about the sales receipts over an extended period of time?

Solution Since

$$\begin{aligned}\lim_{t\to\infty}(5000 - 4000e^{-t}) &= \lim_{t\to\infty} 5000 - \lim_{t\to\infty}(4000e^{-t}) \\ &= \lim_{t\to\infty} 5000 - (4000)\lim_{t\to\infty} e^{-t} \\ &= 5000 - (4000)\cdot 0 = 5000 \qquad \text{[from Equation (5)]}\end{aligned}$$

it follows that the daily sales receipts will level off toward a limiting value of \$5000 over a long period of time. The graph of the sale, Formula (6), is shown in Figure 7.

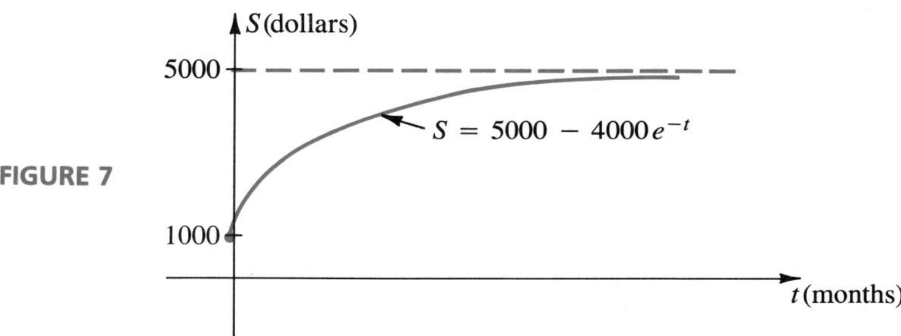

FIGURE 7

EXERCISE SET 14.1

1. (a) Use a calculator to complete the following table of values for $f(x) = (\frac{2}{7})^{0.4x}$. Round your values to two decimal places.

x	−4	−3	−2	−1	0	1	2	3	4
$(\frac{2}{7})^{0.4x}$									

(b) Use the values obtained in (a) to help sketch the graph of $y = (\frac{2}{7})^{0.4x}$.

2. (a) Use a calculator to complete the following table of values for $f(x) = (0.6)^{-0.3x}$. Round your values to two decimal places.

x	−4	−3	−2	−1	0	1	2	3	4
$(0.6)^{-0.3x}$									

(b) Use the values obtained in (a) to help sketch the graph of $y = (0.6)^{-0.3x}$.

3. (a) Use a calculator to complete the following table of values for $f(x) = 5e^x$. Round your values to two decimal places.

x	−4	−3	−2	−1	0	1	2	3	4
$5e^x$									

(b) Use the values obtained in (a) to help sketch the graph of $y = 5e^x$.

4. (a) Use a calculator to complete the following table of values for $f(x) = 2e^{-0.3x}$. Round your values to two decimal places.

x	−4	−3	−2	−1	0	1	2	3	4
$2e^{-0.3x}$									

(b) Use the values obtained in (a) to help sketch the graph of $y = 2e^{-0.3x}$.

In Exercises 5–16 use a calculator; sketch the graph of the given function, and specify the domain and range.

5. $f(x) = e^{2x}$

6. $f(x) = e^{-x}$

7. $f(x) = (\frac{1}{4})^x$

8. $f(x) = (\frac{1}{4})^{-x}$

9. $f(x) = 1 - 2^x$

10. $f(x) = 1 - 2^{-x}$

11. $f(x) = 3^x/2$

12. $f(x) = (3^{-x})/4$

13. $f(x) = 0.2e^{-x}$

14. $f(x) = 1 + e^{-x}$

15. $f(x) = 20e^{0.4x}$

16. $f(x) = 50e^{-0.2x}$

17. Graph the functions $f(x) = 6^x$ and $g(x) = 6^{-x}$ in the same coordinate system.

18. Referring to the Tale of the Servant and the Wheat, after how many days was the servant's reward:
(a) one wagon load
(b) one field
(c) one principality
(d) all the wheat in India

19. Consider a species of bacteria that reproduces every 20 minutes. Beginning with a culture of 1000 bacteria, write the formula $f(t)$ that expresses how many bacteria will be present t minutes later. How many will be present after:
(a) 1 hour (b) 2 hours
(c) 3 hours (d) 4 hours

20. Repeat Exercise 19 for a species of bacteria that reproduces very 40 minutes. (Round off answers to two decimal places.)

21. Suppose you deposit 1 cent in a savings account the first day, and every day thereafter you deposit double that amount. Find the amount deposited on day n as a function of n. What is the deposit on the 15th day? On the 21st day? On the 31st day? (Express your answer as a power of 2.)

22. Suppose you deposit 1 cent in a savings account the first day, and every 2 days thereafter you deposit triple that amount. Find the amount deposited on day n as a function of n. What is the deposit on the 15th day? On the 21st day? On the 31st day? (Express your answer as a power of 3.)

23. Use Appendix Table C.5 to help find

$$\lim_{x\to 0} \frac{e^x - 1}{x}$$

24. Use Appendix Table C.5 to help find

$$\lim_{x\to 0} \frac{1 - e^{-x}}{x}$$

25. Use Appendix Table C.5 to help find

$$\lim_{x\to 0}\left(\frac{5}{e^x} + 2\right)$$

26. Use Appendix Table C.5 to help find

$$\lim_{x\to 0} \frac{x}{e^x}$$

In Exercises 27–30 use Formula (5) to help find the limit, if it exists.

27. $\lim_{x\to\infty} (4 + 2e^{-x})$

28. $\lim_{x\to\infty}\left(\frac{1}{x} - 5e^{-x}\right)$

29. $\lim_{x\to\infty} \frac{7}{1 + 5e^{-x}}$

30. $\lim_{x\to\infty} (x + 2e^{-x})$

31. **(Marketing)** Suppose the daily sales S (in dollars) of a new product varies with time according to the formula

$$S = 1000 - 700e^{-t}$$

where t is the time elapsed (in months) after the product is introduced.
(a) What can be said about the sales receipts over an extended period of time?
(b) Use graph paper and Appendix Table C.5 to make an accurate graph of the sales formula.

32. **(Ecology)** Suppose the number of rainbow trout in a lake varies according to the formula

$$P = \frac{2400}{1 + 5e^{-t}}$$

where t is the time elapsed (in months after the lake is stocked).
(a) How many trout were in the lake initially?
(b) Over an extended period of time, what will happen to the trout population?

33. In a region beset by killer bees, control measures are established to try to keep the population of bees from growing without bound. It has been found that t months after the measures have been instituted, the number of bees is given by

$$P = \frac{5000}{5 + e^{-2t}}$$

(a) How many bees were in the region when the control measures were started?
(b) Over an extended period what will happen to the bee population? [*Hint:* $e^{-2t} = (e^{-t})^2$.]

14.2 LOGARITHMIC FUNCTIONS

In this section we study an important class of funtions called **logarithmic functions**.

TABLE 1

x	-5	-4	-3	-2	-1	-0.5	-0.3	0	0.3	0.5	1	2	3	4	5
2x	$\frac{1}{32}$	$\frac{1}{16}$	$\frac{1}{8}$	$\frac{1}{4}$	$\frac{1}{2}$	$\cong 0.71$	$\cong 0.81$	1	$\cong 1.23$	1.41	2	4	8	16	32

Consider the table of values of 2^x shown in Table 1 and the graph of $y = 2^x$ shown in Figure 1 of Section 14.1. Note that 2^x is always positive, and we have stated (without proof) that the graph of $y = 2^x$ is a continuous curve that is increasing. Therefore, if we pick any positive value of y, for example, $y = 12$, then we can see from the graph in Figure 1 that there is a value of x, for example, x_0, such that $y = 12 = 2^{x_0}$. In the same way, we can see graphically that for any positive real value of y, for example, y_0, there is corresponding value of x, x_0, such that

$$y_0 = 2^{x_0}$$

We define the exponent x_0 to be the **logarithm to the base 2 of y_0**, which is written symbolically as $\log_2 y_0$. Thus,

$$\log_2 8 = 3 \qquad \text{since } 2^3 = 8$$

$$\log_2 \frac{1}{32} = -5 \qquad \text{since } 2^{-5} = \frac{1}{32}$$

Although this discussion is limited to base 2, any other base a greater than 1 will give an analogous result. Thus,

$$\begin{aligned} x &= 3^y && \text{is equivalent to} && y = \log_3 x \\ x &= 10^y && \text{is equivalent to} && y = \log_{10} x \\ x &= e^y && \text{is equivalent to} && y = \log_e x \end{aligned}$$

In the following discussion $\log_a x$ implies that $a > 1$. We now define the logarithm function:

Logarithm of x to the Base a ($a > 1$)

$$\log_a x$$

is the exponent to which a must be raised to obtain x. In other words,

$$y = \log_a x \quad \text{and} \quad x = a^y$$

are equivalent statements.

Thus, $\log_{10} 1000 = 3$ since 10 must be raised to the third power to obtain 1000. Similarly,

$$\begin{aligned} \log_2 16 &= 4 && (\text{since } 2^4 = 16) \\ \log_{10} 10{,}000 &= 4 && (\text{since } 10^4 = 10{,}000) \\ \log_{10} \tfrac{1}{10} &= -1 && (\text{since } 10^{-1} = \tfrac{1}{10}) \\ \log_4 64 &= 3 && (\text{since } 4^3 = 64) \\ \log_{10} 1 &= 0 && (\text{since } 10^0 = 1) \\ \log_{16} 4 &= \tfrac{1}{2} && (\text{since } 16^{1/2} = 4) \end{aligned}$$

Of special importance are logarithms using base 10 (called **common logarithms**) and logarithms to the base e (called **natural logarithms**). These logarithms have their own special notation:

$$\log x \text{ denotes } \log_{10} x$$
$$\ln x \text{ denotes } \log_e x$$

EXAMPLE 1

Logarithm Statement	Equivalent Exponent Statement
$\log 100 = 2$	$10^2 = 100$
$\log \frac{1}{100} = -2$	$10^{-2} = \frac{1}{100}$
$\log 1 = 0$	$10^0 = 1$
$\ln 1 = 0$	$e^0 = 1$
$\ln \frac{1}{e^5} = -5$	$e^{-5} = \frac{1}{e^5}$
$\ln e = 1$	$e^1 = e$

Computing ln x and log x with a Calculator

On a calculator with log x and ln x keys (Figure 1), the following sequence of keystrokes can be used to evaluate log x and ln x:

[x] [log x]

[x] [ln x]

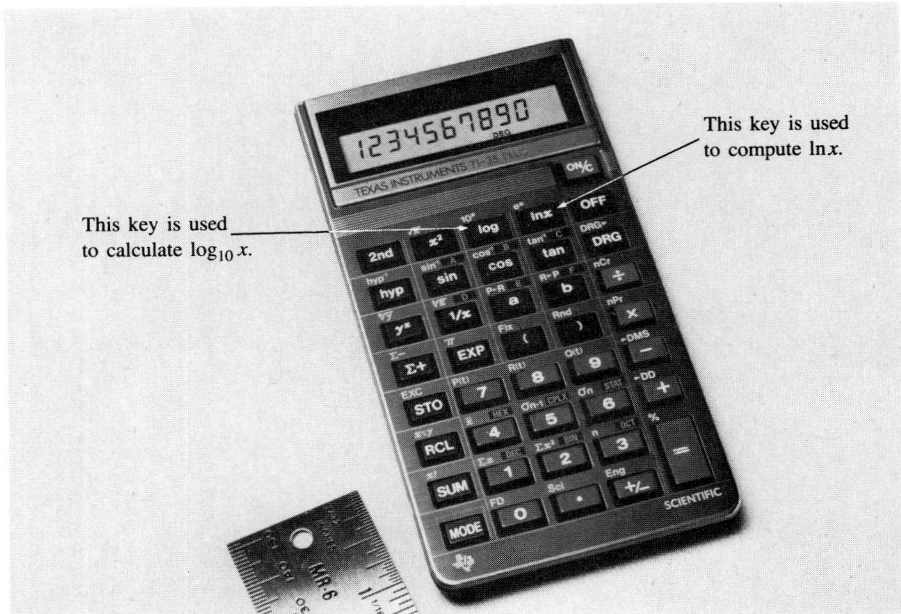

FIGURE 1

For example,

Keystrokes	Result
[2] [ln x]	[0.6931471806] = ln 2
[.3] [ln x]	[−1.203972804] = ln 0.3
[12] [log x]	[1.079181246] = log 12
[.02] [log x]	[−1.698970004] = log 0.02

The actual number of digits displayed will depend on the calculator.

For convenience we have included a brief table of natural logarithms in Appendix Table C.6.

EXAMPLE 2 Use a calculator or Appendix Table C.6 to graph $y = \ln x$.

Solution The table is constructed by rounding off the values in Appendix Table C.6 to one decimal place. From the table we obtain the graph in Figure 2.

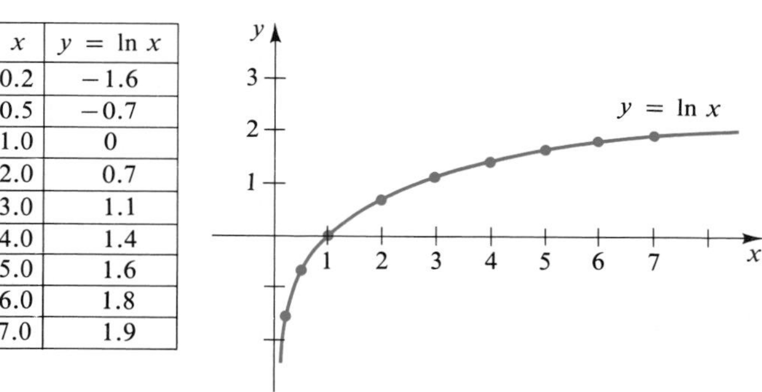

x	$y = \ln x$
0.2	−1.6
0.5	−0.7
1.0	0
2.0	0.7
3.0	1.1
4.0	1.4
5.0	1.6
6.0	1.8
7.0	1.9

FIGURE 2

The graph in Figure 2 is typical of all logarithmic functions; that is, for every base $a > 1$, the graph of $y = \log_a x$ has the general shape shown in Figure 3.

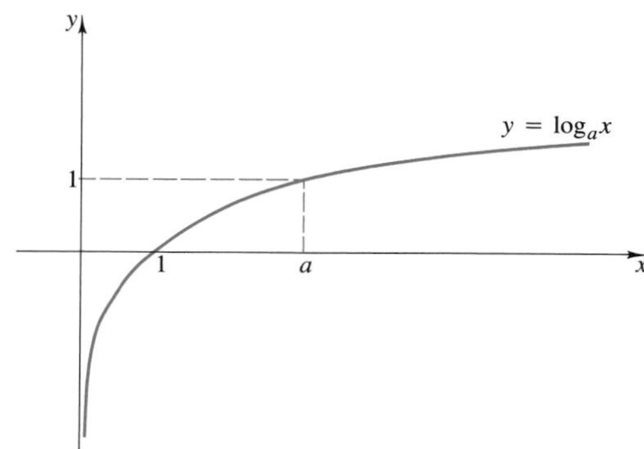

FIGURE 3

The basic facts about logarithmic functions are summarized here (recall that $a > 1$):

1. For every base a, the graph of $y = \log_a x$ crosses the x-axis at $x = 1$. This is because $a^0 = 1$, and consequently $\log_a 1 = 0$.
2. For every base a, the graph of $y = \log_a x$ has $y = 1$ at $x = a$. This is because $a^1 = a$, and consequently $\log_a a = 1$.
3. $\text{Log}_a x$ is positive if $x > 1$ and negative if $0 < x < 1$.
4. As x approaches zero from the right, the graph of $y = \log_a x$ approaches the negative y-axis but never touches it.
5. As x increases, y increases and in fact approaches infinity but does so very slowly.
6. $\text{Log}_a x$ is defined only for positive values of x.
7. For every base $a > 1$, the logarithm function $f(x) = \log_a x$ is continuous at every positive real number x. (This fact is shown in Section 14.4.)

EXAMPLE 3 Graph the equations $y = 2^x$ and $y = \log_2 x$ on the same coordinate system.

Solution Observe that the graphs in Figure 4 are mirror images of one another about the line $y = x$. To see why this is so, recall that $y = \log_2 x$ and $x = 2^y$ are equivalent equations and thus have the same graph. But the graphs of $x = 2^y$ and $y = 2^x$ are mirror images of one another about the line $y = x$ since each is obtained from the other by interchanging x and y (see Supplementary Exercise 26 in Chapter 1). Thus, $y = \log_2 x$ and $y = 2^x$ have graphs that are mirror images about $y = x$.

FIGURE 4

x	-3	-2	-1	0	1	2	3
2^x	$\frac{1}{8}$	$\frac{1}{4}$	$\frac{1}{2}$	1	2	4	8

x	$\frac{1}{8}$	$\frac{1}{4}$	$\frac{1}{2}$	1	2	4	8
$\log_2 x$	-3	-2	-1	0	1	2	3

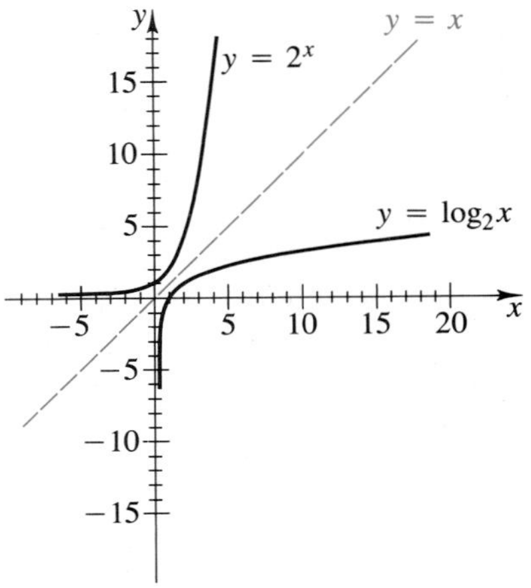

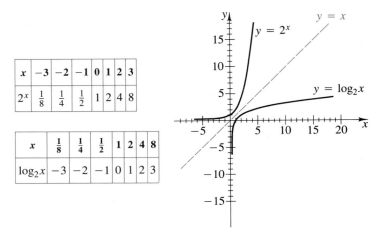

Geometric Relationship between Logarithmic and Exponential Functions In general, the graphs of

$$y = \log_a x \quad \text{and} \quad y = a^x$$

are reflections of one another about the line $y = x$ (Figure 5).

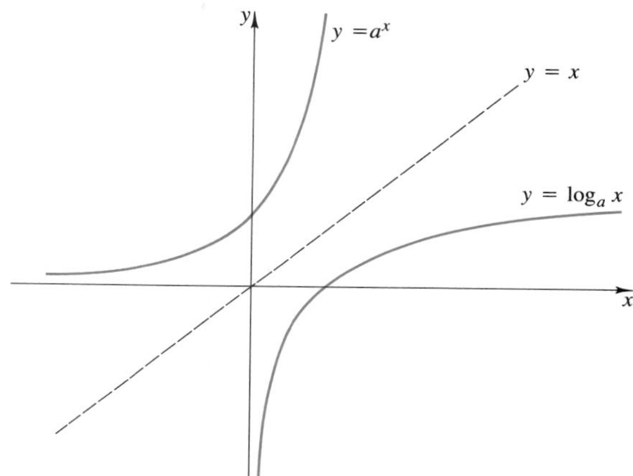

FIGURE 5

Historically, logarithms were developed to simplify numerical calculations. This was made possible by properties of logarithms that can be used to convert multiplication problems to addition problems, division problems to subtraction problems, and extraction of roots to multiplication problems. Although the advent of digital computers has greatly reduced the value of logarithms as a computational tool, the basic properties of logarithms that yesterday made them a valuable tool for numerical calculations today make them a useful tool for a variety of applications. Table 2 reviews some of these properties.

TABLE 2

Properties of Logarithms	Examples
1. $\log_a(yz) = \log_a y + \log_a z$	$\log(5 \cdot 6) = \log 5 + \log 6$
2. $\log_a \frac{y}{z} = \log_a y - \log_a z$	$\log \frac{5}{6} = \log 5 - \log 6$
3. $\log_a y^z = z \log_a y$	$\log 6^5 = 5 \log 6$
4. $\log_a a^y = y$	$\log 10^x = x$ $\ln e^x = x$
5. $\log_a a = 1$	$\log 10 = 1$
6. $\log_a 1 = 0$	$\log 1 = 0$
7. If $\log_a x = \log_a y$, then $x = y$	If $\ln x = \ln 0.01$, then $x = 0.01$
8. $x = e^{\ln x}$, $x > 0$	$4 = e^{\ln 4}$

These properties hold because logarithms are exponents. We prove Property 1 and leave the proofs of the other properties as an exercise.

Let

$$\log_a y = u \quad \text{and} \quad \log_a z = v$$

These statements can be written in exponential form as

$$a^u = y \quad \text{and} \quad a^v = z$$

Then

$$yz = a^u a^v = a^{u+v}$$

Writing this last equation in logarithmic form, we have

$$\log_a(yz) = u + v$$

from which it follows that

$$\log_a(yz) = \log_a y + \log_a z$$

EXAMPLE 4 Using $\ln 2 = 0.693$ and $\ln 3 = 1.099$ (where needed), determine:

(a) $\ln 6$ (b) $\ln \frac{2}{3}$ (c) $\ln \frac{3}{2}$

(d) $\ln \sqrt{6e}$ (e) $\ln e^2$ (f) $\log 10^{-5}$

Solution (a) $\ln 6 = \ln(2 \cdot 3) = \ln 2 + \ln 3 = 0.693 + 1.099 = 1.792$ (Property 1)

Solution (b) $\ln \frac{2}{3} = \ln 2 - \ln 3 = 0.693 - 1.099 = -0.406$ (Property 2)

Solution (c) $\ln \frac{3}{2} = \ln 3 - \ln 2 = 1.099 - 0.693 = 0.406$ (Property 2)

Solution (d) $\ln \sqrt{6e} = \ln (6e)^{1/2} = \frac{1}{2} \ln (6e)$ (Property 3)

$= \frac{1}{2}(\ln 6 + \ln e)$ (Property 1)

$= \frac{1}{2}(\ln 2 + \ln 3 + \ln e) = \frac{1}{2}(0.693 + 1.099 + 1)$

$= 1.396$

Solution (e) $\ln e^2 = 2 \ln e = 2$

Solution (f) $\log 10^{-5} = -5 \log 10 = -5$

EXAMPLE 5 Solve the equation $e^t = 3$ for t.

Solution

$e^t = 3$

$\ln e^t = \ln 3$ (take ln of both sides since the base of e^t is e)

$t = \ln 3$ (Property 4)

$= 1.0986$ (Appendix Table C.6 or a calculator)

EXAMPLE 6 Solve the equation $90 = 100e^{-0.5t}$ for t.

Solution

$$
\begin{aligned}
90 &= 100e^{-0.5t} \\
0.9 &= e^{-0.5t} && \text{(divide by 100)} \\
\ln 0.9 &= \ln e^{-0.5t} && \text{(take ln of both sides)} \\
&= -0.5t && \text{(Property 4)} \\
-0.1054 &= -0.5t && \text{(Appendix Table C.6 or a calculator)} \\
t &= \frac{0.1054}{0.5} = 0.2108
\end{aligned}
$$

The following example uses Property 7.

EXAMPLE 7 Solve the equation for x:

$$\log x = 3 \log 2 - \tfrac{1}{2} \log 9 + \log 5$$

Solution

$$
\begin{aligned}
\log x &= 3 \log 2 - \frac{1}{2} \log 9 + \log 5 \\
&= \log 2^3 - \log 9^{1/2} + \log 5 && \text{(Table 2, Property 3)} \\
&= \log 8 - \log 3 + \log 5 && \text{(properties of exponents)} \\
&= \log \frac{8}{3} + \log 5 && \text{(Table 2, Property 2)} \\
&= \log\left(\frac{8}{3} \cdot 5\right) = \log \frac{40}{3} && \text{(Table 2, Property 1)} \\
x &= \frac{40}{3} && \text{(Table 2, Property 7)}
\end{aligned}
$$

EXAMPLE 8 **MEASUREMENT OF EARTHQUAKE INTENSITY** Magnitudes of earthquakes are measured using the Richter scale. On this scale the magnitude R of an earthquake is given by

$$R = \log \frac{I}{I_0}$$

where I_0 is a standard intensity used for comparison and I is the intensity of the earthquake being measured. Thus, a magnitude of $R = 3$ on the Richter scale means

$$3 = \log \frac{I}{I_0}$$

$$\log 1000 = \log \frac{I}{I_0}$$

THE LOMA PRIETA EARTHQUAKE OF 1989

On October 17, 1989, at 5:04 P.M. a major earthquake with a reading of 7.1 on the Richter scale struck the San Francisco area. Its epicenter was located 75 miles south of San Francisco between Santa Cruz and San Jose. About 63 people died, and damage was estimated to be approximately 6 billion dollars. The earthquake forced the cancellation of the third game of baseball's World Series which was due to start at Candlestick Park in San Francisco at 5:30 P.M.

This earthquake was the fourth worst in U.S. history, as far as loss of life is concerned, since the 1906 San Francisco earthquake (700 dead, 8.3 on the Richter scale), the 1933 Long Beach earthquake (120 dead, 6.3 on the Richter scale), and the 1964 Alaskan earthquake (114 dead, 8.5 on the Richter scale).

$$\frac{I}{I_0} = 1000$$
$$I = 1000I_0$$

which states that the intensity of the earthquake measured is $10^3 = 1000$ times as great as the standard. Similarly, an earthquake registering 8 on the Richter scale has an intensity $10^8 = 100{,}000{,}000$ times that of the standard earthquake. Based on past experience, the following damages can be expected:

Reading	Damage
2.0	Not noticed
4.5	Some damage in a very limited area
6.0	Hazardous serious damage with destruction of buildings in a limited area
7.0	Felt over a wide area with significant damage
8.0	Great damage
8.7	Maximum recorded

EXERCISE SET 14.2

In Exercises 1 and 2 evaluate the given expression without using tables or a calculator.

1. (a) $\log 1000$ (b) $\log \frac{1}{1000}$
   (c) $\log 0.00001$ (d) $\log 100^{20}$
   (e) $\log \sqrt{1000}$ (f) $\log_3 81$

2. (a) $\log 10{,}000$ (b) $\log \frac{1}{10{,}000}$
   (c) $\log_{1/2} 4$ (d) $\ln e^3$
   (e) $\ln 1/e^{2/3}$ (f) $\log 10{,}000^{1/4}$

In Exercises 3–6 use a calculator to evaluate the given quantities. Round your answers to four decimal places.

3. (a) $\ln 2$ (b) $\ln 1.7$
   (c) $\ln \sqrt{7}$ (d) $\ln \frac{9}{16} - \ln 3$
   (e) $\ln 12.82^3$ (f) $\ln(\sqrt{3} + e)$

4. (a) $\ln 8$ (b) $\ln(1/3.2)$
   (c) $\ln(\sqrt{5} - 2)$ (d) $\ln \frac{2}{3} + \ln \frac{9}{5}$
   (e) $\ln 153$ (f) $\ln \pi$

5. (a) $\log 6$ (b) $\log 2.9$
(c) $\log \sqrt{17}$ (d) $\log(\frac{1}{5} + \frac{5}{3})$
(e) $\log(1/0.23)$ (f) $\log(1 + \sqrt{5})$

6. (a) $\log 1.3^4$ (b) $\log 1/19$
(c) $\log(4\sqrt{6})$ (d) $\log 1.5^3 - \log 3.6^3$
(e) $\log(1/0.017)$ (f) $\log(5\sqrt{7} + 9)$

In Exercises 7 and 8 write each expression in exponential form.

7. (a) $\log 100 = 2$ (b) $\log \frac{1}{100} = -2$
(c) $\ln(1/e) = -1$ (d) $\ln 1 = 0$

8. (a) $\log \sqrt{10} = \frac{1}{2}$ (b) $\log 0.0001 = -4$
(c) $\ln(1/\sqrt{e}) = -\frac{1}{2}$ (d) $\ln e^{2/3} = \frac{2}{3}$

In Exercises 9 and 10 write each expression in logarithmic form.

9. (a) $10{,}000 = 10^4$ (b) $\frac{1}{1000} = 10^{-3}$
(c) $e^{-2} = 1/e^2$ (d) $e^1 = e$

10. (a) $0.0001 = 10^{-4}$ (b) $e^0 = 1$
(c) $e^{-3/2} = 1/e^{3/2}$ (d) $2^{-4} = \frac{1}{16}$

In Exercises 11 and 12 write the given expression without using logarithms.

11. (a) $\ln e^7$ (b) $\ln \sqrt[3]{e}$
(c) $\ln e^{-0.6t}$ (d) $\ln(e^{0.8}/e^{0.2})$

12. (a) $3^{\log_3 6}$ (b) $\ln \sqrt[5]{e^2}$
(c) $\ln e^{-0.3t^2}$ (d) $\ln(e^{0.3}e^{0.4}/e^{0.2})$

13. Given that $\log 2 = 0.3010$ and $\log 3 = 0.4771$, find:
(a) $\log 6$ (b) $\log \frac{2}{3}$ (c) $\log 9$
(d) $\log \sqrt{3}$ (e) $\log \frac{1}{2000}$ (f) $\log 54$

14. Given that $\ln 2 = 0.6931$ and $\ln 5 = 1.6094$, find:
(a) $\ln 10$ (b) $\ln \frac{5}{2}$ (c) $\ln 50$
(d) $\ln \sqrt[3]{100}$ (e) $\ln \frac{1}{250}$ (f) $\ln \sqrt{1/500}$

In Exercises 15–20 sketch the graph of the given function.

15. $f(x) = \log_3 x$ **16.** $f(x) = \ln x$

17. $f(x) = \log_2(4x)$ **18.** $f(x) = \log_3(x/9)$

19. $f(x) = \ln(2x)$ **20.** $f(x) = \ln(1 + x)$

In Exercises 21–42 solve for x.

21. $\log_3 x = 2$ **22.** $\log_2(4x) = 3$

23. $4^x = 5$ **24.** $3^{-x} = 4$

25. $\ln x = 3$ **26.** $\ln(3x) = 4$

27. $\ln(2x - 1) = 1$

28. $\ln(x + 1) - \ln(x - 1) = \ln 2$

29. $\ln x - \ln 5 = 1$ **30.** $e^x = 3$

31. $e^{2x} = 4$ **32.** $e^{x+1} = 2$

33. $\ln x^2 = 6$ **34.** $\ln x^3 = -6$

35. $e^{x-1} = 4$ **36.** $x + 2 = \ln 3$

37. $\log x = 3 \log 3 + 2 \log 5 - 3 \log 4$

38. $\log_3 3x + \log_3(3 - x) = \log_3 4$

39. $\ln 2x - \ln 3 = \ln 15$

40. $\log x - \frac{1}{2} \log 4 = 2 \log 5$

41. $\log_2(x + 2) + \log_2(x - 2) = \log_2 4$

42. $\log_2(x + 2) - \log_2(x - 2) = \log_2 4$

43. In each part obtain a numerical value for t (to four decimal places). Use a calculator where necessary.
(a) $e^t = 3$ (b) $500 = 200e^t$ (c) $80 = 100e^{2t}$

44. Follow the directions of Exercise 43.
(a) $105 - 35e^{-0.2t} = 35$
(b) $e^{-0.04t} = 0.03$
(c) $3 = 2 \ln t$

45. Establish:
(a) Property 2 of Table 2
(b) Property 3 of Table 2
(c) Property 4 of Table 2
(d) Property 5 of Table 2

46. **(Advertising models)** Natural logarithms arise in certain models for advertising response. Suppose a model relating sales to advertising is described by the equation

$$N = 3000 + 200 \ln (x + 1)$$

where N = number of units sold
x = amount spent for advertising (in thousands of dollars)

(a) How many units would be sold with no advertising expenditure?
(b) Find the number of units sold with an expenditure of \$39,000 (round off to the nearest integer).
(c) According to the model, how much advertising money must be spent to sell 5000 units?

47. **(Measurement of sound)** The loudness of a sound is measured by using the decibel scale (named after Alexander Graham Bell, the inventor of the telephone). On this scale, the loudness L of a sound of intensity I is given (in decibels) by

$$L = 10 \log \frac{I}{I_0}$$

where I_0 is the intensity of a sound that is considered barely audible by the human ear.

(a) Human speech has an approximate intensity of $10^6 I_0$. Find its loudness in decibels.

(b) Suppose a passage on a record has an intensity of $10{,}000 I_0$. Find its loudness in decibels.

48. **(Measurement of acidity)** The pH of a fluid measures its acidity and is defined by

$$\text{pH} = -\log\,[\text{H}^+]$$

where the symbol $[\text{H}^+]$, ranging from 0 to 14, measures the concentration of hydrogen ions in the fluid. A neutral solution such as distilled water has a pH of 7. A solution with a pH of less than 7 is said to be acidic and one with a pH greater than 7 is said to be basic.

(a) How much greater is the $[\text{H}^+]$ concentration in vinegar, which has a pH of 3, than in human saliva, which has a pH of 6?

(b) Cola has a pH of approximately 2.0, whereas tomato juice has a pH of approximately 4.0. Which is more acidic? By how many times?

(c) Household ammonia has a pH of approximately 11.9, whereas ordinary milk has a pH of approximately 6.9. Which is more acidic? By how many times?

14.3 *e* AND e^x AS LIMITS AND APPLICATIONS TO COMPOUND INTEREST

In Chapter 9 we encounter the number e and see that the function e^{rt} occurs naturally in the calculation of interest compounded continuously. In Section 14.1 we discuss the graph of this function. So far, however, we have not given a proper definition of e. We have given a decimal approximation to e but have not shown where it comes from. Now that we have some knowledge of limits, we can do better. First we define e properly.

Definition of *e*

$$e = \lim_{u \to \infty} \left(1 + \frac{1}{u}\right)^u \tag{1}$$

TABLE 1

u	1	2	5	10	100	1000	10,000	100,000
$1 + 1/u$	2	1.5	1.2	1.1	1.01	1.001	1.0001	1.00001
$(1 + 1/u)^u$	2	2.25	2.48832	2.59374	2.70481	2.71692	2.71815	2.71827

Consider Table 1, found by using a calculator to five decimal places. This is the basis for claiming that $e \cong 2.71828182845\ldots$.

It can also be shown that for every real number x,

$$e^x = \lim_{u \to \infty} \left(1 + \frac{x}{u}\right)^u \tag{2}$$

EXAMPLE 1 Find $\lim_{u\to\infty} (1 - 2/u)^u$.

Solution

$$\lim_{u\to\infty}\left(1 - \frac{2}{u}\right)^u = \lim_{u\to\infty}\left(1 + \frac{(-2)}{u}\right)^u = e^{-2} \qquad \text{[by Formula (2)]}$$

$$= \frac{1}{e^2}$$

Application to Continuous Compounding of Interest

In Section 9.2 we see that if a principal of P dollars is invested at an annual interest rate j, compounded m times a year, then after t years (n conversion periods, where $n = mt$) the investment will have grown to a "compound amount"

$$S = P\left(1 + \frac{j}{m}\right)^{mt} \tag{3}$$

The formula for continuous (or instantaneous) compound interest can be obtained from Equation (3) by letting $m \to \infty$. That is, we may think of interest as compounding infinitely many times per year. When we do that, we see that after t years at nominal rate j, a principal P invested at continuous compound interest will have grown to the amount

$$S = \lim_{m\to\infty} P\left(1 + \frac{j}{m}\right)^{mt} \tag{4}$$

To evaluate this limit, note first that $j/m = jt/mt$ and second that $mt \to \infty$. Therefore Equation (4) is equivalent to

$$S = P \lim_{mt\to\infty}\left(1 + \frac{jt}{mt}\right)^{mt} \tag{5}$$

Now let $u = mt$. Then

$$S = P\lim_{u\to\infty}\left(1 + \frac{jt}{u}\right)^u$$

$$= Pe^{jt} \qquad \text{[by Formula (2)]}$$

Therefore,

$$S = Pe^{jt} \tag{6}$$

This is a mathematical *proof* of Formula (7) of Section 9.2.

EXAMPLE 2 **DOUBLING TIME** The length of time it takes for an investment to double is called its **doubling time**. Find the doubling time for an amount invested at 8% interest per year, assuming the interest is

(a) simple interest

(b) compounded semiannually

(c) compounded continuously

Solution (a) Let P be the principal and $j = .08$. For simple interest, the amount after t years is

$$S = P + Pjt$$

We want to find the value of t that yields $S = 2P$. Thus we solve

$$2P = P + P(.08)t$$

for t.

$$P = P(.08t)$$
$$1 = .08t$$
$$\frac{1}{.08} = t$$

Thus, $t = 12.5$ years.

Solution (b) If interest is compounded semiannually, then the amount after t years is $S = P(1 + j/2)^{2t}$ [Equation (3) with $m = 2$]. Again, we want to find the value of t that makes $S = 2P$. Thus, setting $j = .08$, we have

$$2P = P\left(1 + \frac{.08}{2}\right)^{2t}$$
$$2 = (1.04)^{2t}$$

To solve for t, we take the natural logarithm of both sides,

$$\ln 2 = 2t \ln 1.04$$
$$\frac{\ln 2}{2 \ln 1.04} = t$$

With the aid of a calculator, we find that $t = 8.84$ years (that is, 8 years, 10 months).

Solution (c) After t years at continuous 8% compound interest, the compound amount is $S = Pe^{.08t}$ [by Equation (6)]. Again, we solve for the value of t that makes $S = 2P$:

$$2P = Pe^{.08t}$$
$$2 = e^{.08t}$$

Now, again, to solve for t we must take the natural logarithm of both sides,

$$\ln 2 = .08t$$
$$\frac{\ln 2}{.08} = t$$

Using a calculator, we find $t = 8.66$ years (that is, 8 years, 8 months).

EXERCISE SET 14.3

In Exercises 1–4 use a calculator.

1. Complete the following table, to five decimal places, to verify $\lim_{u\to\infty}(1 + 1/u)^u = e$.

u	2	20	200	2000	20,000
$1 + 1/u$					
$(1 + 1/u)^u$					

2. (a) Complete the following table, to five decimal places, to determine $\lim_{v\to 0^+}(1 + v)^{1/v}$.

v	1	0.1	0.01	0.001	0.0001
$1 + v$					
$(1 + v)^{1/v}$					

(b) How are the answers to Exercises 1 and 2a related? Can you give a reason why this relation holds?

3. Complete the following table, to five decimal places, to determine e^2.

u	5	50	500	5000	50,000
$1 + 2/u$					
$(1 + 2/u)^u$					

4. Complete the following table, to five decimal places, to determine e^{-1}.

u	1	0.1	0.01	0.001	0.0001
$1 - u$					
$(1 - u)^{1/u}$					

In Exercises 5–8 find the indicated limit; use Formula (2).

5. $\lim_{u\to\infty}(1 + 3/u)^u$ **6.** $\lim_{u\to\infty}(1 - 1/u)^u$

7. $\lim_{u\to\infty}(1 + 1/u)^{-u}$ **8.** $\lim_{u\to\infty}[1 + 1/(2u)]^u$

In Exercises 9–12 let $v = 1/u$ (or $u = 1/v$) and evaluate the indicated limits.

9. $\lim_{u\to 0}(1 + 2u)^{1/u}$ **10.** $\lim_{u\to 0}(1 + u)^{-1/u}$

11. $\lim_{u\to 0}(1 - u)^{1/u}$ **12.** $\lim_{u\to 0}(1 + u)^{1/(2u)}$

13. Suppose \$20,000 is invested at 10% compounded continuously. What is the value of the investment after 5 years?

14. How much money should be invested at 8% interest compounded continuously to yield \$10,000 in 5 years?

15. How long will it take an investment at 8% interest compounded quarterly to:
(a) double in value (b) triple in value

16. How long will it take an investment at 7% interest compounded continuously to:
(a) double in value (b) triple in value

17. Find the doubling time for an investment at 6% interest if the interest is:
(a) simple
(b) compounded semiannually
(c) compounded continuously

18. (a) Show that the continuous compound interest formula, (6), can be rewritten as

$$jt = \ln\frac{S}{P}$$

(b) Suppose \$10,000 invested at continuous compound interest grows to \$11,000 in 1 year. What is the annual interest rate j?

(c) Suppose \$10,000 is invested at 12% compounded continuously. How long will it take for the investment to reach a value of \$15,000?

19. Assume that \$10,000 is invested at 8% per year. How long will it take for the value to reach \$15,000 if the investment is:
(a) simple
(b) compounded quarterly
(c) compounded continuously

14.4 DERIVATIVES OF EXPONENTIAL AND LOGARITHMIC FUNCTIONS

Our primary objective in this section is to obtain the derivatives of the natural logarithm function ln x and the natural exponential function e^x. As we see in this section, and in later sections, these functions and their derivatives have numerous important applications.

Derivative of the Exponential Function

To obtain the derivative of $f(x) = e^x$ we use the definition of the derivative. Thus,

$$
\begin{aligned}
f'(x) &= \lim_{h\to 0} \frac{f(x+h) - f(x)}{h} && \text{(derivative definition)} \\
&= \lim_{h\to 0} \frac{e^{x+h} - e^x}{h} && [f(x+h) = e^{x+h}] \\
&= \lim_{h\to 0} \frac{e^x e^h - e^x}{h} && \text{(property of exponents)} \\
&= \lim_{h\to 0} e^x \cdot \frac{e^h - 1}{h} && \text{(factoring)}
\end{aligned}
$$

Because the factor e^x does not involve h, it remains *constant* as $h \to 0$. Therefore, since we are allowed to move a multiplicative constant through a limit sign, we can write

$$f'(x) = e^x \lim_{h\to 0} \frac{e^h - 1}{h} \tag{1}$$

Although we omit the formal proof, it is not hard to see (Table 1) with the aid of a calculator that

$$\lim_{h\to 0} \frac{e^h - 1}{h} = 1$$

TABLE 1

	Approaching 0 from the left					
h	-1	-0.5	-0.25	-0.1	-0.01	-0.001
$\frac{e^h - 1}{h}$	0.632	0.787	0.885	0.952	0.995	0.9995
	Approaching 0 from the right					
h	1	0.5	0.25	0.1	0.01	0.001
$\frac{e^h - 1}{h}$	1.718	1.297	1.136	1.052	1.005	1.0005

(See Exercise 23 of Section 14.1.) Thus Equation (1) becomes

$$f'(x) = e^x \cdot 1 = e^x$$

The exponential function e^x is differentiable everywhere, and

$$\frac{d}{dx}[e^x] = e^x \tag{2}$$

In other words, the derivative of the natural exponential function is the exponential function itself. As with other functions we may apply the chain rule to (2) to obtain a more general rule.

If u is a differentiable function of x, then

$$\frac{d}{dx}[e^u] = e^u \frac{du}{dx} \tag{3}$$

That is, $\frac{d}{dx}[e^{u(x)}] = u'(x)e^{u(x)}$.

Since a differentiable function is continuous (see Section 12.6), we conclude that the exponential function $f(x) = e^x$ is continuous everywhere.

EXAMPLE 1 Find the derivative of each of the following functions, using Equation (3).

(a) $f(x) = e^{8x}$ (b) $f(x) = 4e^{3x} + x^e$ (c) $f(x) = e^{x^3}$ (d) $f(x) = 4x^5e^x$

Solution (a) Using Equation (3) we have

$$\frac{d}{dx}[e^{8x}] = e^{8x}\frac{d}{dx}(8x) = 8e^{8x}$$

Solution (b)

$$\frac{d}{dx}[4e^{3x} + x^e] = 4\frac{d}{dx}[e^{3x}] + \frac{d}{dx}[x^e]$$

and using (3) we have

$$\begin{aligned}\frac{d}{dx}[4e^{3x} + x^e] &= 4e^{3x}\frac{d}{dx}[3x] + ex^{e-1}\\ &= 12e^{3x} + ex^{e-1}\end{aligned}$$

Note that x^e is differentiated as a power, whereas e^x is differentiated as an exponential.

Solution (c) Using (3) we obtain

$$\frac{d}{dx}[e^{x^3}] = e^{x^3}\frac{d}{dx}(x^3) = 3x^2e^{x^3}$$

Solution (d) By the product rule,

$$\begin{aligned}\frac{d}{dx}[4x^5e^x] &= 4x^5\frac{d}{dx}[e^x] + e^x\frac{d}{dx}[4x^5]\\ &= 4x^5e^x + 20x^4e^x\end{aligned}$$

EXAMPLE 2 Show that $f(x) = e^x$ has no relative maximum or minimum and that e^x is an increasing function. Also show that the graph of e^x is concave up for all x.

Solution To determine relative extrema, set $f'(x) = 0$ and solve for the critical points. Observe that $f'(x) = e^x > 0$ for all x. Therefore e^x has no relative maximum or minimum values, and since its derivative is positive for all x, it is an increasing function. Since $f''(x) = e^x$ is positive for all x, e^x is concave up for all x.

EXAMPLE 3 What is the instantaneous rate of change of

$$h(x) = \frac{x}{1 + e^x}$$

with respect to x when $x = 0$?

Solution By the quotient rule,

$$\begin{aligned} h'(x) &= \frac{(1 + e^x) \cdot \frac{d}{dx}(x) - x \cdot \frac{d}{dx}(1 + e^x)}{(1 + e^x)^2} \\ &= \frac{(1 + e^x) \cdot 1 - x \cdot e^x}{(1 + e^x)^2} \\ &= \frac{1 + e^x - xe^x}{(1 + e^x)^2} \end{aligned}$$

Thus the instantaneous rate of change when $x = 0$ is

$$h'(0) = \frac{1 + e^0 - 0e^0}{(1 + e^0)^2} = \frac{2}{(2)^2} = \frac{1}{2}$$

EXAMPLE 4 **HYDROCARBON EMISSIONS** The rate at which hydrocarbons are emitted by an automobile engine depends on the speed of the automobile. Suppose the emission rate $R(x)$ in milligrams per minute is related to the speed x in miles per hour by

$$R(x) = xe^{-x/30}$$

At what speed is the emission rate $R(x)$ a maximum?

Solution Since x represents speed, we must have $x \geq 0$; so the problem is to maximize $R(x)$ over the interval $[0, +\infty)$. We first find the critical values of $R(x)$. Using the product rule for derivatives, we obtain

$$R'(x) = x\left(-\frac{1}{30}e^{-x/30}\right) + e^{-x/30} = e^{-x/30}\left(1 - \frac{x}{30}\right)$$

We want to solve $R'(x) = 0$. Since $e^{-x/30} > 0$, we want to solve

$$1 - \frac{x}{30} = 0$$

Solving gives $x = 30$ as the only critical value. To determine the nature of this value, we apply the first derivative test as follows. Observe that $e^{-x/30}$ is always positive. If $x < 30$, then $\frac{x}{30} < 1$ so $1 - \frac{x}{30} > 0$. If $x > 30$, then $\frac{x}{30} > 1$ so $1 - \frac{x}{30} < 0$. Thus, $R'(x)$ is positive for $x < 30$ and negative for $x > 30$. By the first derivative test, $R(x)$ has a relative maximum and an absolute maximum at

$x = 30$. Therefore the maximum rate of hydrocarbon emission occurs when the automobile is driven at 30 miles per hour.

The Standard Normal Probability Curve

One of the most important curves in the study of probability and statistics is the standard normal probability curve (see Chapter 8) given by the equation

$$y = \frac{1}{\sqrt{2\pi}} e^{-(1/2)x^2} \tag{4}$$

In the exercises we ask you to show that this curve has the shape illustrated in Figure 1 (see Exercise 51).

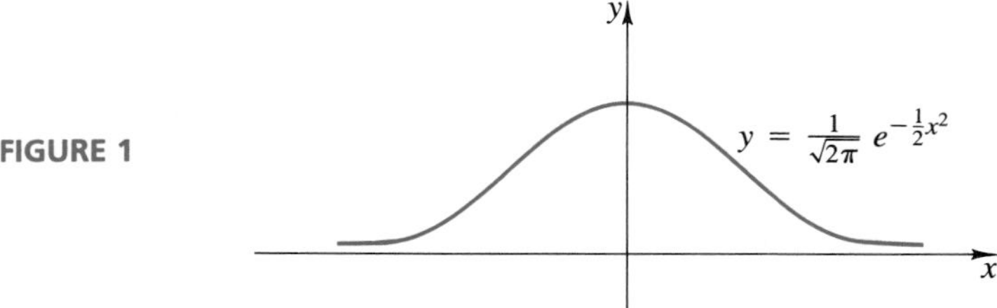

FIGURE 1

The standard normal probability curve has several other common names: the **bell curve** because of its shape and the **Gaussian curve** in honor of its discoverer, Karl Friedrich Gauss.*

The standard normal probability curve is a member of a larger family of curves, the normal probability curves. The curves in this family have equations of the form

$$y = \frac{1}{\sqrt{2\pi}\,\sigma} e^{-\frac{1}{2}\left(\frac{x-\mu}{\sigma}\right)^2} \tag{5}$$

where μ and σ are constants and σ must be positive (see Section 8.6). The standard normal curve (4) is the special case of form (5) where $\mu = 0$ and $\sigma = 1$ (verify this). All normal probability curves have a bell shape; however, as illustrated in Figure 2, the peak of a normal probability curve occurs at the point $x = \mu$, and the curve becomes more compressed as σ increases; the inflection points occur at $x = \mu \pm \sigma$. (See Exercise 51 for the special case when $\mu = 0$ and $\sigma = 1$.)

**Karl Friedrich Gauss* (1777–1855) was a German mathematician and scientist. Sometimes called the "prince of mathematicians," Gauss is regarded by many as the greatest mathematician who ever lived. Gauss was a child prodigy, and stories abound about his childhood feats of mathematics. It is reported that before he was 3 years old he corrected a long tedious calculation in his father's payroll, doing the computations in his head! It is also said that in his first arithmetic class he added the integers from 1 to 100 in a few seconds while his classmates toiled for the entire class period. At age 19, he solved a problem that baffled Euclid—inscribing a regular polygon with 17 sides in a circle, using only a straightedge and compass. Gauss's genius was brought to the attention of Karl Wilhelm Ferdinand, Duke of Brunswick, who then subsidized his education. Gauss's contributions to mathematics were diverse and brilliant—he discovered the Gaussian curve, founded the modern study of number theory, and made major discoveries in geometry. In addition to his work in mathematics he did important work in physics and astronomy.

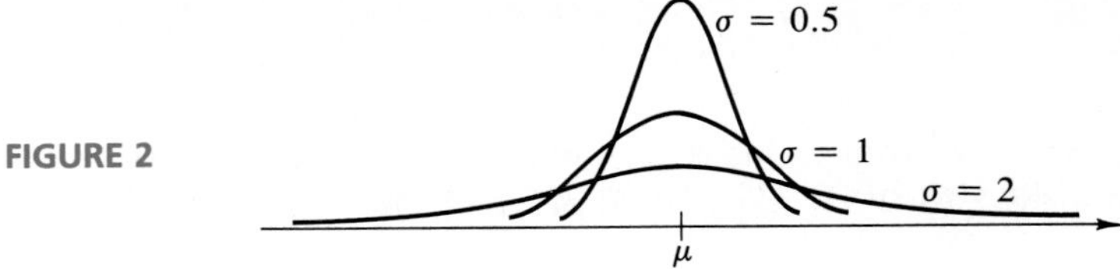

FIGURE 2

Normal probability curves arise in many applications, but of special interest is the fact that when an examination is given to a large number of students (e.g., the SAT examinations), the graph that shows the number of students obtaining each possible score is usually approximately a normal probability curve with μ equal to the average score on the examination.

Normal probability curves are discussed in some detail in Section 8.6 (see also Section 17.7).

Derivative of a^x

Finally, we consider arbitrary exponential functions

$$f(x) = a^x$$

where a is an arbitrary positive constant. To differentiate this function we first express it in terms of the natural exponential function. Since, by Property 8 of Table 2 in Section 14.2, $a = e^{\ln a}$, it follows that

$$a^x = (e^{\ln a})^x$$

By applying Property 5 of Table 3 in Section 14.1 to the right-hand side,

$$a^x = e^{x \ln a} \tag{6}$$

Therefore,

$$\begin{aligned} \frac{d}{dx}[a^x] &= \frac{d}{dx}[e^{x \ln a}] \\ &= e^{x \ln a} \frac{d}{dx}[x \ln a] \qquad \text{[by Formula (3)]} \\ &= e^{x \ln a} \ln a \\ &= (e^{\ln a})^x \ln a \\ &= a^x \ln a \end{aligned}$$

We have therefore established the following rule:

If a is an arbitrary positive constant, then

$$\frac{d}{dx}[a^x] = a^x \ln a \tag{7}$$

The chain rule gives us the more general rule:

If a is an arbitrary positive constant and u is a differentiable function of x, then

$$\frac{d}{dx}[a^{u(x)}] = a^{u(x)}u'(x) \ln a \tag{8}$$

Since a^x is differentiable, it is continuous. Also note that for $a > 1$, $\frac{d}{dx}(a^x) > 0$; therefore, for $a > 1$, a^x is an increasing function. For $a < 1$, $\frac{d}{dx}(a^x) < 0$ and a^x is a decreasing function. (See Example 2 of this section and Example 1 of Section 14.1.)

EXAMPLE 5 Find the derivative of each of the following functions:

(a) $f(x) = 27^x$ (b) $f(x) = x^2 2^x$ (c) $f(x) = 5^{x^2+1}$

Solution (a) By Rule (7),

$$\frac{d}{dx}[27^x] = 27^x \ln 27$$

Solution (b) By the product rule,

$$\begin{aligned}\frac{d}{dx}[x^2 2^x] &= x^2 \frac{d}{dx}[2^x] + 2^x \frac{d}{dx}[x^2] \\ &= x^2 2^x \ln 2 + 2^x\, 2x \\ &= x 2^x(x \ln 2 + 2)\end{aligned}$$

Solution (c) By Rule (8),

$$\frac{d}{dx}[5^{x^2+1}] = 5^{x^2+1} 2x \ln 5$$

Derivative of the Logarithmic Function

It can be shown that the natural logarithm function

$$g(x) = \ln x \qquad x > 0 \tag{9}$$

is differentiable. To compute its derivative we proceed as follows. As discussed in Section 14.2, we can rewrite (9) as

$$x = e^{g(x)} \tag{10}$$

so that

$$\begin{aligned}\frac{d}{dx}[x] &= \frac{d}{dx}[e^{g(x)}] \\ 1 &= e^{g(x)} \cdot g'(x) && \text{[from Equation (3)]} \\ 1 &= x \cdot g'(x) && \text{[from Equation (10)]} \\ g'(x) &= \frac{1}{x}\end{aligned}$$

In other words:

> The natural logarithm function $\ln x$ is differentiable for all $x > 0$, and
>
> $$\frac{d}{dx}[\ln x] = \frac{1}{x} \tag{11}$$

When we combine Formula (11) with the chain rule, we obtain the more general rule:

> If $u(x)$ is a differentiable function of x, then whenever $u(x) > 0$,
>
> $$\frac{d}{dx}[\ln u(x)] = \frac{u'(x)}{u(x)} \tag{12}$$
>
> or, more simply,
>
> $$\frac{d}{dx}[\ln u] = \frac{1}{u}\frac{du}{dx} \tag{13}$$

EXAMPLE 6 Find the derivative of each of the following functions:

(a) $f(x) = \ln(3x), \quad x > 0$

(b) $f(x) = \ln(5x^2 + 7x - 8), \quad 5x^2 + 7x - 8 > 0$

(c) $f(x) = \ln \sqrt{4x - 3}, \quad x > \frac{3}{4}$ (d) $f(x) = \ln \frac{x^2 + 1}{5x + 3}, \quad x \geq 0$

In the solutions that follow you should observe how important it is to know and use the fundamental properties of logarithms shown in Table 2 of Section 14.2. We urge you to review that table now; we refer to these properties of logarithms by number.

Solution (a) Let $u = 3x$. Then,

$$\begin{aligned}\frac{d}{dx}[\ln(3x)] &= \frac{d}{dx}[\ln u]\\ &= \frac{1}{u}\frac{du}{dx}\\ &= \frac{1}{u} \cdot 3\\ &= \frac{3}{3x}\\ &= \frac{1}{x}\end{aligned}$$

As an alternative solution, observe that

$$\ln(3x) = \ln 3 + \ln x \qquad \text{(Property 1 of logarithms)}$$

Thus,

$$\frac{d}{dx}\ln(3x) = \frac{d}{dx}[\ln 3] + \frac{d}{dx}[\ln x]$$

$$= 0 + \frac{1}{x} \qquad \text{(ln 3 is constant)}$$

$$= \frac{1}{x}$$

Solution (b) Let $u = 5x^2 + 7x - 8$. Then,

$$\frac{d}{dx}[\ln(5x^2 + 7x - 8)] = \frac{d}{dx}[\ln u] = \frac{1}{u}\frac{du}{dx}$$

$$= \frac{10x + 7}{5x^2 + 7x - 8}$$

Solution (c)

$$\frac{d}{dx}[\ln\sqrt{4x - 3}] = \frac{d}{dx}[\ln(4x - 3)^{1/2}]$$

$$= \frac{d}{dx}\left[\frac{1}{2}\ln(4x - 3)\right] \qquad \text{(Property 3 of logarithms)}$$

$$= \frac{1}{2}\frac{d}{dx}[\ln u] \qquad (\text{let } u = 4x - 3)$$

$$= \frac{1}{2u}\frac{du}{dx}$$

$$= \frac{1}{2}\cdot\frac{4}{4x - 3} = \frac{2}{4x - 3}$$

Solution (d) Let

$$u = \frac{x^2 + 1}{5x + 3}$$

Then,

$$\ln u = \ln(x^2 + 1) - \ln(5x + 3) \qquad \text{(Property 2 of logarithms)}$$

Thus,

$$\frac{d}{dx}\ln u = \frac{d}{dx}\ln(x^2 + 1) - \frac{d}{dx}\ln(5x + 3)$$

$$= \frac{\frac{d}{dx}(x^2 + 1)}{x^2 + 1} - \frac{\frac{d}{dx}(5x + 3)}{5x + 3}$$

$$= \frac{2x}{x^2 + 1} - \frac{5}{5x + 3}$$

EXAMPLE 7 A certain bacteria colony consisting initially of 25,000 bacteria is treated with a drug, and after t hours the number of bacteria present in the colony is

$$N(t) = 25{,}000 + 5000\left[\frac{t}{100} - \ln(1 + 2t)\right]$$

(a) What is the instantaneous rate of change in this number after 1 hour? After 2 hours? After 3 hours? Interpret the results.
(b) After how many hours is $N(t)$ a minimum?

Solution (a) The instantaneous rate of change in the number of bacteria present is the derivative

$$N'(t) = 5000\left[\frac{1}{100} - \frac{2}{1 + 2t}\right]$$

For $t = 1$ the instantaneous rate of change is

$$\begin{aligned} N'(1) &= 5000\left(\frac{1}{100} - \frac{2}{3}\right) \\ &= 5000\left(\frac{-197}{300}\right) \\ &\cong -3283 \end{aligned}$$

Since the derivative is negative, the number of bacteria present is *decreasing* at the rate of 3283 bacteria per hour, after 1 hour.

For $t = 2$, the instantaneous rate of change is

$$\begin{aligned} N'(2) &= 5000\left(\frac{1}{100} - \frac{2}{5}\right) = 5000\left(\frac{-39}{100}\right) \\ &= -1950 \end{aligned}$$

Thus after 2 hours, the number of bacteria present is decreasing at the rate of 1950 bacteria per hour.

For $t = 3$,

$$\begin{aligned} N'(3) &= 5000\left(\frac{1}{100} - \frac{2}{7}\right) = 5000\left(\frac{-193}{700}\right) \\ &\cong -1379 \end{aligned}$$

Thus after 3 hours, the number of bacteria present is decreasing at the rate of 1379 bacteria per hour.

Solution (b) To find the critical points of $N(t)$ we set $N'(t)$ equal to zero:

$$N'(t) = 5000\left(\frac{1}{100} - \frac{2}{1 + 2t}\right) = 0$$

We find that the only critical point is $t = 199/2 = 99.5$.

Observe that as time progresses, the rate at which the colony is decreasing in number is itself decreasing. The number of bacteria present continues to decline until time $t = 99.5$ hours, but by ever-diminishing amounts. After 99.5 hours, the number of bacteria in the colony increases! (The effect of the drug has worn off.) Thus, $N(t)$ is a minimum after 99.5 hours.

EXAMPLE 8 In Table C.7 of the Appendix we find values of $(1 + i)^n$ for some values of i and integral values of n. This table is used in computing the compound value of $1 invested at interest rate i per period, after n compounding periods. Suppose we invest $1000 at 8% per year compounded quarterly and want to know how fast the compound amount

$$S = 1000(1 + .02)^{4t}$$

is growing after t years. To be specific, let us compute this instantaneous rate of increase after 5 years. Now,

$$\frac{dS}{dt} = 1000 \frac{d}{dt}[(1.02)^{4t}]$$

We want to find dS/dt when $t = 5$. Using Formula (8) we get

$$\begin{aligned}\frac{dS}{dt} &= 1000(1.02)^{4t}(4) \ln 1.02 \\ &= 4000 \ln 1.02\ (1.02)^{4t} \\ &= (79.2105)(1.02)^{4t} \qquad (\ln 1.02 = 0.0198)\end{aligned}$$

When $t = 5$, this becomes

$$\begin{aligned}\frac{dS}{dt} &= (79.2105)(1.02)^{20} \\ &= (79.2105)(1.4859) \\ &= 117.7026\end{aligned}$$

Thus $dS/dt = \$117.70$, so that at the fifth anniversary of the investment it is growing instantaneously at the rate of $117.70 per year.

EXERCISE SET 14.4

In Exercises 1–36 find dy/dx.

1. $y = e^{4x}$

2. $y = e^{-x}$

3. $y = 3e^{5x^2}$

4. $y = e^{\sqrt{x}}$

5. $y = 2x - e^{2x} + 3e^{4x}$

6. $y = 4e^{x^3}$

7. $y = x^3e^x$

8. $y = e^x/x$

9. $y = (e^x)^4$

10. $y = 1/(1 + e^x)$

11. $y = \frac{1}{5}xe^{x^2}$

12. $y = x^4e^{-x}$

13. $y = (e^x + x)^6$

14. $y = \sqrt{e^x + 1}$

15. $y = \ln(4x)$

16. $y = \ln x^3$

17. $y = \ln(7x^3 - 4)$

18. $y = \ln(3x^2 - 4)^{11}$

19. $y = \ln(1/x)$

20. $y = \ln[(x - 1)/(x + 1)]$

21. $y = \ln \sqrt[3]{4x + 5}$

22. $y = \ln \sqrt{2x^3 + 1}$

23. $y = x^2 \ln x$

24. $y = (x^2 + x) \ln(x + 1)$

25. $y = (\ln x)/x^3$

26. $y = 1/\ln x$

27. $y = \ln e^x$

28. $y = \ln(xe^x)$

29. $y = \ln e^{x^2}$

30. $y = \ln e^{-x}$

31. $y = e^{\ln x^2}$

32. $y = e^{\ln x}$

33. $y = x^6 2^x$

34. $y = 3^x/x^2$

35. $y = 3^{x^2}$

36. $y = 2^{\sqrt{x}}$

37. Find the equation of the tangent line to $y = e^x$ at $(0, 1)$.

38. Find the equation of the tangent line to $y = \ln x$ at the point where $x = 1$.

39. Let $y = e^x \ln(x + 1)$. Find the instantaneous rate of change of y with respect to x when $x = 0$.

40. Let $w = t \ln t$. Find the instantaneous rate of change of w with respect to t when $t = e$.

In Exercises 41–44 determine the intervals over which the given function is increasing, and those over which it is decreasing, and sketch the graph of $f(x)$.

41. $f(x) = e^x$ **42.** $f(x) = e^{-x}$

43. $f(x) = \ln x$ **44.** $f(x) = x - \ln x$

In Exercises 45–48 find the maximum and minimum values of f on the stated interval. If a maximum or minimum does not exist, state that in your answer.

45. $f(x) = e^{-x}$ on $[0, 2]$ **46.** $f(x) = \ln x$ on $[1, 2]$

47. $f(x) = xe^{-x}$ on $(-\infty, +\infty)$

48. $f(x) = \dfrac{\ln x}{x}$ on $(0, +\infty)$

In Exercises 49 and 50 locate and classify the critical values as relative maxima, relative minima, or neither, and sketch the curve.

49. $f(x) = e^{x^2}$ **50.** $f(x) = x \ln x$

51. (a) Use a calculator to help graph the standard normal probability curve (see Figure 1):

$$y = \frac{1}{\sqrt{2\pi}} e^{-(1/2)x^2}$$

(b) Find the maximum, minimum, and inflection points, and sketch the curve.

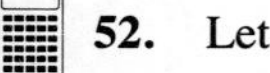

52. Let

$$y = \frac{1}{\sqrt{2\pi}\sigma} e^{-\frac{1}{2}\left(\frac{x-\mu}{\sigma}\right)^2}$$

(a) Find y if $\sigma = 1$, $\mu = 2$, and $x = 4$.
(b) Find y if $\sigma = 2$, $\mu = 1$, and $x = -3$.
(c) Show that the maximum value of y occurs when $x = \mu$.

53. Assume that an object is moving along the x-axis with known position (in inches) given by

$$s(t) = e^{t^2}$$

after t seconds. Find the position, velocity, and acceleration at time $t_0 = -1$.

54. A certain bacteria colony consisting initially of 40,000 bacteria is treated with a drug, and after t hours the number of bacteria present in the colony is

$$N(t) = 40{,}000 + 4000\left[\frac{t}{50} - \ln(1 + 3t)\right]$$

What is the instantaneous rate of change in this number after 1 hour? After 2 hours? After 3 hours? Interpret the results.

55. **(Ecology)** A lake is stocked with steelhead trout. If the lake can support a maximum of 5500 such trout, then according to a standard model (the *inhibited growth model*) the number $P(t)$ of trout in the lake after t months will have a formula of the form

$$P(t) = \frac{5500}{1 + 10e^{-kt}}$$

where k is a constant.
(a) Estimate the constant k if the trout population after 6 months is 800 (use Appendix Table C.5).
(b) Find dP/dt.
(c) How rapidly is the trout population growing after 10 months?

56. **(Learning)** It is determined by experimentation that in a certain assembly line operation, a typical experienced worker can produce at most 100 units per day. If a new worker is placed on the assembly line, then according to a standard model (the *Hullian* or *learning model*) the worker's daily output $P(t)$ after t days of training will have a formula of the form

$$P(t) = 100(1 - e^{-kt})$$

where k is a constant that varies with the individual.
(a) Estimate the constant k for an individual who produces 50 units per day after 1 day of training.
(b) Find dP/dt.
(c) Find the rate at which $P(t)$ is increasing with time after 5 days of training.

57. **(Experimental psychology)** Suppose a physical stimulus of measurable strength S produces a measurable sensation (response) R. According to the psychophysical Weber–Fechner law, R and S are related by an equation of the form

$$R(S) = a \ln S + b$$

that is, the response R is a linear function of the logarithm of the stimulus S.
(a) Let S_0 be the threshold level for S—the lowest level at which the stimulus may be perceived. Mathematically, $R(S_0) = 0$. Show that $b = -a \ln S_0$ and con-

sequently, $R = a \ln(S/S_0)$. Compare this result with the equation for earthquake intensity given in Example 8 of Section 14.2, and explain the similarity.

(b) Show that $dR/dS = a/S$. This form of the Weber–Fechner law states that the rate of change in sensation is inversely proportional to the strength of the stimulus. The rate of perception (response) decreases as the intensity of the stimulus increases. This law has been successfully applied in studying human response to physical pressure, light intensity, sound, smell, taste, and other stimuli. In medicine it applies in the study of the relationship between drug dosage and response.

(c) Suppose a given stimulus-response relationship satisfies the relation $dR/dS = 3.8/S$ and that the threshold value of S is 11.6. Working backward from (b) to (a), find a function $R(S)$ that satisfies this relation.

58. A certain bacteria colony consisting initially of 4000 bacteria grows in such a way that after t hours the number of bacteria present is $N(t) = 4000 \cdot 2^{4t}$. Find the instantaneous rate of change of $N(t)$ after 10 hours.

14.5 EXPONENTIAL GROWTH AND DECAY MODELS

Many physical quantities increase or decrease with time in proportion to the amount of the quantity present. Some typical examples are human population, certain kinds of investment interest, radioactivity, alcohol level in the blood after drinking, and bacteria in a culture. In this section we use the derivative to study the growth of such quantities.

Exponential Growth

A physical quantity is said to have an **exponential growth model** if at each instant of time its rate of growth is proportional to the amount of the quantity present.

Let us consider a positive quantity with an exponential growth model. At each instant of time we let $Q = Q(t)$ denote the amount of the quantity present at time t, where t represents the *time elapsed* from some initial observation. For example, if time is measured in seconds, $t = 1$ means 1 second after the initial observation, $t = 7.6$ means 7.6 seconds after the initial observation, and $t = 0$ is the value of t at the initial observation.

Because the quantity $Q(t)$ has an exponential growth model, its rate of growth dQ/dt is proportional to the amount Q present at each instant. Thus dQ/dt and Q are related by the equation

$$\frac{dQ}{dt} = kQ \tag{1}$$

where k is a constant of proportionality. Equation (1) is an example of a **differential equation**, that is, an equation involving the derivative of an unknown function. Differential equations are examined more fully in Chapter 18.

It is easy to see that one solution of the differential equation (1) is

$$Q(t) = Q_0 e^{kt} \tag{2}$$

where Q_0 is an arbitrary constant. For on differentiating both sides of (2), we obtain

$$\begin{aligned}\frac{dQ(t)}{dt} &= \frac{d}{dt}(Q_o e^{kt}) \\ &= Q_0 \frac{d}{dt}(e^{kt}) \qquad (Q_0 \text{ is constant}) \\ &= Q_o e^{kt}\frac{d}{dt}(kt) \\ &= Q_0 e^{kt} \cdot k \\ &= k(Q_0 e^{kt}) \\ &= kQ(t) \qquad \text{[from Formula (2)]}\end{aligned}$$

That is,

$$\frac{dQ}{dt} = kQ$$

which means that the function (2) is a solution of the differential equation. We prove in Chapter 18 that Equation (2) is the only solution to differential equation (1).

The constant Q_0 in the solution (2) has a natural interpretation when the differential equation (1) is used to model an exponential growth quantity. The value of Q at the initial time $t = 0$ is, from (2),

$$Q(0) = Q_0 e^{k \cdot 0} = Q_0 e^0 = Q_0 \cdot 1 = Q_0 \tag{3}$$

Thus the constant Q_0 in Formula (2) is *the amount of the quantity present initially*.

The constant k in Formula (1) is called the **growth constant** since it determines how rapidly $Q(t)$ grows with time. At each instant the quantity $Q(t)$ grows at a rate given by

$$\text{rate of growth} = kQ \tag{4}$$

where Q is the amount present at time t. Thus, the larger the value of k, the more rapid the increase in $Q(t)$ with time.

As an illustration, Figure 1 compares the graphs of two quantities with exponential models that have the same initial value but different growth constants.

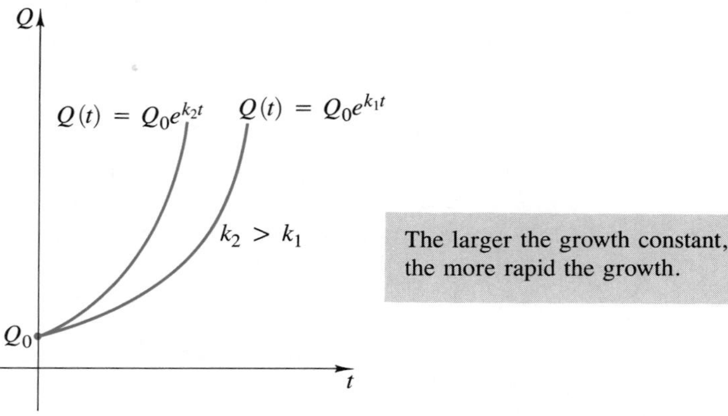

FIGURE 1

EXAMPLE 1 Suppose that a quantity has an exponential growth model described by the formula

$$Q(t) = 60e^{0.05t} \tag{5}$$

where t is measured in hours. For this model, the amount (in units) present initially ($t = 0$) is

$$Q(0) = 60$$

and the growth constant is $k = 0.05$. After 1 hour ($t = 1$) the amount of the quantity present to two decimal places is, from (5),

$$\begin{aligned} Q(1) &= 60e^{0.05} \\ &= 60(1.0513) \cong 63.08 \text{ units} \qquad \text{(Appendix Table C.5 or calculator)} \end{aligned}$$

and after 5 hours ($t = 5$) the amount present is

$$\begin{aligned} Q(5) &= 60e^{0.25} \\ &= 60(1.2840) \cong 77.04 \text{ units} \qquad \text{(Appendix Table C.5 or calculator)} \end{aligned}$$

The graph of the function $Q(t)$ is shown in Figure 2.

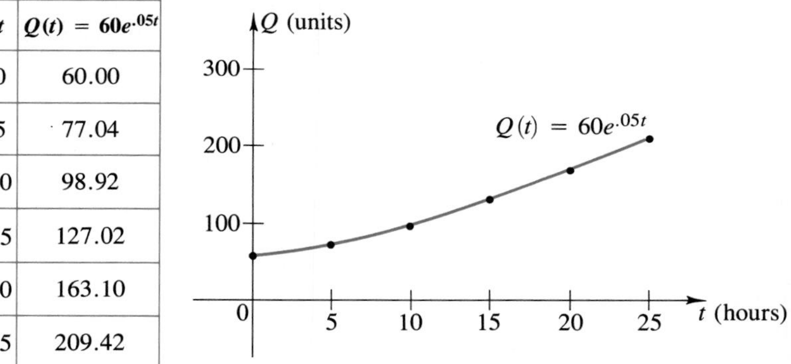

t	$Q(t) = 60e^{.05t}$
0	60.00
5	77.04
10	98.92
15	127.02
20	163.10
25	209.42

FIGURE 2

Exponential Decay

A physical quantity is said to have an **exponential decay model** if at each instant of time its rate of decrease (or decay) is proportional to the amount of the quantity present.

If Q has an exponential decay model, then

$$\frac{dQ}{dt} = -kQ \tag{6}$$

where the minus sign indicates that Q **decreases** at a rate proportional to Q. This differential equation is satisfied by the function (verify)

$$Q(t) = Q_0e^{-kt} \tag{7}$$

where Q_0 is the amount present initially and t is the time elapsed. We call k the *decay* constant. According to Equation (6), at each instant the quantity Q **decays** at the rate

$$\text{rate of decay} = kQ \tag{8}$$

The larger the value of k, the more rapid the decrease in $Q(t)$ with time. By way of illustration, Figure 3 compares the graph of two quantities with exponential decay models that have the same initial value but different decay rates.

FIGURE 3

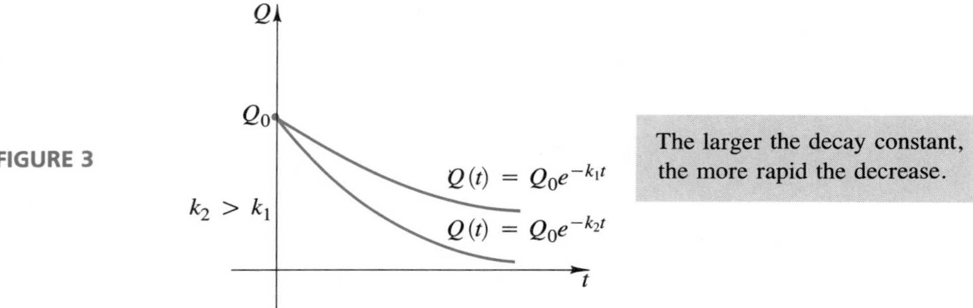

EXAMPLE 2 Suppose that a quantity has an exponential decay model described by the formula

$$Q(t) = 100e^{-0.1t}$$

where t is measured in minutes. For this model the amount present initially $(t = 0)$ is

$$Q(0) = 100 \text{ units}$$

and the decay constant is $k = 0.1$. The graph of $Q(t)$ is shown in Figure 4.

FIGURE 4

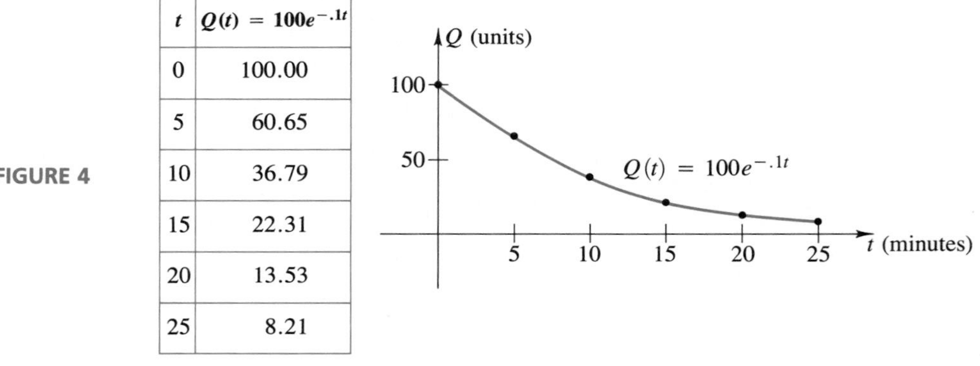

t	$Q(t) = 100e^{-.1t}$
0	100.00
5	60.65
10	36.79
15	22.31
20	13.53
25	8.21

Remark Note that the minus sign in Equation (6) is not part of the decay constant. The decay constant is the positive number k.

Remark In exponential growth or decay models, the constant k is often expressed as a percentage and is called the **growth (decay) rate**. A growth or decay rate of 2% would mean $k = 0.02$, and a growth or decay rate of 300% would mean $k = 3$. (In calculus the growth rate of a function has a slightly different meaning from its use here. However, the terms *growth rate* and *decay rate* are so common in exponential modeling problems that we do not hesitate to use them.)

We now turn to some practical examples of growth and decay models.

Global Population Growth

In 1798 the controversial English social philosopher Thomas Mathus* published a book entitled *An Essay on the Principle of Population as It Affects the Future Improvement of Society*. In this work Malthus proposed that human population has an exponential growth model and that eventually population must exceed the food supply. History has shown the Malthusian model of population growth to be inaccurate for technologically developed countries. However, many modern demographers (population scientists) feel that exponential growth models are suitable for lesser developed countries and for the global population as a whole, at least over a period of 30 or 40 years.

EXAMPLE 3 According to data published by the United Nations, the world population at the beginning of 1975 was approximately 4 billion. Assuming an exponential growth model and a growth rate of approximately 2% per year, estimate the world population for the year 2000.

Solution Let us measure time t in years and population $Q(t)$ in billions. If we take the beginning of 1975 as the initial observation, then the initial value of $Q(t)$ is

$$Q_0 = 4 \qquad \text{(billion)}$$

Since the growth rate is 2% per year, we have $k = 0.02$, so the equation

$$Q(t) = Q_0 e^{kt}$$

becomes

$$Q(t) = 4e^{0.02t}$$

By the year 2000 the time t elapsed from the initial observation in 1975 will be $t = 25$ (years). Thus from the computation above, the population by the year 2000 will be

$$Q(25) = 4e^{0.02(25)} = 4e^{0.5} = 4(1.6487) \cong 6.59$$

which is approximately 6.6 billion. In Figure 5 we show the graph of global population based on this model; Figure 6 shows a bar chart of world population by continent (adapted with permission from *Brittanica Atlas* © 1982, Encyclopaedia Brittanica, Inc.).

---

* *Thomas Robert Malthus* (1766–1834), English social philosopher, was born to a prosperous middle-class family. His theory that population always tends to outrun food supply led him to advocate limits on human reproduction. Malthus spent much of his life as a professor of history and economics for the East India Company's Haileybury College.

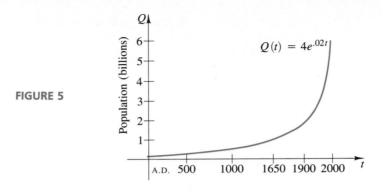

FIGURE 5

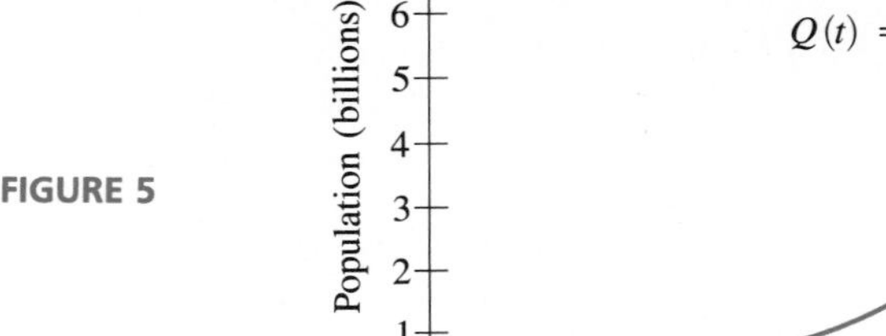

FIGURE 6

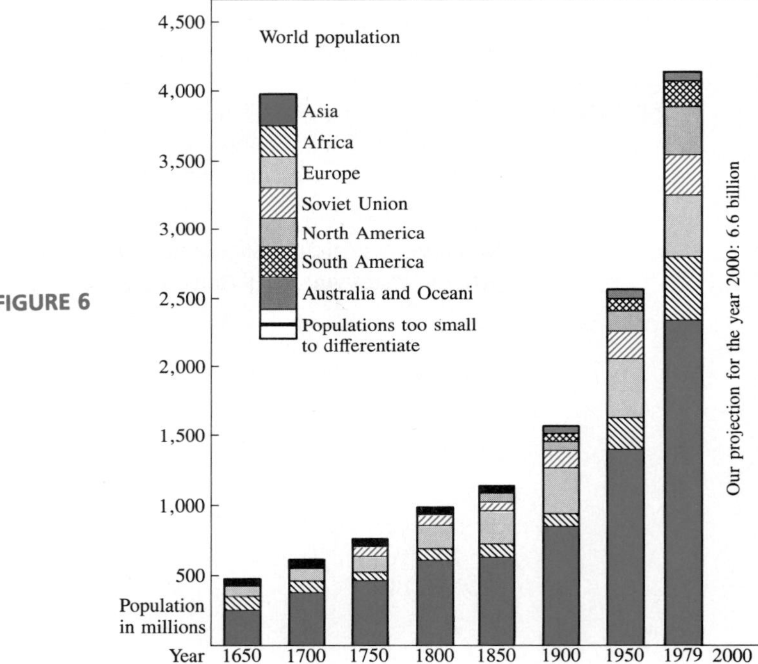

EXAMPLE 4 The following statement appears in the *Encyclopedia Americana**: "Average annual growth of world population is about 2% per year which means doubling of the world's population every 35 years." Show that this statement is correct, assuming an exponential growth model.

Solution Let Q_0 be the world population at any point in time. Since $k = 0.02$, the population at any later time will be

$$Q(t) = Q_0e^{kt} = Q_0e^{0.02t} \tag{9}$$

*Americana Corporation, *Encyclopedia Americana,* vol. 22 (New York: Grolier, 1981).

The population will have doubled when $Q(t) = 2Q_0$. Thus from (9) the time t required for doubling will satisfy

$$2Q_0 = Q_0e^{0.02t}$$

Dividing through by Q_0 yields $2 = e^{0.02t}$ and taking the natural logarithm of both sides gives

$$\ln 2 = \ln e^{0.02t} = 0.02t \ln e = 0.02t$$

Thus the doubling time is

$$t = \frac{1}{0.02} \ln 2$$

or from Appendix Table C.6

$$t = \frac{1}{0.02}(0.6931) = 34.655$$

or approximately 35 years.

It is interesting to note that the doubling time of 35 years does not depend on the quantity Q_0 present initially. At any point in time the population will double in the subsequent 35 years (assuming a 2% growth rate). Thus, with a continued 2% growth rate, the global population of 4 billion in 1975 will double to 8 billion by the year 2010 and will double again to 16 billion by 2045. Therefore, between 2010 and 2045 the Earth will add to its *existing* population 8 billion new people—twice the present global population!

It was not accidental in the previous example that the time period required for doubling was independent of the initial population. This is a property of all exponential growth models. To see this, let Q be any quantity with an exponential growth model. From any initial value Q_0 the time required to double will satisfy

$$2Q_0 = Q_0e^{kt}$$

or on dividing by Q_0, $2 = e^{kt}$ so that

$$\begin{aligned} \ln 2 &= \ln e^{kt} \\ &= kt \ln e = kt \\ t &= \frac{\ln 2}{k} \end{aligned}$$

which is independent of Q_0. We call the quantity

$$T = \frac{\ln 2}{k} \tag{10}$$

the **doubling time** for the growth model. In the case of an exponential decay model the time required for Q to reduce by half is given by the same formula (Exercise 16). For such models we will call T the **halving time** or sometimes the **half-life**. Doubling and halving times are illustrated graphically in Figure 7.

FIGURE 7

Q, $8Q_0$, $4Q_0$, $2Q_0$, Q_0, T, $2T$, $3T$, t

Exponential growth model with doubling time T

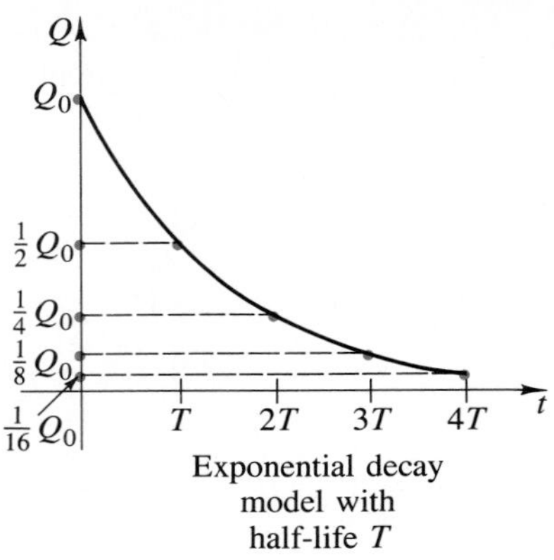

Exponential decay model with half-life T

Radioactive Decay Radioactive elements continually undergo a process of disintegration called **radioactive decay**. It is a physical fact that the rate of decay is proportional to the amount of the element present. As a consequence, the quantity $Q(t)$ of any radioactive substance that has an exponential decay model is given by

$$Q(t) = Q_0 e^{-kt} \tag{11}$$

where Q_0 is the amount present initially, t is the time elapsed, and k is the decay rate.

EXAMPLE 5 Potassium 42 has a decay rate of approximately 5.5% per hour.

(a) If 1000 grams of potassium 42 is present initially, how much will be left after 2 hours?

(b) What is the half-life of potassium 42?

Solution (a) From the given data $Q_0 = 1000$ and $k = 0.055$, so from (11) the quantity left after $t = 2$ hours is

$$Q(2) = 1000e^{-(0.055)2} = 1000e^{-0.11} = 1000(0.8958) = 895.8 \text{ grams}$$

Solution (b) From Formula (10) the half-life T is

$$T = \frac{\ln 2}{k} = \frac{0.6931}{0.055} \cong 12.6 \text{ hours}$$

EXAMPLE 6 Carbon 14 is a radioactive carbon isotope with a half-life of 5750 years. What is the decay rate?

Solution Formula (10) can be written as

$$k = \frac{\ln 2}{T}$$

Substituting the half-life $T = 5750$ and the value $\ln 2 = 0.6931$ yields the decay rate

$$k = \frac{0.6931}{5750} = 0.0001205$$

or approximately $k = 0.012\%$ per year.

Carbon Dating of Fossils

When the nitrogen in the Earth's upper atmosphere is bombarded by cosmic radiation, the radioactive element carbon 14 is produced. This carbon 14 combines with oxygen to form carbon dioxide, which is ingested by plants, which in turn are eaten by animals. In this way all living plants and animals absorb quantities of radioactive carbon 14. In 1947 the American nuclear scientist W. F. Libby proposed the theory that the percentage of carbon 14 in the atmosphere and in living tissues of plants is the same.* When a plant or animal dies, the carbon 14 in the tissue begins to decay. Thus, the age of a plant or animal fossil can be estimated by determining how much of its carbon 14 content has decayed. Libby won the Nobel prize in 1960 for his discovery.

In the years 1950 and 1951 a research team from the Texas Memorial Museum unearthed charred bison bones and the so-called Folsom points, which were projectile tips probably used for darts. It was clear from the evidence that the bones came from a bison cooked by the makers of the points. Thus, by carbon 14 dating of the bones, the research team was able to establish that hunters ("Folsom man") roamed North America some 10,000 years ago. The following example illustrates how such calculations are made by anthropologists.

EXAMPLE 7 **CARBON DATING AND THE "FOLSOM POINTS"†** Assuming that chemical tests show the Folsom bison bones to have lost 70% of their carbon 14, estimate the age of the bones.

Solution From Example 6 the decay rate for carbon 14 is $k = 0.012\%$ per year, so the amount $Q(t)$ of carbon 14 that remains after t years from an initial amount Q_0 is

$$Q(t) = Q_0 e^{-0.00012t}$$

Since the bones lost 70% of their carbon 14, the amount that remains is 30% of the original quantity Q_0 (or $0.3Q_0$). Thus, to find the age t we solve

$$\begin{aligned}
0.3Q_0 &= Q_0 e^{-0.00012t} \\
0.3 &= e^{-0.00012t} \\
\ln 0.3 &= \ln e^{-0.00012t} \\
\ln 0.3 &= -0.00012t \\
-1.2040 &= -0.00012t \qquad \text{(Appendix Table C.6 or a calculator)} \\
t &= \frac{1.2040}{0.00012} = 10{,}033
\end{aligned}$$

Thus, the bison bones are approximately 10,033 years old.

---

* Radiocarbon Dating, *American Scientist*, **44** (1956): 98–112.

† The data in this problem are based on results reported by E. H. Sellards, Age of Folsom Man, *Science*, **115** (1952): 98.

Estimating Future U.S. Trade Exports

Past data on international U.S. exports of goods and services (based on figures reported by the U.S. Department of Commerce) is shown in Figure 8. The shape of the graph strongly suggests the possibility of describing U.S. trade exports by an exponential growth model.

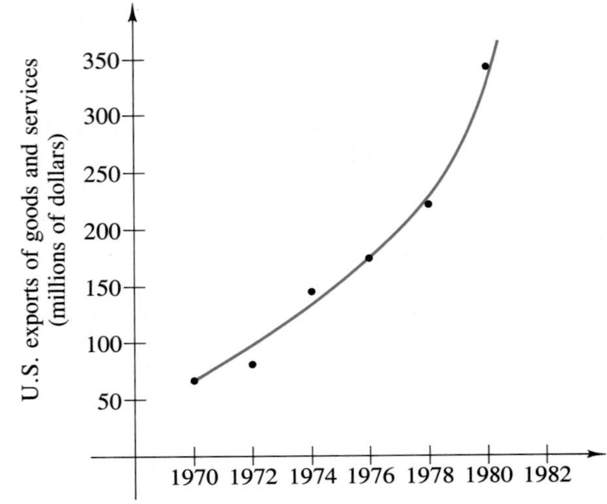

Year	International U.S. exports (in millions of dollars)
1970	65
1972	80
1974	150
1976	170
1978	220
1980	340

FIGURE 8

EXAMPLE 8 Assuming an exponential growth model, find:

(a) the growth constant k, using the data from the years 1970 and 1980

(b) the estimated value of exports for the year 2000

Solution (a) Let $Q = Q(t)$ denote the total annual U.S. exports t years after 1970. Then $t = 0$ represents the year 1970. The value Q_0 of exports for this initial year is

$$Q_0 = 65 \qquad \text{(million dollars)}$$

This means that the value of exports (in millions of dollars) after t years will be

$$Q(t) = Q_0 e^{kt} = 65e^{kt} \tag{12}$$

Since the year 1980 corresponds to $t = 10$ $(1980 - 1970 = 10)$ and since the value of exports for 1980 is $Q(10) = 340$, it follows from (12) with $t = 10$ and the table in Figure 8 that

$$340 = 65e^{10k}$$

To solve for k, we take the natural logarithm of both sides of the equation:

$$\ln 340 = \ln(65e^{10k}) = \ln 65 + \ln e^{10k} = \ln 65 + 10k$$

$$10k = \ln 340 - \ln 65 = \ln \frac{340}{65}$$

$$k = \frac{1}{10} \ln \frac{340}{65} \cong 0.165$$

Thus, the growth rate is approximately 16.5% per year.

SHERLOCK HOLMES AND THE CASE OF THE HOLLYWOOD MURDER

Holmes: Curious case, eh, Watson.
Watson: Righto, Holmes, but open and shut.
Holmes: How so, Watson?
Watson: It's obvious, Holmes. Bigwig, the Hollywood producer, told McHam that he was a second-rate actor and threw him off the set. McHam went to Bigwig's house that night, the two argued, McHam stabbed him and left the bloke dead as a carp. Neighbors heard the fight and saw McHam leaving Bigwig's house at 11:00 P.M.—no doubt about it—McHam's the fiend.
Holmes: Good thinking, Watson, but dead wrong. McHam is innocent!
Watson: I don't understand.
Holmes: Elementary, my dear Watson—it's a simple application of exponential functions. The coroner measured the body temperature to be 95.2°F at 6:00 A.M. precisely. Exactly 1 hour later he found the temperature to be 94°F. Moreover, the room thermostat was set to 70°F. So you see, McHam is innocent.
Watson: As usual, I don't get it.
Holmes: Watson, sometimes you are annoyingly dull. It boils down to the formula

$$T = (T_0 - A)e^{-kt} + A$$

which is called Newton's law of cooling. In this formula T_0 is the original body temperature, t is the elapsed time, A is the room temperature, k is a decay constant, and T is the temperature at time t. Using this formula and the available information, I deduced that Bigwig was killed between 3:24 A.M. and 3:25 A.M.—well after McHam left the house.
Watson: Amazing.
Holmes: Nonsense, it's simple deductive reasoning. Why not work Exercise 17 and I'll lead you through the details.
Watson: Hrumpf.

Solution (b) Using the value $k = 0.165$ obtained in (a), it follows from (12) that

$$Q(t) = 65e^{0.165t}$$

Since the year 2000 corresponds to $t = 30$ (2000 − 1970 = 30), we obtain

$$Q(30) = 65e^{0.165(30)} = 65e^{4.95} \cong 9176$$

Thus, the estimated value of exports for the year 2000 is 9176 million dollars.

It should be emphasized that when a mathematical model based on past data is used to predict future results, there is an underlying assumption that the conditions and assumptions on which the model is based will continue to apply. In the case of U.S. exports, the government has been attempting to reduce the trade deficit in various ways; thus it is reasonable to expect that the values predicted by the model will ultimately deviate significantly from the actual values.

A NUCLEAR DISASTER

On April 25, 1986, a major nuclear disaster occurred at the Chernobyl reactor near the city of Kiev in the Soviet Union. There was a fire and an apparent meltdown of the reactor core. The resulting explosion spewed radioactive materials into the atmosphere, thereby contaminating the soil of nearby farms. One of the radioactive substances deposited on the soil was cesium 137, which has a half-life of 37 years. It takes approximately 7 half-lives for radioactive material to decay to a level that is considered safe for farming. Thus, unless the surface soil is stripped, it will take $7 \cdot 37 = 259$ years before safe farming can be resumed in the area.

EXERCISE SET 14.5

1. Consider the exponential growth model given by

$$Q = 500e^{0.02t}$$

(a) Find the growth constant.
(b) What is the initial value of Q?
(c) Use Appendix Table C.5 to complete the following chart.

t	1	5	20	225
Q				

(d) At what rate is Q growing initially? [Use Formula (4).]
(e) At what rate is Q growing when $t = 3$? [Use Formula (4).]

2. Consider the exponential decay model given by

$$Q = 3000e^{-0.005t}$$

(a) Find the decay constant.
(b) What is the initial value of Q?
(c) Use Appendix Table C.5 to complete the following chart.

t	10	30	200	3000
Q				

(d) At what rate is Q decaying initially? [Use Formula (8).]
(e) At what rate is Q decaying when $Q = 100$? [Use Formula (8).]

3. For the growth model in Exercise 1, use Appendix Table C.5 to complete the following chart.

Q	500	1250	2300
t			

4. For the decay model in Exercise 2, use Appendix Table C.5 to complete the following chart.

Q	3000	1500	600
t			

5. Consider the exponential growth model given by

$$Q = Q_0e^{0.4t}$$

where t is in seconds.
(a) How long does it take for Q to double in value?
(b) How long to quadruple in value?

6. Consider the exponential decay model given by

$$Q = Q_0e^{-0.6t}$$

where t is in years.
(a) How long does it take for Q to decay to half its value?
(b) How long does it take to decay to one-fourth its value?

7. (a) Find the growth rate of an exponential growth model with a doubling time of 5 years.
(b) Find the decay rate of an exponential decay model with a half-life of 100 hours.

8. Radium 228 has a half-life of 6.7 years. How much time is required for 80% of the radium to decay?

9. The decay rate of krypton 85 is 6.3% per year. What is its half-life?

10. Find a formula relating the growth rate and *tripling time* for an exponential model.

11. **(Global population growth)** From the United Nations data in Example 3, the world population at the beginning of 1975 was approximately 4 billion and growing exponentially with a growth rate of approximately 2% per year.

(a) Using this data, complete the following chart.

	1975	1980	1985	1990	1995	2000	2005	2010
World population (in billions)								

(b) Sketch a graph of these data.

12. **(Anesthesiology)** Sodium pentobarbital, commonly used for surgical anesthesia, is absorbed by the body organs at a rate proportional to its concentration in the bloodstream. As a result, the bloodstream concentration follows an exponential decay model. Assume that a surgical patient requires 25 milligrams of sodium pentobarbital in the bloodstream per kilogram of body weight to maintain a proper level of anesthesia for surgery. How many milligrams of the drug must be administered to a 60-kilogram patient to maintain a proper level of anesthesia for a $\frac{1}{2}$-hour operation, assuming the drug bloodstream concentration has a decay rate of 14% per hour?

13. **(Carbon dating)** Fossil remains of a human skeleton have one-twentieth the original carbon 14 content. Use the method of Example 7 to estimate the age of the fossil.

14. **(Nuclear energy supply)** The power supply for an experimental lunar sensor uses a radioisotope whose power output P in watts decreases with time according to the model

$$P = 40e^{-0.006t}$$

where t is in days.

(a) What is the power output after 200 days of operation?
(b) What is the percentage decrease in power output over this 200-day period?
(c) What is the half-life of the power supply?
(d) If the sensing device requires 16 watts of power to operate, how long will it stay in operation?

15. **(Modeling business expansion)** To plan for future expansion, a manufacturer wants to estimate the projected sales volume for the year 1990. The firm's past sales record is shown in Figure 9. Assume an exponential growth model and follow the method of Example 8 to find:

(a) the growth rate k, using the data for the year 1980
(b) the estimated sales volume for the year 1990

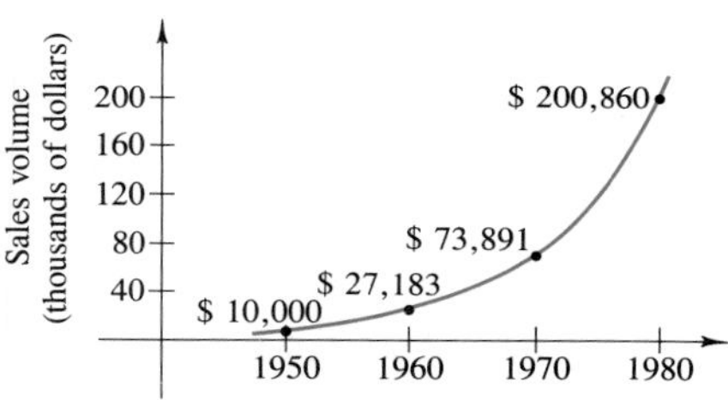

FIGURE 9

16. Show that for an exponential decay model, the halving time is given by Formula (10).

17. **(Sherlock Holmes and the Case of the Hollywood Murder)**

(a) Use Newton's Law of Cooling with $T = 94$, $T_0 = 95.2$, $A = 70$, and $t = 1$ to determine e^{-k}. (Round to six decimal places on a calculator.)
(b) Use Newton's Law of Cooling with $T = 98.6$, $T_0 = 95.2$, $A = 70$, and e^{-k} as determined in part (a) to compute t, the time before (if t is negative) 6:00 A.M. Verify that Bigwig was killed between 3:24 and 3:25 A.M. (Round to six decimal places on a calculator.)

KEY IDEAS FOR REVIEW

- **Exponential function** $f(x) = a^x \qquad a > 0, a \neq 1$

- e An irrational number that is approximately 2.71828. . . .

- $\log_a x$ The exponent y to which the base $a > 0$, $a \neq 1$, must be raised to obtain x.

- **Properties of logarithms**
  $\log_a 1 = 0$
  $\log_a a = 1$
  $\log_a xy = \log_a x + \log_a y$
  $\log_a \frac{x}{y} = \log_a x - \log_a y$
  $\log_a x^k = k \log_a x$
  $\log 10^x = x$
  $\ln e^x = x$
  If $\log_a x = \log_a y$, then $x = y$.
  $e^{\ln x} = x,\ x > 0$
  $10^{\log x} = x,\ x > 0$

- $e = \lim_{u \to \infty} \left(1 + \frac{1}{u}\right)^u$

- $e^x = \lim_{u \to \infty} \left(1 + \frac{x}{u}\right)^u$

- $\frac{d}{dx}[e^{u(x)}] = u'(x)e^{u(x)}$

- $\frac{d}{dx}[a^{u(x)}] = a^{u(x)}u'(x) \ln a$

- $\frac{d}{dx}[\ln u(x)] = \frac{u'(x)}{u(x)}$

- **Exponential growth model** A situation in which the rate of growth of a physical quantity at each instant of time is proportional to the amount of the quantity present.

- **Exponential decay model** A situation in which the rate of decay of a physical quantity at each instant of time is proportional to the amount of the quantity present.

- $Q = Q_0e^{kt}$ The formula for exponential growth.

- $Q = Q_0e^{-kt}$ The formula for exponential decay.

- **Doubling time and halving time (half-life)** $T = \frac{\ln 2}{k}$

SUPPLEMENTARY EXERCISES

1. Sketch the graph of $f(x) = e^{-2x}$. Specify the domain and range.

2. Sketch the graph of $f(x) = 1 - 2^{-x}$. Can you see what happens to the values of $f(x)$ as x becomes larger and larger?

3. A certain bacteria colony doubles in size every hour. If there are 1000 bacteria present initially, write the formula $f(t)$ that expresses the number of bacteria present t hours later. How many will be present after (a) 5 hours and (b) 12 hours?

4. Determine all values of x at which $f(x) = 1/(1 - e^{-x})$ is continuous.

5. Sketch the graph of $f(x) = \ln(1 + x^2)$.

6. Determine all values of x at which $f(x) = \ln(1 - x^2)$ is continuous.

In Exercises 7–10 use the values $\log 2 = 0.3010$, $\log 3 = 0.4771$, and $\log 5 = 0.6990$ to find the value of the expressions. Express your answer to four decimal places.

7. $\log 75$
8. $\log 20$
9. $\log \sqrt{7.5}$
10. $\log 0.3$

In Exercises 11–15 solve for x.

11. $\log_2 x^2 = 4$

12. $\ln(3x - 2) = 2$

13. $e^{x^2} = 4$

14. $3^{1-x} = 4$

15. $\ln \dfrac{x - 2}{4} - \ln x = \ln 5$

In Exercises 16 and 17 use a calculator to determine which number is greater.

16. 2^{π}, π^2
17. 3^{π}, π^3

18. The period T (in seconds) of a simple pendulum of length L (in feet) is given by the formula

$$T = 2\pi\sqrt{\frac{L}{g}}$$

Using common logarithms, find the approximate value of T if $L = 4.72$ feet, $g = 32.2$, and $\pi = 3.14$.

19. The area of a triangle whose sides are a, b, and c in length is given by the formula

$$A = \sqrt{s(s - a)(s - b)(s - c)}$$

where $s = \frac{1}{2}(a + b + c)$. Use logarithms to find the approximate area of a triangle whose sides (in feet) are 12.86, 13.72, and 20.3.

20. The number N of radios that an assembly line worker can assemble daily after t days of training is given by

$$N = 60 - 60e^{-0.04t}$$

After how many days of training does the worker assemble 40 radios daily?

21. The population P of a certain city t years from now is given by

$$P = 20{,}000e^{0.05t}$$

How many years from now will the population be 50,000?

22. Suppose the population of a certain country is increasing exponentially with a growth rate of 3% per year. If the population in 1985 is 55 million people, what will it be in the year 2015?

23. A substance is known to have a decay rate of 6% per hour. Approximately how many hours are required for the remaining quantity to be half of the original quantity?

24. Thorium has a half-life of 18.9 days. How long does it take for 65% of the thorium present to decay?

25. The half-life of actinium is 21.7 years. What is its decay rate?

26. Fossil remains of a human skeleton found in a cave contained 30% of the original carbon 14 content. Estimate the age of the fossil.

27. Approximately how much money should a 35-year-old woman invest now at continuous compound interest of 10% per year to obtain the sum of $20,000 on her retirement at age 65?

In Exercises 28 and 29 find the indicated limit.

28. $\displaystyle\lim_{u\to\infty}\left(1 + \frac{2}{3u}\right)^{3u}$

29. $\displaystyle\lim_{u\to\infty}\left(1 - \frac{2}{3u}\right)^{(1/3)u}$ [*Hint:* Let $v = \frac{1}{3}u$]

30. How long will it take an investment at 10% annual interest compounded quarterly to double in value?

31. Suppose \$10,000 is invested at 10% annual interest compounded continuously. How long will it take for the investment to grow to \$18,000?

In Exercises 32–37 find dy/dx.

32. $y = e^x/x^2$ **33.** $y = e^{-x}/(1 - e^x)$

34. $y = \ln(4x^2)$ **35.** $y = \ln\sqrt{x^2 - x}$

36. $y = \ln e^{x^2}$ **37.** $e^{\ln\sqrt{x}}$

38. Find the equation of the tangent line to $y = e^{x^2}\ln x$ at the point where $x = 1$.

39. Let $y = \ln[(1 - x)/(x^2 + 1)]$. Find the instantaneous rate of change of y with respect to x when $x = 0$.

40. Let $f(x) = e^x/x$.
(a) Determine the intervals over which $f(x)$ is increasing and those over which it is decreasing.
(b) Locate and classify the critical values of $f(x)$ as relative maxima, relative minima, or neither.
(c) Where is the curve concave up? Concave down? Locate all inflection points.
(d) Sketch the graph of $f(x)$.

41. Follow the directions of Exercise 40 for $f(x) = (\ln x)/x$.

42. A certain bacteria colony consisting initially of 50,000 bacteria is treated with a drug, and after t hours the number of bacteria present is

$$N(t) = 50{,}000 + 2000\left[\frac{t}{20} - \ln(1 + 5t)\right]$$

(a) Find the instantaneous rate of change of $N(t)$ after 5 hours.
(b) After how many hours is the number of bacteria a minimum?

CHAPTER TEST

1. Solve $\ln(x^2 - 1) = 4$ for x.

2. A person on an assembly line produces P items per day after t days of training, where

$$P = 400(1 - e^{-t})$$

How many days of training will it take this person to be able to produce 300 items per day?

3. The number of bacteria in a culture after t hours is given by $Q(t) = Q_0e^{0.01t}$. If there are 400 bacteria present initially, how many bacteria will be present after 2 *days*?

4. Thallium 208 has a decay rate of approximately 22.35% per minute.
(a) If 2000 grams of thallium 208 is present initially, how many grams will be left after 4 minutes?
(b) What is the half-life of thallium 208?

5. Suppose that a certain substance has a decay rate of 5% per hour. Approximately how many hours are required for the remaining quantity to be one-fourth the original quantity?

6. Find $\lim_{t\to\infty}\left(1 - \frac{1}{2t}\right)^t$

7. In each part find dy/dx.
(a) $y = \dfrac{e^{x^2} - x}{\ln x^3}$ (b) $y = \ln\dfrac{x}{x^2 + 1}$

8. Let $f(x) = xe^{-x}$.
(a) Determine the intervals over which $f(x)$ is increasing and those over which it is decreasing.
(b) Locate and classify the critical values of $f(x)$ as relative maxima, relative minima, or neither.
(c) Where is the curve concave up? Concave down? Locate all inflection points.
(d) Sketch the graph of $f(x)$.

9. Find the equation of the tangent line to $y = e^{x^2-1}\ln(x^2 + 1)$ at the point where $x = 0$.

10. Suppose \$15,000 is invested 12% annual interest compounded continuously. How long does it take for the investment to grow to \$22,000?

11. Suppose a certain bacteria colony consisting initially of 10,000 bacteria grows in such a way that the number of bacteria present after t hours is $N(t) = 10{,}000(2^{3t})$. Find the instantaneous rate of charge of $N(t)$ after 4 hours.

12. Assume that an object moves along the x-axis with known position (in feet) given by

$$s(t) = t^2e^{-t}$$

after t seconds. Find the position, velocity, and acceleration at time $t_0 = 3$.

15

The Integral

Traditionally, calculus is divided into two main areas, differential calculus and integral calculus. As we have seen in Chapters 13 and 14, the fundamental mathematical tool used in differential calculus is the derivative. We are now about to embark on a study of integral calculus, in which the fundamental mathematical tool is called the integral. Just as the notion of derivative has a geometric basis (the slope of a tangent line), the notion of integral also has a geometric basis, the area between a curve and the x-axis. The following sections show that there is a close relationship between the derivative and the integral; this relationship is established by the fundamental theorem of calculus.

15.1 ANTIDERIVATIVES AND INDEFINITE INTEGRALS

As described in Chapter 12, the instantaneous rate of change of a function f is given at each point by its derivative f'. In this section we study the reverse problem: given the derivative f' at each point, can we find f itself? For example, given the instantaneous speed of an object at each instant, can we determine its position at each instant? Or, given a marginal cost function, can we determine the cost function itself? We begin with an example.

EXAMPLE 1 Let us find $F(x)$ given that

$$F'(x) = 2x \tag{1}$$

From our experience with derivatives we can guess the answer

$$F(x) = x^2$$

almost immediately. However, there are other answers as well; for example,

$$F(x) = x^2 + 1 \qquad F(x) = x^2 - \frac{1}{3} \qquad F(x) = x^2 + \sqrt{2}$$

In fact, for any constant C, the function

$$F(x) = x^2 + C \tag{2}$$

will have derivative $F'(x) = 2x$.

The process of obtaining a function from its derivative is called **antidifferentiation** or **integration**, and a function F such that $F'(x) = f(x)$ is called an **antiderivative** of f. Thus we have just shown that any function of the form $F(x) = x^2 + C$ is an antiderivative of $f(x) = 2x$. You might reasonably ask whether there are any antiderivatives of $f(x) = 2x$ that we missed; that is, antiderivatives that cannot be obtained by substituting a value for C in Equation (2). The following shows that the answer is no; we have indeed obtained all the antiderivatives:

THE EQUAL DERIVATIVE THEOREM If $F'(x) = G'(x)$ at each point of an interval, then $F(x) - G(x)$ is constant over that interval.

Although we omit the proof, the principle is easy to visualize; since $F'(x) = G'(x)$ at each point, the tangents to the graphs of F and G have the same slope and hence are parallel at each x (Figure 1). Thus the graphs of F and G are themselves "parallel"; that is, $F(x)$ and $G(x)$ differ by a constant.

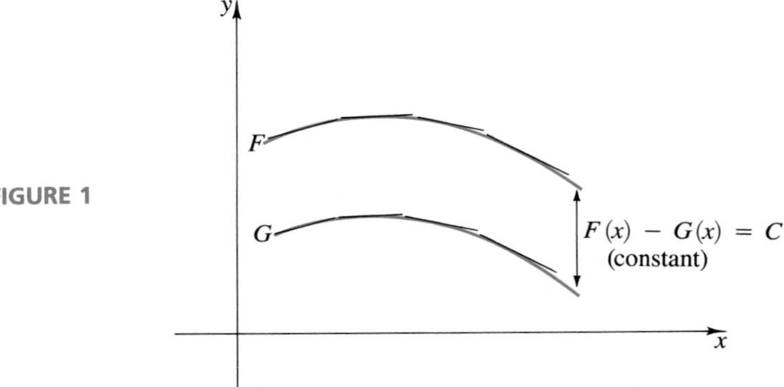

FIGURE 1

In light of the equal derivative principle, if $F(x)$ is any antiderivative of $f(x) = 2x$, then $F(x) - x^2$ is constant since $F(x)$ and x^2 both have derivative $2x$. Thus for some constant C,

$$F(x) - x^2 = C \quad \text{or} \quad F(x) = x^2 + C$$

which proves that (2) describes *all* antiderivatives of $2x$.

In general, if $F(x)$ is any antiderivative of $f(x)$, then the most general antiderivative of $f(x)$ is specified by the formula $F(x) + C$, and we write

$$\int f(x)\, dx = F(x) + C$$

where we read $\int f(x)\, dx$ as "the integral of $f(x)$."

For example, we can write

$$\int 2x\,dx = x^2 + C$$

Just as the symbol d/dx represents the operation of differentiation, so the symbol $\int dx$ represents the operation of *antidifferentiation*. The symbol $\int$ is called an **integral sign**, and the symbol $\int f(x)\,dx$ is called the **indefinite integral** of $f(x)$. The arbitrary constant C is called the **constant of integration**, and the function $f(x)$ is called the **integrand**.

Since integration is the reverse of differentiation, every differentiation formula has a companion integration formula. For example, if k is a constant, then we can write:

Differentiation Formula	Corresponding Integration Formula	
$\dfrac{d}{dx}[kx] = k$	$\displaystyle\int k\,dx = kx + C$	(3)
$\dfrac{d}{dx}\left[\dfrac{x^{k+1}}{k+1}\right] = x^k \quad (k \neq -1)$	$\displaystyle\int x^k\,dx = \frac{x^{k+1}}{k+1} + C \quad (k \neq -1)$	(4)
$\dfrac{d}{dx}[e^x] = e^x$	$\displaystyle\int e^x\,dx = e^x + C$	(5)

EXAMPLE 2 From Formula (3)

$$\int 2\,dx = 2x + C$$

$$\int dx = \int 1\,dx = 1x + C = x + C$$

Integration formula (4) can be stated as: *To integrate a power of x (other than −1), increase the power of x by 1 and divide by the increased power.*

EXAMPLE 3 From Formula (4)

$$\int x^2\,dx = \frac{x^3}{3} + C$$

$$\int x^3\,dx = \frac{x^4}{4} + C$$

$$\int \sqrt{x}\,dx = \int x^{1/2}\,dx = \frac{x^{1+(1/2)}}{1 + 1/2} + C = \frac{2}{3}x^{3/2} + C$$

$$\int x^{-5}\,dx = \frac{x^{-5+1}}{-5+1} + C = -\frac{x^{-4}}{4} + C$$

The following rules are obtained by reversing the corresponding differentiation rules of Section 12.7.

The Constant Times a Function Rule If k is a constant, then

$$\int kf(x)\,dx = k\int f(x)\,dx \tag{6}$$

In words, a *multiplicative constant* can be moved past an integral sign.

The Sum and Difference Rules

$$\int [f(x) + g(x)]\,dx = \int f(x)\,dx + \int g(x)\,dx \tag{7}$$

$$\int [f(x) - g(x)]\,dx = \int f(x)\,dx - \int g(x)\,dx \tag{8}$$

In words, the integral of a sum (or difference) is the sum (or difference) of the integrals.

EXAMPLE 4

$$\int 3x^7\,dx = 3\int x^7\,dx = 3\frac{x^8}{8} + C = \frac{3}{8}x^8 + C$$

$$\int (x + e^x)\,dx = \int x\,dx + \int e^x\,dx = \frac{x^2}{2} + e^x + C$$

$$\begin{aligned}\int (2x^3 - 4x^2 + 1)\,dx &= \int 2x^3\,dx - \int 4x^2\,dx + \int 1\,dx\\ &= 2\int x^3\,dx - 4\int x^2\,dx + \int 1\,dx\\ &= 2\left(\frac{x^4}{4}\right) - 4\left(\frac{x^3}{3}\right) + x + C\\ &= \frac{1}{2}x^4 - \frac{4}{3}x^3 + x + C\end{aligned}$$

The result in any integration problem can be checked by differentiating the result to obtain the integrand. Thus, the last result in Example 4 can be checked as follows:

$$\begin{aligned}\frac{d}{dx}\left(\frac{1}{2}x^4 - \frac{4}{3}x^3 + x + C\right) &= \frac{1}{2}(4x^3) - \frac{4}{3}(3x^2) + 1\\ &= 2x^3 - 4x^2 + 1\end{aligned}$$

EXAMPLE 5

$$\int \left(\frac{1}{\sqrt{x}} + \frac{2}{x^3}\right) dx = \int \frac{1}{\sqrt{x}}\, dx + \int \frac{2}{x^3}\, dx$$

$$= \int x^{-1/2}\, dx + 2\int x^{-3}\, dx$$

$$= \frac{x^{1/2}}{1/2} + 2\left(\frac{x^{-2}}{-2}\right) + C$$

$$= 2\sqrt{x} - \frac{1}{x^2} + C$$

Since x is just a variable, any other variable can be substituted for it in the integration formulas, as shown in the next example.

EXAMPLE 6

$$\int t^3\, dt = \frac{t^4}{4} + C$$

$$\int e^u\, du = e^u + C$$

Integration of 1/*x*

Formula (4) is used to integrate all powers of x with the exception of x^{-1}. This contrasts with the situation in differentiation where one power rule works for *all* powers of x. In integration, a separate rule must be found for the antiderivative of x^{-1}. Fortunately, one is readily available.

In Section 14.4 we show that if $x > 0$, then

$$\frac{d}{dx}[\ln x] = \frac{1}{x}$$

This leads to the integration formula

$$\int \frac{1}{x}\, dx = \ln x + C$$

which holds if $x > 0$. However, for many applications, the requirement that x be positive is too restrictive. The following formula is valid for all nonzero values of x:

$$\int \frac{1}{x}\, dx = \ln|x| + C \tag{9}$$

Proof To prove Formula (9), it is sufficient to prove that for all $x \neq 0$,

$$\frac{d}{dx}[\ln|x|] = \frac{1}{x}$$

where $|x|$ is the absolute value of x, defined in Section 11.1.

Case 1 Consider $x > 0$. Then $\ln|x| = \ln x$, and

$$\frac{d}{dx}[\ln|x|] = \frac{d}{dx}[\ln x] = \frac{1}{x}$$

Case 2 Consider $x < 0$. Then $|x| = -x$, so $\ln|x| = \ln(-x)$, and

$$\frac{d}{dx}[\ln|x|] = \frac{d}{dx}[\ln(-x)] = \frac{1}{-x}\cdot(-1) \qquad \text{(by the chain rule)} = \frac{1}{x}$$

Therefore, regardless of whether $x > 0$ or $x < 0$,

$$\frac{d}{dx}[\ln|x|] = \frac{1}{x}$$

Restating this last result as an integral, we have

$$\int \frac{1}{x}\,dx = \int x^{-1}\,dx = \ln|x| + C \qquad x \neq 0$$

which establishes Equation (9).

EXAMPLE 7

$$\int\left(x^3 - \frac{1}{x}\right)dx = \int x^3\,dx - \int \frac{1}{x}\,dx = \frac{x^4}{4} - \ln|x| + C$$

Determining the Constant of Integration

In applied problems, there are often conditions that determine a specific value for the constant of integration. These conditions are often called **boundary conditions**.

EXAMPLE 8 A manufacturer determines that the marginal cost in dollars is given by

$$M(x) = x^2 + 3x$$

Find the cost function $C(x)$, assuming that the fixed cost (cost when $x = 0$ units are produced) is \$30.

Solution Since the marginal cost $M(x) = x^2 + 3x$ is the derivative of the cost function, we have

$$C'(x) = x^2 + 3x$$

Thus the unknown cost function $C(x)$ is an antiderivative of $x^2 + 3x$; consequently, it can be determined by integration:

$$C(x) = \int (x^2 + 3x)\, dx$$
$$= \int x^2\, dx + 3 \int x\, dx$$
$$= \frac{1}{3}x^3 + \frac{3}{2}x^2 + K$$

(We have used K for the constant of integration to avoid confusion with the cost C.) Since $C = \$30$ when $x = 0$, we have $C(0) = 30$. Thus,

$$\frac{1}{3}(0) + \frac{3}{2}(0) + K = 30$$

or $K = 30$. Thus the cost function is

$$C(x) = \frac{1}{3}x^3 + \frac{3}{2}x^2 + 30$$

EXERCISE SET 15.1

In Exercises 1–24 find the indefinite integral.

1. $\int 5\, dx$
2. $\int 3\, dx$
3. $\int t^5\, dt$
4. $\int x^8\, dx$
5. $\int x^{1/4}\, dx$
6. $\int t^{2/3}\, dt$
7. $\int 20\, x^3\, dx$
8. $\int 15\, x^4\, dx$
9. $\int 6/x^2\, dx$
10. $\int 3/t^5\, dt$
11. $\int 2e^u\, du$
12. $\int 7e^x\, dx$
13. $\int 50/x\, dx$
14. $\int 30/t\, dt$
15. $\int \sqrt{x}\, dx$
16. $\int (x - 1)\, dx$
17. $\int (8x + 6x^2)\, dx$
18. $\int (3A^2 + 2A - 1)\, dA$
19. $\int (3x^2 - 5x^{3/4} + 4)\, dx$
20. $\int (t^2 + e^t - 1/t)\, dt$
21. $\int (2x^{2/3} - 3e^x + 2/x)\, dx$
22. $\int (3x^2 - 2x^{3/2} + 4/x^3)\, dx$
23. $\int \left(2x^3 + \sqrt{x} + \frac{1}{x^2} + 5\right) dx$
24. $\int \left(\frac{1}{2x} + 3e^x + \frac{1}{\sqrt{x}}\right) dx$

In Exercises 25–30 find f from the given information.

25. $f'(x) = x + 2$, $f(3) = 5$
26. $f'(x) = x - 5$, $f(2) = 4$
27. $f'(x) = x^2 + 3$, $f(-1) = 2$
28. $f'(x) = x^2 - 5$, $f(0) = 3$
29. $f'(x) = 2 + 3e^x$, $f(0) = 5$
30. $f'(x) = 2/x$, $f(1) = -3$
31. Given that the point $(-2, 1)$ lies on the curve $y = f(x)$ and given that $f'(x) = 3x^2$, find $f(x)$.
32. Given that the point $(0, 8)$ lies on the curve $w = g(t)$ and given that $dw/dt = 7e^t$, find $g(t)$.
33. Suppose the rate of change of a certain population $P(t)$ with respect to time is given by
$$P'(t) = 25{,}000 + 4t^{2/5}$$
and at $t = 0$, the population is $P(0) = 50{,}000$.
(a) Find an expression for $P(t)$.
(b) What will the population be when $t = 20$?

34. An automobile moves along a straight track in such a way that its velocity $v(t)$ after t seconds is $v(t) = \sqrt{t}$ (feet per second). Find the distance traveled by the car after 5 seconds ($t = 5$). [*Hint:* Let $s(t)$ denote the distance traveled after t seconds. Thus $s(0) = 0$ and $ds/dt = v(t)$.]

35. **(Marginal analysis)** A company determines that its marginal cost is

$$C'(x) = x^2 - 3x$$

Find the cost function $C(x)$ if the fixed cost (cost of producing zero units) is \$1000.

36. (a) Show by differentiation that any function of the form $F(x) = x + C$ is an antiderivative of the constant function 1.

(b) From (a) and the equal derivative principle, the functions of the form $F(x) = x + C$ should have "parallel" graphs. Graph $F(x)$ in the cases $C = -1$, $C = 0$, $C = 1$, and $C = 2$.

15.2 THE DEFINITE INTEGRAL—AN INTRODUCTION

As we show in this section, many applied problems are mathematically equivalent to the problem of finding the **area under a curve** $y = f(x)$ over an interval $[a, b]$. By that we mean the area between the curve $y = f(x)$ and the x-axis and between the vertical lines $x = a$ and $x = b$, assuming $f(x) \geq 0$ for all $a \leq x \leq b$ (Figure 1).

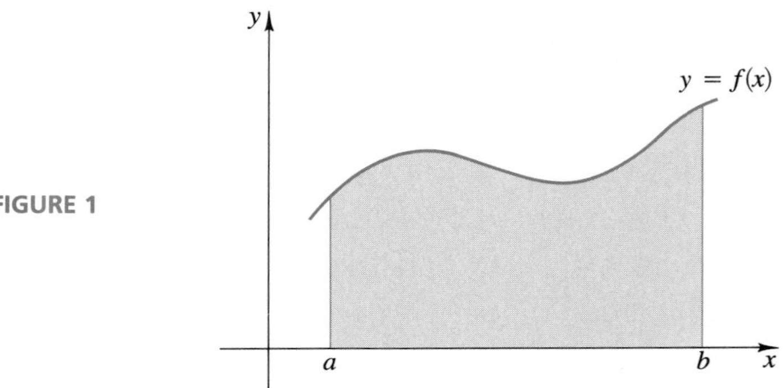

FIGURE 1

Let f be a continuous function whose graph does not dip below the x-axis over an interval $[a, b]$. Surprising as it may seem, the key to finding the area under $y = f(x)$ over the fixed interval $[a, b]$ is to study first how the area under this curve *varies* as we change the right endpoint b. For this purpose, we replace the fixed right endpoint b by a variable endpoint x, and we denote by $A(x)$ the area under the curve over the interval $[a, x]$ (Figure 2). We call $A(x)$ the **area function** for f starting from a.

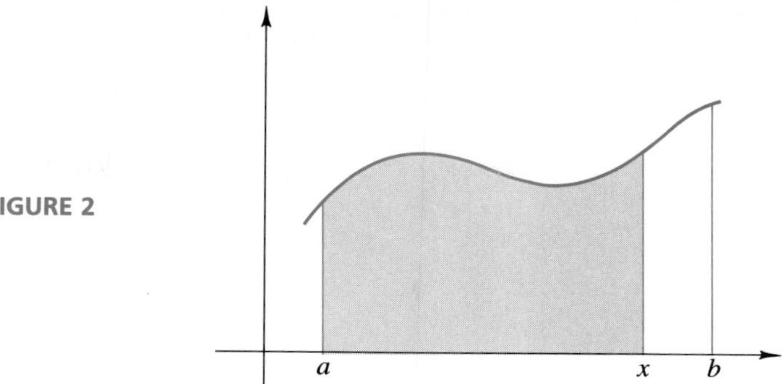

FIGURE 2

EXAMPLE 1 Let f be the constant function defined by $f(x) = 2$. Since the graph of f is the horizontal line $y = 2$ (Figure 3), the area under the graph of f over the interval $[0, x]$ is the area of a rectangle of height 2 and base x. Thus the area function for f starting from 0 is

$$A(x) = 2x$$

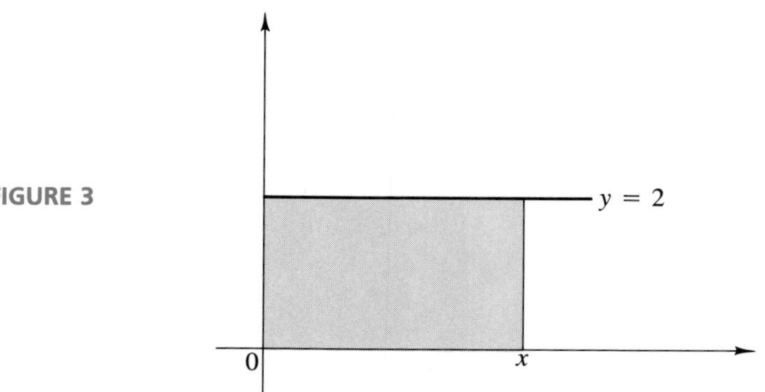

FIGURE 3

EXAMPLE 2 Let $f(x) = 3x$, and find the area function $A(x)$ starting from $x = 0$.

Solution The graph of f is the straight line $y = 3x$. The area under the graph of f over the interval $[0, x]$ is the area of the triangle shown in Figure 4. Since the triangle has base x and height $3x$, its area is

$$A(x) = \frac{1}{2} \cdot x \cdot 3x = \frac{3}{2}x^2$$

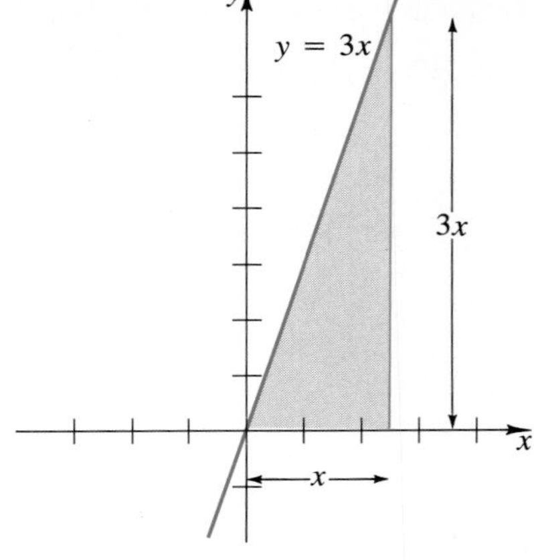

FIGURE 4

EXAMPLE 3 Let

$$f(x) = x - 1$$

The graph of f is the line $y = x - 1$. As indicated in Figure 5, the area under this graph over the interval $[1, x]$ is the area of a triangle with base $x - 1$ and height $x - 1$. Thus the area function for f starting from 1 is

$$A(x) = \frac{1}{2}(x - 1)(x - 1) = \frac{1}{2}(x^2 - 2x + 1) = \frac{1}{2}x^2 - x + \frac{1}{2}$$

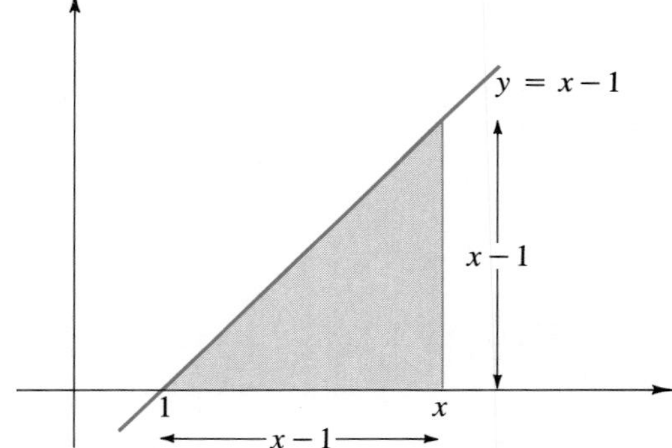

FIGURE 5

The Derivative of $A(x)$

In these three examples we obtain explicit formulas for the area functions because the curves are simple enough that we can use area formulas from geometry. However, even for many simple curves there are no formulas from elementary geometry to help us determine the area function. For example, no basic geometry formula will help us find the area under the curve $y = x^2$ over the interval $[0, x]$ (Figure 6). What is surprising, however, is that the *derivative* of the area function is always easy to obtain. To see why, let us examine the derivatives of the area functions in the examples above:

FIGURE 6

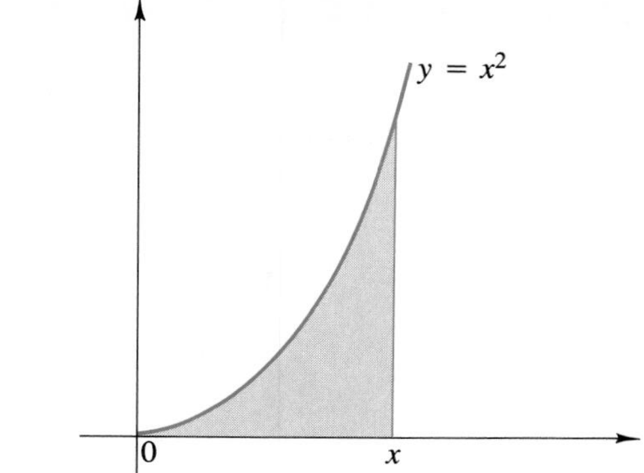

	Function	Area Function	Derivative of Area Function
Example 1	$f(x) = 2$	$A(x) = 2x$	$A'(x) = 2$
Example 2	$f(x) = 3x$	$A(x) = \frac{3}{2}x^2$	$A'(x) = 3x$
Example 3	$f(x) = x - 1$	$A(x) = \frac{1}{2}x^2 - x + \frac{1}{2}$	$A'(x) = x - 1$

In each case, the derivative of the area function turns out to be the same as the original function f. This is not accidental; it is a consequence of the following major result:

THE FUNDAMENTAL THEOREM OF CALCULUS—PART I Suppose the function f is continuous and nonnegative throughout the interval $[a, b]$. Then the area function $A(x)$ is differentiable everywhere on the interval $[a, b]$. Moreover, for all x in $[a, b]$,

$$A'(x) = f(x) \tag{1}$$

Although we do not present a proof here, we do give a plausibility argument: Let $A(x)$ denote the area function for f, starting from a. To determine $A'(x)$ we use the definition of derivative:

$$A'(x) = \lim_{h \to 0} \frac{A(x + h) - A(x)}{h}$$

For arbitrary $h > 0$, $A(x + h) - A(x)$ is the difference between the area from a to $x + h$ and the area from a to x shown in Figure 7a. Thus $A(x + h) - A(x)$ is the shaded area in Figure 7b.

It is plausible that this area could be obtained by taking a rectangle of the same width with height equal to $f(x^*)$ for some x^* in the interval $[x, x + h]$, as shown in Figure 7b. Then,

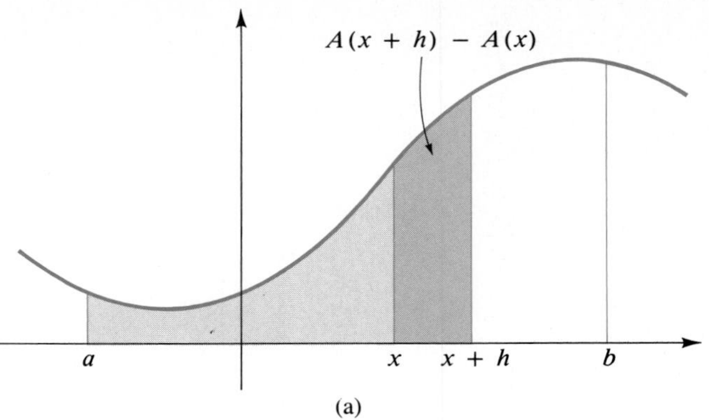

(a)

FIGURE 7

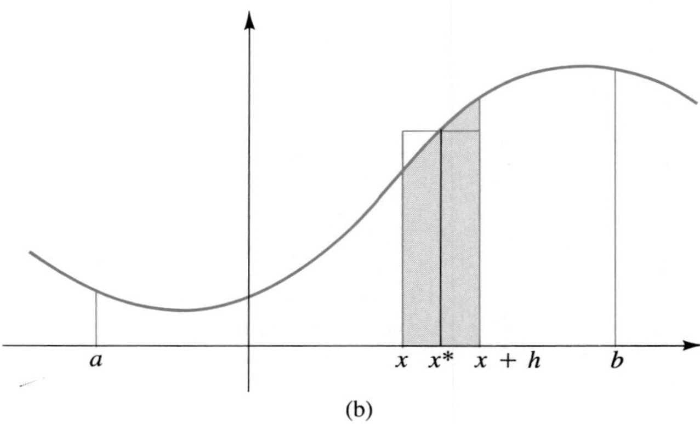

(b)

$$A(x + h) - A(x) = f(x^*) \cdot h$$
$$\frac{A(x + h) - A(x)}{h} = f(x^*)$$

and

$$\lim_{h \to 0^+} \frac{A(x + h) - A(x)}{h} = \lim_{h \to 0^+} f(x^*) \tag{2}$$

As $h \to 0^+$, $x^* \to x$ because x^* is between x and $x + h$. The function f is continuous; hence,

$$\lim_{h \to 0^+} f(x^*) = f(x) \tag{3}$$

A similar argument will show that

$$\lim_{h \to 0^-} f(x^*) = f(x) \tag{4}$$

Equations (2), (3), and (4) together tell us that

$$\lim_{h \to 0} \frac{A(x + h) - A(x)}{h} = f(x)$$

Therefore, $A'(x) = f(x)$.

EXAMPLE 4 Find the area under $y = x^2$ over the interval $[0, x]$, as shown in Figure 8.

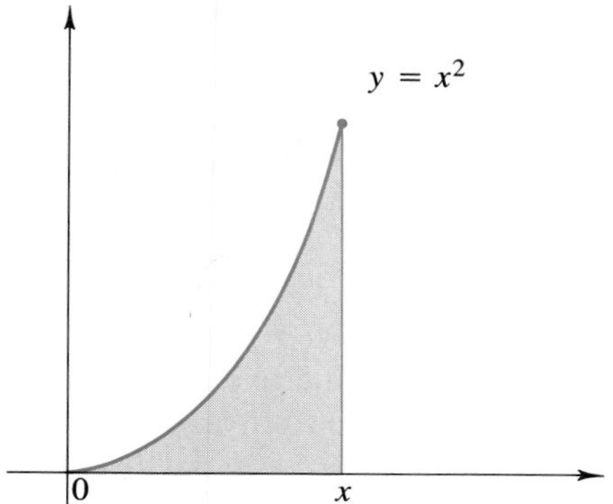

FIGURE 8

Solution From the fundamental theorem of calculus, the area $A(x)$ satisfies

$$A'(x) = x^2$$

which states that $A(x)$ is an antiderivative of x^2. Thus,

$$A(x) = \int x^2\,dx = \frac{x^3}{3} + C \tag{5}$$

where the constant of integration is not yet determined. However, if $x = 0$, then the interval $[0, x]$ reduces to a point, in which case the area under the curve over the interval $[0, x]$ is 0. Thus, $A(x) = 0$ if $x = 0$; that is, $A(0) = 0$. Thus, from (5),

$$0 = 0 + C \quad \text{or} \quad C = 0$$

Therefore, the area is

$$A(x) = \frac{x^3}{3}$$

EXAMPLE 5 Find the area under the curve $y = x^2$ over the interval $[0, 2]$ (Figure 9).

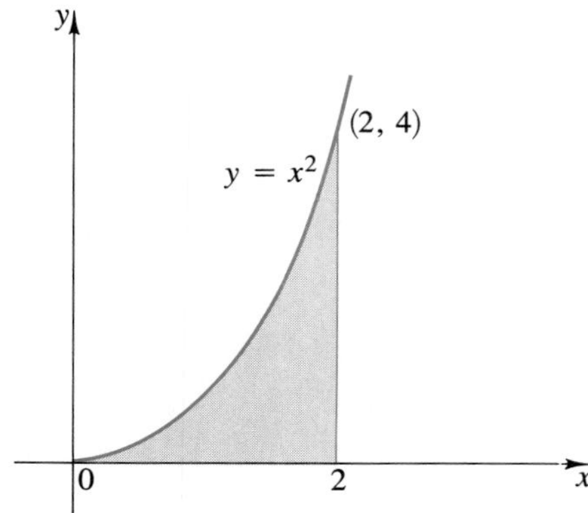

FIGURE 9

Solution From Example 4 the area under the curve over $[0, x]$ is $A(x) = x^3/3$. Letting $x = 2$ yields the area under the curve over $[0, 2]$. This area is

$$A(2) = \frac{(2)^3}{3} = \frac{8}{3} \qquad \text{(square units)}$$

The fundamental theorem of calculus, part I, tells us that the area function $A(x)$ is *an* antiderivative. But there are, of course, infinitely many antiderivatives of f. Example 4 shows how we may use the condition

$$A(a) = 0 \tag{6}$$

to determine the "constant of integration" and thereby single out the area function $A(x)$ from among all the antiderivatives of $f(x)$.

It turns out that *any* antiderivative of $f(x)$ can be used to calculate the area under the curve $y = f(x)$, which is the essence of the next important theorem.

THE FUNDAMENTAL THEOREM OF CALCULUS—PART II Suppose f is continuous and nonnegative throughout the interval $[a, b]$. Let $A(x)$ denote the area under the curve $y = f(x)$ over the interval $[a, x]$, where $a \leq x \leq b$. If $F(x)$ is *any* antiderivative of $f(x)$ over $[a, b]$, then for all x in $[a, b]$,

$$A(x) = F(x) - F(a) \tag{7}$$

To see why this formula holds, let $F(x)$ denote any antiderivative of $f(x)$ over $[a, b]$. Then, for all x in $[a, b]$, $F'(x) = f(x) = A'(x)$. Thus, $F(x)$ and $A(x)$ have the same derivative over $[a, b]$. From the equal derivative theorem, $F(x)$ and $A(x)$ differ only by a constant amount over $[a, b]$. That is, there is some constant C such that for all x in $[a, b]$,

$$A(x) - F(x) = C$$

so

$$A(x) = F(x) + C \tag{8}$$

Since (8) holds for all x in $[a, b]$, it must hold when $x = a$. Thus,

$$\begin{aligned} A(a) &= F(a) + C \\ 0 &= F(a) + C \qquad \text{[from Equation (6)]} \end{aligned}$$

and

$$C = -F(a) \tag{9}$$

Substituting (9) into (8) we obtain

$$A(x) = F(x) - F(a)$$

We have thus shown that if $f(x)$ is a nonnegative continuous function on the interval $[a, b]$, then the area function $A(x)$ is an antiderivative of $f(x)$ and the area under the curve $y = f(x)$ over the interval $[a, b]$ is

$$F(b) - F(a)$$

where F is *any* antiderivative of f over $[a, b]$. The difference $F(b) - F(a)$ can be denoted by

$$\int_a^b f(x)\,dx$$

and is called the **definite integral** of f from a to b. In Section 15.4 we will give a more formal definition of the definite integral. It can be shown that the fundamental theorem of calculus can be generalized as follows:

If $f(x)$ is any continuous function (not necessarily nonnegative) on $[a, b]$, then $f(x)$ has an antiderivative F on $[a, b]$ and

$$\int_a^b f(x)\,dx = F(b) - F(a) \tag{10}$$

The numbers a and b in (10) are called the **limits of integration**, and the expression on the right side of (10) is often written as

$$F(x)\Big|_a^b$$

which means, subtract $F(a)$ from $F(b)$. For example,

$$x^2\Big|_1^3 = (3)^2 - (1)^2 = 8$$

With this notation, (10) can be written as

$$\int_a^b f(x)\,dx = F(x)\Big|_a^b$$

EXAMPLE 6 (a) Evaluate the definite integral $\int_2^3 x^3\,dx$.
(b) Evaluate the definite integral $\int_{-1}^3 x^3\,dx$.

Solution (a)

$$\int x^3\,dx = \frac{x^4}{4} + C$$

Since any antiderivative of x^3 will suffice to evaluate the given definite integral, we take $C = 0$, so that $F(x) = x^4/4$. Thus,

$$\int_2^3 x^3\,dx = \frac{x^4}{4}\Big|_2^3 = \frac{3^4}{4} - \frac{2^4}{4} = \frac{81}{4} - \frac{16}{4} = \frac{65}{4}$$

The computations in this example can be arranged more compactly by writing

$$\int_2^3 x^3\,dx = \int x^3\,dx\Big|_2^3 = \frac{x^4}{4}\Big|_2^3 = \frac{81}{4} - \frac{16}{4} = \frac{65}{4}$$

Solution (b)

$$\int_{-1}^{3} x^3\,dx = \frac{x^4}{4}\bigg|_{-1}^{3} = \frac{81}{4} - \frac{1}{4} = \frac{80}{4} = 20$$

EXAMPLE 7 Use the definite integral to find the area under $y = x^2$ over the interval $[0, 2]$.

Solution The area A is

$$A = \int_0^2 x^2\,dx = \int x^2\,dx\,\bigg|_0^2 = \frac{x^3}{3}\bigg|_0^2 = \frac{8}{3} - 0 = \frac{8}{3}$$

This agrees with the result obtained in Example 4 in which we solve this same problem (less efficiently) using the area function.

EXAMPLE 8 Find the area under the curve $y = e^x$ over the interval $[0, 2]$ (Figure 10).

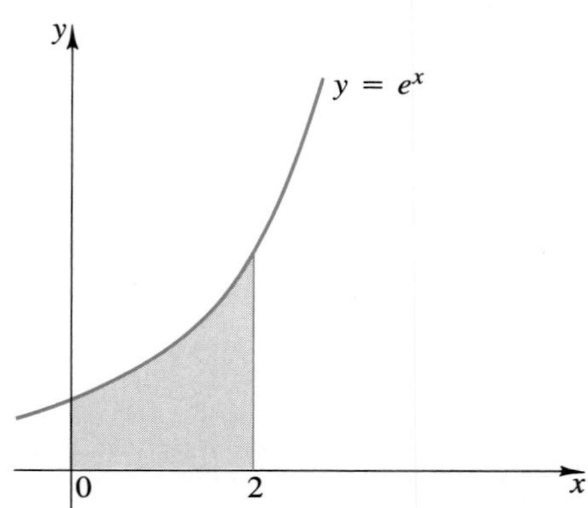

FIGURE 10

Solution The area A we want is

$$A = \int_0^2 e^x\,dx = \int e^x\,dx\,\bigg|_0^2 = e^x\bigg|_0^2 = e^2 - e^0 = e^2 - 1$$

EXERCISE SET 15.2

In Exercises 1–10 evaluate the definite integral.

1. $\int_{-1}^{2} 3\,dx$
2. $\int_2^5 dx$
3. $\int_0^3 x^3\,dx$
4. $\int_{-1}^{2} x^5\,dx$
5. $\int_0^2 e^t\,dt$
6. $\int_{-1}^{0} e^s\,ds$
7. $\int_1^3 1/x\,dx$
8. $\int_2^8 1/t\,dt$
9. $\displaystyle\int_3^5 \frac{1}{x^2}\,dx$
10. $\displaystyle\int_1^2 \frac{1}{x^2}\,dx$

In Exercises 11–16 give a definite integral whose value is the shaded area; then calculate the area from your integral. (Use appendix tables, where needed.)

11.

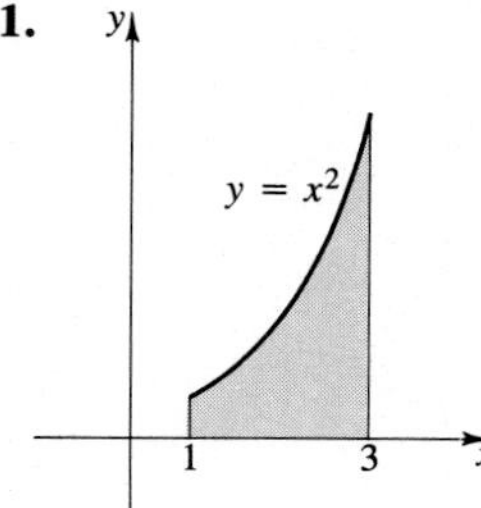

12.

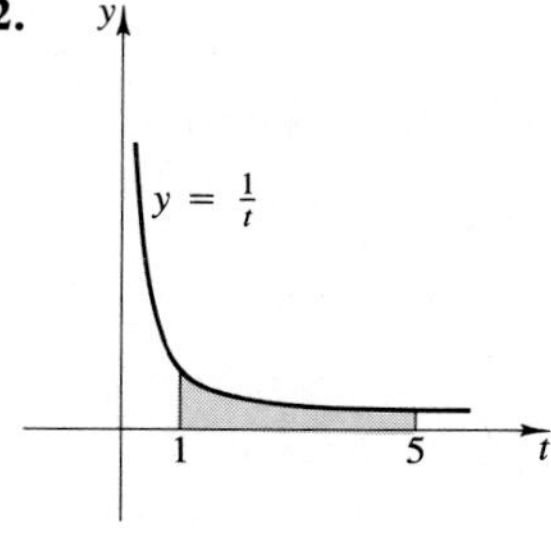

13.

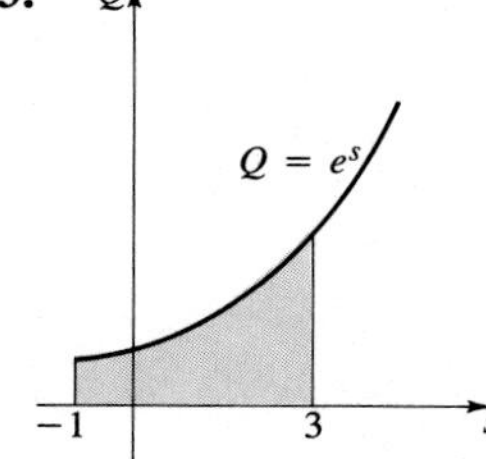

14.

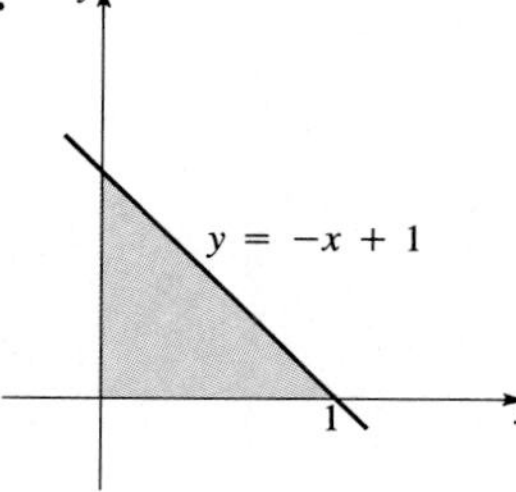

15.

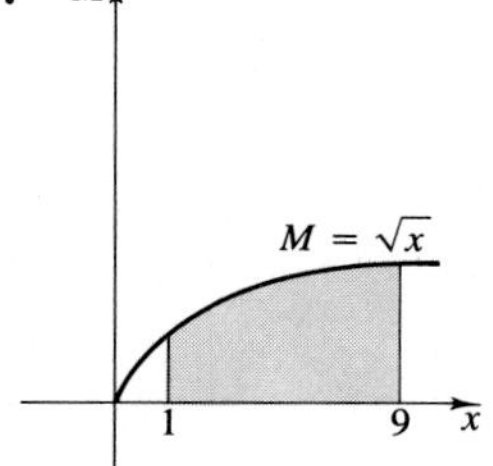

16.

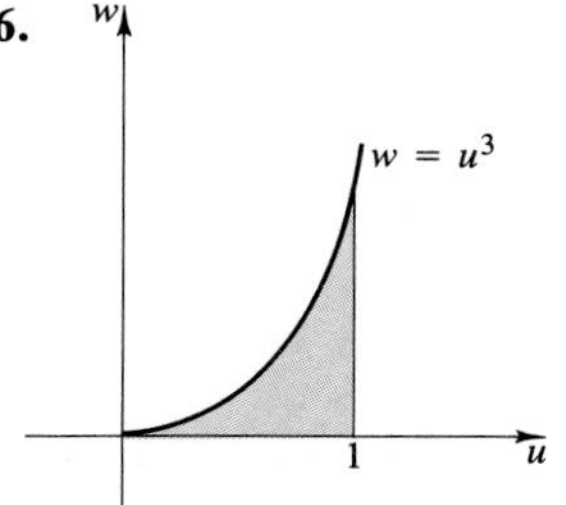

In Exercises 17–30 use a definite integral to find the area under the given curve over the indicated interval.

17. $y = 3, [-1, 2]$

18. $y = 4, [2, 5]$

19. $y = 3x, [1, 4]$

20. $y = 2x, [2, 5]$

21. $y = x^2, [-2, 2]$

22. $y = x^2 - 2, [2, 4]$

23. $y = 3 - x^2, [-1, 1]$

24. $y = 4 + 2x^2, [0, 2]$

25. $y = \sqrt{x}, [1, 4]$

26. $y = 3x - x^2, [0, 3]$

27. $y = e^x, [0, 1]$

28. $y = e^x, [0, 5]$

29. $y = 2/x, [1, 2]$

30. $y = 3/x, [2, 5]$

*In Exercises 31 and 32 for the area function **A(x)** of the shaded region,*

(a) Find $A(x)$ using part I of the fundamental theorem of calculus (the method of Example 4).

(b) Find $A'(x)$, and verify the result of the theorem.

31.

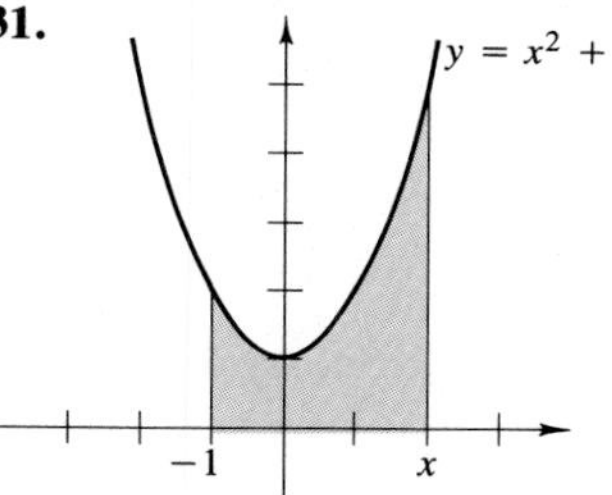

32.

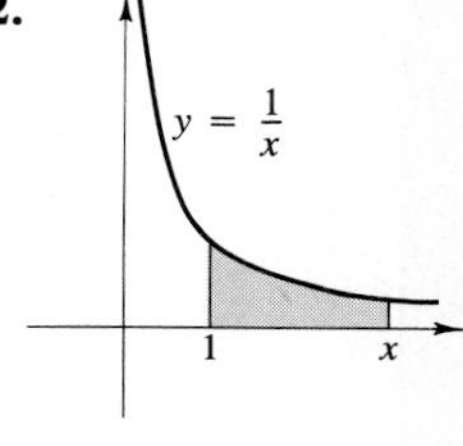

33. Find the area of the shaded region in Figure 11. [*Hint:* The shaded area forms part of a rectangle. Find the area of the remaining portion of the rectangle first.]

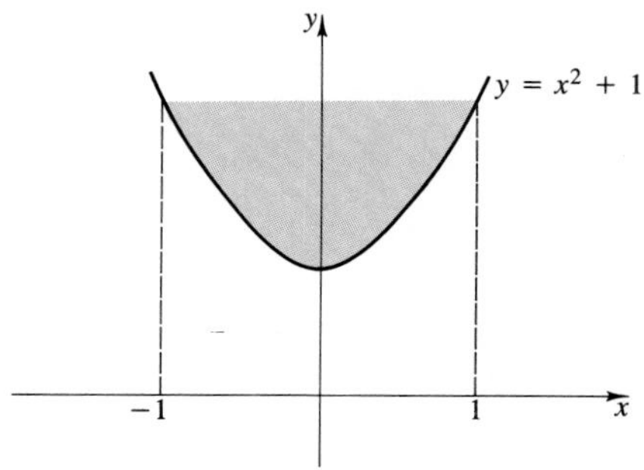

FIGURE 11

15.3 PROPERTIES OF INTEGRALS

In this section we are concerned with formulas that summarize some basic properties of definite integrals. We begin with six properties that follow directly from the corresponding properties of indefinite integrals described in Section 15.1.

Property 1 $\displaystyle\int_a^b k\,dx = k(b - a)$

Property 2 $\displaystyle\int_a^b x^n\,dx = \begin{cases} \dfrac{b^{n+1} - a^{n+1}}{n + 1} & \text{if } n \neq -1 \\ \ln|b| - \ln|a| & \text{if } n = -1 \text{ and } a \text{ and } b \text{ have same sign} \end{cases}$

Property 3 $\displaystyle\int_a^b e^x\,dx = e^b - e^a$

Property 4 $\displaystyle\int_a^b k\,f(x)\,dx = k\int_a^b f(x)\,dx$

Property 5 $\displaystyle\int_a^b [f(x) + g(x)]\,dx = \int_a^b f(x)\,dx + \int_a^b g(x)\,dx$

Property 6 $\displaystyle\int_a^b [f(x) - g(x)]\,dx = \int_a^b f(x)\,dx - \int_a^b g(x)\,dx$

EXAMPLE 1

$$\begin{aligned}\int_2^5 \left(1 + 3x^2 - \frac{1}{x}\right) dx &= \int_2^5 dx + 3\int_2^5 x^2\,dx - \int_2^5 \frac{1}{x}\,dx \\ &= x\,\Big|_2^5 + 3\frac{x^3}{3}\,\Big|_2^5 - \ln|x|\,\Big|_2^5 \\ &= (5 - 2) + 125 - 8 - (\ln|5| - \ln|2|) \\ &= 3 + 117 + \ln 2 - \ln 5 \\ &= 120 + \ln\frac{2}{5}\end{aligned}$$

The next three properties are concerned with the endpoints of the interval of integration.

Property 7 $\displaystyle\int_a^a f(x)\,dx = 0$

To prove Property 7 we need only note that

$$\int_a^a f(x)\,dx = F(a) - F(a) = 0$$

Property 8 $\displaystyle\int_a^b f(x)\,dx = \int_a^c f(x)\,dx + \int_c^b f(x)\,dx$

To obtain Property 8 we need only observe that if F is an antiderivative of f, then

$$\begin{aligned}\int_a^c f(x)\,dx + \int_c^b f(x)\,dx &= [F(c) - F(a)] + [F(b) - F(c)] \\ &= F(b) - F(a) \\ &= \int_a^b f(x)\,dx\end{aligned}$$

When f is a nonnegative function on $[a, b]$ and $a \leq c \leq b$ (Figure 1), then Property 8 states that the area under $y = f(x)$ from a to b is the area from a to c plus the area from c to b.

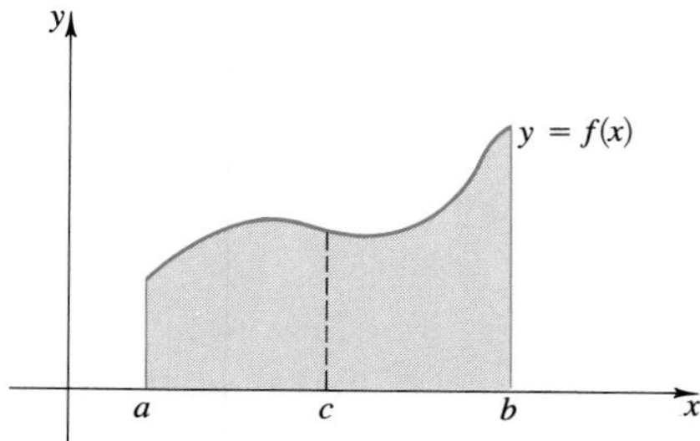

FIGURE 1

EXAMPLE 2

(a) $\displaystyle\int_3^3 (x^4 - 11x^2 + 15)\,dx = 0$

(b) $\displaystyle\int_{-2}^3 x^2\,dx = \int_{-2}^1 x^2\,dx + \int_1^3 x^2\,dx$

Integrating Continuous Piecewise-Defined Functions

Although the definition of the definite integral given in Section 15.2 requires that the integrand be continuous everywhere in the interval of integration, Property 8 can be used to extend that definition to cover piecewise continuous functions as well, as shown in Examples 3, 4, and 5.

EXAMPLE 3 Let $f(x) = |x|$ as shown in Figure 2. To integrate this function over $[-1, 2]$ we partition the interval into two subintervals $[-1, 0]$ and $[0, 2]$, and using Property 8 integrate as follows:

$$\begin{aligned}\int_{-1}^{2} |x|\, dx &= \int_{-1}^{0} |x|\, dx + \int_{0}^{2} |x|\, dx \\ &= \int_{-1}^{0} (-x)\, dx + \int_{0}^{2} x\, dx \\ &= -\frac{x^2}{2}\Big|_{-1}^{0} + \frac{x^2}{2}\Big|_{0}^{2} \\ &= \left[0 - \left(-\frac{1}{2}\right)\right] + \left[\frac{4}{2} - 0\right] \\ &= \frac{1}{2} + 2 \\ &= \frac{5}{2}\end{aligned}$$

FIGURE 2

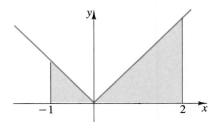

The reason we must use two separate integrals is that there is no single expression that will serve as an antiderivative of $|x|$ over the intervals $[-1, 0)$ and $(0, 2]$.

> **Property 9** If f is continuous everywhere on the closed interval $[a, b]$ and $g(x) = f(x)$ everywhere on the open interval (a, b), then
>
> $$\int_a^b g(x)\, dx = \int_a^b f(x)\, dx$$

EXAMPLE 4 Let $p(x)$ denote the "postage function," introduced in Example 5 of Section 12.5 (Figure 3). We can use Property 9 to integrate $p(x)$ between successive integers:

$$\begin{aligned}\int_2^3 p(x)\, dx &= \int_2^3 75\, dx \qquad \text{because } p(x) = 75 \text{ for } 2 \le x < 3 \\ &= 75(3 - 2) \\ &= 75\end{aligned}$$

Note that $p(3) \neq 75$, but that does not affect the value of the integral since

$$\int_3^3 p(x)\, dx = 0$$

FIGURE 3

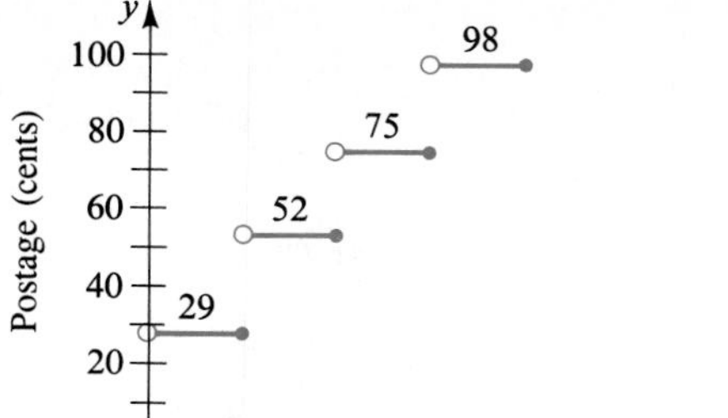

EXAMPLE 5 Evaluate $\int_0^4 p(x)\,dx$, where $p(x)$ denotes the postage function.

Solution Referring to Figure 3 we see that $p(x)$ has discontinuities within the interval $[0, 4]$ at $x = 0, 1, 2, 3$, and 4. Thus with the help of Property 8,

$$\int_0^4 p(x)\,dx = \int_0^1 p(x)\,dx + \int_1^2 p(x)\,dx + \int_2^3 p(x)\,dx + \int_3^4 p(x)\,dx$$

Using Property 9 on each of the integrals separately, we obtain

$$\int_0^4 p(x)\,dx = \int_0^1 29\,dx + \int_1^2 52\,dx + \int_2^3 75\,dx + \int_3^4 98\,dx$$

Then, from Property 1,

$$\int_0^4 p(x)\,dx = 29(1-0) + 52(2-1) + 75(3-2) + 98(4-3)$$

$$= 254$$

This answer may be understood by observing that the area under the four "steps" shown in Figure 3 is 254 square units.

Areas Below the *x*-axis

In the previous section we see that the definite integral $\int_a^b f(x)\,dx$ of a function f that is continuous and nonnegative on $[a, b]$ can be interpreted as the area under the graph of f over the interval $[a, b]$. However, if you will take a moment to reread the definition of the definite integral, you will see that we do not require f to be nonnegative on $[a, b]$; indeed, the defining equation

$$\int_a^b f(x)\,dx = F(b) - F(a)$$

makes perfectly good sense, even if f assumes both positive and negative values on $[a, b]$. For example, the function $f(x) = 1 - x$ has both positive and negative values for x in the interval $[0, 2]$ (Figure 4), yet we can still write

$$\int_0^2 (1 - x)\, dx = \int (1 - x)\, dx \Big|_0^2 = \left(x - \frac{x^2}{2}\right)\Big|_0^2 = \left[2 - \frac{4}{2}\right] - [0] = 0$$

What does the integral $\int_a^b f(x)\, dx$ represent when $f(x)$ assumes both positive and negative values on $[a, b]$? Clearly, the integral *cannot* represent the total area between the graph of f and the interval $[a, b]$ since we see above that

$$\int_0^2 (1 - x)\, dx = 0$$

yet from Figure 4 the total area bounded by $f(x) = 1 - x$, the x-axis, $x = 0$, and $x = 2$ is not zero.

FIGURE 4

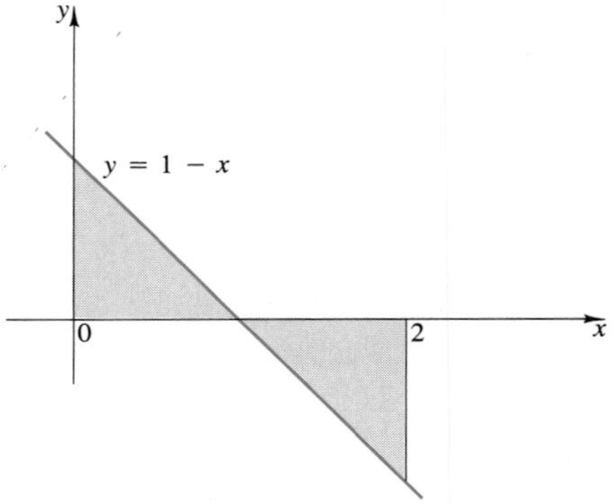

We now try to find the meaning of a definite integral whose integrand has negative values.

Suppose that $g(x) \geq 0$ on $[a, b]$ and $y = f(x)$ is the reflection (or "mirror image") of the curve $y = g(x)$ about the x-axis (Figure 5). As shown in Figure 5a, the functions f and g are related by $f(x) = -g(x)$ for each x in $[a, b]$. Thus, to maintain the validity of Property 4,

$$\int_a^b f(x)\, dx = \int_a^b -g(x)\, dx = -\int_a^b g(x)\, dx \tag{1}$$

Since $\int_b^a g(x)\, dx$ represents the area A between $y = g(x)$ and the interval $[a, b]$, it follows from (1) that

$$\int_a^b f(x)\, dx = -A \tag{2}$$

By symmetry, A is also the area between $y = f(x)$ and the interval $[a, b]$ (Figure 5b), so Equation (2) yields the following result:

FIGURE 5

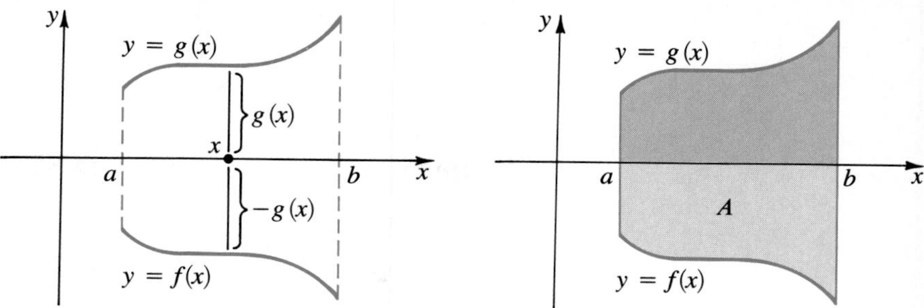

Property 10 If $f(x) \le 0$ for all x in $[a, b]$, then

$$\int_a^b f(x)\,dx$$

represents the *negative* of the area between $y = f(x)$ and the interval $[a, b]$.

To interpret $\int_a^b f(x)\,dx$ in the case where $f(x)$ has both positive and negative values on $[a, b]$, consider the curve $y = f(x)$ in Figure 6. According to Property 8 for definite integrals, we can evaluate $\int_a^b f(x)\,dx$ by writing

$$\int_a^b f(x)\,dx = \int_a^c f(x)\,dx + \int_c^b f(x)\,dx \tag{3}$$

FIGURE 6

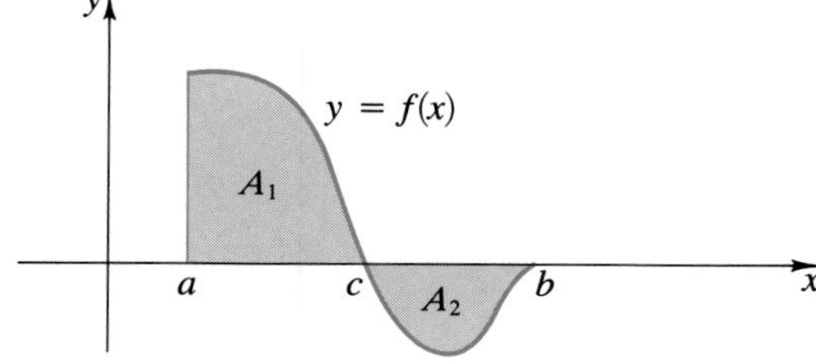

Consider the areas of A_1 and A_2 shown in Figure 6. Since $f(x) \ge 0$ on the interval $[a, c]$, it follows that

$$\int_a^c f(x)\,dx = A_1$$

and since $f(x) \le 0$ on the interval $[c, b]$, it follows that

$$\int_c^b f(x)\,dx = -A_2$$

Thus, (3) can be written as

$$\int_a^b f(x)\,dx = A_1 + (-A_2) = A_1 - A_2 \tag{4}$$

This illustrates the following result:

Property 11 (A Geometric Interpretation of the Definite Integral) If f is a continuous (or piecewise continuous) function on $[a, b]$, then the definite integral $\int_a^b f(x)\,dx$ represents the **net area** between the curve $y = f(x)$ and the x-axis, that is, the area above the x-axis minus the area below the x-axis.

EXAMPLE 6 We have seen that

$$\int_0^2 (1 - x)\,dx = 0$$

The explanation for this result should now be clear from Figure 4. The area of the triangular region above the x-axis is equal to the triangular area below the x-axis.

EXAMPLE 7 We have

$$\int_{-2}^{3} (x^2 - 4)\,dx = \int (x^2 - 4)\,dx \Big|_{-2}^{3} = \left(\frac{x^3}{3} - 4x\right)\Big|_{-2}^{3}$$

$$= \left(\frac{27}{3} - 12\right) - \left(-\frac{8}{3} + 8\right)$$

$$= -3 - \frac{16}{3} = -\frac{25}{3}$$

The negative value for the answer indicates that the portion of the area between the curve $y = x^2 - 4$ and the interval $[-2, 3]$ lying below the x-axis is greater than the portion that lies above the x-axis (Figure 7).

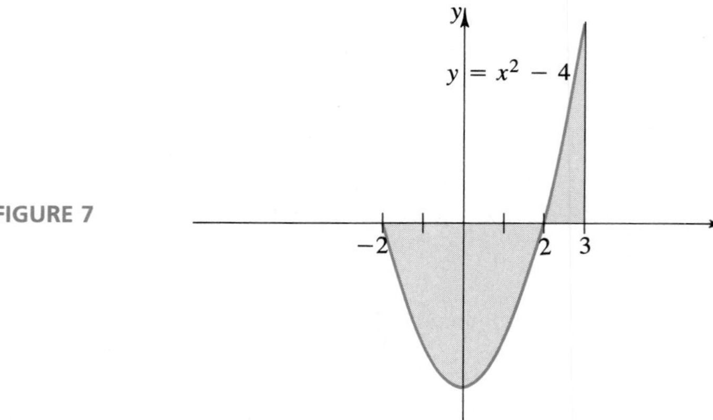

FIGURE 7

EXAMPLE 8 Show that the result obtained in Example 7 is the area above the x-axis under the curve minus the area below.

Solution We must calculate separately the areas above and below the x-axis and then subtract.

The area above the x-axis is

$$\int_2^3 (x^2 - 4)\,dx = \int (x^2 - 4)\,dx\,\Big|_2^3 = \left(\frac{x^3}{3} - 4x\right)\Big|_2^3$$

$$= \left(\frac{27}{3} - 12\right) - \left(\frac{8}{3} - 8\right)$$

$$= -3 - \left(-\frac{16}{3}\right) = \frac{7}{3}$$

The *negative* of the area below the x-axis is

$$\int_{-2}^2 (x^2 - 4)\,dx = \int (x^2 - 4)\,dx\,\Big|_{-2}^2 = \left(\frac{x^3}{3} - 4x\right)\Big|_{-2}^2$$

$$= \left(\frac{8}{3} - 8\right) - \left(-\frac{8}{3} + 8\right)$$

$$= \frac{16}{3} - 16 = -\frac{32}{3}$$

Thus the area below the x-axis is $\frac{32}{3}$ and the area above minus the area below is

$$\frac{7}{3} - \frac{32}{3} = -\frac{25}{3}$$

This agrees with the result in Example 7.

To determine the area between a curve defined by a continuous function f and the x-axis we proceed as follows:

Step 1 Find the values of x at which the curve crosses the x-axis. That is, set $f(x)$ equal to zero and solve for x. Also, sketch the graph of $f(x)$.

Step 2 Find the areas that lie above the x-axis and those that lie below the x-axis.

Step 3 Add the areas lying above the x-axis to those lying below.

Figure 8 illustrates this procedure. In this figure the total area between the curve $y = f(x)$ and the x-axis, from a to b, is

$$A = A_1 + A_2 + A_3$$

$$= \int_a^{c_1} f(x)\,dx - \int_{c_1}^{c_2} f(x)\,dx + \int_{c_2}^{b} f(x)\,dx$$

(Note: on $[c_1, c_2]$, $A_2 = -\int_{c_1}^{c_2} f(x)\,dx$.)

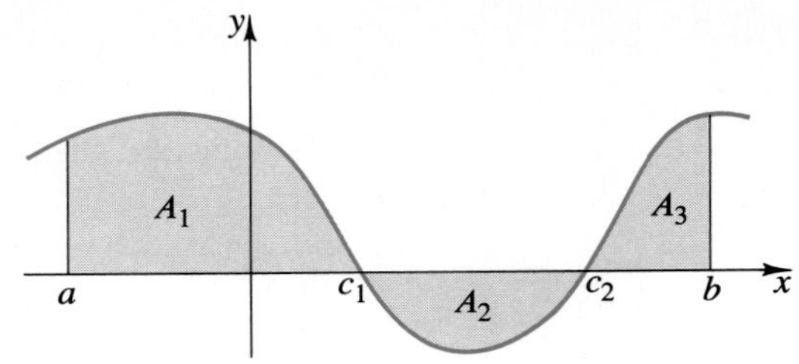

FIGURE 8

EXAMPLE 9 Find the area between the curve $y = x^2 - x$ and the x-axis from $x = 0$ to $x = 3$.

Solution Figure 9 shows the graph of the curve.

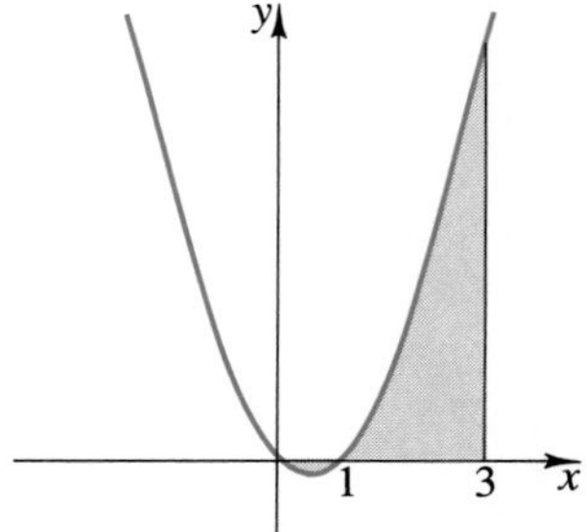

FIGURE 9

Step 1 Set $f(x) = x^2 - x = 0$ and solve for x:

$$x(x - 1) = 0$$
$$x = 0 \qquad x = 1$$

Thus, the curve crosses the x-axis at $x = 0$ and $x = 1$.

Step 2 The area below the x-axis is

$$A_1 = -\int_0^1 (x^2 - x)\, dx = -\left(\frac{x^3}{3} - \frac{x^2}{2}\right)\Bigg|_0^1$$
$$= -\left[\left(\frac{1}{3} - \frac{1}{2}\right) - 0\right] = \frac{1}{6}$$

The area above the x-axis is

$$A_2 = \int_1^3 (x^2 - x)\, dx$$
$$= \frac{x^3}{3} - \frac{x^2}{2}\Bigg|_1^3$$
$$= \frac{27}{3} - \frac{9}{2} - \left(\frac{1}{3} - \frac{1}{2}\right) = \frac{28}{6}$$

Step 3 The area between the curve and the x-axis is then

$$A_1 + A_2 = \frac{1}{6} + \frac{28}{6} = \frac{29}{6}$$

EXERCISE SET 15.3

1. (a) Evaluate the definite integral $\int_2^3 (3 - x)\, dx$.
   (b) Use the formula for the area of a triangle (area $= \frac{1}{2} \cdot$ base $\cdot$ height) and the graph of $y = 3 - x$ in Figure 10 to check the result in (a).

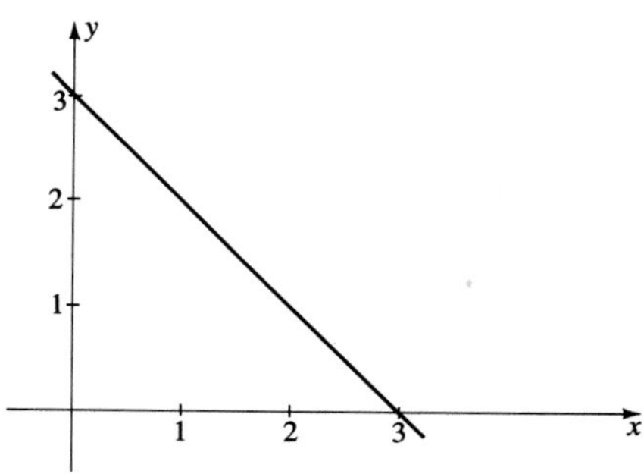

FIGURE 10

2. (a) Evaluate the definite integral $\int_0^2 \frac{1}{2}x\, dx$.
   (b) Use the formula for the area of a triangle (area $= \frac{1}{2} \cdot$ base $\cdot$ height) and the graph of $y = \frac{1}{2}x$ (Figure 11) to check the result in (a).

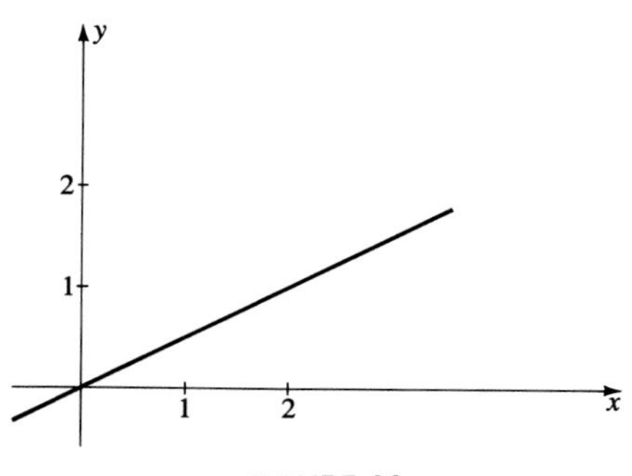

FIGURE 11

3. (a) Evaluate the definite integral $\int_0^3 (2 - x)\, dx$.
   (b) Use the formula for the area of a triangle (area $= \frac{1}{2} \cdot$ base $\cdot$ height) and the graph of $y = 2 - x$ (Figure 12) to check the result in (a).

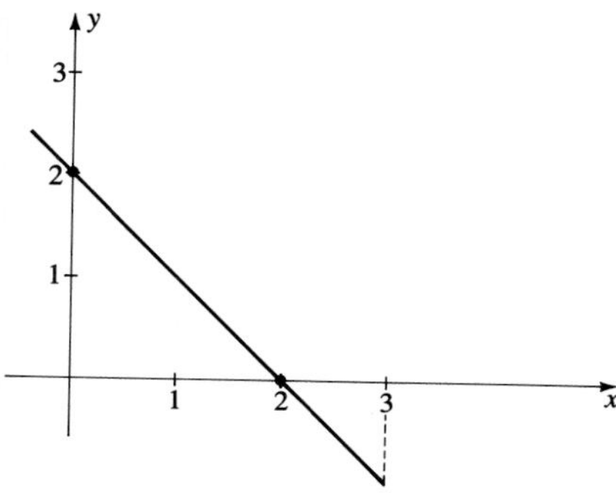

FIGURE 12

4. (a) Evaluate the definite integral $\int_0^1 (3 - x)\, dx$.
   (b) Use the formula for the area of a trapezoid [area $= \frac{1}{2}$(sum of the bases) $\cdot$ altitude] and the graph in Figure 13 to check the result of (a).

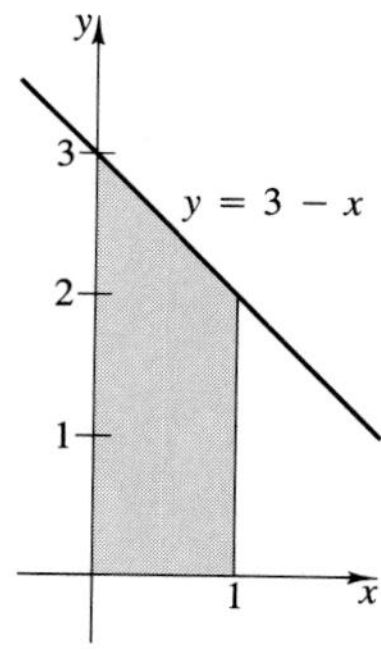

FIGURE 13

5. (a) Evaluate $\int_0^1 x^2\, dx + \int_1^2 x^2\, dx$.
   (b) Write the sum in (a) as a single definite integral and evaluate this integral.

6. (a) Evaluate $\int_{-1}^2 e^x\, dx + \int_2^3 e^x\, dx$.
   (b) Write the sum in (a) as a single integral and evaluate this integral.

7. (a) Evaluate $\int_1^3 1/x\, dx + \int_3^5 1/x\, dx$.
   (b) Write the sum in (a) as a single integral and evaluate this integral.

8. Suppose $\int_2^8 f(x)\,dx = 24$ and $\int_6^8 f(x)\,dx = 19$. Use Property 8 to find $\int_2^6 f(x)\,dx$.

9. Suppose $\int_{-1}^{10} f(x)\,dx = 30$ and $\int_{-1}^{3} f(x)\,dx = 8$. Use Property 8 to find $\int_3^{10} f(x)\,dx$.

In Exercises 10–22 evaluate the definite integral.

10. $\int_2^4 3x\,dx$
11. $\int_2^5 5x^2\,dx$
12. $\int_0^1 (2x^2 + 3x^4)\,dx$
13. $\int_1^2 (3x^3 - 7x^2)\,dx$
14. $\int_{-1}^2 (4x^3 - 2x^2 + 2)\,dx$
15. $\int_9^{16} 5\sqrt{x}\,dx$
16. $\int_0^1 (2\sqrt{s} + s^2 - 2)\,ds$
17. $\int_0^4 (2e^t - 4t^3)\,dt$
18. $\int_3^3 x^{11}\,dx$
19. $\int_7^7 \sqrt[3]{x+1}\,dx$
20. $\int_1^8 (2u^{-1/3} + 2u^3)\,du$
21. $\displaystyle\int_1^2 \left(3x - \frac{2}{3x^2} + \frac{4}{x}\right) dx$
22. $\int_1^8 (x^{1/5} - x^{1/3})\,dx$

In Exercises 23–26 evaluate the given integral using the method of Example 3. In Exercises 23–25 check your answer by drawing the graph and finding the area of the appropriate triangle.

23. $\int_{-2}^2 |x|\,dx$
24. $\int_{-1}^3 |2x|\,dx$
25. $\int_1^4 |x - 3|\,dx$
26. $\int_0^3 |x - 2|\,dx$

In Exercises 27–34 find the areas of the shaded regions.

27.

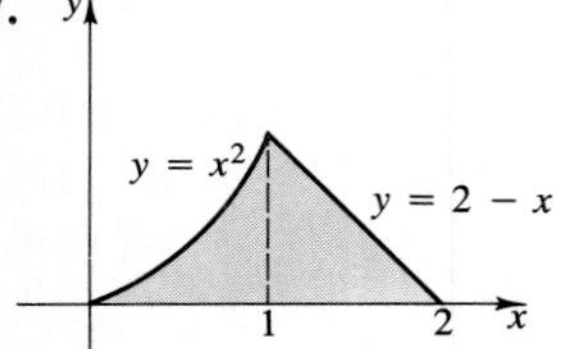

28.

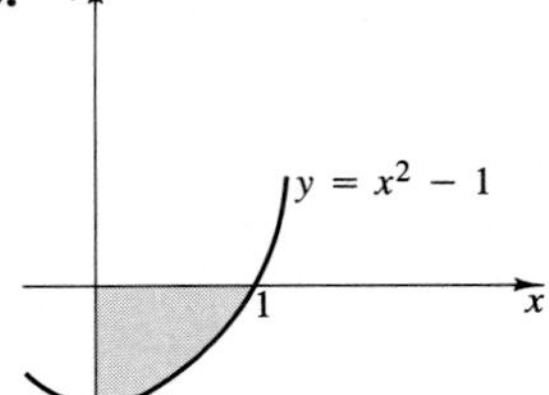

29.

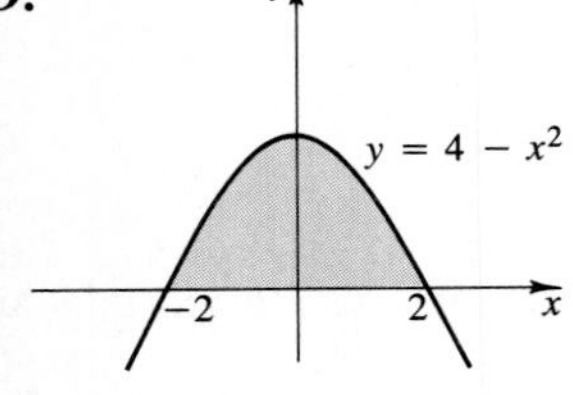

30.

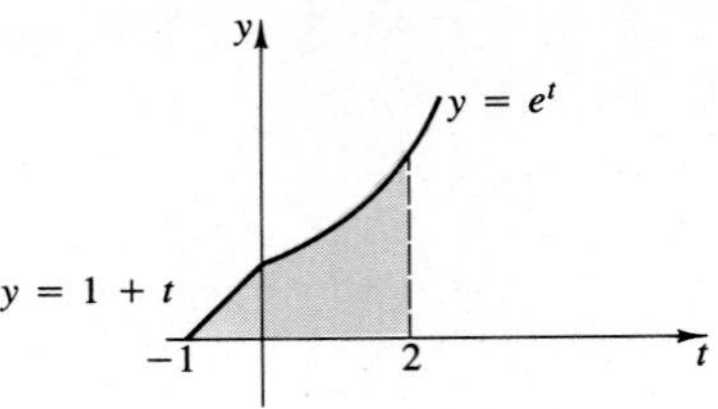

31.

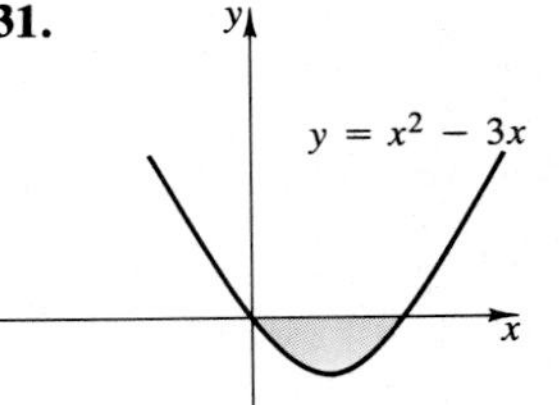

32.

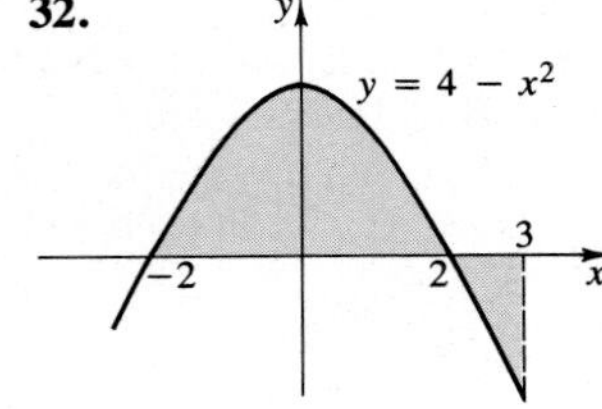

33.

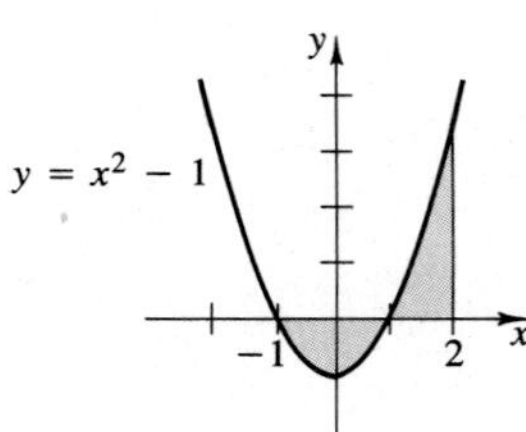

34.

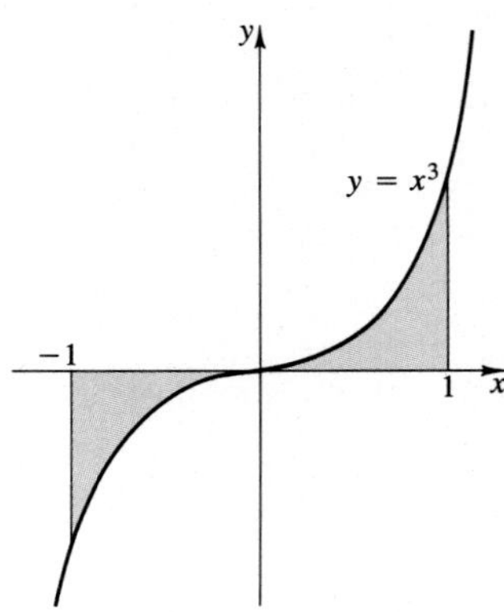

15.4 THE DEFINITE INTEGRAL AS A LIMIT OF RIEMANN SUMS

In our earlier development of the definite integral, $\int_a^b f(x)\,dx$, we have taken an approach that was motivated by the desire to keep the exposition simple and easy to follow. This approach required us to oversimplify the situation in at least two essential ways:

1. We have *assumed* the existence of the area function $A(x)$ and that it behaves according to our intuitive expectations. Actually, we have not even attempted an adequate *definition* of "area."
2. Our use of $\int_a^b f(x)\,dx$ required that $f(x)$ have an antiderivative. We have not considered cases in which we cannot find an antiderivative of f. To provide an integral in such cases we must take a completely different approach to the definition of $\int_a^b f(x)\,dx$.

More than a century ago, Bernhard Riemann* developed a rigorous approach to the definite integral that removes these objections. Riemann's approach simultaneously *defines* both "area"—between a curve $y = f(x)$ and the x-axis—and "integral" and assures that they are equal, without any reference to antidifferentiation. We can explain his approach through the following (simplified) procedure.

Consider an arbitrary function $f(x)$ that is continuous and nonnegative over the closed interval $[a, b]$. The graph of such a function over $[a, b]$ will not go below the x-axis. Divide the interval $[a, b]$ into a fixed number, say, n, of subintervals of equal length. For example, in Figure 1 we have divided $[a, b]$ into six such subintervals. Using these subintervals we can decompose the region under the curve $y = f(x)$ over $[a, b]$ into vertical strips, as indicated in Figure 1. Each vertical strip has width

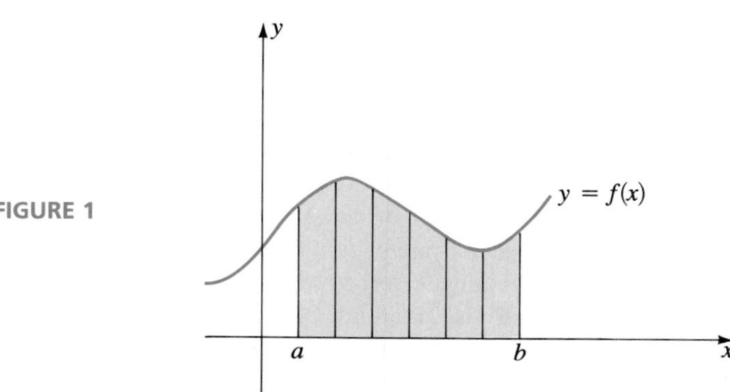

FIGURE 1

*_Georg Friedrich Bernhard Riemann_ (1826–1866) was a German mathematician. Bernhard Riemann, as he is commonly known, was the son of a Protestant minister. He obtained his elementary education from his father and showed brilliance in mathematics at an early age. In college, at Göttingen University, he studied theology and philology but eventually transferred to mathematics and studied under Gauss. In 1862 Riemann contracted pleuritis and was seriously ill for the rest of his life. He died in 1866 at the age of 39. Riemann's early death was unfortunate since his mathematical work was brilliant and of fundamental importance. His work in geometry was used by Albert Einstein some 50 years later in formulating the theory of relativity.

$$\Delta x = \frac{b - a}{n} \tag{1}$$

(Read Δx as "delta x"; Δ is the Greek letter *delta*. In this notation the symbol is "Δx.")

In general, there is no simple formula for the areas of these strips because of their curved upper boundaries. However, we can *approximate* each strip by a rectangle whose base is Δx and whose height is the value of $f(x)$ at a point $\overline{x}_i$ chosen in the corresponding subinterval (Figure 2).

FIGURE 2

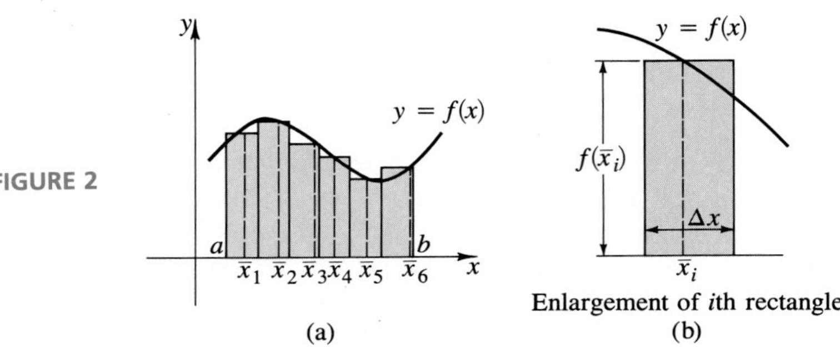

(a)

Enlargement of ith rectangle
(b)

We let A_1, A_2, A_3, and so on, denote the areas of the rectangles so formed. Then the area of the ith rectangle (Figure 2b) is

$$A_i = f(\overline{x}_i)\,\Delta x \tag{2}$$

Adding the areas of all such rectangles we obtain the **Riemann sum**

$$\begin{aligned} R_n &= A_1 + A_2 + \cdots + A_n \\ &= f(\overline{x}_1)\,\Delta x + f(\overline{x}_2)\,\Delta x + \cdots + f(\overline{x}_n)\,\Delta x \\ &= [f(\overline{x}_1) + f(\overline{x}_2) + \cdots + f(\overline{x}_n)]\,\Delta x \end{aligned} \tag{3}$$

Each such Riemann sum serves as an approximation to the area under $y = f(x)$ over $[a, b]$.

Now we come to the crucial observation. As we let the number of subintervals become larger and larger (that is, $n \to \infty$), the widths of all the approximating rectangles decrease ($\Delta x \to 0$), and the discrepancy between the Riemann sum R_n and the area under $y = f(x)$ over $[a, b]$ decreases to 0 (Figure 3). Thus Riemann *defines the area* under the curve $y = f(x)$ over $[a, b]$ to be $\lim\limits_{n\to\infty} R_n$, in cases where this limit exists. He also *defines the definite integral* for continuous functions $f(x)$ to be the same limit.

$$\int_a^b f(x)\,dx = \lim_{n\to\infty} R_n \tag{4}$$

In this way the definite integral is defined as a limit without any reference to a picture or to antidifferentiation.

FIGURE 3

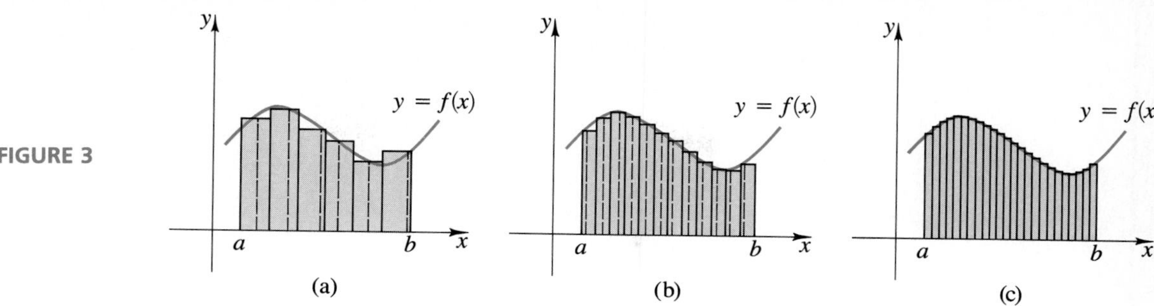

EXAMPLE 1 Calculate $\int_0^2 (x + 1)\,dx$ by Riemann's definition.

Solution We begin by dividing the interval $[0, 2]$ into n equal subintervals. Each such subinterval will have width

$$\Delta x = \frac{2-0}{n} = \frac{2}{n}$$

(See Figure 4.)

FIGURE 4

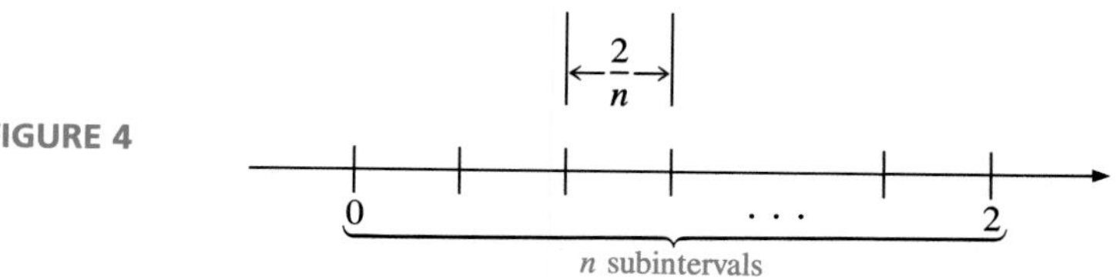

Over each subinterval we construct a rectangle of height $f(x)$, using any point $\overline{x}_i$ in the subinterval. To be specific, we may choose $\overline{x}_i$ to be the right endpoint of the interval (Figure 5).

FIGURE 5

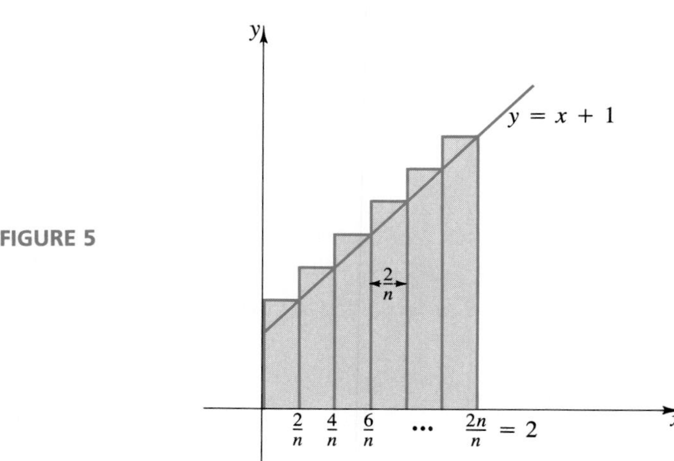

TABLE 1

i	$\overline{x}_i$	$f(\overline{x}_i)$	Δx	A_i
1	$\frac{2}{n}$	$\frac{2}{n} + 1$	$\frac{2}{n}$	$\left(\frac{2}{n} + 1\right)\frac{2}{n}$
2	$\frac{4}{n}$	$\frac{4}{n} + 1$	$\frac{2}{n}$	$\left(\frac{4}{n} + 1\right)\frac{2}{n}$
3	$\frac{6}{n}$	$\frac{6}{n} + 1$	$\frac{2}{n}$	$\left(\frac{6}{n} + 1\right)\frac{2}{n}$
$\vdots$	$\vdots$	$\vdots$	$\vdots$	$\vdots$
n	$\frac{2n}{n}$	$\frac{2n}{n} + 1$	$\frac{2}{n}$	$\left(\frac{2n}{n} + 1\right)\frac{2}{n}$

To calculate the Riemann sum R_n we use Table 1 and Formula (3). We obtain

$$\begin{aligned} R_n &= \left[\left(\frac{2}{n} + 1\right) + \left(\frac{4}{n} + 1\right) + \left(\frac{6}{n} + 1\right) + \cdots + \left(\frac{2n}{n} + 1\right)\right]\frac{2}{n} \\ &= \left[\left(\frac{2}{n} + \frac{4}{n} + \frac{6}{n} + \cdots + \frac{2n}{n}\right) + (1 + 1 + \cdots + 1)\right]\frac{2}{n} \\ &= \left[\frac{2}{n}(1 + 2 + 3 + \cdots + n) + n\right]\frac{2}{n} \end{aligned} \tag{5}$$

For example, when we take 10 subintervals ($n = 10$), Formula (5) yields ($\Delta x = \frac{2}{10} = 0.2$):

$$\begin{aligned} R_{10} &= [0.2(1 + 2 + \cdots + 10) + 10](0.2) \\ &= [0.2(55) + 10](0.2) \\ &= 4.2 \end{aligned}$$

In this case the area of the *shaded* region in Figure 5 is equal to 4.2. This number is not the actual area under the curve $y = x + 1$ over $[0, 2]$. To get a better approximation to that area we could take more subintervals. For example, taking $n = 100$, Formula (5) yields

$$\begin{aligned} R_{100} &= [0.02(1 + 2 + \cdots + 100) + 100](0.02) \\ &= [101 + 100](0.02) \\ &= 4.02 \end{aligned}$$

With the help of a calculator or computer we can calculate R_n for larger and larger values of n. For example, $R_{1000} = 4.0020$ and $R_{10,000} = 4.0002$. From this we might guess that $\lim_{n\to\infty} R_n = 4$. To actually *prove* that the limit is 4 we need the first of the following two identities.

Two Identities For every positive integer n,

$$1 + 2 + 3 + \cdots + n = \frac{n(n+1)}{2} \qquad (6)$$

$$1^2 + 2^2 + 3^2 + \cdots + n^2 = \frac{n(n+1)(2n+1)}{6} \qquad (7)$$

We apply the first identity (6) to Equation (5) and obtain

$$\begin{aligned} R_n &= \left[\frac{2}{n} \cdot \frac{n(n+1)}{2} + n\right]\frac{2}{n} \\ &= [(n+1) + n]\frac{2}{n} \\ &= (2n+1)\left(\frac{2}{n}\right) \\ &= 4 + \frac{2}{n} \end{aligned}$$

From this it is clear that $\lim_{n\to\infty} R_n = 4$.

We can easily compare this result with our previous method of evaluating a definite integral, as follows:

$$\begin{aligned} \int_0^2 (x+1)\,dx &= \left(\frac{1}{2}x^2 + x\right)\Big|_0^2 \\ &= (2+2) - 0 \\ &= 4 \end{aligned}$$

Riemann's method is *not* easier than antidifferentiation! Its main purpose is to provide a rigorous mathematical foundation for the definite integral, but it is also important in many advanced applications. We will see some of these applications in the next chapter.

The fundamental theorem of calculus can now be restated as:

FUNDAMENTAL THEOREM OF CALCULUS

PART I If $f(x)$ is continuous on a closed interval $[a, b]$, then $f(x)$ has an antiderivative $F(x)$.

PART II If $\Delta x = \dfrac{b-a}{n}$ and $\overline{x}_i$ is any point in the interval $[x_{i-1}, x_i]$, then

$$\begin{aligned} \int_a^b f(x)\,dx &= \lim_{n\to\infty} [f(\overline{x}_1) + f(\overline{x}_2) + \cdots + f(\overline{x}_n)\,\Delta x \\ &= F(b) - F(a) \end{aligned}$$

THE VOLUME OF A SPHERE

The definite integral as a limit of Riemann sums can be used to compute the volume of solids of revolution. We illustrate by computing the volume of a sphere.

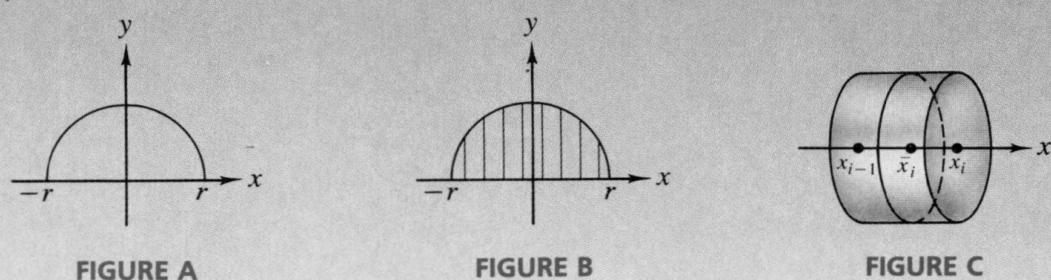

FIGURE A FIGURE B FIGURE C

As shown in Figure A, a sphere of radius r can be obtained by rotating the upper half of the circle $x^2 + y^2 = r^2$ about the x-axis. Note that the equation of the upper half of this circle is $y = f(x) = \sqrt{r^2 - x^2}$. Now divide the interval from $-r$ to r into a fixed number, say n, of subintervals of equal length. Using these subintervals, the region under the upper half of the circle is divided into vertical strips as shown in Figure B. Select a point $\bar{x}_i$ in each subinterval $[x_{i-1}, x_i]$. When the circle is rotated about the x-axis, each vertical strip will generate a disk whose radius is approximately $f(\bar{x}_i)$ and whose thickness is $\Delta x \left(= \dfrac{2r}{n}\right)$. See Figure C. Since the volume of a disk is

$$\pi \cdot (\text{radius})^2 \cdot (\text{thickness})$$

the volume of the disk shown in Figure C is approximately

$$\pi[f(\bar{x}_i)]^2 \, \Delta x$$

Adding up the volumes of the disks from 1 to n yields an approximate value for the volume of the sphere:

$$\begin{aligned} V &\cong \pi[f(\bar{x}_1)]^2 \, \Delta x + \pi[f(\bar{x}_2)]^2 \, \Delta x + \cdots + \pi[f(\bar{x}_n)]^2 \, \Delta x \\ &= \pi\{[f(\bar{x}_1)]^2 + [f(\bar{x}_2)]^2 + \cdots + [f(\bar{x}_n)]^2\} \, \Delta x \end{aligned} \tag{a}$$

As n approaches ∞ and as Δx approaches 0, the Riemann sum in (a) simultaneously approaches the volume of the sphere and the integral $\pi \int_{-r}^{r} [f(x)]^2 \, \Delta x$. Thus

$$\begin{aligned} V &= \pi \int_{-r}^{r} [f(x)]^2 \, dx \\ &= \pi \int_{-r}^{r} [r^2 - x^2] \, dx \\ &= \pi \left[r^2 x - \frac{x^3}{3} \right] \Bigg|_{-r}^{r} \\ &= \frac{4}{3} \pi r^3 \end{aligned}$$

Hence, the volume of a sphere is $\frac{4}{3}\pi r^3$.

EXERCISE SET 15.4

1. Let $f(x) = 2x$ over the interval $I = [0, 8]$.
   (a) Divide the interval I into 4 subintervals of equal length (that is, $n = 4$). In each of these 4 subintervals take $\bar{x}_i$ to be the right endpoint and calculate the resulting Riemann sum R_n (use the method of Example 1). Draw a graph illustrating the procedure.
   (b) Repeat part (a) using $n = 8$ instead of $n = 4$.

2. Repeat Exercise 1, but in each subinterval choose $\bar{x}_i$ to be the left endpoint instead of the right endpoint.

3. Repeat Exercise 1, but in each subinterval choose $\bar{x}_i$ to be the midpoint.

4. Compare the answers in Exercises 1, 2, and 3 with the actual value of $\int_0^8 2x\,dx$. Show that in this case the integral equals one of the Riemann sums and is the average of the other two.

5. Evaluate $\int_0^1 3x\,dx$ by Riemann's method using the following procedure. Let n be a general positive integer. For the function $f(x) = 3x$ over the interval $I = [0, 1]$, take $\bar{x}_i$ to be the right endpoint in every subinterval and find the formula for the resulting Riemann sum R_n. Apply identity (6); then find $\lim_{n\to\infty} R_n$. Compare your answer with $\int_0^1 3x\,dx$.

6. Repeat Exercise 5, taking $\bar{x}_i$ to be the left endpoint in each subinterval. Compare this answer with that of Exercise 5.

7. Let $f(x) = x^2 + 1$ over the interval $I = [0, 4]$. Follow the instructions (a) and (b) of Exercise 1.

8. Repeat Exercise 7, but in each subinterval choose $\bar{x}_i$ to be the left endpoint.

9. Repeat Exercise 7, but in each subinterval choose $\bar{x}_i$ to be the midpoint.

10. Compare the answers in Exercises 7, 8, and 9 with the actual value of $\int_0^4 (x^2 + 1)\,dx$. Show that in this case, the integral equals none of the Riemann sums and is not the average of any two of them.

11. Evaluate $\int_0^1 x^2\,dx$ by Riemann's method in the following way. Let n be a general positive integer. For the function $f(x) = x^2$ over the interval $I = [0, 1]$, take $\bar{x}_i$ to be the right endpoint in every subinterval (as in Exercise 7) and find the formula for the resulting Riemann sum R_n. Apply identity (7); then find $\lim_{n\to\infty} R_n$. Compare your answer with $\int_0^1 x^2\,dx$.

12. Repeat Exercise 11, taking $\bar{x}_i$ to be the left endpoint in each subinterval. Compare this answer with that of Exercise 11.

13. Evaluate $\int_0^2 (2x + 3)\,dx$ by Riemann's method.

14. Evaluate $\int_1^3 (x - 1)\,dx$ by Riemann's method.

15. Evaluate $\int_0^3 x^2\,dx$ by Riemann's method.

16. Evaluate $\int_0^2 (x^2 - 3)\,dx$ by Riemann's method.

KEY IDEAS FOR REVIEW

- **Antiderivative of $f(x)$** A function $F(x)$ such that $F'(x) = f(x)$.
- **Integration (antidifferentiation)** The process of finding an antiderivative.
- **Indefinite integral** The symbol $\int f(x)\,dx$ denoting an antiderivative of $f(x)$.
- **Basic integration formulas**

$$\int k\,dx = kx + C$$

$$\int x^k\,dx = \frac{x^{k+1}}{k+1} + C \qquad (k \neq -1)$$

$$\int e^x\,dx = e^x + C$$

$$\int \frac{1}{x}\,dx = \ln|x| + C$$

Rules of integration

$$\int kf(x)\,dx = k\int f(x)\,dx \qquad \text{(constant times a function rule)}$$

$$\int [f(x) + g(x)]\,dx = \int f(x)\,dx + \int g(x)\,dx \qquad \text{(sum rule)}$$

$$\int [f(x) - g(x)]\,dx = \int f(x)\,dx - \int g(x)\,dx \qquad \text{(difference rule)}$$

Definite integral $\int_a^b f(x)\,dx.$

Properties of the definite integral

$$\int_a^b kf(x)\,dx = k\int_a^b f(x)\,dx$$

$$\int_a^b [f(x) + g(x)]\,dx = \int_a^b f(x)\,dx + \int_a^b g(x)\,dx$$

$$\int_a^b [f(x) - g(x)]\,dx = \int_a^b f(x)\,dx - \int_a^b g(x)\,dx$$

$$\int_a^b f(x)\,dx = \int_a^c f(x)\,dx + \int_c^b f(x)\,dx$$

$$\int_a^a f(x)\,dx = 0$$

Geometric interpretation of the definite integral

$$\int_a^b f(x)\,dx = \begin{bmatrix}\text{area under } y = f(x)\\ \text{above } [a, b]\end{bmatrix} - \begin{bmatrix}\text{area above } y = f(x)\\ \text{below } [a, b]\end{bmatrix}$$

Riemann sum

$$\begin{aligned} R_n &= A_1 + A_2 + \cdots + A_n \\ &= [f(\overline{x}_1) + f(\overline{x}_2) + \cdots + f(\overline{x}_n)]\,\Delta x \end{aligned}$$

Fundamental theorem of calculus If $f(x)$ is continuous on $[a, b]$, then $f(x)$ has an antiderivative $F(x)$ and

$$\int_a^b f(x)\,dx = \lim_{n\to\infty} R_n = F(b) - F(a)$$

SUPPLEMENTARY EXERCISES

In Exercises 1–4 find the indefinite integral.

1. $\int (2x - x^2)\,dx$ **2.** $\int (3e^x + 4/x)\,dx$

3. $\int (2/t^4 - t^{-1/2})\,dt$ **4.** $\int (\sqrt{x} + 1/\sqrt{x} - x^{2/3})\,dx$

5. If $f'(x) = x + e^x$ and $f(-2) = 3$, find $f(x)$.

6. Given that the point (1, 3) lies on the curve $y = f(x)$ and that the slope of the tangent line to the graph of $f(x)$ at (x, y) is $2x^3$, find the function $f(x)$.

In Exercises 7–10 evaluate the definite integral.

7. $\int_{-3}^{4} (2x + x^2)\,dx$ **8.** $\int_1^3 (1/t^2 + e^t)\,dt$

9. $\int_1^2 (2e^s - 3/s)\,ds$ **10.** $\int_4^9 (2x^{-1/2} + x^2 + e^x)\,dx$

In Exercises 11–14 calculate the area of the shaded region by evaluating a definite integral.

11.

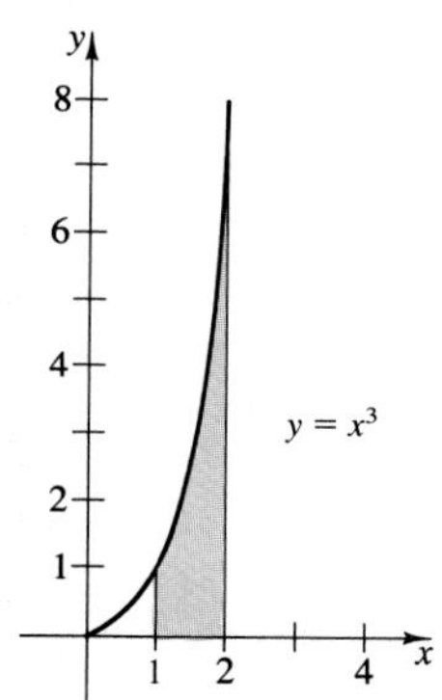

12.

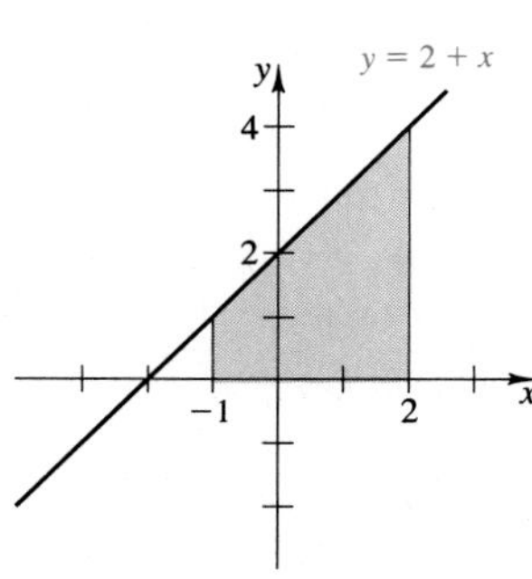

13.

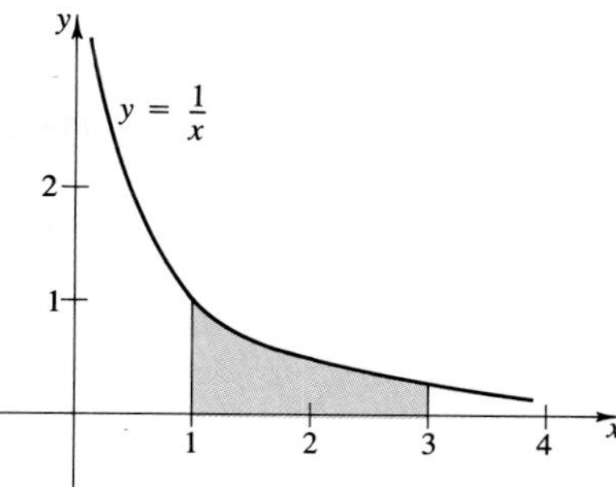

14.

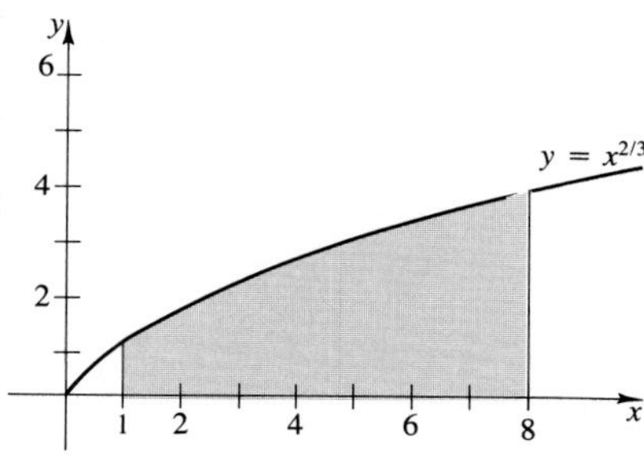

In Exercises 15–18 find the area under the curve over [a, b].

15. $y = 2x - x^2$, [1, 2] **16.** $y = x^{2/3}$, [0, 8]

17. $y = 2e^x$, [1, 4] **18.** $y = 1/x$, [2, 5]

In Exercises 19–22 compute the area of the shaded region.

19.

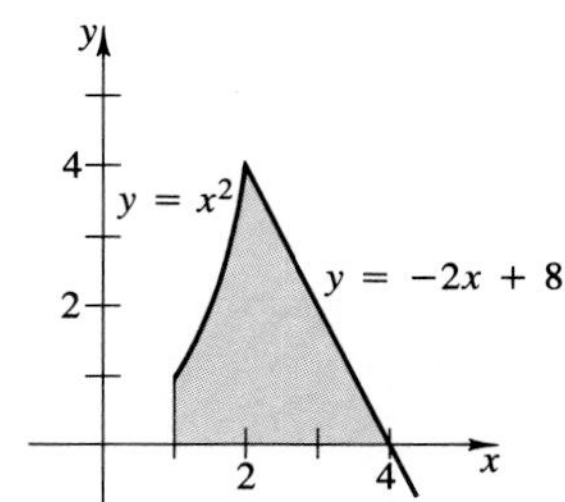

20.

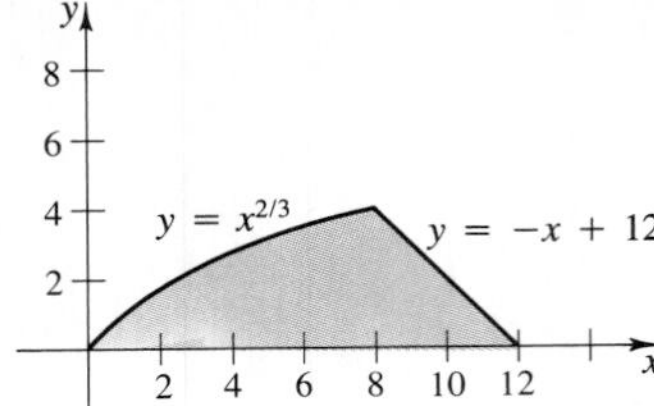

21.

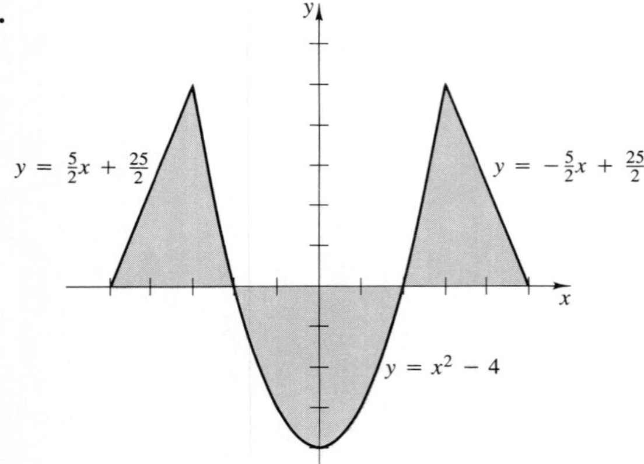

22.

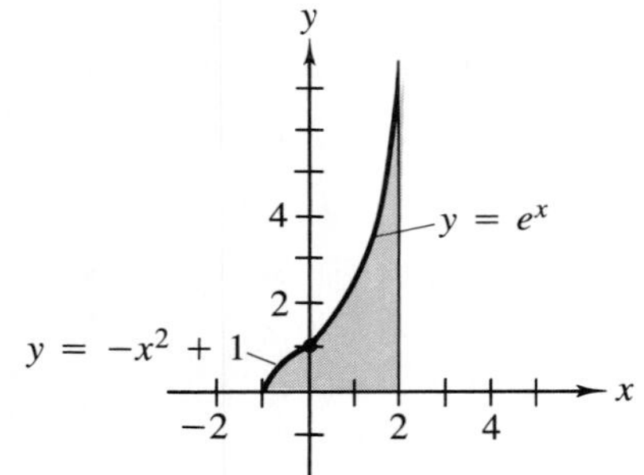

(Optional) *For Exercises 23 and 24 let $f(x) = x - 1$ and let $I = [0, 4]$ and divide I into 4 subintervals.*

23. Calculate the Riemann sum R_n by taking $\overline{x}_i$ as the right endpoint of the subinterval.

24. Calculate the Riemann sum R_n by taking $\overline{x}_i$ as the left endpoint of the subinterval.

25. **(Optional)** Evaluate $\int_2^4 (2x + 1)\, dx$ by Riemann's method.

26. **(Optional)** Evaluate $\int_{-1}^4 (1 - x^2)\, dx$ by Riemann's method.

27. **(Optional)** Evaluate $\int_2^4 (2x - 1)\, dx$ by Riemann's method.

28. **(Optional)** Evaluate $\int_{-1}^2 (1 - x^2)\, dx$ by Riemann's method.

(Optional) *For Exercises 29–34 we define a new function, the* ***greatest integer function****, by*

$$[x] = \text{the greatest integer} \le x$$

For example, $[3.04] = 3$, $[-3.04] = -4$, $[5] = 5$, $[0.85] = 0$, $[-0.12] = -1$, $[0] = 0$, and $[-1] = -1$. Or, in general,

$$[x] = n \qquad \text{for } n \le x < n + 1$$

The graph of this function is illustrated in Figure 1. Evaluate the following integrals using the method of Example 5 in Section 15.3.

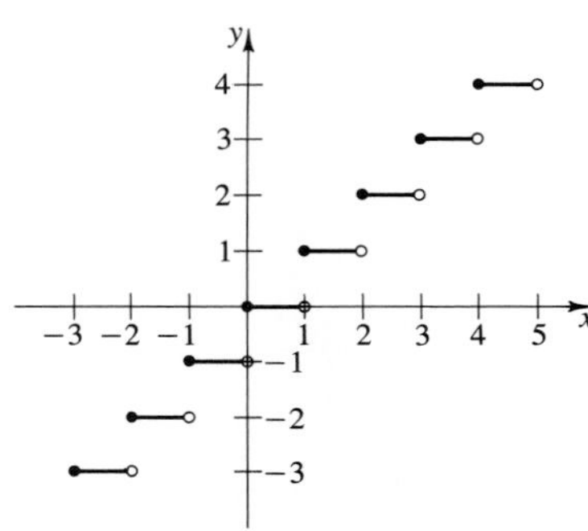

FIGURE 1

29. $\int_0^1 [x]\, dx$ **30.** $\int_1^3 [x]\, dx$ **31.** $\int_0^4 [x]\, dx$

32. $\int_2^5 [x]\, dx$ **33.** $\int_{-2}^2 [x]\, dx$ **34.** $\int_{-3}^1 2\,[x]\, dx$

35. **(Marginal analysis)** A sporting goods manufacturer determines that its marginal cost is

$$C'(x) = x^2 - \tfrac{1}{2}x^{1/2}$$

Find the cost function $C(x)$ if the fixed cost is \$5000.

CHAPTER TEST

1. Compute $\int(2e^x - 3/x + x^2)\,dx$
2. Compute $\int(2x^{-1/3} - 4/x^2 + 3e^x - \frac{3}{4})\,dx$
3. Evaluate $\int_1^8 (3x^{-1/3} - 2x^2 + 4/x)\,dx$
4. Evaluate $\int_1^9 (2e^t - 3/t + \sqrt{t})\,dt$
5. Calculate the area of the shaded region.

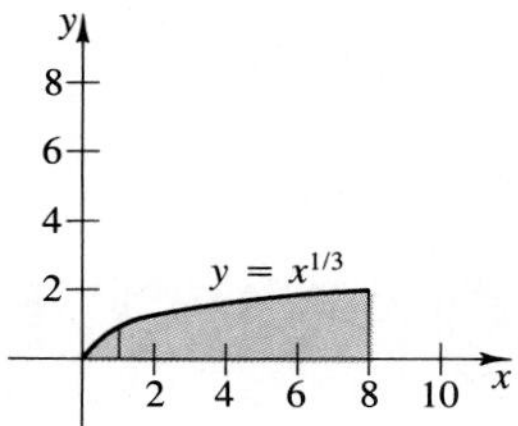

6. Find the area of the shaded region.

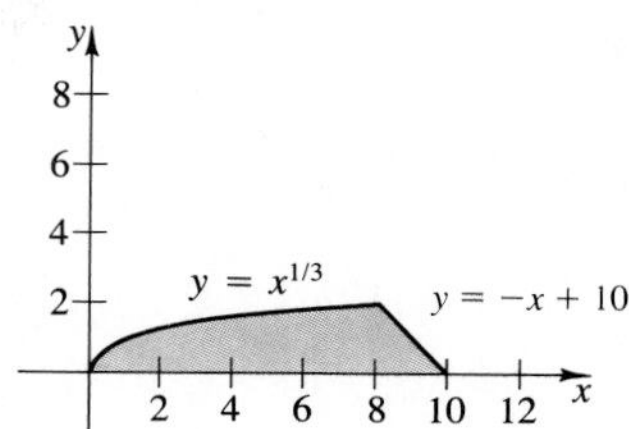

7. Find the area under the curve $y = x^{2/3} + \frac{1}{4}$ over $[1, 8]$.

16

Techniques of Integration

Our procedure for integrating a function requires that we first find an antiderivative of that function. In general, this is intrinsically a more complicated task than differentiating a function. For example, our formulas enable us to easily differentiate the functions

$$f(x) = x^2 e^x \quad \text{and} \quad g(x) = e^{-x^2}$$

With the help of the product rule we find

$$\begin{aligned} f'(x) &= x^2 \frac{d}{dx}(e^x) + e^x \frac{d}{dx}(x^2) \\ &= x^2 e^x + 2xe^x \end{aligned}$$

Using the chain rule we obtain

$$g'(x) = e^{-x^2} \frac{d}{dx}(-x^2) = -2xe^{-x^2}$$

Unfortunately, integrating these functions is not so simple. To find their antiderivatives we must "undifferentiate" them. This is not as easy a task as you might expect. Indeed, to antidifferentiate the first function requires an entirely new technique, whereas to antidifferentiate the second is hopeless!

16.1 THE METHOD OF SUBSTITUTION

The technique of substitution presented in this section can often be used to reduce an unfamiliar integration problem to a familiar one. We first summarize the three basic integration formulas studied so far. For reasons that will be clear to you shortly, we use u as the variable of integration in these formulas rather than x.

Integration Formulas

$$\int u^k\,du = \frac{u^{k+1}}{k+1} + C \qquad (k \neq -1) \tag{1}$$

$$\int \frac{1}{u}\,du = \ln|u| + C \tag{2}$$

$$\int e^u\,du = e^u + C \tag{3}$$

These formulas are obtained by antidifferentiation. For example,

$$\int u^3\,du = \frac{u^4}{4} + C$$

because

$$\frac{d}{du}\left[\frac{u^4}{4} + C\right] = u^3$$

Similarly, we can show that

$$\int x\left(\frac{1}{2}x^2 + 4\right)^{80} dx = \frac{1}{81}\left(\frac{1}{2}x^2 + 4\right)^{81} + C$$

since by the chain rule

$$\frac{d}{dx}\left[\frac{1}{81}\left(\frac{1}{2}x^2 + 4\right)^{81}\right] = \left(\frac{1}{2}x^2 + 4\right)^{80} \cdot \frac{1}{2} \cdot 2x = x\left(\frac{1}{2}x^2 + 4\right)^{80}$$

This antidifferentiation problem is solved by reversing the chain rule. We now use this approach in general and thus obtain another integration formula.

Substitution

Suppose $f(x)$ and $g(x)$ are two known functions and that $f(x)$ is differentiable. Let $F(x)$ be an antiderivative of $f(x)$ so that $F'(x) = f(x)$. Letting $u = g(x)$, we recall the chain rule from Section 12.8:

$$\begin{aligned}\frac{d}{dx}[F(g(x))] &= \frac{d}{dx}[F(u)] \\ &= \frac{dF}{du}\cdot\frac{du}{dx} \\ &= F'(g(x))g'(x) \\ &= f(g(x))g'(x) \qquad [\text{since } F'(u) = f(u)]\end{aligned}$$

We now obtain the integration formula

$$\int f(g(x))g'(x)\,dx = F(g(x)) + C \tag{4}$$

We can simplify the procedure by defining for $u = g(x)$ the **differential** of u:

$$du = g'(x)\,dx \tag{5}$$

The differential in x is the symbol dx.

Equation (4) can now be rewritten as

$$\int f(u)\,du = F(u) + C$$

where F is any antiderivative of f. (Note that the variable of integration is now u. There are no x's in the integrand.)

EXAMPLE 1 To illustrate the basic idea of integration by substitution, consider the problem above:

$$\int x\left(\frac{1}{2}x^2 + 4\right)^{80} dx$$

Let us set $u = g(x) = \frac{1}{2}x^2 + 4$ so that

$$du = g'(x)\,dx = x\,dx \qquad \text{[by Equation (5)]}$$

Thus our integral can be written as

$$\begin{aligned}\int x\left(\frac{1}{2}x^2 + 4\right)^{80} dx &= \int u^{80}\,du\\ &= \frac{1}{81}u^{81} + C \qquad \text{[from Equation (1)]}\\ &= \frac{1}{81}\left(\frac{1}{2}x^2 + 4\right)^{81} + C \qquad \left(\text{substituting } u = \tfrac{1}{2}x^2 + 4\right)\end{aligned}$$

EXAMPLE 2 Evaluate $\int 2x\sqrt{x^2 + 5}\,dx$.

Solution Let $u = x^2 + 5$. Then

$$du = \frac{d}{dx}[x^2 + 5]\,dx = 2x\,dx$$

Thus, after making these substitutions we have

$$\begin{aligned}\int 2x\sqrt{x^2 + 5}\,dx &= \int \sqrt{u}\,du\\ &= \int u^{1/2}\,du\\ &= \frac{2}{3}u^{3/2} + C \qquad \text{[from Equation (1)]}\\ &= \frac{2}{3}(x^2 + 5)^{3/2} + C \qquad (\text{substituting } u = x^2 + 5)\end{aligned}$$

We check by showing that the derivative of our answer is the integrand:

$$\begin{aligned}\frac{d}{dx}\left[\frac{2}{3}(x^2+5)^{3/2}+C\right] &= \frac{2}{3}\cdot\frac{3}{2}(x^2+5)^{3/2-1}\cdot(2x)+0\\ &= (x^2+5)^{1/2}(2x)\\ &= 2x\sqrt{x^2+5}\end{aligned}$$

EXAMPLE 3 Evaluate $\displaystyle\int \frac{x}{x^2+1}\,dx$.

Solution Let $u = x^2 + 1$. Then

$$du = 2x\,dx$$

In the given integral we do not have $2x\,dx$, only $x\,dx$. We obtain $x\,dx$ by dividing both sides of the last equation by 2:

$$x\,dx = \frac{1}{2}\,du$$

Making these substitutions we have

$$\begin{aligned}\int \frac{x}{x^2+1}\,dx &= \int \frac{1}{x^2+1}\,x\,dx = \int \frac{1}{u}\left(\frac{1}{2}\,du\right)\\ &= \frac{1}{2}\int \frac{1}{u}\,du\\ &= \frac{1}{2}\ln|u| + C \qquad \text{[from Formula (2)]}\\ &= \frac{1}{2}\ln|x^2+1| + C \qquad (\text{substituting } u = x^2+1)\\ &= \frac{1}{2}\ln(x^2+1) + C \qquad (\text{since } x^2+1>0)\\ &= \ln\sqrt{x^2+1} + C\end{aligned}$$

We check the answer as follows:

$$\begin{aligned}\frac{d}{dx}\left[\frac{1}{2}\ln(x^2+1)+C\right] &= \frac{1}{2}\cdot\frac{1}{x^2+1}\,\frac{d}{dx}(x^2+1)+0\\ &= \frac{2x}{2(x^2+1)}\\ &= \frac{x}{x^2+1}\end{aligned}$$

In this example you may have wondered how we "knew" to let $u = x^2 + 1$ rather than $u = 2x$, $u = 1/(1 + x^2)$, or some other choice. The answer is that we *did not* really know in advance that the substitution $u = x^2 + 1$ would work; substitution is a "trial and error" technique. Sometimes we might have to try several different substitutions before we find one that works or we reach the conclusion that there is no substitution that works. Again, note that if the method of substitution is used, the resulting integrand is a function *only* of the new variable u.

EXAMPLE 4 Evaluate

$$\int_0^1 \frac{3x^2}{(x^3 + 9)^2}\,dx$$

Solution We first evaluate the indefinite integral

$$\int \frac{3x^2}{(x^3 + 9)^2}\,dx$$

Suppose we try the substitution $u = x^3$; then

$$du = 3x^2\,dx$$

This yields

$$\int \frac{3x^2}{(x^3 + 9)^2}\,dx = \int \frac{1}{(u + 9)^2}\,du$$

which is not an integral we know how to evaluate directly. Let us try a different substitution, for example, $u = x^3 + 9$, so that

$$du = 3x^2\,dx$$

Then

$$\begin{aligned}\int \frac{3x^2}{(x^3 + 9)^2}\,dx &= \int \frac{1}{u^2}\,du = \int u^{-2}\,du \\ &= -u^{-1} + C = -\frac{1}{u} + C \\ &= -\frac{1}{x^3 + 9} + C\end{aligned}$$

Therefore

$$\begin{aligned}\int_0^1 \frac{3x^2}{(x^3 + 9)^2}\,dx &= -\frac{1}{x^3 + 9}\bigg|_0^1 = \left(-\frac{1}{10}\right) - \left(-\frac{1}{9}\right) \\ &= \frac{1}{9} - \frac{1}{10} = \frac{1}{90}\end{aligned}$$

EXAMPLE 5 Evaluate $\int e^{x^3} x^2\,dx$.

Solution Let $u = x^3$; then $\quad du = 3x^2\,dx$

and $x^2\,dx = \frac{1}{3}\,du$. Thus,

$$\int e^{x^3} x^2\,dx = \int e^u\left(\frac{1}{3}\,du\right)$$

$$= \frac{1}{3}\int e^u \, du$$

$$= \frac{1}{3}e^u + C$$

$$= \frac{1}{3}e^{x^3} + C$$

We check our answer as follows:

$$\frac{d}{dx}\left[\frac{1}{3}e^{x^3} + C\right] = \frac{1}{3}e^{x^3} \cdot 3x^2 + 0 = e^{x^3} \cdot x^2$$

EXAMPLE 6 Evaluate

$$\int \frac{e^x}{1 + e^x} \, dx$$

Solution Let $u = 1 + e^x$ so that

$$du = e^x \, dx$$

Thus

$$\begin{aligned} \int \frac{e^x}{1 + e^x} \, dx &= \int \frac{du}{u} \\ &= \ln |u| + C \\ &= \ln |1 + e^x| + C \\ &= \ln (1 + e^x) + C \qquad \text{(since } e^x > 0 \text{ for all } x\text{)} \end{aligned}$$

We check our answer as follows:

$$\frac{d}{dx}[\ln (1 + e^x) + C] = \frac{1}{1 + e^x} e^x + 0 = \frac{e^x}{1 + e^x}$$

EXAMPLE 7 Evaluate

$$\int_1^e \frac{\ln x}{x} \, dx$$

Solution In the indefinite integral

$$\int \frac{\ln x}{x} \, dx$$

let $u = \ln x$ so that

$$du = \frac{1}{x} \, dx$$

Thus

$$\int \frac{\ln x}{x} \, dx = \int \ln x \left(\frac{1}{x} \, dx\right) = \int u \, du = \frac{u^2}{2} + C$$

$$= \frac{(\ln x)^2}{2} + C$$

Therefore

$$\int_1^e \frac{\ln x}{x}\,dx = \left.\frac{(\ln x)^2}{2}\right|_1^e = \frac{(\ln e)^2}{2} - \frac{(\ln 1)^2}{2}$$

$$= \frac{1}{2} - 0 = \frac{1}{2}$$

EXERCISE SET 16.1

In Exercises 1–24 evaluate the given integral by the method of substitution. Check each answer by differentiating the result and comparing it with the integrand.

1. $\int 3x^2(x^3 + 5)^{10}\,dx$
2. $\int \frac{3x^2}{x^3 + 1}\,dx$
3. $\int \frac{2x}{(x^2 - 1)^{15}}\,dx$
4. $\int x^3(x^4 + 1)\,dx$
5. $\int \frac{x}{(x^2 + 4)^3}\,dx$
6. $\int_0^4 \frac{1}{x + 2}\,dx$
7. $\int_0^1 \frac{1}{2 - x}\,dx$
8. $\int 3t^2\sqrt{t^3 + 5}\,dt$
9. $\int t^3\sqrt{t^4 + 2}\,dt$
10. $\int_0^3 2xe^{x^2}\,dx$
11. $\int e^{-2x}\,dx$
12. $\int (3x^2 + 2x)e^{x^3+x^2}\,dx$
13. $\int_1^e \frac{(\ln x)^2}{x}\,dx$
14. $\int \frac{\ln 3x}{x}\,dx$
15. $\int_0^1 e^{3t}\,dt$
16. $\int_0^1 x(x^2 - 1)^{23}\,dx$
17. $\int \sqrt{3x - 2}\,dx$
18. $\int \frac{1}{(4x + 5)^{10}}\,dx$
19. $\int e^x\sqrt{1 + e^x}\,dx$
20. $\int \frac{e^x}{(e^x + 8)^3}\,dx$
21. $\int \frac{x^2}{\sqrt{x^3 + 1}}\,dx$
22. $\int \frac{e^x}{\sqrt{1 + e^x}}\,dx$
23. $\int (x^3 + 1)^{10}x^2\,dx$
24. $\int (3x^2 + 2x)(x^3 + x^2 + 1)^{10}\,dx$

16.2 INTEGRATION BY PARTS

In the list of basic integral formulas there is the conspicuous absence of a product rule. No rule gives the integral of $f(x)g(x)$ in terms of the integrals of $f(x)$ and $g(x)$—as you might expect from having learned the product rule for differentiation. In Section 16.1 we have seen that certain products can be integrated by the method of substitution. Another useful important technique is "integration by parts," which is derived from the product rule of differentiation.

THEOREM (INTEGRATION BY PARTS) If $f(x)$ and $g(x)$ are differentiable functions, then

$$\int f(x)g'(x)\,dx = f(x)g(x) - \int g(x)f'(x)\,dx \tag{1}$$

Alternatively, letting $u = f(x)$ and $v = g(x)$ and using differential notation $du = f'(x)\,dx$ and $dv = g'(x)\,dx$, we have

$$\int u\,dv = uv - \int v\,du \tag{2}$$

Proof The differential of $f(x)g(x)$ is

$$\begin{aligned} d[f(x)g(x)] &= [f(x)g(x)]'\,dx \\ &= [f(x)g'(x) + f'(x)g(x)]\,dx \qquad \text{(by the product rule)} \\ &= f(x)g'(x)\,dx + g(x)f'(x)\,dx \end{aligned}$$

or

$$d\,(uv) = u\,dv + v\,du \tag{3}$$

where u and v are functions of x. Integrating both sides of (3) with respect to x, we obtain

$$\int d\,(uv) = \int u\,dv + \int v\,du$$

or

$$uv = \int u\,dv + \int v\,du \qquad \left(\text{since } \int d\,(uv) = uv\right)$$

or

$$\int u\,dv = uv - \int v\,du$$

which is Equation (2).

To use the method of integration by parts for evaluating an integral $\int h(x)\,dx$, we try to rewrite $h(x)\,dx$ as $u\,dv$ where $\int v\,du$ is easier to evaluate then $\int u\,dv$.

EXAMPLE 1 Evaluate $\int xe^x\,dx$.

Solution Let

$$u = x \quad \text{and} \quad dv = e^x\,dx$$

so that

$$du = dx \quad \text{and} \quad v = \int e^x\,dx = e^x$$

Thus,

$$\begin{aligned} \int xe^x\,dx &= \int u\,dv \\ &= uv - \int v\,du \qquad \text{[by Equation (2)]} \\ &= xe^x - \int e^x\,dx \qquad \left(\text{Note that we can easily determine } \int e^x\,dx.\right) \\ &= xe^x - e^x + C \end{aligned}$$

We check our answer as follows:

$$\begin{aligned} \frac{d}{dx}[xe^x - e^x + C] &= x\frac{d}{dx}[e^x] + e^x\frac{dx}{dx} - \frac{d}{dx}[e^x] + 0 \\ &= xe^x + e^x - e^x \\ &= xe^x \end{aligned}$$

Warning As with substitution, integration by parts is a "trial and error" method; its success hinges on making the right choice for u and dv. For example, in the evaluation of $\int xe^x\,dx$ in Example 1, had we chosen $u = e^x$ and $dv = x\,dx$ so that

$$du = e^x\,dx \quad \text{and} \quad v = \int x\,dx = \frac{x^2}{2}$$

we would have obtained from Equation (2)

$$\int xe^x\,dx = \int u\,dv$$

$$= uv - \int v\,du$$

$$= \frac{x^2}{2}e^x - \int \frac{x^2}{2}e^x\,dx$$

However, the new integral $\int (x^2/2)e^x\,dx$ is more complicated than the original, so this choice of u and dv would be of no help.

EXAMPLE 2 Evaluate $\int x \ln x\,dx$.

Solution Let $u = \ln x$ and $dv = x\,dx$ so that

$$du = \frac{1}{x}\,dx \quad \text{and} \quad v = \int x\,dx = \frac{x^2}{2}$$

Thus

$$\int x \ln x\,dx = \int u\,dv$$

$$= uv - \int v\,du$$

$$= \frac{x^2}{2}\ln x - \int \left(\frac{x^2}{2}\right)\left(\frac{1}{x}\right) dx$$

$$= \frac{x^2}{2}\ln x - \frac{1}{2}\int x\,dx$$

$$= \frac{x^2}{2}\ln x - \frac{1}{2}\left(\frac{x^2}{2}\right) + C$$

$$= \frac{x^2}{2}\ln x - \frac{x^2}{4} + C$$

We check our answer as follows:

$$\frac{d}{dx}\left[\frac{x^2}{2}\ln x - \frac{x^2}{4} + C\right] = \left(\frac{x^2}{2}\right)\left(\frac{1}{x}\right) + \frac{2x}{2}\ln x - \frac{2x}{4} + 0$$

$$= \frac{x}{2} + x\ln x - \frac{x}{2}$$

$$= x \ln x$$

EXAMPLE 3 Evaluate $\int x^2 e^{3x}\,dx$.

Solution Let $u = x^2$ and $dv = e^{3x}\,dx$. Then $du = 2x\,dx$ and $v = \int e^{3x}\,dx = e^{3x}/3$. Thus

$$\int x^2 e^{3x}\,dx = \int u\,dv$$

$$= uv - \int v\, du$$

$$= (x^2)\left(\frac{e^{3x}}{3}\right) - \int \left(\frac{e^{3x}}{3}\right)(2x)\, dx$$

$$= \frac{1}{3}x^2e^{3x} - \frac{2}{3}\int xe^{3x}\, dx \qquad (4)$$

The last integral is one that must be evaluated "by parts," but it has the advantage of being simpler than the original. We now evaluate it.

Let $u = x$ and $dv = e^{3x}$. Then $du = dx$ and $v = e^{3x}/3$. Thus

$$\int xe^{3x}\, dx = \int u\, dv$$

$$= uv - \int v\, du$$

$$= x\left(\frac{e^{3x}}{3}\right) - \int \frac{e^{3x}}{3}\, dx$$

$$= \frac{1}{3}xe^{3x} - \frac{1}{3}\frac{1}{3}e^{3x} + C \qquad (5)$$

Substituting Equation (5) into (4) we have

$$\int x^2e^{3x}\, dx = \frac{1}{3}x^2e^{3x} - \frac{2}{3}\left[\frac{1}{3}xe^{3x} - \frac{1}{9}e^{3x} + C\right]$$

$$= \frac{1}{3}x^2e^{3x} - \frac{2}{9}xe^{3x} + \frac{2}{27}e^{3x} + C$$

You may check this answer as we have done in Examples 1 and 2.

Remark Example 3 shows that in using integration by parts we must be patient. The method often leaves a second integral that has to be evaluated. We may have to use integration by parts three or more times in the same problem.

Helpful Hints for Integration by Parts

1. $\int P(x)e^x\, dx$, where $P(x)$ is a polynomial.
   Use integration by parts. Let $u = P(x)$, $dv = e^x\, dx$.
2. $\int P(x) \ln x\, dx$, where $P(x)$ is a polynomial.
   Use integration by parts. Let $u = \ln x$, $dv = P(x)\, dx$.
3. $\displaystyle\int \frac{(\ln x)^k}{x}\, dx$

   Use simple substitution. Let $u = \ln x$, $du = \dfrac{1}{x}\, dx$. Integration by parts will not work.

EXAMPLE 4 Evaluate $\int \ln x\, dx$, where $x > 0$.

Solution At first this integral does not seem appropriate for the method of integration by parts because no product is indicated. Nevertheless, we let

$$u = \ln x \quad \text{and} \quad dv = dx$$

$$du = \frac{1}{x}\, dx \quad \text{and} \quad v = x$$

Therefore,

$$\begin{aligned}\int \ln x\, dx &= \int u\, dv = uv - \int v\, du \\ &= (\ln x)(x) - \int x \cdot \left(\frac{1}{x}\right) dx \\ &= x \ln x - \int dx \\ &= x \ln x - x + C\end{aligned}$$

We check our answer as follows:

$$\begin{aligned}\frac{d}{dx}[x \ln x - x + C] &= x \cdot \frac{d}{dx}[\ln x] + (\ln x)\frac{d}{dx}[x] - 1 + 0 \\ &= \frac{x}{x} + \ln x - 1 \\ &= \ln x\end{aligned}$$

EXERCISE SET 16.2

In Exercises 1–20 evaluate the given integral by the method of integration by parts.

1. $\int xe^{-x}\, dx$
2. $\int 3xe^{2x}\, dx$
3. $\int x^2 \ln x\, dx$
4. $\int x^2 e^x\, dx$
5. $\int xe^{-x/2}\, dx$
6. $\int 2x/e^x\, dx$ [*Hint:* $1/e^x = e^{-x}$]
7. $\int x \ln 2x\, dx$
8. $\int \sqrt{x} \ln x\, dx$
9. $\int (\ln x)^2\, dx$
10. $\int x (\ln x)^2\, dx$
11. $\displaystyle\int \frac{\ln x}{x^2}\, dx$
12. $\displaystyle\int \frac{\ln x}{\sqrt{x}}\, dx$
13. $\int x^3 \ln x\, dx$
14. $\int (x^2 + 2)e^x\, dx$
15. $\int (x^2 + 2x) \ln x\, dx$
16. $\int \ln x^2\, dx$ [*Hint:* Use properties of logarithms.]
17. $\int \ln \sqrt{x}\, dx$ [*Hint:* Use properties of logarithms.]
18. $\displaystyle\int \frac{x}{e^x}\, dx$
19. $\int x^3 e^x\, dx$
20. $\int x^3 e^{x^2}\, dx$

In Exercises 21–24 evaluate the given integral by any method.

21. $\int e^{2x} e^{3x}\, dx$
22. $\displaystyle\int \frac{(\ln x)^4}{x}\, dx$
23. $\displaystyle\int \frac{(\ln 2x)^{1/2}}{x}\, dx$
24. $\displaystyle\int \frac{\ln x}{x^{3/2}}\, dx$

16.3 USING A TABLE OF INTEGRALS

So far we have seen only a few of the many techniques of integration, and there remain many integrals that we are unable to evaluate with this limited knowledge. For our purposes it is not appropriate to devote more time and energy to the more advanced techniques, many of which require trigonometric formulas. Instead, there are convenient tables of integrals readily available. It is customary for standard calculus textbooks to include a table of integrals. One popular textbook has a table of more than 140 integrals; another has a table of more than 200. You may try to use these tables when you encounter an integral that you do not know how to evaluate.

Our table of integrals, which appears at the end of this section, includes 38 formulas. Like most such tables, it is divided into categories; our categories are:

Basic forms (you should have already memorized these)

Integrals involving $au + b$

Integrals involving $\sqrt{au + b}$

Integrals involving $u^2 \pm a^2$, $a > 0$

Integrals involving exponential and logarithmic functions

The table is arranged according to the *form* of the integrand; you must match the integral with one of the forms in the table. Sometimes the integrand requires a little algebraic manipulation before it matches one of the forms in the table. In all these formulas, u may be a differentiable function of x. Thus one form may be used for many different integrals.

EXAMPLE 1 The same form

$$\int u^n\,du = \frac{u^{n+1}}{n+1} + C \qquad n \neq -1$$

[Formula (4) in the table] is used in each of the following:

(a) $$\int x^3\,dx = \frac{x^4}{4} + C$$

(Here, $u = x$ and $n = 3$.)

(b) $$\int 3x^2(x^3 + 10)^5\,dx = \frac{(x^3 + 10)^6}{6} + C$$

(Here, $u = x^3 + 10$, $du = 3x^2\,dx$, and $n = 5$.)

(c) $$\int \frac{(\ln x)^7}{x}\,dx = \frac{1}{8}(\ln x)^8 + C$$

(Here, $u = \ln x$, $du = 1/x\,dx$, and $n = 7$.)

We now give some examples showing how the integral table is used in practice.

EXAMPLE 2 Use the integral table to evaluate

(a) $\int \frac{4x}{3x-5}\,dx$ (b) $\int \frac{x^2}{(2x+3)^2}\,dx$ (c) $\int \frac{1}{2x^2+5x-3}\,dx$

Solution (a)

$$\int \frac{4x\,dx}{3x-5} = 4\int \frac{x\,dx}{3x-5}$$

The latter integral matches Formula (10) when $a = 3$ and $b = -5$. Thus

$$\int \frac{4x}{3x-5}\,dx = 4\left[\frac{x}{3} - \left(\frac{-5}{9}\right)\ln|3x-5|\right] + C$$

$$= \frac{4x}{3} + \frac{20}{9}\ln|3x-5| + C$$

Solution (b) This integral matches Formula (13) with $a = 2$ and $b = 3$. Thus,

$$\int \frac{x^2}{(2x+3)^2}\,dx = \frac{1}{8}\left[(2x+3) - \frac{9}{2x+3} - 6\ln|2x+3|\right] + C$$

Solution (c)

$$\int \frac{1}{2x^2+5x-3}\,dx = \int \frac{dx}{(2x-1)(x+3)}$$

After the denominator has been factored, we see that this integral matches Formula (17), with $a = 2$, $b = -1$, $c = 1$, and $d = 3$. Thus,

$$\int \frac{1}{2x^2+5x-3}\,dx = \frac{1}{(-1)(1)-(2)(3)}\ln\left|\frac{x+3}{2x-1}\right| + C$$

$$= -\frac{1}{7}\ln\left|\frac{x+3}{2x-1}\right| + C$$

$$= \frac{1}{7}\ln\left|\frac{2x-1}{x+3}\right| + C$$

Do you believe these results? If you want to be sure an answer is correct, just differentiate it and compare the result with the original integrand.

EXAMPLE 3 Use the integral table to evaluate

(a) $\int \frac{1}{\sqrt{x^2-4}}\,dx$ (b) $\int \frac{1}{x^2-5}\,dx$ (c) $\int \frac{1}{x(4-2x)}\,dx$

Solution (a) This integral matches Formula (29) when $a = 2$. Thus

$$\int \frac{1}{\sqrt{x^2-4}}\,dx = \ln|x + \sqrt{x^2-4}| + C$$

EARLY TECHNIQUES OF INTEGRATION

Two of the early techniques of integration were developed by the brothers Jakob and Johann Bernoulli, members of an extraordinary mathematical family (see also page 325).

In 1576 the Protestant Bernoulli family fled Catholic Spanish Holland to Basel, Switzerland, to escape religious persecution. By marrying into wealthy Basel families the Bernoullis became wealthy merchants. The Bernoulli family produced 12 mathematicians in five generations, several of whom made great contributions to mathematics. The first Bernoulli to become famous was Jakob (1654–1705), who popularized the work of Leibniz on calculus. He suggested the name *integral* to Leibniz and greatly extended the work of Newton and Leibniz. In 1699 Jakob evaluated the integral

$$\int \frac{a^2}{a^2 - x^2}\, dx \qquad (a \text{ is a constant})$$

by making the substitution

$$x = a\left(\frac{b^2 - u^2}{b^2 + u^2}\right)$$

leading to the integral

$$-\int \frac{a}{u}\, du$$

which can be easily evaluated.

Jakob's brother Johann (1667–1748) was also a great mathematician, even more prolific than Jakob and instrumental in spreading the knowledge of calculus throughout Europe. In 1702 Johann observed that

$$\frac{a^2}{a^2 - x^2} = \frac{a}{2}\left(\frac{1}{a + x} + \frac{1}{a - x}\right)$$

so

$$\int \frac{a^2}{a^2 - x^2}\, dx = \frac{a}{2}\int \frac{1}{a + x}\, dx + \frac{a}{2}\int \frac{1}{a - x}\, dx$$

The first integral on the right side can be easily evaluated by letting $u = a + x$; the second by letting $u = a - x$. This approach, now referred to as the *method of partial fractions,* had also been developed independently by Leibniz in 1702.

Jakob and Johann were extremely competitive men. Johann attempted to steal some of his brother's ideas and even threw his own son out of the house for having won a prize from the French Academy of Sciences, a prize that Johann had hoped to win.

Solution (b) This integral matches Formula (27) when $a^2 = 5$ and $a = \sqrt{5}$. Thus

$$\int \frac{1}{x^2 - 5}\,dx = \frac{1}{2\sqrt{5}} \ln\left|\frac{x - \sqrt{5}}{x + \sqrt{5}}\right| + C$$

Solution (c) From Formula (14) with $a = -2$ and $b = 4$, we obtain

$$\int \frac{1}{x(4 - 2x)}\,dx = \frac{1}{4} \ln\left|\frac{x}{4 - 2x}\right| + C$$

EXAMPLE 4 Evaluate

$$\int \frac{1}{\sqrt{4x^2 + 9}}\,dx$$

Solution This integral does not exactly match any of the forms in the table. However, if it were not for the factor of 4 multiplying the x^2, it would match Formula (29). We remedy this as follows:

$$\int \frac{1}{\sqrt{4x^2 + 9}}\,dx = \int \frac{1}{\sqrt{4(x^2 + \frac{9}{4})}}\,dx = \frac{1}{2}\int \frac{1}{\sqrt{x^2 + \frac{9}{4}}}\,dx$$

We now apply Formula (29) with $a^2 = \frac{9}{4}$ to obtain

$$\int \frac{1}{\sqrt{4x^2 + 9}}\,dx = \frac{1}{2}\int \frac{1}{\sqrt{x^2 + \frac{9}{4}}}\,dx = \frac{1}{2} \ln\left|x + \sqrt{x^2 + \tfrac{9}{4}}\right| + C$$

EXAMPLE 5 Evaluate each of the following integrals:
(a) $\int 5^x\,dx$ (b) $\int 2x^3e^x\,dx$ (c) $\int x^2e^{3x}\,dx$

Solution (a) By Formula (33),

$$\int 5^x\,dx = \frac{5^x}{\ln 5} + C$$

Solution (b) We use Formula (34) twice and then Formula (32):

$$\begin{aligned}
\int 2x^3e^x\,dx &= 2\int x^3e^x\,dx \\
&= 2(x^3e^x - 3\int x^2e^x\,dx) && \text{[by Formula (34)]} \\
&= 2x^3e^x - 6(x^2e^x - 2\int xe^x\,dx) && \text{[by Formula (34)]} \\
&= 2x^3e^x - 6x^2e^x + 12\int xe^x\,dx \\
&= 2x^3e^x - 6x^2e^x + 12[e^x(x - 1)] + C && \text{[by Formula (32)]} \\
&= 2e^x(x^3 - 3x^2 + 6x - 6) + C
\end{aligned}$$

Solution (c) Let $u = 3x$. Then $du = 3\,dx$ and $x^2 = (u/3)^2$, so

$$\begin{aligned}
\int x^2\,e^{3x}\,dx &= \int \frac{u^2}{9}\cdot e^u \cdot \frac{1}{3}\,du \\
&= \frac{1}{27}\int u^2e^u\,du
\end{aligned}$$

We can now apply Formula (34) to obtain

$$\begin{aligned}
\int x^2 e^{3x}\,dx &= \frac{1}{27}\left(u^2e^u - 2\int ue^u\,du\right) \\
&= \frac{1}{27}u^2e^u - \frac{2}{27}[e^u(u-1)] + C \qquad \text{[by Formula (32)]} \\
&= \frac{1}{27}u^2e^u - \frac{2}{27}ue^u + \frac{2}{27}e^u + C \\
&= \frac{e^u}{27}(u^2 - 2u + 2) + C \\
&= \frac{e^{3x}}{27}(9x^2 - 6x + 2) + C
\end{aligned}$$

Observe that Formula (34) is used twice in succession in solving Example 5(b). Formula (34), as well as Formulas (24) and (26), are called *reduction formulas*. Instead of leading directly to the answer in one step, they reduce the integral to a formula involving a simpler integral. These formulas are intended to be used over again as many times as necessary to reduce the given integral to one that no longer requires reduction. For that reason they are also called *recursion formulas*.

It should be noted that there are limitations to our integral table. For example, the integral

$$\int \frac{dx}{1+x^2} \tag{a}$$

cannot be evaluated using our table. They may seem quite surprising, since its near-twin,

$$\int \frac{dx}{1-x^2}$$

is readily evaluated using Formula (28) in the table. Similarly,

$$\int \sqrt{1-x^2}\,dx \tag{b}$$

cannot be found from our table even though both $\int\sqrt{1+x^2}\,dx$ and $\int\sqrt{x^2-1}\,dx$ may be found using Formula (30). What is even more surprising is the *reason* for the absence of these formulas. Although we would never guess it, the formulas for (a) and (b) require inverse trigonometric functions. It would be inappropriate to list these formulas (and others) without presenting a detailed analysis of the calculus of trigonometric functions and their inverses. Such an analysis is entirely outside the scope of this book.

Unfortunately, no table of integrals can be exhaustive. One popular mathematical handbook has a table of nearly 400 integrals and still does not include every possible integral.

EXERCISE SET 16.3

In Exercises 1–34 evaluate each given integral by using the table of integrals found at the conclusion of the exercise set. In each exercise tell which formula applies and what substitution(s) you are using.

1. $\int \frac{x}{3x+6}\,dx$

2. $\int \frac{1}{11x-8}\,dx$

3. $\int \frac{1}{t(3t-2)^2}\,dt$

4. $\int \frac{t}{(t+4)^2}\,dt$

5. $\int \frac{x^2}{(5x+2)^2}\,dx$

6. $\int \frac{1}{x^2(5x+2)}\,dx$

7. $\int \frac{1}{(x+1)(x-3)}\,dx$

8. $\int \frac{1}{2x^2-x-1}\,dx$

9. $\int (3s-7)\,ds$

10. $\int s\sqrt{2s+3}\,ds$

11. $\int \frac{3x}{2x-1}\,dx$

12. $\int \frac{\sqrt{2x+1}}{3x}\,dx$

13. $\int x^2\sqrt{3x+4}\,dx$

14. $\int \sqrt{x^2-9}\,dx$

15. $\int \frac{1}{\sqrt{w^2-9}}\,dw$

16. $\int \sqrt{w^2+1}\,dw$

17. $\int 2^x\,dx$

18. $\int x3^x\,dx$

19. $\int x^2 \ln x\,dx$

20. $\int x \ln x^2\,dx$

21. $\int \ln(x+3)\,dx$

22. $\int 5ze^{4z}\,dz$

23. $\int \frac{1}{x \ln x}\,dx$

24. $\int \frac{1}{4x^2-9}\,dx$

25. $\int \frac{x^2}{x-1}\,dx$

26. $\int \frac{1}{x(3x-1)}\,dx$

27. $\int \frac{r^2}{\sqrt{7r+3}}\,dr$

28. $\int \frac{r}{\sqrt{7r-3}}\,dr$

29. $\int \frac{x}{x^2+x-6}\,dx$

30. $\int \sqrt{5x+1}\,dx$

31. $\int \sqrt{1+4x^2}\,dx$

32. $\int \frac{1}{4-9x^2}\,dx$

A BRIEF TABLE OF INTEGRALS

Basic Forms

1. $\int du = u + c$

2. $\int af(u)\,du = a\int f(u)\,du$

3. $\int [f(u) \pm g(u)]\,du = \int f(u)\,du \pm \int g(u)\,du$

4. $\int u^n\,du = \frac{u^{n+1}}{n+1} + C \quad (\text{if } n \neq -1)$

5. $\int \frac{1}{u}\,du = \ln|u| + C$

6. $\int e^u\,du = e^u + C$

7. $\int f(x)g'(x)\,dx = f(x)g(x) - \int g(x)f'(x)\,dx \quad \left(\int u\,dv = uv - \int v\,du\right)$

Integrals Involving $au + b$

8. $\int (au + b)^n \, du = \frac{1}{n+1}\frac{1}{a}(au + b)^{n+1} + C \qquad (n \neq -1)$

9. $\int \frac{1}{au + b} \, du = \frac{1}{a} \ln|au + b| + C$

10. $\int \frac{u}{au + b} \, du = \frac{u}{a} - \frac{b}{a^2} \ln|au + b| + C$

11. $\int \frac{u^2}{au + b} \, du = \frac{1}{a^3}\left[\frac{1}{2}(au + b)^2 - 2b(au + b) + b^2 \ln|au + b|\right] + C$

12. $\int \frac{u}{(au + b)^2} \, du = \frac{1}{a^2}\left[\frac{b}{au + b} + \ln|au + b|\right] + C$

13. $\int \frac{u^2}{(au + b)^2} \, du = \frac{1}{a^3}\left[(au + b) - \frac{b^2}{au + b} - 2b \ln|au + b|\right] + C$

14. $\int \frac{1}{u(au + b)} \, du = \frac{1}{b} \ln\left|\frac{u}{au + b}\right| + C$

15. $\int \frac{1}{u^2(au + b)} \, du = -\frac{1}{bu} + \frac{a}{b^2} \ln\left|\frac{au + b}{u}\right| + C$

16. $\int \frac{1}{u(au + b)^2} \, du = \frac{1}{b(au + b)} + \frac{1}{b^2} \ln\left|\frac{u}{au + b}\right| + C$

17. $\int \frac{1}{(au + b)(cu + d)} \, du = \frac{1}{bc - ad} \ln\left|\frac{cu + d}{au + b}\right| + C \qquad (\text{if } bc - ad \neq 0)$

18. $\int \frac{u}{(au + b)(cu + d)} \, du = \frac{1}{bc - ad}\left[\frac{b}{a} \ln|au + b| - \frac{d}{c} \ln|cu + d|\right] + C \qquad (\text{if } bc - ad \neq 0)$

Integrals Involving $\sqrt{au + b}$

19. $\int \sqrt{au + b} \, du = \frac{2}{3a}(au + b)^{3/2} + C$

20. $\int u\sqrt{au + b} \, du = \frac{2(3au - 2b)(au + b)^{3/2}}{15a^2} + C$

21. $\int u^2\sqrt{au + b} \, du = \frac{2}{105a^3}(15a^2u^2 - 12abu + 8b^2)(au + b)^{3/2} + C$

22. $\int \frac{u}{\sqrt{au + b}} \, du = \frac{2}{3a^2}(au - 2b)\sqrt{au + b} + C$

23. $\int \frac{u^2}{\sqrt{au + b}} \, du = \frac{2}{15a^3}(3a^2u^2 - 4abu + 8b^2)\sqrt{au + b} + C$

24. $\int \frac{u^n}{\sqrt{au + b}} \, du = \frac{2u^n\sqrt{au + b}}{a(2n + 1)} - \frac{2bn}{a(n + 1)} \int \frac{u^{n-1}}{\sqrt{au + b}} \, du \qquad n \geq 2$

25. $\int \frac{1}{u\sqrt{au + b}} \, du = \frac{1}{\sqrt{b}} \ln\left|\frac{\sqrt{au + b} - \sqrt{b}}{\sqrt{au + b} + \sqrt{b}}\right| + C \qquad (\text{if } b > 0)$

26. $\int \frac{\sqrt{au + b}}{u} \, du = 2\sqrt{au + b} + b \int \frac{1}{u\sqrt{au + b}} \, du$

Integrals Involving $u^2 \pm a^2$, $a > 0$

27. $\displaystyle\int \frac{1}{u^2 - a^2}\,du = \frac{1}{2a}\ln\left|\frac{u-a}{u+a}\right| + C$

28. $\displaystyle\int \frac{1}{a^2 - u^2}\,du = \frac{1}{2a}\ln\left|\frac{u+a}{u-a}\right| + C$

29. $\displaystyle\int \frac{1}{\sqrt{u^2 \pm a^2}}\,du = \ln|\,u + \sqrt{u^2 \pm a^2}\,| + C$

30. $\displaystyle\int \sqrt{u^2 \pm a^2}\,du = \frac{u}{2}\sqrt{u^2 \pm a^2} \pm \frac{a^2}{2}\ln|\,u + \sqrt{u^2 \pm a^2}\,| + C$

Integrals Involving Exponential and Logarithmic Functions

31. $\displaystyle\int e^{au}\,du = \frac{1}{a}e^{au} + C$

32. $\displaystyle\int ue^{au}\,du = \frac{e^{au}}{a^2}(au - 1) + C$

33. $\displaystyle\int a^u\,du = \frac{a^u}{\ln a} + C \qquad (\text{for } a > 0,\ a \neq 1)$

34. $\displaystyle\int u^n e^u\,du = u^n e^u - n\int u^{n-1}e^u\,du$

35. $\displaystyle\int u^n a^u\,du = \frac{u^n a^u}{\ln a} - \frac{n}{\ln a}\int u^{n-1}a^u\,du \qquad (\text{for } a > 0)$

36. $\displaystyle\int \ln u\,du = u\ln u - u + C$

37. $\displaystyle\int u^n \ln u\,du = \frac{u^{n+1}}{(n+1)^2}[(n+1)\ln u - 1] + C$

38. $\displaystyle\int \frac{1}{u\ln u}\,du = \ln|\ln u| + C$

16.4 NUMERICAL APPROXIMATION OF INTEGRALS (OPTIONAL)

In contrast with the previous section, which is concerned exclusively with antidifferentiation, the present section focuses on evaluating *definite* integrals. Unlike an indefinite integral, a definite integral is a number. When we evaluate a definite integral, we seek not a function but a single *numerical* value. When the definite integral $\int_a^b f(x)\,dx$ is difficult or impossible to evaluate exactly, we need other methods. Consider the following example.

EXAMPLE 1 Evaluate $\displaystyle\int_0^1 \sqrt{1 - x^2}\,dx$.

Solution Our usual procedure for evaluating a definite integral has been to first find an antiderivative $F(x)$ of the integrand $f(x)$ and then to evaluate $F(b) - F(a)$ where a and b are the limits of integration. However [see (b) of Section 16.3],

$$\int \sqrt{1 - x^2}\,dx$$

is one of those antiderivatives we cannot obtain using the techniques available to us in this text. Thus we must rely on another approach.

This problem is easily solved if we look at it in another way. Recall that $\int_0^1 \sqrt{1 - x^2}\,dx$ represents the area under the curve $y = \sqrt{1 - x^2}$ from $x = 0$ to $x = 1$. Figure 1 shows this region. Thus the desired integral is the area of one-quarter of a circle with radius 1. The total area of this circle is π. Therefore,

$$\int_0^1 \sqrt{1 - x^2}\,dx = \frac{\pi}{4}$$

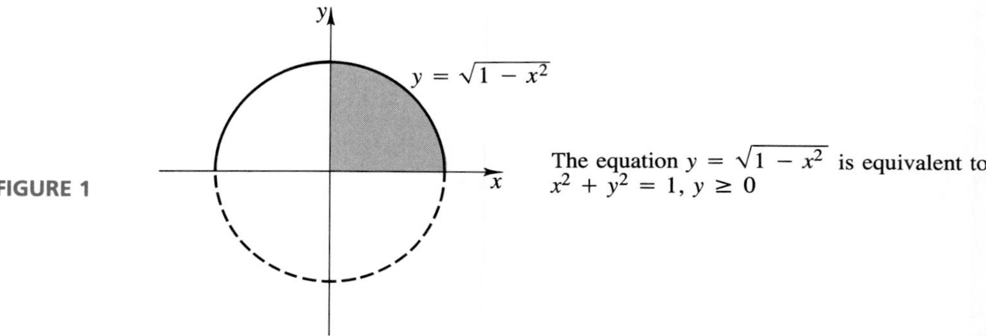

FIGURE 1

The equation $y = \sqrt{1 - x^2}$ is equivalent to $x^2 + y^2 = 1$, $y \geq 0$

The method to be described in this section makes use of the fact that a definite integral may be interpreted as an area. We cannot expect, in general, to be as lucky as we were in Example 1—where we just happened to already know a formula for the desired area. We therefore develop an alternative method for approximating this area.

Approximate Integration—Trapezoidal Approximations (Optional)

Our problem is to calculate

$$\int_a^b f(x)\,dx \tag{1}$$

There are times when we cannot find an antiderivative of $f(x)$ or times when it is inconvenient to do so. There are, however, a number of methods that will generally not give us the exact value of (1) but will give us a good approximation to this value. For this reason we refer to these methods as **approximate integration methods**. Be assured, however, that these approximations can be made quite accurate—especially with the aid of a computer. We concentrate on one such approximate integration method—the method of **trapezoidal approximations**.

Consider a function $f(x)$ that is continuous on $[a, b]$. Subdivide $[a, b]$ into n subintervals of equal length $\Delta x = (b - a)/n$. Define $x_0, x_1, \ldots, x_n$ and $y_0, y_1, \ldots, y_n$ by

$$
\begin{array}{lll}
x_0 = a & \text{and} & y_0 = f(x_0) \\
x_1 = a + \Delta x & \text{and} & y_1 = f(x_1) \\
x_2 = a + 2\Delta x & \text{and} & y_2 = f(x_2) \\
\vdots & & \vdots \\
x_n = a + n\Delta x = b & \text{and} & y_n = f(x_n)
\end{array}
$$

The numbers $x_0, x_1, x_2, \ldots, x_n$ are the endpoints of the subintervals in Figure 2a, and the numbers $y_0, y_1, y_2, \ldots, y_n$ are the heights of the curve above these endpoints. Connect the points (x_i, y_i) by straight-line segments to form trapezoids, as shown in Figure 2a. This figure also shows the resulting trapezoidal approximation to the area under the curve $y = f(x)$ over $[a, b]$ when $n = 8$.

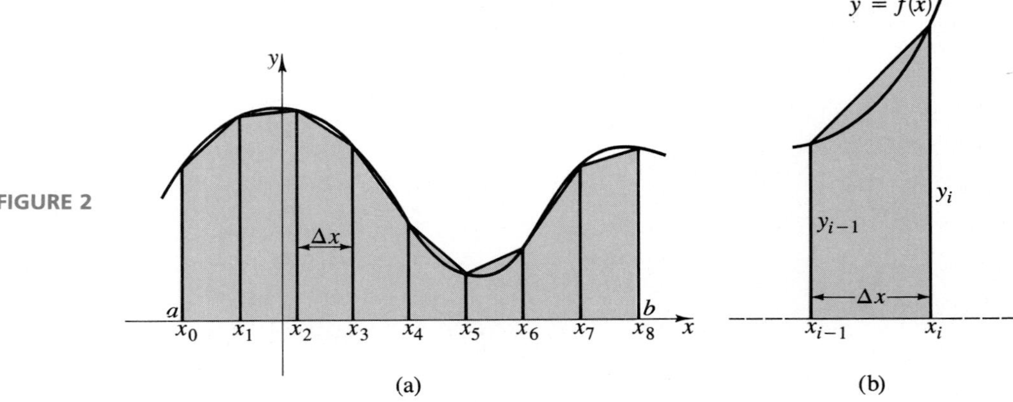

FIGURE 2

A typical trapezoid is shown in Figure 2b. Recall that the area of a trapezoid is one-half the sum of its "bases" times its altitude. Thus the area of the ith trapezoid shown in Figure 2b is $\frac{1}{2}(y_{i-1} + y_i)\Delta x$. The sum of the areas of all the trapezoids is

$$
\begin{aligned}
T_n &= \frac{1}{2}(y_0 + y_1)\Delta x + \frac{1}{2}(y_1 + y_2)\Delta x + \frac{1}{2}(y_2 + y_3)\Delta x + \cdots + \frac{1}{2}(y_{n-1} + y_n)\Delta x \\
&= \frac{\Delta x}{2}[(y_0 + y_1) + (y_1 + y_2) + (y_2 + y_3) + \cdots + (y_{n-1} + y_n)]
\end{aligned}
$$

The trapezoidal approximation

$$
T_n = \frac{\Delta x}{2}(y_0 + 2y_1 + 2y_2 + \cdots + 2y_{n-1} + y_n) \tag{2}
$$

is therefore a reasonable approximation of $\int_a^b f(x)\,dx$. It can be shown that the maximum error involved in using the trapezoidal rule is E_n, where

$$|E_n| \le \frac{M(b-a)^3}{12n^2} \tag{3}$$

Here M is a number such that $|f''(x)| \le M$ for $a \le x \le b$ and

$$T_n - |E_n| \le \int_a^b f(x)\,dx \le T_n + |E_n|^* \tag{4}$$

EXAMPLE 2 (a) Evaluate $\int_0^2 x^2\,dx$ by using the trapezoid approximation with $n = 4$ and determine the maximum error incurred by using this approximation.
(b) Repeat part (*a*) with $n = 8$.

Solution (a) $n = 4$. First we let $\Delta x = (2 - 0)/4 = \frac{1}{2}$. Then

$$\begin{aligned} x_0 &= 0 & y_0 &= 0^2 = 0 \\ x_1 &= \frac{1}{2} & y_1 &= \left(\frac{1}{2}\right)^2 = \frac{1}{4} \\ x_2 &= 1 & y_2 &= (1)^2 = 1 \\ x_3 &= \frac{3}{2} & y_3 &= \left(\frac{3}{2}\right)^2 = \frac{9}{4} \\ x_4 &= 2 & y_4 &= (2)^2 = 4 \end{aligned}$$

and

$$\begin{aligned} T_4 &= \frac{\Delta x}{2}[y_0 + 2y_1 + 2y_2 + 2y_3 + y_4] \\ &= \frac{1/2}{2}\left[0 + 2\left(\frac{1}{4}\right) + 2(1) + 2\left(\frac{9}{4}\right) + 4\right] \\ &= \frac{1}{4}\left[\frac{1}{2} + 2 + \frac{9}{2} + 4\right] \\ &= \left(\frac{1}{4}\right)(11) = 2.75 \end{aligned}$$

For $n = 4$, the maximum error in this approximation is

$$E_4 \le \frac{M(2-0)^3}{12 \cdot 16} \qquad \text{[by (3)]}$$

Since $f(x) = x^2$, $f'(x) = 2x$ and $f''(x) = 2 \le 2$; therefore, $M = 2$ and

$$|E_4| \le \frac{2(8)}{12 \cdot 16} = 0.0833$$

Hence, by (4) we have

---

* The error is only determinable if $f''(x)$ is defined and has an upper bound on $[a, b]$.

$$2.75 - 0.0833 \le \int_0^2 x^2\,dx \le 2.75 + 0.0833$$

$$2.6667 \le \int_0^2 x^2\,dx \le 2.8333$$

By comparison recall that

$$\int_0^2 x^2\,dx = \left.\frac{x^3}{3}\right|_0^2 = \frac{8}{3} = 2.6667$$

Solution (b) $n = 8$. Now we let $\Delta x = (2-0)/8 = \frac{1}{4}$, then

$$\begin{aligned}
x_0 &= 0 & y_0 &= 0\\
x_1 &= \frac{1}{4} & y_1 &= \frac{1}{16}\\
x_2 &= \frac{1}{2} & y_2 &= \frac{1}{4}\\
x_3 &= \frac{3}{4} & y_3 &= \frac{9}{16}\\
x_4 &= 1 & y_4 &= 1\\
x_5 &= \frac{5}{4} & y_5 &= \frac{25}{16}\\
x_6 &= \frac{3}{2} & y_6 &= \frac{9}{4}\\
x_7 &= \frac{7}{4} & y_7 &= \frac{49}{16}\\
x_8 &= 2 & y_8 &= 4
\end{aligned}$$

$$\begin{aligned}
T_8 &= \frac{\Delta x}{2}[y_0 + 2y_1 + 2y_2 + 2y_3 + 2y_4 + 2y_5 + 2y_6 + 2y_7 + y_8]\\
&= \frac{1/4}{2}\left[0 + \frac{2}{16} + \frac{2}{4} + \frac{18}{16} + 2 + \frac{50}{16} + \frac{18}{4} + \frac{98}{16} + 4\right]\\
&= 2.6875
\end{aligned}$$

Again, $M = 2$; therefore,

$$|E_8| \le \frac{M(b-a)^3}{12n^2} = \frac{2(2)^3}{(12)(64)} = 0.0208 \qquad \text{[by (3)]}$$

and

$$2.6875 - 0.0208 \le \int_0^2 x^2\,dx \le 2.6875 + 0.0208$$

$$2.6667 \le \int_0^2 x^2\,dx \le 2.7083$$

Note that E_n decreases as n increases and that T_8 is a better approximation to the integral than is T_4.

EXERCISE SET 16.4

In Exercises 1–12 use the given value of n to evaluate the given integral using the trapezoidal approximation. A calculator will be helpful.

1. $\int_0^2 (3x + 5)\,dx,\ n = 4$

2. $\int_1^7 x^2\,dx,\ n = 6$

3. $\int_{-1}^1 (1 - x^2)\,dx,\ n = 4$

4. $\int_{-3}^3 x^3\,dx,\ n = 6$

5. $\int_1^5 \frac{1}{x}\,dx,\ n = 8$

6. $\int_0^4 \frac{1}{x + 1}\,dx,\ n = 8$

7. $\int_0^4 e^x\,dx,\ n = 4$

8. $\int_1^3 2^x\,dx,\ n = 4$

9. $\int_0^5 \sqrt{1 + x^3}\,dx,\ n = 5$

10. $\int_0^2 \sqrt{1 + x^2}\,dx,\ n = 4$

11. $\int_0^4 \frac{1}{x + 1}\,dx,\ n = 8$

12. $\int_0^3 \frac{1}{1 + x^2}\,dx,\ n = 3$

KEY IDEAS FOR REVIEW

- **The method of substitution** See page 712.
- **Integration by parts** See page 718.
- **Tables of integrals** See pages 728–730.
- **Trapezoidal approximation (optional)**

$$T_n = \frac{\Delta x}{2}[y_0 + 2y_1 + 2y_2 + \cdots + 2y_{n-1} + y_n]$$

approximates

$$\int_a^b f(x)\,dx \qquad \text{where } \Delta x = \frac{b - a}{n}$$

$|E_n|$ = maximum error incurred when using the trapezoidal rule:

$$|E_n| \le \frac{M(b - a)^3}{12n^2} \qquad \text{where } |f''(x)| \le M \text{ for } a \le x \le b$$

SUPPLEMENTARY EXERCISES

In Exercises 1–8 evaluate the given integral by the method of substitution.

1. $\int (3 - 2x)^7\,dx$

2. $\int \frac{1}{8t + 3}\,dt$

3. $\int e^{5-3x}\,dx$

4. $\int \frac{e^{\sqrt{t}}}{\sqrt{t}}\,dt$

5. $\int_0^1 \frac{s}{3s^2 + 1}\,ds$

6. $\int \frac{4t^3 + 5}{t^4 + 5t}\,dt$

7. $\int \frac{1}{x \ln x}\,dx$

8. $\int \frac{1}{x(\ln x)^2}\,dx$

In Exercises 9–16 evaluate the given integral by the method of integration by parts or by substitution.

9. $\int x^2 e^x \, dx$

10. $\int (t + 2)e^t \, dt$

11. $\int x^2 e^{-x} \, dx$

12. $\int \frac{(\ln x)^3}{x} \, dx$

13. $\int_1^e t^3 \ln t \, dt$

14. $\int_1^e \frac{\ln x}{\sqrt{x}} \, dx$

15. $\int \frac{\sqrt{\ln x}}{x} \, dx$

16. $\int_1^e \frac{\ln t}{t^2} \, dt$

In Exercises 17–24 evaluate the given integral by using the table of integrals found at the end of Exercise Set 16.3.

17. $\int \frac{1}{\sqrt{x^2 + 25}} \, dx$

18. $\int \frac{x}{(2x - 5)^2} \, dx$

19. $\int \frac{1}{1 - 9x^2} \, dx$

20. $\int te^{-t/3} \, dt$

21. $\int 3t\sqrt{2t + 1} \, dt$

22. $\int \frac{1}{(3t - 2)(t + 4)} \, dt$

23. $\int x^2\sqrt{5x + 8} \, dx$

24. $\int \frac{1}{x^2 + 2x + 1} \, dx$

CHAPTER TEST

In Exercises 1–3 evaluate the given integral by the method of substitution.

1. $\int \frac{x + 1}{x^2 + 2x} \, dx$

2. $\int (2 + e^{3x})e^{3x} \, dx$

3. $\int \frac{(1 + \ln t)^2}{t} \, dt$

In Exercises 4–6 evaluate the given integrals by the method of integration by parts.

4. $\int \frac{x + 4}{e^x} \, dx$

5. $\int \sqrt{t} \ln t \, dt$

6. $\int \ln (2 + 3x) \, dx$

In Exercises 7–10 evaluate the given integral by using the table of integrals at the end of Exercise Set 16.3.

7. $\int \frac{\sqrt{2 + 9x}}{x} \, dx$

8. $\int t^2 e^{5t} \, dt$

9. $\int \frac{x^2}{(3x + 1)^2} \, dx$

10. $\int \frac{1}{x^2(3x + 4)} \, dx$

17

Applications of Integration

17.1 INTRODUCTION

Integration has a wide variety of applications including situations seemingly remote from the notion of area under a curve. In this chapter we encounter a few of them and see how the integral enables us to solve problems in business, economics, physics, biology, ecology, and probability theory.

The basic question that arises is how to tell when the definite integral is called for in a particular application. The answer involves two methods of defining the definite integral that are discussed in Sections 15.2 and 15.4 and that are shown to be equivalent by the fundamental theorem of calculus.

One method involves the definition of $\int_a^b f(x)\,dx$ for $f(x)$, a continuous function, in terms of Riemann sums (Section 15.4). We repeat this definition for convenience.

If $f(x)$ is continuous on $[a, b]$ and the interval $[a, b]$ is subdivided into n equal subintervals each of width $\Delta x = (b - a)/n$ and $\overline{x}_i$ is an arbitrary point in the subinterval $[x_{i-1}, x_i]$, then

$$\int_a^b f(x)\,dx = \lim_{n\to\infty} [f(\overline{x}_1) + f(\overline{x}_2) + \cdots + f(\overline{x}_n)]\,\Delta x \tag{1}$$

This definition of $\int_a^b f(x)\,dx$ is used in Sections 17.2 through 17.4. Since $\overline{x}_i$ is an arbitrary point in $[x_{i-1}, x_i]$, we select $\overline{x}_i = x_i$, the right endpoint of the interval.

The other method of defining the definite integral of a continuous function $f(x)$ (Section 15.2) is

$$\int_a^b f(x)\,dx = F(b) - F(a) \tag{2}$$

where $F(x)$ is any antiderivative of the integrand $f(x)$. This method is used in Section 17.5.

In both methods we make use of the area interpretation of the definite integral when appropriate.

17.2 AREA BETWEEN TWO CURVES

Suppose we want to compute the area A of the region bounded by the curves $y = f(x)$ and $y = g(x)$ over the interval $[a, b]$ with $f(x) \geq g(x)$, where $f(x)$ and $g(x)$ are continuous functions (Figure 1).

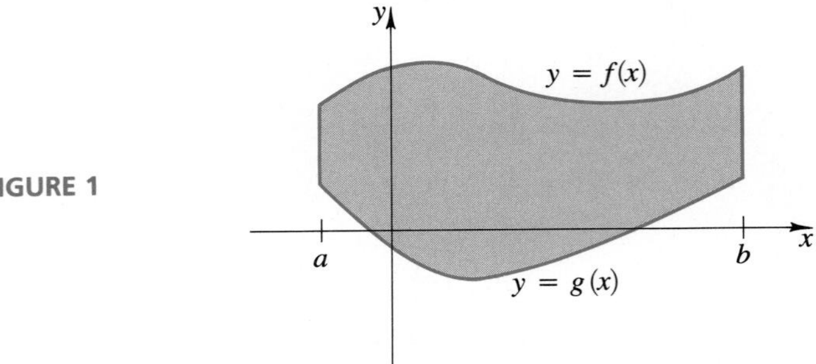

FIGURE 1

To approximate this area, we partition the interval $[a, b]$ into n equal subintervals each of width $\Delta x = (b - a)/n$, where $x_i = a + i\,\Delta x$, $x_0 = a$, and $x_n = b$. We construct rectangles with bases Δx and heights $f(\overline{x}_i) - g(\overline{x}_i) \geq 0$, where $x_{i-1} \leq \overline{x}_i \leq x_i$. In fact, for convenience we select $\overline{x}_i = x_i$ (Figure 2). Then the area A is approximately the sum of the areas of the rectangles:

$$A \cong [f(x_1) - g(x_1)]\,\Delta x + [f(x_2) - g(x_2)]\,\Delta x + \cdots + [f(x_n) - g(x_n)]\,\Delta x$$

Since $f(x) - g(x)$ is continuous, using Riemann's approach we have

$$\begin{aligned} A &= \lim_{n\to\infty} \{[f(x_1) - g(x_1)] + [f(x_2) - g(x_2)] + \cdots + [f(x_n) - g(x_n)]\}\,\Delta x \\ &= \int_a^b [f(x) - g(x)]\,dx \qquad \text{[by Equation (1) of Section 17.1]} \end{aligned}$$

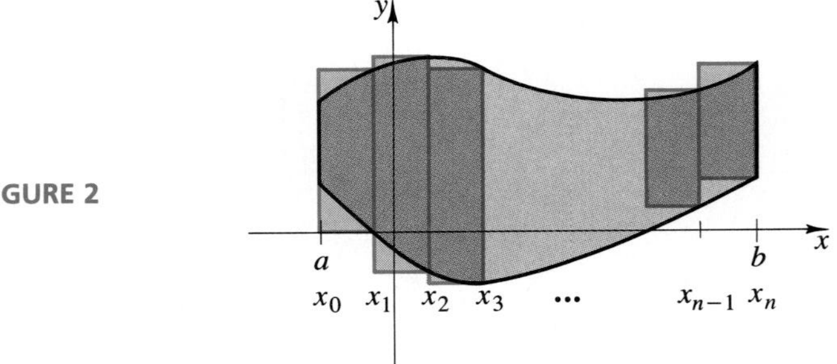

FIGURE 2

We therefore obtain the following result:

Area Between Curves If f and g are continuous functions with $f(x) \geq g(x)$ on $[a, b]$, then the area A between $y = f(x)$ and $y = g(x)$ over $[a, b]$ is

$$A = \int_a^b [f(x) - g(x)]\,dx \tag{1}$$

EXAMPLE 1 Find the area of the region enclosed by the curves $y = x^2$ and $y = x$.

Solution We begin by sketching the curves to obtain a clear picture of the region involved (Figure 3). Next we determine the points of intersection, if any, by observing that any such point (x, y) must satisfy both $y = x^2$ and $y = x$ so that

$$\begin{aligned} x &= x^2 \\ x^2 - x &= 0 \\ x(x - 1) &= 0 \end{aligned}$$

which yields $x = 0$ and $x = 1$. Thus the interval with which we are concerned is $[0, 1]$, and the area is

$$\int_0^1 (x - x^2)\, dx = \left(\frac{x^2}{2} - \frac{x^3}{3}\right)\bigg|_0^1 = \left(\frac{1}{2} - \frac{1}{3}\right) - 0 = \frac{1}{6}$$

FIGURE 3

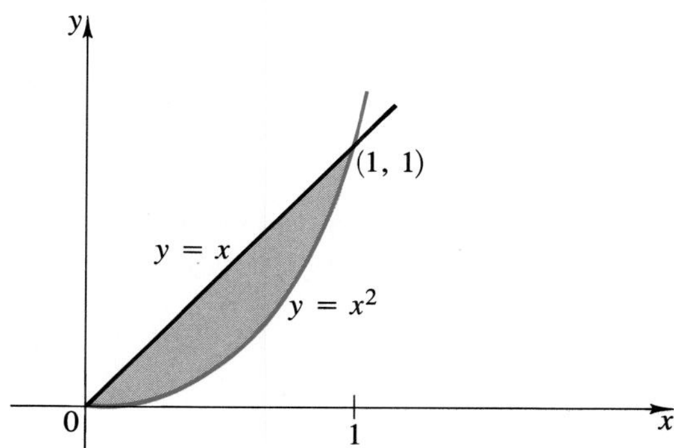

EXAMPLE 2 Find the area of the region bounded by the curve $y = x^2 - 2$ and the line $y = x$.

Solution We first sketch the curves in Figure 4 and determine the points of intersection. We seek x and y that simultaneously satisfy

$$y = x \quad \text{and} \quad y = x^2 - 2$$

FIGURE 4

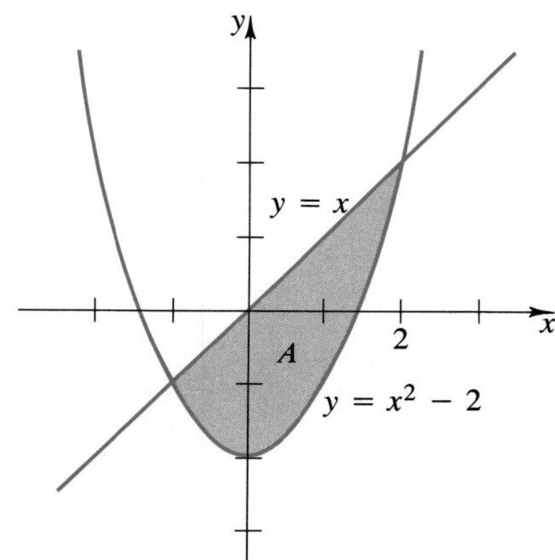

From the equation $x = x^2 - 2$, we have

$$\begin{aligned} 0 &= x^2 - x - 2 \\ &= (x - 2)(x + 1) \end{aligned}$$

from which we obtain $x = -1$ and $x = 2$. Thus, by Formula (1) the area is

$$\begin{aligned} A &= \int_{-1}^{2} [x - (x^2 - 2)]\, dx \\ &= \int_{1}^{2} (x - x^2 + 2)\, dx \\ &= \left(\frac{x^2}{2} - \frac{x^3}{3} + 2x\right)\Bigg|_{-1}^{2} \\ &= \left(2 - \frac{8}{3} + 4\right) - \left[\frac{1}{2} - \left(-\frac{1}{3}\right) - 2\right] \\ &= \frac{9}{2} \end{aligned}$$

Suppose we wish to find the area between two curves that cross one another at some point c in the interval $[a, b]$ over which we want the area, as shown in Figure 5.

FIGURE 5

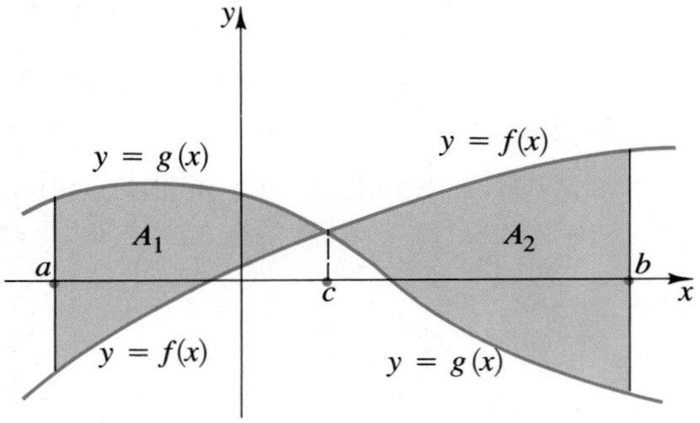

In the interval $[a, c]$, $g(x) \geq f(x)$, whereas in the interval $[c, b]$, $f(x) \geq g(x)$. Thus, the total area between these two curves is

$$\begin{aligned} A &= A_1 + A_2 \\ &= \int_a^c [g(x) - f(x)]\, dx + \int_c^b [f(x) - g(x)]\, dx \end{aligned}$$

The additional step in solving this problem is to find the points at which the curves cross one another. We will now have to evaluate two integrals rather than just one.

EXAMPLE 3 Find the area bounded by the curves $y = \sqrt{x}$ and $y = x^2$ over the interval $[0, 2]$.

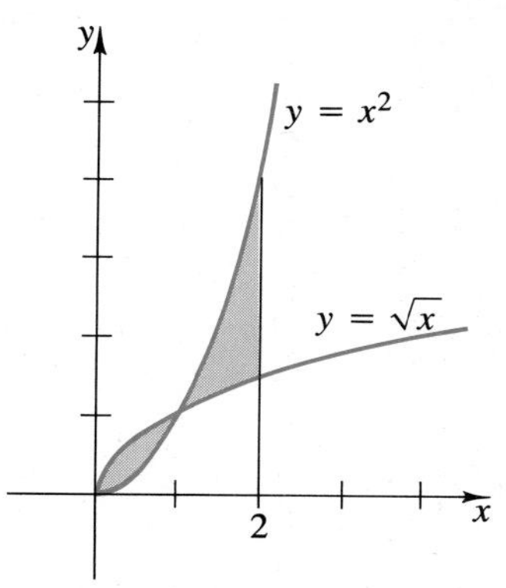

FIGURE 6

Solution We first sketch the curves in Figure 6 and determine the points of intersection of the two curves. To determine the x-coordinate of the point of intersection we solve the equation $\sqrt{x} = x^2$. That is, squaring both sides,

$$\begin{aligned} x &= x^4 \\ x^4 - x &= 0 \\ x(x^3 - 1) &= 0 \\ x = 0 \quad &\text{or} \quad x = 1 \end{aligned}$$

Note that $\sqrt{x} \geq x^2$ if $0 \leq x \leq 1$, and $x^2 \geq \sqrt{x}$ if $1 \leq x \leq 2$. Hence,

$$\begin{aligned} A &= \int_0^1 \left(\sqrt{x} - x^2\right) dx + \int_1^2 (x^2 - \sqrt{x})\, dx \\ &= \left(\frac{2}{3}x^{3/2} - \frac{1}{3}x^3\right)\Bigg|_0^1 + \left(\frac{1}{3}x^3 - \frac{2}{3}x^{3/2}\right)\Bigg|_1^2 \\ &= \left(\frac{2}{3} - \frac{1}{3}\right) + \left\{\left[\frac{8}{3} - \frac{2}{3}\left(2\sqrt{2}\right)\right] - \left(\frac{1}{3} - \frac{2}{3}\right)\right\} \\ &= \frac{1}{3} + \frac{8}{3} - \frac{4\sqrt{2}}{3} + \frac{1}{3} \\ &= \frac{10 - 4\sqrt{2}}{3} \end{aligned}$$

EXERCISE SET 17.2

In Exercises 1–8 find the area of the shaded region.

1.

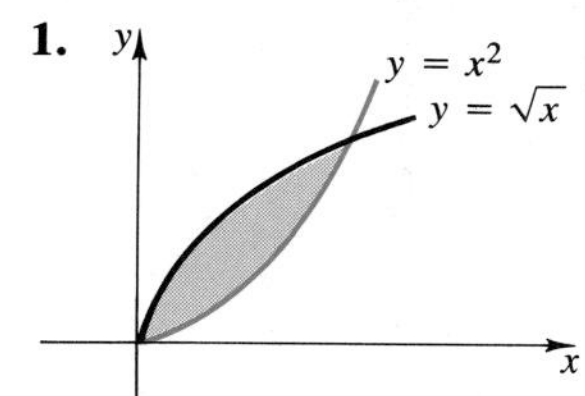

2.

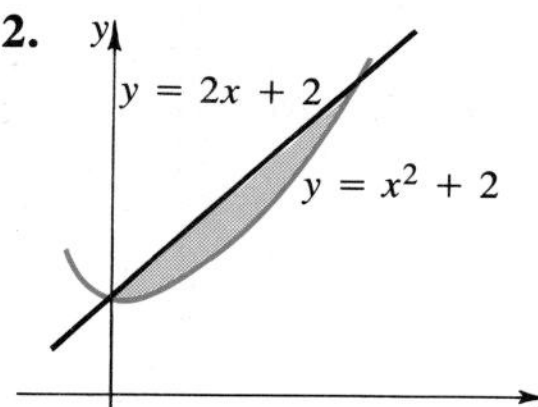

4.

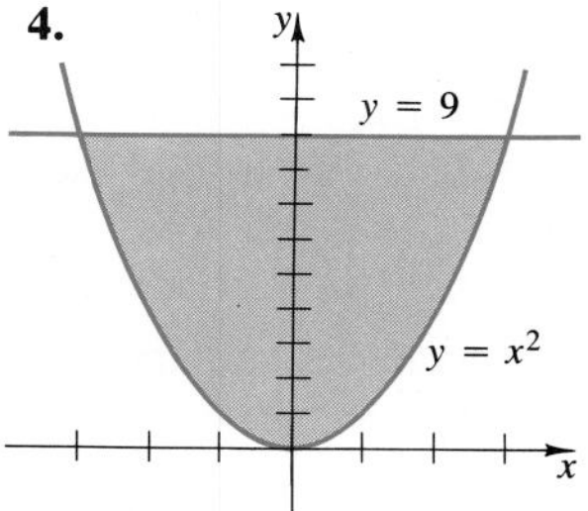

5.

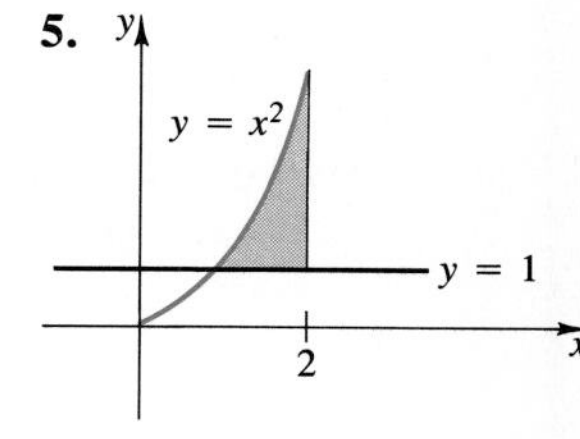

3.

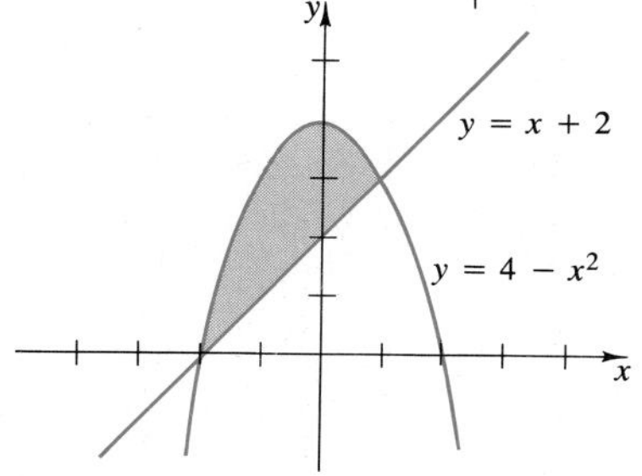

6.

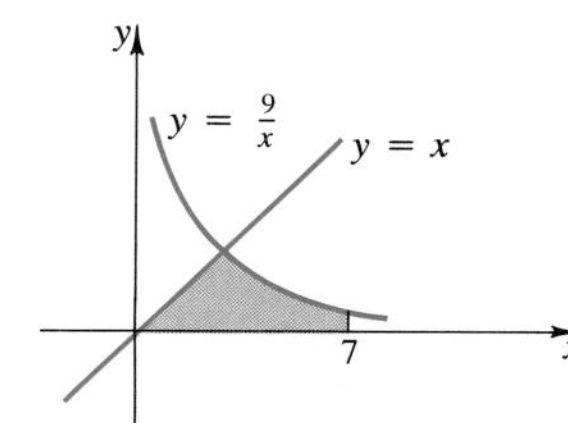

7.

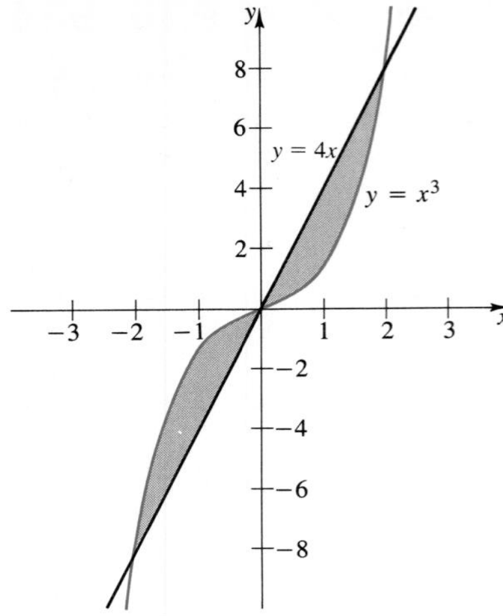

8.

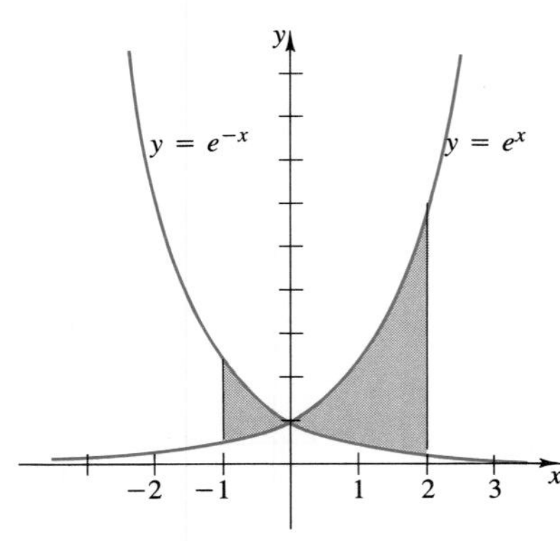

9. Find the area of the region enclosed by the curves $y = 3x^2$ and $y = 3x$.

10. Find the area of the region enclosed by the curves $y = x^2$ and $y = x + 2$.

11. Find the area of the region enclosed by the curves $y = 5/x$ and $y = 6 - x$.

12. Find the area of the region enclosed by the curves $y = x^2 - x$ and $y = x - x^2$.

13. Find the area of the "triangular" region in the first quadrant bounded by the y-axis, the line $y = 3$, and the curve $y = \sqrt{x}$.

14. Find the total area of the region bounded by the curve $y = x^2 - 4x$, the line $x = -3$, and the x-axis.

15. Find the area of the region enclosed by the curves $y = x^2$ and $y = 2x + 3$.

16. Find the area of the region enclosed by the curves $y = 9 - x^2$ and $y = 5$.

17. Find the area of the region enclosed by the curves $y = -\sqrt{x}$ and $y = -x^2$.

17.3 AVERAGE VALUE OF A FUNCTION

If we wish to calculate the **arithmetic average** or **mean** of the n numbers

$$y_1, y_2, \ldots, y_n$$

then we can apply the formula

$$\bar{y} = \frac{y_1 + y_2 + \cdots + y_n}{n} \tag{1}$$

as done in Section 8.2. Suppose we wish to find the "average" value of a continuous function $y = f(x)$ over the interval $[a, b]$. For example, we may wish to find the average temperature $T(x)$ as time goes through one 24-hour period. A reasonable way to do this would be to take temperature readings every hour, record these results as temperatures, $T(x_1), T(x_2), \ldots, T(x_{24})$, and then calculate their arithmetic average. We would have

$$\bar{T} \cong \frac{T(x_1) + T(x_2) + \cdots + T(x_{24})}{24} \tag{2}$$

Of course, our calculation would be improved if we took more temperature readings, for example, every half hour instead of every hour. In fact, the *best* answer for the average temperature would be found by taking infinitely many readings:

$$\bar{T} = \lim_{n\to\infty} \frac{T(x_1) + T(x_2) + \cdots + T(x_n)}{n} \tag{3}$$

This formula leads to an integral, as follows. Imagine dividing the interval $[a, b] = [0, 24]$ into n equal subintervals, each of length $\Delta x = (b - a)/n = 24/n$. Then $n = 24/\Delta x$, so Formula (3) can be written as

$$\begin{aligned}\bar{T} &= \lim_{n\to\infty} \frac{T(x_1) + T(x_2) + \cdots + T(x_n)}{24/\Delta x} \\ &= \lim_{n\to\infty} \frac{\Delta x[T(x_1) + T(x_2) + \cdots + T(x_n)]}{24} \\ &= \frac{1}{24} \lim_{n\to\infty} \Delta x[T(x_1) + T(x_2) + \cdots + T(x_n)]\end{aligned}$$

The limit on the right is just

$$\frac{1}{24}\int_0^{24} T(x)\, dx$$

by Equation (1) of Section 17.1.

This procedure generalizes to arbitrary functions, and we have the following definition:

Definition The **average** or **mean value** of an integrable function $f(x)$ over a closed interval $[a, b]$ is

$$\bar{f} = \frac{1}{b-a}\int_a^b f(x)\, dx \tag{4}$$

EXAMPLE 1 Suppose during one day (time $x = 0$ to $x = 24$ hours) the temperature $T(x)$ follows the pattern

$$T(x) = 52 - 0.2x + 0.06x^2$$

Find the average temperature during that day.

Solution

$$\begin{aligned}\bar{T} &= \frac{1}{24}\int_0^{24} (52 - 0.2x + 0.06x^2)\, dx \\ &= \frac{1}{24}(52x - 0.1x^2 + 0.02x^3)\Big|_0^{24} \\ &= \frac{1}{24}[52(24) - 0.1(576) + (0.02)(13{,}824)] \\ &= \frac{1}{24}[1248 - 57.6 + 276.48] \\ &\cong 61.12^\circ\end{aligned}$$

THE INTEGRAL IN MEDICINE: CAT SCANS

A recent and important application of mathematics in medicine is the area of computerized axial tomography (CAT), a new innovation in X-ray technology. A full description of the mathematical theory is too involved to describe here; nevertheless, we can give an indication of what motivates the theory.

The output of an ordinary X-ray machine is a two-dimensional "photograph" of the object (or person!) under consideration; we call it a *radiograph*. The brightness of each point in the radiograph is an indication of the total amount of mass in the photographed object between the X-ray source and the point on the radiograph (see Figure A).

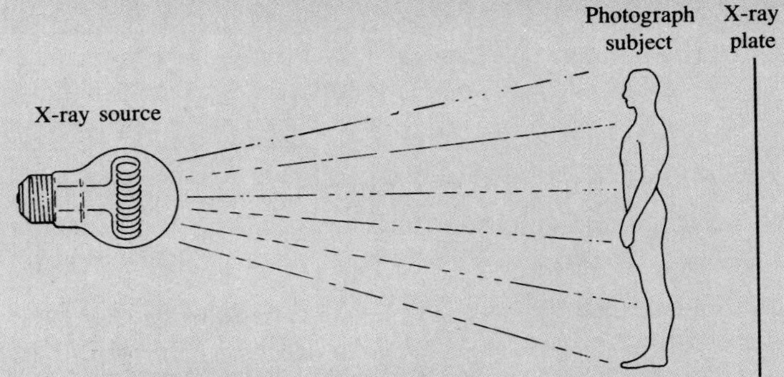

An obvious defect of such a photographic technique is the problem of shadowing. For example, consider the two different objects being X-rayed in Figure B; the single thick slab will obviously produce a radiograph identical to that of the two parallel thin slabs.

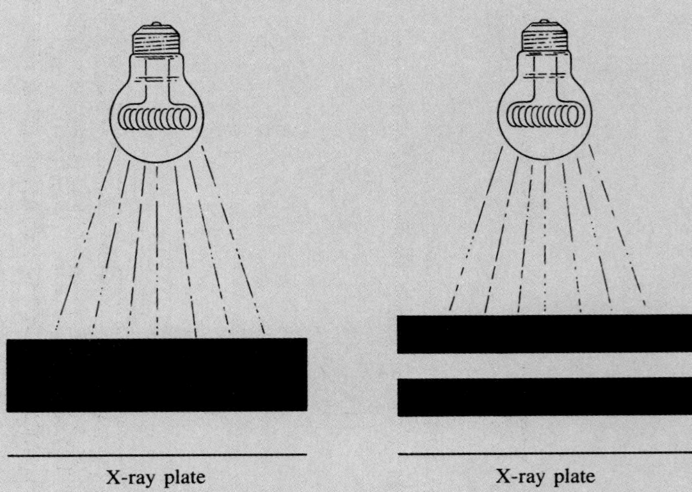

One's immediate reaction is to simply take the picture from a different angle; rotating the X-ray machine by 90°, for example, will certainly give us radiographs distinguishing the two objects. But in the day-to-day use of X-rays, the situation is far more complex. First of all, the object under study (interior tumor, broken bone, etc.) is usually hidden from sight, and thus it is difficult to know beforehand which angle to photograph from. Secondly, the *geometry*, or physical contours, of the object is far more complicated than our simple model. An X-ray of a human chest involves bone material, cartilage, lung tissue, heart tissue, and more, all casting lighter and deeper shadows onto the radiograph. And it is often the case that the material that one is most interested in studying leaves the lightest photographic imprint: this is true for clinical use of X-rays for investigation of lung disorders, for example.

Nevertheless, the idea of rotating the angle from which the radiograph is taken is the key to understanding the new X-ray machines. Instead of taking one or a few high-intensity radiographs from different angles, CAT scan machines take thousands of low-grade radiographs from many different angles, with overall total dosage roughly comparable to that of conventional radiographs. There are other differences between conventional radiographs and CAT scans. The X-ray source for CAT scans is focused down to a narrow pencil beam; and at the receptor end, rather than have a photographic plate record the intensity of the resultant X-ray beam, there are electronic sensors.

This is where calculus plays a role. The observer knows the intensity of the initial X-ray beam (this is calibrated) and has sensors which record the intensity of the beam after it has passed through the object being scanned. The difference in intensity is a measure of the density of material lying along the X-ray path. If $f(x, y, z)$ represents the density of the scanned object, then the data from the CAT scan essentially gives us the integral of f along the ray path. A main mathematical problem of CAT scan technology is to use this information about the function f to identify it uniquely. Great strides have been made toward understanding this problem, but much more needs to be done and it remains the focus of much current research.

EXAMPLE 2 Find the average value of the function $y = \ln x$ as x varies from 1 to 5.

Solution

$$\bar{y} = \frac{1}{5-1}\int_1^5 \ln x\, dx$$

To perform the integration we use integration by parts or Formula (36) in the integral table (Section 16.3). We then have

$$\begin{aligned}
\bar{y} &= \frac{1}{4}(x \ln x - x)\Big|_1^5 \\
&= \frac{1}{4}[(5 \ln 5 - 5) - (1 \ln 1 - 1)] \\
&= \frac{1}{4}[5 \ln 5 - 4] \\
&\cong 1.01
\end{aligned}$$

EXERCISE SET 17.3

In Exercises 1–8 find the average value of the given function over the given interval.

1. $f(x) = x^2 + x$, over $[0, 6]$
2. $f(x) = \sqrt{1 + x}$, over $[3, 8]$
3. $f(x) = e^x$, over $[-1, 1]$
4. $f(x) = xe^x$, over $[0, 2]$
5. $f(x) = 1/x$, over $[2, 10]$
6. $f(x) = (3x - 4)^2$, over $[1, 5]$
7. $f(x) = 1/(x^2 - 4)$, over $[-1, 1]$. Use integral tables.
8. $f(x) = 1/(x^2 - x - 6)$, over $[0, 2]$. Use integral tables.
9. The number of bacteria in a certain colony after t hours is known to be $B = 1000e^{0.85t}$. Find the average number of bacteria present during the first 4 hours.
10. The amount (in pounds) of a certain radioactive substance present after t months is $A = 600e^{-0.02t}$. Find the average amount present during the first 10 years.
11. Find the average slope of the curve $y = x^3$ over the interval $[-1, 1]$.
12. A car, starting from rest, has velocity $v(t) = 5t + 2t^2$ feet/second after t seconds. Find its average velocity over the first 6 seconds.
13. Suppose that x hours after midnight the temperature at a certain location is
$$T = 30 - \frac{1}{12}(x - 6)^2$$
Find the average temperature between midnight and 6:00 A.M.
14. Suppose that after an advertising campaign, daily sales receipts s (in dollars) of a new product vary with time according to the formula
$$s = 1000 - 200e^{-t}$$
where t is the elapsed time in months. Find the average daily sales in the first month.
15. Suppose the supply equation for a commodity is
$$P = 4 + \frac{20}{s}$$
where P = price per unit (in dollars), and s = the number of units produced weekly. Find the average price per unit as the weekly supply changes from 10 to 15 units.
16. The demand equation for a certain commodity is
$$p(x) = 800 - 2x^2$$
where $p(x)$ is the market price per unit (in dollars) when there are x units available on the market. As the number of available units increases steadily from 9 to 15 units, what is the average market price?
17. When \$1000 is invested at 12% interest, compounded continuously, the value of the investment after t years is
$$A = 1000e^{0.12t}$$
What is the average value of the investment over the first 5 years?

17.4 CONSUMERS' SURPLUS AND PRODUCERS' SURPLUS

In this section we explore some mathematical relationships between the market price of a commodity and the supply and demand for that commodity. In this context the definite integral has several significant applications.

Demand, or Sales Level

For our purposes, an industry is composed of all the producers of a certain commodity. The market for the commodity consists of all the consumers of the commodity, possibly including the industry itself. We let

p = (variable) **market price** of commodity
x = number of units of commodity demanded by market when unit price is p (1)

In other words, x represents the number of units that will be sold when the price is p. It is sometimes referred to as the **sales level** at a given price p.

An equation relating p and x is called a **demand equation**. In this chapter we write it as

$$p = f(x) \qquad \text{a "price function" of demand}$$

Demand functions are generally decreasing, since a decrease in price usually causes an increase in demand for the commodity and an increase in price usually causes a decrease in demand. Figure 1 shows a typical demand curve.

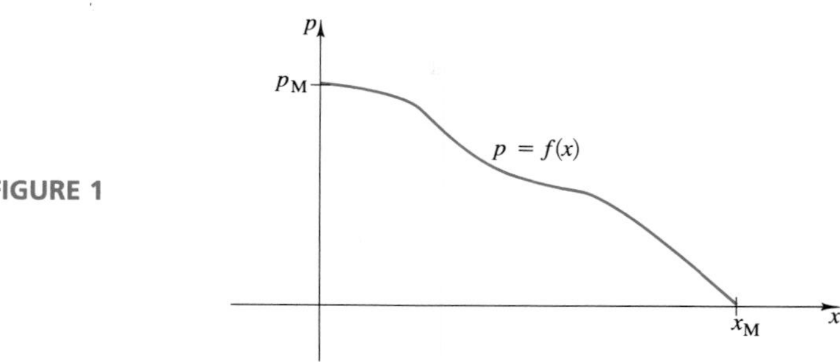

FIGURE 1

The price p_M (maximum price) is "too high"; at this price there is no demand for the product. The demand x_M represents the maximum sales level. The market will not accept more than x_M units of the commodity, even if the industry were to give them away.

Not every consumer is willing to pay the same price. For a given price p in the interval $(0, p_M)$ (Figure 1), there is some demand (sales level) x. Conversely, corresponding to every sales level x there is a price $p = f(x)$ that some consumers will pay.

Ideal Total Revenue

Suppose that the industry could judge perfectly each consumer's "price"—the highest price each customer is willing to pay. Suppose further that the industry sells to every customer at that price. Since revenue at a given price p equals the product of the price p and the number x of units sold, we have

$$R = px$$

Under these ideal conditions, when the sales level is x, the industry's *revenue* from sales of the next Δx units would be

$$\Delta R \cong p \cdot \Delta x = f(x)\, \Delta x \tag{2}$$

Therefore, from the producer's point of view, **the ideal total revenue** IR_p is the sum of all these revenues, ΔR, as x varies from 0 to x_M.

If the interval $[0, x_M]$ is divided into n equal subintervals each of width $\Delta x = (x_M - 0)/n = x_M/n$, then

$$IR_p \cong \Delta R_1 + \Delta R_2 + \cdots + \Delta R_n \cong f(x_1)\, \Delta x + f(x_2)\, \Delta x + \cdots + f(x_n)\, \Delta x$$

Hence,

$$IR_p = \lim_{n \to \infty} [f(x_1) + f(x_2) + \cdots + f(x_n)] \, \Delta x = \int_0^{x_M} f(x) \, dx$$

by Equation (1) of Section 17.1. Thus,

$$IR_p = \int_0^{x_M} f(x) \, dx \tag{3}$$

Since $f(x) \geq 0$, this integral can be considered to represent the area under the curve $p = f(x)$ (see Section 15.2). Therefore, *for the producer, the industry's ideal total revenue is the total area under the demand curve*. The area interpretation allows us to picture some of the following results.

Consumers' Surplus

No industry can judge perfectly each consumer's "price." Instead, the industry arrives at a *market price* p_0 in some other way. This market price is in the open interval $(0, p_M)$ and corresponds to some sales level x_0. The industry's total revenue is then

$$R = p_0 \cdot x_0 \tag{4}$$

In Figure 2 this revenue is represented by the lightly shaded rectangle of base x_0 and height p_0. Since the area is less than the total area under the demand curve, then the actual revenue (4) is less than the ideal total revenue (3).

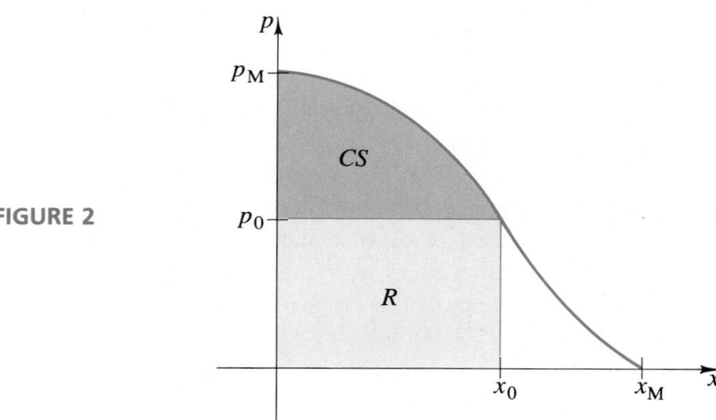

FIGURE 2

When the industry sets the market price at p_0, it loses the revenue it could have gained by selling its commodity at higher prices to those customers willing to pay them. This loss in revenue is represented by the heavily shaded region in Figure 2. Since this industry loss is also the consumers' gain, it is called the **consumers' surplus at market price** p_0. Since this value can be represented as an area, it is representable as an integral.

Definition 1 If $p = f(x)$ is the demand equation for a commodity for which the market price is p_0, then the **consumers' surplus $CS(p_0)$ at market price p_0** is

$$CS(p_0) = \int_0^{x_0} [f(x) - p_0]\, dx \tag{5}$$

where x_0 is the sales level corresponding to price p_0. $CS(p_0)$ represents the total savings to all consumers who paid the price p_0 but would have been willing to pay a higher price.

EXAMPLE 1 Suppose the demand equation for a commodity is $p = f(x) = 120 - 2x - x^2$.

(a) Find the ideal total revenue for the industry.

(b) Find the actual revenue and the consumers' surplus when the sales level is $x_0 = 6$, and interpret the result.

(c) Find the actual revenue and the consumers' surplus when the market price is $p_0 = \$105$, and interpret the results.

Solution (a) We first sketch the demand equation

$$p = 120 - 2x - x^2 = (12 + x)(10 - x)$$

in Figure 3a. We see that the maximum price is $p_M = 120$ and that x_M is 10.

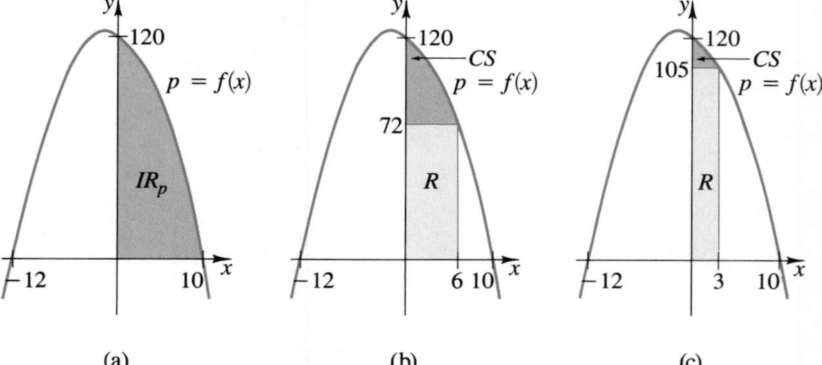

FIGURE 3

The ideal total revenue is the area under the demand curve, the shaded region in Figure 3a. Using Formula (3),

$$\begin{aligned} IR_p &= \int_0^{10} f(x)\, dx = \int_0^{10} (120 - 2x - x^2)\, dx \\ &= \left(120x - x^2 - \frac{x^3}{3}\right)\Bigg|_0^{10} \\ &= 1200 - 100 - \frac{1000}{3} \\ &= \$766.67 \end{aligned}$$

Solution (b) When $x_0 = 6$, $p_0 = 120 - 2(6) - (6)^2 = 120 - 48 = 72$. Thus, by Formula (4) the actual revenue is

$$R = p_0 \cdot x_0 = 72 \cdot 6 = \$432$$

From Formula (5) the consumers' surplus is

$$\begin{aligned} CS &= \int_0^6 [(120 - 2x - x^2) - 72]\, dx = \int_0^6 (48 - 2x - x^2)\, dx \\ &= \left(48x - x^2 - \frac{x^3}{3}\right)\Bigg|_0^6 = 288 - 36 - 72 \\ &= \$180 \end{aligned}$$

This result means that when the sales level is 6 units, the total savings to all the customers willing to pay more than \$72 per unit is \$180. Both R and CS are shown in Figure 3b.

Solution (c) When $p_0 = 105$, we find x_0 by solving the equation

$$\begin{aligned} 105 &= 120 - 2x - x^2 \\ 0 &= 15 - 2x - x^2 \\ &= (5 + x)(3 - x) \end{aligned}$$

so that $x = -5$ or 3. Since x cannot be negative, we conclude that $x_0 = 3$. Using Formula (4) we find the actual revenue:

$$R = p_0 \cdot x_0 = 105 \cdot 3 = \$315$$

Using Formula (5) we find the consumers' surplus:

$$\begin{aligned} CS &= \int_0^3 ((120 - 2x - x^2) - 105)\, dx \\ &= \left(15x - x^2 - \frac{x^3}{3}\right)\Bigg|_0^3 = 45 - 9 - 9 \\ &= \$27 \end{aligned}$$

This means that when the market price is \$105 per unit, the *total* savings to all the consumers willing to pay more than that is only \$27. Both R and CS are shown in Figure 3c.

Supply, or Production Level

Continuing with our analysis of an industry producing a commodity, let x and p be as defined in Equation (1) and let

$$y = \text{number of units of commodity supplied by industry when unit price is } p \tag{6}$$

The value of y is often called the **production level** corresponding to price p, since it represents the number of units the industry is willing to produce when the market price is p.

An equation relating p and y is called a **supply equation**. We usually write it as

$$p = g(y) \qquad \text{a "price function" of supply}$$

Supply functions are generally increasing, since an increase in market price usually stimulates producers to increase production. A typical supply curve is shown in Figure 4. Here, the price p_m (minimum price) is "too low"; at this price producers are unwilling to supply the commodity. The production level y_M represents the maximum possible supply. No matter what the price, producers are unable to produce more than y_M units of the commodity.

FIGURE 4

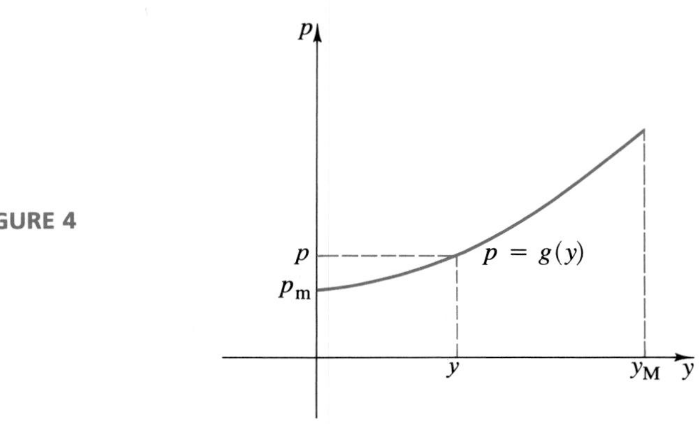

Producers' Surplus

For a given market price of p_0 dollars per unit, some producers can produce the commodity at a *lower* cost. Thus, when the industry sets the market price at p_0 dollars, those producers actually gain by selling at the market price.

Suppose each consumer could judge perfectly each producer's lowest acceptable price and pay each producer only that price. Then, when the production level is at y, the industry's revenue from producing the next Δy units would be

$$\Delta R \cong p \cdot \Delta y = g(y)\,\Delta y$$

Subdividing the interval $[0, y_M]$ into n equal subdivisions each of width $\Delta y = (y_M - 0)/n = y_M/n$, from the consumers' point of view the industry's *ideal total revenue* IR_c would be the sum of all these revenues ΔR as y varies from 0 to y_M. Thus,

$$IR_c \cong \Delta R_1 + \Delta R_2 + \cdots + \Delta R_n \cong g(y_1)\,\Delta y + g(y_2)\,\Delta y + \cdots + g(y_n)\,\Delta y$$

Letting $n \to \infty$, we have

$$IR_c = \int_0^{y_M} g(y)\,dy \tag{7}$$

by Equation (1) of Section 17.1

Again since $g(y) \geq 0$, *the industry's ideal total revenue (from the point of view of the consumer) can be considered as the area under the supply curve*. It may be seen as the lightly shaded area in Figure 5.

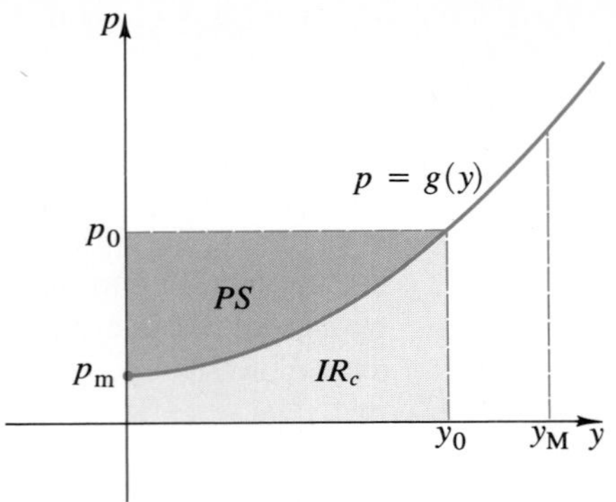

FIGURE 5

When the industry sells all its commodity at the market price p_0, with corresponding production level y_0, its total revenue is

$$R = p_0 \cdot y_0 \tag{8}$$

In Figure 5 this total revenue is represented by the entire shaded rectangle. The heavily shaded region in Figure 5 represents the total excess income realized by the producers who would be willing to supply the commodity at a price less than p_0. This excess is called the **producers' surplus at market price p_0**. Since area can be represented as an integral, we have:

> **Definition 2** If $p = g(y)$ is the supply equation for a commodity for which the market price is p_0, then the **producers' surplus $PS(p_0)$ at market price p_0** is
>
> $$PS(p_0) = \int_0^{y_0} [p_0 - g(y)]\, dy \tag{9}$$
>
> where y_0 is the production level corresponding to price p_0. $PS(p_0)$ represents the total excess revenue earned by all producers who sold at price p_0 but would have been willing to sell at a lower price.

EXAMPLE 2 Suppose the supply equation for a commodity is $p = g(y) = 8 + 0.2y + 0.1y^2$ (in thousands). Assume that the maximum possible supply is $y_M = 10$.

(a) Find the ideal total revenue from the consumer's viewpoint.

(b) Find the actual revenue and the producers' surplus when the production level is $y_0 = 5$, and interpret the results.

(c) Find the actual revenue and the producers' surplus when the market price is $p_0 = 9.5$, and interpret the results.

Solution (a) We first sketch the supply equation $p = g(y) = 8 + 0.2y + 0.1y^2$ as in Figure 6a.

FIGURE 6

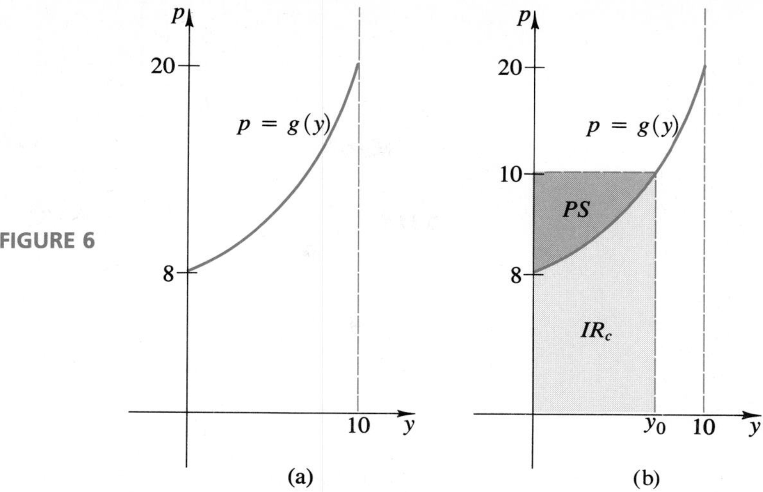

The ideal total revenue is the area of the lightly shaded region in Figure 6b:

$$
\begin{aligned}
IR_c &= \int_0^{10} g(y)\,dy = \int_0^{10} (8 + 0.2y + 0.1y^2)\,dy \\
&= \left(8y + 0.1y^2 + \frac{0.1}{3}y^3\right)\Bigg|_0^{10} \\
&= 80 + 10 + \frac{100}{3} = 123.333 \qquad \text{(thousands)} \\
&= \$123{,}333
\end{aligned}
$$

Solution (b) When $y_0 = 5$, the price is $p_0 = 8 + 0.2(5) + 0.1(5)^2 = 11.5$. By Formula (8) the actual total revenue is

$$
\begin{aligned}
R &= p_0 \cdot y_0 = (11.5)(5) = 57.5 \qquad \text{(thousands)} \\
&= \$57{,}500
\end{aligned}
$$

From Formula (9) the producers' surplus is

$$
\begin{aligned}
PS &= \int_0^5 [11.5 - (8 + 0.2y + 0.1y^2)]\,dy \\
&= \left(3.5y - 0.1y^2 - \frac{0.1}{3}y^3\right)\Bigg|_0^5 \\
&= 17.5 - 2.5 - \frac{12.5}{3} \\
&= 10.833 \qquad \text{(thousands)} \\
&\cong \$10{,}833
\end{aligned}
$$

This means that when the sales level is at 5 units, the total excess revenue realized by all the producers willing to produce for less than \$11,500 per unit is \$10,833.

Solution (c) When $p_0 = 9.5$, we find y_0 by solving the equation

$$\begin{aligned} 9.5 &= 8 + 0.2y + 0.1y^2 \\ 0 &= -1.5 + 0.2y + 0.1y^2 \\ &= y^2 + 2y - 15 = (y + 5)(y - 3) \end{aligned}$$

Thus $y_0 = -5$ or 3. Since y_0 cannot be negative, we obtain $y_0 = 3$. Using Formula (9), the producers' surplus is

$$\begin{aligned} PS &= \int_0^3 [9.5 - (8 + 0.2y + 0.1y^2)]\,dy \\ &= \left(1.5y - 0.1y^2 - \frac{0.1}{3}y^3\right)\Bigg|_0^3 \\ &= 4.5 - 0.9 - 0.9 = 2.7 \qquad \text{(thousands)} \\ &= \$2700 \end{aligned}$$

This means that when the market price is at \$9500, the total excess revenue realized by all the producers willing to produce the commodity for less than this price is \$2700.

Equilibrium

In general the market price for a commodity is not set arbitrarily but is determined by the tendency for supply and demand to balance. Suppose we know both the demand function $p = f(x)$ and the supply function $p = g(y)$. The market is said to be in **equilibrium** when $x = y$, that is, when the demand equals the supply. The market price p_0 for which this occurs is called the **equilibrium price**. The demand at the equilibrium price is called the **market demand** x_0, and the supply at the equilibrium price is called the **market supply** y_0. To solve algebraically for the equilibrium price, observe that the condition $x = y$ means that

$$p = f(x) = g(x) \qquad \text{(letting } x = y\text{)} \tag{10}$$

If we solve Equation (10) for x_0, then $p_0 = f(x_0)$ and $y_0 = x_0$.

EXAMPLE 3 Find the equilibrium price for a commodity that has demand equation $p = 50 - 3x$ (in dollars) and supply equation $p = 18 + 2y$ (in dollars). Also find the market demand and supply.

Solution Letting $x = y$,

$$\begin{aligned} p = 50 - 3x &= 18 + 2x \\ 32 &= 5x \\ x_0 &= \frac{32}{5} = 6.40 \end{aligned}$$

Therefore, the equilibrium price $p_0 = f(x_0)$ is

$$50 - 3(6.40) = \$30.80 \text{ per unit}$$

the market demand is $x_0 = 6.40$ units, and the market supply is $y_0 = 6.40$ units.

For a market in equilibrium, all the relevant features can be illustrated by using a remarkably simple graphing technique. We rewrite the supply equation $p = g(y)$ as $p = g(x)$ (since $x = y$). We then graph both the demand equation $p = f(x)$ and the supply equation in the same coordinate system, as in Figure 7.

FIGURE 7

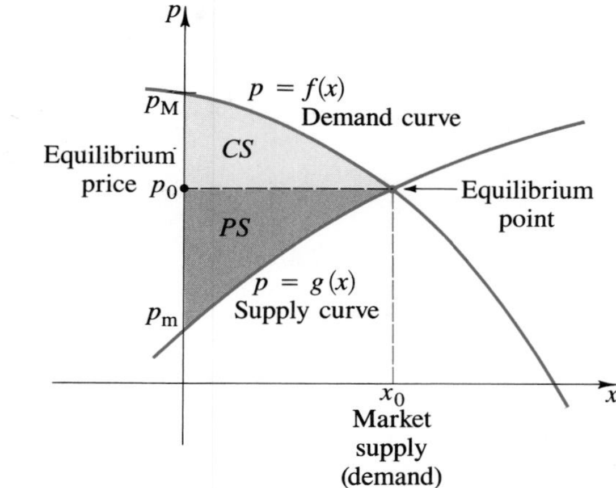

The point where the supply curve intersects the demand curve is called the **equilibrium point**. The p-coordinate of this point, p_0, is the equilibrium price, because at this price the demand equals the supply. The x-coordinate of the equilibrium point, x_0, is both the market demand and the market supply.

The total revenue of the industry is the area of the rectangle having the origin and the equilibrium point at opposite corners. The consumers' surplus and the producers' surplus are the areas of the regions indicated in Figure 7.

EXAMPLE 4 Consider a market with demand equation $p = 4 - \frac{2}{3}x$ and supply equation $p = 1 + y^2/9$, where p is measured in dollars and x and y represent thousands of units. Assuming that the market is in equilibrium, find the market price, supply, and demand and also find the industry revenue, consumers' surplus, and producers' surplus.

Solution We begin by finding the equilibrium point. We replace y by x and solve for the intersection of the supply and demand curves:

$$\begin{aligned} 1 + \frac{x^2}{9} &= 4 - \frac{2}{3}x \\ 9 + x^2 &= 36 - 6x \\ x^2 + 6x - 27 &= 0 \\ (x + 9)(x - 3) &= 0 \\ x = -9 \quad \text{or} \quad x &= 3 \end{aligned}$$

Since x cannot be negative, we have $x_0 = 3$. Thus the market supply and demand is 3000 units. To find the equilibrium price we substitute $x_0 = 3$ into either the demand equation or the supply equation:

$$p_0 = 1 + \frac{3^2}{9} = 1 + 1 = 2$$

Thus the market price is \$2 per unit. Next we sketch the graph (Figure 8).

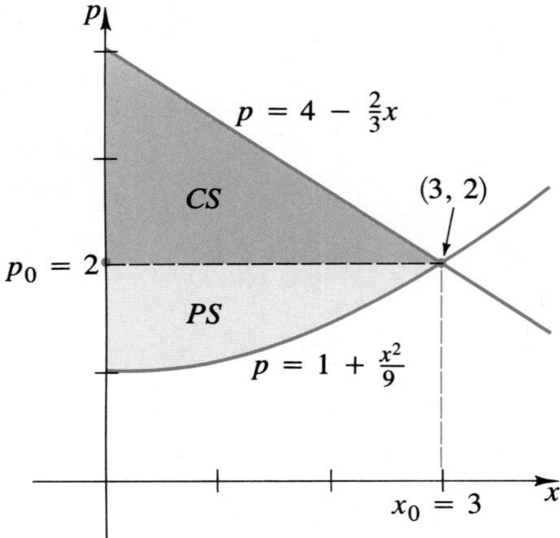

FIGURE 8

Using Figure 8 or Formula (4) we see that the industry revenue is $R = p_0 \cdot x_0 = 2 \cdot 3 = 6$ (in thousands), so $R = \$6000$. Using Figure 8 or Formula (5) we obtain the consumers' surplus:

$$\begin{aligned} CS &= \int_0^3 \left[\left(4 - \frac{2}{3}x\right) - 2\right] dx \\ &= \int_0^3 \left(2 - \frac{2}{3}x\right) dx = \left(2x - \frac{1}{3}x^2\right)\Big|_0^3 \\ &= 6 - 3 = 3 \end{aligned}$$

Using Figure 8 or Formula (9) we obtain the producers' surplus:

$$\begin{aligned} PS &= \int_0^3 \left[2 - \left(1 + \frac{x^2}{9}\right)\right] dx \\ &= \int_0^3 \left(1 - \frac{x^2}{9}\right) dx = \left(x - \frac{x^3}{27}\right)\Big|_0^3 \\ &= 3 - 1 = 2 \end{aligned}$$

Therefore, the consumers' surplus is \$3000 and the producers' surplus is \$2000.

EXERCISE SET 17.4

In Exercises 1–4 follow these steps for the demand equation given.

(a) *Sketch the graph of the demand equation, and find p_M and x_M.*

(b) *Find the ideal total revenue for the industry from the producers' point of view.*

(c) *For the given sales level x_0, find the actual revenue and the consumers' surplus. Sketch a graph and interpret your results.*

(d) *For the given market price p_0, find the actual revenue and the consumers' surplus. Sketch a graph and interpret your results.*

1. $p = 300 - 6x$; for part (c), $x_0 = 30$; for part (d), $p_0 = \$240$

2. $p = 1500 - 15x^2$; for part (c), $x_0 = 6$; for part (d), $p_0 = \$1125$

3. $p = 300 - x - 2x^2$; for part (c), $x_0 = 9$; for part (d), $p_0 = \$245$

4. $p = \sqrt{100 - x}$; for part (c), $x_0 = 64$; for part (d), $p_0 = \$5$

In Exercises 5–8 follow these steps for the given supply equation with maximum supply y_M.

(a) *Sketch the graph of the supply equation, and find p_m.*

(b) *Find the ideal total revenue of the industry from the consumers' point of view.*

(c) *For the given production level y_0, find the actual revenue and the producers' surplus. Sketch a graph and interpret your results.*

(d) *For the given market price p_0, find the actual revenue and the producers' surplus. Sketch a graph and interpret your results.*

5. $p = 6y + 1$, $y_M = 7$; for part (c), $y_0 = 4$; for part (d), $p_0 = \$19$

6. $p = 2 + 9y^2/100$, $y_M = 10$; for part (c), $y_0 = 8$; for part (d) $p_0 = \$4.25$

7. $p = \sqrt{y + 4}$, $y_M = 96$; for part (c), $y_0 = 60$; for part (d), $p_0 = \$9$

8. $p = e^{y/5}$, $y_M = 25$; for part (c), $y_0 = 20$; for part (d), $p_0 = \$e^3$

In Exercises 9–12 both a demand equation and a supply equation are given for a market. Assume that the market is in equilibrium.

(a) *Find the market price.*

(b) *Find the supply and demand values.*

(c) *Find the industry revenue.*

(d) *Find the consumers' surplus.*

(e) *Find the producers' surplus.*

(f) *Sketch an illustrative graph by letting $x = y$ and graphing $p = f(x)$ and $p = g(x)$ in the same coordinate system.*

9. $p = 12 - \frac{4}{5}x$, $p = 2 + \frac{6}{5}y$

10. $p = 30 - 2x$, $p = \frac{1}{2}y^2$

11. $p = 20 - x^2$, $p = 3y^2 + 4$

12. $p = 16/\sqrt{x + 1}$, $p = \sqrt{y + 1}$

17.5 INTEGRATING RATES OF CHANGE

Recall from Section 15.2 that the defining equation for the definite integral is

$$\int_a^b f(x)\,dx = F(b) - F(a)$$

where F is any antiderivative of the integrand f. In particular, if we integrate the derivative $f'(x)$ of a function $f(x)$, we obtain

$$\int_a^b f'(x)\,dx = f(b) - f(a) \tag{1}$$

since $f(x)$ is an antiderivative of the integrand $f'(x)$. This result has a useful geometric interpretation. The right side of (1) represents the change in the value of $f(x)$ as x varies from a to b (Figure 1), and the left side is the integral from a to b of the rate of change of f with respect to x. Thus, restating Equation (1):

> The definite integral $\int_a^b f'(x)\,dx$ of the rate of change of f with respect to x yields the change in the value of f as x varies from a to b.

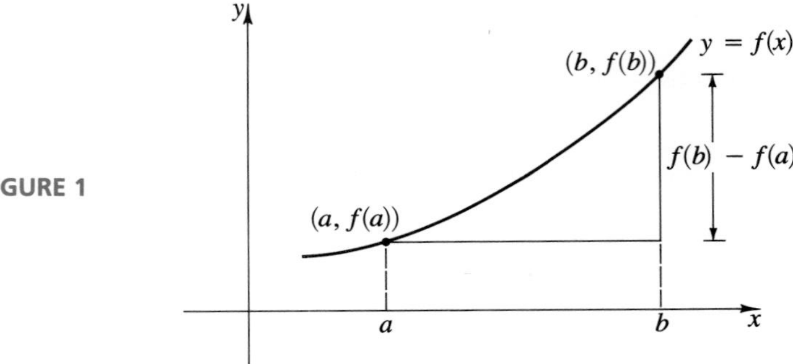

FIGURE 1

EXAMPLE 1 Let $f(x) = x^3$. Find $\int_1^2 f'(x)\,dx$.

Solution From (1),

$$\int_1^2 f'(x)\,dx = f(2) - f(1) = 2^3 - 1^3 = 7$$

As a check,

$$\int_1^2 f'(x)\,dx = \int_1^2 3x^2\,dx = 3 \cdot \frac{x^3}{3}\bigg|_1^2 = x^3\bigg|_1^2 = 2^3 - 1^3 = 7$$

Remark It is easy to forget the order in which the subtraction in (1) should be performed. To help remember it, note that x varies from an initial value of $x = a$ to a final value of $x = b$, so $f(a)$ is the *initial value* of f and $f(b)$ the *final value*. With this terminology, (1) can be stated as:

$$\int_a^b f'(x)\,dx = \text{final value of } f - \text{initial value of } f$$

Applications to Marginal Analysis

Recall from Section 13.2 that marginal cost (MC), marginal revenue (MR), and marginal profit (MP) are the derivatives of the cost function, revenue function, and profit function, respectively. That is,

$$\begin{aligned} \text{MC} &= \text{marginal cost} = C'(x) \\ \text{MR} &= \text{marginal revenue} = R'(x) \\ \text{MP} &= \text{marginal profit} = P'(x) \end{aligned}$$

Thus, if we integrate MC, MR, or MP over an interval $[a, b]$, we obtain from (1):

$$\begin{aligned} \int_a^b (\text{MC})\,dx &= \int_a^b C'(x)\,dx = C(b) - C(a) \\ &= \text{change in cost as } x \text{ varies from } a \text{ to } b \end{aligned}$$

$$\int_a^b (\text{MR})\,dx = \int_a^b R'(x)\,dx = R(b) - R(a)$$

$$= \text{change in revenue as } x \text{ varies from } a \text{ to } b$$

$$\int_a^b (\text{MP})\,dx = \int_a^b P'(x)\,dx = P(b) - P(a)$$

$$= \text{change in profit as } x \text{ varies from } a \text{ to } b$$

EXAMPLE 2 A manufacturer determines that the firm's marginal cost and marginal revenue functions are

$$\text{MC} = C'(x) = 100 - 0.1x$$
$$\text{MR} = R'(x) = 100 + 0.1x$$

(a) Find the change in revenue that results when the sales level increases from 20 to 30 units.

(b) Find the revenue resulting from the sale of 30 units.

(c) If the fixed cost (cost of producing $x = 0$ units) is \$400, find the cost of producing 30 units.

Solution (a) As the sales level increases from $x = 20$ to $x = 30$, the revenue changes an amount

$$\int_{20}^{30} (\text{MR})\,dx = \int_{20}^{30} (100 + 0.1x)\,dx = (100x + 0.05x^2)\Big|_{20}^{30}$$
$$= 3045 - 2020 = 1025$$

Thus the revenue increases by \$1025.

Solution (b) If we assume that the revenue from the sale of $x = 0$ units is zero, then the revenue resulting from the sale of $x = 30$ units can be viewed as the change in revenue that results when sales increase from $x = 0$ to $x = 30$. This change is given by

$$\int_{0}^{30} (\text{MR})\,dx = \int_{0}^{30} (100 + 0.1x)\,dx = (100x + 0.05x^2)\Big|_{0}^{30}$$
$$= 3045 - 0 = 3045$$

Therefore, the revenue received from selling 30 units is \$3045.

Solution (c) As the production level increases from $x = 0$ to $x = 30$, the cost changes an amount

$$\int_{0}^{30} (\text{MC})\,dx = \int_{0}^{30} (100 - 0.1x)\,dx = (100x - 0.05x^2)\Big|_{0}^{30}$$
$$= 2955 - 0 = 2955$$

Thus the total cost for manufacturing 30 units will be the fixed cost plus the added cost as the production increases from $x = 0$ to $x = 30$ units; that is,

$$\text{total cost} = \$400 + \$2955 = \$3355$$

Applications to Motion

Both velocity and acceleration are rates of change. The velocity of an object is the rate of change of its position coordinate along an x-axis with respect to time. Thus,

$$v(t) = \frac{dx}{dt} = x'(t)$$

Similarly, acceleration is the rate of change in velocity:

$$a(t) = \frac{dv}{dt} = v'(t)$$

Thus,

$$x(t) = \int v(t)\,dt \quad \text{and} \quad v(t) = \int a(t)\,dt \tag{2}$$

If $v(t)$ is positive for $a \leq t \leq b$, then from Equation (1),

$$\begin{aligned} \int_a^b v(t)\,dt &= \int_a^b x'(t)\,dt = x(b) - x(a) \\ &= \text{distance traveled from time } t = a \text{ to } t = b \end{aligned} \tag{3}$$

Remark In general, the distance traveled between time $t = a$ and $t = b$ is $\int_a^b |v(t)|\,dt$. However, in our examples $v(t) \geq 0$ for all values of t; therefore, the distance traveled is $\int_a^b v(t)\,dt$.

EXAMPLE 3 An automobile moves along a straight road in such a way that its velocity $v(t)$ after t seconds is $v(t) = 3t^{1/2}$ feet per second. How far does the automobile travel during the first 100 seconds?

Solution The velocity $v(t)$ is the rate of change of position of the car with respect to time. Also, the total distance traveled is the change in position of the car from time $t = 0$ to time $t = 100$. Thus, by Equation (3)

$$\begin{aligned} \text{distance traveled} &= \int_0^{100} v(t)\,dt \\ &= \int_0^{100} 3t^{1/2}\,dt = 3 \cdot \frac{2}{3} t^{3/2} \Big|_0^{100} \\ &= 2000 - 0 \\ &= 2000 \text{ feet} \end{aligned}$$

EXAMPLE 4 Suppose an object is propelled along an x-axis with velocity $v(t) = \sqrt{1 + 3t}$ feet per second at time t. How far does it travel during the time interval from $t = 1$ to $t = 5$?

Solution The distance traveled is the change in x from $t = 1$ to $t = 5$. Moreover, $x'(t) = v(t)$. Thus, by Equation (3),

$$\begin{aligned}
\text{distance traveled} &= \int_1^5 v(t)\,dt \\
&= \int_1^5 \sqrt{1+3t}\,dt \\
&= \left(\frac{1}{3}\right)\cdot\left(\frac{2}{3}\right)(1+3t)^{3/2}\Big|_1^5 \\
&= \left(\frac{2}{9}\right)(16^{3/2}-4^{3/2}) = \left(\frac{2}{9}\right)(64-8) = \frac{112}{9} \\
&= 12\frac{4}{9}\text{ feet}
\end{aligned}$$

Motion of Free-Falling Bodies When an object is dropped from a height and allowed to fall freely under the influence of gravity alone, its acceleration is constant:

$$a(t) = 32 \text{ feet per second per second}$$

By knowing this acceleration we can use Equation (2) to help find the velocity and position of the object at time t.

EXAMPLE 5 When a stone is dropped from a cliff, how far does it fall in the first 4 seconds?

Solution Imagine the stone falling along a y-axis directed downward (Figure 2). We want to find the change in y from time $t = 0$ to time $t = 4$. Thus we want

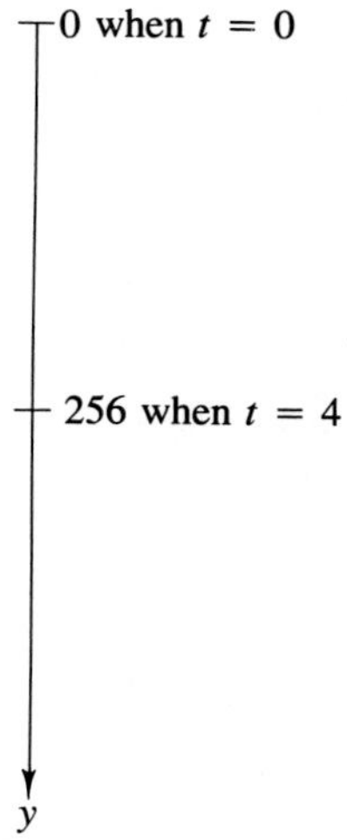

$$\text{distance traveled} = \int_0^4 v(t)\,dt \tag{4}$$

But $v(t) = \int a(t)\,dt$ by Equation (2). Substituting $a = 32$ we have

$$v(t) = \int 32\,dt = 32t + C$$

where $C = 0$ since $v(0) = 0$. Substituting this expression into Equation (4) gives us

$$\begin{aligned}
\text{distance traveled} &= \int_0^4 32t\,dt \\
&= 16t^2\Big|_0^4 \\
&= 256\text{ feet}
\end{aligned}$$

0 when $t = 0$

256 when $t = 4$

y

FIGURE 2

Application to Biology

EXAMPLE 6 An experimental drug changes the average subject's body temperature at a rate

$$r(t) = -0.003t^2 + 0.01t$$

where
$$T(t) = \text{temperature at time } t$$
$$r(t) = T'(t)$$

with $r(t)$ in degrees Fahrenheit per hour and t the number of hours elapsed after administration of the drug. How much of a temperature change will occur between the second and fifth hours ($t = 2$ to $t = 5$)?

Solution Since $r(t)$ is the *rate* at which temperature changes with time, the total change in temperature from $t = 2$ to $t = 5$ is by Equation (1)

$$T(5) - T(2) = \int_2^5 T'(t)\,dt = \int_2^5 r(t)\,dt = \int_2^5 (-0.003t^2 + 0.01t)\,dt$$
$$= (-0.001t^3 + 0.005t^2)\Big|_2^5$$
$$= 0 - 0.012 = -0.012$$

Thus the temperature decreases 0.012°F during the period $t = 2$ to $t = 5$.

Application to Analysis of Natural Resources

Suppose that $R(t)$ is the known rate at which a natural resource (e.g., coal, copper, zinc, oil, or gas) is being consumed. If we pick a reference point $t = 0$ in time and let $Q(t)$ be the amount of the product consumed during the time interval $[0, t]$, then

$$R(t) = \text{rate of consumption} = Q'(t)$$

Since $Q(0) = 0$ (why?), it follows from (1) that the amount of the resource consumed during a time interval $[0, T]$ is

$$Q(T) = Q(T) - Q(0) = \int_0^T Q'(t)\,dt = \int_0^T R(t)\,dt$$

EXAMPLE 7 In the beginning of 1975 ($t = 0$), zinc was being consumed at a rate of 478,850 short tons per year with the consumption rate increasing at 4.5% per year. Assuming an exponential growth model for the rate of consumption, we have*

$$R(t) = Q'(t) = 478{,}850e^{0.045t}$$

Estimate the total amount of zinc used from January 1, 1975, to December 31, 1984.

Solution From Equation (1), the total consumption between $t = 0$ (January 1, 1975) and $t = 10$ (December 31, 1984), as shown in the time line of Figure 3, is

---

* *The World Almanac* (New York: Newspaper Enterprise Association, 1981).

$$Q(10) = \int_0^{10} R(t)\,dt = \int_0^{10} 478{,}850e^{0.045t}\,dt$$
$$= \frac{478{,}850}{0.045}e^{0.045t}\Big|_0^{10}$$
$$= 10{,}641{,}111(e^{0.45} - e^0)$$
$$\cong 10{,}641{,}111(0.5683) \qquad \text{(from Appendix Table C.5)}$$
$$\cong 6{,}047{,}343 \text{ short tons}$$

FIGURE 3

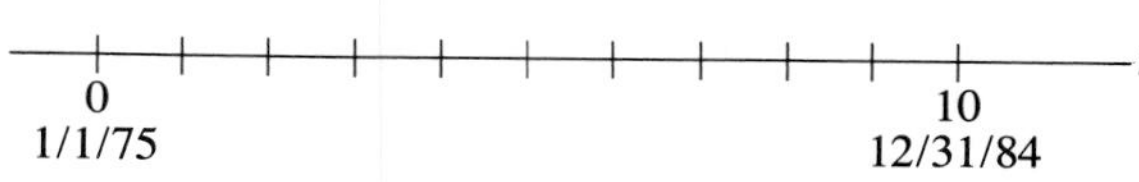

EXERCISE SET 17.5

1. Use Formula (1) to evaluate $\int_3^5 f'(x)\,dx$, where $f(x) = x^2$. Check your result by integrating directly.

2. In each part, use Formula (1) to compute $\int_1^3 f'(x)\,dx$.
(a) $f(x) = 1/(x^2 + 3)$ (b) $f(x) = \ln x$

3. In each part, use Formula (1) to compute $\int_0^2 f'(x)\,dx$.
(a) $f(x) = x^3$ (b) $f(x) = e^{x^2/2}$

4. Use the data in the graph below to find $\int_1^3 f'(x)\,dx$.

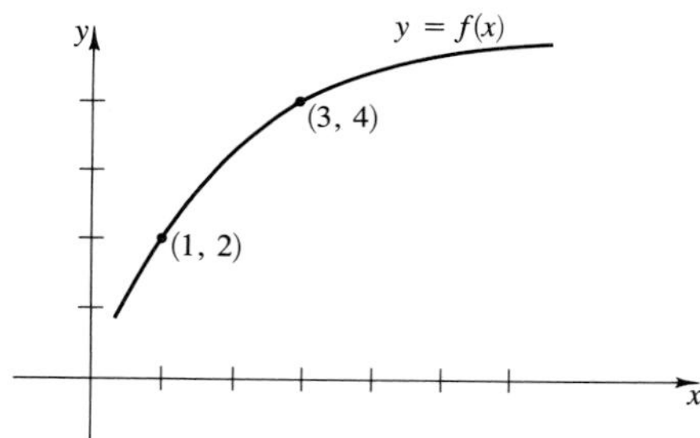

5. **(Marginal analysis)** A manufacturer determines that the firm's marginal cost and marginal revenue functions are
$$\text{MC} = C'(x) = 200 - 0.4x$$
$$\text{MR} = R'(x) = 200 + 0.2x$$
(a) Find the change in revenue that results when the sales level is increased from 10 to 50 units.
(b) Find the revenue resulting from the sale of 70 units.
(c) If the fixed cost is $1000, find the cost of producing 70 units.

6. **(Marginal analysis)** Consider the manufacturing problem of Exercise 5.
(a) Find the total profit from manufacturing and selling 70 units. [*Hint:* Don't forget to take the fixed cost into account.]
(b) Find the profit from manufacturing and selling k units.

7. **(Marginal analysis)** Suppose the marginal cost function (in dollars) for a small manufacturing company producing x units per day is
$$C'(x) = 50 + 0.4x$$
If the company is currently producing 10 units per day, how much more will it cost the company to produce 20 units per day?

8. **(Marginal analysis)** Suppose the marginal revenue function for the sale of x units is given (in dollars) by $R'(x) = 100 + 4x - x^2$.
(a) Find the revenue that results from the sale of 15 units.
(b) Find the total revenue function.
(c) Find the demand equation $p = f(x)$ relating the number of units sold x to the price p. [*Hint:* $R(x) = xp$.]
(d) When the price is $76 per unit, what is the demand?

9. **(Consumption of natural resources)** In 1975, the rate of energy consumption in the United States was approximately 72 quadrillion Btu per year (*Time Magazine,* May 1977). Assuming that the rate of consumption was growing exponentially with a growth rate of 4% per year, estimate the total amount of energy used from January 1, 1976 ($t = 0$), to December 31, 1985 ($t = 10$).

10. **(Consumption of natural resources)** In 1974, U.S. natural gas reserves were being consumed at a rate of 22 million cubic feet per year (U.S. Transportation Department data).

(a) Assuming that the rate of consumption is growing exponentially with a growth rate of 2% per year, estimate the amount of gas that will be used from 1974 ($t = 0$) to 1994 ($t = 20$).

(b) The U.S. Geological Survey estimates U.S. gas reserves at 655 trillion cubic feet. Starting from January 1, 1974, as $t = 0$, how long will U.S. gas reserves last at the rate of consumption in (a)?

11. **(Motion)** A stone dropped from the top of a building falls so that its speed after t seconds is $v(t) = 32t$ feet per second.

(a) How far will the stone fall in 25 seconds?

(b) How long will it take for the stone to fall 400 feet?

12. **(Motion)** An object moves along an x-axis in such a way that at time t (in seconds) its velocity is $v = t + 2t^3$ units per second. If it starts at the point $x = 1$, where will the object be 2 seconds later?

13. **(Motion)** An object moves along an x-axis in such a way that at time t (in seconds) its acceleration (in feet per second per second) is

$$a = 12t^2 + 2t$$

Suppose that at time $t = 0$ the object is at the point $x(0) = 2$ and has velocity $v(0) = 1$. Find its velocity and position 3 seconds later.

14. **(Motion)** A stone is dropped from a height of 576 feet. How fast will it be falling at the moment when it hits the ground?

15. **(Ecology)** A processing plant begins dumping sewage into a stream at the rate of

$$R(t) = 300t^2 + 3t$$

gallons per day, where t is the number of days elapsed after dumping begins.

(a) What is the amount of sewage dumped between the fifth and tenth days inclusive ($t = 5$ to $t = 10$)?

(b) Assuming no sewage is being dumped at time $t = 0$, what is the total amount of sewage dumped during the first 5 days ($t = 0$ to $t = 5$)?

16. **(Salvage value)** The dollar resale value $V(t)$ of an industrial machine t years after its purchase (known as its **salvage value**) declines more rapidly in the first few years than later. For a machine depreciating at the rate

$$V'(t) = -\frac{4500}{t + 5}$$

(a) find the total loss in salvage value over the first 3 years.

(b) find the total loss in salvage value over the next 3 years.

17. **(Advertising)** As the result of a 5-day sales campaign, a company's rate of sales $r(t)$ is expected to grow according to the curve shown below, where $r(t)$ is in dollars per day and t is the number of days elapsed from the start of the campaign. Find the increase in total sales during the campaign period.

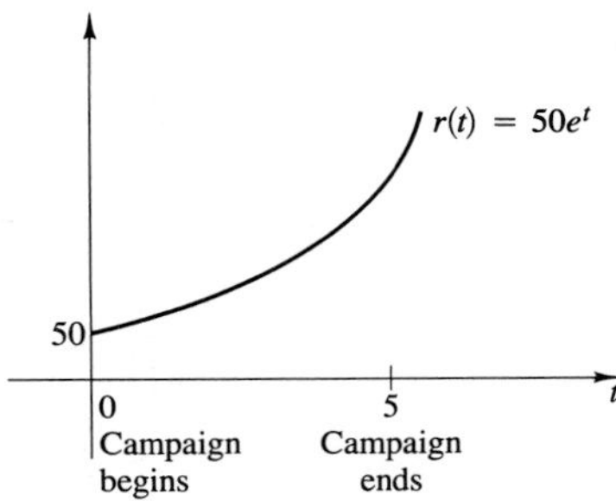

18. **(Biology)** An astronaut's reaction time changes at a rate

$$r(t) = 0.004t^2 - 0.02t$$

where $r(t)$ is in milliseconds per hour and t is the number of hours that the astronaut is weightless. How much of an increase in reaction time will occur between the fifth and tenth hours of weightlessness?

17.6 IMPROPER INTEGRALS

In a number of important applications we need to extend the definition of the definite integral to allow for integration over infinite intervals. We start with the following example.

EXAMPLE 1 The rate of decay of 1000 grams of potassium 42 is

$$r(t) = \frac{dA}{dt} = 55e^{-0.055t}$$

where t is measured in hours. (See Example 5 of Section 14.5.)

(a) How much of the initial 1000 grams will disintegrate by time $t = T$, $T = 10$ hours, $T = 100$ hours, and $T = 1000$ hours?

(b) What is the amount that will eventually disintegrate?

Solution (a) The amount $A(T)$ of the original 1000 grams that will disintegrate by time T is

$$\begin{aligned} A(T) &= \int_0^T r(t)\,dt = \int_0^T 55e^{-0.055t}\,dt \\ &= \frac{55}{-0.055}e^{-0.055t}\Big|_0^T \\ &= -1000e^{-0.055T} + 1000e^0 \\ &= 1000(1 - e^{-0.055T}) \end{aligned}$$

If

$$\begin{aligned} T &= 10 \qquad & A(10) &= 1000(1 - e^{-0.55}) \cong 423.05 \text{ grams} \\ T &= 100 \qquad & A(100) &= 1000(1 - e^{-5.5}) \cong 995.91 \text{ grams} \\ T &= 1000 \qquad & A(1000) &= 1000(1 - e^{-55}) \cong 1000 \text{ grams} \end{aligned}$$

Solution (b) The amount A of the 1000 grams that will eventually disintegrate can be defined as

$$\begin{aligned} \lim_{T\to\infty} 1000(1 - e^{0.055T}) &= 1000(1 - 0) \\ &= 1000 \text{ grams} \end{aligned}$$

This result indicates that

$$\lim_{T\to\infty} \int_0^T r(t)\,dt$$

is meaningful, so the amount that will eventually disintegrate is 1000 grams of potassium 42.

EXAMPLE 2 Find the area of the region bounded by the curve $y = 2/x^2$, the x-axis, and the line $x = 1$ (Figure 1). Because the region extends over an *infinite* interval, we cannot obtain its area by integrating in the usual way. To obtain the area, we first compute the area of the region bounded by the curve $y = 2/x^2$, the x-axis, the

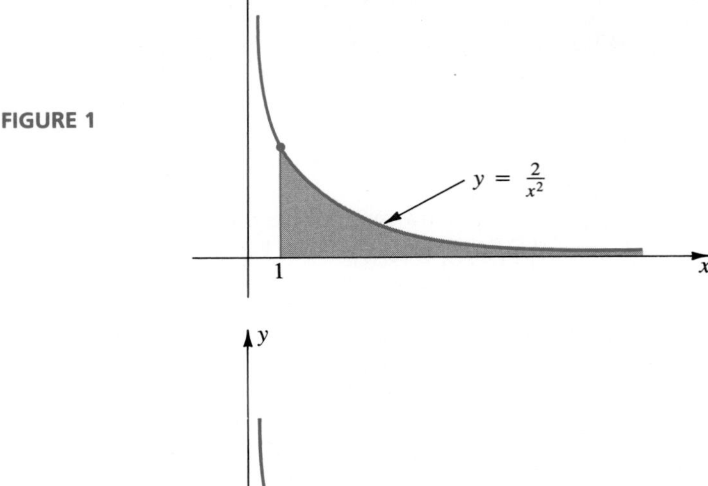

FIGURE 1

FIGURE 2

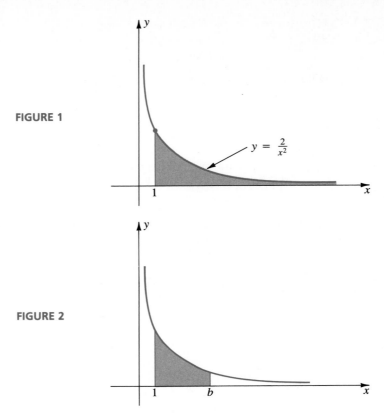

line $x = 1$, and the line $x = b$ (Figure 2). Once we have found this area, we let b approach infinity to obtain the entire area shown in Figure 1.

The area of the region shown in Figure 2 is

$$\int_1^b \frac{2}{x^2}\,dx = 2\int_1^b \frac{1}{x^2}\,dx$$

$$= -\frac{2}{x}\bigg|_1^b = -\frac{2}{b} - \left(-\frac{2}{1}\right) = 2 - \frac{2}{b}$$

If we now let b approach infinity, we can obtain the area A under the curve in Figure 1. This area is

$$A = \lim_{b\to\infty}\int_1^b \frac{2}{x^2}\,dx = \lim_{b\to\infty}\left(2 - \frac{2}{b}\right)$$

$$= 2 - \lim_{b\to\infty}\frac{2}{b} = 2 - 0 = 2$$

It is usual to write

$$\int_1^\infty \frac{2}{x^2}\,dx$$

as an abbreviation for

$$\lim_{b\to\infty}\int_1^b \frac{2}{x^2}\,dx$$

In this notation, the result in Example 1 can be summarized as

$$\left[\text{area under } y = \frac{2}{x^2} \text{ over } [1, \infty)\right] = \int_1^\infty \frac{2}{x^2}\,dx = 2$$

In general, we state the following definition:

Definition 1 If $f(x)$ is continuous for $x \geq a$, then the **improper integral** of $f(x)$ from a to ∞, denoted by

$$\int_a^\infty f(x)\,dx$$

is defined by

$$\int_a^\infty f(x)\,dx = \lim_{b\to\infty}\int_a^b f(x)\,dx \tag{1}$$

provided this limit exists.

If the limit in (1) exists as a finite number, then the improper integral has a finite value and we say that the integral **converges**. Otherwise, we say that the integral diverges.

EXAMPLE 3 Evaluate $\int_1^\infty 1/\sqrt{x}\,dx$.

Solution From Equation (1)

$$\begin{aligned}
\int_1^\infty \frac{1}{\sqrt{x}}\,dx &= \lim_{b\to\infty}\int_1^b \frac{1}{\sqrt{x}}\,dx \\
&= \lim_{b\to\infty}\int_1^b x^{-1/2}\,dx \\
&= \lim_{b\to\infty} \left.\frac{x^{1/2}}{1/2}\right|_1^b \\
&= \lim_{b\to\infty} \left. 2\sqrt{x}\right|_1^b \\
&= \lim_{b\to\infty}(2\sqrt{b} - 2)
\end{aligned}$$

Since $\sqrt{b}$ increases as b does, the last limit is infinite. Thus, the integral diverges to $+\infty$:

$$\int_1^\infty \frac{1}{\sqrt{x}}\,dx = +\infty$$

We occasionally need to integrate a function over the infinite interval $(-\infty, b]$ instead of $[a, \infty)$. We use the following definition:

Definition 2 If $f(x)$ is continuous for $x \geq b$, then the **improper integral** of $f(x)$ from $-\infty$ to b is defined by

$$\int_{-\infty}^{b} f(x)\,dx = \lim_{a\to -\infty} \int_{a}^{b} f(x)\,dx \tag{2}$$

If the limit on the right side exists as a finite number, then we say that the improper integral **converges**. Otherwise, we say that it **diverges**.

EXAMPLE 4 Evaluate $\int_{-\infty}^{0} e^x\,dx$.

Solution From Definition 2,

$$\begin{aligned}\int_{-\infty}^{0} e^x\,dx &= \lim_{a\to -\infty} \int_{a}^{0} e^x\,dx \\ &= \lim_{a\to -\infty} e^x \Big|_{a}^{0} \\ &= \lim_{a\to -\infty} (e^0 - e^a) \\ &= 1 - \lim_{a\to -\infty} e^a \\ &= 1 - 0 \qquad \text{[from Equation (4) in Section 14.1]} \\ &= 1\end{aligned}$$

Therefore, $\int_{-\infty}^{0} e^x\,dx = 1$, and we say that the improper integral converges to 1.

EXAMPLE 5 Evaluate $\int_{-\infty}^{1} x\,dx$.

Solution By Equation (2),

$$\begin{aligned}\int_{-\infty}^{1} x\,dx &= \lim_{a\to -\infty} \int_{a}^{1} x\,dx \\ &= \lim_{a\to -\infty} \frac{1}{2}x^2 \Big|_{a}^{1} \\ &= \lim_{a\to -\infty} \left(\frac{1}{2} - \frac{1}{2}a^2\right) \\ &= \frac{1}{2} - \frac{1}{2}\lim_{a\to -\infty} a^2 \\ &= -\infty\end{aligned}$$

Therefore, the improper integral diverges.

Finally, we also need to be able to integrate a function over the entire real line $(-\infty, \infty)$. For that we use the following definition.

Definition 3 If $f(x)$ is continuous for all real numbers, then the **improper integral** of $f(x)$ from $-\infty$ to ∞ is

$$\int_{-\infty}^{\infty} f(x)\,dx = \int_{-\infty}^{0} f(x)\,dx + \int_{0}^{\infty} f(x)\,dx \tag{3}$$

where the integrals on the right side are as defined in Definitions 1 and 2.

If *both* limits on the right side of Equation (3) exist, then we say that the improper integral **converges** to their sum. If one or both of the limits diverges, then we say that the improper integral **diverges** and has no value.

EXAMPLE 6 Evaluate $\int_{-\infty}^{\infty} e^x\,dx$.

Solution By Definition 3,

$$\int_{-\infty}^{\infty} e^x\,dx = \int_{-\infty}^{0} e^x\,dx + \int_{0}^{\infty} e^x\,dx$$

Evaluating the integrals on the right side, one at a time, we see that

$$\int_{-\infty}^{0} e^x\,dx = 1$$

as shown in Example 4. Now, by Equation (1),

$$\begin{aligned}\int_{0}^{\infty} e^x\,dx &= \lim_{b\to\infty} \int_{0}^{b} e^x\,dx \\ &= \lim_{b\to\infty} e^x \Big|_0^b \\ &= \lim_{b\to\infty} (e^b - e^0) \\ &= \left(\lim_{b\to\infty} e^b\right) - 1 \\ &= \infty\end{aligned}$$

Therefore, $\int_0^{\infty} e^x\,dx$ diverges, and we conclude that $\int_{-\infty}^{\infty} e^x\,dx$ diverges.

EXAMPLE 7 **ECOLOGY** Suppose that t hours after the rupture of a pipeline carrying a dangerous chemical the chemical is leaking at a rate of

$$r(t) = \frac{1000}{(1 + 2t)^2} \qquad \text{gallons per hour}$$

If this were allowed to continue indefinitely, how much of the chemical would eventually leak out?

Solution Let $A(t)$ denote the amount of the chemical leaked out at time t. Since $r(t) = dA/dt$, the total accumulated amount leaked out at time t_0 is

$$A(t_0) = A(t_0) - A(0) = \int_0^{t_0} r(t)\,dt \qquad \text{(since } A(0) = 0\text{)}$$

For the solution to our problem, we want $A(\infty)$, or

$$\begin{aligned}\int_0^{\infty} r(t)\,dt &= \lim_{b\to\infty} \int_0^b \frac{1000}{(1+2t)^2}\,dt \\ &= 1000 \lim_{b\to\infty} \left.\frac{-1/2}{1+2t}\right|_0^b \\ &= 1000 \lim_{b\to\infty}\left[\frac{-1}{2(1+2b)} - \left(-\frac{1}{2}\right)\right] \\ &= 1000\left(0 + \frac{1}{2}\right) \\ &= 500 \text{ gallons}\end{aligned}$$

EXERCISE SET 17.6

In Exercises 1–16 evaluate each improper integral if it converges.

1. $\int_3^{\infty} \frac{1}{x^3}\,dx$
2. $\int_1^{\infty} \frac{1}{x^{3/2}}\,dx$
3. $\int_0^{\infty} e^{-3x}\,dx$
4. $\int_3^{\infty} \frac{2x}{x^2-1}\,dx$
5. $\int_0^{\infty} xe^{-x^2}\,dx$
6. $\int_2^{\infty} \frac{1}{\sqrt[3]{x}}\,dx$
7. $\int_0^{\infty} \frac{x}{\sqrt{x^2+3}}\,dx$
8. $\int_3^{\infty} \frac{1}{(x+2)^2}\,dx$
9. $\int_{-\infty}^{0} \frac{1}{(x-4)^2}\,dx$
10. $\int_{-\infty}^{-1} \frac{1}{x^4}\,dx$
11. $\int_{-\infty}^{-1} x^4\,dx$
12. $\int_{-\infty}^{3} \frac{1}{\sqrt{4-x}}\,dx$
13. $\int_{-\infty}^{-3} \frac{1}{(x+2)^{2/3}}\,dx$
14. $\int_{-\infty}^{-1} \frac{1}{x^{2/3}}\,dx$
15. $\int_4^{\infty} \frac{x}{x^2-3}\,dx$
16. $\int_0^{\infty} \frac{x}{(x^2+3)^{3/2}}\,dx$
17. $\int_{-\infty}^{\infty} e^{2x}\,dx$
18. $\int_{-\infty}^{\infty} xe^{-x^2}\,dx$
19. $\int_{-\infty}^{\infty} e^{-|x|}\,dx$
20. $\int_{-\infty}^{\infty} \frac{x}{x^2+1}\,dx$
21. A company purchases a new machine that should last indefinitely and that after t years should produce a net income of
$$r(t) = 20{,}000e^{-t/12}$$
dollars per year. Find the total anticipated income from this machine.
22. Suppose that t minutes after an explosion a toxic substance is dispersed into the atmosphere at the rate of
$$r(t) = \frac{1000}{(1+4t)^2}$$
ounces per hour. How many ounces of the substance are eventually dispersed into the atmosphere?

17.7 APPLICATIONS OF INTEGRATION IN PROBABILITY THEORY

In this section, we touch briefly on applications of integration to problems in probability. A more detailed study of this topic is usually given in a separate probability course.*

Recall from Section 8.1 that a random variable X is called **finite discrete** if it can take on only finitely many values, and it is called **continuous** if its possible values form an entire interval of numbers.

In Chapter 8 we see that probabilities for finite discrete random variables can be obtained by adding probabilities of appropriate events. However, for continuous random variables, probabilities cannot usually be computed in this way because there are infinitely many sample points. Instead, probabilities are obtained by integration as follows: If X is a continuous random variable, then by using experimental data or theoretical considerations, we try to obtain a function $f(x)$, called a **probability density function** for X, with the property that the area under $y = f(x)$ over the interval $[a, b]$ is equal to the probability $P(a \le X \le b)$ (Figure 1). A probability density function is usually referred to as the **pdf of a distribution**.

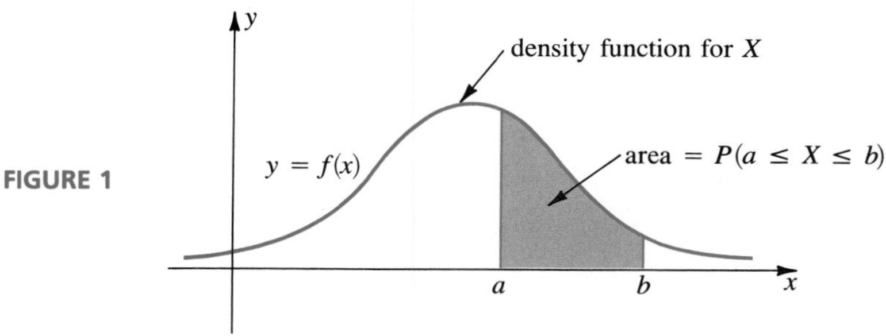

FIGURE 1

We have encountered some important pdf's already, namely, those of the Gaussian or normal curves of Section 8.6 and the chi-square curves of Section 8.8.

In order for a function f to be a pdf for a random variable X, it must satisfy two conditions:

1. $f(x) \ge 0$ for all x.
2. The total area under $y = f(x)$ is equal to 1; that is, $\int_{-\infty}^{\infty} f(x)\, dx = 1$.

In addition if $f(x)$ is the pdf of a continuous random variable X, then for any interval $[a, b]$

$$P(a \le X \le b) = \int_b^a f(x)\, dx$$

It can be shown that for any real number c and continuous random variable X, $P(X = c) = 0$. Therefore,

* For a discussion on probability, written at this same level, see Bernard Kolman and Charles G. Denlinger, *Applied Calculus,* Chap. 11 (Orlando, Fla.: Harcourt Brace Jovanovich, 1989).

$$P(a \le X \le b) = P(a < X < b) = P(a \le X < b) = P(a < X \le b)$$
$$= \int_a^b f(x)\,dx \qquad (1)$$

Since $f(x)$ is continuous, $\int_a^b f(x)\,dx = F(b) - F(a)$, where F is an antiderivative of f. If $F(x)$ is chosen as $\int_{-\infty}^{x} f(x)\,dx$, then $F(x)$ is called the **distribution function** for the random variable X and

$$F(x) = P(X \le x)$$

Moreover,

$$\begin{aligned} F(b) - F(a) &= P(a < X \le b) \\ &= P(a \le X \le b) \qquad \text{[by (1)]} \end{aligned}$$

As we have just noted, $P(a \le X \le b)$ can be computed by using either the distribution function or the pdf of X. In the examples considered in this text the pdf will be given.

Remark $P(a \le X \le b)$ is equal to the area under $y = f(x)$ over $[a, b]$. Conditions 1 and 2 ensure that the probabilities computed in Equation (1) are nonnegative and less than or equal to 1.

EXAMPLE 1 Suppose a random variable X has a pdf defined by

$$f(x) = \begin{cases} 1 - \frac{1}{2}x & 0 \le x \le 2 \\ 0 & \text{elsewhere} \end{cases}$$

(See Figure 2.)

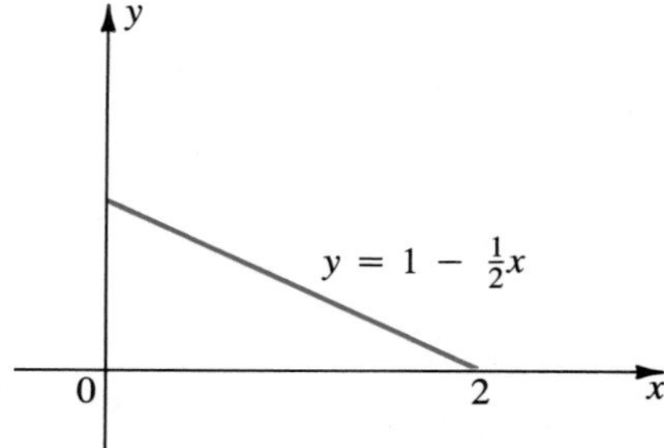

FIGURE 2

(a) Show that $f(x)$ satisfies conditions 1 and 2 of a pdf.

(b) Find $P(1 \le X \le 2)$.

(c) Find $P(X \le 1)$. (d) Find $P(\frac{1}{2} \le X)$.

Solution (a) We are given that

$$f(x) = \begin{cases} 1 - \frac{1}{2}x & 0 \le x \le 2 \\ 0 & \text{elsewhere} \end{cases}$$

1. Since $1 - \frac{1}{2}x \ge 0$ when $0 \le x \le 2$ and is zero elsewhere, $f(x)$ satisfies condition 1.

2. $$\int_{-\infty}^{\infty} f(x)\,dx = \int_{-\infty}^{0} 0\,dx + \int_{0}^{2}\left(1 - \frac{1}{2}x\right)dx + \int_{2}^{\infty} 0dx \quad \text{(additive property of the integral)}$$

$$= 0 + \int_{0}^{2}\left(1 - \frac{1}{2}x\right)dx + 0 = \left(x - \frac{x^2}{4}\right)\Big|_{0}^{2}$$

$$= 1$$

so $f(x)$ satisfies condition 2.

Solution (b)

$$P(1 \le X \le 2) = \int_{1}^{2}\left(1 - \frac{1}{2}x\right)dx = \left(x - \frac{x^2}{4}\right)\Big|_{1}^{2} = 1 - \frac{3}{4} = \frac{1}{4}$$

Solution (c) The pdf of X assumes nonzero values only in the interval $[0, 2]$.

$$P(X \le 1) = \int_{\infty}^{1} f(x)\,dx = \int_{-\infty}^{0} 0\,dx + \int_{0}^{1}\left(1 - \frac{1}{2}x\right)dx = \left(x - \frac{x^2}{4}\right)\Big|_{0}^{1}$$

$$= \left(\frac{3}{4} - 0\right) = \frac{3}{4}$$

Solution (d) Because $f(x) = 0$ for $x > 2$, it follows that

$$P\left(\frac{1}{2} \le X\right) = \int_{1/2}^{2}\left(1 - \frac{1}{2}x\right)dx + \int_{2}^{\infty} 0dx$$

$$= \left(x - \frac{x^2}{4}\right)\Big|_{1/2}^{2} = 1 - \frac{7}{16} = \frac{9}{16}$$

In the remainder of this section we study pdf's for two types of random variables: uniformly distributed random variables and exponentially distributed random variables. If $f(x)$ is said to be defined on an interval $[a, b]$, we assume that $f(x) = 0$ outside of this interval.

Uniformly Distributed Random Variables

Let x represent a number chosen "at random" from the interval $[0, 10]$ and let us try to find $P(2 \le X \le 4)$, that is, the probability that this number falls in the subinterval $[2, 4]$ (Figure 3).

FIGURE 3

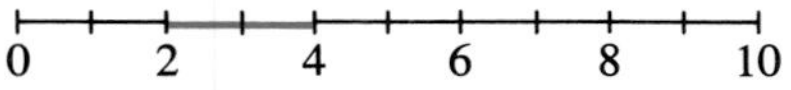

Because the interval $[2, 4]$ occupies 2 of the 10 units in the interval $[0, 10]$, it is intuitively clear that

$$P(2 \le X \le 4) = \frac{\text{length of } [2, 4]}{\text{length of } [0, 10]} = \frac{2}{10} = \frac{1}{5}$$

More generally, if we ask for the probability that X lies in a subinterval $[a, b]$ of $[0, 10]$, then we obtain in a similar way

$$P(a \le X \le b) = \frac{\text{length of subinterval}}{\text{length of entire interval}} = \frac{b - a}{10}$$

(See Figure 4.)

FIGURE 4

$$0 \quad a \quad b \quad 10 \qquad (b - a)$$

A pdf for this random variable is

$$f(x) = \begin{cases} \frac{1}{10} & 0 \le x \le 10 \\ 0 & \text{elsewhere} \end{cases}$$

(Figure 5a). This is indeed a pdf for X because $f(x) \ge 0$ for all x and

$$\int_{-\infty}^{\infty} f(x)\,dx = \int_{-\infty}^{0} 0\,dx + \int_{0}^{10} \frac{1}{10}\,dx + \int_{10}^{\infty} 0\,dx = 1$$

Moreover if $[a, b]$ is a subinterval of $[0, 10]$, then

$$\int_a^b f(x)\,dx = \int_a^b \frac{1}{10}\,dx = \frac{1}{10}x\Big|_a^b = \frac{1}{10}b - \frac{1}{10}a = \frac{b-a}{10} = P(a \le X \le b)$$

(See Figure 5b.)

FIGURE 5

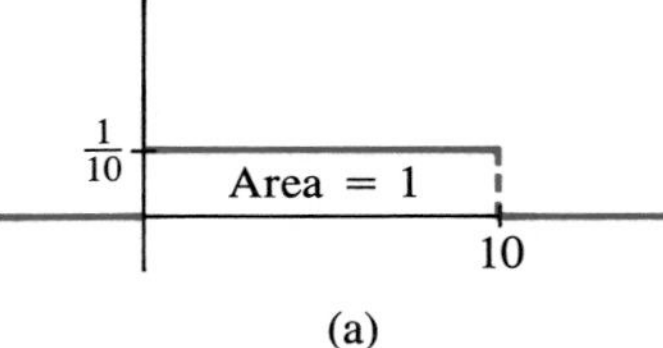

(a)

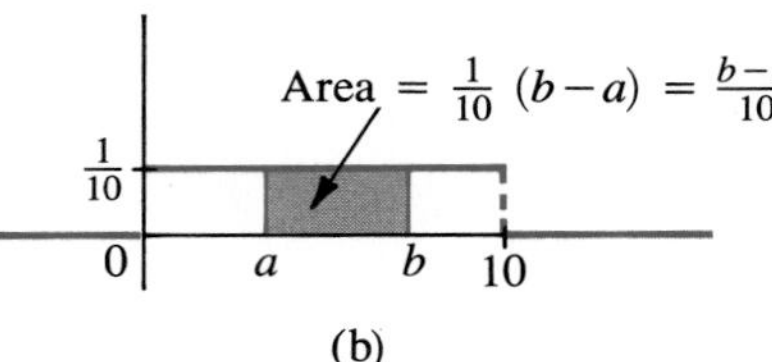

(b)

To generalize this example, if X is a number chosen at random from an interval $[c, d]$, then X is said to be **uniformly distributed** over $[c, d]$. The pdf for X is

$$f(x) = \begin{cases} \dfrac{1}{d-c} & c \le x \le d \\ 0 & \text{elsewhere} \end{cases} \tag{2}$$

EXAMPLE 2 Due to variations in air traffic and weather, the 10:00 A.M. flight from Philadelphia to Chicago takes off (at random) between 10:00 A.M. and 10:15 A.M. What is the probability that a passenger will experience a delay in takeoff of 10 minutes or more?

Solution If X is the delay (in minutes), then X is a random number between 0 and 15, inclusive. Thus, X is uniformly distributed over the interval $[0, 15]$, and its pdf is

$$f(x) = \begin{cases} \dfrac{1}{15} & 0 \le x \le 15 \\ 0 & \text{elsewhere} \end{cases}$$

Thus, the probability of a delay of 10 minutes or more is

$$P(10 \le X \le 15) = \int_{10}^{15} \frac{1}{15}\,dx = \frac{1}{15}x\Big|_{10}^{15} = 1 - \frac{10}{15} = \frac{1}{3}$$

The Mean of a Continuous Random Variable

In Section 8.2 we define the mean $E(X)$ of a discrete random variable. This definition is now extended to continuous random variables as follows:

Definition If $f(x)$ is the pdf of a continuous random variable X, the **mean** or **expected value of X, $E(X)$,** is

$$\text{mean of } X = E(X) = \int_{-\infty}^{\infty} xf(x)\,dx \tag{3}$$

provided the improper integral converges.

EXAMPLE 3 Consider the uniform distribution of Example 2. What is the mean of this distribution?

Solution Since

$$f(x) = \begin{cases} \dfrac{1}{15} & 0 \le x \le 15 \\ 0 & \text{elsewhere} \end{cases}$$

by Equation (3),

$$\begin{aligned} \text{mean} &= \int_{-\infty}^{\infty} xf(x)\,dx = \int_{-\infty}^{0} 0\,dx + \int_{0}^{15} x\left(\frac{1}{15}\right)dx + \int_{15}^{\infty} 0\,dx \\ &= \int_{0}^{15} \frac{x}{15}\,dx = \left.\frac{x^2}{30}\right|_0^{15} = \frac{225}{30} = \frac{15}{2} = 7.5 \end{aligned}$$

This result implies that the expected delay in takeoff is 7.5 minutes.

Exponentially Distributed Random Variables

Another important pdf in applications is

$$f(x) = \begin{cases} 0 & x < 0 \\ ke^{-kx} & x \ge 0 \end{cases} \tag{4}$$

where $x \ge 0$ and k is a positive constant (Figure 6).

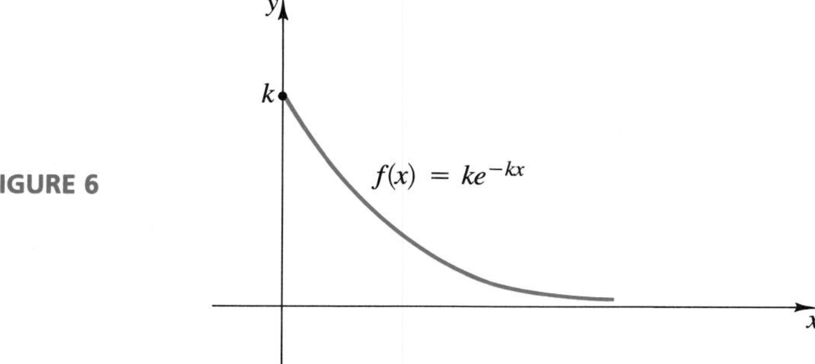

FIGURE 6

A random variable X with such a pdf is said to be **exponentially distributed**. Examples of exponentially distributed random variables are the length of life of a manufactured item such as a lightbulb, resistor, or television; the distance between two successive cars on a highway; and the duration of a phone call.

EXAMPLE 4 (a) Show that the pdf of the exponential distribution satisfies conditions 1 and 2 of a pdf.

(b) Find $E(X)$.

Solution (a) When $x \geq 0$, $f(x) > 0$ since $ke^{-kx} > 0$ for $k > 0$. Also $f(x) = 0$ for $x < 0$. Therefore, condition 1 is satisfied.

$$\begin{aligned}\int_{-\infty}^{\infty} f(x)\,dx &= \int_{-\infty}^{0} 0dx + \int_{0}^{\infty} ke^{-kx}\,dx \\ &= \int_{0}^{\infty} ke^{-kx}\,dx = \lim_{b\to\infty}\int_{0}^{b} ke^{-kx}\,dx \\ &= \lim_{b\to\infty}(-e^{-kx})\Big|_{0}^{b} \\ &= \lim_{b\to\infty}(-e^{-kb} + e^{0}) \\ &= \lim_{b\to\infty}(1 - e^{-kb}) \\ &= 1\end{aligned}$$

Therefore condition 2 is satisfied.

Solution (b)

$$\begin{aligned}E(X) &= \int_{-\infty}^{\infty} xf(x)\,dx = \int_{-\infty}^{0} 0\,dx + \int_{0}^{\infty} xke^{-kx}\,dx \\ &= \lim_{b\to\infty} k\int_{0}^{b} xe^{-kx}\,dx \\ &= \lim_{b\to\infty} k\left(-\frac{xe^{-kx}}{k} - \frac{e^{-kx}}{k^2}\right)\Bigg|_{0}^{b} \qquad \text{[Integral Formula (32) in Section 16.3]} \\ &= \lim_{b\to\infty}\left(-be^{-kb} - \frac{e^{-kb}}{k} + 0 + \frac{1}{k}\right)\end{aligned}$$

It can be shown that $\lim_{b\to\infty} be^{-kb} = 0$. Using this fact,

$$E(X) = -0 - 0 - 0 + \frac{1}{k} = \frac{1}{k}$$

Equivalently,

$$k = \frac{1}{E(X)} \tag{5}$$

EXAMPLE 5 A random variable X is exponentially distributed and has a mean value of 1.

(a) Find its pdf.

(b) Verify that the total area under the pdf is 1.

Solution (a) From Equation (5),

$$k = \frac{1}{\text{expected value of } X} = \frac{1}{1} = 1$$

Thus, from (4), the pdf is

$$f(x) = \begin{cases} e^{-x} & x \geq 0 \\ 0 & \text{elsewhere} \end{cases}$$

Solution (b) The total area under the curve $y = f(x)$ for $x \geq 0$ is given by the improper integral:

$$A = \int_{-\infty}^{\infty} f(x)\,dx = \int_{-\infty}^{0} 0dx + \int_{0}^{\infty} e^{-x}\,dx = \lim_{b\to\infty} \int_{0}^{b} e^{-x}\,dx$$

$$= \lim_{b\to\infty} -e^{-x}\Big|_{0}^{b} = \lim_{b\to\infty} (1 - e^{-b})$$

$$= 1$$

EXAMPLE 6 Company records show the expected length of the firm's business calls to be 5 minutes. Assuming that the length of a call has an exponential distribution, find:

(a) the probability that a call will be 3 minutes or less

(b) the probability that a call will be longer than 3 minutes

Solution (a) If X is the length of a call (in minutes), then from Equation (5)

$$k = \frac{1}{\text{expected value of } X} = \frac{1}{5} = 0.2$$

so the pdf for X is $f(x) = 0.2e^{-0.2t}$, where $x \geq 0$. The probability that a call will be 3 minutes or less is

$$P(-\infty \leq X \leq 3) = \int_{-\infty}^{0} 0dx + \int_{0}^{3} 0.2e^{-0.2x}\,dx$$

$$= -e^{-0.2x}\Big|_{0}^{3}$$

$$= -e^{-0.6} + e^{0}$$

$$= 1 - e^{-0.6}$$

$$= 1 - 0.5488 \qquad \text{(from Appendix Table C.5)}$$

$$= 0.4512$$

In other words, approximately 45.12% of all calls will be 3 minutes or less.

Solution (b) The probability that a call will be longer than 3 minutes is

$$P(3 < X < \infty) = \int_{3}^{\infty} 0.2e^{-0.2x}\,dx$$

However, rather than evaluate this improper integral, we can take advantage of our work in (a) by noting that

$$P(3 < X < \infty) = 1 - P(0 \le X \le 3)$$

(Why?) Therefore,

$$P(3 < X < \infty) = 1 - 0.4512 = 0.5488$$

EXERCISE SET 17.7

1. Let $f(x) = \frac{1}{4}x$, where $1 \le x \le 3$ and zero elsewhere.
(a) Show that f satisfies conditions 1 and 2 of a pdf.
(b) Let X be a continuous random variable whose pdf is f. Find $P(1 \le X \le 2)$.
(c) For the random variable in (b), find $P(X \ge 2)$.
(d) For the random variable in (b), find $P(X \le \frac{3}{2})$.

2. Let $f(x) = \frac{3}{2}x^2$, where $-1 \le x \le 1$ and zero elsewhere.
(a) Show that f satisfies conditions 1 and 2 of a pdf.
(b) Let X be a continuous random variable whose pdf is f. Find $P(-\frac{1}{2} \le X \le \frac{1}{2})$.
(c) For the random variable in (b), find $P(X \le 0)$.
(d) For the random variable in (b), find $P(X \ge 0)$.

3. A number X is chosen at random from the interval $[5, 25]$.
(a) Find a pdf for X.
(b) What is the probability that X satisfies $6 \le X \le 12$?
(c) What is the probability that $X \ge 5$?
(d) What is $E(X)$?

4. Let X be uniformly distributed over the interval $[1, 6]$. Find
(a) $P(2 \le X \le 3)$
(b) $P(2 \le X)$
(c) $P(X \le 4)$
(d) $P(X \ge 1)$
(e) Find $E(X)$

5. Let X be the random variable in Exercise 4. Find the value of k such that $P(2 \le X \le k) = .5$.

6. The voltage in a 120-volt line varies randomly between 115 and 125 volts.
(a) What is the probability that the voltage will be between 118 and 123 volts?
(b) If a photographic enlarger works improperly when the voltage is 118 volts or lower, what percentage of the time will the enlarger work properly?
(c) What is the expected voltage?

7. The 9:00 A.M. train from Philadephia to New York leaves at random between 9:00 and 9:10 A.M.
(a) What is the probability that a passenger boarding at 9:00 A.M. will wait longer than 5 minutes for the train to leave?
(b) What is the probability that the wait will be less than 10 minutes?
(c) If a person reaches the station at 9:07 A.M., what are the chances that the passenger will still catch the train?
(d) What is the expected waiting time?

8. Suppose a commuter train runs every half hour and you arrive at the station at random.
(a) What is the probability you will wait at least 10 minutes?
(b) What is the probability you will wait 15 minutes or less?
(c) What is the probability you will wait at least 2 minutes, but no longer than 5 minutes?
(d) What is the expected waiting time?

9. Let X be an exponentially distributed random variable with an expected value of 2.
(a) Find the pdf for X.
(b) Find $P(X \le 1)$.
(c) Find $P(1 \le X \le 4)$.
(d) Use the result in (b) to find $P(X > 1)$.
(e) Find $P(X > 1)$ by evaluating an appropriate improper integral.

10. Let X be an exponentially distributed random variable with an expected value of 5.
(a) Find the pdf for X.
(b) Find $P(X \le 5)$.
(c) Find $P(0 \le X \le 3)$.
(d) Use the result in (b) to find $P(X > 5)$.
(e) Find $P(X > 5)$ by evaluating an appropriate improper integral.

11. Assume the expected wait in a dentist's office is 20 minutes and the waiting time is exponentially distributed.
(a) What is the probability a patient will wait 10 minutes or less before seeing the dentist?
(b) What percentage of the patients wait longer than 10 minutes?

12. Suppose the expected distance between successive cars on a bridge is 150 feet and the distance is an exponentially distributed random variable.
(a) What is the probability that the distance between two successive cars will be no more than 75 feet?
(b) What is the probability that the distance will be between 30 and 75 feet?

13. A certain brand of light bulb is claimed to have an expected life of 1000 hours. Assuming the life of such a bulb to be exponentially distributed, find:
(a) the probability that a bulb will last 500 hours or less
(b) the probability that a bulb will last more than 500 hours
(c) the percentage of bulbs that will last between 1000 and 2000 hours

KEY IDEAS FOR REVIEW

Area between curves

$$\begin{bmatrix} \text{area between } y = f(x) \text{ and } y = g(x) \\ \text{over } [a, b] \text{ when } f(x) \geq g(x) \end{bmatrix} = \int_a^b [f(x) - g(x)]\, dx$$

Average value of an integrable function $f(x)$ over $[a, b]$

$$\bar{f} = \frac{1}{b - a} \int_a^b f(x)\, dx$$

Demand or sales level The number of units of the commodity demanded by the market at a certain unit price p.

Demand equation An equation relating the price p and the demand x.

Supply or production level The number y of units of the commodity supplied by the industry when the unit price is p.

Supply equation An equation relating the price p and the supply y.

Industry's ideal total revenue $IR_C = \int_0^{y_M} g(y)\, dy$

Consumers' surplus $CS(p_0) = \int_0^{x_0} [f(x) - p_0]\, dx$

Producers' surplus $PS(p_0) = \int_0^{y_0} [p_0 - g(y)]\, dy$

Equilibrium price Price for which demand equals supply.

Formula for integrating a rate of change $\int_a^b f'(x)\, dx = f(b) - f(a)$

Improper integral $\int_a^{\infty} f(x)\, dx = \lim_{b \to \infty} \int_a^b f(x)\, dx$

$$\int_{-\infty}^{b} f(x)\, dx = \lim_{a \to -\infty} \int_a^b f(x)\, dx$$

$$\int_{-\infty}^{\infty} f(x)\, dx = \int_{-\infty}^{0} f(x)\, dx + \int_0^{\infty} f(x)\, dx$$

Probability density function for a random variable X A function $f(x)$ satisfying:

1. $f(x) \geq 0$ for all x
2. Area under $y = f(x)$ is equal to 1.

In addition, $P(a \leq X \leq b) = \int_a^b f(x)\,dx$.

Expected value of a random variable X with pdf $f(x)$

$$E(X) = \int_{-\infty}^{\infty} xf(x)\,dx$$

Uniformly distributed random variable X is uniformly distributed on $[a, b]$ if its pdf and mean are, respectively,

$$f(x) = \frac{1}{b-a} \quad \text{where } a \leq x \leq b \text{ and is zero elsewhere}$$

$$E(X) = \frac{a+b}{2}$$

Exponentially distributed random variable X is exponentially distributed if its pdf and mean are, respectively,

$$f(x) = ke^{-kx} \quad \text{where } x \geq 0 \text{ and is zero elesewhere}$$

$$E(X) = \frac{1}{k}$$

SUPPLEMENTARY EXERCISES

1. Find the area of the region bounded by $y = x^2$ and $y = x^3$.
2. Find the area of the region bounded by $y = x^2$, $y = x^3$, $x = 0$, and $x = 2$.
3. If the temperature of a heated rod 10 inches long is given by
$$T(x) = e^{-x}$$
where x is a position on the rod, what is the average temperature of the rod?
4. What is the average value of $f(x) = x/(x^2 + 1)^2$ on the interval $[1, 3]$?
5. If the rate of change of temperature T with respect to time is
$$T'(t) = e^t - t$$
find the change in temperature from $t = 0$ to $t = 3$.
6. Evaluate if possible. $\displaystyle\int_1^{\infty} \frac{1}{(1+x)^3}\,dx$
7. Evaluate if possible. $\displaystyle\int_{-\infty}^{-1} \frac{1}{(1-x)^2}\,dx$
8. Evaluate if possible. $\displaystyle\int_{-\infty}^{\infty} e^{-t}\,dt$
9. Evaluate if possible. $\displaystyle\int_2^{\infty} \frac{1}{x^3}\,dx$
10. Evaluate if possible. $\displaystyle\int_5^{\infty} \frac{1}{\sqrt{x-1}}\,dx$

11. The pdf of the standard normal distribution is

$$f(x) = \frac{1}{\sqrt{2\pi}} e^{-x^2/2} \qquad -\infty < x < \infty$$

Since

$$P(a < x < b) = \int_a^b \frac{1}{\sqrt{2\pi}} e^{-x^2/2}\, dx$$

cannot be evaluated exactly, numerical integration methods such as the trapezoidal approximation must be used. However it is not difficult to determine the mean $E(X)$ for this distribution. Find $E(X)$.

12. Let

$$f(x) = \begin{cases} 0 & x < 0 \\ \frac{1}{2} e^{-x/2} & x \ge 0 \end{cases}$$

be the pdf of an exponential distribution.
(a) Find $P(0 \le X \le 1)$.
(b) Find $P(0 \le X)$.
(c) Find $E(X)$.

13. Let

$$f(x) = \begin{cases} \frac{1}{25} & 0 \le x \le 25 \\ 0 & \text{elsewhere} \end{cases}$$

be the pdf of a uniform distribution.
(a) Find $P(0 \le X \le 10)$.
(b) Find $P(5 \le X \le 20)$.
(c) Find $E(X)$.

14. Let

$$f(x) = \begin{cases} \frac{2}{21} x & 2 \le x \le 5 \\ 0 & \text{elsewhere} \end{cases}$$

(a) Show that $f(x)$ satisfies the properties of a pdf.
If $f(x)$ is the pdf of a random variable x, then
(b) find $P(2 \le X \le 5)$.
(c) find $P(2 \le X \le 3)$.
(d) find $E(X)$.

CHAPTER TEST

1. Find the area of the region bounded by $y = 2x$ and $y = x^2$.

2. What is the average value of $f(x) = x \ln x$ on the interval $[1, 4]$?

3. If the price of stock X has varied throughout the day according to the equation

$$P(t) = \frac{100t}{1 + t^2} \qquad 0 \le t \le 5 \qquad (t \text{ in hours})$$

what was the average price of the stock during this time period?

4. If the rate of change of the cost of stock Y was

$$r(t) = C'(t) = 50(1 - t^2) \qquad 0 \le t \le 1$$

what was the change in the cost of this stock within the first half hour of trading $(0 \le t \le \frac{1}{2})$?

5. Evaluate.

$$\int_1^\infty \frac{1}{x^{1/3}}\, dx$$

if possible.

6. A random variable X has a uniform distribution with pdf

$$f(x) = \begin{cases} \frac{1}{15} & 0 \le x \le 15 \\ 0 & \text{elsewhere} \end{cases}$$

(a) What is $P(1 \le X \le 2)$?
(b) What is $E(X)$?

18

Differential Equations

18.1 INTRODUCTION

An equation that involves an unknown function and one or more of its derivatives is called a **differential equation.** Some examples are

$$\frac{dy}{dx} = 2x \qquad \left(\frac{dy}{dx}\right)^2 - y = e^x \qquad \frac{dQ}{dt} = 2Q$$

In the first two equations, $y = f(x)$ is an unknown function of x, and in the third, $Q = f(t)$ is an unknown function of t.

Differential equations arise in a variety of important applications, especially those involving rates of change. In this chapter we discuss some applied problems and show how to solve the resulting equations.

A function that satisfies a differential equation is called a **solution** of that equation.

EXAMPLE 1 Show that the function $y = e^x$ is a solution of the differential equation

$$\frac{dy}{dx} - y = 0 \tag{1}$$

Solution We must show that when $y = e^x$ is subtracted from its derivative, the result is equal to 0. Since

$$\frac{dy}{dx} = e^x$$

it follows that

$$\frac{dy}{dx} - y = e^x - e^x = 0$$

so $y = e^x$ is a solution of (1). ▰

In general, a differential equation will have infinitely many solutions. For example,

$$y = Ce^x \tag{2}$$

is a solution of Equation (1) for any value of C since

$$\frac{dy}{dx} = \frac{d}{dx}(Ce^x) = Ce^x$$

so that

$$\frac{dy}{dx} - y = Ce^x - Ce^x = 0$$

It is shown in Section 18.3 that *every* solution of (1) is obtainable by substituting a numerical value for C in (2). For this reason Equation (2) is called the **general solution** of Equation (1). Solutions obtained by substituting numerical values for C are called **particular solutions**. Thus,

$$y = 2e^x \qquad y = -3e^x \qquad y = \sqrt{2}e^x$$

are particular solutions of (1).

In many applications we are not interested in finding all solutions of a differential equation, but rather a particular solution satisfying certain restrictions dictated by physical considerations. A condition that specifies the value of y for some specific value of x is called an **initial condition**, and the problem of solving a differential equation subject to an initial condition is called an **initial-value problem**.

EXAMPLE 2 Show that the function $y = Cx^3 - x^2$ is a solution of the differential equation

$$x\frac{dy}{dx} - 3y = x^2 \tag{3}$$

for all values of C, and find a particular solution satisfying the initial condition $y = 1$ when $x = 1$.

Solution Let $y = Cx^3 - x^2$. Then

$$\frac{dy}{dx} = \frac{d}{dx}(Cx^3 - x^2) = 3Cx^2 - 2x$$

and it follows that

$$\begin{aligned} x\frac{dy}{dx} - 3y &= x(3Cx^2 - 2x) - 3(Cx^3 - x^2) \\ &= 3Cx^3 - 2x^2 - 3Cx^3 + 3x^2 \\ &= x^2 \end{aligned}$$

Thus, $y = Cx^3 - x^2$ satisfies Equation (3).

To find a particular solution such that $x = 1$ when $y = 1$, we substitute these values in the general solution $y = Cx^3 - x^2$ to obtain

$$1 = C - 1 \quad \text{or} \quad C = 2$$

Thus, $y = 2x^3 - x^2$ is the desired particular solution.

Let us now consider how differential equations can arise. The simplest differential equations occur in integration problems. For example, if we want to evaluate

$$\int x^2\,dx$$

then we must find a function $y = f(x)$ such that

$$\frac{dy}{dx} = x^2$$

As we know from our study of integration, the solution is

$$y = \int x^2\,dx = \frac{x^3}{3} + C$$

In general, evaluating the integral

$$\int f(x)\,dx$$

is equivalent to solving the differential equation

$$\frac{dy}{dx} = f(x)$$

EXAMPLE 3 Solve the differential equation

$$\frac{dy}{dx} = x + 1$$

Solution

$$y = \int (x + 1)\,dx = \frac{x^2}{2} + x + C$$

EXAMPLE 4 **DEPRECIATION** An automobile initially worth \$10,000 depreciates in such a way that when the automobile is x years old, the rate of depreciation (in dollars per year) is

$$200x - 2000 \qquad (0 \le x \le 10) \tag{4}$$

(a) Find the value of the automobile after x years.

(b) Find the value of the automobile after 5 years.

Solution (a) If $V(x)$ is the value after x years, then (4) is the derivative of $V(x)$, since it is the rate of change in the value. Thus, $V(x)$ satisfies

$$\frac{dV}{dx} = 200x - 2000$$

To solve this differential equation for V, we integrate:

$$V = \int (200x - 2000)\,dx = 100x^2 - 2000x + C \tag{5}$$

To find C we use the fact that the automobile is initially worth \$10,000, so $V = 10{,}000$ when $x = 0$. Substituting these values in (5) yields

$$10{,}000 = C$$

Therefore the value after x years is

$$V = 100x^2 - 2000x + 10{,}000 \qquad (0 \le x \le 10)$$

Solution (b) At the end of 5 years ($x = 5$) the value of the automobile is

$$V = 100(5)^2 - 2000(5) + 10{,}000 = \$2500$$

The graph of the value of the automobile is shown in Figure 1.

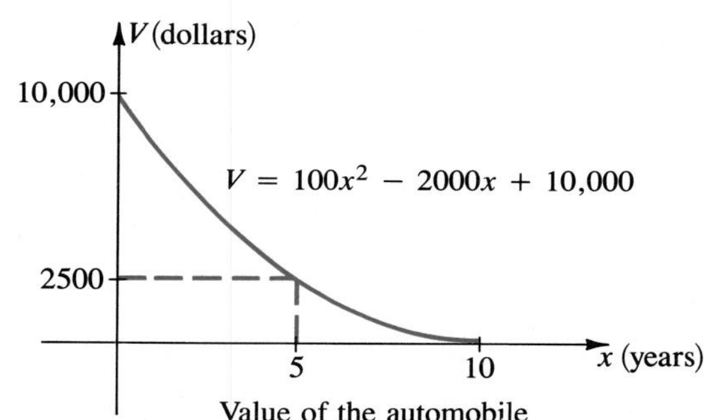

FIGURE 1 Value of the automobile

EXAMPLE 5 **MARGINAL ANALYSIS** Recall from Section 13.2 that the marginal cost MC, marginal revenue MR, and marginal profit MP are the derivatives of the cost function, revenue function, and profit function, respectively. That is,

$$\text{MC} = \text{marginal cost} = \frac{dC}{dx}$$

$$\text{MR} = \text{marginal revenue} = \frac{dR}{dx}$$

$$\text{MP} = \text{marginal profit} = \frac{dP}{dx}$$

where x represents the number of units produced. A manufacturer determines that the firm's marginal revenue function (in dollars per unit) is

$$\frac{dR}{dx} = 100 + 0.1x$$

Assuming that no revenue is received when no units are sold, find:

(a) the revenue from selling x units

(b) the revenue from selling 10 units

Solution (a) The revenue $R(x)$ satisfies the differential equation

$$\frac{dR}{dx} = 100 + 0.1x$$

Integrating yields

$$R = \int (100 + 0.1x)\,dx = 100x + 0.05x^2 + C \qquad (6)$$

To find C we use the fact that $R = 0$ when $x = 0$. Substituting these values in (6) yields $C = 0$. Therefore,

$$R = 100x + 0.05x^2 \qquad \text{dollars} \qquad (7)$$

Solution (b) Substituting $x = 10$ in Equation (7) yields

$$R = 100(10) + 0.05(10)^2 = \$1005$$

which is the revenue from selling 10 units.

Remark This problem can also be solved using the method of Example 2 in Section 17.5.

EXAMPLE 6 **MEDICINE** An experimental drug changes the average subject's body temperature at a rate equal to

$$-0.001t + 0.03\sqrt{t} \qquad \text{degrees Fahrenheit per hour}$$

where t is the number of hours elapsed after administration of the drug. Find the subject's temperature $T(t)$, t hours after the drug is administered if the initial temperature is 98.6°F.

Solution Since dT/dt is the rate at which the temperature changes, $T(t)$ satisfies the differential equation

$$\frac{dT}{dt} = -0.001t + 0.03\sqrt{t}$$

Integrating yields

$$T = \int (-0.001t + 0.03\sqrt{t})\,dt = -0.0005t^2 + 0.02t^{3/2} + C \qquad (8)$$

Since the initial temperature (at $t = 0$) is 98.6°F, we have

$$T = 98.6 \qquad \text{when } t = 0$$

Substituting these values in (8) yields $C = 98.6$, so (8) becomes

$$T = -0.0005t^2 + 0.02t^{3/2} + 98.6$$

which is the desired formula.

Remark This problem can also be solved using the method of Example 6 in Section 17.5.

EXERCISE SET 18.1

In Exercises 1–6 verify that y is a solution of the differential equation.

1. $dy/dx = 3x^2$; $y = x^3 + 2$
2. $dy/dx + 4y = 0$; $y = Ce^{-4x}$ (C any constant)
3. $dy/dt - 3y = e^{3t}$; $y = te^{3t}$
4. $(dy/dx)^2 - 4y = 8$; $y = x^2 - 2$
5. $dy/dx = \dfrac{x}{y}$; $y = \sqrt{x^2 + C}$ (C any constant)
6. $dy/dx = 2xy$; $y = Ce^{x^2}$ (C any constant)

In Exercises 7 and 8 verify that (a), (b), and (c) are solutions of the given differential equation.

7. $y'' - 2y' = 3y$; (a) $y = e^{3x}$ (b) $y = e^{-x}$ (c) $y = C_1e^{3x} + C_2e^{-x}$
8. $6y'' = -y' + 2y$; (a) $y = e^{x/2}$ (b) $y = e^{-2x/3}$ (c) $y = C_1e^{x/2} + C_2e^{-2x/3}$

In Exercises 9–12 verify that y is a solution of the differential equation and determine C so the given initial condition is satisfied.

9. $dy/dx = 4x^3$; $y = x^4 + C$; $y = 2$ when $x = 1$
10. $dy/dx = -3y$; $y = Ce^{-3x}$; $y = 4$ when $x = 0$
11. $2x\, dy/dx = x^2 + 1$; $y = \frac{1}{4}x^2 + \frac{1}{2}\ln x + C$; $y = 1$ when $x = 1$
12. $dy/dt = t/y^2$; $y = (\frac{3}{2}t^2 + C)^{1/3}$; $y = 2$ when $t = 0$

In Exercises 13–18 solve the differential equation.

13. $dy/dx = 3 - x$
14. $dy/dx = 5\sqrt{x}$
15. $dy/dx = x^{1/5} - 2x^{1/3}$
16. $dy/dx = e^{2x}$
17. $2x\, dy/dx = x^2 + 1$
18. $dy/dx = 1/\sqrt{x^2 - 9}$ [*Hint:* Use a table of integrals.]
19. **(Depreciation)** A yacht initially worth \$20,000 depreciates so that the rate of depreciation (in dollars per year) is

$$100x - 2000 \qquad (0 \le x \le 10)$$

   (a) Find the value of the yacht after x years.
   (b) Find the value of the yacht after 10 years.

20. **(Depreciation)** An industrial machine initially worth \$35,000 depreciates in such a way that after t years its rate of depreciation is $24t^2 + 80t - 4000$ (in dollars per year), where $0 \le t \le 9$.
   (a) Find the value of the machine after t years.
   (b) Find the value of the machine after 5 years.

21. **(Marginal analysis)** A manufacturer determines that the firm's marginal profit function is

$$\frac{dP}{dx} = -2x + 600 \qquad \text{(in dollars per unit)}$$

   Find the profit function $P(x)$ from selling x units if no profit is made when no units are sold.

22. **(Marginal analysis)** A manufacturer determines that the marginal cost function is

$$\frac{dC}{dx} = x^2 + x \qquad \text{(in dollars per unit)}$$

   Find the cost function $C(x)$ for manufacturing x units if the fixed cost (cost of producing zero units) is \$500.

23. **(Medicine)** An experimental antihistamine drug produces drowsiness as a side effect. As a result, a subject's reaction time T (in milliseconds) to a certain stimulus increases at a rate

$$\frac{dT}{dt} = e^{-t} \qquad \text{milliseconds per minute}$$

   where t is the time elapsed after the drug is administered. Find the reaction time t minutes after the drug is administered if the reaction time is 500 milliseconds initially.

24. **(Medicine)** A patient's body eliminates a drug from the blood stream at the rate of

$$-\frac{40}{1 + t} \qquad \text{milliliters per hour}$$

   t hours after the drug has been administered. If a patient is given 100 milliliters of the drug, how much of it remains in the bloodstream after 8 hours?

18.2 SEPARABLE DIFFERENTIAL EQUATIONS

We have already seen that when a differential equation can be written in the form

$$\frac{dy}{dx} = f(x)$$

then the solution can be obtained by integrating $f(x)$. We now show how to solve differential equations that can be written in the form

$$\frac{dy}{dx} = f(x)g(y) \tag{1}$$

where the right side is a product of a function of x and a function of y. Such equations are called **separable**.

EXAMPLE 1 Which of the following are separable differential equations?

(a) $\dfrac{dy}{dx} = x^3y^2$ (b) $\dfrac{dy}{dx} + e^xy = 0$ (c) $\dfrac{dy}{dx} = x + y$

Solution Equation (a) is separable because the right side is the product of a function of x and a function of y.

Equation (b) is also separable because it can be rewritten as

$$\frac{dy}{dx} = -e^xy$$

where the right side is the product of a function of x and a function of y.

Equation (c) is not separable because the right side is a sum (not a product) of a function of x and a function of y.

Using the definition of the differential (see Section 16.1), it can be shown that Equation (1) can be written in the alternate form

$$\frac{1}{g(y)}\,dy = f(x)\,dx \tag{2}$$

In this form the x variables are on one side and the y variables are on the other; we say that the variables have been **separated**. The solutions of (1) can now be obtained by integrating both sides of (2):

$$\int \frac{dy}{g(y)} = \int f(x)\,dx$$

EXAMPLE 2 Find all solutions of the separable differential equation $dy/dx = 4x^3e^{-y}$.

Solution Separating variables we obtain

$$e^y\,dy = 4x^3\,dx$$

Integrating both sides yields

$$\int e^y\,dy = \int 4x^3\,dx$$

or

$$e^y + C_1 = x^4 + C_2$$

or

$$e^y = x^4 + C_2 - C_1 \tag{3}$$

where C_1 and C_2 are arbitrary constants of integration. If we combine these into a single arbitrary constant $C = C_2 - C_1$, then Equation (3) can be written as

$$e^y = x^4 + C \tag{4}$$

In general, when integrating both sides of an equation we need only one constant of integration.

Whenever possible, we want to express y directly as a function of x. In this case we can achieve this by taking the natural logarithm of both sides of Equation (4) to obtain $\ln e^y = \ln (x^4 + C)$ or

$$y = \ln (x^4 + C) \tag{5}$$

which is the general solution of the given differential equation.

The procedure for solving separable differential equations can be summarized as follows:

Solving Separable Differential Equations

Step 1 Separate the variables.
Step 2 Integrate both sides.
Step 3 Combine the constants of integration into a single arbitrary constant.
Step 4 If possible, express y directly as a function of x.

EXAMPLE 3 Solve the initial-value problem $dy/dx = x/y^2$, where $y = 3$ if $x = 0$.

Solution Separating variables, we obtain

$$y^2\,dy = x\,dx$$

and integrating both sides yields

$$\int y^2\,dy = \int x\,dx$$

or

$$\frac{y^3}{3} = \frac{x^2}{2} + C \tag{6}$$

Substituting the initial condition $x = 0$, $y = 3$ in (6) yields

$$\frac{27}{3} = 9 = C$$

Equation (6) becomes $y^3/3 = x^2/2 + 9$. Solving for y,

$$y = \left(\frac{3}{2}x^2 + 27\right)^{1/3} \tag{7}$$

which is the solution to the given initial-value problem.

Sometimes the solution to a differential equation is an equation relating x and y, which cannot be solved to express y as an explicit function of x. The following example illustrates this possibility.

TRAVELING TO THE MOON

As we know, when an object is tossed straight up into the air, gravity will tend to pull the object back toward the Earth. Isaac Newton discovered that the acceleration due to the force of gravity (neglecting air resistance) equals 32 feet per second squared near the surface of the Earth. The acceleration decreases as the distance from the Earth increases.

The acceleration a due to the earth's gravitational attraction acting on a rocket aimed at the moon can be computed to be

$$a = -\frac{97{,}280}{r^2} \tag{a}$$

where r is the distance from the rocket to the center of the Earth. The minus sign in Equation (a) means that the acceleration is reducing the initial velocity v_0 at which the rocket leaves the Earth. What should v_0 be so that we can reach the moon? We follow the discussion of Morris Kline in his excellent book *Mathematics and the Physical World* (New York: Dover Publications, 1981). Since

$$a = \frac{dv}{dt} \tag{b}$$

we can rewrite (a) as

$$\frac{dv}{dt} = -\frac{97{,}280}{r^2}$$

Equation (b) is a differential equation that is not of the type discussed in this chapter. This equation can be solved as follows. Multiply both sides of Equation (b) by v:

$$v\frac{dv}{dt} = -\frac{97{,}280}{r^2}v \tag{c}$$

Since $v = dr/dt$, we rewrite (c) by substituting dr/dt for v on the right side:

$$v\frac{dv}{dt} = -\frac{97{,}280}{r^2}\frac{dr}{dt}$$

$$v\,dv = -\frac{97{,}280}{r^2}\,dr$$

The equation is now separable. Integrating both sides we have

$$\frac{v^2}{2} = \frac{97{,}280}{r} + C \tag{d}$$

When $t = 0$, $v = v_0$ (the initial velocity) and $r = 4000$ miles (the radius of the Earth), since the rocket leaves the surface of the Earth. Substituting these values in Equation (d) we have

$$\frac{v_0{}^2}{2} = \frac{97{,}280}{4000} + C$$

so

$$C = \frac{v_0{}^2}{2} - \frac{97{,}280}{4000}$$

Hence,

$$\frac{v^2}{2} = \frac{v_0{}^2}{2} + (97{,}280)\left(\frac{1}{r} - \frac{1}{4000}\right)$$

or

$$v^2 = v_0{}^2 + 2(97{,}280)\left(\frac{1}{r} - \frac{1}{4000}\right) \tag{e}$$

To determine the initial velocity v_0 so that the rocket will reach the moon, we will make the simplifying assumption that the moon's gravity does not affect the velocity of the rocket. Thus, when the rocket reaches the moon (240,000 miles from the center of the Earth), $v = 0$. Substituting $r = 240{,}000$ and $v = 0$ in (e) we solve for v_0, obtaining $v_0 \cong 6.9159$ miles per second, or $v_0 \cong 24{,}897$ miles per hour. Thus, the rocket should leave the Earth with an initial velocity of approximately 24,897 miles per hour.

EXAMPLE 4 Solve the initial-value problem $dy/dx = 8y/(y + 1)$, where $y = 1$ when $x = 1$.

Solution Separating variables we obtain

$$\frac{y + 1}{y}\, dy = 8\, dx$$

and integrating both sides yields

$$\int \left(1 + \frac{1}{y}\right) dy = \int 8\, dx$$

$$y + \ln|y| = 8x + C \tag{8}$$

Substituting the initial conditions yields

$$1 + \ln|1| = 8(1) + C$$
$$1 + 0 = 8 + C$$
$$-7 = C \tag{9}$$

Substituting Equation (9) into Equation (8), the solution is the equation

$$y + \ln|y| = 8x - 7 \tag{10}$$

We cannot solve this equation for y. In this particular case, however, x can be written as a function of y:

$$x = \frac{1}{8}(y + \ln|y| + 7) \tag{11}$$

Sometimes we cannot solve the solution equation for either x or y. In such cases the solution equation determines a function implicitly but not explicitly.

EXERCISE SET 18.2

In Exercises 1–16 solve the separable differential equation.

1. $dy/dx = 3y$
2. $dy/dx = 2 + y$
3. $dy/dx = x/y$
4. $dy/dx = xy$
5. $3y^2\, dy/dx = e^x$
6. $dy/dx = x/(6y^2)$
7. $dy/dx = x\sqrt{y}$
8. $dy/dx = x/\sqrt{y}$
9. $dy/dx = 2xe^{-y}$
10. $dy/dx = 3y^2$
11. $x\, dy/dx = x^2 + 1$
12. $dy/dx = 2y/x$
13. $e^y\sqrt{x}\, dy/dx = 1 + x$
14. $dy/dx = e^{x+y}$
15. $dQ/dt = 1/Q^2$
16. $dy/dt = 3t^2/y^4$

In Exercises 17–24 solve the initial-value problem.

17. $dy/dx = -2x$; $y = 3$ when $x = 0$
18. $dy/dx = e^{-y}$; $y = 1$ when $x = 0$
19. $dy/dx = 4y^{-2}$; $y = 8$ when $x = 1$
20. $dy/dx = 4yx^3$; $y = e$ when $x = 0$
21. $dy/dx = xy + y$; $y = 2$ when $x = 0$
22. $dy/dx = y^2e^{-x}$; $y = 1$ when $x = 0$
23. $(x^2 + 3)\, dy/dx = x$; $y = 5$ when $x = 1$
24. $(x^2 + 1)\, dy/dx = x$; $y = 4$ when $x = 0$

18.3 APPLICATIONS OF SEPARABLE DIFFERENTIAL EQUATIONS

Differential equations play a central role in applied mathematics. In problems in which the rate of change in a function is known to bear a mathematical relation to the value of the function and its input variable, the natural mathematical tool to use is a differential equation. The typical application discussed in this section is best viewed as an initial-value problem.

When a differential equation is used to solve a real-world problem, it is said to be a **mathematical model** of the problem. The model must reflect the essential aspects of the problem so that the solution to the mathematical model may be regarded as a reliable solution to the real-world problem.

The following steps will help you solve typical word problems in this section.

Step 1 Read the problem completely—from beginning to end, slowly. Then read it again. On the second reading pick out the key features of the problem.

Step 2 Look for one variable whose rate of change relative to another variable is given in relation to one or both of the variables. Express that relation as a differential equation.

Step 3 Solve the differential equation. (For problems of this section, it is separable.)

Step 4 The solution obtained in Step 3 involves one or more constants. Look to the problem for information that can be substituted into this solution to enable you to evaluate the constants.

Step 5 Finish by substituting the values of the constants into the solution obtained in Step 3.

Step 6 Read the problem again to see if any additional questions remain to be answered.

Exponential Growth or Decay Models

A quantity $Q > 0$ is said to have an **exponential growth or decay model** if the rate of change in the quantity Q is always proportional to the amount of the quantity present. That is,

$$\frac{dQ}{dt} = kQ \tag{1}$$

If $k > 0$, we call Equation (1) a **growth model**, and if $k < 0$, we call Equation (1) a **decay model**.

In Section 14.5 we see that one solution to the differential equation (1) is

$$Q = Q_0 e^{kt} \tag{2}$$

where $Q_0 = Q(0)$. In that section, we could not prove that this is the *only* solution, but we can now do so, as Example 1 shows.

EXAMPLE 1 Solve the initial-value problem

$$\frac{dQ}{dt} = kQ \qquad \text{where } Q = Q_0 \text{ if } t = 0 \tag{3}$$

Solution This is a separable differential equation in which the variables are Q and t rather than y and x as before. But the solution procedure is the same.

Separating variables, we obtain

$$\frac{1}{Q}\, dQ = k\, dt$$

and integrating yields

$$\int \frac{1}{Q}\,dQ = \int k\,dt$$

or

$$\ln|Q| = kt + K$$

Since $Q > 0$ we can drop the absolute-value sign and write

$$\ln Q = kt + K$$

We can solve for Q in terms of t by writing

$$e^{\ln Q} = e^{kt+K}$$

or

$$Q = e^{kt}e^{K}$$

Since e^K is constant we can denote it more simply by C, so the solutions of the differential equation are

$$Q = Ce^{kt} \tag{4}$$

To find C we substitute the initial condition $t = 0$, $Q = Q_0$ into (4). This yields

$$Q_0 = Ce^0 = C$$

When we substitute this value of C into Equation (4), we see that the solution of the initial-value problem (3) is

$$Q = Q_0e^{kt}$$

Therefore, the solution given by (2) is unique.

EXAMPLE 2 **EFFECTS OF RADIATION** When complex molecules present in the cells of living tissues are exposed to a beam of radiation, they may be irreversibly altered. When this happens, they are said to be "damaged." Let N denote the number of undamaged molecules present after receiving a dose D of the radiation, and let $N_0 = N(0)$, the number of undamaged molecules present immediately before exposure to the radiation. Suppose research indicates that the rate of change in N relative to D is proportional to N. Write a differential equation that expresses N as a function of D and solve.

Solution The rate of change in N relative to D is proportional to N. That is,

$$\frac{dN}{dD} = kN \tag{5}$$

Note that k must be negative, since the number N of undamaged molecules decreases as the dosage D increases. As shown in Example 1, the solution to this differential equation is

$$N = N_0e^{kD} \tag{6}$$

Simple Bounded Exponential Growth Models

When a quantity Q grows according to an exponential growth model, its size increases indefinitely (Figure 1). However, few quantities can grow in this way because eventually the resources needed for growth, such as space and raw materials, become depleted. In this section we develop more realistic growth models in which there is an upper limit on the size of Q.

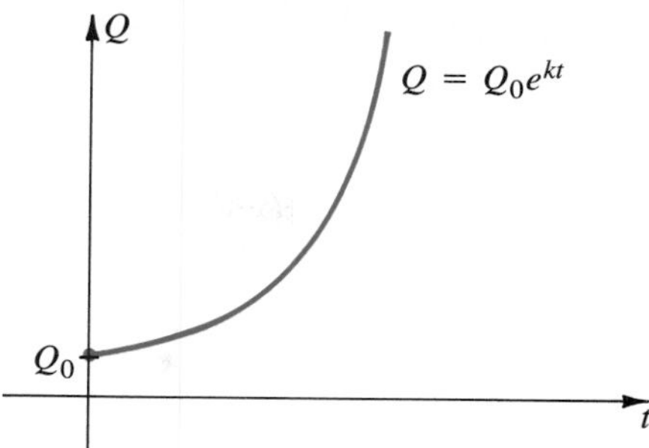

FIGURE 1

Simple Bounded Exponential Growth Models A quantity Q is said to have a **simple bounded exponential growth model** if there is an upper limit L on the size of Q and the rate of growth of Q at each instant is proportional to the difference between L and Q.

Suppose a quantity Q has a simple bounded exponential growth model and we know the amount of Q present at time $t = 0$. To set up an initial-value problem whose solution tells us the value of Q at each instant t, we proceed as follows.

The statement of the problem tells us that we want to solve the initial-value problem

$$\frac{dQ}{dt} = k(L - Q) \qquad Q(0) = Q_0 \tag{7}$$

where k is some constant of proportionality and L is also a constant. Separating variables, Equation (7) becomes

$$\frac{1}{L - Q}\, dQ = k\, dt$$

Integrating, we have

$$\int \frac{1}{L - Q}\, dQ = \int k\, dt$$

$$-\ln|L - Q| = kt + C$$

Since Q is always less than L, $L - Q > 0$ and we may drop the absolute-value signs and obtain

$$-\ln(L - Q) = kt + C$$

$$e^{-\ln(L-Q)} = e^{kt+C}$$

$$\frac{1}{L - Q} = e^{kt}e^{C} \tag{8}$$

We can eliminate the constant C by using the initial condition $Q(0) = Q_0$. Substituting $t = 0$ into Equation (8) we obtain

$$\frac{1}{L - Q_0} = e^0e^C = e^C$$

Therefore, (8) becomes

$$\frac{1}{L - Q} = e^{kt}\frac{1}{L - Q_0}$$

Inverting this equation we obtain

$$L - Q = \frac{L - Q_0}{e^{kt}}$$

$$-Q = \frac{L - Q_0}{e^{kt}} - L$$

$$Q = L - (L - Q_0)e^{-kt} = L - Le^{-kt} + Q_0e^{-kt}$$

Finally, we write

$$Q = L(1 - e^{-kt}) + Q_0e^{-kt} \tag{9}$$

In many problems $Q_0 = 0$ (see Exercise 7), in which case the last term in Equation (9) drops out.

EXAMPLE 3 **TERMINAL VELOCITY** When an object falls through the atmosphere, as when a sky diver jumps from an airplane, its velocity is affected by two factors. First, gravity imparts a constant acceleration $g = 32$ feet per second per second. Second, air friction causes an upward force that acts in opposition to gravity. This actually places a limit on the velocity of the falling object so that the velocity tends toward a constant M, called the **terminal velocity**. It is known that the acceleration of the object at time t is proportional to the difference between M and the velocity at time t.

(a) Write a differential equation that expresses this relation.

(b) How fast is the object falling after t seconds? [*Hint:* Set up an initial-value problem, and solve.]

Solution (a) We are told in the problem that the acceleration satisfies the equation

$$a(t) = k[M - v(t)] \tag{10}$$

where k is a constant of proportionality. But acceleration is the derivative of velocity. Thus Equation (10) becomes

$$\frac{dv}{dt} = k[M - v(t)]$$

Solution (b) We want to solve the initial-value problem

$$\frac{dv}{dt} = k[M - v] \qquad \text{with } v(0) = 0 \tag{11}$$

(The condition $v(0) = 0$ comes from the assumption that the object is merely released, not thrown, downward.) The solution to Equation (11) is found by using Equation (9):

$$v = M(1 - e^{-kt}) \tag{12}$$

Some concrete examples may be found in the exercises.

Logistic Growth Models

Logistic Growth Models A quantity Q is said to have a **logistic growth model** if there is an upper limit L on the size of Q, and the rate of growth at each instant is proportional to both Q and the difference between Q and the upper limit L.

It follows that for Q to have a logistic growth model, the rate of growth dQ/dt must be proportional to both Q and the difference $L - Q$. Thus Q must satisfy the differential equation

$$\frac{dQ}{dt} = kQ(L - Q) \tag{13}$$

where k is a constant of proportionality.

Note the effect of the "inhibiting factor" $L - Q$ in (13); as Q approaches its upper limit L, the factor $L - Q$ approaches zero, which in turn causes the growth rate dQ/dt to level off and approach zero.

We now show how to solve Equation (13). Separating the variables, we obtain

$$\frac{1}{Q(L - Q)}\, dQ = k\, dt$$

and integrating yields

$$\int \frac{1}{Q(L - Q)}\, dQ = \int k\, dt$$

The integral on the left can be evaluated using Formula 14 in the integral table (see Section 16.3) with $u = Q$, $b = L$, and $a = -1$. Thus, we obtain

$$\frac{1}{L} \ln \left| \frac{Q}{L - Q} \right| = kt + C \tag{14}$$

We can drop the absolute-value signs since $Q > 0$ and $L - Q > 0$ and rewrite Equation (14) as

$$\frac{1}{L} \ln \frac{Q}{L - Q} = kt + C$$

To solve for Q we write

$$\begin{aligned} \ln \frac{Q}{L - Q} &= Lkt + LC \\ e^{\ln[Q/(L-Q)]} &= e^{Lkt+LC} \\ \frac{Q}{L - Q} &= e^{Lkt}e^{LC} \end{aligned}$$

Since e^{LC} is a constant, we denote it more simply by N to obtain

$$\begin{aligned}\frac{Q}{L-Q} &= Ne^{Lkt}\\ Q &= Ne^{Lkt}(L-Q)\\ &= NLe^{Lkt} - QNe^{Lkt}\\ Q + QNe^{Lkt} &= NLe^{Lkt}\\ Q(1+Ne^{Lkt}) &= NLe^{Lkt}\end{aligned}$$

To simplify the formula for Q further, we divide the numerator and denominator by Ne^{Lkt}. This yields

$$Q\left(\frac{1}{Ne^{Lkt}} + 1\right) = L$$

or

$$Q = \frac{L}{(1/N)e^{-Lkt} + 1}$$

Finally, if we let $C = 1/N$ and interchange the order of the terms in the denominator, we obtain

$$Q = \frac{L}{1 + Ce^{-Lkt}} \tag{15}$$

which is sometimes called the **logistic equation**. If $Q = Q_0$ is the amount of the quantity present when $t = 0$, then it can be shown that the graph of the logistic equation has the shape shown in Figure 2.

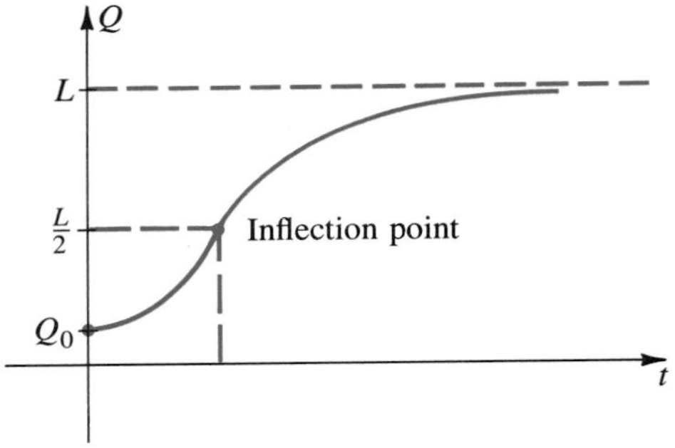

FIGURE 2

Differentiating the right side of Equation (13) we obtain

$$Q'' = k(L - 2Q)Q'$$

(Verify this.) Thus, when $Q = L/2$, we have an inflection point. This is the turning point in the growth rate; at values less than $Q = L/2$ the growth rate is increasing, and for Q greater than $L/2$ it is decreasing.

The following examples are typical applications of logistic growth models to population growth constrained by environmental factors.

EXAMPLE 4 **ECOLOGY** A wildlife preserve contains a sufficient number of nesting areas to support a maximum population of 330 whooping cranes. Suppose the preserve is stocked initially with 30 cranes and that 66 cranes are counted 24 months later.

Assuming that a logistic growth model applies, find an expression for the number Q of whooping cranes in the preserve t months after it is stocked.

Solution A formula for Q is given by the logistic Equation (15). Since $L = 330$ is the limiting value for Q, we obtain

$$Q = \frac{330}{1 + Ce^{-330kt}} \tag{16}$$

Since $Q = 30$ when $t = 0$,

$$30 = \frac{330}{1 + Ce^{0}} = \frac{330}{1 + C}$$

Thus,

$$1 + C = \frac{330}{30} = 11$$

so $C = 10$ and Equation (16) becomes

$$Q = \frac{330}{1 + 10e^{-330kt}} \tag{17}$$

To find k, we use the fact that $Q = 66$ when $t = 24$, so

$$66 = \frac{330}{1 + 10e^{-330k(24)}} = \frac{330}{1 + 10e^{-7920k}}$$

This can be rewritten as

$$1 + 10e^{-7920k} = \frac{330}{66} = 5$$

or

$$e^{-7920k} = \frac{4}{10} = 0.4$$

Taking the natural logarithm of both sides yields $-7920k = \ln 0.4$:

$$-7920k = -0.9163 \qquad \text{(from Appendix Table C.6)}$$

$$k = \frac{0.9163}{7920} = 0.0001157$$

Substituting this value into Equation (17) yields

$$Q = \frac{330}{1 + 10e^{-330(0.0001157)t}}$$

or

$$Q = \frac{330}{1 + 10e^{-0.03818t}} \tag{18}$$

which is the desired formula.

EXAMPLE 5 Use Formula (18) to determine how many months are required for the whooping crane population to reach 110.

Solution We want to find the value of t for which $Q = 110$. From (18) this value of t satisfies

$$110 = \frac{330}{1 + 10e^{-0.03818t}}$$

or

$$1 + 10e^{-0.03818t} = \frac{330}{110} = 3$$

or

$$e^{-0.03818t} = \frac{2}{10} = 0.2$$

Taking the natural logarithm of both sides yields $-0.03818t = \ln 0.2 = -1.6094$ (from Appendix Table C.6) or

$$t = \frac{1.6094}{0.03818} = 42.15 \cong 42$$

Thus, approximately 42 months are required for the population to reach 110 whooping cranes.

Logistic growth models and logistic curves are also used to model the spread of epidemics, learning processes, and information transmission through advertising or rumors. Such applications are explored in the exercises.

EXERCISE SET 18.3

In Exercises 1–4 set up a differential equation, and solve.

1. A substance is known to dissolve in water at a rate proportional to the undissolved amount remaining in the water. If 10 pounds of the substance are put into water, 1 pound will dissolve in 30 minutes. How much will remain after 2 hours? After how many hours will only 2 pounds remain?

2. Suppose the buying power of the dollar is decreasing at a rate that is proportional to the buying power at time t (in years). If a dollar today is worth 92 cents 1 year from today, how much is it worth t years from today? After how many years will it be worth only 50 cents?

3. In 1982 the population of the world was 4.6 billion. If the rate of growth of the population Q is $dQ/dt = 0.02Q$, find the world population in the year 2082.

4. A certain type of radio depreciates at a rate that is always proportional to its value at the time. One of these radios that sells for \$300 is worth \$250 two years later. How much is it worth when it is 10 years old?

Exercises 5–12 use simple bounded exponential growth models.

5. **(Terminal velocity)** An object is dropped from an airplane at time $t = 0$ (in seconds) with initial downward velocity $v(0) = 0$. If it reaches 20% of its terminal velocity in 5 seconds, how long will it take for it to reach 50% of its terminal velocity?

6. **(Terminal velocity)** An object dropped from an airplane reaches half its terminal velocity in 10 seconds. One second later it is falling at a rate of 200 feet per second. What is its terminal velocity?

7. **(Learning)** In many learning situations, the amount of information absorbed by a person increases rapidly at first and then tapers off as the person reaches the limit of his or her learning ability. Thus, if Q measures the amount of information learned and t the amount of time expended in

the learning process, the graph of Q versus t will be similar to the learning curve in Figure 3. For such models, Q satisfies

$$\frac{dQ}{dt} = k(L - Q)$$

where k is a constant of proportionality. Show that the particular solution of the given differential equation, which satisfies the initial condition $Q = 0$ when $t = 0$, is

$$Q = L - Le^{-kt}$$

The graph of this equation, which is shown in Figure 3, is called a **learning curve** because it arises in various learning situations.

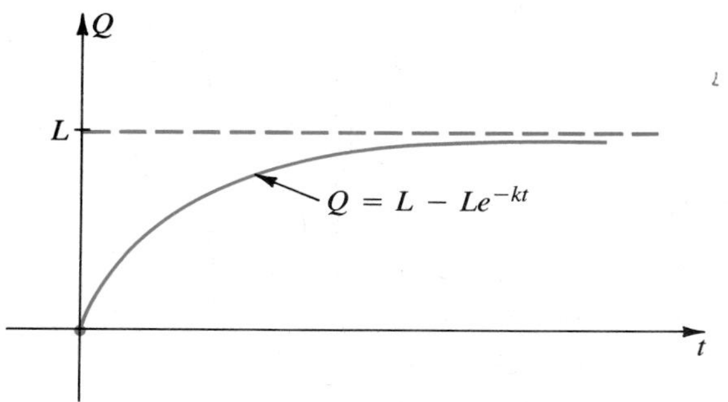

FIGURE 3

8. Suppose an automobile assembly line worker can process

$$Q = 50 - 50e^{-0.2t}$$

engines per hour after t months on the job.
(a) How many engines can the worker process per hour after 1 month on the job? (Use Appendix Table C.5.)
(b) After extensive experience, approximately how many engines will the worker be able to process per hour?
(c) Use graph paper and Appendix Table C.5 to make an accurate graph of Q versus t.

9. (a) Use the results of Equations (7) through (9) to find the solution of

$$\frac{dQ}{dt} = 2(10 - Q)$$

(b) Find the particular solution that satisfies the initial condition, $Q = 0$ when $t = 0$.
(c) For the function in (b), find $\lim_{t \to \infty} Q$.

10. **(Learning)** A new assembly line worker produces 30 units of work the first day on the job and by the end of the fifteenth day produces 50 units per day. Her efficiency improves in such a way that if Q represents her daily production quantity, then on her t th day of work, $dQ/dt = k(90 - Q)$.
(a) Solve an initial-value problem to determine the formula for Q. (Use the results of Example 3.)
(b) How many units of work does she produce daily after 30 days on the job? After 60 days?
(c) After extensive experience, how many units of work will she be able to produce daily? After how many days of work will her daily production be within 1 unit of that figure?

11. **(Newton's law of cooling)** Consider an object that has been heated to an initial temperature T_0, then placed into a cooler medium with constant temperature M and allowed to cool so that its temperature at later time t is T. Newton's law of cooling says that the rate of cooling dT/dt is proportional to the difference between the temperature of the object and that of the surrounding medium. That is, $dT/dt = k(M - T)$. (See also the Sherlock Holmes problem in Chapter 14, page 667.) Use the method of Example 3 to obtain a formula for the temperature T of the object at time t.

12. **(Cooling)** A cup of coffee at 190°F is placed in a room of constant temperature 70°F, and in 5 minutes the temperature of the coffee drops to 150°F.
(a) Set up and solve an initial-value problem whose solution yields T, the temperature of the coffee t minutes after it is placed in the room.
(b) What is the temperature of the coffee 10 minutes after being placed in the room?
(c) How long does it take the coffee to cool to 100°F?

In Exercises 13–19 use logistic growth models.

13. **(Ecology)** A lake can support a maximum population of 4800 rainbow trout. Suppose the lake is stocked with 800 trout initially, and there are 2400 trout 8 months later. Assuming that a logistic growth model applies, find:
(a) a formula for the number of trout in the lake t months after it is stocked
(b) the number of trout in the lake 16 months after it is stocked
(c) the number of months required for the trout population to reach 2000

14. **(Spread of epidemics)** In a town with a population of 8100, the Asian flu causes an epidemic that spreads according to a logistic growth model. At the time the disease is discovered, there are 100 people affected, and 20 days later 1000 are affected.
(a) Find a formula for the number of people affected t days after the disease is discovered.
(b) Find the number of people affected 30 days after the disease is discovered.
(c) Find the number of days required for half the town's population to be affected.

15. **(Advertising)** An advertising firm determines that the market size for its new product is 96,000 people. At the time an advertising campaign begins, 4000 people in the market are already aware of the product, and 2 months later 12,000 people are aware of it. Assume that the number of people who become aware of the product follows a logistic growth model.
(a) Find a formula for the number of people who are aware of the product t months after the advertising campaign begins.
(b) Find the number of people who are aware of the product 6 months after the advertising campaign begins.
(c) If the advertising campaign is to stop when three-fourths of the market is aware of the product, how long will the campaign have to run?

16. **(Spread of a rumor)** The rate at which a rumor spreads through a community is proportional to both the number of people who have heard the rumor and the number who have not. Suppose that in a community of 2000 people, 10 people hear the rumor initially and 100 have heard it 1 day later.
(a) Find a formula for the number of people who have heard the rumor t days after it began.
(b) Find the number of people who have heard the rumor 10 days after it began.
(c) How long will it take for 500 people to hear the rumor?

17. It was noted in this section that a logistic curve

$$Q = \frac{L}{1 + Ce^{-Lkt}}$$

has an inflection point when $Q = L/2$ (see Figure 2). By solving the equation

$$\frac{L}{2} = \frac{1}{1 + Ce^{-Lkt}}$$

for t, show that this inflection point occurs when

$$t = \frac{\ln C}{Lk}$$

18. (a) Use the result of Exercise 17 to find the value of t at which the logistic curve

$$Q = \frac{10}{1 + 2e^{-10t}}$$

has an inflection point.
(b) Make a sketch of the curve that shows the initial value (at $t = 0$), the maximum limiting value L, and the inflection point.

19. Follow the directions of Exercise 18 for the curve

$$Q = \frac{2000}{1 + 200e^{-5t}}$$

Exercises 20 and 21 require methods similar to those of Example 4.

20. **(Birth and immigration)** Another modification of the exponential growth model applied to a population is to allow individuals to enter and leave the population in such a way that the net immigration flow is constant. Then the differential equation for a population P with birth rate k and immigration flow i is

$$\frac{dP}{dt} = kP + i$$

(We assume the death rate is incorporated into the calculation of the birth rate.)
(a) Show that the solution to this differential equation is $P = Ce^{kt} - i/k$.
(b) Suppose $P = P_0$ when $t = 0$. Solve for the constant C and show that the solution is

$$P = P_0e^{kt} - \left(\frac{i}{k}\right)(1 - e^{kt})$$

21. **(Birth and immigration)** Consider a small community with an annual net birth rate of 3% and a net immigration flow of 200 persons per year.
(a) If the population is presently 50,000, what will it be in t years? In 10 years?
(b) Solve for t in the equation of Exercise 20(b).
(c) Use the answer to (b) to find out how long it will take the population of 50,000 to double under these conditions.

KEY IDEAS FOR REVIEW

- **Differential equation** An equation involving an unknown function and its derivatives.
- **Solution of a differential equation** A function that satisfies the differential equation.
- **General solution of a differential equation** A formula involving an arbitrary constant from which all solutions of the differential equation can be obtained by substituting numerical values for the constant.
- **Particular solution of a differential equation** A solution obtained by substituting a numerical value for the arbitrary constant in the general solution.
- **Initial condition** A condition that specifies the value of a solution, $y = f(x)$, for some specific value of x.
- **Separable differential equation** One that can be written in the form

$$\frac{dy}{dx} = f(x)g(y)$$

- **Exponential growth model** Q has an exponential growth model if it satisfies

$$\frac{dQ}{dt} = kQ$$

- **Simple bounded exponential model** Q has a simple bounded exponential growth model if it satisfies

$$\frac{dQ}{dt} = k(L - Q)$$

- **Logistic growth model** Q has a logistic growth model if it satisfies

$$\frac{dQ}{dt} = kQ(L - Q)$$

- **Logistic equation**

$$Q = \frac{L}{1 + Ce^{-Lkt}}$$

SUPPLEMENTARY EXERCISES

1. Solve the differential equation

$$\frac{dy}{dx} = \frac{x^2}{y^2}$$

given $y = 1$ when $x = 0$.

2. A firm's marginal profit function (in dollars per unit) is

$$\frac{dP}{dx} = -x^2 + 1000$$

Find the profit function obtained from selling x units if no profit is made when no units are sold.

In Exercises 3–6 solve the given differential equation with the specified initial condition.

3. $dy/dx = (y + 1)/x$, if $y = 0$ when $x = 1$
4. $\sqrt{x}\, dy/dx = 1/y$, if $y = 1$ when $x = 1$

5. $dQ/dt = 5Q$, if $Q = 1$ when $t = 0$

6. $dQ/dt = 5(1 - Q)$, if $Q = 0$ when $t = 5$

7. The acceleration a of a freely falling body is $a = -32$ feet per second per second, and the initial velocity is 0. (Neglect air friction.)
   (a) Find the velocity v as a function of t.
   (b) Find the velocity at $t = 1$.

8. Solve the differential equation $dQ/dt = 5Q(10 - Q)$, if $Q = 2$ when $t = 0$.

9. A hot bath is drawn. The temperature of the water is 120°F. The room is at constant temperature 70°F. In 10 minutes the temperature of the bath drops to 110°F. Set up and solve an initial-value problem whose solution yields T, the temperature t minutes after the bath is drawn.

10. Solve the Sherlock Holmes problem in Section 14.5, page 667.

CHAPTER TEST

1. Solve the differential equation

$$\frac{dy}{dx} = xe^x \qquad \text{if } y = 1 \text{ when } x = 0$$

2. Solve the differential equation

$$\frac{dQ}{dt} = Qt \qquad \text{if } Q = 1 \text{ when } t = 0$$

3. Solve the differential equation

$$\frac{dy}{dt} = y(4 - y) \qquad \text{if } y = 2 \text{ when } t = 0$$

4. The marginal cost (in dollars) is given as

$$\frac{dC}{dx} = x^2 - 2x + 1000$$

   (a) Find the cost as a function of x if the cost is \$500 when no units are sold.
   (b) What is the cost of manufacturing 20 units?

5. If the velocity of an object moving along a straight line is

$$v(t) = \frac{ds}{dt} = 16t - 64$$

what is the position of the particle at time t if $s = 128$ feet when $t = 0$?

19

Functions of Several Variables

19.1 FUNCTIONS OF TWO VARIABLES; COORDINATE SYSTEMS IN THREE DIMENSIONS

The only functions studied so far have been functions of one independent variable. However, many quantities in applied problems depend on more than one independent variable. For example, the familiar formula for the area of a rectangle,

$$A = lw \tag{1}$$

relates the area to the two variables: length l and width w. The volume V of a rectangular box,

$$V = lwh \tag{2}$$

is expressed in terms of the three variables: length, width, and height. More generally, we can conceive of quantities that depend on four, five, or even more independent variables. For simplicity, we limit the discussion in this section to quantities depending on two independent variables x and y.

> A dependent variable z is a **function of the independent variables x and y**:
>
> $$z = f(x, y) \tag{3}$$
>
> (read "z equals f of x and y") if f is a rule associating a unique value of z with each ordered pair of numbers (x, y); the value of z associated with the pair (x, y) is $f(x, y)$.

In Equation (1), $A = lw$ implies that $A = f(l, w)$. In general there can be any number of independent variables; for example, from Equation (2), $V = g(l, w, h)$.

A function of two variables can be viewed as an input–output machine, just as is the case with a function of one variable. With the two variables, however, the input is an ordered *pair* (x, y) of numbers (Figure 1).

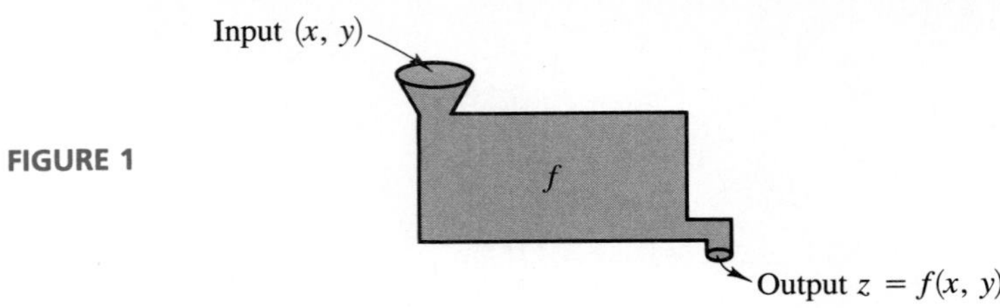

FIGURE 1

EXAMPLE 1 If the function f is given by $f(x, y) = 2x + 3y + 4$, then f associates the number $2x + 3y + 4$ with the ordered pair (x, y). Thus,

$$\begin{aligned} f(1, 0) &= 2(1) + 3(0) + 4 = 6 \\ f(0, 1) &= 2(0) + 3(1) + 4 = 7 \\ f(0, 0) &= 2(0) + 3(0) + 4 = 4 \\ f(-1, 3) &= 2(-1) + 3(3) + 4 = 11 \end{aligned}$$

Viewed as an input–output relation, this can be seen in tabular form:

Input	(1, 0)	(0, 1)	(0, 0)	(−1, 3)
Output	6	7	4	11

EXAMPLE 2 Suppose $z = f(x, y)$, where $f(x, y) = x^2y^3$. Find the values of z that f associates with the ordered pairs (4, 2) and $(\sqrt{2}, -3)$.

Solution Substituting $x = 4$ and $y = 2$ in the formula $z = f(x, y) = x^2y^3$ yields

$$z = f(4, 2) = 4^2 \cdot 2^3 = 128$$

and substituting $x = \sqrt{2}$ and $y = -3$ yields

$$z = f(\sqrt{2}, -3) = (\sqrt{2})^2 \cdot (-3)^3 = -54$$

EXAMPLE 3 **BUSINESS** An auto service station sells two brands of all-weather tires. Brand A costs the station \$60 each, and brand B costs them \$75 each. The demand for one brand is affected by its selling price as well as by the selling price of the competing brand. When the selling prices of brand A and brand B are p dollars and q dollars, respectively, their respective demand functions are

$$\begin{aligned} x &= 150 - 4p + 3q \qquad \text{brand A tires per month} \\ y &= 180 + 4p - 5q \qquad \text{brand B tires per month} \end{aligned}$$

Find the monthly revenue function $R(p, q)$, cost function $C(p, q)$, and profit function $P(p, q)$.

Solution Since the station sells x tires of brand A monthly at p dollars each and y tires of brand B monthly at q dollars each, its monthly revenue function is

$$\begin{aligned} R(p, q) &= xp + yq \\ &= (150 - 4p + 3q)p + (180 + 4p - 5q)q \\ &= 150p + 180q - 4p^2 - 5q^2 + 7pq \end{aligned}$$

Similarly, its monthly cost function is

$$\begin{aligned} C(p, q) &= 60x + 75y \\ &= 60(150 - 4p + 3q) + 75(180 + 4p - 5q) \\ &= 22{,}500 + 60p - 195q \end{aligned}$$

Therefore, the monthly profit function is

$$\begin{aligned} P(p, q) &= R(p, q) - C(p, q) \\ &= (150p + 180q - 4p^2 - 5q^2 + 7pq) - (22{,}500 + 60p - 195q) \\ &= 90p + 375q - 4p^2 - 5q^2 + 7pq - 22{,}500 \end{aligned}$$

The **domain** of a function $f(x, y)$ is the set of all ordered pairs (x, y) of real numbers that can be used as inputs to f. It may be viewed geometrically as a set of points (x, y) in the xy plane, which is two-dimensional. The **range** of $f(x, y)$ is the set of all real numbers z that can be outputs of f. It is a subset of the real number line, which is one-dimensional. Since two dimensions are required for the domain and a third dimension for the range, the graph of the equation $z = f(x, y)$ is a three-dimensional concept.

Coordinate Systems in Three Dimensions

A **rectangular coordinate system** in three dimensions (also called a **Cartesian coordinate system**) is formed by three mutually perpendicular coordinate lines that intersect at their origins; these lines are called the ***x*-axis**, the ***y*-axis**, and the ***z*-axis** (Figure 2).

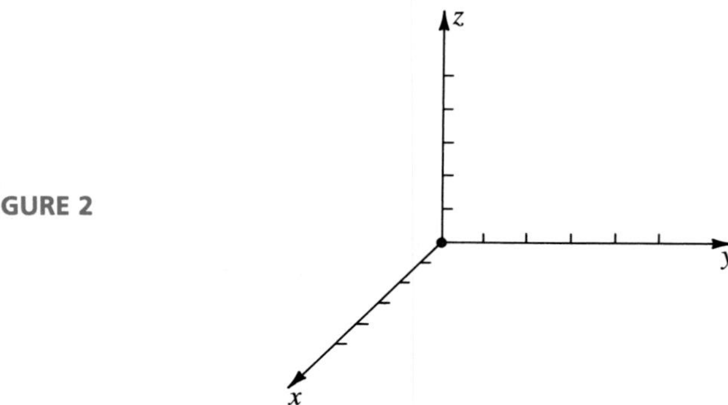

FIGURE 2

The point of intersection of the axes is called the **origin** of the coordinate system. Each pair of coordinate axes determines a plane, called a **coordinate plane**. These are referred to as the ***xy*-plane**, the ***xz*-plane**, and the ***yz*-plane**. To each point P in three-dimensional space we can assign an ordered triple of numbers (a, b, c), called the **coordinates of *P***. We can do this by passing three planes through P perpendicular to the coordinate axes and recording the coordinates a, b, and c of the intersections of the planes with the x-, y-, and z-axes, respectively (Figure 3).

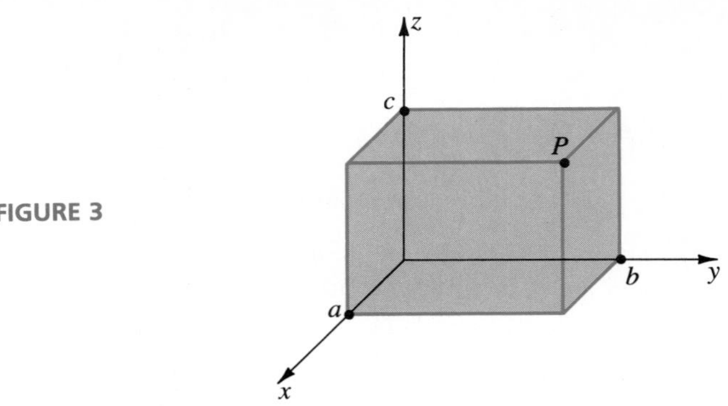

FIGURE 3

Observe that the three coordinates of P may be interpreted as:

a = x-coordinate of P = distance of P from the yz plane (positive or negative according to whether P is in front of, or behind, the yz plane)

b = y-coordinate of P = distance of P from the xz plane (positive or negative according to whether P is to the right of, or left of, the xz plane)

c = z-coordinate of P = distance of P from the xy plane (positive or negative according to whether P is above or below the xy plane)

It is evident from this construction that each point P in three dimensions has a unique set of coordinates (a, b, c), and, conversely, each ordered triple of numbers (a, b, c) determines a unique point P. Frequently, we denote the point P with coordinates (a, b, c) by $P(a, b, c)$.

EXAMPLE 4 In Figure 4 we have plotted the points $P(7, 5, 9)$ and $Q(7, 8, -8)$.

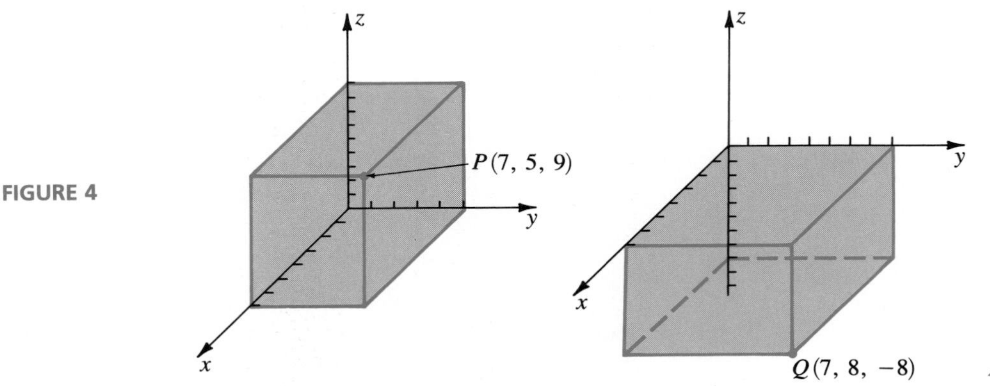

FIGURE 4

Graphs of Functions

Recall that the graph of a function $f(x)$ of one variable consists of all points in the plane whose coordinates (x, y) satisfy the equation $y = f(x)$. In general, the graph of $f(x)$ is some curve in the plane. Analogously, we define the **graph of a function $f(x, y)$** of two variables to be the set of all points in three dimensions whose coordinates (x, y, z) satisfy the equation $z = f(x, y)$. In general, the graph of $f(x, y)$ will be some **surface**.

EXAMPLE 5 The graph of the function

$$f(x, y) = \sqrt{x^2 + y^2}$$

consists of all points (x, y, z) that satisfy the equation

$$z = \sqrt{x^2 + y^2} \tag{4}$$

Some typical points on the graph are (3, 4, 5), (0, 0, 0), and (−2, 0, 2). (You should verify that the coordinates of these points satisfy Equation (4).) Although we will omit the details, it can be shown that the graph of Equation (4) is the cone in Figure 5.

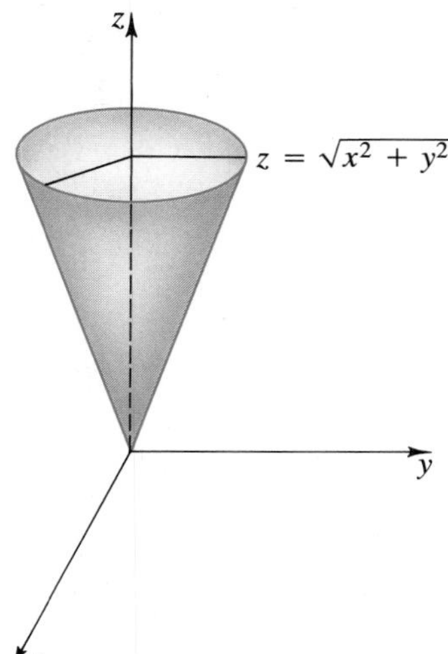

FIGURE 5

It is not usually possible to graph $z = f(x, y)$ by plotting points, because too many points are needed to obtain the shape of the surface. Techniques for graphing surfaces are studied in more advanced treatments of calculus, and are not considered in this text. For our purposes it is sufficient for you to understand the following interpretation of the graph of a function $f(x, y)$:

An Interpretation of the Graph of *f*(*x*, *y*) If the graph of $z = f(x, y)$ is the surface S and if $f(x, y) > 0$, then the value of z that f assigns to the ordered pair (x, y) is equal to the vertical distance between the surface S and the point $(x, y, 0)$ in the xy-plane (Figure 6). If $f(x, y) < 0$, then the surface will lie below the xy-plane and z will be negative.

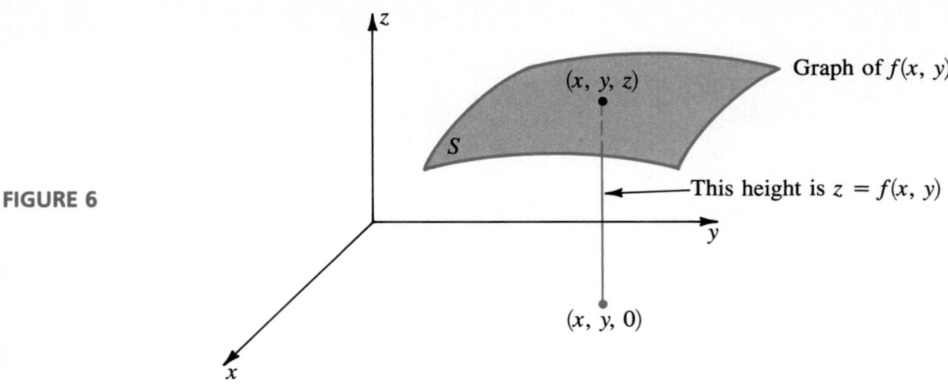

FIGURE 6

EXAMPLE 6 In Example 5 we note that the graph of the function $f(x, y) = \sqrt{x^2 + y^2}$ is the cone in Figure 5. Because

$$f(3, 4) = \sqrt{3^2 + 4^2} = 5$$

it follows that the height of the cone over the point (3, 4) is $z = 5$ units (Figure 7).

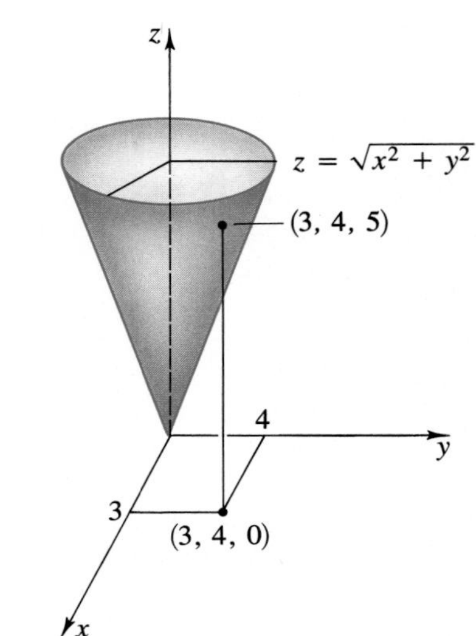

FIGURE 7

COMPUTER GRAPHICS

Computer technology has made great progress in the last few years in the graphics area. Computers are now used to effectively sketch the graph of a function $z = f(x, y)$. Many of these programs allow the user to vary a number of parameters, such as:

1. The bounds on x and y.
2. The number of points plotted (the grid points).
3. The point from which we view the surface. Changing the viewpoint allows us to picture different portions of the surface.
4. The scale along the x- and y-axes. This enables us to make peaks more or less prominent.
5. The frame. Graphs can be sketched with or without a frame around them.

Below we show two computer-generated graphs of the function

$$z = 10\left(x^3 + xy^4 - \frac{x}{5}\right)e^{-(x^2+y^2)} + 0.3e^{-[(x-1.225)^2+y^2]}$$

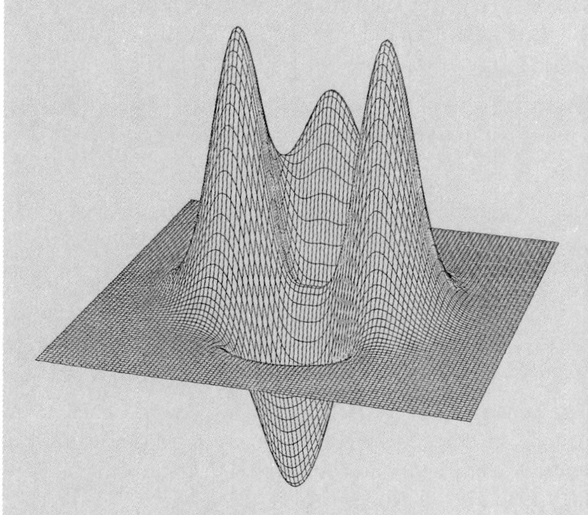

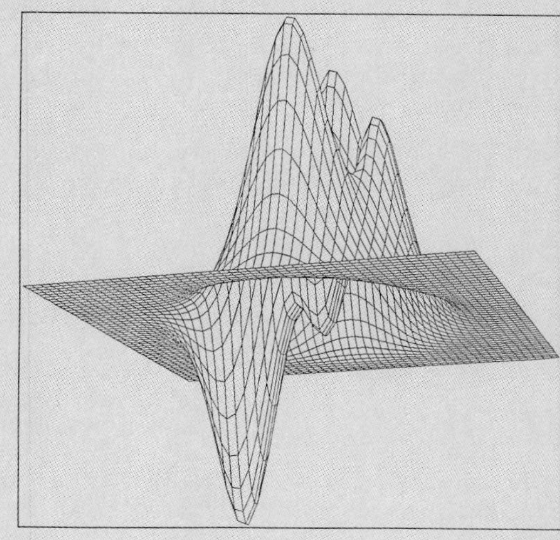

EXERCISE SET 19.1

1. Let $f(x, y) = 2xy^2 - x + 3$. Find:
(a) $f(1, 0)$ (b) $f(0, 1)$
(c) $f(-1, 4)$ (d) $f(0, 0)$

2. Let $z = f(x, y)$, where $f(x, y) = 3x^2y^2 - x + y - 3$. Find the value of z that f associates with the ordered pair:
(a) $(1, 0)$ (b) $(0, 1)$
(c) $(-2, 2)$ (d) $(-3, -2)$

3. Let $g(x, y) = xe^y + ye^x$. Find:
(a) $g(0, 1)$ (b) $g(1, 0)$
(c) $g(1, 1)$ (d) $g(-1, 2)$

4. Let $z = h(x, y) = xye^x + \ln y$. Find the value of z that h associates with the ordered pair:
(a) $(0, 1)$ (b) $(0, e)$
(c) $(1, e)$ (d) $(-1, e)$

5. Let $f(x, y) = (x^2 - y^2)/(x + y)$. Find:
(a) $f(\sqrt{2}, 0)$ (b) $f(\sqrt{3}, \sqrt{3})$
(c) $f(1, 2)$ (d) $f(-4, 1)$

6. Let $f(x, y) = 2x^2 - 3y$. Find:
(a) $f(0, 0)$ (b) $f(a, b)$
(c) $f(0, r)$ (d) $f(s, 1)$

7. Let P be the point with coordinates (x, y, z). Describe the condition(s) the coordinates must satisfy if:
(a) P lies in the xy-plane
(b) P lies in the yz-plane
(c) P lies in the xz-plane

8. Let P be the point with coordinates (x, y, z). Describe the condition(s) the coordinates must satisfy if:
(a) P lies on the x-axis
(b) P lies on the y-axis
(c) P lies on the z-axis

9. Copy Figure 4 and label all corners of the boxes with their coordinates.

10. Plot the following points in a rectangular coordinate system:
(a) $(0, 2, 3)$ (b) $(1, -1, 2)$
(c) $(4, 3, 5)$ (d) $(2, 0, -1)$

11. Find the coordinates of the points A through H in Figure 8.

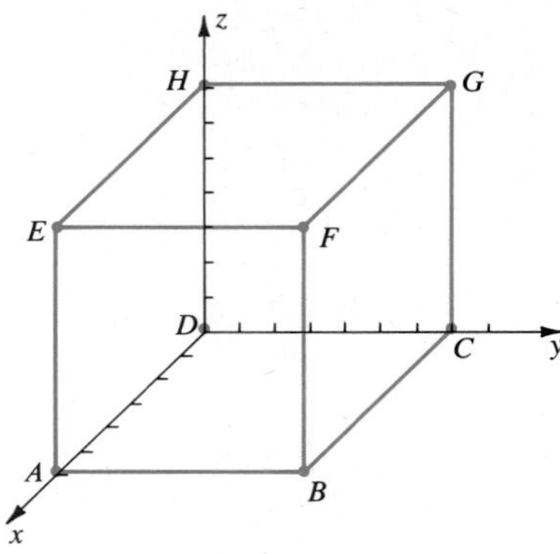

FIGURE 8

12. Find the height of the surface $z = x^2 + y^2$ above the xy-plane at the point:
(a) $(1, 1)$ (b) $(2, 3)$ (c) $(0, 0)$

13. Find the height of the surface $z = 4x + y + 4$ above the xy-plane at the point:
(a) $(2, 1)$ (b) $(0, 0)$ (c) $(-1, 2)$

14. **(Business)** A computer manufacturer finds that the profit P (in millions of dollars) is given by

$$P(x, y) = 3 + 5x + 2y$$

where x and y are the amounts (in millions of dollars) spent on research and development, respectively. Find:
(a) $P(2, 4)$ (b) $P(4, 2)$

15. **(Business)** In Exercise 14, how much must be spent on research to achieve \$16 million profit if \$3 million is spent on development?

16. The function $f(x, y) = 30x^{2/3}y^{1/3}$ gives the number of gallons of film developer that can be produced with x units of labor and y units of capital. Find:
(a) $f(8, 27)$ (b) $f(27, 64)$

17. **(Revenue, cost, and profit)** A company manufactures two types of stereo speakers per year: x thousand deluxe quality and y thousand budget quality. The annual revenue and cost functions (in thousands of dollars) for the company are

$$R(x, y) = 250x + 100y$$
$$C(x, y) = 2x^2 - 3xy + 5y^2 + 150x + 50y + 950$$

(a) Find the profit function $P(x, y)$.
(b) Find the annual revenue, cost, and profit when the company produces 8000 deluxe and 10,000 budget speakers per year.
(c) Find the revenue, cost, and profit functions when the company is producing an equal number of the two types of speakers.
(d) Find the revenue, cost, and profit functions when the company is producing twice as many budget speakers as deluxe speakers.

18. (Demand and revenue) The university store sells official varsity warm-up suits in two competing brands: brand A selling at p dollars each and brand B selling at q dollars each. The demand for one brand is affected by its price as well as by the price of its competitor. The monthly demand equations for brand A and brand B are, respectively,

$$x = 90 - 3p + 2q$$
$$y = 80 + 4p - 5q$$

(a) Find the (combined) monthly revenue function $R(p, q)$ for the warm-up suits.
(b) Find the (combined) monthly revenue when brand A sells for \$10 and brand B sells for \$12.

19.2 PARTIAL DERIVATIVES

Let $f(x, y)$ be a function of two variables, and let us introduce a dependent variable

$$z = f(x, y)$$

We are interested in studying the rate at which z changes as we vary the independent variables x and y. Of particular importance are the rate of change of z with respect to x when the variable y remains constant and the rate of change of z with respect to y when the variable x remains constant. This leads us to the following definitions:

If $z = f(x, y)$, then the **partial derivative of z with respect to x** is the derivative of z with respect to x with y treated as a constant. This partial derivative is denoted by

$$\frac{\partial z}{\partial x} \quad \text{or} \quad \frac{\partial f}{\partial x} \quad \text{or} \quad f_x$$

If $z = f(x, y)$, then the **partial derivative of z with respect to y** is the derivative of z with respect to y with x treated as a constant. This partial derivative is denoted by

$$\frac{\partial z}{\partial y} \quad \text{or} \quad \frac{\partial f}{\partial y} \quad \text{or} \quad f_y$$

The symbols $\partial z/\partial x$ and $\partial z/\partial y$ are used instead of dz/dx and dz/dy to indicate that we are differentiating a function that has more than one variable.

EXAMPLE 1 Let

$$z = f(x, y) = x^2 + xy + y^2$$

Find the partial derivative of z with respect to x and the partial derivative of z with respect to y.

Solution To find the partial derivative of z with respect to x, we treat y as a constant and differentiate z with respect to x, obtaining

$$\begin{aligned}\frac{\partial z}{\partial x} &= \frac{\partial}{\partial x}(x^2 + xy + y^2)\\ &= \frac{\partial}{\partial x}(x^2) + y\frac{\partial}{\partial x}(x) + \frac{\partial}{\partial x}(y^2)\\ &= 2x + y + 0\\ &= 2x + y\end{aligned}$$

Alternatively, we can express this result in the notation

$$f_x(x, y) = 2x + y$$

To find the partial derivative of z with respect to y we treat x as a constant and differentiate z with respect to y, obtaining

$$\begin{aligned}\frac{\partial z}{\partial y} &= \frac{\partial}{\partial y}(x^2 + xy + y^2)\\ &= \frac{\partial}{\partial y}(x^2) + x\frac{\partial}{\partial y}(y) + \frac{\partial}{\partial y}(y^2)\\ &= 0 + x + 2y\\ &= x + 2y\end{aligned}$$

Alternatively, we can express this result in the notation

$$f_y(x, y) = x + 2y$$

If $z = f(x, y)$, then the values of the partial derivatives when $x = x_0$ and $y = y_0$ are denoted by

$$\left.\frac{\partial z}{\partial x}\right|_{(x_0, y_0)} \quad \text{and} \quad \left.\frac{\partial z}{\partial y}\right|_{(x_0, y_0)}$$

or

$$f_x(x_0, y_0) \quad \text{and} \quad f_y(x_0, y_0)$$

EXAMPLE 2 Let

$$z = f(x, y) = x^2 + xy + y^2$$

Then from the previous example,

$$\frac{\partial z}{\partial x} = f_x(x, y) = 2x + y$$

$$\frac{\partial z}{\partial y} = f_y(x, y) = x + 2y$$

Thus, if $x = 1$ and $y = 3$, we obtain

$$\left.\frac{\partial z}{\partial x}\right|_{(1,3)} = f_x(1, 3) = 2(1) + 3 = 5$$

With y held constant, this represents the instantaneous rate of change of z with respect to x when $x = 1$, $y = 3$.

Similarly,

$$\left.\frac{\partial z}{\partial y}\right|_{(1,3)} = f_y(1, 3) = 1 + 2(3) = 7$$

This value represents the instantaneous rate of change of z with respect to y when $x = 1$, $y = 3$, and x remains constant.

EXAMPLE 3 Suppose the cost C (in cents) of manufacturing one cardboard box of paper clips is given by $C = 3x + 10y + 2$, where

$x =$ the unit cost of the cardboard (in cents)
$y =$ the unit cost of the metal (in cents)

Then

$$\frac{\partial C}{\partial x} = 3 \quad \text{and} \quad \frac{\partial C}{\partial y} = 10$$

The first result tells that if the cost y of the metal remains constant, then the manufacturing cost C will increase at a constant rate of 3 cents a box for each 1-cent increase in the cost of the cardboard. Similarly, if the cost x of the cardboard remains fixed, then the cost C will increase at a constant rate of 10 cents per box for each 1-cent increase in the cost of the metal.

Geometric Interpretation of Partial Derivatives

Recall that in a two-dimensional rectangular coordinate system, the set of points satisfying $x = a$ forms a line parallel to the y-axis, and those satisfying $y = b$ form a line parallel to the x-axis (Figure 1). Analogously, in a three-dimensional rectangular coordinate system, the set of points satisfying $x = a$ forms a vertical plane parallel to the yz-plane, and those satisfying $y = b$ form a vertical plane parallel to the xz-plane (Figure 2).

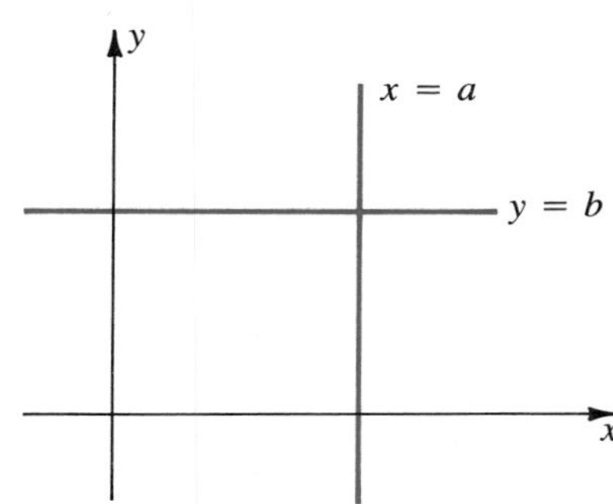

FIGURE 1

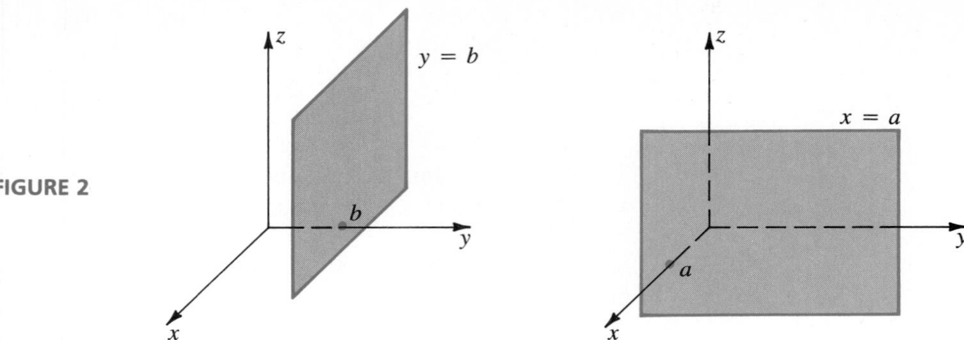

FIGURE 2

We can now give a geometric interpretation of the partial derivatives of a function of two variables. If $z = f(x, y)$ and y is held constant, for example, $y = y_0$, then the point (x, y, z) lies on the curve that is the intersection of the surface $z = f(x, y)$ and the vertical plane $y = y_0$ (Figure 3).

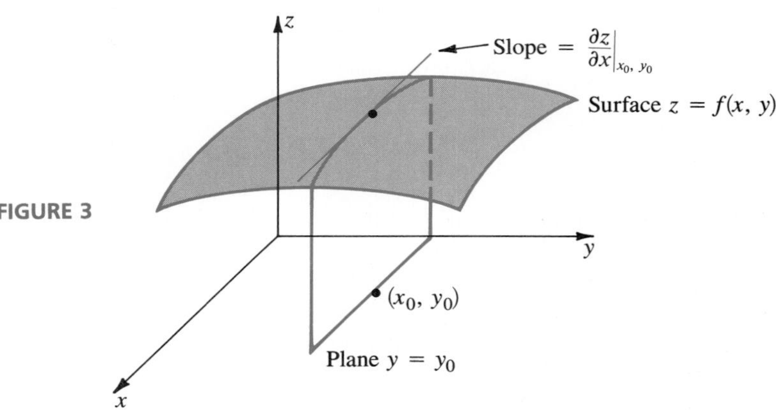

FIGURE 3

Thus,

$$\left.\frac{\partial z}{\partial x}\right|_{(x_0, y_0)}$$

is the slope of the tangent to this curve at the point (x_0, y_0). Similarly,

$$\left.\frac{\partial z}{\partial y}\right|_{(x_0, y_0)}$$

is the slope of the tangent to the intersection of the surface $z = f(x, y)$ and the vertical plane $x = x_0$ at the point (x_0, y_0) (Figure 4).

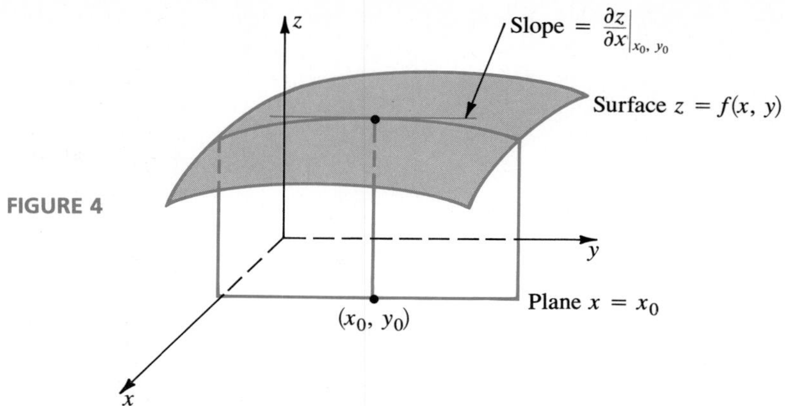

FIGURE 4

Second-Order Partial Derivatives

If $z = f(x, y)$, then

$$f_x = \frac{\partial z}{\partial x} \quad \text{and} \quad f_y = \frac{\partial z}{\partial y}$$

are both functions of x and y, which means we can form partial derivatives of them, thereby obtaining what are called the **second-order partial derivatives:**

$$\frac{\partial}{\partial x}\left(\frac{\partial z}{\partial x}\right) = \frac{\partial^2 z}{\partial x^2} \quad \text{or} \quad (f_x)_x = f_{xx}$$

$$\frac{\partial}{\partial y}\left(\frac{\partial z}{\partial x}\right) = \frac{\partial^2 z}{\partial y\,\partial x} \quad \text{or} \quad (f_x)_y = f_{xy}$$

$$\frac{\partial}{\partial x}\left(\frac{\partial z}{\partial y}\right) = \frac{\partial^2 z}{\partial x\,\partial y} \quad \text{or} \quad (f_y)_x = f_{yx}$$

$$\frac{\partial}{\partial y}\left(\frac{\partial z}{\partial y}\right) = \frac{\partial^2 z}{\partial y^2} \quad \text{or} \quad (f_y)_y = f_{yy}$$

Observe that in this notation, f_{xy} means that we first differentiate with respect to x, then with respect to y, while

$$f_{yx} = \frac{\partial^2 z}{\partial x\,\partial y} = \frac{\partial}{\partial x}\left(\frac{\partial z}{\partial y}\right)$$

means that we first differentiate with respect to y and then with respect to x.

EXAMPLE 4 If

$$f(x, y) = x^2e^y + ye^x + xy^2$$

find $f_{xx}, f_{xy}, f_{yx}, f_{yy}$.

Solution First we form the partial derivatives:

$$f_x = 2xe^y + ye^x + y^2$$
$$f_y = x^2e^y + e^x + 2xy$$

Then we obtain the second-order partial derivatives:

$$f_{xx} = \frac{\partial}{\partial x}(f_x) = \frac{\partial}{\partial x}(2xe^y + ye^x + y^2) = 2e^y + ye^x$$

$$f_{yy} = \frac{\partial}{\partial y}(f_y) = \frac{\partial}{\partial y}(x^2e^y + e^x + 2xy) = x^2e^y + 2x$$

$$f_{xy} = \frac{\partial}{\partial y}(f_x) = \frac{\partial}{\partial y}(2xe^y + ye^x + y^2) = 2xe^y + e^x + 2y$$

$$f_{yx} = \frac{\partial}{\partial x}(f_y) = \frac{\partial}{\partial x}(x^2e^y + e^x + 2xy) = 2xe^y + e^x + 2y$$

In this example it turns out that

$$f_{xy} = f_{yx}$$

Although this equality does not always hold, it is true for many of the commonly encountered functions.

EXERCISE SET 19.2

In Exercises 1–8 find f_x, f_y, and f_{yx}.

1. $f(x, y) = x^3 + xy + 2y^2$
2. $f(x, y) = x^2y - y^2x + y^2$
3. $f(x, y) = xe^y + y^2e^x - y^3$
4. $f(x, y) = xe^{3x+2y}$
5. $f(x, y) = e^{xy}$
6. $f(x, y) = y/x - x/y$
7. $f(x, y) = \ln(xy)$
8. $f(x, y) = (x + y)\ln(xy)$

In Exercises 9–12 calculate $f_x(x_0, y_0)$, $f_y(x_0, y_0)$, and $f_{xy}(x_0, y_0)$.

9. $f(x, y) = xy^2 + 2xy + y^3$; $(x_0, y_0) = (-1, 2)$
10. $f(x, y) = x^3 + xy^3 - x^2y + y^4$; $(x_0, y_0) = (1, 2)$
11. $f(x, y) = xe^y + y^2e^x$; $(x_0, y_0) = (0, 1)$
12. $f(x, y) = x\ln(xy)$; $(x_0, y_0) = (1, 1)$

In Exercises 13–18 calculate f_{xx}, f_{yy}, f_{xy}, and f_{yx}.

13. $f(x, y) = 2x^2 + xy - y^2$
14. $f(x, y) = 3x^2 - 2xy + y^4x^2$
15. $f(x, y) = 2x + 5y$
16. $f(x, y) = e^{xy}$
17. $f(x, y) = xe^{x+y}$
18. $f(x, y) = y\ln x$

In Exercises 19–22 evaluate f_{xx}, f_{yy}, f_{xy}, and f_{yx} at the given point (x_0, y_0).

19. $f(x, y) = 3x^2y + yx - 2y^2$; $(x_0, y_0) = (-1, 1)$
20. $f(x, y) = 2x^2y + y^3$; $(x_0, y_0) = (-2, 3)$
21. $f(x, y) = e^{2x+3y}$; $(x_0, y_0) = (0, 1)$
22. $f(x, y) = y\ln x$; $(x_0, y_0) = (2, 3)$
23. Let $f(x, y) = \ln(x^2 + y^2)$. Show that $f_{xx} + f_{yy} = 0$.
24. **(Business)** A manufacturer's cost C depends on the number of employees x and on the cost of materials and overhead y (in thousands of dollars).
    (a) If the relationship is
    $$C(x, y) = 20 + 3x^2 + 4y$$
    find $\partial C/\partial x$ and $\partial C/\partial y$.
    (b) Evaluate $\partial C/\partial x$ and $\partial C/\partial y$ when $x = 10$ and $y = 3$.
    (c) In words, explain the meaning of the values obtained in (b).

25. **(Ecology)** The pollution index I for a certain lake is defined to be

$$I = \ln x + \ln y + 3xy$$

where x is the number of milligrams of detergent and y is the number of milligrams of metallic salts in a standard sample.

(a) Find $I_x(1, 2)$ and $I_y(1, 2)$.

(b) In words, explain the meaning of the values obtained in (a).

26. **(Biomedical engineering)** In certain fields of engineering design it becomes important to know how the average person's body surface area varies with height and weight (e.g., the body surface area affects the amount of moisture evaporating in a fixed time period). A commonly used empirical formula relating body surface area A (square meters) to weight W (kilograms) and height H (meters) is

$$A = 2.024W^{0.425}H^{0.725}$$

(a) Find $\partial A/\partial W$ and $\partial A/\partial H$ when $W = 91$ and $H = 2$. Use the approximations

$$91^{0.425} \cong 6.80 \qquad 91^{-0.575} \cong 0.075,$$
$$2^{0.725} \cong 1.65 \qquad 2^{-0.275} \cong 0.826$$

(b) Give a physical interpretation of the results in (a).

19.3 APPLICATIONS OF PARTIAL DERIVATIVES TO OPTIMIZATION

In Section 13.4 we considered the problem of maximizing or minimizing a function $f(x)$ of one variable. We now consider the problem of maximizing or minimizing functions of two variables.

Recall that for functions of one variable we distinguish between relative maxima and absolute maxima. A function $f(x)$ is said to have an *absolute maximum* at $x = a$ if $f(a)$ is the largest value for $f(x)$, and it is said to have a *relative maximum* at $x = a$ if $f(a)$ is the largest value for $f(x)$ in the "immediate vicinity" of $x = a$; similarly for absolute minima and relative minima (Figure 1).

FIGURE 1

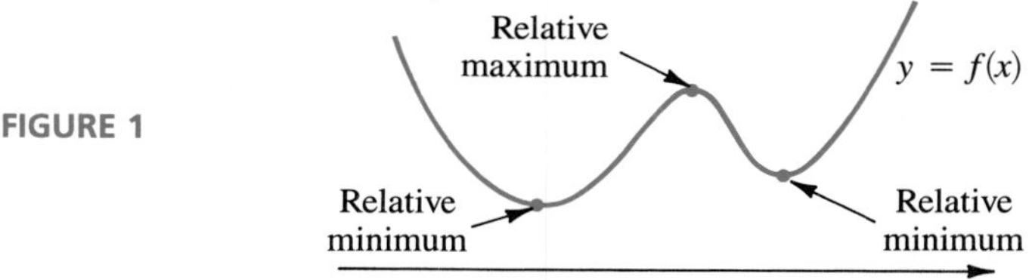

A relative maximum for f may or may not be an absolute maximum. Similarly, a relative minimum may or may not be an absolute minimum.

For functions of two variables, there are analogous notions. The function $f(x, y)$ is said to have an **absolute maximum** at (a, b) if $f(a, b)$ is the largest value that $f(x, y)$ can have, and f is said to have a **relative maximum** at (a, b) if $f(a, b)$ is the largest value for $f(x, y)$ in the immediate vicinity of the point (a, b). The terms *absolute minimum* and *relative minimum* are defined similarly. (The term *immediate vicinity* is admittedly vague, but your intuitive interpretation of this term is sufficient for our purposes.) See Figure 2.

FIGURE 2

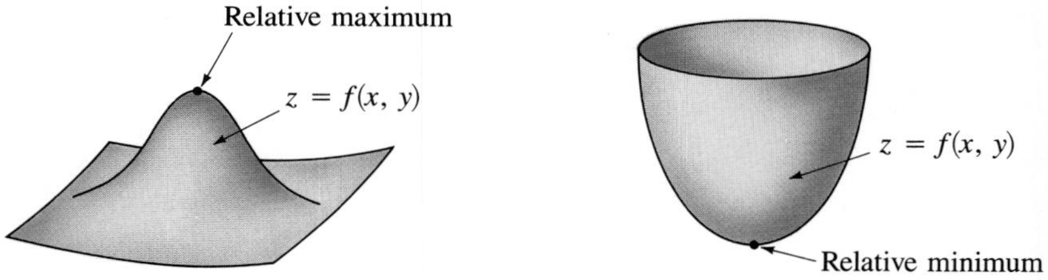

Relative maximum

$z = f(x, y)$

$z = f(x, y)$

Relative minimum

For functions of one variable, we see that if $f(x)$ has a relative maximum or a relative minimum at $x = a$ and if f is differentiable at $x = a$, then $f'(a) = 0$; that is, $y = f(x)$ has a horizontal tangent at every relative maximum or relative minimum where f is differentiable. We called a point $x = a$ a *critical point* of $f(x)$ if $f'(a) = 0$. Locating the critical points of f helps us to find the relative maxima and relative minima of f.

There is an analogous concept for functions of two variables. A point (a, b) is called a **critical point** of $f(x, y)$ if both

$$f_x(a, b) = 0 \quad \text{and} \quad f_y(a, b) = 0$$

To be technically correct, we should also include as critical points those points where the partial derivatives do not exist, but in this textbook we do not consider points of nondifferentiability.

It should be evident from Figure 3 that if $f(x, y)$ has a relative maximum or relative minimum at (a, b) and if f has partial derivatives at that point, then

$$f_x(a, b) = 0 \quad \text{and} \quad f_y(a, b) = 0$$

FIGURE 3

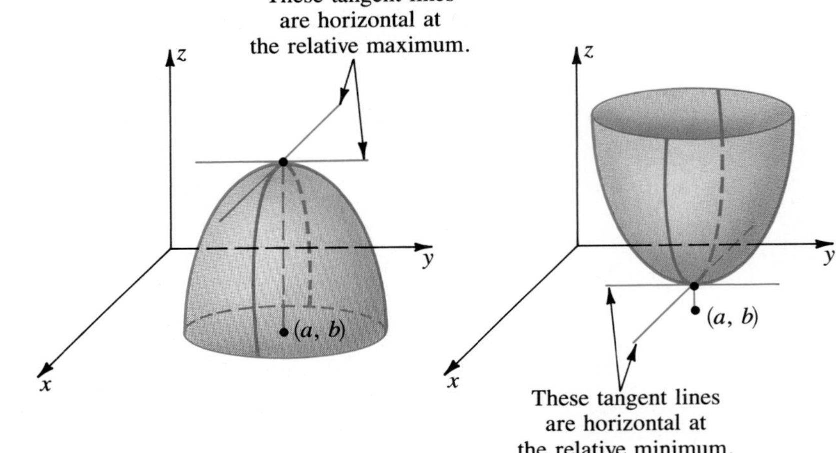
These tangent lines
are horizontal at
the relative maximum.
z
y
x
(a, b)
z
y
x
(a, b)
These tangent lines
are horizontal at
the relative minimum.

In other words, (a, b) is a critical point of f. Thus, we have the following rule:

Finding Relative Maxima and Minima of *f*(*x*, *y*) The relative maxima and relative minima of a function $f(x, y)$ with partial derivatives must occur at critical points.

EXAMPLE 1 Find the critical points of

$$f(x, y) = 2x^2 + xy + y^2 + 3y - 2x$$

Solution

$$f_x(x, y) = 4x + y - 2$$
$$f_y(x, y) = x + 2y + 3$$

To find the critical points we must solve the system of equations

$$\begin{aligned} 4x + y - 2 &= 0 \\ x + 2y + 3 &= 0 \end{aligned} \quad \text{or} \quad \begin{aligned} 4x + y &= 2 \\ x + 2y &= -3 \end{aligned}$$

This system of *linear equations* can be solved by the methods discussed in Section 1.4 to obtain $x = 1$ and $y = -2$. Thus, $(1, -2)$ is the only critical point of f.

EXAMPLE 2 Find the critical points of

$$f(x, y) = 2xy + x^2 - y^3$$

Solution

$$\begin{aligned} f_x(x, y) &= 2y + 2x \\ f_y(x, y) &= 2x - 3y^2 \end{aligned}$$

To find the critical points we must solve the system of equations

$$\begin{aligned} 2y + 2x &= 0 \\ 2x - 3y^2 &= 0 \end{aligned}$$

These are not both linear equations, so the methods of Section 1.4 do not apply. To solve them, we solve the first equation for y in terms of x, obtaining

$$y = -x \tag{1}$$

Then substituting this value of y in the second equation, we obtain

$$2x - 3x^2 = 0$$

or

$$x(2 - 3x) = 0$$

or $x = 0$ and $x = \frac{2}{3}$. From Equation (1) the corresponding y values are $y = 0$ and $y = -\frac{2}{3}$, which yields two critical points: $(0, 0)$ and $(\frac{2}{3}, -\frac{2}{3})$.

At a critical point (x_0, y_0) a function $f(x, y)$ *may* have a relative maximum or relative minimum but need not have either one. There are other possibilities. At a critical point (x_0, y_0) the function $f(x, y)$ may have a **saddle point**, as illustrated in Figure 4a, or it may have a graph as shown in Figure 4b.

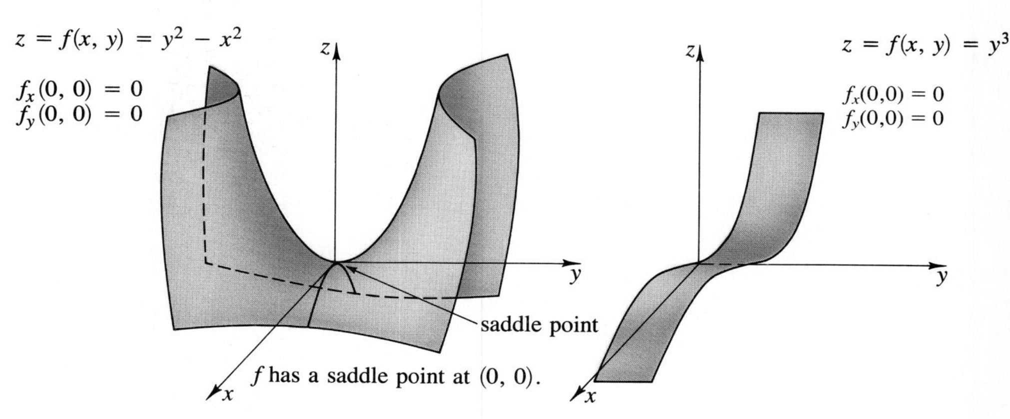

FIGURE 4

For functions of one variable, we can classify critical points as relative maxima or minima by using the second derivatives test. The analogous test for functions of two variables is more complicated.

Second Derivatives Test for Relative Maxima and Minima Let (a, b) be a critical point of $f(x, y)$ and let

$$M = f_{xx}(a, b) \cdot f_{yy}(a, b) - [f_{xy}(a, b)]^2 \tag{2}$$

then:

(a) If $M > 0$, then f has a relative maximum or relative minimum at (a, b). In this case:
  (i) If $f_{xx}(a, b) < 0$, then f has a relative maximum at (a, b).
  (ii) If $f_{xx}(a, b) > 0$, then f has a relative minimum at (a, b).

(b) If $M < 0$, then f has neither a relative maximum nor a relative minimum at (a, b).

(c) If $M = 0$, then the test does not apply. Another test must be used.

The proof or even an intuitive motivation for this test is beyond the scope of this textbook. However, we do give some examples.

EXAMPLE 3 In Example 1 we show that $(1, -2)$ is a critical point of

$$f(x, y) = 2x^2 + xy + y^2 + 3y - 2x$$

To apply the second derivatives test we must calculate

$$f_{xx}(1, -2) \qquad f_{yy}(1, -2) \qquad f_{xy}(1, -2)$$

But

$$f_x(x, y) = 4x + y - 2$$
$$f_y(x, y) = x + 2y + 3$$

and

$$f_{xx}(x, y) = 4 \qquad \text{so } f_{xx}(1, -2) = 4$$
$$f_{xy}(x, y) = 1 \qquad \text{so } f_{xy}(1, -2) = 1$$
$$f_{yy}(x, y) = 2 \qquad \text{so } f_{yy}(1, -2) = 2$$

Thus,

$$M = f_{xx}(1, -2) f_{yy}(1, -2) - [f_{xy}(1, -2)]^2$$
$$= (4)(2) - (1)^2 = 7$$

Since $M > 0$ and $f_{xx}(1, -2) = 4 > 0$, if follows that $f(x, y)$ has a relative minimum at $(1, -2)$.

EXAMPLE 4 In Example 2 we show $(0, 0)$ to be a critical point of

$$f(x, y) = 2xy + x^2 - y^3$$

We leave it for you to show that

$$f_{xx}(x, y) = 2 \qquad f_{yy}(x, y) = -6y \qquad f_{xy}(x, y) = 2$$

Thus at the critical point (0, 0),

$$\begin{aligned} M &= f_{xx}(0, 0) f_{yy}(0, 0) - [f_{xy}(0, 0)]^2 \\ &= (2)(0) - (2)^2 \\ &= -4 \end{aligned}$$

Since $M < 0$, there is neither a relative maximum nor a relative minimum at (0, 0). We leave it for you to show that at the other critical point $(\frac{2}{3}, -\frac{2}{3})$, $f(x, y)$ has a relative minimum value. This surface has no relative maximum values.

Lines of Best Fit: The Method of Least Squares

Suppose that as the result of gathering data from an experiment, we have n "data points" $(x_1, y_1), (x_2, y_2), \ldots, (x_n, y_n)$. Suppose also that we make the assumption that the data "should" all fall on the same straight line. The nature of the experiment may preclude that from actually happening (see Section 1.5). In such cases we often want to obtain the line

$$y = mx + b \tag{3}$$

which "best" fits the data points. We have already seen one way of doing this in Section 1.5, using the principle of least squares (see Figures 4 and 5 of Section 1.5).

The **principle of least squares** suggests that the line that "best fits" the points $(x_1, y_1), (x_2, y_2), \ldots, (x_n, y_n)$ is the one that minimizes the sum of the squares of the vertical distances from the data points to the line.

As shown in Figure 5, for four points this line is one that minimizes

$$d_1^2 + d_2^2 + d_3^2 + d_4^2$$

FIGURE 5

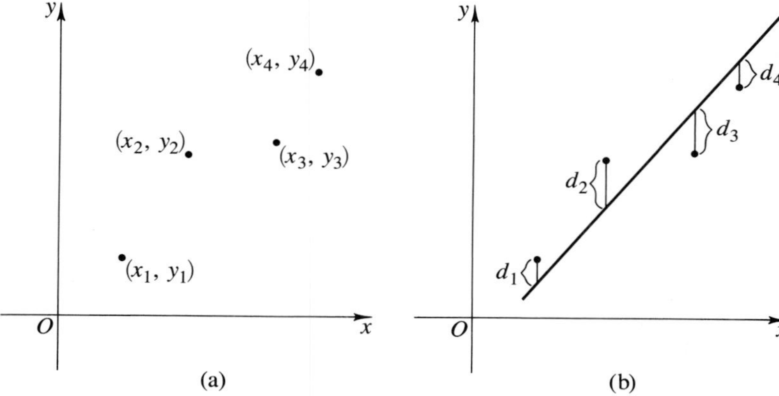

In Section 1.5, we state the formulas that provide the correct line, but we do not explain the logic behind them. With the aid of partial differentiation we now explain the formulas through a valid derivation. The derivation is quite general.

As an aid in understanding it, you should follow Examples 5 and 6 of Section 1.5 in parallel with the derivation.

Consider a general line $y = mx + b$, and for each point (x_i, y_i), let $d_i =$ the vertical distance from (x_i, y_i) to the line. As shown in Figure 5,

$$d_i = (mx_i + b) - y_i \tag{4}$$

for each $i = 1, 2, \ldots, n$. Let us define a function

$$F(m, b) = d_1^2 + d_2^2 + \cdots + d_n^2 \tag{5}$$

$$= (mx_1 + b - y_1)^2 + (mx_2 + b - y_2)^2 + \cdots + (mx_n + b - y_n)^2 \tag{6}$$

Our problem is to find those values of m and b that minimize $F(m, b)$. We begin by finding the critical points of $F(m, b)$. We find the partial derivatives of F and set them equal to 0:

$$F_m(m, b) = 2(mx_1 + b - y_1)x_1 + 2(mx_2 + b - y_2)x_2 + \cdots + 2(mx_n + b - y_n)x_n \tag{7}$$

and

$$F_b(m, b) = 2(mx_1 + b - y_1) + 2(mx_2 + b - y_2) + \cdots + 2(mx_n + b - y_n) \tag{8}$$

The critical points are found by solving $F_m(m, b) = 0$ and $F_b(m, b) = 0$ simultaneously. Setting $F_m(m, b) = 0$ yields

$$2(mx_1 + b - y_1)x_1 + 2(mx_2 + b - y_2)x_2 + \cdots + 2(mx_n + b - y_n)x_n = 0 \tag{9}$$

Setting $F_b(m, b) = 0$ yields

$$2(mx_1 + b - y_1) + 2(mx_2 + b - y_2) + \cdots + 2(mx_n + b - y_n) = 0 \tag{10}$$

To simplify Equations (9) and (10) we first divide both sides by 2 and then collect like terms:

$$\begin{aligned} A &= x_1 + x_2 + \cdots + x_n \\ B &= y_1 + y_2 + \cdots + y_n \\ C &= x_1^2 + x_2^2 + \cdots + x_n^2 \\ D &= x_1 y_1 + x_2 y_2 + \cdots + x_n y_n \end{aligned} \tag{11}$$

Then the critical points of F are found by solving the system of two linear equations in the two unknowns m and b [obtained from Equations (12) and (13)]:

$$m(x_1^2 + x_2^2 + \cdots + x_n^2) + b(x_1 + x_2 + \cdots + x_n) - (x_1 y_1 + x_2 y_2 + \cdots + x_n y_n) = 0 \tag{12}$$

$$m(x_1 + x_2 + \cdots + x_n) + b(1 + 1 + \cdots + 1) - (y_1 + y_2 + \cdots + y_n) = 0 \tag{13}$$

which can be written as

$$\begin{aligned} Cm + Ab &= D \\ Am + nb &= B \end{aligned} \tag{14}$$

When we solve this system simultaneously (see Exercise 29), we obtain

$$m = \frac{nD - AB}{nC - A^2} \tag{15}$$

and

$$b = \frac{CB - AD}{nC - A^2} \tag{16}$$

EXAMPLE 5 Consider the four data points $(-2, -1)$, $(-1, 1)$, $(1, 3)$, and $(3, 3)$. We compute the values of m and b for which the line $y = mx + b$ is the best least-squares fit to these data points as follows.

Step 1 Calculate d_1, d_2, d_3, and d_4 in terms of m and b, using Equation (4):

$$\begin{aligned} d_1 &= (-2m + b) + 1 \\ d_2 &= (-m + b) - 1 \\ d_3 &= (m + b) - 3 \\ d_4 &= (3m + b) - 3 \end{aligned}$$

Step 2 Calculate $F(m, b)$ using Equation (6):

$$F(m, b) = (-2m + b + 1)^2 + (-m + b - 1)^2 + (m + b - 3)^2 + (3m + b - 3)^2$$

Step 3 Find $F_m(m, b)$ and $F_b(m, b)$ as in Equations (7) and (8):

$$\begin{aligned} F_m(m, b) &= 2(-2m + b + 1)(-2) + 2(-m + b - 1)(-1) + 2(m + b - 3)(1) \\ &\quad + 2(3m + b - 3)(3) \\ &= 8m - 4b - 4 + 2m - 2b + 2 + 2m + 2b - 6 + 18m + 6b - 18 \\ &= 30m + 2b - 26 \\ F_b(m, b) &= 2(-2m + b + 1) + 2(-m + b - 1) + 2(m + b - 3) \\ &\quad + 2(3m + b - 3) \\ &= -4m + 2b + 2 - 2m + 2b - 2 + 2m + 2b - 6 + 6m + 2b - 6 \\ &= 2m + 8b - 12 \end{aligned}$$

Step 4 Find the critical points of $F(m, b)$. That is, solve

$$\begin{aligned} 30m + 2b - 26 &= 0 \\ 2m + 8b - 12 &= 0 \end{aligned}$$

Multiplying the first equation by -4 yields

$$\begin{aligned} -120m - 8b + 104 &= 0 \\ 2m + 8b - 12 &= 0 \end{aligned}$$

then adding the two equations, we obtain

$$\begin{aligned} -118m + 92 &= 0 \\ m &= \frac{46}{59} \end{aligned}$$

The first equation is equivalent to $2b = -30m + 26$, or $b = -15m + 13$, from which it follows that

$$b = -15\left(\frac{46}{59}\right) + 13$$

$$= \frac{77}{59}$$

Thus the only critical point of $F(m, b)$ is $(\frac{46}{59}, \frac{77}{59})$.

Step 5 To be certain that $(\frac{46}{59}, \frac{77}{59})$ minimizes F rather than maximizes it, we use the second derivatives test:

$$F_{mm} = 30 \qquad F_{mb} = 2 \qquad F_{bb} = 8$$

From Equation (2),

$$M = F_{mm}F_{bb} - (F_{mb})^2 = (30)(8) - 2^2 > 0$$

Since $F_{mm} > 0$ also, we conclude that F has a relative minimum at $(\frac{46}{59}, \frac{77}{59})$.

Step 6 Finally, the line of best least-squares fit is

$$y = \frac{46}{59}x + \frac{77}{59}$$

EXAMPLE 6 Use Equations (11), (15), and (16) to solve Example 5.

Solution The points are $(-2, -1)$, $(-1, 1)$, $(1, 3)$, and $(3, 3)$. From Table 1 we obtain the values for A, B, C, and D.

TABLE 1

x_i	y_i	x_i^2	x_iy_i
−2	−1	4	2
−1	1	1	−1
1	3	1	3
3	3	9	9
$A = 1$	$B = 6$	$C = 15$	$D = 13$

Equations (15) and (16) provide the values of m and b directly. Since $n = 4$,

$$m = \frac{4(13) - 1(6)}{4(15) - 1^2} = \frac{52 - 6}{60 - 1} = \frac{46}{59}$$

$$b = \frac{15(6) - 1(13)}{4(15) - 1^2} = \frac{90 - 13}{60 - 1} = \frac{77}{59}$$

These answers agree with those obtained in Example 5.

EXERCISE SET 19.3

In Exercises 1–6 find the critical points of f.

1. $f(x, y) = 2x^2 + 5xy - y^2$
2. $f(x, y) = x^2 + y^2 - 4x + 6y + 2$
3. $f(x, y) = 3x^2 + y^2 + 3x - 2y + 3$
4. $f(x, y) = x^2 + xy + 2y + 2x - 3$
5. $f(x, y) = x^3 + y^3 - 3xy$
6. $f(x, y) = x^3 + y^2 + xy$

In Exercises 7–16 find the relative maxima and minima, if any. Use the second derivatives test.

7. $f(x, y) = x^2 + xy + y^2 - 4x - 5y$
8. $f(x, y) = x^2 + 2x + y^2 + 6y + 8$
9. $f(x, y) = x^2 + y^2$
10. $f(x, y) = x^2 - y^2$
11. $f(x, y) = -2x^2 - 2xy - y^2 + 4x + 2y$
12. $f(x, y) = 2x^3 - 6xy + 6x + 3y^2 - 18y$
13. $f(x, y) = x^2 + y^2 - 3xy$
14. $f(x, y) = -x^2 + 6x - y^2 + 2y + 12$
15. $f(x, y) = x^3 + 2xy - y^3$
16. $f(x, y) = x^2 - 4xy + y^2 + 5$

17. **(Business)** A company manufactures two items, I and II, that sell for \$40 and \$60, respectively. The cost of producing x units of I and y units of II is

$$600 + 4x + 12y + 0.02(6x^2 + 6y^2)$$

Assuming that profit = revenue − cost, find the values of x and y that maximize the profit.

18. **(Business)** A department store's daily profit P (in dollars) depends on the number of salespeople x and on the number of departments y. The relationship is

$$P = 8000 - (4 - x)^2 - (8 - y)^2$$

(a) What values of x and y will maximize the profit?
(b) What is the maximum profit?

19. **(Business)** An electronics firm finds that if it spends x million dollars on research and y million dollars on development, then its total expenditure $E(x, y)$ (in millions of dollars) is given by

$$E(x, y) = 2000 + \frac{1}{2}x^2 - 50x + \frac{1}{2}y^3 - \frac{75}{2}y$$

How much money should be spent on research and how much on development to minimize the total expenditure?

20. Bob Jones finds that the profit $P(x, y)$ from selling x TV's and y computers each week is

$$P(x, y) = 4000 - 2x^2 + xy - y^2 + 49y$$

What values of x and y will maximize his profit?

21. Find the dimensions of a closed rectangular box of least surface area and having a volume of 50 cubic inches. [*Hint:* Let the dimensions be x, y, and z and begin by expressing V in terms of x and y alone.]

22. A rectangular chest of volume 22 cubic feet is to be constructed as follows. The top is to be made of material costing 80 cents per square foot, the ends and sides costing 40 cents per square foot, and the bottom 30 cents per square foot. Let x and y denote the length and width of the box. Find the cost function $C(x, y)$ and determine the values x and y that minimize the cost of the materials. What is the minimum cost? Also, what is the height of the box?

23. Consider the four data points (1, 3), (2, 1), (3, 4), and (4, 3). Use the methods of Example 5 to find the line $y = mx + b$, which is the best least-squares fit to these data points.
(a) Calculate d_1, d_2, d_3, and d_4 in terms of m and b, using Equation (4).
(b) Calculate $F(m, b)$, using Equation (6).
(c) Find $F_m(m, b)$ and $F_b(m, b)$, as in Equations (7) and (8).
(d) Find the critical point of $F(m, b)$.
(e) Use the second derivatives test to determine whether this critical point does indeed minimize $F(m, b)$.
(f) Write the desired equation $y = mx + b$.
(g) Compare your results with the answer to Example 5 of Section 1.5.

24. With the same four data points given in Exercise 23, use the method of Example 6 to obtain the line of best fit.

25. With the data points (−3, 2), (−1, 0), (1, 1), and (2, −2), use the method of Example 5 to find the line of best fit.

26. With the data points of Exercise 25, use the method of Example 6 to obtain the line of best fit.

27. With the data points $(-4, 2)$, $(-1, 3)$, $(0, 2)$, $(2, 3)$, and $(3, 4)$, use the method of Example 5 to find the line of best fit.

28. With the data points of Exercise 27, use the method of Example 6 to obtain the line of best fit.

29. Solve the system (14) and thereby derive Equations (15) and (16).

19.4 OPTIMIZATION USING LAGRANGE MULTIPLIERS

In this section we present an amazingly effective technique for maximizing or minimizing a function of two or three variables given an additional equation relating the variables. We can only present the "how to" of the method; the "why it works" is beyond the scope of this course. We will apply this technique only to rational functions in two or three variables.

The method of Lagrange multipliers is designed to handle problems known as "constrained optimization problems" for which a maximum or a minimum exists. We first summarize the form of these problems and then show how they may be solved.

Constrained Optimization Problem—Two Variables Find the points (x_0, y_0) that maximize or minimize the function

$$f(x, y)$$

subject to the constraint

$$g(x, y) = 0$$

EXAMPLE 1 Find the point (x_0, y_0) on the line $3x + 4y = 15$ for which $x^2 + y^2$ is minimum. In this example we want to minimize the function

$$f(x, y) = x^2 + y^2$$

subject to the constraint

$$3x + 4y - 15 = 0$$

Thus, in this example, $g(x, y) = 3x + 4y - 15$. (The solution is given in Example 3.)

Functions of Three Variables

There are many occasions in which we require functions of more than two variables. For example, consider a rectangular box with length l, width w, and height h (Figure 1).

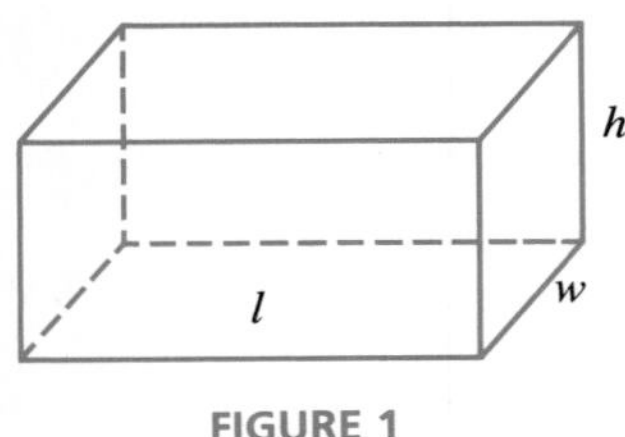

FIGURE 1

The volume of this box is a function of the three variables,

$$V = V(l, w, h) = lwh$$

Similarly, the surface area is a function of the three variables l, w, and h:

$$A = A(l, w, h) = 2lw + 2hw + 2lh$$

Each of the functions $V = V(l, w, h)$ and $A = A(l, w, h)$ given above, have three first partial derivatives:

$$\frac{\partial V}{\partial l} = wh \qquad \frac{\partial V}{\partial w} = lh \qquad \frac{\partial V}{\partial h} = lw$$

and

$$\frac{\partial A}{\partial l} = 2w + 2h \qquad \frac{\partial A}{\partial w} = 2l + 2h \qquad \frac{\partial A}{\partial h} = 2w + 2l$$

Constrained Optimization Problem—Three Variables Find the points (x_0, y_0, z_0) that maximize or minimize the function

$$f(x, y, z)$$

subject to the constraint

$$g(x, y, z) = 0$$

EXAMPLE 2 Find the dimensions of the closed rectangular box of largest volume that can be made of 150 square inches of (surface) material (Figure 1).

In Example 2 we want to maximize the function

$$f(l, w, h) = lwh$$

subject to the constraint

$$2lw + 2hw + 2lh = 150$$

Thus, in this example,

$$g(x, y, z) = 2lw + 2hw + 2lh - 150$$

(The solution is given in Example 4.)

The Method of Lagrange Multipliers

Method of Lagrange Multipliers—Two Variables If, among all points (x, y) satisfying the constraint $g(x, y) = 0$, the function $f(x, y)$ takes on its greatest or least value at (x_0, y_0), then there is some number λ such that

$$f_x(x_0, y_0) = \lambda g_x(x_0, y_0) \tag{1}$$

and

$$f_y(x_0, y_0) = \lambda g_y(x_0, y_0) \tag{2}$$

The number λ in Equations (1) and (2) is called a **Lagrange multiplier**.

EXAMPLE 3 **SOLUTION TO EXAMPLE 1** We want to minimize the function

$$f(x, y) = x^2 + y^2 \tag{3}$$

subject to the constraint

$$g(x, y) = 3x + 4y - 15 = 0 \tag{4}$$

Solution We begin by calculating partial derivatives:

$$f_x(x, y) = 2x \qquad g_x(x, y) = 3$$
$$f_y(x, y) = 2y \qquad g_y(x, y) = 4$$

By Lagrange's method, we seek a number λ such that Equations (1) and (2) are satisfied. By Equation (1)

$$2x = \lambda 3 \quad \text{or} \quad x = \frac{3}{2}\lambda \tag{5}$$

and by Equation (2)

$$2y = \lambda 4 \quad \text{or} \quad y = 2\lambda \tag{6}$$

Substituting (5) and (6) into (4) we see that we must have

$$3\left(\frac{3}{2}\lambda\right) + 4(2\lambda) - 15 = 0$$

$$\frac{9}{2}\lambda + 8\lambda - 15 = 0$$

$$\frac{25}{2}\lambda = 15$$

$$\lambda = \frac{6}{5} \tag{7}$$

Substituting (7) into (5) and (6) we obtain

$$x = \frac{3}{2}\lambda = \frac{3}{2}\left(\frac{6}{5}\right) = \frac{9}{5}$$

$$y = 2\lambda = 2\left(\frac{6}{5}\right) = \frac{12}{5}$$

Thus the only point on the line $3x + 4y = 15$ at which $f(x, y)$ can be maximum or minimum is $(\frac{9}{5}, \frac{12}{5})$.

Observe that $f(x, y) = x^2 + y^2$ cannot be maximum at $(\frac{9}{5}, \frac{12}{5})$ because larger values of x and y satisfying $g(x, y) = 0$ will make $f(x, y)$ larger. For example, $g(5, 0) = 0$ and $f(5, 0) > f(\frac{9}{5}, \frac{12}{5}) = 9$. It can be shown that the minimum value is $f(\frac{9}{5}, \frac{12}{5}) = 9$.

Method of Lagrange Multipliers—Three Variables If, among all points (x, y, z) satisfying the constraint $g(x, y, z) = 0$, the function $f(x, y, z)$ takes on its greatest or least value at (x_0, y_0, z_0), then there is some number λ such that

$$f_x(x_0, y_0, z_0) = \lambda g_x(x_0, y_0, z_0) \tag{8}$$
$$f_y(x_0, y_0, z_0) = \lambda g_y(x_0, y_0, z_0) \tag{9}$$
$$f_z(x_0, y_0, z_0) = \lambda g_z(x_0, y_0, z_0) \tag{10}$$

EXAMPLE 4 **SOLUTION TO EXAMPLE 2** We want to maximize the function

$$f(l, w, h) = lwh \tag{11}$$

subject to the constraint

$$2lw + 2hw + 2lh - 150 = 0$$

Solution We can simplify this constraint by dividing both sides by 2. We then have

$$g(l, w, h) = lw + hw + lh - 75 = 0 \tag{12}$$

Calculating partial derivatives, we have

$$\begin{aligned} f_l(l, w, h) &= wh & g_l(l, w, h) &= w + h \\ f_w(l, w, h) &= lh & g_w(l, w, h) &= l + h \\ f_h(l, w, h) &= lw & g_h(l, w, h) &= w + l \end{aligned}$$

By Lagrange's method, we seek a number λ such that

$$wh = \lambda(w + h) \tag{13}$$
$$lh = \lambda(l + h) \tag{14}$$
$$lw = \lambda(w + l) \tag{15}$$

Mutliplying both sides of (13) by l, and both sides of (14) by w, we get

$$lwh = l\lambda(w + h) \quad \text{and} \quad wlh = w\lambda(l + h) \tag{16}$$

The left sides of these equations are equal. Therefore,

$$\begin{aligned} l\lambda(w + h) &= w\lambda(l + h) \\ lh\lambda &= wh\lambda \end{aligned} \tag{17}$$

Since we want to maximize $f(l, w, h)$, we know from Equation (11) that $h \neq 0$ and we know from Equation (13) that $\lambda \neq 0$. Therefore we can cancel h and λ from both sides of (17), obtaining

$$l = w \tag{18}$$

When we work with Equations (14) and (15) in the same manner, we obtain

$$w = h \tag{19}$$

Equations (18) and (19) imply

$$l = w = h$$

and substituting into (12), we obtain

$$\begin{aligned} w^2 + w^2 + w^2 &= 75 \\ 3w^2 &= 75 \\ w^2 &= 25 \\ w &= 5 \text{ inches} \end{aligned} \tag{20}$$

Therefore, the volume is maximized by taking

$$l = w = h = 5 \text{ inches}$$

EXERCISE SET 19.4

In Exercises 1–8 use the method of Lagrange multipliers to find all points where the given function may have a maximum or minimum value, subject to the given constraint. Find the value of the function at the resulting points.

1. $f(x, y) = x^2 + y^2$, $x + 3y = 5$
2. $f(x, y) = 4x + 3y$, $x^2 + y^2 = 1$
3. $f(x, y) = xy$, $3x^2 + y^2 = 24$
4. $f(x, y) = xy^2$, $x^2 + y^2 = 3$
5. $f(x, y, z) = x + 2y + 3z$, $2x^2 + 2y^2 + 2z^2 = 7$
6. $f(x, y, z) = x^2 + y^2 + z^2$, $x + y + z = 15$
7. $f(x, y, z) = xyz$, $x^2 + y^2 + z^2 = 1$
8. $f(x, y, z) = 2x + y + 12z$, $xyz = 9$

In Exercises 9–16 use the method of Lagrange multipliers.

9. Find two numbers whose sum is 100 and whose product is maximum.
10. Find two numbers whose product is 100 and whose sum is minimum.
11. A piece of wire of length L is bent into the shape of a rectangle. Show that to enclose the greatest possible area, the rectangle must be a square.
12. Use the method of Lagrange multipliers to solve Exercise 21 of Section 19.3.
13. Find the point on the line $3x + 5y = 7$ for which $x^2 + y^2$ is minimum.
14. Find the point on the line $2x - y = 4$ for which $(x - 1)^2 + (y - 3)^2$ is minimum.
15. A manufacturer has an order of 400 units, which it will produce in its two factories, A and B. The total cost function (in thousands) for manufacturing a units in factory A and b units in factory B is

$$C(a, b) = 3a^2 + ab + 2b^2 + 10$$

To minimize the cost function, how many of the units should be manufactured in factory A and how many in factory B?

16. Find the dimensions of a closed rectangular box of surface area 54 square feet, whose volume is maximum.
17. The U.S. Postal Service has this restriction on mailing a fourth-class parcel: the girth (perimeter of one end) plus the length cannot exceed 108 inches. Find the maximum volume of a permissible rectangular parcel.
18. Find the dimensions of the *open* rectangular box of maximum volume that can be made from 192 square inches of material.

In Exercises 19–22 the function $U(x, y)$ or $U(x, y, z)$ is called a utility (or "satisfaction") function. Suppose a consumer has a monthly budget of B dollars to spend on x items each priced at p_x dollars and y items each priced at p_y dollars and, in the three-variable case, z items each priced at p_z dollars. The consumer wants to maximize the utility function (satisfaction) while staying within the budget. That is, the consumer wants to

maximize $U(x, y)$, subject to the condition
$xp_x + yp_y \le B$

or

maximize $U(x, y, z)$, subject to
$xp_x + yp_y + zp_z \le B$

Assume that such a maximum exists and determine how to allocate x and y (or x, y, and z) so that the maximum is achieved.

19. $U(x, y) = x^2y^2$; $p_x = \$3$, $p_y = \$4$, $B = \$240$
20. $U(x, y) = x^2 + y^2 - xy$; $p_x = \$5$, $p_y = \$8$, $B = \$267$
21. $U(x, y, z) = xy^2z$; $p_x = \$3$, $p_y = \$1$, $p_z = \$4$, $B = \$48$
22. $U(x, y, z) = xyz^{3/2}$; $p_x = \$2$, $p_y = \$4$, $p_z = \$3$, $B = \$70$

KEY IDEAS FOR REVIEW

- **Graph of a function $f(x, y)$** The graph of the equation $z = f(x, y)$.
- **Partial derivative of $z = f(x, y)$ with respect to x (denoted by $\partial z/\partial x$, $\partial f/\partial x$, or f_x)** The derivative of z with respect to x, with y treated as a constant.

- **Partial derivative of $z = f(x, y)$ with respect to y (denoted by $\partial z/\partial y$, $\partial f/\partial y$, or f_y)** The derivative of z with respect to y, with x treated as a constant.
- **Notation** $\left.\dfrac{\partial z}{\partial x}\right|_{(x_0, y_0)}$ [also written $f_x(x_0, y_0)$] The value of $\dfrac{\partial z}{\partial x} = f_x$ at (x_0, y_0).
- **Second-order partial derivatives**

$$\frac{\partial^2 z}{\partial x^2} = \frac{\partial}{\partial x}\left(\frac{\partial z}{\partial x}\right) \qquad \frac{\partial^2 z}{\partial y^2} = \frac{\partial}{\partial y}\left(\frac{\partial z}{\partial y}\right)$$

$$\frac{\partial^2 z}{\partial x\, \partial y} = \frac{\partial}{\partial x}\left(\frac{\partial z}{\partial y}\right) \qquad \frac{\partial^2 z}{\partial y\, \partial x} = \frac{\partial}{\partial y}\left(\frac{\partial z}{\partial x}\right)$$

- **Relative maximum** $f(x, y)$ has a relative maximum at (a, b) if $f(a, b)$ is the largest value for $f(x, y)$ in the immediate vicinity of (a, b).
- **Relative minimum** $f(x, y)$ has a relative minimum at (a, b) if $f(a, b)$ is the smallest value for $f(x, y)$ in the immediate vicinity of (a, b).
- **Absolute maximum** $f(x, y)$ has an absolute maximum at (a, b) if $f(a, b)$ is the largest value that $f(x, y)$ can have.
- **Absolute minimum** $f(x, y)$ has an absolute minimum at (a, b) if $f(a, b)$ is the smallest value that $f(x, y)$ can have.
- **Critical point of $f(x, y)$** A point (a, b) where $f_x(a, b) = 0$ and $f_y(a, b) = 0$.
- **Test for relative maxima and minima** See page 822.
- **Lines of best fit: The method of least squares** See page 823.
- **Method of Lagrange multipliers** See pages 829 and 830.

SUPPLEMENTARY EXERCISES

1. Let $f(x, y) = (x^2 - y^2)/(x^2 + y^2)$.
(a) For what point(s) (x, y) is $f(x, y)$ not defined?
(b) Find $f(1, 2)$, $f(-2, 1)$, and $f(3, 2)$.

2. Let $f(x, y) = x^2 - 2xy + y^3 + 18x$.
(a) Find f_x, f_y, f_{xx}, f_{yy}, f_{xy}, and f_{yx}.
(b) Evaluate $f_x(0, 1)$, $f_{xx}(0, 1)$, and $f_{xy}(0, 1)$.
(c) Find f_{xxy}.

Hint: $$f_{xxy} = \frac{\partial}{\partial y}\left(\frac{\partial}{\partial x}\left(\frac{\partial f}{\partial x}\right)\right)$$

3. Let $f(x, y) = x^{3/2} + xy^2 - x^2y + y^3$.
(a) Find f_x, f_y, f_{xx}, f_{yy}, f_{xy}, and f_{yx}.
(b) Evaluate $f_x(4, 1)$, $f_y(4, -1)$, and $f_{yx}(3, 2)$.
(c) Find f_{xyx}.

Hint: $$f_{xyx} = \frac{\partial}{\partial x}\left(\frac{\partial}{\partial y}\left(\frac{\partial f}{\partial x}\right)\right)$$

4. Find the critical points of

$$f(x, y) = 3x^3 + y^2 - 9x + 4y + 5$$

5. Find the critical points of

$$f(x, y) = x^2 + 3xy + 6y^2 + x - 3y + 2$$

6. Test the critical points obtained in Exercise 4 to determine if they yield relative extrema.

7. Test the critical points obtained in Exercise 5 to determine if they yield relative extrema.

In Exercises 8–11 find the relative extrema of f, if any exist.

8. $f(x, y) = 3x^2 + 2xy - y^2$

9. $f(x, y) = x^3 + y^3 - 3xy$

10. $f(x, y) = -x^3 - y^3 + 6xy$

11. $f(x, y) = x^3 + y^3 + 18xy$

12. A closed rectangular box having a surface area of 24 square meters is to be constructed. Find the dimensions that will yield a box of maximum volume.

13. When a corn farmer spends x hundred dollars for labor and y hundred dollars for fertilizers, the farm's profit (in thousands of dollars per acre) is given by

$$P(x, y) = 6x - x^2 - y^2 + 4y - 4$$

How much money should the farmer spend for labor and fertilizers to maximize the profit? What is the maximum profit?

14. **(Economics)** A manufacturer has a monopoly on two types of solar engines: A and B. Market research indicates that if x engines of model A are made, they can all be sold at $150 - 3x$ dollars each, and, if y engines of type B are made, they can all be sold at $160 - 4y$ dollars each. The total cost (in dollars) of manufacturing x engines of type A and y engines of type B is $46x + 30y + 2xy$. How many engines of each type should be made to maximize the profit? What is the maximum profit?

15. Find the line which is the best least-squares fit to the following data points: $(0, -3)$, $(1, -2)$, $(2, 3)$, $(2, 4)$, $(3, 3)$.

16. Find the line which is the best least-squares fit to the following data points: $(3, 2)$, $(4, 4)$, $(5, 6)$, $(6, 7)$, $(7, 8)$, $(8, 12)$.

In Exercises 17–20 use the method of Lagrange multipliers to find the extreme values.

17. Minimize $f(x, y) = x^2 + y^2$ subject to $x + y = 20$.

18. Maximize $f(x, y) = xy + 4$ subject to $x^2 + y^2 = 2$.

19. Find the extreme values of $x - y$ subject to $x^2 + 3y^2 = 3$.

20. Minimize $f(x, y, z) = x^2 + y^2 + z^2$ subject to $x + y + z = 8$.

CHAPTER TEST

1. Let $f(x, y) = x^2 - 2xy + y^2 - 2$. Find $f(-1, 2)$, $f(0, 1)$, and $f(-3, -2)$.

2. Let $f(x, y) = x^2 - 2xy + y^3 - 2$.
   (a) Find f_x, f_y, f_{xy}, and f_{yy}.
   (b) Find $f_x(-1, 2)$, $f_y(-1, 2)$, $f_{xy}(-1, 2)$, and $f_{yy}(-1, 2)$.

3. (a) Find the critical points of $f(x, y) = x^2 - 3xy + y^2 + 100$.
   (b) Test the critical points to determine whether they yield a relative maximum, relative minimum, or neither.

4. An automobile manufacturer determines that the profit P (in millions of dollars) is related to the advertising expenditures by

$$P(x, y) = -3x^2 - 2y^2 + x + 2y + 4xy + 150$$

where x is the amount (in millions of dollars) spent on television advertising and y is the amount (in millions of dollars) spent on magazine advertising. Find the values of x and y that maximize the profit, and determine the maximum profit.

5. Find the line $y = mx + b$ that is the best least-squares fit to the following data points:

$$(-1, 0) \qquad (1, 3) \qquad (2, 2) \qquad (3, 4)$$

6. Use the method of Lagrange's multipliers to find two numbers whose sum is 64 and whose product is a maximum.

A

Appendix A
Algebra Review

A.1 BASIC PROPERTIES OF THE REAL NUMBERS

You should recall the following rule for multiplying positive and negative real numbers.

> **Rule of Signs** The product of two numbers that have the same sign is positive; the product of two numbers that have opposite signs is negative.

Thus,

$$\begin{aligned} (2)(3) &= 6 \\ (-2)(-4) &= 8 \\ (-3)(5) &= -15 \\ (6)(-7) &= -42 \end{aligned}$$

Moreover, the negative, $-a$, of a number a is the product $(-1)a$. Thus,

$$-3 = (-1)(3)$$

This leads to important properties of signs, shown in Table 1.

TABLE 1

Property	Example
$-(-a) = a$	$-(-5) = 5$
$(-a)b = a(-b) = -(ab)$	$(-3)(4) = (3)(-4) = -(3 \cdot 4) = -12$
$(-a)(-b) = ab$	$(-4)(-5) = 4 \cdot 5 = 20$
$-(a + b) = (-a) + (-b) = -a - b$	$-(2 + 5) = (-2) + (-5) = -2 - 5 = -7$
$-(a - b) = (-a) + b$	$-(4 - 1) = (-4) + 1 = -3$
$\frac{-a}{b} = \frac{a}{-b} = -\frac{a}{b}$	$\frac{-3}{4} = \frac{3}{-4} = -\frac{3}{4}$

Table 2 summarizes the rules for adding, subtracting, multiplying, dividing, and simplifying fractions.

TABLE 2

Rule	Explanation	Examples
$\frac{ac}{bc} = \frac{a}{b}$	To simplify a fraction, cancel common factors from the numerator and denominator.	$\frac{6}{8} = \frac{3 \cdot 2}{4 \cdot 2} = \frac{3}{4}$; $\frac{34}{51} = \frac{2 \cdot 17}{3 \cdot 17} = \frac{2}{3}$
$\frac{a}{c} + \frac{b}{c} = \frac{a+b}{c}$ $\frac{a}{c} - \frac{b}{c} = \frac{a-b}{c}$	To add or subtract fractions with the same denominator, add or subtract their numerators, and keep the denominator the same.	$\frac{3}{5} + \frac{4}{5} = \frac{3+4}{5} = \frac{7}{5}$ $\frac{8}{15} - \frac{17}{15} = \frac{8-17}{15} = \frac{-9}{15} = -\frac{3}{5}$
$\frac{a}{b} + \frac{c}{d} = \frac{ad+cb}{bd}$ $\frac{a}{b} - \frac{c}{d} = \frac{ad-cb}{bd}$	To add or subtract fractions with different denominators, change them to fractions with the same denominator and then add or subtract.	$\frac{3}{4} + \frac{2}{3} = \frac{3 \cdot 3 + 2 \cdot 4}{12} = \frac{17}{12}$ $\frac{5}{6} - \frac{4}{3} = \frac{5 \cdot 3 - 4 \cdot 6}{18} = \frac{-9}{18} = -\frac{1}{2}$
$\frac{a}{b} \cdot \frac{c}{d} = \frac{ac}{bd}$	To multiply fractions, multiply their numerators and multiply their denominators.	$\frac{1}{2} \cdot \frac{2}{3} = \frac{2}{6} = \frac{1}{3}$ $-\frac{2}{4} \cdot \frac{4}{5} = -\frac{8}{20} = -\frac{2}{5}$
$\frac{a}{b} \div \frac{c}{d} = \frac{a}{b} \cdot \frac{d}{c} = \frac{ad}{bc}$	To divide one fraction by another, invert the divisor and multiply.	$\frac{5}{2} \div \frac{1}{2} = \frac{5}{2} \cdot \frac{2}{1} = 5$ $\frac{1}{2} \div \frac{2}{3} = \frac{1}{2} \cdot \frac{3}{2} = \frac{3}{4}$ $\frac{3}{7} \div -4 = \frac{3}{7} \cdot \frac{1}{-4} = -\frac{3}{28}$

EXAMPLE 1 Evaluate $\frac{3/2}{5/3}$.

Solution

$$\frac{3/2}{5/3} = \frac{3}{2} \div \frac{5}{3} = \frac{3}{2} \cdot \frac{3}{5} = \frac{9}{10}$$

Percentages

Percentages are merely an alternate way of expressing decimals. The expression ***n* percent** means $n/100$. Thus,

$$3\% = \frac{3}{100} = .03 \quad \text{and} \quad 7\tfrac{1}{4}\% = \frac{7.25}{100} = .0725$$

The following are some rules of conversion:

1. To write a percent as a decimal, move the decimal point two places to the left and remove the % sign.
2. To write a percent as a fraction, remove the percent sign and divide by 100.
3. To write a decimal as a percent, move the decimal point two places to the right and add the % sign.
4. To write a fraction as a percent, express the fraction as a decimal and follow rule 3.

EXAMPLE 2

$$8\tfrac{1}{2}\% = 8.5\% = .085$$

$$12.5\% = \frac{12.5}{100} = \frac{125}{1000} = \frac{1}{8}$$

$$48\% = \frac{48}{100} = \frac{12}{25}$$

$$2.7 = 270\%$$

$$.003 = .3\%$$

$$\frac{3}{8} = .375 = 37.5\%$$

EXAMPLE 3 Find 16% of 200.

Solution 16% of 200 is (.16) times 200, and (.16)(200) = 32.

EXAMPLE 4 If your mortgage payment is \$300 per month, what percentage of your \$800 per month take-home pay is spent on the mortgage?

Solution The percentage is

$$\frac{300}{800} = \frac{3}{8} = .375 = 37.5\%$$

EXAMPLE 5 If you purchase a \$1000 savings certificate whose annual yield is 8.17%, how much will the certificate be worth after 1 year?

Solution Interest would be 8.17% of \$1000 = (.0817)(1000) = \$81.70, so the certificate would be worth \$1000 + \$81.70 = \$1081.70.

EXERCISE SET A.1

In Exercises 1–28 perform the operations and simplify your answer.

1. $2 - (3 - 5)$
2. $5 - 3(-4)$
3. $6 + 7(-8 + 6)$
4. $(8 - 5)(4 - 7)$
5. $(12 - 8)/(13 - 5)$
6. $1 - [1 - (1 - 2)]$
7. $3 - [4(-1 - 3)]$
8. $-[3 - (11 - 20)(-2)]$
9. $\frac{1}{3} \cdot \frac{2}{3}$
10. $\frac{1}{3} + \frac{2}{3}$
11. $\frac{1}{3} \div \frac{2}{3}$
12. $\frac{1}{3} - \frac{2}{3}$
13. $\frac{1}{4} + \frac{1}{5}$
14. $\frac{1}{4} \cdot \frac{1}{5}$
15. $\frac{2}{5} \div 3$
16. $5\frac{1}{2} + \frac{1}{3}$
17. $\frac{1}{2} + \frac{3}{4}$
18. $\frac{2}{35} - \frac{5}{6}$
19. $\frac{2}{35} - \frac{5}{7}$
20. $\frac{2}{5} \div \frac{5}{7}$
21. $\frac{2}{5} \div \frac{7}{5}$
22. $(\frac{5}{2} \div \frac{2}{3}) \cdot \frac{1}{5}$
23. $\dfrac{\frac{2}{3}}{\frac{5}{4}}$
24. $\dfrac{\frac{2}{5}}{\frac{3}{10}}$
25. $3 + \dfrac{\frac{2}{3}}{2}$
26. $\dfrac{2 - \frac{1}{3}}{3 + \frac{1}{4}}$
27. $\dfrac{\frac{1}{2} - \frac{3}{8}}{\frac{1}{3} + \frac{1}{4}}$
28. $\dfrac{\frac{3}{5} - \frac{1}{10}}{1 + \frac{1}{2}}$
29. Write the following percentages as fractions, and reduce:
(a) 12% (b) 2% (c) .02%
(d) 90% (e) 225% (f) 38%

30. Write the following percentages as decimals:
(a) 18% (b) 7% (c) 100%
(d) 80% (e) .03% (f) 300%

31. Write the following decimals as percentages:
(a) .03 (b) .56 (c) 1.1
(d) .015 (e) 2.8 (f) .0003

32. Write the following fractions as percentages:
(a) $\frac{1}{5}$ (b) $\frac{3}{8}$ (c) $\frac{1}{25}$
(d) $\frac{9}{4}$ (e) $\frac{19}{200}$ (f) $\frac{5}{12}$

33. (a) Find 18% of 297.
(b) Find .03% of 7015.
(c) What percent of 18 is 9?
(d) What percent of 35 is 12?
(e) What percent of 9 is 18?

34. If you earn $40.00 per year interest on an $800.00 bank deposit, what is the interest rate?

35. Three people decide to rent a house together and contribute to the rent according to their income. It is decided that David will pay $\frac{1}{4}$ and Jim will pay $\frac{3}{7}$ of the rent. What fraction will the third person pay?

36. At the beginning of the first day of September, the price of common stock in the ABC Power Company was $21 per share. Over the next 30 days the price changed as follows: it went up $\frac{1}{4}$ (dollars) each day for the first 11 days, stayed the same for 15 days, dropped $\frac{1}{2}$ each day for the next 3 days, and went up $2\frac{1}{8}$ on the last day. What was the price of the stock at the close of September?

37. A merchant bought calculators for $12.00 each and she wants to sell them for 12.5% more than she paid. What will be the selling price of each calculator?

38. If you earn $13,500 per year and get a 7% raise, what will your new annual salary be?

A.2 ALGEBRAIC NOTATION AND THE BASIC RULES OF ALGEBRA

Algebra derives much of its power from the precision and brevity of its notation. How often have we heard that a picture is worth a thousand words! For example, the compact formula

$$A = LW \tag{1}$$

represents the area of a rectangle of length L and width W. An important feature of algebraic notation is its generality. Formula (1) is equally applicable for any length L and any width W.

In algebra we distinguish between two kinds of quantities, constants and variables. A **constant** is a quantity whose value does not change, and a **variable** is a quantity that can assume various real values. Frequently, but not always, variables are denoted by letters near the end of the alphabet, such as

$$u \quad v \quad w \quad x \quad y \quad z$$

and constants by letters near the beginning of the alphabet, such as

$$a \quad b \quad c \quad d \quad \ldots$$

To make more symbols available, constants and variables are sometimes denoted by capital letters or letters with subscripts. Thus,

$$a_1 \quad a_2 \quad a_3 \quad b_1 \quad b_4 \quad P_3 \quad P_0$$

are all regarded as different symbols.

Two constants that occur often in mathematical applications are π, which is approximately equal to 3.14159, and e, which is approximately equal to 2.71828.

EXAMPLE 1 The formula $C = 2\pi r$ relates the circumference C and radius r of a circle. The quantities C and r are variables since their values vary with the size of the circle, and 2π is a constant.

Table 1 shows the special symbols used to compare the relative sizes of quantities.

TABLE 1

Symbol	Meaning	Examples
$=$	is equal to	$8 + 2 = 4 + 6$; $x + 1 = 1 + x$
$\neq$	is not equal to	$5 \neq 7$; $x + 1 \neq x$
$\cong$	is approximately equal to	$\pi \cong 3.14$; $e \cong 2.718$
$>$	is greater than	$1 > 0$; $4 > 3$; $-1 > -5$
$\geq$	is greater than or equal to	$x^2 \geq 0$; $x \geq -3$
$<$	is less than	$4 < 5$; $-1 < 0$; $-10 < -7$
$\leq$	is less than or equal to	$x \leq 3$; $x + y \leq 5000$

EXAMPLE 2 To vote in a U.S. election, a citizen must be of age $x \geq 18$.

From previous work, you should be familiar with the following basic properties of real numbers:

(a) The commutative laws:

$$x + y = y + x \qquad xy = yx$$

(b) The associative laws:

$$x + (y + z) = (x + y) + z \qquad x(yz) = (xy)z$$

(c) The distributive laws:

$$\begin{aligned} x(y + z) &= xy + xz \\ (x + y)z &= xz + yz \\ x(y - z) &= xy - xz \\ (x - y)z &= xz - yz \end{aligned}$$

(d) Properties of 0:

$$x + 0 = x \qquad 0x = 0 \qquad x + (-x) = 0$$

(e) Properties of 1:

$$1x = x \qquad x\left(\frac{1}{x}\right) = 1 \quad \text{where } x \neq 0$$

(f) Cancellation properties:

If $x + y = x + z$, then $y = z$.
If $xy = xz$ and $x \neq 0$, then $y = z$.

(g) Nonnegative squares:

If x is any real number, then $x^2 \geq 0$.

To evaluate an algebraic expression such as

$$2(x - 3)(y + x) - \frac{y}{2(x + 1)} + y$$

for particular values of x and y, such as $x = 2$ and $y = 3$, proceed as follows:

Step	Example
To evaluate an algebraic expression for particular values of x and y:	$2(x-3)(y+x)-\dfrac{y}{2(x+1)}+y$ when $x=2$ and $y=3$
1. Substitute the given values.	**1.** $2(2-3)(2+3)-\dfrac{3}{2(2+1)}+3$
2. Perform all operations within parentheses.	**2.** $2(-1)(5)-\dfrac{3}{2(3)}+3$
3. Perform multiplication and division.	**3.** $-10-\dfrac{1}{2}+3$
4. Perform addition and subtraction.	**4.** $-\dfrac{15}{2}$

EXERCISE SET A.2

1. Given that $x = 2$, $y = -3$, and $z = 4$, find each of the following:

(a) $x(3y - z)$ (b) $5x - \dfrac{3}{y}$

(c) $\dfrac{z}{x} + 4xy$ (d) $\dfrac{1}{x} - \dfrac{1}{y}$

In Exercises 2–9 express each of the following in algebraic notation, using variables for the unknown numbers:

2. three fourths of a number

3. five less than six times a number

4. one half of the sum of three numbers

5. one number subtracted from three times another

6. 6% of a number

7. 8% of the amount by which a number exceeds 50

8. Area equals length times width.

9. The area of a circle is π times the square of its radius.

In Exercises 10–18 insert the proper symbol (>, <) between the numbers to make a true statement.

10. 5, 7 **11.** −2, −3 **12.** 0, 5

13. $\frac{1}{2}, \frac{1}{3}$ **14.** $-\frac{1}{2}, -\frac{1}{3}$ **15.** π, 2

16. $\frac{4}{5}, \frac{2}{3}$ **17.** $\frac{5}{3}, \frac{5}{4}$ **18.** $\frac{4}{9}, \frac{5}{12}$

In Exercises 19–32 determine whether the equation is true or false.

19. $2(x + y) = 2x + y$

20. $3 \cdot (xy) = (3x) \cdot (3y)$

21. $a/3 + b/3 = (a + b)/6$

22. $1/x + 1/y = 1/(x + y)$

23. $(4a + b)/12 = (a + b)/3$

24. $1 + 1/x = (x + 1)/x$

25. $x^2 + 2x = x(x + 1) + x$

26. $(x/2)^2 = x^2/2$

27. $(-x)^2 = x^2$

28. $(a \div b) \div c = a \div (b \div c)$

29. $a - (b + c) = (a - b) + c$

30. $x(y + z) = xy + xz$

31. $x(y - z) = yx - zx$

32. $(x - y)z = zx - zy$

A.3 SOLVING EQUATIONS; LINEAR EQUATIONS

If the value 2 is substituted for x in the equation

$$3x - 2 = 4$$

the two sides become equal. We say that $x = 2$ is a **solution** of the equation.

The following operations are helpful for finding the solutions of an equation:

(a) The same quantity may be added to or subtracted from both sides of an equation.

(b) Both sides of an equation may be multiplied (or divided) by the same *nonzero* quantity.

EXAMPLE 1 Solve for x:

$$\begin{aligned} 4x - 7 &= 5 \\ 4x &= 12 \qquad \text{(adding 7 to both sides)} \\ x &= 3 \qquad \text{(dividing both sides by 4)} \end{aligned}$$

We have solved for x.

EXAMPLE 2 Solve for x:

$$\begin{aligned} 2x - 7 &= 6x - 3 \\ 2x - 4 &= 6x \qquad \text{(adding 3 to both sides)} \\ -4 &= 4x \qquad \text{(subtracting } 2x \text{ from both sides)} \\ -1 &= x \qquad \text{(dividing both sides by 4)} \end{aligned}$$

We have solved for x.

After we have obtained what we believe to be a solution of an equation, we can check to see whether it really is a solution by substituting it directly into the original equation.

EXAMPLE 3 **EXAMPLE 1 REVISITED** To check that $x = 3$ is a solution of $4x - 7 = 5$, we substitute this value:

$$\begin{aligned} 4(3) - 7 &= 5 \\ 12 - 7 &= 5 \\ 5 &= 5 \end{aligned}$$

It checks.

For certain equations we are interested in solving for one variable in terms of another. For example, the equation

$$C = 2\pi r$$

gives the circumference C of a circle of radius r. If we want to express the radius of the circle in terms of the circumference, we divide $C = 2\pi r$ by 2π to obtain

$$r = \frac{C}{2\pi}$$

EXAMPLE 4 Solve for x in terms of y:

$$x = 4 + \frac{3x}{y}$$

$$x - \frac{3x}{y} = 4 \qquad \left(\text{subtracting } \frac{3x}{y} \text{ from both sides}\right)$$

$$x\left(1 - \frac{3}{y}\right) = 4 \qquad (\text{by the distributive law})$$

$$x\left(\frac{y-3}{y}\right) = 4 \qquad (\text{combining terms})$$

$$x = \frac{4y}{y-3} \qquad \left(\text{multiplying both sides by } \frac{y}{y-3}\right)$$

Linear Equations A **linear equation in x** is an equation that can be written in the form

$$ax + b = 0 \qquad (\text{where } a \neq 0)$$

Some examples are

$$\begin{aligned} 3x + 4 &= 0 \qquad (a = 3,\ b = 4) \\ 5x - 2 &= 0 \qquad (a = 5,\ b = -2) \\ 2x &= 3 \qquad (\text{rewrite as } 2x - 3 = 0) \end{aligned}$$

EXAMPLE 5 Solve $5x + 7 = 0$.

Solution

$$5x = -7 \qquad (\text{subtracting 7 from both sides})$$

$$x = -\frac{7}{5} \qquad (\text{dividing both sides by 5})$$

EXAMPLE 6 Solve $3x - 2(2x - 5) = 2(x + 3) - 8$.

Solution

$$\begin{aligned} 3x - 4x + 10 &= 2x + 6 - 8 \qquad (\text{distributive law}) \\ -x + 10 &= 2x - 2 \qquad (\text{simplified}) \\ 10 &= 3x - 2 \qquad (\text{added } x \text{ to both sides}) \\ 12 &= 3x \qquad (\text{added 2 to both sides}) \\ 4 &= x \qquad (\text{divided both sides by 3}) \end{aligned}$$

EXAMPLE 7 Solve

$$\frac{2x+5}{2} + \frac{x-4}{3} = 2$$

Solution To clear the equation of fractions, we multiply both sides by 6:

$$6\left(\frac{2x+5}{2}\right) + 6\left(\frac{x-4}{3}\right) = (6)(2)$$

$$\begin{aligned} 3(2x + 5) + 2(x - 4) &= 12 \\ 6x + 15 + 2x - 8 &= 12 \\ 8x + 7 &= 12 \\ 8x &= 5 \\ x &= \frac{5}{8} \end{aligned}$$

EXAMPLE 8 Solve

$$\frac{2x - 3}{x + 2} = \frac{5}{3}$$

Solution

$$\begin{aligned} 3(2x - 3) &= (x + 2)5 && \text{[multiplied both sides by } 3(x + 2)\text{]} \\ 6x - 9 &= 5x + 10 \\ x - 9 &= 10 && \text{(subtracted } 5x \text{ from both sides)} \\ x &= 19 \end{aligned}$$

EXERCISE SET A.3

In Exercises 1–12 find all solutions and check your answers.

1. $5x = -30$
2. $\frac{1}{3}x + 1 = 5$
3. $4x - 5 = 8x + 7$
4. $3(2x - 7) = 2x + 7$
5. $3 - 4(1 - 2x) = 7$
6. $2x - 3(2 + x) = 4(2x - 3) + 1$
7. $4x - 3(x + 2) = 6 + x$
8. $(5x + 4) - (3x - 7) = 1$
9. $1/x^2 = 9/x$, $[x^2 = x \cdot x]$
10. $\dfrac{1}{x - 3} = 5$
11. $\dfrac{5}{x + 1} = \dfrac{3}{x + 2}$
12. $3 - \dfrac{1}{x} = 6$

In Exercises 13–22 solve for x.

13. $xy = 1$
14. $xy = 2 + xy^2$
15. $Ax + By = C$
16. $PV = xRT$
17. $y = mx + b$
18. $y = 11x + 4$
19. $y - x = y + x$
20. $x/y = a/b$
21. $y^2 + 3xy - 4 = 0$
22. $s = \dfrac{a}{1 - x}$
23. Which of the following equations are linear?
    (a) $3x + 4 = 2$
    (b) $3x + 4 = x - 2$
    (c) $2x + 1 = x(2 - x)$
    (d) $\dfrac{x}{x + 1} + 2 = x - 3$
    (e) $\frac{2}{3}x + 1 = \frac{1}{2}x - 2$
    (f) $x^2 = 9$
    (g) $\dfrac{3}{x} + x = 2$
    (h) $x = 0$

In Exercises 24–38 solve for the variable that appears.

24. $2a + 1 - 6a = a - 3$
25. $10m + 4 - 2m = 3m + 7 + 6m$
26. $8 - (2x - 5) = 12 - 3x$

27. $\frac{1}{3} - \frac{x-1}{2} = 1$

28. $\frac{x+1}{3} - \frac{x}{4} = \frac{1}{2}$

29. $\frac{1}{3}x - \frac{5}{2} = \frac{3}{2}x + \frac{2}{3}$

30. $\frac{2}{5}x - \frac{3}{7} = \frac{1}{7}x - \frac{2}{5}$

31. $\frac{2a-3}{3} + \frac{5-3a}{7} = 4$

32. $5 - \frac{2c-3}{3} = \frac{c-2}{3}$

33. $\frac{3a}{10} - \frac{5}{2} = \frac{a}{6} - \frac{1}{2}$

34. $2.3x - 2.4 = 1.6 - 1.7x$

35. $2.5x - 13.5 = 3.7 - 1.8x$

36. $\frac{3-4x}{3} = \frac{9}{5} - \frac{2x-3}{5}$

37. $\frac{1}{2a} + \frac{1}{a} = \frac{1}{6}$

38. $\frac{x-1}{7-x} = \frac{3}{4}$

39. The formula for temperature conversion is

$$F = \frac{9}{5}C + 32$$

where F is degrees Fahrenheit and C is degrees Celsius.

(a) Solve this equation for C.
(b) What is the Fahrenheit temperature when the Celsius temperature is 10°?
(c) What is the Celsius temperature when the Fahrenheit temperature is 86°?
(d) Is there any temperature at which the Fahrenheit degrees equal the Celsius degrees? Use an equation to find out.

40. The formula for simple interest is $I = Prt$, where I is the interest earned by an investment of P dollars at an interest rate r per year for t years.
(a) Solve this equation for r.
(b) Solve this equation for t.
(c) If \$800 is invested at 7% per year for 6 years, find the interest earned.
(d) If an initial investment of \$2000 has grown in value to \$3020 while being invested at simple interest for 6 years, what is the interest rate per year?
(e) If you borrow \$4000 and agree to pay it back in one lump sum, together with 9% per year interest, how much do you have to pay back after 3 years and 6 months?

41. The formula for the total sum of money S that you will have after t years of investing P dollars at simple interest rate r per year is $S = P + Prt$.
(a) Solve this formula for P.
(b) Solve this formula for t.
(c) How much money must you invest today at 7% simple interest per year to have a total amount of \$1000 in 10 years?

A.4 SOLVING LINEAR INEQUALITIES

If the value 2 is substituted for x in the inequality

$$3x - 2 < 8 \tag{1}$$

we obtain $3(2) - 2 = 4 < 8$, which is a true statement. We say that $x = 2$ **satisfies** inequality (1). However, if 5 is substituted for x in (1), we obtain $3(5) - 2 = 13 \not< 8$. We say that $x = 5$ does **not satisfy** inequality (1).

In general, we can solve linear inequalities by proceeding as follows:

> **Solving Linear Inequalities** The same operations can be performed with inequalities as with equations, except that multiplication or division by a *negative* number reverses the inequality sign.

EXAMPLE 1 Solve for x: $4x - 7 \leq 5$.

Solution

$$\begin{aligned} 4x - 7 &\le 5 \\ 4x &\le 12 && \text{(added 7 to both sides)} \\ x &\le 3 && \text{(divided both sides by 4)} \end{aligned}$$

The solution is the set of all real numbers $x \le 3$.

EXAMPLE 2 Solve for x: $-4x - 7 \le 5$.

Solution

$$\begin{aligned} -4x - 7 &\le 5 \\ -4x &\le 12 && \text{(added 7 to both sides)} \\ x &\ge -3 && \text{(divided both sides by } -3) \end{aligned}$$

The solution set consists of all real numbers $x \ge -3$.

EXAMPLE 3 Solve for x: $2x + 7 > 6x + 3$.

Solution

$$\begin{aligned} 2x + 7 &> 6x + 3 \\ 2x + 4 &> 6x && \text{(added } -3 \text{ to both sides)} \\ 4 &> 4x && \text{(added } -2x \text{ to both sides)} \\ 1 &> x && \text{(divided both sides by 4)} \end{aligned}$$

The solution set consists of all $x > 1$.

EXERCISE SET A.4

In Exercises 1–20 solve the given inequality.

1. $x + 4 < 8$

2. $x - 3 \ge 2$

3. $2y < -1$

4. $2r + 5 < 9$

5. $\frac{1}{2}y - 2 \le 2$

6. $x + 5 < 4$

7. $x + 5 \ge -1$

8. $3x < 6$

9. $3x - 2 > 4$

10. $\frac{3}{2}x + 1 \ge 4$

11. $x + 3 < -3$

12. $2 < a + 3$

13. $2x \ge 0$

14. $3x - 1 \ge 2$

15. $3 \le 2x + 1$

16. $x - 2 \le 5$

17. $-5 > b - 3$

18. $-\frac{1}{2}y \ge 4$

19. $4x + 3 \le 11$

20. $4 \ge 3b - 2$

A.5 EXPONENTS AND RADICALS

Positive integral exponents indicate repeated multiplications:

$$3 \cdot 3 \cdot 3 \cdot 3 \cdot 3 = 3^5$$
$$7 \cdot 7 \cdot 7 \cdot 7 = 7^4$$

Here the numbers 5 and 4 are called **exponents** or **powers**; the 3 and 7 are called **bases**. Thus for every natural number n:

$$x^n = \underbrace{x \cdot x \cdot x \cdot \cdots \cdot x}_{n \text{ factors}} \qquad (1)$$

For example,

$$\begin{aligned} x^1 &= x \\ x^2 &= x \cdot x \\ x^3 &= x \cdot x \cdot x \\ &\vdots \\ x^7 &= x \cdot x \cdot x \cdot x \cdot x \cdot x \cdot x \\ &\vdots \end{aligned}$$

Observe that when we increase the exponent on x by 1, we are multiplying by x:

$$x^8 = x^7 \cdot x$$

Conversely, when we decrease the exponent on x by 1, we are dividing by x:

$$x^{14} = x^{15} \div x$$

This pattern suggests that we define a zero exponent by

$$x^0 = x^1 \div x^1 = 1$$

and negative integer exponents by

$$\begin{aligned} x^{-1} &= x^0 \div x = 1 \div x = \frac{1}{x} \\ x^{-2} &= x^{-1} \div x = \frac{1}{x} \div x = \frac{1}{x} \cdot \frac{1}{x} = \frac{1}{x^2} \\ x^{-3} &= x^{-2} \div x = \frac{1}{x^2} \div x = \frac{1}{x^2} \cdot \frac{1}{x} = \frac{1}{x^3} \\ &\vdots \end{aligned}$$

To summarize

$$x^0 = 1$$

$$x^{-n} = \frac{1}{x^n} \quad \text{if } n \text{ is a natural number, } x \neq 0 \qquad (2)$$

EXAMPLE 1

$$\begin{aligned} 5^3 \cdot 5^4 &= (5 \cdot 5 \cdot 5)(5 \cdot 5 \cdot 5 \cdot 5) = 5^7 \\ \pi^2 \cdot \pi^4 &= (\pi \cdot \pi)(\pi \cdot \pi \cdot \pi \cdot \pi) = \pi^6 \\ 4^5 \div 4^3 &= \frac{4 \cdot 4 \cdot 4 \cdot 4 \cdot 4}{4 \cdot 4 \cdot 4} = 4^2 \\ e^6 \div e^2 &= \frac{e \cdot e \cdot e \cdot e \cdot e \cdot e}{e \cdot e} = e^4 \end{aligned}$$

$$(4^3)^2 = 4^3 \cdot 4^3 = (4 \cdot 4 \cdot 4)(4 \cdot 4 \cdot 4) = 4^6$$
$$(2 \cdot 7)^3 = (2 \cdot 7)(2 \cdot 7)(2 \cdot 7) = (2 \cdot 2 \cdot 2) \cdot (7 \cdot 7 \cdot 7) = 2^3 \cdot 7^3$$
$$\left(\frac{2}{7}\right)^3 = \frac{2}{7} \cdot \frac{2}{7} \cdot \frac{2}{7} = \frac{2^3}{7^3}$$

Table 1 summarizes the results of integer exponents.

TABLE 1

Rule	Example
$x^m \cdot x^n = x^{m+n}$	$2^3 \cdot 2^{-5} = 2^{-2} = \frac{1}{2^2} = \frac{1}{4}$
$x^m \div x^n = x^{m-n},\ x \neq 0$	$2^4 \div 2^{-2} = 2^{4-(-2)} = 2^6 = 64$
$(x^m)^n = x^{mn}$	$(2^{-3})^2 = 2^{-6} = \frac{1}{2^6} = \frac{1}{64}$
$(xy)^m = x^m y^m$	$6^3 = (2 \cdot 3)^3 = 2^3 \cdot 3^3 = 8 \cdot 27 = 216$
$\left(\frac{x}{y}\right)^m = \frac{x^m}{y^m},\ y \neq 0$	$\left(\frac{3}{4}\right)^3 = \frac{3^3}{4^3} = \frac{27}{64}$

EXAMPLE 2

(a) $\dfrac{3^4 \cdot 7^3}{3^9 \cdot 7^2} = 3^{4-9}7^{3-2} = 3^{-5}7^1 = \dfrac{7}{3^5} = \dfrac{7}{243}$

(b) $\dfrac{x^{-3}x^{-4}}{x^5} = \dfrac{x^{-7}}{x^5} = x^{-7-5} = x^{-12} = \dfrac{1}{x^{12}}$

(c) $\dfrac{a^3b^5}{a^{-4}b^2} = a^{3-(-4)}b^{5-2} = a^7b^3$

(d) $\dfrac{(2x^2)^3(3y^4)^{-2}}{(5xy^{-2})^2(xy^3)} = \dfrac{2^3x^63^{-2}y^{-8}}{5^2x^2y^{-4}xy^3} = \dfrac{(8/9)x^6y^{-8}}{25x^3y^{-1}} = \dfrac{8}{225}x^3y^{-7} = \dfrac{8x^3}{225y^7}$

(e) $\left(\dfrac{x}{3y^2}\right)^{-4} = \left(\dfrac{3y^2}{x}\right)^4 = \dfrac{3^4y^8}{x^4} = \dfrac{81y^8}{x^4}$

Beware

$-x^2$ does not equal $(-x)^2$.

$(x + y)^n$ does not equal $x^n + y^n$ unless $n = 1$.

$3x^n$ does not equal $(3x)^n$ unless $n = 1$.

A number r is called a **square root** of the number a if

$$r^2 = a$$

and r is called a **cube root** of a if

$$r^3 = a$$

In general, if n is a positive integer, then r is called an nth root of a if

$$r^n = a$$

It can be shown that every positive number has two square roots, one root being the negative of the other. For example, 5 and -5 are square roots of 25 since

$$(5)^2 = 25 \qquad \text{and} \qquad (-5)^2 = 25$$

Negative numbers do not have real square roots. (Why?)

Every real number has exactly one real cube root. For example,

$$2 \text{ is a cube root of } 8 \text{ since } 2^3 = 8$$
$$-3 \text{ is a cube root of } -27 \text{ since } (-3)^3 = -27$$

The positive square root of a is denoted by the symbol $\sqrt{a}$, and the real cube root of a number a is denoted by $\sqrt[3]{a}$. Thus

$$\sqrt{25} = 5 \qquad \text{and} \qquad \sqrt[3]{-27} = -3$$

In general, a positive number a has exactly one positive nth root when n is even and exactly one real nth root when n is odd. In each case the root described is denoted by the symbol $\sqrt[n]{a}$, which is called a **radical**.

EXAMPLE 3 $\sqrt[4]{625} = 5$ (not -5 or ± 5)

$\sqrt[4]{-625}$ is not a real number

$\sqrt[3]{8} = 2$

$\sqrt[3]{-8} = -2$

$\sqrt{8}$ is not an integer, but since $2^2 = 4$ and $3^2 = 9$, $\sqrt{8}$ is some real number between 2 and 3

$\sqrt{-8}$ is not a real number

$\sqrt[5]{32} = 2$

Note $\sqrt[n]{0} = 0$ for every n.

If m and n are natural numbers and $x \geq 0$, then we define **rational exponents** by

$$x^{1/n} = \sqrt[n]{x}$$

and

$$x^{m/n} = (x^m)^{1/n} = \sqrt[n]{x^m}$$

or

$$x^{m/n} = (x^{1/n})^m = (\sqrt[n]{x})^m$$

It can be shown that the last two definitions are equivalent. If x is negative, $x^{1/n}$ is defined as above only if n is an odd integer. We define negative rational exponents by

$$x^{-(m/n)} = \frac{1}{x^{m/n}} \qquad (x \neq 0)$$

It can be proved that the rules of exponents given in Table 1 remain valid when the exponents are allowed to be arbitrary rational numbers.

EXAMPLE 4

$$16^{1/2} = \sqrt{16} = 4$$

$$8^{2/3} = (\sqrt[3]{8})^2 = 2^2 = 4 \quad \text{or} \quad 8^{2/3} = \sqrt[3]{8^2} = \sqrt[3]{64} = 4$$

$$9^{-3/2} = \frac{1}{9^{3/2}} = \frac{1}{(\sqrt{9})^3} = \frac{1}{3^3} = \frac{1}{27}$$

$$\sqrt{5}\,\sqrt[3]{5} = 5^{1/2}5^{1/3} = 5^{(1/2)+(1/3)} = 5^{5/6} = \sqrt[6]{5^5} = \sqrt[6]{3125}$$

EXAMPLE 5

$$\left(\frac{x^{-4}}{4y^6}\right)^{-1/2} = \frac{x^2}{4^{-1/2}y^{-3}} = 4^{1/2}x^2y^3 = 2x^2y^3$$

$$\frac{(125x^4)^{2/3}}{xy^2} = \frac{(5^3)^{2/3}(x^4)^{2/3}}{xy^2} = \frac{5^2x^{(8/3)-1}}{y^2} = \frac{25x^{5/3}}{y^2}$$

EXAMPLE 6 Simplify each expression, expressing the answer in terms of positive exponents only.

(a) $\left(\frac{9a^{1/3}}{a^{1/2}}\right)^{1/2}$ (b) $(3x^{1/3}y^{-4/3})^3$ (c) $(6S^{1/5})(2S^{-2/3})$

Solution

(a) $\left(\frac{9a^{1/3}}{a^{1/2}}\right)^{1/2} = \frac{9^{1/2}a^{1/6}}{a^{1/4}} = 3a^{(1/6)-(1/4)} = 3a^{-1/12} = \frac{3}{a^{1/12}}$

Solution

(b) $(3x^{1/3}y^{-4/3})^3 = 27x^1y^{-4} = \frac{27x}{y^4}$

Solution

(c) $(6S^{1/5})(2S^{-2/3}) = 12S^{(1/5)+(-2/3)} = 12S^{-7/15} = \frac{12}{S^{7/15}}$

So far, we have defined expressions of the form x^k, where k is a rational number. For expressions like

$$x^{\pi} \qquad x^{\sqrt{2}} \qquad x^{-\sqrt{5}}$$

where k is irrational, the precise definition requires calculus. However, for practical purposes an irrational exponent can always be approximated to any degree of accuracy by a rational exponent. For example, in an expression like

$$1024^{\sqrt{2}}$$

we might approximate $\sqrt{2}$ as

$$\sqrt{2} \cong 1.4 = \frac{14}{10}$$

to obtain

$$1024^{\sqrt{2}} \cong (1024)^{14/10} = (\sqrt[10]{1024})^{14} = 2^{14} = 16{,}384$$

The accuracy of this approximation can be improved by using more decimal places in the approximation of $\sqrt{2}$.

It can be shown that the rules of rational exponents listed in Table 1 are also valid for irrational exponents.

In expressions such as $2^{(x-1)^2/3}$ the exponent of the base 2 is $(x - 1)^2/3$. Such an expression is evaluated by first evaluating the exponent $(x - 1)^2/3$ and then raising 2 to that power.

EXAMPLE 7 Evaluate $2^{(x-1)^2/3}$ when $x = 4$.

Solution We first evaluate $(x - 1)^2/3$ when $x = 4$:

$$2^{(4-1)^2/3} = 2^{3^2/3} = 2^3 = 8$$

EXAMPLE 8 Evaluate e^{x^2} when $x = 2$.

Solution

$$e^{x2} = e^{2^2} = e^4 \cong (2.71828)^4 \cong 54.598$$

EXERCISE SET A.5

In Exercises 1–15 evaluate the expressions.

1. $(\frac{1}{2})^4$
2. 3^{-3}
3. $(-2)^3$
4. $(-2)^4$
5. $-(2^4)$
6. $1/2^{-5}$
7. $(-\frac{1}{3})^{-2}$
8. $(2^{10}2^{-6})/4^2$
9. $9^3 3^{-4} 3^0$
10. $\dfrac{3^3 4^2}{3^5 4^{-2}}$
11. $27^{2/3}$
12. $16^{-3/4}$
13. $(\frac{1}{8})^{1/3}$
14. $(-1/125)^{2/3}$
15. $(0.01)^{-5/2}$

In Exercises 16–21 evaluate the expression, if possible.

16. $\sqrt[3]{64}$
17. $\sqrt[5]{-32}$
18. $\sqrt[4]{-16}$
19. $-\sqrt[4]{16}$
20. $\sqrt[4]{81}$
21. $\sqrt[3]{-1000}$

In Exercises 22–25 find two consecutive integers so that the value of the expression is between them.

22. $\sqrt{7}$
23. $\sqrt{21}$
24. $\sqrt[4]{50}$
25. $\sqrt[3]{-5}$

26. Use the rules of exponents to prove the following properties of radicals:

$$\sqrt[n]{xy} = \sqrt[n]{x} \cdot \sqrt[n]{y}$$

$$\sqrt[n]{\frac{x}{y}} = \frac{\sqrt[n]{x}}{\sqrt[n]{y}}$$

$$\sqrt[m]{\sqrt[n]{x}} = \sqrt[mn]{x}$$

27. Use the properties in Exercise 26 to simplify the expressions. Express your answers with radicals and positive integer exponents only.
    (a) $\sqrt[4]{32}$ (b) $\sqrt{6}/\sqrt{3}$
    (c) $\sqrt{x^6}$ (d) $\sqrt{9x^5y^7}$
    (e) $\sqrt[3]{81x^7y^{12}}$ (f) $\sqrt[20]{a^{40}}$

In Exercises 28–54 simplify and express the answers using positive exponents only.

28. $(3a^2b^{-3})(5a^4b^6)$

29. $\dfrac{36x^4y^3z^2}{-12x^6yz^{-1}}$

30. $(4x^2y^{-3})^{-2}$

31. $(2a^3b)^{-4}(8a/b)^3$

32. $\dfrac{(2x^2)^3(3y^4)^2}{(5xy)^3(x^2y^2)^5}$

33. $(2^{-2})^5(p^2)^{-1}$

34. $(2a^0b^{-1})(ab^2)^{-2}$

35. $(abc^4d^3xyz)^0(a/b)^3$

36. $[(2x^2y)^4(3xy^2)^{-5}]^{-2}$

37. $\left[\dfrac{(3x^3y)^5}{(4xy^7)^{-3}}\right]^4$

38. $a^{3/2}a^{-5/2}$

39. $(x^5)^{2/3}$

40. $\left(\dfrac{a^{2/3}}{a^{-1/3}}\right)^3$

41. $(x^{-3/2}\sqrt{y})^2$

42. $(x^6y^{15}z^{-9})^{2/3}$

43. $\sqrt{x}\sqrt[3]{x}\sqrt[4]{x}$

44. $\sqrt{a\sqrt[3]{a}}$

45. $(4a^2b^4)^{1/2}(8a^3b^2)^{1/3}$

46. $\left(\dfrac{x^{4/3}y^{-1/2}}{x^{2/3}y^{-3/2}}\right)^2$

47. $(8x\sqrt{x})^{2/3}$

48. $\left(\dfrac{64a^3b^7}{27bc^6}\right)^{-4/3}$

49. $(a^{3/5}b^{-7/10}c^{-1/2})^{20}c^{10}$

50. $\dfrac{\sqrt{7x}}{\sqrt[4]{49x}}$

51. $\dfrac{10\sqrt{x^5y^2}}{\sqrt{x^3y}}$

52. $\sqrt[3]{\dfrac{a^8}{24b^{10}}}$

53. $\sqrt[3]{x^{-6}y^{12}}$

54. $\sqrt[4]{\pi^2/e^5}$

55. $(3\pi^2/e)^{3/2}$

56. Evaluate $e^{(x-2)^2/4}$ when $x = 6$ (let $e \cong 2.71828$).

57. Evaluate $\pi^{-(x-3)^{2/3}}$ when $x = 6$ (let $\pi \cong 3.14159$).

58. By substituting nonzero values for x and y, show that:
(a) $(x + y)^2$ does not equal $x^2 + y^2$
(b) $\sqrt{x + y}$ does not equal $\sqrt{x} + \sqrt{y}$
(c) $\sqrt{x^2 + y^2}$ does not equal $x + y$

A.6 POLYNOMIALS AND FACTORING

A **polynomial in x** is an algebraic expression of the form

$$p(x) = a_nx^n + a_{n-1}x^{n-1} + \cdots + a_1x + a_0 \tag{1}$$

where $a_0, a_1, \ldots, a_n$ are constants and n is a nonnegative integer. The constants $a_0, a_1, a_2, \ldots, a_n$ are called the **coefficients** of the polynomial, and the expressions

$$a_0,\ a_1x,\ \ldots,\ a_{n-1}x^{n-1},\ a_nx^n$$

are called the **terms**. Note that a polynomial does not involve divisions by variables, fractional or negative exponents on variables, or variables inside radicals. The following expressions are *not* polynomials in x:

$$\frac{2}{x} \qquad x^{1/3} + 1 \qquad 5x^{-2} + 3x + 2 \qquad \sqrt{x^2} + 2x - 3$$

Note also, that constants are polynomials; this is the special case where the coefficients $a_n, a_{n-1}, \ldots, a_1$ are all zero in (1).

Polynomials are classified according to their degree. By definition the **degree** of a polynomial in x is the highest power of x with a nonzero coefficient. The constant zero is not assigned any degree.

In a similar manner a **polynomial in x and y** is an algebraic expression obtained by adding terms of the form ax^ry^s, where a is a constant and r and s are nonnegative integers. The degree of the term ax^ry^s, $a \neq 0$, is $r + s$; therefore, the degree of $-3x^4y^5$ is 9. The degree of a polynomial in x and y is the degree of the highest degree term that has a nonzero coefficient. For example, the polynomial

$$2x^2y^2 - 3x^3y^4 + xy^4 - x^2 + 2y$$

has degree 7. The terms cx^ry^s and dx^ty^u are said to be **like terms** if $r = t$ and $s = u$. Thus, $2x^2y^3$ and $-3x^2y^3$ are like terms.

When *adding* or *subtracting* polynomials, we are merely combining like terms. Two things should be remembered:

1. Only like terms can be combined.
2. To remove a pair of parentheses, *each* term inside the parentheses must be multiplied by the number outside the parentheses.

EXAMPLE 1

(a) $(4x + 2) + (6x - 3) = 4x + 6x + 2 - 3 = 10x - 1$

(b) $(2x^2 + x - 1) - (x^3 + 4x^2 - 5) = 2x^2 + x - 1 - x^3 - 4x^2 + 5$
$= -x^3 - 2x^2 + x + 4$

(c) $(3xy - 4z) + (3x - xy + z) = 3xy - 4z + 3x - xy + z$
$= 2xy + 3x - 3z$

(d) $(x^3 - 2x^2 + x - 5) - 3(x^2 - 2x + 1) = x^3 - 2x^2 + x - 5 - 3x^2 + 6x - 3$
$= x^3 - 5x^2 + 7x - 8$

(e) $2x - 3y - [4x - 2(x - y)] = 2x - 3y - [4x - 2x + 2y]$
$= 2x - 3y - [2x + 2y]$
$= 2x - 3y - 2x - 2y$
$= -5y$

Beware

$2(x + 3)$ does not equal $2x + 3$.

$x - (y - 2)$ does not equal $x - y - 2$.

To *multiply* polynomials, we multiply each term of the first factor by each term of the second factor, and then simplify by combining like terms.

EXAMPLE 2

(a) $2x(3x^2 + 7) = 2x(3x^2) + 2x(7) = 6x^3 + 14x$

(b) $(x^3 + 2x^2)(4x^2 - 2) = (x^3)(4x^2) + x^3(-2) + (2x^2)(4x^2) + (2x^2)(-2)$
$= 4x^5 + 8x^4 - 2x^3 - 4x^2$

(c) $(x + y)(x - y) = x^2 - xy + yx - y^2 = x^2 - y^2$

(d) $(3a - 5b)(a + 2b) = (3a)(a) + (-5b)(a) + (3a)(2b) + (-5b)(2b)$
$= 3a^2 - 5ab + 6ab - 10b^2$
$= 3a^2 + ab - 10b^2$

Sometimes multiplication is best done by placing one polynomial under the other, then multiplying each term of the bottom polynomial by each term of the top, aligning like terms, and adding. For example,

$$\begin{array}{r}
5x^3 + 4x^2 - x + 7 \\
x^2 - 3x + 4 \\
\hline
5x^5 + 4x^4 - x^3 + 7x^2 \qquad\qquad\quad \\
- 15x^4 - 12x^3 + 3x^2 - 21x \qquad \\
20x^3 + 16x^2 - 4x + 28 \\
\hline
5x^5 - 11x^4 + 7x^3 + 26x^2 - 25x + 28
\end{array}$$

This shows that

$$(5x^3 + 4x^2 - x + 7) \cdot (x^2 - 3x + 4) = 5x^5 - 11x^4 + 7x^3 + 26x^2 - 25x + 28$$

Factoring Polynomials

Factoring a polynomial is the process of expressing it as a product of polynomials of lower degree. When we write

$$x^2 + 3x + 2 = (x + 2)(x + 1)$$

we have factored $x^2 + 3x + 2$ into the **factors** $x + 2$ and $x + 1$. Factoring is an extremely helpful tool used often in simplifying complicated algebraic expressions.

It is natural to consider factoring in conjunction with multiplication since each is the reverse of the other. Table 1 shows some common products and their factorizations.

TABLE 1

Multiplying	Factoring
(a) *Removing parentheses* $3x^2(2x - 5) = 6x^3 - 15x^2$	(a) *Removing a common factor* $6x^3 - 15x^2 = 3x^2(2x - 5)$
(b) *Sum times difference* $(x + y)(x - y) = x^2 - y^2$	(b) *Difference of two squares* $x^2 - y^2 = (x + y)(x - y)$ $x^2 - 4 = (x + 2)(x - 2)$
(c) *Square of a binomial* $(x + y)^2 = x^2 + 2xy + y^2$ $(x - y)^2 = x^2 - 2xy + y^2$	(c) *Perfect square* $x^2 + 2xy + y^2 = (x + y)^2$ $x^2 + 10x + 25 = (x + 5)^2$ $x^2 - 2xy + y^2 = (x - y)^2$ $x^2 - 6x + 9 = (x - 3)^2$

EXAMPLE 3

(a) $3x^2y + 17xy = xy(3x + 17)$ (removing a common factor)

(b) $3a^3 - 12a = 3a(a^2 - 4)$ (removing a common factor)
$= 3a(a + 2)(a - 2)$ (difference of two squares)

(c) $x^2 + 6x + 9 = (x + 3)^2$ (perfect square)

The next example shows how to factor a second-degree polynomial

$$x^2 + bx + c$$

that is not a perfect square or a difference of two squares.

EXAMPLE 4 Factor $x^2 + 2x - 15$.

Solution The objective is to find values of d and e such that

$$x^2 + 2x - 15 = (x + d)(x + e)$$

Multiplying out the right side and collecting like terms yields

$$x^2 + 2x - 15 = x^2 + (d + e)x + de$$

Thus, d and e must be chosen so that $de = -15$ and $d + e = 2$. By *trial and error*, $d = -3$ and $e = 5$, so

$$x^2 + 2x - 15 = (x + d)(x + e) = (x - 3)(x + 5)$$

EXERCISE SET A.6

In Exercises 1–12 determine which expressions are polynomials and find the degree.

1. $8x^2y$

2. $\sqrt{3}\,x$

3. x^3y^2/z

4. $(z^2xy^2)z$

5. $\sqrt{7x^2y}$

6. $(14x^5y^8)^{3/2}$

7. $7x^2y - 3/x$

8. $8 + 9x - 7x^4$

9. 1597

10. $5^{2/3}xy^4 - 11z$

11. $\sqrt{x^2 + x^4 + 1}$

12. $x^3 + x^{-3}$

In Exercises 13–27 perform the indicated operations and simplify.

13. $(7x^2 - 11x + 5) - (3x^2 + 4x - 8)$

14. $(2x + 5)(x - 4)$

15. $(x^3 + x + 1) + 3x(2x + \frac{1}{3})$

16. $(xy + 4y^2)(x - 3y)$

17. $x(4x^3 - 7x^2 + 3) - (x + 1)(3x^2 + 7)$

18. $(a + b)^2$

19. $(a + b)^3$

20. $(a + b + c) + (2a - b)$

21. $(a + b + c)(2a - b)$

22. $(x^3 - 3x^2 + 7) - 2x[x^2 + 4x - 3(x^2 - 1)]$

23. $(\frac{1}{2}x - \frac{1}{3}y)(2x + 3y)$

24. $(x^2 + 3xy - 2y^2) + 2y(1 - 3x + 4y)$

25. $(x^3 - y^3)(x^3 + y^3)$

26. $(2x^3 + 3x^2 - 7)(x^2 - 5x + 2)$

27. $(x + y + z)(x + y - z)$

In Exercises 28–50 factor the given algebraic expression.

28. $9x^2 - 9x$

29. $4ab + 4b^3$

30. $2x(x - 1) + 3(x - 1)$

31. $1 + x + x^2 + x^3$

32. $x^2 + 14x + 49$

33. $x^2 - 8x + 16$

34. $x^2 - 1$

35. $b^2 - 25$

36. $x^2 - 8x - 33$

37. $x^2 + x - 12$

38. $x^2 - 11xy + 24y^2$

39. $x^2y - y^3$

40. $16a^2 - 9b^2$

41. $4a^2 + 4a - 3$

42. $x^2 - 13x - 30$

43. $2x^4 - 8x^2$

44. $x^4 - 13x^2 + 36$

45. $xy^3 - x^3y$

46. $6x^2 - 7x - 3$

47. $9x^2 + 30x + 25$

48. $x^4 - 81$

49. $x^2 - 2$

50. $4b - 4b^2 + b^3$

A.7 RATIONAL EXPRESSIONS

A **rational expression** is a quotient of two polynomials. Some examples are

$$\frac{3x + 1}{x^2 + 2x - 5} \qquad \frac{1}{x} \qquad x^3 + 2 = \frac{x^3 + 2}{1} \qquad \frac{xy + z}{x^2 - z^3}$$

Since the variables in a rational expression assume real values, the rules for real numbers in Table 2 of Section A.1 can be used to combine and simplify rational expressions.

Rules for Rational Expressions

1. To simplify a rational expression, cancel all common factors by dividing the numerator and denominator by the common factors.

EXAMPLE 1

(a) $\frac{14xy^2}{8xy} = \frac{2 \cdot 7 \cdot x \cdot y \cdot y}{2 \cdot 4 \cdot x \cdot y} = \frac{7y}{4}$

(b) $\frac{3x^2 + 2x - 1}{x + 1} = \frac{(3x - 1)(x + 1)}{x + 1} = \frac{3x - 1}{1} = 3x - 1$

(c) $\frac{5x^2 - 8x - 4}{x^2 - 4} = \frac{(x - 2)(5x + 2)}{(x + 2)(x - 2)} = \frac{5x + 2}{x + 2}$

(d) $\frac{4x + y}{2x^2}$ cannot be reduced. Although x is a factor of the denominator, it is not a factor of the *entire* numerator, so we cannot cancel x's.

2. To multiply rational expressions, multiply their numerators, multiply their denominators, and then cancel common factors.

EXAMPLE 2

(a) $\frac{x^2 + x}{x + 1} \cdot \frac{x^2 + 1}{x} = \frac{(x^2 + x)(x^2 + 1)}{x(x + 1)} = \frac{x(x + 1)(x^2 + 1)}{x(x + 1)} = x^2 + 1$

(b) $\frac{10x}{9y} \cdot \frac{4yz}{25x} = \frac{(10x)(4yz)}{(9y)(25x)} = \frac{8z}{45}$

(c) $\frac{3x^2 + 6x}{4y} \cdot \frac{y^3 - y^2}{x + 2} = \frac{3x(x + 2)y^2(y - 1)}{4y(x + 2)} = \frac{3xy(y - 1)}{4}$

3. To divide two rational expressions, invert the divisor and then multiply.

EXAMPLE 3

(a) $\frac{9a^2x^3}{6cy} \div \frac{12a^3c}{2xy} = \frac{9a^2x^3}{6cy} \cdot \frac{2xy}{12a^3c} = \frac{x^4}{4ac^2}$

(b) $\frac{x^2 + 8x + 15}{x + 2} \div \frac{x^2 - 25}{x^2 - 5x - 14} = \frac{x^2 + 8x + 15}{x + 2} \cdot \frac{x^2 - 5x - 14}{x^2 - 25}$

$$= \frac{(x+5)(x+3)}{x+2} \cdot \frac{(x-7)(x+2)}{(x+5)(x-5)}$$
$$= \frac{(x+3)(x-7)}{x-5}$$

4. To add or subtract rational expressions with the same denominator, add or subtract the numerators and keep the denominator the same.

EXAMPLE 4

(a) $\frac{x^2-1}{x+3} + \frac{3x^2+x-4}{x+3} = \frac{(x^2-1)+(3x^2+x-4)}{x+3}$
$= \frac{4x^2+x-5}{x+3}$

(b) $\frac{5xy+2}{x^2y} - \frac{xy-1}{x^2y} = \frac{(5xy+2)-(xy-1)}{x^2y} = \frac{4xy+3}{x^2y}$

5. Rational expressions A/B and C/D can be added and subtracted according to the formulas

$$\frac{A}{B} + \frac{C}{D} = \frac{AD+CB}{BD} \quad \text{and} \quad \frac{A}{B} - \frac{C}{D} = \frac{AD-CB}{BD} \tag{1}$$

In words, to add or subtract rational expressions with different denominators, convert them to expressions with the same denominator and then add or subtract.

EXAMPLE 5

(a) $\frac{x}{x+1} + \frac{3}{x} = \frac{x^2+3(x+1)}{(x+1)\cdot x} = \frac{x^2+3x+3}{x^2+x}$

(b) $\frac{5}{x+2} - \frac{4}{x-2} = \frac{5(x-2)-4(x+2)}{(x+2)(x-2)} = \frac{x-18}{x^2-4}$

When the denominators B and D in (1) have common factors, the common denominator BD is more complicated than necessary. The following example illustrates some alternative procedures for obtaining a common denominator in such cases.

EXAMPLE 6

(a) $\frac{1}{x} + \frac{3}{xy} = \frac{1}{x}\cdot\frac{y}{y} + \frac{3}{xy} = \frac{y}{xy} + \frac{3}{xy} = \frac{y+3}{xy}$

(b) $\frac{4a}{a-b} + \frac{b}{b-a} = \frac{4a}{a-b} + \frac{-b}{a-b} = \frac{4a-b}{a-b}$

(c) $\frac{x}{x+1} - \frac{3}{x^2+x} = \frac{x}{x+1}\cdot\frac{x}{x} - \frac{3}{x^2+x} = \frac{x^2}{x^2+x} - \frac{3}{x^2+x} = \frac{x^2-3}{x^2+x}$

(d) $\frac{a}{bc} + \frac{c}{ab} = \frac{a^2}{abc} + \frac{c^2}{abc} = \frac{a^2+c^2}{abc}$

(e) $$\frac{3}{x^2-1}-\frac{3}{x^2-2x+1}=\frac{3}{(x-1)(x+1)}-\frac{2}{(x-1)^2}$$
$$=\frac{3(x-1)}{(x-1)^2(x+1)}-\frac{2(x+1)}{(x-1)^2(x+1)}$$
$$=\frac{3(x-1)-2(x+1)}{(x-1)^2(x+1)}$$
$$=\frac{3x-3-2x-2}{(x-1)^2(x+1)}$$
$$=\frac{x-5}{(x-1)^2(x+1)}$$

We conclude this section with some warnings about common errors.

Don't Make These Common Mistakes

(a) $\frac{a+c}{b+c}$ does *not* equal $\frac{a}{b}$. The c's cannot be canceled because they are not factors of the numerator and denominator.

(b) $\frac{ax}{bx+cy}$ does *not* equal $\frac{a}{b+cy}$. The x in the denominator is not a factor of the whole denominator, and it cannot be canceled.

(c) $\frac{x}{a}+\frac{y}{b}$ does *not* equal $\frac{x+y}{a+b}$. The correct addition is

$$\frac{x}{a}+\frac{y}{b}=\frac{xb}{ab}+\frac{ya}{ab}=\frac{xb+ya}{ab}$$

(d) $\frac{a}{b+c}$ does *not* equal $\frac{a}{b}+\frac{a}{c}$. The correct addition is

$$\frac{a}{b}+\frac{a}{c}=\frac{ac+ab}{bc}$$

EXERCISE SET A.7

1. Which of the following are rational expressions?
   (a) $3x/5$ (b) $(4x-7)/(x^3+3x)$
   (c) $(3x^2-1)/\sqrt{x}$ (d) 489
   (e) $1/(x-2)$ (f) $\sqrt{(1+x)(1-x)}$

In Exercises 2–17 simplify each rational expression by canceling common factors.

2. $\frac{12ab}{4b^2}$

3. $\frac{xy^2z^3}{x^2y^3z^4}$

4. $\frac{x(x+y)}{x^2-y^2}$

5. $\frac{2x^2+2xy}{3x+3y}$

6. $\frac{4xy+y^2}{8x^2+xy}$

7. $\frac{3a^2+a}{12a+4}$

8. $\frac{x^2+2xy+y^2}{x^2-y^2}$

9. $\frac{x^2-9}{2x^2+6x}$

10. $\frac{x^2+4x+4}{x^2+x-2}$

11. $\frac{9-x^2}{x^2-6x+9}$

12. $\dfrac{5x^2 - 20x}{x^2 + x - 20}$

13. $\dfrac{2x^2 + 9x - 5}{x^2 + 7x + 10}$

14. $\dfrac{4x + 3}{4x^2 - x - 3}$

15. $\dfrac{2x^2 - x - 28}{4x^2 + 28x + 49}$

16. $\dfrac{16a^3 - a^3b^2}{a^3b + 4a^2b}$

17. $\dfrac{a^2 - 9}{a^4 - 81}$

18. True or false (for all values of a, b, and x)?

(a) $\dfrac{1}{x} + \dfrac{2}{3} = \dfrac{3}{x + 3}$

(b) $\dfrac{1}{x} + \dfrac{2}{3} = \dfrac{3 + 2x}{3x}$

(c) $\dfrac{a(a + b)}{a(a + b)^2} = \dfrac{0}{a + b} = 0$

(d) $\dfrac{x + 2}{x + 3} = \dfrac{2}{3}$

(e) $\dfrac{3x}{x + 4} = \dfrac{3}{4}$

(f) $\dfrac{4}{x - 2} = \dfrac{4}{x} - \dfrac{4}{2} = \dfrac{4}{x} - 2$

(g) $\dfrac{x - 2}{4} = \dfrac{x}{4} - \dfrac{2}{4} = \dfrac{x}{4} - \dfrac{1}{2}$

(h) $\dfrac{x^2 + 2x + 1}{x^2 + 2x + 3} = \dfrac{1}{3}$

(i) $\dfrac{x^2 - 1}{(x - 1)^2} = \dfrac{x^2 - 1}{x^2 - 1} = 1$

(j) $\dfrac{1}{x} - \dfrac{x - 3}{x} = \dfrac{4 - x}{x}$

In Exercises 19–57 perform the indicated operations and simplify. In your answers, you may leave the numerator and denominator in factored form.

19. $\dfrac{1}{x^2} \cdot \dfrac{x}{3}$

20. $\dfrac{1}{x^2} \cdot 3$

21. $\dfrac{y}{2x} \div \dfrac{y - 2}{4x^2}$

22. $\dfrac{5x^2}{2y^2} \cdot \dfrac{25y}{5x^7}$

23. $\left(\dfrac{2x}{3y}\right)^3 \div \left(\dfrac{y}{x}\right)^2$

24. $\dfrac{x + 5}{x^2 - 9} \cdot (x + 3)$

25. $\dfrac{x + 2}{4x^2} \div \dfrac{x + 2}{8x}$

26. $\dfrac{4a^2}{a + b} \cdot \dfrac{a^2 - b^2}{2}$

27. $\dfrac{x + 12}{x + 7} \div \dfrac{x^2 - 4}{x^2 - 49}$

28. $(x^2 - 4) \cdot \dfrac{2x - 3}{x - 2}$

29. $\dfrac{4x + 8y}{3x - 3y} \div \dfrac{2x + 4y}{x - y}$

30. $\dfrac{a^2 + a}{a^2} \div \dfrac{a^3 + a^2}{a^3}$

31. $\dfrac{x + 8}{x^2 - 16} \cdot \dfrac{x^2 - 2x - 8}{x^2 + 7x - 8}$

32. $\dfrac{x^2 + 2x - 8}{x^2 - 3x + 2} \cdot \dfrac{x^2 - 1}{x^2 + 5x + 4}$

33. $\dfrac{4x^2 + 4x + 1}{x - 3} \div \dfrac{2x + 1}{x^2 + 6x - 27}$

34. $\dfrac{x^2 + 3xy + 2y^2}{x^2 - 2xy} \cdot \dfrac{x^2 + xy - 6y^2}{x^2 + 4xy + 3y^2}$

35. $\dfrac{x}{y} + \dfrac{2}{x}$

36. $\dfrac{3}{x} + \dfrac{4}{x^2}$

37. $\dfrac{11x}{3y} - \dfrac{5x}{6y}$

38. $\dfrac{2}{x} - \dfrac{1}{5}$

39. $\dfrac{3x}{y} - 5$

40. $\dfrac{x}{5a^2} - \dfrac{2}{15a}$

41. $\dfrac{3}{x} - \dfrac{y}{x + 2}$

42. $\dfrac{1}{x + 3} - \dfrac{1}{x + 2}$

43. $\dfrac{5}{a + b} - \dfrac{2}{a - b}$

44. $\frac{x}{y^2} - \frac{1}{y} + \frac{x^3}{y^3}$

45. $\frac{1}{ab} + \frac{1}{ac} + \frac{1}{bc}$

46. $\frac{3}{2x - 3y} + \frac{2}{3y - 2x}$

47. $\frac{a}{a^2 - 4} - \frac{1}{a + 2}$

48. $\frac{3x}{x^2 - 1} - \frac{2x}{x^2 - 2x + 1}$

49. $\frac{x - 3}{9x^2} + \frac{3}{2x} - \frac{x - 1}{6x^3}$

50. $\frac{1}{x - 1} + 2 - \frac{1}{x + 1}$

51. $\frac{2x - 3}{x^2 - 8x + 15} + \frac{x - 1}{2x^2 - 5x - 3}$

52. $\frac{x - 3}{x^2 - 3x + 2} - \frac{x + 1}{x^2 + 2x - 3}$

53. $\frac{x^2}{x^2 + 2x + 1} + \frac{x}{6} - \frac{1}{3x + 3}$

54. $\left(\frac{1}{x} + \frac{1}{2}\right) \div \left(\frac{1}{x} - \frac{1}{2}\right)$

55. $\frac{1}{x - y}\left(\frac{x}{y} - \frac{y}{x}\right)$

56. $\left(x - \frac{1}{x}\right) \div (x + 1)$

57. $1 \div \left\{1 - \left[1 \div \left(1 + \frac{1}{x}\right)\right]\right\}$

A.8 QUADRATIC EQUATIONS

A **quadratic equation** in x is an equation that is expressible in the form

$$ax^2 + bx + c = 0 \tag{1}$$

where a, b, and c are real numbers, and $a \neq 0$. Some examples are

$$x^2 - 3x + 2 = 0$$
$$x^2 - 1 = 0$$
$$4x^2 = 4x - 1 \quad \text{(rewrite as } 4x^2 - 4x + 1 = 0\text{)}$$

If $b = 0$ in (1), there is no x term, and the equation can be solved by the methods illustrated in the next two examples.

EXAMPLE 1 Solve $3x^2 - 18 = 0$.

Solution We isolate x^2, obtaining

$$3x^2 = 18$$
$$x^2 = 6$$

Thus, we have the two solutions $x = \sqrt{6}$ and $x = -\sqrt{6}$.

EXAMPLE 2 Solve $x^2 + 9 = 0$.

Solution The equation can be rewritten as

$$x^2 = -9$$

which has no solutions in the real number system, since $x^2 \geq 0$ for all real values of x.

If $b \neq 0$ but the left side of the equation

$$ax^2 + bx + c = 0$$

can be factored easily, then this equation can be solved by the **method of factoring**. This technique is based on the property of real numbers, which states:

If $AB = 0$, then $A = 0$ or $B = 0$ (or both).

The converse is also true. That is, the product of two numbers is zero if and only if at least one of the numbers is zero.

To see how this result is used to solve quadratic equations, consider the equation

$$x^2 - 2x - 3 = 0$$

By factoring the left side, this equation can be rewritten as

$$(x - 3)(x + 1) = 0$$

Thus, the solutions are those values of x for which $x - 3 = 0$ or $x + 1 = 0$. This yields the two solutions $x = 3$ and $x = -1$.

EXAMPLE 3 Solve by factoring:

(a) $x^2 + 7x = 0$ (b) $x^2 - 5x + 6 = 0$ (c) $x^2 - 2x + 1 = 0$

Solution

(a) Factoring the left side of $x^2 + 7x = 0$ yields

$$x(x + 7) = 0$$

Thus, $x = 0$ or $x + 7 = 0$, so the solutions are $x = 0$ and $x = -7$.

Solution

(b) Factoring the left side of $x^2 - 5x + 6 = 0$ yields

$$(x - 2)(x - 3) = 0$$

Thus, $x - 2 = 0$ or $x - 3 = 0$, so the solutions are $x = 2$ and $x = 3$.

Solution

(c) Factoring the left side of $x^2 - 2x + 1 = 0$ yields

$$(x - 1)(x - 1) = 0$$

Thus, $x - 1 = 0$, so the only solution is $x = 1$.

EXAMPLE 4 Solve $2y^2 = 30 - 4y$.

Solution We begin by bringing all terms to the left side:

$$2y^2 + 4y - 30 = 0$$

Dividing by 2 and factoring we have

$$y^2 + 2y - 15 = 0$$
$$(y + 5)(y - 3) = 0$$

Thus, $y + 5 = 0$ or $y - 3 = 0$, which yields the solutions $y = -5$ and $y = 3$.

The Quadratic Formula The solutions of a quadratic equation

$$ax^2 + bx + c = 0$$

can *always* be obtained by the **quadratic formula**

$$x = \frac{-b \pm \sqrt{b^2 - 4ac}}{2a} \tag{2}$$

This formula is not "magic" but has a logical derivation from still another method of solving a quadratic equation called **completing the square**. Consult any standard algebra textbook for a discussion of this method and a subsequent derivation of the quadratic formula.

EXAMPLE 5 Solve by the quadratic formula:

(a) $x^2 - 7x + 12 = 0$ (b) $x^2 - 2x - 1 = 0$

(c) $x^2 - 6x = 5$ (d) $x^2 + x + 1 = 0$

Solution

(a) Substituting $a = 1$, $b = -7$, $c = 12$ in the quadratic formula

$$x = \frac{-b \pm \sqrt{b^2 - 4ac}}{2a}$$

we obtain

$$x = \frac{-(-7) \pm \sqrt{(-7)^2 - 4(1)(12)}}{2(1)} = \frac{7 \pm \sqrt{1}}{2} = \frac{7 \pm 1}{2}$$

Using the + yields the solution $x = 4$, and using the $-$ yields the solution $x = 3$.

Solution

(b) Substituting $a = 1$, $b = -2$, $c = -1$ in the quadratic formula yields

$$x = \frac{-(-2) \pm \sqrt{(-2)^2 - 4(1)(-1)}}{2(1)}$$
$$= \frac{2 \pm \sqrt{8}}{2} = \frac{2 \pm 2\sqrt{2}}{2}$$
$$= 1 \pm \sqrt{2}$$

Thus, the solutions are $x = 1 + \sqrt{2}$ and $x = 1 - \sqrt{2}$.

Solution

(c) We first rewrite the equation as

$$x^2 - 6x - 5 = 0$$

Substituting $a = 1$, $b = -6$, $c = -5$ in the quadratic formula yields

$$x = \frac{-(-6) \pm \sqrt{(-6)^2 - 4(1)(-5)}}{2(1)}$$

$$= \frac{6 \pm \sqrt{56}}{2} = \frac{6 \pm 2\sqrt{14}}{2}$$

$$= 3 \pm \sqrt{14}$$

Solution

(d) Substituting $a = 1$, $b = 1$, $c = 1$ in the quadratic formula yields

$$x = \frac{-1 \pm \sqrt{1^2 - 4(1)(1)}}{2}$$

$$= \frac{-1 \pm \sqrt{-3}}{2}$$

But $\sqrt{-3}$ is not a real number. Thus in the real number system the equation has no solution.

EXERCISE SET A.8

In Exercises 1–6 solve by the method of Example 1.

1. $x^2 - 4 = 0$ **2.** $2x^2 - 18 = 0$

3. $x^2 = 0$ **4.** $x^2 + 4 = 0$

5. $2x^2 - 14 = 0$ **6.** $3x^2 - 5 = 0$

In Exercises 7–22 solve by factoring.

7. $x^2 - 100 = 0$ **8.** $x^2 = 9$

9. $x^2 - 4x + 3 = 0$ **10.** $x^2 + 3x - 18 = 0$

11. $6y^2 + y - 12 = 0$ **12.** $6x - 5 = x^2$

13. $2y^2 = 3 - 5y$ **14.** $x^2 = 8 - 2x$

15. $6x^2 + 5x = 4$ **16.** $4x^2 - 49 = 0$

17. $2x^2 - 7x - 4 = 0$ **18.** $6x^2 = 7x + 5$

19. $6y^2 - y - 2 = 0$ **20.** $y^2 - 4y = 21$

21. $10y^2 - 3y = 1$ **22.** $6y^2 - 6y = 7y - 6$

In Exercises 23–32 solve by the quadratic formula.

23. $x^2 - 2x - 2 = 0$ **24.** $2x^2 + 2x - 1 = 0$

25. $x^2 + x - 1 = 0$ **26.** $2x^2 - 4x - 3 = 0$

27. $3x^2 + 4x = 4$ **28.** $2x^2 - 20x - 6 = 0$

29. $x^2 + 2x + 3 = 0$ **30.** $x^2 - 2x - 4 = 0$

31. $-6y = 3 - y^2$ **32.** $2y^2 = 3y - 2$

B

Appendix B Logic

Suppose your friend tells you that if she studies for the final exam, then she will pass the finite math course. Two weeks later you meet her in the bookstore and find out that she did indeed pass the course. Can you safely conclude that she did study for the exam?

The field of logic helps us to answer such questions. More specifically, **logic** is the science that provides rules for determining the truth or falsity of statements and deals with methods of reasoning. Logical reasoning is used in mathematics and science as well as in everyday life to solve a wide variety of problems. In this appendix we review the basic ideas of logic.

Statements

A **statement** or **proposition** is a declarative sentence that is either true or false, but not both.

EXAMPLE 1 The following are statements:

(a) $5 + 7 = 12$.

(b) London is the capital of England.

(c) It will rain in London tomorrow.

(d) Mars has no plant life.

(e) $5 + 7 = 13$.

Statements (a) and (b) are true, while statement (e) is false. The truth or falsity of statement (c) will be determined tomorrow. Statement (d) is either true or false, but not both, although at the time of this writing we are unable to ascertain its truth value. When an appropriate spacecraft is able to visit Mars, we will find out if the statement is true or false.

EXAMPLE 2 The following are not statements:

(a) Don't smoke here.

(b) Do you speak French?

(c) $4 + 2x = 5$.

Sentence (a) is a command, not a declarative sentence. Sentence (b) is a question, not a declarative sentence. While (c) is true for some values of x, such as $x = \frac{1}{2}$, it is false for all other values of x.

Logical Connectives and Compound Statements

Just as the letters $x, y, z, w, \ldots$ are used in mathematics to denote variables that can be replaced by real numbers, in logic we use the letters $p, q, r, \ldots$ to denote propositional variables, that is, variables that can be replaced by statements. For example, we can write

p: It is cool outside.

q: I will wear a sweater.

Statements or propositional variables can be combined by logical connectives to form new statements called **compound statements**. For example, we could combine the above statements into the following compound statement by using the connective *and:*

p *and* q: It is cool outside *and* I will wear a sweater.

The truth or falsity of a compound statement is determined by the truth or falsity of the statements being combined and by the type of connective used. We now define the following connectives: *not*, *and*, *or*, *if . . . then*, *if and only if*.

The **negation** of the statement p is the statement *not* p, denoted by $\sim p$. Strictly speaking, *not* is not a connective, since it does not join two statements to produce a compound statement. The statement *not* p is true when p is false, and it is false when p is true. This information is summarized in Table 1, called a **truth table**. More specifically, a truth table gives the truth values of a compound statement in terms of the possible truth values of the statements being combined.

TABLE 1

p	$\sim p$
T	F
F	T

Observe that Table 1 defines the connective *not*.

EXAMPLE 3 Give the negation of the following statements:

(a) p: It is raining.

(b) q: 3 is a positive number.

(c) r: $2 + 5 = 3$

Solution

(a) $\sim p$: It is *not* raining.

(b) $\sim q$: 3 is *not* a positive number.

(c) $\sim r$: $2 + 5 \neq 3$.

LOGIC; OR, THE ART OF THINKING

Describing his reading at the age of sixteen, Benjamin Franklin in his *Autobiography* recalled: "And I read about this Time Locke on Human Understanding, and the Art of Thinking by Messrs. du Port Royal." The work was said to have revolutionized the teaching of logic, and as the translator wrote in this preface "is so full of fine Reflections for the common Use of Life, and so differently handled from the Scholastical Manner, that it has been every where well received, and translated into all Languages." The authors illustrated their points with piquant examples, an explanatory method which Franklin later used most effectively.

In 1733 he gave the volume to the Library Company. On the title page is his signature, "B. Franklin's," in the smooth round handwriting of his youth. It may well be the earliest Franklin signature extant.

Printed for W. Taylor in Pater-Noster-Row

Given by B. Franklin

LOGIC;

OR, THE

Art of Thinking:

CONTAINING

(Besides the COMMON RULES)

MANY

New Observations,

That are of great Use in forming an Exactness of Judgment.

IN FOUR PARTS.

I. Consisting of Reflections upon the Ideas, or first Operation of the Mind.
II. Of the Reflections Men have made upon their Judgments.
III. Of Reasoning.
IV. Of Method; or the clearest Manner of demonstrating any Truth.

Done from the New *French* Edition.

By Mr. *OZELL.*

LONDON:

Printed for WILLIAM TAYLOR, at the *Ship* in *Pater-noster-row*, MDCCXVII.

TABLE 2

p	q	$p \wedge q$
T	T	T
T	F	F
F	T	F
F	F	F

The **conjunction** of the statements of p and q is the compound statement p *and* q, written as $p \wedge q$, where the connective *and* is denoted by $\wedge$. The conjunction of p and q is defined to be true when both p and q are true and false otherwise. The truth table in Table 2 gives the truth values of $p \wedge q$ in terms of the truth values of p and q.

EXAMPLE 4 Form the conjunction of p and q and use Table 2 to determine its truth or falsity.

(a) p: $3 + 2 = 5$.
q: $(7)(8) = 56$.

(b) p: Paris is the capital of England.
q: $2 < 3$.

Solution

(a) $p \wedge q$: $3 + 2 = 5$ *and* $(7)(8) = 56$. From Table 2, since both p and q are true, $p \wedge q$ is true.

(b) $p \wedge q$: Paris is the capital of England *and* $2 < 3$. From Table 2, since p is false and q is true, $p \wedge q$ is false.

TABLE 3

p	q	$p \vee q$
T	T	T
T	F	T
F	T	T
F	F	F

Observe, as shown in Example 4(b), that in logic we can form the conjunction of any two statements, whether or not they are related.

The **disjunction** of the statements p and q is the compound statement p *or* q, written as $p \vee q$, where the connective *or* is denoted by $\vee$. The truth table in Table 3 defines the truth values of $p \vee q$ in terms of the truth values of p and q. Observe that $p \vee q$ is false only when both p and q are false.

EXAMPLE 5 Form the disjunction of p and q and use Table 3 to determine its truth or falsity.

(a) p: $3 + 2 = 5$.
q: $(7)(8) = 56$.

(b) p: Paris is the capital of England.
q: $2 < 3$.

(c) p: $5 + 3 > 10$.
q: Boston is in Pennsylvania.

Solution

(a) $p \vee q$: $3 + 2 = 5$ *or* $(7)(8) = 56$. From Table 3, since both p and q are true, $p \vee q$ is true.

(b) $p \vee q$: Paris is the capital of England *or* $2 < 3$. From Table 3, since p is false and q is true, $p \vee q$ is true.

(c) $p \vee q$: $5 + 3 > 10$ *or* Boston is in Pennsylvania. From Table 3, since both p and q are false, $p \vee q$ is false.

In everyday language we would not be likely to form the disjunction of two unrelated statements. However, as shown in Examples 5(b) and (c), this can be done in logic. The connective *or* is used in two different ways in everyday English. When we make the true statement "I left my keys at work or I left them home," we have statements p: "I left my keys at work" and q: "I left them (the keys) home." In this case either p is true or q is true, both cannot be true. Thus, *or* is being used in an *exclusive* sense. On the other hand, when we say, "He will become rich or he will become famous," we again have a disjunction of two statements p: "He will become rich" and q: "He will become famous." In this case, one possibility could occur or *both* possibilities could occur. Thus, *or* is being used in an *inclusive* sense. In mathematics and science and in this book, *or* is always used in an inclusive sense unless stated otherwise.

If p and q are statements, the compound statement *if p then q*, written as $p \to q$, where the connective *if . . . then* is denoted by $\to$, is called a **conditional statement**. Statement p is called the **antecedent**, while statement q is called the **consequent**.

EXAMPLE 6 Write the conditional $p \to q$ in words.

(a) p: It is raining.
q: I will get wet.

(b) p: $2 + 3 = 6$.
q: It is raining.

Solution

(a) If it is raining, then I will get wet.

(b) If $2 + 3 = 6$, then it is raining.

TABLE 4

p	q	$p \to q$
T	T	T
T	F	F
F	T	T
F	F	T

Observe that in logic the conditional statement $p \to q$ can be formed when there is no cause-and-effect relationship between p and q, as in Example 6(b). However, in ordinary English a conditional statement is used only when there is a causal relationship between p and q. The truth table in Table 4 defines the truth values of $p \to q$ in terms of the truth values of p and q.

Observe that when the antecedent p is false, the conditional statement $p \to q$ is defined to be true no matter what the truth value of the consequent q. Thus, the conditional statement in Example 6(b), "If $2 + 3 = 6$, then it is raining," is true whether or not it is raining because the antecedent "$2 + 3 = 6$" is false.

EXAMPLE 7 Write the conditional statement $p \to q$ in words and determine its truth or falsity.

(a) p: $3 + 2 = 5$.
q: $(7)(8) = 56$.

(b) p: Paris is the capital of England.
q: $2 < 3$.

(c) p: $5 + 3 > 10$.
q: Boston is in Pennsylvania.

Solution

(a) $p \rightarrow q$: If $3 + 2 = 5$, then $(7)(8) = 56$. From Table 4, since both p and q are true, $p \rightarrow q$ is true.

(b) $p \rightarrow q$: If Paris is the capital of England, then $2 < 3$. From Table 4, since p is false and q is true, $p \rightarrow q$ is true.

(c) $p \rightarrow q$: If $5 + 3 > 10$, then Boston is in Pennsylvania. From Table 4, since both p and q are false, $p \rightarrow q$ is true.

In English the following are represented by $p \rightarrow q$:

q, if p

p only if q

p is a sufficient condition for q

q is a necessary condition for p

q, provided that p

With each conditional statement $p \rightarrow q$ we define three other conditional statements: the **converse**, $q \rightarrow p$; the **contrapositive**, $\sim q \rightarrow \sim p$; and the **inverse**, $\sim p \rightarrow \sim q$.

EXAMPLE 8 Give the converse, contrapositive, and inverse of the conditional statement

If I have money, then I will go to the movies.

Solution

Converse: If I will go to the movies, then I have money.

Contrapositive: If I will not go to the movies, then I do not have money.

Inverse: If I do not have money, then I will not go to the movies.

TABLE 5

p	q	$p \leftrightarrow q$
T	T	T
T	F	F
F	T	F
F	F	T

If p and q are statements, then the compound statement p *if and only if* q, written as $p \leftrightarrow q$, where the connective *if and only if* is denoted by $\leftrightarrow$, is called a **biconditional**. The truth table in Table 5 gives the truth values of $p \leftrightarrow q$

EXAMPLE 9 Give the biconditional $p \leftrightarrow q$ in words and determine its truth or falsity.

(a) p: $3 + 2 = 5$
q: $(7)(8) = 56$.

(b) p: Paris is the capital of England.
q: $2 < 3$.

(c) p: $5 + 3 > 10$.
q: Boston is in Pennsylvania.

Solution

(a) $p \leftrightarrow q$: $3 + 2 = 5$ if and only if $(7)(8) = 56$. From Table 5, since both p and q are true, $p \leftrightarrow q$ is true.

(b) $p \leftrightarrow q$: Paris is the capital of England if and only if $2 < 3$. From Table 5, since p is false and q is true, $p \leftrightarrow q$ is false.

(c) $p \leftrightarrow q$: $5 + 3 > 10$ if and only if Boston is in Pennsylvania. From Table 5, since both p and q are false, $p \leftrightarrow q$ is true.

In English the biconditional statement $p \leftrightarrow q$ is also stated as

p if and only if q

p is necessary and sufficient for q

Truth Tables

In general a compound statement may consist of component statements, each of which in turn is a compound statement. For example, the statement

$$(\sim p \vee q) \rightarrow (p \wedge \sim q) \tag{1}$$

is a compound statement whose component statements are $(\sim p \vee q)$ and $(p \wedge \sim q)$. Its truth or falsity depends upon the truth values of its propositional variables p and q. We use a truth table to determine the truth values of compound statements as follows.

For a compound statement with two propositional variables p and q, form a table whose first two columns are labeled p and q. In these two columns list the four possible truth values for p and q. The succeeding columns are labeled with the components of the given statements. The final column is labeled with the given statement. The truth values of these components are determined from Tables 1 to 5.

EXAMPLE 10 Determine the truth values of the compound statement.

(a) $\sim(p \wedge q)$

(b) $\sim(\sim p \vee q)$

Solution

(a) We construct Table 6.

TABLE 6

(1) p	(2) q	(3) $p \wedge q$	(4) $\sim(p \wedge q)$
T	T	T	F
T	F	F	T
F	T	F	T
F	F	F	T
		↑ Using columns 1 and 2 and Table 2	↑ Using column 3 and Table 1

The truth values of $\sim(p \wedge q)$ are shown in column 4.

Solution

(b) We construct Table 7.

TABLE 7

(1) p	(2) q	(3) $\sim p$	(4) $\sim p \vee q$	(5) $\sim(\sim p \vee q)$
T	T	F	T	F
T	F	F	F	T
F	T	T	T	F
F	F	T	T	F
		↑ Using column 1 and Table 1	↑ Using columns 3 and 2 and Table 3	↑ Using column 4 and Table 1

The truth values of $\sim(\sim p \vee q)$ are shown in column 5.

EXAMPLE 11 Determine the truth values of the compound statement

$$(\sim p \vee q) \rightarrow [(p \wedge \sim q) \vee p]$$

Solution We construct Table 8.

TABLE 8

(1) p	(2) q	(3) $\sim p$	(4) $\sim p \vee q$	(5) $\sim q$	(6) $p \wedge \sim q$	(7) $(p \wedge \sim q) \vee p$	(8) $(\sim p \vee q) \rightarrow [(p \wedge \sim q) \vee p]$
T	T	F	T	F	F	T	T
T	F	F	F	T	T	T	T
F	T	T	T	F	F	F	F
F	F	T	T	T	F	F	F
		↑ Using column 1 and Table 1	↑ Using columns 3 and 2 and Table 3	↑ Using column 2 and Table 1	↑ Using columns 1 and 5 and Table 2	↑ Using columns 6 and 1 and Table 3	↑ Using columns 4 and 7 and Table 4

The truth values of the given statement are shown in column 8.

A compound statement is a **tautology** if it is true regardless of the truth values of the propositional variables. A compound statement is a **contradiction** if it is false regardless of the truth values of the propositional variables.

EXAMPLE 12 Use truth tables to determine whether or not the following statements are tautologies, contradictions, or neither.

(a) $p \wedge \sim p$

(b) $p \vee \sim p$

(c) $(p \to q) \leftrightarrow (q \to p)$

(d) $\sim(p \wedge q) \leftrightarrow (\sim p \vee \sim q)$

Solution

(a) We construct Table 9.

TABLE 9

p	$\sim p$	$p \wedge \sim p$
T	F	F
F	T	F

Therefore, $p \wedge \sim p$ is a contradiction.

Solution

(b) We construct Table 10.

TABLE 10

p	$\sim p$	$p \vee \sim p$
T	F	T
F	T	T

Therefore, $p \vee \sim p$ is a tautology.

Solution

(c) We construct Table 11.

TABLE 11

p	q	$p \to q$	$q \to p$	$(p \to q) \leftrightarrow (q \to p)$
T	T	T	T	T
T	F	F	T	F
F	T	T	F	F
F	F	T	T	T

Therefore, $(p \to q) \leftrightarrow (q \to p)$ is neither a tautology nor a contradiction.

Solution

(d) We construct Table 12.

TABLE 12

p	q	$p \wedge q$	$\sim(p \wedge q)$	$\sim p$	$\sim q$	$\sim p \vee \sim q$	$\sim(p \wedge q) \leftrightarrow (\sim p \vee \sim q)$
T	T	T	F	F	F	F	T
T	F	F	T	F	T	T	T
F	T	F	T	T	F	T	T
F	F	F	T	T	T	T	T

Therefore, $\sim(p \wedge q) \leftrightarrow (\sim p \vee \sim q)$ is a tautology.

Table 13 gives a list of common tautologies.

TABLE 13

Some common tautologies	
$p \vee \sim p$	
$\sim(\sim p) \leftrightarrow p$	
$(p \wedge p) \leftrightarrow p$	
$(p \vee p) \leftrightarrow p$	
$(p \wedge q) \leftrightarrow (q \wedge p)$	
$(p \vee q) \leftrightarrow (q \vee p)$	
$\sim(p \wedge q) \leftrightarrow (\sim p \vee \sim q)$	DeMorgan's laws
$\sim(p \vee q) \leftrightarrow (\sim p \wedge \sim q)$	DeMorgan's laws
$\sim(p \rightarrow q) \leftrightarrow (p \wedge \sim q)$	
$\sim(p \leftrightarrow q) \leftrightarrow [(p \wedge \sim q) \vee (q \wedge \sim p)]$	

Relations between Statements

If R and S are compound statements, then R **logically implies** S when $R \rightarrow S$ is a tautology. It follows from the truth table for the conditional (see Table 4) that $R \rightarrow S$ fails to be a tautology only when R is true and S is false. Thus, we need only check that when R is true, S must be true also. Therefore, R logically implies S means that when R is true, S must be true.

EXAMPLE 13 Show that $[p \wedge (p \rightarrow q)]$ logically implies q.

Solution To prove that $[p \wedge (p \rightarrow q)] \rightarrow q$ is a tautology, we construct the truth table in Table 14.

TABLE 14

			R	S	$R \rightarrow S$
p	q	$p \rightarrow q$	$p \wedge (p \rightarrow q)$	q	$[p \wedge (p \rightarrow q)] \rightarrow q$
T	T	T	T	T	T
T	F	F	F	F	T
F	T	T	F	T	T
F	F	T	F	F	T

Since the conditional is a tautology, we conclude that $[p \wedge (p \rightarrow q)]$ logically implies q.

Observe in Example 13 that the only case that actually needed to be checked was the case where $[p \wedge (p \rightarrow q)]$ is true (the shaded portion of the truth table.) Thus, Table 14 could have been written as shown in Table 15.

TABLE 15

			R	S	$R \rightarrow S$
p	q	$p \rightarrow q$	$p \wedge (p \rightarrow q)$	q	$[p \wedge (p \rightarrow q)] \rightarrow q$
T	T	T	T	T	T
T	F	F	F		
F	T	T	F		
F	F	T	F		

We only need to check the cases for which R is true.

EXAMPLE 14 Show that $[q \wedge (p \to q)]$ does not logically imply p.

Solution To prove that $[q \wedge (p \to q)] \to p$ is not a tautology, we construct the truth table in Table 16.

TABLE 16

			R	S	$R \to S$
p	q	$p \to q$	$q \wedge (p \to q)$	p	$[q \wedge (p \to q)] \to p$
T	T	T	T	T	T
T	F	F	F		
F	T	T	T	F	F
F	F	T	F		

Since we have found a case where $[q \wedge (p \to q)]$ is true but p is false, we conclude that $[q \wedge (p \to q)]$ does not logically imply p.

If R and S are compound statements, then R and S are **logically equivalent** if $R \leftrightarrow S$ is a tautology. This means that in any compound statement containing R, we may replace R by S without changing the logical meaning of the original statement. For example, since $\sim(\sim p) \leftrightarrow p$ is a tautology (see Table 13), $\sim(\sim p)$ is logically equivalent to p, and therefore we may replace $\sim(\sim p)$ by p wherever it occurs.

EXAMPLE 15 Show that the conditional $p \to q$ and its contrapositive $\sim q \to \sim p$ are logically equivalent.

Solution We prove that $(p \to q) \leftrightarrow [(\sim q) \to (\sim p)]$ is a tautology. See Table 17.

TABLE 17

		R			S	$R \leftrightarrow S$
p	q	$p \to q$	$\sim q$	$\sim p$	$\sim q \to \sim p$	$(p \to q) \leftrightarrow [(\sim q) \to (\sim p)]$
T	T	T	F	F	T	T
T	F	F	T	F	F	T
F	T	T	F	T	T	T
F	F	T	T	T	T	T

This truth table establishes the desired result.

EXAMPLE 16 Show that the inverse and converse of $p \to q$ are logically equivalent.

Solution The inverse of $p \to q$ is $\sim p \to \sim q$. By Example 15 it is logically equivalent to its contrapositive, $\sim(\sim q) \to \sim(\sim p)$, which in turn is logically equivalent to $q \to p$, the converse of $p \to q$.

EXAMPLE 17 Show that $[(p \to q) \wedge (q \to p)]$ and $(p \leftrightarrow q)$ are logically equivalent.

Solution We construct the truth table of $[(p \to q) \wedge (q \to p)] \leftrightarrow (p \leftrightarrow q)$ in Table 18.

TABLE 18

				R	S	$R \leftrightarrow S$
p	q	$p \to q$	$q \to p$	$[(p \to q) \wedge (q \to p)]$	$p \leftrightarrow q$	$[(p \to q) \wedge (q \to p)] \leftrightarrow (p \leftrightarrow q)$
T	T	T	T	T	T	T
T	F	F	T	F	F	T
F	T	T	F	F	F	T
F	F	T	T	T	T	T

This truth table establishes the desired result.

Some of the tautologies in Table 13 lead to the logically equivalent statements in Table 19.

TABLE 19

Some common logically equivalent statements	
$\sim(\sim p)$	is logically equivalent to p
$p \wedge p$	is logically equivalent to p
$p \vee p$	is logically equivalent to p
$p \wedge q$	is logically equivalent to $q \wedge p$
$p \vee q$	is logically equivalent to $q \vee p$
$\sim(p \wedge q)$	is logically equivalent to $\sim p \vee \sim q$
$\sim(p \vee q)$	is logically equivalent to $\sim p \wedge \sim q$
$\sim(p \to q)$	is logically equivalent to $p \wedge \sim q$
$\sim(p \leftrightarrow q)$	is logically equivalent to $(p \wedge \sim q) \vee (q \wedge \sim p)$

Argument An argument is a set of compound statements $P_1, P_2, \ldots, P_r$ called **premises** and a compound statement C called the **conclusion**. If $[P_1 \wedge P_2 \wedge \cdots \wedge P_r]$ logically implies C, we say that the argument is **valid**. Otherwise, it is said to be **invalid**. From the definition of logical implication, we see that an argument is valid if when the conjunction of all the premises is true, then the conclusion is true; or an argument is valid if when all the premises are true, then the conclusion is true. Arguments are frequently written as

$$\begin{array}{l} P_1 \\ P_2 \\ \cdot \\ \cdot \\ \cdot \\ \underline{P_r \quad} \\ \therefore C \end{array}$$

where the symbol $\therefore$ means *therefore*.

EXAMPLE 18 Determine whether the following arguments are valid.

(a) $p \to q$
$\underline{p}$
$\therefore q$

(b) q
$\underline{p \to q}$
$\therefore p$

Solution

(a) We need to determine whether $[p \land (p \to q)]$ logically implies q. In fact we only need to determine whether the conclusion is true when the premises are true. From Table 20 we see that the argument is valid.

TABLE 20

		P_1	P_2	C
p	q	$p \to q$	p	q
T	T	T	T	T
T	F	F	T	
F	T	T	F	
F	F	T	F	

We only need consider the cases where all the premises are true

Solution

(b) We need to determine whether $[q \land (p \to q)]$ logically implies p. We only need to determine whether the conclusion is true when the premises are true. From Table 21 we see that the argument is invalid.

TABLE 21

	P_1	P_2	C
p	q	$p \to q$	p
T	T	T	T
T	F	F	
F	T	T	F
F	F	T	

In this case the conclusion is false when the premises are true. Therefore, the argument is invalid.

EXAMPLE 19 Determine whether the following argument is valid.

If taxes go up, then I will not be able to buy a house.
Taxes go up.

∴ I will not be able to buy a house.

Solution Let p and q denote the following statements.

p: Taxes go up.

q: I will not be able to buy a house.

Then the given argument takes the form

$$\begin{array}{l} p \to q \\ \underline{p\qquad} \\ \therefore q \end{array}$$

which was seen to be valid in Example 18(a).

EXAMPLE 20 Determine whether the following argument is valid.

I will be happy or I will be rich.
I will be happy.
∴ I will be rich.

Solution Let p and q denote the following statements.

p: I will be happy.
q: I will be rich.

Then the given argument takes the form

$$\begin{array}{l} p \vee q \\ \underline{p\qquad} \\ \therefore q \end{array}$$

To determine whether the given argument is valid we need to determine whether $[(p \vee q) \wedge p]$ logically implies q, or more simply to determine whether the conclusion is true when the premises are true. See Table 22.

TABLE 22

		P_1	P_2	C	
p	q	$p \vee q$	p	q	
T	T	T	T	T	The argument is not valid because in this case, the conclusion is false when the premises are true.
T	F	T	T	F	
F	T	T	F		
F	F	F	F		

Hence, the given argument is not valid.

Some examples of valid and invalid arguments are given in Table 23.

TABLE 23

Valid arguments		Invalid arguments
$p \to q$ p $\therefore q$	This argument is referred to as *modus ponens*	$p \to q$ q $\therefore p$
$p \to q$ $\sim q$ $\therefore \sim p$	This argument is referred to as *modus tollens*	$p \lor q$ p $\therefore \sim q$
$p \lor q$ $\sim p$ $\therefore q$		$p \to q$ $\sim p$ $\therefore q$
$\sim p \lor q$ $\sim q$ $\therefore \sim p$		$\sim(p \land q)$ p $\therefore q$
$p \to q$ $q \to r$ $\therefore p \to r$		$\sim(p \to q)$ $\sim q$ $\therefore \sim p$

Finally, observe that we have not considered statements of the following forms:

(a) *Every* dog is an animal.

(b) *All* apples are of the Macintosh variety.

(c) *There exists* an integer whose square is zero.

(d) *There is* an integer whose square is negative.

Each of these statements contains one of the **quantifiers** *every, all, there exists,* and *there is*. Statement (a) is, of course, true. Statement (b) is false since there is at least one other variety of apples, such as Stayman apples. Statement (c) is true since $0^2 = 0$. Statement (d) is false since the square of every integer is nonnegative. For further discussion on this topic refer to Hugues Leblanc and William A. Wisdom, *Deductive Logic,* 2d ed. (Boston: Allyn and Bacon, 1976).

Exercise Set

1. Which of the following are statements?
(a) Will you go to New York?
(b) I will go to New York today.
(c) $x^3 - x + 1 = 5$.
(d) Go to the bank.
(e) I will not study next Monday.

2. Give the negations of the following statements.
(a) $3 + 4 \geq 2$.
(b) 8 is an odd integer.
(c) You will buy some stock next week.
(d) It is not going to rain tomorrow.

3. In each of the following, form the conjunction and the disjunction of p and q.
(a) p: $7 = (5)(2)$. q: $3 = 2 + 1$.
(b) p: I will study tonight. q: I will pass the exam tomorrow.
(c) p: I will eat dinner. q: I will watch TV.

4. Determine the truth or falsity of each of the following compound statements.
(a) $3 + 4 \geq 2$ and 8 is an odd integer.
(b) 3 is an odd integer and $3 + 4 \geq 2$.
(c) The world is round and dogs are animals.
(d) The world is flat and dogs are animals.

5. Determine the truth or falsity of each of the following compound statements.
(a) $3 + 4 \geq 2$ or 8 is an odd integer.
(b) 3 is an odd integer or $3 + 4 \geq 2$.
(c) The world is round or dogs are animals.
(d) The world is flat or dogs are animals.

6. Suppose that

p: IBM stock goes up.

q: John will sell his IBM stock.

are both true statements. Determine the truth values of the following statements.
(a) If IBM stock goes up, then John will not sell his IBM stock.
(b) If IBM stock goes down, then John will sell his IBM stock.
(c) If IBM stock goes down, then John will not sell his IBM stock.
(d) If IBM stock goes up, then John will sell his IBM stock.

7. Write the conditional statement $p \to q$ in words and determine its truth or falsity.
(a) p: 2 is a positive number; q: $2^2 = 4$.
(b) p: England is in Asia; q: France is in Europe.
(c) p: $3 < -1$; q: $4 + 5 = 9$.

8. Write the biconditional statement $p \leftrightarrow q$ in words and determine its truth or falsity.
(a) p: $\frac{8}{2} = 4$; q: $3 < 0$.
(b) p: Texas is west of Mississippi; q: Philadelphia is in Pennsylvania.
(c) p: $6^2 = 25$; q: The earth has no moons.

9. Suppose that

p: Michael takes finite math.

q: Susan takes English.

are both true statements. Determine the truth values of the following statements.
(a) If Michael takes finite math, then Susan takes English.
(b) If Michael does not take finite math, then Susan does not take English.
(c) If Michael does not take finite math, then Susan takes English.
(d) If Michael takes finite math, then Susan does not take English.
(e) Michael does not take finite math if and only if Susan does not take English.
(f) Michael takes finite math if and only if Susan takes English.

10. State the converse of the following conditional statements.
(a) If $1 + 3 = 5$, then Boston is in Pennsylvania.
(b) If it is sunny, then it is hot.
(c) If I am rich, then I am not happy.
(d) If you discover a cure for cancer, then you will win the Nobel prize.

11. State the contrapositive of the following conditional statements.
(a) If $2 < 3$, then Paris is the capital of England.
(b) If it snows, then you cannot drive to work.
(c) If I have money, then I will go dancing.
(d) If I don't study, then I won't pass the exam.

12. Suppose that "I will eat" is a true statement and "I will go to the theater" is a false statement. Determine the truth value of the following statements.
(a) I will eat if and only if I will go to the theater.
(b) If I go to the theater, then I will not eat.
(c) If I will eat, then I will not go to the theater.
(d) I will eat or I will go to the theater.
(e) I will eat and I will not go to the theater.

13. Suppose that "I will write the paper" and "I will watch TV" are both false statements. Determine the truth value of the following statements.
(a) I will write the paper and I will not watch TV.
(b) I will write the paper or I will watch TV.
(c) If I will write the paper, then I will watch TV.
(d) If I will watch TV, then I will write the paper.
(e) If I will not watch TV, then I will not write the paper.

In Exercises 14–19 construct a truth table for each statement and determine whether the statement is a tautology, contradiction, or neither.

14. (a) $(p \wedge \sim q) \vee (p \wedge q)$
(b) $(p \vee q) \wedge (\sim p \wedge \sim q)$

15. (a) $(\sim p \vee q) \rightarrow p$
(b) $(p \wedge \sim p) \rightarrow q$

16. (a) $(p \vee \sim p) \rightarrow (q \wedge \sim q)$
(b) $(p \rightarrow q) \leftrightarrow (q \rightarrow p)$

17. (a) $(p \rightarrow q) \leftrightarrow (\sim q \rightarrow \sim p)$
(b) $(p \rightarrow q) \vee (q \rightarrow p)$

18. (a) $\sim(p \vee \sim q) \leftrightarrow (\sim p \wedge q)$
(b) $\sim(\sim p \wedge q) \rightarrow (p \wedge \sim q)$

19. (a) $[(p \rightarrow q) \wedge (q \rightarrow r)] \rightarrow (p \rightarrow r)$ [*Hint*: There are eight cases for the possible combinations of truth values for p, q, and r.]

In Exercises 20–23 determine whether R logically implies S.

20. R: $\sim p \vee q$; S: p

21. R: $\sim p \vee q$; S: q

22. R: $p \vee \sim p$; S: $q \wedge \sim p$

23. R: $(p \rightarrow q) \wedge (q \rightarrow r)$; S: $p \rightarrow r$

In Exercises 24–27 determine whether R and S are logically equivalent.

24. R: $\sim p \vee q$; S: $\sim(p \wedge \sim q)$

25. R: $\sim(p \vee \sim q)$; S: $\sim p \vee q$

26. R: $p \wedge \sim p$; S: $q \wedge \sim q$

27. R: $(p \rightarrow q) \wedge (q \rightarrow r)$; S: $p \rightarrow r$

In Exercises 28–33 determine whether the given arguments are valid.

28. (a) p
$\underline{\sim p \vee q}$
$\therefore q$

(b) p
$\underline{\sim q \vee p}$
$\therefore q$

29. (a) $\sim(p \vee q)$
$\underline{p}$
$\therefore p \vee \sim q$

(b) $p \rightarrow q$
$\underline{\sim q}$
$\therefore \sim p$

30. $p \rightarrow q$
$\underline{q \rightarrow r}$
$\therefore \sim p \vee r$ [*Hint*: The truth table will have eight cases giving the truth values of p, q, and r.]

31. (a) $p \rightarrow q$
$\underline{\sim p}$
$\therefore q$

(b) $\sim(p \wedge q)$
$\underline{p}$
$\therefore \sim q$

32. (a) $p \wedge (q \rightarrow p)$
$\underline{\sim q \vee p}$
$\therefore p \wedge q$

(b) $(q \rightarrow p) \rightarrow (p \wedge q)$
$\underline{\sim p}$
$\therefore q$

33. $\sim p \vee q$
$\underline{q \rightarrow r}$
$\therefore \sim p \vee r$
[*Hint*: The truth table will have eight cases giving the truth values of p, q, and r.]

In Exercises 34–37 determine whether the following arguments are valid.

34. I will study or I will go to the movies.
I did not study.
∴ I went to the movies.

35. If the tax rate is lowered, I will buy a car.
I did not buy a car.
∴ The tax rate was not lowered.

36. If it is a sunny day, I will take a walk.
I took a walk.
∴ It was a sunny day.

37. I will take a walk or I will watch TV.
I did not watch TV.
∴ I did not take a walk.

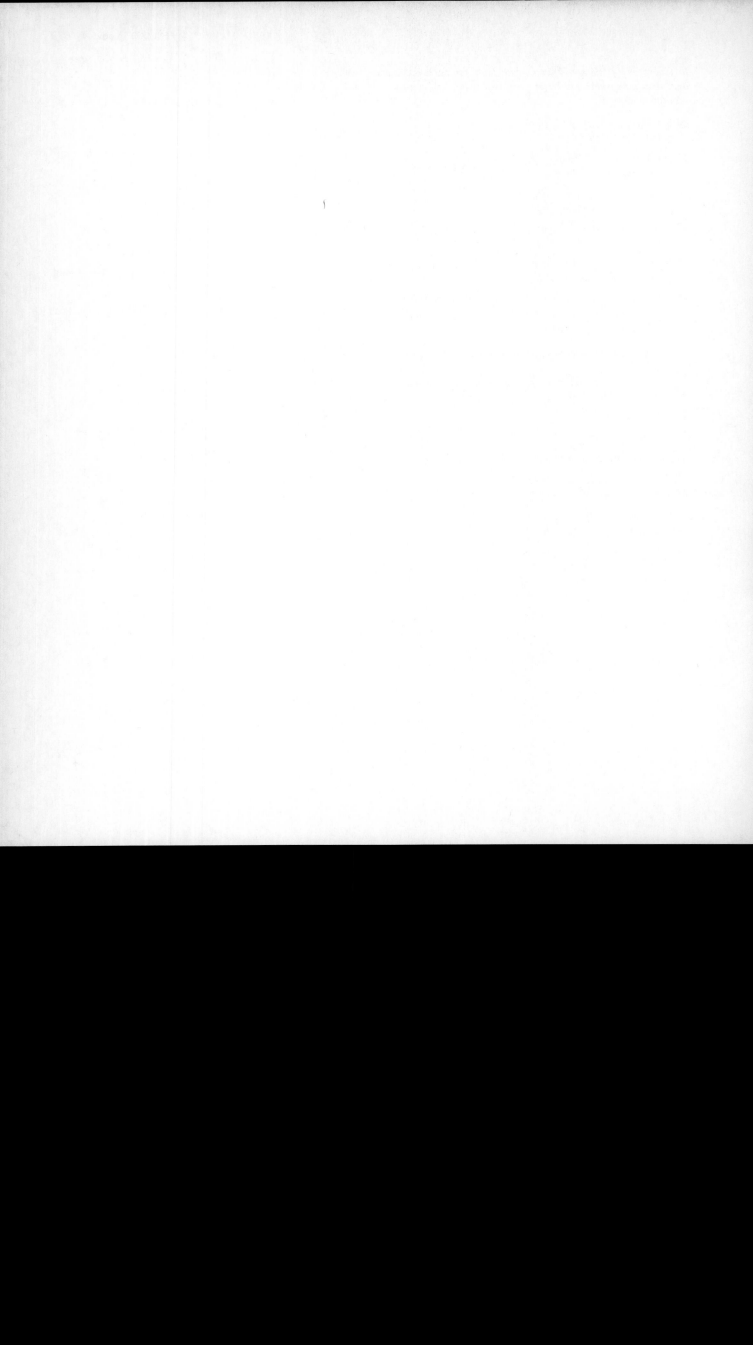

Appendix C
Tables

TABLE C-1 Areas under the Standard Normal Curve

The table below gives the shaded area shown in this figure.

z	0	1	2	3	4	5	6	7	8	9
−3.4	.0003	.0003	.0003	.0003	.0003	.0003	.0003	.0003	.0003	.0002
−3.3	.0005	.0005	.0005	.0004	.0004	.0004	.0004	.0004	.0004	.0003
−3.2	.0007	.0007	.0006	.0006	.0006	.0006	.0006	.0005	.0005	.0005
−3.1	.0010	.0009	.0009	.0009	.0008	.0008	.0008	.0008	.0007	.0007
−3.0	.0013	.0013	.0013	.0012	.0012	.0011	.0011	.0011	.0010	.0010
−2.9	.0019	.0018	.0018	.0017	.0016	.0016	.0015	.0015	.0014	.0014
−2.8	.0026	.0025	.0024	.0023	.0023	.0022	.0021	.0021	.0020	.0019
−2.7	.0035	.0034	.0033	.0032	.0031	.0030	.0029	.0028	.0027	.0026
−2.6	.0047	.0045	.0044	.0043	.0041	.0040	.0039	.0038	.0037	.0036
−2.5	.0062	.0060	.0059	.0057	.0055	.0054	.0052	.0051	.0049	.0048
−2.4	.0082	.0080	.0078	.0075	.0073	.0071	.0069	.0068	.0066	.0064
−2.3	.0107	.0104	.0102	.0099	.0096	.0094	.0091	.0089	.0087	.0084
−2.2	.0139	.0136	.0132	.0129	.0125	.0122	.0119	.0116	.0113	.0110
−2.1	.0179	.0174	.0170	.0166	.0162	.0158	.0154	.0150	.0146	.0143
−2.0	.0228	.0222	.0217	.0212	.0207	.0202	.0197	.0192	.0188	.0183
−1.9	.0287	.0281	.0274	.0268	.0262	.0256	.0250	.0244	.0239	.0233
−1.8	.0359	.0351	.0344	.0336	.0329	.0322	.0314	.0307	.0301	.0294
−1.7	.0446	.0436	.0427	.0418	.0409	.0401	.0392	.0384	.0375	.0367
−1.6	.0548	.0537	.0526	.0516	.0505	.0495	.0485	.0475	.0465	.0455
−1.5	.0668	.0655	.0643	.0630	.0618	.0606	.0594	.0582	.0571	.0559

(Table continued on next page)

z	0	1	2	3	4	5	6	8	8	9
−1.4	.0808	.0793	.0778	.0764	.0749	.0735	.0721	.0708	.0694	.0681
−1.3	.0968	.0951	.0934	.0918	.0901	.0885	.0869	.0853	.0838	.0823
−1.2	.1151	.1131	.1112	.1093	.1075	.1056	.1038	.1020	.1003	.0985
−1.1	.1357	.1335	.1314	.1292	.1271	.1251	.1230	.1210	.1190	.1170
−1.0	.1587	.1562	.1539	.1515	.1492	.1469	.1446	.1423	.1401	.1379
−0.9	.1841	.1814	.1788	.1762	.1736	.1711	.1685	.1660	.1635	.1611
−0.8	.2119	.2090	.2061	.2033	.2005	.1977	.1949	.1922	.1894	.1867
−0.7	.2420	.2389	.2358	.2327	.2296	.2266	.2236	.2206	.2177	.2148
−0.6	.2743	.2709	.2676	.2643	.2611	.2578	.2546	.2514	.2483	.2451
−0.5	.3085	.3050	.3015	.2981	.2946	.2912	.2877	.2843	.2810	.2776
−0.4	.3446	.3409	.3372	.3336	.3300	.3264	.3228	.3192	.3156	.3121
−0.3	.3821	.3783	.3745	.3707	.3669	.3632	.3594	.3557	.3520	.3483
−0.2	.4207	.4168	.4129	.4090	.4052	.4013	.3974	.3936	.3897	.3859
−0.1	.4602	.4562	.4522	.4483	.4443	.4404	.4364	.4325	.4286	.4247
−0.0	.5000	.4960	.4920	.4880	.4840	.4801	.4761	.4721	.4681	.4641
0.0	.5000	.5040	.5080	.5120	.5160	.5199	.5239	.5279	.5319	.5359
0.1	.5398	.5438	.5478	.5517	.5557	.5596	.5636	.5675	.5714	.5753
0.2	.5793	.5832	.5871	.5910	.5948	.5987	.6026	.6064	.6103	.6141
0.3	.6179	.6217	.6255	.6293	.6331	.6368	.6406	.6443	.6480	.6517
0.4	.6554	.6591	.6628	.6664	.6700	.6736	.6772	.6808	.6844	.6879
0.5	.6915	.6950	.6985	.7019	.7054	.7088	.7123	.7157	.7190	.7224
0.6	.7257	.7291	.7324	.7357	.7389	.7422	.7454	.7486	.7517	.7549
0.7	.7580	.7611	.7642	.7673	.7704	.7734	.7764	.7794	.7823	.7852
0.8	.7881	.7910	.7939	.7967	.7995	.8023	.8051	.8078	.8106	.8133
0.9	.8159	.8186	.8212	.8238	.8264	.8289	.8315	.8340	.8365	.8389
1.0	.8413	.8438	.8461	.8485	.8508	.8531	.8554	.8577	.8599	.8621
1.1	.8643	.8665	.8686	.8708	.8729	.8749	.8770	.8790	.8810	.8830
1.2	.8849	.8869	.8888	.8907	.8925	.8944	.8962	.8980	.8997	.9015
1.3	.9032	.9049	.9066	.9082	.9099	.9115	.9131	.9147	.9162	.9177
1.4	.9192	.9207	.9222	.9236	.9251	.9265	.9279	.9292	.9306	.9319
1.5	.9332	.9345	.9357	.9370	.9382	.9394	.9406	.9418	.9429	.9441
1.6	.9452	.9463	.9474	.9484	.9495	.9505	.9515	.9525	.9535	.9545
1.7	.9554	.9564	.9573	.9582	.9591	.9599	.9608	.9616	.9625	.9633
1.8	.9641	.9649	.9656	.9664	.9671	.9678	.9686	.9693	.9699	.9706
1.9	.9713	.9719	.9726	.9732	.9738	.9744	.9750	.9756	.9761	.9767
2.0	.9772	.9778	.9783	.9788	.9793	.9798	.9803	.9808	.9812	.9817
2.1	.9821	.9826	.9830	.9834	.9838	.9842	.9846	.9850	.9854	.9857
2.2	.9861	.9864	.9868	.9871	.9875	.9878	.9881	.9884	.9887	.9890
2.3	.9893	.9896	.9898	.9901	.9904	.9906	.9909	.9911	.9913	.9916
2.4	.9918	.9920	.9922	.9925	.9927	.9929	.9931	.9932	.9934	.9936
2.5	.9938	.9940	.9941	.9943	.9945	.9946	.9948	.9949	.9951	.9952
2.6	.9953	.9955	.9956	.9957	.9959	.9960	.9961	.9962	.9963	.9964
2.7	.9965	.9966	.9967	.9968	.9969	.9970	.9971	.9972	.9973	.9974
2.8	.9974	.9975	.9976	.9977	.9977	.9978	.9979	.9979	.9980	.9981
2.9	.9981	.9982	.9982	.9983	.9984	.9984	.9985	.9985	.9986	.9986
3.0	.9987	.9987	.9987	.9988	.9988	.9989	.9989	.9989	.9990	.9990
3.1	.9990	.9991	.9991	.9991	.9992	.9992	.9992	.9992	.9993	.9993
3.2	.9993	.9993	.9994	.9994	.9994	.9994	.9994	.9995	.9995	.9995
3.3	.9995	.9995	.9995	.9996	.9996	.9996	.9996	.9996	.9996	.9997
3.4	.9997	.9997	.9997	.9997	.9997	.9997	.9997	.9997	.9997	.9998

TABLE C.2 Table of Binomial Probabilities

		p													
n	*x*	.01	.05	.10	.20	.30	.40	.50	.60	.70	.80	.90	.95	.99	*x*
2	0	.980	.902	.810	.640	.490	.360	.250	.160	.090	.040	.010	.002	.000	0
	1	.020	.095	.180	.320	.420	.480	.500	.480	.420	.320	.180	.095	.020	1
	2	.000	.002	.010	.040	.090	.160	.250	.360	.490	.640	.810	.902	.980	2
3	0	.970	.857	.729	.512	.343	.216	.125	.064	.027	.008	.001	.000	.000	0
	1	.029	.135	.243	.384	.441	.432	.375	.288	.189	.096	.027	.007	.000	1
	2	.000	.007	.027	.096	.189	.288	.375	.432	.441	.384	.243	.135	.029	2
	3	.000	.000	.001	.008	.027	.064	.125	.216	.343	.512	.729	.857	.970	3
4	0	.961	.815	.656	.410	.240	.130	.062	.026	.008	.002	.000	.000	.000	0
	1	.039	.171	.292	.410	.412	.346	.250	.154	.076	.026	.004	.000	.000	1
	2	.001	.014	.049	.154	.265	.346	.375	.346	.265	.154	.049	.014	.001	2
	3	.000	.000	.004	.026	.076	.154	.250	.346	.412	.410	.292	.171	.039	3
	4	.000	.000	.000	.002	.008	.026	.062	.130	.240	.410	.656	.815	.961	4
5	0	.951	.774	.590	.328	.168	.078	.031	.010	.002	.000	.000	.000	.000	0
	1	.048	.204	.328	.410	.360	.259	.156	.077	.028	.006	.000	.000	.000	1
	2	.001	.021	.073	.205	.309	.346	.312	.230	.132	.051	.008	.001	.000	2
	3	.000	.001	.008	.051	.132	.230	.312	.346	.309	.205	.073	.021	.001	3
	4	.000	.000	.000	.006	.028	.077	.156	.259	.360	.410	.328	.204	.048	4
	5	.000	.000	.000	.000	.002	.010	.031	.078	.168	.328	.590	.774	.951	5
6	0	.941	.735	.531	.262	.118	.047	.016	.004	.001	.000	.000	.000	.000	0
	1	.057	.232	.354	.393	.303	.187	.094	.037	.010	.002	.000	.000	.000	1
	2	.001	.031	.098	.246	.324	.311	.234	.138	.060	.015	.001	.000	.000	2
	3	.000	.002	.015	.082	.185	.276	.312	.276	.185	.082	.015	.002	.000	3
	4	.000	.000	.001	.015	.060	.138	.234	.311	.324	.246	.098	.031	.001	4
	5	.000	.000	.000	.002	.010	.037	.094	.187	.303	.393	.354	.232	.057	5
	6	.000	.000	.000	.000	.001	.004	.016	.047	.118	.262	.531	.735	.941	6
7	0	.932	.698	.478	.210	.082	.028	.008	.002	.000	.000	.000	.000	.000	0
	1	.066	.257	.372	.367	.247	.131	.055	.017	.004	.000	.000	.000	.000	1
	2	.002	.041	.124	.275	.318	.261	.164	.077	.025	.004	.000	.000	.000	2
	3	.000	.004	.023	.115	.227	.290	.273	.194	.097	.029	.003	.000	.000	3
	4	.000	.000	.003	.029	.097	.194	.273	.290	.227	.115	.023	.004	.000	4
	5	.000	.000	.000	.004	.025	.077	.164	.261	.318	.275	.124	.041	.002	5
	6	.000	.000	.000	.000	.004	.017	.055	.131	.247	.367	.372	.257	.066	6
	7	.000	.000	.000	.000	.000	.002	.008	.028	.082	.210	.478	.698	.932	7
8	0	.923	.663	.430	.168	.058	.017	.004	.001	.000	.000	.000	.000	.000	0
	1	.075	.279	.383	.336	.198	.090	.031	.008	.001	.000	.000	.000	.000	1
	2	.003	.051	.149	.294	.296	.209	.109	.041	.010	.001	.000	.000	.000	2
	3	.000	.005	.033	.147	.254	.279	.219	.124	.047	.009	.000	.000	.000	3
	4	.000	.000	.005	.046	.136	.232	.273	.232	.136	.046	.005	.000	.000	4
	5	.000	.000	.000	.009	.047	.124	.219	.279	.254	.147	.033	.005	.000	5
	6	.000	.000	.000	.001	.010	.041	.109	.209	.296	.294	.149	.051	.003	6
	7	.000	.000	.000	.000	.001	.008	.031	.090	.198	.336	.383	.279	.075	7
	8	.000	.000	.000	.000	.000	.001	.004	.017	.058	.168	.430	.663	.923	8
9	0	.914	.630	.387	.134	.040	.010	.002	.000	.000	.000	.000	.000	.000	0
	1	.083	.299	.387	.302	.156	.060	.018	.004	.000	.000	.000	.000	.000	1
	2	.003	.063	.172	.302	.267	.161	.070	.021	.004	.000	.000	.000	.000	2
	3	.000	.008	.045	.176	.267	.251	.164	.074	.021	.003	.000	.000	.000	3
	4	.000	.001	.007	.066	.172	.251	.246	.167	.074	.017	.001	.000	.000	4

n	x	p .01	.05	.10	.20	.30	.40	.50	.60	.70	.80	.90	.95	.99	x
	5	.000	.000	.001	.017	.074	.167	.246	.251	.172	.066	.007	.001	.000	5
	6	.000	.000	.000	.003	.021	.074	.164	.251	.267	.176	.045	.008	.000	6
	7	.000	.000	.000	.000	.004	.021	.070	.161	.267	.302	.172	.063	.003	7
	8	.000	.000	.000	.000	.000	.004	.018	.060	.156	.302	.387	.299	.083	8
	9	.000	.000	.000	.000	.000	.000	.002	.010	.040	.134	.387	.630	.914	9
10	0	.904	.599	.349	.107	.028	.006	.001	.000	.000	.000	.000	.000	.000	0
	1	.091	.315	.387	.268	.121	.040	.010	.002	.000	.000	.000	.000	.000	1
	2	.004	.075	.194	.302	.233	.121	.044	.011	.001	.000	.000	.000	.000	2
	3	.000	.010	.057	.201	.267	.215	.117	.042	.009	.001	.000	.000	.000	3
	4	.000	.001	.011	.088	.200	.251	.205	.111	.037	.006	.000	.000	.000	4
	5	.000	.000	.001	.026	.103	.201	.246	.201	.103	.026	.001	.000	.000	5
	6	.000	.000	.000	.006	.037	.111	.205	.251	.200	.088	.011	.001	.000	6
	7	.000	.000	.000	.001	.009	.042	.117	.215	.267	.201	.057	.010	.000	7
	8	.000	.000	.000	.000	.001	.011	.044	.121	.233	.302	.194	.075	.004	8
	9	.000	.000	.000	.000	.000	.002	.010	.040	.121	.268	.387	.315	.091	9
	10	.000	.000	.000	.000	.000	.000	.001	.006	.028	.107	.349	.599	.904	10
11	0	.895	.569	.314	.086	.020	.004	.000	.000	.000	.000	.000	.000	.000	0
	1	.099	.329	.384	.236	.093	.027	.005	.001	.000	.000	.000	.000	.000	1
	2	.005	.087	.213	.295	.200	.089	.027	.005	.001	.000	.000	.000	.000	2
	3	.000	.014	.071	.221	.257	.177	.081	.023	.004	.000	.000	.000	.000	3
	4	.000	.001	.016	.111	.220	.236	.161	.070	.017	.002	.000	.000	.000	4
	5	.000	.000	.002	.039	.132	.221	.226	.147	.057	.010	.000	.000	.000	5
	6	.000	.000	.000	.010	.057	.147	.226	.221	.132	.039	.002	.000	.000	6
	7	.000	.000	.000	.002	.017	.070	.161	.236	.220	.111	.016	.001	.000	7
	8	.000	.000	.000	.000	.004	.023	.081	.177	.257	.221	.071	.014	.000	8
	9	.000	.000	.000	.000	.001	.005	.027	.089	.200	.295	.213	.087	.005	9
	10	.000	.000	.000	.000	.000	.001	.005	.027	.093	.236	.384	.329	.099	10
	11	.000	.000	.000	.000	.000	.000	.000	.004	.020	.086	.314	.569	.895	11
12	0	.886	.540	.282	.069	.014	.002	.000	.000	.000	.000	.000	.000	.000	0
	1	.107	.341	.377	.206	.071	.017	.003	.000	.000	.000	.000	.000	.000	1
	2	.006	.099	.230	.283	.168	.064	.016	.002	.000	.000	.000	.000	.000	2
	3	.000	.017	.085	.236	.240	.142	.054	.012	.001	.000	.000	.000	.000	3
	4	.000	.002	.021	.133	.231	.213	.121	.042	.008	.001	.000	.000	.000	4
	5	.000	.000	.004	.053	.158	.227	.193	.101	.029	.003	.000	.000	.000	5
	6	.000	.000	.000	.016	.079	.177	.226	.177	.079	.016	.000	.000	.000	6
	7	.000	.000	.000	.003	.029	.101	.193	.227	.158	.053	.004	.000	.000	7
	8	.000	.000	.000	.001	.008	.042	.121	.213	.231	.133	.021	.002	.000	8
	9	.000	.000	.000	.000	.001	.012	.054	.142	.240	.236	.085	.017	.000	9
	10	.000	.000	.000	.000	.000	.002	.016	.064	.168	.283	.230	.099	.006	10
	11	.000	.000	.000	.000	.000	.000	.003	.017	.071	.206	.377	.341	.107	11
	12	.000	.000	.000	.000	.000	.000	.000	.002	.014	.069	.282	.540	.886	12

TABLE C.3 Table of 5% Critical Levels for χ^2-Curves

v Degrees of freedom	1	2	3	4	5	6	7	8	9	10
c 5% critical level	3.84	5.99	7.81	9.49	11.1	12.6	14.4	15.5	16.9	18.3

TABLE C.4 Table of Square Roots and Cube Roots

n	$\sqrt{n}$	$\sqrt{10n}$	$\sqrt[3]{n}$	n	$\sqrt{n}$	$\sqrt{10n}$	$\sqrt[3]{n}$
1.0	1.000	3.162	1.000	5.5	2.345	7.416	1.765
1.1	1.049	3.317	1.032	5.6	2.366	7.483	1.776
1.2	1.095	3.464	1.063	5.7	2.387	7.550	1.786
1.3	1.140	3.606	1.091	5.8	2.408	7.616	1.797
1.4	1.183	3.742	1.119	5.9	2.429	7.681	1.807
1.5	1.225	3.873	1.145	6.0	2.449	7.746	1.817
1.6	1.265	4.000	1.170	6.1	2.470	7.810	1.827
1.7	1.304	4.123	1.193	6.2	2.490	7.874	1.837
1.8	1.342	4.243	1.216	6.3	2.510	7.937	1.847
1.9	1.378	4.359	1.239	6.4	2.530	8.000	1.857
2.0	1.414	4.472	1.260	6.5	2.550	8.062	1.866
2.1	1.449	4.583	1.281	6.6	2.569	8.124	1.876
2.2	1.483	4.690	1.301	6.7	2.588	8.185	1.885
2.3	1.517	4.796	1.320	6.8	2.608	8.246	1.895
2.4	1.549	4.899	1.339	6.9	2.627	8.307	1.904
2.5	1.581	5.000	1.357	7.0	2.646	8.367	1.913
2.6	1.612	5.099	1.375	7.1	2.665	8.426	1.922
2.7	1.643	5.196	1.392	7.2	2.683	8.485	1.931
2.8	1.673	5.292	1.409	7.3	2.702	8.544	1.940
2.9	1.703	5.385	1.426	7.4	2.720	8.602	1.949
3.0	1.732	5.477	1.442	7.5	2.739	8.660	1.957
3.1	1.761	5.568	1.458	7.6	2.757	8.718	1.966
3.2	1.789	5.657	1.474	7.7	2.775	8.775	1.975
3.3	1.817	5.745	1.489	7.8	2.793	8.832	1.983
3.4	1.844	5.831	1.504	7.9	2.811	8.888	1.992
3.5	1.871	5.916	1.518	8.0	2.828	8.944	2.000
3.6	1.897	6.000	1.533	8.1	2.846	9.000	2.008
3.7	1.924	6.083	1.547	8.2	2.864	9.055	2.017
3.8	1.949	6.164	1.560	8.3	2.881	9.110	2.025
3.9	1.975	6.245	1.574	8.4	2.898	9.165	2.033
4.0	2.000	6.325	1.587	8.5	2.915	9.220	2.041
4.1	2.025	6.403	1.601	8.6	2.933	9.274	2.049
4.2	2.049	6.481	1.613	8.7	2.950	9.327	2.057
4.3	2.074	6.557	1.626	8.8	2.966	9.381	2.065
4.4	2.098	6.633	1.639	9.9	2.983	9.434	2.072
4.5	2.121	6.708	1.651	9.0	3.000	9.487	2.080
4.6	2.145	6.782	1.663	9.1	3.017	9.539	2.088
4.7	2.168	6.856	1.675	9.2	3.033	9.592	2.095
4.8	2.191	6.928	1.687	9.3	3.050	9.644	2.103
4.9	2.214	7.000	1.698	9.4	3.066	9.695	2.110
5.0	2.236	7.071	1.710	9.5	3.082	9.747	2.118
5.1	2.258	7.141	1.721	9.6	3.098	9.798	2.125
5.2	2.280	7.211	1.732	9.7	3.114	9.849	2.133
5.3	2.302	7.280	1.744	9.8	3.130	9.899	2.140
5.4	2.324	7.348	1.754	9.9	3.146	9.950	2.147

TABLE C.5 Table of Exponentials and Their Reciprocals

x	e^x	e^{-x}
0.00	1.0000	1.0000
0.01	1.0101	0.9900
0.02	1.0202	0.9802
0.03	1.0305	0.9704
0.04	1.0408	0.9608
0.05	1.0513	0.9512
0.06	1.0618	0.9418
0.07	1.0725	0.9324
0.08	1.0833	0.9231
0.09	1.0942	0.9139
0.10	1.1052	0.9048
0.11	1.1163	0.8958
0.12	1.1275	0.8869
0.13	1.1388	0.8781
0.14	1.1503	0.8694
0.15	1.1618	0.8607
0.16	1.1735	0.8521
0.17	1.1853	0.8437
0.18	1.1972	0.8353
0.19	1.2092	0.8270
0.20	1.2214	0.8187
0.21	1.2337	0.8106
0.22	1.2461	0.8025
0.23	1.2586	0.7945
0.24	1.2712	0.7866
0.25	1.2840	0.7788
0.26	1.2969	0.7711
0.27	1.3100	0.7634
0.28	1.3231	0.7558
0.29	1.3364	0.7483
0.30	1.3499	0.7408
0.35	1.4191	0.7047
0.40	1.4918	0.6703
0.45	1.5683	0.6376
0.50	1.6487	0.6065
0.55	1.7333	0.5769
0.60	1.8221	0.5488
0.65	1.9155	0.5220
0.70	2.0138	0.4966
0.75	2.1170	0.4724
0.80	2.2255	0.4493
0.85	2.3396	0.4274
0.90	2.4596	0.4066
0.95	2.5857	0.3867
1.0	2.7183	0.3679
1.1	3.0042	0.3329
1.2	3.3201	0.3012
1.3	3.6693	0.2725
1.4	4.0552	0.2466
1.5	4.4817	0.2231
1.6	4.9530	0.2019
1.7	5.4739	0.1827
1.8	6.0496	0.1653
1.9	6.6859	0.1496
2.0	7.3891	0.1353
2.1	8.1662	0.1225
2.2	9.0250	0.1108
2.3	9.9742	0.1003
2.4	11.023	0.0907
2.5	12.182	0.0821
2.6	13.464	0.0743
2.7	14.880	0.0672
2.8	16.445	0.0608
2.9	18.174	0.0550
3.0	20.086	0.0498
3.1	22.198	0.0450
3.2	24.533	0.0408
3.3	27.113	0.0369
3.4	29.964	0.0334
3.5	33.115	0.0302
3.6	36.598	0.0273
3.7	40.447	0.0247
3.8	44.701	0.0224
3.9	49.402	0.0202
4.0	54.598	0.0183
4.1	60.340	0.0166
4.2	66.686	0.0150
4.3	73.700	0.0136
4.4	81.451	0.0123
4.5	90.017	0.0111
4.6	99.484	0.0101
4.7	109.95	0.0091
4.8	121.51	0.0082
4.9	134.29	0.0074
5	148.41	0.0067
6	403.43	0.0025
7	1096.6	0.0009
8	2981.0	0.0003
9	8103.1	0.0001
10	22026	0.00005
11	59874	0.00002
12	162,754	0.000006
13	442,413	0.000002
14	1,202,604	0.0000008
15	3,269,017	0.0000003

TABLE C.6 Table of Natural Logarithms

n	$\log_e n$	n	$\log_e n$	n	$\log_e n$
		4.5	1.5041	9.0	2.1972
0.1	−2.3026	4.6	1.5261	9.1	2.2083
0.2	−1.6094	4.7	1.5476	9.2	2.2192
0.3	−1.2040	4.8	1.5686	9.3	2.2300
0.4	−0.9163	4.9	1.5892	9.4	2.2407
0.5	−0.6931	5.0	1.6094	9.5	2.2513
0.6	−0.5108	5.1	1.6292	9.6	2.2618
0.7	−0.3567	5.2	1.6487	9.7	2.2721
0.8	−0.2231	5.3	1.6677	9.8	2.2824
0.9	−0.1054	5.4	1.6864	9.9	2.2925
1.0	0.0000	5.5	1.7047	10	2.3026
1.1	0.0953	5.6	1.7228	11	2.3979
1.2	0.1823	5.7	1.7405	12	2.4849
1.3	0.2624	5.8	1.7579	13	2.5649
1.4	0.3365	5.9	1.7750	14	2.6391
1.5	0.4055	6.0	1.7918	15	2.7081
1.6	0.4700	6.1	1.8083	16	2.7726
1.7	0.5306	6.2	1.8245	17	2.8332
1.8	0.5878	6.3	1.8405	18	5.8904
1.9	0.6419	6.4	1.8563	19	2.9444
2.0	0.6931	6.5	1.8718	20	2.9957
2.1	0.7419	6.6	1.8871	25	3.2189
2.2	0.7885	6.7	1.9021	30	3.4012
2.3	0.8329	6.8	1.9169	35	3.5553
2.4	0.8755	6.9	1.9315	40	3.6889
2.5	0.9163	7.0	1.9459	45	3.8067
2.6	0.9555	7.1	1.9601	50	3.9120
2.7	0.9933	7.2	1.9741	55	4.0073
2.8	1.0296	7.3	1.9879	60	4.0943
2.9	1.0647	7.4	2.0015	65	4.1744
3.0	1.0986	7.5	2.0149	70	4.2485
3.1	1.1314	7.6	2.0281	75	4.3175
3.2	1.1632	7.7	2.0142	80	4.3820
3.3	1.1939	7.8	2.0541	85	4.4427
3.4	1.2238	7.9	2.0669	90	4.4998
3.5	1.2528	8.0	2.0794	95	4.5539
3.6	1.2809	8.1	2.0919	100	4.6052
3.7	1.3083	8.2	2.1041		
3.8	1.3350	8.3	2.1163		
3.9	1.3610	8.4	2.1282		
4.0	1.3863	8.5	2.1401		
4.1	1.4110	8.6	2.1518		
4.2	1.4351	8.7	2.1633		
4.3	1.4586	8.8	2.1748		
4.4	1.4816	8.9	2.1861		

TABLE C.7 i = rate of interest per period n = number of periods $i = \frac{1}{4}\%$

n	$(1+i)^n$	$(1+i)^{-n}$	$s_{\overline{n}\vert i}$	$a_{\overline{n}\vert i}$	$\dfrac{1}{s_{\overline{n}\vert i}}$	$\dfrac{1}{a_{\overline{n}\vert i}}$
1	1.0025 0000	0.9975 0623	1.0000 0000	0.9975 0623	1.0000 0000	1.0025 0000
2	1.0050 0625	0.9950 1869	2.0025 0000	1.9925 2492	0.4993 7578	0.5018 7578
3	1.0075 1877	0.9925 3734	3.0075 0625	2.9850 6227	0.3325 0139	0.3350 0139
4	1.0100 3756	0.9900 6219	4.0150 2502	3.9751 2446	0.2490 6445	0.2515 6445
5	1.0125 6266	0.9875 9321	5.0250 6258	4.9627 1766	0.1990 0250	0.2015 0250
6	1.0150 9406	0.9851 3038	6.0376 2523	5.9478 4804	0.1656 2803	0.1681 2803
7	1.0176 3180	0.9826 7370	7.0527 1930	6.9305 2174	0.1417 8928	0.1442 8928
8	1.0201 7588	0.9802 2314	8.0703 5110	7.9107 4487	0.1239 1035	0.1264 1035
9	1.0227 2632	0.9777 7869	9.0905 2697	8.8885 2357	0.1100 0462	0.1125 0462
10	1.0252 8313	0.9753 4034	10.1132 5329	9.8638 6391	0.0988 8015	0.1013 8015
11	1.0278 4634	0.9729 0807	11.1385 3642	10.8367 7198	0.0897 7840	0.0922 7840
12	1.0304 1596	0.9704 8187	12.1663 8277	11.8072 5384	0.0821 9370	0.0846 9370
13	1.0329 9200	0.9680 6171	13.1967 9872	12.7753 1555	0.0757 7595	0.0782 7595
14	1.0355 7448	0.9656 4759	14.2297 9072	13.7409 6314	0.0702 7510	0.0727 7510
15	1.0381 6341	0.9632 3949	15.2653 6520	14.7042 0264	0.0655 0777	0.0680 0777
16	1.0407 5882	0.9608 3740	16.3035 2861	15.6650 4004	0.0613 3642	0.0638 3642
17	1.0433 6072	0.9584 4130	17.3442 8743	16.6234 8133	0.0576 5587	0.0601 5587
18	1.0459 6912	0.9560 5117	18.3876 4815	17.5795 3250	0.0543 8433	0.0568 8433
19	1.0485 8404	0.9536 6700	19.4336 1727	18.5331 9950	0.0514 5722	0.0539 5722
20	1.0512 0550	0.9512 8878	20.4822 0131	19.4844 8828	0.0488 2288	0.0513 2288
21	1.0538 3352	0.9489 1649	21.5334 0682	20.4334 0477	0.0464 3947	0.0489 3947
22	1.0564 6810	0.9465 5011	22.5872 4033	21.3799 5488	0.0442 7278	0.0467 7278
23	1.0591 0927	0.9441 8964	23.6437 0843	22.3241 4452	0.0422 9455	0.0447 9455
24	1.0617 5704	0.9418 3505	24.7028 1770	23.2659 7957	0.0404 8121	0.0429 8121
25	1.0644 1144	0.9394 8634	25.7645 7475	24.2054 6591	0.0388 1298	0.0413 1298
26	1.0670 7247	0.9371 4348	26.8289 8619	25.1426 0939	0.0372 7312	0.0397 7312
27	1.0697 4015	0.9348 0646	27.8960 5865	26.0774 1585	0.0358 4736	0.0383 4736
28	1.0724 1450	0.9324 7527	28.9657 9880	27.0098 9112	0.0345 2347	0.0370 2347
29	1.0750 9553	0.9301 4990	30.0382 1330	27.9400 4102	0.0332 9093	0.0357 9093
30	1.0777 8327	0.9278 3032	31.1133 0883	28.8678 7134	0.0321 4059	0.0346 4059
31	1.0804 7773	0.9255 1653	32.1910 9210	29.7933 8787	0.0310 6449	0.0335 6449
32	1.0831 7892	0.9232 0851	33.2715 6983	30.7165 9638	0.0300 5569	0.0325 5569
33	1.0858 8687	0.9209 0624	34.3547 4876	31.6375 0262	0.0291 0806	0.0316 0806
34	1.0886 0159	0.9186 0972	35.4406 3563	32.5561 1234	0.0282 1620	0.0307 1620
35	1.0913 2309	0.9163 1892	36.5292 3722	33.4724 3126	0.0273 7533	0.0298 7533
36	1.0940 5140	0.9140 3384	37.6205 6031	34.3864 6510	0.0265 8121	0.0290 8121
37	1.0967 8653	0.9117 5445	38.7146 1171	35.2982 1955	0.0258 3004	0.0283 3004
38	1.0995 2850	0.9094 8075	39.8113 9824	36.2077 0030	0.0251 1843	0.0276 1843
39	1.1022 7732	0.9072 1272	40.9109 2673	37.1149 1302	0.0244 4335	0.0269 4335
40	1.1050 3301	0.9049 5034	42.0132 0405	38.0198 6336	0.0238 0204	0.0263 0204
41	1.1077 9559	0.9026 9361	43.1182 3706	38.9225 5697	0.0231 9204	0.0256 9204
42	1.1105 6508	0.9004 4250	44.2260 3265	39.8229 9947	0.0226 1112	0.0251 1112
43	1.1133 4149	0.8981 9701	45.3365 9774	40.7211 9648	0.0220 5724	0.0245 5724
44	1.1161 2485	0.8959 5712	46.4499 3923	41.6171 5359	0.0215 2855	0.0240 2855
45	1.1189 1516	0.8937 2281	47.5660 6408	42.5108 7640	0.0210 2339	0.0235 2339
46	1.1217 1245	0.8914 9407	48.6849 7924	43.4023 7047	0.0205 4022	0.0230 4022
47	1.1245 1673	0.8892 7090	49.8066 9169	44.2916 4137	0.0200 7762	0.0225 7762
48	1.1273 2802	0.8870 5326	50.9312 0842	45.1786 9463	0.0196 3433	0.0221 3433
49	1.1301 4634	0.8848 4116	52.0585 3644	46.0635 3580	0.0192 0915	0.0217 0915
50	1.1329 7171	0.8826 3457	53.1886 8278	46.9461 7037	0.0188 0099	0.0213 0099

$i = \frac{1}{4}\%$

n	$(1+i)^n$	$(1+i)^{-n}$	$s_{\overline{n}\|i}$	$a_{\overline{n}\|i}$	$\frac{1}{s_{\overline{n}\|i}}$	$\frac{1}{a_{\overline{n}\|i}}$
51	1.1358 0414	0.8804 3349	54.3216 5449	47.8266 0386	0.0184 0886	0.0209 0886
52	1.1386 4365	0.8782 3790	55.4574 5862	48.7048 4176	0.0180 3184	0.0205 3184
53	1.1414 9026	0.8760 4778	56.5961 0227	49.5808 8953	0.0176 6906	0.0201 6906
54	1.1443 4398	0.8738 6312	57.7375 9252	50.4547 5265	0.0173 1974	0.0198 1974
55	1.1472 0484	0.8716 8391	58.8819 3650	51.3264 3656	0.0169 8314	0.0194 8314
56	1.1500 7285	0.8695 1013	60.0291 4135	52.1959 4669	0.0166 5858	0.0191 5858
57	1.1529 4804	0.8673 4178	61.1792 1420	53.0632 8847	0.0163 4542	0.0188 4542
58	1.1558 3041	0.8651 7883	62.3321 6223	53.9284 6730	0.0160 4308	0.0185 4308
59	1.1587 1998	0.8630 2128	63.4879 9264	54.7914 8858	0.0157 5101	0.0182 5101
60	1.1616 1678	0.8608 6911	64.6467 1262	55.6523 5769	0.0154 6869	0.0179 6869
61	1.1645 2082	0.8587 2230	65.8083 2940	56.5110 7999	0.0151 9564	0.0176 9564
62	1.1674 3213	0.8565 8085	66.9728 5023	57.3676 6083	0.0149 3142	0.0174 3142
63	1.1703 5071	0.8544 4474	68.1402 8235	58.2221 0557	0.0146 7561	0.0171 7561
64	1.1732 7658	0.8523 1395	69.3106 3306	59.0744 1952	0.0144 2780	0.0169 2780
65	1.1762 0977	0.8501 8848	70.4839 0964	59.9246 0800	0.0141 8764	0.0166 8764
66	1.1791 5030	0.8480 6831	71.6601 1942	60.7726 7631	0.0139 5476	0.0164 5476
67	1.1820 9817	0.8459 5343	72.8392 6971	61.6186 2974	0.0137 2886	0.0162 2886
68	1.1850 5342	0.8438 4382	74.0213 6789	62.4624 7355	0.0135 0961	0.0160 0961
69	1.1880 1605	0.8417 3947	75.2064 2131	63.3042 1302	0.0132 9674	0.0157 9674
70	1.1909 8609	0.8396 4037	76.3944 3736	64.1438 5339	0.0130 8996	0.0155 8996
71	1.1939 6356	0.8375 4650	77.5854 2345	64.9813 9989	0.0128 8902	0.0153 8902
72	1.1969 4847	0.8354 5786	78.7793 8701	65.8168 5774	0.0126 9368	0.0151 9368
73	1.1999 4084	0.8333 7442	79.9763 3548	66.6502 3216	0.0125 0370	0.0150 0370
74	1.2029 4069	0.8312 9618	81.1762 7632	67.4815 2834	0.0123 1887	0.0148 1887
75	1.2059 4804	0.8292 2312	82.3792 1701	68.3107 5146	0.0121 3898	0.0146 3898
76	1.2089 6191	0.8271 5523	83.5851 6505	69.1379 0670	0.0119 6385	0.0144 6385
77	1.2119 8532	0.8250 9250	84.7941 2797	69.9629 9920	0.0117 9327	0.0142 9327
78	1.2150 1528	0.8230 3491	86.0061 1329	70.7860 3411	0.0116 2708	0.0141 2708
79	1.2180 5282	0.8209 8246	87.2211 2857	71.6070 1657	0.0114 6511	0.0139 6511
80	1.2210 9795	0.8189 3512	88.4391 8139	72.4259 5169	0.0113 0721	0.0138 0721
81	1.2241 5070	0.8168 9289	89.6602 7934	73.2428 4458	0.0111 5321	0.0136 5321
82	1.2272 1108	0.8148 5575	90.8844 3004	74.0577 0033	0.0110 0298	0.0135 0298
83	1.2302 7910	0.8128 2369	92.1116 4112	74.8705 2402	0.0108 5639	0.0133 5639
84	1.2333 5480	0.8107 9670	93.3419 2022	75.6813 2072	0.0107 1330	0.0132 1330
85	1.2364 3819	0.8087 7476	94.5752 7502	76.4900 9548	0.0105 7359	0.0130 7359
86	1.2395 2928	0.8067 5787	95.8117 1321	77.2968 5335	0.0104 3714	0.0129 3714
87	1.2426 2811	0.8047 4600	97.0512 4249	78.1015 9935	0.0103 0384	0.0128 0384
88	1.2457 3468	0.8027 3915	98.2938 7060	78.9043 3850	0.0101 7357	0.0126 7357
89	1.2488 4901	0.8007 3731	99.5396 0527	79.7050 7581	0.0100 4625	0.0125 4625
90	1.2519 7114	0.7987 4046	100.7884 5429	80.5038 1627	0.0099 2177	0.0124 2177
91	1.2551 0106	0.7967 4859	102.0404 2542	81.3005 6486	0.0098 0004	0.0123 0004
92	1.2582 3882	0.7947 6168	103.2955 2649	82.0953 2654	0.0096 8096	0.0121 8096
93	1.2613 8441	0.7927 7973	104.5537 6530	82.8881 0628	0.0095 6446	0.0120 6446
94	1.2645 3787	0.7908 0273	105.8151 4972	83.6789 0900	0.0094 5044	0.0119 5044
95	1.2676 9922	0.7888 3065	107.0796 8759	84.4677 3966	0.0093 3884	0.0118 3884
96	1.2708 6847	0.7868 6349	108.3473 8681	85.2546 0315	0.0092 2957	0.0117 2957
97	1.2740 4564	0.7849 0124	109.6182 5528	86.0395 0439	0.0091 2257	0.0116 2257
98	1.2772 3075	0.7829 4388	110.8923 0091	86.8224 4827	0.0090 1776	0.0115 1776
99	1.2804 2383	0.7809 9140	112.1695 3167	87.6034 3967	0.0089 1508	0.0114 1508
100	1.2836 2489	0.7790 4379	113.4499 5550	88.3824 8346	0.0088 1446	0.0113 1446

$i = \frac{1}{2}\%$

n	$(1+i)^n$	$(1+i)^{-n}$	$s_{\overline{n}\rvert i}$	$a_{\overline{n}\rvert i}$	$\frac{1}{s_{\overline{n}\rvert i}}$	$\frac{1}{a_{\overline{n}\rvert i}}$
1	1.0050 0000	0.9950 2488	1.0000 0000	0.9950 2488	1.0000 0000	1.0050 0000
2	1.0100 2500	0.9900 7450	2.0050 0000	1.9850 9938	0.4987 5312	0.5037 5312
3	1.0150 7513	0.9851 4876	3.0150 2500	2.9702 4814	0.3316 7221	0.3366 7221
4	1.0201 5050	0.9802 4752	4.0301 0013	3.9504 9566	0.2481 3279	0.2531 3279
5	1.0252 5125	0.9753 7067	5.0502 5063	4.9258 6633	0.1980 0997	0.2030 0997
6	1.0303 7751	0.9705 1808	6.0755 0188	5.8963 8441	0.1645 9546	0.1695 9546
7	1.0355 2940	0.9656 8963	7.1058 7939	6.8620 7404	0.1407 2854	0.1457 2854
8	1.0407 0704	0.9608 8520	8.1414 0879	7.8229 5924	0.1228 2886	0.1278 2886
9	1.0459 1058	0.9561 0468	9.1821 1583	8.7790 6392	0.1089 0736	0.1139 0736
10	1.0511 4013	0.9513 4794	10.2280 2641	9.7304 1186	0.0977 7057	0.1027 7057
11	1.0563 9583	0.9466 1487	11.2791 6654	10.6770 2673	0.0886 5903	0.0936 5903
12	1.0616 7781	0.9419 0534	12.3355 6237	11.6189 3207	0.0810 6643	0.0860 6643
13	1.0669 8620	0.9372 1924	13.3972 4018	12.5561 5131	0.0746 4224	0.0796 4224
14	1.0723 2113	0.9325 5646	14.4642 2639	13.4887 0777	0.0691 3609	0.0741 3609
15	1.0776 8274	0.9279 1688	15.5365 4752	14.4166 2465	0.0643 6436	0.0693 6436
16	1.0830 7115	0.9233 0037	16.6142 3026	15.3399 2502	0.0601 8937	0.0651 8937
17	1.0884 8651	0.9187 0684	17.6973 0141	16.2586 3186	0.0565 0579	0.0615 0579
18	1.0939 2894	0.9141 3616	18.7857 8791	17.1727 6802	0.0532 3173	0.0582 3173
19	1.0993 9858	0.9095 8822	19.8797 1685	18.0823 5624	0.0503 0253	0.0553 0253
20	1.1048 9558	0.9050 6290	20.9791 1544	18.9874 1915	0.0476 6645	0.0526 6645
21	1.1104 2006	0.9005 6010	22.0840 1101	19.8879 7925	0.0452 8163	0.0502 8163
22	1.1159 7216	0.8960 7971	23.1944 3107	20.7840 5896	0.0431 1380	0.0481 1380
23	1.1215 5202	0.8916 2160	24.3104 0322	21.6756 8055	0.0411 3465	0.0461 3465
24	1.1271 5978	0.8871 8567	25.4319 5524	22.5628 6622	0.0393 2061	0.0443 2061
25	1.1327 9558	0.8827 7181	26.5591 1502	23.4456 3803	0.0376 5186	0.0426 5186
26	1.1384 5955	0.8783 7991	27.6919 1059	24.3240 1794	0.0361 1163	0.0411 1163
27	1.1441 5185	0.8740 0986	28.8303 7015	25.1980 2780	0.0346 8565	0.0396 8565
28	1.1498 7261	0.8696 6155	29.9745 2200	26.0676 8936	0.0333 6167	0.0383 6167
29	1.1556 2197	0.8653 3488	31.1243 9461	26.9330 2423	0.0321 2914	0.0371 2914
30	1.1614 0008	0.8610 2973	32.2800 1658	27.7940 5397	0.0309 7892	0.0359 7892
31	1.1672 0708	0.8567 4600	33.4414 1666	28.6507 9997	0.0299 0304	0.0349 0304
32	1.1730 4312	0.8524 8358	34.6086 2375	29.5032 8355	0.0288 9453	0.0338 9453
33	1.1789 0833	0.8482 4237	35.7816 6686	30.3515 2592	0.0279 4727	0.0329 4727
34	1.1848 0288	0.8440 2226	36.9605 7520	31.1955 4818	0.0270 5586	0.0320 5586
35	1.1907 2689	0.8398 2314	38.1453 7807	32.0353 7132	0.0262 1550	0.0312 1550
36	1.1966 8052	0.8356 4492	39.3361 0496	32.8710 1624	0.0254 2194	0.0304 2194
37	1.2026 6393	0.8314 8748	40.5327 8549	33.7025 0372	0.0246 7139	0.0296 7139
38	1.2086 7725	0.8273 5073	41.7354 4942	34.5298 5445	0.0239 6045	0.0289 6045
39	1.2147 2063	0.8232 3455	42.9441 2666	35.3530 8900	0.0232 8607	0.0282 8607
40	1.2207 9424	0.8191 3886	44.1588 4730	36.1722 2786	0.0226 4552	0.0276 4552
41	1.2268 9821	0.8150 6354	45.3796 4153	36.9872 9141	0.0220 3631	0.0270 3631
42	1.2330 3270	0.8110 0850	46.6065 3974	37.7982 9991	0.0214 5622	0.0264 5622
43	1.2391 9786	0.8069 7363	47.8395 7244	38.6052 7354	0.0209 0320	0.0259 0320
44	1.2453 9385	0.8029 5884	49.0787 7030	39.4082 3238	0.0203 7541	0.0253 7541
45	1.2516 2082	0.7989 6402	50.3241 6415	40.2071 9640	0.0198 7117	0.0248 7117
46	1.2578 7892	0.7949 8907	51.5757 8497	41.0021 8547	0.0193 8894	0.0243 8894
47	1.2641 6832	0.7910 3390	52.8336 6390	41.7932 1937	0.0189 2733	0.0239 2733
48	1.2704 8916	0.7870 9841	54.0978 3222	42.5803 1778	0.0184 8503	0.0234 8503
49	1.2768 4161	0.7831 8250	55.3683 2138	43.3635 0028	0.0180 6087	0.0230 6087
50	1.2832 2581	0.7792 8607	56.6451 6299	44.1427 8635	0.0176 5376	0.0226 5376

$i = \frac{1}{2}\%$

n	$(1+i)^n$	$(1+i)^{-n}$	$s_{\overline{n}\vert i}$	$a_{\overline{n}\vert i}$	$\frac{1}{s_{\overline{n}\vert i}}$	$\frac{1}{a_{\overline{n}\vert i}}$
51	1.2896 4194	0.7754 0902	57.9283 8880	44.9181 9537	0.0172 6269	0.0222 6269
52	1.2960 9015	0.7715 5127	59.2180 3075	45.6897 4664	0.0168 8675	0.0218 8675
53	1.3025 7060	0.7677 1270	60.5141 2090	46.4574 5934	0.0165 2507	0.0215 2507
54	1.3090 8346	0.7638 9324	61.8166 9150	47.2213 5258	0.0161 7686	0.0211 7686
55	1.3156 2887	0.7600 9277	63.1257 7496	47.9814 4535	0.0158 4139	0.0208 4139
56	1.3222 0702	0.7563 1122	64.4414 0384	48.7377 5657	0.0155 1797	0.0205 1797
57	1.3288 1805	0.7525 4847	65.7636 1086	49.4903 0505	0.0152 0598	0.0202 0598
58	1.3354 6214	0.7488 0445	67.0924 2891	50.2391 0950	0.0149 0481	0.0199 0481
59	1.3421 3946	0.7450 7906	68.4278 9105	50.9841 8855	0.0146 1392	0.0196 1392
60	1.3488 5015	0.7413 7220	69.7700 3051	51.7255 6075	0.0143 3280	0.0193 3280
61	1.3555 9440	0.7376 8378	71.1188 8066	52.4632 4453	0.0140 6096	0.0190 6096
62	1.3623 7238	0.7340 1371	72.4744 7507	53.1972 5824	0.0137 9796	0.0187 9796
63	1.3691 8424	0.7303 6190	73.8368 4744	53.9276 2014	0.0135 4337	0.0185 4337
64	1.3760 3016	0.7267 2826	75.2060 3168	54.6543 4839	0.0132 9681	0.0182 9681
65	1.3829 1031	0.7231 1269	76.5820 6184	55.3774 6109	0.0130 5789	0.0180 5789
66	1.3898 2486	0.7195 1512	77.9649 7215	56.0969 7621	0.0128 2627	0.0178 2627
67	1.3967 7399	0.7159 3544	79.3547 9701	56.8129 1165	0.0126 0163	0.0176 0163
68	1.4037 5785	0.7123 7357	80.7515 7099	57.5252 8522	0.0123 8366	0.0173 8366
69	1.4107 7664	0.7088 2943	82.1553 2885	58.2341 1465	0.0121 7206	0.0171 7206
70	1.4178 3053	0.7053 0291	83.5661 0549	58.9394 1756	0.0119 6657	0.0169 6657
71	1.4249 1968	0.7017 9394	84.9839 3602	59.6412 1151	0.0117 6693	0.0167 6693
72	1.4320 4428	0.6983 0243	86.4088 5570	60.3395 1394	0.0115 7289	0.0165 7289
73	1.4392 0450	0.6948 2829	87.8408 9998	61.0343 4222	0.0113 8422	0.0163 8422
74	1.4464 0052	0.6913 7143	89.2801 0448	61.7257 1366	0.0112 0070	0.0162 0070
75	1.4536 3252	0.6879 3177	90.7265 0500	62.4136 4543	0.0110 2214	0.0160 2214
76	1.4609 0069	0.6845 0923	92.1801 3752	63.0981 5466	0.0108 4832	0.0158 4832
77	1.4682 0519	0.6811 0371	93.6410 3821	63.7792 5836	0.0106 7908	0.0156 7908
78	1.4755 4622	0.6777 1513	95.1092 4340	64.4569 7350	0.0105 1423	0.0155 1423
79	1.4829 2395	0.6743 4342	96.5847 8962	65.1313 1691	0.0103 5360	0.0153 5360
80	1.4903 3857	0.6709 8847	98.0677 1357	65.8023 0538	0.0101 9704	0.0151 9704
81	1.4977 9026	0.6676 5022	99.5580 5214	66.4699 5561	0.0100 4439	0.0150 4439
82	1.5052 7921	0.6643 2858	101.0558 4240	67.1342 8419	0.0098 9552	0.0148 9552
83	1.5128 0561	0.6610 2346	102.5611 2161	67.7953 0765	0.0097 5028	0.0147 5028
84	1.5203 6964	0.6577 3479	104.0739 2722	68.4530 4244	0.0096 0855	0.0146 0855
85	1.5279 7148	0.6544 6248	105.5942 9685	69.1075 0491	0.0094 7021	0.0144 7021
86	1.5356 1134	0.6512 0644	107.1222 6834	69.7587 1135	0.0093 3513	0.0143 3513
87	1.5432 8940	0.6479 6661	108.6578 7968	70.4066 7796	0.0092 0320	0.0142 0320
88	1.5510 0585	0.6447 4290	110.2011 6908	71.0514 2086	0.0090 7431	0.0140 7431
89	1.5587 6087	0.6415 3522	111.7521 7492	71.6929 5608	0.0089 4837	0.0139 4837
90	1.5665 5468	0.6383 4350	113.3109 3580	72.3312 9958	0.0088 2527	0.0138 2527
91	1.5743 8745	0.6351 6766	114.8774 9048	72.9664 6725	0.0087 0493	0.0137 0493
92	1.5822 5939	0.6320 0763	116.4518 7793	73.5984 7487	0.0085 8724	0.0135 8724
93	1.5901 7069	0.6288 6331	118.0341 3732	74.2273 3818	0.0084 7213	0.0134 7213
94	1.5981 2154	0.6257 3464	119.6243 0800	74.8530 7282	0.0083 5950	0.0133 5950
95	1.6061 1215	0.6226 2153	121.2224 2954	75.4756 9434	0.0082 4930	0.0132 4930
96	1.6141 4271	0.6195 2391	122.8285 4169	76.0952 1825	0.0081 4143	0.0131 4143
97	1.6222 1342	0.6164 4170	124.4426 8440	76.7116 5995	0.0080 3583	0.0130 3583
98	1.6303 2449	0.6133 7483	126.0648 9782	77.3250 3478	0.0079 3242	0.0129 3242
99	1.6384 7611	0.6103 2321	127.6952 2231	77.9353 5799	0.0078 3115	0.0128 3115
100	1.6466 6849	0.6072 8678	129.3336 9842	78.5426 4477	0.0077 3194	0.0127 3194

$i = \frac{3}{4}\%$

n	$(1+i)^n$	$(1+i)^{-n}$	$s_{\overline{n}\rvert i}$	$a_{\overline{n}\rvert i}$	$\frac{1}{s_{\overline{n}\rvert i}}$	$\frac{1}{a_{\overline{n}\rvert i}}$
1	1.0075 0000	0.9925 5583	1.0000 0000	0.9925 5583	1.0000 0000	1.0075 0000
2	1.0150 5625	0.9851 6708	2.0075 0000	1.9777 2291	0.4981 3200	0.5056 3200
3	1.0226 6917	0.9778 3333	3.0225 5625	2.9555 5624	0.3308 4579	0.3383 4579
4	1.0303 3919	0.9705 5417	4.0452 2542	3.9261 1041	0.2472 0501	0.2547 0501
5	1.0380 6673	0.9633 2920	5.0755 6461	4.8894 3961	0.1970 2242	0.2045 2242
6	1.0458 5224	0.9561 5802	6.1136 3135	5.8455 9763	0.1635 6891	0.1710 6891
7	1.0536 9613	0.9490 4022	7.1594 8358	6.7946 3785	0.1396 7488	0.1471 7488
8	1.0615 9885	0.9419 7540	8.2131 7971	7.7366 1325	0.1217 5552	0.1292 5552
9	1.0695 6084	0.9349 6318	9.2747 7856	8.6715 7642	0.1078 1929	0.1153 1929
10	1.0775 8255	0.9270 0315	10.3443 3940	9.5995 7958	0.0966 7123	0.1041 7123
11	1.0856 6441	0.9210 9494	11.4219 2194	10.5206 7452	0.0875 5094	0.0950 5094
12	1.0938 0690	0.9142 3815	12.5075 8636	11.4349 1267	0.0799 5148	0.0874 5148
13	1.1020 1045	0.9074 3241	13.6013 9325	12.3423 4508	0.0735 2188	0.0810 2188
14	1.1102 7553	0.9006 7733	14.7034 0370	13.2430 2242	0.0680 1146	0.0755 1146
15	1.1186 0259	0.8939 7254	15.8136 7923	14.1369 9495	0.0632 3639	0.0707 3639
16	1.1269 9211	0.8873 1766	16.9322 8183	15.0243 1261	0.0590 5879	0.0665 5879
17	1.1354 4455	0.8807 1231	18.0592 7394	15.9050 2492	0.0553 7321	0.0628 7321
18	1.1439 6039	0.8741 5614	19.1947 1849	16.7791 8107	0.0520 9766	0.0595 9766
19	1.1525 4009	0.8676 4878	20.3386 7888	17.6468 2984	0.0491 6740	0.0566 6740
20	1.1611 8414	0.8611 8985	21.4912 1897	18.5080 1969	0.0465 3063	0.0540 3063
21	1.1698 9302	0.8547 7901	22.6524 0312	19.3627 9870	0.0441 4543	0.0516 4543
22	1.1786 6722	0.8484 1589	23.8222 9614	20.2112 1459	0.0419 7748	0.0494 7748
23	1.1875 0723	0.8421 0014	25.0009 6336	21.0533 1473	0.0399 9846	0.0474 9846
24	1.1964 1353	0.8358 3140	26.1884 7059	21.8891 4614	0.0381 8474	0.0456 8474
25	1.2053 8663	0.8296 0933	27.3848 8412	22.7187 5547	0.0365 1650	0.0440 1650
26	1.2144 2703	0.8234 3358	28.5902 7075	23.5421 8905	0.0349 7693	0.0424 7693
27	1.2235 3523	0.8173 0380	29.8046 9778	24.3594 9286	0.0335 5176	0.0410 5176
28	1.2327 1175	0.8112 1966	31.0282 3301	25.1707 1251	0.0322 2871	0.0397 2871
29	1.2419 5709	0.8051 8080	32.2609 4476	25.9758 9331	0.0309 9723	0.0384 9723
30	1.2512 7176	0.7991 8690	35.5029 0184	26.7750 8021	0.0298 4816	0.0373 4816
31	1.2606 5630	0.7932 3762	34.7541 7361	27.5683 1783	0.0287 7352	0.0362 7352
32	1.2701 1122	0.7873 3262	36.0148 2991	28.3556 5045	0.0277 6634	0.0352 6634
33	1.2796 3706	0.7814 7158	37.2849 4113	29.1371 2203	0.0268 2048	0.0343 2048
34	1.2892 3434	0.7756 5418	38.5645 7819	29.9127 7621	0.0259 3053	0.0334 3053
35	1.2989 0359	0.7698 8008	39.8538 1253	30.6826 5629	0.0250 9170	0.0325 9170
36	1.3086 4537	0.7641 4896	41.1527 1612	31.4468 0525	0.0242 9973	0.0317 9973
37	1.3184 6021	0.7584 6051	42.4613 6149	32.2052 6576	0.0235 5082	0.0310 5082
38	1.3283 4866	0.7528 1440	43.7798 2170	32.9580 8016	0.0228 4157	0.0303 4157
39	1.3383 1128	0.7472 1032	45.1081 7037	33.7052 9048	0.0221 6893	0.0296 6893
40	1.3483 4861	0.7416 4796	46.4464 8164	34.4469 3844	0.0215 3016	0.0290 3016
41	1.3584 6123	0.7361 2701	47.7948 3026	35.1830 6545	0.0209 2276	0.0284 2276
42	1.3686 4969	0.7306 4716	49.1532 9148	35.9137 1260	0.0203 4452	0.0278 4452
43	1.3789 1456	0.7252 0809	50.5219 4117	36.6389 2070	0.0197 9338	0.0272 9338
44	1.3892 5642	0.7198 0952	51.9008 5573	37.3587 3022	0.0192 6751	0.0267 6751
45	1.3996 7584	0.7144 5114	53.2901 1215	38.0731 8136	0.0187 6521	0.0262 6521
46	1.4101 7341	0.7091 3264	54.6897 8799	38.7823 1401	0.0182 8495	0.0257 8495
47	1.4207 4971	0.7038 5374	56.0999 6140	39.4861 6775	0.0178 2532	0.0253 2532
48	1.4314 0533	0.6986 1414	57.5207 1111	40.1847 8189	0.0173 8504	0.0248 8504
49	1.4421 4087	0.6934 1353	58.9521 1644	40.8781 9542	0.0169 6292	0.0244 6292
50	1.4529 5693	0.6882 5165	60.3942 5732	41.5664 4707	0.0165 5787	0.0240 5787

$i = \frac{3}{4}\%$

n	$(1+i)^n$	$(1+i)^{-n}$	$s_{\overline{n}\rvert i}$	$a_{\overline{n}\rvert i}$	$\frac{1}{s_{\overline{n}\rvert i}}$	$\frac{1}{a_{\overline{n}\rvert i}}$
51	1.4638 5411	0.6831 2819	61.8472 1424	42.2495 7525	0.0161 6888	0.0236 6888
52	1.4748 3301	0.6780 4286	63.3110 6835	42.9276 1812	0.0157 9503	0.0232 9503
53	1.4858 9426	0.6729 9540	64.7859 0136	43.6006 1351	0.0154 3546	0.0229 3546
54	1.4970 3847	0.6679 8551	66.2717 9562	44.2685 9902	0.0150 8938	0.0225 8938
55	1.5082 6626	0.6630 1291	67.7688 3409	44.9316 1193	0.0147 5605	0.0222 5605
56	1.5195 7825	0.6580 7733	69.2771 0035	45.5896 8926	0.0144 3478	0.0219 3478
57	1.5309 7509	0.6531 7849	70.7966 7860	46.2428 6776	0.0141 2496	0.0216 2496
58	1.5424 5740	0.6483 1612	72.3276 5369	46.8911 8388	0.0138 2597	0.0213 2597
59	1.5540 2583	0.6434 8995	73.8701 1109	47.5346 7382	0.0135 3727	0.0210 3727
60	1.5656 8103	0.6386 9970	75.4241 3693	48.1733 7352	0.0132 5836	0.0207 5836
61	1.5774 2363	0.6339 4511	76.9898 1795	48.8073 1863	0.0129 8873	0.0204 8873
62	1.5892 5431	0.6292 2592	78.5672 4159	49.4365 4455	0.0127 2795	0.0202 2795
63	1.6011 7372	0.6245 4185	80.1564 9590	50.0610 8640	0.0124 7560	0.0199 7560
64	1.6131 8252	0.6198 9266	81.7576 6962	50.6809 7906	0.0122 3127	0.0197 3127
65	1.6252 8139	0.6152 7807	83.3708 5214	51.2962 5713	0.0119 9460	0.0194 9460
66	1.6374 7100	0.6106 9784	84.9961 3353	51.9069 5497	0.0117 6524	0.0192 6524
67	1.6497 5203	0.6061 5170	86.6336 0453	52.5131 0667	0.0115 4286	0.0190 4286
68	1.6621 2517	0.6016 3940	88.2833 5657	53.1147 4607	0.0113 2716	0.0188 2716
69	1.6745 9111	0.5971 6070	89.9454 8174	53.7119 0677	0.0111 1785	0.0186 1785
70	1.6871 5055	0.5927 1533	91.6200 7285	54.3046 2210	0.0109 1464	0.0184 1464
71	1.6998 0418	0.5883 0306	93.3072 2340	54.8929 2516	0.0107 1728	0.0182 1728
72	1.7125 5271	0.5839 2363	95.0070 2758	55.4768 4880	0.0105 2554	0.0180 2554
73	1.7253 9685	0.5795 7681	96.7195 8028	56.0564 2561	0.0103 3917	0.0178 3917
74	1.7383 3733	0.5752 6234	98.4449 7714	56.6316 8795	0.0101 5796	0.0176 5796
75	1.7513 7486	0.5709 7999	100.1833 1446	57.2026 6794	0.0099 8170	0.0174 8170
76	1.7645 1017	0.5667 2952	101.9346 8932	57.7693 9746	0.0098 1020	0.0173 1020
77	1.7777 4400	0.5625 1069	103.6991 9949	58.3319 0815	0.0096 4328	0.0171 4328
78	1.7910 7708	0.5583 2326	105.4769 4349	58.8902 3141	0.0094 8074	0.0169 8074
79	1.8045 1015	0.5541 6701	107.2680 2056	59.4443 9842	0.0093 2244	0.0168 2244
80	1.8180 4398	0.5500 4170	109.0725 3072	59.9944 4012	0.0091 6821	0.0166 6821
81	1.8316 7931	0.5459 4710	110.8905 7470	60.5403 8722	0.0090 1790	0.0165 1790
82	1.8454 1691	0.5418 8297	112.7222 5401	61.0822 7019	0.0088 7136	0.0163 7136
83	1.8592 5753	0.5378 4911	114.5676 7091	61.6201 1930	0.0087 2847	0.0162 2847
84	1.8732 0196	0.5338 4527	116.4269 2845	62.1539 6456	0.0085 8908	0.0160 8908
85	1.8872 5098	0.5298 7123	118.3001 3041	62.6838 3579	0.0084 5308	0.0159 5308
86	1.9014 0536	0.5259 2678	120.1873 8139	63.2097 6257	0.0083 2034	0.0158 2034
87	1.9156 6590	0.5220 1169	122.0887 8675	63.7317 7427	0.0081 9076	0.0156 9076
88	1.9300 3339	0.5181 2575	124.0044 5265	64.2499 0002	0.0080 6423	0.0155 6423
89	1.9445 0865	0.5142 6873	125.9344 8604	64.7641 6875	0.0079 4064	0.0154 4064
90	1.9590 9246	0.5104 4043	127.8789 9469	65.2746 0918	0.0078 1989	0.0153 1989
91	1.9737 8565	0.5066 4063	129.8380 8715	65.7812 4981	0.0077 0190	0.0152 0190
92	1.9885 8905	0.5028 6911	131.8118 7280	66.2841 1892	0.0075 8657	0.0150 8657
93	2.0035 0346	0.4991 2567	133.8004 6185	66.7832 4458	0.0074 7382	0.0149 7382
94	2.0185 2974	0.4954 1009	135.8039 6531	67.2786 5467	0.0073 6356	0.0148 6356
95	2.0336 6871	0.4917 2217	137.8224 9505	67.7703 7685	0.0072 5571	0.0147 5571
96	2.0489 2123	0.4880 6171	139.8561 6377	68.2584 3856	0.0071 5020	0.0146 5020
97	2.0642 8814	0.4844 2850	141.9050 8499	68.7428 6705	0.0070 4696	0.0145 4696
98	2.0797 7030	0.4808 2233	143.9693 7313	69.2236 8938	0.0069 4592	0.0144 4592
99	2.0953 6858	0.4772 4301	146.0491 4343	69.7009 3239	0.0068 4701	0.0143 4701
100	2.1110 8384	0.4736 9033	148.1445 1201	70.1746 2272	0.0067 5017	0.0142 5017

$i = 1\%$

n	$(1+i)^n$	$(1+i)^{-n}$	$s_{\overline{n}\vert i}$	$a_{\overline{n}\vert i}$	$\frac{1}{s_{\overline{n}\vert i}}$	$\frac{1}{a_{\overline{n}\vert i}}$
1	1.0100 0000	0.9900 9901	1.0000 0000	0.9900 9901	1.0000 0000	1.0100 0000
2	1.0201 0000	0.9802 9605	2.0100 0000	1.9703 9506	0.4975 1244	0.5075 1244
3	1.0303 0100	0.9705 9015	3.0301 0000	2.9409 8521	0.3300 2211	0.3400 2211
4	1.0406 0401	0.9609 8034	4.0604 0100	3.9019 6555	0.2462 8109	0.2562 8109
5	1.0510 1005	0.9514 6569	5.1010 0501	4.8534 3124	0.1960 3980	0.2060 3980
6	1.0615 2015	0.9420 4524	6.1520 1506	5.7954 7647	0.1625 4837	0.1725 4837
7	1.0721 3535	0.9327 1805	7.2135 3521	6.7281 9453	0.1386 2828	0.1486 2828
8	1.0828 5671	0.9234 8322	8.2856 7056	7.6516 7775	0.1206 9029	0.1306 9029
9	1.0936 8527	0.9143 3982	9.3685 2727	8.5660 1758	0.1067 4036	0.1167 4036
10	1.1046 2213	0.9052 8695	10.4622 1254	9.4713 0453	0.0955 8208	0.1055 8208
11	1.1156 6835	0.8963 2372	11.5668 3467	10.3676 2825	0.0864 5408	0.0964 5408
12	1.1268 2503	0.8874 4923	12.6825 0301	11.2550 7747	0.0788 4879	0.0888 4879
13	1.1380 9328	0.8786 6260	13.8093 2804	12.1337 4007	0.0724 1482	0.0824 1482
14	1.1494 7421	0.8699 6297	14.9474 2132	13.0037 0304	0.0669 0117	0.0769 0117
15	1.1609 6896	0.8613 4947	16.0968 9554	13.8650 5252	0.0621 2378	0.0721 2378
16	1.1725 7864	0.8528 2126	17.2578 6449	14.7178 7378	0.0579 4460	0.0679 4460
17	1.1843 0443	0.8443 7749	18.4304 4314	15.5622 5127	0.0542 5806	0.0642 5806
18	1.1961 4748	0.8360 1731	19.6147 4757	16.3982 6858	0.0509 8205	0.0609 8205
19	1.2081 0895	0.8277 3992	20.8108 9504	17.2260 0850	0.0480 5175	0.0580 5175
20	1.2201 9004	0.8195 4447	22.0190 0399	18.0455 5297	0.0454 1531	0.0554 1531
21	1.2323 9194	0.8114 3017	23.2391 9403	18.8569 8313	0.0430 3075	0.0530 3075
22	1.2447 1586	0.8033 9621	24.4715 8598	19.6603 7934	0.0408 6372	0.0508 6372
23	1.2571 6302	0.7954 4179	25.7163 0183	20.4558 2113	0.0388 8584	0.0488 8584
24	1.2697 3465	0.7875 6613	26.9734 6485	21.2433 8726	0.0370 7347	0.0470 7347
25	1.2824 3200	0.7797 6844	28.2431 9950	22.0231 5570	0.0354 0675	0.0454 0675
26	1.2952 5631	0.7720 4796	29.5256 3150	22.7952 0366	0.0338 6888	0.0438 6888
27	1.3082 0888	0.7644 0392	30.8208 8781	23.5596 0759	0.0324 4553	0.0424 4553
28	1.3212 9097	0.7568 3557	32.1290 9669	24.3164 4316	0.0311 2444	0.0411 2444
29	1.3345 0388	0.7493 4215	33.4503 8766	25.0657 8530	0.0298 9502	0.0398 9502
30	1.3478 4892	0.7419 2292	34.7848 9153	25.8077 0822	0.0287 4811	0.0387 4811
31	1.3613 2740	0.7345 7715	36.1327 4045	26.5422 8537	0.0276 7573	0.0376 7573
32	1.3749 4068	0.7273 0411	37.4940 6785	27.2695 8947	0.0266 7089	0.0366 7089
33	1.3886 9009	0.7201 0307	38.8690 0853	27.9896 9255	0.0257 2744	0.0357 2744
34	1.4025 7699	0.7129 7334	40.2576 9862	28.7026 6589	0.0248 3997	0.0348 3997
35	1.4166 0276	0.7059 1420	41.6602 7560	29.4085 8009	0.0240 0368	0.0340 0368
36	1.4307 6878	0.6989 2495	43.0768 7836	30.1075 0504	0.0232 1431	0.0332 1431
37	1.4450 7647	0.6920 0490	44.5076 4714	30.7995 0994	0.0224 6805	0.0324 6805
38	1.4595 2724	0.6851 5337	45.9527 2361	31.4846 6330	0.0217 6150	0.0317 6150
39	1.4741 2251	0.6783 6967	47.4122 5085	32.1630 3298	0.0210 9160	0.0310 9160
40	1.4888 6373	0.6716 5314	48.8863 7336	32.8346 8611	0.0204 5560	0.0304 5560
41	1.5037 5237	0.6650 0311	50.3752 3709	33.4996 8922	0.0198 5102	0.0298 5102
42	1.5187 8989	0.6584 1892	51.8789 8946	34.1581 0814	0.0192 7563	0.0292 7563
43	1.5339 7779	0.6518 9992	53.3977 7936	34.8100 0806	0.0187 2737	0.0287 2737
44	1.5493 1757	0.6454 4546	54.9317 5715	35.4554 5352	0.0182 0441	0.0282 0441
45	1.5648 1075	0.6390 5492	56.4810 7472	36.0945 0844	0.0177 0505	0.0277 0505
46	1.5804 5885	0.6327 2764	58.0458 8547	36.7272 3608	0.0172 2775	0.0272 2775
47	1.5962 6344	0.6264 6301	59.6263 4432	37.3536 9909	0.0167 7111	0.0267 7111
48	1.6122 2608	0.6202 6041	61.2226 0777	37.9739 5949	0.0163 3384	0.0263 3384
49	1.6283 4834	0.6141 1921	62.8348 3385	38.5880 7871	0.0159 1474	0.0259 1474
50	1.6446 3182	0.6080 3882	64.4631 8218	39.1961 1753	0.0155 1273	0.0255 1273

$i = 1\%$

n	$(1+i)^n$	$(1+i)^{-n}$	$s_{\overline{n}\mid i}$	$a_{\overline{n}\mid i}$	$\frac{1}{s_{\overline{n}\mid i}}$	$\frac{1}{a_{\overline{n}\mid i}}$
51	1.6610 7814	0.6020 1864	66.1078 1401	39.7981 3617	0.0151 2680	0.0251 2680
52	1.6776 8892	0.5960 5806	67.7688 9215	40.3941 9423	0.0147 5603	0.0247 5603
53	1.6944 6581	0.5901 5649	69.4465 8107	40.9843 5072	0.0143 9956	0.0243 9956
54	1.7114 1047	0.5843 1336	71.1410 4688	41.5686 6408	0.0140 5658	0.0240 5658
55	1.7285 2457	0.5785 2808	72.8524 5735	42.1471 9216	0.0137 2637	0.0237 2637
56	1.7458 0982	0.5728 0008	74.5809 8192	42.7199 9224	0.0134 0824	0.0234 0824
57	1.7632 6792	0.5671 2879	76.3267 9174	43.2871 2102	0.0131 0156	0.0231 0156
58	1.7809 0060	0.5615 1365	78.0900 5966	43.8486 3468	0.0128 0573	0.0228 0573
59	1.7987 0960	0.5559 5411	79.8709 6025	44.4045 8879	0.0125 2020	0.0225 2020
60	1.8166 9670	0.5504 4962	81.6696 6986	44.9550 3841	0.0122 4445	0.0222 4445
61	1.8348 6367	0.5449 9962	83.4863 6655	45.5000 3803	0.0119 7800	0.0219 7800
62	1.8532 1230	0.5396 0358	85.3212 3022	46.0396 4161	0.0117 2041	0.0217 2041
63	1.8717 4443	0.5342 6097	87.1744 4252	46.5739 0258	0.0114 7125	0.0214 7125
64	1.8904 6187	0.5289 7126	89.0461 8695	47.1028 7385	0.0112 3013	0.0212 3013
65	1.9093 6649	0.5237 3392	90.9366 4882	47.6266 0777	0.0109 9667	0.0209 9667
66	1.9284 6015	0.5185 4844	92.8460 1531	48.1451 5621	0.0107 7052	0.0207 7052
67	1.9477 4475	0.5134 1429	94.7744 7546	48.6585 7050	0.0105 5136	0.0205 5136
68	1.9672 2220	0.5083 3099	96.7222 2021	49.1669 0149	0.0103 3889	0.0203 3889
69	1.9868 9442	0.5032 9801	98.6894 4242	49.6701 9949	0.0101 3280	0.0201 3280
70	2.0067 6337	0.4983 1486	100.6763 3684	50.1685 1435	0.0099 3282	0.0199 3282
71	2.0268 3100	0.4933 8105	102.6831 0021	50.6618 9539	0.0097 3870	0.0197 3870
72	2.0470 9931	0.4884 9609	104.7099 3121	51.1503 9148	0.0095 5019	0.0195 5019
73	2.0675 7031	0.4836 5949	106.7570 3052	51.6340 5097	0.0093 6706	0.0193 6706
74	2.0882 4601	0.4788 7078	108.8246 0083	52.1129 2175	0.0091 8910	0.0191 8910
75	2.1091 2847	0.4741 2949	110.9128 4684	52.5870 5124	0.0090 1609	0.0190 1609
76	2.1302 1975	0.4694 3514	113.0219 7530	53.0564 8638	0.0088 4784	0.0188 4784
77	2.1515 2195	0.4647 8726	115.1521 9506	53.5212 7364	0.0086 8416	0.0186 8416
78	2.1730 3717	0.4601 8541	117.3037 1701	53.9814 5905	0.0085 2488	0.0185 2488
79	2.1947 6754	0.4556 2912	119.4767 5418	54.4370 8817	0.0083 6983	0.0183 6983
80	2.2167 1522	0.4511 1794	121.6715 2172	54.8882 0611	0.0082 1885	0.0182 1885
81	2.2388 8237	0.4466 5142	123.8882 3694	55.3348 5753	0.0080 7179	0.0180 7179
82	2.2612 7119	0.4422 2913	126.1271 1931	55.7770 8666	0.0079 2851	0.0179 2851
83	2.2838 8390	0.4378 5063	128.3883 9050	56.2149 3729	0.0077 8887	0.0177 8887
84	2.3067 2274	0.4335 1547	130.6722 7440	56.6484 5276	0.0076 5273	0.0176 5273
85	2.3297 8997	0.4292 2324	132.9789 9715	57.0776 7600	0.0075 1998	0.0175 1998
86	2.3530 8787	0.4249 7350	135.3087 8712	57.5026 4951	0.0073 9050	0.0173 9050
87	2.3766 1875	0.4207 6585	137.6618 7499	57.9234 1535	0.0072 6418	0.0172 6418
88	2.4003 8494	0.4165 9985	140.0384 9374	58.3400 1520	0.0071 4089	0.0171 4089
89	2.4243 8879	0.4124 7510	142.4388 7868	58.7524 9030	0.0070 2056	0.0170 2056
90	2.4486 3267	0.4083 9119	144.8632 6746	59.1608 8148	0.0069 0306	0.0169 0306
91	2.4731 1900	0.4043 4771	147.3119 0014	59.5652 2919	0.0067 8832	0.0167 8832
92	2.4978 5019	0.4003 4427	149.7850 1914	59.9655 7346	0.0066 7624	0.0166 7624
93	2.5228 2869	0.3963 8046	152.2828 6933	60.3619 5392	0.0065 6673	0.0165 6673
94	2.5480 5698	0.3924 5590	154.8056 9803	60.7544 0982	0.0064 5971	0.0164 5971
95	2.5735 3755	0.3885 7020	157.3537 5501	61.1429 8002	0.0063 5511	0.0163 5511
96	2.5992 7293	0.3847 2297	159.9272 9256	61.5277 0299	0.0062 5284	0.0162 5284
97	2.6252 6565	0.3809 1383	162.5265 6548	61.9086 1682	0.0061 5284	0.0161 5284
98	2.6515 1831	0.3771 4241	165.1518 3114	62.2857 5923	0.0060 5503	0.0160 5503
99	2.6780 3349	0.3734 0832	167.8033 4945	62.6591 6755	0.0059 5936	0.0159 5936
100	2.7048 1383	0.3697 1121	170.4813 8294	63.0288 7877	0.0058 6574	0.0158 6574

$i = 1\frac{1}{4}\%$

n	$(1+i)^n$	$(1+i)^{-n}$	$s_{\overline{n}\rvert i}$	$a_{\overline{n}\rvert i}$	$\frac{1}{s_{\overline{n}\rvert i}}$	$\frac{1}{a_{\overline{n}\rvert i}}$
1	1.0125 0000	0.9876 5432	1.0000 0000	0.9876 5432	1.0000 0000	1.0125 0000
2	1.0251 5625	0.9754 6106	2.0125 0000	1.9631 1538	0.4968 9441	0.5093 9441
3	1.0379 7070	0.9634 1833	3.0376 5625	2.9265 3371	0.3292 0117	0.3417 0117
4	1.0509 4534	0.9515 2428	4.0756 2695	3.8780 5798	0.2453 6102	0.2578 6102
5	1.0640 8215	0.9397 7706	5.1265 7229	4.8178 3504	0.1950 6211	0.2075 6211
6	1.0773 8318	0.9281 7488	6.1906 5444	5.7460 0992	0.1615 3381	0.1740 3381
7	1.0908 5047	0.9167 1593	7.2680 3762	6.6627 2585	0.1375 8872	0.1500 8872
8	1.1044 8610	0.9053 9845	8.3588 8809	7.5681 2429	0.1196 3314	0.1321 3314
9	1.1182 9218	0.8942 2069	9.4633 7420	8.4623 4498	0.1056 7055	0.1181 7055
10	1.1322 7083	0.8831 8093	10.5816 6637	9.3455 2591	0.0945 0307	0.1070 0307
11	1.1464 2422	0.8722 7746	11.7139 3720	10.2178 0337	0.0853 6839	0.0978 6839
12	1.1607 5452	0.8615 0860	12.8603 6142	11.0793 1197	0.0777 5831	0.0902 5831
13	1.1752 6395	0.8508 7269	14.0211 1594	11.9301 8466	0.0713 2100	0.0838 2100
14	1.1899 5475	0.8403 6809	15.1963 7988	12.7705 5275	0.0658 0515	0.0783 0515
15	1.2048 2918	0.8299 9318	16.3863 3463	13.6005 4592	0.0610 2646	0.0735 2646
16	1.2198 8955	0.8197 4635	17.5911 6382	14.4202 9227	0.0568 4672	0.0693 4672
17	1.2351 3817	0.8096 2602	18.8110 5336	15.2299 1829	0.0531 6023	0.0656 6023
18	1.2505 7739	0.7996 3064	20.0461 9153	16.0295 4893	0.0498 8479	0.0623 8479
19	1.2662 0961	0.7897 5866	21.2967 6893	16.8193 0759	0.0469 5548	0.0594 5548
20	1.2820 3723	0.7800 0855	22.5629 7854	17.5993 1613	0.0443 2039	0.0568 2039
21	1.2980 6270	0.7703 7881	23.8450 1577	18.3696 9495	0.0419 3749	0.0544 3749
22	1.3142 8848	0.7608 6796	25.1430 7847	19.1305 6291	0.0397 7238	0.0522 7238
23	1.3307 1709	0.7514 7453	26.4573 6695	19.8820 3744	0.0377 9666	0.0502 9666
24	1.3473 5105	0.7421 9707	27.7880 8403	20.6242 3451	0.0359 8665	0.0484 8665
25	1.3641 9294	0.7330 3414	29.1354 3508	21.3572 6865	0.0343 2247	0.0468 2247
26	1.3812 4535	0.7239 8434	30.4996 2802	22.0812 5299	0.0327 8729	0.0452 8729
27	1.3985 1092	0.7150 4626	31.8808 7337	22.7962 9925	0.0313 6677	0.0438 6677
28	1.4159 9230	0.7062 1853	33.2793 8429	23.5025 1778	0.0300 4863	0.0425 4863
29	1.4336 9221	0.6974 9978	34.6953 7659	24.2000 1756	0.0288 2228	0.0413 2228
30	1.4516 1336	0.6888 8867	36.1290 6880	24.8889 0623	0.0276 7854	0.0401 7854
31	1.4697 5853	0.6803 8387	37.5806 8216	25.5692 9010	0.0266 0942	0.0391 0942
32	1.4881 3051	0.6719 8407	39.0504 4069	26.2412 7418	0.0256 0791	0.0381 0791
33	1.5067 3214	0.6636 8797	40.5385 7120	26.9049 6215	0.0246 6786	0.0371 6786
34	1.5255 6629	0.6554 9429	42.0453 0334	27.5604 5644	0.0237 8387	0.0362 8387
35	1.5446 3587	0.6474 0177	43.5708 6963	28.2078 5822	0.0229 5111	0.0354 5111
36	1.5639 4382	0.6394 0916	45.1155 0550	28.8472 6737	0.0221 6533	0.0346 6533
37	1.5834 9312	0.6315 1522	46.6794 4932	29.4787 8259	0.0214 2270	0.0339 2270
38	1.6032 8678	0.6237 1873	48.2629 4243	30.1025 0133	0.0207 1983	0.0332 1983
39	1.6233 2787	0.6160 1850	49.8662 2921	30.7185 1983	0.0200 5365	0.0325 5365
40	1.6436 1946	0.6084 1334	51.4895 5708	31.3269 3316	0.0194 2141	0.0319 2141
41	1.6641 6471	0.6009 0206	53.1331 7654	31.9278 3522	0.0188 2063	0.0313 2063
42	1.6849 6677	0.5934 8352	54.7973 4125	32.5213 1874	0.0182 4906	0.0307 4906
43	1.7060 2885	0.5861 5656	56.4823 0801	33.1074 7530	0.0177 0466	0.0302 0466
44	1.7273 5421	0.5789 2006	58.1883 3687	33.6863 9536	0.0171 8557	0.0296 8557
45	1.7489 4614	0.5717 7290	59.9156 9108	34.2581 6825	0.0166 9012	0.0291 9012
46	1.7708 0797	0.5647 1397	61.6646 3721	34.8228 8222	0.0162 1675	0.0287 1675
47	1.7929 4306	0.5577 4219	63.4354 4518	35.3806 2442	0.0157 6406	0.0282 6406
48	1.8153 5485	0.5508 5649	65.2283 8824	35.9314 8091	0.0153 3075	0.0278 3075
49	1.8380 4679	0.5440 5579	67.0437 4310	36.4755 3670	0.0149 1563	0.0274 1563
50	1.8610 2237	0.5373 3905	68.8817 8989	37.0128 7575	0.0145 1763	0.0270 1763

$i = 1\frac{1}{4}\%$

n	$(1+i)^n$	$(1+i)^{-n}$	$s_{\overline{n}\vert i}$	$a_{\overline{n}\vert i}$	$\frac{1}{s_{\overline{n}\vert i}}$	$\frac{1}{a_{\overline{n}\vert i}}$
51	1.8842 8515	0.5307 0524	70.7428 1226	37.5435 8099	0.0141 3571	0.0266 3571
52	1.9078 3872	0.5241 5332	72.6270 9741	38.0677 3431	0.0137 6897	0.0262 6897
53	1.9316 8670	0.5176 8229	74.5349 3613	38.5854 1660	0.0134 1653	0.0259 1653
54	1.9558 3279	0.5112 9115	76.4666 2283	39.0967 0776	0.0130 7760	0.0255 7760
55	1.9802 8070	0.5049 7892	78.4224 5562	39.6016 8667	0.0127 5145	0.0252 5145
56	2.0050 3420	0.4987 4461	80.4027 3631	40.1004 3128	0.0124 3739	0.0249 3739
57	2.0300 9713	0.4925 8727	82.4077 7052	40.5930 1855	0.0121 3478	0.0246 3478
58	2.0554 7335	0.4865 0594	84.4378 6765	41.0795 2449	0.0118 4303	0.0243 4303
59	2.0811 6676	0.4804 9970	86.4933 4099	41.5600 2419	0.0115 6158	0.0240 6158
60	2.1071 8135	0.4745 6760	88.5745 0776	42.0345 9179	0.0112 8993	0.0237 8993
61	2.1335 2111	0.4687 0874	90.6816 8910	42.5033 0054	0.0110 2758	0.0235 2758
62	2.1601 9013	0.4629 2222	92.8152 1022	42.9662 2275	0.0107 7410	0.0232 7410
63	2.1871 9250	0.4572 0713	94.9754 0034	43.4234 2988	0.0105 2904	0.0230 2904
64	2.2145 3241	0.4515 6259	97.1625 9285	43.8749 9247	0.0102 9203	0.0227 9203
65	2.2422 1407	0.4459 8775	99.3771 2526	44.3209 8022	0.0100 6268	0.0225 6268
66	2.2702 4174	0.4404 8173	101.6193 3933	44.7614 6195	0.0098 4065	0.0223 4065
67	2.2986 1976	0.4350 4368	103.8895 8107	45.1965 0563	0.0096 2560	0.0221 2560
68	2.3273 5251	0.4296 7277	106.1882 0083	45.6261 7840	0.0094 1724	0.0219 1724
69	2.3564 4442	0.4243 6817	108.5155 5334	46.0505 4656	0.0092 1527	0.0217 1527
70	2.3858 9997	0.4191 2905	110.8719 9776	46.4696 7562	0.0090 1941	0.0215 1941
71	2.4157 2372	0.4139 5462	113.2578 9773	46.8836 3024	0.0088 2941	0.0213 2941
72	2.4459 2027	0.4088 4407	115.6736 2145	47.2924 7431	0.0086 4501	0.0211 4501
73	2.4764 9427	0.4037 9661	118.1195 4172	47.6962 7093	0.0084 6600	0.0209 6600
74	2.5074 5045	0.3988 1147	120.5960 3599	48.0950 8240	0.0082 9215	0.0207 9215
75	2.5387 9358	0.3938 8787	123.1034 8644	48.4889 7027	0.0081 2325	0.0206 2325
76	2.5705 2850	0.3890 2506	125.6422 8002	48.8779 9533	0.0079 5910	0.0204 5910
77	2.6026 6011	0.3842 2228	128.2128 0852	49.2622 1761	0.0077 9953	0.0202 9953
78	2.6351 9336	0.3794 7879	130.8154 6863	49.6416 9640	0.0076 4436	0.0201 4436
79	2.6681 3327	0.3747 9387	133.4506 6199	50.0164 9027	0.0074 9341	0.0199 9341
80	2.7014 8494	0.3701 6679	136.1187 9526	50.3866 5706	0.0073 4652	0.0198 4652
81	2.7352 5350	0.3655 9683	138.8202 8020	50.7522 5389	0.0072 0356	0.0197 0356
82	2.7694 4417	0.3610 8329	141.5555 3370	51.1133 3717	0.0070 6437	0.0195 6347
83	2.8040 6222	0.3566 2547	144.3249 7787	51.4699 6264	0.0069 2881	0.0194 2881
84	2.8391 1300	0.3522 2268	147.1290 4010	51.8221 8532	0.0067 9675	0.0192 9675
85	2.8746 0191	0.3478 7426	149.9681 5310	52.1700 5958	0.0066 6808	0.0191 6808
86	2.9105 3444	0.3435 7951	152.8427 5501	52.5136 3909	0.0065 4267	0.0190 4267
87	2.9469 1612	0.3393 3779	155.7532 8945	52.8529 7688	0.0064 2041	0.0189 2041
88	2.9837 5257	0.3351 4843	158.7002 0557	53.1881 2531	0.0063 0119	0.0188 0119
89	3.0210 4948	0.3310 1080	161.6839 5814	53.5191 3611	0.0061 8491	0.0186 8491
90	3.0588 1260	0.3269 2425	164.7050 0762	53.8460 6036	0.0060 7146	0.0185 7146
91	3.0970 4775	0.3228 8814	167.7638 2021	54.1689 4850	0.0059 6076	0.0184 6076
92	3.1357 6085	0.3189 0187	170.8608 6796	54.4878 5037	0.0058 5272	0.0183 5272
93	3.1749 5786	0.3149 6481	173.9966 2881	54.8028 1518	0.0057 4724	0.0182 4724
94	3.2146 4483	0.3110 7636	177.1715 8667	55.1138 9154	0.0056 4425	0.0181 4425
95	3.2548 2789	0.3072 3591	180.3862 3151	55.4211 2744	0.0055 4366	0.0180 4366
96	3.2955 1324	0.3034 4287	183.6410 5940	55.7245 7031	0.0054 4541	0.0179 4541
97	3.3367 0716	0.2996 9666	186.9365 7264	56.0242 6698	0.0053 4941	0.0178 4941
98	3.3784 1600	0.2959 9670	190.2732 7980	56.3202 6368	0.0052 5560	0.0177 5560
99	3.4206 4620	0.2923 4242	193.6516 9580	56.6126 0610	0.0051 6391	0.0176 6391
100	3.4634 0427	0.2887 3326	197.0723 4200	56.9013 3936	0.0050 7428	0.0175 7428

$i = 1\frac{1}{2}\%$

n	$(1+i)^n$	$(1+i)^{-n}$	$s_{\overline{n}\vert i}$	$a_{\overline{n}\vert i}$	$\frac{1}{s_{\overline{n}\vert i}}$	$\frac{1}{a_{\overline{n}\vert i}}$
1	1.0150 0000	0.9852 2167	1.0000 0000	0.9852 2167	1.0000 0000	1.0150 0000
2	1.0302 2500	0.9706 6175	2.0150 0000	1.9558 8342	0.4962 7792	0.5112 7792
3	1.0456 7838	0.9563 1699	3.0452 2500	2.9122 0042	0.3283 8296	0.3433 8296
4	1.0613 6355	0.9421 8423	4.0909 0338	3.8543 8465	0.2444 4479	0.2594 4479
5	1.0772 8400	0.9282 6033	5.1522 6693	4.7826 4497	0.1940 8932	0.2090 8932
6	1.0934 4326	0.9145 4219	6.2295 5093	5.6971 8717	0.1605 2521	0.1755 2521
7	1.1098 4491	0.9010 2679	7.3229 9419	6.5982 1396	0.1356 5616	0.1515 5616
8	1.1264 9259	0.8877 1112	8.4328 3911	7.4859 2508	0.1185 8402	0.1335 8402
9	1.1433 8998	0.8745 9224	9.5593 3169	8.3605 1732	0.1046 0982	0.1196 0982
10	1.1605 4083	0.8616 6723	10.7027 2167	9.2221 8455	0.0934 3418	0.1084 3418
11	1.1779 4894	0.8489 3323	11.8632 6249	10.0711 1779	0.0842 9384	0.0992 9384
12	1.1956 1817	0.8363 8742	13.0412 1143	10.9075 0521	0.0766 7999	0.0916 7999
13	1.2135 5244	0.8240 2702	14.2368 2960	11.7315 3222	0.0702 4036	0.0852 4036
14	1.2317 5573	0.8118 4928	15.4503 8205	12.5433 8150	0.0647 2332	0.0797 2332
15	1.2502 3207	0.7998 5150	16.6821 3778	13.3432 3301	0.0599 4436	0.0749 4436
16	1.2689 8555	0.7880 3104	17.9323 6984	14.1312 6405	0.0557 6508	0.0707 6508
17	1.2880 2033	0.7763 8526	19.2013 5539	14.9076 4931	0.0520 7966	0.0670 7966
18	1.3073 4064	0.7649 1159	20.4893 7572	15.6725 6089	0.0488 0578	0.0638 0578
19	1.3269 5075	0.7536 0747	21.7967 1636	16.4261 6837	0.0458 7847	0.0608 7847
20	1.3468 5501	0.7424 7042	23.1236 6710	17.1686 3879	0.0432 4574	0.0582 4574
21	1.3670 5783	0.7314 9795	24.4705 2211	17.9001 3673	0.0408 6550	0.0558 6550
22	1.3875 6370	0.7206 8763	25.8375 7994	18.6208 2437	0.0387 0332	0.0537 0332
23	1.4083 7715	0.7100 3708	27.2251 4364	19.3308 6145	0.0367 3075	0.0517 3075
24	1.4295 0281	0.6995 4392	28.6335 2080	20.0304 0537	0.0349 2410	0.0499 2410
25	1.4509 4535	0.6892 0583	30.0630 2361	20.7196 1120	0.0332 6345	0.0482 6345
26	1.4727 0953	0.6790 2052	31.5139 6896	21.3986 3172	0.0317 3196	0.0467 3196
27	1.4948 0018	0.6689 8574	32.9866 7850	22.0676 1746	0.0303 1527	0.0453 1527
28	1.5172 2218	0.6590 9925	34.4814 7867	22.7267 1671	0.0290 0108	0.0440 0108
29	1.5399 8051	0.6493 5887	35.9987 0085	23.3760 7558	0.0277 7878	0.0427 7878
30	1.5630 8022	0.6397 6243	37.5386 8137	24.0158 3801	0.0266 3919	0.0416 3919
31	1.5865 2642	0.6303 0781	39.1017 6159	24.6461 4582	0.0255 7430	0.0405 7430
32	1.6103 2432	0.6209 9292	40.6882 8801	25.2671 3874	0.0245 7710	0.0395 7710
33	1.6344 7918	0.6118 1568	42.2986 1233	25.8789 5442	0.0236 4144	0.0386 4144
34	1.6589 9637	0.6027 7407	43.9330 9152	26.4817 2849	0.0227 6189	0.0377 6189
35	1.6838 8132	0.5938 6608	45.5920 8789	27.0755 9458	0.0219 3363	0.0369 3363
36	1.7091 3954	0.5850 8974	47.2759 6921	27.6606 8431	0.0211 5240	0.0361 5240
37	1.7347 7663	0.5764 4309	48.9851 0874	28.2371 2740	0.0204 1437	0.0354 1437
38	1.7607 9828	0.5679 2423	50.7198 8538	28.8050 5163	0.0197 1613	0.0347 1613
39	1.7872 1025	0.5595 3126	52.4806 8366	29.3645 8288	0.0190 5463	0.0340 5463
40	1.8140 1841	0.5512 6232	54.2678 9391	29.9158 4520	0.0184 2710	0.0334 2710
41	1.8412 2868	0.5431 1559	56.0819 1232	30.4589 6079	0.0178 3106	0.0328 3106
42	1.8688 4712	0.5350 8925	57.9231 4100	30.9940 5004	0.0172 6426	0.0322 6426
43	1.8968 7982	0.5271 8153	59.7919 8812	31.5212 3157	0.0167 2465	0.0317 2465
44	1.9253 3302	0.5193 9067	61.6888 6794	32.0406 2223	0.0162 1038	0.0312 1038
45	1.9542 1301	0.5117 1494	63.6142 0096	32.5523 3718	0.0157 1976	0.0307 1976
46	1.9835 2621	0.5041 5265	65.5684 1398	33.0564 8983	0.0152 5125	0.0302 5125
47	2.0132 7910	0.4967 0212	67.5519 4018	33.5531 9195	0.0148 0342	0.0298 0342
48	2.0434 7829	0.4893 6170	69.5652 1929	34.0425 5365	0.0143 7500	0.0293 7500
49	2.0741 3046	0.4821 2975	71.6086 9758	34.5246 8339	0.0139 6478	0.0289 6478
50	2.1052 4242	0.4750 0468	73.6828 2804	34.9996 8807	0.0135 7168	0.0285 7168

$i = 1\frac{1}{2}\%$

n	$(1+i)^n$	$(1+i)^{-n}$	$s_{\overline{n}\rceil i}$	$a_{\overline{n}\rceil i}$	$\frac{1}{s_{\overline{n}\rceil i}}$	$\frac{1}{a_{\overline{n}\rceil i}}$
51	2.1368 2106	0.4679 8491	75.7880 7046	35.4676 7298	0.0131 9469	0.0281 9469
52	2.1688 7337	0.4610 6887	77.9248 9152	35.9287 4185	0.0128 3287	0.0278 3287
53	2.2014 0647	0.4542 5505	80.0937 6489	36.3829 9690	0.0124 8537	0.0274 8537
54	2.2344 2757	0.4475 4192	82.2951 7136	36.8305 3882	0.0121 5138	0.0271 5138
55	2.2679 4398	0.4409 2800	84.5295 9893	37.2714 6681	0.0118 3018	0.0268 3018
56	2.3019 6314	0.4344 1182	86.7975 4292	37.7058 7863	0.0115 2106	0.0265 2106
57	2.3364 9259	0.4279 9194	89.0995 0606	38.1338 7058	0.0112 2341	0.0262 2341
58	2.3715 3998	0.4216 6694	91.4359 9865	38.5555 3751	0.0109 3661	0.0259 3661
59	2.4071 1308	0.4154 3541	93.8075 3863	38.9709 7292	0.0106 6012	0.0256 6012
60	2.4432 1978	0.4092 9597	96.2146 5171	39.3802 6889	0.0103 9343	0.0253 9343
61	2.4798 6807	0.4032 4726	98.6578 7149	39.7835 1614	0.0101 3604	0.0251 3604
62	2.5170 6609	0.3972 8794	101.1377 3956	40.1808 0408	0.0098 8751	0.0248 8751
63	2.5548 2208	0.3914 1669	103.6548 0565	40.5722 2077	0.0096 4741	0.0246 4741
64	2.5931 4442	0.3856 3221	106.2096 2774	40.9578 5298	0.0094 1534	0.0244 1534
65	2.6320 4158	0.3799 3321	108.8027 7215	41.3377 8618	0.0091 9094	0.0241 9094
66	2.6715 2221	0.3743 1843	111.4348 1374	41.7121 0461	0.0089 7386	0.0239 7386
67	2.7115 9504	0.3687 8663	114.1063 3594	42.0808 9125	0.0087 6376	0.0237 6376
68	2.7522 6896	0.3633 3658	116.8179 3098	42.4442 2783	0.0085 6033	0.0235 6033
69	2.7935 5300	0.3579 6708	119.5701 9995	42.8021 9490	0.0083 6329	0.0233 6329
70	2.8354 5629	0.3526 7692	122.3637 5295	43.1548 7183	0.0081 7235	0.0231 7235
71	2.8779 8814	0.3474 6495	125.1992 0924	43.5023 3678	0.0079 8727	0.0229 8727
72	2.9211 5796	0.3423 3000	128.0771 9738	43.8446 6677	0.0078 0779	0.0228 0779
73	2.9649 7533	0.3372 7093	130.9983 5534	44.1819 3771	0.0076 3368	0.0226 3368
74	3.0094 4996	0.3322 8663	133.9633 3067	44.5142 2434	0.0074 6473	0.0224 6473
75	3.0545 9171	0.3273 7599	136.9727 8063	44.8416 0034	0.0073 0072	0.0223 0072
76	3.1004 1059	0.3225 3793	140.0273 7234	45.1641 3826	0.0071 4146	0.0221 4146
77	3.1469 1674	0.3177 7136	143.1277 8292	45.4819 0962	0.0069 8676	0.0219 8676
78	3.1941 2050	0.3130 7523	146.2746 9967	45.7949 8485	0.0068 3645	0.0218 3645
79	3.2420 3230	0.3084 4850	149.4688 2016	46.1034 3335	0.0066 9036	0.0216 9036
80	3.2906 6279	0.3038 9015	152.7108 5247	46.4073 2349	0.0065 4832	0.0215 4832
81	3.3400 2273	0.2993 9916	156.0015 1525	46.7067 2265	0.0064 1019	0.0214 1019
82	3.3901 2307	0.2949 7454	159.3415 3798	47.0016 9720	0.0062 7583	0.0212 7583
83	3.4409 7492	0.2906 1531	162.7316 6105	47.2923 1251	0.0061 4509	0.0211 4509
84	3.4925 8954	0.2863 2050	166.1726 3597	47.5786 3301	0.0060 1784	0.0210 1784
85	3.5449 7838	0.2820 8917	169.6652 2551	47.8607 2218	0.0058 9396	0.0208 9396
86	3.5981 5306	0.2779 2036	173.2102 0389	48.1386 4254	0.0057 7333	0.0207 7333
87	3.6521 2535	0.2738 1316	176.8083 5695	48.4124 5571	0.0056 5584	0.0206 5584
88	3.7069 0723	0.2697 6666	180.4604 8230	48.6822 2237	0.0055 4138	0.0205 4138
89	3.7625 1084	0.2657 7996	184.1673 8954	48.9480 0234	0.0054 2984	0.0204 2984
90	3.8189 4851	0.2618 5218	187.9299 0038	49.2098 5452	0.0053 2113	0.0203 2113
91	3.8762 3273	0.2579 8245	191.7488 4889	49.4678 3696	0.0052 1516	0.0202 1516
92	3.9343 7622	0.2541 6990	195.6250 8162	49.7220 0686	0.0051 1182	0.0201 1182
93	3.9933 9187	0.2504 1369	199.5594 5784	49.9724 2055	0.0050 1104	0.0200 1104
94	4.0532 9275	0.2467 1300	203.5528 4971	50.2191 3355	0.0049 1273	0.0199 1273
95	4.1140 9214	0.2430 6699	207.6061 4246	50.4622 0054	0.0048 1681	0.0198 1681
96	4.1758 0352	0.2394 7487	211.7202 3459	50.7016 7541	0.0047 2321	0.0197 2321
97	4.2384 4057	0.2359 3583	215.8960 3811	50.9376 1124	0.0046 3186	0.0196 3186
98	4.3020 1718	0.2324 4909	220.1344 7868	51.1700 6034	0.0045 4268	0.0195 4268
99	4.3665 4744	0.2290 1389	224.4364 9586	51.3990 7422	0.0044 5560	0.0194 5560
100	4.4320 4565	0.2256 2944	228.8030 4330	51.6247 0367	0.0043 7057	0.0193 7057

$i = 1\frac{3}{4}\%$

n	$(1+i)^n$	$(1+i)^{-n}$	$s_{\overline{n}\mid i}$	$a_{\overline{n}\mid i}$	$\frac{1}{s_{\overline{n}\mid i}}$	$\frac{1}{a_{\overline{n}\mid i}}$
1	1.0175 0000	0.9828 0098	1.0000 0000	0.9828 0098	1.0000 0000	1.0175 0000
2	1.0353 0625	0.9658 9777	2.0175 0000	1.9486 9875	0.4956 6295	0.5131 6295
3	1.0534 2411	0.9492 8528	3.0528 0625	2.8979 8403	0.3275 6746	0.3450 6746
4	1.0718 5903	0.9329 5851	4.1062 3036	3.8309 4254	0.2435 3237	0.2610 3237
5	1.0906 1656	0.9169 1254	5.1780 8939	4.7478 5508	0.1931 2142	0.2106 2142
6	1.1097 0235	0.9011 4254	6.2687 0596	5.6489 9762	0.1595 2256	0.1770 2256
7	1.1291 2215	0.8856 4378	7.3784 0831	6.5346 4139	0.1355 3059	0.1530 3059
8	1.1488 8178	0.8704 1157	8.5075 3045	7.4050 5297	0.1175 4292	0.1350 4292
9	1.1689 8721	0.8554 4135	9.6564 1224	8.2604 9432	0.1035 5813	0.1210 5813
10	1.1894 4449	0.8407 2860	10.8253 9945	9.1012 2291	0.0923 7534	.0.1098 7534
11	1.2102 5977	0.8262 6889	12.0148 4394	9.9274 9181	0.0832 3038	0.1007 3038
12	1.2314 3931	0.8120 5788	13.2251 0371	10.7395 4969	0.0756 1377	0.0931 1377
13	1.2529 8950	0.7980 9128	14.4565 4303	11.5376 4097	0.0691 7283	0.0866 7283
14	1.2749 1682	0.7843 6490	15.7095 3253	12.3220 0587	0.0636 5562	0.0811 5562
15	1.2972 2786	0.7708 7459	16.9844 4935	13.0928 8046	0.0588 7739	0.0763 7739
16	1.3199 2935	0.7576 1631	18.2816 7721	13.8504 9677	0.0546 9958	0.0721 9958
17	1.3430 2811	0.7445 8605	19.6016 0656	14.5950 8282	0.0510 1623	0.0685 1623
18	1.3665 3111	0.7317 7990	20.9446 3468	15.3268 6272	0.0477 4492	0.0652 4492
19	1.3904 4540	0.7191 9401	22.3111 6578	16.0460 5673	0.0448 2061	0.0623 2061
20	1.4147 7820	0.7068 2458	23.7016 1119	16.7528 8130	0.0421 9122	0.0596 9122
21	1.4395 3681	0.6946 6789	25.1163 8938	17.4475 4919	0.0398 1464	0.0573 1464
22	1.4647 2871	0.6827 2028	26.5559 2620	18.1302 6948	0.0376 5638	0.0551 5638
23	1.4903 6146	0.6709 7817	28.0206 5490	18.8012 4764	0.0356 8796	0.0531 8796
24	1.5164 4279	0.6594 3800	29.5110 1637	19.4606 8565	0.0338 8565	0.0513 8565
25	1.5429 8054	0.6480 9632	31.0274 5915	20.1087 8196	0.0322 2952	0.0497 2952
26	1.5699 8269	0.6369 4970	32.5704 3969	20.7457 3166	0.0307 0269	0.0482 0269
27	1.5974 5739	0.6259 9479	34.1404 2238	21.3717 2644	0.0292 9079	0.0467 9079
28	1.6254 1290	0.6152 2829	35.7378 7977	21.9869 5474	0.0279 8151	0.0454 8151
29	1.6538 5762	0.6046 4697	37.3632 9267	22.5916 0171	0.0267 6424	0.0442 6424
30	1.6828 0013	0.5942 4764	39.0171 5029	23.1858 4934	0.0256 2975	0.0431 2975
31	1.7122 4913	0.5840 2716	40.6999 5042	23.7698 7650	0.0245 7005	0.0420 7005
32	1.7422 1349	0.5739 8247	42.4121 9955	24.3438 5897	0.0235 7812	0.0410 7812
33	1.7727 0223	0.5641 1053	44.1544 1305	24.9079 6951	0.0226 4779	0.0401 4779
34	1.8037 2452	0.5544 0839	45.9271 1527	25.4623 7789	0.0217 7363	0.0392 7363
35	1.8352 8970	0.5448 7311	47.7308 3979	26.0072 5100	0.0209 5082	0.0384 5082
36	1.8674 0727	0.5355 0183	49.5661 2949	26.5427 5283	0.0201 7507	0.0376 7507
37	1.9000 8689	0.5262 9172	51.4335 3675	27.0690 4455	0.0194 4257	0.0369 4257
38	1.9333 3841	0.5172 4002	53.3336 2365	27.5862 8457	0.0187 4990	0.0362 4990
39	1.9671 7184	0.5083 4400	55.2669 6206	28.0946 2857	0.0180 9399	0.0355 9399
40	2.0015 9734	0.4996 0098	57.2341 3390	28.5942 2955	0.0174 7209	0.0349 7209
41	2.0366 2530	0.4910 0834	59.2357 3124	29.0852 3789	0.0168 8170	0.0343 8170
42	2.0722 6624	0.4825 6348	61.2723 5654	29.5678 0136	0.0163 2057	0.0338 2057
43	2.1085 3090	0.4742 6386	63.3446 2278	30.0420 6522	0.0157 8666	0.0332 8666
44	2.1454 3019	0.4661 0699	65.4531 5367	30.5081 7221	0.0152 7810	0.0327 7810
45	2.1829 7522	0.4580 9040	67.5985 8386	30.9662 6261	0.0147 9321	0.0322 9321
46	2.2211 7728	0.4502 1170	69.7815 5908	31.4164 7431	0.0143 3043	0.0318 3043
47	2.2600 4789	0.4424 6850	72.0027 3637	31.8589 4281	0.0138 8836	0.0313 8836
48	2.2995 9872	0.4348 5848	74.2627 8425	32.2938 0129	0.0134 6569	0.0309 6569
49	2.3398 4170	0.4273 7934	76.5623 8298	32.7211 8063	0.0130 6124	0.0305 6124
50	2.3807 8893	0.4200 2883	78.9022 2468	33.1412 0946	0.0126 7391	0.0301 7391

$i = 1\frac{3}{4}\%$

n	$(1+i)^n$	$(1+i)^{-n}$	$s_{\overline{n}\vert i}$	$a_{\overline{n}\vert i}$	$\frac{1}{s_{\overline{n}\vert i}}$	$\frac{1}{a_{\overline{n}\vert i}}$
51	2.4224 5274	0.4128 0475	81.2380 1361	33.5540 1421	0.0123 0269	0.0298 0269
52	2.4648 4566	0.4057 0492	83.7054 6635	33.9597 1913	0.0119 4665	0.0294 4665
53	2.5079 8046	0.3987 2719	86.1703 1201	34.3584 4632	0.0116 0492	0.0291 0492
54	2.5518 7012	0.3918 6947	88.6782 9247	34.7503 1579	0.0112 7672	0.0287 7672
55	2.5965 2785	0.3851 2970	91.2301 6259	35.1354 4550	0.0109 6129	0.0284 6129
56	2.6419 6708	0.3785 0585	93.8266 9043	35.5139 5135	0.0106 5795	0.0281 5795
57	2.6882 0151	0.3719 9592	96.4686 5752	35.8859 4727	0.0103 6606	0.0278 6606
58	2.7352 4503	0.3655 9796	99.1568 5902	36.2515 4523	0.0100 8503	0.0275 8503
59	2.7831 1182	0.3593 1003	101.8921 0405	36.6108 5526	0.0098 1430	0.0273 1430
60	2.8318 1628	0.3531 3025	104.6752 1588	36.9639 8552	0.0095 5336	0.0270 5336
61	2.8813 7306	0.3470 5676	107.5070 3215	37.3110 4228	0.0093 0172	0.0268 0172
62	2.9317 9709	0.3410 8772	110.3884 0522	37.6521 3000	0.0090 5892	0.0265 5892
63	2.9831 0354	0.3352 2135	113.3202 0231	37.9873 5135	0.0088 2455	0.0263 2455
64	3.0353 0785	0.3294 5587	116.3033 0585	33.3168 0723	0.0085 9821	0.0260 9821
65	3.0884 2574	0.3237 8956	119.3386 1370	38.6405 9678	0.0083 7952	0.0258 7952
66	3.1424 7319	0.3182 2069	122.4270 3944	38.9588 1748	0.0081 6813	0.0256 6813
67	3.1974 6647	0.3127 4761	125.5695 1263	39.2715 6509	0.0079 6372	0.0254 6372
68	3.2534 2213	0.3073 6866	128.7669 7910	39.5789 3375	0.0077 6597	0.0252 6597
69	3.3103 5702	0.3020 8222	132.0204 0124	39.8810 1597	0.0075 7459	0.0250 7459
70	3.3682 8827	0.2968 8670	135.3307 5826	40.1779 0267	0.0073 8930	0.0248 8930
71	3.4272 3331	0.2917 8054	138.6990 4653	40.4696 8321	0.0072 0985	0.0247 0985
72	3.4872 0990	0.2867 6221	142.1262 7984	40.7564 4542	0.0070 3600	0.0245 3600
73	3.5482 3607	0.2818 3018	145.6134 8974	41.0382 7560	0.0068 6750	0.0243 6750
74	3.6103 3020	0.2769 8298	149.1617 2581	41.3152 5857	0.0067 0413	0.0242 0413
75	3.6735 1098	0.2722 1914	152.7720 5601	41.5874 7771	0.0065 4570	0.0240 4570
76	3.7377 9742	0.2675 3724	156.4455 6699	41.8550 1495	0.0063 9200	0.0238 9200
77	3.8032 0888	0.2629 3586	160.1833 6441	42.1179 5081	0.0062 4285	0.0237 4285
78	3.8697 6503	0.2584 1362	163.9865 7329	42.3763 6443	0.0060 9806	0.0235 9806
79	3.9374 8592	0.2539 6916	167.8563 3832	42.6303 3359	0.0059 5748	0.0234 5748
80	4.0063 9192	0.2496 0114	171.7938 2424	42.8799 3474	0.0058 2093	0.0233 2093
81	4.0765 0378	0.2453 0825	175.8002 1617	43.1252 4298	0.0056 8828	0.0231 8828
82	4.1478 4260	0.2410 8919	179.8767 1995	43.3663 3217	0.0055 5936	0.0230 5936
83	4.2204 2984	0.2369 4269	184.0245 6255	43.6032 7486	0.0054 3406	0.0229 3406
84	4.2942 8737	0.2328 6751	188.2449 9239	43.8361 4237	0.0053 1223	0.0228 1223
85	4.3694 3740	0.2288 6242	192.5392 7976	44.0650 0479	0.0051 9375	0.0226 9375
86	4.4459 0255	0.2249 2621	196.9087 1716	44.2899 3099	0.0050 7850	0.0225 7850
87	4.5237 0584	0.2210 5770	201.3546 1971	44.5109 8869	0.0049 6636	0.0224 6636
88	4.6028 7070	0.2172 5572	205.8783 2555	44.7282 4441	0.0048 5724	0.0223 5724
89	4.6834 2093	0.2135 1914	210.4811 9625	44.9417 6355	0.0047 5102	0.0222 5102
90	4.7653 8080	0.2098 4682	215.1646 1718	45.1516 1037	0.0046 4760	0.0221 4760
91	4.8487 7496	0.2062 3766	219.9299 9798	45.3578 4803	0.0045 4690	0.0220 4690
92	4.9336 2853	0.2026 9057	224.7787 7295	45.5605 3860	0.0044 4882	0.0219 4882
93	5.0199 6703	0.1992 0450	229.7124 0148	45.7597 4310	0.0043 5327	0.0218 5327
94	5.1078 1645	0.1957 7837	234.7323 6850	45.9555 2147	0.0042 6017	0.0217 6017
95	5.1972 0324	0.1924 1118	239.8401 8495	46.1479 3265	0.0041 6944	0.0216 6944
96	5.2881 5429	0.1891 0190	245.0373 8819	46.3370 3455	0.0040 8101	0.0215 8101
97	5.3806 9699	0.1858 4953	250.3255 4248	46.5228 8408	0.0039 9480	0.0214 9480
98	5.4748 5919	0.1826 5310	255.7062 3947	46.7055 3718	0.0039 1074	0.0214 1074
99	5.5706 6923	0.1795 1165	261.1810 9866	56.8850 4882	0.0038 2876	0.0213 2876
100	5.6681 5594	0.1764 2422	266.7517 6789	47.0614 7304	0.0037 4880	0.0212 4880

$i = 2\%$

n	$(1+i)^n$	$(1+i)^{-n}$	$s_{\overline{n}\vert i}$	$a_{\overline{n}\vert i}$	$\frac{1}{s_{\overline{n}\vert i}}$	$\frac{1}{a_{\overline{n}\vert i}}$
1	1.0200 0000	0.9803 9216	1.0000 0000	0.9803 9216	1.0000 0000	1.0200 0000
2	1.0404 0000	0.9611 6878	2.0200 0000	1.9415 6094	0.4950 4950	0.5150 4950
3	1.0612 0800	0.9423 2233	3.0604 0000	2.8838 8327	0.3267 5467	0.3467 5467
4	1.0824 3216	0.9238 4543	4.1216 0800	3.8077 2870	0.2426 2375	0.2626 2375
5	1.1040 8080	0.9057 3081	5.2040 4016	4.7134 5951	0.1921 5839	0.2121 5839
6	1.1261 6242	0.8879 7138	6.3081 2096	5.6014 3089	0.1585 2581	0.1785 2581
7	1.1486 8567	0.8705 6018	7.4342 8338	6.4719 9107	0.1345 1196	0.1545 1196
8	1.1716 5938	0.8534 9037	8.5829 6905	7.3254 8144	0.1165 0980	0.1365 0980
9	1.1950 9257	0.8367 5527	9.7546 2843	8.1622 3671	0.1025 1544	0.1225 1544
10	1.2189 9442	0.8203 4830	10.9497 2100	8.9825 8501	0.0913 2653	0.1113 2653
11	1.2433 7431	0.8042 6304	12.1687 1542	9.7868 4805	0.0821 7794	0.1021 7794
12	1.2682 4179	0.7884 9318	13.4120 8973	10.5753 4122	0.0745 5960	0.0945 5960
13	1.2936 0663	0.7730 3253	14.6803 3152	11.3483 7375	0.0681 1835	0.0881 1835
14	1.3194 7876	0.7578 7502	15.9739 3815	12.1062 4877	0.0626 0197	0.0826 0197
15	1.3458 6834	0.7430 1473	17.2934 1692	12.8492 6350	0.0578 2547	0.0778 2547
16	1.3727 8571	0.7284 4581	18.6392 8525	13.5777 0931	0.0536 5013	0.0736 5013
17	1.4002 4142	0.7141 6256	20.0120 7096	14.2918 7188	0.0499 6984	0.0699 6984
18	1.4282 4625	0.7001 5937	21.4123 1238	14.9920 3125	0.0467 0210	0.0667 0210
19	1.4568 1117	0.6864 3076	22.8405 5863	15.6784 6201	0.0437 8177	0.0637 8177
20	1.4859 4740	0.6729 7133	24.2973 6980	16.3514 3334	0.0411 5672	0.0611 5672
21	1.5156 6634	0.6597 7582	25.7833 1719	17.0112 0916	0.0387 8477	0.0587 8477
22	1.5459 7967	0.6468 3904	27.2989 8354	17.6580 4820	0.0366 3140	0.0566 3140
23	1.5768 9926	0.6341 5592	28.8449 6321	18.2922 0412	0.0346 6810	0.0546 6810
24	1.6084 3725	0.6217 2149	30.4218 6247	18.9139 2560	0.0328 7110	0.0528 7110
25	1.6406 0599	0.6095 3087	32.0302 9972	19.5234 5647	0.0312 2044	0.0512 2044
26	1.6734 1811	0.5975 7928	33.6709 0572	20.1210 3576	0.0296 9923	0.0496 9923
27	1.7068 8648	0.5858 6204	35.3443 2383	20.7068 9780	0.0282 9309	0.0482 9309
28	1.7410 2421	0.5743 7455	37.0512 1031	21.2812 7236	0.0269 8967	0.0469 8967
29	1.7758 4469	0.5631 1231	38.7922 3451	21.8443 8466	0.0257 7836	0.0457 7836
30	1.8113 6158	0.5520 7089	40.5680 7921	22.3964 5555	0.0246 4992	0.0446 4992
31	1.8475 8882	0.5412 4597	42.3794 4079	22.9377 0152	0.0235 9635	0.0435 9635
32	1.8845 4059	0.5306 3330	44.2270 2961	23.4683 3482	0.0226 1061	0.0426 1061
33	1.9222 3140	0.5202 2873	46.1115 7020	23.9885 6355	0.0216 8653	0.0416 8653
34	1.9606 7603	0.5100 2817	48.0338 0160	24.4985 9172	0.0208 1867	0.0408 1867
35	1.9998 8955	0.5000 2761	49.9944 7763	24.9986 1933	0.0200 0221	0.0400 0221
36	2.0398 8734	0.4902 2315	51.9943 6719	25.4888 4248	0.0192 3285	0.0392 3285
37	2.0806 8509	0.4806 1093	54.0342 5453	25.9694 5341	0.0185 0678	0.0385 0678
38	2.1222 9879	0.4711 8719	56.1149 3962	26.4406 4060	0.0178 2057	0.0378 2057
39	2.1647 4477	0.4619 4822	58.2372 3841	26.9025 8883	0.0171 7114	0.0371 7114
40	2.2080 3966	0.4528 9042	60.4019 8318	27.3554 7924	0.0165 5575	0.0365 5575
41	2.2522 0046	0.4440 1021	62.6100 2284	27.7994 8945	0.0159 7188	0.0359 7188
42	2.2972 4447	0.4353 0413	64.8622 2330	28.2347 9358	0.0154 1729	0.0354 1729
43	2.3431 8936	0.4267 6875	67.1594 6777	28.6615 6233	0.0148 8993	0.0348 8993
44	2.3900 5314	0.4184 0074	69.5026 5712	29.0799 6307	0.0143 8794	0.0343 8794
45	2.4378 5421	0.4101 9680	71.8927 1027	29.4901 5987	0.0139 0962	0.0339 0962
46	2.4866 1129	0.4021 5373	74.3305 6447	29.8923 1360	0.0134 5342	0.0334 5342
47	2.5363 4352	0.3942 6836	76.8171 7576	30.2865 8196	0.0130 1792	0.0330 1792
48	2.5870 7039	0.3865 3761	79.3535 1927	30.6731 1957	0.0126 0184	0.0326 0184
49	2.6388 1179	0.3789 5844	81.9405 8966	31.0520 7801	0.0122 0396	0.0322 0396
50	2.6915 8803	0.3715 2788	84.5794 0145	31.4236 0589	0.0118 2321	0.0318 2321

$i = 2\%$

n	$(1+i)^n$	$(1+i)^{-n}$	$s_{\overline{n}\|i}$	$a_{\overline{n}\|i}$	$\frac{1}{s_{\overline{n}\|i}}$	$\frac{1}{a_{\overline{n}\|i}}$
51	2.7454 1979	0.3642 4302	87.2709 8948	31.7878 4892	0.0114 5856	0.0314 5856
52	2.8003 2819	0.3571 0100	90.0164 0927	32.1449 4992	0.0111 0909	0.0311 0909
53	2.8563 3475	0.3500 9902	92.8167 3746	32.4950 4894	0.0107 7392	0.0307 7392
54	2.9134 6144	0.3432 3433	95.6730 7221	32.8382 8327	0.0104 5226	0.0304 5226
55	2.9717 3067	0.3365 0425	98.5865 3365	33.1747 8752	0.0101 4337	0.0301 4337
56	3.0311 6529	0.3299 0613	101.5582 6432	33.5046 9365	0.0098 4656	0.0298 4656
57	3.0917 8859	0.3234 3738	104.5894 2961	33.8281 3103	0.0095 6120	0.0295 6120
58	3.1536 2436	0.3170 9547	107.6812 1820	34.1452 2650	0.0092 8667	0.0292 8667
59	3.2166 9685	0.3108 7791	110.8348 4257	34.4561 0441	0.0090 2243	0.0290 2243
60	3.2810 3079	0.3047 8227	114.0515 3942	34.7608 8668	0.0087 6797	0.0287 6797
61	3.3466 5140	0.2988 0614	117.3325 7021	35.0596 9282	0.0085 2278	0.0285 2278
62	3.4135 8443	0.2929 4720	120.6792 2161	35.3526 4002	0.0082 8643	0.0282 8643
63	3.4818 5612	0.2872 0314	124.0928 0604	35.6398 4316	0.0080 5848	0.0280 5848
64	3.5514 9324	0.2815 7170	127.5746 6216	35.9214 1486	0.0078 3855	0.0278 3855
65	3.6225 2311	0.2760 5069	131.1261 5541	36.1974 6555	0.0076 2624	0.0276 2624
66	3.6949 7357	0.2706 3793	134.7486 7852	36.4681 0348	0.0074 2122	0.0274 2122
67	3.7688 7304	0.2653 3130	138.4436 5209	36.7334 3478	0.0072 2316	0.0272 2316
68	3.8442 5050	0.2601 2873	142.2125 2513	36.9935 6351	0.0070 3173	0.0270 3173
69	3.9211 3551	0.2550 2817	146.0567 7563	37.2485 9168	0.0068 4665	0.0268 4665
70	3.9995 5822	0.2500 2761	149.9779 1114	37.4986 1929	0.0066 6765	0.0266 6765
71	4.0795 4939	0.2451 2511	153.9774 6937	37.7437 4441	0.0064 9446	0.0264 9446
72	4.1611 4038	0.2403 1874	158.0570 1875	37.9840 6314	0.0063 2683	0.0263 2683
73	4.2443 6318	0.2356 0661	162.2181 5913	38.2196 6975	0.0061 6454	0.0261 6454
74	4.3292 5045	0.2309 8687	166.4625 2231	38.4506 5662	0.0060 0736	0.0260 0736
75	4.4158 3546	0.2264 5771	170.7917 7276	38.6771 1433	0.0058 5508	0.0258 5508
76	4.5041 5216	0.2220 1737	175.2076 0821	38.8991 3170	0.0057 0751	0.0257 0751
77	4.5942 3521	0.2176 6408	179.7117 6038	39.1167 9578	0.0055 6447	0.0255 6447
78	4.6861 1991	0.2133 9616	184.3059 9558	39.3301 9194	0.0054 2576	0.0254 2576
79	4.7798 4231	0.2092 1192	188.9921 1549	39.5394 0386	0.0052 9123	0.0252 9123
80	4.8754 3916	0.2051 0973	193.7719 5780	39.7445 1359	0.0051 6071	0.0251 6071
81	4.9729 4794	0.2010 8797	198.6473 9696	39.9456 0156	0.0050 3405	0.0250 3405
82	5.0724 0690	0.1971 4507	203.6203 4490	40.1427 4663	0.0049 1110	0.0249 1110
83	5.1738 5504	0.1932 7948	208.6927 5180	40.3360 2611	0.0047 9173	0.0247 9173
84	5.2773 3214	0.1894 8968	213.8666 0683	40.5255 1579	0.0046 7581	0.0246 7581
85	5.3828 7878	0.1857 7420	219.1439 3897	40.7112 8999	0.0045 6321	0.0245 6321
86	5.4905 3636	0.1821 3157	224.5268 1775	40.8934 2156	0.0044 5381	0.0244 5381
87	5.6003 4708	0.1785 6036	230.0173 5411	41.0719 8192	0.0043 4750	0.0243 4750
88	5.7123 5402	0.1750 5918	235.6177 0119	41.2470 4110	0.0042 4416	0.0242 4416
89	5.8266 0110	0.1716 2665	241.3300 5521	41.4186 6774	0.0041 4370	0.0241 4370
90	5.9431 3313	0.1682 6142	247.1566 5632	41.5869 2916	0.0040 4602	0.0240 4602
91	6.0619 9579	0.1649 6217	253.0997 8944	41.7518 9133	0.0039 5101	0.0239 5101
92	6.1832 3570	0.1617 2762	259.1617 8523	41.9136 1895	0.0038 5859	0.0238 5859
93	6.3069 0042	0.1585 5649	265.3450 2093	42.0721 7545	0.0037 6868	0.0237 6868
94	6.4330 3843	0.1554 4754	271.6519 2135	42.2276 2299	0.0036 8118	0.0236 8118
95	6.5616 9920	0.1523 9955	278.0849 5978	42.3800 2254	0.0035 9602	0.0235 9602
96	6.6929 3318	0.1494 1132	284.6466 5898	42.5294 3386	0.0035 1313	0.0235 1313
97	6.8267 9184	0.1464 8169	291.3395 9216	42.6759 1555	0.0034 3242	0.0234 3242
98	6.9633 2768	0.1436 0950	298.1663 8400	42.8195 2505	0.0033 5383	0.0233 5383
99	7.1025 9423	0.1407 9363	305.1297 1168	42.9603 1867	0.0032 7729	0.0232 7729
100	7.2446 4612	0.1380 3297	312.2323 0591	43.0983 5164	0.0032 0274	0.0232 0274

$i = 2\frac{1}{4}\%$

n	$(1+i)^n$	$(1+i)^{-n}$	$s_{\overline{n}\vert i}$	$a_{\overline{n}\vert i}$	$\frac{1}{s_{\overline{n}\vert i}}$	$\frac{1}{a_{\overline{n}\vert i}}$
1	1.0225 0000	0.9779 9511	1.0000 0000	0.9779 9511	1.0000 0000	1.0225 0000
2	1.0455 0625	0.9564 7444	2.0225 0000	1.9344 6955	0.4944 3758	0.5169 3758
3	1.0690 3014	0.9354 2732	3.0680 0625	2.8698 9687	0.3259 4458	0.3484 4458
4	1.0930 8332	0.9148 4335	4.1370 3639	3.7847 4021	0.2417 1893	0.2642 1893
5	1.1176 7769	0.8947 1232	5.2301 1971	4.6794 5253	0.1912 0021	0.2137 0021
6	1.1428 2544	0.8750 2427	6.3477 9740	5.5544 7680	0.1575 3496	0.1800 3496
7	1.1685 3901	0.8557 6946	7.4906 2284	6.4102 4626	0.1335 0025	0.1560 0025
8	1.1948 3114	0.8369 3835	8.6591 6186	7.2471 8461	0.1154 8462	0.1379 8462
9	1.2217 1484	0.8185 2161	9.8539 9300	8.0657 0622	0.1014 8170	0.1239 8170
10	1.2492 0343	0.8005 1013	11.0757 0784	8.8662 1635	0.0902 8768	0.1127 8768
11	1.2773 1050	0.7828 9499	12.3249 1127	9.6491 1134	0.0811 3649	0.1036 3649
12	1.3060 4999	0.7656 6748	13.6022 2177	10.4147 7882	0.0735 1740	0.0960 1740
13	1.3354 3611	0.7488 1905	14.9082 7176	11.1635 9787	0.0670 7686	0.0895 7686
14	1.3654 8343	0.7323 4137	16.2437 0788	11.8959 3924	0.0615 6230	0.0840 6230
15	1.3962 0680	0.7162 2628	17.6091 9130	12.6121 6551	0.0567 8852	0.0792 8852
16	1.4276 2146	0.7004 6580	19.0053 9811	13.3126 3131	0.0526 1663	0.0751 1663
17	1.4597 4294	0.6850 5212	20.4330 1957	13.9976 8343	0.0489 4039	0.0714 4039
18	1.4925 8716	0.6699 7763	21.8927 6251	14.6676 6106	0.0456 7720	0.0681 7720
19	1.5261 7037	0.6552 3484	23.3853 4966	15.3228 9590	0.0427 6182	0.0652 6182
20	1.5605 0920	0.6408 1647	24.9115 2003	15.9637 1237	0.0401 4207	0.0626 4207
21	1.5956 2066	0.6267 1538	26.4720 2923	16.5904 2775	0.0377 7572	0.0602 7572
22	1.6315 2212	0.6129 2457	28.0676 4989	17.2033 5232	0.0356 2821	0.0581 2821
23	1.6682 3137	0.5994 3724	29.6991 7201	17.8027 8955	0.0336 7097	0.0561 7097
24	1.7057 6658	0.5862 4668	31.3674 0338	18.3890 3624	0.0318 8023	0.0543 8023
25	1.7441 4632	0.5733 4639	33.0731 6996	18.9623 8263	0.0302 3599	0.0527 3599
26	1.7833 8962	0.5607 2997	34.8173 1628	19.5231 1260	0.0287 2134	0.0512 2134
27	1.8235 1588	0.5483 9117	36.6007 0590	20.0715 0376	0.0273 2188	0.0498 2188
28	1.8645 4499	0.5363 2388	38.4242 2178	20.6078 2764	0.0260 2525	0.0485 2525
29	1.9064 9725	0.5245 2213	40.2887 6677	21.1323 4977	0.0248 2081	0.0473 2081
30	1.9493 9344	0.5129 8008	42.1952 6402	21.6453 2985	0.0236 9934	0.0461 9934
31	1.9932 5479	0.5016 9201	44.1446 5746	22.1470 2186	0.0226 5280	0.0451 5280
32	2.0381 0303	0.4906 5233	46.1379 1226	22.6376 7419	0.0216 7415	0.0441 7415
33	2.0839 6034	0.4798 5558	48.1760 1528	23.1175 2977	0.0207 5722	0.0432 5722
34	2.1308 4945	0.4692 9641	50.2599 7563	23.5868 2618	0.0198 9655	0.0423 9655
35	2.1787 9356	0.4589 6960	52.3908 2508	24.0457 9577	0.0190 8731	0.0415 8731
36	2.2278 1642	0.4488 7002	54.5696 1864	24.4946 6579	0.0183 2522	0.0408 2522
37	2.2779 4229	0.4389 9268	56.7974 3506	24.9336 5848	0.0176 0643	0.0401 0643
38	2.3291 9599	0.4293 3270	59.0753 7735	25.3629 9118	0.0169 2753	0.0394 2753
39	2.3816 0290	0.4198 8528	61.4045 7334	25.7828 7646	0.0162 8543	0.0387 8543
40	2.4351 8897	0.4106 4575	63.7861 7624	26.1935 2221	0.0156 7738	0.0381 7738
41	2.4899 8072	0.4016 0954	66.2213 6521	26.5951 3174	0.0151 0087	0.0376 0087
42	2.5460 0528	0.3927 7216	68.7113 4592	26.9879 0390	0.0145 5364	0.0370 5364
43	2.6032 9040	0.3841 2925	71.2573 5121	27.3720 3316	0.0140 3364	0.0365 3364
44	2.6618 6444	0.3756 7653	73.8606 4161	27.7477 0969	0.0135 3901	0.0360 3901
45	2.7217 5639	0.3674 0981	76.5225 0605	28.1151 1950	0.0130 6805	0.0355 6805
46	2.7829 9590	0.3593 2500	79.2442 6243	28.4744 4450	0.0126 1921	0.0351 1921
47	2.8456 1331	0.3514 1809	82.0272 5834	28.8258 6259	0.0121 9107	0.0346 9107
48	2.9096 3961	0.3436 8518	84.8728 7165	29.1695 4777	0.0117 8233	0.0342 8233
49	2.9751 0650	0.3361 2242	87.7825 1126	29.5056 7019	0.0113 9179	0.0338 9179
50	3.0420 4640	0.3287 2608	90.7576 1776	29.8343 9627	0.0110 1836	0.0335 1836

$i = 2\frac{1}{4}\%$

n	$(1+i)^n$	$(1+i)^{-n}$	$s_{\overline{n}\mid i}$	$a_{\overline{n}\mid i}$	$\frac{1}{s_{\overline{n}\mid i}}$	$\frac{1}{a_{\overline{n}\mid i}}$
51	3.1104 9244	0.3214 9250	93.7996 6416	30.1558 8877	0.0106 6102	0.0331 6102
52	3.1804 7852	0.3144 1810	96.9101 5661	30.4703 0687	0.0103 1884	0.0328 1884
53	3.2520 3929	0.3074 9936	100.0906 3513	30.7778 0623	0.0099 9094	0.0324 9094
54	3.3252 1017	0.3007 3287	103.3426 7442	31.0785 3910	0.0096 7654	0.0321 7654
55	3.4000 2740	0.2941 1528	106.6678 8460	31.3726 5438	0.0093 7489	0.0318 7489
56	3.4765 2802	0.2876 4330	110.0679 1200	31.6602 9768	0.0090 8530	0.0315 8530
57	3.5547 4990	0.2813 1374	113.5444 4002	31.9416 1142	0.0088 0712	0.0313 0712
58	3.6347 3177	0.2751 2347	117.0991 8992	32.2167 3489	0.0085 3977	0.0310 3977
59	3.7165 1324	0.2690 6940	120.7339 2169	32.4858 0429	0.0082 8268	0.0307 8268
60	3.8001 3479	0.2631 4856	124.4504 3493	32.7489 5285	0.0080 3533	0.0305 3533
61	3.8856 3782	0.2573 5801	128.2505 6972	33.0063 1086	0.0077 9724	0.0302 9724
62	3.9730 6467	0.2516 9487	132.1362 0754	33.2580 0573	0.0075 6795	0.0300 6795
63	4.0624 5862	0.2461 5635	136.1092 7221	33.5041 6208	0.0073 4704	0.0298 4704
64	4.1538 6394	0.2407 3971	140.1717 3083	33.7449 0179	0.0071 3411	0.0296 3411
65	4.2473 2588	0.2354 4226	144.3255 9477	33.9803 4405	0.0069 2878	0.0294 2878
66	4.3428 9071	0.2302 6138	148.5729 2066	34.2106 0543	0.0067 3070	0.0292 3070
67	4.4406 0576	0.2251 9450	152.9158 1137	34.4357 9993	0.0065 3955	0.0290 3955
68	4.5405 1939	0.2202 3912	157.3564 1713	34.6560 3905	0.0063 5500	0.0288 5500
69	4.6426 8107	0.2153 9278	161.8969 3651	34.8714 3183	0.0061 7677	0.0286 7677
70	4.7471 4140	0.2106 5309	166.5396 1758	35.0820 8492	0.0060 0458	0.0285 0458
71	4.8539 5208	0.2060 1769	171.2867 5898	35.2881 0261	0.0058 3816	0.0283 3816
72	4.9631 6600	0.2014 8429	176.1407 1106	35.4895 8691	0.0056 7728	0.0281 7728
73	5.0748 3723	0.1970 5065	181.1038 7705	35.6866 3756	0.0055 2169	0.0280 2169
74	5.1890 2107	0.1927 1458	186.1787 1429	35.8793 5214	0.0053 7118	0.0278 7118
75	5.3057 7405	0.1884 7391	191.3677 3536	36.0678 2605	0.0052 2554	0.0277 2554
76	5.4251 5396	0.1843 2657	196.6735 0941	36.2521 5262	0.0050 8457	0.0275 8457
77	5.5472 1993	0.1802 7048	202.0986 6337	36.4324 2310	0.0049 4808	0.0274 4808
78	5.6720 3237	0.1763 0365	207.6458 8329	36.6087 2675	0.0048 1589	0.0273 1589
79	5.7996 5310	0.1724 2411	213.3179 1567	36.7811 5085	0.0046 8784	0.0271 8784
80	5.9301 4530	0.1686 2993	219.1175 6877	36.9497 8079	0.0045 6376	0.0270 6376
81	6.0635 7357	0.1649 1925	225.0477 1407	37.1147 0004	0.0044 4350	0.0269 4350
82	6.2000 0397	0.1612 9022	231.1112 8763	37.2759 9026	0.0043 2692	0.0268 2692
83	6.3395 0406	0.1577 4105	237.3112 9160	37.4337 3130	0.0042 1387	0.0267 1387
84	6.4821 4290	0.1542 6997	243.6507 9567	37.5880 0127	0.0041 0423	0.0266 0423
85	6.6279 9112	0.1508 7528	250.1329 3857	37.7388 7655	0.0039 9787	0.0264 9787
86	6.7771 2092	0.1475 5528	256.7609 2969	37.8864 3183	0.0038 9467	0.0263 9467
87	6.9296 0614	0.1443 0835	263.5380 5060	38.0307 4018	0.0037 9452	0.0262 9452
88	7.0855 2228	0.1411 3286	270.4676 5674	38.1718 7304	0.0036 9730	0.0261 9730
89	7.2449 4653	0.1380 2724	277.5531 7902	38.3099 0028	0.0036 0291	0.0261 0291
90	7.4079 5782	0.1349 8997	284.7981 2555	38.4448 9025	0.0035 1126	0.0260 1126
91	7.5746 3688	0.1320 1953	292.2060 8337	38.5769 0978	0.0034 2224	0.0259 2224
92	7.7450 6621	0.1291 1445	299.7807 2025	38.7060 2423	0.0033 3577	0.0258 3577
93	7.9193 3020	0.1262 7331	307.5257 8645	38.8322 9754	0.0032 5176	0.0257 5176
94	8.0975 1512	0.1234 9468	315.4451 1665	38.9557 9221	0.0031 7012	0.0256 7012
95	8.2797 0921	0.1207 7719	323.5426 3177	39.0765 6940	0.0030 9078	0.0255 9078
96	8.4660 0267	0.1181 1950	331.8223 4099	39.1946 8890	0.0030 1366	0.0255 1366
97	8.6564 8773	0.1155 2029	340.2883 4366	39.3102 0920	0.0029 3868	0.0254 3868
98	8.8512 5871	0.1129 7828	348.9448 3139	39.4231 8748	0.0028 6578	0.0253 6578
99	9.0504 1203	0.1104 9221	357.7960 9010	39.5336 7968	0.0027 9489	0.0252 9489
100	9.2540 4630	0.1080 6084	366.8465 0213	39.6417 4052	0.0027 2594	0.0252 2594

$i = 2\frac{1}{2}\%$

n	$(1+i)^n$	$(1+i)^{-n}$	$s_{\overline{n}\|i}$	$a_{\overline{n}\|i}$	$\frac{1}{s_{\overline{n}\|i}}$	$\frac{1}{a_{\overline{n}\|i}}$
1	1.0250 0000	0.9756 0976	1.0000 0000	0.9756 0976	1.0000 0000	1.0250 0000
2	1.0506 2500	0.9518 1440	2.0250 0000	1.9274 2415	0.4938 2716	0.5188 2716
3	1.0768 9063	0.9285 9941	3.0756 2500	2.8560 2356	0.3251 3717	0.3501 3717
4	1.1038 1289	0.9059 5064	4.1525 1563	3.7619 7421	0.2408 1788	0.2658 1788
5	1.1314 0821	0.8838 5429	5.2563 2852	4.6458 2850	0.1902 4686	0.2152 4686
6	1.1596 9342	0.8622 9687	6.3877 3673	5.5081 2536	0.1565 4997	0.1815 4997
7	1.1886 8575	0.8412 6524	7.5474 3015	6.3493 9060	0.1324 9543	0.1574 9543
8	1.2184 0290	0.8207 4657	8.7361 1590	7.1701 3717	0.1144 6735	0.1394 6735
9	1.2488 6297	0.8007 2836	9.9545 1880	7.9708 6553	0.1004 5689	0.1254 5689
10	1.2800 8454	0.7811 9840	11.2033 8177	8.7520 6393	0.0892 5876	0.1142 5876
11	1.3120 8666	0.7621 4478	12.4834 6631	9.5142 0871	0.0801 0596	0.1051 0596
12	1.3448 8882	0.7435 5589	13.7955 5297	10.2577 6460	0.0724 8713	0.0974 8713
13	1.3785 1104	0.7254 2038	15.1404 4179	10.9831 8497	0.0660 4827	0.0910 4827
14	1.4129 7382	0.7077 2720	16.5189 5284	11.6909 1217	0.0605 3652	0.0855 3652
15	1.4482 9817	0.6904 6556	17.9319 2666	12.3813 7773	0.0557 6646	0.0807 6646
16	1.4845 0562	0.6736 2493	19.3802 2483	13.0550 0266	0.0515 9899	0.0765 9899
17	1.5216 1826	0.6571 9506	20.8647 3045	13.7121 9772	0.0479 2777	0.0729 2777
18	1.5596 5872	0.6411 6591	22.3863 4871	14.3533 6363	0.0446 7008	0.0696 7008
19	1.5986 5019	0.6255 2772	23.9460 0743	14.9788 9134	0.0417 6062	0.0667 6062
20	1.6386 1644	0.6102 7094	25.5446 5761	15.5891 6229	0.0391 4713	0.0641 4713
21	1.6795 8185	0.5953 8629	27.1832 7405	16.1845 4857	0.0367 8733	0.0617 8733
22	1.7215 7140	0.5808 6467	28.8628 5590	16.7654 1324	0.0346 4661	0.0596 4661
23	1.7646 1068	0.5666 9724	30.5844 2730	17.3321 1048	0.0326 9638	0.0576 9638
24	1.8087 2595	0.5528 7535	32.3490 3798	17.8849 8583	0.0309 1282	0.0559 1282
25	1.8539 4410	0.5393 9059	34.1577 6393	18.4243 7642	0.0292 7592	0.0542 7592
26	1.9002 9270	0.5262 3472	36.0117 0803	18.9506 1114	0.0277 6875	0.0527 6875
27	1.9478 0002	0.5133 9973	37.9120 0073	19.4640 1087	0.0263 7687	0.0513 7687
28	1.9964 9502	0.5008 7778	39.8598 0075	19.9648 8866	0.0250 8793	0.0500 8793
29	2.0464 0739	0.4886 6125	41.8562 9577	20.4535 4991	0.0238 9127	0.0488 9127
30	2.0975 6758	0.4767 4269	43.9027 0316	20.9302 9259	0.0227 7764	0.0477 7764
31	2.1500 0677	0.4651 1481	46.0002 7074	21.3954 0741	0.0217 3900	0.0467 3900
32	2.2037 5694	0.4537 7055	48.1502 7751	21.8491 7796	0.0207 6831	0.0457 6831
33	2.2588 5086	0.4427 0298	50.3540 3445	22.2918 8094	0.0198 5938	0.0448 5938
34	2.3153 2213	0.4319 0534	52.6128 8531	22.7237 8628	0.0190 0675	0.0440 0675
35	2.3732 0519	0.4213 7107	54.9282 0744	23.1451 5734	0.0182 0558	0.0432 0558
36	2.4325 3532	0.4110 9372	57.3014 1263	23.5562 5107	0.0174 5158	0.0424 5158
37	2.4933 4870	0.4010 6705	59.7339 4794	23.9573 1812	0.0167 4090	0.0417 4090
38	2.5556 8242	0.3912 8492	62.2272 9664	24.3486 0304	0.0160 7012	0.0410 7012
39	2.6195 7448	0.3817 4139	64.7829 7906	24.7303 4443	0.0154 3615	0.0404 3615
40	2.6850 6384	0.3724 3062	67.4025 5354	25.1027 7505	0.0148 3623	0.0398 3623
41	2.7521 9043	0.3633 4695	70.0876 1737	25.4661 2200	0.0142 6786	0.0392 6786
42	2.8209 9520	0.3544 8483	72.8398 0781	25.8206 0683	0.0137 2876	0.0387 2876
43	2.8915 2008	0.3458 3886	75.6608 0300	26.1664 4569	0.0132 1688	0.0382 1688
44	2.9638 0808	0.3374 0376	78.5523 2308	26.5038 4945	0.0127 3037	0.0377 3037
45	3.0379 0328	0.3291 7440	81.5161 3116	26.8330 2386	0.0122 6751	0.0372 6751
46	3.1138 5086	0.3211 4576	84.5540 3443	27.1541 6962	0.0118 2676	0.0368 2676
47	3.1916 9713	0.3133 1294	87.6678 8530	27.4674 8255	0.0114 0669	0.0364 0669
48	3.2714 8956	0.3056 7116	90.8595 8243	27.7731 5371	0.0110 0599	0.0360 0599
49	3.3532 7680	0.2982 1576	94.1310 7199	28.0713 6947	0.0106 2348	0.0356 2348
50	3.4371 0872	0.2909 4221	97.4843 4879	28.3623 1168	0.0102 5806	0.0352 5806

$i = 2\frac{1}{2}\%$

n	$(1+i)^n$	$(1+i)^{-n}$	$s_{\overline{n}\rvert i}$	$a_{\overline{n}\rvert i}$	$\frac{1}{s_{\overline{n}\rvert i}}$	$\frac{1}{a_{\overline{n}\rvert i}}$
51	3.5230 3664	0.2838 4606	100.9214 5751	28.6461 5774	0.0099 0870	0.0349 0870
52	3.6111 1235	0.2769 2298	104.4444 9395	28.9230 8072	0.0095 7446	0.0345 7446
53	3.7013 9016	0.2701 6876	108.0556 0629	29.1932 4948	0.0092 5449	0.0342 5449
54	3.7939 2491	0.2635 7928	111.7569 9645	29.4568 2876	0.0089 4799	0.0339 4799
55	3.8887 7303	0.2571 5052	115.5509 2136	29.7139 7928	0.0086 5419	0.0336 5419
56	3.9859 9236	0.2508 7855	119.4396 9440	29.9648 5784	0.0083 7243	0.0333 7243
57	4.0856 4217	0.2447 5956	123.4256 8676	30.2096 1740	0.0081 0204	0.0331 0204
58	4.1877 8322	0.2387 8982	127.5113 2893	30.4484 0722	0.0078 4244	0.0328 4244
59	4.2924 7780	0.2329 6568	131.6991 1215	30.6813 7290	0.0075 9307	0.0325 9307
60	4.3997 8975	0.2272 8359	135.9915 8995	30.9086 5649	0.0073 5340	0.0323 5340
61	4.5097 8449	0.2217 4009	140.3913 7970	31.1303 9657	0.0071 2294	0.0321 2294
62	4.6225 2910	0.2163 3179	144.9011 6419	31.3467 2836	0.0069 0126	0.0319 0126
63	4.7380 9233	0.2110 5541	149.5236 9330	31.5577 8377	0.0066 8790	0.0316 8790
64	4.8565 4464	0.2059 0771	154.2617 8563	31.7636 9148	0.0064 8249	0.0314 8249
65	4.9779 5826	0.2008 8557	159.1183 3027	31.9645 7705	0.0062 8463	0.0312 8463
66	5.1024 0721	0.1959 8593	164.0962 8853	33.1605 6298	0.0060 9398	0.0310 9398
67	5.2299 6739	0.1912 0578	169.1986 9574	32.3517 6876	0.0059 1021	0.0309 1021
68	5.3607 1658	0.1865 4223	174.4286 6314	32.5383 1099	0.0057 3300	0.0307 3300
69	5.4947 3449	0.1819 9241	179.7893 7971	32.7203 0340	0.0055 6206	0.0305 6206
70	5.6321 0286	0.1775 5358	185.2841 1421	32.8978 5698	0.0053 9712	0.0303 9712
71	5.7729 0543	0.1732 2300	190.9162 1706	33.0710 7998	0.0052 3790	0.0302 3790
72	5.9172 2806	0.1689 9805	196.6891 2249	33.2400 7803	0.0050 8417	0.0300 8417
73	6.0651 5876	0.1648 7615	202.6063 5055	33.4049 5417	0.0049 3568	0.0299 3568
74	6.2167 8773	0.1608 5478	208.6715 0931	33.5658 0895	0.0047 9222	0.0297 9222
75	6.3722 0743	0.1569 3149	214.8882 9705	33.7227 4044	0.0046 5358	0.0296 5358
76	6.5315 1261	0.1531 0389	221.2605 0447	33.8758 4433	0.0045 1956	0.0295 1956
77	6.6948 0043	0.1493 6965	227.7920 1709	34.0252 1398	0.0043 8997	0.0293 8997
78	6.8621 7044	0.1457 2649	234.4868 1751	34.1709 4047	0.0042 6463	0.0292 6463
79	7.0337 2470	0.1421 7218	241.3489 8795	34.3131 1265	0.0041 4338	0.0291 4338
80	7.2095 6782	0.1387 0457	248.3827 1265	34.4518 1722	0.0040 2605	0.0290 2605
81	7.3898 0701	0.1353 2153	255.5922 8047	34.5871 3875	0.0039 1248	0.0289 1248
82	7.5745 5219	0.1320 2101	262.9820 8748	34.7191 5976	0.0038 0254	0.0288 0254
83	7.7639 1599	0.1288 0098	270.5566 3966	34.8479 6074	0.0036 9608	0.0286 9608
84	7.9580 1389	0.1256 5949	278.3205 5566	34.9736 2023	0.0035 9298	0.0285 9298
85	8.1569 6424	0.1225 9463	286.2785 6955	35.0962 1486	0.0034 9310	0.0284 9310
86	8.3608 8834	0.1196 0452	294.4355 3379	35.2158 1938	0.0033 9633	0.0283 9633
87	8.5699 1055	0.1166 8733	302.7964 2213	35.3325 0671	0.0033 0255	0.0283 0255
88	8.7841 5832	0.1138 4130	311.3663 3268	35.4463 4801	0.0032 1165	0.0282 1165
89	9.0037 6228	0.1110 6468	320.1504 9100	35.5574 1269	0.0031 2353	0.0281 2353
90	9.2288 5633	0.1083 5579	329.1542 5328	35.6657 6848	0.0030 3809	0.0280 3809
91	9.4595 7774	0.1057 1296	338.3831 0961	35.7714 8144	0.0029 5523	0.0279 5523
92	9.6960 6718	0.1031 3460	347.8426 8735	35.8746 1604	0.0028 7486	0.0278 7486
93	9.9384 6886	0.1006 1912	357.5387 5453	35.9752 3516	0.0027 9690	0.0277 9690
94	10.1869 3058	0.0981 6500	367.4772 2339	36.0734 0016	0.0027 2126	0.0277 2126
95	10.4416 0385	0.0957 7073	377.6641 5398	36.1691 7089	0.0026 4786	0.0276 4786
96	10.7026 4395	0.0934 3486	388.1057 5783	36.2626 0574	0.0025 7662	0.0275 7662
97	10.9702 1004	0.0911 5596	398.8084 0177	36.3537 6170	0.0025 0747	0.0275 0747
98	11.2444 6530	0.0889 3264	409.7786 1182	36.4426 9434	0.0024 4034	0.0274 4034
99	11.5255 7693	0.0867 6355	421.0230 7711	36.5294 5790	0.0023 7517	0.0273 7517
100	11.8137 1635	0.0846 4737	432.5486 5404	36.6141 0526	0.0023 1188	0.0273 1188

$i = 3\%$

n	$(1+i)^n$	$(1+i)^{-n}$	$s_{\overline{n}\rvert i}$	$a_{\overline{n}\rvert i}$	$\frac{1}{s_{\overline{n}\rvert i}}$	$\frac{1}{a_{\overline{n}\rvert i}}$
1	1.0300 0000	0.9708 7379	1.0000 0000	0.9708 7379	1.0000 0000	1.0300 0000
2	1.0609 0000	0.9425 9591	2.0300 0000	1.9134 6970	0.4926 1084	0.5226 1084
3	1.0927 2700	0.9151 4166	3.0909 0000	2.8286 1135	0.3235 3036	0.3535 3036
4	1.1255 0881	0.8884 8705	4.1836 2700	3.7170 9840	0.2390 2705	0.2690 2705
5	1.1592 7407	0.8626 0878	5.3091 3581	4.5797 0719	0.1883 5457	0.2183 5457
6	1.1940 5230	0.8374 8426	6.4684 0988	5.4171 9144	0.1545 9750	0.1845 9750
7	1.2298 7837	0.8130 9151	7.6624 6218	6.2302 8296	0.1305 0635	0.1605 0635
8	1.2667 7008	0.7894 0923	8.8923 3605	7.0196 9219	0.1124 5639	0.1424 5639
9	1.3047 7318	0.7664 1673	10.1591 0613	7.7861 0892	0.0984 3386	0.1284 3386
10	1.3439 1638	0.7440 9391	11.4638 7931	8.5302 0284	0.0872 3051	0.1172 3051
11	1.3842 3387	0.7224 2128	12.8077 9569	9.2526 2411	0.0780 7745	0.1080 7745
12	1.4257 6089	0.7013 7988	14.1920 2956	9.9540 0399	0.0704 6209	0.1004 6209
13	1.4685 3371	0.6809 5134	15.6177 9045	10.6349 5533	0.0640 2954	0.0940 2954
14	1.5125 8972	0.6611 1781	17.0863 2416	11.2960 7314	0.0585 2634	0.0885 2634
15	1.5579 6742	0.6418 6195	18.5989 1389	11.9379 3509	0.0537 6658	0.0837 6658
16	1.6047 0644	0.6231 6694	20.1568 8130	12.5611 0203	0.0496 1085	0.0796 1085
17	1.6528 4763	0.6050 1645	21.7615 8774	13.1661 1847	0.0459 5253	0.0759 5253
18	1.7024 3306	0.5873 9461	23.4144 3537	13.7535 1308	0.0427 0870	0.0727 0870
19	1.7535 0605	0.5702 8603	25.1168 6844	14.3237 9911	0.0398 1388	0.0698 1388
20	1.8061 1123	0.5536 7575	26.8703 7449	14.8774 7486	0.0372 1571	0.0672 1571
21	1.8602 9456	0.5375 4928	28.6764 8572	15.4150 2414	0.0348 7178	0.0648 7178
22	1.9161 0341	0.5218 9250	30.5367 8030	15.9369 1664	0.0327 4739	0.0627 4739
23	1.9735 8651	0.5066 9175	32.4528 8370	16.4436 0839	0.0308 1390	0.0608 1390
24	2.0327 9411	0.4919 3374	34.4264 7022	16.9355 4212	0.0290 4742	0.0590 4742
25	2.0937 7793	0.4776 0557	36.4592 6432	17.4131 4769	0.0274 2787	0.0574 2787
26	2.1565 9127	0.4636 9473	38.5530 4225	17.8678 4242	0.0259 3829	0.0559 3829
27	2.2212 8901	0.4501 8906	40.7096 3352	18.3270 3147	0.0245 6421	0.0545 6421
28	2.2879 2768	0.4370 7675	42.9309 2252	18.7641 0823	0.0232 9323	0.0532 9323
29	2.3565 6551	0.4243 4636	45.2188 5020	19.1884 5459	0.0221 1467	0.0521 1467
30	2.4272 6247	0.4119 8676	47.5754 1571	19.6004 4135	0.0210 1926	0.0510 1926
31	2.5000 8035	0.3999 8715	50.0026 7818	20.0004 2849	0.0199 9893	0.0499 9893
32	2.5750 8276	0.3883 3703	52.5027 5852	20.3887 6553	0.0190 4662	0.0490 4662
33	2.6523 3524	0.3770 2625	55.0778 4128	20.7657 9178	0.0181 5612	0.0481 5612
34	2.7319 0530	0.3660 4490	57.7301 7652	21.1318 3668	0.0173 2196	0.0473 2196
35	2.8138 6245	0.3553 8340	60.4620 8181	21.4872 2007	0.0165 3929	0.0465 3929
36	2.8982 7833	0.3450 3243	63.2759 4427	21.8322 5250	0.0158 0379	0.0458 0379
37	2.9852 2668	0.3349 8294	66.1742 2259	22.1672 3544	0.0151 1162	0.0451 1162
38	3.0747 8348	0.3252 2615	69.1594 4927	22.4924 6159	0.0144 5934	0.0444 5934
39	3.1670 2698	0.3157 5355	72.2342 3275	22.8082 1513	0.0138 4385	0.0438 4385
40	3.2620 3779	0.3065 5684	75.4012 5973	23.1147 7197	0.0132 6238	0.0432 6238
41	3.3598 9893	0.2976 2800	78.6632 9753	23.4123 9997	0.0127 1241	0.0427 1241
42	3.4606 9589	0.2889 5922	82.0231 9645	23.7013 5920	0.0121 9167	0.0421 9167
43	3.5645 1677	0.2805 4294	85.4838 9234	23.9819 0213	0.0116 9811	0.0416 9811
44	3.6714 5227	0.2723 7178	89.0484 0911	24.2542 7392	0.0112 2985	0.0412 2985
45	3.7815 9584	0.2644 3862	92.7198 6139	24.5187 1254	0.0107 8518	0.0407 8518
46	3.8950 4372	0.2567 3653	96.5014 5723	24.7754 4907	0.0103 6254	0.0403 6254
47	4.0118 9503	0.2492 5876	100.3965 0095	25.0247 0783	0.0099 6051	0.0399 6051
48	4.1322 5188	0.2419 9880	104.4083 9598	25.2667 0664	0.0095 7777	0.0395 7777
49	4.2562 1944	0.2349 5029	108.5406 4785	25.5016 5693	0.0092 1314	0.0392 1314
50	4.3839 0602	0.2281 0708	112.7968 6729	25.7297 6401	0.0088 6549	0.0388 6549

$i = 3\%$

n	$(1+i)^n$	$(1+i)^{-n}$	$s_{\overline{n}\vert i}$	$a_{\overline{n}\vert i}$	$\frac{1}{s_{\overline{n}\vert i}}$	$\frac{1}{a_{\overline{n}\vert i}}$
51	4.5154 2320	0.2214 6318	117.1807 7331	25.9512 2719	0.0085 3382	0.0385 3382
52	4.6508 8590	0.2150 1280	121.6961 9651	26.1662 3999	0.0082 1718	0.0382 1718
53	4.7904 1247	0.2087 5029	126.3470 8240	26.3749 9028	0.0079 1471	0.0379 1471
54	4.9341 2485	0.2026 7019	131.1374 9488	26.5776 6047	0.0076 2558	0.0376 2558
55	5.0821 4859	0.1967 6717	136.0716 1972	26.7744 2764	0.0073 4907	0.0373 4907
56	5.2346 1305	0.1910 3609	141.1537 6831	26.9654 6373	0.0070 8447	0.0370 8447
57	5.3916 5144	0.1854 7193	146.3883 8136	27.1509 3566	0.0068 3114	0.0368 3114
58	5.5534 0098	0.1800 6984	151.7800 3280	27.3310 0549	0.0065 8848	0.0365 8848
59	5.7200 0301	0.1748 2508	157.3334 3379	27.5058 3058	0.0063 5593	0.0363 5593
60	5.8916 0310	0.1697 3309	163.0534 3680	27.6755 6367	0.0061 3296	0.0361 3296
61	6.0683 5120	0.1647 8941	168.9450 3991	27.8403 5307	0.0059 1908	0.0359 1908
62	6.2504 0173	0.1599 8972	175.0133 9110	28.0003 4279	0.0057 1385	0.0357 1385
63	6.4379 1379	0.1553 2982	181.2637 9284	28.1556 7261	0.0055 1682	0.0355 1682
64	6.6310 5120	0.1508 0565	187.7017 0662	28.3064 7826	0.0053 2760	0.0353 2760
65	6.8299 8273	0.1464 1325	194.3327 5782	28.4528 9152	0.0051 4581	0.0351 4581
66	7.0348 8222	0.1421 4879	201.1627 4055	28.5950 4031	0.0049 7110	0.0349 7110
67	7.2459 2868	0.1380 0853	208.1976 2277	28.7330 4884	0.0048 0313	0.0348 0313
68	7.4633 0654	0.1339 8887	215.4435 5145	28.8670 3771	0.0046 4159	0.0346 4159
69	7.6872 0574	0.1300 8628	222.9068 5800	28.9971 2399	0.0044 8618	0.0344 8618
70	7.9178 2191	0.1262 9736	230.5940 6374	29.1234 2135	0.0043 3663	0.0343 3663
71	8.1553 5657	0.1226 1880	238.5118 8565	29.2460 4015	0.0041 9266	0.0341 9266
72	8.4000 1727	0.1190 4737	246.6672 4222	29.3650 8752	0.0040 5404	0.0340 5404
73	8.6520 1778	0.1155 7998	255.0672 5949	29.4806 6750	0.0039 2053	0.0339 2053
74	8.9115 7832	0.1122 1357	263.7192 7727	29.5928 8107	0.0037 9191	0.0337 9191
75	9.1789 2567	0.1089 4521	272.6308 5559	29.7018 2628	0.0036 6796	0.0336 6796
76	9.4542 9344	0.1057 7205	281.8097 8126	29.8075 9833	0.0035 4849	0.0335 4849
77	9.7379 2224	0.1026 9131	291.2640 7469	29.9102 8964	0.0034 3331	0.0334 3331
78	10.0300 5991	0.0997 0030	301.0019 9693	30.0099 8994	0.0033 2224	0.0333 2224
79	10.3309 6171	0.0967 9641	311.0320 5684	30.1067 8635	0.0032 1510	0.0332 1510
80	10.6408 9056	0.0939 7710	321.3630 1855	30.2007 6345	0.0031 1175	0.0331 1175
81	10.9601 1727	0.0912 3990	332.0039 0910	30.2920 0335	0.0030 1201	0.0330 1201
82	11.2889 2079	0.0885 8243	342.9640 2638	30.3805 8577	0.0029 1576	0.0329 1576
83	11.6275 8842	0.0860 0236	354.2529 4717	30.4665 8813	0.0028 2284	0.0328 2284
84	11.9764 1607	0.0834 9743	365.8805 3558	30.5500 8556	0.0027 3313	0.0327 3313
85	12.3357 0855	0.0810 6547	377.8569 5165	30.6311 5103	0.0026 4650	0.0326 4650
86	12.7057 7981	0.0787 0434	390.1926 6020	30.7098 5537	0.0025 6284	0.0325 6284
87	13.0869 5320	0.0764 1198	402.8984 4001	30.7862 6735	0.0024 8202	0.0324 8202
88	13.4795 6180	0.0741 8639	415.9853 9321	30.8604 5374	0.0024 0393	0.0324 0393
89	13.8839 4865	0.0720 2562	429.4649 5500	30.9324 7936	0.0023 2848	0.0323 2848
90	14.3004 6711	0.0699 2779	443.3489 0365	31.0024 0714	0.0022 5556	0.0322 5556
91	14.7294 8112	0.0678 9105	457.6493 7076	31.0702 9820	0.0021 8508	0.0321 8508
92	15.1713 6556	0.0659 1364	472.3788 5189	31.1362 1184	0.0021 1694	0.0321 1694
93	15.6265 0652	0.0639 9383	487.5502 1744	31.2002 0567	0.0020 5107	0.0320 5107
94	16.0953 0172	0.0621 2993	503.1767 2397	31.2623 3560	0.0019 8737	0.0319 8737
95	16.5781 6077	0.0603 2032	519.2720 2568	31.3226 5592	0.0019 2577	0.0319 2577
96	17.0755 0559	0.0585 6342	535.8501 8645	31.3812 1934	0.0018 6619	0.0318 6619
97	17.5877 7076	0.0568 5769	552.9256 9205	31.4380 7703	0.0018 0856	0.0318 0856
98	18.1154 0388	0.0552 0164	570.5134 6281	31.4932 7867	0.0017 5281	0.0317 5281
99	18.6588 6600	0.0535 9383	588.6288 6669	31.5468 7250	0.0016 9886	0.0316 9886
100	19.2186 3198	0.0520 3284	607.2877 3269	31.5989 0534	0.0016 4667	0.0316 4667

$i = 3\frac{1}{2}\%$

n	$(1+i)^n$	$(1+i)^{-n}$	$s_{\overline{n}\rvert i}$	$a_{\overline{n}\rvert i}$	$\frac{1}{s_{\overline{n}\rvert i}}$	$\frac{1}{a_{\overline{n}\rvert i}}$
1	1.0350 0000	0.9661 8357	1.0000 0000	0.9661 8357	1.0000 0000	1.0350 0000
2	1.0712 2500	0.9335 1070	2.0350 0000	1.8996 9428	0.4914 0049	0.5264 0049
3	1.1087 1788	0.9019 4271	3.1062 2500	2.8016 3698	0.3219 3418	0.3569 3418
4	1.1475 2300	0.8714 4223	4.2149 4288	3.6730 7921	0.2372 5114	0.2722 5114
5	1.1876 8631	0.8419 7317	5.3624 6588	4.5150 5238	0.1864 8137	0.2214 8137
6	1.2292 5533	0.8135 0064	6.5501 5218	5.3285 5302	0.1526 6821	0.1876 6821
7	1.2722 7926	0.7859 9096	7.7794 0751	6.1145 4398	0.1285 4449	0.1635 4449
8	1.3168 0904	0.7594 1156	9.0516 8677	6.8739 5554	0.1104 7665	0.1454 7665
9	1.3628 9735	0.7337 3097	10.3684 9581	7.6076 8651	0.0964 4601	0.1314 4601
10	1.4105 9876	0.7089 1881	11.7313 9316	8.3166 0532	0.0852 4137	0.1202 4137
11	1.4599 6972	0.6849 4571	13.1419 9192	9.0015 5104	0.0760 9197	0.1110 9197
12	1.5110 6866	0.6617 8330	14.6019 6164	9.6633 3433	0.0684 8395	0.1034 8395
13	1.5639 5606	0.6394 0415	16.1130 3030	10.3027 3849	0.0620 6157	0.0970 6157
14	1.6186 9452	0.6177 8179	17.6769 8636	10.9205 2028	0.0565 7073	0.0915 7073
15	1.6753 4883	0.5968 9062	19.2956 8088	11.5174 1090	0.0518 2507	0.0868 2507
16	1.7339 8604	0.5767 0591	20.9710 2971	12.0941 1681	0.0476 8483	0.0826 8483
17	1.7946 7555	0.5572 0378	22.7050 1575	12.6513 2059	0.0440 4313	0.0790 4313
18	1.8574 8920	0.5383 6114	24.4996 9130	13.1896 8173	0.0408 1684	0.0758 1684
19	1.9225 0132	0.5201 5569	26.3571 8050	13.7098 3742	0.0379 4033	0.0729 4033
20	1.9897 8886	0.5025 6588	28.2796 8181	14.2124 0330	0.0353 6108	0.0703 6108
21	2.0594 3147	0.4855 7090	30.2694 7068	14.6979 7420	0.0330 3659	0.0680 3659
22	2.1315 1158	0.4691 5063	32.3289 0215	15.1671 2484	0.0309 3207	0.0659 3207
23	2.2061 1448	0.4532 8563	34.4604 1373	15.6204 1047	0.0290 1880	0.0640 1880
24	2.2833 2849	0.4379 5713	36.6665 2821	16.0583 6760	0.0272 7283	0.0622 7283
25	2.3632 4498	0.4231 4699	38.9498 5669	16.4815 1459	0.0256 7404	0.0606 7404
26	2.4459 5856	0.4088 3767	41.3131 0168	16.8903 5226	0.0242 0540	0.0592 0540
27	2.5315 6711	0.3950 1224	43.7590 6024	17.2853 6451	0.0228 5241	0.0578 5241
28	2.6201 7196	0.3816 5434	46.2906 2734	17.6670 1885	0.0216 0265	0.0566 0265
29	2.7118 7798	0.3687 4815	48.9107 9930	18.0357 6700	0.0204 4538	0.0554 4538
30	2.8067 9370	0.3562 7841	51.6226 7728	18.3920 4541	0.0193 7133	0.0543 7133
31	2.9050 3148	0.3442 3035	54.4294 7098	18.7362 7576	0.0183 7240	0.0533 7240
32	3.0067 0759	0.3325 8971	57.3345 0247	19.0688 6547	0.0174 4150	0.0524 4150
33	3.1119 4235	0.3213 4271	60.3412 1005	19.3902 0818	0.0165 7242	0.0515 7242
34	3.2208 6033	0.3104 7605	63.4531 5240	19.7006 8423	0.0157 5966	0.0507 5966
35	3.3335 9045	0.2999 7686	66.6740 1274	20.0006 6110	0.0149 9835	0.0499 9835
36	3.4502 6611	0.2898 3272	70.0076 0318	20.2904 9381	0.0142 8416	0.0492 8416
37	3.5710 2543	0.2800 3161	73.4578 6930	20.5705 2542	0.0136 1325	0.0486 1325
38	3.6960 1132	0.2705 6194	77.0288 9472	20.8410 8736	0.0129 8214	0.0479 8214
39	3.8253 7171	0.2614 1250	80.7249 0604	21.1024 9987	0.0123 8775	0.0473 8775
40	3.9592 5972	0.2525 7247	84.5502 7775	21.3550 7234	0.0118 2728	0.0468 2728
41	4.0978 3381	0.2440 3137	88.5095 3747	21.5991 0371	0.0112 9822	0.0462 9822
42	4.2412 5799	0.2357 7910	92.6073 7128	21.8348 8281	0.0107 9828	0.0457 9828
43	4.3897 0202	0.2278 0590	96.8486 2928	22.0626 8870	0.0103 2539	0.0453 2539
44	4.5433 4160	0.2201 0231	101.2383 3130	22.2827 9102	0.0098 7768	0.0448 7768
45	4.7023 5855	0.2126 5924	105.7816 7290	22.4954 5026	0.0094 5343	0.0444 5343
46	4.8669 4110	0.2054 6787	110.4840 3145	22.7009 1813	0.0090 5108	0.0440 5108
47	5.0372 8404	0.1985 1968	115.3509 7255	22.8994 3780	0.0086 6919	0.0436 6919
48	5.2135 8898	0.1918 0645	120.3882 5659	23.0912 4425	0.0083 0646	0.0433 0646
49	5.3960 6459	0.1853 2024	125.6018 4557	23.2765 6450	0.0079 6167	0.0429 6167
50	5.5849 2686	0.1790 5337	130.9979 1016	23.4556 1787	0.0076 3371	0.0426 3371

$i = 3\frac{1}{2}\%$

n	$(1+i)^n$	$(1+i)^{-n}$	$s_{\overline{n}\rvert i}$	$a_{\overline{n}\rvert i}$	$\frac{1}{s_{\overline{n}\rvert i}}$	$\frac{1}{a_{\overline{n}\rvert i}}$
51	5.7803 9930	0.1729 9843	136.5828 3702	23.6286 1630	0.0073 2156	0.0423 2156
52	5.9827 1327	0.1671 4824	142.3632 3631	23.7957 6454	0.0070 2429	0.0420 2429
53	6.1921 0824	0.1614 9589	148.3459 4958	23.9572 6043	0.0067 4100	0.0417 4100
54	6.4088 3202	0.1560 3467	154.5380 5782	24.1132 9510	0.0064 7090	0.0414 7090
55	6.6331 4114	0.1507 5814	160.9468 8984	24.2640 5323	0.0062 1323	0.0412 1323
56	6.8653 0108	0.1456 6004	167.5800 3099	24.4097 1327	0.0059 6730	0.0409 6730
57	7.1055 8662	0.1407 3433	174.4453 3207	24.5504 4760	0.0057 3245	0.0407 3245
58	7.3542 8215	0.1359 7520	181.5509 1869	24.6864 2281	0.0055 0810	0.0405 0810
59	7.6116 8203	0.1313 7701	188.9052 0085	24.8177 9981	0.0052 9366	0.0402 9366
60	7.8780 9090	0.1269 3431	196.5168 8288	29.9447 3412	0.0050 8862	0.0400 8862
61	8.1538 2408	0.1226 4184	204.3949 7378	25.0673 7596	0.0048 9249	0.0398 9249
62	8.4392 0793	0.1184 9453	212.5487 9786	25.1858 7049	0.0047 0480	0.0397 0480
63	8.7345 8020	0.1144 8747	220.9880 0579	25.3003 5796	0.0045 2513	0.0395 2513
64	9.0402 9051	0.1106 1591	229.7225 8599	25.4109 7388	0.0043 5308	0.0393 5308
65	9.3567 0068	0.1068 7528	238.7628 7650	25.5178 4916	0.0041 8826	0.0391 8826
66	9.6841 8520	0.1032 6114	248.1195 7718	25.6211 1030	0.0040 3031	0.0390 3031
67	10.0231 3168	0.0997 6922	257.8037 6238	25.7208 7951	0.0038 7892	0.0388 7892
68	10.3739 4129	0.0963 9538	267.8268 9406	25.8172 7489	0.0037 3375	0.0387 3375
69	10.7370 2924	0.0931 3563	278.2008 3535	25.9104 1052	0.0035 9453	0.0385 9453
70	11.1128 2526	0.0899 8612	288.9378 6459	26.0003 9664	0.0034 6095	0.0384 6095
71	11.5017 7414	0.0869 4311	300.0506 8985	26.0873 3975	0.0033 3277	0.0383 3277
72	11.9043 3624	0.0840 0300	311.5524 6400	26.1713 4275	0.0032 0973	0.0382 0973
73	12.3209 8801	0.0811 6232	323.4568 0024	26.2525 0508	0.0030 9160	0.0380 9160
74	12.7522 2259	0.0784 1770	335.7777 8824	26.3309 2278	0.0029 7816	0.0379 7816
75	13.1985 5038	0.0757 6590	348.5300 1083	26.4066 8868	0.0028 6919	0.0378 6919
76	13.6604 9964	0.0732 0376	361.7285 6121	26.4798 9244	0.0027 6450	0.0377 6450
77	14.1386 1713	0.0707 2827	375.3890 6085	26.5506 2072	0.0026 6390	0.0376 6390
78	14.6334 6873	0.0683 3650	389.5276 7798	26.6189 5721	0.0025 6721	0.0375 6721
79	15.1456 4013	0.0660 2560	404.1611 4671	26.6849 8281	0.0024 7426	0.0374 7426
80	15.6757 3754	0.0637 9285	419.3067 8685	26.7487 7567	0.0023 8489	0.0373 8489
81	16.2243 8835	0.0616 3561	434.9825 2439	26.8104 1127	0.0022 9894	0.0372 9894
82	16.7922 4195	0.0595 5131	451.2069 1274	26.8699 6258	0.0022 1628	0.0372 1628
83	17.3799 7041	0.0575 3750	467.9991 5469	26.9275 0008	0.0021 3676	0.0371 3676
84	17.9882 6938	0.0555 9178	485.3791 2510	26.9830 9186	0.0020 6025	0.0370 6025
85	18.6178 5881	0.0537 1187	503.3673 9448	27.0368 0373	0.0019 8662	0.0369 8662
86	19.2694 8387	0.0518 9553	521.9852 5329	27.0886 9926	0.0019 1576	0.0369 1576
87	19.9439 1580	0.0501 4060	541.2547 3715	27.1388 3986	0.0018 4756	0.0368 4756
88	20.6419 5285	0.0484 4503	561.1986 5295	27.1872 8489	0.0017 8190	0.0367 8190
89	21.3644 2120	0.0468 0679	581.8406 0581	27.2340 9168	0.0017 1868	0.0367 1868
90	22.1121 7595	0.0452 2395	603.2050 2701	27.2793 1564	0.0016 5781	0.0366 5781
91	22.8861 0210	0.0436 9464	625.3172 0295	27.3230 1028	0.0015 9919	0.0365 9919
92	23.6871 1568	0.0422 1704	648.2033 0506	27.3652 2732	0.0015 4273	0.0365 4273
93	24.5161 6473	0.0407 8941	671.8904 2073	27.4060 1673	0.0014 8834	0.0364 8834
94	25.3742 3049	0.0394 1006	696.4065 8546	27.4454 2680	0.0014 3594	0.0364 3594
95	26.2623 2856	0.0380 7735	721.7808 1595	27.4835 0415	0.0013 8546	0.0363 8546
96	27.1815 1006	0.0367 8971	748.0431 4451	27.5202 9387	0.0013 3682	0.0363 3682
97	28.1328 6291	0.0355 4562	775.2246 5457	27.5558 3948	0.0012 8995	0.0362 8995
98	29.1175 1311	0.0343 4359	803.3575 1748	27.5901 8308	0.0012 4478	0.0362 4478
99	30.1366 2607	0.0331 8221	832.4750 3059	27.6233 6529	0.0012 0124	0.0362 0124
100	31.1914 0798	0.0320 6011	862.6116 5666	27.6554 2540	0.0011 5927	0.0361 5927

$i = 4\%$

n	$(1 + i)^n$	$(1 + i)^{-n}$	$s_{\overline{n}\rvert i}$	$a_{\overline{n}\rvert i}$	$\dfrac{1}{s_{\overline{n}\rvert i}}$	$\dfrac{1}{a_{\overline{n}\rvert i}}$
1	1.0400 0000	0.9615 3846	1.0000 0000	0.9615 3846	1.0000 0000	1.0400 0000
2	1.0816 0000	0.9245 5621	2.0400 0000	1.8860 9467	0.4901 9608	0.5301 9608
3	1.1248 6400	0.8889 9636	3.1216 0000	2.7750 9103	0.3203 4854	0.3603 4854
4	1.1698 5856	0.8548 0419	4.2464 6400	3.6298 9522	0.2354 9005	0.2754 9005
5	1.2166 5290	0.8219 2711	5.4163 2256	4.4518 2233	0.1846 2711	0.2246 2711
6	1.2653 1902	0.7903 1453	6.6329 7546	5.2421 3686	0.1507 6190	0.1907 6190
7	1.3159 3178	0.7599 1781	7.8982 9448	6.0020 5467	0.1266 0961	0.1666 0961
8	1.3685 6905	0.7306 9021	9.2142 2626	6.7327 4487	0.1085 2783	0.1485 2783
9	1.4233 1181	0.7025 8674	10.5827 9531	7.4353 3161	0.0944 9299	0.1344 9299
10	1.4802 4428	0.6755 6417	12.0061 0712	8.1108 9578	0.0832 9094	0.1232 9094
11	1.5394 5406	0.6495 8093	13.4863 5141	8.7604 7671	0.0741 4904	0.1141 4904
12	1.6010 3222	0.6245 9705	15.0258 0546	9.3850 7376	0.0665 5217	0.1065 5217
13	1.6650 7351	0.6005 7409	16.6268 3768	9.9856 4785	0.0601 4373	0.1001 4373
14	1.7316 7645	0.5774 7508	18.2919 1119	10.5631 2293	0.0546 6897	0.0946 6897
15	1.8009 4351	0.5552 6450	20.0235 8764	11.1183 8743	0.0499 4110	0.0899 4110
16	1.8729 8125	0.5339 0818	21.8245 3114	11.6522 9561	0.0458 2000	0.0858 2000
17	1.9479 0050	0.5133 7325	23.6975 1239	12.1656 6885	0.0421 9852	0.0821 9852
18	2.0258 1652	0.4936 2812	25.6454 1288	12.6592 9697	0.0389 9333	0.0789 9333
19	2.1068 4918	0.4746 4242	27.6712 2940	13.1339 3940	0.0361 3862	0.0761 3862
20	2.1911 2314	0.4563 8695	29.7780 7858	13.5903 2634	0.0335 8175	0.0735 8175
21	2.2787 6807	0.4388 3360	31.9692 0172	14.0291 5995	0.0312 8011	0.0712 8011
22	2.3699 1879	0.4219 5539	34.2479 6979	14.4511 1533	0.0291 9881	0.0691 9881
23	2.4647 1554	0.4057 2633	36.6178 8858	14.8568 4167	0.0273 0906	0.0673 0906
24	2.5633 0416	0.3901 2147	39.0826 0412	15.2469 6314	0.0255 8683	0.0655 8683
25	2.6658 3633	0.3751 1680	41.6459 0829	15.6220 7994	0.0240 1196	0.0640 1196
26	2.7724 6978	0.3606 8923	44.3117 4462	15.9827 6918	0.0225 6738	0.0625 6738
27	2.8833 6858	0.3468 1657	47.0842 1440	16.3295 8575	0.0212 3854	0.0612 3854
28	2.9987 0332	0.3334 7747	49.9675 8298	16.6630 6322	0.0200 1298	0.0600 1298
29	3.1186 5145	0.3206 5141	52.9662 8630	16.9837 1463	0.0188 7993	0.0588 7993
30	3.2433 9751	0.3083 1867	56.0849 3775	17.2920 3330	0.0178 3010	0.0578 3010
31	3.3731 3341	0.2964 6026	59.3283 3526	17.5884 9356	0.0168 5535	0.0568 5535
32	3.5080 5875	0.2850 5794	62.7014 6867	17.8735 5150	0.0159 4859	0.0559 4859
33	3.6483 8110	0.2740 9417	66.2095 2742	18.1476 4567	0.0151 0357	0.0551 0357
34	3.7943 1634	0.2635 5209	69.8579 0851	18.4111 9776	0.0143 1477	0.0543 1477
35	3.9460 8899	0.2534 1547	73.6522 2486	18.6646 1323	0.0135 7732	0.0535 7732
36	4.1039 3255	0.2436 6872	77.5983 1385	18.9082 8195	0.0128 8688	0.0528 8688
37	4.2680 8986	0.2342 9685	81.7022 4640	19.1425 7880	0.0122 3957	0.0522 3957
38	4.4388 1345	0.2252 8543	85.9703 3626	19.3678 6423	0.0116 3192	0.0516 3192
39	4.6163 6599	0.2166 2061	90.4091 4971	19.5844 8484	0.0110 6083	0.0510 6083
40	4.8010 2063	0.2082 8904	95.0255 1570	19.7927 7388	0.0105 2349	0.0505 2349
41	4.9930 6145	0.2002 7793	99.8265 3633	19.9930 5181	0.0100 1738	0.0500 1738
42	5.1927 8391	0.1925 7493	104.8195 9778	20.1856 2674	0.0095 4020	0.0495 4020
43	5.4004 9527	0.1851 6820	110.0123 8169	20.3707 9494	0.0090 8989	0.0490 8989
44	5.6165 1508	0.1780 4635	115.4128 7696	20.5488 4129	0.0086 6454	0.0486 6454
45	5.8411 7568	0.1711 9841	121.0293 9204	20.7200 3970	0.0082 6246	0.0482 6246
46	6.0748 2271	0.1646 1386	126.8705 6772	20.8846 5356	0.0078 8205	0.0478 8205
47	6.3178 1562	0.1582 8256	132.9453 9043	21.0429 3612	0.0075 2189	0.0475 2189
48	6.5705 2824	0.1521 9476	139.2632 0604	21.1951 3088	0.0071 8065	0.0471 8065
49	6.8333 4937	0.1463 4112	145.8337 3429	21.3414 7200	0.0068 5712	0.0468 5712
50	7.1066 8335	0.1407 1262	152.6670 8366	21.4821 8462	0.0065 5020	0.0465 5020

$i = 4\frac{1}{2}\%$

n	$(1+i)^n$	$(1+i)^{-n}$	$s_{\overline{n}\rvert i}$	$a_{\overline{n}\rvert i}$	$\dfrac{1}{s_{\overline{n}\rvert i}}$	$\dfrac{1}{a_{\overline{n}\rvert i}}$
1	1.0450 0000	0.9569 3780	1.0000 0000	0.9569 3780	1.0000 0000	1.0450 0000
2	1.0920 2500	0.9157 2995	2.0450 0000	1.8726 6775	0.4889 9756	0.5339 9756
3	1.1411 6613	0.8762 9660	3.1370 2500	2.7489 6435	0.3187 7336	0.3637 7336
4	1.1925 1860	0.8385 6134	4.2781 9113	3.5875 2570	0.2337 4365	0.2787 4365
5	1.2461 8194	0.8024 5105	5.4707 0973	4.3899 7674	0.1827 9164	0.2277 9164
6	1.3022 6012	0.7678 9574	6.7168 9166	5.1578 7248	0.1488 7839	0.1938 7839
7	1.3608 6183	0.7348 2846	8.0191 5179	5.8927 0094	0.1247 0147	0.1697 0147
8	1.4221 0061	0.7031 8513	9.3800 1362	6.5958 8607	0.1066 0965	0.1516 0965
9	1.4860 9514	0.6729 0443	10.8021 1423	7.2687 9050	0.0925 7447	0.1375 7447
10	1.5529 6942	0.6439 2768	12.2882 0937	7.9127 1818	0.0813 7882	0.1263 7882
11	1.6228 5305	0.6161 9874	13.8411 7879	8.5289 1692	0.0722 4818	0.1172 4818
12	1.6958 8143	0.5896 6386	15.4640 3184	9.1185 8078	0.0646 6619	0.1096 6619
13	1.7721 9610	0.5642 7164	17.1599 1327	9.6828 5242	0.0582 7535	0.1032 7535
14	1.8519 4492	0.5399 7286	18.9321 0937	10.2228 2528	0.0528 2032	0.0978 2032
15	1.9352 8244	0.5167 2044	20.7840 5429	10.7395 4573	0.0481 1381	0.0931 1381
16	2.0223 7015	0.4944 6932	22.7193 3673	11.2340 1505	0.0440 1537	0.0890 1537
17	2.1133 7681	0.4731 7639	24.7417 0689	11.7071 9143	0.0404 1758	0.0854 1758
18	2.2084 7877	0.4528 0037	26.8550 8370	12.1599 9180	0.0372 3690	0.0822 3690
19	2.3078 6031	0.4333 0179	29.0635 6246	12.5932 9359	0.0344 0734	0.0794 0734
20	2.4117 1402	0.4146 4286	31.3714 2277	13.0079 3645	0.0318 7614	0.0768 7614
21	2.5202 4116	0.3967 8743	33.7831 3680	13.4047 2388	0.0296 0057	0.0746 0057
22	2.6336 5201	0.3797 0089	36.3033 7795	13.7844 2476	0.0275 4565	0.0725 4565
23	2.7521 6635	0.3633 5013	38.9370 2996	14.1477 7489	0.0256 8249	0.0706 8249
24	2.8760 1383	0.3477 0347	41.6891 9631	14.4954 7837	0.0239 8703	0.0689 8703
25	3.0054 3446	0.3327 3060	44.5652 1015	14.8282 0896	0.0224 3903	0.0674 3903
26	3.1406 7901	0.3184 0248	47.5706 4460	15.1466 1145	0.0210 2137	0.0660 2137
27	3.2820 0956	0.3046 9137	50.7113 2361	15.4513 0282	0.0197 1946	0.0647 1946
28	3.4296 9999	0.2915 7069	53.9933 3317	15.7428 7351	0.0185 2081	0.0635 2081
29	3.5840 3649	0.2790 1502	57.4230 3316	16.0218 8853	0.0174 1461	0.0624 1461
30	3.7453 1813	0.2670 0002	61.0070 6966	16.2888 8854	0.0163 9154	0.0613 9154
31	3.9138 5745	0.2555 0241	64.7523 8779	16.5443 9095	0.0154 4345	0.0604 4345
32	4.0899 8104	0.2444 9991	68.6662 4524	16.7888 9086	0.0145 6320	0.0595 6320
33	4.2740 3018	0.2339 7121	75.7562 2628	17.0228 6207	0.0137 4453	0.0587 4453
34	4.4663 6154	0.2238 9589	77.0302 5646	17.2467 5796	0.0129 8191	0.0579 8191
35	4.6673 4781	0.2142 5444	81.4966 1800	17.4610 1240	0.0122 7045	0.0572 7045
36	4.8773 7846	0.2050 2817	86.1639 6581	17.6660 4058	0.0116 0578	0.0566 0578
37	5.0968 6049	0.1961 9921	91.0413 4427	17.8622 3979	0.0109 8402	0.0559 8402
38	5.3262 1921	0.1877 5044	96.1382 0476	18.0499 9023	0.0104 0169	0.0554 0169
39	5.5658 9908	0.1796 6549	101.4644 2398	18.2296 5572	0.0098 5567	0.0548 5567
40	5.8163 6454	0.1719 2870	107.0303 2306	18.4015 8442	0.0093 4315	0.0543 4315
41	6.0781 0094	0.1645 2507	112.8466 8760	18.5661 0949	0.0088 6158	0.0538 6158
42	6.3516 1548	0.1574 4026	118.9247 8854	18.7235 4975	0.0084 0868	0.0534 0868
43	6.6374 3818	0.1506 6054	125.2764 0402	18.8742 1029	0.0079 8235	0.0529 8235
44	6.9361 2290	0.1441 7276	131.9138 4220	19.0183 8305	0.0075 8071	0.0525 8071
45	7.2482 4843	0.1379 6437	138.8499 6510	19.1563 4742	0.0072 0202	0.0522 0202
46	7.5744 1961	0.1320 2332	146.0982 1353	19.2883 7074	0.0068 4471	0.0518 4471
47	7.9152 6849	0.1263 3810	153.6726 3314	19.4147 0884	0.0065 0734	0.0515 0734
48	8.2714 5557	0.1208 9771	161.5879 0163	19.5356 0654	0.0061 8858	0.0511 8858
49	8.6436 7107	0.1156 9158	169.8593 5720	19.6512 9813	0.0058 8722	0.0508 8722
50	9.0326 3627	0.1107 0965	178.5030 2828	19.7620 0778	0.0056 0215	0.0506 0215

$i = 5\%$

n	$(1+i)^n$	$(1+i)^{-n}$	$s_{\overline{n}\|i}$	$a_{\overline{n}\|i}$	$\frac{1}{s_{\overline{n}\|i}}$	$\frac{1}{a_{\overline{n}\|i}}$
1	1.0500 0000	0.9523 8095	1.0000 0000	0.9523 8095	1.0000 0000	1.0500 0000
2	1.1025 0000	0.9070 2948	2.0500 0000	1.8594 1043	0.4878 0488	0.5378 0488
3	1.1576 2500	0.8638 3760	3.1525 0000	2.7232 4803	0.3172 0856	0.3672 0856
4	1.2155 0625	0.8227 0247	4.3101 2500	3.5459 5050	0.2320 1183	0.2820 1183
5	1.2762 8156	0.7835 2617	5.5256 3125	4.3294 7667	0.1809 7480	0.2309 7480
6	1.3400 9564	0.7462 1540	6.8019 1281	5.0756 9207	0.1470 1747	0.1970 1747
7	1.4071 0042	0.7106 8133	8.1420 0845	5.7863 7340	0.1228 1982	0.1728 1982
8	1.4774 5544	0.6768 3936	9.5491 0888	6.4632 1276	0.1047 2181	0.1547 2181
9	1.5513 2822	0.6446 0892	11.0265 6432	7.1078 2168	0.0906 9008	0.1406 9008
10	1.6288 9463	0.6139 1325	12.5778 9254	7.7217 3493	0.0795 0457	0.1295 0457
11	1.7103 3936	0.5846 7929	14.2067 8716	8.3064 1422	0.0703 8889	0.1203 8889
12	1.7958 5633	0.5568 3742	15.9171 2652	8.8632 5164	0.0628 2541	0.1128 2541
13	1.8856 4914	0.5303 2135	17.7129 8285	9.3935 7299	0.0564 5577	0.1064 5577
14	1.9799 3160	0.5050 6795	19.5986 3199	9.8986 4094	0.0510 2397	0.1010 2397
15	2.0789 2818	0.4810 1710	21.5785 6359	10.3796 5804	0.0463 4229	0.0963 4229
16	2.1828 7459	0.4581 1152	23.6574 9177	10.8377 6956	0.0422 6991	0.0922 6991
17	2.2920 1832	0.4362 9669	25.8403 6636	11.2740 6625	0.0386 9914	0.0886 9914
18	2.4066 1923	0.4155 2065	28.1323 8467	11.6895 8690	0.0355 4622	0.0855 4622
19	2.5269 5020	0.3957 3396	30.5390 0391	12.0853 2086	0.0327 4501	0.0827 4501
20	2.6532 9771	0.3768 8948	33.0659 5410	12.4622 1034	0.0302 4259	0.0802 4259
21	2.7859 6259	0.3589 4236	35.7192 5181	12.8211 5271	0.0279 9611	0.0779 9611
22	2.9252 6072	0.3418 4987	38.5052 1440	13.1630 0258	0.0259 7051	0.0759 7051
23	3.0715 2376	0.3255 7131	41.4304 7512	13.4885 7388	0.0241 3682	0.0741 3682
24	3.2250 9994	0.3100 6791	44.5019 9887	13.7986 4179	0.0224 7090	0.0724 7090
25	3.3863 5494	0.2953 0277	47.7270 9882	14.0939 4457	0.0209 5246	0.0709 5246
26	3.5556 7269	0.2812 4073	51.1134 5376	14.3751 8530	0.0195 6432	0.0695 6432
27	3.7334 5632	0.2678 4832	54.6691 2645	14.6430 3362	0.0182 9186	0.0682 9186
28	3.9201 2914	0.2550 9364	58.4025 8277	14.8981 2726	0.0171 2253	0.0671 2253
29	4.1161 3560	0.2429 4632	62.3227 1191	15.1410 7358	0.0160 4551	0.0660 4551
30	4.3219 4238	0.2313 7745	66.4388 4750	15.3724 5103	0.0150 5144	0.0650 5144
31	4.5380 3949	0.2203 5947	70.7607 8988	15.5928 1050	0.0141 3212	0.0641 3212
32	4.7649 4147	0.2098 6617	75.2988 2937	15.8026 7667	0.0132 8042	0.0632 8042
33	5.0031 8854	0.1998 7254	80.0637 7084	16.0025 4921	0.0124 9004	0.0624 9004
34	5.2533 4797	0.1903 5480	85.0669 5938	16.1929 0401	0.0117 5545	0.0617 5545
35	5.5160 1537	0.1812 9029	90.3203 0735	16.3741 9429	0.0110 7171	0.0610 7171
36	5.7918 1614	0.1726 5741	95.8363 2272	16.5468 5171	0.0104 3446	0.0604 3446
37	6.0814 0694	0.1644 3563	101.6281 3886	16.7112 8734	0.0098 3979	0.0598 3979
38	6.3854 7729	0.1566 0536	107.7095 4580	16.8678 9271	0.0092 8423	0.0592 8423
39	6.7047 5115	0.1491 4797	114.0950 2309	17.0170 4067	0.0087 6462	0.0587 6462
40	7.0399 8871	0.1420 4568	120.7997 7424	17.1590 8635	0.0082 7816	0.0582 7816
41	7.3919 8815	0.1352 8160	127.8397 6295	17.2943 6796	0.0078 2229	0.0578 2229
42	7.7615 8756	0.1288 3962	135.2317 5110	17.4232 0758	0.0073 9471	0.0573 9471
43	8.1496 6693	0.1227 0440	142.9933 3866	17.5459 1198	0.0069 9333	0.0569 9333
44	8.5571 5028	0.1168 6133	151.1430 0559	17.6627 7331	0.0066 1625	0.0566 1625
45	8.9850 0779	0.1112 9651	159.7001 5587	17.7740 6982	0.0062 6173	0.0562 6173
46	9.4342 5818	0.1059 9668	168.6851 6366	17.8800 6650	0.0059 2820	0.0559 2820
47	9.9059 7109	0.1009 4921	178.1194 2185	17.9810 1571	0.0056 1421	0.0556 1421
48	10.4012 6965	0.0961 4211	188.0253 9294	18.0771 5782	0.0053 1843	0.0553 1843
49	10.9213 3313	0.0915 6391	198.4266 6259	18.1687 2173	0.0050 3965	0.0550 3965
50	11.4673 9979	0.0872 0373	209.3479 9572	18.2559 2546	0.0047 7674	0.0547 7674

$i = 6\%$

n	$(1+i)^n$	$(1+i)^{-n}$	$s_{\overline{n}\vert i}$	$a_{\overline{n}\vert i}$	$\frac{1}{s_{\overline{n}\vert i}}$	$\frac{1}{a_{\overline{n}\vert i}}$
1	1.0600 0000	0.9433 9623	1.0000 0000	0.9433 9623	1.0000 0000	1.0600 0000
2	1.1236 0000	0.8899 9644	2.0600 0000	1.8333 9267	0.4854 3689	0.5454 3689
3	1.1910 1600	0.8396 1928	3.1836 0000	2.6730 1195	0.3141 0981	0.3741 0981
4	1.2624 7696	0.7920 9366	4.3746 1600	3.4651 0561	0.2285 9149	0.2885 9149
5	1.3382 2558	0.7472 5817	5.6370 9296	4.2123 6379	0.1773 9640	0.2373 9640
6	1.4185 1911	0.7049 6054	6.9753 1854	4.9173 2433	0.1433 6263	0.2033 6263
7	1.5036 3026	0.6650 5711	8.3938 3765	5.5823 8144	0.1191 3502	0.1791 3502
8	1.5938 4807	0.6274 1237	9.8974 6791	6.2097 9381	0.1010 3594	0.1610 3594
9	1.6894 7896	0.5918 9846	11.4913 1598	6.8016 9227	0.0870 2224	0.1470 2224
10	1.7908 4770	0.5583 9478	13.1807 9494	7.3600 8705	0.0758 6796	0.1358 6796
11	1.8982 9856	0.5267 8753	14.9716 4264	7.8868 7458	0.0667 9294	0.1267 9294
12	2.0121 9647	0.4969 6936	16.8699 4120	8.3838 4394	0.0592 7703	0.1192 7703
13	2.1329 2826	0.4688 3902	18.8821 3767	8.8526 8296	0.0529 6011	0.1129 6011
14	2.2609 0396	0.4423 0096	21.0150 6593	9.2949 8393	0.0475 8491	0.1075 8491
15	2.3965 5819	0.4172 6506	23.2759 6988	9.7122 4899	0.0429 6276	0.1029 6276
16	2.5403 5168	0.3936 4628	25.6725 2808	10.1058 9527	0.0389 5214	0.0989 5214
17	2.6927 7279	0.3713 6442	28.2128 7976	10.4772 5969	0.0354 4480	0.0954 4480
18	2.8543 3915	0.3503 4379	30.9056 5255	10.8276 0348	0.0323 5654	0.0923 5654
19	3.0255 9950	0.3305 1301	33.7599 9170	11.1581 1649	0.0296 2086	0.0896 2086
20	3.2071 3547	0.3118 0473	36.7855 9120	11.4699 2122	0.0271 8456	0.0871 8456
21	3.3995 6360	0.2941 5540	39.9927 2668	11.7640 7662	0.0250 0455	0.0850 0455
22	3.6035 3742	0.2775 0510	43.3922 9028	12.0415 8172	0.0230 4557	0.0830 4557
23	3.8197 4966	0.2617 9726	46.9958 2769	12.3033 7898	0.0212 7848	0.0812 7848
24	4.0489 3464	0.2469 7855	50.8155 7735	12.5503 5753	0.0196 7900	0.0796 7900
25	4.2918 7072	0.2329 9863	54.8645 1200	12.7833 5616	0.0182 2672	0.0782 2672
26	4.5493 8296	0.2198 1003	59.1563 8272	13.0031 6619	0.0169 0435	0.0769 0435
27	4.8223 4594	0.2073 6795	63.7057 6568	13.2105 3414	0.0156 9717	0.0756 9717
28	5.1116 8670	0.1956 3014	68.5281 1162	13.4061 6428	0.0145 9255	0.0745 9255
29	5.4183 8790	0.1845 5674	73.6397 9832	13.5907 2102	0.0135 7961	0.0735 7961
30	5.7434 9117	0.1741 1013	79.0581 8622	13.7648 3115	0.0126 4891	0.0726 4891
31	6.0881 0064	0.1642 5484	84.8016 7739	13.9290 8599	0.0117 9222	0.0717 9222
32	6.4533 8668	0.1549 5740	90.8897 7803	14.0840 4339	0.0110 0234	0.0710 0234
33	6.8405 8988	0.1461 8622	97.3431 6471	14.2302 2961	0.0102 7293	0.0702 7293
34	7.2510 2528	0.1379 1153	104.1837 5460	14.3681 4114	0.0095 9843	0.0695 9843
35	7.6860 8679	0.1301 0522	111.4347 7987	14.4982 4636	0.0089 7386	0.0689 7386
36	8.1472 5200	0.1227 4077	119.1208 6666	14.6209 8713	0.0083 9483	0.0683 9483
37	8.6360 8712	0.1157 9318	127.2681 1866	14.7367 8031	0.0078 5743	0.0678 5743
38	9.1542 5235	0.1092 3885	135.9042 0578	14.8460 1916	0.0073 5812	0.0673 5812
39	9.7035 0749	0.1030 5552	145.0584 5813	14.9490 7468	0.0068 9377	0.0668 9377
40	10.2857 1794	0.0972 2219	154.7619 6562	15.0462 9687	0.0064 6154	0.0664 6154
41	10.9028 6101	0.0917 1905	165.0476 8356	15.1380 1592	0.0060 5886	0.0660 5886
42	11.5570 3267	0.0865 2740	175.9505 4457	15.2245 4332	0.0056 8342	0.0656 8342
43	12.2504 5463	0.0816 2962	187.5075 7724	15.3061 7294	0.0053 3312	0.0653 3312
44	12.9854 8191	0.0770 0908	199.7580 3188	15.3831 8202	0.0050 0606	0.0650 0606
45	13.7646 1083	0.0726 5007	212.7435 1379	15.4558 3209	0.0047 0050	0.0647 0050
46	14.5904 8748	0.0685 3781	226.5081 2462	15.5243 6990	0.0044 1485	0.0644 1485
47	15.4659 1673	0.0646 5831	241.0986 1210	15.5890 2821	0.0041 4768	0.0641 4768
48	16.3938 7173	0.0609 9840	256.5645 2882	15.6500 2661	0.0038 9765	0.0638 9765
49	17.3775 0403	0.0575 4566	272.9584 0055	15.7075 7227	0.0036 6356	0.0636 6356
50	18.4201 5427	0.0542 8836	290.3359 0458	15.7618 6064	0.0034 4429	0.0634 4429

$i = 7\%$

n	$(1+i)^n$	$(1+i)^{-n}$	$s_{\overline{n}\rvert i}$	$a_{\overline{n}\rvert i}$	$\frac{1}{s_{\overline{n}\rvert i}}$	$\frac{1}{a_{\overline{n}\rvert i}}$
1	1.0700 0000	0.9345 7944	1.0000 0000	0.9345 7944	1.0000 0000	1.0700 0000
2	1.1449 0000	0.8734 3873	2.0700 0000	1.8080 1817	0.4830 9179	0.5530 9179
3	1.2250 4300	0.8162 9788	3.2149 0000	2.6243 1604	0.3110 5167	0.3810 5167
4	1.3107 9601	0.7628 9521	4.4399 4300	3.3872 1126	0.2252 2812	0.2952 2812
5	1.4025 5173	0.7129 8618	5.7507 3901	4.1001 9744	0.1738 9069	0.2438 9069
6	1.5007 3035	0.6663 4222	7.1532 9074	4.7665 3966	0.1397 9580	0.2097 9580
7	1.6057 8148	0.6227 4974	8.6540 2109	5.3892 8940	0.1155 5322	0.1855 5322
8	1.7181 8618	0.5820 0910	10.2598 0257	5.9712 9851	0.0974 6776	0.1674 6776
9	1.8384 5921	0.5439 3374	11.9779 8875	6.5152 3225	0.0834 8647	0.1534 8647
10	1.9671 5136	0.5083 4929	13.8164 4796	7.0235 8154	0.0723 7750	0.1423 7750
11	2.1048 5195	0.4750 9280	15.7835 9932	7.4986 7434	0.0633 5690	0.1333 5690
12	2.2521 9159	0.4440 1196	17.8884 5127	7.9426 8630	0.0559 0199	0.1259 0199
13	2.4098 4500	0.4149 6445	20.1406 4286	8.3576 5074	0.0496 5085	0.1196 5085
14	2.5785 3415	0.3878 1724	22.5504 8786	8.7454 6799	0.0443 4494	0.1143 4494
15	2.7590 3154	0.3624 4602	25.1290 2201	9.1079 1401	0.0397 9462	0.1097 9462
16	2.9521 6375	0.3387 3460	27.8880 5355	9.4466 4860	0.0358 5765	0.1058 5765
17	3.1588 1521	0.3165 7439	30.8402 1730	9.7632 2299	0.0324 2519	0.1024 2519
18	3.3799 3228	0.2958 6392	33.9990 3251	10.0590 8691	0.0294 1260	0.0994 1260
19	3.6165 2754	0.2765 0833	37.3789 6479	10.3355 9524	0.0267 5301	0.0967 5301
20	3.8696 8446	0.2584 1900	40.9954 9232	10.5940 1425	0.0243 9293	0.0943 9293
21	4.1405 6237	0.2415 1309	44.8651 7678	10.8355 2733	0.0222 8900	0.0922 8900
22	4.4304 0174	0.2257 1317	49.0057 3916	11.0612 4050	0.0204 0577	0.0904 0577
23	4.7405 2986	0.2109 4688	53.4361 4090	11.2721 8738	0.0187 1393	0.0887 1393
24	5.0723 6695	0.1971 4662	58.1766 7076	11.4693 3400	0.0171 8902	0.0871 8902
25	5.4274 3264	0.1842 4918	63.2490 3772	11.6535 8318	0.0158 1052	0.0858 1052
26	5.8073 5292	0.1721 9549	68.6764 7036	11.8257 7867	0.0145 6103	0.0845 6103
27	6.2138 6763	0.1609 3037	74.4838 2328	11.9867 0904	0.0134 2573	0.0834 2573
28	6.6488 3836	0.1504 0221	80.6976 9091	12.1371 1125	0.0123 9193	0.0823 9193
29	7.1142 5705	0.1405 6282	87.3465 2927	12.2776 7407	0.0114 4865	0.0814 4865
30	7.6122 5504	0.1313 6712	94.4607 8632	12.4090 4118	0.0105 8640	0.0805 8640
31	8.1451 1290	0.1227 7301	102.0730 4137	12.5318 1419	0.0097 9691	0.0797 9691
32	8.7152 7080	0.1147 4113	110.2181 5426	12.6465 5532	0.0090 7292	0.0790 7292
33	9.3253 3975	0.1072 3470	118.9334 2506	12.7537 9002	0.0084 0807	0.0784 0807
34	9.9781 1354	0.1002 1934	128.2587 6481	12.8540 0936	0.0077 9674	0.0777 9674
35	10.6765 8148	0.0936 6294	138.2368 7835	12.9476 7230	0.0072 3396	0.0772 3396
36	11.4239 4219	0.0875 3546	148.9134 5984	13.0352 0776	0.0067 1531	0.0767 1531
37	12.2236 1814	0.0818 0884	160.3374 0202	13.1170 1660	0.0062 3685	0.0762 3685
38	13.0792 7141	0.0764 5686	172.5610 2017	13.1934 7345	0.0057 9505	0.0757 9505
39	13.9948 2041	0.0714 5501	185.6402 9158	13.2649 2846	0.0053 8676	0.0753 8676
40	14.9744 5784	0.0667 8038	199.6351 1199	13.3317 0884	0.0050 0914	0.0750 0914
41	16.0226 6989	0.0624 1157	214.6095 6983	13.3941 2041	0.0046 5962	0.0746 5962
42	17.1442 5678	0.0583 2857	230.6322 3972	13.4524 4898	0.0043 3591	0.0743 3591
43	18.3443 5475	0.0545 1268	247.7764 9650	13.5069 6167	0.0040 3590	0.0740 3590
44	19.6284 5959	0.0509 4643	266.1208 5125	13.5579 0810	0.0037 5769	0.0737 5769
45	21.0024 5176	0.0476 1349	285.7493 1084	13.6055 2159	0.0034 9957	0.0734 9957
46	22.4726 2338	0.0444 9859	306.7517 6260	13.6500 2018	0.0032 5996	0.0732 5996
47	24.0457 0702	0.0415 8747	329.2243 8598	13.6916 0764	0.0030 3744	0.0730 3744
48	25.7289 0651	0.0388 6679	353.2700 9300	13.7304 7443	0.0028 3070	0.0728 3070
49	27.5299 2997	0.0363 2410	378.9989 9951	13.7667 9853	0.0026 3853	0.0726 3853
50	29.4570 2506	0.0339 4776	406.5289 2947	13.8007 4629	0.0024 5985	0.0724 5985

$i = 8\%$

n	$(1 + i)^n$	$(1 + i)^{-n}$	$s_{\overline{n}\mid i}$	$a_{\overline{n}\mid i}$	$\frac{1}{s_{\overline{n}\mid i}}$	$\frac{1}{a_{\overline{n}\mid i}}$
1	1.0800 0000	0.9259 2593	1.0000 0000	0.9259 2593	1.0000 0000	1.0800 0000
2	1.1664 0000	0.8573 3882	2.0800 0000	1.7832 6475	0.4807 6923	0.5607 6923
3	1.2597 1200	0.7938 3224	3.2464 0000	2.5770 9699	0.3080 3351	0.3880 3351
4	1.3604 8896	0.7350 2985	4.5061 1200	3.3121 2684	0.2219 2080	0.3019 2080
5	1.4693 2808	0.6805 8320	5.8666 0096	3.9927 1004	0.1704 5645	0.2504 5645
6	1.5868 7432	0.6301 6963	7.3359 2904	4.6228 7966	0.1363 1539	0.2163 1539
7	1.7138 2427	0.5834 9040	8.9228 0336	5.2063 7006	0.1120 7240	0.1920 7240
8	1.8509 3021	0.5402 6888	10.6366 2763	5.7466 3894	0.0940 1476	0.1740 1476
9	1.9990 0463	0.5002 4897	12.4875 5784	6.2468 8791	0.0800 7971	0.1600 7971
10	2.1589 2500	0.4631 9349	14.4865 6247	6.7100 8140	0.0690 2949	0.1490 2949
11	2.3316 3900	0.4288 8286	16.6454 8746	7.1389 6426	0.0600 7634	0.1400 7634
12	2.5181 7012	0.3971 1376	18.9771 2646	7.5360 7802	0.0526 9502	0.1326 9502
13	2.7196 2373	0.3676 9792	21.4952 9658	7.9037 7594	0.0465 2181	0.1265 2181
14	2.9371 9362	0.3404 6104	24.2149 2030	8.2442 3698	0.0412 9685	0.1212 9685
15	3.1721 6911	0.3152 4170	27.1521 1393	8.5594 7869	0.0368 2954	0.1168 2954
16	3.4259 4264	0.2918 9047	30.3242 8304	8.8513 6916	0.0329 7687	0.1129 7687
17	3.7000 1805	0.2702 6895	33.7502 2569	9.1216 3811	0.0296 2943	0.1096 2943
18	3.9960 1950	0.2502 4903	37.4502 4374	9.3718 8714	0.0267 0210	0.1067 0210
19	4.3157 0106	0.2317 1206	41.4462 6324	9.6035 9920	0.0241 2763	0.1041 2763
20	4.6609 5714	0.2145 4821	45.7619 6430	9.8181 4741	0.0218 5221	0.1018 5221
21	5.0338 3372	0.1986 5575	50.4229 2144	10.0168 0316	0.0198 3225	0.0998 3225
22	5.4365 4041	0.1839 4051	55.4567 5516	10.2007 4366	0.0180 3207	0.0980 3207
23	5.8714 6365	0.1703 1528	60.8932 9557	10.3710 5895	0.0164 2217	0.0964 2217
24	6.3411 8074	0.1576 9934	66.7647 5922	10.5287 5828	0.0149 7796	0.0949 7796
25	6.8484 7520	0.1460 1790	73.1059 3995	10.6747 7619	0.0136 7878	0.0936 7878
26	7.3963 5321	0.1352 0176	79.9544 1515	10.8099 7795	0.0125 0713	0.0925 0713
27	7.9880 6147	0.1251 8682	87.3507 6836	10.9351 6477	0.0114 4810	0.0914 4810
28	8.6271 0639	0.1159 1372	95.3388 2983	11.0510 7849	0.0104 8891	0.0904 8891
29	9.3172 7490	0.1073 2752	103.9659 3622	11.1584 0601	0.0096 1854	0.0896 1854
30	10.0626 5689	0.0993 7733	113.2832 1111	11.2577 8334	0.0088 2743	0.0888 2743
31	10.8676 6944	0.0920 1605	123.3458 6800	11.3497 9939	0.0081 0728	0.0881 0728
32	11.7370 8300	0.0852 0005	134.2135 3744	11.4349 9944	0.0074 5081	0.0874 5081
33	12.6760 4964	0.0788 8893	145.9506 2044	11.5138 8837	0.0068 5163	0.0868 5163
34	13.6901 3361	0.0730 4531	158.6266 7007	11.5869 3367	0.0063 0411	0.0863 0411
35	14.7853 4429	0.0676 3454	172.3168 0368	11.6545 6822	0.0058 0326	0.0858 0326
36	15.9681 7184	0.0626 2458	187.1021 4797	11.7171 9279	0.0053 4467	0.0853 4467
37	17.2456 2558	0.0579 8572	203.0703 1981	11.7751 7851	0.0049 2440	0.0849 2440
38	18.6252 7563	0.0536 9048	220.3159 4540	11.8288 6899	0.0045 3894	0.0845 3894
39	20.1152 9768	0.0497 1341	238.9412 2103	11.8785 8240	0.0041 8513	0.0841 8513
40	21.7245 2150	0.0460 3093	259.0565 1871	11.9246 1333	0.0038 6016	0.0838 6016
41	23.4624 8322	0.0426 2123	280.7810 4021	11.9672 3457	0.0035 6149	0.0835 6149
42	25.3394 8187	0.0394 6411	304.2435 2342	12.0066 9867	0.0032 8684	0.0832 8684
43	27.3666 4042	0.0365 4084	329.5830 0530	12.0432 3951	0.0030 3414	0.0830 3414
44	29.5559 7166	0.0338 3411	356.9496 4572	12.0770 7362	0.0028 0152	0.0828 0152
45	31.9204 4939	0.0313 2788	386.5056 1738	12.1084 0150	0.0025 8728	0.0825 8728
46	34.4740 8534	0.0290 0730	418.4260 6677	12.1374 0880	0.0023 8991	0.0823 8991
47	37.2320 1217	0.0268 5861	452.9001 5211	12.1642 6741	0.0022 0799	0.0822 0799
48	40.2105 7314	0.0248 6908	490.1321 6428	12.1891 3649	0.0020 4027	0.0820 4027
49	43.4274 1899	0.0230 2693	530 3427 3742	12.2121 6341	0.0018 8557	0.0818 8557
50	46.9016 1251	0.0213 2123	573.7701 5642	12.2334 8464	0.0017 4286	0.0817 4286

D

Appendix D Answers to Odd-numbered Exercises

CHAPTER ONE

Exercise Set 1.1, page 8

1. (a) terminating, repeating, rational (b) repeating, rational (c) irrational (d) repeating, rational
(e) terminating, repeating, rational

3. (b) (d) (f) (e) (c) (a)
$-5\ -4\ -3\ -2\ -1\ 0\ 1\ 2\ 3\ 4\ 5\ 6$

5. 4 **7.** -2 **9.** -5

11. $-6\ -5\ -4\ -3\ -2\ -1\ 0\ 1\ 2\ 3\ 4\ 5\ 6\ 7\ 8\ 9$

13. $-2\ -1\ 0\ 1\ 2\ 3\ 4\ 5\ 6\ 7$

15. $10 > 9.99$ **17.** $a \geq 0$ **19.** $x > 0$ **21.** $b \leq -4$ **23.** $b \geq 5$ **25.** $3 < 5$
27. $-3 < -2$ **29.** $\frac{1}{2} > -\frac{1}{4}$ **31.** $-3.4 < 4.2$ **33.** $a < 0$ **35.** $e > f$ **37.** $0 < e$

39.

	Quadrant			
Coordinates	**I**	**II**	**III**	**IV**
$(-1, 3)$		✓		
$(2, 7)$	✓			
$(4, -8)$				✓
$(0, 2)$				
$(-\sqrt{2}, -\pi)$			✓	
$(4, 0)$				

41.

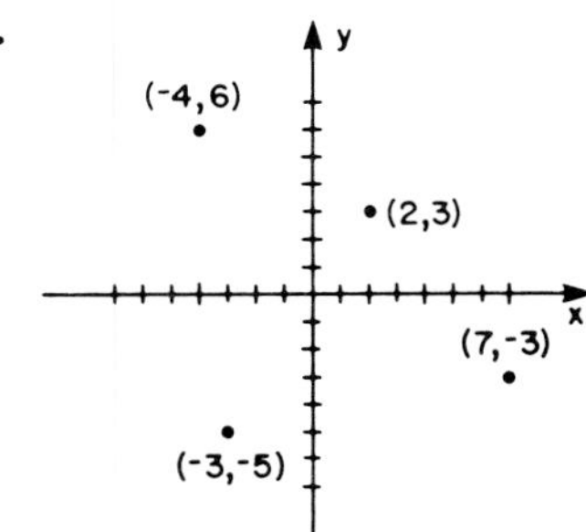

43.

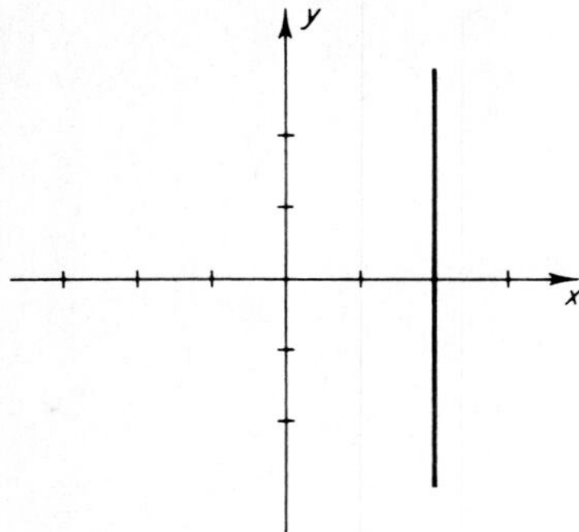

45. $(-2, 9)$

Exercise Set 1.2, page 14

1. (a), (b), (d) **3.** (a) $7/3$ (b) -3 (c) $-3/2$ (d) 3

5.

x	-3	-2	-1	0	1	2	3
$y = 1 - x^2$	-8	-3	0	1	0	-3	-8

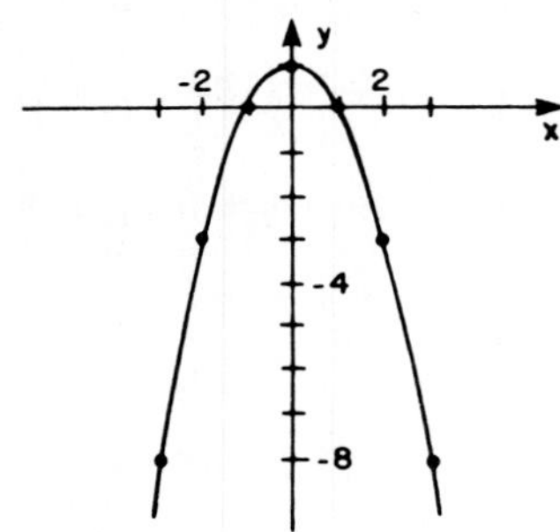

	x-intercept	y-intercept
7.	4	$-\frac{4}{3}$
9.	-1	1
11.	1	-1
13.	None	1
15.	± 4	None

17. x-intercept: $-\frac{3}{2}$; y-intercept: -3

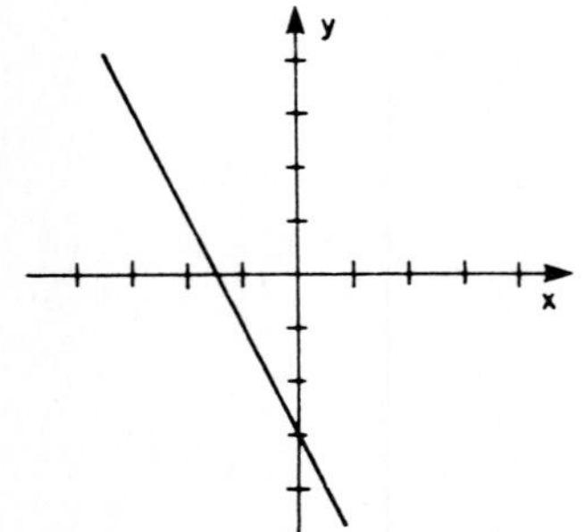

19. x-intercept: $\frac{5}{2}$; y-intercept: $\frac{5}{3}$

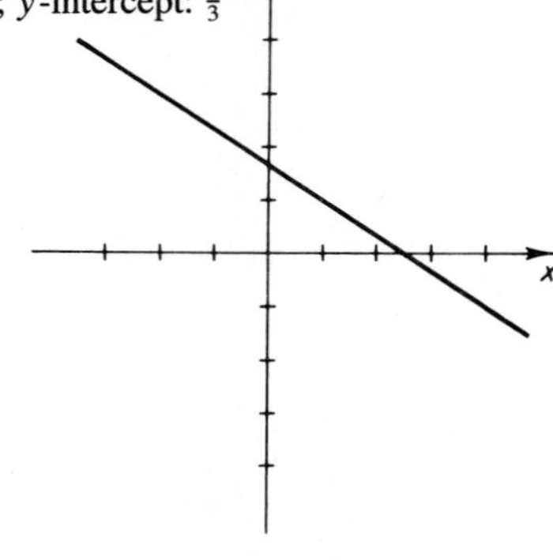

21. x-intercept: -1; y-intercept: 1

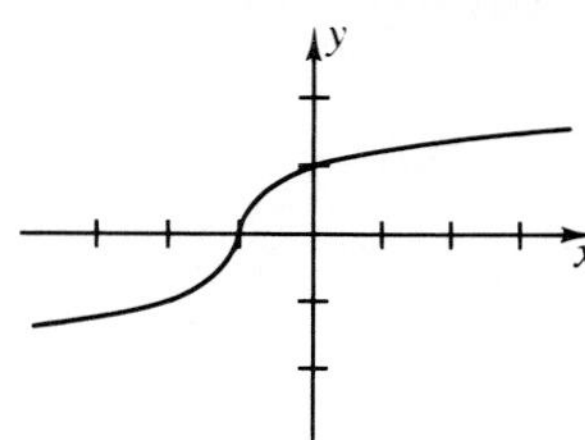

23. x-intercept: $\pm\dfrac{\sqrt{3}}{2}$; y-intercept: 3

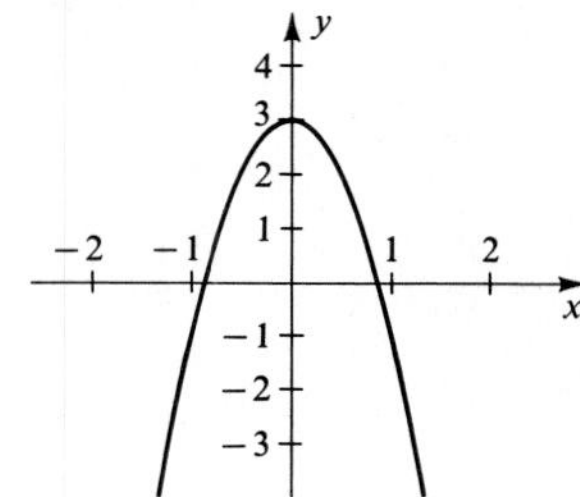

25. x-intercept: none; y-intercept: none

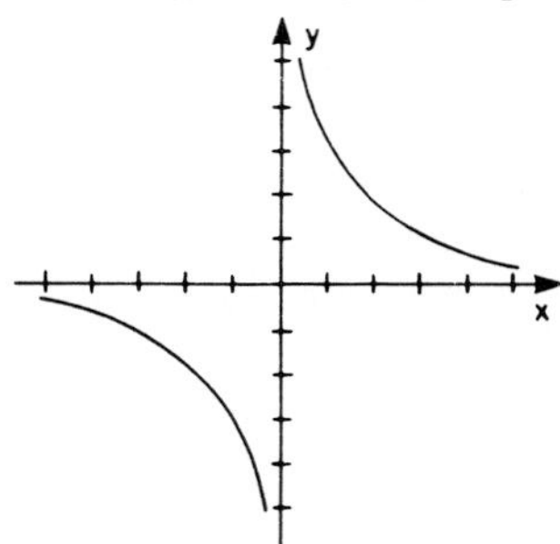

27. 1

29.

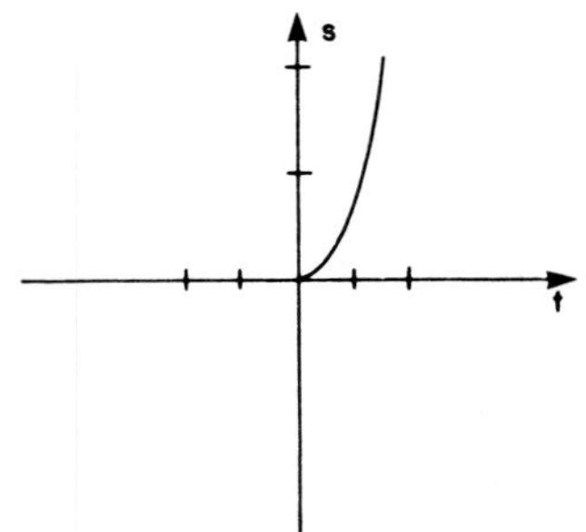

Exercise Set 1.3, page 23

1. (a) $-\frac{1}{5}$ (b) 2 (c) $\frac{2}{3}$ (d) 0 (e) $\frac{519}{919}$
3. (a) $\frac{3}{2}$ (b) -8 **5.** 2 **7.** (a) parallel (b) not parallel (c) not parallel (same line)
9. (a) $y = 2x + 4$ (b) $y = -\frac{4}{3}x - 2$ (c) $y = 3$
11. (a) $y = 3 + (x - 2)$ (b) $y = 2 - \frac{2}{3}x$ (c) $y = 3 - 3(x - 3)$ (d) $y = 3$
(e) $y = -4 + 6(x - 2)$ (f) $y = \frac{2}{3} - 4(x - \frac{1}{2})$
13. (a) $y = 3 - \frac{1}{5}(x - 2)$ (b) $y - 1 = \frac{10}{3}(x - \frac{3}{2})$
15. (a) $y = x + 1$ (b) $y = \frac{1}{5}(-4x + 2)$ **17.** $y = -2x + 4$
19. (a) $x - \frac{1}{3}y = -7; A = 1, B = -\frac{1}{3}, C = -7$ (b) $y = 2; A = 0, B = 1, C = 2$
(c) $4x - y = 5; A = 4, B = -1, C = 5$ (d) $x - 3y = -6; A = 1, B = -3, C = -6$
21. (a)

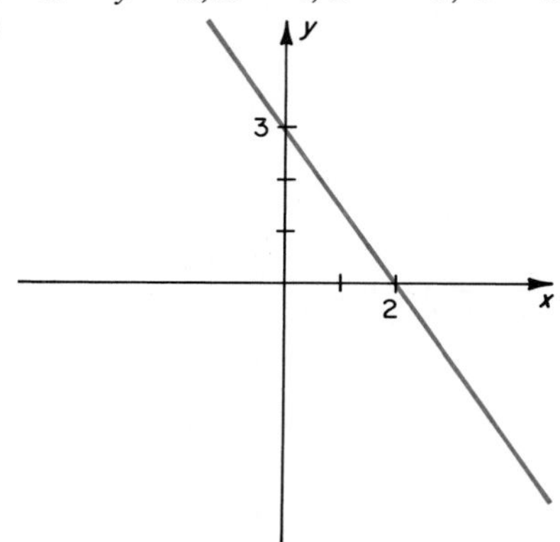

(b)

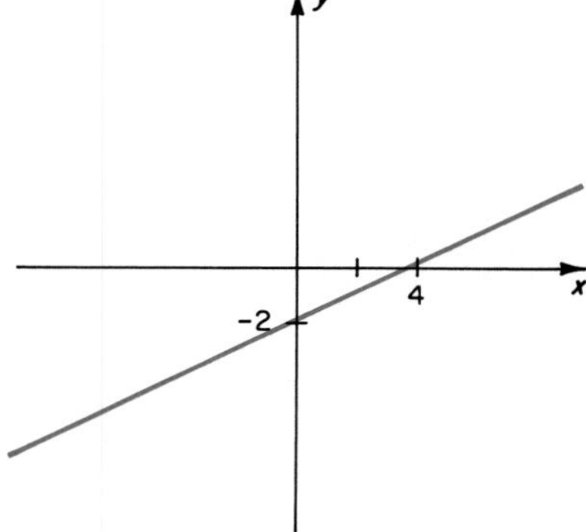

(c)

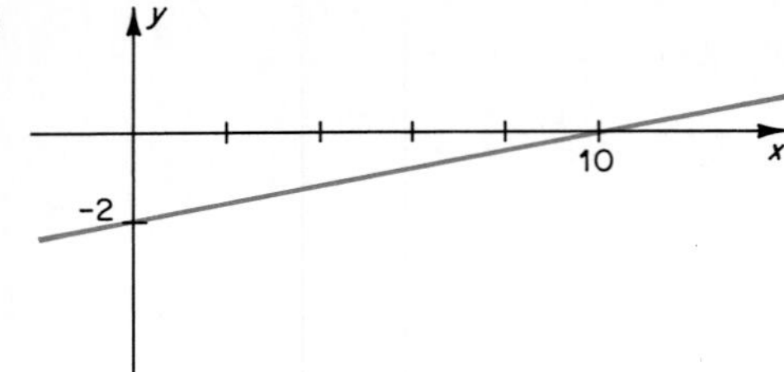

(d)

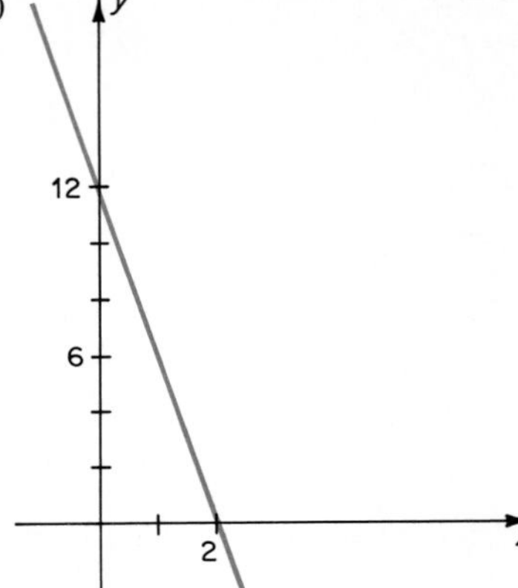

23. (a)

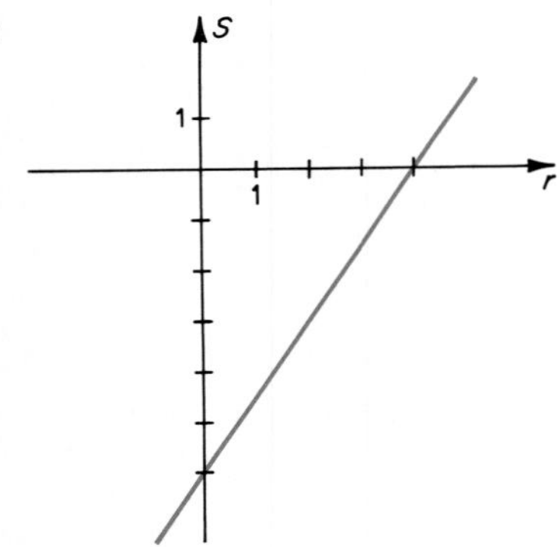

(b)

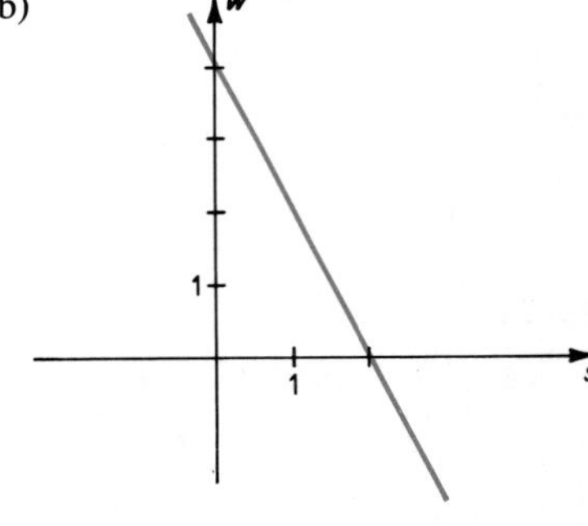

(c)

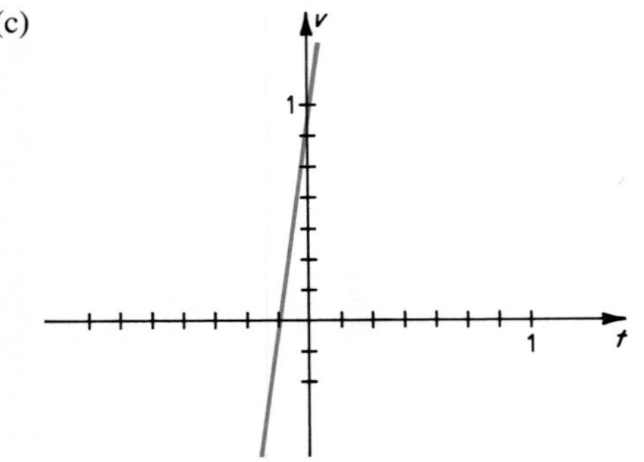

25. $y = 2x - 2$ **29.** $P = 2x + 5000$

Exercise Set 1.4, page 29

1. $(-1, 2)$ **3.** (a) $x = -2, y = 2$ (b)

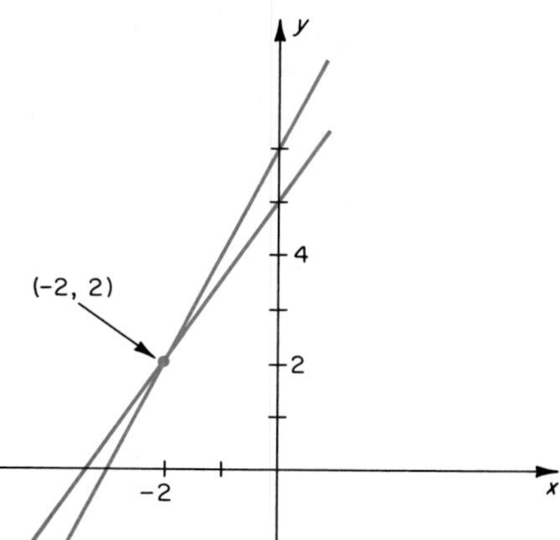

5. (a)

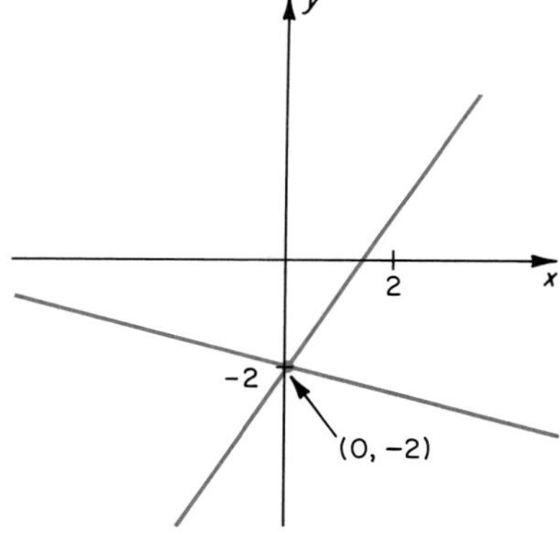

(b)

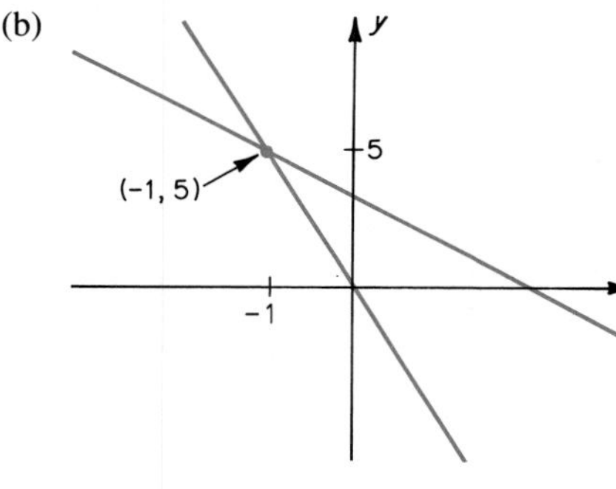

(c)

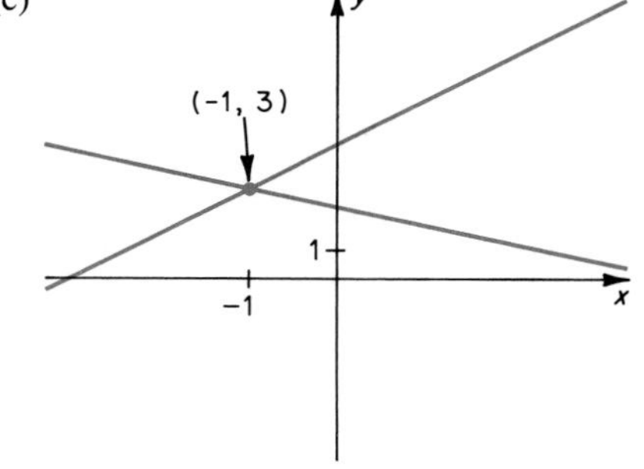

(d)

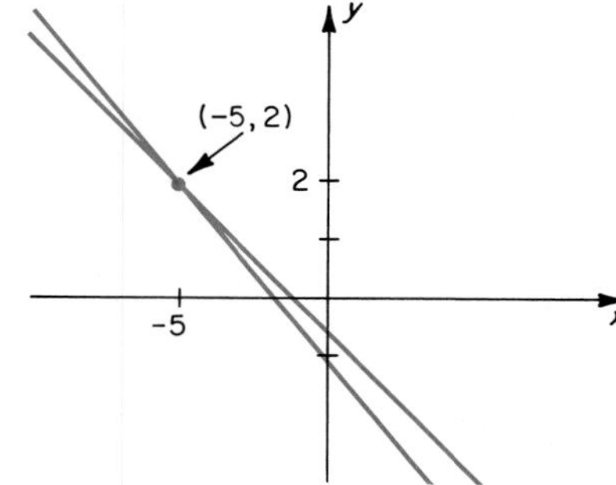

7. (a)

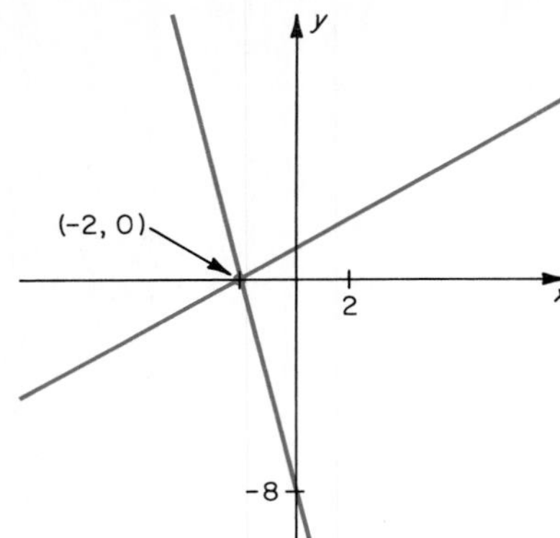

(b)

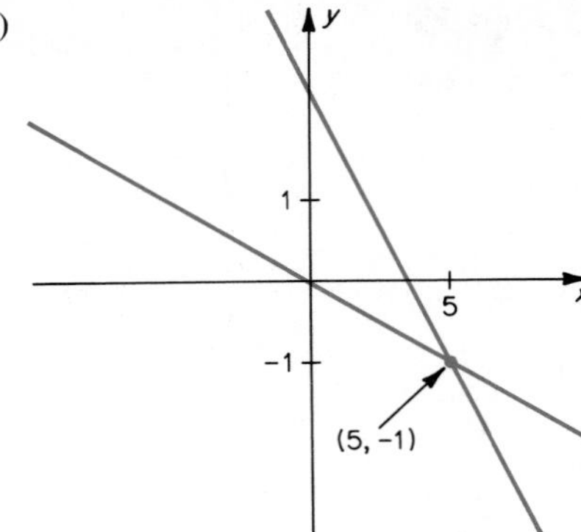

(c)

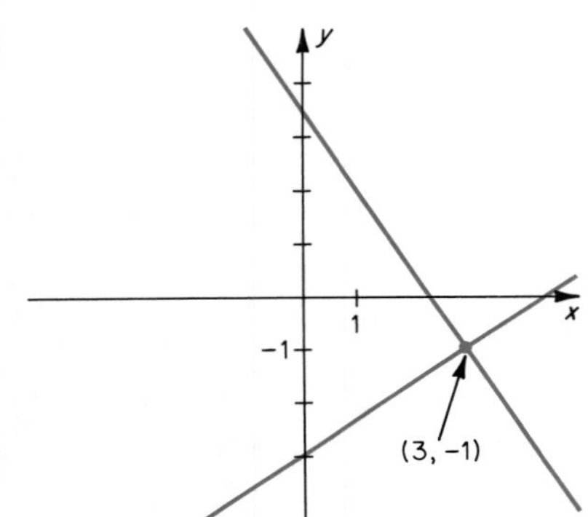

(d)

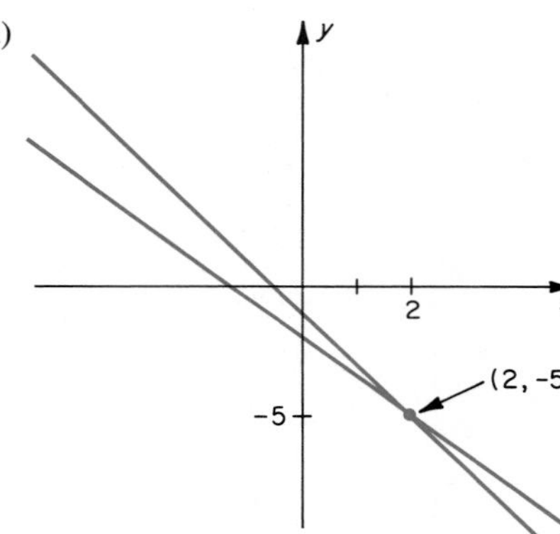

9. (a) $x = -3$, $y = 3$ (b) $x = 5$, $y = 0$ (c) $x = 5$, $y = 4$ (d) $x = 4$, $y = 2$

11. (b) the lines are parallel (c)

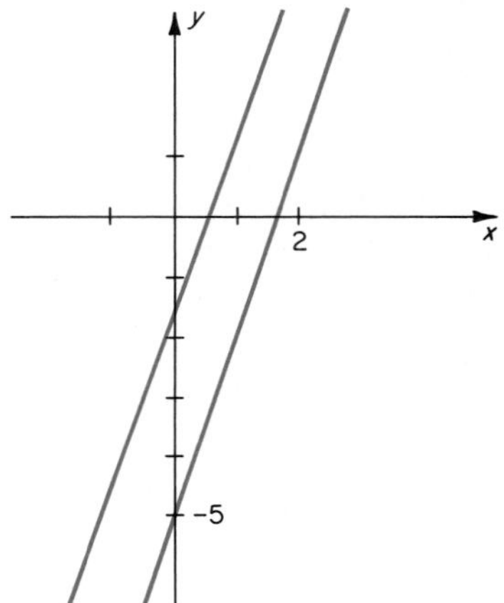

13. (b), (c) **15.** $x = \dfrac{ed - bf}{ad - bc}$, $y = \dfrac{af - ce}{ad - bc}$ **17.** 3 dimes and 5 nickels

19. \$4000 invested in type *A* bonds and \$2000 invested in type *B* bonds

Exercise Set 1.5, page 40

1. (a) $S = 12{,}000 + 840t$ (b) \$16,200 (c) \$14,940 **3.** \$6000
5. (a) $V = 24{,}000 - 1920t$ (b) \$10,560 (c) 9.9 (approximately) (d) 12.5 **7.** \$60,000
9. (a) $N = \frac{64}{3} + \frac{64}{3}t$ (b)

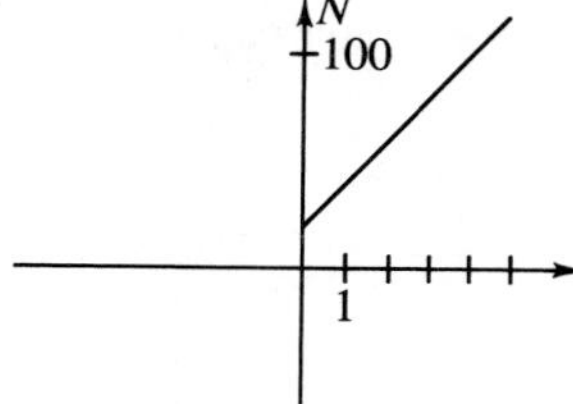

(c) 213,333

13. $y = \frac{17}{10}x + \frac{3}{10}$ **17.** (a) $y = 6.177x + 59.190$ (b) 127.1 (c) to simplify the arithmetic
19. 1100 meals **21.** 4000 newspapers
23. (a) equilibrium price = \$15; 4000 compact discs sold (b)

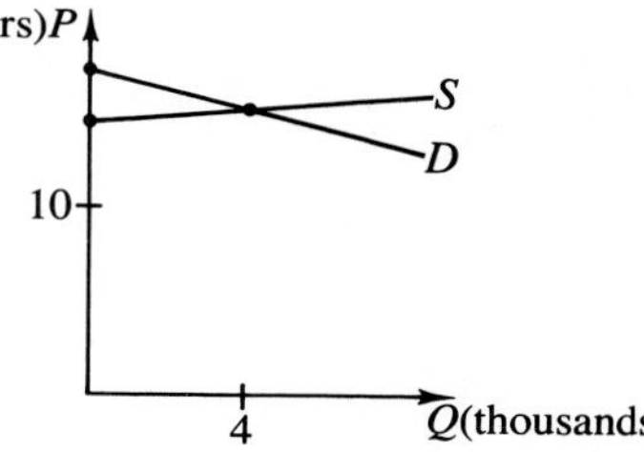

25. (a) equilibrium price = \$5 (b) yes, 9 units (c) yes, 6 units

Exercise Set 1.6, page 47

1. (a)

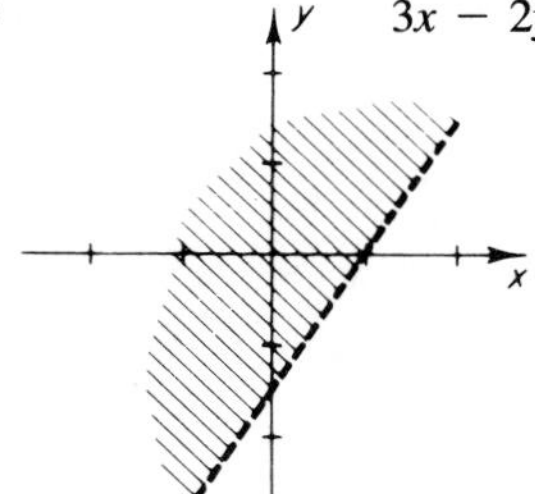

(b)

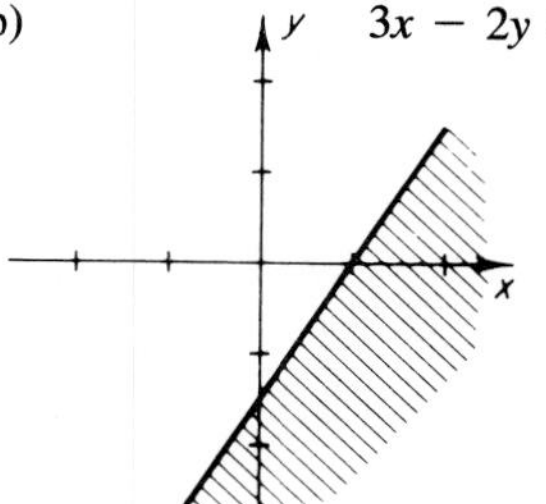

3. (a)

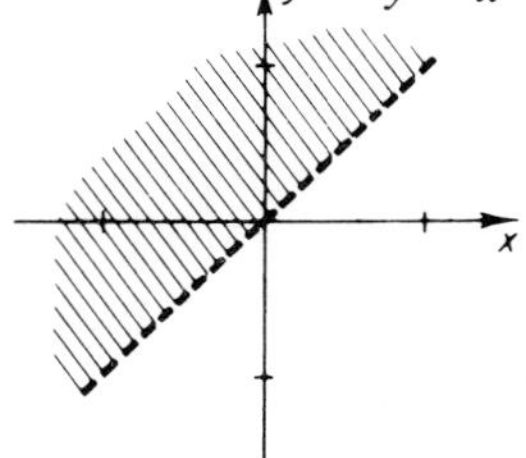

(b)

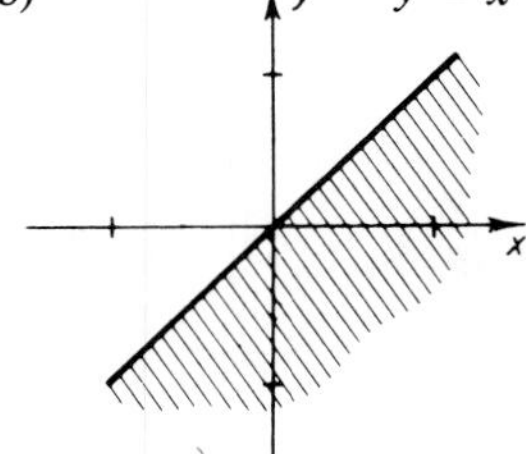

5. (a)

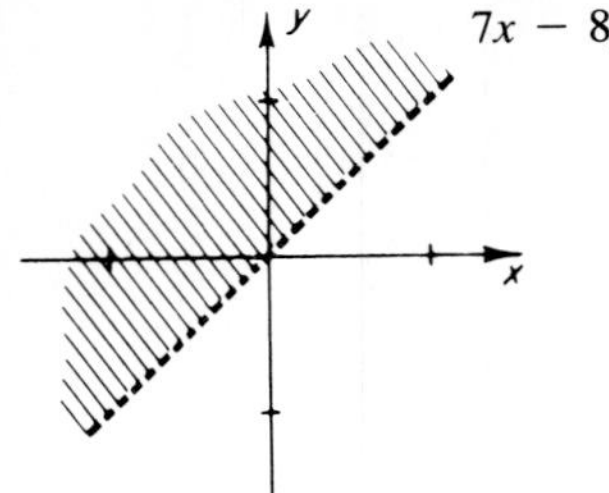

(b)

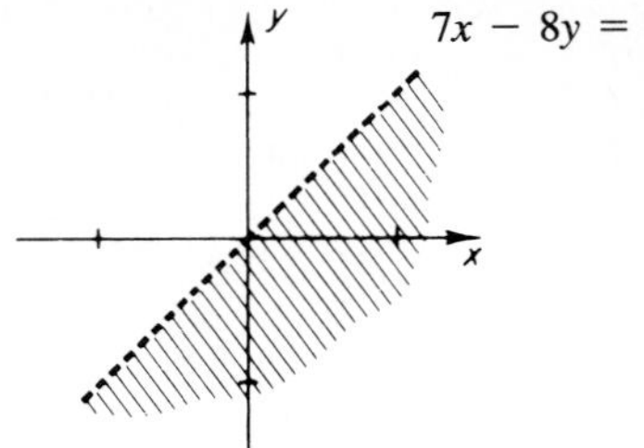

7.

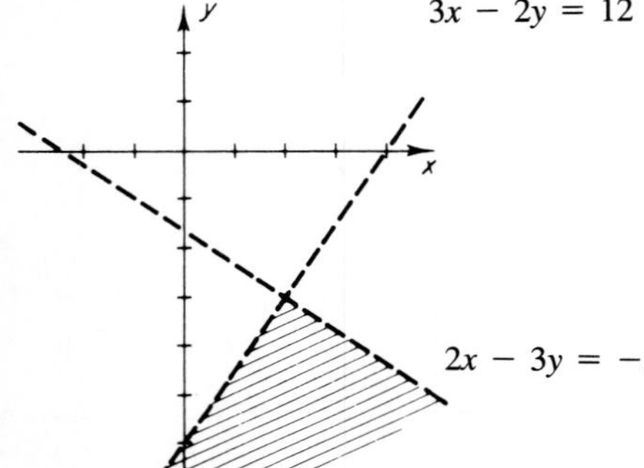

9.

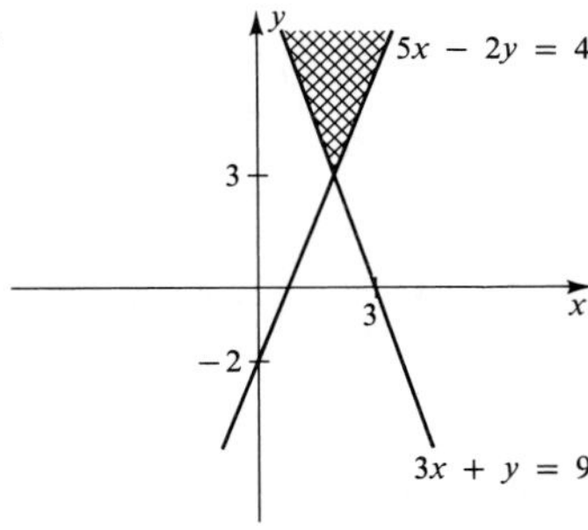

11.

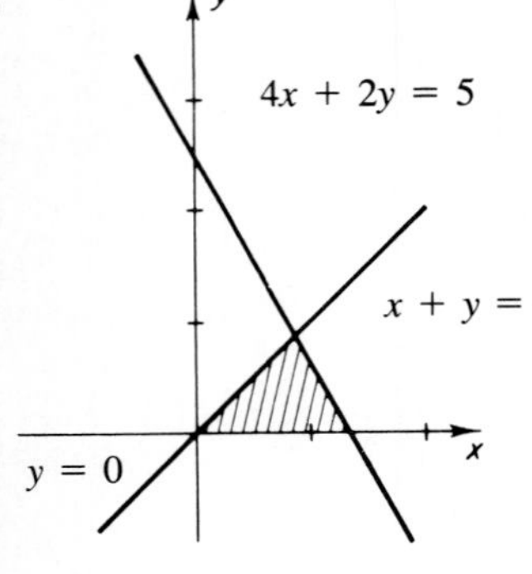

13. no solution

15.

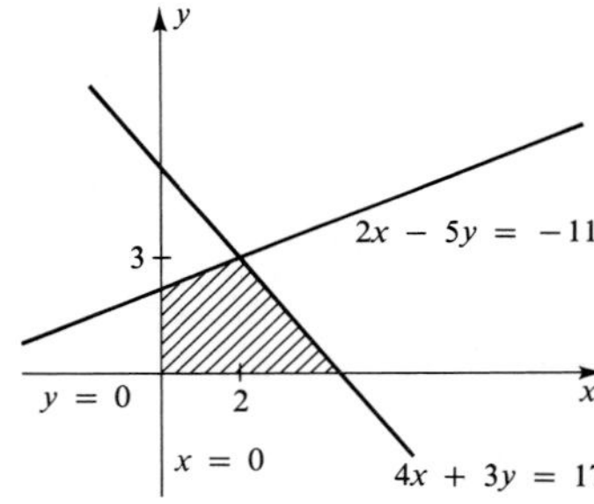

17. $y - x \geq 0$ **19.** $3y - 2x \leq 6$ **21.** $y + x \geq 0$; $y - x \geq -3$

Supplementary Exercises, page 49

	x-intercept	y-intercept
1.	3, −3	−9
3.	2, −2	−4

5.
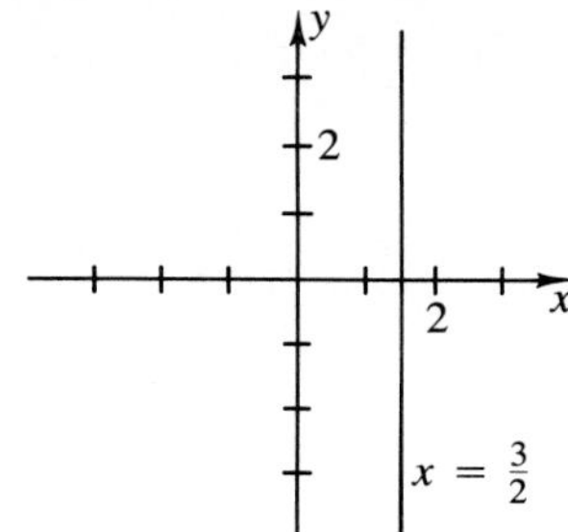

7.
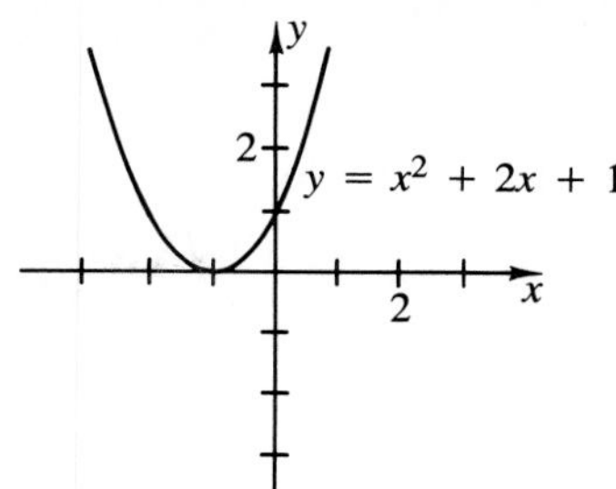

9.
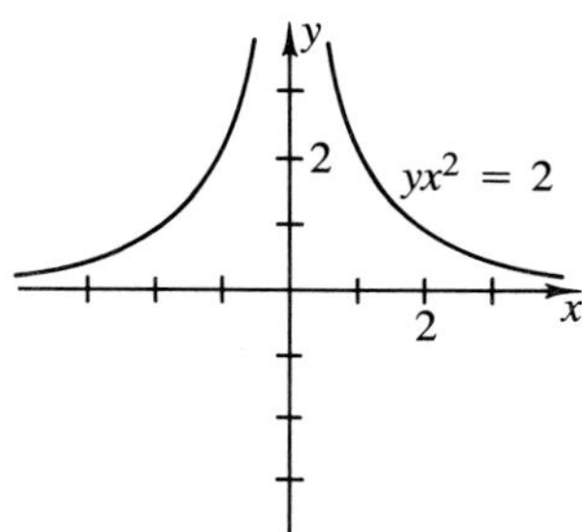

11.
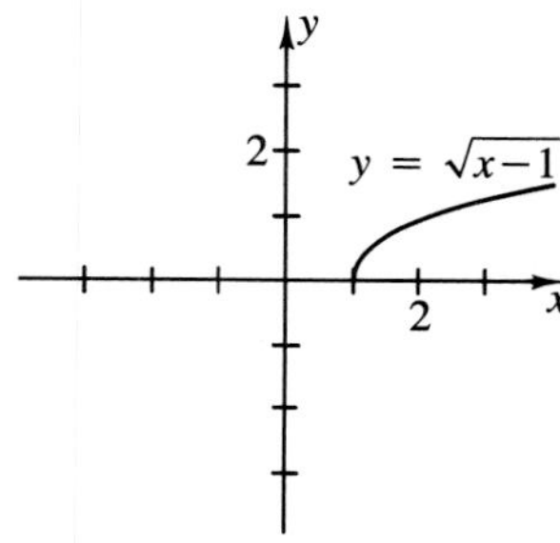

13. $y = 4x + 11$ **15.** $y = 1$ **17.** $y = \frac{4}{5}x + \frac{13}{5}$ **19.** x-intercept: a; y-intercept: b **23.** $(\frac{5}{2}, 5)$
27. $x = 2, y = -3$ **29.** $x = 0, y = 0$ **31.** 8% per year **33.** \$760 million **35.** $y = 1.8x - 0.6$
37. equilibrium price = \$.40; 68,000 pounds of bananas sold **39.** $x + 2y \geq 1; x - 2y \geq 6$

Chapter Test, page 51

1. x-intercepts: 3 and -3, y-intercept: 27
2.
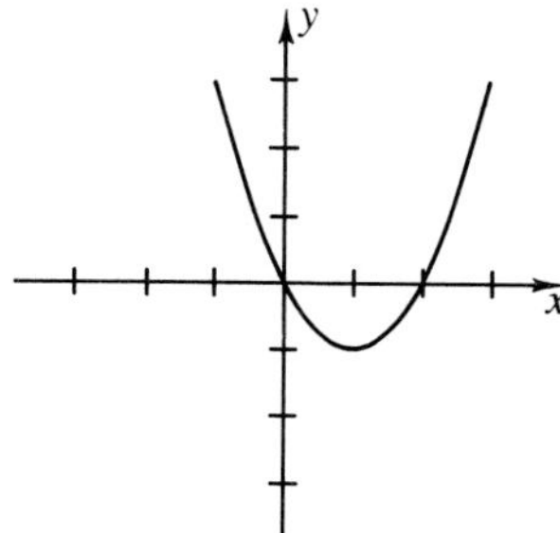

3. $3x + 2y = 3$ **4.** $3x + 2y = 6$ **5.** $x = \frac{3}{4}, y = \frac{19}{8}$ **6.** 6 years **7.** 50,000 pounds of pollutant
8. break-even point = 2,000 pens; cost = revenue = \$1600
9.
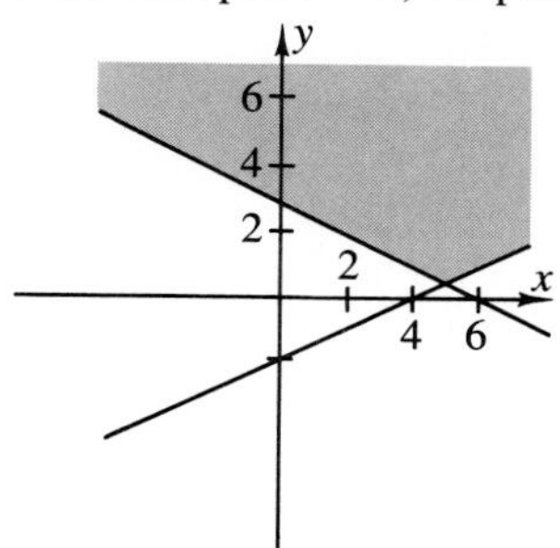

CHAPTER TWO

Exercise Set 2.1, page 59

1. (a), (d) **3.** (b), (d) **5.** (b), (c)

7. (a)
$$\begin{aligned} 4x_1 - 2x_2 &= 7 \\ -2x_1 + 3x_2 + 5x_3 &= -4 \\ 3x_1 + 2x_2 - 5x_3 &= 3 \\ 2x_1 + x_2 + 3x_3 &= 4 \end{aligned}$$

(b)
$$\begin{aligned} 2x_1 + 3x_2 &= 5 \\ 4x_1 + 2x_2 &= 1 \end{aligned}$$

(c)
$$\begin{aligned} 3x_1 + 2x_2 + 4x_3 + 2x_4 &= 5 \\ 2x_1 + 3x_2 + x_3 + 3x_4 &= 1 \end{aligned}$$

(d)
$$\begin{aligned} x_1 &= 1 \\ x_2 &= 2 \\ x_3 &= 3 \end{aligned}$$

11.
$$\begin{aligned} 2x_1 + 2x_2 &= 12 \\ 3x_1 + x_2 &= 16 \\ 2x_1 + 4x_2 &= 14 \end{aligned}$$

13. inconsistent since the three lines do not intersect at a single point

Exercise Set 2.2, page 69

1. (a), (c)

3. (a) $x_1 = 3,\ x_2 = -2,\ x_3 = 4$
(b) $x_1 = 3 - 2t,\ x_2 = 6 - 3t,\ x_3 = -2 - 4t,\ x_4 = t$, where t is arbitrary
(c) $x_1 = 1 - 2s - 5t,\ x_2 = s,\ x_3 = -2 - 4t,\ x_4 = 3 - 2t,\ x_5 = t$, where s and t are arbitrary
(d) system is inconsistent

5. (a) $\begin{bmatrix} 1 & 0 & -3 & 6 & 0 \\ 0 & 1 & -1 & 3 & 0 \\ 0 & 0 & 0 & 0 & 1 \end{bmatrix}$ (b) $\begin{bmatrix} 1 & 0 & 0 & 0 & 2 \\ 0 & 1 & 2 & 1 & 3 \\ 0 & 0 & 0 & 0 & 0 \end{bmatrix}$ (c) $\begin{bmatrix} 1 & 0 & 0 & 2 & 2 \\ 0 & 0 & 1 & 3 & -3 \\ 0 & 0 & 0 & 0 & 0 \end{bmatrix}$

7. (a) $x = 6,\ y = 7,\ z = 2$ (b) $x_1 = 0,\ x_2 = 0,\ x_3 = 0$
(c) $x_1 = 3 - s + t,\ x_2 = -5 + s - 5t,\ x_3 = -2 - t,\ x_4 = s,\ x_5 = t$, where s and t are arbitrary
(d) $x_1 = 9 - 5s,\ x_2 = 4 - 2s,\ x_3 = 0,\ x_4 = s$, where s is arbitrary

9. (a) $x_1 = -10,\ x_2 = 2,\ x_3 = 3,\ x_4 = 0$ (b) $x_1 = 1,\ x_2 = 1$ (c) $x_1 = -1,\ x_2 = 2,\ x_3 = 2$
(d) $x_1 = \frac{7}{2} - \frac{1}{2}s,\ x_2 = s,\ x_3 = 4$, where s is arbitrary

11. no solution if $a = -2$; exactly one solution if a is neither -2 nor 2; infinitely many solutions if $a = 2$

17. 3 vessels of type A, 4 vessels of type B, 5 vessels of type C

19. three 12-inch sets, eight 16-inch sets, and five 19-inch sets

Exercise Set 2.3, page 78

1. (a) 2×3 (b) 3×2 (c) 1×2 (d) 1×1 **3.** $u = 5,\ v = 2,\ w = 0$

5. (a) 3×4 (d) 3×3

11.

	Under $10,000	Over $10,000
Republican	205	192
Democrat	317	128
Independent	96	63

13.

	Store 1	Store 2	Store 3
Produce	2	1	2
Meats	2	1	6
Canned goods	1	2	2

Exercise Set 2.4, page 87

1. (a) 3×2 (e) 4×4 (f) 3×2 (g) 3×2

3. $\begin{bmatrix} 40 & -38 & -58 \\ 40 & 32 & 4 \end{bmatrix}$ (c) $\begin{bmatrix} -144 & 157 \\ -84 & 78 \end{bmatrix}$

11. $A = \begin{bmatrix} 3 & -1 & 2 & 1 & 1 \\ -2 & 1 & 4 & 1 & -2 \\ 0 & 3 & 6 & 3 & 1 \end{bmatrix} \quad X = \begin{bmatrix} x_1 \\ x_2 \\ x_3 \\ x_4 \\ x_5 \end{bmatrix} \quad B = \begin{bmatrix} 7 \\ -8 \\ 9 \end{bmatrix}$

15. $PM = \begin{bmatrix} 53 \\ 22 \end{bmatrix}$, where the first and second rows give, respectively, the daily production of 3- and 10-speed bikes

19. (a) \$1.40

(b) $AB = $

	East coast	West coast	
	\$1.40	\$1.70	Business calculators
	\$1.90	\$2.30	Standard calculators

Exercise Set 2.5, page 97

1. yes, (b)

3. (a) $\begin{bmatrix} \frac{1}{2} & -1 & \frac{1}{2} \\ \frac{1}{2} & 0 & -\frac{1}{2} \\ -\frac{1}{2} & 1 & \frac{1}{2} \end{bmatrix}$ (b) $\begin{bmatrix} \frac{1}{3} & \frac{1}{3} & -\frac{2}{3} \\ \frac{1}{6} & -\frac{1}{3} & \frac{7}{6} \\ -\frac{1}{6} & \frac{1}{3} & -\frac{1}{6} \end{bmatrix}$ (c) not invertible

5. (a) $\begin{bmatrix} 3 & 2 & 0 & 6 \\ -2 & -1 & 0 & -3 \\ -2 & -3 & -1 & -6 \\ 2 & 2 & 1 & 4 \end{bmatrix}$ (b) $\begin{bmatrix} 0 & -\frac{1}{3} & \frac{1}{3} \\ 0 & \frac{1}{3} & \frac{2}{3} \\ 1 & -\frac{1}{3} & -\frac{8}{3} \end{bmatrix}$ (c) not invertible

7. $x = 4,\ y = 2,\ z = -1$ **9.** $x_1 = -1,\ x_2 = 3,\ x_3 = 2$ **11.** $x = 1,\ y = 0,\ z = -2$

13. (a) $x_1 = \frac{27}{25},\ x_2 = \frac{17}{25},\ x_3 = \frac{13}{25}$ (b) $x_1 = 1,\ x_2 = 1,\ x_3 = 2$ (c) $x_1 = -\frac{11}{25},\ x_2 = \frac{19}{25},\ x_3 = -\frac{34}{25}$

19. $\begin{bmatrix} -10 \\ 14 \end{bmatrix}$ **21.** $\begin{bmatrix} 5 & 0 \\ -4 & 22 \end{bmatrix}$ **25.** $A^{-1} = \begin{bmatrix} 2 & -1 \\ -5 & 3 \end{bmatrix}$

(a) $x = 2,\ y = -5$

(b) $x = -15,\ y = 41$

Supplementary Exercises, page 99

1. $k \neq 4$

3. (a) $\left[\begin{array}{ccc|c} 1 & 0 & 0 & \frac{58}{16} \\ 0 & 1 & 0 & -\frac{33}{16} \\ 0 & 0 & 1 & -\frac{35}{16} \end{array}\right]$ (b) $\left[\begin{array}{cccc|c} 1 & 0 & 0 & -\frac{3}{32} & \frac{77}{32} \\ 0 & 1 & 0 & \frac{47}{32} & -\frac{31}{32} \\ 0 & 0 & 1 & -\frac{5}{16} & \frac{11}{16} \end{array}\right]$

5. (a) inconsistent system

(b) $x = 3,\ y = -1,\ z = 2$ (c) $x_1 = \frac{59}{11} + t,\ x_2 = -\frac{25}{11} + \frac{2}{11}t,\ x_3 = -\frac{21}{11} - \frac{19}{11}t,\ x_4 = t$

(d) $x_1 = 1 - \frac{4}{3}s + \frac{1}{6}t,\ x_2 = \frac{1}{3}s - \frac{8}{3}t,\ x_3 = s,\ x_4 = t$, where s and t are arbitrary.

9. $a = -4$, no solution

$a = 4$, infinitely many solutions

$a \neq \pm 4$, only one solution

11. (a) $\begin{bmatrix} 14 & -1 & 10 \\ 21 & 37 & -2 \end{bmatrix}$ (b) impossible (c) impossible (d) impossible **17.** not invertible

19. $\begin{bmatrix} -3 & \frac{1}{2} & \frac{5}{2} \\ \frac{5}{2} & -\frac{1}{4} & -\frac{7}{4} \\ 1 & 0 & -1 \end{bmatrix}$ **21.** $x = -1,\ y = 2,\ z = 3$ **23.** $x_1 = 3,\ x_2 = 4,\ x_3 = -2$

25. $\begin{bmatrix} \frac{1}{14} & \frac{3}{14} \\ \frac{2}{7} & -\frac{1}{7} \end{bmatrix}$ **27.** 2 ounces of food A, 3 ounces of food B, and 3 ounces of food C

29. 2 cc of Chem 1, 1 cc of Chem 2, and 4 cc of Chem 3

Chapter Test, page 101

1. no solution if $a = -5$, one solution if $a \neq \pm 5$, infinitely many solutions if $a = 5$

2. (a) $x_1 = \frac{7}{2} - 2s$, $x_2 = s$, $x_3 = -\frac{5}{2} - t$, $x_4 = \frac{3}{2} + t$, $x_5 = t$, where s and t are arbitrary
(b) inconsistent system

3. (a) impossible (b) impossible (c) impossible (d) $\begin{bmatrix} 10 & 6 \\ -15 & -9 \end{bmatrix}$ (e) $[1]$

4. $x = \frac{2}{3}$, $x = -1$ **5.** (a) $\begin{bmatrix} \frac{1}{2} & \frac{1}{2} & -\frac{1}{2} \\ \frac{1}{2} & -\frac{1}{2} & \frac{1}{2} \\ -\frac{1}{2} & \frac{1}{2} & \frac{1}{2} \end{bmatrix}$ (b) $x = \frac{1}{2}$, $y = \frac{13}{2}$, $z = -\frac{5}{2}$ **6.** A = 2, B = 3, C = 3

CHAPTER THREE

Exercise Set 3.1, page 108

1. (a) $18x + 12y$
(b) $18x + 12y \leq 1500$
$x \geq 60$
$y \geq 30$
(c) $z = 120x + 60y$
(d) $z = 110x + 70y$

3. minimize $z = G + \frac{3}{2}T$, where
$4G + 10T \geq 100$
$4G + 2T \geq 60$
$G \geq 0$
$T \geq 0$

5. minimize $z = 8M + 12N$, where
$M + N \geq 7$
$2M + N \geq 10$
$M \geq 0$
$N \geq 0$

7. maximize $z = 5A + 3B$, where
$10A + 5B \leq 450$
$6A + 12B \leq 480$
$9A + 9B \leq 450$
$A \geq 0$
$B \geq 0$

9. maximize $z = 0.3x + 0.4y$, where
$5x + 6y \leq 12{,}000$
$5x + 3y \leq 9{,}000$
$x \geq 0$
$y \geq 0$

11. minimize $z = 10{,}000A + 12{,}000B$, where
$8A + 8B \leq 48$
$100A + 200B \geq 1000$
$A \geq 0$
$B \geq 0$

13. maximize $z = 25A + 35S$, where
$A + S \leq 120$
$32A + 8S \leq 160$
$12A + 24S \leq 1200$
$A \geq 0$
$S \geq 0$

15. minimize $z = 0.5R + 0.6L$, where
$R + L \geq 100$
$0.01R + 0.01L \leq 4$
$0.03R + 0.01L \leq 6$
$R \geq 0$
$L \geq 0$

17. minimize $z = 1000H + 600L$, where
$H + L \leq 30$
$H \geq 3$
$H \leq 10$
$L \geq 18$

19. (a) maximize $z = 3x + y$, where
$3x + y \leq 6$
$x + y \leq 4$
$x \geq 0$
$y \geq 0$
(b) maximize $z = 4x + y$, where
$3x + y \leq 6$
$x + y \leq 4$
$x \geq 0$
$y \geq 0$

21. maximize $z = 10x_1 + 9x_2 + 11x_3 + 8x_4 + 7x_5 + 12x_6 + 16x_7 + 10x_8 + 13x_9 + 10x_{10} + 9x_{11} + 15x_{12}$, where

$$\begin{array}{lll} x_1 + x_2 + x_3 + x_4 \le 150 & x_1 + x_5 + x_9 \ge 180 & x_1 \ge 0,\ x_2 \ge 0,\ x_3 \ge 0 \\ x_5 + x_6 + x_7 + x_8 \le 300 & x_2 + x_6 + x_{10} \ge 175 & x_4 \ge 0,\ x_5 \ge 0,\ x_6 \ge 0 \\ x_9 + x_{10} + x_{11} + x_{12} \le 275 & x_3 + x_7 + x_{11} \ge 170 & x_7 \ge 0,\ x_8 \ge 0,\ x_9 \ge 0 \\ & x_4 + x_8 + x_{12} \ge 200 & x_{10} \ge 0,\ x_{11} \ge 0,\ x_{12} \ge 0 \end{array}$$

23. maximize $z = 20{,}000 + 13x \times 10y$, where

$$\begin{aligned} x + y &\ge 50 \\ x &\le 150 \\ y &\le 150 \\ x &\ge 0 \\ y &\ge 0, \end{aligned}$$

x = no. of acres of tomatoes, y = no. of acres of corn

Exercise Set 3.2, page 121

1. (b), (c) **3.** (8, 4) **5.** $x = 9, y = 0$ **7.** $x = \frac{7}{3}, y = 6$

9. (a) $x = 1$ and $y = 2$, or $x = \frac{7}{3}$ and $y = 6$, or any point on the line segment connecting (1, 2) and $(\frac{7}{3}, 6)$
(b) $x = 0, y = 6$

11. 40 tables of type A, 10 tables of type B

13. 0 containers from the Jones Corp., 2000 containers from the Jackson Corp.

15. 0 planes of type A, 5 planes of type B **17.** 20 acres of soybeans, 0 acres of alfalfa

19. 100 gallons of R, 0 gallons of L **21.** 3 minutes of H, 18 minutes of L

23. (a) 1 gram of Curine I and 3 grams of Curine II, or 2 grams of Curine I and 0 grams of Curine II, or any point on the line segment connecting (1, 3) and (2, 0)
(b) 2 grams of Curine I, 0 grams of Curine II

Supplementary Exercises, page 122

1. (a) **3.** (3, 1) **5.** 320 pounds of yellow cake, 0 pounds of white cake

7. $37,500 invested in U.S. Treasury bonds, $12,500 invested in corporate bonds; maximum annual interest = $6375

9. 60 acres of crop A, 40 acres of crop B

11. 0.1 gallon for large mower (rider), 0.5 gallon for small mower

Chapter Test, page 123

1. (b), (d) **2.** (a) (6, 1) (b) (2, 5)

3. minimum value of z is 8 for $x = 4, y = 0$ **4.** 150 8-bit computers, 250 16-bit computers

5. 120 large containers, 260 small containers

CHAPTER FOUR

Exercise Set 4.1, page 128

1. (a) **3.** no

5. (a) maximize $z = 5x + 7y$
subject to

$$\begin{aligned} 4x - 6y + u &= 9 \\ 2x + 7y + v &= 3 \\ 5x - 8y + w &= 2 \\ x &\ge 0 \\ y &\ge 0 \\ u &\ge 0 \\ v &\ge 0 \\ w &\ge 0 \end{aligned}$$

(b) maximize $z = -x_1 + x_2 - x_3 + x_4$
subject to

$$\begin{aligned} 2x_1 + 3x_3 + x_4 + v &= 8 \\ x_1 - 2x_2 + x_4 + w &= 6 \\ x_1 &\ge 0 \\ x_2 &\ge 0 \\ x_3 &\ge 0 \\ x_4 &\ge 0 \\ v &\ge 0 \\ w &\ge 0 \end{aligned}$$

7. (a) maximize $z = 10x + 12y$
subject to
$$\begin{aligned} 0.2x + 0.4y + v &= 30 \\ 0.2x + 0.2y + w &= 20 \\ x &\geq 0 \\ y &\geq 0 \\ v &\geq 0 \\ w &\geq 0 \end{aligned}$$

(b) $x = 50,\ y = 50,\ v = 0,\ w = 0$

9. (a) $A = \begin{bmatrix} 1 & 0 & -1 \\ 0 & 1 & -1 \\ 1 & -1 & 0 \end{bmatrix}$ $B = \begin{bmatrix} 1 \\ 2 \\ 3 \end{bmatrix}$ $C = [4 \quad -2 \quad 7]$ $X = \begin{bmatrix} x \\ y \\ t \end{bmatrix}$

(b) $A = [1 \quad 3]$ $B = [5]$ $C = [1 \quad -1]$ $X = \begin{bmatrix} x_1 \\ x_2 \end{bmatrix}$

11. maximize $z = 6x + 7y + 9t$
subject to
$$\begin{aligned} 2x - 4y + 5t &\leq 3 \\ 7x + y + 3t &\leq 9 \\ x &\geq 0 \\ y &\geq 0 \\ t &\geq 0 \end{aligned}$$

Exercise Set 4.2, page 138

1. maximum value of z is 7 for $x = 0,\ y = 1$

3. maximum value of z is 18 for $x = 3,\ y = 0$

5. maximum value of z is 0 for $x = 0,\ y = 0$

9. (a) maximize $z = 2x + 5y$
subject to
$$\begin{aligned} 3x + 2y + v &= 2 \\ 2x + 5y + w &= 8 \\ x &\geq 0 \\ y &\geq 0 \\ v &\geq 0 \\ w &\geq 0 \end{aligned}$$

(b)
$$\begin{array}{cccccc} x & y & v & w & z & \\ \hline -2 & -5 & 0 & 0 & 1 & 0 \\ 3 & 2 & 1 & 0 & 0 & 2 \\ 2 & 5 & 0 & 1 & 0 & 8 \end{array}$$

(c)
$$\begin{bmatrix} 0 & 0 & 0 & 1 & 1 & 8 \\ \frac{11}{5} & 0 & 1 & -\frac{2}{5} & 0 & -\frac{6}{5} \\ \frac{2}{5} & 1 & 0 & \frac{1}{5} & 0 & \frac{8}{5} \end{bmatrix}$$

(d)
$$\begin{bmatrix} 0 & 0 & 0 & 1 & 1 & 8 \\ 0 & -\frac{11}{2} & 1 & -\frac{3}{2} & 0 & -10 \\ 1 & \frac{5}{2} & 0 & \frac{1}{2} & 0 & 4 \end{bmatrix}$$

11. (a) $x,\ v,\ z$ (b) $y,\ v,\ z$ (c) $x_3,\ v_1,\ z$ (d) $x,\ y,\ v,\ z$

Exercise Set 4.3, page 153

1. $x = 0,\ y = 0,\ v = 12,\ w = 6$

$$\begin{array}{c} \begin{array}{cccccc} x & y & v & w & z & \end{array} \\ \left[\begin{array}{cccccc} -4 & -3 & 0 & 0 & 1 & 0 \\ 2 & 3 & 1 & 0 & 0 & 12 \\ -3 & 2 & 0 & 1 & 0 & 6 \end{array}\right] \end{array}$$

3. $x = 0,\ y = 0,\ u = 4,\ v = 6,\ w = 1$

$$\begin{array}{c} \begin{array}{ccccccc} x & y & u & v & w & z & \end{array} \\ \left[\begin{array}{ccccccc} -8 & -6 & 0 & 0 & 0 & 1 & 0 \\ 1 & 1 & 1 & 0 & 0 & 0 & 4 \\ 1 & 3 & 0 & 1 & 0 & 0 & 6 \\ -1 & 1 & 0 & 0 & 1 & 0 & 1 \end{array}\right] \end{array}$$

5. (a)

$$\begin{array}{rcl} 5x_3 + 7u \qquad + 6w + z &=& 40 \\ x_1 - \quad 2x_3 + u \qquad + 2w \qquad &=& 6 \\ x_2 + 4x_3 + u \qquad - w \qquad &=& 14 \\ 3x_3 \qquad + v + 3w \qquad &=& 12 \end{array}$$

(b) $x_1 = 6,\ x_2 = 14,\ x_3 = 0,\ u = 0,\ v = 12,\ w = 0$

(c) $z = 40$

7. (a), (c)

9. $y,\ x,\ 4$

11. $\left[\begin{array}{ccccccc} 0 & 0 & 27 & 4 & 8 & 1 & 540 \\ 1 & 0 & 3 & \frac{1}{2} & 1 & 0 & 50 \\ 0 & 1 & -4 & -\frac{3}{2} & 0 & 0 & 30 \end{array}\right]$

13. $\left[\begin{array}{ccccccccc} 0 & 0 & 0 & 0 & \frac{4}{3} & \frac{23}{3} & 0 & 1 & 360 \\ 0 & 0 & 0 & 1 & -\frac{2}{3} & \frac{5}{3} & -2 & 0 & 80 \\ 0 & 1 & 0 & 0 & \frac{1}{2} & 3 & \frac{5}{2} & 0 & 390 \\ 0 & 0 & 1 & 0 & 0 & -1 & 0 & 0 & 150 \\ 1 & 0 & 0 & 0 & \frac{1}{6} & \frac{1}{3} & \frac{1}{2} & 0 & 30 \end{array}\right]$

15. $z = \frac{144}{7}$ for $x = \frac{60}{7},\ y = \frac{8}{7},\ t = 0$ **17.** $z = \frac{12}{5}$ for $x_1 = \frac{6}{5},\ x_2 = 0,\ x_3 = 0$

21. 0 gallons of L and E, 125 gallons of R

Exercise Set 4.4, page 159

1. $z = -\frac{26}{3}$ for $x = \frac{8}{3},\ y = 1$ **3.** $z = -\frac{16}{3}$ for $x = 0,\ y = \frac{8}{3},\ t = 0$

5. $z = 8$ for $x_1 = 0,\ x_2 = 0,\ x_3 = 8$ **7.** $z = 6$ for $x_1 = 2,\ x_2 = 0$

9. $z = \frac{158}{11}$ for $x_1 = \frac{18}{11},\ x_2 = \frac{10}{11}$ **11.** 7 ounces of M, 0 ounces of N

Supplementary Exercises, page 160

1. (a) maximize $z = 3x + 5y$
subject to

$$\begin{array}{r} 2x - \ \ y \le 4 \\ -3x + 2y \le 2 \\ x \ge 0 \\ y \ge 0 \end{array}$$

(b) maximize $z = -4x_1 + 3x_2 - x_3$
subject to

$$\begin{array}{r} 5x_1 - 3x_2 + x_3 \le 2 \\ x_1 - 2x_2 \qquad \le 4 \\ x_1 \ge 0 \\ x_2 \ge 0 \\ x_3 \ge 0 \end{array}$$

3.

Corner points	Basic feasible solutions				$z = 3x + 4y$
(0, 0)	$x = 0$	$y = 0$	$v = 3$	$w = 7$	0
(1, 0)	$x = 1$	$y = 0$	$v = 0$	$w = 5$	3
(0, 7)	$x = 0$	$y = 7$	$v = 10$	$w = 0$	28
(2, 3)	$x = 2$	$y = 3$	$v = 0$	$w = 0$	18

The optimal solution is $x = 0$, $y = 7$, $z = 28$.

5. (a) x_2, u, v (b) $x_2 = 40$, $u = 20$, $v = 12$

7. $x_1 = 8$, $x_3 = 4$, $v = 15$, $z = 120$

9.
$$\begin{array}{c} \\ z \\ t \\ x \\ v \end{array} \begin{array}{c} \begin{array}{cccccccc} x & y & t & u & v & w & z & \end{array} \\ \left[\begin{array}{cccccccc} 0 & 9 & 0 & 0 & 0 & 1 & 1 & 105 \\ 0 & -\frac{44}{3} & 1 & -\frac{2}{3} & 0 & -\frac{10}{3} & 0 & \frac{10}{3} \\ 1 & -\frac{29}{3} & 0 & -\frac{23}{3} & 0 & -\frac{4}{3} & 0 & \frac{10}{3} \\ 0 & \frac{5}{3} & 0 & \frac{2}{3} & 1 & \frac{1}{3} & 0 & \frac{5}{3} \end{array}\right] \end{array}$$

11. $x_1 = 2$, $x_2 = 0$, $x_3 = 4$, $z = 20$

13. maximize $z' = 8y_1 + 15y_2$
subject to
$$\begin{aligned} 2y_1 + 3y_2 &\le 2 \\ y_1 - y_2 &\le 3 \\ y_1 + 4y_2 &\le 1 \\ y_1 &\ge 0 \\ y_2 &\ge 0 \end{aligned}$$

15. make 25 tons of GARDEN. The profit is \$10,000.

Chapter Test, page 162

1.

Corner points	Basic feasible solutions				$z = 2x - 4y$
(0, 0)	$x = 0$	$y = 0$	$u = 8$	$v = 5$	0
(0, 8)	$x = 0$	$y = 8$	$u = 0$	$v = 21$	−32
(3, 2)	$x = 3$	$y = 2$	$u = 0$	$v = 0$	− 2
$(\frac{5}{3}, 0)$	$x = \frac{5}{3}$	$y = 0$	$u = \frac{14}{3}$	$v = 0$	$\frac{10}{3}$

The optimal solution is $x = \frac{5}{3}$, $y = 0$, $z = \frac{10}{3}$

2. $x_1 = 1$, $x_2 = \frac{1}{3}$, $x_3 = 0$, $z = 11$

3. maximize $z = 20y_1 + 15y_2 + 12y_3$
subject to
$$\begin{aligned} 2y_1 + 4y_2 + y_3 &\le 3 \\ y_1 - y_2 + y_3 &\le 1 \\ -y_1 + 2y_2 + 3y_3 &\le 4 \\ y_1 &\ge 0 \\ y_2 &\ge 0 \\ y_3 &\ge 0 \end{aligned}$$

4. $x_1 = \frac{5}{4}$, $x_2 = \frac{25}{4}$, $x_3 = 0$, $z = \frac{35}{4}$

5. 6 acres of corn, 0 acres of soybeans, 6 acres of oats. Profit = \$360

6. $\frac{40}{3}$ pounds of CARPET, no GROW. Cost is \$40

CHAPTER FIVE

Exercise Set 5.1, page 167

1. (a) false (b) true (c) true (d) false (e) true (f) false

3. (a) true (b) false (c) true (d) true

5. (a) $\{a, r, d, v, k\}$ (b) $\{m, i, s, p\}$ (c) $\{t, a, b, l, e\}$

7. (a) $\{x \mid x \text{ is a U.S. citizen}\}$ (b) $\{x \mid x \text{ is a U.S. citizen over 40 years of age}\}$ **9.** (b)
11. $\varnothing$, $\{2\}$, $\{5\}$, $\{2, 5\}$ **13.** (a) $\varnothing$, $\{a_1\}$, $\{a_2\}$, $\{a_3\}$, $\{a_1, a_2\}$, $\{a_1, a_3\}$, $\{a_2, a_3\}$, $\{a_1, a_2, a_3\}$ (b) $\varnothing$
15. (a) true (b) false (c) false (d) false
17.

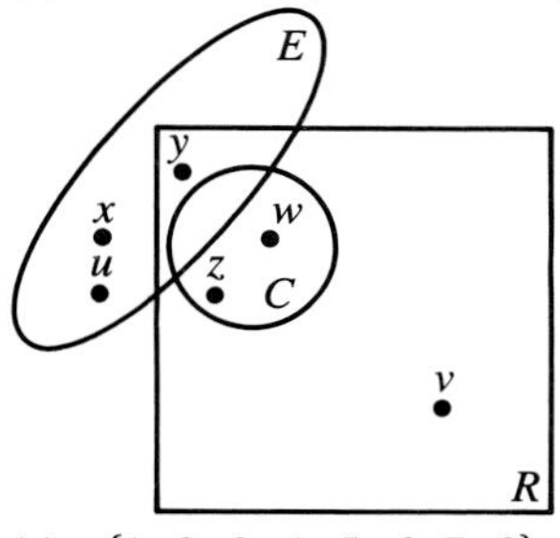

19. (a) $\{1, 2, 3, 4, 5, 6, 7, 9\}$ (b) $\{a, b, c\}$ **21.** no; yes
23. 2^n **25.** $\{1, 2, 3\}$, $\{1, 2, 3, 4\}$, $\{1, 2, 3, 5\}$, $\{1, 2, 3, 6\}$, $\{1, 2, 3, 4, 5\}$, $\{1, 2, 3, 4, 6\}$, $\{1, 2, 3, 5, 6\}$, $\{1, 2, 3, 4, 5, 6\}$

Exercise Set 5.2, page 175

1. (a) $\{3, 7\}$ (b) $\{1\}$ (c) $\{2\}$ (d) $\varnothing$ (e) $\varnothing$ (f) $\varnothing$

3. 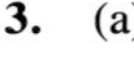

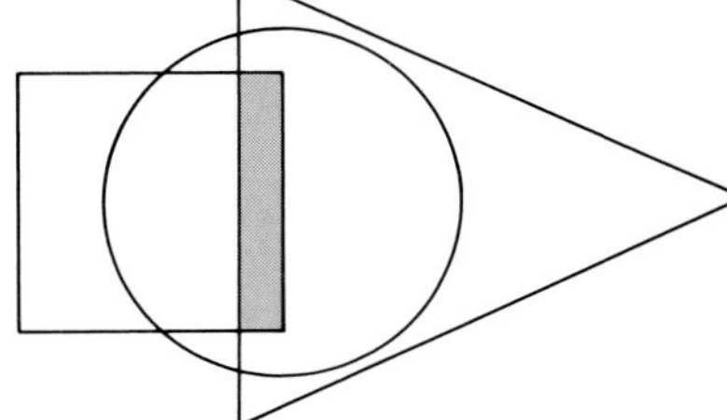

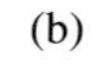

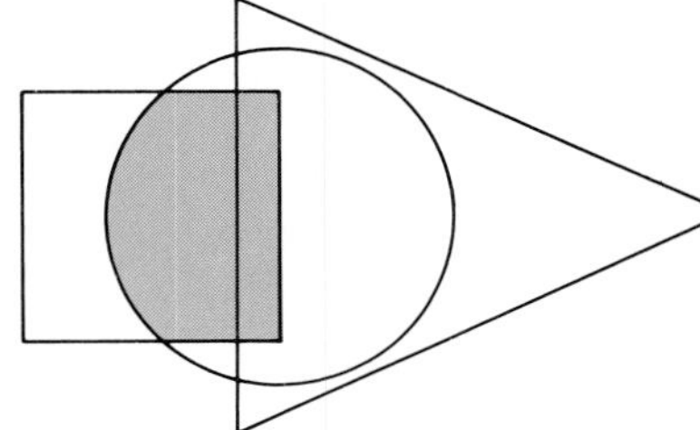

(c)

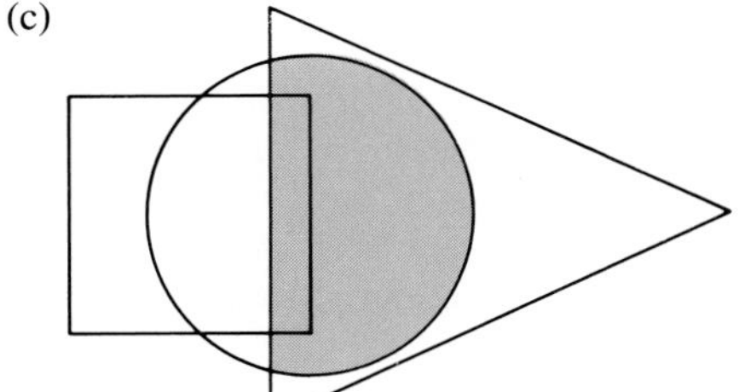

(d)

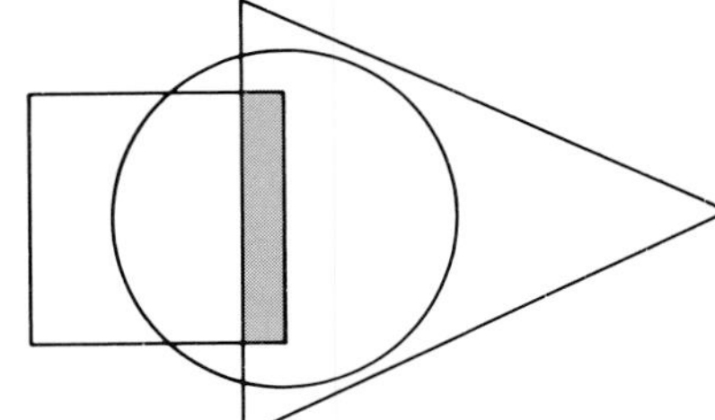

5. (a)

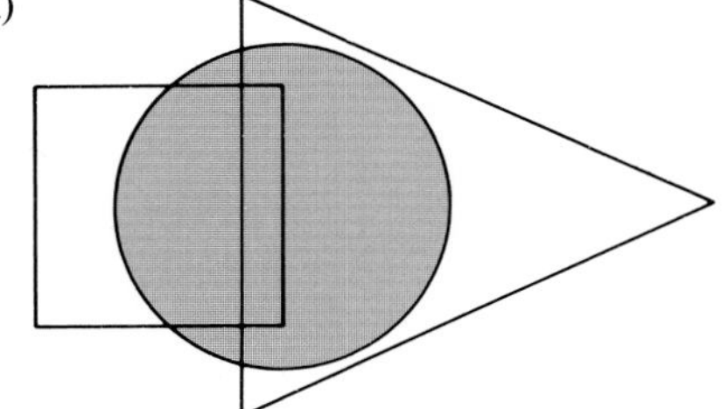

(b)

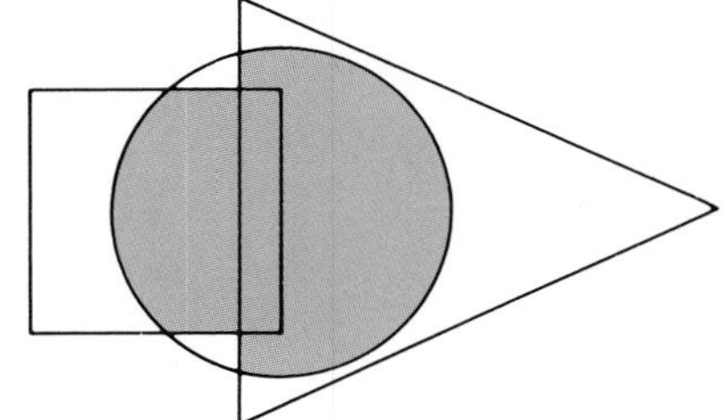

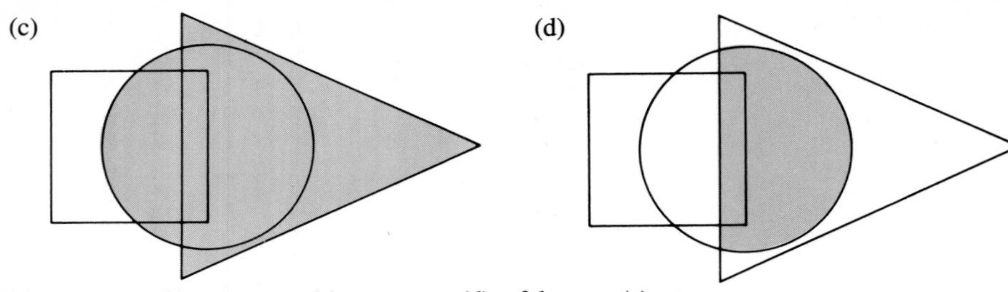

7. (a) true (b) true (c) true (d) false (e) true

9. (a) true (b) true (c) false (d) false

19. $C \cap (T \cup R)$

21. (a) all male administrative employees (b) all employees (c) all female administrative technical employees (d) either female or administrative or technical employees (e) all male administrative employees working for the company at least 5 years

23. (a) those who drive cars with engines that are more than 250 horsepower
(b) those who drive cars with engines that are more than 200 horsepower
(c) those who are male and over 25 years of age and who drive cars with engines that are more than 200 horsepower
(d) those who are female or over 20 years of age or who drive cars with engines that are more than 200 horsepower
(e) those who drive cars with engines that are more than 250 horsepower and are either over 20 years of age or female

25.

Blood type	Antigen A	Antigen B	Antigen Rh
A^-	yes	no	no
A^+	yes	no	yes
B^-	no	yes	no
B^+	no	yes	yes
AB^-	yes	yes	no
AB^+	yes	yes	yes
O^-	no	no	no
O^+	no	no	yes

Exercise Set 5.3, page 182

1. (a) $\{b, c, e, g\}$ (b) $\{b, c, e, f, g, h\}$ (c) $\{b, c, e, g\}$ (d) U (e) $\varnothing$

3. (a) the set of stocks traded on the New York Stock Exchange that have not paid a dividend for at least one of the past 10 years.
(b) the set of stocks traded on the New York Stock Exchange that have a price-to-earnings ratio of more than 12
(c) the set of stocks traded on the New York Stock Exchange that have not paid a dividend for at least one of the past 10 years and that have a price-to-earnings ratio of more than 12
(d) the set of stocks traded on the New York Stock Exchange that either have not paid a dividend for at least one of the past 10 years or have a price-to-earnings ratio of more than 12

5. U = the set of all integers; $A' = \{x \mid x \text{ is an integer satisfying } x \leq 4\}$

7. U = the set of all letters in the English alphabet; $C' = \{a, e, i, o, u\}$

9. (a)

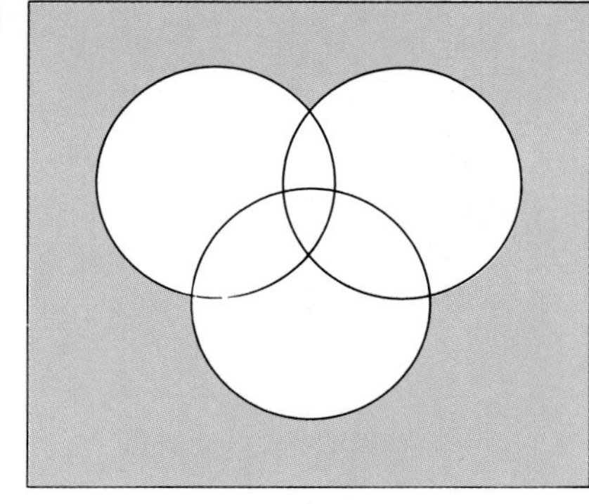

(b)

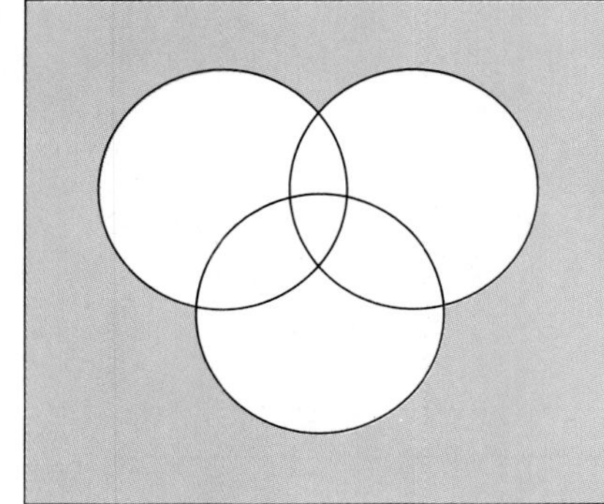

(c)

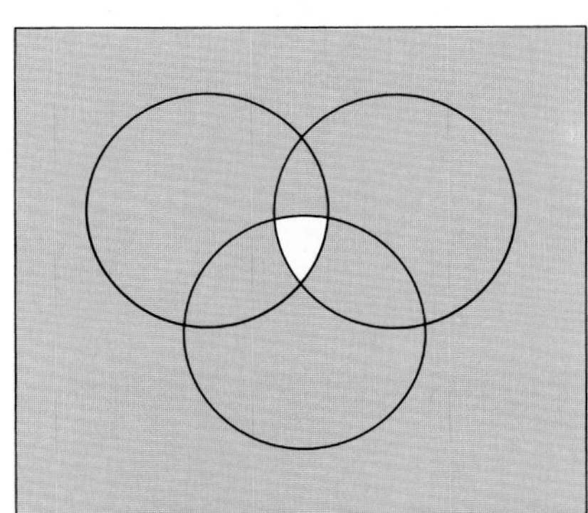

(d)

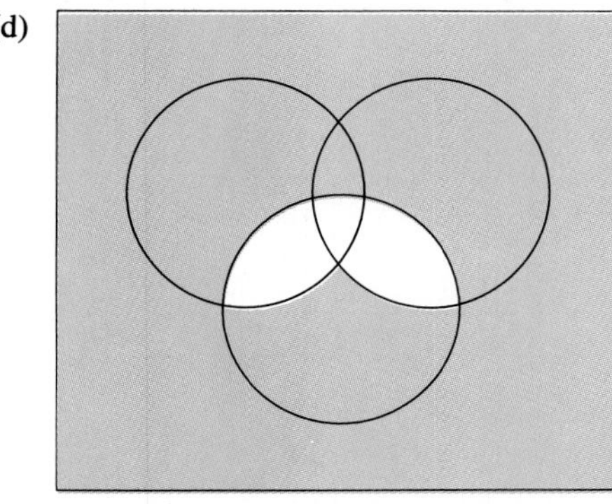

13. (a) $(A \cup B \cup C \cup D)' = A' \cap B' \cap C' \cap D'$ $(A \cap B \cap C \cap D)' = A' \cup B' \cup C' \cup D'$

(b) $(A_1 \cup A_2 \cup \cdots \cup A_n)' = A_1' \cap A_2' \cap \cdots \cap A_n'$ $(A_1 \cap A_2 \cap \cdots \cap A_n)' = A_1' \cup A_2' \cup \cdots \cup A_n'$

15. (a) x (b) y (c) none (d) none

17. (a)

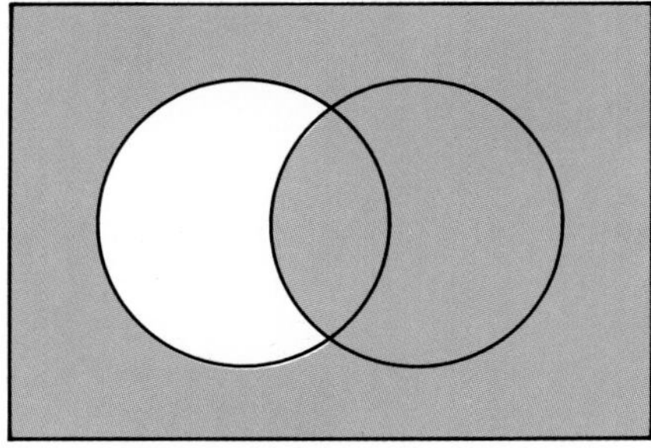

(b)

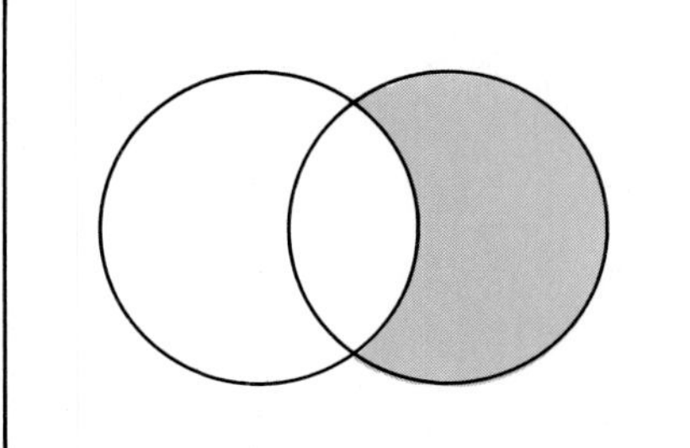

(c)

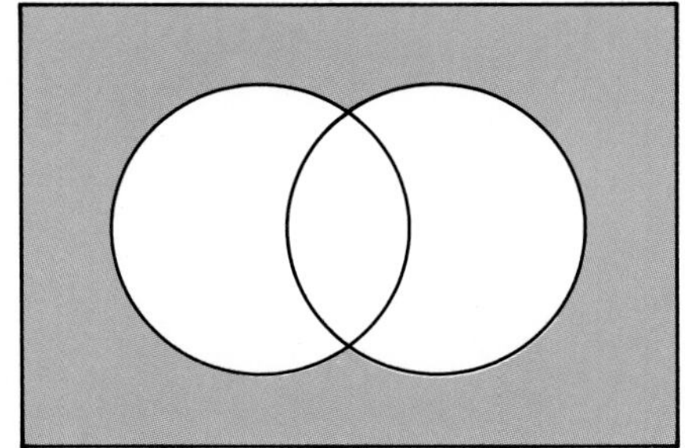

19. (a)

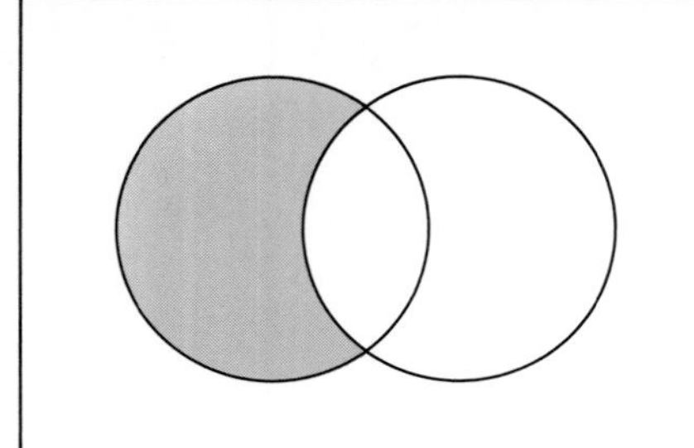

(b)

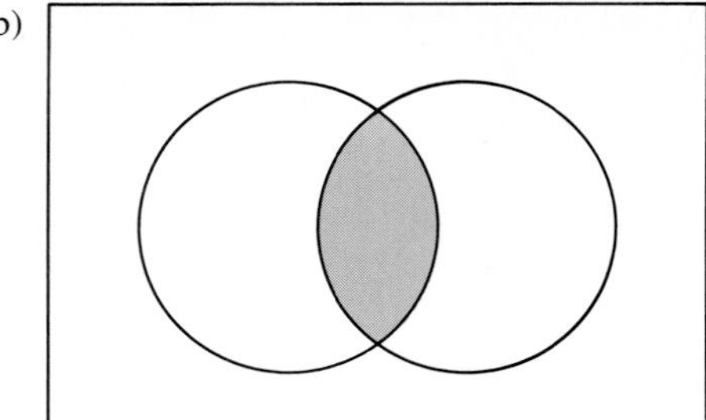

(c)

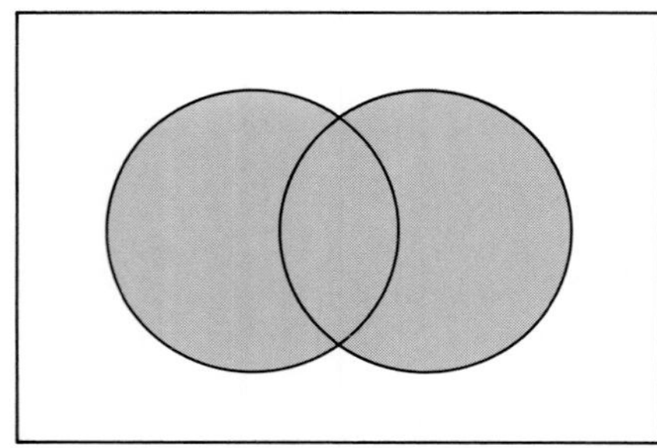

21. (a)

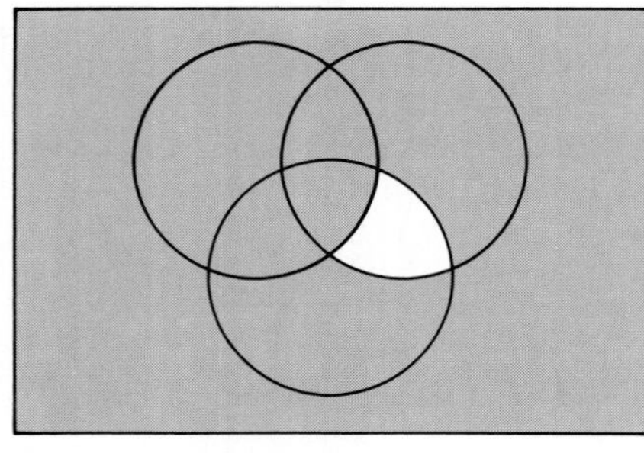

(b)

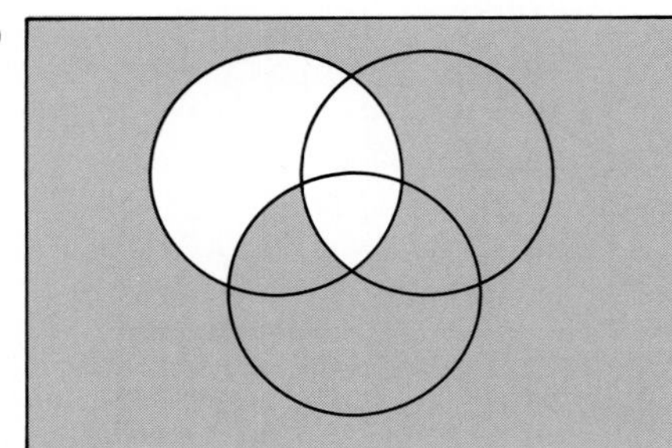

(c)

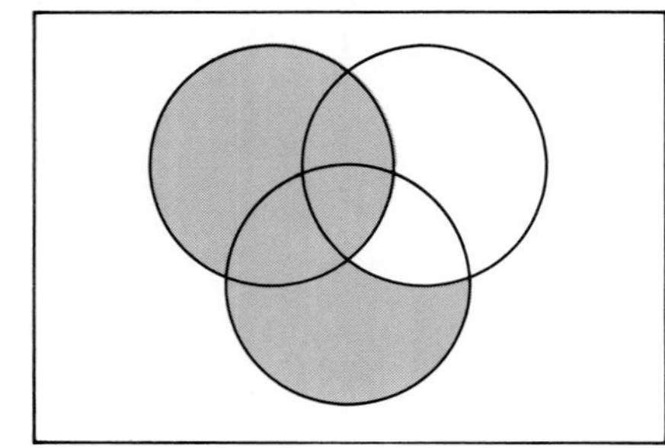

23. $R \cap A \cap (S \cup P) \cap G'$

Supplementary Exercises, page 184

1. (a) $\in$ (b) $\notin$ (c) $\in$ (d) $\notin$ (e) $\subset$ (f) $\subset$

3. (a) true (b) false (c) false (d) true (e) true (f) false

5. true **7.** $\{a, b\}$, $\{a, b, c\}$, $\{a, b, d\}$, $\{a, b, c, d\}$

9. (a) $\{2, 4\}$ (b) $\{1, 2, 4\}$ (c) $\varnothing$ (d) $\{1, 2, 3, 4\}$ (e) $\{1, 2, 3, 4\}$ (f) $\{1, 2, 3, 4\}$

11. (a)

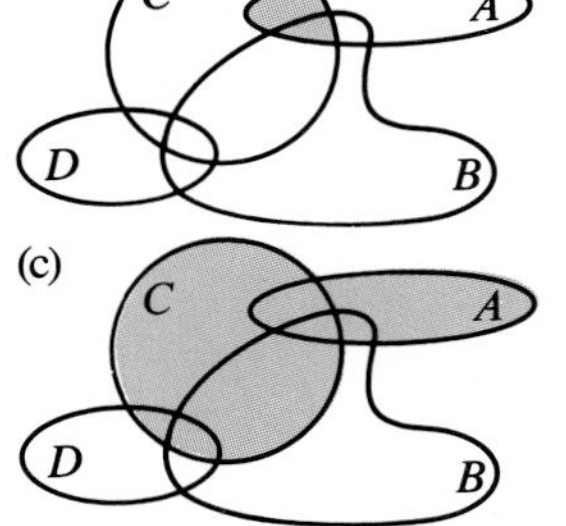

(b)

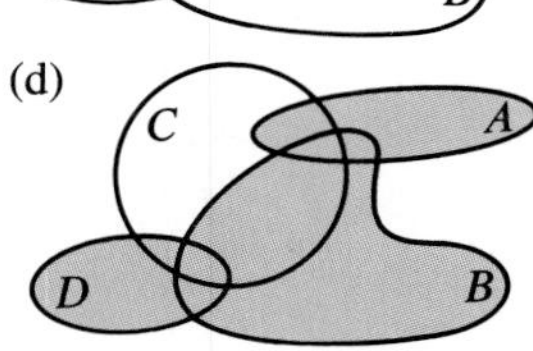

13. (a) false (b) true (c) true (d) false (e) false (f) true

21. $(A \cap B) \cup C$ **23.** (a) $\{1, 4, 7, 8\}$ (b) $\{2, 6, 9\}$ **25.** $\{1, 2, 3, 7, 8\}$

27. (a) $\{x|x$ is an 8-byte machine that supports FORTRAN$\}$
(b) $\{x|x$ is a 16-byte machine that supports FORTRAN and BASIC$\}$
(c) $\{x|x$ is a 16- or 32-byte machine that supports PASCAL$\}$
(d) $\{x|x$ is a 16-byte machine that supports FORTRAN or BASIC$\}$

29. (a) $\varnothing$ (b) $\{1\}$ (c) $\{2, 7\}$ (d) $\{2, 3, 5, 6, 7\}$

31. (a)

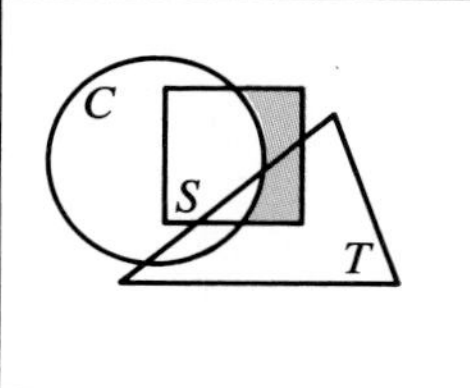

(b)

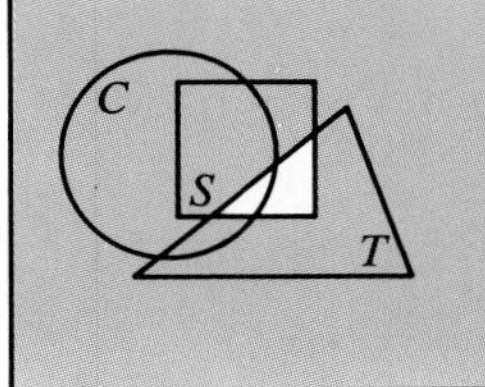

(c)

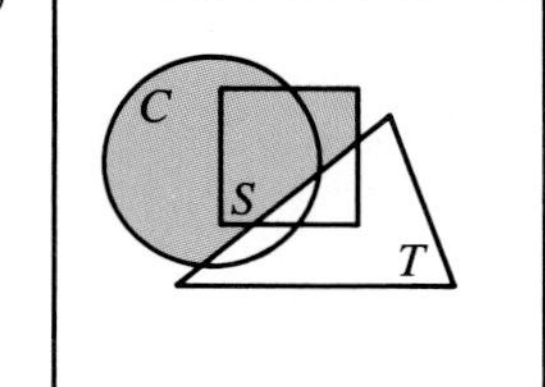

(d)

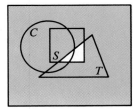

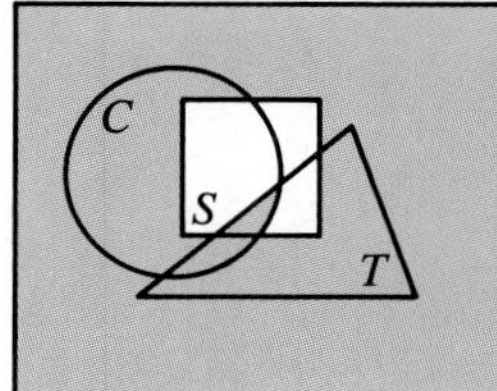

33. $A' \cap B \cap C$

Chapter Test, page 187

1. (a) true (b) false (c) true (d) false **2.** (a) false (b) true (c) false (d) true

3. $\{-3, -2, 0, 2\}$ **4.** $\{-3, -2, -1, 0, 1, 2, 3\}$ **5.** $\{0, 1, 2, 3, 4\}$ **6.** $\{3, 5\}$

7. 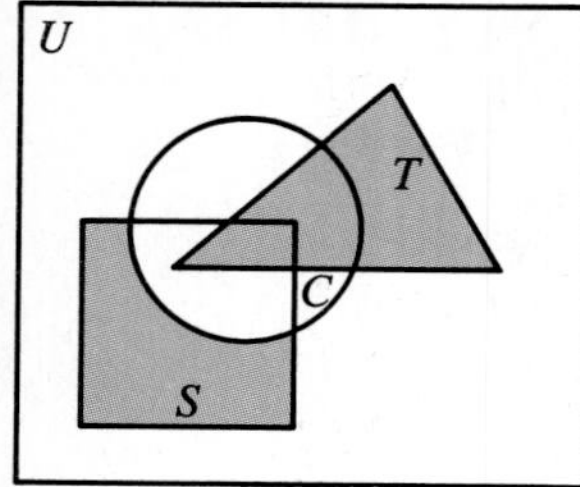

8. 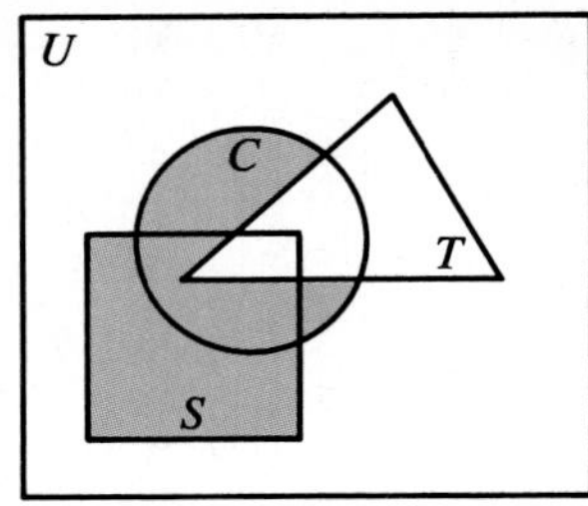

9. $(A \cap C') \cup (A' \cap B \cap C)$ 10. (a) $M \cap (D \cup S) \cap H$ (b) $M' \cap (D \cup E') \cap H'$

CHAPTER SIX

Exercise Set 6.1, page 192

1. (a) $x = 3$ (b) $x = 3, y = 3$ (c) $x = 1, y = 7$ (d) $x = \pm 4$
3. $A \times B = \{(u, q), (u, r), (v, q), (v, r)\}$
   $B \times A = \{(q, u), (q, v), (r, u), (r, v)\}$
   $A \times A = \{(u, u), (u, v), (v, u), (v, v)\}$
   $B \times B = \{(q, q), (q, r), (r, q), (r, r)\}$
5. $A \times B = B \times A = A \times A = B \times B$
   $= \{(-2, -2), (-2, 1), (-2, 4), (1, -2), (1, 1), (1, 4), (4, -2), (4, 1), (4, 4)\}$
7. 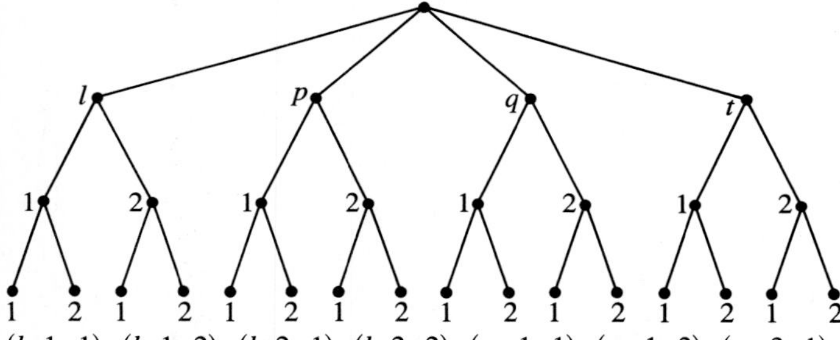

$(l, 1, 1), (l, 1, 2), (l, 2, 1), (l, 2, 2), (p, 1, 1), (p, 1, 2), (p, 2, 1), (p, 2, 2), (q, 1, 1), (q, 1, 2), (q, 2, 1),$
$(q, 2, 2), (t, 1, 1), (t, 1, 2), (t, 2, 1), (t, 2, 2)$
9. 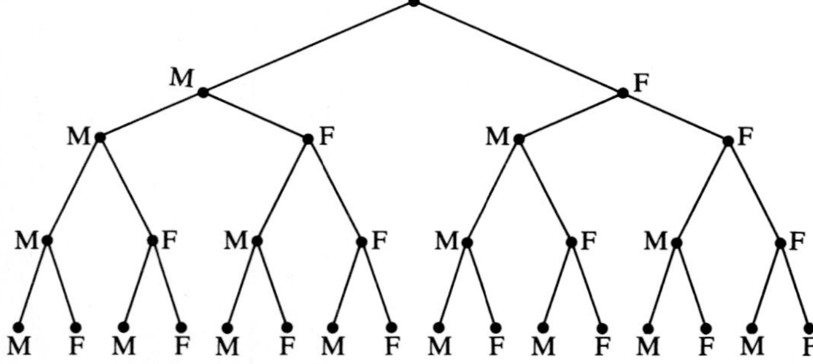

$(M, M, M, M), (M, M, M, F), (M, M, F, M), (M, M, F, F), (M, F, M, M), (M, F, M, F), (M, F, F, M),$
$(M, F, F, F), (F, M, M, M), (F, M, M, F), (F, M, F, M), (F, M, F, F), (F, F, M, M), (F, F, M, F),$
$(F, F, F, M), (F, F, F, F)$

11. (a) $H = \{h, t\}$
(b)

START
h t
h t h t
h t h t h t h t
h t h t h t h t h t h t h t h t

(h, h, h, h), (h, h, h, t), (h, h, t, h), (h, h, t, t), (h, t, h, h), (h, t, h, t), (h, t, t, h), (h, t, t, t), (t, h, h, h), (t, h, h, t), (t, h, t, h), (t, h, t, t), (t, t, h, h), (t, t, h, t), (t, t, t, h), (t, t, t, t)

13. (a) $E = \{a, b, c\}$ $T = \{l, r\}$
(b)

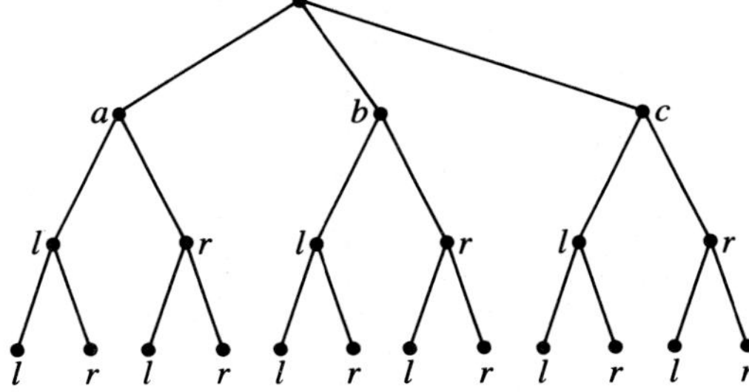

(a, l, l), (a, l, r), (a, r, l), (a, r, r), (b, l, l), (b, l, r), (b, r, l), (b, r, r), (c, l, l), (c, l, r), (c, r, l), (c, r, r)

15. $r = -1, s = 2, t = 3$ **17.** (a) 6 (b) mn **19.** (a) m^2 (b) m^3 (c) m^n

21.

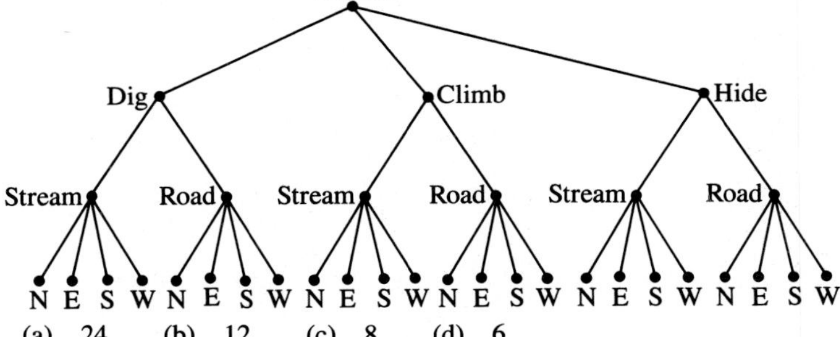

(a) 24 (b) 12 (c) 8 (d) 6

Exercise Set 6.2, page 201

1. (a) 21 (b) 2 (c) 4 (d) 6 (e) infinite (f) 0 **5.** $B = \varnothing$
11. 6 **13.** (a) 36 (b) 30 (c) 33 (d) 32 **17.** (a) 18 **19.** (a) 6^5 (b) 6^n
21. 10^7 **23.** 9 **25.** 878
27. (a) 51 (b) 17 (c) 73 (d) 7.77% (e) 82 (f) 86 **29.** 36,000
31. (a) 3 (b) 6 (c) 26
33. (a) 10 (b) 30

Section 6.3, page 214

1. (a) 720 (b) 5040 (c) 40,320 (d) 1056 (e) 67 (f) 2,162,160 (g) $\frac{1}{2}$
3. (a) 840 (b) 20,058,300 (c) $\frac{1}{6}$ (d) 15 (e) 1
5. (a) 42 (b) 11,880 (c) 120 (d) n **7.** (a) 21 (b) 495 (c) 1 (d) 3,838,380
9. (a) $-\frac{6}{29}$ (b) $\frac{120}{71}$ (c) $\frac{1}{2520}$ **11.** (a) 2, 8 (b) 9 (c) 0, 5
13. 350 **15.** 48 **17.** 2^{32} **19.** 7776; 6^n

21.

abcd	*bacd*	*cabd*	*dabc*
abdc	*badc*	*cadb*	*dacb*
acbd	*bcad*	*cbad*	*dbac*
acdb	*bcda*	*cbda*	*dbca*
adbc	*bdac*	*cdab*	*dcab*
adcb	*bdca*	*cdba*	*dcba*

23. (a) 20 (b)

ab	*ac*	*ad*	*ae*
ba	*bc*	*bd*	*be*
ca	*cb*	*cd*	*ce*
da	*db*	*dc*	*de*
ea	*eb*	*ec*	*ed*

25. 5,200,300 **27.** 1,351,350 **29.** 15,504
31. 720 **33.** 1000 **35.** 720

37.

abc	*abd*	*abe*	*acb*	*acd*	*ace*	*adb*	*adc*	*ade*	*aeb*	*aec*	*aed*
bac	*bad*	*bae*	*bca*	*bcd*	*bce*	*bda*	*bdc*	*bde*	*bea*	*bec*	*bed*
cab	*cad*	*cae*	*cba*	*cbd*	*cbe*	*cda*	*cdb*	*cde*	*cea*	*ceb*	*ced*
dab	*dac*	*dae*	*dba*	*dbc*	*dbe*	*dca*	*dcb*	*dce*	*dea*	*deb*	*dec*
eab	*eac*	*ead*	*eba*	*ebc*	*ebd*	*eca*	*ecb*	*ecd*	*eda*	*edb*	*edc*

39. $C_{26,7} = 657{,}800$ **41.** $C_{9,3} = 84$ **43.** $C_{25,4} = 12{,}650$
45. (a) $6^7 = 279{,}936$ (b) $7 \cdot 5^6$ (c) $C_{7,2} \cdot 5^5$ (d) $5^7 + C_{7,1}5^6 + C_{7,2}5^5$ **49.** 10^9

Exercise Set 6.4, page 221

1. $C_{26,5} \cdot C_{26,2} = 21{,}378{,}500$
3. (a) $C_{52,6} = 20{,}358{,}520$ (b) $C_{13,2} \cdot C_{39,4} = 6{,}415{,}578$
(c) $C_{13,3} \cdot C_{13,3} = 81{,}796$ (d) $C_{13,3} \cdot C_{26,3} = 743{,}600$
5. (a) $C_{9,6} = 84$ (b) $C_{6,4} \cdot C_{3,2} = 45$ (c) 1
7. (a) $C_{50,6} = 15{,}890{,}700$ (b) $C_{10,2} \cdot C_{40,4} = 4{,}112{,}550$ (c) $C_{40,6} = 3{,}838{,}380$
9. (a) 512 (b) 126 (c) 72 **11.** (a) 462 (b) 3360 (c) 10,080
13. (a) $C_{52,9} - C_{39,9}$ (b) $C_{52,9} - C_{13,8} \cdot C_{39,1} - C_{13,9}$
15. (a) $2^{50} - C_{50,2} - C_{50,1} - C_{50,0}$ (b) $2^{50} - C_{50,49} - C_{50,50}$
17. (a) 2517 (b) 137
19. (a) $C_{10,8} = 45$ (b) $C_{10,3} \cdot C_{4,2} \cdot C_{2,1} \cdot C_{11,2} = 79{,}200$ (c) $C_{27,8} = 2{,}220{,}075$
21. choose n so $C_{n,3} \geq 52$; $n = 8$

Exercise Set 6.5, page 227

1. (a) $a^7 + 7a^6b + 21a^5b^2 + 35a^4b^3 + 35a^3b^4 + 21a^2b^5 + 7ab^6 + b^7$
(b) $x^9 + 9x^8y + 36x^7y^2 + 84x^6y^3 + 126x^5y^4 + 126x^4y^5 + 84x^3y^6 + 36x^2y^7 + 9xy^8 + y^9$
(c) $r^{11} + 11r^{10}s + 55r^9s^2 + 165r^8s^3 + 330r^7s^4 + 462r^6s^5 + 462r^5s^6 + 330r^4s^7 + 165r^3s^8 + 55r^2s^9 + 11rs^{10} + s^{11}$
3. (a) $(a - b)^2 = a^2 - 2ab + b^2$
$(a - b)^3 = a^3 - 3a^2b + 3ab^2 - b^3$
$(a - b)^4 = a^4 - 4a^3b + 6a^2b^2 - 4ab^3 + b^4$
$(a - b)^5 = a^5 - 5a^4b + 10a^3b^2 - 10a^2b^3 + 5ab^4 - b^5$
5. (a) $a^8 - 8a^7b + 28a^6b^2 - 56a^5b^3 + 70a^4b^4 - 56a^3b^5 + 28a^2b^6 - 8ab^7 + b^8$
(b) $w^{10} - 10w^9z + 45w^8z^2 - 120w^7z^3 + 210w^6z^4 - 252w^5z^5 + 210w^4z^6 - 120w^3z^7 + 45w^2z^8 - 10wz^9 + z^{10}$
(c) $c^{12} - 12c^{11}d + 66c^{10}d^2 - 220c^9d^3 + 495c^8d^4 - 792c^7d^5 + 924c^6d^6 - 792c^5d^7 + 495c^4d^8 - 220c^3d^9 +$
$66c^2d^{10} - 12cd^{11} + d^{12}$
7. $x^7 + 7x^6 + 21x^5 + 35x^4 + 35x^3 + 21x^2 + 7x + 1$ **9.** $a^3 - 9a^2b + 27ab^2 - 27b^3$

11. $512r^9 - 4608r^8s + 18{,}432r^7s^2 - 43{,}008r^6s^3 + 64{,}512r^5s^4 - 64{,}512r^4s^5 + 43{,}008r^3s^6 - 18{,}432r^2s^7 + 4608rs^8 - 512s^9$
13. (a) $1 + 20p + 190p^2 + 1140p^3 + 4845p^4$ (b) $1 - 20p + 190p^2 - 1140p^3 + 4845p^4$
15. $-480{,}700x^{18}y^7$ **17.** 64 **19.** 128

Supplementary Exercises, page 229

1. (a) $\{(a, 1), (a, 2), (b, 1), (b, 2), (c, 1), (c, 2)\}$ (b) 6 **3.** 35 **5.** 6,480,000 **7.** 420
9. 105 **11.** 120 **13.** 6^5 **15.** 276 **17.** 40 **19.** 55 **21.** 5^{10} **23.** (120)7!
25. $1 + 4x + 6x^2 + 4x^3 + x^4$ **27.** $C_{40,6} = 3{,}838{,}380$ **29.** 4

Chapter Test, page 230

1. 6 **2.** 250 **3.** (a) $7! = 5040$ (b) 288 (c) 576 (d) 240 (e) 144
4. (a) 210 (b) 151,200 **5.** 840 **6.** (a) $C_{52,4} = 270{,}725$ (b) $C_{13,4} = 715$
7. (a) 50 (b) 20 **8.** (a) 243 (b) 40 (c) 51 **9.** $1 - 5x + 10x^2 - 10x^3 + 5x^4 - x^5$
10. 35

CHAPTER SEVEN

Exercise Set 7.1, page 239

1. (a) $\{0, 1, 2, 3, 4, 5, 6, 7, 8, 9, 10\}$ (b) $\{0, 1, 2, \ldots\}$ (c) $\{x | 0 \leq x \leq 100\}$
(d) $\{x | x \geq 0\}$ (e) $\{(h, h, h), (h, h, t), (h, t, h), (t, h, h), (h, t, t), (t, h, t), (t, t, h), (t, t, t)\}$
3. $\varnothing$; S, $\{a, b\}$, $\{a, c\}$, $\{b, c\}$, $\{a\}$, $\{b\}$, $\{c\}$
5. (a) the same number is obtained on both tosses
(b) the first number tossed is 3
(c) a 4 is tossed both times
7. (a) $\{(m, h, d), (m, h, r), (m, h, i), (m, a, d), (m, a, r), (m, a, i), (m, l, d), (m, l, r), (m, l, i), (f, h, d), (f, h, r), (f, h, i), (f, a, d), (f, a, r), (f, a, i), (f, l, d), (f, l, r), (f, l, i)\}$
(b) $\{(m, h, r), (m, a, r), (m, l, r), (f, h, r), (f, a, r), (f, l, r)\}$
(c) $\{(f, h, i), (f, a, i), (f, l, i)\}$
(d) $\{(m, h, d), (m, h, r), (m, h, i), (m, a, d), (m, a, r), (m, a, i), (m, l, d), (m, l, r), (m, l, i), (f, h, d), (f, a, d), (f, l, d)\}$
9. (a) the number tossed is either even or less than 5
(b) the number tossed is divisible by 3 and is less than 5
(c) the number tossed is either divisible by 3 or greater than 4
(d) the number tossed is not divisible by 3 and is less than 5
(e) the number tossed is odd, not divisible by 3, and greater than 4
11. (a) $\{t | t \geq 10\}$ (b) $\{t | 0 \leq t < 15\}$ (c) $\{t | 0 \leq t < 10\}$ (d) $\{t | t \geq 15\}$
(e) $\{t | t \geq 15\}$ (f) $\{t | t \geq 10\}$
13. (a) yes, $E \cap F = \varnothing$ (b) no, $E \cap H \neq \varnothing$ (c) no, $E \cup F \neq S$ (d) yes, $E \cup E' = S$
15. yes, $E \cap F = E \cap G = F \cap G = \varnothing$

Exercise Set 7.2, page 252

1. (a) $\{(h, h)\}, \{(h, t)\}, \{(t, h)\}, \{(t, t)\}$ (b) $\{i\}, \{d\}, \{n\}$
(c) $\{(i, i)\}, \{(i, d)\}, \{(i, n)\}, \{(d, i)\}, \{(d, d)\}, \{(d, n)\}, \{(n, i)\}, \{(n, d)\}, \{(n, n)\}$
(d) $\{a\}, \{b\}, \{c\}, \{d\}, \{e\}$ (e) $\{e\}, \{o\}$ (f) $\{c\}, \{d\}, \{h\}, \{s\}$
(g) $\{0\}, \{1\}, \{2\}, \{3\}, \{4\}, \{5\}, \{6\}, \{7\}, \{8\}, \{9\}, \{10\}$
3. (a) $\frac{2}{3}$ (b) $\frac{2}{3}$ (c) 0 **5.** (a) $\frac{5}{11}$ (b) $\frac{5}{11}$ (c) 1 **7.** (c) and (d)

9. (a) $\{e\}, \{o\}$; equally likely (b) $\{3\}$, {not 3}; not equally likely
(c) {ace}, {not an ace}; not equally likely (d) {black}, {red}; equally likely
(e) $\{a, b\}, \{a, c\}, \{a, d\}, \{b, c\}, \{b, d\}, \{c, d\}$; equally likely

11.

Event	Probability
$\{s_1, s_2, s_3\}$	1
$\{s_1, s_2\}$	$\frac{2}{3}$
$\{s_1, s_3\}$	$\frac{2}{3}$
$\{s_2, s_3\}$	$\frac{2}{3}$
$\{s_1\}$	$\frac{1}{3}$
$\{s_2\}$	$\frac{1}{3}$
$\{s_3\}$	$\frac{1}{3}$
$\varnothing$	0

13. (a) $\frac{3}{8}$ (b) $\frac{7}{8}$ (c) $\frac{1}{2}$ (d) $\frac{1}{2}$ **15.** no
17. (a) $P(\{1\}) = P(\{2\}) = \cdots = P(\{12\}) = \frac{1}{12}$
(b) $P(\{0\}) = P(\{1\}) = \cdots = P(\{9\}) = \frac{1}{10}$ (c) $\frac{7}{120}$
19. $P(2) = \frac{1}{36}$ $P(3) = \frac{2}{36}$ $P(4) = \frac{3}{36}$ $P(5) = \frac{4}{36}$ $P(6) = \frac{5}{36}$ $P(7) = \frac{6}{36}$
$P(8) = \frac{5}{36}$ $P(9) = \frac{4}{36}$ $P(10) = \frac{3}{36}$ $P(11) = \frac{2}{36}$ $P(12) = \frac{1}{36}$
21. $P(\{\text{black}\}) = \frac{2}{3}$ $P(\{\text{red}\}) = \frac{1}{3}$ **23.** (a) $\frac{42}{129}$ (b) $\frac{15}{129}$ **25.** (a) $\frac{80}{350}$
(b) $\frac{23}{70}$ (c) $\frac{26}{280}$

Exercise Set 7.3, page 260

1. (a) .7 (b) 0 (c) .8 **3.** (a) .7 (b) .9 (c) .35 **5.** .1
7. (a) .7 (b) 0 (c) .6 (d) .7 (e) .3 **9.** .96
11. (a) .5 (b) .9 (c) .4 (d) 0 (e) .2 (f) .7 **13.** .4
15. (a) .3 (b) .7 (c) .3
17. (a) 1 to 3 for, 3 to 1 against (b) 3 to 10 for, 10 to 3 against (c) 1 to 17 for, 17 to 1 against (d) 21 to 5 for, 5 to 21 against
19. (a) $\frac{7}{8}$ (b) $\frac{3}{11}$ **21.** (a) $\frac{6}{25}$ (b) $\frac{3}{100}$

Exercise Set 7.4, page 265

1. $\frac{1}{6}$ **3.** (a) $\frac{1}{720}$ (b) $\frac{1}{6}$ (c) $\frac{1}{30}$ **5.** $\frac{21}{128} \cong .164$ **7.** $\frac{247}{256} \cong .965$
9. (a) $\frac{C_{13,5}}{C_{52,5}} \cong .000495$ (b) $\frac{4 \cdot C_{13,5}}{C_{52,5}} \cong .002$ **11.** (a) $\frac{15}{36}$ (b) $\frac{1}{12}$ (c) $\frac{1}{2}$
13. (a) $\frac{C_{20,6} \cdot C_{30,4}}{C_{50,10}} \cong .1034$ (b) $\frac{C_{30,10}}{C_{50,10}} \cong .0029$ (c) $1 - \frac{C_{30,10}}{C_{50,10}} \cong .997$
15. (a) 77 to 25 (b) 220 to 1

Exercise Set 7.5, page 280

1. (a) $\frac{1}{2}$ (b) $\frac{2}{7}$ (c) no; $P(A \cap B) \neq P(A)P(B)$ (d) no; $P(A \cap B) \neq 0$
3. 0.1 **5.** $\frac{2}{33}$ **7.** $\frac{2}{3}$ **9.** (a) $\frac{1}{5}$ (b) $\frac{1}{3}$ **11.** $\frac{2}{25}$
13. (a) $P(A)P(B) = \frac{1}{2} \cdot \frac{1}{2} = \frac{1}{4} = P(A \cap B)$ (b) $P(C)P(D) = \frac{3}{8} \cdot \frac{1}{4} = \frac{3}{32} \neq P(C \cap D) = \frac{1}{8}$
15. (a) $P(E|H) = \frac{917}{926}$, $P(E) = \frac{951}{1000}$ (b) no **19.** (a) .3 (b) .6
21. (a) $P(E) \cdot P(F) = \frac{1}{4} \cdot \frac{3}{13} = \frac{3}{52} = P(E \cap F)$ (b) $P(E) \cdot P(F') = \frac{1}{4} \cdot \frac{10}{13} = \frac{5}{26} = P(E \cap F')$
23. (a) $P(E) \cdot P(F) = \frac{1}{2} \cdot \frac{2}{3} = \frac{1}{3} = P(E \cap F)$ (b) $P(E') \cdot P(F) = \frac{1}{2} \cdot \frac{2}{3} = \frac{1}{3} = P(E' \cap F)$
25. (a) $P(E) \cdot P(F) = \frac{1}{6} \cdot \frac{1}{6} = \frac{1}{36} = P(E \cap F)$ (b) $P(E') \cdot P(F') = \frac{5}{6} \cdot \frac{5}{6} = \frac{25}{36} = P(E' \cap F')$

27. (a) $\frac{336}{720}$ (b) $\frac{576}{720}$ (c) $\frac{336}{720}$ (d) $\frac{672}{720}$ **29.** (a) .60125 (b) .64 (c) $\frac{25}{41}$
31. $(1 - 10^{-10})^{22}$

Exercise Set 7.6, page 288

1. (a) $\frac{9}{22}$ (b) $\frac{9}{22}$ (c) $\frac{2}{11}$ **3.** $\frac{12}{13}$ **5.** $\frac{2}{3}$ **7.** 966 **9.** $\frac{1}{3}$ **11.** $\frac{2}{3}$

Supplementary Exercises, page 291

1. .1 **3.** .5 **5.** $\frac{3}{8}$ **7.** (a) no; $P(A)P(B) \neq P(A \cap B)$ (b) no; $P(A \cap B) \neq 0$
9. no; if A and B are mutually exclusive $P(A \cup B) > 1$. This is impossible.
11. (a) $\frac{1}{9}$ (b) $\frac{1}{6}$ (c) 1 to 8 in favor of
13. (a) 1 to 5 (b) \$5 (c) no; you will lose in the long run. See answer to (b)
15. $\frac{2 \cdot 5!\,3!}{8!} = \frac{1}{28}$ **17.** (a) $\frac{1}{1000}$ (b) $\frac{1}{1000}$ (c) $\frac{1}{1000} \cdot \frac{1}{1000} = \frac{1}{1{,}000{,}000}$ (d) $\frac{1}{1000}$
(e) part (c) requires that a specific number be drawn on two consecutive days; part (d) only requires that the number selected on the second day match the preceding day's selection.
19. (a) $\frac{3750}{6000}$ (b) $\frac{2250}{6000}$ (c) $\frac{5050}{6000}$ (d) $\frac{3000}{5050}$ (e) $\frac{3000}{3750}$
21. $\frac{1}{8}$ **23.** $\frac{11}{20}$ **25.** $\frac{8}{33}$ **27.** 0 **29.** $\frac{5}{11}$

Chapter Test, page 292

1. (a) yes (b) no; a probability cannot be negative (c) no; $P(s) \neq 1$
2. (a) 1 (b) $\frac{1}{2}$ (c) $\frac{1}{3}$ (d) $\frac{1}{4}$ (e) no; $P(A)P(B) \neq P(A \cap B)$ (f) no; $P(A \cap B) \neq 0$
(g) 3 to 1 (h) 1 to 3
3. (a) $\frac{5}{8}$ (b) $\frac{3}{4}$ **4.** $\frac{1}{90}$
5. (a) $\frac{15{,}000}{50{,}000} = \frac{3}{10}$ (b) $\frac{25{,}000}{50{,}000} = \frac{1}{2}$ (c) $\frac{10{,}000}{50{,}000} = \frac{1}{5}$ (d) $\frac{33{,}000}{50{,}000} = \frac{33}{50}$ (e) $\frac{20{,}000}{33{,}000} = \frac{20}{33}$
(f) $\frac{20{,}000}{25{,}000} = \frac{4}{5}$ (g) d_2; because $P(D_2|A) > P(D_1|A)$ and $P(D_2|A) > P(D_3|A)$
6. (a)

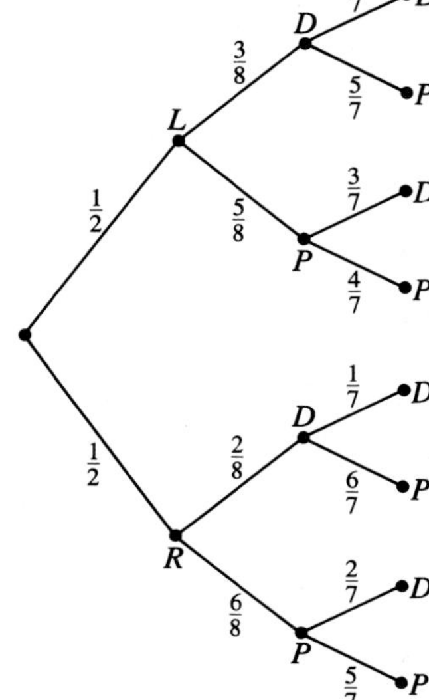

(b) $\frac{31}{52}$ (c) $\frac{2}{5}$

CHAPTER EIGHT

Exercise Set 8.1, page 301

1. (a)

x	1	2	3	4	5	6
$P(X = x)$	$\frac{1}{6}$	$\frac{1}{6}$	$\frac{1}{6}$	$\frac{1}{6}$	$\frac{1}{6}$	$\frac{1}{6}$

(b) $P(X \geq 1) = 1$
(c) $P(X \leq 1) = \frac{1}{6}$

3. (a)

x	0	1	2
$P(X = x)$	$\frac{1}{4}$	$\frac{1}{2}$	$\frac{1}{4}$

(b) $P(X \geq 1) = \frac{3}{4}$
(c) $P(X \leq 1) = \frac{3}{4}$

5. (a)

x	−3	−1	1	3
$P(X = x)$	$\frac{1}{8}$	$\frac{3}{8}$	$\frac{3}{8}$	$\frac{1}{8}$

(b) $P(X \geq 1) = \frac{1}{2}$
(c) $P(X \leq 1) = \frac{7}{8}$

7. (a)

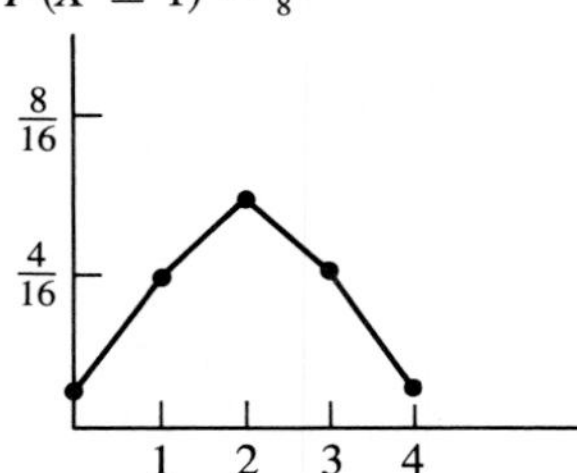

(b)

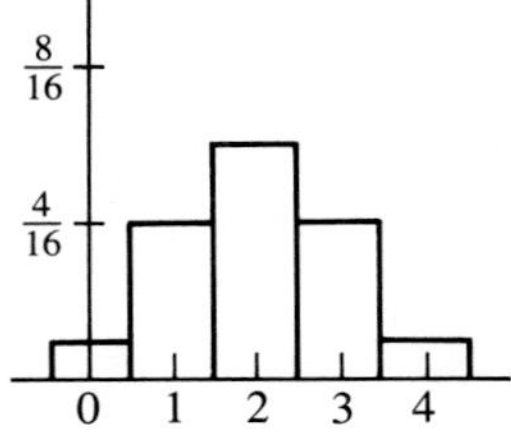

9. (a)

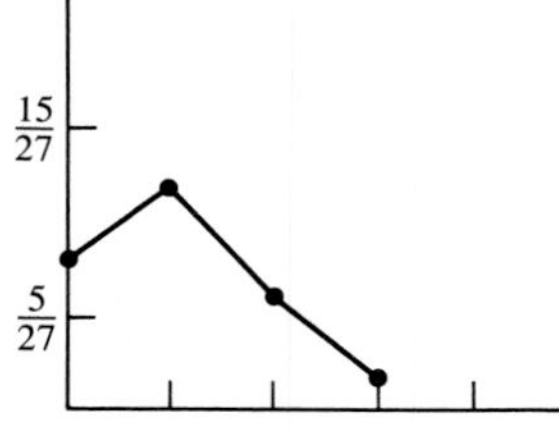

(b)

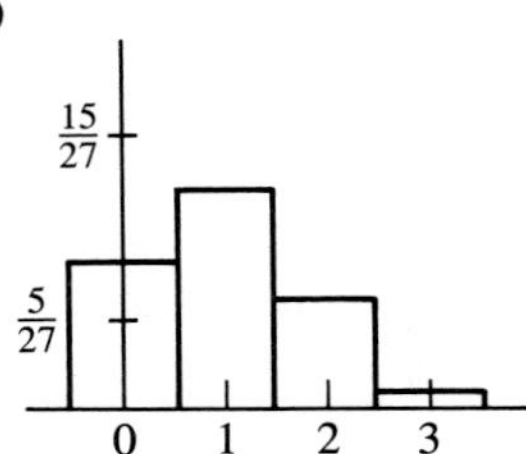

11. (a) infinite discrete
(b) finite discrete
(c) continuous

Exercise Set 8.2, page 310

1. 3.4 **3.** 2.888 **5.** 2 **7.** 1 **9.** (a) 3.6 (b) −0.2 **11.** \$4.60
13. Region A, \$2 million; Region B, \$2.2 million; bid on region B
15. (a) \$240 (b) no; the program costs \$300/wk and saves only \$500 − \$260 = \$240/wk on the average
17. (a) $E(X) \approx -0.17$
$HP = 17\%$
(b) $\dfrac{E(X)}{\text{bet}} = -0.14$
$HP = 14\%$ (c) Game (b): less of an expected loss

Exercise Set 8.3, page 319

1. $E(X) = 2.3$, $\sigma^2 = 0.81$ **3.** $E(X) = 0.224$, $\sigma^2 = 0.067$, $\sigma = 0.260$ **5.** $E(X) = 2$, $\sigma^2 = 1$ **7.** $E(X) = 1$, $\sigma^2 \cong 0.67$
9. Fig. (b) **11.** No. There might be a greater variance in the amount paid to each worker in company A.

Exercise Set 8.4, page 324

1. (a) 1 (b) 2 (c) 1.5 (d) 2.75 **3.** (a) 16 (b) −0.5 (c) 24.1 (d) 9.1
5. (a) at least 0 (b) at least 0 (c) at least $\frac{15}{16}$
7. at least 0.91 **9.** at least $\frac{15}{16}$ **11.** .632 **13.** 20 boards

Exercise Set 8.5, page 333

1. (a) $\frac{256}{625}$ (b) $\frac{256}{625}$ (c) $\frac{96}{625}$ (d) $\frac{16}{625}$ (e) $\frac{1}{625}$ (f) $\frac{624}{625}$ (g) $\frac{113}{625}$
3. (a) 0.042 (b) 0.017 (c) 0.000 (d) 0.083 (e) 0.157 (f) 0.002
5.

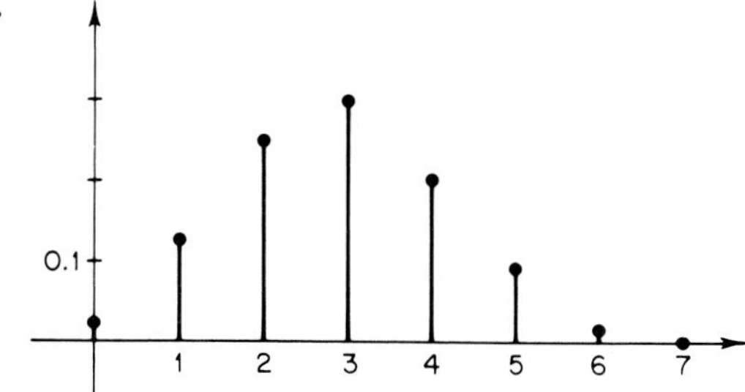

7. (a) 0.107 (b) 0.893 (c) 0.201
9. $\mu = 0.8$, $\sigma^2 = 0.64$ **11.** (a) 0.349 (b) 0.930 (c) 0.001 (d) $\mu = 9$, $\sigma^2 = 0.9$
13. $P(X \geq 2) = 0.916$, $n = 4$
$P(X \geq 2) = 0.989$, $n = 6$
15. 8 heads out of 10. **17.** (a) 0.055 (b) 0.069

Exercise Set 8.6, page 345

1. (a) (b) 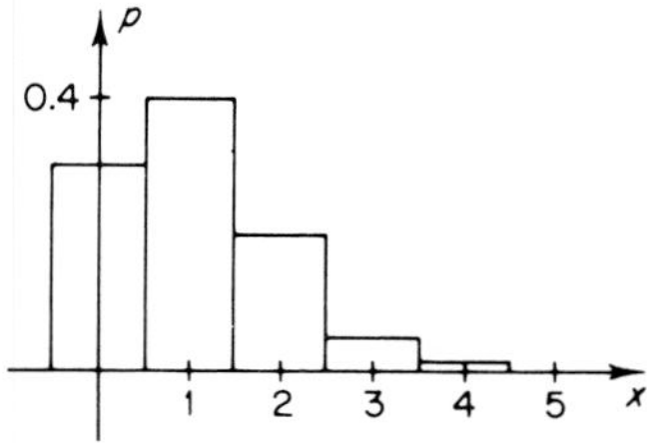 (c) 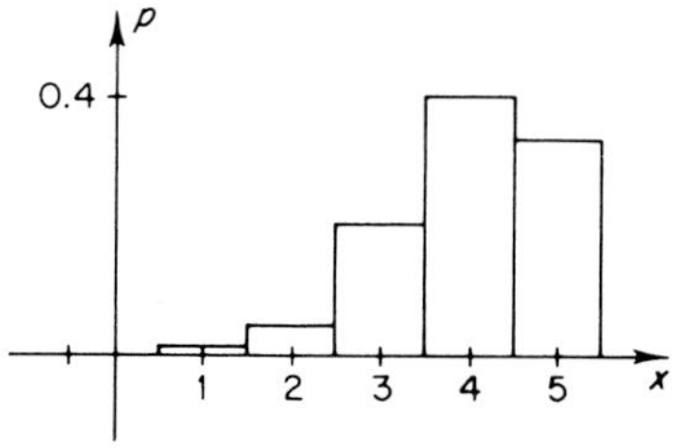 (d)

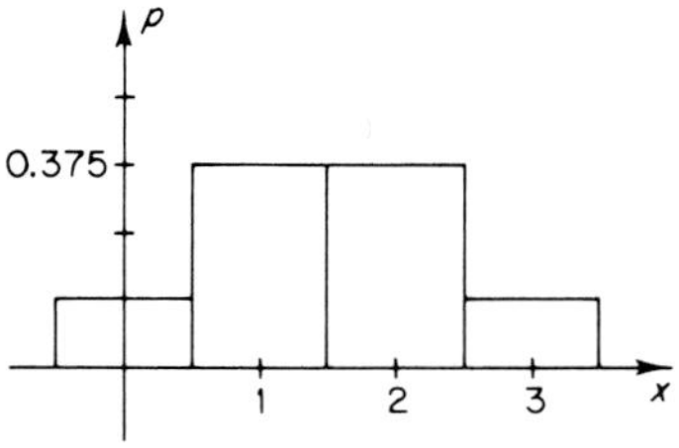

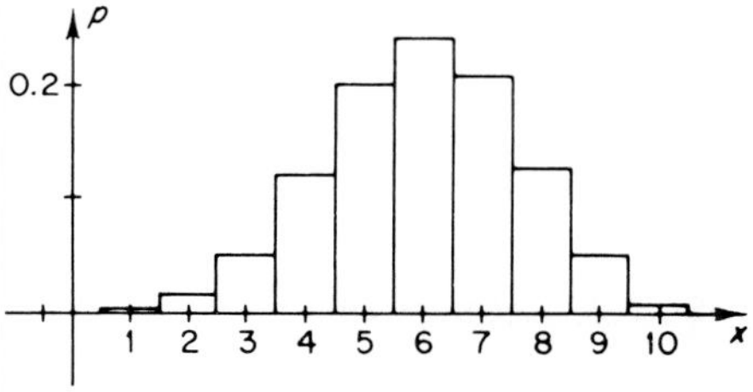

3. (a) $\mu = \frac{3}{2}$, $\sigma^2 = \frac{3}{4}$ (b) $\mu = 1$, $\sigma^2 = 0.8$ (c) $\mu = 250$, $\sigma^2 = \frac{375}{2}$ (d) $\mu = 2250$, $\sigma^2 = 225$
5. (a) 0.8944 (b) 0.2005 (c) 0.3228
7. (a) 1.15 (b) 1.07 (c) −1.92 (d) −2.42

9. (a) $P(Y \leq 10) \cong P(-.5 \leq X \leq 10.5) = .2119$ (b) $P(25 \leq Y \leq 35) \cong P(24.5 \leq X \leq 35.5) = .7445$
(c) $P(Y \geq 155) \cong P(X \geq 154.5) = .2327$
11. (a) 0.9525 (b) 0.0188 (c) 0.8164
13. (a) 0.6915 (b) 0.0928 (c) 0.3085 (d) 0.4649
15. $P(Y < 470) \cong P(-.5 \leq X \leq 469.5) = .0749$
17. (a) $P(Y > 14) \cong P(X \geq 14.5) = .0668$ (b) $P(Y > 70) \cong P(X \geq 70.5) = .0011$
(c) $P(Y > 140) \cong P(X \geq 140.5) \cong 0$
19. 219

Exercise Set 8.7, page 356

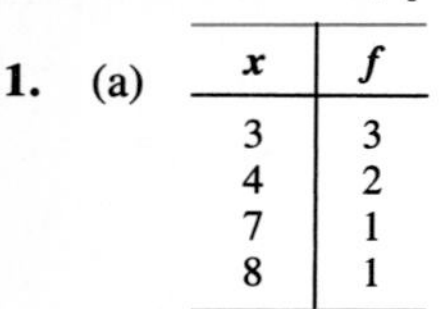

1. (a)

x	f
3	3
4	2
7	1
8	1

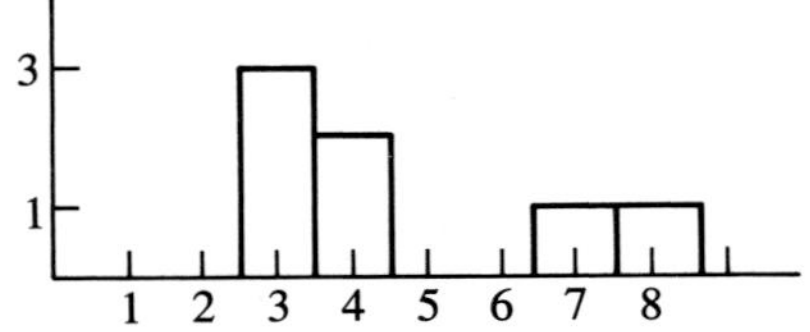

(b) $\bar{x} = 4.57$ (c) median = 4 (d) mode = 3 (e) $s^2 = 4.286$; $s = 2.07$

3. (a)

x	f
0	4
0.3	1
0.5	1
0.8	1
1.2	1
1.5	1
2.3	1

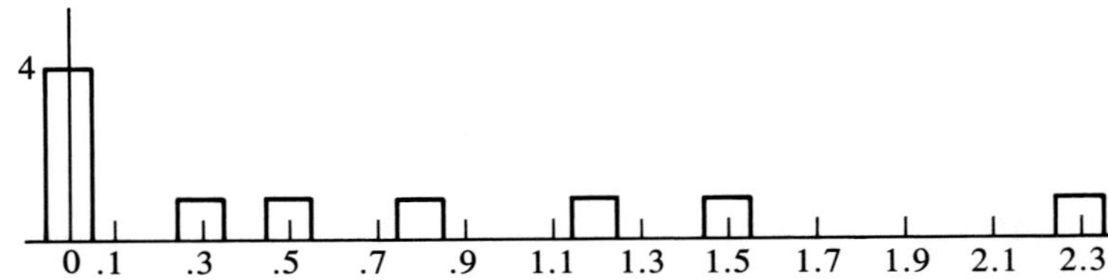

(b) 0.66 (c) 0.4 (d) 0 (e) $s^2 = 0.62$, $s = 0.79$

7. (a)

Market St.		Broad St.	
x	f	x	f
12	1	5	4
15	1	6	3
16	4	7	5
17	2	8	4
18	6	9	5
19	5	10	7
20	9	12	2
21	1		
25	1		

(b) $\bar{x}_M = 18.47$ (c) $\bar{x}_B = 8.13$ (d) yes
9. (a) $\bar{x}_A = 43.375$, $\bar{x}_B = 38.625$ (b) $s_A = 17.74$, $s_B = 26.46$
(c) median $A = 40$, median $B = 30$ (d) mode $A = 40$, mode $B = 10$ (e) A

Exercise Set 8.8, page 365

1. (a) 3.24 (b) 17.64 (c) 4.80 **3.** the data do not support the assumption
5. the data support the conclusion **7.** the data support the assumption
9. model is valid

Supplementary Exercises, page 367

1. 

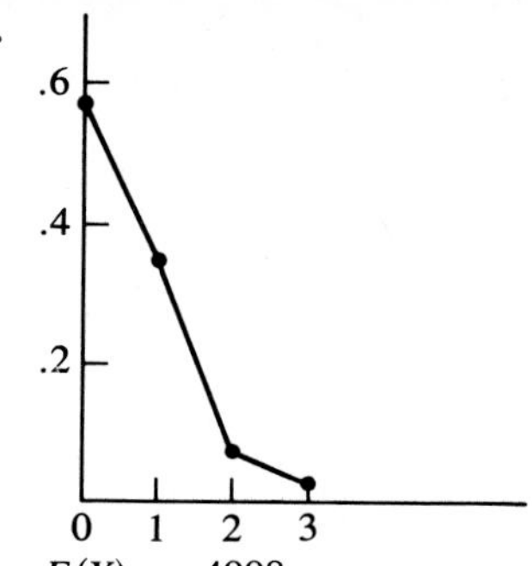

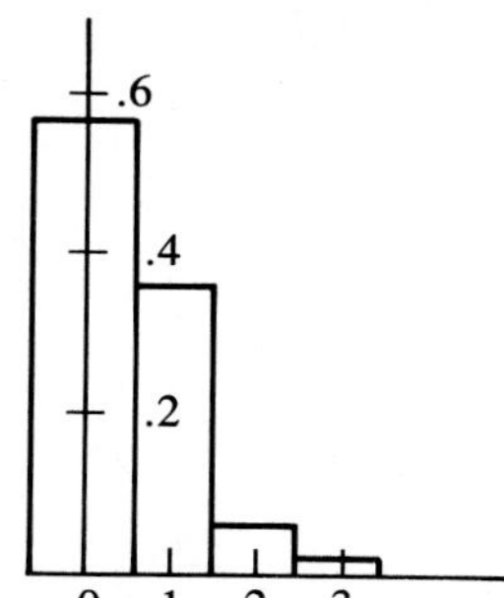

3. $E(X) = .4998$

5. 

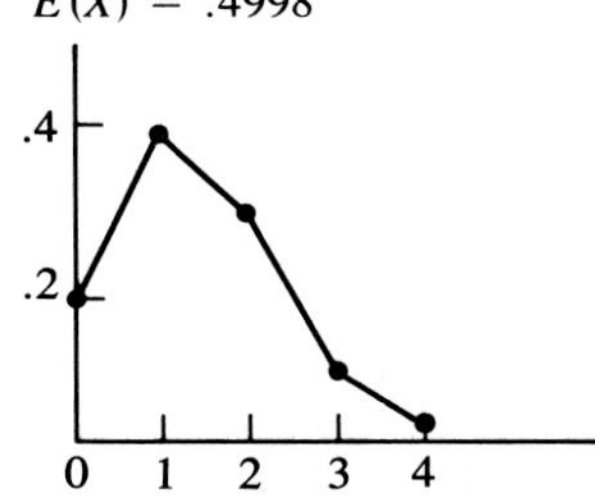

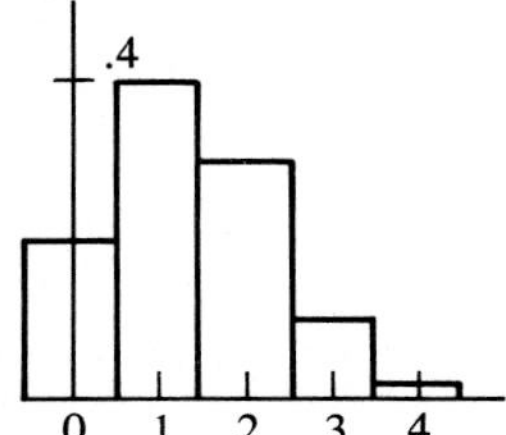

7. $E(X) = \frac{4}{3}$ 9. -0.0526, 5.26% 11. -0.0789, 7.89%

13. B 15. 1.5 17. (a) 0.1608 (b) $\mu = \frac{5}{6}$, $\sigma^2 = \frac{25}{36}$

19. (a) 0.0062 (b) .8351 (c) .1587

21. (a) $\mu = 300$ (b) $\sigma = 15$ (c) .0005 (d) .9143

23. (a)

x	f
38	1
39	2
40	3
42	3
43	3
45	2
46	1
48	4
49	1

(b) $x = 43.4$
(c) $s = 3.53$
(d) median = 43
(e) mode = 48

25. (a) 0.48 (b) 1

27. $\chi^2 = .70$ implies that the viewers regard the three channels to be equally good.

Chapter Test, page 369

1. (a) 0.9703, 0.0294, 0.0003, 1×10^{-6} (b) 0.9409
(c)

1.0
0.8
0.6
0.4
0.2
0 1 2 3

(d) $E(X) = 0.03$, $\text{Var}(X) = 0.0297$ **2.** $\mu = 3.3$, $\text{Var}(X) = 1.21$
3. (a) 0.1317 (b) 0.3292 (c) 0.8683 (d) $\frac{5}{3}$ (e) $\frac{10}{9}$
4. (a) 0.0004 (b) 0.3232 (c) 0.9996 **5.** (a) .5120 (b) .5114
6. (a) 0.5 (b) 0.5 **7.** (a) 34.083 (b) 216.5 (c) 31

CHAPTER NINE

Exercise Set 9.1, page 377

1. (a) $3360 (b) $4080 (c) $3090 (d) $3090 **3.** $4545.45
5. (a) $1023.42 (b) $1023.75 **7.** $1955.99 **9.** $528 **11.** $626.67 **15.** 2.5 years

Exercise Set 9.2, page 391

1. (a) $5618 (b) $5635.80 (c) $5632.46 (d) $5627.54 (e) $5637.48
3. (a) $1397.85 (b) $1402.76 (c) $1395.35
5. (a) $10930.83 (b) $10455.06 (c) $11948.31
7. (a) $3150.85 (b) $3108.61 (c) $3122.99 (d) $3093.92
9. (a) $1536.41 (b) $1551.54 (c) $1554.60 **11.** $1818.87

Exercise Set 9.3, page 396

1. (a) 6.17% (b) 6.14% (c) 6.09% (d) 6.18%
3. 9% per year compounded quarterly **5.** (a) 12.75% (b) 16.18%
7. 3.47% **9.** (a) $384 (b) $16 (c) $400 (d) $8\frac{1}{3}$% (e) $8\frac{1}{3}$% **11.** $1052.63
13. (a) $520.83 (b) $8\frac{1}{3}$%

Exercise Set 9.4, page 406

1. (a) $603 (b) $2,045.61 **3.** $24,500.90 **5.** $5,465,589.59 **7.** $219,199.15
9. (a) $18,850.31 (b) $795.74 **11.** (a) 5.101005 (b) 10.949721 (c) 31.371423
13. (a) $4100.20 (b) $13,677.74 (c) $83,721.17 (d) $2069.02 **15.** $18,893.30
17. $844.13 for television, $55.87 interest **21.** (a) $111.22 (b) $470.73
23. $4769.05

Supplementary Exercises, page 408

1. (a) \$10200 (b) \$200 **3.** 9% compounded continuously **5.** \$378.46 **7.** \$4557.32
9. \$7106.19 **11.** \$4377.83

Chapter Test, page 409

1. (a) \$1150 (b) \$1159.69 (c) \$1161.80 **2.** (a) \$1694.92 (b) \$1674.97 (c) \$1670.60
3. (a) 10.38% (b) 10.52% **4.** (a) \$1930 (b) 7.25%
5. \$2232.11 **6.** \$628.25 **7.** \$14,325.41 **8.** \$17,412.63
9. \$39.98 **10.** \$4516.13

CHAPTER TEN

Exercise Set 10.1, page 418

1. (a) l_{60}(male) = 8,036,408 (b) $l_{30} - l_{40}$(female) = 167,162

3. (a) $\frac{l_{60}}{l_{40}}$ (male) = .8565

(b) $\frac{l_{60}}{l_{40}}$ (female) = .8982 (**Remark:** The male and female mortality tables are not the same. According to these tables, women are expected to live longer than men.)

(c) $\frac{l_{60}}{l_0}$ (female) = .8576 (d) $\frac{l_{60}}{l_0}$ (male) = .8036 [See Remark for 3(b)]

(e) $\frac{l_{80} - l_{85}}{l_{70}}$ (female) = .2463 (f) $\frac{l_{80} - l_{85}}{l_{70}}$ (male) = .2445

5. $P = 10{,}000\frac{l_{45}}{l_{20}}(1.07)^{-25} = \$1737.53$ (male)

7. $P = 10{,}000\frac{l_{45}}{l_{20}}(1.07)^{-25} = \$1763.17$ (female)

Remark: Since women are expected to live longer than men, the company expects to pay benefits on this type of policy more often to a woman than to a man of the same age. Therefore, the premium for a woman is higher than the premium for a man of the same age.

9. $Q = 10{,}000\left[\frac{d_{65}}{l_{65}}(1.07)^{-1} + \frac{d_{66}}{l_{65}}(1.07)^{-2} + \frac{d_{67}}{l_{65}}(1.07)^{-3} + \frac{d_{68}}{l_{65}}(1.07)^{-4}\right] = \$983.55$ (male)

11. Same equation as in Exercise 9—but for female $Q = \$572.00$

Exercise Set 10.2, page 427

1. (b), (c), (f)

3. (a) row 1, column 2; −5 (b) row 1, column 2; 1 (c) row 1, column 2; −4
(d) row 1, column 2; 4

5.

Player II

$$\text{Player I}\quad \begin{matrix} & 3 & 4 \\ 3 & 6 & -7 \\ 4 & -7 & 8 \end{matrix}$$

7.

Player II

$$\text{Player I}\quad \begin{matrix} & \text{Stone} & \text{Scissors} & \text{Paper} \\ \text{Stone} & 0 & 1 & -1 \\ \text{Scissors} & -1 & 0 & 1 \\ \text{Paper} & 1 & -1 & 0 \end{matrix}$$

9. player R shows one finger, player C shows two fingers

13. firms A and B should each use television

15. (a) −2 (b) 2

Exercise Set 10.3, page 441

1. (a) $\frac{17}{12}$ (b) $\frac{33}{25}$ (c) $\frac{4}{3}$

3. $p_1 = \frac{1}{2},\ p_2 = \frac{1}{2},\ q_1 = \frac{3}{4},\ q_2 = \frac{1}{4},\ E = \frac{5}{2}$

5. $p_1 = \frac{4}{7},\ p_2 = \frac{3}{7},\ q_1 = \frac{1}{14},\ q_2 = \frac{13}{14},\ E = \frac{10}{7}$

7. $p_1 = \frac{1}{2},\ p_2 = \frac{1}{2},\ q_1 = \frac{1}{2},\ q_2 = \frac{1}{2},\ E = \frac{1}{2}$

9. $p_1 = \frac{1}{2},\ p_2 = \frac{1}{2},\ p_3 = 0,\ q_1 = \frac{3}{4},\ q_2 = \frac{1}{4},\ q_3 = 0,\ E = \frac{1}{2}$

11. $p_1 = 0,\ p_2 = \frac{4}{7},\ p_3 = \frac{3}{7},\ q_1 = 0,\ q_2 = \frac{1}{14},\ q_3 = \frac{13}{14},\ E = -\frac{4}{7}$

13. Columbus should keep going with probability .627

15. $\frac{2}{3}$ male, $\frac{1}{3}$ female

Exercise Set 10.4, page 448

3. 75% **5.** $\frac{3}{4}$ **7.** 25% of genotype *AA*, 50% of genotype *Aa*, 25% of genotype *aa*

9. 36% red, 48% pink, 16% white

Exercise Set 10.5, page 463

1.

$$\begin{array}{ll} & \text{Next state} \\ & \begin{array}{cc} 1 & 2 \end{array} \\ \begin{array}{r} \text{Present } 1 \\ \text{state } 2 \end{array} & \begin{bmatrix} \frac{2}{5} & \frac{3}{5} \\ \frac{3}{5} & \frac{2}{5} \end{bmatrix} \end{array}$$

3. (a) if the system is in state 1, the probability that at the next observation the system will be in state 2 is $\frac{3}{4}$
(b) $[\frac{1}{4}\ \ \frac{3}{4}]$ (c) $[\frac{2}{5}\ \ \frac{3}{5}]$

5. $[\frac{29}{80}\ \ \frac{51}{80}]$ **7.** $[.722\ \ .278]$ **9.** $[.320\ \ .258\ \ .422]$

11. (a) no power of P has all positive entries (b) P^2 has all positive entries

13. (a) .3 (b) .167

15. 279 spaces at Kennedy, 115 spaces at LaGuardia, 107 spaces at Newark

17. (a) $\begin{bmatrix} \frac{1}{3} & \frac{2}{9} & \frac{4}{9} \\ \frac{5}{8} & \frac{1}{8} & \frac{1}{4} \\ 0 & \frac{5}{8} & \frac{3}{8} \end{bmatrix}$ (b) yes, row sums equal 1

19. (a)

Transition from state 0	Number of times	Transition from state 1	Number of times
0 to 0	8	1 to 0	7
0 to 1	7	1 to 1	6

(b) $\begin{bmatrix} \frac{8}{15} & \frac{7}{15} \\ \frac{7}{13} & \frac{6}{13} \end{bmatrix}$ (c) $[\frac{15}{28}\ \ \frac{13}{28}]$ (d) yes

Supplementary Exercises, page 467

1. $\frac{l_{80}}{l_{50}}(\text{male}) = .3483$

3. $Q = 10{,}000\left[\frac{d_{50}}{l_{50}}(1.07)^{-1} + \frac{d_{51}}{l_{50}}(1.07)^{-2} + \frac{d_{52}}{l_{50}}(1.07)^{-3}\right]$ (female) = \$143.55

5. (b) and (c) **7.** $p_1 = \frac{1}{4},\ p_2 = 0,\ p_3 = \frac{3}{4},\ q_1 = \frac{7}{8},\ q_2 = \frac{1}{8},\ q_3 = 0$

9. 0% **11.** 1 **13.** $[\frac{13}{22},\ \frac{9}{22}]$

CHAPTER TEST, page 468

1. (a) $1 - \frac{l_{60}}{l_{40}}(\text{female}) = .1018$ (b) $1 - \frac{l_{60}}{l_{40}}(\text{male}) = .1435$

(c) the mortality tables indicate that women are expected to live longer than men.

2. $Q = 10{,}000\left[\frac{d_{40}}{l_{40}}(1.06)^{-1} + \frac{d_{41}}{l_{40}}(1.06)^{-2} + \frac{d_{42}}{l_{40}}(1.06)^{-3} + \frac{d_{43}}{l_{40}}(1.06)^{-4}\right]$

$Q(\text{female}) = \$98.31$

$Q(\text{male}) = \$122.60$

Ruth's premium is smaller because it is more likely that she will survive 4 years than it is that Bob will.

3. The saddle point is 3. The value of the game is 3.

4. $p_1 = 0,\ p_2 = \frac{3}{4},\ p_3 = \frac{1}{4},\ q_1 = \frac{3}{4},\ q_2 = \frac{1}{4}$

5. (a)

$$\begin{array}{cc} & \text{II} \\ & \begin{array}{ccc} \text{4D} & \text{3H} & \text{7S} \end{array} \\ \text{I} \begin{array}{c} \text{5H} \\ \text{10C} \\ \text{6S} \end{array} & \begin{bmatrix} -9 & -8 & 12 \\ 14 & 13 & -17 \\ 10 & 9 & -13 \end{bmatrix} \end{array}$$

(b) no

6. 56.25% red; 37.5% pink; 6.25% white

7.

State 2 → (.4) State 1 → (.25) State 1; (.75) State 2

State 2 → (.6) State 2 → (.4) State 1; (.6) State 2

The matrix state two observations later is [.34 .66]

CHAPTER ELEVEN

Exercise Set 11.1, page 478

1. (a) 3 (b) 7 (c) 6 (d) 3 (e) 8

3. (a) 5 (b) 0 (c) 12 (d) -4 (e) $a^2 - 4$ (f) $x^2 - 4x$ (g) $(x + h)^2 - 4$

5. (a) 1 (b) $-\frac{3}{5}$ (c) $-\frac{3}{2}$ (d) $\frac{3}{b-2}$ (e) $\frac{3}{b+5}$ (f) $\frac{3}{b+h-2}$ (g) $\frac{3}{-x-2}$ (h) undefined

7. (a) 5 (b) -4 (c) 1 (d) $x^4 + 1$ (e) $x^2 + 1$ **9.** all real numbers

11. all real numbers except 3 **13.** all real numbers less than or equal to 2

15. range = all real numbers

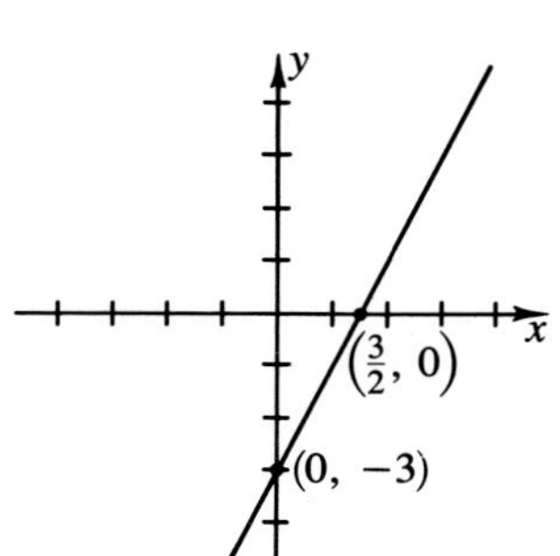

17. range $= \{y \mid y \le 4\}$

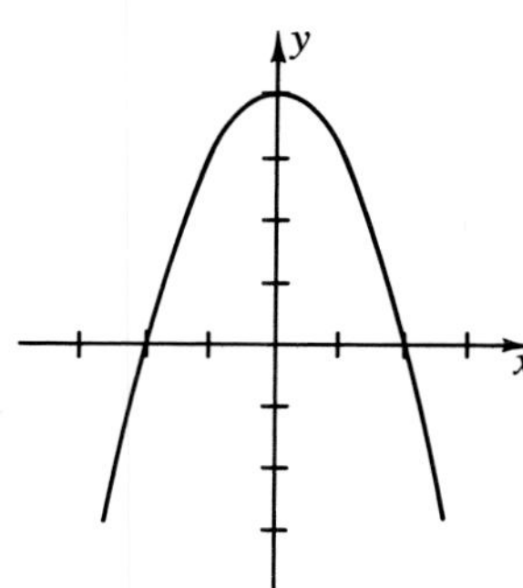

19. range = $\{y \mid y \neq 0\}$

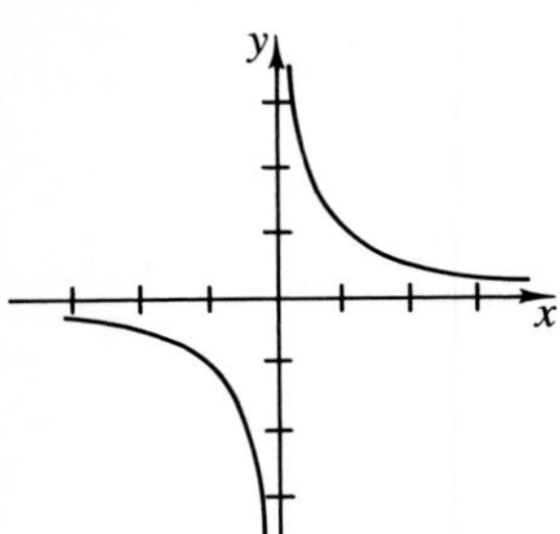

21. range = $\{y \mid y \geq 1\}$

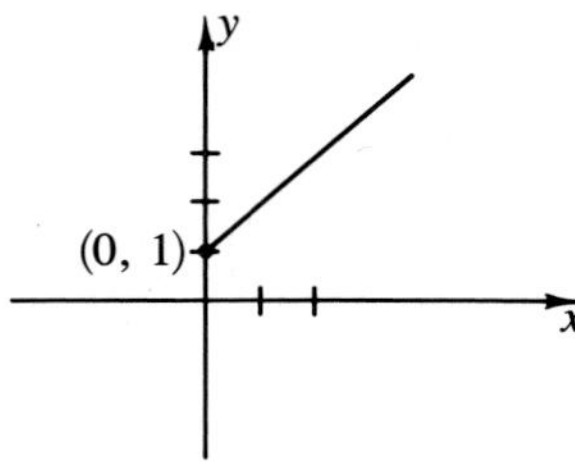

23. range = $\{-1, 1\}$

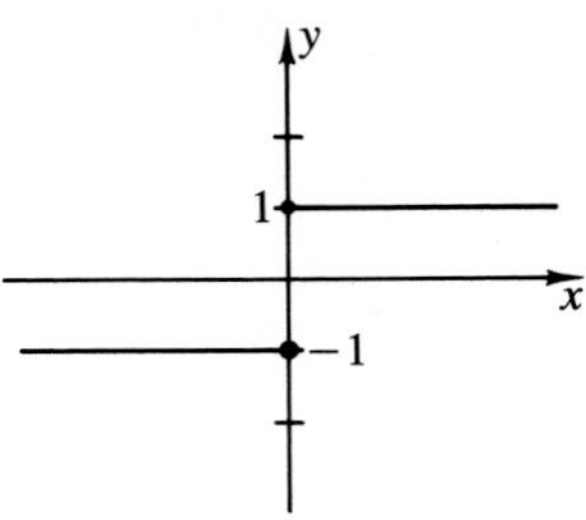

25. no **27.** no **29.** yes

31. (a) $C = 0.5 + 0.3(w - 1)$. $w = 1, 2, \ldots, 10$
(b)

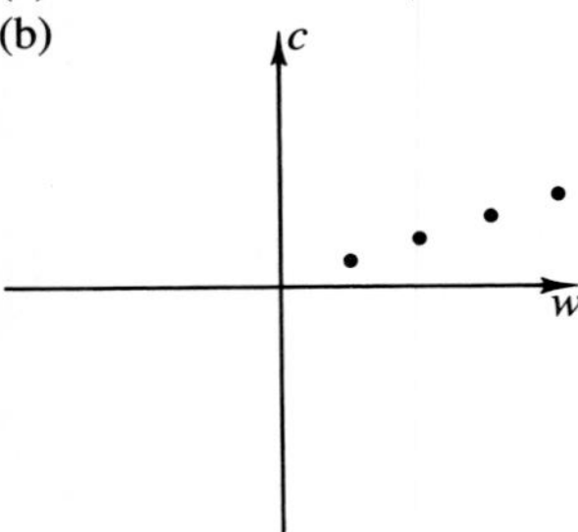

(c) \$2.90

33. (a) $C = 120 + 15r$, $r = 0, 1, \ldots, 6$ (b) $r = 0, 1, \ldots, 6$ (c) \$210

Exercise Set 11.2, page 488

1.

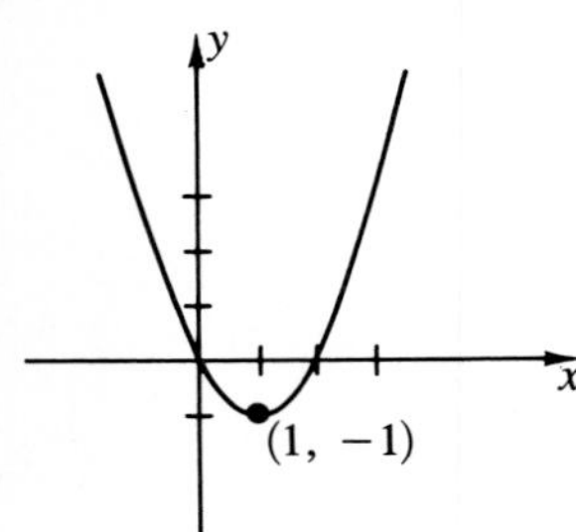

3.

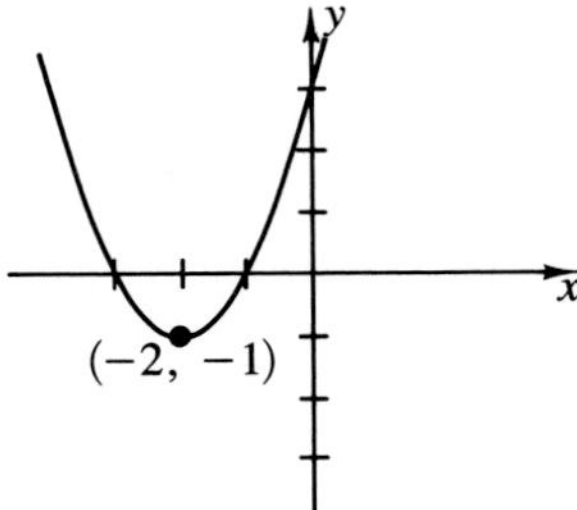

5.

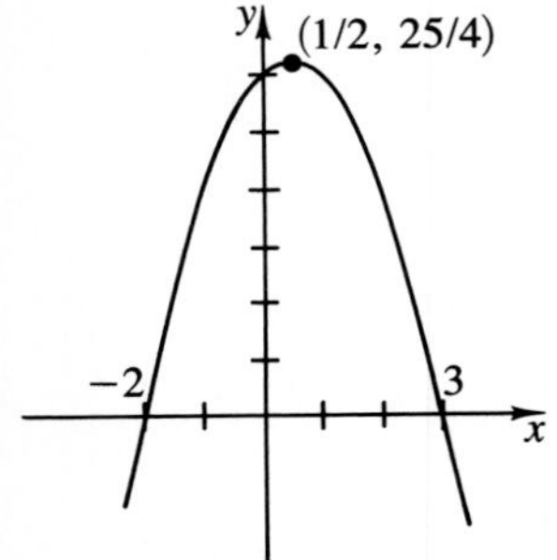

7.

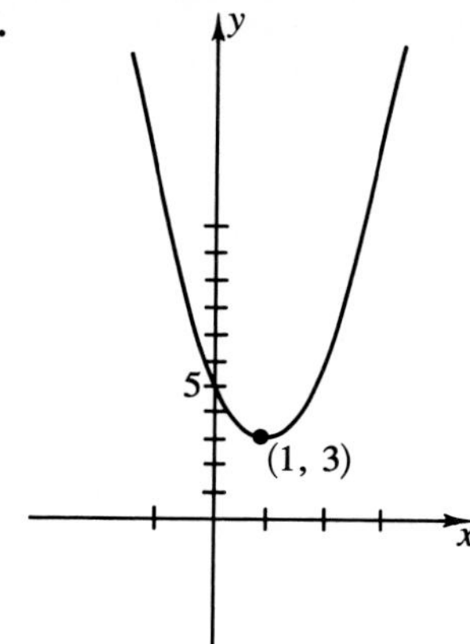

9. $\frac{1}{2}, -2$ **11.** $\frac{3}{2}$ **13.** 1 **15.** 0 **17.** 2

19.

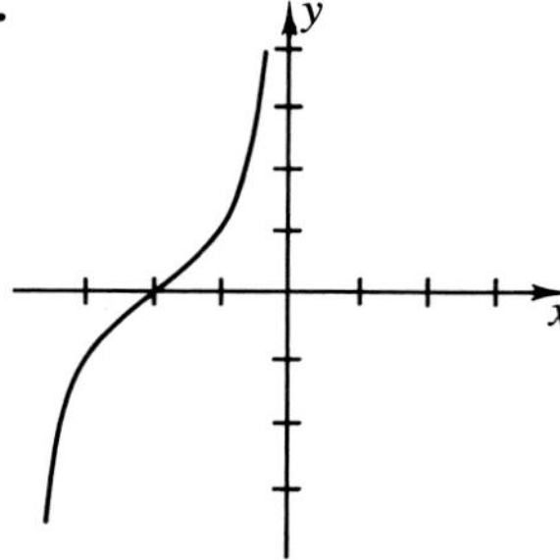

21.

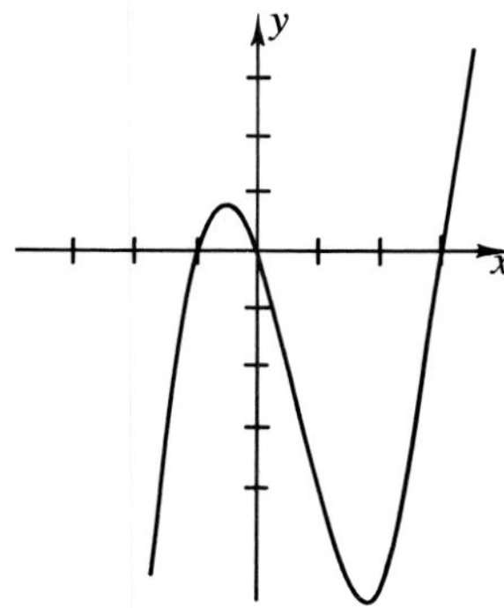

23.

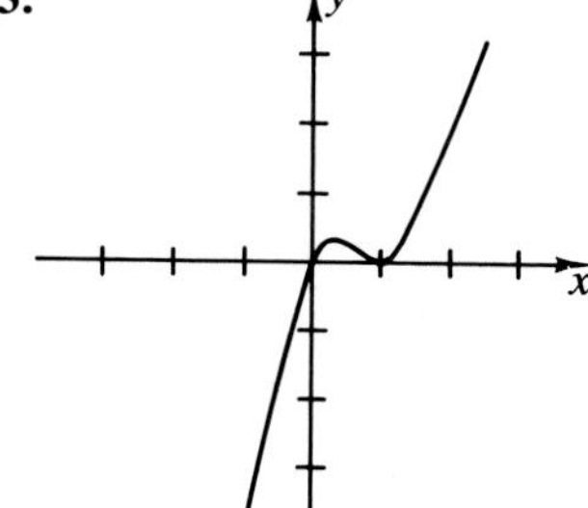

25. (a) \$1000 (b) \$1190 (c) \$5950 (d) \$2,000,000 **27.** $\dfrac{A = w^2 + 49w + 16}{2}$

29. $V = \left(20 - \dfrac{2x}{3}\right)x^2$

31. (a) \$8750 (b) $R(x) = (7 + x)(1250 - 50x)$ (c) $P(x) = 900x - 50x^2$
(d) \$16; attendance = 800; profit = \$4050

Supplementary Exercises, page 490

1. (a) −4 −3 −2 −1 0 1 2 3 4

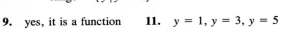

(b) −2 −1 0 1 2

(c) −2 −1 0 1 2 3 4 5 6 7

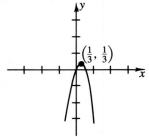

3. (a) −1 (b) 10 (c) 5 (d) $\begin{cases} 4x + 1 & \text{if } x \le 1 \\ 4x^2 + 1 & \text{if } x > 1 \end{cases}$

5.

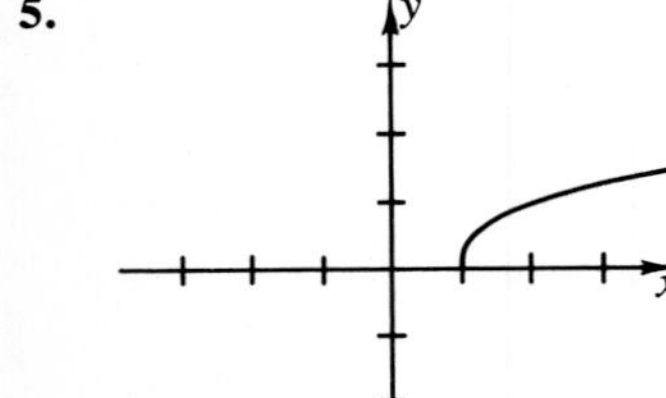

domain = $\{x \mid x \ge 1\}$
range = $\{y \mid y \ge 0\}$

7.

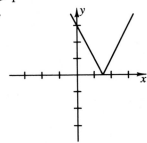

domain = all real numbers
range = $\{y \mid y \ge 0\}$

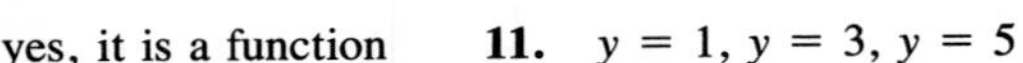

9. yes, it is a function **11.** $y = 1, y = 3, y = 5$ **13.** x-intercepts: 0 and −5; y-intercept: 0

15. $\left(\frac{1}{3}, \frac{1}{3}\right)$

17. $a = 2, b = 4, c = 0$

19. (a) revenue = $pq = p(200 - 10p)$ (b) \$960 (c) \$910 (d) $p = \$10$

Chapter Test, page 491

1. (a) $f(-2) = -3; f(3) = -7$ (b) domain: all x; range: $y \le 1$

(c)

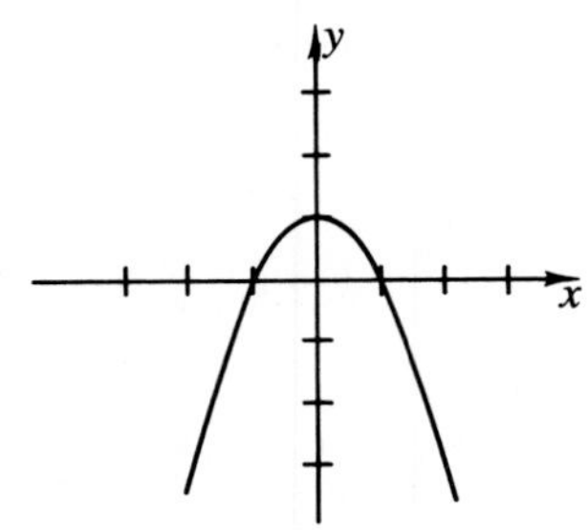

2. (a) x-intercepts: $\frac{1}{2}$, -2; y-intercept: -2 (b) x-intercepts: $-\frac{1}{2}$, -1; y-intercept: 1

3.

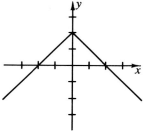

domain: all x
range: $y \leq 2$

4.

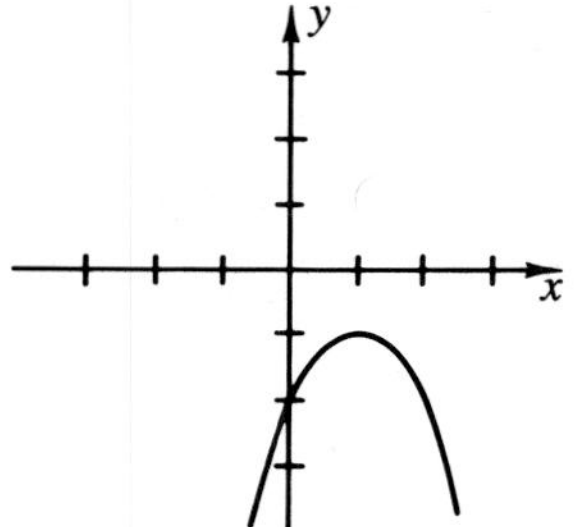

5. $x = 8, x = -2$ **6.** 0

7. (a) $C(s) = \begin{cases} 0.20s & 0 < s < 300 \\ 60 + 0.35(s - 300) & 300 \leq s < 400 \\ 95 + 0.60(s - 400) & 400 \leq s \end{cases}$

(b) \$110

(c)

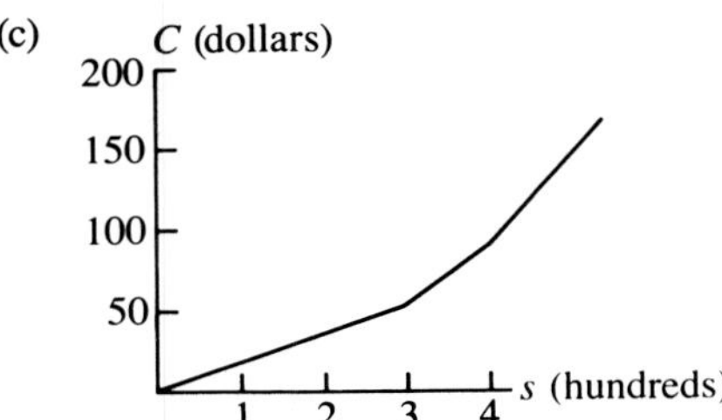

CHAPTER 12

Exercise Set 12.2, page 503

1. (a) 3 (b) 3 (c) 3 (d) The graph of $y = 3x + 4$ is a line whose slope is 3

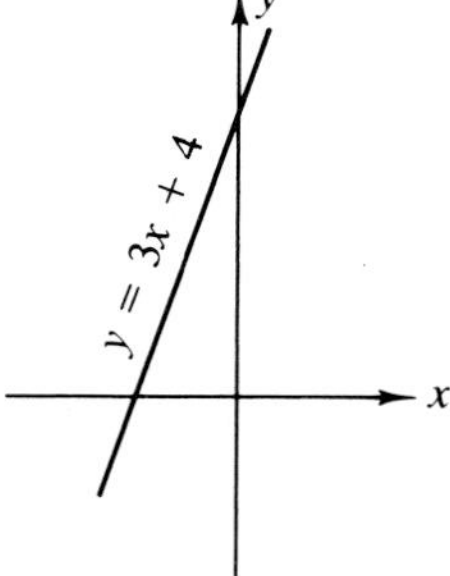

3. −4 **5.** 0 **7.** 19 **9.** $\frac{8}{3}$

11. (a) 432 feet (b) 144 feet/second (c) 96 feet/second (d) 128 feet/second

13. 6 **15.** 24

17. (a) approx. 4202 people/year (b) approx. −3212 people/year (c) more rapidly between 1960 and 1980

19. (a) 1; 0.75; 0.5; 0.25 (b) law of diminishing returns

Exercise Set 12.3, page 514

1. (a) 1 3 (e) 1

(b) 1 3 (f) 1

(c) 1 3 (g) 3

(d) 1 3 (h) 3

3. (a)

x	2.6	2.7	2.8	2.9	2.99	2.999	3.001	3.01	3.1	3.2	3.3	3.4
$f(x) = 2x - 3$	2.2	2.4	2.6	2.8	2.98	2.998	3.002	3.02	3.2	3.4	3.6	3.8

(b) 3

5. (*a*)

x	−1	−0.1	−0.01	−0.001	−0.0001	0.0001	0.001	0.01	0.1	1
$f(x) = \frac{\lvert x \rvert}{2x}$	−0.5	−0.5	−0.5	−0.5	−0.5	0.5	0.5	0.5	0.5	0.5

(*b*) Limit does not exist

7. 6 **9.** 51 **11.** limit does not exist **13.** 0 **15.** 0

17. 4; 4; 4 $\lim_{x \to 1} f(x) = 4$

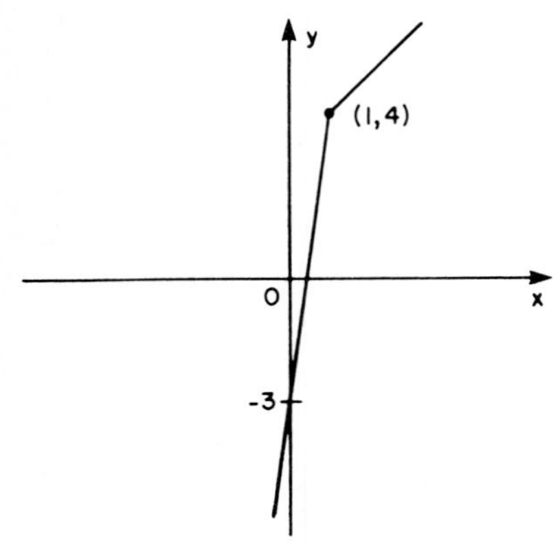

19. $\lim_{t \to b^-} f(t) = c$, $\lim_{t \to b^+} f(t) = 0$, $\lim_{t \to b} f(t)$ does not exist

21.

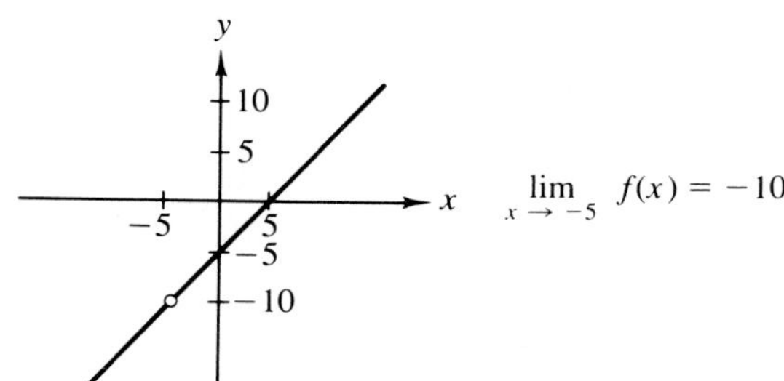

$\lim_{x \to -5} f(x) = -10$

23. $\lim_{x \to 3} \frac{|x-3|}{x-3}$ does not exist

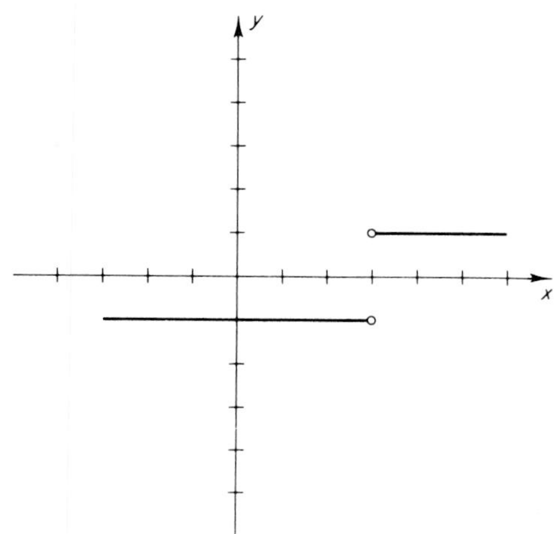

25. 3 **27.** does not exist **29.** $9000

Exercise Set 12.4, page 523

1. $-\infty, \infty$, does not exist **3.** $-\infty, \infty$, does not exist **5.** $\infty, -\infty$, does not exist
7. $-\infty, \infty$, does not exist **9.** $\frac{1}{2}, \frac{1}{2}, \frac{1}{2}$ **11.** 0 **13.** 5 **15.** -3 **17.** $-\infty$ **19.** ∞ **21.** 0
23. 8 **25.** -3 **27.** $-\infty$ **29.** 2 **31.** 1 **33.** $\frac{1}{2}$ **35.** 4 **37.** $y = 1, x = -2$
39. $y = 0, x = 2$ and $x = -1$ **41.** $y = 0, x = -4$

Exercise Set 12.5, page 528

1. (a) yes (b) yes (c) no (d) no (e) no (f) no (g) no (h) yes
3. yes: $f(6) = 12 = \lim_{x \to 6} f(x)$ **5.** yes, all polynomials are continuous **7.** no **9.** no **11.** yes
13. $x = 3$ **15.** no points of discontinuity **17.** no points of discontinuity **19.** $x = 0$
21. no, when the stock falls below 100, it will jump to 500 when the order comes in

Exercise Set 12.6, page 537

1. (a) $-\frac{2}{3}$ (b) 1 (c) 0 (d) -1 **3.** 4 **5.** $\frac{2}{3}$ **7.** -6 **9.** 5 **11.** 24 **13.** -1
15. $10x$ **17.** $2x + 4$ **19.** $\frac{-4}{x^2}$ **21.** (a) 1 (b) 3 (c) -1 (d) 5 **23.** $(3, 13)$; $y = 9x - 14$

25. $D_x(5x + 11) = 5$
if $y = 5x + 11$, then $\frac{dy}{dx} = 5$
if $y = 5x + 11$, then $y' = 5$
$D(5x + 11) = 5$
$\frac{d}{dx}(5x + 11) = 5$

27. $D_x(3/x) = -3/x^2$
if $y = 3/x$, then $y' = -3/x^2$
if $f(x) = 3/x$, then $f'(x) = -3/x^2$
$D(3/x) = -3/x^2$
$\frac{d}{dx}(3/x) = -3/x^2$

29. $D_x(3x - x^3) = 3 - 3x^2$
if $y = 3x - x^3$, then $dy/dx = 3 - 3x^2$
if $y = 3x - x^3$, then $y' = 3 - 3x^2$
$D(3x - x^3) = 3 - 3x^2$
$\frac{d}{dx}(3x - x^3) = 3 - 3x^2$

31. P, R, S, U, W **33.** yes, yes **35.** no, yes **37.** no, no **39.** no, yes
41. (a) (b) yes (c) no

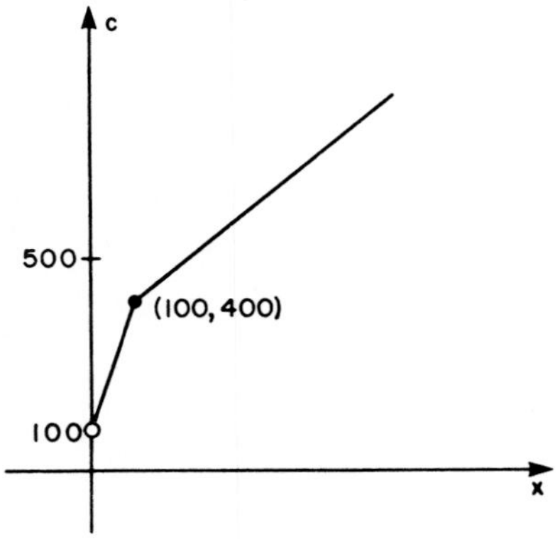

Exercise Set 12.7, page 545

1. $6x^5$ **3.** $12x^3$ **5.** $9x^2 - 4x + 1$ **7.** $\frac{5}{2}x^{-1/2}$ **9.** $-16/x^2$ **11.** $.12x^{-.97}$ **13.** $-4/x^3 + 9x^2$
15. $-12/x^{3/2}$ **17.** $11x^{10}$ **19.** $30x^4 - 16x^3 + 6x - 2$ **21.** $8x^3 + \frac{7}{2}x^{5/2} + \frac{1}{2}x^{-1/2} + 2$ **23.** $6x^5$
25. $\dfrac{100}{(100 - x)^2}$ **27.** $\dfrac{-2x - 3 - 4\sqrt{x}}{2\sqrt{x}\,(2x - 3)^2}$ **29.** $\dfrac{-6x^4 + 8x^3 - 6x^2 + 6x - 2}{(2x^3 + 1)^2}$ **31.** $\dfrac{-2x}{(x^2 + 2)^2}$
33. $\dfrac{-1}{(x + \sqrt{x})^2}\left(1 + \dfrac{1}{2\sqrt{x}}\right)$ **35.** $f'(2) = 10, f'(0) = -2, f'(-2) = -14$
37. (a) -1 (b) 1 (c) -3 (d) 3 **39.** $(3, 13)$ **41.** $(2, -15), (-1, 12)$
43. (a) $6t$ (b) $2x$ (c) $7x^6$ (d) $4\pi r^2$ **45.** (a) $\frac{3}{2}x^{-1/2} + 6x^2$ (b) $-6x^{-3} + 2$

Exercise Set 12.8, page 552

1. $\dfrac{dr}{dt} = \dfrac{dr}{ds}\cdot\dfrac{ds}{dt}$ **3.** (a) $\dfrac{-6}{(x + 5)^3}$ (b) $\dfrac{-6}{(x + 5)^3}$ (c) answer (a) = answer (b)
5. (a) $\dfrac{-6x}{(3x^2 + 4)^2}$ (b) $\dfrac{-6x}{(3x^2 + 4)^2}$ (c) answer (a) = answer (b)
7. $18(9x^2 + 2x)^{17}(18x + 2)$ **9.** $-\frac{1}{2}(2 - x)^{-1/2}$ **11.** $\frac{1}{3}(x^2 - x)^{-2/3}(2x - 1)$ **13.** $-6(2x^2 - 3x)^{-7}(4x - 3)$
15. $(14x^2 - 8x)(7x^3 - 6x^2 + 2)^{-1/3}$ **17.** $\dfrac{15x^2 + 16x}{2\sqrt{3x + 4}}$ **19.** $120x^2(3x^4 + 2)^9(2x^3 - 3)^{19}(5x^4 - 3x + 2)$
21. $\frac{7}{2}[(x + 4)(x + 11)^3]^{-1/2}$ **23.** $\dfrac{-3x^2 - 2}{2\sqrt{x}(x^2 - 2)^2}$ **25.** $\dfrac{x^4 + 2x^3 + 5x^2 - 10x - 7}{(x^2 + x + 1)^2}$
27. $\frac{1}{3}(x^2 + 2x + 1000)^{-2/3}(2x + 2)$ **29.** $-x(x^2 - 9)^{-3/2}$ **31.** $\dfrac{1}{4}\left(\dfrac{x}{1 + x^2}\right)^{-3/4}\dfrac{1 - x^2}{(1 + x^2)^2}$ **33.** 37.04
35. $(2, -\frac{1}{3}), (-2, \frac{7}{3})$ **37.** \$403,225 per day

Supplementary Exercises, page 556

1. (a) 0 (b) $b + a - 1$ (c) 2 **3.** m
5. (a) 26 miles/hour (b) 44 miles/hour (c) 38 miles/hour **7.** 20 gallons/day
9. (a) \$1212 million/liter (b) \$612 million/liter **11.** 1 **13.** ∞ **15.** ∞ **17.** does not exist
19. 0 **21.** ∞ **23.** 3
25. $\lim_{x\to 2^-} f(x) = \lim_{x\to 2^+} f(x) = \lim_{x\to 2} f(x) = 5$

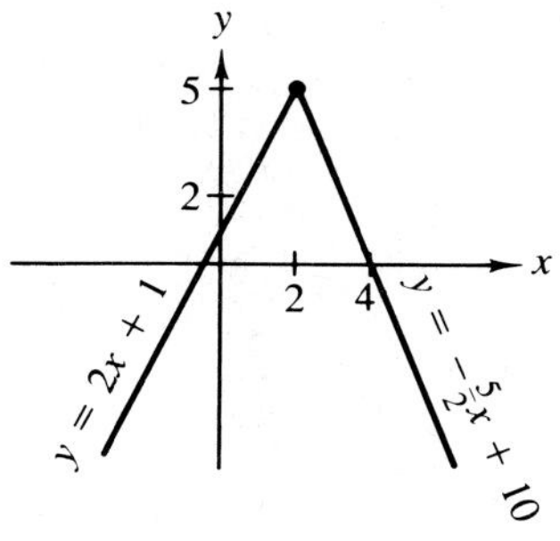

27. $\lim_{x\to 1^-} f(x) = 5$, $\lim_{x\to 1^+} f(x) = -\infty$ **29.** $f(x)$ is continuous at a, b: $f(x)$ is not continuous at c, d **31.** yes

33. (a) $C(x) = \begin{cases} 1.30x & x \le 100 \\ 1.15x & 100 < x \le 150 \\ 1.00x & 150 < x < 200 \\ 0.85x & x \ge 200 \end{cases}$

(b)

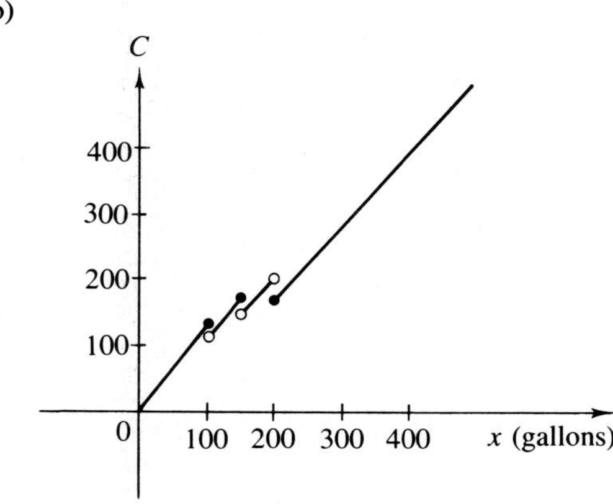

(c) $C(x)$ is discontinuous at (100, 130), (150, 172.5), (200, 170)
35. (a) $\frac{3}{4}$ (b) 0 **37.** -2 **39.** $-\frac{1}{16}$ **41.** $\frac{3}{25}$ **43.** $-2x^{-5/3} + \frac{1}{3}x^{-2/3}$
45. $-\frac{2}{3}x^{-4/3} + \frac{20}{3}x^{1/3} + \frac{4}{3}x^{-5/3}$ **47.** $\dfrac{-x^{1/2}}{(x - x^{1/2})^2}$ **49.** $\dfrac{4x^4 - 3x^2}{\sqrt{x^2 - 1}}$
51. $5(3x^3 + x^2 - 1)^4 (9x^2 + 2x)(x^3 + x^2 - 2x)^2 + 2(x^3 + x^2 - 2x)(3x^2 + 2x - 2)(3x^3 + x^2 - 1)^5$
53. $-4(3x^3 - x^2 + 1)^{-5}(9x^2 - 2x)$ **55.** (a) at $x = 2$, $m = -7$ (b) at $x = -3$, $m = 13$
57. (0, 4), (1, 0), (−1, 0) **59.** $(\frac{2}{3}, 1)$

Chapter Test, page 559

1. (a) -3 (b) -1 **2.** (a) 42 (b) 120 **3.** $2x$ **4.** $\frac{29}{2}$ **5.** (a) $-\infty$ (b) ∞ **6.** 3

7. $\lim_{x\to 4^-} f(x) = \infty$, $\lim_{x\to 4^+} f(x) = -\infty$, $\lim_{x\to 4} f(x)$ does not exist

8. $f(x)$ is not continuous at $x = 0$ since $f(0)$ does not exist

9. (a) $6\left(\dfrac{x^3 - 2x^2}{x^2 + x + 1}\right)^5 \dfrac{(x^2 + x + 1)(3x^2 - 4x) - (x^3 - 2x^2)(2x + 1)}{(x^2 + x + 1)^2}$

(b) $\frac{2}{3}x^5(x^2 - 1)^{-2/3} + 4x^3(x^2 - 1)^{1/3}$

10. $\dfrac{dD}{dt} = \$1{,}328{,}400/\text{month}$

CHAPTER 13

Exercise Set 13.1, page 566

1. $y = -2x + 4$

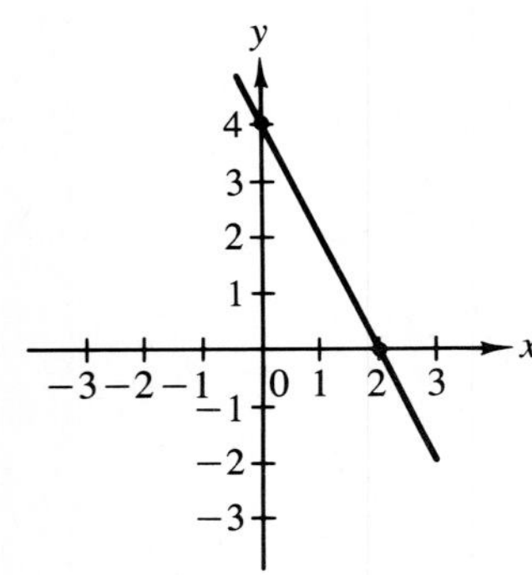

3. $y = 4x - 4$

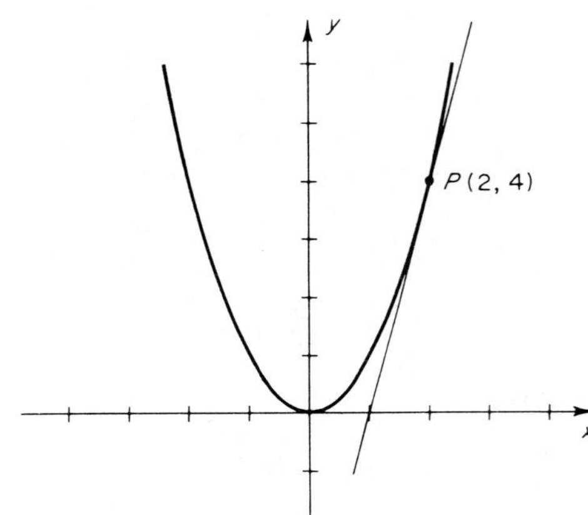

5. $y = -x + 2$

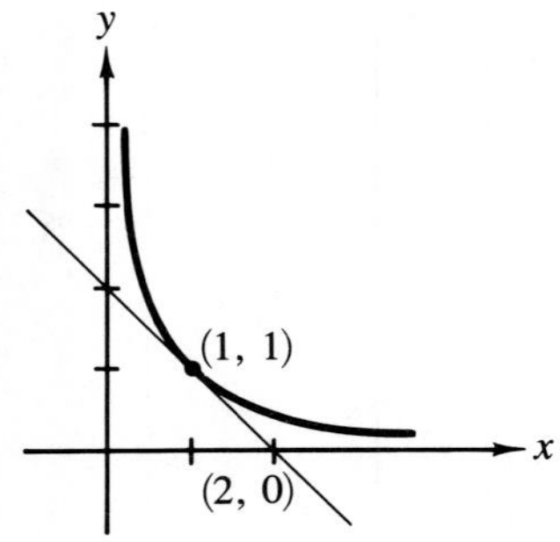

7. $y = 3x + 2$

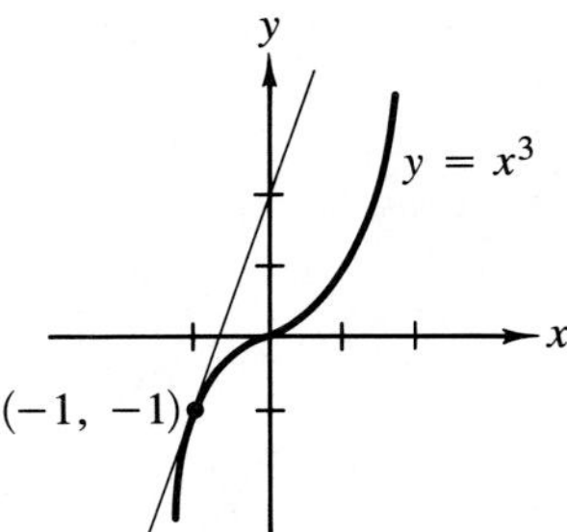

9. $y = \frac{3}{25}x + \frac{1}{25}$ **11.** (a) $4\pi r^2$ (b) π

13. (a) $Q'(x) = -2x + 500$ (b) 460 units per thousand dollars

15. increasing for $x < 0$; decreasing for $x > 0$ **17.** increasing for $x > -2$; decreasing for $x < -2$

19. increasing for all x **21.** decreasing for all $x \neq 0$; $f(0)$ is not defined

23. increasing for $x > 2$; decreasing for $x < 2$
25. increasing for $x < 0$; decreasing for $x > 0$; $f(0)$ is not defined
27. (a) (i) 11 feet to the left of the origin ($x = -11$) (ii) -9 feet per second (iii) it is moving to the left, away from the origin (iv) 6 feet per second per second
(b) (i) 9 feet to the left of the origin ($x = -9$) (ii) 15 feet per second (iii) it is moving to the right, toward the origin (iv) 18 feet per second per second
29. $s(1) = 0$ inch; $v(1) = -4$ inches per second; $a(1) = -2$ inches per second per second
31. $s(2) = \frac{1}{2}$ inch; $v(2) = -\frac{1}{4}$ inch per second; $a(2) = \frac{1}{4}$ inch per second per second
33. (a) $v(t) = 48 - 32t$ feet per second (b) $v(0) = 48$ feet per second
(c) $t = 3$ seconds (d) -48 feet per second
(e) $a(t) = -32$ feet per second per second; the only force acting on the stone is the force due to gravity
(f) $\frac{3}{2}$ seconds (g) 36 feet

Exercise Set 13.2, page 574

1. (a) \$5000 per unit (b) \$3000 per unit (c) $\overline{C}(x)$ is decreasing when $x = 20$; $\overline{C}(x)$ is decreasing when $x = 10$
3. (a) $C'(x) = 50 + \frac{x}{5}$; $R'(x) = \frac{1}{20}$; $P'(x) = \frac{-999}{20} - \frac{x}{5}$ (b) \$55 per unit
(c) \$0.05 per unit (d) $-\$54.95$ per unit (e) $C'(25) = \$55$
5. (a) $R(x) = x(200 - \frac{1}{2}x)$; $P(x) = -\frac{3}{4}x^2 + 200x - 3000$
(b) $R'(x) = 200 - x$; $C'(x) = \frac{1}{2}x$; $P'(x) = -\frac{3}{2}x + 200$ (c) $R'(25) = \$175$ (d) $P'(25) = \$162.50$
7. $R(x) = 200x - \frac{1}{50}x^2$

Exercise Set 13.3, page 578

1. 72π square feet per second **3.** 72π cubic centimeters per second **5.** $\frac{2}{3\pi}$ feet per minute
7. 108 cubic inches per hour **9.** \$790 per hour 11. \$5600 per month 13. 20 units per month
15. \$130 per month

Exercise Set 13.4, page 589

1. absolute maximum value of 18 occurs at $x = 6$; absolute minimum value of 4 occurs at $x = -1$
3. absolute maximum value of 14 occurs at $x = 3$; absolute minimum value of -2 occurs at $x = -1$
5. absolute maximum value of $\frac{1}{4}$ occurs at $x = \frac{1}{2}$; absolute minimum value of -2 occurs at $x = 2$
7. absolute maximum value of 48 occurs at $x = 4$; absolute minimum value of -2 occurs at $x = -1$
9. absolute maximum value of 0 occurs at $x = 0$ and $x = 2$; absolute minimum value of $-\frac{27}{16}$ occurs at $x = \frac{3}{2}$
11. absolute maximum value of 2 occurs at $x = \frac{1}{2}$; absolute minimum value of $\frac{1}{5}$ occurs at $x = 5$
13. 50, 50 **15.** a square with each side 60 feet **17.** 5400 square yards **19.** 11,664 cubic inches
21. \$46,000 per year **23.** 5 units per day **25.** 20 counters **27.** lot size = 100 cameras; 10 orders per year
29. $x = 0.75$ ounce; $xy = 0.1125$ ounce

Exercise Set 13.5, page 602

1. (a) (a, b) and (d, f) (b) (b, d) and (f, g) (c) (c, e) (d) (a, c) and (e, g)
3. (a) b, f (b) d (c) c, e (d) b, d, f
5. (a) b, d, f (b) (a, b) and (f, g) (c) (b, d) and (d, f) (d) (c, d) and (e, g) (e) (a, c) and (d, e)
7. decreases for $x < 3$, increases for $x > 3$, relative minimum at $x = 3$ **9.** decreases for all x
11. decreases for $x < 0$, increases for $x > 0$, relative minimum at $x = 0$
13. decreases for $-3 < x < 1$, increases for $x < -3$ and for $x > 1$; relative maximum at $x = -3$, relative minimum at $x = 1$
15. increases for $x > 0$; no relative maximum or minimum points; $f(x)$ is not defined for $x < 0$

17. 6 **19.** 0 **21.** $\dfrac{2x(x^2-3)}{(x^2+1)^3}$ **23.** concave up for $-\infty < x < \infty$; no inflection points

25. concave up for $x > 2$, concave down for $x < 2$; inflection point at $x = 2$

27. concave up for $x > 3$, concave down for $x < 3$; inflection point at $x = 3$

29. relative maximum at $x = -1$, relative minimum at $x = 1$ **31.** relative minimum at $x = \frac{3}{2}$

33. no relative maxima or minima exist **35.** no relative maxima or minima exist

37.

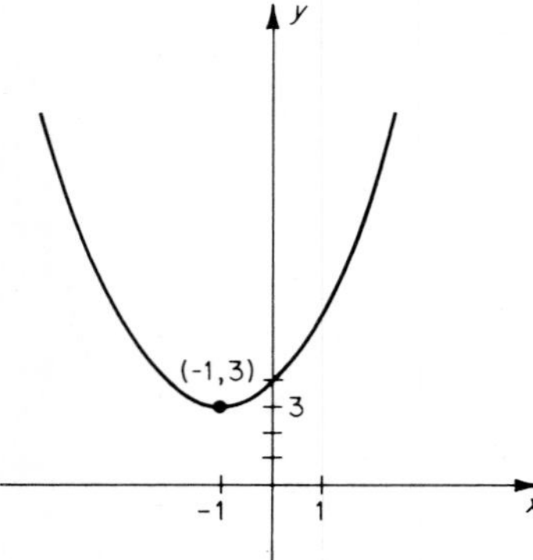

39.

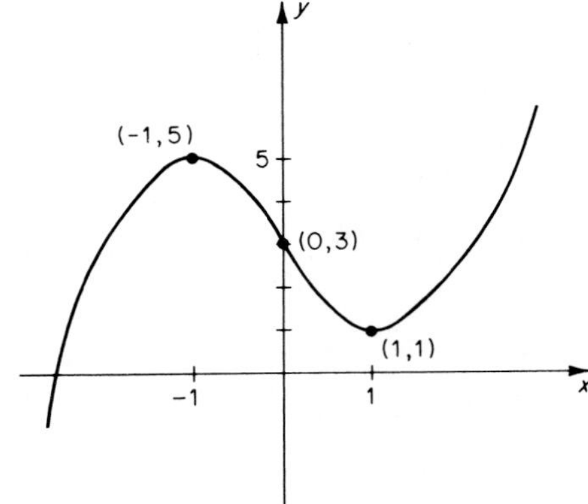

41.

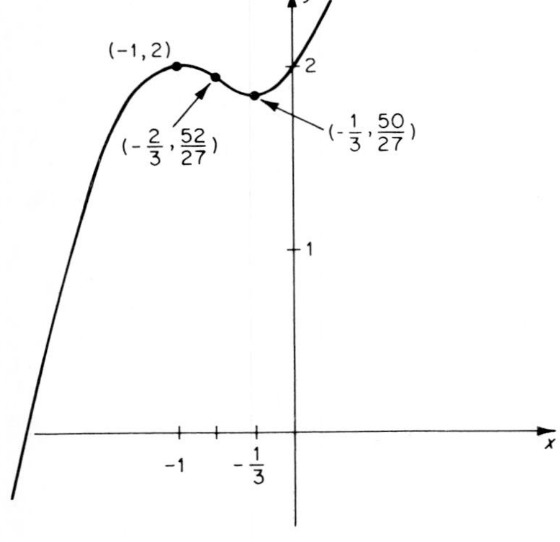

43.

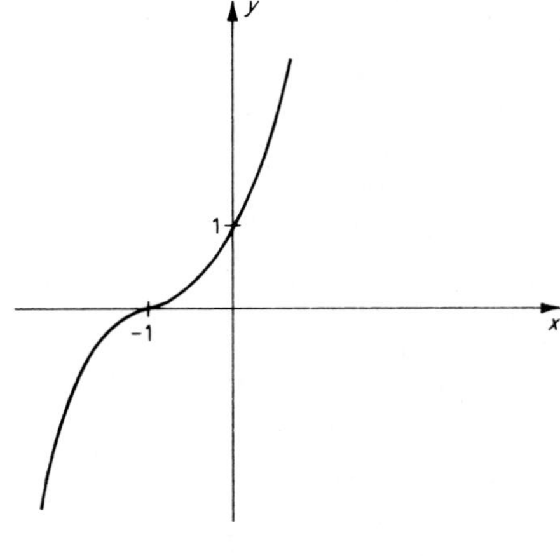

45.

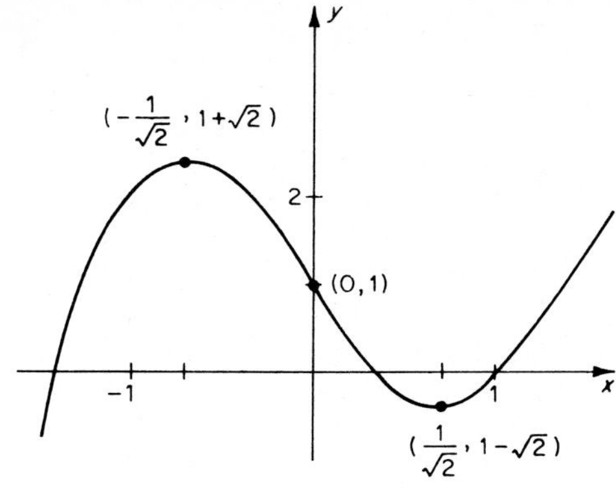

47.

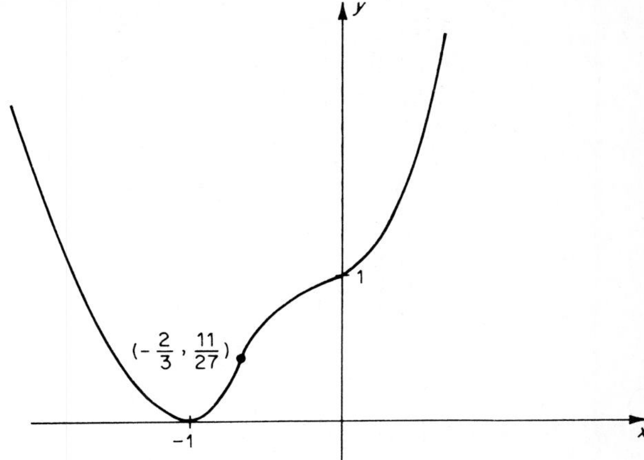

49. at $t = 5$ weeks

Exercise Set 13.6, page 608

1. minimum value of $-\frac{1}{4}$ occurs at $x = \frac{3}{2}$; no maximum value
3. maximum value of $\frac{19}{3}$ occurs at $x = 2$; no minimum value
5. minimum value of 3 occurs at $x = \frac{1}{2}$; no maximum value
7. minimum value of 2 occurs at $x = 1$; no maximum value
9. 2 hours **11.** minimum value of 108 occurs when $x = 6$
13. dimensions 5 inches × 5 inches × 2.5 inches give the minimum surface area of 75 square inches
15. 30 items **17.** 40 miles per hour **19.** $r = h = \dfrac{10}{\sqrt[3]{\pi}}$ **21.** \$15 increase **23.** $D = A$

Exercise Set 13.7, page 618

1. (a) $\Delta x = 4$ (b) $\dfrac{\Delta x}{x} = \dfrac{4}{25}$ (c) $\Delta y = 10$ (d) $\dfrac{\Delta y}{y} = \dfrac{1}{15}$
3. (a) $\dfrac{\Delta x}{x} = \dfrac{2}{3}$ (b) $\dfrac{\Delta y}{y} = \dfrac{16}{9}$ (c) $\dfrac{\Delta y/y}{\Delta x/x} = \dfrac{8}{3}$
5. $E_x(y) = 2$; $f(x)$ is elastic at $x = 3$ **7.** $E_x(y) = -1$; $f(x)$ has unit elasticity at $x = 2$
9. $E_x(y) = \frac{19}{10}$; $f(x)$ is elastic at $x = 3$ **11.** $E_x(y) = -\frac{1}{4}$; $f(x)$ is inelastic at $x = 1$
13. $E_x(y) = 4$; $f(x)$ is elastic at $x = 3$
15. (a) $E_p(x) = \dfrac{-2p^2}{500 - p^2}$ $0 \leq p < \sqrt{500}$ (b) $E_p(x) = -\frac{1}{2}$ at $p = 10$; demand is inelastic
(c) $E_p(x) = -8$ at $p = 20$; demand is elastic (d) $p = \sqrt{\frac{500}{3}} \cong \$12.91$

17. (a) $E_p(x) = \dfrac{-p}{2000 - 2p}$ $0 \le p < 1000$ (b) $E_p(x) = -\frac{1}{8}$ at $p = 200$; demand is inelastic
(c) $E_p(x) = -2$ at $p = 800$; demand is elastic (d) $p = \frac{2000}{3} \cong \$666.67$
19. (a) a price increase brings about an increase in revenue because demand is inelastic [Exercise 15(b)]
(b) a price increase brings about a decrease in revenue because demand is elastic [Exercise 15(c)]
21. at $x = 10$, $E_x(C) = \frac{18}{23}$; cost is inelastic; the cost of producing the eleventh unit will be less than the average cost of the units already produced

at $x = 20$, $E_x(C) = \frac{28}{23}$; cost is elastic; the cost of producing the next unit will be greater than the average cost of the units already produced
23. $E_x(C) = \frac{60}{49}$ when $x = 6$; cost is elastic; the cost of producing the seventh unit will be greater than the average cost of the units already produced

Supplementary Exercises, page 621

1. Slope $= -5$; $y = -5x - 4$ 3. Slope $= \frac{13}{4}$; $y = \frac{13}{4}x - \frac{25}{4}$ 5. $(0, 4)$, $(1, 0)$, $(-1, 0)$
7. critical point is $x = \frac{1}{2}$; $f(x)$ is decreasing for $x < \frac{1}{2}$ and decreasing for $x > \frac{1}{2}$
9. critical points are $x = 0$ and $x = 1$; $f(x)$ is increasing for $x > 1$ and decreasing for $0 < x < 1$ and for $x < 0$
11. maximum value of 12 occurs at $x = 4$; minimum value of $-\frac{1}{4}$ occurs at $x = \frac{1}{2}$
13. maximum value of 2 occurs at $x = 1$; minimum value of $\frac{4}{3}$ occurs at $x = 3$
15. relative minimum at $(1, -1)$; inflection points at $(0, 0)$ and $(\frac{2}{3}, -\frac{16}{27})$

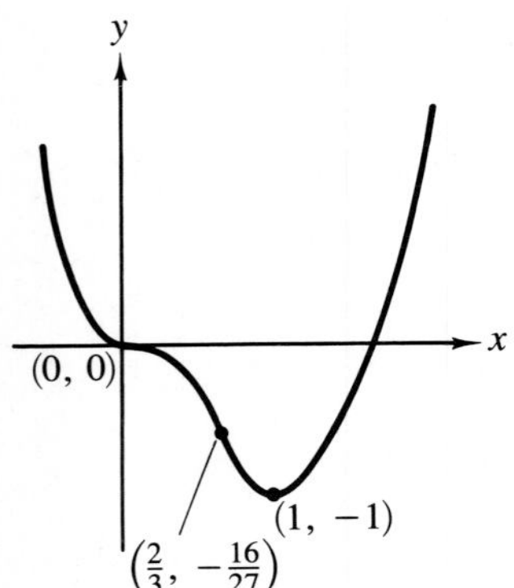

17. 120 grams per week 19. \$1,328,400 per month 21. 40 feet
23. (a) $v(2) = 0$ feet per second (b) $a(2) = -32$ feet per second per second (c) 64 feet
25. 10 shirts at a cost of \$12 per shirt
27. an increase of \$2.50 per ticket will give a maximum revenue of \$2250
29. (a) $E_p(x) = \dfrac{-p^2}{p^2 + 25}$ (b) $E_p(x) = -\frac{4}{5}$ when $p = 10$

(c) there is no price at which the demand has unit elasticity since $\dfrac{p^2}{p^2 + 25} < 1$ for all p

Chapter Test, page 622

1. $y = 12x - 20$
2. relative maximum at $(\frac{1}{2}, \frac{1}{16})$; relative minima at (0, 0) and (1, 0); inflection points at (0.211, 0.028) and (0.789, 0.028)

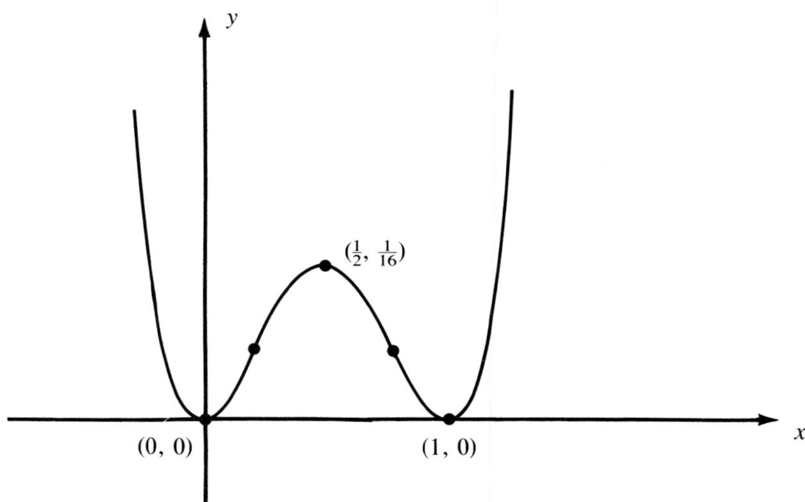

3. (a) object moves to the left when $t < \frac{1}{2}$; to the right when $t > \frac{1}{2}$ (b) $t = \frac{1}{2}$; at this time $x = -\frac{1}{4}$
4. maximum value of 12.375 at $x = 3.5$; minimum value of $-6\sqrt{3} + 1$ at $x = \sqrt{3}$
5. 100 cars; maximum profit is $10,000 6. length = 250 feet; width = 125 feet 7. 50 items
8. $\frac{\sqrt{6}}{12}$ feet per second

CHAPTER 14

Exercise Set 14.1, page 631

1. (a)

x	-4	-3	-2	-1	0	1	2	3	4
$(\frac{2}{7})^{0.4x}$	-7.42	-4.5	-2.72	-1.65	1	0.61	0.37	0.22	0.13

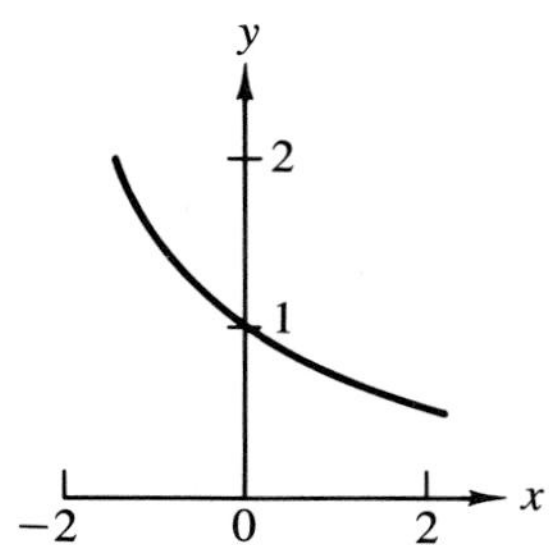

3. (a)

x	−4	−3	−2	−1	0	1	2	3	4
$5e^x$	0.09	0.25	0.68	1.84	5	13.59	36.94	100.43	272.99

(b)

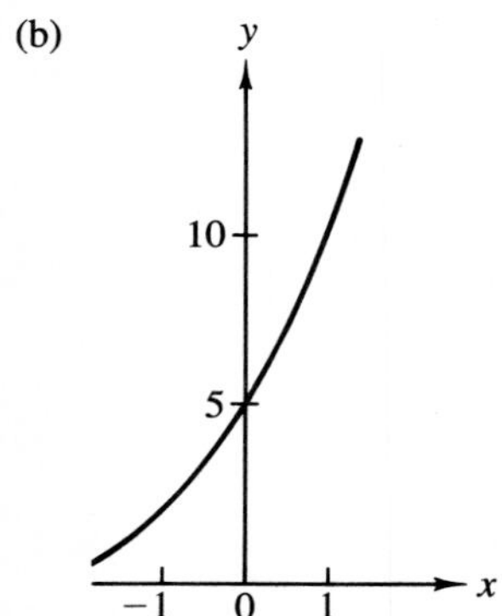

5. domain: all x; range: $y > 0$

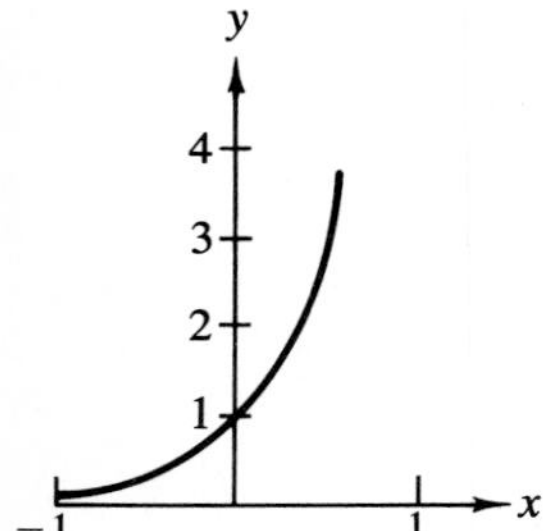

7. domain: all x; range: $y > 0$

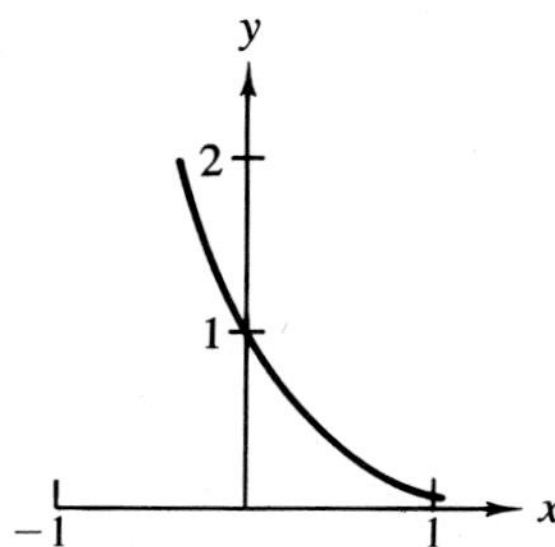

9. domain: all x; range: $y < 1$

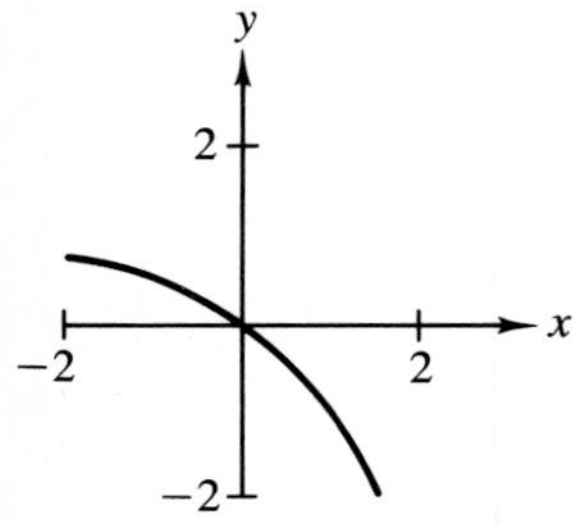

11. domain: all x; range: $y > 0$

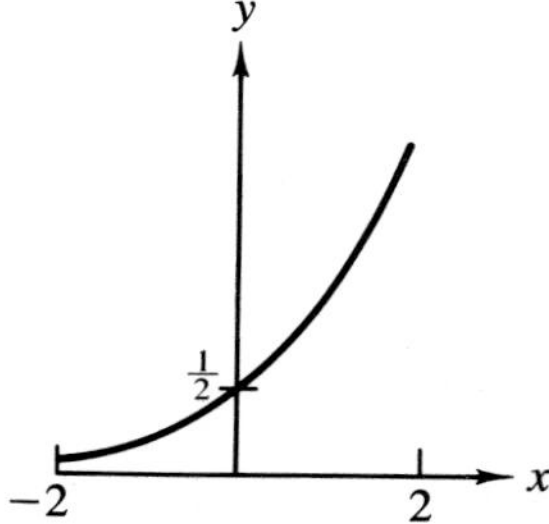

13. domain: all x; range: $y > 0$

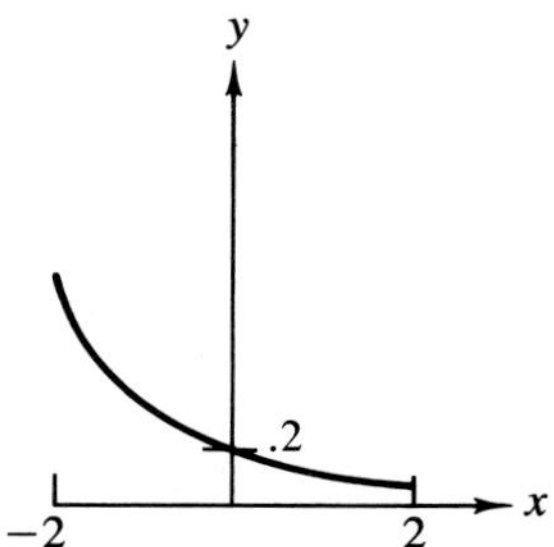

15. domain: all x; range: $y > 0$

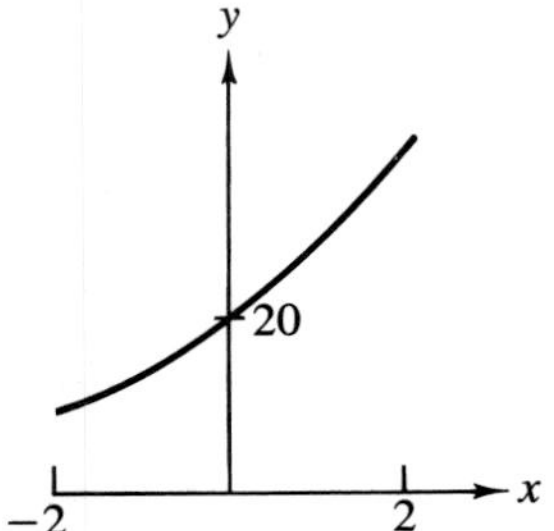

17.

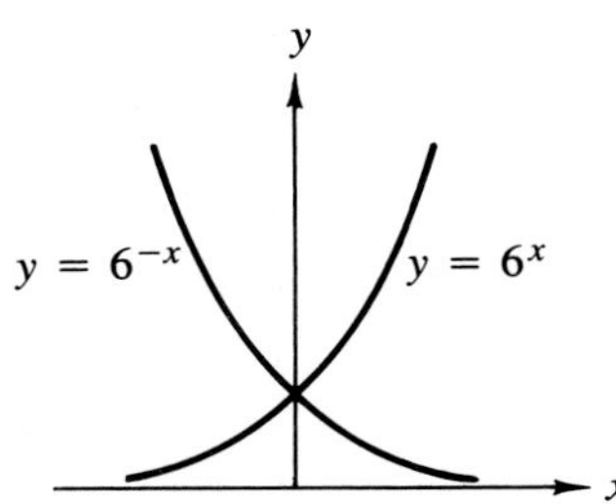

19. (a) 8000 (b) 64,000 (c) 512,000 (d) 4,096,000 **21.** 2^{n-1}, 2^{14}, 2^{20}, 2^{30}
23. 1 **25.** 7 **27.** 4 **29.** 7 **31.** (a) S approaches 1000 (b)

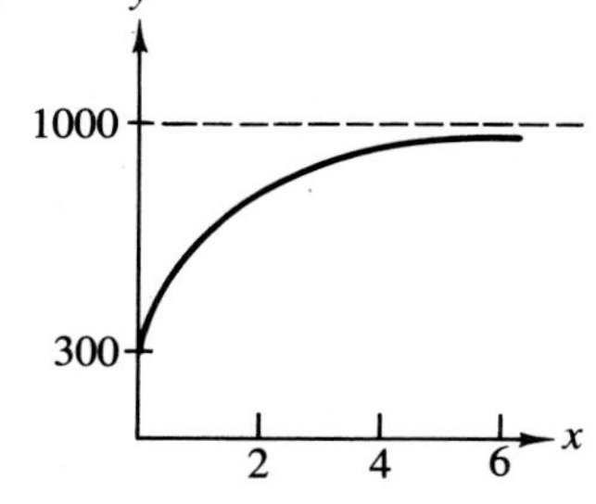

33. (a) $\frac{5000}{6}$ (b) 1000

Exercise Set 14.2, page 640

1. (a) 3 (b) −3 (c) −5 (d) 40 (e) 1.5 (f) 4
3. (a) 0.6931 (b) 0.5306 (c) 0.9730 (d) −1.6740 (e) 7.6530 (f) 1.4930
5. (a) 0.7782 (b) 0.4624 (c) 0.6152 (d) 0.2711 (e) 0.6383 (f) 0.5100
7. (a) $10^2 = 100$ (b) $10^{-2} = 0.01$ (c) $e^{-1} = 1/e$ (d) $e^0 = 1$
9. (a) $\log 10{,}000 = 4$ (b) $\log \frac{1}{1000} = -3$ (c) $\ln 1/e^2 = -2$ (d) $\ln e = 1$
11. (a) 7 (b) $\frac{1}{3}$ (c) $-0.6t$ (d) 0.6
13. (a) 0.7781 (b) −0.1761 (c) 0.9542 (d) 0.2386 (e) −3.3010 (f) 1.7323

15.

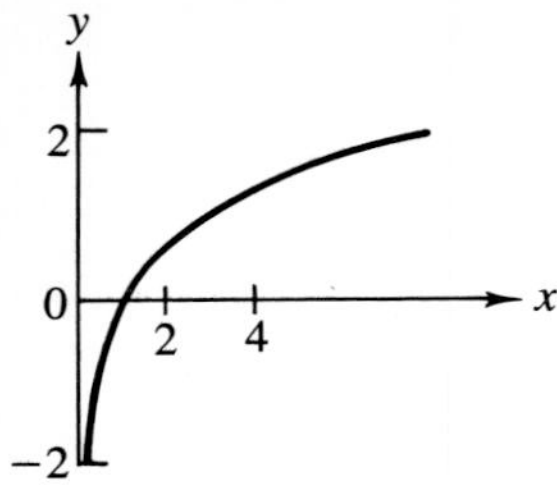

17.

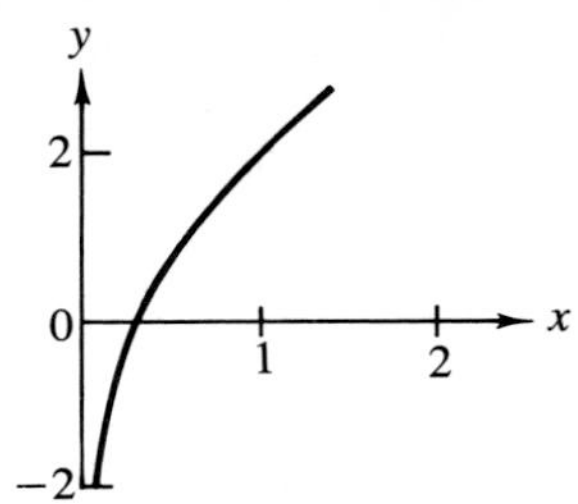

19.

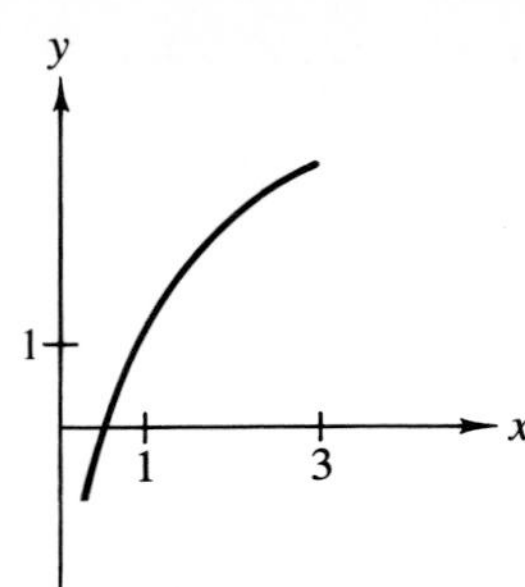

21. 9 **23.** $\dfrac{\ln 5}{\ln 4} \cong 1.161$ **25.** $e^3 \cong 20.09$ **27.** $\dfrac{e+1}{2} \cong 1.86$
29. $5e \cong 13.59$ **31.** $(\ln 4)/2$ **33.** $e^3 \cong 20.09$ **35.** 2.3863 **37.** $\frac{675}{64}$ **39.** 22.5 **41.** 2.8284
43. (a) 1.0986 (b) 0.9163 (c) 0.112 **47.** (a) 60 decibels (b) 40 decibels

Exercise Set 14.3, page 645

1.

u	2	20	200	2000	20,000
$1 + \frac{1}{u}$	1.5	1.05	1.005	1.0005	1.00005
$\left(1 + \frac{1}{u}\right)^u$	2.25	2.6533	2.71152	2.71760	2.71821

3.

u	5	50	500	5000	50,000
$1 + \frac{2}{u}$	1.4	1.04	1.004	1.0004	1.00004
$\left(1 + \frac{2}{u}\right)^u$	5.37824	7.10668	7.35964	7.38610	7.38876

5. e^3 **7.** e^{-1} **9.** e^2 **11.** e^{-1} **13.** \$32,974.43
15. (a) 8.75 years (b) 13.87 years **17.** (a) 16.7 years (b) 11.72 years (c) 11.55 years
19. (a) 6.25 years (b) 5.12 years (c) 5.07 years

Exercise Set 14.4, page 655

1. $4e^{4x}$ **3.** $30xe^{5x^2}$ **5.** $2 - 2e^{2x} + 12e^{4x}$ **7.** $x^3e^x + 3x^2e^x$ **9.** $4e^{4x}$ **11.** $\frac{1}{5}e^{x^2}(2x^2+1)$
13. $6(e^x + x)^5(e^x + 1)$ **15.** $1/x$ **17.** $\dfrac{21x^2}{7x^3 - 4}$ **19.** $-\dfrac{1}{x}$ **21.** $\dfrac{4}{3(4x+5)}$ **23.** $x + 2x \ln x$
25. $\dfrac{1 - 3\ln x}{x^4}$ **27.** 1 **29.** $2x$ **31.** $2x$ **33.** $x^6 \cdot 2^x \cdot \ln 2 + 2^x \cdot 6x^5$ **35.** $2x3^{x^2} \ln 3$ **37.** $y = x + 1$
39. 1

41. increasing for all x **43.** increasing for $x > 0$

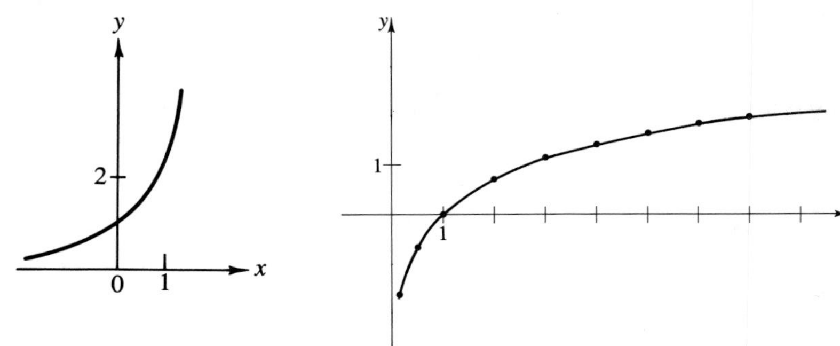

45. absolute maximum value of 1 occurs at $x = 0$; absolute minimum value of e^{-2} occurs at $x = 2$

47. absolute maximum value of e^{-1} occurs at $x = 1$; no absolute minimum value

49. relative minimum at $x = 0$

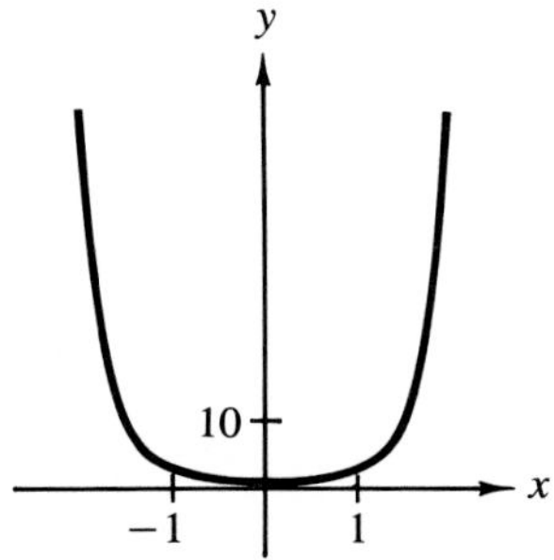

51. (a) relative maximum at $x = 0$; no minimum values; inflection points at $x = -1$ and $x = 1$

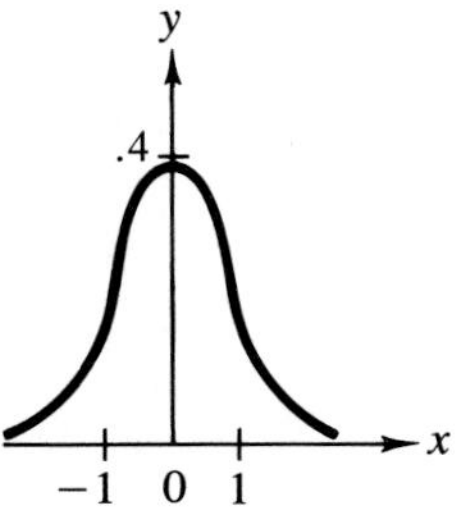

53. $s(-1) = e$ inches; $v(-1) = -2e$ inches/second; $a(-1) = 6e$ inches/second/second

55. (a) $k = 0.0886$ (b) $\dfrac{55{,}000ke^{-kt}}{(1 + 10e^{-kt})^2}$ (c) 76.6 trout/month **57.** (c) $R(S) = 3.8 \ln \dfrac{S}{11.6}$

Exercise Set 14.5, page 668

1. (a) 0.02 (b) 500 (c)

t	1	5	20	225
Q	510.1	552.6	746.9	45,008.6

(d) 10 (e) 10.62

3.

Q	500	1250	2300
t	0	4.58	7.63

5. (a) 1.73 seconds (b) 3.47 seconds **7.** (a) 0.14 (b) 0.0069 **9.** 11 years

11. (a)

t	1975	1980	1985	1995	2000	2005	2010
Q	4	4.42	4.886	5.97	6.59	7.29	8.06

(b)

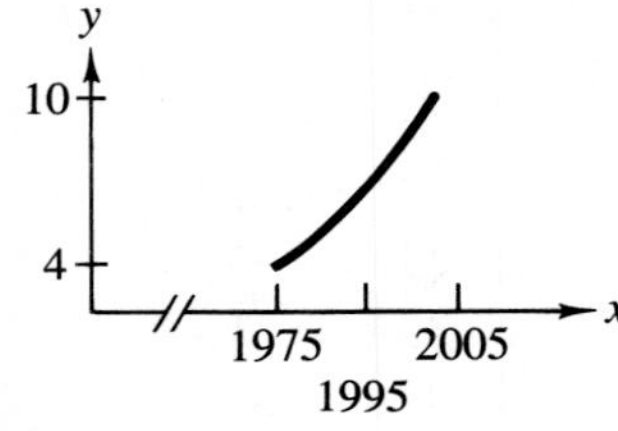

13. Approximately 25,000 years **15.** (a) 10% (b) $545,981 **17.** (a) 0.95238

Supplementary Exercises, page 671

1.

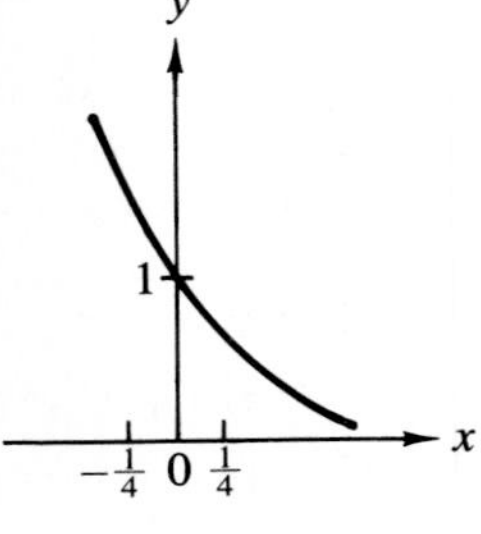

3. (a) 32,000 (b) 4,096,000

5.

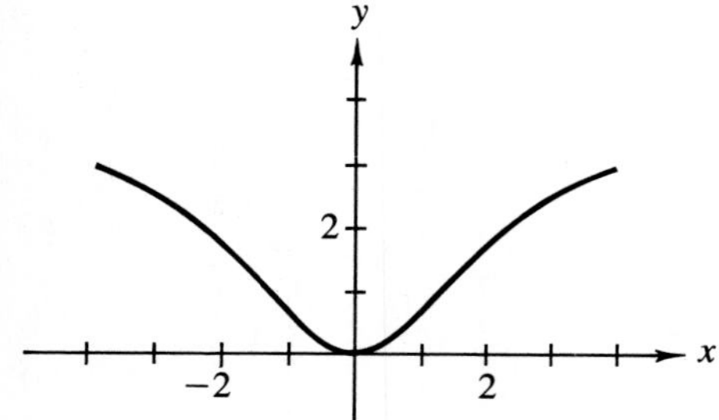

7. 1.8751 **9.** 0.4375 **11.** +4, −4 **13.** ± 1.177 **15.** $-\frac{2}{19}$ **17.** 3^{π}
19. 23.44 square feet **21.** 18.32 years **23.** 11.55 hours **25.** 0.032 **27.** $996 **29.** $e^{-2/9}$
31. 5.9 years **33.** $\dfrac{2 - e^{-x}}{(1 - e^{x})^2}$ **35.** $\dfrac{2x - 1}{2(x^2 - x)}$ **37.** $\dfrac{1}{2\sqrt{x}}$ **39.** −1
41. (a) increasing: $x < e$, decreasing: $x > e$ (b) relative maximum at $x = e$
(c) concave up: $x > e^{3/2}$; concave down: $0 < x < e^{3/2}$; inflection point at $x = e^{3/2}$
(d)

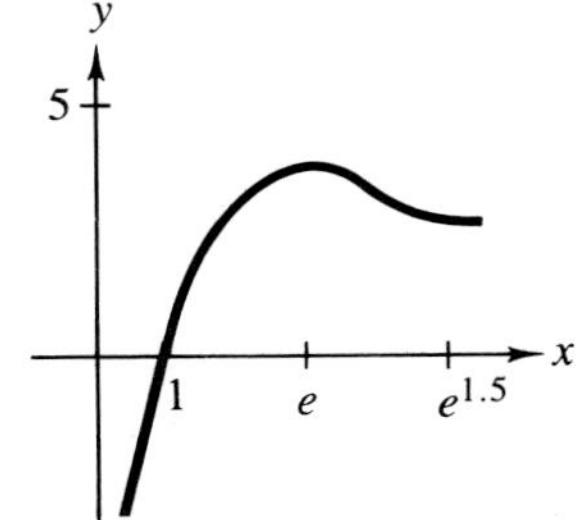

Chapter Test, page 672

1. ±7.46 **2.** 1.39 days **3.** 646.4 bacteria **4.** (a) 4890 grams (b) 3.10 minutes **5.** 27.7
6. $e^{-1/2}$ **7.** (a) $\dfrac{\ln x^3(2xe^{x^2} - 1) - (3/x)(e^{x^2} - x)}{(\ln x^3)^2}$ (b) $\dfrac{1 - x^2}{x(x^2 + 1)}$
8. (a) increasing: $x < 1$, decreasing: $x > 1$ (b) relative maximum at $x = 1$
(c) concave up: $x > 2$ concave down: $x < 2$; inflection point at $x = 2$
(d)

9. $y = 0$ **10.** 3.19 years **11.** Approximately 85 million bacteria per hour
12. $s = 0.448$ feet $v = -0.149$ feet per second $a = -0.050$ feet per second per second

CHAPTER FIFTEEN

Exercise Set 15.1, page 679

1. $5x + C$ **3.** $\frac{1}{6}t^6 + C$ **5.** $\frac{4}{5}x^{5/4} + C$ **7.** $5x^4 + C$ **9.** $-6/x + C$ **11.** $2e^u + C$ **13.** $50 \ln|x| + C$
15. $\frac{2}{3}x^{3/2} + C$ **17.** $4x^2 + 2x^3 + C$ **19.** $x^3 - \frac{20}{7}x^{7/4} + 4x + C$ **21.** $\frac{6}{5}x^{5/3} - 3e^x + 2\ln|x| + C$
23. $\frac{1}{2}x^4 + \frac{2}{3}x^{3/2} - \frac{1}{x} + 5x + C$ **25.** $\frac{1}{2}x^2 + 2x - \frac{11}{2}$ **27.** $\frac{1}{3}x^3 + 3x + \frac{16}{3}$ **29.** $2x + 3e^x + 2$ **31.** $x^3 + 9$
33. (a) $25{,}000t + \frac{20}{7}t^{7/5} + 50{,}000$ (b) 550,189 **35.** $\frac{1}{3}x^3 - \frac{3}{2}x^2 + 1000$

Exercise Set 15.2, page 688

1. 9 **3.** $\frac{81}{4}$ **5.** $e^2 - 1$ **7.** ln 3 **9.** $\frac{2}{15}$ **11.** $\frac{26}{3}$ **13.** $e^3 - 1/e$ **15.** $\frac{52}{3}$ **17.** 9 **19.** 22.5
21. $\frac{16}{3}$ **23.** $\frac{16}{3}$ **25.** $\frac{14}{3}$ **27.** $e - 1$ **29.** 2 ln 2 **31.** (a) $A(x) = \frac{1}{3}x^3 + x + \frac{4}{3}$ (b) $A'(x) = x^2 + 1$
33. $\frac{4}{3}$

Exercise Set 15.3, page 699

1. (a) $\frac{1}{2}$ **3.** (a) $\frac{3}{2}$ **5.** (a) $\frac{8}{3}$ (b) $\int_0^2 x^2\,dx = \frac{8}{3}$ **7.** (a) ln 5 (b) $\int_1^5 1/x\,dx = \ln 5$ **9.** 22
11. 195 **13.** $-\frac{61}{12}$ **15.** $\frac{370}{3}$ **17.** $2e^4 - 258$ **19.** 0 **21.** $\frac{25}{6} + 4 \ln 2$ **23.** 4 **25.** $\frac{5}{2}$ **27.** $\frac{5}{6}$
29. $\frac{32}{3}$ **31.** $\frac{9}{2}$ **33.** $\frac{8}{3}$

Exercise Set 15.4, page 707

1. (a) 80

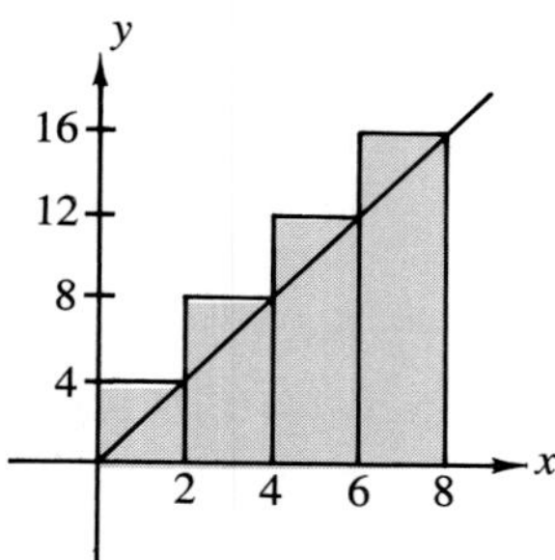

(b) 72

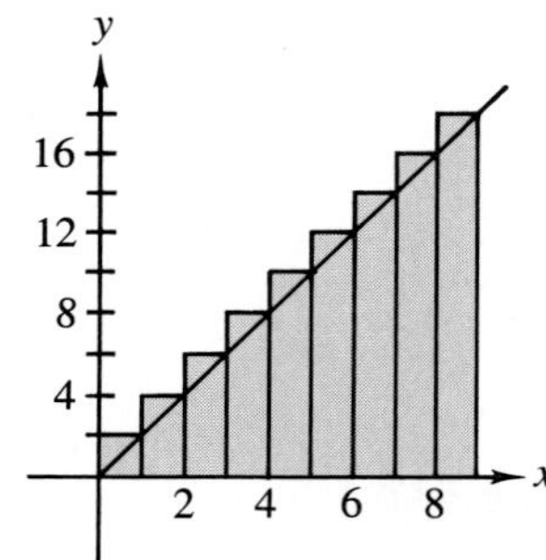

3. (a) 64

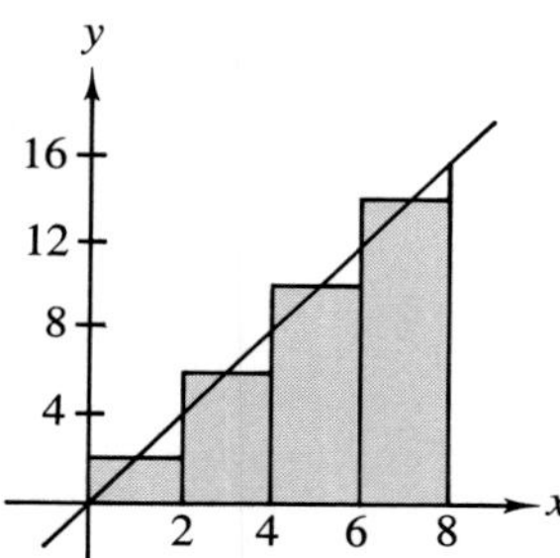

(b) 64

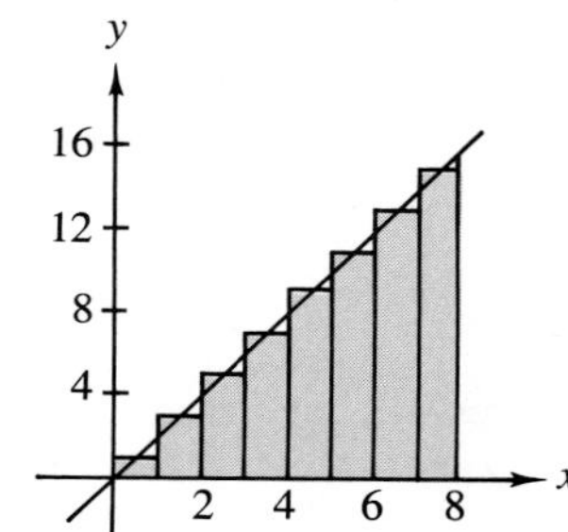

5. $\frac{3}{2}$ **7.** (a) 34

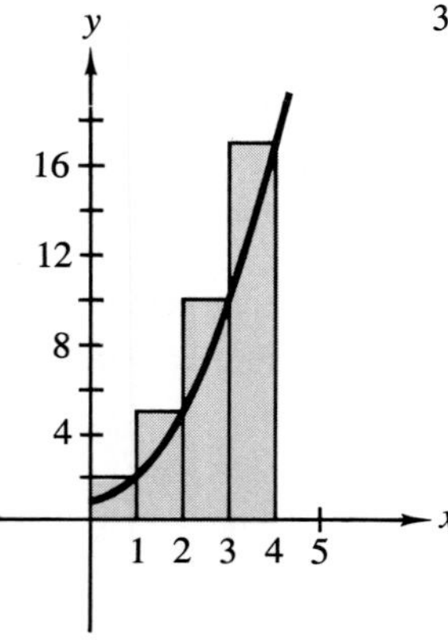

(b) 29.5

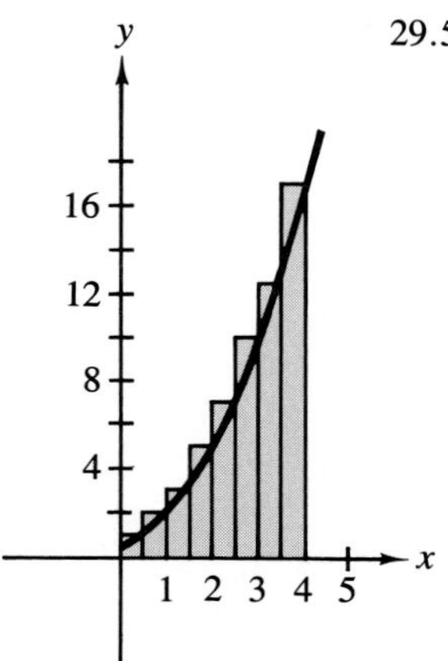

9. (a) 25 (b) 25.25

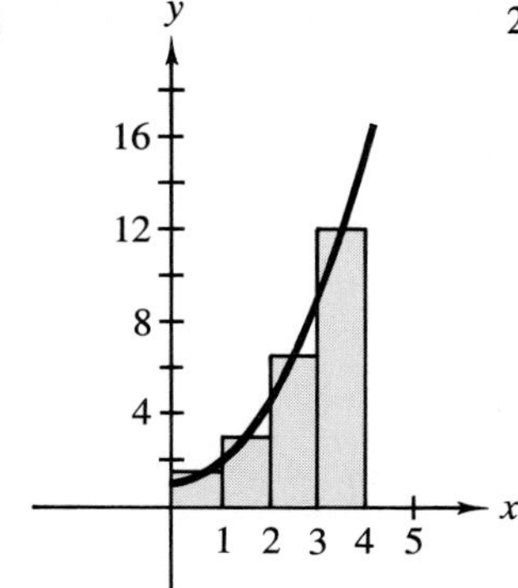

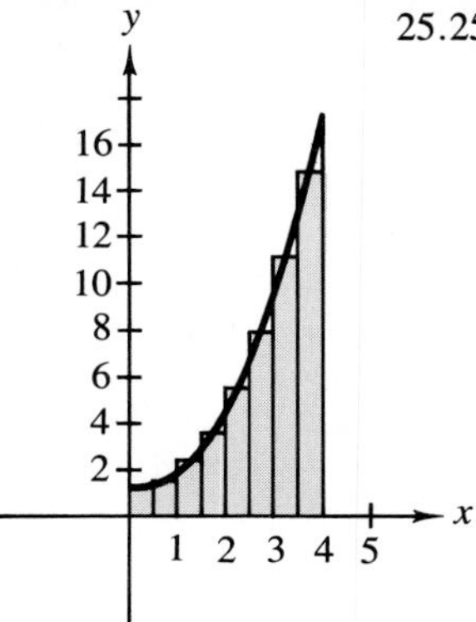

11. $\frac{1}{3}$ **13.** 10 **15.** 9

Supplementary Exercises, page 709

1. $x^2 - \frac{1}{3}x^3 + C$ **3.** $\frac{-2}{3t^3} - 2t^{1/2} + C$ **5.** $f(x) = \frac{x^2}{2} + e^x + 1 - \frac{1}{e^2}$ **7.** $\frac{112}{3}$ **9.** $2e^2 - 3\ln 2 - 2e$
11. $\frac{15}{4}$ **13.** $\ln 3$ **15.** $\frac{2}{3}$ **17.** $2e^4 - 2e$ **19.** $\frac{19}{3}$ **21.** $\frac{76}{3}$ **23.** 6 **25.** 14 **27.** 10
29. 0 **31.** 6 **33.** -2 **35.** $\frac{1}{3}x^3 - \frac{1}{3}x^{3/2} + 5000$

Chapter Test, page 711

1. $2e^x - 3\ln|x| + \frac{1}{3}x^3 + C$ **2.** $3x^{2/3} + \frac{4}{x} + 3e^x - \frac{3}{4}x + C$ **3.** $-\frac{1963}{6} + 4\ln 8$ **4.** $2e^9 - 3\ln 9 + \frac{52}{3} - 2e$
5. 12 **6.** 14 **7.** $\frac{407}{20}$

CHAPTER 16

Section 16.1, page 718

1. $\frac{1}{11}(x^3 + 5)^{11} + C$ **3.** $-\frac{1}{14}(x^2 - 1)^{-14} + C$ **5.** $-\frac{1}{4(x^2 + 4)^2} + C$ **7.** $\ln 2$ **9.** $\frac{1}{6}(t^4 + 2)^{3/2} + C$
11. $-\frac{1}{2}e^{-2x} + C$ **13.** $\frac{1}{3}$ **15.** $\frac{1}{3}(e^3 - 1)$ **17.** $\frac{2}{9}(3x - 2)^{3/2} + C$ **19.** $\frac{2}{3}(1 + e^x)^{3/2} + C$
21. $\frac{2}{3}\sqrt{x^3 + 1} + C$ **23.** $\frac{(x^3 + 1)^{11}}{33} + C$

Section 16.2, page 722

1. $-xe^{-x} - e^{-x} + C$ **3.** $\frac{1}{3}x^3 \ln x - \frac{1}{9}x^3 + C$ **5.** $-2xe^{-x/2} - 4e^{-x/2} + C$ **7.** $\frac{1}{2}x^2 \ln 2x - \frac{1}{4}x^2 + C$
9. $x(\ln x)^2 - 2x\ln x + 2x + C$ **11.** $-\frac{\ln x}{x} - \frac{1}{x} + C$ **13.** $\frac{x^4}{4}\ln x - \frac{x^4}{16} + C$
15. $\left(\frac{x^3}{3} + x^2\right)\ln x - \frac{x^3}{9} - \frac{x^2}{2} + C$ **17.** $\frac{1}{2}(x\ln x - x) + C$ **19.** $x^3e^x - 3x^2e^x + 6xe^x - 6e^x + C$
21. $\frac{1}{5}e^{5x} + C$ **23.** $\frac{2}{3}(\ln 2x)^{3/2} + C$

Section 16.3, page 728

1. $\frac{1}{3}x - \frac{2}{3}\ln|3x+6| + C$ **3.** $-\frac{1}{2(3t-2)} + \frac{1}{4}\ln\left|\frac{t}{3t-2}\right| + C$

5. $\frac{1}{125}\left[5x+2-\frac{4}{5x+2}-4\ln|5x+2|\right]+C$ **7.** $\frac{1}{4}\ln\left|\frac{x-3}{x+1}\right|+C$ **9.** $\frac{3}{2}s^2 - 7s + C$

11. $\frac{3}{2}x + \frac{3}{4}\ln|2x-1| + C$ **13.** $\frac{2}{2835}(135x^2 - 144x + 128)(3x+4)^{3/2} + C$ **15.** $\ln|w + \sqrt{w^2-9}| + C$

17. $2^x/\ln 2 + C$ **19.** $\frac{x^3}{9}[3\ln x - 1] + C$ **21.** $(x+3)\ln(x+3) - x + C$ **23.** $\ln|\ln x| + C$

25. $\frac{1}{2}x^2 + x + \ln(x-1) + C$ **27.** $\frac{2}{5145}(147r^2 - 84r + 72)\sqrt{7r+3} + C$

29. $\frac{1}{5}[3\ln|x+3| + 2\ln|x-2|] + C$ **31.** $x\sqrt{x^2+\frac{1}{4}} + \frac{1}{4}\ln|x+\sqrt{x^2+\frac{1}{4}}| + C$

Section 16.4, page 735

1. 16 **3.** 1.25 **5.** 1.63 **7.** 58 **9.** 23.9 **11.** 2.15

Supplementary Exercises, page 735

1. $-\frac{1}{16}(3-2x)^8 + C$ **3.** $-\frac{1}{3}e^{5-3x} + C$ **5.** $\frac{1}{6}\ln 4$ **7.** $\ln|\ln x| + C$ **9.** $x^2e^x - 2xe^x + 2e^x + C$

11. $-x^2e^{-x} - 2xe^{-x} - 2e^{-x} + C$ **13.** $\frac{1}{16}[3e^4 + 1]$ **15.** $\frac{2(\ln x)^{3/2}}{3} + C$ **17.** $\ln|x + \sqrt{x^2+25}| + C$

19. $\frac{1}{6}\ln\left|\frac{3x+1}{3x-1}\right| + C$ **21.** $\frac{(3t-1)(2t+1)^{3/2}}{5} + C$ **23.** $\frac{2}{13{,}125}(375x^2 - 480x + 512)(5x+8)^{3/2} + C$

Chapter Test, page 736

1. $\frac{1}{2}\ln|x^2+2x| + C$ **2.** $\frac{2}{3}e^{3x} + \frac{1}{6}e^{6x} + C$ **3.** $\frac{1}{3}(1+\ln t)^3 + C$ **4.** $-(x+4)e^{-x} - e^{-x} + C$

5. $\frac{2}{3}t^{3/2}\ln t - \frac{4}{9}t^{3/2} + C$ **6.** $x\ln(2+3x) - x + \frac{2}{3}\ln(2+3x) + C$ **7.** $2\sqrt{2+9x} + \frac{2}{\sqrt{2}}\ln\left|\frac{\sqrt{2+9x}-\sqrt{2}}{\sqrt{2+9x}+\sqrt{2}}\right| + C$

8. $\frac{1}{5}t^2e^{5t} - \frac{2}{25}te^{5t} + \frac{2}{125}e^{5t} + C$ **9.** $\frac{1}{27}[(3x+1) - \frac{1}{3x+1} - 2\ln|3x+1|] + C$ **10.** $-\frac{1}{4x} + \frac{3}{16}\ln\left|\frac{3x+4}{x}\right| + C$

CHAPTER 17

Exercise Set 17.2, page 741

1. $\frac{1}{3}$ **3.** $\frac{9}{2}$ **5.** $\frac{4}{3}$ **7.** 8 **9.** $\frac{1}{2}$ **11.** $12 - 5\ln 5$ **13.** 9 **15.** $\frac{32}{3}$ **17.** $\frac{1}{3}$

Exercise Set 17.3, page 746

1. 15 **3.** $\frac{1}{2}(e - \frac{1}{e})$ **5.** $\frac{1}{8}\ln 5$ **7.** $\frac{1}{8}\ln\frac{1}{9}$ **9.** 8519 **11.** 1 **13.** 29° **15.** $10 + 4\ln\frac{3}{2}$
17. 1370

Exercise Set 17.4, page 757

1. (a)

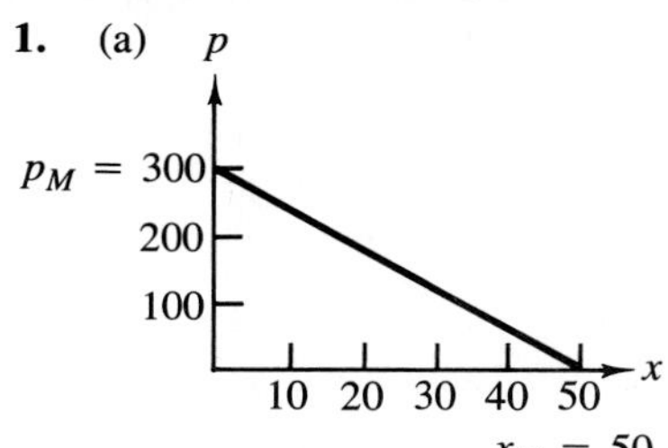

(b) $7500

(c) $3600, $2700

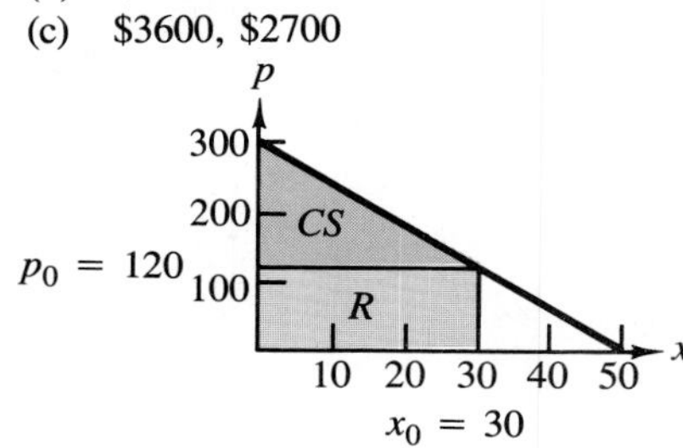

(d) $2400, $300

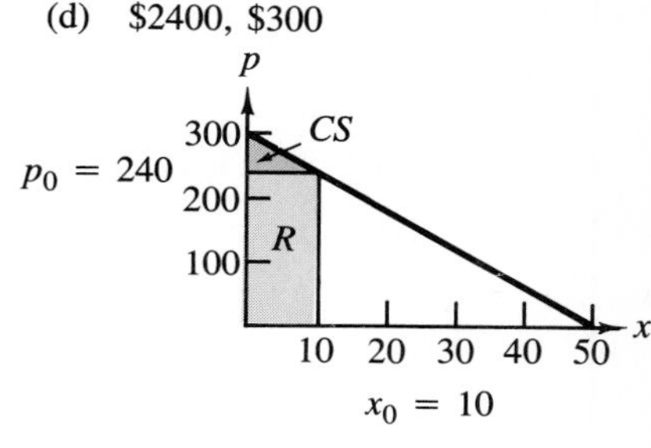

3. (a)

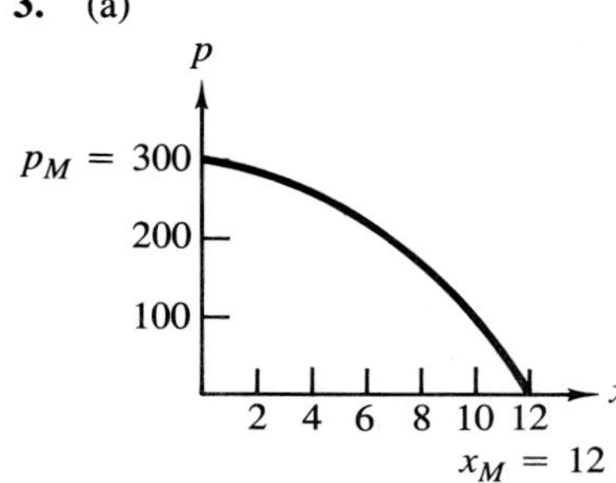

(b) $2376

(c) $1161, $1012.50

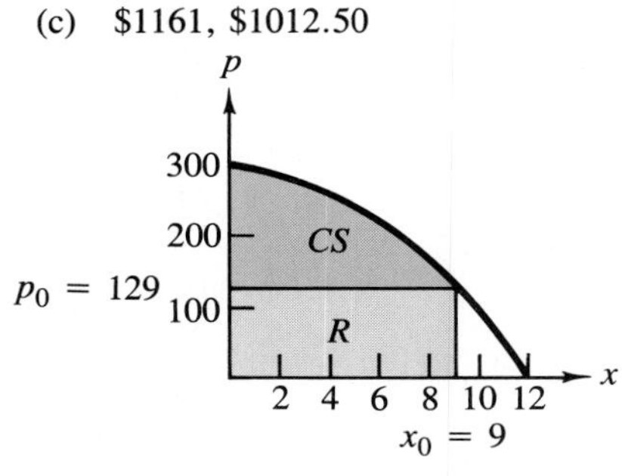

(d) $1225, $179.17

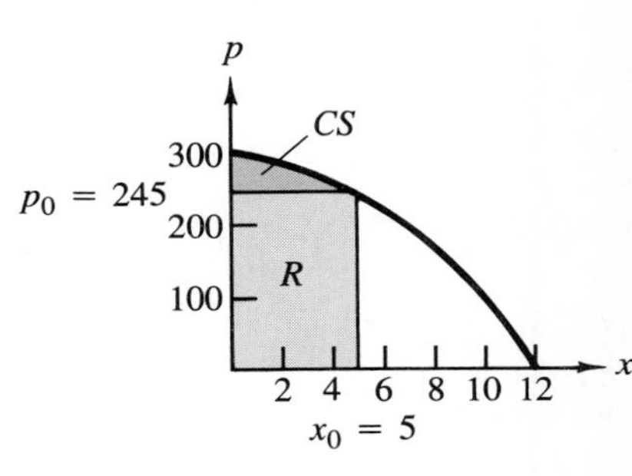

5. (a)

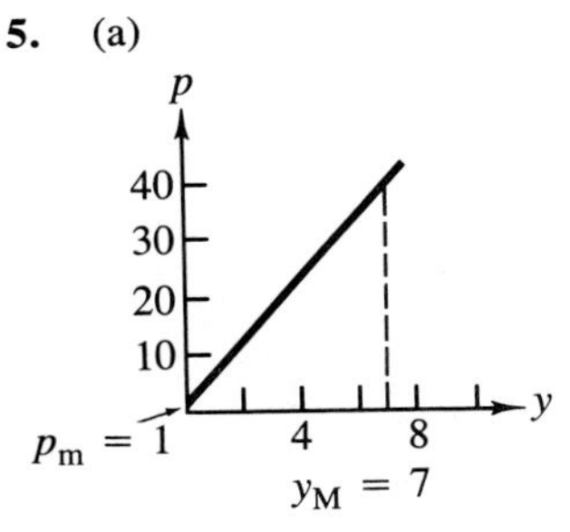

(b) $154

(c) $100, $48

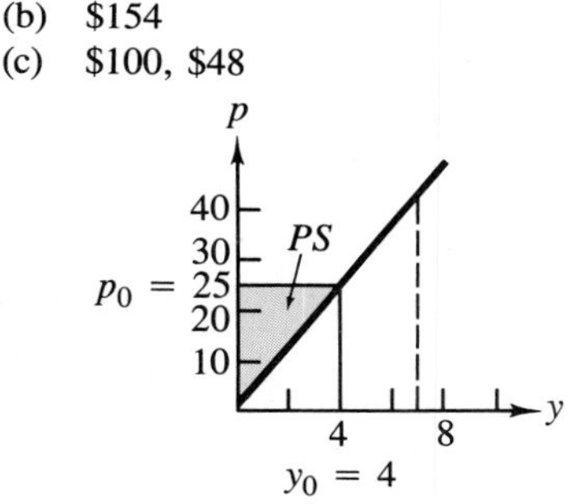

(d) $57, $27

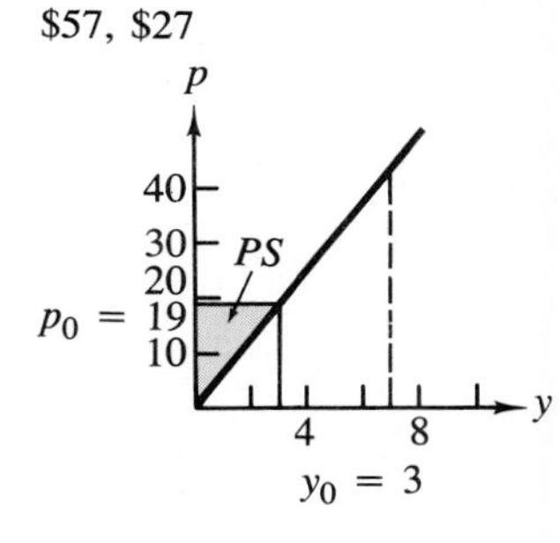

7. (a)

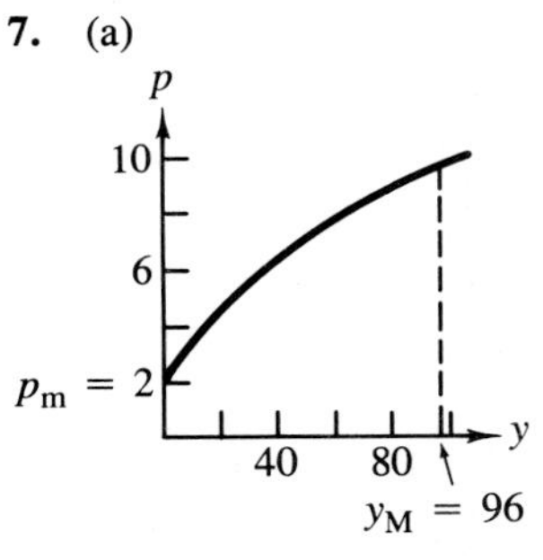

(b) $661.33

(c) $480, $144

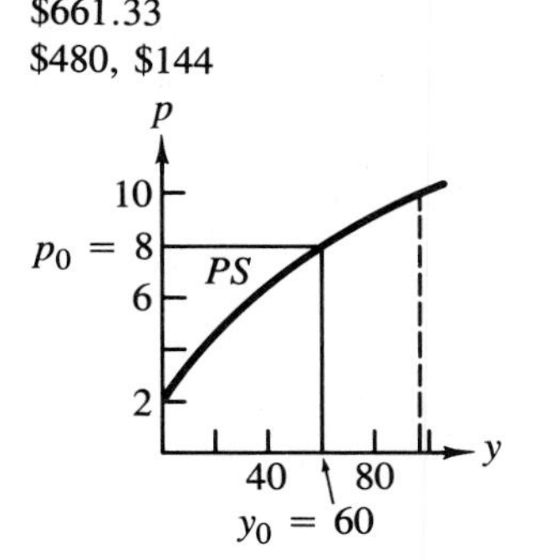

(d) $693, $212.33

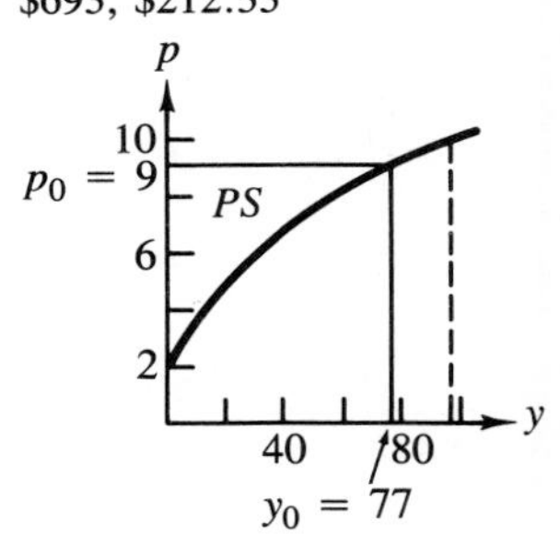

9. (a) \$8 (b) 5 units (c) \$40 (c) \$10 (e) \$15 (f)

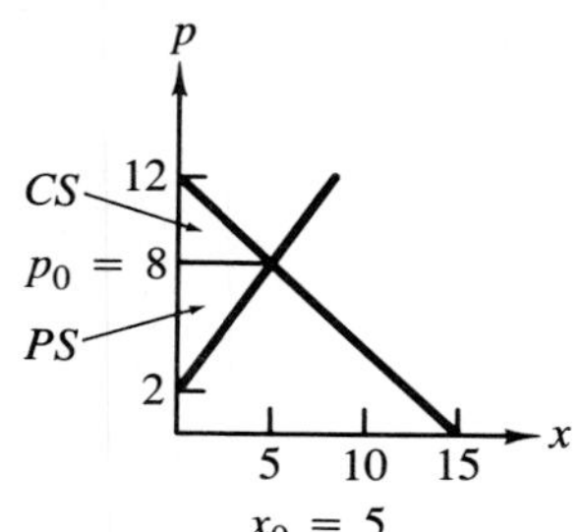

11. (a) \$16 (b) 2 units (c) \$32 (d) $\$\frac{16}{3}$ (e) \$16 (f)

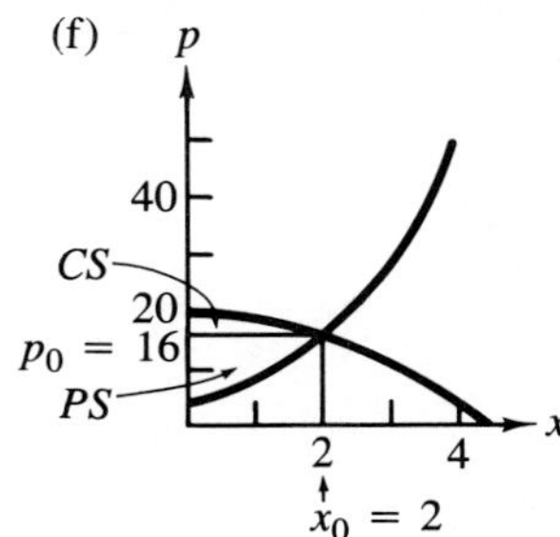

Exercise Set 17.5, page 763

1. 16 **3.** (a) 8 (b) $e^2 - 1$ **5.** (a) \$8240 (b) \$14,490 (c) \$14,020 **7.** \$560
9. 885.28 quadrillion Btu **11.** (a) 10,000 feet (b) 5 seconds **13.** 118 feet/second, 95 feet
15. (a) 87,612.5 gallons (b) 12,537.5 gallons **17.** $50(e^5 - 1)$ dollars

Exercise Set 17.6, page 770

1. $\frac{1}{18}$ **3.** $\frac{1}{3}$ **5.** $\frac{1}{2}$ **7.** diverges **9.** $\frac{1}{4}$ **11.** diverges **13.** diverges **15.** diverges
17. diverges **19.** 2 **21.** \$240,000

Exercise Set 17.7, page 778

1. (b) $\frac{3}{8}$ (c) $\frac{5}{8}$ (d) $\frac{5}{32}$ **3.** (a) $f(x) = \frac{1}{20}$ $5 \le x \le 25$ (b) $\frac{3}{10}$ (c) 1 (d) 15
5. 4.5 **7.** (a) $\frac{1}{2}$ (b) 1 (c) 0.3 (d) 5 minutes
9. (a) $f(x) = 0.5e^{-0.5x}$ $0 \le x < \infty$ (b) 0.3935 (c) 0.4712 (d) 0.6065
11. (a) 0.3935 (b) 60.65% **13.** (a) 0.3935 (b) 0.6065 (c) 23.3%

Supplementary Exercises, page 780

1. $\frac{1}{12}$ **3.** $\frac{1}{10}(1 - e^{-10})$ **5.** $e^3 - \frac{11}{2}$ **7.** $\frac{1}{2}$ **9.** $\frac{1}{8}$ **11.** 0 **13.** (a) $\frac{2}{5}$ (b) $\frac{3}{5}$ (c) 12.5

Chapter Test, page 781

1. $\frac{4}{3}$ **2.** $\frac{8}{3}\ln 4 - \frac{5}{4}$ **3.** 10 ln 26 **4.** $\frac{275}{12}$ **5.** diverges **6.** (a) $\frac{1}{15}$ (b) 7.5

CHAPTER EIGHTEEN

Exercise Set 18.1, page 787

9. $C = 1$ **11.** $C = \frac{3}{4}$ **13.** $y = 3x - \frac{1}{2}x^2 + C$ **15.** $y = \frac{5}{6}x^{6/5} - \frac{3}{2}x^{4/3} + C$ **17.** $y = \frac{1}{4}x^2 + \frac{1}{2}\ln|x| + C$
19. (a) $50x^2 - 2000x + 20{,}000$ (b) \$5000 **21.** $P(x) = -x^2 + 600x$ **23.** $T = -e^{-t} + 501$

Exercise Set 18.2, page 792

1. $y = Ce^{3x}$ **3.** $y = \pm\sqrt{x^2 + C}$ **5.** $y = (e^x + C)^{1/3}$ **7.** $y = (\frac{1}{4}x^2 + C)^2$ **9.** $y = \ln(x^2 + C)$
11. $y = \frac{1}{2}x^2 + \ln|x| + C$ **13.** $y = \ln(2\sqrt{x} + \frac{2}{3}x^{3/2} + C)$ **15.** $Q = (3t + C)^{1/3}$ **17.** $y = -x^2 + 3$
19. $y = (12x + 500)^{1/3}$ **21.** $y = 2e^{(1/2)x^2 + x}$ **23.** $y = \frac{1}{2}\ln(x^2 + 3) + 5 - \ln 2$

Exercise Set 18.3, page 800

1. 6.561 pounds will remain after 2 hours. After 7.64 hours only 2 pounds will remain.
3. 34 billion people **5.** 15.53 seconds **9.** (a) $Q = 10 + Ce^{-2t}$ (b) $Q = 10 - 10e^{-2t}$ (c) 10
11. $T = M + (T_0 - M)e^{-kt}$ **13.** (a) $Q = \dfrac{4800}{1 + 5e^{-[(\ln 5)/8]t}}$ (b) 4000 (c) 6.33 months
15. (a) $Q = \dfrac{96{,}000}{1 + 23^{-1/2 \ln (23/7)t}} = \dfrac{96{,}000}{1 + 23(7/23)^{t/2}}$ (b) 58,239 (c) 7.12 months
19. (a) 1.06 (b)

21. (a) in t years: $(50{,}000 + \frac{200}{0.03})e^{0.03t} - \frac{200}{0.03}$, in 10 years: 69,825
(b) $t = \dfrac{\ln (P + \frac{i}{k}) - \ln (P_0 + \frac{i}{k})}{k}$ (c) 21.08 years

Supplementary Exercises, page 803

1. $y = (x^3 + 1)^{1/3}$ **3.** $y = x - 1$ **5.** $Q = e^{5t}$ **7.** (a) $v = -32t$ (b) -32 feet/second
9. $T = 70 + 50e^{-0.1[\ln (5/4)]t}$ or $T = 70 + 50(\frac{4}{5})^{0.1t}$

Chapter Test, page 804

1. $y = xe^x - e^x + 2$ **2.** $Q = e^{(1/2)t^2}$ **3.** $y = \dfrac{4e^{4t}}{1 + e^{4t}}$
4. (a) $C(x) = \frac{1}{3}x^3 - x^2 + 1000x + 500$ (b) $C(20) = \$22{,}767$ **5.** $s(t) = 8t^2 - 64t + 128$

CHAPTER NINETEEN

Exercise Set 19.1, page 812

1. (a) 2 (b) 3 (c) -28 (d) 3 **3.** (a) 1 (b) 1 (c) $2e$ (d) $-e^2 + 2/e$
5. (a) $\sqrt{2}$ (b) 0 (c) -1 (d) -5 **7.** (a) $z = 0$ (b) $x = 0$ (c) $y = 0$
9.

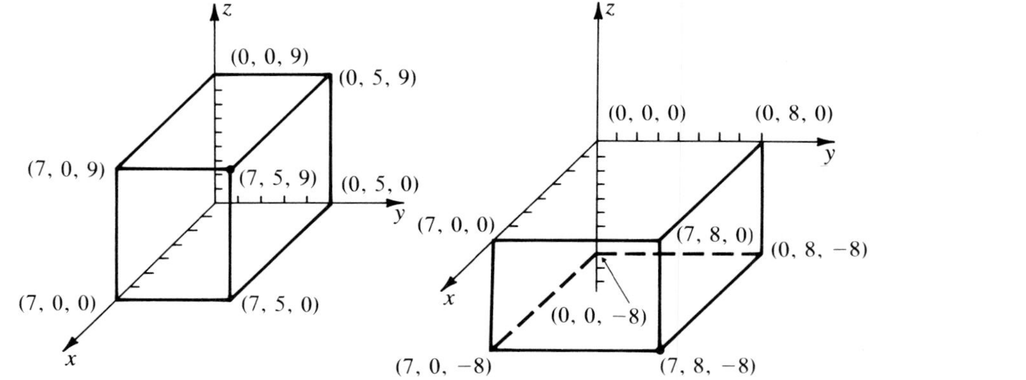

11. A (6, 0, 0) D (0, 0, 0) G (0, 7, 7)
B (6, 7, 0) E (6, 0, 7) H (0, 0, 7)
C (0, 7, 0) F (6, 7, 7)

13. (a) 13 (b) 4 (c) 2 **15.** \$1.4 million **17.** (a) $P(x, y) = -2x^2 + 3xy - 5y^2 + 100x + 50y - 950$
(b) $R = 3000$, $C = 3038$, $P = -38$ (in thousands) (c) $R = 350x$, $C = 4x^2 + 200x + 950$, $P = -4x^2 + 150x - 950$ (d) $R = 450x$, $C = 16x^2 + 250x + 950$, $P = -16x^2 + 200x - 950$

Exercise Set 19.2, page 818

1. $f_x = 3x^2 + y$ $f_y = x + 4y$ $f_{yx} = 1$ **3.** $f_x = e^y + y^2e^x$ $f_y = xe^y + 2ye^x - 3y^2$ $f_{yx} = e^y + 2ye^x$
5. $f_x = ye^{xy}$ $f_y = xe^{xy}$ $f_{yx} = xye^{xy} + e^{xy}$ **7.** $f_x = 1/x$ $f_y = 1/y$ $f_{yx} = 0$
9. $f_x = 8$ $f_y = 6$ $f_{xy} = 6$ **11.** $f_x = e + 1$ $f_y = 2$ $f_{xy} = e + 2$
13. $f_{xx} = 4$ $f_{yy} = -2$ $f_{xy} = f_{yx} = 1$ **15.** $f_{xx} = f_{yy} = f_{xy} = f_{yx} = 0$
17. $f_{xx} = 2e^{x+y} + xe^{x+y}$ $f_{yy} = xe^{x+y}$ $f_{xy} = f_{yx} = e^{x+y} + xe^{x+y}$
19. $f_{xx} = 6$ $f_{yy} = -4$ $f_{xy} = f_{yx} = -5$ **21.** $f_{xx} = 4e^3$ $f_{yy} = 9e^3$ $f_{xy} = f_{yx} = 6e^3$
23. $f_{xx} = \dfrac{-2x^2 + 2y^2}{(x^2 + y^2)^2}$ $f_{yy} = \dfrac{2x^2 - 2y^2}{(x^2 + y^2)^2}$
25. (a) $I_x = 7; I_y = \frac{7}{2}$
(b) the change of the index I with respect to y is half that with respect to x.

Exercise Set 19.3, page 827

1. (0, 0) **3.** $(-\frac{1}{2}, 1)$ **5.** (0, 0) and (1, 1) **7.** relative minimum: $f(1, 2) = -7$ **9.** relative minimum: $f(0, 0) = 0$
11. relative maximum: $f(1, 0) = 2$ **13.** no extrema, saddle point at (0, 0)
15. relative minimum: $f(\frac{2}{3}, -\frac{2}{3}) = -\frac{8}{27}$
17. $x = 150$, $y = 200$ **19.** \$50 million on research, \$5 million on development **21.** $x = y = z = \sqrt[3]{50}$
23. (a) $d_1 = (m + b) - 3$ $d_2 = (2m + b) - 1$ $d_3 = (3m + b) - 4$ $d_4 = (4m + b) - 3$
(b) $(m + b - 3)^2 + (2m + b - 1)^2 + (3m + b - 4)^2 + (4m + b - 3)^2$
(c) $F_m = 60m + 20b - 58$ $F_b = 20m + 8b - 22$ (d) $m = \frac{3}{10}$, $b = 2$ (e) $M > 0$, $F_{mm} > 0$
(f) $y = \frac{3}{10}x + 2$ **25.** $y = -\frac{35}{59}x + \frac{6}{59}$ **27.** $y = \frac{7}{30}x + \frac{14}{5}$

Exercise Set 19.4, page 832

1. $(\frac{1}{2}, \frac{3}{2}); f(\frac{1}{2}, \frac{3}{2}) = \frac{5}{2}$
3. $(2, 2\sqrt{3}), f(2, 2\sqrt{3}) = 4\sqrt{3}$; $(2, -2\sqrt{3}), f(2, -2\sqrt{3}) = -4\sqrt{3}$; $(-2, 2\sqrt{3}), f(-2, 2\sqrt{3}) = -4\sqrt{3}$;
$(-2, -2\sqrt{3}), f(-2, -2\sqrt{3}) = 4\sqrt{3}$ **5.** $(\frac{1}{2}, 1, \frac{3}{2}); f(\frac{1}{2}, 1, \frac{3}{2}) = 7; (-\frac{1}{2}, -1, -\frac{3}{2}); f(-\frac{1}{2}, -1, -\frac{3}{2}) = -7$
7. $\left(\frac{1}{\sqrt{3}}, \frac{1}{\sqrt{3}}, \frac{1}{\sqrt{3}}\right), f\left(\frac{1}{\sqrt{3}}, \frac{1}{\sqrt{3}}, \frac{1}{\sqrt{3}}\right) = \frac{1}{3\sqrt{3}}$; $\left(\frac{1}{\sqrt{3}}, \frac{1}{\sqrt{3}}, -\frac{1}{\sqrt{3}}\right), f\left(\frac{1}{\sqrt{3}}, \frac{1}{\sqrt{3}}, -\frac{1}{\sqrt{3}}\right) = -\frac{1}{3\sqrt{3}}$;
$\left(\frac{1}{\sqrt{3}}, -\frac{1}{\sqrt{3}}, \frac{1}{\sqrt{3}}\right), f\left(\frac{1}{\sqrt{3}}, -\frac{1}{\sqrt{3}}, \frac{1}{\sqrt{3}}\right) = -\frac{1}{3\sqrt{3}}$; $\left(-\frac{1}{\sqrt{3}}, \frac{1}{\sqrt{3}}, \frac{1}{\sqrt{3}}\right), f\left(-\frac{1}{\sqrt{3}}, \frac{1}{\sqrt{3}}, \frac{1}{\sqrt{3}}\right) = -\frac{1}{3\sqrt{3}}$;
$\left(-\frac{1}{\sqrt{3}}, -\frac{1}{\sqrt{3}}, \frac{1}{\sqrt{3}}\right), f\left(-\frac{1}{\sqrt{3}}, -\frac{1}{\sqrt{3}}, \frac{1}{\sqrt{3}}\right) = \frac{1}{3\sqrt{3}}$; $\left(-\frac{1}{\sqrt{3}}, \frac{1}{\sqrt{3}}, -\frac{1}{\sqrt{3}}\right), f\left(-\frac{1}{\sqrt{3}}, \frac{1}{\sqrt{3}}, -\frac{1}{\sqrt{3}}\right) = \frac{1}{3\sqrt{3}}$;
$\left(\frac{1}{\sqrt{3}}, -\frac{1}{\sqrt{3}}, -\frac{1}{\sqrt{3}}\right), f\left(\frac{1}{\sqrt{3}}, -\frac{1}{\sqrt{3}}, -\frac{1}{\sqrt{3}}\right) = \frac{1}{3\sqrt{3}}$;
$\left(-\frac{1}{\sqrt{3}}, -\frac{1}{\sqrt{3}}, -\frac{1}{\sqrt{3}}\right), f\left(-\frac{1}{\sqrt{3}}, -\frac{1}{\sqrt{3}}, -\frac{1}{\sqrt{3}}\right) = -\frac{1}{3\sqrt{3}}$
9. $x = y = 50$ **13.** $(\frac{21}{34}, \frac{35}{34})$ **15.** $a = 150$, $b = 250$ **17.** 11,664 cubic inches
19. $x = 40$, $y = 30$ **21.** $x = 4$, $y = 24$, $z = 3$

Supplementary Exercises, page 833

1. (a) (0, 0) (b) $-\frac{3}{5}, \frac{3}{5}, \frac{5}{13}$ **3.** (a) $f_x = \frac{3}{2}x^{1/2} + y^2 - 2xy$ $f_y = 2xy - x^2 + 3y^2$ $f_{xx} = \frac{3}{4}x^{-1/2} - 2y$ $f_{yy} = 2x + 6y$ $f_{xy} = f_{yx} = 2y - 2x$ (b) $f_x(4, 1) = -4, f_y(4, -1) = -21, f_{xy}(3, 2) = -2$ (c) $f_{xyx} = -2$
5. $(-\frac{7}{5}, \frac{3}{5})$ **7.** minimum **9.** relative minimum: $f(1, 1) = -1$ **11.** relative maximum: $f(-6, -6) = 216$
13. \$300 for labor, \$200 for fertilizer; maximum profit is \$9000 per acre **15.** $y = \frac{5}{2}x - 3$ **17.** $f(10, 10) = 200$
19. $f(\frac{3}{2}, -\frac{1}{2}) = 2$ $f(-\frac{3}{2}, \frac{1}{2}) = -2$

Chapter Test, page 834

1. $-7, -1, -1$ **2.** (a) $f_x = 2x - 2y$ $f_y = -2x + 3y^2$ $f_{xy} = -2$ $f_{yy} = 6y$ (b) $-6, 14, -2, 12$
3. (a) (0, 0) (b) $M = -5$, therefore (0, 0) is a saddle point
4. $x = \frac{3}{2}$, $y = 2$, maximum profit = \$152.75 million **5.** $y = \frac{31}{35}x + \frac{8}{7}$ **6.** $x = y = 32$

APPENDIX A, ALGEBRA REVIEW

Exercise Set A.1, page 837

1. 4 **3.** -8 **5.** $\frac{1}{2}$ **7.** 19 **9.** $\frac{2}{9}$ **11.** $\frac{1}{2}$ **13.** $\frac{9}{20}$ **15.** $\frac{2}{15}$ **17.** $\frac{5}{4}$ **19.** $-\frac{23}{35}$ **21.** $\frac{2}{7}$
23. $\frac{8}{15}$ **25.** $3\frac{1}{3}$ **27.** $\frac{3}{14}$
29. (a) $\frac{3}{25}$ (b) $\frac{1}{50}$ (c) $\frac{1}{5000}$ (d) $\frac{9}{10}$ (e) $\frac{9}{4}$ (f) $\frac{19}{50}$
31. (a) 3% (b) 56% (c) 110% (d) 1.5% (e) 280% (f) 0.03%
33. (a) 53.46 (b) 2.1045 (c) 50% (d) 34.29% (e) 200%
35. $\frac{9}{28}$ **37.** \$13.50

Exercise Set A.2, page 840

1. (a) -26 (b) 11 (c) -22 (d) $\frac{5}{6}$
3. $6x - 5$ **5.** $3x - y$ **7.** $.08(x - 50)$ **9.** $A = \pi r^2$ **11.** $-2 > -3$ **13.** $\frac{1}{2} > \frac{1}{3}$ **15.** $\pi > 2$
17. $\frac{5}{3} > \frac{5}{4}$ **19.** false **21.** false **23.** false **25.** true **27.** true **29.** false **31.** true

Exercise Set A.3, page 843

1. -6 **3.** -3 **5.** 1 **7.** no solutions **9.** $\frac{1}{9}$ **11.** $-\frac{7}{2}$ **13.** $\frac{1}{y}$ **15.** $\dfrac{C - By}{A}$ **17.** $\dfrac{y - b}{m}$
19. 0 **21.** $\dfrac{4 - y^2}{3y}$ **23.** (a), (b), (e), (h) **25.** -3 **27.** $-\frac{1}{3}$ **29.** $-\frac{19}{7}$ **31.** 18 **33.** 15
35. 4 **37.** 9 **39.** (a) $C = \frac{5}{9}(F - 32)$ (b). 50 (c) 30 (d) yes, at -40°C **41.** (a) $P = \dfrac{S}{1 + rt}$
(b) $t = \dfrac{S - P}{Pr}$ (c) \$588.23

Exercise Set A.4, page 845

1. $x < 4$ **3.** $y < -\frac{1}{2}$ **5.** $y \le 8$ **7.** $x \ge -6$ **9.** $x > 2$ **11.** $x < -6$ **13.** $x \ge 0$
15. $x \ge 1$ **17.** $b < -2$ **19.** $x \le 2$

Exercise Set A.5, page 850

1. $\frac{1}{16}$ **3.** -8 **5.** -16 **7.** 9 **9.** 9 **11.** 9 **13.** $\frac{1}{2}$ **15.** 100,000 **17.** -2 **19.** -2
21. -10 **23.** 4, 5 **25.** $-2, -1$
27. (a) $2\sqrt[4]{2}$ (b) $\sqrt{2}$ (c) x^3 (d) $3x^2y^3\sqrt{xy}$ (e) $3x^2y^4\sqrt[3]{xy}$ (f) a^2

29. $\frac{-3y^2z^3}{x^2}$ **31.** $\frac{32}{a^9b^7}$ **33.** $\frac{1}{2^{10}p^2}$ **35.** $\frac{a^3}{b^3}$ **37.** $3^{20}4^{12}x^{72}y^{104}$ **39.** $x^{10/3}$ **41.** $\frac{y}{x^3}$ **43.** $x^{13/12}$
45. $4a^2b^{8/3}$ **47.** $4x$ **49.** $\frac{a^{12}}{b^{14}}$ **51.** $10x\sqrt{y}$ **53.** $\frac{y^4}{x^2}$ **55.** $\frac{3^{3/2}\pi^3}{e^{3/2}}$ **57.** $\pi^{-3} = 0.03225$

Exercise Set A.6, page 854

1. 3 **3.** no **5.** no **7.** no **9.** polynomial of degree zero **11.** no
13. $4x^2 - 15x + 13$ **15.** $x^3 + 6x^2 + 2x + 1$ **17.** $4x^4 - 10x^3 - 3x^2 - 4x - 7$ **19.** $a^3 + 3a^2b + 3ab^2 + b^3$
21. $2a^2 + ab - b^2 + 2ac - bc$ **23.** $x^2 + \frac{5}{6}xy - y^2$ **25.** $x^6 - y^6$ **27.** $x^2 + 2xy + y^2 - z^2$
29. $4b(a + b^2)$ **31.** $(1 + x)(1 + x^2)$ **33.** $(x - 4)^2$ **35.** $(b - 5)(b + 5)$ **37.** $(x - 3)(x + 4)$
39. $y(x - y)(x + y)$ **41.** $(2a + 3)(2a - 1)$ **43.** $2x^2(x - 2)(x + 2)$ **45.** $xy(y - x)(y + x)$ **47.** $(3x + 5)^2$
49. $(x - \sqrt{2})(x + \sqrt{2})$

Exercise Set A.7, page 857

1. (a), (b), (d), (e) **3.** $\frac{1}{xyz}$ **5.** $\frac{2x}{3}$ **7.** $\frac{a}{4}$ **9.** $\frac{x-3}{2x}$ **11.** $\frac{3+x}{3-x}$
13. $\frac{2x-1}{x+2}$ **15.** $\frac{x-4}{2x+7}$ **17.** $\frac{1}{a^2+9}$ **19.** $\frac{1}{3x}$ **21.** $\frac{2xy}{y-2}$ **23.** $\frac{8x^5}{27y^5}$ **25.** $\frac{2}{x}$
27. $\frac{(x+12)(x-7)}{(x-2)(x+2)}$ **29.** $\frac{2}{3}$ **31.** $\frac{x+2}{(x+4)(x-1)}$ **33.** $(2x + 1)(x + 9)$ **35.** $\frac{x^2+2y}{xy}$ **37.** $\frac{17x}{6y}$
39. $\frac{3x-5y}{y}$ **41.** $\frac{3x+6-xy}{x(x+2)}$ **43.** $\frac{3a-7b}{a^2-b^2}$ **45.** $\frac{a+b+c}{abc}$ **47.** $\frac{2}{a^2-4}$ **49.** $\frac{29x^2-9x+3}{18x^3}$
51. $\frac{5x^2-10x+2}{(x-3)(x-5)(2x+1)}$ **53.** $\frac{x^3+8x^2-x-2}{6(x+1)^2}$ **55.** $\frac{x+y}{xy}$ **57.** $x + 1$

Exercise Set A.8, page 862

1. ± 2 **3.** 0 **5.** $\pm\sqrt{7}$ **7.** ± 10 **9.** 1, 3 **11.** $-\frac{3}{2}, \frac{4}{3}$ **13.** $-3, \frac{1}{2}$ **15.** $-\frac{4}{3}, \frac{1}{2}$
17. $-\frac{1}{2}, 4$ **19.** $-\frac{1}{2}, \frac{2}{3}$ **21.** $-\frac{1}{5}, \frac{1}{2}$ **23.** $1 \pm \sqrt{3}$ **25.** $\frac{-1 \pm \sqrt{5}}{2}$ **27.** $-2, \frac{2}{3}$
29. no real solution **31.** $3 \pm 2\sqrt{3}$

APPENDIX B, LOGIC, page 877

1. (b) and (e)
3. (a) $p \wedge q$: $7 = (5)(2)$ and $3 = 2 + 1$
$p \vee q$: $7 = (5)(2)$ or $3 = 2 + 1$
(b) $p \wedge q$: I will study tonight and I will pass the exam tomorrow.
$p \vee q$: I will study tonight or I will pass the exam tomorrow.
(c) $p \wedge q$: I will eat dinner and I will watch TV.
$p \vee q$: I will eat dinner or I will watch TV.
5. (a) true (b) true (c) true (d) true
7. (a) If 2 is a positive number, then $2^2 = 4$. true
(b) If England is in Asia, then France is in Europe. true
(c) If $3 < -1$, then $4 + 5 = 9$. true
9. (a) true (b) true (c) true (d) false (e) true (f) true
11. (a) If Paris is not the capital of England, then $2 \not< 3$.
(b) If you can drive to work, then it will not snow.
(c) If I don't go dancing, then I will not have money.
(d) If I pass the exam, then I will study.

13. (a) false (b) false (c) true (d) true (e) true

15. (a)

p	q	$\sim p$	$\sim p \vee q$	$(\sim p \vee q) \to p$
T	T	F	T	T
T	F	F	F	T
F	T	T	T	F
F	F	T	T	F

neither

(b)

p	q	$\sim p$	$p \wedge \sim p$	$(p \wedge \sim p) \to q$
T	T	F	F	T
T	F	F	F	T
F	T	T	F	T
F	F	T	F	T

tautology

17. (a)

p	q	$p \to q$	$\sim q \to \sim p$	$(p \to q) \leftrightarrow (\sim q \to \sim p)$
T	T	T	T	T
T	F	F	F	T
F	T	T	T	T
F	F	T	T	T

tautology

(b)

p	q	$p \to q$	$q \to p$	$(p \to q) \vee (q \to p)$
T	T	T	T	T
T	F	F	T	T
F	T	T	F	T
F	F	T	T	T

tautology

19.

p	q	r	$p \to q$	$q \to r$	$(p \to q) \wedge (q \to r)$	$p \to r$	$[(p \to q) \wedge (q \to r)] \to (p \to r)$
T	T	T	T	T	T	T	T
T	T	F	T	F	F	F	T
T	F	T	F	T	F	T	T
T	F	F	F	T	F	F	T
F	T	T	T	T	T	T	T
F	T	F	T	F	F	T	T
F	F	T	T	T	T	T	T
F	F	F	T	T	T	T	T

tautology

21.

	S	R	
p	q	$\sim p \vee q$	$R \to S$
T	T	T	T
T	F	F	
F	T	T	T
F	F	T	F

R does not logically imply *S*

23.

p	q	r	$p \to q$	$q \to r$	R $(p \to q) \wedge (q \to r)$	S $p \to r$	$R \to S$
T	T	T	T	T	T	T	T
T	T	F	T	F	F		
T	F	T	F	T	F		
T	F	F	F	T	F		
F	T	T	T	T	T	T	T
F	T	F	T	F	F		
F	F	T	T	T	T	T	T
F	F	F	T	T	T	T	T

R logically implies *S*

25.

p	q	$(p \vee \sim q)$	R $\sim(p \vee \sim q)$	S $\sim p \vee q$	$R \leftrightarrow S$
T	T	T	F	T	F
T	F	T	F	F	T
F	T	F	T	T	T
F	F	T	F	T	F

R and *S* are not logically equivalent.

27.

p	q	r	$p \to q$	$q \to r$	R $(p \to q) \wedge (q \to r)$	S $p \to r$	$R \leftrightarrow S$
T	T	T	T	T	T	T	T
T	T	F	T	F	F	F	T
T	F	T	F	T	F	T	F
T	F	F	F	T	F	F	T
F	T	T	T	T	T	T	T
F	T	F	T	F	F	T	F
F	F	T	T	T	T	T	T
F	F	F	T	T	T	T	T

R and *S* are not logically equivalent

29. (a)

p	q	$\sim(p \vee q)$	$\sim(p \vee q) \wedge p$	$p \vee \sim q$	$[\sim(p \vee q) \wedge p] \to [p \vee \sim q]$
T	T	F	F	T	T
T	F	F	F	T	T
F	T	F	F	F	T
F	F	T	F	T	T

the argument is valid

(b)

p	q	$p \to q$	$\sim q$	$(p \to q) \wedge \sim q$	$[(p \to q) \wedge \sim q] \to \sim p$
T	T	T	F	F	T
T	F	F	T	F	T
F	T	T	F	F	T
F	F	T	T	T	T

the argument is valid

31. (a)

p	q	$p \to q$	$\sim p$	$(p \to q) \land \sim p$	$[(p \to q) \land \sim p] \to q$
T	T	T	F	F	
T	F	F	F	F	
F	T	T	T	T	T
F	F	T	T	T	F

the argument is not valid

(b)

p	q	$\sim(p \land q)$	$\sim(p \land q) \land p$	$[\sim(p \land q) \land p] \to \sim q$
T	T	F	F	
T	F	T	T	T
F	T	T	F	
F	F	T	F	

the argument is valid

33.

p	q	r	$\sim p \lor q$	$q \to r$	$(\sim p \lor q) \land (q \to r)$	$\sim p \lor r$	$[(\sim p \lor q) \land (q \to r)] \to (\sim p \lor r)$
T	T	T	T	T	T	T	T
T	T	F	T	F	F	F	
T	F	T	F	T	F	T	
T	F	F	F	T	F	F	
F	T	T	T	T	T	T	T
F	T	F	T	F	F	T	
F	F	T	T	T	T	T	T
F	F	F	T	T	T	T	T

the argument is valid

35. p: The tax rate is lowered.
q: I will buy a car.

$$\begin{array}{l} p \to q \\ \underline{\sim q} \\ \therefore \sim p \end{array}$$

p	q	$p \to q$	$(p \to q) \land \sim q$	$[(p \to q) \land \sim q] \to \sim p$
T	T	T	F	
T	F	F	F	
F	T	T	F	
F	F	T	T	T

the argument is valid

37. p: I will take a walk.
q: I will watch TV.

$$\begin{array}{l} p \lor q \\ \underline{\sim q} \\ \therefore \sim p \end{array}$$

p	q	$p \lor q$	$(p \lor q) \land \sim q$	$[(p \lor q) \land \sim q] \to \sim p$
T	T	T	F	
T	F	T	T	F
F	T	T	F	
F	F	F	F	

the argument is not valid

Index

A 1
B 2
C 3
D 4
E 5
F 6
G 7
H 8
I 9
J 0

GRAPHS OF FUNCTIONS

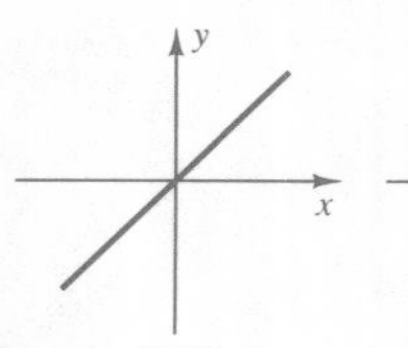

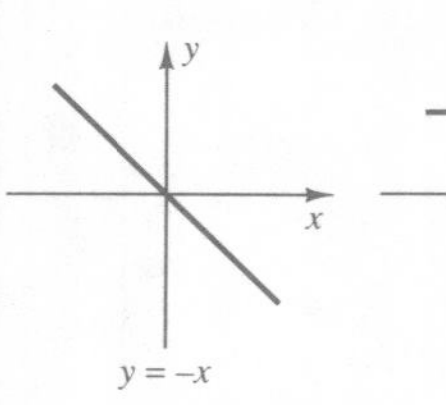

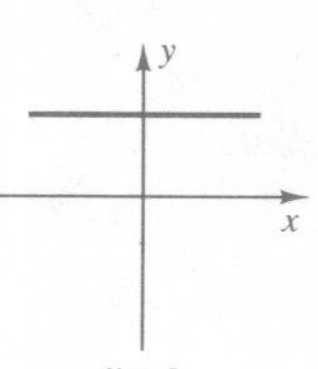

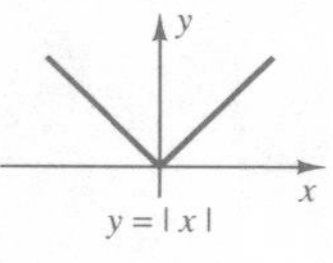

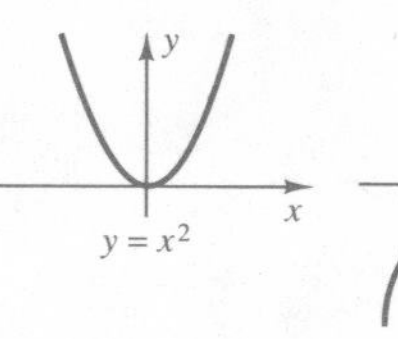

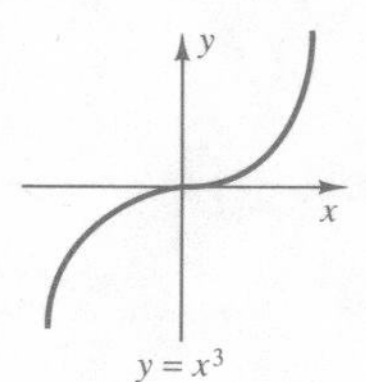

LINEAR INEQUALITIES AND HALF PLANES

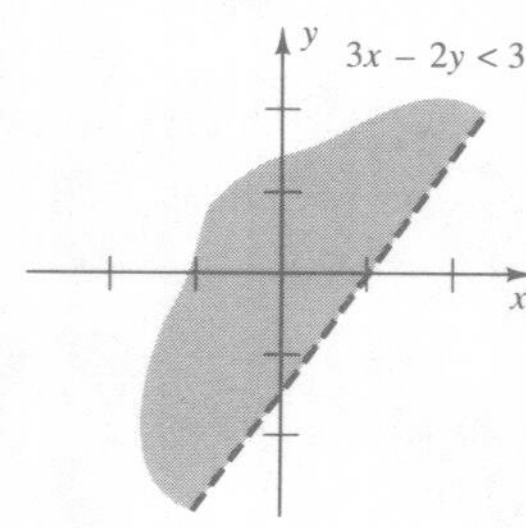

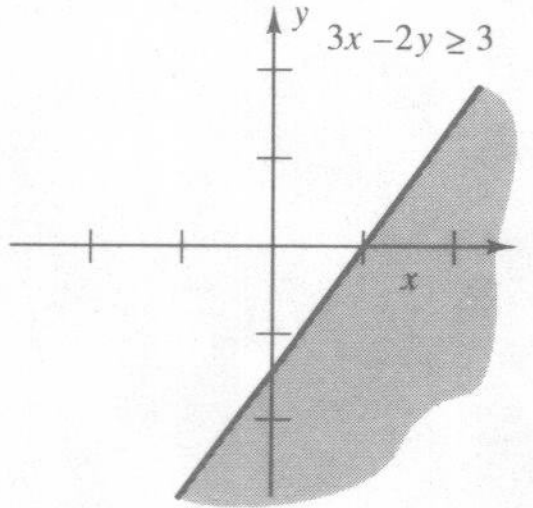

THE EXPONENTIAL FUNCTION

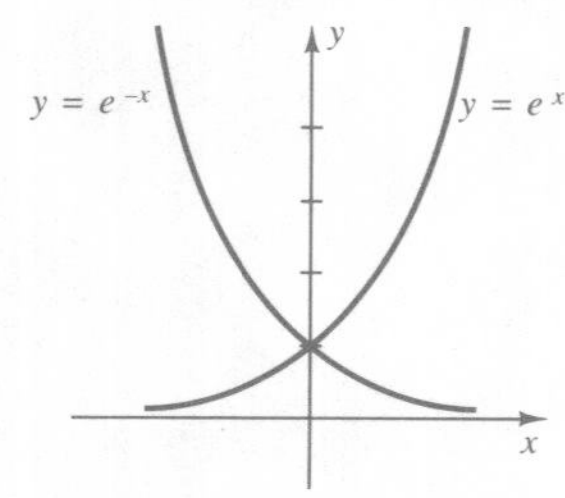

LAWS OF EXPONENTS

$a^m a^n = a^{m+n}$ $\qquad a^0 = 1$

$(a^m)^n = a^{mn}$ $\qquad a^1 = a$

$(ab)^m = a^m b^m$

$\left(\dfrac{a}{b}\right)^m = \dfrac{a^m}{b^m}$ $\qquad a^{-m} = \dfrac{1}{a^m}$

$\dfrac{a^m}{a^n} = a^{m-n}$

LAWS OF LOGARITHMS

$\log_a(xy) = \log_a x + \log_a y$ $\qquad \log_a 1 = 0$

$\log_a a = 1$

$\log_a \dfrac{x}{y} = \log_a x - \log_a y$ $\qquad \log_b x = \dfrac{\log_a x}{\log_a b}$

$\log_a(x^n) = n \log_a x$

$x = a^{\log_a x} \qquad x > 0$

THE NATURAL LOGARITHMIC FUNCTION

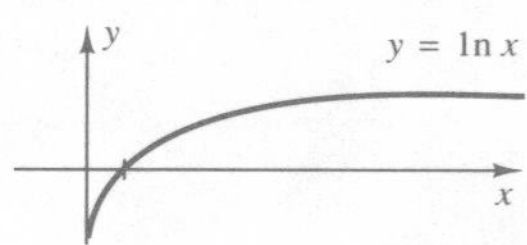